THE
ENCYCLOPEDIA
OF ELDER CARE

THE ENCYCLOPEDIA OF ELDER CARE

MATHY D. MEZEY

EDITOR-IN-CHIEF

**Barbara J. Berkman, Christopher M. Callahan,
Terry T. Fulmer, Ethel L. Mitty, Gregory J. Paveza,
Eugenia L. Siegler, Neville E. Strumpf**

ASSOCIATE EDITORS

Melissa M. Bottrell

MANAGING EDITOR

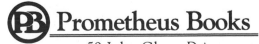
Prometheus Books

59 John Glenn Drive
Amherst, New York 14228-2197

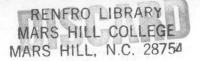
Published 2004 by Prometheus Books

Inquiries should be addressed to
Prometheus Books
59 John Glenn Drive
Amherst, New York 14228–2197
VOICE: 716–691–0133, ext. 207
FAX: 716–564–2711
WWW.PROMETHEUSBOOKS.COM

08 07 06 05 04   5 4 3 2 1

Library of Congress Cataloging-in-Publication Data

The encyclopedia of elder care : the comprehensive resource on geriatric and social care / Mathy D. Mezey, editor-in-chief ; Barbara J. Berkman ... [et al.], associate editors ; Melissa M. Bottrell, managing editor.
    p. cm.
Originally published: New York : Springer Pub., c2001.
Includes bibliographical references and index.
ISBN 1–59102–189–8 (alk. paper)
1. Geriatric nursing—Encyclopedias. 2. Aged—Care—Encyclopedias. 3. Aged—Medical care—Encyclopedias. I. Mezey, Mathy Doval.

RC954.E53 2004
362.198'97—dc22

2004044683

Printed in the United States on acid-free paper

CONTENTS

THE EDITORS

EDITOR-IN-CHIEF

Mathy Doval Mezey, RN, EdD, FAAN, throughout her career as a nurse, teacher, and researcher, has focused on raising the standards of nurses caring for older adults and ensuring that people age in comfort and dignity.

Dr. Mezey received her undergraduate and graduate degrees in nursing from Columbia University. In New York City, she worked for the Visiting Nurse Service of New York and taught at Lehman College, CUNY. From 1981 to 1991, Dr. Mezey was a Professor at the University of Pennsylvania where she directed the geriatric nurse practitioner program and the Robert Wood Johnson Foundation Teaching Nursing Home Project. Beginning in September 1996, Dr. Mezey assumed the position of Director of the John A. Hartford Foundation Institute for Geriatric Nursing in the Division of Nursing at New York University. In this position, she oversees a national initiative to improve geriatric nursing. In addition, she serves as the Independence Foundation Professor of Nursing Education at NYU and also co-chairs the University Geriatrics Committee as well as the Division of Nursing's Scholarship and Development Committee.

Dr. Mezey has a long-standing interest in bioethics research and education. Her research in bioethics examines the decision-making capacity of older adults to execute a health care proxy, and the factors influencing the transfer of nursing home residents to hospitals at the end of life. With Nancy Dubler, she directs the Montefiore Medical Center NYU Nursing Certificate Program in Bioethics and the Medical Humanities. In 1994, Dr. Mezey was a member of the Ethics Task force of the President's Panel on Health Care Reform. She is Trustee Emeritus of Columbia University and sits on the Board of the Visiting Nurse Service of New York. Dr. Mezey is the author of 11 books and nearly 100 chapters and articles. Among her many awards Dr. Mezey has twice received the Geriatric Book-of-the-Year Award by the American Journal of Nursing. She is a Fellow of the Gerontological Society of America and the American Academy of Nursing.

ASSOCIATE EDITORS

Barbara J. Berkman, DSW, is the Helen Rehr/Ruth Fizdale Professor of Health and Mental Health at Columbia University School of Social Work (CUSSW). Dr. Berkman has directed 23 federally and foundation supported research projects focusing on issues in geriatric care, and is currently Principal Investigator and Director of the John A. Hartford Foundation's Geriatric Social Work Faculty Scholar's Program. She has produced over 100 books, chapters, and articles in the areas of social work in geriatric health care. She has been named a fellow of the Gerontological Society of America and of the New York Academy of Medicine. Dr. Berkman received her doctorate from CUSSW and a post-doctoral Kellogg fellowship to study geriatric health care service delivery. She received her MA from the University of Chicago School of Social Service Administration and her BA in Philosophy from the University of Michigan.

Christopher M. Callahan, MD, is the Cornelius and Yvonne Pettinga Scholar in Aging Research and Director of the Indiana University Center for Aging Research. Dr. Callahan is a Scientist in the Regenstrief Institute for Health Care and an Associate Professor of Medicine in the Division of General Internal Medicine and Geriatrics at the Indiana University School of Medicine. He is a Paul B. Beeson Physician

Faculty Scholar in Aging and his primary academic role is in aging research. Dr. Callahan's research focuses on strategies to improve the care of older adults by primary care physicians in primary care settings. His specific research interests include the recognition and treatment of late life depression and dementia. His clinical duties include the care of older adults in ambulatory, inpatient, and long-term care settings.

Terry T. Fulmer, RN, PhD, FAAN, is a Professor of Nursing at New York University Division of Nursing and Director of the New York University Division of Nursing Center for Nursing Research and co-Director for the John A. Hartford Foundation Institute for the Advancement of Geriatric Nursing Practice. She is also the Director of the Consortium of New York Geriatric Education Centers. Dr. Fulmer's program of research focuses on acute care of the elderly and specifically on the subject of elder abuse and neglect. She has received a $1.6 million grant award from the National Institute on Aging in partnership with the National Institute of Nursing for her proposal entitled "Dyadic Vulnerability/Risk Profiling for Elder Neglect." She has written extensively on both subjects, has published over 90 articles and 40 chapters. Two of her nine books have received the American Journal of Nursing Book of the Year Awards. Dr. Fulmer has been elected as a Fellow in the American Academy of Nursing, Gerontological Society of America, and the New York Academy of Medicine. She received her bachelor's degree from Skidmore College and her master's and doctoral degrees from Boston College.

Ethel L. Mitty, EdD, RN, is a Research Associate and Adjunct Assistant Professor in the Division of Nursing and Adjunct Associate Professor at the Robert F. Wagner School for Public Service, both at New York University. After twenty-five years as Director of Nursing in flagship Long-Term Care facilities, Dr. Mitty left administration and conducts bioethics research. Her research has included examination of the Patient Self-Determination Act in hospitals and nursing homes, ethics committees in nursing homes, ascertaining the decisional capacity of nursing home residents to execute a health care proxy, understanding the process and substance of decision making for hospitalization of nursing home residents, and end-of-life planning and care in assisted living facilities. Two areas of special interest to her are cross-cultural perceptions of the major principles of bioethics and health care organization ethics. Dr. Mitty is the author of the Handbook for Directors of Nursing in Long Term Care, and has authored numerous articles on long-term care and bioethics. Dr. Mitty is co-project director on an RO-1 implementation study of hospital transfer decision-making guidelines. Dr. Mitty received her Bachelor of Arts from Queens College, her Masters from Hofstra University, and her Doctorate of Education from Columbia University Teacher's College.

Gregory J. Paveza, MSW, PhD, is currently an Associate Professor in the School of Social Work and one of the founding faculty in the interdisciplinary PhD in Aging Studies Program at the University of South Florida located in Tampa, Florida. Dr. Paveza is a native of Chicago, who received his BA from Lewis College in 1969, his MSW from the University of Hawaii in 1973, and his PhD in Public Health Sciences (Psychiatric Epidemiology) in 1986 from the School of Public Health at the University of Illinois at Chicago. Dr. Paveza has been a clinical social work practitioner, a social service agency administrator, and a health sciences researcher. He is currently an active university researcher and educator. His research interests include issues related to the social consequences of caregiving and Alzheimer's disease including his specific interest in elder mistreatment in these families. Additionally he has a general interest in elder mistreatment in all of its forms and its impact on the broader aging community. He is currently a member of the Institute of Medicine/National Research Council Committee on the Training Needs of Health Professionals to Respond to Family Violence, and a member of the Leadership Council of the Mental Health and Aging Network of the American Society on Aging.

Eugenia L. Siegler, MD, FACP, is Associate Professor of Clinical Medicine in the Division of Geriatrics and Gerontology at the Weill Medical College of Cornell University and Clinical Associate Professor of

Nursing at New York University's Division of Nursing. Her research and clinical interests and publications are in the areas of geriatrics, collaborative practice, and palliative care. She is the editor of two books, one on Geriatric Interdisciplinary Team Training and one on nurse-physician collaboration. Formerly Chief of Geriatrics at the Brooklyn Hospital Center, Dr. Siegler is now Medical Director of the Geriatrics Inpatient Service at the Cornell Campus of New York Presbyterian Hospital. Dr. Siegler's AB is in Biochemical Sciences from Princeton University and her MD is from Johns Hopkins University School of Medicine. She is board certified in internal medicine and has added qualifications in geriatric medicine.

Neville E. Strumpf, PhD, RN, FAAN, is the Interim Dean of the University of Pennsylvania School of Nursing and the Edith Clemmer Steinbright Professor in Gerontology. In addition, she is the Director of the Center for Gerontologic Nursing Science at the School of Nursing. Dr. Strumpf recently completed a $1.8 million NIH/NIA clinical study of hospitalized nursing home residents and has conducted major research aimed at reducing the use of physical restraints on nursing home residents. She is currently engaged in a study of palliative care in nursing homes funded by the Robert Wood Johnson Foundation. Dr. Strumpf is the recipient of many prestigious awards including: Gerontological Nurse of the Year by the American Nurses Association, the Baxter Foundation Episteme Award (along with Dr. Lois Evans) from Sigma Theta Tau, and the Distinguished Alumni Award from New York University. She received her BS in Nursing from the State University of New York at Plattsburgh, her MS from Russell Sage College with a major in Medical-Surgical Nursing/Teaching, and her Ph.D. from New York University.

MANAGING EDITOR

Melissa M. Bottrell, MPH, PhD(c), is an Assistant Research Scientist at the New York University Division of Nursing, a doctoral candidate in bioethics and public policy, and Adjunct Professor at the Robert F. Wagner Graduate School of Public Service at New York University. She has managed and participated in a variety of research projects in geriatric nursing, and bioethics and public policy, including the Nurses Improving Care to Health System's Elders (NICHE) project, issues in transferring nursing home residents to the hospital at the end of life, gerontological nursing content in baccalaureate nursing programs, and hospital advance directives and state's advance directive policy. Her doctoral dissertation examines state level nursing home surveyors' perceptions of palliative care and the problems nursing home palliative care creates for the survey process. In 1992, she received a prestigious California Executive Fellowship to work as a legislative and budget analyst for the State of California. Ms. Bottrell received her Master of Public Health from Boston University's School of Public Health and her Bachelor of Arts in Bioethics and Public Policy from Pomona College.

ASSISTANT MANAGING EDITORS

Kanika Mody, BA, St. George's University School of Medicine, St. George, Grenada.

Abraham Brody, New York University, New York, NY.

ADVISORY BOARD

PUBLISHER'S NOTE

Elder care has become a growing concern for many families, for community and state agencies, and for the many establishments that provide living arrangements for the elderly. "Elder care" includes nursing care, medical attention, often rehabilitation, counseling, social services, and psychological support. It includes the work and expertise of professionals in several fields, as broadly as human living requires. Consequently, to provide work in elder care, service professionals, nurses, and family members need much information from related disciplines. Now, for the first time, a collection of this information has been provided in one source.

It is this *Encyclopedia of Elder Care: The Comprehensive Resource on Geriatric and Social Care*, edited by Mathy D. Mezey, RN, EdD, FAAN, and her team of distinguished associate editors. Dr. Mezey, a noted author of many books and articles, is currently the Director of the John A. Hartford Foundation Institute for the Advancement of Geriatric Nursing Practice. This institution is committed to supporting quality of health care for elderly Americans.

From numerous scientific sources, up-to-date and peer-reviewed, the contents of this unique encyclopedia will serve women and men now engaged in elder care, and also those who are training for the work.

In sum, we are proud, as the publisher, to present the new *Encyclopedia of Elder Care* to the growing number of professionals, support staff, teachers, and students who want to strengthen their knowledge and competence in the care and service of older persons.

Paul Kurtz
Publisher

PREFACE

The Encyclopedia of Elder Care was developed to provide state-of-the-art information on issues of clinical significance to health care providers and others caring for older persons. The issues covered in the encyclopedia include: care of elderly patients with acute and chronic diseases; nursing home care; rehabilitation; home care, including family-based care provision; disease prevention; health promotion and education; social services; case management; assisted living; palliative care, and more. Given the tremendous scope of topics possible for inclusion, the editors used a conceptual index to determine which topics should be included, excluded, or merged into other areas. This conceptual index, which strongly emphasizes interdisciplinary and care perspectives, encompasses the areas of Society, Community, Caregiving/Family, and Patient.

Society: The social and policy issues important in caring for elders:

- Gerontological health care providers
- Financing coverage and costs of health care for elders
- Organizations in aging

Community: The structural supports and circumstances necessary to care for elders:

- Special characteristics of various care locations
- Provider and institutional interactions across the health care continuum

Caregiving and Family: Issues important for family and professional caregivers:

- Caregiver issues
- Technology support for elders and caregivers

Patient: Issues specific to diagnosis, treatment, and disease management:

- Clinical assessment
- Prevention
- Symptom management
- Geriatric syndromes and major diseases of the elderly
- Disease management
- Issues in providing health care for specific racial/ethnic elder populations
- Pharmacology and appropriate prescribing

Because care providers need information about a variety of related issues that might influence their practice, some broader topics are also included. While most textbooks and encyclopedias provide a discipline-specific focus to diagnosis and disease management, we have worked to represent the disciplinary concerns of nursing, social work, and medicine. Dentistry, physical and occupational therapy are also well represented throughout the text. Where possible, interdisciplinary perspectives are provided, for example, the entries on incontinence discuss social management issues in caring for older adults with incontinence in addition to the diagnosis and treatment. In fact, many pieces were written jointly by professionals from more than one discipline. To strengthen this concept further, extensive cross-references are provided to direct the reader to other articles of related interest.

Electronic Resources

With the growing influence of electronic media and Internet resources on information access and publishing in general, we felt that a new print encyclopedia had to provide a bridge between the old and the new. While the content of each entry includes the minimal information a clinician will need to begin to examine a topic, we added to each piece what we hope are helpful steps into the electronic world. The Internet Resources Section includes World Wide Web sites related to the specific topic. The Internet Keywords section includes useful keywords for searching for further information on the World Wide Web and other electronic databases. When more than one entry covers related information—for instance there are multiple entries with respect to skin issues—we have included different Internet Keywords and Internet Resources for each site. However, readers are advised to look to the cross references for each article for additional sources of websites and keywords.

When choosing websites, we tried to suggest highly reputable sites that identify their sources of funding. As such, we have recommended very few commercial websites, such as those funded by pharmaceutical or other for-profit companies. When using the Internet resources, readers are advised to consider not only the keyword that they use to search with but also the power of their electronic search engine and their ability to specify the search parameters. For instance, a search engine like *All the Web* (http://www.ussc.alltheweb. com/) allows one to search easily for "all of the words," "any of the words," or "the exact phrase" that typed into the search box, without needing to construct a Boolean query. A search engine like *Altavista* (http://www.altavista.com/) allows for keyword searches, not only in the title of the page, but also in the text of the individual article. As such, one may be able to search for more specific information without also retrieving mountains of unrelated information. As always, readers are advised to use more specific words to target their search to their exact needs.

In Appreciation

A work of this size and scope could never be completed without the assistance of many individuals. Associate Editors Eugenia Siegler, MD, and Terry Fulmer, RN, PhD, provided special insight in their conceptualization of topic areas, recruiting authors, reading articles for their content and appropriateness, and pinch-hitting for the myriad of last-minute problems. Assistant Managing Editors, Kanika Mody and Abraham Brody maintained the database and provided managerial support necessary to track the progress of nearly 300 articles and to maintain contact with nearly 400 authors. Their involvement went above and beyond what is expected of undergraduate student assistants.

We are indebted to the John A. Hartford Foundation of New York for their financial support and enthusiasm for such a comprehensive work. This support demonstrates the Hartford Foundation's ongoing commitment to interdisciplinary education for health care professionals.

We especially thank Ursula Springer, PhD, and Sheri W. Sussman of Springer Publishing Company. Dr. Springer initially conceived of the need for this volume as part of her ongoing commitment to advancing the field of geriatrics. Sheri W. Sussman offered unwavering technical and psychological support, especially when the project seemed like it just might be too large to manage. Topics that are generally related to the basic sciences and sociocultural aspects of aging but are less directly related to providing health care for elders will continue to be covered in the Springer Publishing Company reference *The Encyclopedia of Aging*, third edition forthcoming in March, 2001.

Mathy D. Mezey, RN, EdD, FAAN
Editor-in-Chief

Melissa M. Bottrell, MPH, PhD(c)
Managing Editor

CONTRIBUTORS

Marsha Aaron, BS
Orthopedic Biomechanics Laboratory
Beth Israel Deaconess Medical Center
Boston, MA

Laila Abdullah, BS
Memory Disorder Clinic
Roskamp Institute
Tampa, FL

Toshiko Abe, RN, PhD
Health Care Allied Science
Tokyo Medical and Dental University Health
*   Allied*
Tokyo, Japan

Kathryn Betts Adams, MSW, LCSW-C
School of Social Work
University of Maryland
Baltimore, MD

Ronald D. Adelman, MD
Medicine
New York Presbyterian
New York, NY

Cathy Alessi
Department of Geriatrics
UCLA/Sepulveda VA Medical Center
Sepulveda, CA

Stacy S. Amano, MA
Department of Medicine
VA Medical Center of Western Los Angeles
Los Angeles, CA

Stanley J. Anderman, MSW, PhD
Center for Psychosocial Study of Health and
*   Illness*
Mailman School of Public Health at Columbia
*   University*
New York, NY

Mary Ann Anderson, RN, PhD
College of Nursing
Quad Cities Regional Program
University of Illinois at Chicago
Moline, IL

Patricia G. Archbold, RN, DNSc, FAAN
Gerontological Nursing
Oregon Health Sciences University
Portland, OR

Harriet Udin Aronow, PhD
UCLA Department of Medicine
Center for Healthy Aging
Santa Monica, CA

Carol D. Austin, PhD
University of Calgary
Faculty of Social Work
Calgary, Canada

Mary Guerriero Austrom, PhD
Psychiatry
University of Indiana School of Medicine
Indianapolis, IN

Elizabeth Ayello, PhD, RN, CS
Division of Nursing
New York University
New York, NY

Susan J. Aziz, MA
WISE Senior Services
Santa Monica, CA

Ronet Bachman, PhD
Criminology
University of Delaware
Newark, DE

Luc Baert, MD, PhD
Department of Urology
University Hospitals Leuven
Leuven, Belgium

Marjorie E. Baker, PhD
Department of Social Work
Wright State University
Dayton, OH

David Baldridge
National Indian Council on Aging
Albuquerque, NM

Lodovico Balducci, MD
Internal Medicine
College of Medicine
Tampa, FL

Amy R. Barlow, RN, MS, CRNP-Adult
School of Nursing
University of Maryland
Baltimore, MD

Virginia W. Barrett, RN, DrPH
Stroud Center
Columbia University
New York, NY

Steven Bartz, MD
Office of Geriatric Medicine
University of Cincinnati Medical Center
Cincinnati, OH

Karen Bassuk, CSW
Independent Consultant
New York, NY

Susan I. Bernatz, PhD
Behavioral Health Medical Group
Beverly Hills, CA

David E. Biegel, PhD, ACSW
Mandel School of Applied Social Sciences
Case Western Reserve University
Cleveland, OH

Joyce Black, RN, PhD
College of Nursing
University of Nebraska Medical Center
Elkhorn, NE

Steven B. Black, MD
Center for Wound Healing at Clarkson
Nebraska Health System
Lincoln, NE

Bennett Blum, MD
Geriatric Division
Park Dietz and Associates, Inc.
Tucson, AZ

Jeffrey Blustein, PhD
Department of Epidemiology and Social
    Medicine
Montefiore Medical Center
Bronx, NY

Jeremy Boal, MD
Henry L. Schwartz Department of
    Geriatric and Adult Development
The Mount Sinai Medical Center
New York, NY

Robert A. Bonomo, MD
Veterans Affairs Medical Center
Cleveland, OH

Enid A. Borden, BA, MA
Meals on Wheels Association of America
Alexandria, VA

Edgar Borgatta, PhD
Institute on Aging
University of Washington
Seattle, WA

Mathias Bostrom, MD
Hospital for Special Surgery
Weill Medical College of Cornell University
New York, NY

Melissa M. Bottrell, MPH
Division of Nursing
New York University
New York, NY

Meg Bourbonniere, MS, RN
School of Nursing
University of Pennsylvania
Philadelphia, PA

Kathryn H. Bowles, PhD, RN
School of Nursing
University of Pennsylvania
Philadelphia, PA

Risa Breckman, MSW
Wright Center on Aging
New York Presbyterian Hospital
New York, NY

Maura Brennan, MD
Department of Medicine
Baystate Medical Center
Springfield, MA

Patrick J. Brennan, MD
Department of Health
Infectious Diseases Division
University of Pennsylvania Medical Center
Philadelphia, PA

Abraham A. Brody
Division of Nursing
New York University
New York, NY

Barbara Brush, PhD, RN
Boston College School of Nursing
Chestnut Hill, MA

Cary Buckner, MD
Department of Neurology
Brooklyn Hospital Center
Brooklyn, NY

Denise Burnette, MSSW, PhD
School of Social Work
Columbia University
New York, NY

Robert N. Butler, MD
International Longevity Center USA, Ltd
New York, NY

K. R. Byju, MD
School of Medicine
Tufts University
Boston, MA

Christopher M. Callahan, MD
Regenstrief Institute for Health Care
Indiana University Center for Aging Research
Indianapolis, IN

Tanya L. Carter, OD
College of Optometry
State University of New York
New York, NY

Barbara Carty, RN, EdD
Division of Nursing
New York University
New York, NY

Christine K. Cassel, MD
Henry L. Schwartz Department of Geriatric and
  Adult Development
Mount Sinai Medical Center
New York, NY

Jean Cassidy, CNM, DrPH
University of Maryland
School of Nursing
Baltimore, MD

Joshua Chodosh, MD
Department of Medicine
UCLA
Los Angeles, CA

Margaret A. Christenson, MPH, OTR,
  FAOTA
Lifease, Inc
New Brighton, MN

Mary Jane Ciccarello, JD
Division of Aging and Adult Services
State of Utah
Salt Lake City, UT

Lissa Clark, MSN, RN
College of Nursing
University of Nebraska Medical Center
Lincoln, NE

Robert L. Clark, PhD
Department of Business Management
North Carolina State University
Raleigh, NC

Peter Cleary
American Federation for Aging Research
New York, NY

Anne Coffman, PT, MS, GCS
Rehabilitation
RehabCare Group
New Berlin, WI

Michael Corcoran, MD
Department of Neurology
University of Maryland School of Medicine
Baltimore, MD

Constance Saltz Corley, PhD, LCSW
School of Social Work
University of Maryland
Baltimore, MD

Barbara Corrigan, MS, RN, CS
Genesis Eldercare
Agawam, MA

Kenneth E. Covinsky, MD, MPH
Division of Geriatrics and the Center on Aging
University of California
San Francisco VA Medical Center
San Francisco, CA

Fiona Crawford, PhD
Roskamp Institute
University of South Florida College of Medicine
Tampa, FL

Craig Curry, MD
Division of Geriatrics
Bellevue Hospital
New York, NY

Sara J. Czaja, PhD
Psychiatry and Behavioral Sciences
University of Miami School of Medicine
Miami, FL

JoAnn Damron-Rodriguez, LCSW,
VA Geriatric Research
*   Education and Clinical Center*
GRECC West Los Angeles
UCLA School of Public Policy and Social
*   Research*
Los Angeles, CA

Carol M. Davis, EdD, PT
Division of Physical Therapy
University of Miami School of Medicine
Coral Gables, FL

Mia Defever, PhD
Policy Forum Center for Health Services and
*   Nursing Research*
School of Public Health
Catholic University of Leuven
Leuven, Belgium

Sabina De Geest, RN, PhD
Institute of Nursing Science
University of Basel
Basel, Switzerland

Dirk De Ridder, MD, PhD, FEBU
Department of Urology
University Hospitals of Leuven
Leuven, Belgium

Eddy Dejaeger, MD, PhD
Department of Geriatrics
University of Leuven
Leuven, Belgium

Richard D. Della Penna, MD
Continuing Care Services
Kaiser Permanente San Diego
San Diego, CA

Cheryl A. Dellasega, PhD, GNP
Division of General Internal Medicine
The Pennsylvania State University
College of Medicine
Hershey, PA

Gretchen J. Diefenbach, PhD
Department of Psychiatry and Behavioral
*   Sciences*
Mental Sciences Institute
University of Texas-Houston Medical School
Houston, TX

Rose Ann DiMaria-Ghalili, PhD, RN, CNSN
West Virginia University School of Nursing
Charleston Division
Charleston, WV

Annemarie Dowling-Castronovo, MA, GNP-CS
Division of Nursing
New York University
New York, NY

Nancy N. Dubler, LLB
Department of Epidemiology and
Social Medicine
Division of Bioethics
Montefiore Medical Center
Bronx, NY

Robert W. Duff, PhD
Department of Social and Behavioral Sciences
University of Portland
Portland, OR

Mitchell L. Dul, OD
College of Optometry
State University of New York
New York, NY

Barbara Edlund, PhD, RN, CS, ANP
College of Nursing
Medical University of South Carolina
Charleston, SC

David J. Ekerdt, PhD
Gerontology Center
University of Kansas Medical Center
Lawrence, KS

Charles A. Emlet, PhD, ACSW
Social Work Program
University of Washington, Tacoma
Tacoma, WA

Steven J. Ersser, PhD, BSC, RGN, Cert TH Ed
School of Nursing and Midwifery
University of Southhampton
Southhampton, England

David V. Espino, MD
Department of Family Practice
University of Texas Health Center
San Antonio, TX

Ronald l. Ettinger, BDS, MDS, DDSc
Department of Prosthodontics
Dows Institute for Dental Research
Iowa City, IA

Lois K. Evans, RN, DNSc, FAAN
Academic Nursing Practice
School of Nursing
University of Pennsylvania
Philadelphia, PA

Georges C.M. Evers, PhD, RN, FEANS
Centre for Health Services and Nursing Research
Catholic University of Leuven
Leuven, Belgium

Linda Farber Post, JD, BSN, MA
Department of Epidemiology and Social
Medicine
Division of Bioethics
Montefiore Medical Center
Bronx, NY

Naomi Feil, BSc, MSW
Validation Training Institute
Cleveland, OH

Gerda G. Fillenbaum, PhD
Center for the Study of Aging and Human
Development
Duke University Medical Center
Durham, NC

Sanford I. Finkel, MD
Council for Jewish Elderly
Chicago, IL

Christine Fitzgerald, RRT, MHS, PhD
Department of Cardiopulmonary and Diagnostic
Imaging
Quinnipiac College
Hamden, CT

Mary FitzGerald, RN, DN, Cert Ed, MN, PhD, FRCNA
The Department of Clinical Nursing
The University of Adelaide
Adelaide, Australia

Sheila FitzSimmons-Scheurer, RN, MA, CS, GNP
Division of Geriatrics and Gerontology
New York Presbyterian Hospital
New York, NY

Ellen Flaherty, RN, MSN
Division of Nursing
New York University
New York, NY

Mary Ann Forciea, MD
Division of Geriatric Medicine
School of Medicine
University of Pennsylvania
Philadelphia, PA

Marquis D. Foreman, RN, PhD, FAAN
Medical-Surgical Nursing
College of Nursing
University of Illinois at Chicago
College of Nursing (M/C802)
Chicago, IL

Barry Fortner, PhD
Rush-Presbyterian-St. Luke's Medical Center
Chicago, IL

Michael Freedman, MD
Geriatric Medicine
NYU Medical Center
New York, NY

Robert Friedland, PhD
National Academy on Aging
Washington, DC

James F. Fries, MD
Department of Immunology and Rheumatology
School of Medicine
Stanford University
Stanford, CA

Terry T. Fulmer, RN, PhD, FAAN
Division of Nursing
New York University
New York, NY

Sandra J. Fulton Picot, PhD, RN, FAAN
Sonya Ziporkin Gershowitz Chair in Gerontology
University of Maryland, Baltimore
Baltimore, MD

Sandy B. Ganz, PT, MS, GCS
Department of Rehabilitation
Amsterdam Nursing Home
New York, NY

Margaretta E. Gennantonio, MD
Office of Geriatric Medicine
University of Cincinnati Medical Center
Cincinnati, OH

Linda K. George, PhD
Center for the Study of Aging and Human Development
Duke University Medical Center
Durham, NC

Kimberly S. Glassman, RN
New York University Medical Center
New York, NY

Ann Goeleven
Department of ENT
University Hospitals
Katholieke Universiteit Leuven
Leuven, Belgium

Robyn L. Golden, MSW, LCSW
Council for Jewish Elderly
Chicago, IL

Marsha E. Goodwin-Beck, RN-C, MSN, MA
Geriatrics and Extended Care
Department of Veterans Affairs
Washington, DC

Elaine S. Gould, MSW
Division of Nursing
New York University
New York, NY

Stuart Green, MD
Department of Rheumatology
Brooklyn Hospital Center
Brooklyn, NY

Lynn B. Greenberg, RD, MS, LD
Nutrition and Food Service
VA Maryland Health Care System
Baltimore, MD

Sherry A. Greenberg, MSN, RN, CS
Division of Nursing
New York University
New York, NY

Michele G. Greene, DrPH
Department of Health and Nutrition Sciences
Brooklyn College
Brooklyn, NY

Robert Greenwood
American Association of Homes and Services for
the Aged
Washington, DC

Margaret M. Grisius, DDS
Department of Oral Medicine
School of Dental Medicine
University of Pennsylvania
Philadelphia, PA

Murray Grossman, MD
Department of Neurology
Hospital of the University of Pennsylvania
Philadelphia, PA

Barry Gurland, MD
Faculty of Medicine and the New York State
Psychiatric Institute
Stroud Center for the Study of Quality of Life
Columbia University
New York, NY

Lisa P. Gwyther, MSW
Family Support Program
Duke Center for Aging
Durham, NC

Kathy M. Haag, RN
Indiana University School of Medicine
Indianapolis, IN

Bernadette B. Hackett, LCSW
The Connecticut VNA
North Haven, CT

Alice H. Hedt, MUA
National Long Term Care Ombudsman Resource
Center
National Long Term Care Ombudsman Center
Washington, DC

Arthur E. Helfand, DPM, DABPPH
Community Health, Aging and Health Policy
Temple University School of Podiatric Medicine
Philadelphia, PA

Lelia B. Helms, PhD, JD
Planning, Policy, and Leadership
College of Education
University of Iowa
Iowa City, IA

Jon Hendricks, PhD
Department of Sociology
Oregon State University
Corvallis, OR

Hugh C. Hendrie, MB, ChB
Department of Psychiatry
Indiana University School of Medicine
Indianapolis, IN

Alice Herb, JD, LLN
Humanities in Medicine
SUNY HSC at Brooklyn
Brooklyn, NY

Dorothy G. Herron, PhD, RN, CS
School of Nursing
University of Maryland
Baltimore, MD

Elizabeth E. Hill-Westmoreland, RN, MS, CS
School of Nursing
University of Maryland at Baltimore
Baltimore, MD

Immanuel K. Ho, MD
Gastroenterology
Brooklyn Hospital Center
Brooklyn, NY

Helen Hoenig, MD, MPH
Physical Medicine and Rehabilitation Services
Veterans Administration Medical Center
Durham, NC

Cynthia Holzer, MD
Office of Geriatric Medicine
University of Cincinnati Medical Center
Cincinnati, OH

Cynthia Hughes Harris, EdD, OTR, FAOTA
School of Allied Health Sciences
Florida A&M University
Tallahassee, FL

Kathryn Hyer, DrPA, MPP
USF Training Academy on Aging
University of Southern Florida
Tampa, FL

Naoki Ikegami, MD
Department of Health Policy and Management
School of Medicine
Keio University
Tokyo, Japan

Pamela Ingham
American Geriatrics Society
New York, NY

Margaret B. Ingraham, BA, MA
Meals on Wheels Association of America
Alexandria, VA

Sharon K. Inouye, MD, MPH
Yale University School of Medicine
New Haven, CT

Yakov Iofel, MD
Department of Medicine
New York Presbyterian Hospital
New York, NY

Nancy S. Jecker, PhD
Department of Medical History and Ethics
University of Washington
Seattle, WA

Elaine Jensen Amella, RN, MSN, PhD, CS-GNP
College of Nursing
Medical University of South Carolina
Charleston, SC

Jerry C. Johnson, MD
Geriatric Medicine Division
School of Medicine
University of Pennsylvania
Philadelphia, PA

Julie E. Johnson, RN, PhD
Orvis School of Nursing
University of Nevada, Reno
Reno, NV

Rebecca A. Johnson, PhD, RN
Sinclair School of Nursing
University of Missouri-Columbia
Columbia, MO

Milla Karev, MD
Office of Geriatric Medicine
University of Cincinnati Medical Center
Cincinnati, OH

Jason H. T. Karlawish, MD
Center for Bioethics and Alzheimer's Disease
Division of Geriatric Medicine
University of Pennsylvania
Philadelphia, PA

Mary Katsikitis, PhD, MAPS
Department of Psychiatry
The University of Adelaide
Adelaide, Australia

Jeanie Kayser-Jones, RN, PhD, FAAN
Department of Physiological Nursing and
*   Medical Anthropology Program*
University of California
San Francisco School of Nursing
San Francisco, CA

John R. Kelly, PhD
University of Illinois at Urbana-Champaign
Champaign, IL

Gary J. Kennedy, MD
Division of Geriatric Psychiatry
Montefiore Medical Center
Bronx, NY

Denis Keohane, MD
Department of Medicine
New York Presbyterian Hospital
New York, NY

Salil Khandwala, MD
Division of Urogynecology and Pelvic
  Reconstruction
Department of Obstetrics and Gynecology
University of Maryland
Baltimore, MD

Mary T. Knapp, MSN, GNP, NHA, FAAN
ZA Consulting
Jenkintown, PA

Rosalind Kopfstein, DSW, MSW
School of Health Sciences
Western Connecticut State University
Danbury, CT

Natalya Kozlova, MD
Section of Geriatric Medicine
Brooklyn Hospital Center
Brooklyn, NY

B. Josea Kramer, PhD
Sepulveda Geriatric Research Education Clinical
  Center
VA Greater Los Angeles Healthcare System
Sepulveda, CA

Betty J. Kramer, PhD
School of Social Work
University of Wisconsin-Madison
Madison, WI

Reto W. Kressig, MD
Geneva University Hospitals
Geneva, Switzerland

John A. Krout, PhD
Gerontology Institute
Ithaca College
Ithaca, NY

Jerome E. Kurent, MD
Center for the Study of Aging
Medical University of South Carolina
Charleston, SC

Lenore H. Kurlowicz, PhD, RN, CS
School of Nursing
University of Pennsylvania
Philadelphia, PA

Mark Lachs, MD, MPH
Department of Medicine
Weill Medical College of Cornell University
New York, NY

Anna Lamnari, MD
Department of Medicine
New York Presbyterian Hospital
New York, NY

Melinda S. Lantz, MD
The Jewish Home and Hospital
New York, NY

Donna Larkin
Joint Commission on Accreditation of Healthcare
  Organizations
Oakbrook Terrace, IL

Beth Latimer, RN, MA, GNP
Division of Nursing
New York University
New York, NY

Sylvie Lauque, RN
Service de Medecine Interne et de Gérontologie
  Clinique
Centre Hospitalier Universitaire de Toulouse
Tolouse, France

M. Powell Lawton, PhD
Eward and Esther Polisher Research Institute
Philadelphia Geriatric Center
Philadelphia, PA

Haejung Lee
College of Medicine Nursing Dept.
Pusan National University
Pusan, South Korea

Mary Lee Wong, MD
Medicine Specialties
Beth Israel Medical Center
New York, NY

Gary L. LeRoy, MD
East Dayton Health Center
Dayton, OH

Felix W. Leung, MD
Division of Gastroenterology
Sepulveda VA Medical Center
Sepulveda, CA

Phoebe S. Liebig, PhD
Ethel Percy Andrus Gerontology Center
University of Southern California
Los Angeles, CA

Jo E. Linder, MD
Portland, ME

Veronica LoFaso, MD
Department of Medicine
New York Presbyterian Hospital
New York, NY

Meridean L. Maas, PhD, RN, FAAN
College of Nursing
University of Iowa
Iowa City, IA

George L. Maddox, PhD
Long Term Care Resources Program
Center for the Study of Aging
Duke University Medical Center
Durham, NC

Richard J. Madonna, OD, MA, FAAO
College of Optometry
State University of New York
New York, NY

Kevin J. Mahoney, PhD
School of Social Work
Boston College
Chestnut Hill, MA

Susan Markey
American Society on Aging
San Francisco, CA

Joanne Marlatt-Otto, MSW
Adult Protection/Elder Rights
Colorado Department of Human Services
Denver, CO

Karen S. Martin, RN, MSN, FAAN
Martin Associates
Omaha, NE

Mary J. Marzullo, RN, BSN
Lifeline, Systems
Larchmont, NY

Edward J. Masoro, PhD
University of Texas Health Science Center
San Antonio, TX

Mary Ann Matteson, PhD, RN, CNS, FAAN
School of Nursing
University of Texas Science Center at San
   Antonio
San Antonio, TX

Cathy McEvoy, PhD
Department of Gerontology
University of South Florida
Tampa, FL

Elizabeth E. McGann, RN, DNS, CS
School of Nursing
Quinnipiac College
Hamden, CT

Eileen M. McGee, RN, MSN
Pine Street Inn Nurses Clinics
Boston, MA

Joan Meehan
American Nurses Association
Washington, DC

Kurt P. Merkelz, MD
Department of Family Practice
University of Texas Health Center
San Antonio, TX

Mathy D. Mezey, RN, EdD, FAAN
Division of Nursing
New York University
New York, NY

Jean-Pierre Michel, MD
Département de Gériatrie
Institutions Universitaires de Gériatrie de
*   Genève*
Genève, Switzerland

Patricia Miller, EdD, OTR, FAOTA
Programs in Occupational Therapy
Columbia University
New York, NY

Lorraine C. Mion, RN, PhD
Department of Nursing
Department of Geriatrics
Mount Sinai Medical Center
New York, NY

Ahmed A. Mirza, MD
Department of Medicine
New York Presbyterian Hospital-Cornell Medical
*   Center*
New York, NY

Ethel Mitty, RN, EdD
Division of Nursing
New York University
New York, NY

Roger D. Mitty, MD
Gastroenterology
St. Elizabeths Medical Center of Boston
Boston, MA

Kanika P. Mody
Division of Nursing
New York University
New York, NY

Abraham Monk, PhD
School of Social Work
Columbia University
New York, NY

Catherine Morency, MS, RN
Gerontology Division
Harvard Medical School
Boston, MA

Amy Morgan, PharmD
Philadelphia College of Pharmacy
University of the Sciences in Philadelphia
Philadelphia, PA

Lynne Morishita, RNCS, GNP, MSN
Department of Family Practice and Community
*   Health*
University of Minnesota
Edina, MN

John E. Morley, MB, BCH
Department of Geriatrics
St. Louis University
St. Louis, MO

Laura Mosqueda, MD
Department of Geriatrics
University of California, Irvine
Orange, CA

Charles P. Mouton, MD
Department of Family Practice
University of Texas Health Center
San Antonio, TX

Michael Mullan, MD, PhD
Roskamp Biologic Psychiatry Research Lab
College of Medicine
Tampa, FL

Michael D. Murray, PharmD, MPH
Regenstrief Institute
Purdue University School of Pharmacy
Indianapolis, IN

Carol M. Musil, PhD
School of Nursing
Case Western Reserve University
Cleveland, OH

Robert J. Myers, LLD
National Commission on Social Security Reform
Washington, DC

Matthias J. Naleppa, PhD
School of Social Work
Virginia Commonwealth University
Richmond, VA

Joseph B. Narus, GNP, MA, RN, CS
Michael Callen-Audre Lorde Community
    Health Center
New York, NY

Robert A. Neimeyer, PhD
Department of Psychology
University of Memphis
Memphis, TN

Catherine O'Keefe, Med, CTRS
Department of Health
PE and Leisure Services
University of South Alabama
Mobile, AL

Corrie J. Odom, PT, DPT, ATC
School of Medicine
Duke University
Durham, NC

Morris A. Okun, PhD
Department of Psychology
Arizona State University
Tempe, AZ

M. Louay Omran, MD
Division of Geriatric Medicine
St. Louis University Health Sciences Center
St. Louis, MO

Kathleen A. Ondus, MSN, RN, CS
Geriatric Services
Southwest General Health Center
Middleburg Heights, OH

Martin Orrell
Psychiatry and Behavioral Sciences
University College of London
London, England

Tom Otwell
American Association of Retired Persons
Washington, DC

Robert M. Palmer, MD, MPH
Department of Geriatrics
Cleveland Clinic Foundation
Cleveland, OH

Maria Pappas-Rogich, DrPH, RN-C
Institute of Technology
State University of New York
Utica, NY

Gregory J. Paveza, MSW, PhD, ACSW
School of Social Work
University of South Florida
Tampa, FL

Alan Pearson, RN, MSc, PhD
Department of Clinical Nursing
School of Nursing
LaTrobe University
Buncloora, Victoria, Australia

David A. Peterson, PhD
Leonard Davis School of Gerontology
Ethel Percy Andrus Gerontology Center
Los Angeles, CA

Kathy Pierce Bradley, EdD, OTR
Department of Occupational Therapy
Medical College of Georgia
Augusta, GA

Patricia P. Pine, PhD
New York State Office for the Aging
Albany, NY

Rosemary Polomano, PhD, RN, FAAN
Center for Nursing Research
Milton S. Hershey Medical Center
Hershey, PA

Carol Porter, PhD, RD, FADA
University of California
San Francisco Medical Center
San Francisco, CA

Valery A. Portnoi, MD
Division of Geriatrics
Beth Israel Medical Center
New York, NY

Lidia Pousada, MD, FACP, AGSF
Division of Geriatrics and Gerontology
Sound Shore Medical Center of Westchester
New York Medical College
New Rochelle, NY

Christopher M. Powers, PhD
Biokinesiology and Physical Therapy
University of Southern California
Los Angeles, CA

Nicholas G. Procter, RN, PhD, MANZCMHN
Division of Health Sciences
University of South Australia
North Terrace, Australia

Charles T. Pu, MD
Department of Geriatrics
Massachusetts General Hospital
Boston, MA

Lauretta Quinn, PhD, RN
Medical Surgical Nursing
College of Nursing
The University of Illinois at Chicago
Chicago, IL

Barrie L. Raik, MD
Department of Medicine
New York Presbyterian Hospital
New York, NY

Phyllis Ramzel, RN, MA
Phyllis Ramzel & Associates, Inc.
Syosset, NY

Amie L. Rang, BA
Social Work Program
University of Washington, Tacoma
Tacoma, WA

Satish S. C. Rao, MD
Department of Internal Medicine, GI Division
School of Medicine
University of Iowa
Iowa City, IA

Michael Reinemer
National Council on Aging
Washington, DC

Barbara Resnick, RN, PhD, CRNP
School of Nursing
University of Maryland
Baltimore, MD

Mohammed Reyazuddin
Gerontological Society of America
Washington, DC

Sandra L. Reynolds, PhD
Gerontology
University of South Florida
Tampa, FL

Carol M. Rhodes, ARNP, MS, CFNP, CCRN, CEN
Family Nurse Practitioner Distance Learning
  Program
SUNY Stony Brook
Stony Brook, NY

Louis B. Rice, MD
Medical Service
Veterans Administration Medical Center,
  Cleveland
Cleveland, OH

Dan Richards, MD
Memory Disorder Clinic
Roskamp Institute
Tampa, FL

Sara E. Rix, PhD
Public Policy Institute
American Association of Retired Persons
Washington, DC

Jeffrey Scott Roberts, PhD
Health Services Research and Development
Serious Mental Illness Treatment Research and
  Evaluation Center
Ann Arbor, MI

Lois Roelofs, Phd, RN
School of Nursing
Trinity Christian College
Palos Heights, IL

Anissa T. Rogers, MA, MSW, PhD
Department of Social and Behavioral Sciences
Social Work Program
University of Portland
Portland, OR

Peri Rosenfeld, PhD
Division of Nursing
New York University
New York, NY

Mona Rosenthal, MPH
Kaiser Permanente Southern California
Pasadena, CA

Brenda Rosenzweig, MD
Division on Aging
Harvard Medical School
Boston, MA

Mary Ann Rosswurm, EdD, RN, FAAN
Program Evaluation and Research
The Drake Center
Cincinnati, OH

Kathy L. Rush, RN, PhD
Honorary Research Associate
University of New Brunswick
Fredericton, New Brunswick, Canada

Carlene Russel, MS, RD, LD
Mercy Health Center, North Iowa
Mason City, IA

Debra Saliba, MD, MPH
Geriatric Research Education and Clinical
    Center
VA Greater Los Angeles Health
    Care System
Los Angeles, CA

Julie M. Sazant, MSW
Division of Nursing
New York University
New York, NY

Alessandra Scalmati, MD, PhD
UJA Montefiore Aging and Memory Center
Bronx, NY

Barbara W. Schneider, BSN, MA
Independent Consultant
Elverson, PA

Edna P. Schwab, MD
Division of Geriatric Medicine
School of Medicine
University of Pennsylvania
Philadelphia, PA

Mary Shelkey, RN, GNP
School of Nursing
Seattle University
Seattle, WA

Andrea Sherman, PhD
Division of Nursing
New York University
New York, NY

Chol Shin, MD, PhD, FCCP
Pulmonary and Critical Care
Korea University
AnSan-Si, South Korea

Heajong Shin, PhD, MW
Department of Social Welfare
College of Social Science
Soonchunhyang University
Shinchang-myun, South Korea

Kyung Rim Shin, RN, EdD, FAAN
College of Nursing Science
Ewha Women's University
Seoul, Korea

Karolyn Siegel, PhD
Mailman School of Public Health
Columbia University
New York, NY

Eugenia L. Siegler, MD, FACP
Division of Geriatrics and Gerontology
Weill Cornell Medical College
Brooklyn, NY

Lori Simon-Rusinowitz, PhD
University of Maryland
College Park, MD

Carole Smyth, RNC, ANP, GNP
UJA Montefiore Aging and Memory Center
Bronx, NY

Rhayun Song, PhD, RN
Nursing Department
College of Medicine
Soonchnhyang University
Con An, South Korea

Mort Soroka, PhD
College of Optometry
State University of New York
New York, NY

Elaine Souder, RN, PhD
College of Nursing
University of Arkansas for Medical Sciences
Little Rock, AR

Aimee Spector
Psychiatry and Behavioral Sciences
University College London
London, England

Ann Marie Spellbring, RN, PhD
Adult Health Nursing
University of Maryland at Baltimore
Baltimore, MD

Melinda Anne Stanley, PhD
Psychiatric and Behavioral Sciences
University of Texas Medical School
Houston Health Science Center
Houston, TX

Alan M. Stark, DDS, ASGD
Department of Oral Medicine
Temple University School of Dentistry
Philadelphia, PA

Els Steeman, RN, MSN
Center for Health Services and Nursing Research
Catholic University of Leuven
Leuven, Belgium

Richard Stein, MD
Department of Medicine
Brooklyn Hospital Center
Brooklyn, NY

Leroy O. Stone, PhD
Analytical Studies Branch
Statistics Canada
Ottawa, Ontario, Canada

Gordon F. Streib, PhD
Department of Sociology
University of Florida
Gainesville, FL

Neville E. Strumpf, PhD, RN, FAAN
School of Nursing
University of Pennsylvania
Philadelphia, PA

Jeannette Takamura, MSW, PhD
U.S. Department of Health and Human Services
Washington, DC

Karen Talerico, PhD
Office of Research Development and Utilization
School of Nursing
Oregon Health Science University
Portland, OR

Catherine J. Tompkins
Association for Gerontology in Higher Education
Washington, DC

John A. Toner, EdD, MPhil
Geriatric Psychiatry Residency and Fellowship
  Programs
Columbia University
The Stroud Center
New York, NY

Colin Torrance, RN, PhD
Department of Nursing
The Alfred Hospital
Prahran, Australia

Mary Grace Umlauf, RN, PhD
School of Nursing
University of Alabama
Tuscaloosa, AL

R. Alexander Vachon, PhD
Formerly Office of Senator Bob Dole
Washington, DC

Glen Van Andel, ReD, CTRS
Department of Physical Education and
*   Recreation*
Calvin College
Grand Rapids, MI

Ben Van Cleyenbreugel, MD
Department of Urology
University Hospitals of Leuven
Leuven, Belgium

Luc Van de Ven, PhD
Department of Geronto-Psychiatry
University Hospitals of Leuven St. Rafael
Leuven, Belgium

Bruno Vellas, MD
Service de Medicine Interne et de Gérontologie
*   Clinique*
Centre Hospitalier Universitaire de Toulouse
Auzeville-Tolosane, France

Maria Vezina, RN, EdD
Department of Nursing Education
The Mount Sinai Hospital
New York, NY

Nancy S. Wadsworth, MSSA, PhDc
Case Western Reserve Geriatric Education
*   Center*
Case Western Reserve University
Cleveland, OH

Martha S. Waite, MSW, LCSW
Interprofessional Team Training and
*   Development*
Geriatric Research Education and Clinical Care
*   Center*
VA Greater Los Angeles Health Care System
Los Angeles, CA

Meredith Wallace, PhDc, RN
Division of Nursing
New York University
New York, NY

Kenneth Walsh, RPN, RGN, Bnurs
Department of Clinical Nursing
The University of Adelaide
Adelaide, Australia

Camille B. Warner, PhDc
Sociology Department
Case Western Reserve University
Cleveland, OH

Renee Warshofsky-Altholz, CSW
Patient Care Services
Beth Israel Medical Center
New York, NY

Terri E. Weaver, Rn, PhD, CS, FAAN
Center for Urban Health Research
School of Nursing
University of Pennsylvania
Philadelphia, PA

Jeanne Y. Wei, MD, PhD
Department of Gerontology
Harvard Medical School
Boston, MA

Thelma Wells, RN, PhD, FAAN, FRCN
Center for Health Science
School of Nursing
University of Wisconsin
Madison, WI

Peter J. Whitehouse, MD, PhD
University Alzheimer Center
Case Western Reserve University
Cleveland, OH

Darryl Wieland, PhD
Geriatric Services
Palmetto Richland Memorial Hospital
Columbia, SC

Joshua M. Wiener, PhD
The Urban Institute
Washington, DC

Stacy Schantz Wilkins, PhD, ABPP
Department of Medicine
School of Medicine
University of California Los Angeles
Los Angeles, CA

Tim J. Wilkinson, MB, ChB, FRACP
Christchurch School of Medicine
Princess Margaret Hospital
Christchurch, New Zealand

Kathy Wilson
American Medical Directors Association
Columbia, MD

Nancy Wilson, LMSW, MA
Huffington Center on Aging
Baylor College of Medicine
Houston, TX

Yukari Yamada, RN, MHSc
Department of Health Policy and Management
Keio University School of Medicine
Tokyo, Japan

Donna L. Yee, PhD
National Asian Pacific Center on Aging
Seattle, WA

Janet Yellowitz, DMD, MPH
Oral Health Care Delivery
School of Dentistry
University of Maryland
Baltimore, MD

Nancy Xiaoshuang Yin, MD
Geriatric Medicine
Weill Medical College of Cornell University
New York, NY

Cora D. Zembrzuski, RN, MSN
The Connecticut VNA
North Haven, CT

Dottie Zoller
International Psychogeriatric Association
Northfield, IL

LIST OF ENTRIES

THE
ENCYCLOPEDIA
OF ELDER CARE

A

ABUSE

See
Elder Neglect
Financial Abuse

ACCESS TO CARE

The social responsibility to provide essential health and social services relies on equitable access to care. Growing demands on the healthcare system predicated predominantly on the increase in the older population, chronic illness, and disability challenge the current delivery system to provide access to acute and long-term care. Inequity in access to health care services, appropriate use of those services, quality of the services provided, and escalating costs of care are four major healthcare policy issues that will compete for attention in the new millennium (Gold, 1998). Perversely, increased focus on the cost of care overlooks widening disparities in access to care.

Health insurance coverage and proximity to providers are the predominant measures of equitable access (Gold, 1998). Residents of inner city and rural areas experience access barriers to healthcare coverage. Most older persons have basic healthcare coverage/benefits through Medicare (97% of elders) and 70% of elders have additional coverage through private insurance. Only 7.8% receive Medicaid, and 2% of elders have no coverage at all. While the number of elders who have Medicaid coverage is small, the numbers vary greatly when broken down by race. Only 5.4% of white elders, as compared to 19.3% African American, and 27.6% Hispanic elders are covered by Medicaid (Centers for Disease Control and Prevention, National Center for Health Statistics, 2000). Integrating benefits through managed care may affect those elders with higher cost chronic conditions.

Differential access must be considered as a contributing factor in health status, service utilization rates, costs of care, treatment trajectories, and intervention outcomes (Newcomer & Benjamin, 1997). Race and ethnicity are important predictors of poor health care access for older persons (Miller et al., 1997; Wallace et al., 1998). The result is often higher morbidity and mortality rates among certain populations. For instance, black women have lower rates of screening mammography resulting in later breast cancer detection, which may contribute to their higher death rates from this disease (Yee & Capitman, 1996).

Defining Access

Defining and measuring access requires differentiating access from related dimensions of healthcare delivery such as service availability, utilization, appropriateness, quality, and satisfaction. Availability simply means that the services exist and there is a possibility of use. Access implies that the available services are approachable and that the means to use them is reachable. Utilization is an essential element of measurement of access. However, the related elements of need and appropriateness of the service must be considered. Finally, quality and satisfaction are related to access. Poor quality services may generate high utilization rates when the care is urgently needed. However, for less needed services, low patient satisfaction may lead to low utilization of available and otherwise accessible services.

The measurement of access can be constructed by asking two questions: What is the utilization rate as measured by the number of encounters of a particular service? What is the estimate of prevailing need within the target population differentiated by age, gender, and racial/ethnic group? Unmet need or inequities in access are measured by calculating the estimate of the condition in the given population minus the number of encounters or utilization rate (Newcomer & Benjamin, 1997). Access evaluation relies on agency documentation of encounters and claims, and regional or national estimates of need.

Access must be defined in specific relationship to a type of service. For older adults, important indicators of access to care include physicians, hospital, rehabilitation, nursing home, other therapeutic programs, skilled/non-skilled home care, specialized living arrangements, and other social services. Accessibility within a community are measured by utilization rates in relationship to the total population in need.

Appropriate utilization of a particular level of care should be assessed at the individual and community level. Data collection and analysis could look at, for example, placement of individuals in nursing homes based on functional level, and emergency room versus community-based care for minority elders (Damron-Rodriguez et al., 1994). In addition to evaluating access in relationship to major type of service or level of care, it can also be evaluated in relationship to a type of intervention, for example, preventive health services such as immunizations and health screenings. Preventive, acute, and chronic care services for the elderly present different challenges of both availability and accessibility.

Acceptability

Despite its accessibility, a service may not be acceptable or preferred. Acceptability connotes usefulness to the population in question. Low utilization rates by minority elderly raise questions about preference for, as well as availability of and access to, other forms of family and community based care for ethnic elders. Structural barriers associated with costs of care, intensity and duration of service, and location preclude drawing the conclusion that preference alone determines minority groups' utilization of care settings (Damron-Rodriguez et al., 1994).

Quality of care and consumer satisfaction are related to acceptability and accessibility. A service may be acceptable or unacceptable to a particular population of elders. If a service is acceptable, then the elders may be very satisfied or only mildly satisfied based on quality and other factors. The technical and professional quality of an acceptable service may be technically very high but yield low consumer satisfaction if it is provided in an unacceptable manner. Thus, access is an essential but not sufficient condition to assure utilization of services.

Recommendations for Appropriate, Accessible, and Acceptable Care

Providers aiming to increase access to services should first assess older adult population characteristics and service needs, designing programs that reduce both structural and cultural barriers.

Appropriate Care: Population characteristics that must be considered in determining appropriate services include health status, functional level (cognitive, physical, and social), and acuity and chronicity of health conditions. Program characteristics include comprehensiveness and intensity of the intervention, the dimensions of assessment, the disciplines involved, the rehabilitation components, and the length of treatment.

Accessible Care: Population characteristics that must be considered in relationship to the accessibility of appropriate service access include income, health care coverage, immigration status, residence, neighborhood, level of disability, and living arrangements. Program characteristics include affordability, desired hours of operation, accommodating location or available transportation, timeliness of service provided, minimal intake procedures and paperwork, and outreach and information. Failure to consider population characteristics can lead to structural barriers that will significantly limit access.

Cultural barriers can make an accessible program or service unacceptable. Population characteristics to consider when creating acceptable services include ethnicity, language, family support systems, education, generation in this country, and acculturation. Program characteristics include cultural and language competence, and family enabling policies.

Less commonly construed as such, accessibility also relates to informal care. An individual may not have access to family care based on conditions embedded in their social support system. For example, a key family caregiver may be ill or otherwise unable to provide personal or instrumental assistance, there may be no living family, or the family may be geographically dispersed. Information and outreach services should be targeted to elders with limited access to informal support.

Access maximization means providing the right services for the right population at the right time and place in a manner that ensures quality and satisfaction. Underlying what providers can do to create better access locally are regional, state, and federal policy initiatives needed to ensure equitable access through adequate healthcare coverage.

JoAnn Damron-Rodriguez

See Also
Future of Care
Information Transfer
Long-Term Care Financing
Medicaid
Medicare
Medicare Managed Care
Risk Assessment and Identification
Support Groups

Internet Key Words
Assessment
Barriers
Healthcare
Utilization

Internet Resources
Agency for Healthcare Research and Quality: Quality Assessment
http://www.ahcpr/qual/

U.S. Centers for Disease Control and Prevention, National Center for Health Statistics
http://www.cdc.gov/nchs

U.S. Department of Health and Human Services
http://www.hhs.gov/policy

ACTIVE LIFE EXPECTANCY

Life expectancy is the average length of life that a particular group may expect to live. For example, at the beginning of the 1990s, the estimated average life expectancy at birth for white males was 72.7 years and for white females, 79.2 years; by contrast, average life expectancy at birth for black males was 64.8 years and for black females 73.5 years. These estimates are computed by knowing the age-specific death rates for a group and applying these death rates to the group for each successive age period. It is possible to estimate the average remaining life for a group at any point in time. For example, for the population of the United States at the beginning of the 1990s, the average remaining life expectancy of persons 65 years old was 17.2 years; for those 75 years old, 10.9 years; for those 85 and older, 6.2 years.

Information on age-specific death rates is essential to understanding changes in the age distribution of the population. Many general demographic studies have indicated that the population of the United States, in fact, of all societies is aging; that is, life expectancy is increasing and more persons are living to older ages. Since health and other problems are associated with older ages, the trend of increasing numbers and proportions of older persons has been of interest not only to demographers and other social scientists, but also to policy makers and social planners. Particularly if extending the period of old age means extending the period that the average person will be *dependent* on others and require care, policy makers and social planners become concerned with what the changes mean in terms of economic and other resources required. Thus, an important question centers on whether dependency increases as life expectancy increases, and this question is focused particularly on the active life expectancy at older ages, i.e., the nondependent period.

Fries and Crapo (1981) advanced a major theoretical formulation that bears on the period of dependency as life expectancy increases. They note that longevity, or the upper limit of life, is not likely to change materially in the foreseeable future. As persons live longer because disease, disorders, and disabilities are reduced or delayed, the problems can only happen in a shorter and shorter period of remaining life. The conclusion then follows that the period of dependency is shortened or compressed as life expectancy is extended. There have been many critiques of this theory and the facts on which the theory presumably rests (Grundy, 1984; Review Symposium, 1982; Schneider & Brody, 1983). Actual studies that bear on the issue require unique data, so the theory cannot be fully tested easily. Among the criticisms, however, a frequently recurring theme is that we are keeping alive longer people who have chronic disabilities. We have remarkable life support systems available. Some persons may be kept alive who may have a high level of dependency and for whom prognosis of recovery of function is small. Katz and his associates (1963) published a study that bears directly on the theory. For example, active life expectancy for women in their study was longer than for men, but the women had longer life expectancy also, and it turned out that the percentage of active life expectancy for men was higher for all age groups. Other aspects of the study were not consistent with the theory.

Active life expectancy as a concept involves information critical to understanding the amount of resources that will be required to maintain an aging population. Many issues are involved in considering active life expectancy, including ethical and moral considerations that define the level of medical and long-term care that is to be provided and the circumstances under which life is to be prolonged. For a general discussion and review of evidence, see Manton and Soldo (1985).

Currently, there is a more general acceptance of the notion that most of the communicable diseases that at one time quickly killed the aged are largely controllable, and thus the vast majority of people will die of degenerative diseases and systemic failures associated with aging. Of the degenerative diseases, heart attacks and strokes frequently lead to a quick death, but most others can involve a prolonged period of disability. Similarly, bodily organ and systemic failures usually are progressive and extended over long periods, so that the notion that one is on a genetic time clock that stops ticking all at once simply does not correspond to reality as it is encountered. This has led to an expectation that a large proportion of the aging population may involve long periods of dependency, and a secondary alternative has become more prominent in the practical consideration about how this can be managed, namely, the notion that older persons should have a more direct part in determining how long and under what circumstances their lives should continue in a dependent status, particularly when factors of quality of life are at issue. The issues of quality of life involve not merely pain, mobility, and the ability to carry out normal functions, but also transition to conditions of mental deterioration, coma, or other vegetative states. For a discussion of the origins of the construct *active life expectancy*, its early use in Canadian research, and the connection of the construct to *disability-free life expectancy* and *quality-adjusted life expectancy*, see Maddox (1994).

EDGAR F. BORGATTA

See Also
Demography of Aging
Health Maintenance
Health Promotion Screening
Morbidity Compression
Multidimensional Functional Assessment

ACTIVITIES

Activities are what people do. They may be special or routine, strenuous or relaxing, solitary or social. Some are commonplace and ordinary, such as walking the dog before breakfast or watching the news before going to bed. Others are extraordinary, such as a once-in-a-lifetime trip to Australia or a family celebration of a 25th wedding anniversary. As a consequence, no article or study can encompass all the activities that older people do in even a week, much less in a year or a lifetime. There are, however, several generalizations that offer a framework for understanding what later-life adults do and what those activities mean.

The first generalization is that older people usually go on doing most of the same activities they have always done. There are, in fact, no "activities of the old," stereotypes notwithstanding. The continuities of later life far outweigh the discontinuities, at least until major trauma forces drastic changes or significant decrements seriously reduce a person's abilities or resources. The general rule, then, is simple: Older people continue most of the activities that contributed to their lives previously. This is especially true for those "core" activities that are accessible, usually in or near the residence, and that form an integral part of daily life: informal interaction and conversation, media use, reading, walking, and performing routine tasks and procedures (Kelly, Steinkamp, & Kelly, 1986). In fact, home-based activities may actually increase for adults over 64 years of age (Iso-ahola, Jackson, & Dunn, 1994).

Most later-life adults seek a "balance" of activities (Kelly, 1987). This balance includes activities that are demanding and demonstrate skill and those that are restful and relaxing; those that involve communication with other people and those that offer withdrawal and disengagement; those that occur regularly during the day and week and those that punctuate periods with change or novelty. This balance changes throughout life as work, family, community, and leisure roles shift in importance and as resources and responsibilities change.

Activities and Aging

Each person builds a repertoire of skills, interests, associations, values, and commitments throughout life. There is no time when that history is abandoned or when we become different persons. Rather, our identities and the ways we cope with life tend to demonstrate considerable continuity over time

(Atchley, 1993). It is hardly surprising, then, that activity patterns and investments also tend to display considerable continuity. Further, we do not at some magical or critical moment define ourselves as different people because of chronological age. Rather, our self-definitions are more ageless than age defined (Kaufman, 1993).

There are no age-designated activities that move prominently to the fore as men and women pass the age of 65, retire, or are widowed. One reason is that older adults, like younger, do so many different things that they continue, at a somewhat slower rate, to engage in a process of adding, subtracting, and substituting activities (Iso-ahola et al., 1994). The demographics of an aging population, however, is one factor in the increased participation in two outdoor activities: walking and golf. Clearly, those formerly labeled "old" by passing the age of 65 do not suddenly cease or transform their activities. Rather, most are now called the "active old" or the "young old" and have become the targets of high levels of recreation and travel marketing.

Development of Activity Study

In 1961, a landmark collection edited by Robert Kleemeier, *Aging and Leisure*, drew attention to the significance of activity in the lives of middle-aged and older adults. Several chapters were based on the Kansas City study in which a variety of activities were found to be associated with changing roles and orientations (Kleemeier, 1961). The multidimensional meanings of the activities were embedded in community and family contexts. Activities, even those designated as leisure, were not segmented and separate from the ongoing flows of life meanings, attachments, and commitments. Rather, they tended to be integrated with the social roles of that life period, including intimate relationships, organizational histories, gender roles, personality development, and economic and cultural resources.

At the same time, Cumming and Henry (1961) proposed the hypothesis that growing old inevitably involves a natural process of disengagement that is indexed by decreased activity. In response, others argued that activity changes are adaptive, that engagement remains significant in aging, and that decrements are selective rather than inevitable (Lemon, Bengtson, & Peterson, 1972; Maddox, 1963). This is consistent with the later Houston study, in which lower rates of participation varied according to the type of activity as well as life circumstances (Gordon, Gaitz, & Scott, 1976). A more recent study found that overall activity levels were significantly lower with age, especially for those 75 and older, but that social and family activities as well as some cultural and organizational activities did not follow that pattern of loss. Rather, the marked declines were concentrated in travel, exercise, sports, and other outdoor or physically demanding activities (Kelly et al., 1986).

Perhaps the pattern that best encompasses both declines and continuities is that called "selective optimization with compensation" (Baltes & Baltes, 1990). Older persons are seen as active agents who strategize and negotiate with their own abilities and resources to maintain as much continuity of satisfaction and meaning as possible (Atchley, 1993). Activities are vehicles of meaning, self-investment, and salient relationships that are central to who we are at any age or in any set of circumstances. There may be some concentration and constriction of activity range in later life, but there is also the possibility of innovation and replacement rather than simple decline and loss (Iso-ahola et al., 1994).

Significance of Activities

There is now a broad consensus that remaining active is a major factor in later-life satisfaction and health (Kelly, 1993). In a longitudinal study, for those with viable health and income, activities outside the home were found to be the main factor in high levels of later-life satisfaction for both men and women (Palmore, 1979). Although it is difficult to ascribe causation, such engagement has consistently been found to be an essential element of an overall lifestyle that is relatively high in quality. Simply having financial and health resources is not enough unless they are employed in activity that is personally and socially involving.

For those moving into a final phase of frailty, a gradual or traumatic loss of health and physical or mental ability includes a constriction of activities, especially those that are physically strenuous or require full mobility. In that period, however, activities that maintain social integration and exercise developed skills continue to be important. Factors in sustaining activity are access, transportation, communication skills, and an atmosphere of inclusion. For residential care institutions, activities that challenge abilities and give a sense of worth and community are especially critical (Voelkl, 1993).

Community-dwelling older adults, however, are not found sitting around waiting for senior golden age programs. Only 15% of respondents aged 65 and older in a massive national survey had visited a senior center even once in the previous year, and less than 10% participated regularly (Cutler & Danigelis, 1993). Rather, older adults usually find their lives more or less filled by ongoing relationships, activities, and responsibilities. They commonly report being too busy to take on new activities. Even in retirement, the other roles and relationships of their lives tend to expand to fill and reconstruct the schedule formerly dominated by employment. Their continuing activity settings tend to be age integrated rather than age designated and age segregated.

All activities, however, are not alike. Filling time is not a major problem for the active old. Ordinary life is not empty, but there are possibilities for extraordinary levels of satisfaction and meaning in later life. Several studies suggest what those activities are likely to be. First, they express lifelong values and give a sense of worth (Kaufman, 1993). Second, they involve regular associations with other people who are valued, especially family and friends. And third, they involve a consistent commitment to activities that utilize skills and challenge abilities (Mannell, 1993). Such activities yield two significant outcomes: a sense of competence in the face of challenge, and a community of those engaged in the common activity.

Several research approaches have identified the factors in an engaged later life. A longitudinal study found that both role shifts and stressful events had significant impacts on activity engagement (Chiriboga & Pierce, 1993). In a community study, particular types of activity were correlated with high levels of life satisfaction: travel and cultural activities for those aged 40 to 64, social and travel activities for those 65 to 74, and family and home-based activities for those 75 and older (Kelly et al., 1986). Of course, such activities are usually associated with higher levels of resources. Another study found that the kinds of activity associated with higher levels of satisfaction are those requiring high levels of involvement, commitment, and skill (Mannell, 1993). These "high-investment" activities, sometimes termed "serious leisure," use skills developed over years of engagement and usually involve communities of common action and identification.

This picture suggests that being entertained or consuming experiences provided by others, however well meaning, does not fill fundamental needs at any age. Certainly, at any age, people need some periods of disengagement from demand. However, they also need activity that challenges and yields communities of common action. Later-life adults still need activity that allows them to identify themselves as persons of worth and ability who are significantly related to others. Such engagement may be found in ongoing sets of relationships and responsibilities, or it may be found in special commitments to challenging activities. Such activities, especially when they continue long-term involvement, become contexts of action and community.

Summary of Later-Life Activity

In summary, the activities of older adults demonstrate the following characteristics:

1. Most older persons are not sitting around with nothing to do. Their routines of ordinary activity usually fill the day and evening.
2. Those most satisfied with their lives are most often engaged in regular activities outside the home that provide challenge and relationships with friends and family.
3. Such activity does not, for the most part, occur in age-segregated settings.
4. Significant activities usually continue to be those that were satisfying in earlier life, although the active old do select, replace, and even start activities. Continuity is demonstrated in identities, established competencies, self-images, values, and relationships.
5. As a consequence, activities most likely to attract older adults build on familiarity, established abilities and identities, communities of action and interaction, and histories of satisfaction. Activity programs are attractive primarily because of their quality rather than any age designation.
6. Conversely, activity programs that require older persons to redefine themselves as old or in any way incompetent or inferior are unlikely to be attractive. The "ageless self" tends to retain values and self-definitions of ability that are not redefined by age alone.

Both personal and social histories are significant in the development of activity interests and abilities. For cohorts who enter later life with higher levels of education and greater financial resources, the range of activities increases. Future cohorts of women and minorities will have experienced fuller opportunities; thus, their interests, resources, and abilities will expand the possibilities for later-life activities. For all, however, the particular history of associations, life chances, circumstances, and crucial events differentiated by gender, race, ethnicity, sexual orientation, and social class constructs a life in which activities are integrated throughout its course.

JOHN R. KELLY

See Also
Active Life Expectancy
Creativity
Leisure Programs
Retirement
Senior Centers
Social Isolation

ACTIVITIES OF DAILY LIVING

Definitions of activities of daily living task categories vary in the literature, thus complicating assessment and treatment descriptions. Globally, such activities have been considered in two major categories: activities of daily living (ADL) and instrumental activities of daily living (IADL). The term ADL collectively describes the performance of functional self-care tasks that may include feeding and eating, dressing, grooming, bathing and showering, toileting, performing acts of sexual expression, and transferring. IADLs specifically refer to cooking, home management, functional communication, and financial management tasks (Maguire, 1996). In the United States, the American Occupational Therapy Association defines ADL as self-maintenance tasks that include grooming, oral hygiene, bathing or showering, toilet hygiene, personal device care, dressing, feeding and eating, medication routine, health maintenance, socialization, functional communication, community mobility, emergency response, and sexual expression. Activities related to home management (clothing care, cleaning, meal preparation and cleanup, shopping, money management, safety procedures, and the care of others) are more specifically classified as work-related tasks. When assessing ADL, health care providers should consider the diversity of terms, the impact on the older adult's performance, and overall safety. ADL serve as real-world tasks that determine the individual's level of performance, independence, productivity, and safety. Developing an operational terminology assists understanding among health care providers and clarifies ADL assessment and treatment concepts (Rogers & Holm, 1998).

ADL Assessment Measures

Methods of obtaining ADL information include self-assessment, assessment of ADL task performance or independence, and determination of the quality of the performance. Qualitative and quantitative approaches may be used in ADL assessment. The clinician's challenge is to use assessment tools that provide an objective description of outcomes that result from ADL intervention. Accepted measures of dependency vary internationally, with cultural and reimbursement system differences. The World Health Organization (WHO) International Classification of Impairments, Disabilities, and Handicaps (ICIDH) provides a model for discussing and organizing descriptions of dependency measures as they correspond to classes of handicap. The ICIDH codes provide an international reference for key areas of functional performance to adjust for international differences. The five ICIDH handicap codes are mobility, communication, self-care, occupation, and social relations. Clinicians selecting an ADL assessment tool should consider (1) comprehensiveness with respect to the ICIDH categories, (2) qualitative and quantitative measurements, (3) performance-based assessment and self-report flexibility, (4) the use of assistive technology and adaptive devices, and (5) statistical significance, such as reliability, validity, and sensitivity (Ottenbacher & Christiansen, 1997). These codes are currently under revision, and interim information about differences between current codes and new standards may be found on the WHO Web site (see "Internet Resources").

Other ADL assessment parameters describe levels of performance. The Health Care Financing Administration (HCFA) defines independence in terms of the levels of assistance needed: "minimal assistance" constitutes 25% assistance, "moderate assistance" 50%, "maximal assistance" 75%, and "total assistance or dependence" 100%. Other definitions of independence may include ordinal scales. Safety, independence, and overall quality of task performance should be included in an ADL evaluation. Measures may also consider the use of assistive technology and the nonphysical assistance needed for task performance. A description of needed technology is important with regard to discharge safety and compliance.

The Three-Part ADL Assessment

Assessing each ADL task involves a three-part process: analyzing the task to be performed, determining the learning capacity of the person, and determining the influence of the individual's environment. The three-point assessment of task, person, and environment allows for an ADL measurement that determines whether the performance is a departure from normal tasks and roles. Analysis of ADL capacity should include assessment of the level of performance needed to successfully complete a defined task. Information is also needed on how an ADL task is performed, when it is performed, the quality of performance, and the amount of assistance needed. A variety of assessment tools exist to measure older adults' ADL performance, and clinicians should carefully match the operational ADL terms used with assessment approaches. Assessment includes clinical reasoning and recognition of the practice setting or patient environment in which tasks are to be performed. ADL tasks that are independently performed may require assistive technology or assistance from others. Tasks that are performed independently but not safely may place the patient at risk for additional disability. Task performance that is not of a quality acceptable to the elder may lead to personal frustration, depression, role restriction, or decreased quality of life.

All ADL assessment tools should consider the elder's learning capacity, including the level of sensorimotor skill, endurance, and cognitive skills. Questions a clinician should ask include the following: Can the individual conceptualize or attend to the ADL task? Can the individual adapt to limitations in performance and learn new methods of ADL performance? Can the individual use assistive technology or devices to perform the ADL task? The elder's ability to perform the task will direct the health care provider in selecting the most successful and cost-efficient level of intervention.

The environment in which the ADL task will be performed must be included in the assessment. Task performance should be matched to intervention strategies that reflect the older adult's actual living conditions. Teaching ADL tasks such as bathing in an enhanced environment will not benefit an individual who returns home to an environment lacking grab bars, equipment, or adaptive plumbing.

Treatment

ADL intervention strategies include restoring function and compensating or adapting for performance limitations (Ottenbacher & Christiansen, 1997). ADL treatments should include approaches to restore function, compensate for disability or impairments, or adapt for permanent inability to perform. ADL treatment involves establishing realistic goals based on consideration of the individual, the demands of the ADL task, and the environment. Treatment approaches that are relevant and medically necessary are available from a variety of clinicians, including occupational therapists, nurses, physical therapists, and caregivers. ADL treatment should consider the client's ability to learn or relearn the ADL task, the complexities of task performance within the individual's abilities, and the techniques or technology needed. ADL treatment considers the value of the task to the individual, the level of performance, and overall safety. Assistive technology may be used, depending on input from the individual, the ability to use such devices, financial resources, and the actual outcome obtained with technology usage. Treatment outcomes are determined by post-intervention assessments that include discharge planning and caregiver education. The ability to perform ADL tasks is critical to independence and quality-of-life issues.

KATHY PIERCE BRADLEY

See Also
Assistive Devices
Cognitive Changes in Aging
Occupational Therapy Assessment and Evaluation
Osteoarthritis
Physical Therapy Services
Rehabilitation
Rheumatoid Arthritis
Self-Care
Stroke/Cerebral Vascular Accident

Internet Key Words
Activities of Daily Living
Assistive Technology

Occupational Performance
Self-Care

Internet Resources
American Occupational Therapy Association, Inc.
http://www.aota.org/

Assistive Technology
http://www.resna.org/resna/webres.htm

Evan Kemp Associates, Inc. (resource for disability
and assistive technology information)
http://www.disability.com/

World Health Organization International Classification of Impairments, Disabilities, and Handicaps
http://www.who.int/icidh

ADULT DAY CARE

Adult day care is a community-based long-term-
care concept introduced to provide elderly individuals and their caregivers with out-of-home support
for part of the day. These programs give older
adults an opportunity for socialization through peer
support and supervised activities and may also offer
specialized care services such as nursing, speech
therapy, physical therapy, and counseling (Cohen-
Mansfield, Besansky, Watson, & Bernhard, 1994).
Adult day care is an opportunity for elderly individuals to enhance and maintain their physical and
mental functioning and well-being longer than if
they were in an institutional setting (Kaye & Kirwin, 1990). Another important function of adult
day care is to provide caregiver respite to prevent
caregiver burnout and enhance family functioning.
It allows caregivers time to work, run errands, or
perform other tasks that might not be completed if
they were supervising their elderly relatives.

Types of Programs

Adult day-care programs vary but are usually one
of three models: social, medical, or mixed. A social
adult day-care program provides socialization, creative and educational activities, meals and nutritional monitoring, supervision by nursing or social
work staff, and medication management in some
programs. However, these programs usually offer
little or no personal care. A medical day-care program is for individuals needing more intensive levels of personal and medical care. Typically, medical
adult day-care programs offer medical, nursing, and
personal care, as well as physical, occupational,
and other forms of therapy. Participants in medical
adult day care require physician orders to receive
medical treatment. The Program of All-Inclusive
Care for the Elderly (PACE) is a well-known example of medical adult day care. PACE's focus is on
the integration of funding and provider financial
risk through capitation, the use of multidisciplinary
case management teams, and the delivery of services through adult day health facilities to assist
nursing home–eligible older adults remain in the
community (Branch, Coulam, & Zimmerman,
1995).

The mixed model of adult day care incorporates concepts and activities that fall under the social and medical models. Distinguishing between
models is often difficult due to service variations.
In addition to the three options, some programs
focus on narrowly defined client populations—
population-specific adult day care—serving individuals with senile dementia. Some adult day-care
centers have intergenerational programming and
joint activities with child day-care or community
programs.

Policies and Standards

Although the National Institute of Adult Day Care
(NIAD) has developed a set of standards and guidelines for adult day-care programs, there are currently no mandatory national standards for such
programs. Significant variations exist from state to
state. Many states require a license to operate an
adult day-care facility. Medicaid certification is required for programs using this funding source for
reimbursement. The NIAD and the Child and Elder

Care Directory Web sites offer an updated overview of programs by state and geographic region.

Program Characteristics and Staffing

Regardless of the type of adult day-care program, services are usually offered up to 12 hours a day. A national survey of adult day-care centers found that programs are generally open 8 hours a day, with approximately 5 hours of formal programming during that time. The typical program has an enrollment of approximately 24 clients, with an average daily attendance of 19. About one-third of the programs reviewed have waiting lists. Many day-care programs offer transportation services to and from the program. Approximately 55% of clients live within 30 minutes of the facility; another 35% travel between 30 and 60 minutes to attend (Conrad, Hanrahan, & Hughes, 1990). Referrals to adult day care come primarily from hospitals, visiting nurses, home health programs, and other human services programs. If adult day-care programs share facilities, they are usually located with nutrition sites, nursing homes, senior centers, or child-care centers (Conrad et al., 1990). Since there are no uniform requirements, staffing may vary significantly between programs. On average, adult day-care programs have approximately four core staff and a total staff of approximately five workers (Conrad et al., 1990). The client-staff ratio is approximately eight to one. Nursing services are usually performed by registered nurses. In medical programs, the nurses administer medications and manage and monitor the personal and medical care of clients as directed by their individual physicians. Licensed social workers direct social work services, offer caregiver support groups, and provide counseling to clients and caregivers. Their tasks may also include outreach, intake and assessment, coordination of services, and assistance with Medicaid applications. However, not all day-care programs have social workers on staff. Specialized therapies are provided by physical, occupational, and speech therapists. Social workers, nurses, recreational therapists, aides, or volunteers may be responsible for the social, creative, and educational activities of a day-care program.

The costs of adult day care vary greatly, depending on the type of program and the services used. The two primary funding sources are Medicaid and participant fees. Program fees range from a few hundred dollars to well over $1,000 a month. Adult day-care programs often receive substantial support from philanthrophy, thus reducing the costs to program participants.

Access to Services

The first step in accessing adult day-care services is to obtain the most current information about such programs in the geographic area. Next, the practitioner and client should review the options and discuss which programs would best fit the client's needs: financial, nutritional, medical, and social support. In nonurban and rural areas, there might be few or no programs available. A visit to the program site should be planned before enrollment, and if possible, the practitioner should accompany the client and caregiver on this visit. If the client is looking at a medical adult day-care program, it is helpful to speak with members of the medical staff and inquire about the appropriateness of the medical care options. Another area to inquire about are the social and recreational activities offered. Although it is difficult to judge the quality of a program by one visit, interactions between staff and program participants may be a good indicator of the facility's atmosphere. Many programs let potential clients spend some time and attend group activities to find out whether they like the program. Another area to appraise is the personnel. Do qualified personnel provide the services? What is the client-staff ratio? What is the overall condition and physical environment of the facility? Are there handrails, signs, and adequate lighting? Does the program offer or coordinate transportation services? Will program personnel come into the home to pick up clients? Once a decision is made, the client may need assistance completing the application. An important role for the practitioner in this context is to review and educate the client about the contractual terms of the adult day-care facility.

Recommendations for the Future

Research indicates a continued underutilization of adult day-care services influenced by costs, transportation, match of services and client needs, and client refusal (Cohen-Mansfield et al., 1994). Suggestions to increase utilization include better marketing, improved transportation, provision of more counseling services, and establishment of quality-of-care standards (Conrad et al., 1990; Kaye & Kirwin, 1990). As with other services, cultural differences exist in the way older adults access and utilize adult day-care programs. The heavy reliance on participant fees may pose a financial barrier to use by older adults from lower economic backgrounds. Thus, additional emphasis should be placed on evaluating service utilization and potential access barriers for individuals from different cultural and economic backgrounds.

MATTHIAS J. NALEPPA

See Also
Caregiver Burden
Caregiver Burnout
Program of All-Inclusive Care for the Elderly
  (PACE)

Internet Key Words
Adult Day Care
Caregiver Respite
Medical Adult Day Care
Mixed Model Adult Day Care
Social Adult Day Care

Internet Resources
Adult Day Services Association
http://www.ncoa.org/nadsa

National Institute of Adult Day Care
http://www.niad.org

Child and Elder Care Directory
http://www.careguide.com

ADULT EDUCATION

Education of older people is primarily an off-shoot from the adult education movement, a movement as old as the nation itself with roots in the social, cultural, and political concerns of the time. Educational programs for older persons today reflect this history and heritage, involving a wide variety of formats and purposes. This instruction includes heterogeneity of sponsors, program types, audiences, and content. No central system supports or monitors these activities, so idiosyncratic patterns have developed around the preferences of administrators and the needs of local communities. This has frequently made the programs responsive to the wishes of the constituent clientele but has neither facilitated development of program categories nor provided much assistance in generating program models that can be replicated in other sites and with other sponsors.

History

One of the earliest comprehensive surveys of educational programs for older people (Donahue, 1955) indicated wide availability of instructional offerings by the mid-1950s but suggested that older people were not the primary audience but were included in programs designed originally for other age groups. Later these programs were modified to make them more appealing and more specifically oriented toward older people.

The 1960s brought the inclusion of programs in a greater number of agencies and institutions. The 1961 White House Conference on Aging gave an impetus to this growth by increasing its visibility and emphasizing the growing social concern for older persons. Instructional efforts for older people took on a social service orientation (Moody, 1976), emphasizing the crisis of adjustment to retirement and the need for outside assistance to overcome the trauma of role change. Program rationales emphasized the needs of older people and the responsibility of social institutions to meet these needs through service programs.

A shift from a social service orientation can be noted in the background paper to the 1971 White House Conference on Aging. In it, McClusky (1971) accurately reflected the new orientation when he emphasized the positive nature of education and the potential of every person, regardless of age. Another major development beginning in the 1970s was the rapid growth of residential education for older persons. Through the Elderhostel Program, weeklong courses of instruction, field visits, and entertainment were provided by colleges and universities across the nation. The response to this offering has proven that a combination of stimulating atmosphere, congenial company, and varied activities is both desired and appreciated by older people.

Another example of a program designed for the well-educated senior can be seen in the Institutes for Retired Professionals. These programs draw retired teachers, professionals, and community leaders into study programs that are administered and taught by older persons.

The 1981 White House Conference on Aging emphasized a theme that is rapidly growing in the United States—self-help. In relating this theme of older people and education, emphasis has been placed on the development of instructional programs that will result in increased problem-solving abilities, especially those that can be used in the workplace or in volunteer roles. The federal government has encouraged employers, state governments, and community organizations to offer instruction that will assist older persons in holding or gaining contributive roles in the community. A number of job placement programs, many using the name Second Careers, are being created in major cities, and instruction in job search skills, self-concept, and specific content is growing.

Technological developments are increasingly being used for instruction of older learners. Networks of computer users (e.g., e-mail distribution lists and list servers), videotapes on exercise or skills development, local access television programming, extensive information on CD-ROM and the Internet, two-way video workshops, and televised classes are all available to adults desiring to learn or pursue a degree but who are not able or interested in traveling to a campus to attend lecture classes.

Implications for the Future

Educational programs for older people in the United States have developed to such a point that we can now begin to gain some perspective on their growth and to identify several implications for the future. Three will be mentioned. First, there now seems to be general acceptance that we are living in a learning society, in which persons of every age will be required to continue to expand their knowledge and skills in order to survive and prosper. Community colleges, public schools, universities, libraries, museums, recreation centers, clubs, senior centers, department stores, and corporations are offering instruction in a broad and rapidly growing variety of topics (Peterson, 1983).

Second, there is now general acceptance that a multipurpose rationale exists for the conduct of educational programs for older people. No longer do the purpose statements of instructional programs deal only with the difficulties and crises of old age; now they frequently emphasize the ability that persons have to grow and develop throughout their lives. Older persons are seen as individuals with potential who can contribute and serve as well as cope and survive. This change in orientation has led to many more programs on self-actualization and growth. Liberal studies, psychological growth, and broadening experience are becoming a greater part of the programs, and lifelong planning is replacing adjustment to retirement.

Finally, as the enrollment of older persons increases, institutions that have simply encouraged older people to participate in their regular programs are beginning to develop special offerings exclusively for them. As this occurs, recognition of the need to accommodate the unique characteristics of older people has developed. Older learners tend to be highly motivated, organized, and determined. It is now apparent that elders often will not accept the bureaucratic procedures of admission, advisement,

and registration that younger students have tolerated. They need adaptations that will make these administrative processes more streamlined and abbreviated. Likewise, they expect educational content and method to be developed to suit their interests and preferences. As institutions begin to understand these needs, major modifications of the usual administrative and instructional procedures will be demanded.

The growth of the number of older students is leading to increased interest in their learning abilities and classroom performance. The continuing development of knowledge about the older person as a learner will be helpful not only in educational institutions but in a variety of community settings where assistance to older people is provided and where their talents (such as serving members of other generations and taking leadership positions in community organizations) are needed for the good of the community.

DAVID PETERSON
Updated by CATHERINE J. TOMPKINS

See Also
Aging Agencies: City and County Level
Creativity

Internet Resource
Elderhostel
http://www.elderhostel.org

ADULT FOSTER CARE HOMES

Since the 1980s, alternative housing options for elders have proliferated in the United States. Known by a variety of terms, these alternative settings provide varying degrees of assistance to the older individual. The geriatric adult foster care (AFC) home, also known as board and care facility or group home, is the oldest form of alternative living arrangement. An AFC provides needed support services while allowing the older adult to remain in the community (Steinhauer, 1982). In the United States, regional variations exist regarding regulation and licensure, and AFCs may be under the jurisdiction of none to several state agencies. Even in states that require licensing, variation exists in the minimum standards, delegated responsibility, and degree of enforcement (Steinhauer, 1982). Since most AFCs are private residences, it is unknown how many AFCs there are currently in the United States, although some estimates are as high as 30,000 facilities.

In general, an AFC home is a private home that is residentially zoned with a nonrelated, live-in caregiver. Typically, AFCs limit the number of residents to no more than five individuals, although larger AFCs exist. Often, an additional state requirement is that only one resident may be bedbound or require maximum assistance with care. The AFC may provide a private bedroom with access to a bathroom on the same floor as the bedroom, but this environmental characteristic is not always present however. Varying by state, the AFC may accept private payment, Medicaid, or both.

Facilities and Caregivers

A live-in caregiver is the basic service requirement of AFCs, regardless of geographic location. AFCs provide a wide range of services, somewhat dependent on whether the AFC is associated with a medical center or social services agency. While all AFCs have a formal caregiver who can provide personal care services, AFC caregivers may also supervise medications, including injections; provide special diets, including tube feedings; provide bladder training, catheter irrigations, and dressing changes.

Formal training of caregivers varies. Reporting on foster homes, Kane and colleagues (1991) and Sherwood and Morris (1983) found that the majority of caregivers in Oregon and Pennsylvania had no formal training. Other investigators found that foster home caregivers who participated in programs linked to medical centers received formal training of up to forty hours (Braun et al., 1988; Oktay & Volland, 1987). Formal training included acquiring skills and knowledge related to personal assistance, home health care, special diets, physical therapy, and psychosocial aspects of aging and illness. Additionally, some researchers reported that up to 70% of the caregivers had some form of health-related employment (Braun et al., 1988).

Physical, environmental, and social characteristics of a facility are important to older adults. Surveys of older adults from Oregon nursing homes and AFCs revealed that characteristics influencing older adults' choice of an AFC were homelike atmosphere, privacy, and flexibility in routines, whereas organized activities and the availability of physical rehabilitation were important to those who chose nursing homes (Nyman et al., 1997; Reinardy & Kane, 1999).

Residents

Most older adults in AFCs are typically older than 55 years, with reported mean ages greater than 70 years. Unlike residents who reside in assisted living facilities, who are generally retired middle class professionals, AFC tenants are more likely to be poor, single, living alone, and often rejected by family members (Oktay & Volland, 1987).

AFC tenants have fewer reported major health problems compared to nursing home residents, but this finding may reflect a reporting bias, since nursing home personnel are more likely to have current medical histories of their residents (Kane et al., 1991). In spite of fewer reported health problems among those residing in AFCs, difficulties with managing activities of daily living (ADL) and instrumental activities of daily living (IADL) do exist. AFC residents requiring assistance with ADL range from 9% to 78% requiring assistance with walking; 4% to 58% requiring assistance with transferring; 12% to 27% requiring assistance with dressing; 30% to 51% requiring assistance with bathing; 4% to 7% requiring assistance with eating; and 5% to 70% requiring assistance with toileting (Braun et al., 1988; Hopp, 1999; Kane et al., 1991; Quinn et al., 1999). The broad ranges in ADL dependency reflect the existence of a subpopulation in AFCs that is very dependent. Although a higher proportion of AFC residents than nursing home residents

are independent in ADL and IADL, these findings demonstrate that significant overlap occurs between the two populations in degree of dependency in ADL and IADL. Not surprisingly, AFC residents are more likely to require assistance with IADL than with ADL. It is not unusual for residents of AFCs to have cognitive and/or behavioral deficits.

Little is known regarding the health outcomes of older adults residing in AFCs. AFCs affiliated with a medical center and formal caregiver training report improved patient well-being, decreased anxiety, improvement in ADL function, and decreased cost of services as compared to patient outcomes in nursing homes (Braun et al., 1988; Oktay & Volland, 1987). In contrast, AFCs lacking affiliation with a medical center report little to no improvement in ADL function (Kane et al., 1991; Sherwood & Morris, 1983). Overall, 49% of AFC residents have increased ADL dependency within the first year of admission to the AFC. Increasing personal care needs influenced whether residents could remain in an AFC. Other factors influencing continued AFC residency included whether the resident was "likable" and able to participate in the home, social functions such as meals, family support, staff trained to provide services, and caregivers who felt that the needed care could be provided. AFC characteristics other than caregiver training associated with positive outcomes included children residing in the AFC, caregivers who involved the older adults in activities, and caregivers who rated the work as satisfying but rated their need to work in the home as somewhat important.

Choosing an Adult Foster Care Home

Older adults and/or family members of older adults who are in need of alternative living arrangements and who are considering an adult foster care home need to examine characteristics of the home and the caregiver. If the AFC is certified or licensed by the state, a potential tenant should review the last survey report, making note of any deficiencies. Safety and physical considerations include safe entry and exit ways, well-lighted entry and parking areas, wheelchair access, maintenance of grounds and building, presence of safety features such as smoke detectors and handrails, and whether a private bedroom is available. The environment should be clean and free of odors with good ventilation and comfortable temperatures. Whereas most AFCs are in residential neighborhoods, the AFC should be easily accessible to family and friends and near the resident's doctor or hospital.

Financial considerations are important, since not all AFCs accept Medicaid payments. Families need to determine if there is a basic fee that covers all services or if additional fees are included for services such as laundry, or amenities such as telephone or cable television. If a contract is signed, families need to know under what circumstances they may terminate the contract, for example, if there is a change in the resident's medical condition. Minimal caregiver training should include first aid skills, safety and fire prevention, and prevention and containment of communicable diseases.

An aging population and longer life expectancy will increase demand for alternative housing care environments, and for caregivers able to provide services to individuals with varying degrees of functional and cognitive impairments and psychosocial needs. Findings that older adults who reside in foster homes have a significant decline within the first year of residency are troubling (Stark et al., 1995). Issues for future research include the lack of knowledge and formal training of caregivers, the numerous terms and lack of a standardized definition of adult foster care that includes services provided, and the lack of federal regulations.

LORRAINE C. MION
KATHLEEN A. ONDUS

See Also
Assisted Living
Consumer-Directed Care
Institutionalization
Nursing Homes

Internet Key Words
Adult Foster Care Homes
Board and Care Facilities
Group Homes

Internet Resources
Assisted Living Network
http://www.alfnet.com

Health Care of Michigan, Resource guide for adult foster care homes
http://www.hcam.org

AARP Research/Adult Foster Care for the Elderly
http://www.research.arrp.org

ADULT PROTECTIVE SERVICES

Adult protective services units are key agencies involved in the identification and treatment of elders who are at risk of being abused or are being abused. Estimates of the prevalence of elder abuse range from 450,000 to 1.8 million cases annually (National Center on Elder Abuse, 1998; Tatara, 1996). Types of elder abuse include physical abuse, sexual abuse, emotional or psychological abuse, neglect, abandonment, financial or material exploitation, and self-neglect. Elder mistreatment occurs in domestic settings and long-term-care facilities as well. The 1997 *Long Term Care Ombudsman Report* indicated that physical abuse was one of the five most frequent types of complaint among the 30,449 complaints received involving board and care homes (Administration on Aging, 1997).

There are no federal statutes or guidelines for the delivery of adult protective services (National Association of Adult Protective Services Administrators, 1994). Each state has some form of legislation addressing protective services for adults or elderly victims, but statutes vary widely. In order to know when, where, and what to report to whom, familiarity with a state's adult protective services laws is essential. A listing of state adult protective services statutes can be found on the National Center for Elder Abuse Web site.

Elder mistreatment is a hidden problem. Victims are reluctant to self-report, usually due to shame, fear, or the physical or mental inability to do so. In all but a few states, reporting of elder mistreatment is mandated by law. In most states, a variety of professionals, including physicians, medical personnel, law-enforcement officers, social workers, nursing home staff, and mental health professionals, are required to report. Many state statutes include penalties for failure to report. Reports are usually made to adult protective services units, which are housed in state or local departments of social services or aging services.

Most reports of elder or adult abuse are made to the reporting agency by telephone. In general, when a report is made, it is first screened to determine whether the victim fits the target population defined in the statute, as well as whether the reported mistreatment is covered by state law. For example, in many states, younger disabled victims as well as the elderly qualify for adult protective services, based on their vulnerability. Some states define "elderly" as age 60 and older; others define it as age 65 and older. Not all states provide services to victims of emotional abuse.

If it is determined that the victim is eligible for adult protective services, a caseworker is assigned to conduct an investigation and make an assessment. In metropolitan agencies, the worker may be an adult protection specialist. Some states have certification programs for professional staff. However, in small, rural areas, the worker conducting the assessment also may have other duties, such as responsibility for child welfare cases.

Because identification and reporting of elder abuse are crucial in initiating treatment, family members, caregivers, professionals, and the elderly need to be familiar with reporting laws in their states, as well as which agencies receive these reports (Quinn & Tomita, 1997). Once a report has been made, the agency receiving the report conducts an investigation to determine the validity of the report and the immediate risk to the victim. Under most state statutes, the confidentiality of good-faith reports is assured. Confidentiality of abuse reporting in the case of long-term-care patients is also protected under the federal Older Americans Act.

Any occurrence of domestic elder mistreatment that appears to involve criminal activity, including physical and sexual assault, armed robbery, and theft, should be reported immediately to the local law-enforcement agency. As with mistreatment that occurs in the community, any criminal act that occurs in a long-term-care facility should be reported to local law enforcement. Adult protective services programs do not have the expertise or the statutory authority to conduct criminal investigations.

Situations of elder mistreatment in long-term-care facilities should be reported to the facility administrator, the state or local agency that regulates the facility, and the local long-term-care ombudsman. In some states, abuse reports involving long-term-care residents should also be made to adult protective services. When there are multiple agencies involved in a long-term-care abuse investigation, coordination and clarification of responsibilities are essential.

Although there are no universal national assessment standards, the process generally includes an evaluation of the types of mistreatment that may have occurred; the victim's immediate risk of further harm; the victim's relationship with the alleged perpetrator; the victim's capacity to make informed decisions, understand the implications of continued maltreatment, and cooperate with the treatment plan; the victim's physical health and any resulting limitations in performing activities of daily living; and environmental hazards and social and emotional supports available to the victim.

Certain basic principles guide the delivery of adult protective services. These principles include:

- The client's right to self-determination and choice

- The use of the least restrictive alternative
- The maintenance of the family unit whenever possible
- The use of community-based services rather than institutions
- The avoidance of blame
- That inadequate or inappropriate services are worse than none

Based on these principles and the assessment findings, as well as available community resources, the adult protective services worker develops a case plan. An essential component of the planning processes is the client's understanding of his or her situation and willingness to accept services designed to reduce further mistreatment. Once services have been initiated, the caseworker may monitor the service providers for a period of time or turn over responsibility to an ongoing case manager, who may be part of the adult protection unit or employed by an outside case management agency.

In situations in which the risk of harm is imminent and the victim lacks the capacity to make informed decisions, a temporary emergency guardianship action may be initiated. As with other aspects of adult protective services, the responsibility for initiating such court procedures depends on state law. Guardianship actions typically involve civil probate procedures.

Treatment provided to victims of elder abuse is dependent on the immediate level of risk to the victim, as well as the victim's capacity to accept or refuse services. An older person who has the capacity to make informed decisions has the right to refuse services, even when exercising that right may increase the risk of further abuse. Elderly persons who lack decisional capacity may have services imposed on them only as the result of a court order.

In states in which adult protective services agencies are permitted by statute to petition the court for guardianship, the caseworker bears primary responsibility for collecting and documenting sufficient information to make a compelling argument to the court. The caseworker may provide information related to the assessment, including an evaluation of the victim's ability to function independently, and a recommendation for the appointment of a guardian. Most states also require a statement from a physician, psychologist, or mental health professional regarding the proposed ward's capacity to make informed decisions. A person's choice to live in a manner that does not conform to accepted community standards of hygiene does not necessarily indicate incapacity. If the proposed ward contests the appointment of a guardian, an attorney will be appointed to represent his or her interests.

Treatment options for adult victims who consent to services include Meals on Wheels, in-home care, medical and mental health treatment, placement, financial management to pay bills and protect assets, and legal services to recover funds and property through civil litigation. In most states, casework counseling is an integral part of adult protective services. It is the ongoing casework relationship that provides emotional support to the victim, as well as monitoring of the other services the client has agreed to receive.

In all treatment options, the wishes and preferences of the mistreated older person take precedence. Treatment must include the least restrictive alternative and support the victim's right to self-determination. Unwanted services cannot be imposed without court intervention based on the victim's incapacity to protect himself or herself from further harm due to a severe mental or physical impairment.

Prevention of the mistreatment of older persons is highly dependent on individual and community awareness. Although many older people may fear becoming victims of violent crime, violent crime is actually rare. However, it is estimated that at least 1 of 20 older Americans will be victims of abuse, exploitation, or neglect each year. This mistreatment is most likely to be committed by a family member, usually an adult child. More than two-thirds of the perpetrators of domestic mistreatment are family members.

For this reason, awareness of the types of abuse and the probability of their occurrence is the primary method of prevention. Older persons should be aware of, and take responsibility for, disease prevention and management of their medical conditions, as well as thoughtful financial planning designed to protect their assets. They need to plan for the time when they will no longer be able to live independently by developing a plan of care and clearly designating decision-making authority to persons who will act in their best interests should they lose the capacity to make informed decisions.

In choosing a paid caregiver, the most careful scrutiny must be undertaken. Home-care agencies should be chosen based on the types of background and history checks conducted by the agency, as well as the level and frequency of supervision provided to home-care staff. Criminal record checks and frank discussions with previous employers should always be conducted.

Since powers of attorney may easily be abused for financial gain, older persons should consider appointing two persons to oversee financial transactions carried out on their behalf. One agent can be a trusted family member or friend, but the other should be an attorney, financial adviser, or disinterested person who does not stand to gain from the estate. In this way, each person acts as a check and balance on the other, thus reducing the possibility of financial exploitation.

Because mistreatment of the elderly may develop slowly over time and is largely hidden, open communication between family members and caregivers is essential. A caregiver may be the first person to identify possible mistreatment and must be assured that the reporting of any suspicions will be taken seriously and kept confidential.

JOANNE MARLATT OTTO

See Also
Crime Victimization
Elder Mistreatment: Overview
Financial Abuse
Elder Neglect

Internet Key Words
Abandonment
Emotional Abuse
Financial Exploitation
Neglect
Physical Abuse
Self-Neglect
Sexual Abuse

Internet Resources
Administration on Aging
http://www.aoa.gov/sheindex.html

Administration on Aging, 1997 National Ombudsman Reporting System Data Tables
http://www.aoa.gov/ltcombudsman/97NORS/default.htm

Caregiver Resources—Information About Services in Your Community
http://www.aoa.gov/elderpage/locator.html

National Association of Adult Protective Services
http://www.NAAPSA.org

National Center on Elder Abuse
http://www.gwjapan.com/NCEA

Wierucka, D., & Goodridge, D. *Institutional elder abuse*
http://www.cjona.org/vulnera.html

ADVANCE DIRECTIVES

Health care professionals, families caring for elders, and elders themselves often face difficult decisions and conflict regarding starting, continuing, or stopping life-sustaining treatments. The situation becomes especially difficult when professional providers have little knowledge of what life-sustaining treatments a patient would want or when no one is appointed (or available) to make health care decisions for an incapacitated elder. Advance directives can help reduce conflict and create peace of mind for elders that their wishes will be followed and that the burden of decision making will be eased for family members and friends.

An advance directive is a document with legal standing that generally gives directions or instructions about what health care treatments or interventions a person would or would not want to receive if he or she were unable to communicate those wishes, or it allows an individual to appoint another person to carry out those wishes. A living will–type advance directive contains an individual's specific written instructions about wanted, limited, or unwanted life-prolonging health care treatments in case the person is incapacitated or unable to communicate. A durable power of attorney for health care (DPAHC) is a type of advance directive that permits an individual to designate *another person*—called a health care agent, surrogate, health care proxy, or attorney-in-fact—to make health care decisions if he or she loses decision-making capacity. Some advance directives combine the elements of a living will and a DPAHC into a single document, which both allows a person to set down broad directions or instructions regarding his or her future medical care and designates a person to see that those wishes are carried out.

An advance directive becomes effective only when an individual is incapacitated or lacks decision-making capacity for a particular health care issue. Most states have detailed statutes that outline the conditions under which an advance directive is legally valid. Living wills are not recognized by statute in New York, Massachusetts, and Michigan. Some states empower a health care agent to make health care decisions on behalf of the patient; other states allow an individual to designate an agent to carry out the intent of a living will. In states that have family-consent laws (laws that designate a legal hierarchy of family members who can make health care decisions), a DPAHC can clarify which individual is authorized by the patient to make health care decisions when there are two family members of equal status in the hierarchy (such as two siblings). Appointing a health care agent is extremely important when an individual would prefer a non–family member to make health care decisions. For example, in the gay community, it is common to appoint a domestic partner or friend rather than a family member as a health care agent under the DPAHC statute.

In practice, many health care providers are unclear about when a living will applies or uncomfortable about deciding when a patient is on a dying trajectory that warrants the triggering of a living will's instructions. Many health care providers say that a living will's directions are either too broad or too narrow to be clearly applicable to every possible situation. Thus, the appointment of a health care agent may be more useful than a living will in treatment decision making, because the agent generally can make decisions about all health treatments, not just decisions to refuse, limit, or withhold life-sustaining treatment.

The Patient Self-Determination Act (PSDA), in effect since 1991, mandates that agencies and institutions receiving Medicare and Medicaid reimbursement provide written information to individuals about their right to participate in medical decision making (as outlined in state law) and to formulate advance directives. Facilities must (1) provide written information to each adult about his or her right under state law to make health care decisions, (2) create written policies for respecting the implementation of such advance directives, (3) ask each person whether an advance directive has been completed, (4) document in medical records the existence of an advance directive, (5) not discriminate with respect to the provision of care based on the presence or absence of an advance directive, (6) comply with applicable state laws regarding advance directives, and (7) educate agency staff and the community about advance directives.

Although surveys of the general public report positive attitudes toward advance directives, only about 15% of community-dwelling elders have completed one. In nursing homes, however, there is evidence that closer to 50% of elders have completed some type of advance directive (High, 1993). Advance-directive completion is more concentrated among whites with higher education and income levels. Evidence suggests that nonwhite patients, patients from lower income groups, and patients with visual or hearing deficits are less likely to receive information about advance directives. All patients, regardless of demographic characteristics, should be approached about advance directives, and steps should be taken to minimize such selective informing. Elders who lack the cognitive capacity to perform some tasks may still be able to designate a health care agent.

All adults should be approached regarding end-of-life care preferences and advance directives. Most elders want to talk about their end-of-life care, and discussions with a trained informant can increase advance-directive completion. Evidence suggests that elders expect health care providers—preferably physicians—to initiate these discussions. Discussions about end-of-life care can begin with questions about what health care or life-sustaining treatment the person might want if he or she were unable to communicate, or who would be expected to make decisions if the person were incapacitated. Providers and patients may find such conversations more useful, and less stressful, if the patients are asked to prioritize their health care goals, thus allowing the professional to infer a preferred pattern of care and more broadly discuss care preferences (Gillick, Berkman, & Cullen, 1999). Forms such as the Values History Form (Center for Health Law and Ethics, 1999) can help patients and providers elucidate care preferences and discuss care situations (Gillick et al., 1999). In settings where nurses, physician assistants, or social workers are not primarily responsible for informing patients about advance directives, such providers nevertheless play a vital role in discussions of end-of-life care preferences. Nurses can approach patients and families after their discussions with physicians about treatment preferences, both to offer support and to further clarify points regarding certain technologies or treatments. Since evidence suggests that patients are more likely to discuss end-of-life preferences and advance planning if the physician is of a similar racial or ethnic background, social workers can help bridge such cultural gaps and mitigate fears and anxiety that may be provoked by such conversations.

End-of-life care discussions should occur either during a regularly scheduled appointment with a primary care provider or during the early stages of an elective admission. In an acute emergency admission, it should take place as soon as possible (Brown, Beck, Boles, & Barrett, 1999). Patients may be more receptive to written information about advance directives from institutions if it is received in a preadmission package to be read at home or as part of the discharge process, when the impact of the institutional admission is still fresh. The mailing of informational materials about advance directives from the primary care provider to the patient's home has also been found to increase the probability of advance-directive completion (Brown et al., 1999). Despite the increasing public discussion of death with dignity, patients may still be unwilling to discuss end-of-life care and advance directives because they are unwilling to discuss their own deaths (Hare, Pratt, & Nelson, 1992). Education and information may not completely counteract the discomfort people feel when they think about dying, so health care professionals need to learn techniques of discussing death realistically and sensitively with patients and their families. These techniques include being aware of patients' spiritual interests and needs and including other professionals, such as clergy, in end-of-life care discussions.

Periodically, and following significant health events, health care providers should review completed advance directives with patients to ascertain whether the information and listed treatment preferences still match their wishes. Patients should be counseled to give copies of the directive to their designated health care agents. Family members and close friends should also get copies, and a copy should be placed in the medical records of the agency or the primary provider's office chart to ensure that the directive is easily accessible in an emergency. In emergency health situations in the community, emergency medical services (EMS) providers may recognize the legality of only those advance-directive cards issued by the state's EMS Program and countersigned by a personal physician. Providers should be aware of individual state law and community standards for emergency situations.

Although nursing home residents are more likely to complete advance directives than community-dwelling elders, there is some concern that residents are completing the form as part of the nursing home admission process rather than as a result of a considered discussion. Each person's right not to complete an advance directive should be respected. Patients should be told that they will not be abandoned by clinicians or the facility and will not receive substandard care if they do not complete an advance directive. That a patient chooses not to formulate a directive should be noted in the medical chart.

Although a written advance directive is preferable, especially in emergency situations, courts view oral advance directives favorably and have enforced or relied on them in adjudicated decisions to forgo life-sustaining treatment. Oral advance directives are more likely to be accepted if such statements (1) were made on serious occasions or were solemn pronouncements, (2) were consistently repeated, (3) were made by a mature person who understood the underlying issues, (4) were consistent with the values demonstrated in other aspects of the person's life, (5) were made shortly before the need for a treatment decision, and (6) addressed with some specificity the actual conditions of the person. Physicians, nurses, and social workers should document such statements when discussing advance directives or end-of-life treatment preferences with patients.

MELISSA M. BOTTRELL

See Also
Palliative Care
Substitute Decision Making

Internet Key Words
Advance Directives
Durable Power of Attorney
Health Care Agent or Proxy
Living Wills
Oral Advance Directives
Patient Self-Determination Act
Surrogate

Internet Resources
Advance Directives International
http://www.adiwills.com/adiread.htm

American Academy of Family Physicians
http://www.aafp.org/patientinfo/directiv.html

Health Care Financing Administration, U.S. Department of Health and Human Services
http://www.medicare.gov/publications/advdir.htm

ADVANCED PRACTICE NURSING

Nearly all nurses care for older adults; therefore, all nurses need preparation in the principles of best practices in geriatrics. Unfortunately, the only nurses who are specifically prepared to care for geriatric clients are geriatric advanced practice (AP) nurses. AP nurses—geriatric nurse practitioners (GNPs), geriatric nurse clinical specialists (GNCs), and geropsychiatric nurse clinicians—are registered nurses who have completed a master's degree program to specialize in geriatric nursing. AP geriatric nurses blend the skills of an experienced nurse with those of a geriatric primary care practitioner. AP geriatric nurses promote the health of older adults, diagnose and manage acute and chronic illnesses and syndromes common in older adults, provide primary care and disease management, and educate patients and families in collaboration with other nursing and health personnel. AP geriatric nurses work in ambulatory settings and clinics, hospitals, nursing homes, and Veterans Administration hospitals and provide home care.

First introduced in the mid-1970s, AP geriatric nurses are educated in 59 academic programs and are certified by the American Nurses Association or state boards of nursing (American Association of Colleges of Nursing, 1998). AP geriatric nurses represent 4.2% of the 161,711 AP nurses currently working (American Association of Colleges of Nursing, 1999). In their master's programs, AP geriatric nurses receive classroom teaching and an average of 690 hours of geriatric practice experience (12 to 16 hours of practice a week over two years) (Mezey, 1995). After graduation, approximately 40% are employed either by nursing homes or by physicians with practices in nursing homes, 35% work in facilities in the inner city, and 20% work in rural areas (Mezey, 1995).

AP geriatric nurses are eligible for reimbursement under Medicare and Medicaid (Omnibus Reconciliation Act of 1997). The reimbursement rate under Medicare is 85% of the rate for comparable physician services. In rural health settings, AP geriatric nurses are reimbursed through the Federal Rural Health Act (Mezey, 1995).

Nurses with geriatric specialization take a comprehensive and rehabilitative approach to care for older patients experiencing chronic and acute illnesses. They develop programs to promote health or to prevent disabilities, such as immunization and fall-prevention programs. In collaboration with physicians and other health professionals, they perform physical, functional, and psychosocial assessments. GNP and geriatric fellowship programs have similar curricula related to geriatric assessment. AP geriatric nurses deliver primary care, diagnose physical and behavioral conditions and diseases associated with aging, and manage commonly occurring acute conditions and stable chronic conditions in collaboration with physicians and the geriatric team. These nurse practitioners also prescribe and manage medications. As of 1999, all states had given nurse practitioners some degree of prescribing authority. In fact, in 11 states, they have prescriptive authority (including controlled substances) independent of any physician involvement (Pearson, 1999). These nurses teach patients, families, and staff; supervise nonprofessional staff; and oversee quality assurance.

There is strong scientific evidence that AP geriatric nurses improve the quality of care and decrease the cost of care in long-term facilities, hospitals, and ambulatory care settings (Paier & Strumpf, 1999). The relative absence of professional nurses and physicians in the nation's 17,000 nursing homes makes AP geriatric nurses particularly valuable in improving the care of the 1.5 million adults who reside in these homes on any given day. For every 100 nursing home beds, the average staffing is only one registered nurse (RN), who is most likely the director of nursing; 1.5 licensed practical nurses (LPNs); and 6.5 nursing assistants. This translates, on average, to one RN for every 49 nursing home patients and a median of 12 minutes of RN time per resident per day. This staffing pattern persists despite a sicker case mix in nursing homes (Paier & Strumpf, 1999).

Numerous evaluations confirm that nurses with advanced preparation in the care of elderly nursing home residents decrease unnecessary hospitalizations and the use of emergency rooms. AP nurses improve admission and ongoing patient assessments, provide better illness prevention and case finding, decrease incontinence, lower the use of psychotropic medications and physical restraints, and generally improve the overall management of chronic and acute health problems. Most of this improvement occurs without incurring additional costs, and in some instances, at reduced costs. These outcomes are in large part attributable to the enhanced teaching and supervision of professional and nonprofessional nursing staff and the decentralization of nursing services, which moves decision making down to the level of the bedside. Geriatric nurse specialists specifically have been shown to ameliorate the most common problems experienced by nursing home residents: urinary incontinence, pressure ulcers, the overuse of physical restraints, and hard-to-manage behavior evidenced by residents with dementia (Paier & Strumpf, 1999).

In hospitals, geriatric clinical nurse experts have been found to (1) significantly reduce morbidity, including preventing or reducing clinical syndromes common to the elderly, such as delirium; (2) shorten hospital length of stay; (3) reduce morbidity following discharge; and (4) reduce emergency room use and readmission after discharge. These outcomes are evident throughout several hospital-based geriatric nurse practice models. The geriatric resource nurse model is designed to improve care by helping the primary bedside nurse develop enhanced skills and knowledge in geriatric nursing. Geriatric assessment and acute care of the elderly (ACE) units redesign the physical environment and use collaborative team decision making. In a transition model, an AP geriatric nurse serves as first contact to an elderly patient on admission, follows the patient and family through the hospital stay, and then serves as the patient's home health nurse during the first four weeks after discharge (Paier & Strumpf, 1999).

In ambulatory settings, AP geriatric nurses have been shown to improve care in Department

of Veterans Affairs facilities, private physicians'
offices, and continuing care retirement communities. Positive outcomes have also been reported
from geriatric nurse specialty practices that address
the common problems afflicting older people living
in the community, such as practices used to assess
and treat elderly patients with urinary incontinence,
elder abuse, and stroke. Home-care agencies are
only now beginning to consider using AP nurses
to deliver primary care to elderly home-bound patients. It is estimated that annually, over 9,000 Medicaid home health agencies deliver services to patients who are primarily elderly (Health Care Financing Administration, 1997). Early evidence of
the efficacy of GNPs in home care, movements
toward capitated home-care payments, and newly
authorized Medicare reimbursement for GNPs are
expected to further spur the use of GNPs in home
care. GNPs have been integral providers in mature
capitated model programs for the frail elderly, such
as PACE demonstration projects and the social
HMOs.

MATHY D. MEZEY

See Also
Geriatric Evaluation and Management Units
Geriatric Interdisciplinary Health Care Teams

Internet Key Words
ACE Units
Advance Practice Nurse (APN)
Ambulatory Care
Geriatric Nurse Practitioner (GNP)
Geropsychiatric Nurse Clinician (GNC)
Hospitals
Nursing Homes
PACE Demonstration Project
Social HMOs

Internet Resources
Gerontological Nursing Interventions Research
  Center
http://www.nursing.uiowa.edu/gnirc/index

National Association of Geriatric Education Centers
http://www.hcoa.org/nagec/

National Conference of the Gerontological Nurse
  Practitioners (NCGNP)
http://www.ncgnp.org/

AGEISM

Ageism is defined as a process of systematic stereotyping and discrimination against people because
they are old, just as racism and sexism accomplish
this for skin color and gender. Older people are
categorized as senile, rigid in thought and manner,
and old-fashioned in morality and skills. In medicine, terms like "crock" and "vegetable" have been
commonly used. Ageism allows the younger generation to see older people as different from themselves; thus, they suddenly cease to identify with
their elders as human beings. This behavior serves
to reduce their own sense of fear and dread of aging.
Stereotyping and myths surrounding old age are
explained in part by a lack of knowledge and insufficient contact with a wide variety of older people.
But another factor comes into play—a deep and
profound dread of growing old.

Ageism is a broader concept than gerontophobia, which refers to a rarer, "unreasonable fear and/
or irrational hatred of older people, whereas ageism
is a much more comprehensive and useful concept"
(Palmore, 1972). This concept and term were introduced in 1968 (Butler, 1969). The underlying psychological mechanism of ageism makes it possible
for individuals to avoid dealing with the reality of
aging, at least for a time. It also becomes possible
to ignore the social and economic plight of some
older persons. Ageism is manifested in a wide range
of phenomena (on both individual and institutional
levels), stereotypes and myths, outright disdain and
dislike, or simply subtle avoidance of contact; discriminatory practices in housing, employment, and
services of all kinds; epithets, cartoons, and jokes.
At times, ageism becomes an expedient method by
which society promotes viewpoints about the aged
in order to relieve itself from responsibility toward
them, and at other times ageism serves a highly
personal objective, protecting younger (usually

middle-aged individuals, often at high emotional cost), from thinking about things they fear (aging, illness, and death).

Ageism, like all prejudices, influences the behavior of its victims (Hausdorff, Levy, & Wei, 1999). The elderly tend to adopt negative definitions about themselves and to perpetuate the various stereotypes directed against them, thereby reinforcing societal beliefs. In a sense, the elderly "collaborate" with the enemy and with stereotypes. Margaret Thaler Singer observed similarities between the Rorschach test findings in members of a sample of healthy aged volunteers in the face of aging and of a sample of American prisoners of war who collaborated with their captors in Korea. Some older people refuse to identify with the elderly and may dress and behave inappropriately in frantic attempts to appear young. Others may underestimate or deny their age.

Ageism can apply to stages of life other than old age. Older persons have many prejudices against the young and the attractiveness and vigor of youth. Angry and ambivalent feelings may flow, too, between the old and the middle-aged. The middle-aged often bear many of the pressures of both young and old, and they experience anger toward both groups. Since the introduction of the concept of ageism, there have been some gains on the part of the elderly. The Age Discrimination and Employment Act of 1967, amended in 1978, ended mandatory retirement in the federal government and advanced it to age 70 in the private sector.

Some of the myths of age include a lack of productivity, disengagement, inflexibility, senility, and loss of sexuality (Stone & Stone, 1997; Bytheway, 1995). There have been some advances in, and more attention to, the productive capabilities of older people, and a better understanding that older persons have desires, capabilities, and satisfaction with regard to sexual activities. The "write-off" of older persons as "senile" because of memory problems, for example, is being replaced by an understanding of the profound and most common forms of what is popularly referred to as "senility," namely, Alzheimer's disease.

Senility is no longer seen as inevitable with age. Rather, it is understood to be a disease or group of diseases. When means of effectively treating dementia are available, ageism will also decline. Reminiscence or life review has helped focus attention on what can be learned from listening to the lives of the old. Indeed, the memoir has become, in the minds of some, the signature genre of our age. Today, old age is in the process of being redefined as a more robust and contributory stage of life. Unfortunately, the underlying dread, fear, and distaste for age remains.

ROBERT N. BUTLER

See Also
Intergenerational Care
Life Review

AGING AFRICAN AMERICANS

Nearly all gerontology professionals provide care to aging African Americans. Yet specialized cultural competency training and an understanding of the unique health care and social services needs of this fast-growing segment of the population are rarely promoted in the literature.

The aging African American population is three times as poor; is considerably more prone to chronic, debilitating physical illnesses; and, with the exception of those 85 and older (who tend to outlive their white counterparts), has a higher mortality and morbidity risk than white elderly (Jackson, Antonucci, & Gibson, 1995). Multiple reasons and contributors intermingle to cause these distressing facts.

Ethnic-Specific Medical Problems

Diseases of the heart, cancer, and HIV and AIDS are the three leading causes of death among African American males. For African American women, strokes replace HIV and AIDS as the third leading cause (Feldman & Fulwood, 1999). Poor diet is a significant contributor to a number of chronic diseases seen in the elderly African American population, such as hypertension, cardiovascular disease, and diabetes. For example, approximately 30% of cancer deaths may be nutritionally related. Many cardiovascular diseases and certain types of cancers are related to saturated fats in foods consumed in the minority community. These and other health management concerns, including hypertension, obesity, and diabetes, have an impact on the level of chronic illness and the high death rate among the African American elderly population (LeRoy, 1997). Hypertension, type II diabetes mellitus, hyperlipidemia (high cholesterol), cardiovascular diseases, and certain cancers (prostate, colon, and breast) have familial (hereditary) etiologies.

Social Structural Inequities

The dual consequences of racism and ageism and the related inequitable economic, educational, political, and social conditions have resulted in a lifetime tainted by impoverishment and multiple disadvantages for many African American elderly (Baker, 1997). These and other related factors contribute to a lack of trust and a reluctance by some aging African Americans to utilize needed services.

Research Concerns and Considerations

Research findings obtained from samples of white elderly subjects are not uniformly generalizable to African American elderly. In terms of general disparate treatment, African American elderly have much in common with others in their cohort. Yet they are not a monolithic entity (Stoller & Gibson, 1997), and intragroup differences must be acknowledged. For example, some African American elderly have supportive children and resourceful extended families, while others in the same cohort are isolated, lonely, and despondent. Many are frail and ill, but some enjoy a measure of affluence and good health (Stoller & Gibson, 1997). Perception of the effects of discrimination, availability of familial resources, personal hardiness, and ability to adapt to life's circumstances all contribute to the aging experience of African American elders. Their overall aging process is shaped by a host of factors, including but not limited to complex social, biological, cultural, and economic variables that influence health, health beliefs, and behaviors (Hooyman & Kiyak, 1999). These factors must be considered if research paradigms and methodologies are to yield valid and reliable data.

Service Utilization and Delivery

Although the need for adequate access to health care is great, low service utilization and limited awareness of and access to services in the African American elderly community are well documented (Feldman & Fulwood, 1999; LeRoy, 1997). There are many possible causes for this. Limited access to medical services, the absence of physicians in impoverished communities, lack of health insurance, and lack of health promotion and disease prevention services (LeRoy, 1997) contribute to the complex and deficient system of service delivery to this vulnerable population. Additionally, many African American elderly may be unaware of available services and how to take advantage of them. Formalized services may not be perceived as accessible or user-friendly to some. Others, particularly those age 75 and older, whose memories bear scar tissue from an antebellum frame of reference (Baker, 1997), may be reluctant to believe that representatives of the health care and social services systems are committed to their best interests. This makes overall intervention and management of quality health and mental health care delivery a challenge for gerontology practitioners.

Effective service delivery to aging African Americans may also be impeded by health care professionals' lack of awareness of the need for enhanced cultural competency training. Although incremental efforts have been made over the past decade, service delivery models have historically not been sensitive to racial and ethnic differences (Jackson et al., 1995). Those minority professionals who have an understanding of and are comfortable working with African American elderly must get involved and offer their voices to all aspects of service delivery, from initial planning through interpretation of outcomes and evaluation efforts.

Professionals serving African American elderly need to understand the manner in which sociodemographic trends, environmental factors, and cultural patterns influence health status as well as help-seeking behaviors among African American elderly. Gerontologists from all disciplines have an obligation to be cognizant of the cumulative effects of socioeconomic forces, including racial injustices and lack of access to adequate employment, housing, transportation, health care, and other resources in communities where indigent African American elderly reside.

Options and Suggestions for Improvement

Specialized training in cultural diversity and an understanding of the unique experience of African American elderly are essential for professional service providers seeking to work effectively with this group of elders. Various initiatives could help enhance service delivery to aging African Americans:

Proactive Health Promotion and Outreach Programs: Programs should be integrated and linked with other community resources used by African American elderly (senior citizen centers, community health centers, churches, and community organizations). Given the historic ability of black churches to mobilize, organize, and influence African Americans in the community, the church would be an excellent liaison between African American elderly and formal systems of care (Baker, 1997).

Culturally Sensitive Communication and Advertising Strategies: Minority-targeted advertisements on television and radio must be utilized to promote health care options, opportunities, and practices on a regular basis. Both written and visual communication should incorporate and reflect compatibility with cultural beliefs. Information should not be written in highly technical language (e.g., professional jargon, undefined acronyms) for an African American or any elderly readership. Visual images should utilize respected role models from the minority community (e.g., clergy, social workers, nurses, physicians, and other allied professionals) to promote and disseminate health care and social services information.

Enhanced Interpersonal Skills: Professionals must:

- *Heighten self-awareness:* Acknowledge and eliminate any biases and stereotypical thinking patterns. Cultural sensitivity should be a way of life.
- *Individualize:* Eliminate stereotypical generalizations. For example, even though their educational opportunities were limited (particularly for those age 75 and over), never assume that elderly African Americans are illiterate. Many who had very little formal education can read, write, and count well, but some cannot. Recognizing individuality is important, because life experiences vary for each African American elder, depending on many lifestyle and individual factors.
- *Show positive regard, courtesy, and respect:* Address African American elderly properly as Mr. or Mrs. (As a means of showing respect, Sir and Ma'am are also commonly used in addressing African American elderly.)
- *Actively listen:* Seek cultural and linguistic interpretations when uncertainty exists. Obtain an understanding of certain pronunciations and the meanings of various concepts (e.g., diabetes mellitus may be understood and referred to as "sugar," losing weight may be termed "falling off"). Ask for clarification and accurate interpretations as needed from the family, church members, or close friends. Withhold diagnostic assessment until there is a clear understanding of the articulated problem.
- *Be genuine:* Demonstrate a genuine sense of caring, kindness, and concern. These qualities are perceived and interpreted positively, even by poorly educated recipients of health and mental health services.

MARJORIE E. BAKER
GARY L. LEROY

Internet Key Words
Barriers
Black/African American Elderly/Aging
Ethnicity
Formal and Informal Services

Health Care
Mental Health

Internet Resources
Healthy People 2010
http://www.health.gov/healthypeople

Institute for Social Research, Research Center for
Group Dynamics, Program for Research on
Black Americans at the University of Michigan
http://www.irs.umich.edu/rcgd/prba

AGING AGENCIES: CITY AND COUNTY LEVEL

Most older adults wish to remain living in their
own homes, but they may need assistance, particularly from local resources, to do so. There are many
options for older adults. A variety of private and
governmental services are available to older adults
at the city and county level. Because local services
vary greatly, however, it is important for providers
to contact the local government for details. Five
sources of support for older adults are typically
available in most areas of the United States: senior
centers, area agencies on aging (AAAs), home
health agencies (HHAs), hospice care, and the Eldercare Locator. Programs and services specifically
designed for older persons who are actual or potential victims of abuse and neglect are provided by
adult protective services.

Senior Centers

Senior centers, in some form, are typically available
at the local level even in rural areas. Approximately
15,000 senior centers exist throughout the United
States (National Council on the Aging [NCOA],
1999), and approximately 15% of older persons
participate in senior centers (Hooyman & Kiyak,
1999). Senior centers are an integral part of the
aging network, serving community needs, assisting
other agencies in serving older adults, and providing opportunities for older adults to develop their

potential as individuals (NCOA, 1990). Centers
may offer activities and programs for seniors to
enjoy recreation and socialize with their peers.
Many senior centers are also congregate meal sites.
Some centers have day programs, and many provide
outreach services through on-site visits from various health care providers (e.g., nurses, podiatrists, nutritionists).

Senior centers are based on the belief that
"aging is a normal developmental process; that human beings need peers with whom they can interact
and who are available as a source of encouragement
and support; and that adults have the right to a
voice in determining matters in which they have a
vital interest" (NCOA, 1990, p. 17). The centers
attempt to provide an environment where older
adults can continue to develop and grow while
forming relationships with others. A variety of formal and informal classes and clubs are offered at
centers. If the senior center does not provide a
needed service, the center may be a resource for
referrals to other community services. For many
older adults and their families, the senior center
serves as the "front door" to community-based services for both the well and the functionally impaired
elderly. Senior centers may be operated by a local
board of directors; they may be part of local, municipal government; or they may be operated as a not-
for-profit agency. The NCOA provides information
on senior centers.

Area Agencies on Aging

AAAs are responsible for planning, coordinating,
evaluating, and monitoring home and community-
based care programs for older adults. Approximately 670 AAAs are run by state, county, or city
governments or as nonprofit or public agencies designated by the state (Binstock, 1991). Each AAA
is responsible for a designated geographical area
known as a planning and service area (PSA). The
AAAs create a plan for their PSA to ensure that
local needs of older persons are being addressed.
AAAs can be an excellent source of information
to service providers and health professionals, as

they are familiar with most, if not all, programs serving the needs of older persons in the local community.

Many health care professionals, especially physicians, are unfamiliar with AAAs and their importance to the continuum of care. AAA staff can provide elders and their families with information and suggestions for local resources based on their specific needs. Through this service, AAAs can act as advocates for local elders at either the individual or the policy level. AAAs have financial responsibility to administer federal, state, and local funds to support locally specific services in their PSA. These services may vary widely by region but may include case management services, transportation, counseling, adult day-care programs, health screening and education, nutritional education, meals, legal assistance, residential repair, physical fitness, recreation, home care, respite care, telephone reassurance, and volunteer services, among others. AAAs monitor the programs they support to ensure that high-quality services are being provided effectively and efficiently (Mid-Florida Area Agency on Aging, 1999). The number of AAAs varies greatly from state to state. Rhode Island has only one AAA, whereas New York has approximately 60. The National Association of Area Agencies on Aging publishes a directory of all AAAs nationwide.

Home Health Agencies

HHAs have become major providers of care for older adults. In 1998, there were 9,655 Medicare-certified HHAs throughout the United States; 66% of clients were age 65 or over (National Association for Home Care, 1999a). Medicare-certified HHAs provide intermittent skilled care. This care is typically provided through the order of a physician for a limited time. There are no exact limitations on the duration of care; it is dependent on the need for skilled services. The goal of the HHA is to help clients remain in their homes safely while receiving the assistance they need to do so. HHAs typically provide a multidisciplinary approach to care that can include skilled, intermittent nursing care; occupational, physical, and speech therapy; respiratory therapy; medical social work; and home health aide services. An in-home assessment is the first step in determining the type and level of client need. A care plan is then designed to address those needs. Nursing services can provide wound care, medication management, and status monitoring, as well as training and support for both the patient and the caregiver. Home health care can be provided either through Medicare, which has stringent eligibility and duration limitations, or through private providers (paid for by the client). Some areas may use HHAs to provide Medicaid-funded long-term custodial care in the home through the federal In-Home Supportive Services Program.

Medicare-certified HHAs can provide social work services to assist seniors with long-range planning, advance directives, short-term counseling, and transition counseling to help people deal with new or changing circumstances; they can also advocate for older persons. Home health aides provide assistance with bathing and grooming, transferring, hygiene, toileting, household chores, meal preparation, exercises, and errands. Home-care agencies can be located by contacting the National Association for Home Care or through the local Yellow Pages. Approximately 30% of home-care agencies throughout the United States are hospital based, so contacting the local hospital may be an effective method of locating HHAs.

Hospice Care

Hospice care is an important component of the local care continuum. Hospice services provide palliative rather than curative care to address the social, emotional, and spiritual needs of terminally ill individuals and their families. Often thought of as a program for those with cancer, hospice care can be provided to any terminally ill individual, regardless of diagnosis, including those with end-stage respiratory disease, heart disease, Alzheimer's disease, AIDS, or amyotrophic lateral sclerosis. Like HHAs, hospice care is typically interdisciplinary in nature and may include nurses, social workers, chaplains, home-care aides, and bereavement counselors to help people cope with their illness and end-of-life issues.

The hospice team specializes in symptom management, providing the patient with the highest possible quality of life. Hospice care can take place in many different settings, including the individual's home, a hospital, or a long-term-care facility. As of December 1997, 2,274 hospice programs had been identified throughout the country, with an additional 400 volunteer hospices (National Association for Home Care, 1999b). Information regarding hospice care at the local level can be obtained through the local Yellow Pages, the National Association for Home Care Web site, or the National Hospice Helpline (1-800-658-8898). This toll-free number can provide referrals to national and international hospice organizations (see the Web site for the National Hospice Organization).

Eldercare Locator

The Eldercare Locator provides the names, telephone numbers, and service information for agencies throughout the United States. This service is staffed by trained professionals who provide the information needed to contact a care provider or agency in the designated area. They provide information on services such as meal programs, home care, transportation, housing alternatives, home repair, recreation, and social activities, as well as legal and other community services (see the Web site). The toll-free number for the Eldercare Locator is 1-800-677-1116, and it is operational Monday through Friday, 9:00 A.M. to 8:00 P.M., eastern standard time. One value of the Eldercare Locator is its ability to research services for family members whose older relatives live far from them.

CHARLES A. EMLET
AMIE RANG

See Also
Adult Protective Services
Aging Agencies: Federal Level

Aging Agencies: State Level
Home Health Care
Hospice
Senior Centers

Internet Key Words
Area Agencies on Aging
Eldercare Locator
Home Health Agencies
Hospice
Senior Centers

Internet Resources
AAA listing
http://www.mfaaa.org/agingnetwork.html#AAA.

Eldercare Locator
http://www.aginginfo.org/elderloc

National Association for Home Care
http://www.nahc.org

National Association of Area Agencies on Aging
http://www.n4a.org

National Council on the Aging
http://www.ncoa.org

National Hospice Organization
http://www.nho.org

AGING AGENCIES: FEDERAL LEVEL

The U.S. Administration on Aging (AoA), an agency of the U.S. Department of Health and Human Services, functions as the federal focal point and advocacy agency for older persons. The AoA administers federally funded programs established under the Older Americans Act and provides support to states and communities in the development of systems of support for older people.

Passed by Congress in 1965, the Older Americans Act has been amended 11 times and was last reauthorized in 1992. Congress intended the act to improve the lives of all older Americans in areas such as income, health, housing, employment, in-home and community services, research, and education. Anyone 60 years of age or older qualifies for programs funded by the act, even though services tend to be targeted at those elders in greatest social and economic need. The act's various titles form the foundation for a broad spectrum of services and providers that have become known as the Aging Network. Implementation of the network over the last 30 years has resulted in an impressive system of services on both the national and local levels that strives to promote independent living, create opportunities for active older persons, and meet the needs of older persons at risk of losing their independence.

The AoA has overall responsibility for administering the Older Americans Act and distributing federal funds in accordance with the act's requirements. In addition, it sets policy for the Aging Network at the state and local levels and funds national grantees that provide research, training, support, and demonstration programs for the Aging Network. The AoA is responsible for numerous functions specified in the act, including effective and visible advocacy for the elderly within the federal government, coordination of research and implementation of programs, provision of technical assistance to states and communities, collection and dissemination of information, and monitoring and evaluation of programs developed pursuant to the act.

The AoA is headed by the assistant secretary for aging (ASA), who is appointed by the president. The ASA has the authority to issue regulations and policies that interpret and implement the Older Americans Act, and it oversees the agency that works in partnership with a network of some 57 state units on aging (SUAs); 680 area agencies on aging (AAAs); 228 Native American, Alaskan, and Hawaiian tribal organizations representing 300 tribes; 5,000 senior centers; and more than 27,000 local service providers. The AoA has nine regional offices throughout the country that provide technical assistance to the states, communicate AoA national policies, and review and monitor the SUAs' plans. The AoA distributes funds authorized by the act to the SUAs on the basis of state plans and

according to a formula that considers the number and percentage of older persons in each state. The SUAs then distribute the funds to AAAs, which contract with local service providers.

The majority of services provided throughout the Aging Network are supported by funds authorized under Title III of the act. Each community offers different services, depending on available resources. Title III funds can be used for information and referral services; supportive services and centers; disease prevention and health promotion; nutrition services, including food distribution and both congregate and home-delivered meals; homemaker, home health aide, and other in-home services for the frail elderly; senior centers and daycare programs; transportation; ombudsman programs for residents of long-term-care facilities; crime prevention and victim assistance programs; translation services for non-English-speaking elders; protective services for abused, neglected, and exploited elders; and legal services.

In addition to services and programs funded under Title III, the AoA administers funds under Title IV of the act, earmarked for its own operations and for providing direct grants and contracts for research, training, and demonstration programs on a national level through the Discretionary Funds Program. The Title IV mandate is aimed, generally, at building knowledge, developing innovative model programs, and training personnel for service in the field of aging.

Title V of the act authorizes funds to subsidize part-time community-service jobs for unemployed, low-income persons age 55 and older. The national organizations that receive most of the money to implement employment programs include the American Association of Retired Persons, Asociación National Pro Personas Mayores, National Center on Black Aged, National Council of Senior Citizens, National Urban League, Green Thumb, National Pacific/Asian Resource Center on Aging, National Indian Council on Aging, U.S. Forest Service, and National Council on the Aging.

Title VI awards annual grants for nutritional and supportive services for American Indian, Alaskan Native, and Native Hawaiian elders. Training and technical assistance are provided to Title VI grantees both electronically and through on-site, telephone, and written consultation; national meetings; and newsletters by AoA staff, Three Feathers Associates, and the Native American Resource Centers. The resource centers receive AoA cooperative agreement grants to serve as focal points for the development and sharing of technical information and expertise to Indian organizations, Title VI grantees, Native American communities, educational institutions, and professionals and paraprofessionals in the field.

Title VII, initiated in the 1992 amendments to the act, oversees the protection of vulnerable elder rights. These activities include state long-term-care ombudsman programs; programs for the prevention of elder abuse, neglect, and exploitation; state elder rights and legal assistance development; and state outreach, counseling, and assistance for insurance and public benefits.

Long-term-care ombudsmen advocate on behalf of individuals and groups of residents in nursing homes, board and care homes, assisted living facilities, and other adult care facilities. They also work to effect systems changes on local, state, and national levels. The AoA provides funds for national support groups, including the National Long-Term Care Ombudsman Resource Center, operated by the National Citizens' Coalition for Nursing Home Reform in conjunction with the National Association of State Units on Aging. The center provides on-call technical assistance and intensive annual training to assist ombudsmen.

The AoA supports a national legal assistance system, primarily through state legal services developers. In addition, AoA's Discretionary Funds Program helps support national legal support projects, as well as centers that provide training and technical assistance to advocates in the field. The support centers collaborate in their efforts through information exchange, training, and liaison with field advocates. Support centers currently funded by the AoA are the National Consumer Law Center, the nation's consumer law expert; the Center for Social Gerontology, a nonprofit research, training, and social policy organization dedicated to promoting the au-

tonomy of older persons and advancing their well-being in society; the American Bar Association Commission on Legal Problems of the Elderly, which examines law-related concerns of older persons; and the American Association of Retired Persons Foundation's National Training Project, which provides training to lawyers and other professionals who advocate for the rights of older persons.

Other public services supported by AoA funds include the Eldercare Locator, the National Aging Information Center, and the National Center on Elder Abuse. The Eldercare Locator is administered by the AoA, together with the National Association of Area Agencies on Aging and the National Association of State Units on Aging. It is a nationwide directory-assistance service designed to help older persons and caregivers locate local support resources. The National Aging Information Center serves as a central source for a wide variety of programs and policies, related materials, and demographic and other statistical data on the health, economic, and social status of older Americans. Its services are free of charge and include access to information, databases, printed materials, statistical information, and a reading room and reference collection. The National Center on Elder Abuse provides information to professionals and the public, offers technical assistance and training to elder abuse agencies and related professionals, conducts short-term research, and assists with elder abuse program and policy development.

Several other federal agencies have programs that provide services to older persons, even though the agencies do not exclusively serve older persons. The Heath Care Financing Administration (HCFA) administers Medicare, Medicaid, and child health insurance programs. Many older Americans receive health care assistance through Medicare and Medicaid. The National Institute on Aging, one of the National Institutes of Health, promotes healthy aging by conducting and supporting biomedical, social, and behavioral research and public education. The Social Security Administration is responsible for several programs that directly affect seniors, including the retirement, survivors, and disability insurance program and the supplemental security income program.

Information on all the agencies mentioned is available on the main Web site of the Administration on Aging (see below). Caregivers, service providers, and others involved with the needs, protection, and advocacy of older persons should be familiar with the vast array of services and programs supported or administered by the AoA and the agencies and organizations with which it collaborates.

MARY JANE CICCARELLO

See Also
Aging Agencies: City and County Level
Aging Agencies: State Level
Older Americans Act
Social Security

Internet Key Words
Administration on Aging
Area Agencies on Aging
Older Americans Act
State Units on Aging

Internet Resources
Administration on Aging
http://www.aoa.gov/

Eldercare Locator
http://www.aoa.gov/elderpage/locator.html

Health Care Financing Administration
http://www.hcfa.gov/

National Aging Information Center
http://www.aoa.gov/NAIC/

National Institute on Aging
http://www.nih.gov/nia

National Senior Citizens Law Center
http://www.nsclc.org/

Social Security Administration
http://www.ssa.gov/

AGING AGENCIES: STATE LEVEL

State services are designed to assist the elderly and their caregivers in maintaining health and adequate finances and remaining safely in the community for as long as possible. Many state services developed pursuant to passage of the federal Older Americans Act. The Act charges states to provide services to the elderly, age 60 and older, and to remove economic, physical, and social barriers to their remaining in the community and out of institutions. Many federally funded services administered by the states have federal guidelines to ensure uniform eligibility nationwide. However, with trends for increased local and state control over the allocation of funds and creation of alternate delivery systems, the same service can vary among states.

The state's role in care for the elderly falls into four categories: research, information dissemination, investigation and licensing, and services. How these functions are provided varies within and among states and within and among state agencies.

Research

States conduct research on topics that affect quality of life and services for their citizens, such as quality of care in nursing homes, cost-effectiveness of managed long-term care, and the effect of taxes, and on new services, such as assisted living. Several states are planning services for the projected increase in the number of elderly in the 21st century.

Dissemination of Information

Many state agencies have toll-free telephone services and Web sites to provide information for the elderly and their caregivers on topics such as finding state services, obtaining preventive services, and contacting investigatory agencies to report fraud and abuse. Almost all states offer a federally funded state health insurance program that provides information and counseling on health insurance, including Medicare and Medicaid, long-term-care insurance, and Medigap insurance policies. State veterans' agencies provide information regarding veterans' benefits (state and federal) such as tax reductions, home-care services, and health care.

Investigation and Licensing

State governments investigate fraud and abuse and control the licensing and certification of individuals (physicians, nurses, social workers, and dentists) and organizations that provide care to elderly residents of the state. Nursing homes and adult homes are licensed, certified, regulated, and surveyed by state agencies. In many states, the office of the long-term-care ombudsman investigates charges of institutional abuse, mistreatment, and neglect.

Services

All states provide adult protective services that investigate reports of and protect the elderly from financial, sexual, physical, and mental abuse. Some states have programs and services for elderly residents that may not be replicated in other states, such as prescription drug programs that subsidize pharmaceuticals, special school and property tax breaks, and recreational benefits. The services may have eligibility criteria based on medical need, age, and income.

Generally, services for the elderly are administered and monitored by state agencies that receive and pass federal and state funds to localities. The agencies that administer these federal funds may differ in different states; programs and jurisdictions vary. Some states provide direct services to senior citizens; others monitor locally delivered services. A local area agency on aging or the state unit on aging can provide information on what services are available in communities and how they can be accessed by clinicians, older adults, and family members.

Several state-operated services support health care needs or provide preventive health programs such as health screening and well-patient clinics. Preventive and rehabilitative mental health services offered at the community level may be state funded and operated.

Many states offer property, sales, and school tax breaks for senior citizens or low-income older adults. Senior citizens may pay reduced admission fees at state parks and recreational sites. Some states also provide recreational or sport events—"senior games" or "senior Olympics"—designed to help older adults remain healthy and competitive.

Many support services provided locally, such as home care to assist older adults with personal hygiene, light housekeeping, meals, and chores, are monitored by the state. A case management or care management professional assigned by the state or local agency assesses the needs of the older person and establishes a plan of care.

Services that may be available outside the home (i.e., community-based services) include nutrition education and counseling; transportation for medical appointments, shopping, and visiting; and other activities. Lunchtime or evening meals can be provided in group (congregate) settings at community sites. Senior centers in urban and rural areas offer support services such as meals, counseling, recreation, and visiting. Social-model adult day centers offer support, management, and nutrition services. Medical-model centers provide rehabilitation and therapy as well as support and nutrition services.

Financial assistance is available through state-monitored programs that help with the costs of housing or home renovation, health care (Medicaid), and prescription drugs and provide state-funded employment opportunities for low-income older adults.

The maze of services can be overwhelming to older adults, particularly those for whom English is a second language, and their families. Thus, it is critical that health care professionals be aware of the services and benefits elders need to maintain a quality life.

PATRICIA P. PINE

See Also
Adult Day Care
Aging Agencies: City and County Level
Aging Agencies: Federal Level
Case Management
Meals On Wheels
Veterans Affairs (VA) Health Care Services for
  Elderly Veterans

Internet Key Words
Information Dissemination
In-Home Care
Investigation and Licensing
Recreation
Research
Services

Internet Resources
New York State Office for the Aging
http://www.aging.state.ny.us

U.S. Administration on Aging
http://www.aoa.dhhs.gov/aoa/pages/state.html

AIDS
See
Human Immunodeficiency Virus (HIV) and
   AIDS

"ALERT" SAFETY SYSTEMS

"Alert" safety systems, or personal response services, as they are known generally, were developed in the 1980s to respond to the emergency needs of the elderly and the physically challenged living alone.

A communicator unit is installed in the person's residence using the existing electrical wiring and the telephone line. A button or other sending mechanism is typically worn around the neck or on the wrist. At the push of the button, an outbound call is made, establishing a connection with a monitoring platform staffed by a person who can respond to the individual in trouble. The qualifications of monitoring platform staff vary according to the company providing the service. Monitoring platforms are available nationwide and typically receive and respond to thousands of signals daily. Historically, monitoring platforms were known as call centers and were located in hospital emergency rooms. With recent technological advances, the trend is toward routing calls to more sophisticated

monitoring platforms designed with computer telephone integration and large databases to meet the needs of the people they support each day.

There are no formal standards for the personal response services industry. However, clinicians, families, and individual clients should consider several factors when choosing a personal response service:

- Voice-to-voice technology allows the older or physically challenged person using the device to have a two-way conversation with monitoring platform staff.
- Waterproof sending mechanisms allow the device to be worn in the bath or shower, where most accidents occur.
- Devices should be lightweight (half an ounce), with an activation area that is easy to depress.
- Persons with limited finger mobility should be able to activate the device. Adaptations include pillow switches, sip and puff switches, and switches that are highly sensitive to movement, which, when placed on any part of the body, such as the eyebrow, can activate a call to the monitoring platform.
- National language lines should be available to meet the needs of non-English-speaking persons.
- The monitoring platform should be equipped with TTY and TTD devices to meet the needs of the deaf and aphasic.
- Response to a call should occur within 30 seconds, to enable emergency response units to save lives.
- Communication ranges should be greater than 300 square feet from the device to the communicator or transmitter.
- A speaker phone enables the person to communicate via the telephone without getting up from a chair or the floor.

Voice-to-voice connection assistance significantly improves the likelihood of a positive outcome. Gurley, Lum, Sande, Lo, and Katz (1996) reported a total mortality of 67% for patients estimated to have been helpless for more than 72 hours, compared with 12% for those who had been helpless for less than 1 hour. These systems may help lengthen an older person's independence. Tinetti and Williams (1997) reported that the inability to get up after a fall was associated with placement in a skilled nursing facility. Further research is needed to demonstrate the cost-effectiveness of this type of service for the older individuals living alone in the community.

Personal response services are currently available through a dozen approved Medicaid and long-term home health care programs nationwide. The service is also available to individuals and families who can pay privately. Costs usually involve an installation fee and a monthly cost of approximately $35 to $45. Referrals for personal response services can be obtained from a local hospital upon discharge, from home-care nurses, or from the local office of the aging.

MARY MARZULLO

See Also
Aging Agencies: City and County Level
Assistive Devices
Burns and Other Safety Issues

Internet Key Words
Emergency Response Service
Medical Alarms, Systems, and Monitoring

Internet Resource
Rehabilitation/adaptive devices
http://www.Abledata.com

ALLERGIC RHINITIS

Atopic diseases are often overlooked or ignored in the elderly because allergies more commonly present in the first three decades of life. Allergies can occur at any age, however, and can persist throughout the aging process. In fact, allergic conditions such as allergic rhinitis, allergic conjunctivitis, pruritus, chronic urticaria, contact dermatitis, food allergy, drug hypersensitivity, and asthma can be presenting first-time complaints in the elderly.

Rhinitis

Rhinitis is prevalent among the elderly and can be caused by allergic rhinitis, polyps, atrophic rhinitis, sinusitis, drug-induced rhinitis, and idiopathic rhinitis. Moreover, the aging process includes normal physiological changes that can aggravate underlying nasal symptoms.

With aging, there is a decrease in total body water content, deterioration of mucus-secreting glands, and reduction of nasal blood flow, all of which contribute to atrophy and drying of the nasal mucosa. Age-related changes include degeneration of collagen and elastic fibers in the dermis, resulting in weakening of the upper and lower lateral nasal cartilages, retraction of the nasal columella, and downward rotation of the nasal tip, which result in increased nasal airflow resistance. The mucociliary system is less effective, and mucus viscosity is increased.

Allergic Rhinitis

Allergic rhinitis is a common disease affecting more than 20 million people in the general population. It is more common in younger than elderly patients, and complaints can be minor to severe. Commonly known as hay fever, allergic rhinitis can cause many symptoms, including nasal congestion, sneezing, runny nose, and itchy eyes, ears, and throat. Nasal congestion can result in mouth breathing, snoring, loss of taste or smell, and "allergic shiners," or dark circles under the eyes. Many elderly patients complain of a postnasal drip. Some patients have symptoms only during certain seasons; others have problem all year long that worsen during certain seasons. Perennial allergens include molds, roaches, dust mites, and household pets. Seasonal allergens include various pollens, including those from trees, grasses, and weeds.

Patients can be diagnosed clinically and treated empirically. However, referral to an allergist for an in vivo skin test or in vitro radioallergosorbent test (RAST) can identify triggers of immediate IgE-mediated reactions and can be confirmatory. Treatment recommendations can then be made.

Causes

The immune system protects the body from dangerous bacteria and viruses while ignoring harmless substances. Allergies occur when the immune system reacts inappropriately to harmless substances such as pollen, animal dander, mold spores, or dust mites. In nonallergic individuals, nothing happens when pollen or cat dander enters their bodies. When these same substances enter the body of an allergic rhinitis sufferer, the immune system is activated. Subsequently, IgE antibody binds to mast cell receptors, resulting in the release of mediators such as histamine, leukotrienes, prostaglandin, and platelet-activating factor. Chemotactic factors are also released and recruit neutrophils, eosinophils, and basophils to produce inflammation and make the eyes and inside of the nose itch, swell, and produce watery mucus.

Management and Treatment

Environmental Control: The first line of defense against allergies is to avoid or minimize exposure to the offending allergen. Allergy testing can help identify the culprit allergens. If the patient is allergic to pollen, windows should be kept closed, and air-conditioning can be used to filter incoming air. The air conditioner filter needs to be changed frequently, but preferably not by the patient. Avoiding unnecessary travel during peak pollination times can be helpful. Weather reports frequently include pollen counts for the area, for the benefit of allergy sufferers.

Allergies to perennial indoor allergens are more challenging to treat in the elderly. Mold, roach, and dust mite avoidance is difficult. Vigorous cleaning in the patient's environment is extremely helpful but not always possible for an elderly person to accomplish. Mold and dust mites proliferate in warm, humid environments; therefore, keeping the home dry, with a humidity of less than 60%, is advised. However, humidity of less than 30% can be irritating to the nose, especially since elderly patients often have dryer and more atrophic mucosa. The elderly may need assistance to fix water leaks, clean their homes, and get rid of pests such as roaches and mice. If the patient has a dust mite allergy, removing carpeting and replacing upholstery with wipeable furniture are useful interventions. Acaricides can be recommended if the elderly refuse to remove carpeting. Benzyl benzoate kills dust mites but does not denature the dust mite allergen; tannic acid is a protein-denaturing agent that does not kill dust mites but denatures the allergen. These products require frequent application, and their use can become expensive and time-consuming. All bedding should be washed in hot water above 135° F (57° C) weekly, and mattresses and pillows should be encased in plastic, impermeable, allergy-proof dust mite covers.

If the patient's pet is the source of or a contributor to the problem (i.e., the inciting allergen is animal dander), avoiding exposure or keeping the pet outdoors or at least off the bed is strongly recommended. Many elderly patients prefer the company of cats, since they do not need to be walked on a daily basis. Unfortunately, cat allergen is produced by the cat's sebaceous glands and becomes airborne when skin scales are shed. Cat allergen particles are small and remain airborne for long periods. Significant amounts of cat allergen can be found in upholstered furniture, carpeting, and bedding. The best recommendation, unfortunately, is for the animal to live elsewhere, which is a difficult situation for many elderly patients, since the pet is often a friend or companion. Even if the animal is removed from the domicile, it may take several months of good household cleaning to reduce cat allergen levels. If the cat cannot be relocated, then bathing the cat weekly, not allowing the animal into the patient's bedroom, and frequent, thorough vacuuming will help. Installing high-efficiency particulate arresting (HEPA) filters on the vacuum cleaner, on air ducts, and in heating systems will also reduce airborne allergens.

Unfortunately, avoidance of allergens is sometimes impossible, for example, removing trees and grass from the patient's surrounding. Mold is ubiquitous both indoors and outdoors; it is almost impossible to completely avoid. It can also be very costly to get rid of all of the patient's rugs and upholstered furniture. Instituting change may be financially unfeasible for elderly on small, fixed incomes. Frail elderly are not physically capable of making certain changes and will require help from the family or other caregiver.

Medication: Three classes of medications are used: antihistamines, decongestants, and anti-inflammatories. Antihistamines block the effect of histamine; they prevent itchy and runny nose, sneezing, and postnasal drip down the back of the throat. Decongestants act by shrinking the dilated blood vessels back to normal size, thereby decreasing swelling and congestion in the nose. Anti-inflammatory medications, such as corticosteroids and mast cell stabilizers, act directly on the immune cells to prevent release of chemical mediators.

Many elderly patients desperate for quick relief go to the nearest pharmacy for an over-the-counter (OTC) solution. Numerous pills, nasal sprays, and eyedrops are available, but nonprescription medications must be used with caution. Most OTCs consist of an antihistamine or a decongestant or a combination of both.

OTCs generally contain older, first-generation antihistamines that are extremely effective at relieving allergy symptoms but can also cause sedation and impaired judgment. Studies show that sedative antihistamines impair driving ability as much as alcohol (Pirisi, 2000). Elderly patients who need to drive should avoid taking such sedative antihistamines. Another problem with first-generation antihistamines is their anticholinergic properties, which produce dryness of the mouth and eyes, blurred vision, constipation, and urinary retention. Since elderly patients are prone to dryness, these medications exacerbate the situation. First-generation antihistamines should be avoided in patients with a history of symptomatic prostatic hyperplasia, bladder neck obstruction, or narrow-angle glaucoma.

Second-generation antihistamines are slightly more costly but are less soporific. Sedating antihistamines in combination with alcohol or other central nervous system depressants (sedatives and tranquilizers) should be avoided because they diminish alertness and cognitive performance. Polypharmacy can be avoided by instructing elderly patients to keep an up-to-date list of their current medications to show their health care providers and pharmacists every time there is a change in medications. Elderly patients should be advised to fill all prescriptions at one pharmacy. Pharmacists often computerize or keep logs of patients' medications, but unfortunately, OTC nonprescription medications are usually not included on such lists.

Since antihistamines have little effect on nasal congestion, OTC pills and nasal sprays often contain decongestants. Decongestant pills do not cause sedation. In fact, the opposite is often true. They can cause jitteriness, irritability, anxiety, insomnia, palpitations, and elevated blood pressure. Rarely, cerebrovascular events and cardiovascular problems, including myocardial infarction and arrhythmia, have been reported. Thus, elderly patients with poorly controlled hypertension may need to avoid decongestants. Decongestant nasal sprays such as oxymetazoline and phenylephrine are extremely effective for short-term use but if used chronically can cause medicamentous rhinitis. Topical decongestant nasal sprays should not be used for more than three or four days; prolonged use can result in increased rebound nasal swelling and congestion.

Many of the newer prescription antihistamines offer much needed-relief with significantly less sedation. Nonsedating prescription antihistamines include fexofenadine and loratadine. Cetirizine, a new antihistamine that is a metabolite of hydroxyzine, is also available but is mildly sedative. Terfenadine and astemizole are nonsedative but have been removed from the U.S. market because they cannot be used with certain medications. Also, serious heart rhythm problems and deaths have been associated with the use of terfenadine or astemizole together with quinine, certain antibiotics, and antifungal medications, as well as when taken with grapefruit juice. Elderly patients, unaware of their removal from the market, may still have them in their medicine cabinets and need to be told to discard them for newer, safer antihistamines. Cetirizine and loratadine are both available in liquid suspension and can be used by elderly patients who have difficulty swallowing pills or have swallowing disorders. Azelastine is the first and only antihistamine available as a nasal spray.

Anti-inflammatory medications are very effective in relieving nasal symptoms of itching, congestion, sneezing, and watery discharge but do little to control eye symptoms. Active ingredients are corticosteroids and mast cell stabilizers. These medications are available as nasal sprays and are often used in conjunction with antihistamine pills. The steroid nasal sprays include fluticasone propionate, triamcinolone acetonide, flunisolide, beclomethasone dipropionate, budesonide, and mometasone furoate monohydrate. The mast cell stabilizer cromolyn was a prescription medication but is now available over the counter as a nasal spray.

The amount of time needed for medications to become effective varies. Antihistamines, because they directly block the chemical mediator histamine, work rapidly and are generally effective within hours. Anti-inflammatory medications work on the immune cells to prevent them from releasing chemical mediators and take longer to have an effect. For example, corticosteroids can take hours to days to become effective, and cromolyn can take two weeks to reach effective levels.

Immunotherapy: If the combination of avoidance techniques and medication fails to adequately control symptoms, "allergy shots" or immunotherapy may be an option to build tolerance to allergens. Increasingly larger doses of allergens are injected over time. Immunotherapy is the only available treatment that has the potential to cure an allergy and has been used in the treatment of seasonal and perennial allergic rhinitis caused by pollens, pet allergens, dust mites, and some molds. Since each patient is allergic to different allergens, immunotherapy must be individualized.

Immunotherapy carries a small risk of systemic reactions that can range from acute rhinoconjunctivitis or urticaria to severe bronchospasm or anaphylaxis. Elderly patients with comorbid heart disease, poorly controlled asthma, or severe chronic obstructive pulmonary disease are at increased risk for severe anaphylactic reactions. The risk-benefit ratio must be carefully weighed for elderly patients.

Immunotherapy is generally covered by Medicare, Medicaid, and private insurance, but coverage may vary from state to state. Precertification or preapproval may be required.

MARY LEE-WONG

See Also
Immunization
Over-the-Counter Drugs and Self-Medication
Polypharmacy: Drug-Drug Interactions
Pruritus (Itching)

ALTERNATIVE/ COMPLEMENTARY MEDICINE

Alternative and complementary therapies are becoming more common in the health care of older individuals in the United States. Controversy over the use of these therapies relates to different conceptions of cause and effect. Many alternative and complementary therapies arose from an Eastern philosophy, in contrast to Western Cartesian and Newtonian thought. Traditional science, or reductionism, has its roots in the early 17th century. René Descartes claimed that the best way to elevate and organize the search for truth was to eliminate that which could not be observed with the five senses. All that could not be seen was to be ignored, and only that which could be measured and experienced was suitable in the scientific search for cause and effect. Later, Sir Isaac Newton developed the theory of gravity, outlined mathematical rules of physics, and described the theories on which modern science is based. In the early 1900s, Albert Einstein suggested another way of viewing reality based on his understanding of the behavior of subatomic particles. Subsequently, quantum physics and systems theory formed the basis for the theoretical foundation of holism—a concept that attempts to describe the outcomes of alternative and complementary therapies.

The extent to which older adults use these therapies and the indications and contraindications for their use in older adults are unknown. Since complementary therapies, on the whole, are aimed at affecting energy flow and restoring homeostasis, many do not qualify for study through randomized controlled trials, so research on their efficacy is taking place by means of case studies and other forms of systematic analysis. Much of the current research is not found in peer-reviewed journals, although a good source is Spencer and Jacobs's *Complementary and Alternative Medicine—An Evidence-Based Approach* (1998). Several new peer-review journals have been started to publish the results of these studies, such as *Alternative Therapies in Health and Medicine*. It is incumbent on allopathic practitioners to stay up-to-date on the breadth of complementary and alternative therapies, their use and contraindications, the training required for effective application, and data supporting their use.

Holism: Based on current knowledge of molecules, atoms, and electron behavior, it is no longer useful to regard humans solely as machines that can be fully understood simply by reducing the whole and analyzing the parts. The challenge of the human organism lies in understanding how it is organized and how the parts interact and exchange information. Atoms and their electrons provide the basis of wave theory, bioelectromagnetism, energy, and the flow of ch'i (Capra, 1996).

Holism focuses on balance and integration of all the interacting elements of the system. Information inherent in the organization of a system gets lost in the separation of the parts (Schwartz & Russek, 1997). The whole is more than simply the sum of the parts. A variety of complementary and alternative therapies have been found to be useful in caring for older people. Generally, test therapies aim to affect the flow of ch'i and, as a result, restore balance or homeostasis in the mind and body and restore information flow, which facilitates the body's natural state of wholeness and healing (Davis, 1997).

Manual Therapies: These therapies include myofascial release, craniosacral therapy, the Rosen method, Rolfing, Hellerwork, soma, neuromuscular therapy, noncontact therapeutic touch, and osteopathic and chiropractic medicine. Except for non-contact therapeutic touch, the manual therapies involve the use of hands on the body-mind surface, thereby stimulating bioelectromagnetic force. Research by Hunt (1996) and Rubik (1995) measured energy flow from the body and suggested that both mechanical and energy forces stimulate response from the tissue.

Myofascial release, developed by Barnes (1990), is an effective manual therapy for older patients with diminished hydration of tissue, myofascial shortening, and cross-linked collagen restriction in their bodies. The practitioner places his or her hands directly on the myofascial tissue of the patient, taking out the stretch, and then gently waits until the tissue responds under the surface of the practitioner's hand. Within 90 to 120 seconds, the tissue begins to move in a flowing manner. The practitioner follows the flow of the tissue with his or her hands so as to increase the length of the tissue with the softening of the myofascia underneath the hands. The cause of this softening, or "melting," of tissue, experienced by practitioners and patients alike, is unknown. It is believed to be the effect of the therapist's energy—the piezoelectric effect—on the polyglycoid layer of the collagen of the myofascia, which increases tissue length and results in a release of trapped energy. The patient then has more freedom to move, gains better posture, and experiences relief of the pain caused by myofascial restriction (Barnes, 1990). Fascial restrictions released in this way over time result in improved balance and strength and help eliminate pain and poor posture. Multiple outcome case studies on myofascial release demonstrate improvements in the quality of life of older people and the prevention of chronic musculoskeletal problems (Barnes, 1990).

Mind-Body Interventions: Such interventions include psychotherapies, support groups, meditation, imagery, hypnosis, dance and music therapy, art therapy, prayer, validation therapy, neurolinguistic psychology, biofeedback, yoga, and t'ai chi. These mind-body interventions demonstrate how movement and verbal and nonverbal communication with the mind and body seem to open up new pathways for thought and therefore unblock energy flow, or ch'i. A growing body of literature examines the effects of t'ai chi on the ability to prevent falls in elderly people and on their quality of life (Wolf et al., 1997).

Movement Awareness Techniques: The Feldenkrais method, the Alexander technique, and the Trager approach are movement awareness techniques. It is postulated that these techniques help people recognize the way they move habitually. By practicing new ways of moving and holding themselves posturally, energy trapped in tissue while maintaining habitual postures is freed.

Traditional Chinese Medicine: Methods include acupuncture, acupressure, polarity reflexology, Touch for Health, Jin Shin Do, and Qi Gong. These approaches within traditional Chinese medicine focus on enhancing the flow of ch'i along body pathways or meridians.

Herbal Approaches: Naturopathy and homeopathy use plants and animal substances to fight off disease, and wise nutrition to combat illness and bring about healing.

Bioelectromagnetics: Thermal applications of nonionizing radiation, such as radio-frequency hyperthermia lasers, radio-frequency surgery, and radio-frequency diathermy, and nonthermal applications of nonionizing radiation are used for bone repair and wound healing. Biomicroelectromagnetics is the term applied to the energy that seems to emanate from the hands of people who are healers (Rubik, 1995). Credible research exists on the use of electromagnetic energy for wound healing and bone repair.

Mind-Body Medicine: Mind-body medicine links traditional linear research methods with complementary and alternative health care practices. The influence of the mind on the body was first introduced by Herbert Benson's research on Tibetan monks. These monks could control their autonomic nervous systems—lower their body temperatures and respiration rates and enter a wakeful hypometabolic physiological state at will (Wallace, Benson, & Wilson, 1971). Ader and Cohen (1991) coined the term psychoneuroimmunology, wherein the mind affects the immune system via the autonomic nervous system and the fluid nervous system

(i.e., the nonadrenergic and noncholinergic nervous system). Pert (1997) articulated the physiological functioning of the fluid nervous system, manifested through the effects of thought on neurotransmitters, neuropeptides, and steroids in the body. This biochemistry differs from the flow of microelectropotentials, or the exchange of energy from the hands of the therapist, but illustrates that the mind and body are inseparable and that the mind communicates with every cell in the body.

Complementary and alternative therapies are energy-based therapies that require belief in the phenomenon of vital flow of energy in the body. We can observe energy at work in the body in many ways; electrocardiograms, electroencephalograms, and electromyograms all measure the energy output from various organs. The piezoelectric effect enables osteoblastic activity that keeps our bones structurally intact. Biomicroelectropotentials, or the exchange of subtle energies in electromagnetic fields that emanate from the hands of healers, are being researched (Rubik, 1995).

Traditional Therapies Applied from a Holistic Approach: In working with older people, massage, exercise, and relaxation can be approached by therapists in a conventional way, wherein the intention is a mechanical effect on a body part, or it can be approached in a holistic way, where the intention is to influence the flow of vital energy and bring about homeostasis. Researchers confirm the importance of having hope and faith in one's physician and practitioners. How this facilitates healing still remains unclear, but to ignore the positive effect of therapeutic presence is to neglect a powerful intervention. How practitioners are with their patients, not just what they do, is important. The exchange of energy with the intention to serve and facilitate healing is critical.

It is believed that complementary and alternative therapies have an effect on patients by way of the energy that emanates from the healer's hands. As we move into the next century, many researchers and practitioners in health care are seriously exploring new ways of viewing reality. What we know about quantum physics and systems theory, the inadequacies of conventional medicine in overcoming chronic illness and autoimmune disease, and the growing tendency of patients and clients to seek out complementary and alternative therapies position us on the verge of a revolution in the linear and materialistic view of reality.

CAROL M. DAVIS

See Also
Over-the-Counter Drugs and Self-Medication

Internet Key Words
Energy Medicine
Holistic
Mind/Body
Piezoelectric Effect
Psychoneuroimmunology

Internet Resources
Aging—Complementary and Alternative Medicine
    Program at Stanford University
http://scrdp.stanford.edu/camps.html

Alternative Medicine Center
http://www.alternativemedicine.net

Alternative Therapies in Health and Medicine
http://www.alternative-therapies.com/

Health World Online newsletter, *Healthy Update*
http://www.healthyupdate.com

National Institutes of Health Center for Complementary and Alternative Medicine
http://nccam.nih.gov./nccam/about/general.shtml

ALZHEIMER'S ASSOCIATION

Alzheimer's disease, the most common form of dementia, is a progressive degenerative disease of the brain. An estimated 4 million Americans have Alzheimer's disease and an estimated 19 million family members consider themselves "caregivers" for persons with Alzheimer's disease. Alzheimer's care takes a unique physical, emotional, and financial toll on families (Ory et al., 1999), and recent research links increased risk of mortality to spousal Alzheimer's care strain (Schulz & Beach, 1999).

Alzheimer's disease is a major cause of death (National Center for Health Statistics, 1998). One in ten persons over age 65 and 30% to 40% of persons over 85 have Alzheimer's disease. Alzheimer's disease progresses over an average of 8 years—for some as many as 20 years—from the onset of symptoms. At later stages, persons with Alzheimer's disease are vulnerable to developing other medical conditions and dying before they would if they did not have Alzheimer's disease. The disease knows no social or economic boundaries and affects men and women almost equally, although more women live to the age of greatest risk. Risk factors include advancing age and a strong family history, although rare familial Alzheimer's disease can begin in the 40s or 50s.

Most persons with Alzheimer's disease are cared for at home, although more than 60% of persons in nursing homes and up to 40% of persons living in assisted living and non–nursing-home residential care have Alzheimer's disease. Alzheimer's disease is devastating to patients and to families, with annual economic value of informal family care estimated at $196 billion in 1997 (Arno et al., 1999). Alzheimer's disease costs American businesses more than $33 billion annually, primarily attributable to lost productivity of family caregivers. The average lifetime cost per patient is estimated to be $174,000, with paid care at home averaging $12,500 per year per patient and nursing home care averaging $42,000 per year (Alzheimer's Association, 1999).

Research

The Alzheimer's Association is the only national voluntary health organization dedicated to research for the causes, cures, treatment, and prevention of Alzheimer's disease and to providing education and support services to affected individuals and their families. The national Alzheimer's Association, headquartered in Chicago and with a public policy office in Washington D.C., operates through a network of more than 200 local and area chapters. Chapters sponsor support groups, publish newsletters, run volunteer telephone helplines, and provide education and support to patients, families, and health and social service professionals caring for persons with Alzheimer's.

Funding biomedical research both through Association funds and at the National Institutes of Health is at the top of the Association's federal agenda. Since 1990, the Association has been successful in boosting federal research funding from $146 million to about $450 million, and the Association itself has funded more than $76 million in research grants since 1982.

The Association's vision is to create a world without Alzheimer's disease while optimizing quality of life for individuals and their families. The organization has moved over time from a sole focus on family support to a broader focus on individuals with Alzheimer's. There is now good evidence of a long latent or preclinical phase of Alzheimer's before symptoms develop, and new evidence that persons with mild cognitive impairment are at high risk of converting to Alzheimer's in three years. With earlier diagnosis, more persons are diagnosed at a point of insight, and their families are looking for support programs that focus on the patient as well as the family.

A primary goal of the Alzheimer's Association is to mobilize worldwide resources, set priorities, and fund select projects for biomedical, social, and behavioral research. In 2000, the Alzheimer's Association led and directed the World Alzheimer Congress, a first world congress joining an international Alzheimer's Research conference with the annual meeting of Alzheimer's Disease International, the international federation of 50 countries' Alzheimer's Societies.

Public Policy

Other goals of the Association are to promote, develop, and disseminate education programs and guidelines for health and social service professionals, to increase public awareness and concern for the impact of Alzheimer's on individuals and families in a diverse society, and to expand access to services, information, and optimal care techniques. Current programs focus on personalized knowledge services through toll-free lines, the Internet and publications, and care coordination services on the local level.

Perhaps the greatest success of this voluntary organization has been its public policy coalitions and extensive federal, state, and local advocacy networks that promote legislation responsive to the needs of individuals with Alzheimer's disease and their families. An annual public policy conference provides opportunities for family advocates from the entire country to meet with elected representatives to discuss a national program to conquer Alzheimer's disease. A state policy clearinghouse tracks long-term care and other legislation at the state and local level that affect Alzheimer's families.

The Chicago office of the national Alzheimer's Association houses the Green-Field Library, publishes research and practice updates for physicians and consumers, coordinates Memory Walks as a national fundraising and awareness program, and hosts an annual education conference for care professionals. A national toll-free hotline (800-272-3900) and Web site (www.alz.org) link families and professionals to local and area chapters and support groups.

LISA P. GWYTHER

See Also
Dementia: Nonpharmacologic Therapy
Dementia: Pharmacologic Therapy
Dementia: Special Care Units
Wandering

Internet Key Words
Alzheimer's Disease
Dementia

Internet Resource
Alzheimer's Association
http://www.alz.org

AMERICAN ASSOCIATION OF HOMES AND SERVICES FOR THE AGING

The American Association of Homes and Services for the Aging (AAHSA) is the national nonprofit organization representing over 5,600 nonprofit nursing homes, continuing care retirement communities, and assisted living, housing, and community service organizations for the elderly.

Through advocacy, grassroots action, and coalition work, AAHSA influences public policy pertaining to health, housing, community, and related services to ensure that aging populations receive the services they need and to protect and enhance the viability of not-for-profit providers. AAHSA also offers members information and assistance in interpreting relevant bills, laws, and regulations.

AAHSA offers timely newsletters, publications, and other communications to keep members up-to-date on congressional and regulatory actions and other trends and issues in the field of aging services.

AAHSA provides education and training in a variety of formats designed to meet the diverse information needs of professionals in the aging services field. AAHSA's annual meeting and exposition is highly acclaimed for its extensive curriculum. The continuing education program offers more than 175 concurrent sessions, special symposia, and intensive workshops. The exposition enables participants to view hundreds of the latest products and services. AAHSA's annual spring conference and exposition, held in Washington, D.C., combines reports on the latest developments in public policy with intensive educational programs. The public policy and educational components are enhanced by visits to members of Congress. Stand-alone seminars focus on topics of major concern to members and permit comprehensive examination of issues vital to the effective management of not-for-profit organizations. The Retirement Housing Professionals Certification Program provides professional recognition and management training in the administrative, property management, and human services aspects of retirement housing.

The AAHSA Development Corporation provides consultation to members on development planning and assists members in obtaining project financing. Its publications and educational programs keep members informed on various capital formation techniques and resources.

ROBERT GREENWOOD

See Also
Assisted Living
Continuing Care Retirement Community
Nursing Homes

Internet Resource
American Association of Homes and Services for the Aging
http://www.aahsa.org

AMERICAN ASSOCIATION OF RETIRED PERSONS

The American Association of Retired Persons (AARP) is the nation's leading organization for people age 50 and older. It serves their needs and interests through information and education, advocacy, and community services provided by a network of local chapters and experienced volunteers throughout the country. The organization also offers members a wide range of special benefits and services, including *Modern Maturity* magazine and the monthly *Bulletin*.

AARP believes that a comprehensive, coordinated long-term-care system needs to be developed that includes universal long-term-care coverage and provides a range of services including in-home assistance, community services, supportive housing, institutional care, rehabilitation services, and transportation. In addition, the long-term-care system should include support for informal care, adequate public financing, private-sector involvement, consumer protections, and national standards for quality. The association's educational and advocacy work on long-term care promotes the advancement of such a system.

TOM OTWELL

Internet Resource
American Association of Retired Persons
http://www.aarp.org

AMERICAN FEDERATION FOR AGING RESEARCH

The American Federation for Aging Research (AFAR) is dedicated to helping scientists further their research in the realms of aging and geriatric medicine. AFAR's mission is to promote healthier aging through biomedical research. The organization hopes to fulfill this mission by supporting biomedical research that promotes healthier aging and furthers our understanding of the aging process and its associated diseases and disorders; building a cadre of new and young scientists to work in aging research and geriatric medicine; offering opportunities for scientists and physicians to exchange new ideas and knowledge about aging; and promoting an awareness among the general public, in the United States and abroad, of the importance of aging research.

PETER CLEARY
KANIKA P. MODY

Internet Resource
American Federation for Aging Research
http://www.Afar.org

AMERICAN GERIATRICS SOCIETY

The American Geriatrics Society (AGS) is the premier professional organization of health care providers dedicated to improving the health and well-being of all older adults. With an active membership of over 6,000 health care professionals, the AGS

has a long history of effecting change in the provision of health care for older adults. In the last decade, the society has become a pivotal force in shaping attitudes, policies, and practices regarding health care for older people. The AGS promotes high-quality, comprehensive, and accessible care for America's older population, including those who are chronically ill and disabled. The organization provides leadership to health care professionals, policy makers, and the public by developing, implementing, and advocating programs in patient care, research, professional and public education, and public policy. In response to the many challenges that a rapidly aging population poses, the AGS has established the Foundation for Health in Aging (FHA). The FHA's goals are to build a bridge between geriatrics health care professionals and the public and to advocate on behalf of older adults and their special needs: wellness and preventive care, self-responsibility and independence, and connections to family and community.

PAMELA INGHAM

Internet Resources
American Geriatrics Society
http://www.americangeriatrics.org
E-mail: info@americangeriatrics.org

Foundation for Health in Aging
http://www.healthinaging.org
E-mail: foundation@americangeriatrics.org

AMERICAN HEALTH CARE ASSOCIATION

The American Health Care Association (AHCA) represents nearly 12,000 not-for-profit and for-profit assisted living, nursing home, and subacute care providers through 50 state organizations. AHCA and its affiliates and member providers advocate for frail elderly and those who, because of social needs, disability, or illness, require services in long-term-care settings. The association represents long-term-care providers to government, business leaders, and the public at large and is an advocate for meaningful change.

AHCA provides education, information, and administrative tools to consumers of long-term care, providers, health care professionals, regulators, and policy makers. Daily and monthly electronic updates on the AHCA Web site and print newsletters and journals (*AHCA Gazette, Hot Issues, Provider Magazine*) distill current clinical and health services research and issues, proposed legislation, and statistical data for all members of its public, professional, and industry constituency.

The annual national conference invites owners and administrators, clinical management staff, directors of nursing, quality assurance directors, compliance officers, and risk managers to special seminars, presentations, and poster sessions on issues affecting long-term care. Many of the sessions are approved for continuing education credit. Regulators and policy makers from the Health Care Financing Administration, Department of Justice, and Office of the Inspector General participate in many of the informational sessions, describe new initiatives, and hear the concerns of providers.

Research is supported, conducted, and disseminated through the Research and Information Service and includes impact assessments of current and proposed public policy; regulatory compliance reports; publication of *Facts and Trends*, an annual compilation of data about residents in assisted living, nursing home, and subacute care facilities; utilization and expenditure reports; and quality initiatives. AHCA supports the industry's need for standardized measures of quality.

Clinical practice guidelines for use by member organizations and others were created by AHCA in collaboration with the American Medical Directors Association. A scholarship program is specifically maintained for student nurses in registered nursing or licensed practical or vocational training programs. The National Council of Assisted Living (NCAL) is a program affiliate of AHCA dedicated to representing the needs and interests of residents of assisted living facilities and their owner-operators. NCAL is a resource for legislative updates and quality guidelines for assisted living, model consumer agreements, and by-state listings of pro-

viders and has an active informational and advocacy role in assisted living.

KANIKA P. MODY

Internet Resource
American Health Care Association
http://www.ahca.org

AMERICAN INDIAN ELDERS

American Indians and Alaskan Natives are collective terms describing over 500 nations, tribes, bands, and native villages. Contemporary American Indian societies vary in culture, lifestyle, health status, health risks, and the experience of aging with chronic disease. Nearly 2 million Native Americans live in the United States, half of whom reside in cities and towns. Although elders are a small proportion of the total American Indian population (6% of rural and reservation inhabitants; 5% of urban inhabitants), the number of elders over age 60 increased 52% between 1980 and 1990. "Elder" refers to a social or physical status, not to chronological age, and the role of an elder is imbued with positive value and with an active engagement in maintaining and transmitting cultural values.

Access to Health Care

Access varies with residential location and government recognition of tribal status. Members of federally recognized tribes living on federal reservations, trust territories, historic Indian areas in Oklahoma, and Alaskan villages are provided free, universal access to health care by the Indian Health Service (IHS), a branch of the U.S. Public Health Service. However, the IHS has insufficient resources for the chronically disabled and for long-term care. The IHS serves only about 40% of the total American Indian population but produces an annual comprehensive report of American Indian and Alaskan Native morbidity and mortality that is often generalized to represent the entire population.

More than half of the total American Indian population, including those over age 65, currently live in urban areas. As with all older Americans, health care is provided on the basis of ability to pay by private agencies or by state and local health departments, sometimes supplemented by Medicaid or Medicare. Eligibility for federally mandated health care through IHS expires after 180 days' residence off-reservation (although eligibility is renewed if reservation residence is reestablished). Virtually all urban American Indian elders were born on or near reservations but migrated after World War II to cities, where they worked, raised their families, retired, and maintain permanent residence. Health information about urban American Indians and Alaskan Natives of all ages generally is available only from isolated cross-sectional studies, which report that these groups have poorer health and functional status, across virtually all measures, than the general population or their reservation counterparts. A study of over 300 elders living in Los Angeles found that they compared unfavorably with their reservation peers and that self-reported poor health status, a marker of morbidity and mortality, was twice the expected frequency.

As for other Americans, the three leading causes of death for American Indian elders age 65 and older are heart disease, malignant neoplasm, and cerebrovascular disease; unlike in the general population, however, diabetes mellitus is the fourth leading cause of death. Autoimmune and other chronic diseases have an earlier onset and higher prevalence rates in American Indian populations. As early as age 45, American Indians share similar health and human services needs as mainstream adults age 65 and older. The implications are that some Native Americans—"elders" in their own communities—may not be eligible for Medicare or other support services when such assistance is first needed.

Chronic Diseases and Intervention Strategies

During the last half of the 20th century, changes in technology, diet, exercise, and lifestyle have in-troduced new risk factors for diseases, in particular, diabetes and its associated complications, hypertension, rheumatoid diseases, and cancer. Most diseases affecting American Indians have preventable risk factors. Intervention programs have focused on tertiary prevention (e.g., education on exercise and diet) that generally recognizes tribal-specific variations in health risks and health behaviors. Successful methods often engage traditional values of respect and authority for older role models and spokespersons, oral transmission of knowledge, nonintrusive guidance, and importance of family and community continuity.

Diabetes and Associated Complications

Early in the 20th century, health records indicated that diabetes was an uncommon disease among American Indians. By the close of the century, type II, non-insulin-dependent diabetes mellitus was endemic; 20.3% of American Indians age 65 and older have been diagnosed with diabetes, compared with 9.3% of the general population age 65 to 74 reported on the National Health and Nutrition Survey (NHANES) II. Mortality attributed to diabetes is 209.6 in 100,000 among American Indians age 65 to 74, compared with 75.7 in 100,000 for the general population. Diabetes is particularly prevalent among Pima (Tohono O'odham) living on the Gila River Reservation in Arizona, where half of all adults over age 35 have diabetes. The considerable tribal variation in incidence and prevalence suggests the involvement of an as yet undiscovered genetic link, variations in lifestyle (still undefined), obesity, or increasing exposure to nontraditional diets and technology. American Indians experience the highest incidence of diabetes-related complications, perhaps as a result of earlier onset or exposure to different or more intense risk factors. Although there are difficulties in comparing the complication rates overall, end-stage renal disease is 6.8 times higher in American Indian patients with diabetes than in whites with diabetes. Most tribes also far exceed the rate of non-Indians for lower-extremity amputations (Valway, Linkins, & Gohde, 1993).

A key diabetes intervention is diet, but as in all societies, food has special social and cultural meanings that go beyond its nutritional value. Urging people to restrict or eliminate foods that are invested with cultural meanings is counterproductive and generally ignored in favor of participating in the social and cultural life of the community.

Culturally specific interpretations of diabetes affect treatment decisions and self-care. Among Seneca, illness is considered to be the result of malevolent aggression, and the patient alone is responsible for health improvement. Seeking available, reasonable, and effective treatment through Western biomedicine does not conflict with traditional values. In contrast, the Sioux and Navajo perspective is that diabetes is a "new" disease or a "white man's disease," reflecting the destruction of their societies and cultures by non-Indians. Because religion and culture are tied to traditional healing methods, the need for nontraditional interventions (e.g., a daily pharmaceutical regime, lifelong dietary restrictions) can be a daily affront.

Although intervention strategies have incorporated elements of traditional lifestyles, modified for contemporary nutrition and activity patterns, behavior-based strategies are difficult to implement, maintain, and evaluate on a long-term basis. Nutrition education based on locally available traditional foods has resulted in minimum weight loss but improved self-care. Fitness programs that promote traditional activities have been enthusiastically embraced. The Zuni fitness program combines traditional dance movements with mild aerobic exercise for less active or impaired older adults and promotes other traditional activities, such as running, for younger, active adults.

Heart Disease and Hypertension

Age-adjusted mortality attributed to heart disease is 7% lower for American Indians than for the general population, but that lower rate masks its seriousness. American Indians age 40 and younger have a three to four times higher cardiovascular mortality than the general population. The IHS reports considerable regional and tribal variation, with tribes of the Northern Plains reporting morbidity and mortality rates as high as or higher than those in the U.S. general population. In Los Angeles, significantly more older American Indian women self-report heart disease and hypertension than older American Indian men. Surprisingly, rates of hypertension are relatively low, despite the high prevalence of obesity and diabetes among American Indians. Nevertheless, these comorbidities represent a serious but generally underestimated health risk that increases with age among both men and women. Other risk factors, such as social and economic stressors, have also been suggested to explain variations in tribal-specific prevalence rates by age and sex.

Awareness and control of hypertension are problematic. In three culturally and geographically diverse regions, only 50% of American Indian adults were aware of their disease, whereas those without hypertension tended to overestimate actual blood pressure (Deprez, Conelehan, & Hart, 1985; Destefano, Conelehan, & Wiant, 1979; Sharlin, Heath, Ford, & Welty, 1993). Studies of American Indian patients with elevated blood pressure consistently note the difficulty in managing hypertension with medication alone, suggesting the need for community-based programs to reduce risk factors (Gillum, Gillum, & Smith, 1984; Sharlin et al., 1993).

Rheumatoid Diseases

Limited archaeological evidence suggests that, unlike diabetes, rheumatoid arthritis may be a New World disease. Genetic studies propose that American Indian populations may be unusually prone to autoimmune rheumatic disease. The prevalence of rheumatoid diseases is higher among many American Indians groups than among Alaskan Natives or whites. Tribal and regional variation is significant. Higher prevalence and incidence rates of rheumatoid arthritis have been documented in Native American cultures such as the Pima (Southwest), Chippewa (Great Lakes), Tlingit, and Yakima (both Northwest). The prevalence of systemic lupus erythematosus is higher than expected among diverse tribes, including Sioux, Arapaho, Crow, Haida, Tlingit, and Tsimshian.

Little attention has been given to educational programs and culturally specific adaptive strategies to cope with rheumatoid diseases among American Indians. Self-reporting instruments to monitor disease progress are problematic in both urban and reservation communities because questions conflict with culture-bound concepts relating to time, autonomy, and communication within a therapeutic setting.

Cancer

American Indians have the lowest cancer survival rate of any U.S. population, with only one-third surviving for five years. As in the general population, the most common cancer site is the lung. In comparison to whites, the incidence of cancers of the gallbladder, kidney, stomach, and cervix is relatively higher in American Indians, as are cancers of the liver and nasopharynx in Native Alaskans. Differences in incidence and in site-specific cancers have been attributed to cultural heterogeneity associated with health behaviors, as well as to environmental factors. The frequency of lung cancer and cardiovascular disease risks have been linked to tribal-specific patterns of cigarette smoking both on and off the reservation. For instance, tribal lung cancer rates in the Southwest, where cigarette smoking is less frequent, are lower than the rates among Northern Plains tribes, where heavier cigarette smoking typically prevails. Environmental and occupational exposure to toxic emissions on and near reservations affects tribes in many states, and prevention of the adverse consequences is the focus of the Agency for Toxic Substances Disease Registry at the U.S. Department of Health and Human Services and the Bureau of Tribal Affairs. Radiation from uranium strip mining, particularly in New Mexico and Arizona, is a risk factor for Navajo. Some have pursued treatment and legal redress, but for others who perceive a culturally valid outcome for disturbing the sacred nature of the land by strip mining, there may be little incentive to seek treatment (Dawson, 1992).

Culturally sensitive health education to reduce cancer risks uses interventions designed for specific communities and trained lay American Indian health educators to deliver the program. Media presentations using American Indian and Alaskan Native women, including elders, as role models have been successful in promoting education about cervical cancer and the need for regular Pap screening (American Indian Health Care Association, 1993; Stillwater, Echavarria, & Lanier, 1995).

Treating the Indian Patient

Successful therapeutic relationships are based on establishing an interest in the patient's well-being and often require a cultural assessment. Clinicians should not underestimate the sophistication of American Indian clients, who may enjoy the same broadcast and print media as other American patients. Traditional etiquette may call for avoidance of eye contact, firm handshakes, and direct questions and responses; these behaviors indicate respect in Indian cultures and should not be mistaken for furtiveness. Silence may indicate responsiveness rather than avoidance or hostility. Conversational silences allow both parties time for reflection to absorb information and to formulate a thorough response; they indicate respect for the serious nature of the business at hand. A calm, accepting, nonjudgmental approach is appreciated in establishing a trusting relationship, which may take more than one visit. Obtaining advance directives is often a difficult and lengthy process, stymied in part by institutional protocols that require written consent to a formal, witnessed document processed by staff with whom patients may not have an established relationship. End-of-life treatment decisions tend to favor natural approaches and accept the inevitability of death. A cultural assessment is a key factor in orienting the discussion on advance directives. Clinicians should be aware of the spiritual healing that patients and families may desire to accompany treatment. When elders can no longer speak for themselves, a family proxy (whose role may reflect the indigenous social structure) usually emerges to express what should be accepted as the authentic wishes of the patient if these have not been previously determined with the provider (Hepburn & Reed, 1995).

B. JOSEA KRAMER

Internet Key Words
American Indian
Health
Native American/Alaskan Native

Internet Resource
Bureau of Indian Affairs
http://www.doi.gov/bureau-indian-affairs.html

AMERICAN MEDICAL DIRECTORS ASSOCIATION

Founded in 1975, the American Medical Directors Association (AMDA) is a national organization representing more than 8,000 medical directors and other physicians who practice in long-term-care settings. AMDA is committed to the continual improvement of the quality of patient care by providing education, advocacy, information, and professional development for medical directors and other physicians. Although the association's name reveals its origins, the reality is that AMDA has always served the interests of both medical directors and attending physicians.

Among AMDA's many accomplishments are its public policies for improved care. These policies include the establishment of the Certified Medical Director (CMD) Program, to demonstrate competence in both clinical medicine and medical direction and administrative responsibilities. It has also worked to improve standards in federal nursing facilities, contributing to the passage of the 1987 Nursing Home Reform Act. To help clinicians more directly, AMDA develops information kits and organizes national symposia to aid in the efforts to improve the long-term care of the elderly.

KATHY WILSON

Internet Resource
American Medical Directors Association
http://www.amda.com

AMERICAN NURSES ASSOCIATION

The American Nurses Association (ANA) is the only full-service professional organization representing the nation's entire registered nurse population. As the largest nursing organization in America, the ANA is the strongest voice for the nursing profession and for workplace advocacy. Dedicated to ensuring that an adequate supply of highly skilled and well-trained nurses is available, the ANA is committed to meeting the needs of nurses as well as health care consumers. The ANA advances the nursing profession by fostering high standards of nursing practice, promoting the economic and general welfare of nurses in the workplace, projecting a positive and realistic view of nursing, and lobbying Congress and regulatory agencies on health care issues affecting nurses and the general public. ANA also publishes *American Nurse* and maintains a Web site.

The three ANA-affiliated organizations are the American Nurses Foundation, the American Academy of Nursing, and the American Nurses Credentialing Center. The American Nurses Foundation was founded in 1955 as the research, education, and charitable affiliate of the ANA. The foundation complements the work of the ANA by raising funds and developing and managing grants to support advances in research, education, and clinical practice. The American Academy of Nursing is an organization of distinguished leaders in nursing who have been recognized for their outstanding contributions to the profession and to health care. The ANA established its certification program in 1973 to provide tangible recognition of professional achievement in defined functional or clinical areas of nursing. To date, more than 150,000 nurses have been certified in 29 specialty areas. The American Nurses Credentialing Center bases its programs on the standards set by the ANA Congress of Nursing Practice.

JOAN MEEHAN

Internet Resource
American Nurses Association
http://www.nursingworld.org

AMERICAN SOCIETY ON AGING

The American Society on Aging (ASA) is committed to improving the quality of life of the elderly by aiding professionals in the areas of education and research. Thousands of professionals throughout the country rely on ASA to keep them on the cutting edge in an aging society. Through renowned educational programming, outstanding publications, and state-of-the-art resources, ASA members tap into the knowledge and experience of the largest network of professionals in the field.

No other organization in the field of aging represents the diversity of settings and professional disciplines reached by ASA. It brings together researchers, practitioners, educators, businesspeople, and policy makers concerned with the physical, emotional, social, economic, and spiritual aspects of aging. ASA is founded on the premise that the complexity of aging in our society can be addressed only as a multidisciplinary whole.

Resources for Education, Training, and Information

The ASA Learning Center is committed to being the premier resource for education, training, and information on aging-related issues. The mission of the Learning Center is to continually improve the knowledge and skills of individuals and organizations in all sectors of society concerned with older adults and their families. It carries out this mission through such programs as the Summer Series on Aging, Web-enhanced teleconferences and computer-based training, and an on-line store and searchable databases that provide one-stop shopping for education and training resources on aging.

Diversity, Cultural Competence, and Personal Growth

ASA is committed to advancing a new standard of professionalism in aging, with diversity and cultural competence at its core. Through special initiatives such as New Ventures in Leadership, Serving El-

ders of Color, and an on-line multicultural aging network, ASA strives to develop leadership in the field of aging that is representative of the racial, ethnic, and cultural diversity of the populations served.

The very nature of work with elders challenges us personally as well as professionally. ASA brings these perspectives together in ways that allow its members to explore opportunities for personal growth. Its largest constituency group (see below) is the Forum on Religion, Spirituality, and Aging, which offers avenues to explore the spiritual side of aging and the quest for meaning in later life. The Creative Aging Institute, conducted in partnership with Elders Share the Arts, melds the arts and creativity with community development and personal enhancement.

Constituency Groups

ASA provides members with the opportunity to join the following constituency groups, which provide newsletters, membership directories, and annual programming in the particular area of interest:

- Business Forum on Aging
- Forum on Religion, Spirituality, and Aging
- Healthcare and Aging Network
- Lesbian and Gay Aging Issues Network
- Lifetime Education and Renewal Network
- Multicultural Aging Network
- Mental Health and Aging Network
- Network on Environments, Services, and Technologies for Maximizing Independence

SUSAN MARKEY

Internet Resource
American Society on Aging
http://www.asaging.org.

AMERICANS WITH DISABILITIES ACT

The Americans With Disabilities Act (ADA) of 1990 (P.L. 101-336) is a civil rights law to promote equal opportunity and greater participation by people with disabilities in employment, services offered by state and local governments and private businesses, and telecommunications. ADA prohibits discrimination on the basis of disability, with protections like those provided under civil rights laws for race, sex, national origin, and religion. However, unlike other civil rights law, ADA requires various proactive measures to ensure access.

History and Theory of Disability Rights Laws

Two historically unprecedented trends of the past 25 years underpin ADA: the development of federal disability rights law and the empowerment of people with disabilities.

The Architectural Barriers Act of 1968 (P.L. 90-480) is the first modern federal disability rights law. Introduced in January 1967 by Senator E. L. Bartlett, its purpose was "[t]o ensure that public buildings financed with federal funds are so designed and constructed as to be accessible to the physically handicapped" (Public Law 90-480). After hearings and amendments, it was signed into law in August 1968 (Katzmann, 1986).

The act was drafted by Bartlett aide Hugh Gallagher. His story not only illustrates a personal struggle with exclusion and accessibility but exemplifies other stories that later convinced Congress of the need for ADA.

In 1952, while in college, Gallagher developed polio and subsequently required use of a wheelchair. In 1963, he went to work for Senator Bartlett. On many occasions he wanted to visit public buildings in Washington, D.C., but most were inaccessible. For example, to enter the National Gallery of Art he needed a small ramp to climb the two-step, 10-inch curb at the museum's entrance on Constitution Avenue.

Gallagher wrote to the National Gallery, asking for a ramp, and was told that one would "destroy the architectural integrity of the building" (H. Gallagher, personal communication, April 3, 1992). Gallagher thought his request was simple and reasonable, and that the National Gallery, as a national museum, was the property of all Americans, not just those who could walk into it. Gallagher eventually got his wish when Senator Bartlett prevailed on the National Gallery's Trustees to install a ramp—which turned out to be nearly invisible.

To solve this access problem generally, Gallagher drafted a bill with Senate Legislative Counsel, one that was "short and simple and that would put in a civil rights context," a mandate that "buildings constructed wholly or in part with federal funds be available to all citizens" (H. Gallagher, personal communication, July 27, 1992).

Although modest—only a page long and with no enforcement provision, the act departed fundamentally from existing laws for the disabled. Previous legislation involved social welfare measures (e.g., providing cash assistance or job training). The Barriers Act was a civil rights law, to promote integration and pointing to constitutional claims of equal protection and due process. It was also the first law built on the theory that disability is not simply a function of an individual's impairment, but an interaction between impairment and environment. Environments—physical and otherwise—can disable or enable a person with a disability. As a corollary, the act was the first disability law that did not require an individual to identify himself as disabled to benefit. Lastly, it expressed new national aspirations for disability policy.

The second disability rights law was Title V of the Rehabilitation Act of 1973 (P.L. 93-112), in particular section 504. Section 504 is a broad guarantee that "No otherwise qualified handicapped individual . . . shall be excluded from the participation in, be denied the benefits of, or be subjected to discrimination under any activity or program receiving Federal financial assistance. . . . " Section 504 was drafted by Senate aides using language from other civil rights laws (Scotch, 1984).

A second trend in the push for disability rights was the birth of the "independent living movement" in the late 1960s—a movement by and for people with disabilities. For example, Ed Roberts, an early leader, also had had polio and required use of a ventilator and wheelchair. In the mid-1960s he entered the University of California at Berkeley over the objections of school officials and later helped found the first "independent living center" in Berkeley in 1972. Today, a nationwide network of such centers provides peer support, advocacy, and services. This movement also reflected the increasing prevalence and complexity of disability—following the success of medical science in keeping people alive from once-fatal conditions but with lifelong disability (Vachon, 1987). For a history of this movement, see Shapiro (1993).

Although these two trends began independently, they married with great force in the mid-1970s when the Carter administration delayed publishing federal regulations to implement section 504. This led to nationwide demonstrations by disabled persons, including sit-ins at federal office buildings, and was perhaps the single most important event in coalescing a "disability community" nationwide and fostering its political education (Bowe, 1986; Scotch, 1984).

Other federal disability rights laws have been enacted subsequently. For a summary, see U.S. Department of Education (1992).

ADA was first proposed by a Reagan-appointed National Council on Disability in a 1986 report entitled, "Toward Independence." Created in 1979, the National Council is an independent federal agency charged with advising the president and Congress on disability policy. The first ADA bill was introduced in Congress in April 1988 but died when the session ended. A substantially revised bill, modeled on section 504 and other civil rights laws, was introduced in May 1989, passed overwhelmingly in 1990 with broad bipartisan support, and was signed into law by President George Bush on July 26, 1990. Bush had spoken out forcefully on disability rights issues as vice-president (Shapiro, 1993).

Specific Provisions of ADA

Preamble: The opening sections describe congressional findings and purposes, including the historical segregation and exclusion experienced by people with disabilities and the nation's disability policy goals, and define disability.

Title I—Employment: Employers may not discriminate against an individual with a disability in hiring, promotion, or other employment benefits if that person is otherwise qualified for a job. Employers must provide "reasonable accommodations" to assist an individual with a disability to meet job requirements, such as job restructuring and adaptive equipment, except where accommodations would be an "undue hardship." Title I only applies to employers with more than 15 employees and is enforced by the Equal Employment Opportunity Commission and private lawsuit.

Title II—Public Services: Title II applies to state and local governments, and has two major subtitles. Subtitle A prohibits discrimination against individuals with disabilities, and requires government facilities and services to be accessible. Subtitle B applies to public transportation—requiring accessible buses, paratransit, or comparable transportation services for individuals who cannot use fixed route bus services, and accessible train cars, bus, and train stations. Title II is enforced by the U.S. Departments of Justice and Transportation, and private lawsuit.

Title III—Public Accommodations and Services Operated by Private Entities: Private entities, such as restaurants, hotels, and retail stores, may not discriminate and must provide auxiliary aids and services to individuals with vision or hearing impairments and other disabilities, unless an undue burden. Physical barriers in existing facilities must be removed, if "readily achievable." If not, other means of providing services must be provided, again if readily achievable. All new construction and alterations must be accessible. Title III is enforced by the U.S. Department of Justice and private lawsuit.

Title IV—Telecommunications: Companies offering telephone services to the general public must offer relay services to individuals who use telecommunications devices for the deaf (TDDs). Title IV is enforced by the Federal Communications Commission.

Older Americans and the Americans With Disabilities Act

Given both the relationship between aging and disability and the aging of the U.S. population, the benefits of ADA can be expected to be especially significant for older Americans and for the nation at large. Although only experience will prove the value of ADA (Vachon, 1992), possible benefits include increased employment, access to public transit, and access to a greater number of restaurants, stores, and service establishments; and greater opportunities to enjoy films, sporting events, and other performances at accessible theatres and stadiums—especially through audio amplification technologies, closed captioning, and wheelchair-accessible seating.

R. ALEXANDER VACHON

See Also

Assistive Devices
Human Immunodeficiency Virus (HIV) and AIDS
Older Americans Act

Internet Resource

U.S. Department of Justice
http://www.usdoj.gov/crt/ada/adahoml.htm

ANEMIA

Anemia in an elderly person (hemoglobin [Hgb] < 12 g/dL or hematocrit [Hct] < 36%) should be viewed as a sign, not a disease. Anemia is not a part of normal aging. Normal parameters for the complete blood count (CBC) indices can be used, although there may be an insignificant age-related rise in the mean corpuscular volume. The first step in treatment is to find the underlying disease that is causing the anemia. The main causes of chronic anemia in the elderly are iron deficiency, vitamin B_{12} or folate deficiency, myelodysplastic syndromes, and anemia of chronic disease. This discussion focuses on the main causes, diagnosis, and treatment of chronic anemia in the elderly.

Diagnosis

Presenting symptoms of anemia may include fatigue, shortness of breath, dyspnea on exertion, anginal chest pain, palpitations, headache, dizziness, change in mental status, syncope, nausea, bowel irregularities, and impotence. The physical signs of anemia include pallor (especially in the mucous membranes), tachycardia, systolic ejection murmur, and a widened pulse pressure. The speed at which the signs and symptoms develop is often a clue to the acuteness of the condition. To rule out an acute gastrointestinal bleed, a rectal exam must be done to characterize the nature of the stool; a guaiac test should be done to check for occult blood. Clinicians should try to elicit a history of melena or other change in bowel habits. Follow-up should include monitoring vital signs and stools for occult blood (on three more separate occasions).

The initial laboratory test is a CBC that not only reports the hemoglobin and hematocrit levels but also provides information on the size and shape of the red blood cells (RBCs). Additional tests that may be helpful in the initial workup are ferritin, iron, total iron binding capacity, vitamin B_{12}, folate, lactate dehydrogenase, indirect bilirubin, and reticulocyte count. Classically, anemia is characterized by the size and appearance of the RBC seen on the peripheral smear (microcytic, normocytic, macrocytic) and by the rate of RBC production, as indicated by the reticulocyte count and reticulocyte index. After the anemia is classified by these indices, the differential diagnosis identifies the disease behind the anemia. Bone marrow examination may be necessary if the diagnosis is not clear from patient history, physical examination, and standard blood tests, and it is required when the anemia is secondary to a malignancy such as leukemia, lymphoma, and multiple myeloma. It is also usually necessary to establish the diagnosis of myelodysplasia.

Microcytic Anemias: A mean corpuscular volume (MCV) less than 80 fL indicates a microcytic anemia. Thalassemia and iron-deficiency, myelodysplastic, and drug- or toxin-induced anemias may all present in this fashion.

Normocytic Anemias: An MCV of greater than 80 and less than 100 fL falls within the normocytic range. The differential diagnosis in this category includes anemia of chronic disease, intrinsic marrow disease (e.g., aplasia or malignancy), and acute blood loss or hemolysis. The finding of a normal MCV may be a confounding factor, since it may represent a combination of microcytic and macrocytic processes or an early stage in the development of the anemia. The range distribution width can be a clue in this situation. A range distribution width greater than 14 indicates that the RBC population is heterogeneous and that a combination of factors may be at work.

Macrocytic Anemias: Anemias that present with an MCV greater than 100 fL include megaloblastic anemia (vitamin B_{12} or folate deficiency) and chronic liver disease.

Treatments

Acute Gastrointestinal Bleed: If an acute bleed is discovered, the patient must be admitted to the hospital; a patient with signs of shock, tachycardia, or orthostatic hypotension should be in a monitored setting, such as an intensive care unit. In general, an elderly patient with a hemoglobin less than 8 g/dL or hematocrit less than 25% should be considered a candidate for a blood transfusion of packed red blood cells. Similarly, if the patient presents with obvious signs of hemorrhage or end-organ damage, such as chest pain or severe dyspnea, with a hemoglobin less than 10 g/dL, an immediate transfusion should be considered (Blinder, 1998). In the very aged (those over age 85) and those with a history of congestive heart failure, transfusions should be administered as half units (125 mL) to run over three to four hours. In addition, 10 mg of furosemide may be given intravenously with each unit to prevent fluid overload.

Iron Deficiency: Iron-deficiency anemias are characterized by microcytosis, low ferritin level (< 20 ng/mL), low iron (< 30 μg/dL), high total iron binding capacity (> 360 μg/dL), high range distribution width (> 14), and low reticulocyte index (< 2%) (Chatta & Lipschitz, 1999). Iron deficiency in the elderly can be caused by inadequate nutrition, achlorhydria, or chronic blood loss (often from the gastrointestinal tract). The finding of iron deficiency therefore warrants a workup of the gas-

trointestinal tract, including upper and lower tract endoscopy, x-rays, or sigmoidoscopy.

Treatment consists of iron supplements taken orally; in emergencies or when iron cannot be given orally, iron dextran may be given intravenously. Oral iron is available in several formulations: ferrous sulfate (usually 325 mg three times a day or 65 mg of elemental iron), ferrous gluconate, ferrous fumarate, and ferrous polysaccharide. Because ferrous sulfate is constipating, a stool softener such as docusate is recommended. Ferrous polysaccharide (Niferex 150 mg) is less constipating and has once-daily dosing, making it a preferred choice in elderly people. Vitamin C administered with the iron can help maintain the iron in its reduced state and improve absorption.

Parenteral iron therapy may be used when the patient is unable to absorb oral iron adequately or in sufficient doses. The amount of iron necessary to replenish the iron stores and restore the hemoglobin to normal levels can be calculated by the following formula: Iron (mg) = 0.3 × body weight (lb) × [100 − (Hgb [g/dL] × 14.8 × 100)] (Freedman, 1998). Iron dextran can be given either intramuscularly or intravenously. A test dose of 0.5 mL should be administered intravenously to ensure that anaphylactic shock does not occur.

Thalassemias: The thalassemias are hereditary disorders characterized by low MCV (often < 70 fL), target cells on peripheral smear, low reticulocyte index, high range distribution width, and normal iron, ferritin, and total iron binding capacity (Freedman, 1998). The form most likely to be encountered in the elderly population is thalassemia minor. No treatment is required, and iron therapy is contraindicated, as it may produce iron overload.

Myelodysplastic Syndromes: The myelodysplastic syndromes are refractory anemia, refractory anemia with ringed sideroblasts, refractory anemia with excess blasts, refractory anemia with excess blasts in transformation, and chronic myelomonocytic leukemia. These are differentiated by high ferritin (> 200 ng/mL), normal or elevated iron levels, and low reticulocyte index. Prior treatment with alkylating agents is a risk factor for developing these syndromes.

Treatment is supportive, with transfusions as necessary. Chemotherapy has not been effective in the treatment of these syndromes. The syndromes may respond to pyridoxine (20 to 200 mg orally once a day). Refractory cases sometimes respond to erythropoietin (dosing described later, under anemia of chronic disease).

Vitamin B_{12} Deficiency: This condition is characterized by macrocytosis, low B_{12} level, high range distribution width, and low reticulocyte index. Borderline low B_{12} levels may be corroborated by finding elevated homocysteine or methylmalonic acid levels. Vitamin B_{12} deficiency may also cause neurological damage, including dementia.

The most efficient way to treat vitamin B_{12} deficiency due to pernicious anemia (autoantibodies to intrinsic factor) is with injections of 1,000 µg of vitamin B_{12} intramuscularly (Fauci et al., 1998). Initially, the injections are given daily for one week, or until a response in hematocrit is seen. Then the maintenance dose is 1,000 µg every month. Potassium and phosphate levels must be monitored during the initial stage of therapy. Alternatively, B_{12} may be given orally at a dose of 1,000 µg a day. Some clinicians recommend a lower dose of 100 µg and monitoring levels for a response. Since the pernicious anemia form of B_{12} deficiency may be associated with gastrointestinal cancer, it is recommended that patients be monitored for this as well.

Folate Deficiency: This condition presents in a similar manner to vitamin B_{12} deficiency. When a macrocytic anemia is diagnosed and a folate deficiency is suspected, it is imperative that both B_{12} and folate levels be checked. Replacing only folate in a patient who is also deficient in vitamin B_{12} can improve the anemia but fails to stop the neurological sequelae of B_{12} deficiency. Treatment is 1 mg of folic acid orally once a day.

Anemia of Chronic Disease: This condition is normally a diagnosis of exclusion. The characteristic findings are a low reticulocyte index (< 2%), normal or reduced iron and total iron binding capacity, normal or increased ferritin, normal range distribution width, and normal bone marrow. Implicated diseases include chronic inflammatory disease such as collagen-vascular disease or malig-

Treatment of Anemia

Type of Anemia	Preferred Treatment
Iron deficiency	Ferrous polysaccharide 150 mg orally once a day
Thalassemia minor	None
Myelodysplasia	Pyridoxine 20 to 200 mg orally once a day; erythropoietin as needed
Vitamin B_{12} deficiency	Maintenance: vitamin B_{12} 1,000 µg intramuscularly every month
Folate deficiency	Folic acid 1 mg orally once a day
Anemia of chronic disease	If refractory, erythropoietin 50 to 100 units/kg subcutaneously 3 times a week
Acute blood loss	Transfusions as needed

nancy, chronic liver disease, chronic renal disease, or endocrine disorders.

A search should be made for correctable nutritional deficiencies (iron, B_{12}, or folate) that may predispose the elderly to infection or exacerbation of comorbidity. Dementia, poverty, or elder abuse can contribute to nutritional inadequacy. Erythropoietin levels should be checked, because even low-normal levels are associated with the development of anemia. In the case of collagen-vascular diseases and malignancy, the anemia often responds to treatment of the underlying illness. In refractory cases and in chronic renal disease, treatment with erythropoietin may be necessary. Erythropoetin is given intravenously to hemodialysis patients and subcutaneously to those not on dialysis. The starting dose is 50 to 100 units per kilogram body weight given three times a week. This dose is then titrated, with a target hematocrit of 30% and a maximum dose of 300 units per kilogram three times a week. Recent findings indicate that some patients can be maintained on once-a-week dosing administered by a home health nurse.

CRAIG C. CURRY
MICHAEL L. FREEDMAN

Internet Key Words
Anemia of Chronic Disease
Folate Deficiency

Iron Deficiency
Myelodysplastic Syndromes
Thallassemia
Vitamin B$_{12}$ Deficiency

ANIMAL-ASSISTED HEALTH CARE

The capacity of animals to assist with the health care of disabled or chronically ill patients has been regarded as vital and powerful. Much of the literature has been confined to clinical anecdotes and case reports or has been limited to the companion aspects alone. The use of animals has been associated with improved emotional and physical health and a reduction in negative attitudes toward the disabled.

Humans have used animals for domestic purposes and as companions for thousands of years. The therapeutic use of animals in health care is more recent. Seventeenth-century Quaker asylums used small animals as part of their social milieu to rehabilitate psychiatric patients. Florence Nightingale, in *Notes on Nursing* (1860/1969), wrote that animals could serve as companions to sick persons with confining or long-term illnesses. Human-animal partnerships in health care settings continue to expand. Increasingly, health care providers are working with animals or encountering older clients in health care settings who rely on these animals for a variety of vital and beneficial functions.

Reports of the therapeutic effects of animals in health care are primarily anecdotal. Empirical research studies are more recent and few in number. Studies suggest, however, that pet ownership or interaction with animals may decrease stress and anxiety and improve cardiovascular health. Better physical health (as assessed by ability to perform activities of daily living) has been associated with owning pets.

Allen and Blascovich (1996) demonstrated the profound psychosocial and economic value of service dogs for people with severe ambulatory disabilities. People with severe, chronic disabilities requiring the use of wheelchairs were randomly assigned to receive trained service dogs within one

month or assigned to a waiting list. Improvements in self-esteem, internal control, psychological well-being, and community social integration were significant within six months of service-dog placement. Increases in school attendance and employment were demonstrated. In addition, recipients of service dogs required markedly fewer hours of paid and informal assistance. This important study identified the cost-effectiveness of service dogs, in addition to the previously described psychosocial and environmental benefits. Other beneficial aspects of human-animal interactions that have been reported include unconditional acceptance, improved sense of well-being, improved emotional health of families, decreased staff stress, increased sense of connection, improved communication skills of withdrawn or isolated persons, and spiritual connectedness (Dossey, 1997).

In modern animal-assisted health care, a variety of animals are used in a multitude of health settings. Although dogs and cats are most commonly associated with these programs, many other species are used, including rabbits, birds, pigs, fish, horses, dolphins, llamas, and even snakes. Breed and species are usually not the most important criteria for such programs. Animals are often chosen for their temperament, tolerance, and energy level, consistent with the health care environment and the focus of the team. The purposes and goals of the program are critical in choosing the animal.

Program Types

Diverse and overlapping terms are used to describe animal-assisted health care programs, such as pet therapy, animal-assisted therapy, and service animals. Although these terms are often used interchangeably, each type is distinct and characterized by different goals. Pet therapy consists of volunteers bringing animals into health care settings. Pet therapy is also called pet visitation or animal-assisted activities. Typically, the animals are not trained, and institutional policies and local health care regulations govern the rules of visitation. Nurses have been active in advocating and instituting pet visitation programs in acute-care settings and nursing homes and for the homebound elderly. The Eden Alternative is a nationally recognized model developed to use live-in pets in long-term-care environments as part of an overall goal to improve the quality of life of nursing home residents.

Animal-assisted therapy (AAT) is more structured and goal-directed than pet therapy. Persons (called handlers) escorting the animals are given training, and animals are screened and certified for health, obedience, sociability, and temperament. Each AAT team has specific goals, and animals are chosen for their ability to assist in accomplishing the psychosocial or physical therapeutic goals for a client. Specific goals might include increasing rapport between a psychotherapist and a client or maximizing mobility and muscle coordination in a physically debilitated client. Visits are structured, and therapeutic outcomes for the client are monitored. Often the AAT human partner is a health care professional.

Service animals are highly trained and legally defined assistance animals. The Americans with Disabilities Act of 1990, a federal civil rights law, defines a service animal as any animal individually trained to do work or perform tasks for the benefit of a person with a disability. The law describes a disabled person as an individual whose physical or mental impairment substantially limits one or more major life activity. The tasks that service dogs are able to perform include guiding persons with impaired vision, alerting persons with hearing impairments to various sounds, pulling wheelchairs, pulling a person into a lying or sitting position, turning switches on or off, retrieving objects, and summoning help. Certain dogs have even been found to sense an impending seizure and alert their owners before the seizure occurs.

Federal laws protect the rights of disabled individuals to be accompanied by their service animals into public places. These animals are closely partnered with their owners, who rely on them to provide vital services. Laws do not restrict the type of service the animals perform, and owners are not required to disclose their disability. Service animals usually wear an identifying harness or vest, but this is not required by law. The rules for interacting with human-animal teams vary when dealing with pet visitation, AAT teams, and service animal teams.

Care Guidelines

Guidelines are available that govern all types of animal-assisted health care programs. Health care facilities are required to adhere to state or federal guidelines regarding the use of animals. Animals require health screening and immunizations by a veterinarian prior to entering a health care facility. For animals that live in a facility, there are regulations outlining the care of the animal. Pet visitation and AAT programs follow the institutional guidelines for dealing with persons who may be allergic or phobic. Staff and residents should be notified ahead of time of animal visits, or visits may be restricted to discrete areas of the health facility. Those who are phobic or allergic or do not wish to participate can then remove themselves from the vicinity. Caution should be exercised when animals are exposed to clients who have disabilities that may cause them to handle the animal roughly or provoke the animal unexpectedly.

Institutional and other regulatory systems (local or state) also regulate infection controls. Handwashing, before and after, is standard when there is contact with an animal. Persons with open wounds or active infectious processes are typically excluded from such programs. Pet visitation and AAT programs have recorded thousands of visits without any substantial risk of zoonoses (diseases transmitted from animal to human) being substantiated in the literature

Strangers, including health care professionals, should always speak to the person before interacting with the animal partner, and should not grab the elbow of visually disabled persons or assist without permission. Such actions may confuse the dog or prevent it from doing its job. Strangers should not talk to, pet, or feed a service animal, as these activities also may distract the animal from its work. It is often necessary to explain to others who complain about an animal's presence that the animal is medically necessary and that federal law protects the right of the person to be accompanied in public places. If a service animal/dog barks or growls, it may be necessary to find out what happened (the dog may have been stepped on). The owner of the service animal/dog may be asked to have the dog lie down, as long as this does not interfere with the animal's work. Despite federal protection, animals acting in a vicious or destructive manner may be excluded from a public setting.

The needs of the animal (adequate water, toileting, and exercise) need to be planned for. Noise, sanitation, staff concerns about the appropriateness of service animals, and cost issues all need to be considered before instituting animal-assistance programs in a health care facility. Costs for the animals vary widely. Pet therapy and AAT team animals are often owned by the handler. Service dogs are the most costly because of their extensive training (up to $15,000). Public funds do not currently reimburse these costs. Service animals are often provided at a nominal fee to disabled individuals by nonprofit animal organizations, but waiting lists can be as long as two years. Many organizations exist that provide information, training, publications, and videos. The Delta Society is a leading international organization for promoting the human-animal bond. It serves as a valuable resource for information about the various types of animal programs.

The use of animals to assist with the care of the older adult offers diverse possibilities and opportunities. Benefits include well-described improvements in psychological well-being and social acceptance and decreased need for paid and unpaid assistance. Animals may be utilized purely as companions for social support, but their usefulness extends to service functions that range from home assistance to facilitation of community activities. The use of service animals within institutions is a recent but important development.

MARY SHELKEY

Internet Key Words
Animal-Assisted Therapy
Animal Therapy
Pet Therapy
Service Animals/Dogs

Internet Resources
Center to Study Human Animal Relationships and Environments
http://www.censhare.umn

Delta Society
http://petsforum.com/deltasociety/default.html

Eden Alternative
http://www.edenalt.com

ANXIETY AND PANIC DISORDERS

Comprehensive care for the elderly includes attention to mental as well as physical health. Although depression among the elderly has been identified as a major problem, anxiety among older adults has received less attention, despite the fact that anxiety disorders are diagnosed more than twice as often as depression in this age group (Regier et al., 1988). Greater awareness about anxiety disorders among professionals working with the elderly will improve the overall care of these patients.

Prevalence of Anxiety Disorders in Older Adults

Six-month prevalence rates of anxiety disorders in the elderly range from 3.5% to 10.2%, with data suggesting that phobias and generalized anxiety disorder (GAD) are most common. Considerably lower rates of obsessive-compulsive disorder, posttraumatic stress disorder, and panic disorder (PD) are seen. It should be noted, however, that even though PD occurs in only a relatively small percentage of older adults with anxiety, these patients are the ones most likely to present for medical treatment. Similarly, symptoms of GAD are described frequently by older patients requesting medical treatment. Older women are generally at higher risk for anxiety disorders than are men.

Recognition of Anxiety Disorders in Older Adults

To recognize anxiety disorders in older adults, it is important to be familiar with the cardinal diagnostic criteria. Phobias, GAD, and PD (American Psychiatric Association, 1994) are the anxiety disorders most likely to be assessed in a medical setting.

Phobia is diagnosed when an individual develops excessive and unrealistic fear of a circumscribed object or situation. Phobia often leads to subsequent avoidance of the feared object or situation. Specific phobias can be related to any number of things, such as animals, heights, or air travel. Social phobia is characterized by anxiety in social situations, usually stemming from fear of embarrassment, criticism, or humiliation.

GAD is characterized by excessive and uncontrollable worry accompanied by several physiological symptoms, including sleep disturbance, irritability, muscle tension, fatigue, restlessness, and difficulty concentrating.

PD is characterized by unexpected panic attacks, which are sudden and intense fear reactions accompanied by physical symptoms such as increased heart rate, perspiration, hyperventilation, chest pain, dizziness, fear of losing control, and fear of dying. Panic attacks usually intensify and peak within 10 to 15 minutes. Many people who develop PD avoid going out or doing things because of fear of triggering a panic attack. When avoidance behavior is significant, *agoraphobia* is also diagnosed.

Most important in the diagnosis of anxiety disorders in older adults is differentiating symptoms from normal aging. Excessive anxiety is not a normal part of aging, but social and developmental changes associated with aging need to be considered in the assessment. For example, new threats to the person's health (e.g., increased risk of physical illness or injury) or financial situation (e.g., loss of income after retirement) may cause an increase in anxiety. To determine whether these fears represent an anxiety disorder, it is important to evaluate the patient's concerns within his or her environmental context. If the fears and worries are "unrealistic" or "excessive," given the known facts of the situation, and cause some disruption in life function (e.g., interpersonal conflict, poor adjustment to retirement), an anxiety disorder may be present. It also is important to differentiate physical symptoms of anxiety from normal physical changes in the aging process. For example, it is easy to disregard physical symptoms of GAD (e.g., problems concentrating, sleep disruption) as signs of normal aging. Therefore, clinicians should assess for the presence of an anxiety disorder when these symptoms are reported.

The typical presentation of anxiety is different in older adults. Older adults focus more on physiological symptoms of anxiety than affective experiences. Patients may be reluctant to report anxious feelings but readily admit to problems with sleep, concentration, fatigue, or aches and pains. Again, clinicians need to assess for anxiety when these physiological complaints are reported. In addition, verbal descriptors used by older adults to report anxiety are unique. In particular, older adults with GAD may deny feeling "worried" but will admit that they "fret" often. Therefore, developing a common language with the patient may help to accurately assess the degree of anxiety.

Older adults are not only more likely to report physical symptoms but also more likely than younger adults to attribute anxiety symptoms to medical causes and to experience medical problems with overlapping symptoms. Since many symptoms of anxiety disorders mimic symptoms of medical disorders (e.g., heart attacks, hyperthyroidism), differential diagnosis requires a thorough history and diagnostic interview, as well as medical tests to rule out biophysical causes. Medical illnesses can also cause stressors that precipitate or exacerbate an anxiety disorder in older adults. Although GAD can be a chronic condition that begins in early adulthood, it can also have a late onset, especially following a stressful event such as a new medical illness. Panic attacks can also develop following a significant illness such as a heart attack if the patient develops a "learned alarm" reaction to perceived physiological changes.

Caring for Older Adults with Anxiety

Once anxiety is recognized as a problem for an older adult, treatment options can include both pharmacotherapy and psychosocial treatments. It is important to emphasize, however, that an individual treatment plan needs to be developed within the context of the patient's living situation (e.g., independent versus assisted living), financial resources, physical mobility, and cognitive capacity. In addition, clinicians must be familiar with the community resources available when a referral is indicated.

Pharmacotherapy: Older adults experiencing anxiety are often treated by their primary care physicians, who prescribe medication—usually benzodiazepines. Lower doses of compounds with short half-lives should be prescribed for as short a duration as possible. Adverse effects of these medications that may be particularly problematic for older adults (e.g., sedation, respiratory depression) need to be monitored closely. Buspirone, beta blockers, and antidepressants are alternatives to reduce tension and have less serious side effects. However, more research is needed before firm conclusions can be drawn about the most efficacious pharmacotherapy for older adults with anxiety.

Psychosocial Treatments: Cognitive-behavioral approaches have received strong empirical support for the treatment of anxiety in older adults (Stanley & Beck, in press), although the most effective treatment depends in part on the presenting complaint. Systematic desensitization is appropriate for treatment of a specific phobia, whereas exposure with response prevention is the treatment of choice for obsessive-compulsive disorder. Specific techniques that are helpful in treating GAD and PD include relaxation training, problem solving, sleep management, and cognitive restructuring (i.e., changing negative thoughts). A referral to a clinical psychologist or licensed social worker may be required to determine the most appropriate psychosocial treatment.

Treatment Adjuncts: In addition to the general therapeutic techniques, education of both the patient and supportive others (e.g., caregivers) is especially important, since older adults tend to be negatively biased about mental health issues. Clinicians can also increase compliance by taking extra steps to facilitate the patient's ability to attend appointments. For example, it may be helpful to provide detailed maps in large print and to offer appointments during low-flow traffic hours or to assist in using public transportation services. Finally, psychosocial treatments for older adults are improved by greater use of visual and written aids to facilitate learning through multisensory pathways.

Anxiety in the elderly is a significant public health problem. Improved awareness of the prevalence of anxiety in this population will lead more clinicians to consider anxiety as a diagnosis and in treatment planning. Attention to the uniqueness of the presentation and management of anxiety disorders will facilitate high-quality care for older adults with anxiety.

GRETCHEN DIEFENBACH

Internet Key Words
Anxiety Treatment
Generalized Anxiety Disorder
Panic Disorder
Phobia
Psychological Assessment

APHASIA, PRIMARY PROGRESSIVE

Aphasia is impairment of a central aspect of language processing, such as comprehension or expression, that is independent of peripheral processes such as hearing or oral speech articulatory movements. Historically, aphasia was thought to have a sudden onset due to a stroke or head trauma. It has become increasingly evident, however, that aphasia can also be primary, that is, due to a neurodegenerative disease. In this situation, the aphasia is progressive, worsening over time until language comprehension or expression is completely lost. Primary progressive aphasia is thus a neurodegenerative syndrome with the cardinal feature of language difficulty.

Core Clinical Features

Like other neurodegenerative conditions, primary progressive aphasia tends to occur in individuals older than 40 years of age. Unlike Alzheimer's disease, however, primary progressive aphasia does not occur more frequently with aging. There is a Poisson distribution, with the mean age of onset about 60 years, but there does not appear to be a distinct gender distribution. There are no apparent predisposing factors for the development of primary progressive aphasia, although there may be a link in some patients with the q21-22 locus of chromosome 17.

The most common primary progressive aphasia syndromes are progressive nonfluent aphasia (PNFA) and progressive fluent aphasia, also known as semantic dementia (SD). PNFA is most notable for extremely effortful speech that is often dysarthric and produced hesitantly at a very slow rate. There is a paucity of grammatical forms, with an overabundance of grammatically simple utterances, and grammatical suffixes are commonly substituted or omitted. Word-finding difficulty is frequent in spontaneous speech, and a confrontation naming deficit is often found. In some PNFA patients, repetition is quite compromised, which may be related to a limitation in short-term memory. These problems in language expression become much less prominent when patients speak highly familiar phrases, recite overlearned sentences, or produce automatic sequences such as counting or reciting the alphabet. PNFA patients also have difficulty with comprehension. This appears to be related mostly to their limited comprehension of grammatical relationships. Comprehension of simple declarative sentences (e.g., The boy chased the girl and the boy is happy) is relatively preserved, in comparison to comprehension of complex sentences with subordinate clauses (e.g., The boy that the girl chased is happy). Single-word comprehension is relatively preserved during the early stages of the disease process. As the condition progresses, the patient is able to produce less speech and eventually becomes mute. Comprehension similarly declines until the patient understands essentially nothing that is said. Reading and writing remain relatively preserved and decline more slowly, lagging up to one or two years behind the patient's oral communication. These visually based modalities become the methods of communication until reading and writing also become impaired. The comprehension and expression of gestures and facial expressions also decline more slowly than oral communication.

Patients with SD have a very different presentation. Their core problem is a profound difficulty understanding single words. They might repeat a word and say: "I'm sure I've heard that word before. What does it mean?" Their speech, though fluent, becomes increasingly empty of content because the meanings of the words they hope to use have been degraded. SD patients often develop increasing difficulty in reading and writing that lags somewhat behind the decline in their oral language. However, their pattern of reading and writing difficulty is quite distinct from that seen in PNFA: SD patients develop a surface dyslexia whereby words are pronounced as they appear (e.g., "choir" pronounced "cho-ire" instead of "quire").

Associated Cognitive Difficulties

Patients with primary progressive aphasia often develop other cognitive difficulties as the underlying disease progresses. Over time, PNFA patients evolve to a frontotemporal form of degeneration. Many of these patients begin to demonstrate a limitation in executive functions, including distractibility, poor planning and mental organization, impaired hypothesis formation and anticipation, and compromised working memory. Patients often become "stuck" repeating the same word or gesture. There may be an associated limitation in initiative and internal motivation, whereby patients sit or lie for hours without any apparent interest in doing anything. Family members frequently report patients' inability to perform any activity beyond the most simple and structured task and describe their difficulty taking a shower or making a sandwich.

Patients with SD may develop a pattern of letter-by-letter reading (as if they are slowly spelling aloud to themselves while reading). This may eventually blend into a form of visual agnosia, with difficulty recognizing line drawings of objects.

Some patients with progressive aphasia can develop a bizarre affect. Families notice disinhibited and socially inappropriate behavior that is entirely out of character, given the patient's premorbid personality. Characteristics of Klüver-Bucy syndrome may emerge, such as hypersexual behavior; hyperoral behavior, such as compulsive food fixations or mouthing inedible materials; and hypervisual behavior that can be manifested as shoplifting small, bright objects. A pattern of meteoric rage responses can develop in reaction to trivial or nonexistent events, and such a response may fade just as rapidly and unexpectedly. Later in the course of these conditions, patients can develop an anterograde memory deficit that limits the acquisition and retention of new material.

Differential Diagnosis

The differential diagnosis of primary progressive aphasia includes other neurological diseases that can cause progressive decline of left hemisphere functioning. For example, a strategically placed arteriovenous malformation, aneurysm, tumor, or other mass lesion can progressively interfere with language functioning. Metabolic disorders such as vitamin B_{12} deficiency or hypothyroidism can mimic some features of primary progressive aphasia. Symptoms of small-vessel ischemic disease and hydrocephalus can be similar to some aspects of primary progressive aphasia. Features of primary progressive aphasia can be seen in infections such as cryptococcal meningitis and subacute conditions such as Creutzfeldt-Jakob disease. Infiltrative diseases such as sarcoidosis and Whipple's disease can have clinical manifestations similar to primary progressive aphasia. A poorly controlled complex partial seizure disorder and exposure to environmental or iatrogenic central nervous system toxins can present as primary progressive aphasia as well.

A careful history and neurological examination can help determine whether any of these conditions—or a common neurodegenerative condition such as Alzheimer's disease—is causing a syndrome resembling primary progressive aphasia. For example, a history of episodic worsening and resolving clinical manifestations suggests an underlying seizure disorder. A disorder in voluntary gaze or axial rigidity can be seen in akinetic-rigid disorders associated with primary progressive aphasia, such as progressive supranuclear palsy and corticobasal degeneration. A gait disorder may be evident in these conditions as well as in hydrocephalus, small-vessel ischemic disease, and vitamin B_{12} deficiency. Progressive weakness and muscle wasting, together with fasciculations, are consistent with an amyotrophic lateral sclerosis (ALS)–dementia syndrome that can present as PNFA and has a poor prognosis. Subtle focal neurological features also are evident in small-vessel ischemic disease, and reduced vibration and proprioception are evident in the distal lower extremities in vitamin B_{12} deficiency. Many of these conditions may not have specific neurological manifestations, and appropriate studies should be obtained to rule them out, including serum studies, electroencephalogram, lumbar puncture, and high-resolution structural imaging studies, such as magnetic resonance imaging. In the case of negative findings on these studies, a functional neuroimaging study demonstrating reduced cerebral activity in the frontal or temporal region of the left hemisphere (in right-handed individuals) is consistent with the diagnosis of primary progressive aphasia.

Autopsy studies on patients presenting with primary progressive aphasia have been few. A summary of the autopsy studies published up to 1997 indicated that PNFA was most likely to be caused by a non-Alzheimer's form of dementia such as Pick's disease, dementia lacking distinctive histopathology, ALS-dementia syndrome, or corticobasal degeneration. SD also may be caused by these conditions, but this form of primary progressive aphasia is more likely to be caused by Alzheimer's disease than is PNFA.

Management

Treatment of a progressive language disorder depends on whether an underlying disease has been identified. In such a situation, treatment of the primary disorder can halt progression of the aphasia and may result in improved language functioning. Treatment of primary progressive aphasia, by comparison, is much more difficult. We are only beginning to understand the underlying pathophysiology of the condition, and management is still relatively primitive. Nevertheless, pharmacological and behavioral interventions are available and are being developed that may benefit patients with primary progressive aphasia.

One pharmacologic approach advocates supplementation of neurotransmitter systems, which are likely to be compromised in these patients. There is some evidence that primary progressive aphasia is related in part to interruption of a fronto-striatothalamofrontal circuit modulated in part by dopamine. Studies of stroke patients indicate that dopamine supplementation can benefit speech fluency in those with a nonfluent form of aphasia. Moreover, dopamine supplementation can help working memory function in some patients with traumatic brain injury, which frequently involves frontal regions. Dopaminergic precursors such as levodopa and dopaminergic agonists such as bromocriptine or pergolide also can be administered to patients with primary progressive aphasia. Clinical observations suggest that in small doses, these medications may have some beneficial effect on the natural history of the disorder and in some cases may improve speech fluency and confrontation naming. Selegiline is a form of monoamine oxidase inhibitor that slows the breakdown of dopamine. In small doses, its activity is selective for the type B form of monoamine oxidase in the central nervous system, so it has fewer systemic side effects. This, too, may help slow the progression of primary progressive aphasia.

Ritalin, a medication used in the management of patients with attention deficit disorder, may help some of the attentional limitations that can emerge in patients with primary progressive aphasia, and it may also have a beneficial effect on the mood of some patients. Aricept is an acetylcholinesterase inhibitor used in Alzheimer's disease to slow the rate of memory decline by slowing the breakdown of the neurotransmitter acetylcholine. Clinical observations suggest that medications such as Aricept can also benefit those with primary progressive aphasia who are having difficulty with anterograde memory. Medications such as trazodone can have a beneficial effect on the dopaminergic system and can help modulate the personality disorder that emerges in some patients with primary progressive aphasia. Patients with comprehension difficulty can become frustrated, agitated, and even paranoid. A mood-stabilizing agent such as valproic acid or carbamazepine may be beneficial, and a small dose of a neuroleptic agent such as olanzapine or chlorpromazine may be necessary. All these medications have side effects that can interfere with their effectiveness or tolerability in individual patients, and it is important to keep these in mind when considering a medication regimen.

Other medications may have some beneficial effect on the underlying condition, even if they have no direct effect on the patient's presenting symptoms. Although there have been no biochemical or medication studies assessing free radical production in primary progressive aphasia, it is fair to assume that factors contributing to neuronal death in Alzheimer's disease are likely to contribute in part to the diseases implicated in primary progressive aphasia. Thus, it would not be unreasonable for patients to take supplemental vitamin E and an anti-inflammatory medication such as ibuprofen. Selegiline may also reduce the production of oxygen free radicals, which can contribute to biochemical inflammation and more rapid neuronal death.

Some behavioral and cognitive interventions can benefit patients with primary progressive aphasia. Clinical observations suggest that traditional cognitive speech therapies have not been very effective. Experimental speech therapies have been attempted in a small number of case studies, with modest success. These have been directed at recruiting alternative cerebral resources to compensate for brain regions that are not functioning optimally. For example, one approach to the treatment of PNFA concentrates on intense practice identifying a verb in a sentence and then identifying the subject of the verb.

Other practical strategies can be initiated to optimize communication in patients with primary progressive aphasia. The use of simple grammatical forms and high-frequency vocabulary terms facilitates comprehension. Supplementation of speech with gestures and speaking slowly can help comprehension as well. Multiple-choice strategies can be initiated to help patients select a response. This avoids difficulties associated with word retrieval and structures the task in a user-friendly fashion that minimizes working memory demands. A vocabulary sheet with frequently used words (paired with appropriate pictures) can be created for the patient. More generally, communication within a familiar, supportive environment (with relevant objects available for demonstration, familiar features that minimize executive resource demands) can be helpful. Although speech comprehension and expression are quite compromised in these patients, written communication may be less severely compromised and can serve as an effective communication method at specific stages of the condition.

(This work was supported in part by funding from USPHS [AG15116, NS35867, and AG17586] and the American Health Assistance Foundation.)

MURRAY GROSSMAN

See Also
Communication Issues for Practitioners
Speech Therapists
Stroke/Cerebral Vascular Accident

Internet Key Words
Frontotemporal Degeneration
Language Disorder
Semantic Dementia

Internet Resources
American Speech-Language-Hearing Association
http://www.asha.org

National Aphasia Association
http://www.aphasia.org

National Institute of Neurological Disorders and Strokes
http://www.ninds.nih.gov

ARTHRITIS
See
Osteoarthritis
Rheumatoid Arthritis

ASIAN AND PACIFIC ISLANDER AMERICAN ELDERS

In the United States, Asian and Pacific Islander (API) Americans are a diverse group of persons who immigrated or whose ancestors immigrated from various countries in Asia and the Pacific Islands. As many as 29 distinct nationality, ethnic, cultural, and language groups are counted by the U.S. Bureau of the Census. The largest Asian ethnic groups include Chinese, Filipinos, Koreans, Japanese, Vietnamese, Asian Indians, Cambodians, Laotians, H'mong, Thai, and Pakistani. The largest Pacific Islander groups include Native Hawaiians, Samoans, Guamanians, and Tongans. These individuals' life experiences have been affected by economic, political, and social policies in areas of the government as diverse as justice, defense, labor, education, and intergovernmental relations. As with all populations, these life experiences have an impact on access to and participation in health and social services programs for individuals and as a group.

Historic Context

Each API ethnic community in the United States has a rich and varied history that is affected by the wave or cohort in which community members arrived. Initial waves of Chinese immigrants were bound by historic policies and practices that re-sulted in their segregation to certain streets of San Francisco; certain mine claims in California, Wyoming, and Nevada; specific low-paid, high-risk jobs to build the transcontinental railroad; and later, certain service-sector jobs that were not competitive with other races. Initial waves of Japanese, Filipinos, and Koreans were allowed limited social and economic roles as farm workers, domestic or food service workers, and day laborers. More recently, relatively large numbers of older adult Chinese, Filipinos, Koreans, Asian Indians, Tongans, and others have arrived as families seek to reunite. Vietnamese, Cambodians, H'mong, and Laotians arrive seeking refuge from political unrest and the effects of war. In the past 40 years, life for many API and other elders of color in the United States has been affected by major social, economic, and political change: reparations paid to Japanese Americans in recognition of their unfair incarceration during World War II; statutory laws to allow APIs to own property and businesses; affirmative efforts to permit all persons fair access to housing and public accommodations; and, in some communities, drivers' tests and election ballots in languages other than English. In some instances, relations within different ethnic groups of APIs have undergone as much change as between each API ethnic group and "Americans."

Demographics

API elders are the fastest growing ethnic population among elders, representing almost 2% of 34 million older adults in the United States. They numbered more than half a million in 1990, and this number is expected to more than double by 2010. API older adults are more likely to be foreign born (70% compared with 9% for non-API elders) and more likely to speak a language other than English at home (80% for APIs, compared with 12% for non-APIs). Yet almost half of foreign-born API elders are naturalized U.S. citizens. Nativity, language, and citizenship differences among API elders are the types of indicators that could help service pro-

viders better understand their API clients. Among Japanese, Chinese, and Vietnamese, for example, differences are pronounced and are further affected by the length of time individuals have been in the United States (e.g., arrival as children or young adults and participation in the U.S. education system).

Program Participation

In community capacity-building assessment meetings between 1997 and 1999 (National Asian Pacific Center on Aging [NAPCA], 1999), elders stated that they were uninformed about, ignored by, and at times unwelcome at mainstream programs for older adults in their communities. Although many API elders (especially in western states) have resided in the same place for decades, paid taxes, and participated in the economic growth of their communities, they are often not visible and do not participate in social and health services programs in their communities, even though data lead us to expect that they would need such help.

API elders are more likely to live in urban and suburban than rural areas. More than two in five API elders live in California, almost one in five live in Hawaii, and almost one in ten live in New York. States such as Illinois, Washington, New Jersey, and Texas each had over 10,000 API elder residents in 1990. Yet there is very little microdata on APIs overall or on any specific API ethnic group in the United States. Considered too small a population to include in analyses of national survey data, API elders are also less likely to participate in national mailed or telephone surveys that are only in English. Many API elders have also learned from life experiences that talking about problems or personal situations to government representatives can result in oppressive and retaliatory actions, such as having benefits curtailed or denied or having family members investigated. Efforts by service providers to understand and respond to the needs of API elders are often problematic and are unsuccessful in establishing rapport, identifying possible interventions that meet stated needs, and achieving successful referrals to available service programs.

Challenges for Service Providers

Recent work consistently indicates that steps must be taken conjointly by service providers and API elder communities to address critical needs and improve the quality of life for API older persons. Communitywide outreach programs and service delivery for infants and young children, AIDS- or HIV-infected individuals, the unemployed, and frail older adults since the late 1970s have met with mixed results and have been largely ineffective and unsuccessful.

Several voluntary community-based organizations that serve specific API ethnic groups have had long-term success in service delivery in their geographic areas. At the same time, many public health and social services programs appear to have abandoned efforts to systematically reach and appropriately serve API elders. Increasing complexity within service systems, scarce resources, and a lack of direction to effectively serve newer, smaller, harder-to-reach communities are among the reasons given for the lack of effort and success in serving API elders. Gathering in forums such as community capacity-building and Medicare outreach meetings to revisit community issues, listen to elders' perspectives, and enjoin problem solvers to sustain conversations until new solutions are developed can be an effective intervention (NAPCA, 1999; Yee, Sanchez, & Shin, 1999).

Addressing language and information barriers is a first step toward increasing understanding among API elders and indicating that service systems want to be responsive to the social, health, and other human services needs of API elders. Quick fixes, such as the use of "black box" communications (i.e., piping the ATT Language Service into medical examination rooms and government offices), may be better than nothing in some cases but could be more harmful in other situations. Anyone who has been frustrated while trying to decipher literally translated instructions (from Japanese or German into English) to assemble furniture or a child's toy or to operate electronic equipment can understand why literal translations from English into an API language cannot be assumed to work

for API older persons. Cultural translation may be needed to ensure communication and rapport between doctor and patient, provider and consumer, outsider and family. Literal interpretation might not be enough to facilitate the breadth and depth of communication needed in such instances.

Health care access issues need to be addressed. Elders need explanations of how service programs and service systems work in ways that make sense to them. Concepts such as prepaid health care or home- and community-based care are not likely to be part of an API elder's vocabulary or life experience. Practitioners must be willing to recognize differences in cultural assumptions about private and public roles. Understanding family roles in decision making among many APIs may mean asking about and being open to ways in which an eldest son, a husband, or a brother-in-law can make decisions rather than the elder patient herself. An API family may need help understanding that "Americans" place importance on end-of-life or insurance issues and that such decisions are made by the individual client. Concurrently, service providers may need to recognize that a patient or client has different priorities. A mutual understanding of differences and a mutual willingness to communicate in the context of such differences are more likely to lead to a negotiated resolution.

Best Practices and Lessons Learned

Anne Fadiman's account (1997) of a harrowing mismatch between the "American" health system and a H'mong family and Sawako Ariyoshi's novel (1984) of a woman's relationship with her frail in-laws in postwar Japan are two excellent portrayals of how help-seeking behavior, problem identification, and problem solving between client and service provider are only the first steps in providing appropriate and good care in cross-cultural situations. Attending to the cross-cultural aspects of each mini-episode of the caregiving relationship is critical to identify appropriate interventions that address the initial problem.

Contact between those whose mission it is to provide services and meet needs and those who need assistance is critical in building relationships. Communication with one another about shared and differing cultural assumptions can build a bridge of cultural competence and make significant differences in the quality of life and health outcomes for API elders.

DONNA L. YEE

Internet Key Words
Asian and Pacific Islanders
Cultural Competence
Language Access
Multiculturalism
Service Access

Internet Resources
American Society on Aging/Multicultural Aging
  Network
http://www.asaging.org

National Asian Pacific Center on Aging
http://www.napca.org.htm

ASPIRATION
See
Swallowing Disorders and Aspiration

ASSESSMENT TOOLS AND DEVICES

Older adults are subject to a wide variety of physical and psychosocial changes that can influence their quality of life and independence. They experience interrelated problems that call for a biopsychosocial approach to address the functional, psychological, social, environmental, and financial aspects of their lives. Assessment can provide clients and caregivers with information that indicates the best course of action for that individual. An assessment may determine eligibility for services and justify the need for interventions, treatments, or further testing (Geron, 1997). George (1997) defined assessment

as "methods of determining the state of individuals on one or more dimensions of abilities, resources or attributes" (p. 32). Assessments may be performed in hospitals, long-term-care facilities, geriatric clinics, assessment centers, or the individual's home.

The assessment process must balance practicality, the possible intrusion into the older person's time, and thoroughness. Although assessments can be completed strictly from clinical interviews, Geron (1997) articulated three reasons for the inclusion of standardized instruments in the assessment process. Instruments should incorporate standardized measures, balance psychometric precision with practicality, and be standardized across users. If these criteria are to be met, providers must be sufficiently trained in specific assessment instruments to understand the hows and whys of each instrument. Physical functioning, cognitive and psychosocial functioning, social support, financial or socioeconomic issues, and environmental concerns should all be assessed.

One important aspect of physical assessment is the determination of independence/dependence in functional status. Activities of daily living (ADL) and instrumental activities of daily living (IADL) are common indicators of functional status. Emlet, Crabtree, Condon, and Treml's (1996) analysis of 13 ADL and IADL assessment instruments included the purpose and clinical utility of each instrument, as well as its completion time, scale construction, reliability, and validity. Kane and Kane (1981), in their classic work *Assessing the Elderly: A Practical Guide to Measurement*, provide an extremely valuable description of various ADL and IADL instruments. Because of the variety of assessment instruments available, it is important to balance the needs of the clients, the discipline, and the agency when choosing an ADL or IADL scale.

Screening tools for cognitive functioning contain questions that assess orientation, memory, language, constructional abilities, and higher cognitive functions. Instruments typically used for this purpose include the Mini-Mental Status Examination (MMSE), the Blessed Informational-Memory-Concentration Test (Fuld, 1978), and the Short Portable

Mental Status Questionnaire (Pfeiffer, 1975). Despite the MMSE's bias around educational level and socioeconomic status of the respondent, Zarit (1997) contends that it is important because it is widely known and many physicians are familiar with it.

In day-to-day work with older clients, professionals need an instrument that is familiar across multiple disciplines. Taking any mental status exam may be threatening to older clients. The older person may not have met the professional before and may be concerned about being regarded as "crazy." Regardless of the instrument chosen, it is important to explain the purpose of the examination. It may be helpful to frame these as "memory questions" or find some means to make the process less threatening. It is particularly important to balance the need for completing the assessment with the individual's anxiety. If trying to answer a question creates too much anxiety for the individual, it may be best to move on, find other ways to test that aspect of cognition, or attempt the question later in the interview.

Kapust and Weintraub (1988) suggest asking the elder to perform specific functional tasks such as preparing tea, finding commonly used objects (keys, tools), using the telephone, and giving a house tour. These tasks assess memory, attention and motivation, language, and visuospatial deficits and help determine functional and cognitive abilities. Furthermore, the tasks are not based on Eurocentric values such as making calculations, repeating phrases, or spelling words backwards, which have no cultural meaning.

Depression is a common psychiatric disorder of old age, viewed as both a constellation of symptoms and a diagnostic category. Screening is critical. Brief measures of depression function as symptom indicators rather than diagnostic tools, and include the Center for Epidemiological Studies Depression Scale (CES-D) (Radloff, 1977), the Beck Depression Scale (Beck, 1961), and the Hamilton Depression Scale. Another instrument, the Geriatric Depression Scale (GDS), specifically designed for use with older populations (Yesavage, 1983), consists of 30 yes/no questions and can be self-adminis-

tered or completed through an interview. A shorter, 15-question version (Sheikh & Yesavage, 1986) is also commonly used. A five-item version of the GDS is reportedly as effective as the 15-question version with a marked reduction in administration time (Hoyl, 1999).

Social support assessment can determine sources of emotional and affective support and identify possible individuals for prevention or intervention activities. Antonucci, Sherman, and Vandewater (1997) inventoried several assessment instruments. The hierarchical mapping technique (Antonucci, 1986) is a qualitative and innovative way for older persons to describe who fits into their social networks according to personal feelings of closeness by placing them inside three concentric circles.

The financial or socioeconomic status of older persons may affect their ability to meet basic needs and their capacity to purchase needed services (Emlet et al., 1996). Financial problems are associated with low income, lack of financial resources, or difficulty in financial management. Emlet et al. (1996) constructed a financial worksheet to assist with financial evaluation. Many elders consider their financial affairs very private. It is important to explain why financial questions are important, even though professionals may feel uncomfortable asking such questions. Approaching the client in a straightforward matter, explaining the purpose, and stating that this information may help determine eligibility for services can reduce the elder's resistance to answering the questions. It may be helpful to make it clear that all information received is confidential.

When possible, assessment of an older person should include the environment. Environmental barriers can hinder accessibility and activity as well as jeopardize safety. Many environmental barriers are architectural in nature. May (1999) provides a useful checklist for home safety. Additional specifications for the homes of persons with disabilities can be found in Rothstein, Roy, and Wolf's *The Rehabilitation Specialist's Handbook* (1991).

A professional service provider can be perceived as compassionate and well meaning or as a "hostile agent of an oppressive and irrational bureaucracy that needs to be placated or cajoled in order to achieve even a little help" (Yee, 1997, p. 27). Although the purpose of assessment is to objectify and quantify need, the client-provider relationship cannot be overemphasized. Taking the time to listen and develop a relationship and being sensitive to clients who tell us what they think we want to hear are important issues when working with elders from other cultures. In many cultures, the process (relationship) is as important as the product (outcome).

Argüelles and Lowenstein (1997) remind us that translating assessment instruments for use with non-English-speaking elders is not a simple undertaking. Individuals from different ethnic and cultural backgrounds may be placed at a disadvantage when using standardized assessment instruments. There are a number of options for addressing this problem. A three-step protocol known as "back translation" can help ensure proper translation of assessment tools. These steps are (1) translating the instrument into another language, (2) translating the instrument back into English by a second individual, and (3) comparing the original and back-translated versions (Argüelles & Lowenstein, 1997). It is also recommended that providers check with the assessment instrument's publisher to determine whether a translated version is available.

Technology is increasingly relevant to geriatric assessment. Telephones are being used to assess individuals. Norton et al. (1999) studied the use of a telephone-modified MMSE and found that it provided a reasonable substitute for more costly face-to-face exams with a population of individuals without major cognitive syndromes. Automated telephone assessments of functional status have been studied and found to be as reliable as personal telephone assessments, but not as reliable as in-home assessments in capturing impairments (Mahoney, Tennstedt, Friedman, & Heeren, 1999).

There are over 200 programs for administering, scoring, and interpreting assessment packages (Zastrow, 1999). Computer programs can transform data into graphic forms that may help clients understand changes in their function, status, or symptoms.

Because of the use of technology and the increasing productivity demands on providers, dehumanization of clients is a risk. Assessment is "conversation with a purpose" (King, 1997, p. 73); the approach is conversational and relaxed, but structured. Providers should use assessment tools but take the time to develop rapport and probe appropriately and not shy away from difficult situations. The goal is to empower the client (Genevay, 1997). Assessors have "power" that is often not recognized. In elder assessment, the assessors should "see us [the client], hear us, and know who we are—in all our complexity, competence and need for autonomy" (Genevay, 1997, p. 18).

CHARLES A. EMLET

See Also
Activities of Daily Living
Assistive Devices
Cognition Instruments
Cognitive Screening Tests: The Mini-Mental Status Exam
Cultural Assessment
Depression Measurement Instruments
Vision Safety

Internet Key Words
Assessment
Cognitive Impairment
Culture
Depression
Functional Status
Social Support
Socioeconomic Status
Technology

Internet Resources
American Geriatrics Society
http://www.americangeriatrics.org/positionpapers/CGA.html

Comprehensive Geriatric Assessment
http://www.agenet.com/kimparticle.html

Geriatric Assessment
http://cpmcnet.columbia.edu/dept/dental/Dental_Educational_Software/

Gerontology_and_Geriatric_Dentistry/
Assessment/indexass.html

ASSISTED LIVING

Assisted living (AL) is a type of congregate residential housing that provides or coordinates personal services, 24-hour supervision and assistance (scheduled and unscheduled), recreational activities, and health-related services. AL is known by some 26 different names, including residential care, board and care, personal care, enriched housing, and domiciliary care, and the distinctions often lie in state statutes, regulations, certification, or licensing; type of accommodation (e.g., shared room or apartment); and the needs that can be met. Currently, more than 600,000 elders are living in approximately 30,000 AL residences in the United States (Assisted Living Federation of America, 1998). AL is for frail elderly and adults with disabilities who need help to live independently or who no longer wish to remain at home. It is unsuitable for those who require 24-hour skilled nursing care or medical monitoring.

An AL residence has a consumer-centered philosophy that promotes wellness and maximizes quality of life, independence, privacy, choice, safety, decision making, and "aging in place" in a homelike environment. Some AL facilities provide a special environment and programs for developmentally disabled or mentally ill residents and those with Alzheimer's disease or other dementias. State regulations set boundaries for the scope of services, but providers (operators) determine what will be offered and which residents (tenants) will be admitted. Private living units, including private bath, are a critical feature of AL. Another feature of privacy is that residents may lock their doors. At least 30 states allow two people to share a unit (apartment) or bedroom, but in some states, only if the two parties choose to do so. Depending on the licensing category, some states allow as many as four people to share a unit (Mollica, 1998).

The number and type of staff vary with the number of residents, their needs, and the services provided. Personal-care staff can be employed by the facility or contracted from an outside agency (e.g., licensed home-care agency). There is no mandatory training curriculum except for staff working in AL facilities caring for residents with dementia or cognitive impairments. Staff must be competent to care for residents and perform their jobs in accordance with state and federal regulations. Relevant staff must know how to respond in an emergency. There must be at least one person available during the night hours. Every AL facility must have an administrator (manager, director, operator) who has overall responsibility for staff performance and resident well-being.

An occupancy or service agreement, executed before or immediately on admission to an AL facility, is based on an assessment of the person's need for services and how these needs can best be met. It must clearly specify the services to be provided, the cost of each service and the payment structure, and relocation criteria should the facility be unable to meet the resident's health and safety needs. All AL facilities monitor the residents' activities to ensure their well-being and provide daily supervision and assistance with instrumental and personal activities of daily living; three meals a day, including therapeutic diets; housekeeping and laundry services; transportation for recreational and shopping trips; and an emergency call system. Qualified staff supervise or assist with self-administration of medication, as permitted by state laws and regulations. Residents who are temporarily incapacitated or recuperating from surgery, injury, or illness, or those who are dying, can remain in the AL facility if it can provide the necessary services and care. Each resident's health care is supervised by a physician of his or her own choosing (Citro & Hermanson, 1999; National Center for Assisted Living, 1998).

Criteria for admission and retention vary widely. Some states simply require that a resident be in stable health and not need 24-hour nursing care; other states' criteria screen out those who are bed-bound, are incontinent, have deep pressure ulcers, need artificial feeding or hydration, or are ventilator dependent. States might have criteria relating to independent ambulation, ability to use the toilet unassisted, and stage of dementia. In some AL facilities, residents can sign a "negotiated risk" contract that allows them to remain in the facility despite functional decline (Mollica, 1998). To some degree, this kind of contract epitomizes resident autonomy and choice, balanced with resident health and safety. Medicare home care can be provided in AL facilities for Medicare beneficiaries who meet eligibility criteria. A registered nurse from a certified home-care agency supervises and monitors the care. Hospice care as a home-care service can also be provided in AL facilities. Generally, this decision is made by the facility and the resident or family and is not contingent on state regulations.

The typical AL resident is female, white, and 83 years old and needs assistance with three activities of daily living (bathing, dressing, and medication administration). Slightly less than half of all residents have some cognitive impairment (e.g., short-term memory loss). Almost two-fifths of all residents use wheelchairs for mobility. Although most residents come to AL facilities from their own homes, slightly under 20% come from nursing homes. After an average stay of 26 months, approximately 45% of residents are discharged to nursing homes and 26% have died. Only 5% of residents leave AL facilities for financial reasons (Citro & Hermanson, 1999).

The AL market is predominantly private pay; there is no entry fee. A growing number of states are paying for the personal-care component for low-income elders by drawing monies from Medicaid; fewer states use the home- and community-based care Medicaid waiver (115c) program. Waiver dollars can be used only if the Medicaid beneficiary would be eligible for nursing home care. Long-term-care insurance is infrequently used or available among AL residents.

Average monthly fees are about $2,000; this is higher than the typical board and care fee but lower than nursing home costs. The actual fee paid by the resident depends on the type of housing (shared vs. private room) and the kind and number of services included in the contract. The AL fee can be constructed as an all-inclusive, flat-rate monthly price; tiered pricing based on the package of services needed (or desired) by the resident; tiered rates based on the resident's acuity level; fee-for-service pricing based on the resident's à la carte selection of services; or some combination of these models. Some services can be purchased on an as-needed basis.

To make AL affordable for an individual whose annual income is less than $25,000, and to slow down if not reduce rising Medicaid costs for nursing home care, several states are developing innovative financing mechanisms to assist developers in constructing affordable AL housing. A growing number of not-for-profit and for-profit nursing homes are converting beds and wings to AL. Many continuing-care retirement communities offer AL either in the tenant's current domicile or by relocation to an AL facility on the premises. Statistics suggest that "light-care" residents who would have been in nursing homes are opting for AL residence. Medicare capitated, managed-care organizations view AL as being well suited for managing rehabilitation and providing a supportive environment for frail managed-care enrollees.

Virtually every state is studying or promulgating regulations and licensure requirements that distinguish AL from other long-term-care and residential models (Mollica, 1998). States have the authority to set provider standards; many have done so with respect to food preparation and fire safety. Standards of care, compliance surveys, and mandatory reporting systems vary among states; there are no uniform requirements. The Joint Commission for the Accreditation of Healthcare Organizations (JCAHO) has developed principles and standards for AL that will be ready for accreditation surveys in 2001. It is unknown at this time whether JCAHO accreditation will be optional or mandatory. Federal quality-of-care standards probably will not be promulgated in the near future, because it is unlikely that the federal government will become a major AL payer. The state role in monitoring and licensure will continue to grow, especially as Medicaid assumes greater responsibility for the costs of care and services.

ETHEL L. MITTY

See Also
Continuing Care Retirement Community

Internet Resources
American Association of Retired Persons
http://www.aarp.org

Assisted Living Federation of America
http://www.alfa.org

National Center for Assisted Living
http://www.ncal.org

Senior Resources
http://www.seniorresource.com

ASSISTED SUICIDE
See
Euthanasia

ASSISTIVE DEVICES

Age-related changes and accompanying diseases often result in functional impairments that lead to a reduction in or loss of independence. The way older people do tasks and where they live can create major challenges. Many of these obstacles can be overcome by using assistive devices or home modifications to compensate for these changes. Absent these adjustments, doing tasks is more difficult, and the home may become unsafe. Many devices are available to help overcome functional limitations, prevent accidents, and promote independence and comfort.

Definition

According to the Technology-Related Assistance for Individuals with Disabilities Act of 1988, an assistive device is an item or product (whether handmade, acquired commercially, modified, or customized) that is used to increase, maintain, or improve functional capabilities of an older person. It may be handled, carried, or in some other way be in direct contact with the person; this includes built-up handles on eating utensils for those with a limited grasp, reachers to assist in retrieving items, and long-handled sponges to assist in bathing.

An assistive device may be placed in or attached to the environment or used within it to compensate for or bypass a potential problem. These interventions include home modifications or adaptations, which can vary in complexity (Cullen & Moran, 1992). For example, offset hinges widen doorways for a person in a wheelchair; alarms can sound so that the caregiver of a person with dementia can be alerted to the person's unsafe movement.

Types of Assistive Devices

Some assistive devices may not be marketed as such. When an item compensates for functional limitation, it is a part of the broad assistive device category.

Some people may require assistance in activities of daily living, such as bathing. Devices that may aid an impaired person include a bath bench, bath lifts from the bottom of the tub, bathmats or appliqués, grab bars, a hand-held shower, and a roll-in shower enclosure. Grooming assistance may be provided by items such as a long-handled comb, shampoo basin, soap dispenser for the shower, or suction denture brush. For toileting, one may use a commode, raised toilet seat, toilet bowl platform that fits between the floor and the toilet, or an incontinence warning device. Devices that aid in dressing include a bracelet fastener, button hook, long-handled shoehorn, dressing stick, elastic cuff extender, elastic shoelaces, sock aid, tactile markers to match clothing, trouser pull-up, zipper pull, and Velcro replacement for zippers and buttons. Built-up handled utensils, mugs or cups with large handles, glass holders, and plate guards are some devices that may aid in eating (Krantz, Christenson, & Lindquist, 1998).

Assistive devices to aid in mobility include bags or baskets for a walker or wheelchair, cane holder, cane ice gripper, offset hinges, door threshold ramp, residential elevator, transfer board, and walker with an integral seat. Positioning, sitting, and rising assistance can be provided by an adjustable bed, bed pull-up, chair rising aid, and grab rail or pole attached to or near the bed. Devices such as a dumbwaiter and shopping cart may be used for aid in lifting and carrying. Dumbwaiters can be installed in homes in a three-foot by three-foot space that is accessible from both floors.

Hand, grasp, or pinch strength or dexterity enhancement can be provided by a number of devices. A universal cuff with a pocket to hold a device, pencil and pen grippers, and window openers are some general aids. For doors, one may need doorknob grippers or lever door handles. For faucets and showers, lever faucet handles and lever shower controls can be useful. An electric cord plug puller or rocker light switcher can provide assistance as well.

Communication assistance is essential as well. Devices that aid in communication include a telephone with a gooseneck receiver holder, amplified telephone, large-number telephone, programmable telephone, and automatic dialer. Vision loss compensation (other than medically prescribed items) can be provided by appliance knobs with large contrasting numbers, large-print books, magnifiers with light, measuring-cup level indicators, talking food scales, and talking money identifiers. For lighting, one may use a light switch dropper, stairway lights, touch-control lamp, and motion-activated lights (Krantz, Christenson, & Lindquist, 1998). Hearing loss compensation (other than medically prescribed items) may require the use of a closed-captioned television, headphones that amplify sound, and a one-to-one pocket talker.

Safety devices that a person may use include nonslip mesh placed between a rug and hard floor, faucet temperature control, and stove element heat timer and sensor. Security devices include an entry security bar for a sliding door, home security system, intruder alarm sensor, and motion-activated exterior lighting system. Medication self-administration aids include an eyedrop dispenser that stabilizes the bottle, pill bottle opener, insulin syringe

holder, a device that clicks for each unit of insulin, and a magnifier that can be placed over bottles to enlarge the increment lines.

Many people require aid in food preparation. One may achieve ease of reach by using kitchen cabinetry pull-out shelves or a reacher or by lowering the shelf height. Ease of grasp may be provided by cookware with easy-grip handles, jar and bottle openers, milk carton holders, and stove knob turners. Items that allow one-handed use may also be helpful. These include cutting boards with spikes and corner guards, one-hand-operated rolling pins and vegetable peelers, rocker food-chopper knives, and saucepan stabilizers.

Recreational activities may also be difficult for an impaired person and require the use of assistive devices. Built-up garden utensils, a garden cart, and a watering wand may be used in indoor and outdoor gardening. For card and board games, one may use a card shuffler, large-number bingo cards, large-number playing cards, and checkers with raised or recessed surfaces. Reading aids include book holders, page turners, and prism glasses. Other recreational devices include fishing poles with automatic casting devices, pool cue bridges, and needlecraft aids, such as a one-handed embroidery hoop, needle threader, and knitting and crocheting aid (Krantz, Christenson, & Lindquist, 1998).

Issues and Strategies in Assistive Device Acceptance and Use

Many factors may influence a person's receptivity to assistive devices and should be considered before introducing or suggesting a product to an older person. If the person does not want the item or feels that it is unnecessary, it will probably be rejected.

Perceived Need: The person should be asked how he or she is now performing a particular task and how successfully. If the individual's actions seem safe, he or she may be reluctant to change. If the person's routine seems unsafe as well as only marginally successful, the problem should be discussed. The inability to see a need for a particular item or change certainly affects acceptance. Denial concerning the nature or progression of a problem may also be a major factor complicating the acceptance of an assistive device. "The most important factor in the eventual utility of assistive technology and interventions is the perspective of the person who will receive the intervention. Elders must be fully informed about the disability, so they can fully participate in decisions regarding the selection of appropriate devices and interventions. Service providers must consider the priorities of elders (and their caregivers)—do they really want or feel the need for certain devices? Have they been given options for devices and the differences between devices?" (Mann & Lane, 1995, p. 23).

Cost: Some older people are reluctant to purchase an item or make a structural change they feel they can do without. Others admit that they need assistive devices, but they cannot afford to buy them.

Knowledge of Availability: Many older people may feel a need for an item but be unaware that it exists because they lack current information on the range of assistive devices available (Mann & Lane, 1995).

Introduction of an Item: An older person might purchase an item based on its successful use by an acquaintance. Some hospitals give the patient a package of assistive devices based on the individual. This approach has met with varied acceptance.

Selection of the Right Product: The person, device, environment, and task need to be assessed for the intricate interaction among these elements.

Training in Use: The person should be able to demonstrate proper use of the device. An occupational therapist can analyze the patient's needs and the product to determine whether modifications need to be made to the product, its location, or the way the person is using the item.

Acceptance: Stressing the convenience aspect of a product makes it more acceptable than stressing the safety or need factor. Many items designed and marketed for convenience are essential everyday tools for older persons with disabilities.

Appearance of the Product: In the past, many products for people with disabilities were designed with little thought for their aesthetics. Most people, regardless of their functional capability, do not want an ugly item or a crude modification in their homes. When all other factors are equal, the more attractive item is likely to be selected.

Help for Caregivers: Some products or home modifications can ease the jobs of caregivers. Absorbed in the needs of the people they are caring for, caregivers may not realize that there are ways to relieve the burden for themselves as well as for the care receivers.

MARGARET CHRISTENSON

See Also
Activities of Daily Living
Hearing Aids
Low Vision
Vision Safety

Internet Resources
Center for Universal Design
http://www.design.ncsu.edu/cud

National Institute on Disability and Rehabilitation Research, U.S. Department of Education
http://www.abledata.com

National Resource Center on Supportive Housing and Home Modification
http://www.homemods.org

Rehabilitation Engineers Society of North America
http://www.resna.org

ASSOCIATION FOR GERONTOLOGY IN HIGHER EDUCATION

The Association for Gerontology in Higher Education (AGHE), an educational unit of the Gerontological Society of America, was established in 1974 as a membership organization of colleges and universities that provides research, education, training, and service programs in the field of aging. Its basic goal is to provide an organizational network to assist faculty and administrators in developing and improving the quality of gerontology and geriatric programs in institutions of higher education. The current membership of AGHE consists of over 300 institutions throughout the United States, Canada, and abroad. AGHE carries out its purposes through the following services and programs.

Meetings: Through annual meetings, workshops, and seminars, information and ideas are shared on curriculum and program development and evaluation, faculty training, and other issues of gerontological education.

Publications: AGHE publications (such as the *AGHExchange* newsletter, *Directory of Educational Programs in Gerontology and Geriatrics*, *Gerontology Program Development and Evaluation*, and *Standards and Guidelines for Gerontology Programs*) provide vehicles for sharing information about educational developments and opportunities, practical assistance, innovative programs, and research related to gerontological education.

Programs and Services: The central office in Washington, D.C., serves as a clearinghouse for information about gerontology programs in higher education for students, faculty, and other interested persons. The *Brief Bibliography Series* (with over 30 titles) points faculty to the best-quality instructional materials in various topical areas of gerontology education. A consultation service assists institutions in locating consultants who can provide individualized guidance on gerontology program development and evaluation. The Database on Gerontology in Higher Education is a computerized listing of over 1,000 gerontology and geriatrics programs in the United States. The newly established Program of Merit is an initiative establishing AGHE as the organizational body that evaluates and recognizes educational programs at all levels (certificate/specialization, minor, undergraduate and graduate degrees) following national standards and guidelines. AGHE administers scholarships and fellowships funded by the AARP Andrus Foundation for graduate and undergraduate students.

Advocacy: A major service of the association is monitoring and advocating the interests of gerontology and geriatrics education among na-

tional leaders, government officials, and the private sector. In addition, AGHE promotes the interests of research and education among the gerontological service community and the interests of gerontology among the higher education community.

Research: The association regularly undertakes analyses, such as the project cosponsored with the Andrus Gerontology Center at the University of Southern California, Core Principles and Outcomes of Gerontology, Geriatrics, and Aging Studies Instruction. These analyses are designed to provide assistance in improving the character and quality of academic programs in gerontology.

AGHE is a nonprofit organization, and the resources for carrying out these activities are derived from members' annual dues, conference registration revenues, sales of publications, and grants and contracts.

CATHERINE J. TOMPKINS

ATROPHIC VAGINITIS

Although many women enjoy the sexual freedom that an empty nest and no longer worrying about contraception may allow, others are alarmed by postmenopausal changes such as atrophic vaginitis (Doress & Siegal, 1987). After menopause, the lack of endogenous estrogen drastically alters the epithelium of the vagina and vulva. Vaginal walls become friable, thin, pale, and smooth, losing their rouge as the amount of adipose tissue and collagen decreases. The introitus shrinks, and the prepuce of the clitoris atrophies. The vaginal canal shortens, narrows, and decreases in elasticity, and healing adhesions may develop (Bachmann, 1994). Sebaceous gland secretion declines, and the amount of vaginal lubrication produced in response to sexual stimulation decreases. In addition, the normally low vaginal pH that inhibits the overgrowth of infectious organisms increases due to lower estrogen levels (Byyny & Speroff, 1996).

These changes in the vulvovaginal epithelium set the stage for inflammation and discomfort. Thinning and friability of the vaginal tissue may cause itching and postcoital bleeding or spotting. Atrophy of the clitoral prepuce leaves the glans unprotected and exposed to increased irritation. Inadequate vaginal lubrication leads to complaints of vaginal dryness—indeed, postmenopausal women may find that a sexual response that took seconds now takes minutes (Kass-Annese, 1997). Decreased lubrication, along with a narrowed introitus and vaginal canal, leads to dyspareunia (painful intercourse). Elevation in pH and the resulting alterations in the vaginal flora can cause a chronic discharge and vulvovaginal burning.

History

Absolutely crucial in the care of postmenopausal women is a careful history, allowing the patient privacy and time for the history to unfold in a relaxed manner. Many older women may feel embarrassed or ashamed bringing up issues of vaginal and sexual health with their health care providers. Therefore, the clinician must be prepared to elicit information on issues that the patient may be relieved to discuss once permission has been granted.

The patient should be questioned specifically regarding dyspareunia, soreness after sexual intercourse, and decreased vaginal lubrication. Anticipation of discomfort can interfere with sexual arousal and cause the patient to avoid sexual interactions with her partner or masturbation. The clinician should inquire about any discharge, burning, itching, dysuria, bleeding, and vaginal dryness and swelling, as well as allergic causes of these symptoms such as changes in soap or toilet tissue. Common symptoms of menopause, such as night sweats, hot flushes, or insomnia, support the diagnosis.

Physical Examination

A complete pelvic examination, including a bimanual exam and Pap smear, should be provided to postmenopausal patients. Choosing a small or Pederson's speculum may aid insertion and avoid pain and irritation of atrophic vaginal tissue. The bimanual exam may be performed gently with just one finger. In diagnosing atrophic vaginitis, the clinician must be sure to rule out infections of other causes, including sexually transmitted infections, which are becoming more common in older women (Barber, 1994).

Treatment Options

The treatment of atrophic vaginitis can be approached in several ways. Treatment in a woman with no contraindications for use and who is highly motivated to prevent heart disease and osteoporosis includes oral or transdermal estrogen replacement therapy (ERT). Topical vaginal cream is an option that efficiently and specifically targets the vulvovaginal epithelium and is safe for most women. Nonetheless, some women may refuse any form of ERT, searching out as-yet unproved "natural" remedies such as herbs or homeopathic medicine. Finally, sexual counseling and advice regarding appropriate lubrication may be all the treatment that some women want or need.

Sexual Activity

Sex, with or without a partner, helps to decrease atrophy by improving circulation to the vulvovaginal tissues (Byyny & Speroff, 1996). Thus, regular sexual activity should be encouraged to maintain optimal vaginal health and sexual function. The subject of masturbation may be broached as a normal, healthy alternative for women. In general, women should be counseled regarding the expected changes in sexual function associated with aging. Women who abstain for some years may find that reestablishing sexual function is difficult (Barber, 1994). Heterosexual women benefit from being informed of expected age-related changes in their male partners as well.

Estrogen Replacement Therapy

Patients considering ERT should be counseled on the risks and benefits and encouraged to make an informed choice. Fears related to cancers and other health problems are common reasons for rejecting ERT and should be addressed squarely. Although two recent studies indicated a greater association between estrogen-progestin hormone replacement therapy and increased risk of breast cancer than the use of estrogen only, these studies were not based on randomized clinical trials. The absence of definitive evidence linking ERT to breast cancer should be emphasized (Byyny & Speroff, 1996), but women should be encouraged to make their own decisions. In addition, women can be informed that compared with any increased risk of breast cancer, the risk of disability and death from heart disease and osteoporosis is much greater (Kass-Annese, 1997). Contraindications to ERT include pregnancy, unexplained vaginal bleeding, active disease or impaired function of the liver, thromboembolic disorder, and breast or any other estrogen-dependent cancer (Hatcher et al., 1998).

Clinicians should be sure to inform women of the increased risk of endometrial hyperplasia and cancer in women with an intact uterus who take estrogen without progestin. A woman without a uterus need not take progestin. Oral therapy for a woman with an intact uterus may consist of estrogen daily for 25 days each month and progestin daily for the last 10 to 14 days of estrogen treatment. Withdrawal bleeding follows the progestin phase. Daily combined estrogen and progestin is another option that avoids withdrawal bleeding in most women (Hatcher et al., 1998). There are several choices for transdermal estrogen patches that are applied once or twice weekly.

Although most women with atrophic vaginitis respond well to oral or transdermal ERT, some women need a supplement of topical estrogen cream (Barber, 1994). Other women may prefer to use vaginal cream to benefit from the localized effects of estrogen while avoiding the systemic effects. Unfortunately, the dosage regimen needed to relieve symptoms while avoiding increased endometrial proliferation and cancer has not been well studied, necessitating progestin use or annual monitoring with endometrial biopsy or transvaginal sonogram. Additionally, estrogen is well absorbed

from the initially atrophic vaginal tissue, but this absorption decreases once the atrophic vaginitis has abated (Byyny & Speroff, 1996). A vaginal ring can be used to introduce estrogen into the vagina to treat urogenital symptoms associated with postmenopausal atrophy. The ring is inserted high into the vagina, where it steadily releases estrogen. It is put into place by the patient and replaced after 90 days (or according to the pharmaceutical manufacturer's instructions). Use of the three-month vaginal ring has been noted to produce a blood level lower than the estimated threshold for endometrial proliferation (Hatcher et al., 1998).

Many sources recommend that a much smaller amount of cream than is listed on the package insert may be effective: one-fourth to one-eighth of an applicator. General recommendations for use of estrogen cream include instructing the patient to rub the cream into the outer tissues as well as placing a small amount of the cream on the outside of the applicator to aid insertion into the vagina. The cream is never to be used for sexual lubrication.

Nonpharmaceutical Treatments

General vulva care may be helpful to many women: keeping the area dry, wearing cotton underwear, and avoiding douches, sprays, and perfumed soaps. A wide variety of suggested sexual lubricants for women with atrophic vaginitis can be found, including egg whites, vitamin E oil, aloe vera gel, Replens, vegetable oil, Astroglide, and K-Y jelly (which may dry before intercourse is completed). Kegel exercises may improve the muscle tone of the vaginal canal. Counseling regarding nonvaginal sexual activities may also be appropriate.

Alternative Treatments

Patients may approach their health care providers with questions regarding the use of scientifically unproved complementary therapies in the treatment of atrophic vaginitis. It is important to note that for women with contraindications for ERT, some herbal remedies may be inappropriate due to their estrogenic effects. Women who want to use these therapies are recommended to visit a certified practitioner of the therapy rather than self-prescribing (Kass-Annese, 1997). Interested clinicians should consult alternative medicine sources to either suggest treatments or investigate treatments their clients may already be using.

AMY R. BARLOW
JEAN CASSIDY
SALIL KHANDWALA

See Also
Alternative/Complementary Medicine
Cancer Treatment
Sexual Health

Internet Resources
American College of Obstetricians and Gynecologists
http://www.acog.org

Iowa Women's Health Center
http://obgyn.uihc.uiowa.edu/patinfo/vulvar/vaginitis.htm

Women's Cancer Network
http://www.wcn.org

AUDIOLOGY ASSESSMENT

One-third of the elderly age 65 and older have hearing impairments (Weinstein, 1998); it is the third most prevalent chronic problem in the elderly after osteoarthritis and hypertension. Two individuals with the same level of hearing loss will react differently. Therefore, it is critical to assess each patient individually for the perceived disabling effects of the hearing loss on social and emotional functioning (Weinstein, 1998).

Patients often self-refer to an audiologist for assessment when the impairment has become noticeable enough to be of concern. The most common complaint is, "I can hear you, but I can't understand you, especially when it is noisy" (Weinstein, 1998). Hearing problems managed in the home without

intervention may become a significant concern upon admission to a nursing home. New residents should be evaluated within two weeks of admission to reduce the likelihood of depression and isolation (Hull, 1995).

The screening process begins with a history and physical exam. The patient should be questioned about when the hearing difficulty began and in what situations it is worse (e.g., social events, dinner table, telephone, work) (Hull, 1995). Indications that hearing should be evaluated include the volume on the TV or radio turned up so high that others complain; noticeable straining to hear conversation; regular misunderstanding of words or asking to have them repeated; lip reading; easy frustration; belief that others do not speak clearly; and tinnitus, dizziness, or ear infections (Warshaw & Moqeeth, 1998). The physical exam should inspect for cerumen impaction, ear infection, and tympanic membrane perforation. Auditory screening tests, such as whispered voice, finger friction, watch tick, and tuning fork, are often inaccurate due to variation in administration and interpretation (Rees, Duckert, & Carey, 1999). The Hearing Handicap Inventory for the Elderly Person, combined with the pure-tone audioscope (a hand-held otoscope with an audiometer), is a simple, quick, cost-effective, and easy-to-administer screening test that has an 83% accuracy in diagnosing a problem (Weinstein, 1998). Given the psychosocial consequences of hearing impairment, the U.S. Preventative Services Task Force recommends that those over 65 have periodic hearing evaluations. Also, the U.S. Public Health Service believes that older adults should be questioned about signs of loss and, if found, referred to an audiologist for a complete assessment (Weinstein, 1998).

Audiologists are trained to assess auditory functioning and hearing sensitivity and uncover peripheral versus central processing problems. An audiologist also provides services such as fitting hearing aids, providing other assistive devices (e.g., pocket talker), and counseling the elderly on their disability in private practice, hospitals, nursing homes, and clinics. Clinical audiology assessment contains four components: pure-tone air conduction, pure-tone bone conduction, speech reception threshold (SRT), and speech discrimination. When working with the elderly, the audiologist should begin with a short practice period to ensure that instructions are understood, be sensitive to physical and emotional tolerance, and allow increased response time and breaks between tasks.

Pure-tone air conduction is plotted on an audiogram consisting of frequency measured in hertz (200 to 8,000) and decibels (−10 to 100). Inability to hear at a decibel level of 70 to 100 indicates profound hearing loss; inability to hear at 25 to 70 decibels indicates mild to moderately severe hearing loss. A sensitivity threshold is obtained with headphones for each frequency and for both ears. Pure-tone bone conduction is obtained by an oscillator placed behind the ear on the mastoid bone, and sensitivity thresholds are measured for 250 to 4,000 Hz. Air conduction loss can represent a loss anywhere in the auditory system; however, bone conduction testing helps define the location of loss because sound transmission bypasses the outer or middle ear. The difference between air and bone conduction delineates the three kinds of hearing loss: conductive, sensorineural, and mixed. The air-bone gap reflects the amount of conductive hearing impairment (air conduction is decreased, and bone conduction is normal). An equal loss of air and bone conduction is sensorineural loss. Mixed loss is a loss in both air and bone conduction but a greater loss in air conduction.

SRT identifies the sensitivity level for speech. Two-syllable words are spoken into each earphone separately to find the lowest level at which 50% of the words can be identified. To determine speech discrimination intelligibility, the tester presents 25 to 50 monosyllabic words into each earphone at comfortably loud levels. It is scored on a scale of 0 to 100%; a score of at least 90% is considered normal.

Further testing should be performed if hearing loss cannot be attributed to presbycusis, ototoxic drugs, or noise trauma (Warshaw & Moqeeth, 1998). Acoustic immittance and brain stem auditory evoked potentials (electrophysiological analysis) can be performed to ascertain whether lesions are central, cochlear, or retrocochlear. Immittance includes acoustic reflex testing and tympanometry. Acoustic reflex testing differentiates retrocochlear from cochlear lesions by testing the stapedius muscle reflex; tympanometry checks the flow of acoustic energy through the middle ear with variation of the air pressure.

In conclusion, medical evaluation by an ear, nose, or throat specialist should be performed in any patient with an audiogram that is not consistent with presbycusis, when there is a large difference between hearing loss and speech discrimination, or when there is sudden or unilateral hearing loss (Warshaw & Moqeeth, 1998).

MARGRETTA E. GENNANTONIO

See Also
Hearing Aids
Hearing Impairment

AUTONOMY

Autonomy is the first of three major principles guiding medical ethics and sets the tone for those principles through its respect for personhood. The two other principles are beneficence/nonmaleficence and justice. Honoring autonomy can be a powerful component in planning elders' future health care and preserving their participation in decision making and adherence to the plan, but it can be an especially difficult concept to put into practice with seniors. In the clinical setting, providers often do not trust elder patients' capacity to make decisions, and elders are often reticent in asserting their autonomy.

Autonomy was best defined in a 1914 landmark case by Justice Benjamin Cardozo: "Every human being of adult years and sound mind has a right to determine what shall be done with his own body." This definition, acknowledging the autonomy of the patient, was the basis for Cardozo's ruling that if surgery is performed without the patient's consent, the surgeon commits an assault and the patient may sue for damages (*Schloendorff v.*

Society of New York Hospital, 211 N.Y. 125, 105 N.E. 92). In short, every adult of sound mind has the right to self-determination—meaning that the patient can consent to or refuse treatment, provided the patient is capable of making the decision. Autonomy and self-determination also encompasses the patient's right to make decisions for the future when he or she is no longer capable. Signing an advance directive such as a living will or a health care proxy documents these future wishes. As such, autonomy lies at the very root of medical care and the provider-patient relationship (Ahronheim, Moreno, & Zuckerman, 2000). Respecting this principle can enhance patient care enormously; violating it can lead to treatment that prolongs life against the patient's wishes.

The notion of autonomy goes back to ancient Greece and can be traced in philosophy, politics, and law throughout history. In the 20th century, the Cardozo decision in *Schloendorff* is a legal beacon for patient autonomy, but it was another half century before medical ethics embraced the principle of autonomy. Paternalism in the best interest of the patient was the prevailing wisdom that informed decision making by physicians. The physician was presumed to know best what would benefit the patient, and the patient was thought to be incapable of understanding the information necessary to make a good decision. It was believed, moreover, that revealing the diagnosis to the patient might be damaging. Autonomy in medical ethics emerged in the 1960s at the time of the civil rights movements in the United States. Authority and claims to power were regularly challenged, and medical technology had become complex enough to raise questions regarding the wisdom of leaving life-or-death decisions to someone other than oneself or one's family (Pellegrino, 1994).

Establishing autonomy as the centerpiece of medical ethics was not easy, and even today, a patient's articulated wishes may be ignored by physicians and other providers. Even providers who honored their patients' decisions felt that patient autonomy threatened the physician-patient relationship and the patient's relationship with nurses and other health care providers. Providers feared that

their own autonomy and freedom to recommend medical treatment would be seriously impaired. Providers failed to recognize that empowering the patient would create a provider-patient relationship of mutual respect, wherein the patient would be a partner in health care planning and decision making.

Mutual respect encompasses the various elements of autonomy that should lead to a trusting provider-patient relationship. First, a patient's ability to exercise autonomy and make sound decisions depends on accurate and comprehensive information. The provider is therefore obligated to make available to the patient all the information necessary to make an informed decision. Such information includes diagnosis, prognosis, treatment options and their respective benefits and burdens, expected outcomes, and any other relevant facts and recommendations (Pellegrino, 1994).

The patient's decision consists of more than just the medical information and recommendations. Autonomy recognizes that each person is unique and possesses his or her own set of religious or other values, traditions, and a sense of what life is about. A medical recommendation is only one consideration—albeit an important one—in making a health care decision. For the physician, nurse, or other clinician, the major consideration is which treatment is most likely to be curative or prolong life (Pellegrino, 1994). The patient, for personal reasons, may decide against the recommended treatment even though this decision may shorten his or her life and may, in the provider's view, not be in the patient's best interest. The provider may encourage the patient to change his or her mind, but ultimately it is the patient' life and the patient's decision how to live that life.

The exercise of autonomy is not an absolute right. Patient autonomy ends where it collides or conflicts with that of another. Mutual respect demands that the provider's integrity and autonomy also be honored. Thus a patient cannot demand that a physician, nurse, or other health care professional provide an inappropriate treatment, intervention, or medication. The professional has the right to refuse to act on a request (Pellegrino, 1994). Similarly, a patient who decides to reject acute treatment cannot demand to be cared for by family or to go home without help if it is deemed unsafe. The patient must choose among reasonable options, which may include discharge home with home care or placement in a long-term-care facility. Continuing to stay at an acute-care facility when the patient no longer wants acute care or when such care is inappropriate is also not a reasonable option. At the same time, a family member or health care agent may not overturn a decision made by the patient, even if the patient has lost decisional capacity. The patient's professed preferences supersede anyone else's decision, even when the proxy or surrogate is authorized to act.

When a patient makes a decision that runs counter to the provider's values, it is the patient's wishes that prevail, but the provider does not have to compromise personal beliefs (Pellegrino, 1994). For example, if a patient refuses artificial hydration and nutrition and the provider feels morally bound to feed the patient, the provider is obligated to honor the patient's wishes but can do so by referring the patient to another physician. In this way, the patient's wishes are honored, and the provider's values are preserved.

Confidentiality of information is another component of patient autonomy (Beauchamp & Childress, 1994). Family members are not entitled to any information about the patient without his or her consent. In an era in which truth-telling in the interest of patient autonomy dominates, a family's desire to shield the patient from the truth or a provider's reluctance to share bad news with the patient can violate the patient's right to keep information private and obstruct his or her ability to make personal decisions. Only if permission is sought from the patient, or if the patient asks the provider to consult with or inform a family member, may the clinician reveal the information. A person's capacity to make a particular decision is critical to autonomy. To make an informed decision, a person must have the ability to understand the nature and consequences of that decision (Ahronheim, Moreno, & Zuckerman, 2000). The patient should be able to repeat previously told statements and indicate understand-

ing of his or her specific decision. The level of capacity needed to make decisions varies with the degree of difficulty in understanding the information required to make a decision. For example, naming a health care agent requires a much lower level of capacity than deciding whether to consent to surgery. In the former, the patient simply decides whom he or she trusts to make decisions when he or she is no longer able to do so. In considering surgery, the patient would need to understand at least what the operation involves, how it will affect him or her, and the extent of recovery—clearly requiring a higher degree of capacity. Therefore, a patient may at the same time be capable of making one decision, such as a health care proxy, but not another, such as consent for surgery. In such a situation, the patient may have the capacity to name a proxy but then has to rely on the proxy to make the more complex decisions.

Patients can manifest waxing and waning capacity. Elderly people particularly may "sundown," being clear and lucid in the morning but losing clarity as the day wears on. In respecting a patient's autonomy, the patient should be consulted and asked to make decisions in the morning, when he or she is able to do so. If the provider is not convinced that the patient has made a conscious, deliberate decision, the provider should confirm the patient's wishes by approaching him or her a second and even a third and fourth time. A consistent response should be persuasive that the patient understands the decision made. For an incapacitated patient who has no advance directive and no surrogate decision maker, beneficence/nonmaleficence becomes the prevailing principle. This means that a clinician who is unable to establish the patient's preferences should be guided in making treatment decisions by the best interest (beneficence) of the patient and, as always, is charged to do no harm (nonmaleficence) (Ahronheim, Moreno, & Zuckerman, 2000).

An elder person in declining health and with diminishing mental capacities is especially vulnerable. His or her control of life and values is threatened when autonomy and right to self-determination are not respected. Empathy for elder patients is essential if the clinician is to understand the fear experienced by such patients when they are admitted to a medical facility, especially in an emergency. Providers must give the elder patient encouragement and assistance in making decisions. Respecting autonomy contributes to the health and welfare of elderly patients, allows maintenance of self-esteem, and may reduce fears regarding the loss of control common upon entrance into the health care system (McWilliam, Brown, Carmichael, & Lehman, 1994).

ALICE HERB

See Also
Advance Directives
Cultural Assessment
Patient-Provider Relationships
Substitute Decision Making
Sundown Syndrome

Internet Key Words
Autonomy
Capacity
Decision Making
Informed Decision
Patient/Provider Relationship
Self-Determination

Internet Resources
American Society of Bioethics and Humanities
http://www.asbh.org

American Society of Law, Medicine, and Ethics
http://www.aslme.org

National Bioethics Advisory Commission
http://www.bioethics.gov

National Reference Center for Bioethics Literature
http://www.georgetown.edu/research/nrcbl

B

BALANCE

Approximately 13% of community-dwelling adults age 65 to 69 (and about 46% of those 85 years or older) complain of balance problems or unsteadiness when walking or changing positions (U.S. Department of Health and Human Services, National Center for Health Statistics, 1996). The purpose of balance is to maintain the body's center of gravity, located anterior to the second sacral vertebra, over the body's base of support. Balance is a complex process that involves the coordination of afferent mechanisms or sensory systems (visual, vestibular, and proprioceptive) and efferent mechanisms or motor systems (muscle strength and flexibility). When the center of gravity extends beyond the body's base of support, instability is experienced. The resulting imbalance is detected by the sensory and motor systems in an attempt to realign the body.

Effects of Aging on Balance

There is an age-related increase in unsteadiness under both static and dynamic conditions. Older adults tend to exhibit greater posterior and anteroposterior sway than lateral sway compared with younger individuals (Woollacott, 1993). These changes are due to a combination of decreased sensory input, slowed motor responses, and musculoskeletal limitations (Dayhoff, Suhrheinrich, Wigglesworth, Topp, & Moore, 1998). Older adults attempt to compensate for these changes by using visual input when proprioceptive feedback is reduced or missing. For example, an older adult often looks down to view the correct placement of the feet when ambulating.

There is a delayed response to balance change in healthy older adults. In some older adults, there is a reversal of the normal distal-to-proximal sequence of muscle activation following a change in balance. A variety of age-related musculoskeletal

changes influences balance. Kyphosis alters the body's balance by moving the center of gravity forward, making it more difficult to maintain standing balance. Although not a normal age change, older adults tend to have decreased muscle strength in the lower extremities. Decreased muscle strength complicates the execution of postural strategies and makes it more difficult for older adults to adjust their center of gravity.

Alterations in Balance Due to Disease

Neurological Conditions: Following a stroke, poor postural control is common and is attributed to vestibular dysfunction and visuospatial impairment (hemianopsia) or neglect (hemi-inattention). On the affected side, the sequence of postural muscle activation changes. Instead of the normal sequence of distal-to-proximal muscle activation, the proximal muscles of the hemiplegic or paretic extremity are activated first, impairing the individual's ability to initiate a quick postural response. Other neurological problems that influence balance include cerebellar degeneration, myelopathy, Parkinson's disease, and peripheral neuropathies.

Peripheral Vestibular Imbalance: Vestibular dysfunction may be either peripheral—involving the semicircular canals, utricle or saccule, vestibular nerve, and nuclei—or central—involving the brain stem and cerebellum. Episodes of vertigo may be due to age-related changes occurring in the vestibular portion of the inner ear, such as degeneration in the ampullary mechanism of the semicircular canal, labyrinthitis or vestibular neuronitis from infection, Meniere's disease, medications, head trauma, or degeneration of the cervical spine. Age-related changes in the vestibular system—a 20% decline in hair cells in the otoliths and a 40% reduction of hair cells in the semicircular canals after the age of 70—have not been directly associated with changes in balance (Connell & Wolf, 1997; Tideiksaar, 1997).

Dizziness (Vertigo):   Dizziness is a common complaint of older adults and can result in imbalance. The causes of dizziness vary but can include peripheral vestibular disorders, strokes, central disorders, drug toxicity, cardiac problems, acute illnesses, or diabetes.

Treatment of Balance Disorders

A comprehensive evaluation should be conducted to determine whether there is a reversible cause of the balance problem, with a special focus on the neurological, sensory, and cardiac systems. Neurological evaluation should include mental status, position and vibration sense, deep tendon reflexes, nystagmus, cerebellar function, and muscle strength. Musculoskeletal evaluation should include range of motion and muscle strength. Cardiac evaluation should include arrhythmia, bruits, valvular disorders, and postural hypotension. Sensory evaluation should include vision and hearing assessments. Medication use should also be evaluated, with consideration given to the need for benzodiazepines, tricyclic antidepressants, anticonvulsants, and hypnotic sedatives—all of which can affect stability.

When no treatable disorder for the balance problems is found, existing function should be supported. Targeted exercise regimens aimed at strengthening lower extremity muscles and improving joint flexibility, and balance training designed to enhance sensory function, can help improve balance (Hu & Woollacott, 1994). Canes, crutches, and walkers increase stability by providing a wider base of support and additional sensory input, and they supplement muscle activity by assisting with both propulsion and deceleration during ambulation. Canes, the least restrictive of the gait aids, are fit properly if the handle of the cane is near the level of the greater trochanter, so that when the cane is grasped, the elbow rests in 20 to 30 degrees of flexion. A cane is typically held on the side opposite the affected lower extremity, and the cane

and affected leg are advanced simultaneously. During the stance phase of the gait cycle, weight is distributed between the cane and the affected leg. Walkers provide the most stable base of support and are indicated for those patients requiring maximal mechanical assistance for ambulation. Users should be advised to keep the walker out in front of them and to avoid the tendency to overstep into the walker. Rolling walkers can be more energy efficient but are difficult to use on carpeted and uneven surfaces and are not as safe as traditional walkers.

BARBARA RESNICK
MICHAEL CORCORAN

See Also
Activities of Daily Living
Gait Assessment Instruments
Gait Disturbances

Internet Key Words
Ambulation
Balance
Functional Performance
Gait

Internet Resources
Neurological Exam, University of Florida Medical
    Education Server
*http://www.medinof.ufl.edu/year1/bcs/clist/
    neuro.htm*

Orthotics and Prosthetics Online Community
http://www.oandp.com

Parkinson's glossary
*http://www.chebucto.ns.ca/health/NSPF/
    glossary.html*

Research Activities Gait Analysis Web page
http://www.ncku.edu.tw/~motion/e_gait.htm

BEHAVIORAL SYMPTOMS
IN PATIENTS WITH DEMENTIA

An estimated 4 million Americans are currently affected with dementia. These older adults exhibit behaviors that caregivers often describe as agitated, disruptive, or disturbing. Behavioral symptoms often lead to caregiver frustration, burnout, and injury; increase the potential for elder mistreatment; and contribute greatly to the high cost of caring for those with Alzheimer's disease and other dementia.

The prevalence of aggressive behavior in community-dwelling older adults with dementia ranges from 55% to 65%; these behaviors are primary predictors of institutionalization. Behavioral symptoms occur in 64% to 91% of all nursing home residents, with physically aggressive behavior occurring in 14% to 51%, verbally aggressive behavior occurring in 25% to 66%, and sexually aggressive behavior occurring in 4% to 8%. Other behavioral symptoms commonly seen in older adults with dementia are pacing, disrobing, wandering, and restlessness. Knowledge about the correlates of behavioral symptoms is limited to the recognition that they occur primarily in the context of personal grooming and are more likely to occur in those with a premorbid history of aggression.

Assessment

It is important to avoid judgmental terms, such as "disruptive" or "disturbing," which focus only on the caregiver's response rather than the potentially unmet needs of the older adult with dementia. Labeling behavior as disruptive may lead to a focus on merely controlling the behavior to minimize caregiver disruption. Caregivers should be encouraged to focus on the actual behavior exhibited and to describe it nonjudgmentally, based on what they observe, using terms such as aggressive, repetitive, or vocal. This is more likely to promote identification of triggers and an understanding of the meaning of the behavior, which in turn promotes effective care strategies targeted at meeting individual needs.

Research has shown that aggressive behavior usually occurs on a continuum, wherein anxiety is exhibited first, progressing to verbal behavior, and, if no effective intervention is implemented, resulting in physically aggressive behavior. Thus, the goal of individualized care is to intervene in the earliest stage of behavior, to prevent aggression. Families or the primary caregiver should be encouraged to keep a behavior log to track each behavior that occurs; the log should note the time the behavior occurred, the environment, the persons present, and any other relevant information. This log can then be used to identify patterns, triggers, early warning signs, and unmet needs during a careful assessment process for targeted intervention. The log can be used as both an assessment tool and an intervention. Families and caregivers will feel like something is being done, and the log may provide data to suggest nondrug strategies. This is important, since drugs may be prescribed to help caregivers feel that something is being done to reduce the behaviors they find distressing. In addition, neuropsychological testing can provide valuable guidelines and information on diagnosis, retained cognitive abilities, and potentially effective strategies for coping with behavioral symptoms.

Treatments

In the past, behavioral symptoms were most frequently treated with physical restraint and psychoactive drugs. A large body of research demonstrates the multiple adverse consequences, and even the possibility of death, with physical restraint use, as well as their lack of efficacy. Physical restraint use has been associated with increased behavioral symptoms in several studies. In some states, the inappropriate use of physical restraints is a crime involving neglect of a care-dependent person. Federal regulations have placed significant restrictions on the use of physical restraints, in response to public outcries about the dangers of their use (Health Care Financing Administration, 1999).

Research on the use of psychoactive drugs to control behavior in older adults with dementia is less sound. Few drug studies demonstrate efficacy for any specific behavior, and older adults with dementia are particularly prone to serious side effects. Unfortunately, caregiver distress, rather than a specific target symptom for which the drug is known to be effective, is frequently the reason that a drug is prescribed. Careful assessment is required prior to initiating psychoactive drug use. The use of psychosocial interventions, the treatment of any comorbid medical problems, and a partial or complete drug holiday are frequently recommended as part of an evaluation before initiating drug treatment. Correction of impaired vision or hearing may reduce aggressive behaviors that are secondary to sensory deficits. Pain and discomfort from acute or chronic medical conditions may masquerade as a variety of symptoms in very impaired older adults. The use of routine, rather than as needed, analgesia is important in treating chronic pain in older adults with dementia, because proper attention to comfort often markedly reduces behavioral symptoms.

Given the state of research on psychoactive drugs and their widespread misuse in the past, only cautious recommendations can be made regarding the use of such drugs for behavioral symptoms. Psychosis, manifested as hallucinations, delusions, or paranoia, is an appropriate indication for antipsychotic drugs that have been shown to be effective for these target symptoms, but there is no clearly established efficacy for agitation. Newer antipsychotic drugs, such as risperidone, may have fewer side effects for older adults with psychoses. Caregivers must have a high suspicion of depression when behavioral symptoms are present. Some professionals recommend a therapeutic trial of a serotonergic antidepressant whenever depression is indicated or suspected. Benzodiazepine drugs are of limited value for behavioral symptoms, and the risk of falls increases with dose and length of drug half-life. Case reports suggest that buspirone or trazodone may be helpful in decreasing behavioral symptoms (American Psychiatric Association, 1997). Case studies report decreased behavioral symptoms with the use of antimanic and antiseizure drugs and decreased aggressive sexual behavior in men with the use of female hormones, but again, the research evidence is insufficient to recommend their use. There is some anecdotal information that cognitive enhancers, such as donepezil, may also reduce behavioral symptoms.

Environmental and psychosocial interventions are the mainstays of safe treatment for behavioral symptoms in older adults with dementia. An environment that is familiar and soothing to the older adult with dementia does much to decrease behavioral symptoms. Lighting, color, and noise can significantly affect behavior, positively or negatively. A proper balance of sensory input without overstimulation may help reduce behaviors. Improved lighting and use of color contrasts can help overcome fear during grooming activities by providing an environment that the older adult can easily distinguish and navigate. Personal-care practices should be evaluated to ensure that caregivers are focusing on individual comfort and needs rather than task completion. Many specific strategies have been developed for altering personal care to minimize behavioral symptoms. Some of these include using one-step directions, offering simple choices during dressing, and altering the form of grooming, such as giving a towel bath rather than a standard shower (Rader & Tornquist, 1995).

Appropriate communication style, content, and complexity can reduce behavioral symptoms in older adults with dementia. Strategies that appear to be helpful are the use of short, grammatically simple sentences, the use of definite nouns, and repetition. Caregivers should pay close attention to nonverbal communication of the older adult. Much can be learned about the older adult's unmet needs through careful attention to facial affect, tone of voice, and gestures. In addition, the caregiver's nonverbal communication of anger or frustration, through a tense affect or high-pitched tone of voice, is associated with increased behaviors.

Planning activities that are meaningful and enjoyable for the older adult can reduce behaviors due to boredom. Activities should be carefully planned to capitalize on preserved abilities; these should be based on previous job tasks or hobbies. Activities should be scheduled to balance stimulation and rest, as fatigue is thought to increase behavioral symptoms. Adult day care or a psychiatric day hospital program, if depression is present, can provide meaningful activities.

Family education and counseling are critical for successful coping with behavior symptoms. Families need to understand that behaviors do not equal a problem but rather express a need (Strumpf, Robinson, Wagner, & Evans, 1998). Education should focus on ways to maximize the older adult's remaining capabilities. Family and caregiver interactions may have negative effects on behavior if they are overly demanding and "test" cognitive abilities. Families may need assistance and education in ways to structure visits so that they are enjoyable for both the older adult and the family. Identifying ways to make caregiving a mutually beneficial experience may help buffer the strain often observed in caregivers. Finally, respite, or time away from caregiving, to prevent burnout is an important part of any intervention plan for an older adult who exhibits behavioral symptoms.

KAREN TALERICO

See Also
Dementia: Nonpharmacological Therapy
Dementia: Pharmacological Therapy
Reality Orientation
Restraints
Validation Therapy

Internet Key Words
Aggression
Agitation
Behaviors
Physical Restraint Use
Psychoactive Drugs
Psychosocial Interventions

Internet Resources
ADEAR Center
http://www.alzheimers.org/

Alzheimer's Association
http://www.alz.org/

Eldercare Locator
http://www.ageinfo.org/elderloc/elderloc.html

BENIGN PROSTATIC HYPERPLASIA

Lower urinary tract symptoms are common in elderly men. Benign prostatic hyperplasia (BPH) is

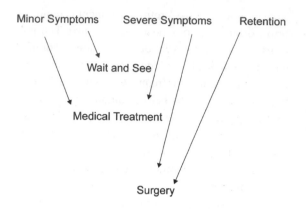

Treatment options for BPH.

the most common finding, but other problems can cause infravesical obstruction, including neoplastic prostatic hyperplasia, urethral stricture, neurological dysfunction, foreign material, pelvic mass, bladder cancer, and bladder stones. The pathophysiology of BPH is related to a combination of mechanical obstruction (due to glandular hypertrophy) and functional obstruction (smooth muscle tone within the prostate).

Symptoms are described as being irritative or obstructive. Irritative symptoms include urgency, frequency, nocturia, and urge incontinence. These symptoms can be reliably evaluated using the International Prostate Symptoms Score. Obstructive symptoms include the weakening of the urinary stream, Valsalva voiding, hesitancy, intermittency, postvoid dribbling, and the sensation of incomplete emptying of the bladder. Severe obstruction can lead to urinary retention, with possible deterioration of renal function. Associated conditions are urinary tract infection, hematuria, and overflow incontinence. Hematuria requires further examination of the entire urinary tract to exclude other conditions such as urological tumors or calculi.

Evaluation includes the patient's general condition, life expectancy, and comorbidities such as mental illness (Petrovich & Baert, 1994). The prostatic evaluation focuses on the exclusion of prostate cancer and on prostate size. Exclusion of prostate cancer is done by assessing the prostate specific antigen (PSA) level, digital transrectal examination, and transrectal ultrasound study. The ultrasound study may also provide information on prostate size that can be used to select appropriate surgical therapy. Prostate size, however, does not correlate well with the severity of obstruction. Cystoscopy provides information about possible disorders in the urethra, prostate, and bladder that could otherwise be mistaken for BPH. An ultrasound study of the kidney may reveal hydroureteronephrosis in cases of chronic overflow incontinence or severe obstruction. Excretory urography is indicated only in selected patients with hematuria, urinary infection, or calculi.

Minimal urodynamic evaluation can consist of uroflowmetry, with measurement of the postvoid residual. Maximum flow rates below 10 mL/second are usually considered to be caused by infravesical obstruction, if there is no sign of detrusor insufficiency. Higher flow rates with symptoms of obstruction may warrant further urodynamic investigation. However, many urologists perform only a minimal urodynamic evaluation and combine these data with the patient's symptoms and the clinical examination to devise a therapeutic plan.

More advanced urodynamic evaluation using videourodynamics and pressure flow studies can show uninhibited bladder contractions, poor detrusor contractility, and elevated detrusor pressure during voiding. To some extent, these examinations allow one to predict the success of therapeutic interventions. There is only a poor correlation, however, between symptoms and urodynamic findings.

Treatment

The treatment of BPH is tailored to the severity of symptoms and the obstruction. If only minor symptomatic discomfort is present in the absence of postvoid residual, a wait-and-see policy is indicated. More severe symptoms may be treated medically. Alpha-blockers such as prazosin, phentolamine, terazosin, doxazosin, alfuzosin, tamsulosin, and indoramin all have a distinct effect on symptoms of BPH and on obstructive parameters (Lepor, 1995). Most alpha-blockers have a hypotensive ef-

fect, and blood pressure monitoring is warranted while the medication is being started. Finasteride is a 5-alpha-reductase blocker that lowers the dihydrotestosterone concentration in prostatic tissue. Its most important effect is reducing prostate size. Currently, the indication is limited to large prostates (> 60 gr). It is important to know that finasteride lowers the PSA level by approximately 50%, thus limiting the usefulness of PSA levels as a cancer screening tool. Androgen ablation by the use of luteinizing hormone–releasing hormone (LHRH) agonists or antiandrogens can be effective, but severe side effects such as gynecomastia, hot flashes, impotence, and loss of libido, as well as the high cost of these medications, limit their use. Bothersome nocturia due to a high nocturnal diuresis may be alleviated by the use of desmopressin spray in selected patients.

Surgical treatment of BPH is indicated in cases of urinary retention, overflow incontinence, hematuria, secondary hydronephrosis, and recurrent urinary tract infection. Transurethral resection (TURP) is the gold standard for the treatment of bladder outflow obstruction caused by BPH and constitutes 95% of all prostatectomy procedures (Riehmann & Bruskewitz, 1994). Over 80% of patients experience regression in urinary symptoms, especially those with urinary retention or severe symptoms. The mortality rate is low (0.2%) and the morbidity rate is about 18%. About 15% of patients do not benefit from the procedure.

Smaller prostate glands (< 20 gr) can be managed by a transurethral incision of the prostate (TUIP). This procedure is easier to perform and to teach than TURP and has a shorter operating time. For larger prostates (> 60 to 80 gr), an open prostatectomy is associated with lower morbidity and mortality because of the lower risk of irrigating fluid absorption and subsequent electrolyte disturbances.

Minimally invasive therapies have been developed for the treatment of BPH in recent years. Balloon dilatation results in a slight improvement of symptoms for a short time. Permanent stenting of the prostatic urethra with a woven metal stent is still under evaluation. Problems of stent migration and encrustation are limiting factors. Microwave hyperthermia and thermotherapy have been shown to be effective in treating BPH, but additional clinical trials are needed (Baert, Ameye, Astrahan, & Petrovich, 1993).

Laser prostatectomy (contact laser or free beam laser) induces coagulation necrosis of the prostate gland. However, laser energy delivery systems are not yet optimal. Chronic catheterization may be needed to allow for adequate sloughing of the prostate until obstruction is relieved. The contact laser induces vaporization of the tissue but is time consuming. Recently, other energy forms such as radiofrequency waves are being used in the treatment of BPH.

Thus far, TURP remains the gold standard of treatment. Following TURP, more than 50% of patients may expect retrograde ejaculation, and 5% to 40% may experience impotence. The incidence of impotence after the newer forms of treatment may be somewhat less, but there are no good data. These complications have to be discussed with the patient preoperatively. Postprostatectomy incontinence is rare (1%), but if present, sphincteric injury, bladder instability, or persistent bladder outlet obstruction should be ruled out.

DIRK DE RIDDER
LUC BAERT

Internet Key Words
Benign Prostatic Hyperplasia
Prostatectomy
Screening
TURP
Urodynamics

Internet Resources
Administration on Aging
http://www.aoa.dhhs.gov/

P\S\L\ Consulting Group, Inc.
http://www.pslgroup.com
http://www.pslgroup.com/ENLARGPROST.HTM

U-Net International
http://www.u-net.net
http://www.personal.u-net.com/pha/

BEREAVEMENT

The loss of significant others, especially a spouse, can be a particularly distressing event for elders. They may be losing important longtime companions and confidants at a time in life when their social networks may already have become quite attenuated. Spousal bereavement places elders at increased risk for physical morbidity, depression, and even mortality.

Clarification of terms is important. The term "bereavement" is usually considered the objective event of the loss. "Grief" or "grieving" is the emotional reaction that follows. "Mourning" refers to the social expression of grief, including mourning rituals.

The Course of Grief

Initial grief reactions typically include shock, numbness, and denial of the loss, followed by disorganization; during this time, individuals may become agitated, confused, and distracted and have somatic complaints. Fatigue and crying are also common. The bereaved often show little interest in daily affairs, loss of appetite, and an inability to concentrate. Bereaved elders may direct anger and aggression against those around them. Individuals may also express guilt over perceived missed opportunities for interaction with the deceased. In the first weeks following the loss, the bereaved may experience visual or auditory hallucinations in which they see or hear the deceased. Such hallucinations are considered a kind of "searching" behavior in which the bereaved seek to find the lost person. The bereaved may need to review memories of the deceased, including events leading up to the loss. This reminiscence process may be an attempt to master the loss and the associated anxiety. The loss may also activate negative self-images of vulnerability and helplessness. Additionally, the bereaved may have a need to idealize the deceased and repress memories of the negative aspects of their relationship (Osterweis, Solomon, & Green, 1984).

During the first several months, these grief reactions gradually abate. By one year following the loss, although a sense of loneliness may persist, most individuals begin the process of reorganization. They accept their loss emotionally and intellectually and realize that they must accommodate to a world without the deceased and begin to reorient themselves to the future rather than the past. Accommodation may require acceptance of a new identity (e.g., as a widow) and the assumption of new responsibilities and tasks (e.g., management of one's finances). By this time, most have gained confidence that they can adequately function without the deceased and carry out expected roles and responsibilities. After the first year, the bereaved may continue to feel increased distress around the anniversary of the loss (i.e., "anniversary reactions"), at the time of marker events (e.g., the birth of a grandchild), or at holidays or family gatherings when the absence of the deceased is poignantly felt.

Theories of grief and mourning are changing. The notion that all bereaved adults pass through a set of predictable stages is increasingly questioned. Within the parameters of normal loss, there may be variability in the intensity and duration of various emotions and behaviors associated with loss. The absence of grief or delayed grief is still typically assumed to warrant professional evaluation (Osterweis, Solomon, & Green, 1984).

Historically, the bereaved was expected to sever emotional ties to the deceased (decathexis). Failing to accomplish this task and maintaining a continuing relationship with the deceased were considered manifestations of unresolved grief and a pathological outcome. Recent adjustments to the theory propose that the survivor maintains an ongoing bond and relationship with the deceased that may change over time. Maintaining such a relationship helps mitigate the sense of loss experienced. Thus, it is not inappropriate or maladaptive for the bereaved to maintain some ongoing tie to the deceased (e.g., through dreams, a felt presence) if they are able to simultaneously invest emotional energy in new relationships and activities (Goulding, 1997).

Risk and Protective Factors

Some factors may increase or protect against older adults' risk for complicated or pathological grief reactions or physical morbidity. If the death was unexpected or sudden, the absence of an opportunity to anticipate the loss and mobilize one's psychological coping resources can enhance the risk of a poor outcome. Conversely, too long a period of preparation can also be maladaptive, especially for family caregivers. For example, when the deceased had a protracted illness, caregivers may have become emotionally and physically exhausted from both the physical demands of caring for the patient and the sustained anxiety of anticipating the loss. Emotional and physical depletion may place caregivers at increased risk for psychological morbidity both before and after the death. Further, when the terminal phase of the illness was protracted and the patient's quality of life compromised, caregivers might have wished for death to come soon or experienced a sense of relief when it finally did occur. These feelings can give rise to considerable guilt and subsequent psychological distress in the postdeath period.

A preexisting psychiatric disorder, especially depression, may recur or intensify, resulting in poor adjustment to the loss. A preexisting chronic illness increases the risk of physical morbidity and greater use of health services by caregivers in the postdeath period. Women appear to be at greater risk of depression following spousal bereavement.

Among the protective factors that have been identified are good social supports in the period following the loss. Social support may facilitate adjustment in a number of ways, including forestalling social withdrawal and the accompanying loneliness, helping the bereaved to manage new tasks that must be assumed, and affording the bereaved a chance to express and work through feelings about the loss. Religious faith can also be of value by offering a belief system that enables the survivor to transcend the loss. Finally, being able to achieve a sense of closure and having no regrets about issues left unsaid or not addressed with the deceased also appear to facilitate adjustment.

Helping the Bereaved Cope with Loss

In coping with loss, the bereaved can benefit from supportive interventions. Yet the anger expressed by the bereaved can drive away valuable sources of support. Individuals against whom anger is directed should be encouraged to take a tolerant attitude toward the bereaved, endure the hostility, and recognize that it is part of the bereaved's reaction to the perceived injustice of the loss. The bereaved will benefit from being able to express anger, both by gaining a greater acceptance of it and by not having to direct it inward. Unfortunately, although there is often adequate support immediately following the loss and at the time of the funeral, family and friends tend to withdraw after the first month or two. Family and friends may hold unrealistic expectations that the bereaved can accept the loss and get on with life. Once the crisis is over, however, the bereaved may need reassurance and sometimes practical assistance in completing new tasks that must be assumed as a result of the loss. Support groups can help the bereaved recognize their feelings and "normalize" positive and negative feelings about the deceased (e.g., anger at them for leaving), thus encouraging verbalization and acceptance of the loss.

Whether depressed bereaved individuals should be treated with medications after the first few months following a loss is debated. Some professionals contend that sadness, lethargy, and other depressive symptoms are normal and adaptive and should not be interfered with. Other professionals argue that grief distress can and should be alleviated with medication. Because depression distorts grief work, it increases the likelihood of poor adaptation (Osterweis, Solomon, & Green, 1984).

In the absence of professional intervention or adequate informal social support, older adults cope with loss in a number of ways. Some elders distract themselves with activities so as not to think about the deceased. Others use alcohol, drugs, or excessive eating to blunt their pain. More adaptive choices include turning to religion as a framework for finding meaning in loss and thereby facilitating acceptance. Religion also may provide support through the personal relationship forged with the divinity or fellow congregates. Churches and synagogues offer supportive services for bereaved members. Greater social involvement in religious organizations following a loss can also help overcome the sense of loneliness that many older widows and widowers experience.

Pathological or Maladaptive Grief

Although a general pattern can be discerned regarding grief reactions experienced by older adults, considerable variability is possible and normal. Thus the question arises of how to differentiate adaptive distress from that which is pathological. Some indications of the latter might include vegetative symptoms such as extreme fatigue, weight loss, sleep disturbances, and agitation that continues beyond the first two to three months. Persistent psychosomatic complaints (especially those mimicking symptoms of the illness that caused the deceased's death) or frequent visits to the physician with complaints of vague or ambiguous symptoms can also signal the need for further evaluation. If distress does not lessen over the first year following the loss, or if there is persistent or excessive anger or self-recriminations and feelings of guilt and worthlessness, professional assessment is usually warranted (Zisook & Schucher, 1996).

Differences Between Bereaved Younger and Older Adults

Older adults seem to suffer less anguish over the loss of a spouse. Serious illness and death may not be seen as "off-time" events by the elderly; they expect illness, disability, and death to strike at any time. Alternatively, younger adults are more likely to see the death of a spouse as untimely, tragic, and unfair. Older adults are more likely to become confused and disoriented by the loss, especially if there is a preexisting cognitive impairment. Elders may suffer more morbidity following spousal death, particularly if there is a preexisting chronic illness that might be exacerbated by the stress of the loss.

Lastly, there may also be a greater tendency of older adults to turn to their religion as a coping resource (Wisocki, 1998).

KAROLYN SIEGEL
STANLEY J. ANDERMAN

Internet Key Words

Bereavement

Death

Dying

Grief

Mourning

Internet Resources

Growth House, Inc.
http://www.growthhouse.org/

National Hospice Organization
http://www.nho.org

TLC Group
http://www.metronet.com/~tlc

BEST PRACTICES IN GERIATRIC CARE

Despite efforts to bolster home and community-based care, hospitalization remains a common event for older people. In 1997 alone, persons 65 years and older accounted for 36% of all hospital stays and 49% of all days of care in hospitals. The average length of a hospital stay was 6.8 days for older people, compared with only 5.5 days for those under 65 (Administration on Aging, 1999). In the future, increasing numbers of older adults will enter hospitals requiring comprehensive treatment for chronic and disabling conditions.

There is substantial evidence that hospitals can be dangerous places for older people, especially those who are frail. Hospitalization exposes people to infections, prolonged bed rest and deconditioning, and treatments that have negative effects on an older person's functional and mental status. Thus, it is essential that personnel caring for older hospital patients have the necessary knowledge and skills to prevent and manage untoward events that place older patients at risk for negative outcomes during and following a hospital stay.

Direct providers of care—nurses and others—require the requisite knowledge and skills to deliver "best practices" to the elderly. In delivering quality care to older adults, issues that should be examined and perfected include holistic assessment, recognition of cultural differences, communication skills, models for information sharing, and decision-making principles and behaviors.

Clinicians working in the acute-care setting need to develop the expertise that will allow them to care for this vulnerable population. Over the past 20 years, a body of knowledge has developed about what constitutes best practices for older adults when they are hospitalized. These best practices include standardized assessment instruments for determining patients at risk for untoward events and validated strategies to manage commonly occurring geriatric syndromes such as sleep disturbances, pressure ulcers, delirium and agitation, urinary incontinence, and falls. In addition, there are best practice models for deploying advanced practice nurses in ways that maximize positive patient outcomes during and following hospitalization.

There are a variety of ways to infuse best practices into the culture of care. One way is to deliver care that is interdisciplinary in nature. Another way is through nurse-driven models, which have proved to be powerful tools for change. Clinical practice protocols can provide care guidance. However, in order to infuse best practices into the care environment, a whole culture shift may be needed. This is where models of best practice should be considered.

The Acute Care of Elders (ACE) Model promotes collaborative team building and the development of nurse-initiated protocols of care (Palmer, Counsell, & Landefeld, 1998). The Geriatric Resource Nurse (GRN) Model integrates geriatric resource nurses, primary care nurses, a geriatric nurse, and a physician specialist on a geriatric care team (Fulmer, 1991). The Quality Cost Model of Transitional Care targets elderly at high risk for poor post-discharge outcomes, and advanced practice nurses provide direct care and coordination of services (Naylor et al., 1999).

Research-based clinical practice protocols can be particularly useful in providing hospital staff with methods of addressing geriatric syndromes in older people. To be useful, however, protocols must be responsive to the needs of the bedside nurse and other clinicians providing care to hospitalized elders. Protocols should be structured to provide guidance on the assessment, prevention, management, and follow-up that can lead to functional decline of the hospitalized elderly.

The Agency for Healthcare Research and Quality (AHRQ) conducts and supports general health services research, including medical effectiveness research; facilitates the development of clinical practice guidelines; and disseminates research findings and guidelines to health care providers, policy makers, and the public (AHRQ, 1999). AHRQ protocols can be improved through their application to specific populations. One example is a program called NICHE: Nurses Improving Care for Health System Elders (Fulmer, Flaherty, & Bottrell, 2000). In contrast to AHRQ protocols, NICHE protocols draw on best practice standards that focus specifically on the needs and special problems of older adults. The NICHE interventions are primarily nurse-driven (whereas AHRQ protocols tend to be geared toward physicians), so they meet a critical need, given that registered nurses (RNs) are so important in hospital care. NICHE assists hospitals and, more recently, nursing homes, improve the knowledge of nurses and other staff regarding best practices in the care of the elderly. Implementation of NICHE has resulted in better staff attitudes toward elders. NICHE provides a variety of tools to help organizations assess staff perceptions regarding the quality of care they provide to older adults, modify nursing care practices to better meet elders' needs, and evaluate the effectiveness of those interventions.

The 14 research-based clinical practice protocols represent the key clinical conditions and circumstances likely to be encountered by a hospital nurse caring for older adults. They reflect current research and national standards, including those promulgated by AHRQ (Abraham, Bottrell, Fulmer, & Mezey, 1999). The protocols address pressure ulcers, urinary incontinence, sleep disturbances, physical restraints, advance directives, pain management, functional assessment, cognitive assessment, eating and feeding difficulties, medication safety, depression, falls, confusion and delirium, and discharge planning.

Although protocols address syndromes experienced by older adults, they are not guidelines for treating diseases. Older adults, unlike their younger counterparts, often enter the hospital with complicated medical problems that overlap and cannot be treated in isolation.

Interdisciplinary collaboration or RN-only model development and selection of protocols can help ensure full adoption within a unit or facility. Professionals are more likely to embrace change when protocols are research based and when their efficacy has been shown. The protocols play a greater role in the provision of care when they are dissected, analyzed, and reworked, based on the opportunities and constraints of a given hospital or specialized care unit.

JULIE M. SAZANT
TERRY T. FULMER
MATHY D. MEZEY

See Also
Clinical Pathways
Deconditioning Prevention
Geriatric Evaluation and Management Units
Home Health Care
Hospital-Based Services
Iatrogenesis
Information Transfer

Internet Key Words
Advance Directives
Assessing Cognitive Function
Assessment of Function
Best Practice
Clinical Pathways
Clinical Practice Protocols

Confusion/Delirium
Depression
Discharge Planning
Eating and Feeding Difficulties
Falls
Hospitals
Medication Safety
Pain Management
Physical Restraints
Pressure Ulcers
Sleep Disturbances
Urinary Incontinence

Internet Resource

John A. Hartford Institute: Best Practices
http://www.nyu.edu/education/nursing/
  hartford.institute/

BODY COMPOSITION

Body composition consists of two parts: lean body mass (LBM) and body fat. LBM is fat-free tissue, excluding the structural lipids in the cell walls and nerve sheaths. Body fat can be found viscerally, within body compartments, and subcutaneously (under the skin). Older persons tend to have less bone density, greater atrophy of muscle and organ tissue, and a higher percentage of body fat. Many epidemiological studies link increased disease risk to percentage of body fat (Bray, 1992). Body mass index (BMI) defines the relationship of body fat to LBM according to weight and height. Persons with very low BMIs are at greater risk for mortality from digestive and pulmonary diseases. Persons with progressively higher BMIs are at progressively higher mortality risk from cardiovascular and gallbladder diseases and diabetes mellitus. Waist-to-hip ratios correlate positively with the incidences of these diseases, as does increasing age itself.

Recent attention has been paid to possible retardation of aging related to calorie deprivation using animal models. Loss of body fat related to decreased caloric intake over a 10-week period lowered blood pressure and had beneficial effects on lipid profiles in middle-aged men while having no

adverse effects on physical or mental performance (Velthuis-te Wieric, van den Berg, Schaafsma, Hendriks, & Brouwer, 1994). Studies examining the effect of extreme leanness in the elderly over long periods would be difficult, if not ethically questionable. Long life and good health depend on much more than calories. To an extent, body fat stores are useful, in that they help one tolerate long periods without food. This is significant for older persons with gastrointestinal conditions or those with high calorie demands associated with cancer. Persons who are initially obese lose only 20% of their LBM during fasting, but more than 50% is lost by those who initially have few fat reserves. Rebuilding LBM can be a difficult process; two-thirds of weight increase is fat, and only one-third is LBM. Undernutrition can be a significant problem in the frail elderly, just as obesity is in those elderly with too much total body fat.

Measurement

Of the many methods for calculating body composition, the most common approach uses height and weight to LBM, which is a function of stature, as is skeletal height. Most epidemiological studies that look at the relationship between body composition and disease risk use BMI as the measure. BMI is calculated by dividing a person's weight in kilograms by the square of his or her height in meters [weight (kg)/height (m)^2]. It can also be calculated using pounds and inches: [weight (lb)/height (in)^2 × 705]. Age-specific guidelines are suggested by Bray (1992), since BMI increases with age. Using standing height measurement to compute BMI in the elderly may give an inaccurate result due to spinal compression and postural changes. Greater accuracy may be achieved in some persons by measuring long bone length, since these bones do not shorten with age or disuse. The formula for estimation of total height in centimeters for women is 84.88 + [1.83 × knee height (cm)] + (−.24 × age in years). For men, the formula is 60.65 + [2.04 × knee height (cm)]. Knee height is measured using the length of the lower leg from the sole of the foot to the anterior of the thigh when the knee and ankle are both bent at 90 degrees (Moore, 1997). A knee height caliper available from Ross Laboratories can also be used.

Other anthropometric methods include skinfold thickness and circumference measurements to estimate subcutaneous fat, assumed to be 50% of total body fat. However, older persons have less subcutaneous and more intra-abdominal fat. Accuracy of such measurements decreases with increasing obesity. Skinfold thickness is measured with calipers over the triceps and biceps, below the scapula, above the iliac crest (waist skinfold), and on the upper thigh. The most complete standards are available for the triceps and scapular measures but are standardized for younger adults. Body circumferences are taken at the mid–upper arm and the calf. Waist-hip ratios are calculated and used frequently in epidemiological studies.

Additional methods of body composition measurement include electrical conductivity and impedence, in which the speed of electrical conductivity correlates to the amount of LBM. Computed tomography can measure visceral and subcutaneous fat deposits, as can photon absorption scans, nuclear magnetic resonance scans, and neutron activation to assess for amounts of calcium, phosphorus, sodium, chlorine, and nitrogen. Body composition may also be estimated by multiplying body potassium by the reciprocal of the average intracellular potassium concentration (total $K^+ \times 0.00833$).

Water displacement (densitometry) measures body density and is based on the fact that lean body tissue contains more fluid than fatty tissue. Thus, a person who contains a greater proportion of LBM displaces more water than an equivalently sized person with a greater proportion of fat. Older people are more frequently dehydrated, however, and may contain less fluid in their lean body tissue. This measurement method is mainly used in research studies, and many sites may not have the necessary facilities or equipment. In addition, this method is more suitable for younger subjects than for frail elderly, due to the demands of water.

Factors Affecting Body Composition

Gender: LBM is about equal between the sexes from birth to puberty, at which time the amount of LBM increases for both sexes, but proportionally more for men than for women. Adult men have 1.4 times the LBM of women. The recommended dietary allowances reflect the two-thirds relationship of LBM between the sexes. In the elderly, the amounts of LBM decline about equally for both sexes, with women maintaining more body fat. Beginning around age 40 years, more bone is lost than is replaced. This is accelerated in women after menopause, when estrogen levels decrease. Demineralization is worsened by prolonged immobility and decreased calcium intake and is greatly accelerated in men and women who have osteoporosis. Muscles also atrophy in bedridden and frail elderly, and because frail older persons are frequently dehydrated, their LBM may become smaller due to fluid loss.

Race: Shorter, lighter races (e.g., Asians) have decreased LBM and decreased heights and weights. North American African American males have increased LBM compared with North American white males, with an average of 1.15 times total body potassium and 1.17 times total body calcium.

Socioeconomic Status: Data indicate that North American African American females in lower socioeconomic groups have more body fat than North American African American or white females in higher socioeconomic groups (Allan, Mayo, & Michel, 1993). Studies of this type use younger adult subjects, however.

Basal Metabolic Rate (BMR):   BMR is the amount of energy required to sustain life over 24 hours, or the number of kilocalories a person requires to remain alive each day. BMR can vary among persons because of multiple causes, including metabolic demand. BMR is more closely related to LBM than to body weight. Starting at age 30 years, BMR decreases 2% for each decade of life. Although this factor should be taken into account when drugs are prescribed for older persons, often it is not. BMR affects the ease with which an older person can gain fat to store extra calories, whether or not food intake increases.

DOROTHY G. HERRON
LYNN B. GREENBERG

See Also
Caloric Intake
Obesity

Internet Key Words
Basal Metabolic Rate
Body Composition
Body Composition Measurement
Body Fat
Body Mass Index
Lean Body Mass
Subcutaneous Fat
Visceral Fat

Internet Resources
Healthy Weight Network
http://www.healthyweightnetwork.com

Mayo Health Oasis
http//www.mayohealth.org

National Institute of Aging
*http//www.nih.gov/health/chip/nia/aging/
    physiologic.html*

BOWEL FUNCTION

The age-related slowdown in gastrointestinal motility, in combination with nervous system, endocrine, and vascular changes, contributes to alteration in patterns of bowel elimination. However, bowel constipation and bowel-related illnesses are not typical of normal aging (Matteson, McConnell, & Linton, 1997); the changes are not in themselves pathological, nor are they causal in diseases such as colon cancer, constipation, or diverticulitis.

Several factors promote normal bowel function in older people, including mobility and exercise, adequate fluid and fiber, and privacy and positioning during bowel elimination. Culture and health belief systems also influence bowel function. Appropriate bowel function management includes assessment of clinical, lifestyle, and psychosocial aspects of an older person.

Assessment

Bowel assessment requires a comprehensive history focusing on past and present health-related behaviors, including bowel habits, laxative use, and frequency and consistency of stool; circumstances resulting in variations, such as anxiety, stress, or spicy food "triggers" that affect elimination; dietary habits; exercise and activity levels; typical fluid intake; past or concurrent diseases, such as colorectal cancer, diverticulitis, and neurological, musculoskeletal, or endocrine disorders; and beliefs and values regarding bowel elimination.

Medication history is important, especially regarding use of aspirin, ibuprofen, anticoagulants such as Coumadin, antacids, antidepressants, antipsychotics, anticholinergics, antidiarrheals, laxatives, and over-the-counter drugs. Older people are the prime consumers of both prescription and over-the-counter medications. Moreover, they frequently experience adverse and toxic reactions to drugs. Many medications carry side effects of intestinal irritation and diminished or increased peristalsis that affect bowel pattern. The clinician must carefully analyze each medication for interactions and side effects.

Physical examination must include abdominal assessment, observation of asymmetry and bulges, auscultation of bowel sounds, palpation for stool, and percussion for densities, masses, or stool; the rectal digital exam looks for rough and bulging surfaces (a possible sign of prostate hypertrophy), asymmetry, and fecal impaction and tests sphincter muscle tone. Diarrhea can mask fecal impaction;

liquid stool oozes around the fecal mask. An abdominal flat plate may be needed to confirm the presence of fecal impaction higher in the intestinal tract.

Diagnostic tests indicated for the assessment of bowel dysfunction include stool for occult blood (annually), barium enema, flexible sigmoidoscopy (every five to seven years after the age of 50), and colonoscopy for those at high risk of colon cancer.

Environment and Body Rhythms

When an individual's body rhythms are disturbed, his or her sense of security and bowel rhythm are challenged. Sleep-wake patterns and eating and activity patterns vary from person to person and give individuals a sense of congruity and constancy. For some, a daily bowel movement is normal; for others, every other day is the norm. Each person has a unique rhythm. Something as simple as skipping a meal or staying up a few additional hours changes how individuals function, including their usual pattern of elimination.

An older person's surroundings might not support regular bowel elimination. The accompanying table illustrates environmental factors that should prompt concern and corrective measures.

Fiber and Fluid

Dietary habits and adequate fluid and fiber intake directly affect bowel habits. The lower the fiber, the higher the risk for breast and colon cancer (Caygill, Charlett, & Hill, 1998; Negri, Franceschi, Parpinel, & LaVecchia, 1998). Fluid intake should range between 1,500 and 2,000 mL daily, barring fluid restriction. Water, as opposed to caffeinated or sugar-based beverages, is recommended. Bottled waters, additions of lemon or lime to water, and juices cut in half with water are all better-quality fluid choices. Sugarless gelatin and ice pops are also alternatives to high-sugar carbonated beverages or beverages containing caffeine, which have a diuretic effect.

Bowel Issues: The Environment and Corrective Interventions

Environmental Issue	Corrective Measure
Lengthy or stair-climbing bedroom-to bathroom distance	Put commode in private area of daytime living space
Lack of privacy during bowel elimination	Use folding dividers to partition space; leave the person, after ensuring safe positioning
Too many changes in daily routine, foods, fluids, and activity patterns	Provide structure, not rigidity, in mealtime, fluids, snacks, and recreation
Changes in sleep-wake pattern	Maintain similar sleeping and waking times, except for special events and activities
Complicated or layered clothing, such as buttons, girdles, and full-length stockings	Keep clothing simple yet comfortable and stylish. Examples: jogging suits, simple underpants and shirts, pants with elasticized waist, Velcro closures
Immobility and dependence on others for toileting	Schedule and pace toileting; ensure that caregivers are aware of the individual's needs; use a bell or call-light system; have mobility evaluated by a physical therapist and teach the person to increase mobility if possible
Lengthy travel without access to toilet	Identify rest stops in advance
Unusual body position and place that requires the individual to use a bedpan, such as spending the night away from home, hospitalization, bedrest, hip or long bone fracture or surgery	Caregivers and nurses must ask the individual about toileting on an hourly basis; allow the individual adequate time after proper positioning, given restrictions; offer sufficient fluids and fiber first, rather than laxatives; reassure the individual that the situation is temporary and that a normal routine will be resumed

Dietary fiber is classified as soluble or insoluble. Fluid binds with insoluble fiber, which is found in foods such as nuts, seeds, some vegetables, and whole-grain products. Once in the large intestinal tract, it expands and facilitates a feeling of fullness and need for defecation. Without fiber, fluid, and activity, constipation is likely. In response, bearing down and straining occur. Some older people experience episodes of dizziness and fainting from the straining because of vagus nerve stimulation.

Recommended dietary fiber intake for adults is 20 to 35 grams per day. A sudden increase in dietary fiber can cause gas, diarrhea, and bloating. Fiber should be added slowly by allowing a few days for adjustment to each graduated increase. Excessive fiber intake can interfere with the absorption of some nutrients. It is also important to drink plenty of water when consuming a lot of fiber. The clinician should recommend the desirable amount of fiber for each patient, particularly for those with cardiac and renal disease.

Fiber intake does not need to be unpalatable or even noticeable. One-half cup of oat or wheat bran or wheat germ may be added to meatloaf, hamburgers, or turkey burgers without compromising flavor. Individuals who do not like vegetables or fruits or who are unable to chew fresh produce may need blenderized fruits and vegetables added to soups, stews, gravies, muffins, and cakes as a portion of the liquid ingredient. For those who enjoy salads, a tossed green salad with 1 to 2 teaspoons of olive oil and spices such as parsley, oregano, chives, and cumin may reduce straining associated with constipation. Whole wheat, rye, or pumpernickel is a better high-fiber alternative to white bread. Starches and refined carbohydrates move slowly through the digestive tract and can cause bloating and flatulence and promote constipation. Eating three dried prunes daily is also recommended.

Mobility

Walking and other activities encourage venous and lymphatic circulation and movement, which carry toxins away from and out of the body. For many, a brisk walk in the morning or evening prompts a bowel movement. The health benefit of regular exercise and activity is clearly identified as a preventive measure for heart disease, hypertension, and osteoporosis; exercise also reduces stress and enhances well-being. Concentrated effort and commitment to ambulate frail older people two or three times daily can be beneficial. The simple formula of "walking and talking" (physical and cognitive stimulation) is recognized as a global principle in enhancing human living.

Psychosocial Variables

Psychosocial effects on a person with bowel dysfunction may include feelings of shame and low self-esteem. The individual internalizes these feelings by withdrawing from family functions, relationships, and social situations, which may lead to depression. Bowel dysfunction may cause changes in lifestyle as well as personal hygiene. The older adult may restrict all outside activities. Routines such as churchgoing may cease. What was once a source of support and comfort may now be a source of anxiety, anger, or guilt. Acknowledgment of these feelings by the caregiver and health professional gives the needed support to the individual.

Managing bowel patterns through a planned program of changes in the environment, nutrition, activity, and medications of an older person contribute to efficient bowel rhythms and a sense of security.

CORA D. ZEMBRZUSKI
BERNADETTE B. HACKETT

See Also
Fecal Incontinence
Gastrointestinal Physiology
Urinary Incontinence Assessment
Urinary Incontinence Treatment

Internet Resources
American Gastroenterologic Society
http://www.gastro.org/index.html

Incontinence on the Internet
http://www.InContiNet.com

International Foundation for Functional Gastrointestinal Disorders
http://www.iffgd.org

BREATHING
See
Dyspnea

BURNS AND OTHER SAFETY ISSUES

Injury is a major cause of morbidity and mortality in the geriatric population. Since most people in this age group live in their own homes, home safety is a serious concern, and prevention of injury is paramount to continuing functionality and quality of life. Falls and their complications are the most significant injuries, causing about half the injury deaths, with motor vehicle accidents and burns from fire and hot water also important. There are a number of safety issues that should concern older adults living in the community.

Counseling elderly persons on specific measures to prevent injuries, in conjunction with efforts to improve mobility, coordination, and sensory function, is the recommended approach to reducing the risk of accidents (State Society on Aging of New York, 1999). Such measures include ensuring proper lighting in all areas of the home, promoting the use of seat belts, installing appropriate assistive devices in the bathroom, removing small area rugs from pathways, installing dead-bolt locks on exterior doors, periodically assessing ability to safely operate an automobile, and preparing for the event of fire or other emergency.

Burns occur in all age groups, but the rate of fatal burns is much greater among persons age 75 and older. Ninety percent of fatal burns result from residential fires. Burns are largely explainable by characteristics of both the individual and the physical environment that interact to produce the burn and associated sequelae. Normal age-related

changes in the skin contribute to poor wound healing and increased risk of infection associated with decreased vascular response.

Cigarette smoking causes almost 50% of all residential fires. The use of space heaters, fireplaces, and heating pads; faulty electrical wires; lack of smoke detectors and fire extinguishers; scalding liquids; and prolonged exposure to the sun can lead to burn injuries in older adults (Kennedy-Malone, Fletcher, & Plank, 2000). Poverty is one of the strongest risk factors for fatality in a residential fire because of the tendency of poor families to dwell in old structures built without adequate fire protection, alarms, exits, and sprinklers. Elderly persons who live in poverty are also more likely to have untreated or inadequately treated chronic health conditions that may compromise their mobility or mental functioning, which compounds their risk for injury due to fire.

A detailed assessment of the burned patient is not within the purview of this entry, but some general principles follow.

History and Physical Examination of the Burn Patient

A complete history of the incident should be obtained, including substances involved, duration of exposure, emergency treatment, and overall condition of the victim before the incident (including mental status, medical diagnoses, medications, prior history of burn injuries). The physical examination should include assessment of vital signs and of the burn area and surrounding tissues. A respiratory system examination may also be warranted. General management of burns requires the assessment of the extent and severity of the burn injury, using the "rule of nines" (Dewar, 1996). The total body area is divided into percentages equal to multiples of nine, with the head and neck, right upper extremity, and left upper extremity counted as 9% each. The anterior chest and abdomen, posterior chest and abdomen, right lower extremity, and left lower extremity each count as 18%. The genitalia are counted as 1%. The percentages are added to determine the extent of the burn injury, which can then be classified by severity, based on the percentage of body surface area (BSA) involved:

- Small (less than 15% BSA)
- Moderate (15% to 49% BSA)
- Large (50% to 69%)
- Massive (70% or greater BSA)

Only second- and third-degree burns are included when calculating the total BSA (Kennedy-Malone et al., 2000).

Treatment of Burns

Immediate treatment for a burn is to apply cold water with wet towels or, if possible, immerse the burned area in cold tap water until the burn is free of pain both in and out of the water. First-degree burns can be treated with cold, wet compresses to limit the extent of the injury and analgesics to reduce pain as needed. Second- and third-degree burns should be irrigated with sterile saline followed by application of a topical antibiotic such as silver sulfadiazine to reduce the incidence of burn wound infection. A sterile occlusive dressing, such as coarse mesh gauze, should then be applied. Dressings should be changed daily if topical antibiotics are not used and twice a day if topical antibiotics are used (Dewar, 1996). For patients with second- and third-degree burns and an infection at the burn site, a sample of the wound should be cultured and the appropriate antibiotic prescribed. Tetanus prophylaxis is recommended for second- and third-degree burns.

Outpatients should be seen 72 hours after the burn injury for examination of the burn site and surrounding tissues for signs of healing or impending infection. The client may need to be evaluated once a week until progressive healing is noted. Older adults who have suffered moderate or major burn injuries require hospitalization. Outpatients being treated for burns may require referral to a physical therapist for wound therapy if the extent of the burn or scarring interferes with the performance of activities of daily living (Kennedy-Malone et al., 2000).

Prevention of Burns

Older adults and their families and caregivers need to take steps to reduce the risk of burns. Important measures include:

- Set the hot water temperature no higher than 110 to 120° F. Always check the water temperature by hand before entering the bath or shower. Taking baths, rather than showers, reduces the risk of scalding from sudden changes in water temperature.
- Establish and practice a primary and alternative evacuation plan in case of a fire or other emergency. Make sure there is a telephone next to the bed.
- If an older person lives alone, consider purchasing an emergency response system. In the event of an emergency, one can simply press a button on a pendant worn around the neck. This causes a communicator unit in the elder's home to dial out to a monitoring platform staffed by individuals who can contact appropriate assistance.
- Place at least one smoke detector on every floor of the home, especially near bedrooms and away from air vents. Check and replace batteries and bulbs according to the manufacturer's instructions. Some fire departments or local government agencies provide assistance in acquiring or installing smoke detectors.
- Use sunscreen with an SPF of at least 15 and wear protective clothing when outside.
- Do not wear loosely fitting or flammable clothing when cooking.
- Check electrical outlets and switches for unsafe wiring conditions.
- Make sure space heaters are used with the proper grounding receptacle and are placed where they cannot be knocked over.
- Check chimneys to make sure they are not clogged with accumulated leaves and other debris. The chimney should be checked and cleaned regularly by a professional.

- Do not sleep with a heating pad turned on. Tucking in the sides of an electric blanket or placing additional coverings on top of it can cause excessive heat buildup, which can start a fire.
- Never smoke in bed or when drowsy. Remove sources of heat or flame from around the bed.
- Keep towels, curtains, and plastic utensils away from the range. Remove any towels hanging on oven handles.
- Make sure lightbulbs are the appropriate wattage and type for lamps or other fixtures to prevent overheating.

MARIA PAPPAS-ROGICH

See Also
"Alert" Safety Systems
Assistive Devices

Internet Key Words
Burns
Home Modification
Safety Risks

Internet Resources
New York State Office for the Aging
http://wwww.aging.state.ny.us/nysofa

Preventive Home Modification
http://www.lifehome.com/artl1006.htm

Seniors: Making Your Home Safer
http://www.lifehome.com/artl0148.htm

C

CALORIC INTAKE

Maintaining energy balance by regulating caloric intake so that it equals calorie expenditure contributes to health and well-being. Excess calories contribute to obesity and more complex health problems. Estimating calorie needs is done using a variety of formulas. The Harris-Benedict equation, a widely accepted formula, uses weight, height, and age to calculate resting calorie needs, also called Resting Energy Expenditure (REE) (Morrison & Hark, 1999).

REE Equation for Males: $66 + [13.7 \times$ weight (kg)] + [$5.0 \times$ height (cm)] $- [6.8 \times$ age] = kcal/day

REE Equation for Females: $655 + [9.7 \times$ weight (kg)] = [$1.8 \times$ height (cm)] $- [4.7 \times$ age] = kcal/day

The equation should be adjusted to 125% of ideal weight for obese individuals. The calculation is then adjusted for activity and injury factors to provide total estimated calorie needs. Total calorie requirements = REE × activity factor × injury factor.

The Harris-Benedict equation provides an estimation of calories needed to maintain weight. For older individuals, maintenance of usual body weight is important for nutritional health. Usual body weight is the preferred standard for most older adults. The usual body weight may require adjustment for amputation or paralysis. Weight for the paraplegic is reduced by 5% to 10% and for a quadriplegic by 10% to 15%. Weight changes due to amputation vary from 0.3% for a hand to 15% for above-the-knee amputation (Niedert, 1998).

Low Calorie Intake

Chronic low calorie intake contributes to weight loss and undernutrition, which have been associated with negative health outcomes, including the development of pressure ulcers, cognitive problems, infections, hip fractures, muscle weakness, edema, and increased mortality. Low body weight and unintentional weight loss are predictive of increased morbidity and mortality, with individuals at the lowest body mass index being at the greater risk. Weight loss is an indicator of declining nutritional status and can be identified before clinical signs and symptoms of overt malnutrition occur.

Weight loss also has a negative impact on physical functioning. A weight loss of 3% or more per year is associated with increased risk of inability to perform activities of daily living, such as walking, standing, and bathing. Loss of 5% in one month significantly impacts nutritional health. A recent loss of 10% body weight predicts mortality in the elderly. Because of the significant negative consequences of weight loss, nursing home regulations require routine monitoring of residents' weights and appropriate interventions to prevent weight loss (Morley & Evans, 1999). Identification of weight loss can help ensure appropriate and timely nutritional interventions.

Weight loss may occur for numerous reasons, including cancer, depression, dementia, medica-

TABLE 1   Activity Factor

Confined to bed	1.2
Ambulatory	1.3
Normal activity	1.5–1.75
Extremely active	2.0

TABLE 2   Injury Factor

Postoperative (no complications)	1.0–1.05
Long fracture	1.15–1.3
Wound healing	1.2–1.6
Pulmonary disease	1.3
Sepsis	1.75–1.85

tion, age-related changes in taste and smell, functional disability, and disorders of the gastrointestinal tract. In approximately 25% of cases, no cause of weight loss can be found, despite extensive evaluation.

Factors that may impede a person's ability to consume food include chewing problems, impaired mobility, changes in cognition, and overly strict diets, such as low salt or low fat. Therapeutic diets for the elderly should be liberalized whenever possible to maximize nutritional intake and prevent excessive weight loss and functional decline (Morley & Evans, 1999). Additionally, depression needs to be managed and anorexigenic drugs eliminated.

Elderly individuals have fragile homeostatic mechanisms that easily become imbalanced because of reduced physiological reserves. Hospitalized older individuals are at risk of weight loss due to the unfamiliar setting, nothing-by-mouth status for tests or surgery, loss of strength and independence, and inadequate assistance with meals.

The goals of nutrition screening and interventions include maintenance of optimal nutritional status; maintenance of a reasonable weight or weight restoration, if underweight; oral or enteral supplementation, if food intake is inadequate to meet nutritional need; adequate protein and fluid intake; appropriate restorative interventions to facilitate food intake (e.g., swallowing evaluation, assistive feeding devices); evaluation of cognitive and emotional status and implementation of behavior strategies to enhance cognition and address feeding problems; evaluation of oral health status and provision of appropriate dental hygiene; assessment of medications and modification of prescriptions to improve food intake, if possible; referral to social services intervention programs to meet specific identified needs (e.g., home-delivered meals, housekeeping assistance, medications management, transportation assistance); exercise to promote appetite and increase stamina; and moderation or abstinence in alcoholic beverage intake.

Meals and snacks should be supplemented with commercial medical nutritional products if the older person is unable to obtain adequate calories from food.

TABLE 3   Making Foods Calorie- or Nutrient-Dense

Milk and Dairy Products	Fruit	Vegetables
Switch to whole milk Use half and half or double-strength milk (fluid milk mixed with powdered milk) for cereal, in beverages, in cooking Add cheese to sandwiches and casseroles	Blend ice cream with soft fruits Roll fruit in nuts or dip in chocolate	Stuff vegetables with cottage cheese, cream, pimento cheese, or other cheese Use full-fat salad dressings

Meat and Meat Substitutes		Bread or Grain Products
Use ground nuts or seeds in place of bread crumbs Increase use of meat, egg, or chicken salad for sandwich fillings Add deviled eggs as a side dish		Stuff biscuits, rolls, muffins, or bread sticks with cheese before baking Add nuts and dried fruit to cereals, quick breads, cookies, cake, or other baked products

		Fats
		Increase fat content of the diet as tolerated through the use of butter, margarine, mayonnaise, peanut butter, and oils.

Excess Calorie Intake

Obesity among older persons is an important contributor to disability. Obesity may be associated with hypertension, diabetes mellitus, cardiovascular disease, and osteoarthritis. It can have profound functional and psychosocial consequences. As older adults suffer functional limitation, activity levels and consequent energy expenditure are further reduced. Decreased function and increased dependency may be considered failure to thrive as a result of obesity.

Quality of life worsens as weight increases. Depression is common in obese older persons. Obesity treatment in the older population needs to focus on interventions that address psychosocial concerns (Jensen & Rogers, 1998).

Lifestyle Changes

Incorporating lifestyle changes or implementing prescribed modifications in diet is a complex process, which most people do not understand, thus interfering with successful implementation. Lifestyle and diet changes are motivated by knowledge of a problem. At this beginning stage, the opportunity exists for the health professional to provide information to clarify the problem and gradually introduce what diet or lifestyle changes are needed. In addition to gaining knowledge, the individual must believe society, family, and friends will support their change efforts. If this basic belief is not there, most individuals will not venture into change. If there is evidence that change will be supported and understood, then the individual is ready to consider change.

Exploring a person's values, beliefs, priorities in life, and the overall effect the old behavior can have on all of these areas is an effective way to facilitate change. Establishing clear, specific, short term, and measurable goals gives a clear sense about the behaviors that are needed to implement the prescribed lifestyle or diet modification.

The change process has the potential of recycling back to questioning the need for change and testing of old behaviors. Coping skills and stress management skills are essential components of the education process to facilitate the prescribed modifications in diet.

Benefits can be achieved with small changes in lifestyle, including a moderate reduction in calories while increasing physical activity. A modest first goal of a 10% to 20% weight reduction with a loss of one-half to one pound a week and maintenance of weight loss is reasonable. Unrealistic weight-loss goals are likely to result in failure.

Dietary interventions for obese older persons should focus on good nutrition using the Food Guide Pyramid as a guideline. A diet limited to 30% fat that provides high fiber, adequate protein (1 gram per kilogram per day), and adequate fluids is the basic recommendation. Portion control and avoidance of high-calorie snacks may be adequate to result in weight loss when accompanied by exercise or resistance training to improve muscle strength. Resistance exercises two to three times a week can be done using resistance machines, elastic bands, or simple weight-lifting devices. This type of exercise is essential to preserve or increase muscle mass during weight loss and is a benefit for older individuals who must lose weight (Evans & Cyr-Campbell, 1997).

Normal Ranges

In adults, anthropometric indicators of energy balance include the following measurements: height, weight, waist circumference, middle upper arm circumference, various skinfold thickness and other extremity circumference measures, body mass index (BMI; weight in kilograms divided by height in meters squared), waist-to-hip ratio, and percent body fat. Bioelectrical impedance devices, underwater weighing, or other technologies can be used to estimate lean and fat mass. BMI is widely used to estimate body composition and health risk. However, it does not reflect body composition for a physically fit person with a low body fat proportion and large lean body mass with a BMI suggestive of obesity (White, 1999).

Healthy older adults should have a BMI between 22 and 27. The healthiest BMI is approximately 24 for adults older than 70 years of age, since this is associated with the least health care utilization and lowest risk of mortality. A low BMI reflects protein energy malnutrition and is associated with poor health outcomes. Functional impairment occurs with both extremes of BMI (i.e., below 24 and above 32) (Niedert, 1998; White, 1999).

A healthy weight for older adults is slightly heavier than the weight recommended for younger adults. If weight reduction is indicated, weight should be lost slowly and accompanied with progressive resistive exercises so that lean body mass is not compromised and protein, vitamin and mineral intakes remain adequate (Evans & Campbell, 1993).

CARLENE RUSSELL

See Also
Body Composition
Obesity

Internet Key Words
BMI
Calorie Needs
Diet Interventions
Healthy Weight
Obesity
Overweight
Progressive Resistance Training
Undernutrition
Weight Loss

Internet Resources
American Dietetic Association
http://www.eatright.org

Council for Nutrition
http://www.LTCnutrition.com

CANCER TREATMENT

Cancer is the second most common cause of death and a major cause of morbidity for persons age 65 and over in the United States. Approximately 50% of all cancer patients are age 65 and over, and this percentage is likely to increase with the expansion of the older population (Redmond & Aapro, 1997; Yancik & Ries, 1998).

Causes

The association of cancer and age may be accounted for by two non–mutually exclusive explanations. First, carcinogenesis is a time-consuming process. The development of cancer requires that a person live several years from the original exposure to the initiating factor. Second, to some extent, cellular aging mimics carcinogenesis and primes the aging cells to the effects of late-stage carcinogens (promoters). Of special interest is so-called proliferative senescence, a process during which, paradoxically, the aging cells become more prone to neoplastic transformation at the same time they lose the ability to replicate. The enhanced susceptibility of older individuals to late-stage carcinogens suggests that older individuals may benefit from chemoprevention.

Biology

In at least five types of cancer, the prognosis changes with patient age (see the accompanying table). In studying the mechanism through which age may influence the biology of cancer, it is helpful to distinguish a "seed effect," involving changes intrinsic to the tumor cells, from a "soil effect," involving changes in the physiology of the older person. For example, the worse prognosis for acute myelogenous leukemia (AML) in older individuals may be accounted for mainly by changes related to the malignant cells; in the case of large cell lymphoma, the soil effect appears predominant. The prognosis of large cell lymphoma may worsen in the aged because the circulating concentration of interleukin-6 increases with age. In the case of breast cancer, both the seed and the soil effects may contribute to produce a tumor with a more indolent course.

Despite the influence of age on prognosis, chronological age should not determine the treatment of individual patients. Even though the prognosis for AML becomes worse with age, as many as 65% of patients over age 60 may obtain complete

Change in Prognosis Based on Patient Age

Cancer	Change in Prognosis with Age	Mechanism
Acute myelogenous leukemia	Worse	Increased resistance to chemotherapy due to increased prevalence of glycoprotein P–mediated multidrug resistance Increased prevalence of unfavorable chromosomal abnormalities Involvement of the pluripotent hemopoietic stem cells by the neoplastic process
Large cell non-Hodgkin's lymphoma	Worse	Increased circulating levels of interleukin-6
Breast cancer	Improved	Increased prevalence of hormone receptor–rich well-differentiated tumor cells Decreased cell proliferation Decreased circulating concentrations of estrogen Decreased mononuclear cell reaction to the primary tumor, with decreased production of growth-stimulating cytokines
Ovarian cancer	Worse	Unknown
Non–small cell lung cancer	Improved	Unknown

remission of their disease, and a number of them may be cured. Although 60% to 80% of older women present with hormone-responsive breast cancer, the remaining 20% to 40% present with aggressive disease in need of aggressive cytotoxic chemotherapy (Balducci, Silliman, & Baekey, 1998).

Prevention

Primary prevention of cancer involves the elimination of environmental carcinogens and the administration of substances that may offset carcinogenesis (chemoprevention). The feasibility of chemoprevention has been shown in randomized controlled trials; cis-retinoic acid prevents second malignancies of the head and neck area, and the selective estrogen receptor modulator tamoxifen reduces the incidence of breast cancer by 40% in women at risk of developing the disease (Balducci et al., 1998). Other substances undergoing active investigation include raloxifene for breast cancer and nonsteroidal anti-inflammatory drugs for colon cancer. Seemingly, older individuals benefit at least as much as younger individuals from primary prevention of cancer, because they may be more susceptible to environmental carcinogens. In the case of chemoprevention, however, it is important to establish whether the reduction in cancer-related mortality and morbidity outweighs the complications of the treatment.

Secondary prevention of cancer involves screening of asymptomatic persons at risk. The positive predictive value of screening tests, such as mammography or fecal occult blood test, may improve with the age of the population due to increased prevalence of cancer. At the same time, the benefits of screening may be lessened due to the decreased life expectancy of the aged and the effects of previous screening, which may have eliminated the majority of "prevalence" cases. There is no clear evidence that screening mammography reduces the breast cancer–related mortality in women over 70 or that screening for colorectal cancer reduces the cancer-related mortality in persons over 80. In the absence of randomized and controlled studies, it appears reasonable to institute some type of screening program for any person with a life expectancy of at least three years, as the benefits of screening first become apparent three years after the beginning of the screening program. Biennial mammography and fecal occult blood testing and annual physical breast exams are probably adequate screening for persons over age 70 (Yancik & Ries, 1998).

Treatment

Treatment of cancer involves local forms of treatment, such as surgery and radiotherapy, and systemic forms of treatment, including hormonal therapy, cytotoxic chemotherapy, and biological therapy.

Surgery

The risk of surgical mortality and complications increases in persons age 70 and older. In the majority of cases, however, the increased mortality is accounted for by the mortality related to emergency surgery, including obstruction and perforation of the large bowel. Thus, surgical mortality in older individuals may be lessened by a widespread screening program for cancer of the large bowel. Recent data show that both general anesthesia and cancer-related procedures may be safely performed in people who are 100 years old and older.

Radiation Therapy

Radiation therapy may provide valid palliation and occasional cure of cancer in persons of all ages. At least three large studies conducted in Italy, France, and the United States showed that patients over age 70 tolerate doses of chemotherapy comparable to those used in younger individuals, with similar risk of complications.

Cytotoxic Chemotherapy

Because aging may be associated with changes in the pharmacokinetics and pharmacodynamics of cytotoxic drugs and with increased susceptibility of normal tissues to the complications of treatment, the dosages of cytotoxic agents may need to be modified. Pharmacokinetic changes in absorption,

volume distribution, metabolism, and excretion occur with age. With the development of new oral drugs, changes in drug absorption may become more relevant in the use of chemotherapeutic agents. Changes in the volume concentrations of water-soluble agents occur as a result of a decline in lean body weight, a decline in albumin concentration, and anemia. The effects of anemia on volume concentration have recently been recognized. Many cytotoxic agents are heavily bound to red blood cells. In the presence of anemia, the concentration of free agent in the circulation increases.

A marked decrease in the activity of type I hepatic reactions, involving the P450 cytochrome system, has been reported in frail elderly patients. Also, these reactions are the main sites of drug interactions, which are particularly likely in older individuals who take several medications each day. Abnormalities in these reactions may reduce the activation and deactivation of drugs such as the oxaphosphorines (cyclophosphamide and ifosfamide). Unfortunately, no reliable clinical tests are available to assess these reactions.

Because the decline in glomerular filtration rate (GFR) is one of the most consistent physiological changes in aging, the renal excretion of drugs is reduced in the majority of older individuals. Although it is possible that other excretory mechanisms compensate for decreased renal function, it is prudent to use caution with drugs excreted though the kidneys (e.g., methotrexate, bleomycin, carboplatin) or drugs that give origin to active and toxic metabolites excreted through the kidneys. Idarubicin and daunorubicin are metabolized to idarubicinol and daunorubicinol, respectively. These alcohols are largely responsible for the activity of the drugs and are excreted through the kidneys. Cytarabine in high doses leads to accumulation in the circulation of ara-uridine, a neurotoxic metabolite eliminated mainly through the kidneys.

In addition to the pharmacokinetic changes that occur with age, pharmacodynamic changes such as altered intracellular metabolism of drugs and decreased ability to repair DNA occur. In elders, the ability of cells to catabolize fluorinated pyrimidines is reduced due to decreased concentration of the enzyme dihydropyrimidine dehydroxilase. Also, the ability of normal cells to eliminate *cis*-platin–induced DNA adducts is reduced. With age, the toxicity of normal tissues is altered, and the risk and severity of myelodepression, mucositis, central and peripheral neurotoxicity, and cardiotoxicity also appear to increase.

The risk of chemotherapy-related complications may be ameliorated by a number of simple provisions. Measures that may mitigate the toxicity of antineoplastic chemotherapy in older persons include adjusting the dose of chemotherapy to the patient's GFR, maintaining the hemoglobin levels greater than or equal to 12 g/dL, and using myelopoietic growth factors (granulocyte or granulocyte-macrophage colony-stimulating factor) on patients age 70 and older receiving moderately toxic chemotherapy (CHOP, cyclophosphamide-doxorubicin). Interventions involving aggressive fluid resuscitation in the presence of mucositis may lower the toxicity of antineoplastic chemotherapy in older people (Cova, Beretta, & Balducci, 1998).

Although the initial dose of chemotherapy should be adjusted to GFR, subsequent doses should be escalated or de-escalated according to the degree of toxicity observed, because alternative mechanisms of drug excretion may play a role in some older individuals. The maintenance of hemoglobin levels at 12 g/dL or greater may reduce the toxicity of chemotherapy and at the same time minimize fatigue in older individuals. Prophylactic use of hemopoietic growth factors is recommended, as grade 3 and 4 myelotoxicity occurs in more than 50% of lymphoma patients age 70 and older treated with CHOP, and the treatment-related mortality among these patients is 5% to 30%.

A number of drug antidotes may render chemotherapy more manageable in older individuals (Lichtman, 1998). These include amifostine, which may function as a "pan-protector" to reduce the risk of myelodepression, cardiotoxicity, neurotoxicity, and nephrotoxicity; desrazoxane for protection of the myocardium from anthracyclines; and interleukin-11, which reduces the risk of thrombocytopenia and platelet transfusions.

In addition to drug antidotes, patient selection is extremely important to minimize the complications of chemotherapy. In general, frail patients are not good candidates for aggressive chemotherapy but may receive valuable palliation from drugs with limited toxicity, including gemcitabine, vinorelbine, capecitabine, and taxanes. Among the nonfrail elderly, those dependent in one or more instrumental activities of daily living require particularly close attention, as they may be at increased risk for chemotherapy-related complications.

Cancer in older persons is an increasingly common problem, and they may benefit from both primary and secondary cancer prevention. Surgical and radiation treatments are generally safe in older individuals. Chemotherapy may be safe and effective in properly selected patients.

LODOVICO BALDUCCI

See Also
Health Maintenance
Health Promotion Screening

Internet Key Words
Aging
Biology
Cancer
Cytotoxic Chemotherapy
Prevention
Treatment

Internet Resources
National Cancer Institute
http://www.nci.nih.gov

Oncolink
http://www.oncolink.upenn.edu

Women's Cancer Network
http://www.wcn.org/

CARDIOVASCULAR FITNESS
See
Deconditioning Prevention
Exercise and the Cardiovascular Response

CAREGIVER BURDEN

Family caregivers provide 80% of all unpaid non-professional care, numbering more than 25 million persons. Although most are middle-aged female relatives (e.g., daughters, daughters-in-law), other female and male relatives and nonrelatives also function as informal caregivers. Some caregivers report positive outcomes from the caregiving experience, but many more report negative consequences (Picot, Debanne, Namazi, & Wykle, 1997). Researchers who study caregiving label these negative perceptions and consequences "caregiver burden."

Caregiver burden sometimes describes tasks performed by caregivers to compensate for elders' functional and behavioral deficits. Investigators also use the term to describe the tension, stress, strain, or upset that caregivers feel in response to these deficits. Most commonly, caregiver burden is defined as the extent to which caregivers perceive that caregiving has had an adverse effect on their emotional, social, financial, and physical functioning (Zarit, Reever, & Bach-Peterson, 1980). This is the definition used here, with the addition of the adverse consequence of spiritual distress (Picot et al., 1997).

Caregivers who are experiencing caregiver burden often do not seek assistance for themselves. Rather, caregivers may be introduced to health providers when they accompany elders for their health care visits. Caregivers report that health care providers rarely inquire about their health status and concerns, and when caregivers do articulate the negative effects of caregiving, health providers do not always recognize the request for assistance. Thus, informal caregivers are often referred to as the hidden or secondary patients.

Identifying which caregivers are vulnerable to caregiver burden is difficult, because each caregiver perceives her situation uniquely. Pearlin, Mullan, Semple, and Skaff's (1990) conceptual model of caregiver stress suggests that practitioners assess for the background or context of stress, primary and secondary stressors, mediators of stress, and outcomes of stress to determine whether caregiver

burden is a likely result. Background factors to assess include socioeconomic status, caregiving history (e.g., the relationship of the caregiver to the elder, past relationship quality, family and network composition, and health problems of the elder), and the availability, accessibility, and acceptability of community resources. Assessment of the cultural meaning of caregiving and the caregiver's role is important, because culture influences the acknowledgment and meaning of problems, responses to problems, and accepted health regimens (Picot, Stuckey, Humphrey, Smyth, & Whitehouse, 1996).

Primary stressors originate from the functional and behavioral deficits of the elder, and particular attention must be given to the number, duration, frequency, and unrelenting nature of care demands. The quickest way to assess these care demands is to administer a standard measure of physical and instrumental activities of daily living (Duke University Center for the Study of Aging and Human Development, 1978; Katz, Ford, Moskowitz, Jackson, & Jaffe, 1963). It is also important to assess the elder's resistance to the caregiver's help. The Texas Research Institute of Mental Sciences Behavioral Problem Checklist can be used to assess the range of behavioral problems exhibited by dementia and nondementia patients living in private homes and the corresponding responses of caregivers (Niederehe, 1988; Picot, 1995). Burnout may result when caregiving demands exceed the caregiver's resources (e.g., time, physical, financial); when there are continual conflicts between the elder's, the caregiver's, and the family's needs; and when there is no progress.

Caregivers who provide care to elders with predominantly physical deficits, including incontinence, are often prone to physical health problems and limitations on their social and recreational activities. Caregivers of cognitively impaired or mentally ill elders who exhibit disruptive or unpredictable behaviors requiring vigilance, or whose illness is associated with stigma (e.g., AIDS, cancer) are prone to depression, anger, hostility, grief, fatigue, and social relationship strain. Both care situations threaten the finances of the caregiver.

Secondary stressors, an outgrowth of primary stressors, consist of role and intrapsychic strains.

Role strains often arise from conflict about the care management of the elder within the context of the family, including employment and financial aspects of care provision. As care demands increase, health professionals must assess for any reduction in the caregiver's social and recreational activities. Social isolation accentuates both caregiver burden and depression, placing the elder and caregiver at risk for elder abuse and mistreatment. Intrapsychic strains include lowered self-esteem and perceptions of competence and gain. Caregivers who have competing roles (e.g., employee, parent, wife); who are young, female, and white; who have low self-esteem and low sense of mastery; and who have little knowledge of the condition contributing to the elder's disability (e.g., dementia) appear to be most at risk for caregiver burden resulting from secondary stressors.

Mediators of stress outcomes include coping and social support. The coping measure suggested by Pearlin et al. (1990) offers strategies related to the management of stress, the meaning of the caregiving situation, and the resulting symptoms of stress. In assessing social support, it is crucial to avoid assuming that the availability or size of a social support network translates into caregiver satisfaction or absence of burden.

Global assessment of the caregiver's self-perceived health—physical, mental, social, and spiritual—can be assessed with Robinson's (1983) 13-item caregiver strain index. As the care demands of the elder change, necessitating changes in caregiving tasks, health care professionals should periodically reassess the caregiver's health status. Assessment of the physical health consequences of caregiving should include a global assessment of the self-perceived health of caregivers (i.e., excellent, good, fair, or poor) in comparison to that of their peers, as well as attention to the worsening of chronic condition indicators, such as increasing blood pressure, blood glucose, hemoglobin A1-c (control of blood glucose over several months), and dyslipidemia; compromised immune system responses; and increased doctor visits, medication use, and substance abuse. Measures that include symptoms of specific organ systems, such as the Cornell Medical Index, help the practitioner target specific areas of concern for monitoring and intervention.

Mental health evaluations should include stress symptoms of denial, anger, social withdrawal, sleeplessness, irritability, and lack of concentration. A caregiver reporting the following symptoms for at least two weeks may be depressed: low or irritable mood; feelings of worthlessness, self-reproach, or excessive guilt; suicidal thinking or attempts; motor retardation, agitation, or disturbed sleep; fatigue and loss of energy; loss of interest or pleasure in usual activities; difficulty thinking or concentrating; and changes in appetite or weight. The most commonly used measure of depression is the Center for Epidemiologic Studies Depression Scale.

The social health of the caregiver can be evaluated by listing who is both available and reliable in the caregiver's social support network to provide respite, empathetic listening, or help with care tasks (Pearlin et al., 1990). Spiritual health often helps buffer the development of other adverse health consequences for caregivers, but spiritual distress can occur when caregivers are separated from their religious ties (many caregivers are homebound) or their beliefs are challenged by the caregiving situation and the elder's suffering (Picot et al., 1997). Queries about caregivers' ability to attend religious services, receive visits from fellow worshippers in the home, and read or hear religious messages, and questions about the importance of religion in their lives, can gauge the spiritual health of caregivers.

Day-to-day strategies to lessen caregiver stress and caregiver burden include acquiring the knowledge and skills to take care of the elder's problems and working out a system that balances caregiving with other roles and health promotion needs (Nkongho & Archbold, 1996; Schultz, 2000). To maintain their own health, caregivers may need assistance identifying and coordinating acceptable family and professional helpers and planing for anticipated changes, including transferring care of the elder to someone else or to an institution. Caregivers may need support or counseling to accept that transferring care responsibilities can have positive consequences for both the caregiver and the elder (Fink & Picot, 1995).

Health care professionals must be empathetic listeners as they conduct a careful assessment of caregivers' backgrounds, stressors, mediators, and outcomes. Effective strategies to avoid or minimize caregiver burden should be planned with the caregiver and should be responsive to each caregiver's unique situation.

SANDRA J. FULTON PICOT
LORRIE L. POWEL

See Also
Caregiver Burnout
Caregiving Benefits
Caregiving Relationships
Elder Mistreatment: Overview

Internet Key Words
Caregiver Burden
Caregiver Burnout
Caregiver Health
Caregiver Quality of Life
Caregiver Strain
Caregiver Stress

Internet Resources
American Association of Retired Persons: Health and Wellness and Explore Health
*http://www.aarp.org/indexes/
   health.html#caregiving*

Guide to Internet Resources Related to Aging
http://www.aarp.org/cyber/sd13.htm#top

Internet Resources on Aging: Caregiving
*http://www-cpr.maxwell.syr.edu/gerontologist/ira/
   june97/june97.htm*

National Family Caregivers' Association
http://www.nfcacares.org

CAREGIVER BURNOUT

Health professionals and informal caregivers of serious and chronically ill patients can recognize

burnout by being sensitive to the signs that define burnout:

- A general loss of energy that is often conceptualized as exhaustion from the constant drain on personal resources and a response to unrelieved stress (Felton, 1998; Muldary, 1983).
- A state of physical, emotional, and mental exhaustion resulting from the demands of intense involvement with people over a long period (Pines & Aronson, 1988).

Burnout is viewed as a state of mind that frequently afflicts individuals who work with other people and give much more than they get back in return. It is a syndrome that includes physical, emotional, and mental fatigue; feelings of helplessness and hopelessness; and a lack of interest in and enthusiasm for work and life in general (Maslach & Leiter, 1997). Burnout can be observed among teachers, police officers, social workers, mental health workers, health care professionals, and informal caregivers. Among professional caregivers, burnout syndrome is related to a loss of idealism, decreased commitment to helping those in need, emotional detachment from and negative attitudes toward patients, feelings of powerlessness over the conditions of their work, and mental and physical exhaustion (Melchior et al., 1997; Muldary, 1983).

Effects of Burnout

Burnout usually occurs after less than two years of caregiving (Muldary, 1983). Caregivers experiencing burnout focus more on negative events of the past or negative prospects for the future and narrow their attention to events outside themselves. They give less attention to immediate circumstances and become increasingly preoccupied with internal thoughts. This narrowing attention reduces their ability to process information efficiently and solve problems as they arise. As a result, there is increased risk of errors in judgment, neglect of caregiving activities, and accidents (Felton, 1998).

Burnout results from the caregiver's inefficient appraisal of circumstances or lack of adequate coping resources, thus impairing the caregiver's ability to respond to continually changing conditions and demands (Muldary, 1983). Such a caregiver may respond inadequately to the care recipient's needs. As caregiving enthusiasm and commitment decline, overall motivation to provide quality health care decreases (Muldary, 1983). Health care becomes just an activity that is performed (grudgingly) by a demoralized caregiver, and only to the extent necessary to avoid serious adverse consequences. Simultaneously, loss of empathy, caring, and respect for care recipients emerges; this is communicated by avoiding eye contact and performing rigidly technical caregiving activities without any other human interactions. Reflected in the language of the caregivers, the dehumanizing attitude makes care recipients seem more like objects, problems, or disease entities than individuals. Dehumanization results in neglect of the care recipient.

Burnout often impairs the control and coping mechanisms used to regulate emotional expression (Muldary, 1983). Thus, burned-out caregivers often express anger and anxiety. Many burned-out caregivers believe that things would be better if they terminated their caregiving activities or used alcohol and drugs. They become unwilling to cooperate with others and tend to spend more time alone.

Predisposing Factors

Although burnout is a highly variable syndrome that cannot be predicted by any single factor, certain people are at more risk for burnout than others. People who have a high degree of empathy for others, are sensitive to the needs of others, have compulsive personalities, and lack self-confidence are more vulnerable to burnout (Muldary, 1983). Individuals with compulsive personalities like to ensure predictability and control in their world; imperfect or unpredictable caregiving situations provoke stress. Individuals who lack self-confidence or are submissive tend to subordinate their own needs to those of others. Such persons are

unlikely to say no; they avoid making personal demands to keep from displeasing others and must feel that they are needed and appreciated (Muldary, 1983). Thus, when positive outcomes are not readily apparent—as is often the case in long-term caregiving situations—emotional stress accumulates over time and leads to burnout.

Interventions

Individual Coping Strategies: The basic principle in managing burnout is to encourage individuals to do whatever they can to change a given situation, or to train them to cope more effectively when it is not possible to change the situation (Muldary, 1983). Caregivers need to be aware of the physical and psychological signals of chronic stress and their responses to it by consciously monitoring themselves.

There are two modes of coping strategies: emotion focused and problem focused (Lazarus, 1991). Emotion-focused coping strategies are used by family caregivers to make themselves feel better without changing the source of stress. The most common emotion-focused strategy is overconsumption of alcohol and use of mood-altering substances. However, these strategies have only short-term effects. In other cases, caregivers attempt to detach themselves from the caring situation, but because of their familial relationship, distancing may be difficult. Relaxation techniques such as tension relaxation, letting go, sensory awareness, and yoga are useful emotion-focused coping activities for reducing muscle tension and achieving a feeling of well-being.

Problem-focused strategies are efforts to master stressful situations. Individuals may go directly to the source of stress and seek change it through interpersonal communication. Family caregivers can be trained to communicate more effectively and use functional coping strategies in the therapeutic environment. A change in perspective might also be required in managing burnout. Some caregivers believe that they are entirely responsible for their care recipients' well-being; others feel that what they do does not make any difference in the situation. Direct action for personal change involves learning how to do some things differently.

Family Therapy: When working with families of burned-out caregivers, a focus on context, relationship, and meaning can be helpful. Many family therapists attend to the structure and process of the family's interaction with the environment, the intergenerational family system, and the current family system (Hartman, 1995).

The focus on the family-environment interface attempts to alter the environment. Family members need to look at their resources and draw on the strengths of their relationships. This may help family members make better use of conditions in their environment and gain control and power in their lives. The focus on the intergenerational family system stems from the belief that people and their current families are shaped in major ways by powerful intergenerational forces. The third focus is on the current family system—its structure, communication, and organization. The therapist tries to create a context in which family members gain a different perspective of themselves and their world. The therapist has to be personally authentic and connected while trying to influence the family's transactional patterns. By giving attention to the needs, feelings, and words of every member of the family, the family therapist may help establish a relationship that can change the burnout environment (Catherall, 1999).

Caregiver's Bill of Rights

Health professionals can help family caregivers manage burnout by enhancing their coping strategies, providing family therapy, and acknowledging caregivers' rights and responsibilities. According to the "Caregiver's Bill of Rights," caregivers have the right to seek information about providing better caregiving activities, as well as protecting their own health, spirits, and relationships (Torres-Standovik, 1999). The bill of rights makes family caregivers understand that they are allowed to pursue better caregiving environments with the help and support from other family members and health profession-

als. It also enables them to find alternatives if home caregiving is no longer physically, financially, or emotionally feasible. Advocating caregivers' rights allows family caregivers to ask for outside help that mitigates burnout and encourages effective caregiving.

Caregivers need to recognize their own needs, such as adequate information about the nature of the care receiver's condition; coping skills; assistance with psychological, financial, and medical services; emotional support; and long-term planning. Health care professionals also need to collaborate with caregivers' families to elicit their expertise, resources, and strengths.

HAEJUNG LEE
RHAYUN SONG
HEAJONG SHIN

See Also
Caregiver Burden
Caregiving Benefits
Caregiving Relationships

Internet Key Words
Burnout
Caregiving
Coping Strategies

Internet Resources
The Caregiver's Handbook
http://www.medsupport.org/caregiverguide.htm

Caregiver Survival Resources
http://www.caregiver911.com

Family Caregiver Alliance
http://www.caregiver.org

National Family Caregivers' Association
http://www.nfcacares.org

CAREGIVING BENEFITS

Family Based Caregiving

Most of the assistance and care provided to older adults with physical or cognitive disabilities is provided by family members, including spouses, adult children, and other relatives. Given the complexities of family caregiving, geriatric care clinicians will provide services to older clients who are themselves caregivers as well as those elders who receive care. Family members initiate referrals to social service and health care agencies, frequently accompany older adults with cognitive or physical impairments to medical appointments, and seek services for themselves to cope with the challenges of caregiving. Optimal care planning for older adults entails active engagement of the family caregiver in assessment, care management, and intervention. Geriatric professionals in all disciplines who have a contextual understanding of the processes and experiences of family caregiving will be better suited to engage and intervene with these later life families.

Research over the past two decades has sought to better understand the experience of family caregivers through documentation of the challenges, strains, and burdens frequently inherent in the caregiving role. Recently, there is growing interest in the study of the more positive aspects of the caregiving experience. A variety of terms have been used to describe the positive aspects of caregiving including satisfactions, pleasures, rewards, enjoyment, growth, meaning, uplifts, enhanced relationships, and gratification (Kramer, 1997). These positive aspects of caregiving are broadly conceptualized as "gain," which is defined as "the extent to which the caregiving role is appraised to enhance an individual's life space and be enriching" (Kramer, 1997, p. 219).

Caregiver gain includes any positive affective or practical return that is experienced as a direct result of being a caregiver, and it is an aspect of care provision that they value and seem to want to talk about. In a study of 644 caregivers in the United Kingdom, 95% identified something good and/or rewarding about providing care (Bamford et al., 1998). Benefits perceived by the caregivers included heightened care receiver well-being; a sense of satisfaction or achievement; feeling appreciated; receiving companionship; reciprocating past help; and enabling the care receiver to remain at home. In a qualitative study of 94 caregivers in the United

States, 90% of caregivers reported that they valued positive aspects of caregiving and relationships (Farran et al., 1991). Caregivers valued feelings of confidence in providing quality care, the care receiver's love for them, the positive relationship that they experienced with the elder, memories and accomplishments, and family and social relationships.

Evidence suggests that gain is also an important predictor of caregiver well-being. In a conceptual model of caregiver adaptation (Figure 1), the appraisal of role gain is hypothesized to play an important role in understanding caregiver well being (Kramer, 1997). Documentation of the intervening mechanisms of appraisals of gain and strain add to our understanding of the negative versus positive indicators of psychological well-being (Lawton et al., 1991). Given that the well-being of the care receiver is often directly dependent upon the well-being of the caregiver, geriatric clinicians and service providers are advised to work to enhance caregiver gain.

Gain has been correlated with motivation for providing care, better quality of the relationship, satisfaction with social activities, action oriented types of coping, positive affect (Kramer, 1997), and religiosity (Picot et al., 1997). In addition, studies have suggested that some persons are more likely to report caregiver gain. For example, two studies found that African-American caregivers expressed greater caregiving rewards and satisfaction than Caucasian caregivers (Lawton et al., 1992; Picot et al., 1997). Caregivers who were older and less educated were also more likely to report caregiver gain (Kramer, 1997). While caution is advised in generalizing these findings, given the need for replication studies, these results suggest potential areas of assessment to explore when working with the family caregiver, and potential domains for intervention.

Assessment

Multiple areas of inquiry are identified by the conceptual model of caregiver adaptation (Figure 1) and should be assessed when working with family caregivers. Assessment of caregiver appraisals of

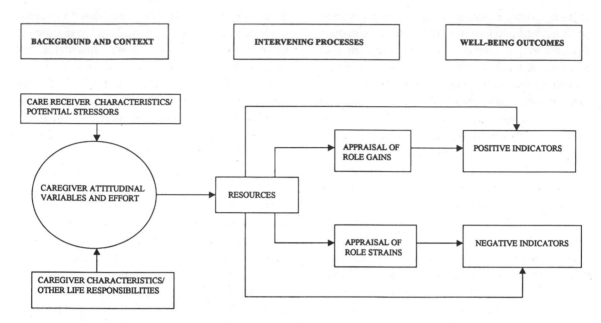

FIGURE 1   Conceptual model of caregiver adaptation. Reprinted from *The Gerontologist* (1997), volume 37, number 2, page 229. Copyright © The Gerontological Society of America. Reproduced by permission of the publisher.

role gain and role strain should go hand in hand. Assessment tools for examining caregiving strains are reviewed elsewhere in this volume (see Caregiver Burden and Caregiver Burnout). Several measurement approaches have been used to evaluate caregiver gain, including qualitative inquiry, single-item indicators, dichotomous or frequency ratings, and multi-item scales in which caregivers are asked to rate several items according to their frequency or occurrence (Kramer, 1997). Different measures will be more appropriate under particular circumstances. Exceedingly lengthy measures are inadvisable when numerous other variables are being assessed simultaneously.

A simple way to initiate assessment of both gain and strain is to ask the question, "How has caregiving affected you?" Other questions to explore with the caregiver include: "What are the most pleasant aspects of caregiving?"; "What keeps you going or gives you hope?"; "What are the good things in your life?" (Farran et al., 1991). Recognition of the ways individuals feel enriched by caregiving in the assessment process, and validating

their experience of and capacity for growth may allow the geriatric practitioner to more appropriately validate feelings and experiences, and to empower the caregiver. Caregivers who find meaning, growth, and reward in caregiving may be happier in the caregiving role. As such, caregivers who are unable to articulate any gain in caregiving may be at greater risk for poorer well-being.

Enhancing Gain

Much attention has been given to developing interventions that might reduce the strain experienced by family caregivers. Interventions and direct services to family caregivers should also seek to enhance the sense of gain that is experienced.

First, professionals should encourage caregivers to talk about their growth and the positive outcomes they have experienced as a result of caregiving. Help them to articulate the ways their life has been enhanced. A time limited, continuous writing journal exercise can be used with individuals or groups. To engage caregivers in a reflective writing process that dislodges left brain control of expression, the writing is made continuous and somewhat repetitive. Caregivers are told to choose from several phrases to facilitate their writing. For example, "I would never have chosen for _____ to become ill, but from this I have learned _____" or "I would never have chosen to experience the challenge of caregiving for _____, but from this has come _____." They are instructed to write out the entire phrase, filling in the blanks briefly, and then writing out the whole phrase again while filling in the second blank with a new thought each time. Caregivers then share their insights with the service provider or with other caregivers in a group setting.

Second, professionals should seek to identify the caregiver's strengths and affirm the competence of the caregiver. Caregivers should understand that they are providing a valuable service to the care receiver and identify the ways in which they are excelling as a caregiver.

Finally, geriatric practitioners could teach caregivers to use action-oriented types of coping, work to enhance social activities, and provide or encourage spiritual counseling.

BETTY J. KRAMER

See Also
Caregiver Burden
Caregiver Burnout
Caregiving Relationships
Coping with Chronic Illness

Internet Key Words
Caregiver Well-Being
Caregiver Rewards
Caregiver Satisfactions
Caregiver Gain
Positive Aspects

CAREGIVING RELATIONSHIPS

A number of social factors have influenced family-based care, including the aging population, changes in the health care system, the increase in multigenerational households, and a shift from acute to chronic illnesses in older adults. When a family becomes involved in the care of one of its members, it takes on a new, supportive function. Consequently, a number of structural and functional changes occur within the larger family system that affect many aspects of family life, including family relationships. Structural aspects of family relationships refer to the composition of the family social network and the participation of its members, whereas functional aspects refer to the family's ability to garner different types of support for specific needs (Li, Seltzer, & Greenberg, 1997). Caregiving relationships fall within several domains, including relationships between the caregiver and the care recipient, those among different caregivers, and those between formal (paid provider, physician, nurse) and family-based care providers.

Several factors must be considered when making a decision to undertake family-based care. The relationships between the caregiver and the care recipient and among the caregivers themselves affect this decision in many ways. The decision also depends on the available resources, which include tangible resources, such as money or the available space in the caregiver's home, and social resources, such as the number and proximity of family members who can help with family-based care. A less tangible resource, but equally important, is the family's ability to discuss and negotiate a plan of family-based care among its members, including the care recipient when possible. For example, the family members must determine where the care recipient will live and how the responsibilities of care will be divided, if that is an option. If the care recipient requires a high level of care, providing respite to the primary caregivers is important.

Central to the issue of family caregiving are the prior relationships of the care recipient with the caregivers and the relationships between or among the caregivers. When the relationship of the caregiver to the care recipient has been good, the provision of family-based care may be seen as another demonstration of caring, affection, and support. When the relationship with the care recipient has been characterized by periods of ambivalence or conflict, the caregiving relationship is likely to reflect the usual style of relating. Thus, the prior relationship becomes the basis on which the caregiver–care recipient relationship changes, usually from a relationship with reciprocal support and exchange to one of increased dependency. This shift is often initially uncomfortable for both the caregiver and the care recipient.

Similarly, the relationships among multiple family caregivers represent not only their current ways of interacting and solving problems but also the "family ledger" of who owes what to whom. Family caregiving often brings to the forefront prior issues and unresolved conflicts, with a backdrop of stress and emotional reactivity, particularly if the decision about family-based caregiving is sudden. In contrast, watching the gradual deterioration of a family member due to a debilitating illness is difficult, but it allows the family to begin considering options for the care of that member. Because family caregiving is a time of change for all involved, it is also an opportunity for families to strengthen relationships and develop new ways to solve problems.

Women are somewhat more likely to take on the role of primary caregiver. Since women generally outlive men, more women than men may expect to provide family-based care to their spouses. Adult children generally recognize that caregiving responsibility is likely; their chances of doing so are increased if they are married. The adult child with the fewest competing responsibilities (e.g., career, spouse, children) may be the designated caregiver in a family of multiple siblings, although sometimes the most able family member emerges to assume the caregiver role. Parents are less likely to anticipate caregiving to their older children, but as life spans increase, more parents may outlive their children and help in their care when they are ill. Such caregiving to adult children may be extremely burdensome both physically and emotionally. Caregiving to siblings is somewhat less common, in part because individuals are likely to turn to their children first for caregiving help. In addition, geographic distance between adult siblings may make caregiving impractical and even impair the sibling relationships that undergird family-based caregiving.

A number of background factors influence the type of care and the extent to which family-based care is offered to older persons. The gender of the caregiver likely influences the types of caregiving help provided. Women are somewhat more likely to assist with housekeeping, meal preparation, and personal care, whereas men tend to offer transportation assistance, make household repairs, and manage financial affairs. Factors such as racial and ethnic background, family traditions and history, and religion also influence decisions to care for family members at home and the degree to which family caregivers feel burdened. Historically, persons of color have been more likely to care for family members at home due to a strong sense of family, extended networks of helpers made up of family and friends, financial considerations, and past discrimination by formal health services. There is a higher incidence of caregiving among Asian American, African American, and Hispanic households than white households, and these families are more likely to care for more than one person. Additionally, caregivers in these minority groups are more likely to live with the care recipient and to receive help from others. Thus, caregiving for older persons is not universally distressing, particularly if it is considered a culturally acceptable, rewarding, or expected event.

Other family-based care relationships reflect nontraditional family structures, such as friends, domestic partners, and extended families. Recent social changes have made domestic partners eligible for health care and other benefits that extend the support base to many individuals. Single persons—divorced, widowed, or never married—must rely on their extended families and broader social network, as well as paid help, to provide assistance. In caregiving situations, networks of extended family and friends often deal with the same issues and encounter similar stresses in arranging caregiving responsibilities and tasks.

One form of family-based caregiving that has gained recent attention focuses on grandparents and grandchildren. Increasing numbers of children live in homes maintained by grandparents, with or without one or more parents present. These grandparents' involvement in the daily care of grandchildren is markedly different from that of grandparents who provide day care or baby-sitting or have more traditional grandparenting roles. In situations in which the grandparents take on parenting responsibilities and become surrogate parents to grandchildren, they are likely to have complex relationships with the child's parents.

Where caregiving falls in the life course has an impact on the effects of caregiving (Moen, Robison, & Dempster-McClain, 1995). In addition to the relationships with those directly involved in caregiving, other aspects of the caregiver's life are affected. The type of care necessary and the available supports influence the degree to which caregiving demands spill over into other settings, such as one's work. The Family and Medical Leave Act of 1993, which provides up to three months of leave for the care of family members, recognizes and provides support for such caregivers. In some circumstances, such as debilitating diseases of long duration such as Alzheimer's or Parkinson's disease, the caregiving relationship may be extensive and long-standing; in other cases, caregiving may not extend beyond several months or a year. Maintaining existing relationships with family and friends is important to the caregiver's health, yet the activities of caregiving deplete the time and energy necessary to sustain these relationships. Family caregivers must be assisted to find time and respite to engage in rewarding noncaregiving activities that can prevent burnout and alleviate the burden of caregiving responsibilities.

Due to the importance of family caregiving, health care professionals should work to establish partnerships with family caregivers. Nurses, physicians, social workers, and other health professionals can collaborate with informal caregivers to achieve their mutual goals focused on the care recipient.

Professionals must also take the lead in supporting the caregiver, the relationship between the caregiver and the care recipient, and the relationships among the other caregivers as well.

CAROL M. MUSIL
CAMILLE B. WARNER

See Also
Caregiver Burden
Caregiver Burnout
Caregiving Benefits

Internet Resources
Caregiver Survival Resources
http://www.caregiver911.com

Caregiving Online
http://www.caregiving.com

Family Caregiver Alliance
http://www.caregiver.org

National Family Caregivers Association
http://www.nfcacares.org

CASE MANAGEMENT

Case management (also known as care management, care coordination, and service coordination) has become a familiar service and function in the provision of health and social services for older adults. Various models of case management have emerged in primary, acute, home, and long-term care, and most recently in managed care. Case management practice varies. Depending on the agency and program, case managers may be nurses, social workers, physicians, or other health care professionals.

History

Case management was a core service in the community-based long-term-care demonstration projects implemented in the early 1970s. A primary goal of these projects was to delay or avoid premature or inappropriate nursing home placement. Waivered Medicaid funds were made available to support community-based care plans. In programs funded by Medicaid waivers, each client was assigned a case manager. Subsequent expansion of community-based care has been fueled by two major trends. First, funding for home and community-based services has increased. Second, federal and state legislation mandates case managers to arrange and monitor community-based long-term-care services for program clients. Service coordination is necessary to overcome the obstacles encountered in highly fragmented delivery systems.

Driven by the introduction of diagnosis-related groups (DRGs), case management has also emerged in acute care, as pressure for timely and appropriate discharge has increased. In this context, case managers utilize care maps and critical pathways to monitor patients' progress and develop discharge plans to settings that meet patients' needs. Acute-care case managers focus on quality, cost-effective, timely, and appropriate service provision.

Definition

The National Advisory Committee on Long Term Care Case Management defines case management as a "coordinating service that helps frail elders and others with functional impairments and their families identify and secure cost effectively administered services appropriate to the consumers' needs" (Connecticut Community Care, Inc., 1994, p. 5). In managed care, case management is "active oversight of healthcare delivery [that] will ensure individuals receive appropriate and quality service in a cost-effective manner" (Murer & Lenhoff-Brick, 1997, p. 3).

Many program-specific definitions of case management exist. In community-based long-term care, the case management process includes outreach, screening, assessment, care planning, plan implementation, monitoring, and reassessment. In managed care, cost containment pervades and

guides the case management process. A case manager may not be assigned to every case, but case managers working in managed-care settings have the authority to approve or deny services based on cost and benefit caps. Hence, in any given setting or program, case management reflects characteristics of the target population as well as the structure of the program's funding. As such, there is considerable variation in implementation of the component case management activities. Caseload size varies considerably, reflecting the diversity of programs, target populations, and settings where case managers are employed.

Components

Outreach attempts to identify persons likely to need case management, given the complexity of their circumstances. Programs target diverse populations. Target populations are specified and operationally defined in policy and program regulations. In managed care, potentially high-cost cases are identified before initiating service delivery. *Screening* is a preliminary assessment of the client's circumstances and resources to determine presumptive eligibility. Standardized protocols are used to determine whether a client's status and situation meet the criteria of the program's target population and how service costs should be monitored. Outreach and screening are important gatekeeping mechanisms that affect the accuracy of the program's targeting efforts, the effectiveness of cost-containment mechanisms, and program operation and management.

Comprehensive assessment is a "method for collecting in-depth information about a person's social situation and physical, mental and psychological function which allows identification of the person's problems and care needs" (Schneider & Weiss, 1982, p. 12). Typically, comprehensive assessment focuses on several domains: physical health, cognitive functioning, emotional status, ability to perform activities of daily living, social supports, physical environment, and financial resources. Many programs utilize standardized multi-dimensional assessment instruments.

Care planning builds on information collected during the assessment process. It requires clinical judgment, knowledge of community resources, creativity, and sensitivity. Care planning is a key resource-allocation process. The care plan specifies services, providers, frequency of service delivery, and costs. Critical considerations in care plan development are the willingness and availability of informal caregivers. Balance between formal and informal services is a major consideration in the care planning process. Clients and caregivers participate in the process.

Service arrangement involves contacting formal and informal providers to arrange services specified in the care plan. Case managers often negotiate for services with providers when making referrals to other agencies. When case managers have the authority to purchase services on their clients' behalf, they may order services directly from providers.

Case managers systemically *monitor* changes in clients' situations and alter care plans to meet clients' current needs. Ongoing monitoring, combined with timely modification of care plans, helps ensure that program expenditures reflect current client needs and are not based on outdated assessment data.

Reassessment involves determining whether changes in the client's situation have occurred since the last assessment. Systematic and regularly scheduled reassessments also assist in evaluating the attainment of outcomes specified in the care plan.

Role Conflict and System Constraints

Case management goals and tasks are theoretically oriented toward both the client and the service delivery system. As such, case managers are simultaneously client advocates and delivery system agents. While working to ensure that care plans are appropriate for individual clients, case managers also are charged with containing costs and designing care plans within financial parameters. The nature and extent of case manager advocacy activities can vary considerably. In community-based long-term-care settings, case managers are a single point of contact in a complex delivery system. Role conflict for case managers often centers on the balance between client-centered and system-focused orientations. In the context of managed care, pervasive cost-containment goals may reduce a case manager's sensitivity to potential ethical and role conflicts.

Program Context

Case management services are constrained by four program characteristics: financing and reimbursement, targeting criteria, gatekeeping mechanisms, and organizational structure. Financing and reimbursement may include client cost sharing, waivers, fee for service, and capitation. Targeting criteria define the population to be serviced—for example, those who are at risk of nursing home placement, are functionally impaired, have a frail support system, or have unmet needs. Gatekeeping mechanisms, designed to control the number and type of services clients receive, include a cap on the cost of individual care plans, authorization power, and provider risk sharing. Organizational auspices refers to the agency that employs case managers (e.g., provider agency, freestanding case management agency, hospital, nursing home, day care, managed care). Despite disagreement about what to call it, as well as considerable variation in its implementation, case management will continue to have a key role and function in elder care.

CAROL D. AUSTIN

Internet Key Words
Care Coordination
Case Management
Home and Community-Based Services
Service Coordination

Internet Resources
Case Management Resource Guide
http://www.cmrg.com

National Association of Professional Geriatric
   Case Managers
http://www.caremanager.org

CATARACTS

The lens of the eye accounts for approximately
30% of its focusing power and is the only part of
the eye that can change shape to allow focus on
objects at different distances. Loss of the lens's
ability to change shape, presbyopia, is an age-related change. A cataract is opacity in the normally
transparent lens. It is not a "film" or "growth" over
the eye but a loss of clarity that can occur at any
age and for a variety of reasons, although normal
aging of the lens is the most common cause. Virtually everyone who lives long enough will develop
a cataract. There is no medication that can prevent
or reverse a cataract once it forms.

Symptoms vary with the type and density of
the cataract and with the patient's activity level
and visual needs. Slowly progressive cataracts and
nuclear or cortical cataracts (affecting the central
nucleus or the outer cortex of the lens) may be
less visually disturbing than a rapidly progressing
posterior subcapsular cataract (typically a dense,
focal change on the visual axis). A bedridden 83-
year-old is typically less affected by a cataract than
an active 63-year-old with work-related visual demands. Clinicians have to be aware that cataracts
do not affect each person the same way.

The presenting complaint may be blurry vision
at a distance, near, or both; inability to see street
signs while driving; difficulty with small print; or
a "film" or "fog" over the eye, indicative of cloud-
ing of the lens. Since a cloudy lens scatters light
more than a clear lens, patients are often bothered
by glare from bright lights. Another interesting phenomenon is "second sight," in which the changes
in the lens increase myopia or nearsightedness,
allowing the patient to see better up close without
glasses; some patients are able to read without
glasses or bifocals.

Assessment of a cataract patient begins with
a thorough eye examination, keeping in mind that
cataract is not the only cause of visual change in
the elderly. Macular degeneration, glaucoma, and
other eye diseases are potential causes of vision
loss in the geriatric population that can be ruled
out by a comprehensive eye examination.

It is vital for the patient to understand that
cataracts do not improve with time and will worsen
(although at different rates in different people) and
that surgery will be needed when spectacles or other
visual aids are no longer helpful. Timing of surgery
is the heart of care for cataract patients. Many can be
helped with appropriate visual aids such as frequent
changes in spectacle prescription, magnifiers, con-
trast-enhancing lenses, tints, and glare-reducing
lenses, which may improve the patient's function
and delay the need for surgery temporarily or
permanently.

Cataract surgery should be recommended
when the patient cannot perform some or all activities of daily living because of the cataract. This is
a subjective decision; the timing differs from patient to patient. There is no need to wait for the
lens to "ripen," as in the past, nor is there any
absolute visual acuity level that must be reached
before surgery is undertaken. The choice to have
cataract surgery should be made by the patient after
all options have been discussed, the prognosis without surgery is understood, and the risks and benefits
of the type of surgery recommended by the surgeon
are fully explained and understood.

Current techniques in cataract surgery make
it among the safest and most common surgical procedures performed in the United States. Understanding the procedure can make it easier to decide
on surgery when it is indicated.

Typically, the 10- to 15-minute procedure is
done on an outpatient basis at an ambulatory or
outpatient surgery center. The patient is usually
able to resume nonstrenuous, everyday activities
within 24 hours. Incision type, use of stitches or
"no-stitch" procedures, and use of postoperative
patching depend on the surgeon, type of cataract,
presence or absence of concurrent eye disease, and
the eye's response during surgery. Preoperative assessment includes a complete examination and
measurements of the eye to determine the appro-

priate power of the intraocular lens (IOL) that will replace the cataractous lens. A physical examination, laboratory work, and an electrocardiogram are often required. The extent of the preoperative physical is dependent on the patient's health, most recent physical examination, and requirements of the center where the procedure will be done—all of which should be clearly specified prior to the surgery.

The vast majority of cataract surgeries currently performed involve removal of the opacified lens and insertion of an IOL in its place. The IOL replaces the focusing power of the lens and obviates the need for the thick cataract glasses or contact lenses of the past. Current IOLs may be "foldable," allowing them to be placed into the eye through a smaller incision, which is associated with faster rates of healing and fewer complications; this has led to the development of the self-healing, "no-stitch" type of cataract procedure. Clear cornea surgery—incision through the avascular cornea—is associated with the fastest healing and fewest complications.

Although IOLs replace the focusing power of the lens, most IOLs are fixed-focus lenses; they cannot adjust to different planes of focus, as can the normal lens. Patients often require reading glasses after surgery, and they need to be advised of this. Multifocal IOLs, recently approved by the Food and Drug Administration, focus objects at a variety of distances and reduce the need for glasses following cataract surgery.

A common misconception is that cataracts are removed by laser. Although this type of surgery is being studied, lasers are not currently used in cataract removal. However, lasers are used in the postoperative management of cataract patients.

The patient may be given antibiotics the day before surgery to reduce the risk of intraoperative infection. Nothing can be taken by mouth except clear liquids after midnight the night before the procedure. Upon arrival at the surgery center, the patient is administered several topical eyedrops for antibiosis and to dilate the pupil. Immediately before the surgery, the eye is anesthetized. General anesthesia is rarely used. Typically, local injection around the eye provides deep anesthesia and arrests eye movements. Risks are minimal, but there is a chance of damage to the structures around the eye. Many cataract surgeons have recently begun using topical anesthesia for cataract surgery; it has the least risk, causes the least patient discomfort and eliminates the apprehension associated with injection. However, the patient feels some sensation during the procedure and can move the eye. A skillful surgeon and good preoperative instructions to the patient are strongly recommended.

The eye may or may not be patched, depending on the surgeon's preference and the type of procedure performed. Postoperative medications include an antibiotic and a steroid to reduce inflammation, often as a combination drop.

The patient may be moderately uncomfortable the first day and night after surgery. Pain is much less of a factor, however, with the advent of small incision procedures. Intense pain is unexpected and should be reported to the surgeon at once. The first postoperative visit is the day after surgery; if there are no complications, the next follow-up visit is one to two weeks following surgery. Antibiotic-steroid drops are gradually tapered, depending on the degree of inflammation. Glasses may be prescribed four to six weeks following surgery—sooner with the less invasive procedures. The patient must be alert for and report any significant decrease in vision, pain, redness, discharge, flashes of light, or new "floaters" (black spots in front of the eye) throughout the postoperative period.

Posterior capsular opacification is a common postoperative complication. In most cataract surgeries, the lens capsule is left in place to support the IOL. This can serve as a site for the growth of lens tissue that clouds the vision; the patient reports feeling that the cataract has grown back. Patients must be assured that once a cataract is removed, it does not come back. Making a clear opening in the posterior capsule using the YAG laser (YAG laser capsulotomy) easily treats this capsular change, takes a few minutes, and is done in the surgeon's office.

Cataracts need not be the debilitating eye condition they once were. Appropriate education and use of visual aids can markedly improve function prior to the need for surgery. Current surgical techniques have minimized postoperative discomfort and reduced recovery time. Laser surgical techniques may further improve outcome. The IOL frees the postoperative cataract patient from the cumbersome cataract glasses or contact lenses of the past. The future promises the use of multifocal IOLs that function more like the natural lens of the eye.

RICHARD J. MADONNA

See Also
Eye Care Providers
Vision Changes and Care

Internet Key Words
Cataract
Cataract Surgery
Intraocular Lens (IOL)
"No-Stitch"
Visual Aids
YAG Laser Capsulotomy

Internet Resources
American Academy of Ophthalmology
http://www.eyenet.org

American Optometric Association
http://aoanet.org

On-line medical consultation
http://www.mediconsult.com

CEREBRAL VASCULAR ACCIDENT
See
Stroke/Cerebral Vascular Accident

CHEST PAIN: NONCARDIAC CAUSES

Chest pain is one of the most frequent complaints encountered by health care providers in both acute and nonacute settings. The character of chest pain is a major determinant in emergency department triage and hospital admission (Summers et al., 1999). The positive or negative sequelae associated with the prognosis weigh heavily on initial physical assessment and management. Failure to recognize serious heart disease can be fatal, yet many patients with chest pain have normal physical examinations. These include patients with coronary artery disease and with gastrointestinal, pulmonary, and psychiatric disorders. The initial approach is to rule out myocardial ischemia and other coronary etiologies. Once this is done, evaluation of further differentials should be pursued. The focus here is on noncardiac chest pain and the most common differentials: pulmonary embolus, generalized anxiety disorder, costochondritis, and herpes zoster.

Pulmonary Embolus

Pulmonary embolism is an occlusion of a portion of the pulmonary vascular bed by an embolus. Pulmonary emboli are the most common cause of pulmonary disease in hospitalized individuals. They are the third leading cause of death in the United States, accounting for 100,000 deaths per year, and they contribute to another 100,000 deaths per year. Approximately 50% of the deaths occur within two hours of the embolic event as a result of not being detected and treated.

Chest pain occurs in 80% to 90% of cases. Classically, the pain is described as pleuritic, but it can be of any character or at any location on the chest. Common symptoms include dyspnea, usually of acute onset; hemoptysis; palpitations; and anxiety.

Often, the patient is tachypneic. Pulmonary examination may identify wheezing, consolidation, and friction rub, or it may be normal. Atrial arrhythmias, in particular atrial fibrillation, may be found, as may signs of congestive heart failure (CHF). The lower extremities should be examined for signs of deep vein thrombosis that may be the source of the pulmonary embolism. The clinician should also inquire about recent surgeries, especially orthopedic. Symptoms vary, and no one symptom is diagnostic. Therefore, a high index of suspicion must be maintained to pursue and rule out diagnoses.

Hypoxia, evidenced by arterial blood gas (PO_2) less than 60, is often a good indicator, as is an elevated A-a gradient. Electrocardiography to rule out cardiac origins of chest pain and a chest x-ray to rule out other causes of chest pain and dyspnea should be done. A V/Q scan is necessary to look for areas of ventilation and perfusion mismatch. If this is nonconclusive, pulmonary angiography, the gold standard in diagnosis of pulmonary embolism, is required. Doppler studies of the lower extremities should be performed to evaluate for the presence of thrombii (Summers et al., 1999).

Porcine muscosal unfractionated heparin (UFH) is the initial therapy for pulmonary embolism. Intravenous UFH is given by intravenous (IV) bolus, followed by infusion to achieve and maintain an activatcd partial thromboplastin timc (APTT) prolongation, which is equivalent to 1.5 to 2.0 times baseline. Warfarin is started at the same time as IV UFH. UFH is continued for 5 days, and stopped when the international ration (INR) is 2.0 for at least 24 to 48 hours. Warfarin dosing is adjusted to maintain an INR between two and three. For patients who have failed anticoagulation treatment (i.e., repeat pulmonary embolism or development of pulmonary embolism while on anticoagulants), a mechanical filter may be needed.

The role of thrombolytic therapy for venous thromboembolism remains uncertain. There are a few case studies that demonstrate some efficacy, however to date no well-designed randomized clinical trial has shown that the benefits of thrombolysis exceed the risks for a well-defined subgroup of patients with acute pulmonary embolism. The largest multinational registry of patients in whom acute pulmonary embolism had been diagnosed found that 3% of 311 patients who underwent thrombolysis for acute pulmonary embolism suffered an intracranial hemorrhage. Due to the potential risks of thrombolysis for pulmonary embolism, treatment remains controversial as investigators continue to study other methods of treatment (Dipiro et al., 1999).

Patient education should focus on etiology, medications, deep breathing and relaxation tech-niques, signs and symptoms of decompensation, and the importance of patient compliance.

Gastroesophageal Reflux Disease

Gastroesophageal reflux disease (GERD) is a chronic condition characterized by a motility disorder, with an incompetent lower esophageal sphincter (LES) retrograde flow of stomach contents into the esophagus. It also can be caused by delayed emptying and impaired esophageal clearance. GERD is the most prevalent condition originating in the gastrointestinal tract (Richter, 1997). GERD is estimated to affect between 25% and 35% of the United States population (Scott & Gelhot, 1999). Clinical manifestations of GERD present with various symptoms, including pyrosis, early satiety, abdominal fullness, bloating, dysphagia, odynophagia, and belching.

Atypical symptoms associated with GERD have been found in about 50% of patients with noncardiac chest pain, 75% of patients with chronic hoarseness, and between 70% and 80% of patients with asthma (Katz, 1998).

Diagnosis of GERD may be accomplished by a classical history of symptoms, including pyrosis and regurgitation, and a trial of therapy can be initiated without further diagnostic tests. Treatment includes dietary modifications, smoking cessation, elevation of the head of the bed, and avoidance of late or large evening meals. In addition, pharmacological management may be instituted, including H2 antagonists, prokinetic agents, and proton pump inhibitors. Further diagnostic investigation may be warranted if esophageal symptoms do not respond to medical management. These include radiographs, upper GI endoscopy, Berstein test, and a 24-hour esophageal pH monitor. Surgery may be a final alternative, with a Nissen fundoplication utilized because quite frequently it has a positive outcome.

Patient follow-up to determine effectiveness of treatment and adjustments of regimen are essential. Patient education is focused on increasing signs and symptoms of decompensation, etiologies of disease, lifestyle modifications, and medications. Patients should be advised that medical management for GERD is long term, and that compliance is necessary for maximal benefit.

Generalized Anxiety Disorder

Patients with generalized anxiety disorder (GAD) experience gradual, persistent anxiety without the traditional specific symptoms found in panic disorders, phobias, or obsessive-compulsive disorders. They exhibit varying degrees of uneasiness, apprehensiveness, or worry about multiple real or projected problems. The intensity of worry is out of proportion to the situation. Fatigue, restlessness, irritability, muscle tension, and sleep disturbance are usually present. Symptoms vary with each individual but are often punctuated by bursts of autonomic activity that include chest pain, palpitations, tachycardia, hyperventilation, dyspnea, globus hystericus, indigestion, paresthesia, and sweating. Chest pain is often described as acute and sharp over the precordial area.

As soon as the patient is medically cleared, diagnosis focuses on further testing. Instruments used to confirm or refute the diagnosis of GAD include the Zung Anxiety Self-Assessment Scale, Sheehan Patient Rated Anxiety Scale, Hamilton Anxiety and Depression Scales, and Covi Anxiety and Rashkin Depression Scales (McGlynn & Metcalf, 1992).

Treatment modalities include psychotherapy, pharmacotherapy, and aggressive patient education. Psychotherapy should address modifications in both cognitive and unconscious behaviors. Behavior therapy is used to help alleviate and prevent recurrent maladaptive behaviors associated with some episodes of anxiety. Supportive counseling and family therapy are also of benefit. Other techniques that may be helpful include relaxation therapy, biofeedback, visualization, meditation, exercise, and desensitization. If anxiety is severe enough to warrant medication, three classes of medications are commonly used. Azipirones are well tolerated,

although the mechanism of action often takes two to four weeks. Benzodiazepines are effective and act promptly. Tricyclic antidepressant medications have also been used with good effectiveness. The clinician needs to proceed cautiously when prescribing potential addictive substances that need judicious monitoring.

Successful management of patients with GAD lies in the degree and effectiveness of patient education. As with all disease states, the more the patient understands and gets involved with the illness, the better the prognosis. When educated about the causes of GAD, and its symptoms and medications, the patient can become a proactive partner in the health care process.

Costochondritis

Costochondritis is an inflammation of unknown etiology of the costal cartilage in the costochondral and costosternal articulations. It is one of the most common causes of anterior chest pain, affects woman more than men, and is usually seen in patients over 40 years of age. Costochondritis tends to affect the third, fourth, and fifth costochondral joints. Focal tenderness is often present on palpation. The onset of chest pain from chostochrondritis, described as sharp and nonradiating to the jaw, neck, or upper extremities, usually occurs in the afternoon, versus the more traditional early morning chest pain associated with myocardial infarction. The onset of pain in the parasternal area may be acute or gradual, with possible radiation to the arms, shoulders, and entire chest. Often pain is aggravated by sneezing, coughing, deep inspiration, upper trunk movement, and reaching. Pain is intensified with palpation of the anterior chest wall, with localized point tenderness over costochondral junctions. Usually no swelling or crepitus is noted; if present, however, it is known as Tietze's syndrome. Erythema and predominant pain over the left anterior wall may also be present.

An electrocardiogram (EKG) and chest radiograph are useful in ruling out other diseases; other tests are usually not beneficial. Diagnosis is traditionally made based on physical examination. Costochondritis may coexist with other disease processes, particularly atherosclerotic heart disease.

Nonsteriodal anti-inflammatory agents (NSAIDs) and local heat are the cornerstones of therapy. Analgesics may be considered for pain management, and, depending on severity, a lidocaine-corticosteroid preparation injected into joints may offer some relief. Refractory cases may require surgery for resection of the involved cartilage. There are no special diet recommendations. Activity is as tolerated, with avoidance of aggravating maneuvers affecting the pectoralis major muscles during repetitive activity. Proper posture alignment and stretching exercises on a regular basis can alleviate symptoms and decrease incidence.

Patient education is focused on etiology and on signs and symptoms of decompensation, which include deterioration of physical state and of the treatment regimen. The patient should be advised that improvement is often gradual, occurring over several weeks, and that recurrence is possible.

Varicella Infection: Herpes Zoster

Herpes zoster is the reactivation of varicella-zoster virus (VSV) of the Herpesviridae family. After primary infection, the VSV becomes latent in the dorsal root ganglia and can remain so indefinitely, ranging from years to the lifetime of the individual. When the immune system suffers insult or wanes with age or other disease, VSV reemerges as "shingles." Viral reactivation occurs in 60% to 90% of individuals who had the primary infection of chickenpox as children. Incubation ranges from 10 to 20 days. The infectious period begins 48 hours before onset of clinical symptoms and persists through the period of new lesion formation, usually 3 to 5 days.

If the virus emerges on the C6 through C8 dermatomes, the prodrome will include anterior radiating chest pain, myalgia, and often fever. The prodrome precedes the defining rash by 2 to 3 days.

Chest pain is described as a sharp, often stabbing sensation with radiation if the virus is infecting the correlating dermatomes. Classically, the rash begins with an erythematous base, with scattered small vesicles often seen on the trunk or face initially. The rash usually erupts along unilateral dermatomes. The lesions are pruritic and painful. Constitutional symptoms may also be present, including low-grade fever, anorexia, myalgia, fatigue, or headache. The rash, with multiple stages of lesions, presents in a random and often linear pattern. Excoriated areas may also be present as a result of scratching the pruritic lesions.

If chest pain precedes the appearance of lesions, an EKG may be done to rule out cardiac origin. A trial of sublingual nitroglycerine may be employed to see if the agent relieves pain. If it does not, the etiology of the chest pain is probably not cardiac in nature. Further diagnosis is obtained from history, symptoms, and clinical interpretation. Serum varicella titers and enzyme-linked immunosorbent assay may be performed, as well as Tzanck smears or skin biopsy if the etiology remains unclear.

Treatment focuses on symptom relief. Antipruritic measures, including oatmeal baths, baking soda, calamine lotion, and antihistamines, are of some benefit. Acetaminophen or ibuprofen is suggested for fevers greater than 101.5°F. Antiviral therapy may be instituted for those with more severe symptoms. It is best to begin antivirals within 36 to 72 hours of onset of symptoms for best results. More commonly prescribed antivirals include oral acyclovir 800 mg five times per day and famciclovir 500 mg every 8 hours. Creatinine clearance must be monitored and medications adjusted accordingly.

The patient is advised to clean lesions with a mild soap, and pat them dry with a clean towel. No dressings are needed for the rash. The vesicles last for 2 to 3 days, then crust over and slowly resolve over 3 to 4 weeks (Berg, 1993). Approximately 20% of individuals experience postherpetic neuralgia (Ragozzino, 1982), which is most effectively treated with tricyclic antidepressants (Volmink, 1996). The patient needs reassurance that the disease is generally benign and has a limited course. Patient education is focused on etiology of disease, increasing sign and symptoms of decompensation (including those of infection) and medications.

CAROL M. RHODES

See Also
Anxiety and Panic Disorders
Depression Measurement Instruments
Gastrointestinal Physiology
Heartburn

Internet Key Words
Costochondritis
Gastroesophageal Reflux Disease
Generalized Anxiety Disorder
Herpes Zoster
Pulmonary Embolus
Shingles

Internet Resources
Chest Pain When It's Not Your Heart, The Mayo Clinic
http://www.mayohealth.org/mayo/9810/htm/ cp.htm

Emergency Medicine and Primary Care
http://www.EMBBS.com

Generalized Anxiety Disorder, National Institute of Mental Health
http://www.nimh.nih.gov/anxiety/anxiety/gad/ index.htm

Shingles (Herpes Zoster), National Institute of Neurological Disorders and Stroke
http://www.ninds.nih.gov/patients/Disorder/ SHINGLES/shingles.htm

Updated Guidelines for the Diagnosis and Treatment of Gastroesophageal Reflux Disease, American Journal of Gastroenterology
http://www-east.elsevier.com/ajg/issues/9406/ ajg1123fla.htm

CHRONIC ILLNESS

More than 100 million people currently experience chronic medical or mental illness in the United States (Hoffman, Rice, & Sung, 1996). This is projected to increase to 134 million by the year 2020 (Institute for Health and Aging, 1996). Miller (1992) defines chronic illness as "an altered health state that will not be cured by surgical procedure or a short course of medical therapy." Strauss (1975) defines chronic illness as an impairment or deviation that has one or more of the following characteristics: is permanent; leaves residual disabilities; causes nonreversible pathological alterations; necessitates special rehabilitation; or may involve a long period of supervision, observation, and care.

People can be born with health problems or can develop illnesses at any age that require care, but chronic disorders and diseases become increasingly prevalent with age. "Almost 100 million people in the United States have some form of chronic condition, from minor ailments to severely disabling illnesses and impairments" (Institute for Health and Aging, 1996). Arthritis, hypertension, heart disease, and hearing impairment are among the most prevalent chronic conditions in older, non-institutionalized persons in the United States (Cobbs, Duthie, & Murphy, 1999).

Chronic illness is expensive. In 1990, the total cost of chronic illness reached $659 billion. Approximately one-third of this was attributed to indirect medical costs of premature death and lost productivity. These costs do not include the lost productivity of caregivers (Institute for Health and Aging, 1996). Chronic illness necessitates chronic care. Unfortunately, the health care delivery system for the elderly does not adequately provide for those living with chronic disease. Inadequacies include lack of coordination among medical consultants and specialists, shortage of geriatricians and gerontological nurses and social workers, lack of standardized training for health care professionals in geriatric assessment and care, insufficient training in interdisciplinary team collaboration, and health care financing programs insufficient to support individuals with chronic disease. For example, Medicare will not pay for a home health aide for a person with Alzheimer's disease, but when the patient becomes impoverished, these services can be covered by Medicaid.

Confronted with declining health and failing capabilities, older people experience overwhelming fear of losing the most intimate and significant elements of life: independence, friends and family, status upon retirement, physical comfort, and life itself. These fears, and related anxiety and frustration, can cause multiple syndromes such as depression, restlessness, and insomnia. Some people become easily tearful; others complain incessantly or even intimidate family or caregivers in an effort to relieve the feelings of helplessness. Still others withdraw. These patients need help regaining emotional control and need reassurance that someone will care for them when they are no longer able to care for themselves. A variety of system responses can help meet the needs of patients requiring chronic care.

Provide Prevention and Early-Detection Services: Primary prevention refers to health promotion practices that can delay dependency, such as nutritional support, smoking cessation, home safety, exercise, and immunizations. Secondary prevention refers to screening for diseases such as diabetes, hypertension, breast cancer, colon cancer, glaucoma, and depression. Screening can detect problems early and prevent worsening conditions. Tertiary prevention includes patient support, education, monitoring, and rehabilitation in an effort to forestall further health and functional decline in individuals already diagnosed with a chronic illness, such as chronic obstructive pulmonary disease or coronary artery disease.

Develop a Team Approach: Patients with chronic illnesses require an interdisciplinary team to adequately respond to their needs. The team should comprise a physician, nurse, social worker, and specialists as needed (e.g., occupational therapist, physical therapist, nutritionist, psychiatrist, psychologist, and pharmacist). Participation of the patient and family in health care discussions and planning is critical to providing care and monitoring the illness. They are important and unique contributors to the team.

Educate Health Professionals: Professionals need to appreciate the multidimensional impact of chronic disease on daily living. Creative programs and curricula need to be developed and implemented to achieve this awareness. Project DOCC, an innovative training program for physicians designed and offered by chronically ill patients and families, has been instituted at numerous medical centers nationally. For example, parents of chronically ill children teach pediatric trainees about the issues of chronic disease by giving "pediatric grand rounds" and facilitating house calls to their homes (Shaw, 1999).

Conduct a Thorough Assessment: The degree and intensity of involvement by the professional team depend on the seriousness of the illness and its impact on functionality. A variety of assessment issues should guide the development of intervention strategies.

1. *Disease onset, diagnosis, and certainty:* Was the disease onset rapid or gradual? Was it life-threatening, debilitating, or merely annoying? What is the impact of the diagnosis? (The responses can range from feeling sad and frustrated to feeling overwhelmed and suicidal.) Is the diagnosis definitive, or is there a possibility for better or worse news?
2. *Disease etiology:* Is there a genetic component, or does the patient feel responsible for and perhaps guilty about the disease occurrence (e.g., emphysema or lung cancer in smokers)?
3. *Disease course, pain, and familiarity:* Will the patient stabilize, improve, or worsen with time? What is the potential for the patient's increased self-reliance? Should the onset of significant pain be anticipated? What is the patient's and family's understanding of the nature and course of the illness? What preparation do the patient and family need to provide future care?
4. *Other acute and chronic illnesses:* Is the patient already suffering from acute health problems or adapting to life with a chronic illness? If so, what is the anticipated effect of the additional health problems on the patient's overall well-being?
5. *Functional status:* What is the patient's overall functional ability as determined by assessing activities of daily living and instrumental activities of daily living?

6. *Restrictions:* What areas of life will be changed, including mobility, diet, cognitive function, employment, and self-care?
7. *Interventions:* Will the patient require surgery, medications, physical therapy, occupational therapy, psychotherapy, or a new diet regimen?
8. *Service site:* Will care be provided at home or in an outpatient or inpatient setting?
9. *Cognitive and emotional status:* Do the patient and caregiver understand the disease and associated problems and have the emotional capacity to adapt to lifestyle changes and re·trictions?
10. *Social stigma:* Is the illness something the patient experiences shame about (e.g., HIV)?
11. *Spiritual status:* Has the illness affected the patient's spiritual life and faith? What is there about the patient's spiritual life that the health care professional needs to know to be helpful (Narayanasamy, 1996)?
12. *Social supports:* Does the patient have a network of friends and family to provide emotional comfort and assistance with new routines? Are family members able to provide the needed assistance? If yes, will they be successful at it?
13. *Environmental obstacles:* Would structural changes in the home design or fixtures improve the patient's quality of life? Safety?
14. *Financial status:* Are financial resources sufficient to cover needed health care services, such as medications, personal care, or social support? To pay for an aide? To retrofit the home for optimal functioning?
15. *Community resources:* Would medical day care, respite services, or friendly visitors be helpful?

Provide Multipronged Intervention: Interventions should optimize the patient's functional independence, improve social connections, and advance the patient's comfort and well-being.

1. *Medical and surgical interventions:* The core of the treatment plan is the assurance that the best medical and surgical practices are being utilized. The nurse frequently has a pivotal role in caring for the patient and coordinating and monitoring medical services.
2. *Education and disease monitoring:* Patient and family members usually need education about the illness and require training in the use of surgical supplies and equipment and medication administration. Education is frequently provided by a nurse, is usually didactic and experiential, and may be ongoing, requiring reminders and review.
3. *Rehabilitation:* To achieve restoration within the limits of a disability and to prevent further deterioration of a chronic condition, realistic short- and long-term rehabilitation goals are set, based on the patient's medical problems, physical limitations, potential for rehabilitation, and participation readiness. Slow, methodical, step-by-step movement toward the short-term goals achieves the long-range goals.
4. *Environmental adaptations:* User-friendly environments for those with disabilities include simple and low-cost interventions such as grab bars, improved lighting, and wheelchair-accessible counters.
5. *Social services and mental health services:* Patients need to understand the chronic illness and how it will affect their self-concept. Social workers can help patients accept a functional loss, grieve that loss, and learn to utilize remaining capabilities. Ascertaining a person's pre-illness personality can be helpful in understanding his or her adjustment to illness and treatment adherence.

Clinicians need to discuss financial solutions with the patient and family and look within the immediate community and beyond for needed resources. For example, pharmaceutical companies may be able to provide some medications at no charge to the patient. Social workers and nurses are often key in recognizing new or unresolved problems and bringing them to the attention of team members for discussion and resolution.

Social supports are key to preventing loneliness, depression, and premature cognitive decline.

Clinicians can help preserve present relationships by helping caregivers find respite and support services, identifying family counseling services when needed, and helping patients expand their social networks through programs and services offered through religious institutions, unions, and disease-specific organizations. A growing number of these organizations have Web sites and chat rooms.

RISA BRECKMAN
RONALD ADELMAN

See Also
Coping with Chronic Illness
Health Maintenance

Internet Resources
Family Caregiver Alliance
http://www.caregiver.org

Robert Wood Johnson Foundation, Chronic Care in America: A 21st Century Challenge
http://www.rwjf.org/library/chrcare

Robert Wood Johnson Foundation and the National Institute for Computer-Assisted Reporting
http://www.chronicnet.org/

CHRONIC OBSTRUCTIVE PULMONARY DISEASE

Chronic obstructive pulmonary disease (COPD) is the fifth leading cause of death in the United States and is one of the major medical disorders in the elderly (peak prevalence 65 to 74 years of age) (Adair, 1999; Witta, 1997). Approximately 12 million Americans have chronic bronchitis, and 2 million patients have emphysema. The disease course is usually slow and progressive; it often does not come to medical attention until the advanced stages.

There are two major types of COPD—emphysematous and chronic bronchitis—but the disease often presents with mixed features (Petty, 1998; Phillips & Hnatiuk, 1998). Emphysematous patients, known as "pink puffers," are thin, maintain a near-normal PaO_2, and look adequately oxygen-

ated. This form of COPD is closely associated with cigarette smoking and takes a long time for airflow obstruction to occur. Chronic bronchitis is defined by daily excessive mucus production or sputum for at least three consecutive months in two consecutive years and is also associated with cigarette smoking. Typical patients exhibit a productive cough, retain carbon dioxide with a lower PaO_2, have a tendency toward cyanosis, and are slightly overweight. This form of COPD is more easily treatable and has a better prognosis and survival rate than the emphysematous form.

The most typical symptom of COPD is a productive cough with or without shortness of breath, followed by the insidious onset of dyspnea on exertion. Weight loss, psychiatric disturbances such as depression and feelings of hopelessness, memory loss, easy fatigability, and morning headache are common symptoms in the advanced stage of COPD.

Eighty to 90% of COPD cases in the United States are associated with cigarette smoking, making COPD one of the major preventable diseases. Measures to prevent the deterioration of elderly patients with COPD include influenza and pneumococcal vaccination, pulmonary rehabilitation, and supplemental oxygen for patients with hypoxemia. Replacement therapy for persons at risk for emphysema because of a genetic deficiency syndrome (α_1-antitrypsin) usually occurs by 40 to 50 years of age. Preventing pneumonia in patients with COPD is one of the most important measures associated with morbidity and mortality. Yet influenza and pneumococcal vaccines are underused in the United States. Only 20% to 30% of high-risk patients receive annual influenza vaccines, and about 10% receive the pneumococcal vaccine (usually given only once). Indications for both vaccinations are high-risk patients with chronic lung or cardiovascular diseases, diabetes, renal disease, hemoglobinopathy, or immunosuppression, and persons older than 65 years (Tregonning & Langley, 1999).

The diagnostic feature of COPD is the finding of airflow limitation on spirometry. Important prognostic information can be obtained from a pulmonary function test. Predictors of poor survival in COPD include low forced expiratory volume in 1 second (FEV_1), the presence of cor pulmonale, age greater than 65 years, low body weight, and residence at high altitudes (Adair, 1999).

The only therapies with a proven impact on outcome are smoking cessation and oxygen therapy. Smoking cessation is associated with a slower rate of decline in lung function, demonstrating that it is never too late to stop smoking.

The two most commonly used brochodilators are inhaled anticholinergics and β_2-agonists for acutely improved airflow in patients with COPD. Long-term effects of these agents have not been determined. Anticholinergic agents are the first choice for regular maintenance therapy. Ipratropium is the inhaled anticholinergic bronchodilator currently approved in the United States. Compared with inhaled β_2-agonists such as albuterol, inhaled ipratropium is marginally longer acting, remarkably devoid of significant side effects, and extremely well tolerated, as systemic anticholinergic effects are rare. The most common side effects are cough and dry mouth. Ipratropium is a quaternary ammonium salt that causes bronchodilation by inhibiting vagus-mediated bronchoconstriction. It is slower in onset and should be used on a regular basis rather than as needed or as rescue therapy. Ipratropium increases FEV_1 significantly better than albuterol in patients with COPD. If the patient's response to ipratropium therapy is inadequate, an inhaled β_2-agonist can be added. Oral β-agonists are available, but the metered-dose inhaler (MDI) is the preferred method of administration for patients with COPD, as it provides more rapid relief and has fewer side effects. The recommended dosage is about two puffs per usage. Inhaled β_2-agonists are safe and are recommended on an as-needed basis rather than for regular use.

The MDI is often used ineffectively, especially by the elderly. Instructions in the proper use of this device are essential. The combination of ipratropium and inhaled β_2-agonists acts rapidly and is more effective than either agent alone in improving lung compliance. The combination does not always show greater bronchodilating effect than that obtained by the maximal dose of either agent alone, but it has fewer side effects because of less frequent dosing.

The use of theophylline in the treatment of COPD is controversial. Theophylline and its analogues have both bronchodilator and nonbronchodilator effects that appear to benefit some patients. These effects include increased contractility and increased hypoxic drive to respiration. Elderly patients are vulnerable to serious toxic effects of theophylline such as cardiac arrhythmia, seizures, insomnia, headache, anxiety, and tremors, which may occur when an acceptable blood level is exceeded (range of 5 to 20 µg/mL).

Corticosteroids clearly have a beneficial effect on airway inflammation, such as in asthma, but they are not as effective in COPD. Systemic steroids significantly improve baseline FEV_1 in 10% of patients with COPD but are associated with serious side effects, particularly in elderly COPD patients, who are prone to hyperglycemia, osteoporosis, myopathy, cataracts, hypertension, peptic ulcer disease, adrenal insufficiency, and a range of mood and behavioral changes. Inhaled corticosteroids may produce fewer side effects than systemic steroids but are generally less effective.

Oxygen therapy improves both quality of life and survival in hypoxemic patients with COPD. The measurement of PaO_2 is essential to prescribing oxygen therapy. Supplemental oxygen is required under the following conditions: if PaO_2 is 55 mm Hg or less in a sitting; if there are more than two hours during which the patient's SaO_2 is less than 90%; or if PaO_2 falls to 55 mm Hg or lower during exercise.

Empirical antibiotics are of limited benefit, even though a mixture of *Streptococcus pneumoniae*, *Haemophilus influenzae*, *Moraxella catarrhalis*, and *Chlamydia* species can be recovered from the lower respiratory tracts of patients who have COPD. The group that benefits most from empirical antibiotics includes those who are elderly, have a worsening PFT, and have purulent sputum.

The goal of pulmonary rehabilitation in the elderly is to improve their highest possible level of functional capacity and to prevent or treat reversible diseases. It is a multidisciplinary program incorporating exercise, patient education, psychosocial and nutritional interventions, smoking cessation, optimization of medication, and respiratory therapy counseling (Dow & Mest, 1997). Pulmonary rehabilitation enhances the sense of well-being by alleviating symptoms, maximizing functional capacity, increasing exercise tolerance, and improving quality of life. Such a program should be instituted as early as possible and continued as a long-term treatment of patients with COPD. Furthermore, the importance of environmental control, such as the avoidance of irritants, should not be underestimated.

CHOL SHIN

See Also
Chest Pain: Noncardiac Causes
Cough

Internet Key Words
Chronic Bronchitis
COPD
Elderly

Internet Resources
American Association for Respiratory Care
http://www.aarc.org/

University of Iowa Healthcare, Virtual Hospital, COPD with Pulmonary Hypertension and Cor Pulmonale
http://www.vh.org/Providers/TeachingFiles/ PulmonaryCoreCurric/COPD/COPD.html

CLINICAL PATHWAYS

Collaboration among professionals is an essential element of today's health care environment. Clinical pathways, or care guidelines, are tools that promote collaboration among clinicians who plan and implement care for a variety of patients. Pathways are often created by interdisciplinary teams, including physicians, nurses, social workers, therapists, and nutritionists, for patients in hospital or community settings such as home care or primary care clinics. Patients with chronic illness benefit from disease management programs that provide guidelines for the entire continuum of care for the duration of the illness (Zitter, 1997). Elderly patients benefit when pathways incorporate elder-specific care into the plan.

Pathway development involves the selection of a specific case type, such as uncomplicated exacerbation of heart failure or ischemic stroke. Defining the membership of the interdisciplinary team is essential. In the development phase, teams that are physician led and have a strong clinical focus are more successful than those that do not involve expert practitioners. The team begins the development process by reviewing literature that supports a best-practice approach to patient care and then identifies the outcomes that patients should achieve before hospital discharge. For pathways that span a continuum of care, outcomes to be achieved by the first postdischarge provider visit or at the end of home-care services are usually identified. Within integrated health systems, home health nurses and nurses in other community settings commonly participate in designing the continuum-of-care pathway.

Once the outcomes have been identified, each provider determines the details of interdisciplinary care that must occur daily to help patients achieve those outcomes. For example, it is imperative that elderly patients' mobility be maintained in the hospital setting. Pathways help cue staff that patients should be mobilized out of bed early in the postoperative course and thus ensure that elders are not delayed in their clinical progress.

Implementation of pathways is an area that merits attention and involves a commitment from several disciplines (Dykes & Wheeler, 1997). Pathway drafts should be shared with a wider audience than the development team. Commonly, members of the development team are responsible for sharing the draft with colleagues in their respective disciplines and eliciting their feedback on the details or processes of care. This sharing among colleagues promotes buy-in from caregivers and can provide additional information to the development team. Physicians, particularly, need to share the draft with other physicians caring for the patient case type.

Although all care is individualized, a general guide to overall care is useful and ensures that the best standard of care is applied to all patients. For example, the Agency for Healthcare Research and Quality heart failure guidelines are commonly used to support processes of care for patients with heart failure. Infusing a pathway with those best practices ensures that the recommended details of care are not overlooked for a population of patients. Promoting a pathway as a general guideline for best practice helps alleviate clinicians' concerns regarding "cookbook" care. Pathways can serve as communication tools for a wide variety of caregivers, including medical students and house staff. Helping physicians and clinicians understand early in the process that a pathway can alleviate redundancies and repetitive communication reinforces the value of the product to caregivers.

Pathways require periodic evaluation to ensure that the processes of care are based on best practice and to identify patients who are not meeting outcomes. This type of ongoing evaluation is often accomplished through case managers or clinical nurse specialists working with a specific patient population. Patients with chronic conditions, such as stroke, can benefit from pathway systems that include close communication among providers at all sites of care (Newlin, Gibbs, Lonowski, & Meyers, 1996).

Pathways can incorporate specific elder assessments, thus focusing all clinicians on the needs of hospitalized elders. Being alert to the possibility of confusion, unsteady gait, or incontinence can help clinicians better plan for possible problems before they become complications. Paynter, Ambrose, and Dolan (1997) incorporated geriatric evaluation and management into hospital-based pathways, with the guidance of geriatric experts on pathway teams. These tools provide specific details to maintain the functional status of frail elders yet provide a framework for disease-specific care management.

Hospitals with computerized information systems can customize specific diagnosis-related group (DRG)–based pathways with elder-specific needs. Those facilities with paper pathways can create a version for common elder patient groups, such as those with pneumonia, or tailor their pathways to elders when their data show that the majority of patients are in the elder age group.

Patient pathways are designed to guide patients and their families through the hospital recovery period. They are written in patient-friendly language and may include graphic elements or icons to depict hospital events. A patient pathway developed for stroke patients at the University of Washington Academic Medical Center in Seattle provided an opportunity to communicate care details to a multicultural patient and family population (Nemeth, Hendricks, Salaway, & Garcia, 1998).

Postdischarge continuum pathways are essential to keep elders from experiencing unnecessary hospital readmission. Although patients are expected to reach such clinical milestones as ambulating or being able to eat their usual diet by the time of discharge, elders often require additional care in their homes. Home health nurses are significant contributors to ensuring continuation of the plan, particularly with regard to assisting elders with activities of daily living or teaching self-medication skills. A continuum pathway that is focused on the specific needs of an elder heart failure patient, for example, may emphasize initial skills during hospitalization such as learning medication routines, self-managing diet, and monitoring weight daily. The home-care component of the continuum pathway may focus on continued assessment of the patient's adherence to medication and diet through detailed cardiopulmonary examination and reinforcement of weight monitoring. It would also teach and reinforce more individualized components of self-care to help the patient live with the chronic condition. Such specificity allows home-care nurses to focus their visits to help patients achieve outcomes within appropriate periods and can help ensure that care details are not overlooked (Huggins & Phillips, 1998).

Clinical pathways offer clinicians an opportunity to outline key processes that must occur for patients to move safely through the continuum of care. Pathways for elders should address all sites at which care is delivered, from hospitalization to rehabilitation to long-term care to home. This continuum approach promotes communication and collaboration among caregivers and ensures that elder-specific care is delivered with a focus on quality.

KIMBERLY S. GLASSMAN

Internet Key Words
Care Paths

Internet Resources
Case Management Society of America
http://www.ccm.org

Case Manager Resource List
http://www.cmsa.org/resources/sites.html

Center for Case Management
http://www.cfcm.com

COGNITION INSTRUMENTS

Owing to an aging population, dementia is more prevalent. The incidence of dementia increases with age, doubling every five years after age 65 until age 85, when the incidence reaches 50%. Early detection can yield substantial treatment, due to new agents that slow cognitive deterioration. Testing instruments can help clinicians screen, detect, and monitor cognitive deterioration.

Dementia is a progressive, global, cognitive impairment involving memory, language, visual-spatial ability, executive function, and praxy. Of dementia's many varieties, Alzheimer's disease (AD) is the most common. Although many dementias are irreversible, some are treatable and even partially reversible, such as B_{12} deficiency and ETOH abuse. In the early stages of dementia, a range of etiologies prompts various clinical presentations, each demonstrating a distinct set of deficits. For example, AD affects the temporal lobe first and impairs memory; vascular dementia (VaD) provokes a variety of symptoms depending on the affected regions of the brain. Parkinson's disease and other subcortical dementias have an early impact on the frontal lobe. Specific instruments are required to differentiate these conditions. Not all instruments assess functioning with equal accuracy, sensitivity, and reliability.

Ideally, primary care providers should screen persons over age 65 for cognitive deficits annually. However, any trained health care professional can administer the instruments discussed here. With limited practitioner time, instruments must be sensitive, specific, valid, reliable, easy to administer, and brief. The most commonly used instruments, indications for their use, and criteria to determine when further testing is indicated are described here.

Preparation for Testing

To interpret tests meaningfully and assess a dementia's etiology, the clinician must obtain from the patient (and, if possible, a collateral) a detailed clinical history that is sufficiently thorough to identify sensory impairments that can alter test performance. Collateral information is helpful when evaluating Activities of Daily Living (ADL) and Instrumental ADL (IADL). Absent a collateral to corroborate history, the clinician must carefully observe the person's grooming, hygiene, appropriateness of clothing to the weather, state of nourishment, capacity to understand and follow medical recommendations, medication adherence, and regular attendance to appointments, tests, and follow-ups.

Accurate evaluation of a patient's performance requires understanding the patient's cultural and ethnic background, primary language, education level, employment history, and overall premorbid functioning. Education level, for instance, can alter the cut-off points of certain instruments. Some tests display greater cultural biases than others, yielding uneven accuracy levels across populations.

Depression and pain can have an impact on performance and test results. Clinicians need to be meticulous and cautious: even with substantial impairment, a patient can maintain social skills and present surprisingly well—at least, superficially.

Screening Instruments

All cognitive impairment screening instruments evolved with research and testing among many institutionalized and community-based populations. Key attributes and limitations of the most commonly used instruments are discussed in this section.

The Functional Activities Questionnaire

The Functional Activities Questionnaire (FAQ) (Pfeffer, 1982) is a standardized assessment of a patient's level of functioning by an informant who knows the patient best. It includes ten complex tasks, such as writing checks; assembling business papers; shopping alone; playing a game; pursuing a hobby; using a stove; preparing a meal; tracking events; discussing a book, magazine, or TV show; remembering appointments, family occasions, and medications; and traveling by car or bus from the patient's neighborhood. The informant rates the patient's ability to complete each task on a Likert-type scale; 0 indicates normalcy; 3 indicates dependency. Additional responses include "never did the activity, but could now" (0 points) and "never did and would have difficulty now" (1 point). Sums of all ten activity-ratings total 0 to 30; a score of 9 or above indicates impairment. Importantly, a clinician must differentiate impairments due to physical-sensory disability from those due to cognitive deficit (i.e., a patient who cannot shop because of blindness or arthritis, as compared to confusion or impaired memory). Some informants will be able to fill out this questionnaire by themselves, while others will need some help from the clinician. In this case, no more than 2 to 3 minutes should be required.

Blessed

The Blessed Orientation-Memory-Concentration Test (Blessed) (Costa, 1996) is a six-item instrument to measure short-term recall, concentration, spatial ability, and orientation. The test requires verbal ability. Although brief and reliable, the test does not measure cognitive abilities, as does the Mini-Mental Status Exam (MMSE). Questions regarding orientation have higher weightings than do questions addressing concentration, memory, and

mathematics. Scores range from 0 to 28, with a score of 10 or above indicating dementia.

Clock Drawing Test

The Clock Drawing Test (CDT) (Freedman, 1994) is a brief, widely used, and easy-to-administer instrument to assess executive control and temporal-parietal abilities. The CDT is reliable and less influenced by language, culture, and education, given that most people use clocks. The test's main limitation stems from its many versions, each varying slightly in scoring criteria and instructions. While each version accurately assesses certain functions, different scoring methods impede comparisons across studies. Yet, the CDT's fundamental concept remains easy to understand and sound. A clinician asks a patient to draw a clock set at a specific time. Instructions differ in setting the clock at different times, in drawing free hand versus completing a predrawn circle, and in copying a clock that the examiner has already drawn. Scoring also varies by the number of details evaluated. Practitioners should use one version of CDT consistently to facilitate follow-up comparisons.

Memory-Impairment Screen

The Memory-Impairment Screen (MIS) (Buschke, 1999) is a brief, reliable, and valid test for early dementia and memory impairment. The MIS is part of a longer, more complex screen with high discriminative validity. The MIS tests only memory, proving most accurate for patients with a suspected diagnosis of AD and least accurate for persons with suspected frontal lobe deficits, subcortical dementias, and intact memory. The MIS tests free and cued recall of four words that the patient has been asked to learn. A clinician determines the patient's score, on a scale of 0 to 8, by making the following calculation: $(2 \times \text{Free Recall}) + \text{Obtained Cued Recall}$. The cutoff score of 4 has a sensitivity of 0.69 for mild dementia and 0.92 for moderate dementia. When analyses are restricted only to AD dementia, the sensitivity for mild dementia is 0.79 and 0.95 for moderate impairment.

Other Tests

Several other instruments are available to test for cognitive deficits. A full battery of tests typically requires several hours to administer, score, and interpret, and should be administered only by trained practitioners. Most clinicians caring for the elderly need to be familiar with only a few instruments and know when to refer a patient to a specialist. Some of the more complex tests are relevant primarily for their usefulness in clinical research. Among such tests is the Dementia Rating Scale (DRS) (Gardner et al., 1981) that consists of 36 tasks, comprising 5 subscales that measure attention, initiation/perseveration, construction, conceptualization, and memory. The DRS's total score cut-off is 123, with separate cut-off scores for each subcategory indicating specific deficits. For example, a patient with impaired scores in the initiation/perseveration and conceptualization subscales, but with intact memory, would suggest a frontal deficit and make a diagnosis of AD unlikely.

The Boston Naming Test (Kaplan et al., 1983) assesses naming difficulties associated with aphasia common to cortical dementias. The Trail-Making Test (Reintan & Wolfson, 1985) measures attentional capacity, with particular sensitivity to frontal lobe deficits and early detection of subcortical dementia.

No instrument for screening cognitive deficits is perfect or complete. Some are more affected by primary language, education, and cultural background; others are less standardized or measure only specific functions. Testing with the MMSE, MIS, and CDT will generate a reasonably thorough set of results. Together, these three tests require 10 to 15 minutes to complete, are easily interpreted, and facilitate a superior screen for cortical and subcortical dementia than does any single instrument. Most importantly, a clinician must determine when to refer a patient for further testing. If a patient with a high level of education and premorbid functioning complains of cognitive decline, but scores above the cut-off threshold in a screening test, the patient should be referred for further testing. Moreover, any person showing a discrepancy between functional capacity and screening performance should see a specialist.

ALESSANDRA SCALMATI
CAROLE SMYTH

See Also
Cognitive Screening Tests: The Mini-Mental Status Exam
Depression Measurement Instruments
Pain: Chronic

COGNITIVE CHANGES IN AGING

Cognitive changes are a hallmark of the aging process, and the impact of those changes can be observed in almost all aspects of an older person's life. The speed of learning to program a new VCR, the likelihood of remembering the instructions for a new medication, and the ease of comprehending a document comparing five Medicare programs diminish with aging. Even in the absence of dementing illnesses, most older adults will experience some degree of cognitive decline, primarily in memory efficiency, by the seventh decade of life.

Cognitive Aging

Many aspects of cognition begin to decline measurably by middle age in the average adult. Information-processing speed peaks in the 20s and continually declines thereafter, becoming marked in the 70s and 80s, contributing to poorer memory performance, difficulty comprehending spoken and written communications, less effective reasoning ability, and impaired learning (Salthouse, 1996). Cognitive slowing is particularly evident with complex tasks or those that require multiple steps for completion.

Aging is also associated with decreased working memory functioning—the ability to hold some information in consciousness while manipulating other information. Working memory is essential to comprehension; for example, ambiguous sentences are held in working memory until later sentences clarify them. Decline in working memory also contributes to poorer memory performance, especially when the person must learn or recall under conditions that tax processing ability. Noisy environments—visually, auditorily, or informationally noisy—place demands on working memory that can interfere with comprehension, learning, and remembering.

Although many aspects of cognition decline with advancing age, not all cognitive capabilities follow this pattern. Most individuals who remain mentally active will continue to acquire general knowledge throughout life, which supports much of the success in day-to-day cognitive activities. With cognitive slowing and deficits in working memory, greater reliance may be placed on the automatic processes supported by accumulated general knowledge. Rather than attempting to understand a complex, new soufflé recipe, an older woman may make the same chocolate cake she has baked for 45 years. Much of daily life involves tasks performed with minimal conscious processing; these automatic activities are relatively well preserved in later life. Knowledge and wisdom acquired throughout one's lifetime continue to have important positive influences on a person's daily activities and decisions. Even though most people will experience frustrating memory lapses from time to time as they grow older, the majority of people will not suffer significant cognitive problems in their daily activities.

Self-Reports

Caregivers call upon their clients and patients to provide self-reports. The older adult has to comprehend questions, accurately retrieve the answers, and truthfully report symptoms, names and schedules of medications, family histories, dietary intakes, moods, and so forth. Unfortunately, self-reports can be unreliable sources of critical information, more so with age-related cognitive decline (Schwartz et al., 1999).

Consider a 78-year-old woman who is asked, "How have you felt during the past month?" If she interprets this question as asking whether she has been sick or experienced any severe pain, she might answer "I've been feeling fine." If she interprets the question as asking whether she has had any minor aches and pains, she might answer "Not so good—I've had better days than this." If she interprets the question as asking whether she has been sad or depressed, her answer might be totally different. Even a well-structured question like "How many pieces of fruit do you eat on an average day?" may elicit different answers if the half grapefruit eaten at breakfast and the other half eaten at lunch are considered as one piece of fruit or two. Comprehension can be increased by using simply worded questions, examples, and by allowing the person to read the question as well as hear it. Speech is sometimes too fast for an older adult's limited working memory, thereby interfering with comprehension. Auditory presentations tax working memory. Older adults are more likely than younger adults to endorse the final alternative of a set of auditorily presented choices, but this difference is not observed when choices are presented visually.

At any age, it is difficult to remember the frequency of events, such as visits to the doctor; this memory task becomes more difficult with aging. Greater accuracy may be achieved by asking the person to name the doctor and the purpose of each visit. Written response alternatives, such as lists of diseases or symptoms, rather than open-ended questions can also improve recall for topics such as family medical histories.

Care-Related Information

Although considerable time and resources are spent presenting information to people of all ages regarding their care, much of this information is soon forgotten. The problem is compounded in old age, particularly if the information is presented audi- torily. Rapid rate and cadence of speech, failing to make eye contact to judge comprehension, failing to signal the most important parts of the message, using jargon, and providing too much information all have greater negative effects on memory for older adults than for younger adults. Memory can be aided by asking the person to repeat back the important points and by providing a clear and sim- ple written summary (Morrow et al., 1988). Older care recipients should be encouraged to take notes and then read them back to check accuracy. When in doubt about the care recipient's memory, a surro- gate should be included in the discussion. Under high-stress situations, even people with good mem- ory ability may have difficulty comprehending and remembering instructions.

Cognitive factors also influence medication adherence behaviors. Research suggests that the young-old (ages 60 to 75) tend to have relatively high rates of medication adherence, but that the older-old (over age 75) are more likely to forget their medications (Morrell et al., 1997). Cognitive interventions, such as providing the older adult with a pill organizer and a chart outlining the proper medication regimen, are effective in reducing omis- sion errors, probably because they reduce the mem- ory demands of complex regimens.

Communicating with Older Adults

Speakers produce between two and three words per second, requiring listeners to comprehend at the same pace. Spoken communications can strain the working memory capacity of older adults. Unlike written communications in which the reader can go back and reread a passage, the listener cannot get back words spoken even a few seconds ago. Com- munication with older adults may benefit from us- ing sentences with fewer syntactic complexities (such as embedded clauses) and with clearly pre- sented major points. Speakers can improve the qual- ity of communication by providing examples and summaries of the important points, without having to talk down to or demean the listener. Reduced

auditory acuity makes it more difficult to distin- guish spoken sounds and increases reliance on non- verbal components of the communication. Older adults make great use of their knowledge to fill in what was not well heard. They also rely on the normal intonations and pitch changes of spoken communications to comprehend what is being said. However, customary prosodic patterns can be dis- torted in unexpected ways. For example, computer- generated speech, such as that used in telephone menus, lacks the prosody of normal speech. Live speakers who have given the same information to scores of patients may begin to sound like robots; speech is fast and lacks both intonation and pitch changes, which cue important content. Conversely, the speaker must be wary of exaggerating the slower rate of speech or use of prosodic features when talking with older adults, in order to avoid patronizing "elderspeak."

In summary, older adults' slowed cognitive processing and their reduced working memory re- sources can contribute to difficulties in compre- hending and remembering important care-related information. However, if the information is pre- sented in such a way that the older adult can use his or her many years of knowledge to process it, most age-related cognitive deficits should be minimal.

CATHY L. MCEVOY

See Also

Internet Key Words
Cognitive Aging
Cognitive Slowing
Communicating with the Elderly
Memory for Medical Information

Survey Memory
Working Memory

COGNITIVE SCREENING TESTS: THE MINI-MENTAL STATUS EXAM

Cognitive Screening Tests for Dementia

Accurate identification and management of cognitive deficits is a common challenge in the primary care of older patients. Studies document the lack of dementia detection (Froehlich et al., 1998), suggesting that physicians do not routinely identify patients with moderate or even severe degrees of dementia. Failure to recognize impairment in memory and judgment may limit the effectiveness of the clinical encounter, increasing the likelihood of inaccurate and incomplete histories (Larson et al., 1984). Recommended therapies may be misunderstood or forgotten. In light of these concerns, screening is recommended for cognitive impairment in elderly primary care patients (Rubenstein et al., 1988; National Institute of Health Consensus Development Conference, 1988) as is use of a "quantitative mental status examination" (Costa et al., 1996).

Several cognitive testing options are available to the health care professional. While screening tests identify individuals at greater risk for impairment, detailed neuropsychological tests are required for definitive diagnosis when considering dementia. The focus of this article is on the MMSE screening test for dementia, and its strengths and limitations.

While dementia is often missed without quantitative screening, the results of screening should be supplemented with a careful functional history. A history of specific functional decline is further support of cognitive impairment without which the results of cognitive screening may simply reflect educational level or cultural background. Quantitative screening provides a measure to gauge severity and identify clinical changes over time.

The opportunity for cognitive assessment begins with initial patient contact. Attention to physical appearance and affect are important windows to mental functioning. Details of the history may provide important clues. Maintaining normal conversation, fluency, and attention are necessary elements of normal cognition. However, evidence of these abilities alone is not sufficient for excluding the presence of cognitive impairment. In fact, social skills are often preserved in the early period of cognitive decline. Without some objective quantitative measure, the clinician may miss even significant cognitive impairment.

The criteria for a good cognitive screening test require that it be relatively brief, easy to use, and psychometrically sound. A good screening test should be sensitive enough to miss relatively few cases and yet be accurate enough to limit the number of false positives, test characteristics referred to as sensitivity and specificity, respectively. Finally, when several equivalent options are available to the clinician, it is desirable to choose a widely used test.

Orientation questions, while having good specificity (> 90%) have poor sensitivity (50%) and are therefore inadequate as the only method for screening. A useful and quick screen for cognitive impairment consists of using recall of three items at one minute, serial sevens, and the clock-drawing test (description follows). Abnormal results should prompt further investigation.

The Mini-Mental Status Examination (MMSE)

The MMSE (Folstein et al., 1975) is the most commonly used quantitative instrument to screen for moderate or more severe impairment. It includes 30 items that cover seven domains, including orientation to time and place, registration, attention, calculation, short-term recall, language, and construction. While proven reliable, its validity is less secure unless adjusted scoring methods are used to address the potential for educational bias. The MMSE is relatively insensitive to early or mild forms of dementia.

Cognitive impairment is often unrecognized by practitioners without the use of some quantitative screening tool (Larsen, 1984). Professional organizations and consensus groups have recommended screening for cognitive impairment in elderly primary care patients and these recommendations include the use of a "quantitative mental status examination" (Costa et al., 1996). It is useful to administer the MMSE or a similar widely accepted instrument during an older person's initial visit. The psychometric properties of the MMSE are such that repeated testing can detect changes in cognition over time for those who are already impaired. It is also appropriate to readminister the MMSE after an older person has a change in health status (e.g., post hospitalization) or a significant functional status change (e.g., a fall). In the hospital, the MMSE is useful for distinguishing between states of delirium and dementia.

Content of the MMSE

Orientation: is measured with 10 items. Patients are questioned on their knowledge of the correct date, month, year, day of the week, season, state, county, city, name of the building, and floor, and receive one point for each correct response.

Registration: To test registration, patients are asked to immediately repeat 3 words. The score is based upon the number of correct words repeated (0 to 3). The number of times required to correctly repeat all 3 words can be recorded but only the first trial is counted towards the total score. Repeated trials may be necessary to ensure registration of all 3 items in order to correctly test the capacity for recall of those items. There is no standard approach with respect to the 3 words used; the assessor may choose any 3 words, which are typically objects and conceptually unrelated to one another.

Attention and Calculation: The subject is asked to count backwards by 7s starting with 100 and is given points for correct subtraction whether or not an error preceded the correct subtraction, e.g., 100, 92, 85, 78, 71 earns 4 points as does 100, 93, 86, 79, 71. The patient is asked to spell "world" backwards with correct points for each letter in its correct position; D R L O W = 3 points. If a patient is able to provide any correct serial 7s then that

score is used instead of the score for spelling "world" backwards. If no subtractions are correct then the latter is used as the score for "calculation."

Short-Term Recall: To test this cognitive domain the patient is asked to repeat the 3 words provided earlier (an approximate 5 minute delay) to test whether the subject has incorporated this new information. If items are missed, the examiner can provide clues. The ability to recall only with clues, while not part of the formal scoring, suggests problems with information retrieval, while the ability to encode new information appears to be intact.

Language Capacity: Is tested by naming, comprehension of the spoken word, repetition, comprehension of reading, and writing. The patient is asked to identify by name a pen and a wristwatch. Repeating "no ifs ands or buts" after directed to do so by the examiner tests repetition. Subjects should not be penalized for problems with articulation, such as dysarthria. A 3-step command, "Take this paper in your right hand, fold it in half, and place it on the floor," is stated only once and tests comprehension of spoken language. Asking the subject to read and follow the direction, "Close your eyes," tests comprehension of written language. Writing a complete sentence is another test of written language. Proper spelling and punctuation is not required. Sensory impairments should be noted and will lower the maximum possible score, by eliminating the possible score for those areas limited by physical impairment.

Construction: The patient is shown two intersecting pentagons and then directed to draw them. This activity tests capacity for construction and specifically measures visual-spatial and executive functions, the latter of which enables planning and carrying out a sequence of steps such as is required in dressing or cooking. Each pentagon must clearly have five sides and the intersection must occur with an angle of each pentagon.

Scoring

The total number of correct answers is summed for a possible maximal score of 30. Refusal to answer a question, however, presents a scoring dilemma.

Epidemiological research suggests scoring refusals as missed answers, given that a refusal most likely represents an inability to answer the question (Fillenbaum et al., 1988). Interpretations of particular cut-points have been influenced by a number of studies, which indicate that scores vary with age and education. Age and education-corrected norms, based on data from the National Institute of Mental Health Epidemiological Catchment Area Program surveys, conducted between 1980 and 1984, suggest specific cutoffs that can be used as criteria for initiating further neuropsychological testing (Malloy et al., 1997). For individuals with a college education, a cutoff score of 29 is suggested. For individuals with 5 to 8 or 9 to 12 years of education, suggested cutoff scores are 23 and 27, respectively.

The Clock-Drawing Test

The Clock-Drawing Test directs the patient to place numbers within a circle (drawn by the examiner) and then to draw the hands showing typically "ten minutes past eleven o'clock." Although this is a conventionally used time, any time that requires abstraction is acceptable. For example, "two forty-five" but not "ten minutes to eleven" would be acceptable. One can see with the latter example that a patient can draw hands to the ten and the eleven in a concrete fashion, i.e., without abstraction. It is a sensitive test of parietal lobe dysfunction and can provide information about executive skills (planning) and spatial orientation, and can identify focal intracranial processes. Although the clock test can supplement information obtained from the MMSE and adds prognostic information (Ferucci et al., 1996), it can be misleading if used as the sole measure of mental function.

Conclusion

Cognitive assessment can be anxiety provoking, potentially demeaning, and even insulting if not framed or introduced properly. Opening statements such as, "I'm going to ask you a series of questions that I routinely ask all of my patients. Some ques-

tions will seem easy; others will be more difficult. By the way, how far did you get in school?" can be helpful.

A determination of cognitive impairment should prompt further investigation for potential reversible causes, although these are relatively infrequent. Additional benefits of screening include the increased likelihood of detecting undiagnosed physical problems in searching for a cause of cognitive impairment (Larson et al., 1987). Vision and hearing deficits can mimic cognitive impairment, resulting in scores below cutoffs, and treatment effects can be dramatic. Medications frequently cause mental impairment. Depression often mimics dementia and is very amenable to treatment. Understanding the extent and specific nature of cognitive deficits enables the clinician to make informed decisions about needed social supports, to be sensitized to the extent of caregiver burden, and to make more effective and appropriate treatment decisions while addressing related conditions.

JOSHUA CHODOSH

See Also
Cognition Instruments
Cognitive Changes in Aging
Dementia: Overview
Depression in Dementia
Genetic Factors in Alzheimer's Disease
Mental Capacity Assessment
Vascular and Lewy Body Dementia

Internet Key Words
Cognition
Dementia
Depression
Mini-Mental State
MMSE
Screening

COMMUNICATION ISSUES FOR PRACTITIONERS

Communication that produces understanding between older adults and health care providers is central to high-quality care. All communication requires a two-way receptive and expressive process that can be impaired by age-related sensory and physical losses. Communication works best in a "shared reality." It is frequently and erroneously taken for granted that we see the same thing at the same time. For example, elders with dementia are in a different reality. Families, patients, and health care providers with differing goals are all in different realities. Communication interchange is facilitated when expectations are fulfilled. Preconceived ideas, cultural expectations, or unanticipated and confusing patient behaviors make continued conversation difficult. Comfort level with a conversation is also critical to understanding. Sensitive topics such as sexuality, urinary incontinence, depression, elder abuse, terminal care, advance directives, cognitive impairments, and death may cause different kinds and intensities of discomfort for elders, providers, and family members. Lack of confidence, diminished ability, lack of control, sociocultural differences, environment, or ageist views can inhibit open communication and leave essential components of care unaddressed (Gould & Mariano, 1999). Effective communication with older adults requires the personal communication skills of openness, honesty, respect, clarity, directness, and assertiveness (Gastel, 1994; Marshall & Houseman, 2000).

Active listening is both an attitude and a skill. Older adults may not have a good listening attitude or skills because of physical difficulties or anxiety. Professionals as well as elders may be poor listeners because of cultural differences or long-standing patterns of behavior. Good listening requires concern, centering, connecting, concentrating, capturing, and clarifying (Marshall & Houseman, 2000). Listening is an acquired skill; an individual has to be predisposed to hear what another has to say. Health care professionals need to value what elderly patients have to say, even when it at first appears confusing or irrelevant. It is the responsibility of the professional—not the elder—to clarify what is unclear. Using closed rather than open-ended questions can focus and help direct both listening and responding.

Speakers produce between two and three words per second, requiring listeners to comprehend at the same pace, which can strain the working memory capacity of older adults. Enhancing conversation with nonverbal components and using simple sentences can help older adults use their knowledge of conversation standards and information from intonation and pitch to fill gaps that result from distorted comprehension. Rapid and monotone speech that lacks intonation and pitch, such as that generated by computers in telephone menus or by health care professionals who repeat the same information regularly, lacks the changes that cue important content. However, health care professionals must be wary of exaggerated or overly slow speech in order to avoid patronizing elders (McEvoy, 2000).

Physical, cognitive, psychological, and sociocultural barriers to communication can be generated by the older adult, the health care professional, and the environmental and organizational setting in which the patient and provider find themselves.

Physical barriers consist of functional deficits in hearing, vision, speech, and movement. Often dismissed as a normal part of aging, they can significantly affect receptive and expressive functioning. These deficits do not work like an on-off electrical switch. Physical capacity spans a continuum; changes over time and multiple conditions that exacerbate isolation of the elderly can increase problems. In addition, pain can inhibit communication. Pain can be so debilitating that it becomes the focus of attention for an elder and greatly inhibits the ability to listen. The importance of identifying physical health barriers cannot be overemphasized. All too often, communication difficulties due to sensory or physical impairments are mistaken for "confusion" or dismissed as a normal part of aging.

Techniques for communicating with hearing impaired older adults begin with getting the person's attention. Facing the older adult and keeping light on the speaker's face rather than backlighting are important. After asking whether the person has a hearing aid, it may be necessary to check that it is in position and working properly. Speech should be slow and clear but not overemphasized. Low-pitched tones are heard more easily than higher tones. Background noise needs to be eliminated. A misunderstood sentence should be rephrased rather than repeated. Additional nonverbal approaches to communication such as gestures, diagrams, and written materials help comprehension. Having the elderly person repeat essential facts ensures understanding. The elder should be cued in advance when the topic of conversation is about to change (League for the Hard of Hearing, 1996). With the patient's permission, a sign can be posted above the bed, on the door, on the gurney, or outside the medical record to alert health care personnel of the hearing impairment. Doing this, however, may be subject to state regulations about patient privacy and confidentiality.

Techniques for communicating with the visually impaired elderly always begin with announcing or telling the elder that a person is entering or leaving the room. Activities should be narrated and precise directions given, such as "right" and "left" rather than general terms like "over there." Vision aids are helpful and ensure that there is adequate light. The elder may be wearing eyeglasses, but they should be checked for cleanliness. Unusual posture may be due to vision impairments such as a lack of peripheral vision rather than avoidance behavior. For print material, 14-point black type on a white background is helpful. Respectful touch that is sensitive to cultural preferences can enhance communication (The Light House, Inc., 1999). As with a hearing impaired patient, a sign can be posted with the patient's consent, as long as state regulations are followed.

Techniques for communicating with the speech impaired elderly depend on recognition of the speech impairment to assuage embarrassment and frustration. Alternative mechanisms for communication (e.g., written materials, signs, cards), gestures, and body language can augment communication.

Barriers caused by cognitive impairment can make communication a frustrating and exhausting experience. The most common behavioral responses from cognitively impaired elders are confusion, embarrassment, and aggression. Because verbal ability is often reduced, nonverbal forms of communication are increased. In addition, understanding the impact of external influences such as alcohol and drug dependency, pharmacological reactions from multidrug use, and metabolic and nutritional deficiencies on cognitive ability helps health care professionals create realistic expectations and strategies for both short- and long-term communication with cognitively impaired older adults.

Effective communication with cognitively impaired elderly requires realistic expectations. Trying to force reality orientation on the severely demented in nursing homes or expecting a cognitively impaired elder to remember a complicated drug regimen is unrealistic. One direction at a time elicits more comprehension and a successful execution. Alerting the elder to a change in subject helps him or her focus. Written instructions with pictures can help any patient, and certainly cognitively impaired older adults, achieve greater understanding. Many "disturbing" behaviors and "incomprehensible" statements are logical for the cognitively impaired person, but unfortunately, they are not logical to us (Feil, 1993). Confidence and trust gained through verbal and physical reassurance and validation can create the bonding needed for good communication; otherwise, incomprehension can result in aggression or withdrawal. Disruptive verbal or nonverbal behavior does not necessarily mean what it appears on the surface; rather, it might represent fear, expression of loss, unmet physical or emotional needs, or an attempt to maintain past behavior patterns.

Psychological barriers that impede effective communication derive not only from elderly people but also from health care professionals. Issues of power and control, and histories of personal loss and painful relationships and events—our personal baggage—are significant psychological barriers. Additional psychological conditions such as depression, which is prevalent in older adults and not a normal part of aging, or fixations and obsessive behaviors can foster withdrawal or inhibit productive communication (Rader & Tornquist, 1995).

Techniques for circumventing psychological barriers depend on awareness of issues, personal baggage, and psychiatric conditions. It is important to recognize and respect what an elderly person brings to the table. However, a clinician's expression of sympathy should not legitimize withdrawal or aggressive behavior.

Sociocultural barriers to communication between an older adult and health care provider can be the most difficult to circumvent, in part because they are the most elusive. Culture (beliefs, values, customs), socioeconomic status, education, and assimilation all influence the effectiveness of communication with the elderly. However, language is the medium through which we communicate many of our cultural differences. In some cultures, language creates reality as well as reflects it. Some people will not say "cancer" because they are convinced that uttering the word will bring on the disease. Working with language interpreters is never easy. Family members or caregivers who double as interpreters are especially difficult, because their own needs and frustrations often interfere with their ability to translate the words of the elders and health care professionals. Family interpreters should be directed to "translate," not "interpret," and need reassurance that they will be asked for their views after the patient interview.

Techniques for communicating with socioculturally diverse elderly can be greatly enhanced by using key words in the language of the older patient to foster trust and comfort. Understanding the cultural beliefs and value systems of elderly patients and their families—particularly regarding respect, nutrition, pain, and death—will help in understanding their perspective and expectations. Appreciating key customs and rituals will prevent awkward social or communication errors (Goodman, 1989). All this requires a suspension of stereotypes and prejudice.

Common ground is essential for successful communication with the elderly, and health care professionals must make the effort to find it. It is imperative to understand the elder's point of view, as all people want to be seen as individuals. It is an injustice to communicate in the same manner with a healthy 68-year-old and an 89-year-old with multiple chronic conditions. Each of us looks at the world through a unique prism. For the elderly, as for all of us, physical limitations, cognitive and psychological status, and sociocultural backgrounds influence point of view—the way we perceive, understand, and communicate with the world. Age tempers this view. Health care workers view the world through a professional prism that sometimes filters out significant parts of a person's life. Good communication means looking at the world through the patient's prism, which requires empathy and respect.

ELAINE S. GOULD
BETH LATIMER

See Also
Cognitive Changes in Aging
Cultural Assessment
Hearing Aids
Hearing Impairment
Low Vision
Vision Changes and Care

Internet Key Words
Active Listening
Communication
Sensory Loss
Validation

Internet Resources
Alzheimer's Association
http://www.alz.org

League for the Hard of Hearing
http://www.lhh.org

Lighthouse, NYC
http://www.lighthouse.org/index_main.htm

COMMUNITY ASSESSMENT

Comprehensive Geriatric Assessment (CGA) is "a multi-disciplinary evaluation in which the multiple problems of older persons are uncovered, described, and explained, if possible; and in which the resources and strengths of the person are catalogued, need for services assessed, and a coordinated care plan developed to focus interventions of the person's problems" (NIH, 1988). A cornerstone of successful community assessment, CGA in the home setting provides a unique picture of the older person.

Assessment

A brief summary of what is typically covered in the comprehensive community assessment of an older individual includes:

- Physical health
  Medical history, review of systems, self-reported chronic conditions, self-perceived health status
  Physical examination (vision, hearing, nutritional status, oral health, skin, gait and balance performance)
  Lab studies
  Examination of home pharmacy, medication, and other daily supplement use
  Health maintenance compliance
- Psychological (mental) health
  Depression screening
  Cognitive screening
  Self-perceived quality of life
- Social health
  Social support network (size, quality of relationship)
  Socioeconomic status
- Environmental health
  Safety
  Accessibility to services
- Assessment of function
  Basic activities of daily living (BADLs)
  Instrumental activities of daily living (IADLs)
  Advanced activities of daily living (AADLs)

Results from in-home CGA studies reveal a high prevalence of suboptimally treated health problems (Stuck et al., 1995). The most common include hearing deficit (65% of clients), musculoskeletal problems (63%), arthritis (61%), hyperten-

sion (58%), cataracts (50%), and unsafe environments (46%). Less frequent, but with major potential health consequences, are other vision deficits (35%), urinary incontinence (32%), osteoporosis (31%), depression (26%), anemia (23%), arrhythmia (23%), postural hypotension (23%), and gait and balance disorder (23%).

For some aspects of assessment, the home is not an ideal setting. Too much or too little light and ambient noise interfere with parts of the examination, low beds are uncomfortable for the nurse assessor, and clutter or small living spaces make it difficult to assess gait and balance. Some clients feel awkward disrobing other than in an examination room.

Recommendations

Community assessment becomes an intervention only when linked with a care plan. The plan can take different forms depending on the client or patient population. For frail elderly, community CGA is the basis of medical management and recommendations for placement, or services needed to substitute for nursing home placement. For well elderly, CGA is a risk appraisal method with recommendations made to the older person for what he or she can do to maintain or improve health and prevent functional decline. CGA can improve post-hospital outcomes, manage chronic disease, supplement regular primary care for the elderly, and prevent functional decline (Rubenstein et al., 1997).

As the front end of a system of care, community assessment is supported by a comprehensive set of community resources, educational materials, and negotiation strategies. Each health problem uncovered may have several alternative management responses. For example, an elder's gait and balance disorder may need to be presented to her as a serious health risk (of falling). Recommendations could be to discontinue use of high heeled sandals and remove throw rugs in her home (self care), see the podiatrist for foot care (community resource), and see her primary care physician regarding the increasing pain in her hips (physician referral).

Adherence

The next step in the care system requires turning recommendations into action. In clinic and hospital based CGA, the best health outcomes resulted from situations where the clinician doing the CGA also carried out the plan (Stuck et al., 1993) in contrast to typical community assessment where recommendations are made to be implemented by the older person, family, or primary care provider.

One advantage of community assessment, reported by CGA program nurses, was that the relationship with the older person was on more equal terms (Stuck et al., 1995). The nurse remains the expert, but the older person becomes more partner than patient. In addition, common self-care recommendations—taught, modeled, and reinforced in the home environment—have a greater likelihood of being followed. In one study, nurse practitioners made 5,694 specific recommendations to 202 clients over three years. Fifty-one percent involved a self-care activity; 20%, referral to a non-physician professional or community service; and 29%, referral to a physician (typically the client's own primary care provider). Referrals to physicians had the highest level of full or partial adherence (70%), self-care recommendations were followed about 60% of the time, and community referrals about 50% (Alessi et al., 1997). Adherence to recommendations made to the clients appeared to follow a pattern related to the degree of habit change involved, perceived seriousness of the problem, and familiarity with recommended behavior.

Clinician Competencies

The clinician doing community assessment should have the knowledge and skill to complete all the components of the CGA, including familiarity with a variety of self-care and community resources,

knowing how and when to seek consultation from colleagues, when to refer to physician care (and when not to), and how to use principles of adult learning to promote behavior change for good health behaviors and outcomes.

An interdisciplinary clinical team is extremely important to successful community assessment. The team may include a nurse practitioner or community nurse, with backup from a geriatrician, and consultation as needed from a social worker. Depending on the available resources and the skills of the nurse, consultation from a physical therapist, nutritionist, and pharmacist can be helpful. The intervention is more cost effective if only one person is in the field (i.e., the nurse); obtaining adherence is more focused if only one person interacts with the client.

Time Frame

Behavior change and improvement in health status (or prevention of functional decline) takes place over months and years. Repeated assessments yield new problems and recommendations. Reinforcement of positive change is a continual process. In a randomized controlled trial (Stuck et al., 1995), after three years of annual in-home CGA and quarterly home visits by a nurse practitioner, intervention group subjects (n = 215) had significantly fewer permanent nursing home placements, and fewer nursing home days, than the control group (n = 199). In addition, the participants were more independent in daily chores and activities. They also had increased physician visits (which, while increasing the costs of the program may have also led to concomitant reduction in nursing home days) and used more community services that promoted socialization, such as senior transportation and special community college programs. There were no significant differences between the intervention and control groups in the use of in-home supportive or personal care services.

Community assessment programs are slow in gaining acceptance. As for any intervention that is primarily diagnostic and preventive, payment sources are elusive and enrollment is a multifaceted issue. Nevertheless, growing evidence of cost-effectiveness of CGA makes it a promising model for further development and wider implementation.

HARRIET UDIN ARONOW

See Also
Geriatric Interdisciplinary Health Care Teams
Health Maintenance
Home Health Care
Primary Care Principles

Internet Key Words
Comprehensive Geriatric Assessment
In-home Assessment

Internet Resources
1997 World Congress of Gerontology
http://www.cas.finders.edu.au/iag/proceedings/proc0027.htm

Merck Manual of Geriatrics
http://www.merck.com/pubs/mm_geriatrics/

Center for Healthy Aging
http://www.centerforhealthyaging.org

CONFUSION/DELIRIUM: RISK, DIAGNOSIS, ASSESSMENT, AND INTERVENTIONS

Delirium is an acute reversible disorder characterized by disturbance of consciousness with reduced ability to focus, sustain, or shift attention; change in cognition; or a perceptual disturbance that occurs over a short period and tends to fluctuate over the course of the day (American Psychiatric Association, 1994). This disorder is frequently misdiagnosed as dementia or mental illness (commonly, schizophrenia), independently contributes to poor outcomes of acute care (Inouye, Rushing, Foreman, Palmer, & Pompei, 1998), and is frequently unrecognized. Moreover, agitated behaviors associated with delirium can lead to the use of physical and chemical restraints that heighten the risk of functional loss and serious complications.

Prevalence in hospitalized patients ranges from 10% to 80%; higher rates are associated with greater physiological instability and severity of illness, older age, general surgery, and intensive care unit stays. Patients may be discharged while still exhibiting symptoms of delirium. Persistent memory deficits are not uncommon months after the initial episode of delirium. Many individuals suffer recurrences (Levkoff et al., 1994) and are more likely to become demented (Rockwood et al., 1994).

Each year, delirium complicates hospital stays for more than 2.3 million older people, involves more than 17.5 million inpatient days, and accounts for more than $4 billion (1994 dollars) of Medicare expenditures (Inouye et al., 1999). Substantial additional costs after hospital discharge are associated with the need for institutionalization, rehabilitation, and home care. Thus, prevention, prompt and early diagnosis, and effective management of delirium are critical.

Risk: A multifactorial model of delirium that includes the complex interrelationships of predisposing factors (baseline host vulnerability) and precipitating factors (acute insults) is clinically useful in identifying and quantifying risk for delirium and in providing direction for minimizing such risk (Inouye & Charpentier, 1996). Predisposing factors include existing cognitive impairment, severe physical illness, multisensory impairment, stroke, Parkinson's disease, and dehydration. Precipitating factors include elements of and treatments for the acute illness, for example, malnutrition, polypharmacy, use of physical restraints and other immobilizing devices, and iatrogenic events.

Diagnosis: Underdiagnosis by health care professionals is attributed to semantic ambiguity, variation in the behavioral manifestations of delirium, similarity to and coexistence with dementia, failure of nurses and physicians to routinely use standardized methods of detection and screening, and a sense of being unable to influence the course and outcomes of delirium once it is diagnosed. However, evidence suggests that modest educational interventions with health professionals can increase the detection of delirium and thereby improve the outcomes for patients (Rockwood et al., 1994). Standards for surveillance of delirium include screening on admission to the hospital and screening every eight hours as an element of the standard nursing assessment. Additionally, when there is evidence of new inattention, unusual or inappropriate behavior or speech, or noticeable changes in the way the patient utilizes information and makes decisions, and the patient reports feeling "mixed up" or muddled, assessment should be repeated. Information from family and friends is important in the diagnosis of delirium; usual behavior can be differentiated from that which is not normal for the patient.

Assessment Instruments: Instruments to screen for or diagnose delirium include Folstein's Mini–Mental State Examination (Folstein, Folstein, & McHugh, 1975), Inouye's Confusion Assessment Method (the most frequently used instrument in research and clinical practice), Vermeersch's Clinical Assessment of Confusion— Form A, Albert's Delirium Symptom Interview, Trzepacz's Delirium Rating Scale, Neelon and Champagne's NEECHAM Confusion Scale, O'Keefe's Delirium Assessment Scale, and Breitbart's Memorial Delirium Assessment Scale. Each instrument has advantages and disadvantages; the instrument used should depend in part on the purpose and the patient population. (See Robertsson [1998] for a detailed review of instruments.)

Etiology: Imbalance in the cholinergic and dopaminergic neurotransmitter systems is most commonly implicated in the neuropathogenesis of delirium (Trzepacz, 1998). The multifactorial origin of the imbalance includes pharmacological agents (both intoxication and withdrawal), especially those used for sedative-hypnotic purposes; dehydration, with and without electrolyte disturbances; hypoxia; infection, especially upper respiratory and urinary tract; metabolic disturbances; and nutritional deficiencies (Foreman et al., 1999b).

Interventions: Of the few studies on the effectiveness of systematic interventions to prevent or treat delirium, most have found only modest benefits. Primary prevention is accomplished through risk reduction, identification of patients at risk, and targeted strategies to minimize or eliminate the occurrence of precipitating factors. Treatment should correct or eliminate the underlying causes while providing symptomatic and supportive care; effective solutions require multifactorial interventions.

The Elder Life Program (Inouye et al., 1999) is a multicomponent intervention consisting of six standardized protocols for managing specific risk factors for delirium: cognitive impairment, sleep deprivation, immobility, visual impairment, hearing impairment, and dehydration. Data suggest that this strategy results in significant reductions in the number and duration of episodes of delirium in hospitalized older patients. However, the intervention has no significant effect on the severity of delirium or on recurrence rates. Once delirium has occurred, the intervention is less effective and efficient. Thus, primary prevention is the most effective treatment strategy.

Pharmacotherapy for symptomatic and supportive therapy should include somatic interventions, such as haloperidol, thioridazine, or atypical neuroleptics; pain management that avoids the use of meperidine hydrochloride; and nonpharmacological interventions to promote sleep. Medication reviews (including over-the-counter and home or folk remedies) should be used to minimize the use of problematic medications—those with anticholinergic side effects, sedative-hypnotics, H_2 blockers— and to detect drug-drug and drug-nutrient interactions.

A therapeutic and protective environment is created by promoting meaning and orientation; balancing sensory stimulation with quiet, low-stimulus periods; incorporating normal diurnal variations in lighting and activity; promoting continuity of caregivers; providing appropriate individual cognitive stimulation; and including family and friends in caregiving.

MARQUIS D. FOREMAN
SHARON K. INOUYE

See Also
Cognition Instruments
Dehydration

Dementia: Overview
Mental Capacity Assessment
Over-the-Counter Drugs and Self-Medication
Parkinsonism
Polypharmacy Management
Stroke/Cerebral Vascular Accident

Internet Key Words
Acute Care
Confusion
Delirium
Dementia
Hospitalized Elderly

CONSERVATORSHIP
See
Guardianship and Conservatorship

CONSUMER-DIRECTED CARE

Consumer direction is a philosophy and orientation whereby informed consumers make key choices about the home and community-based services they receive. It is most frequently found in programs offering personal assistance services (PAS) under Medicaid or state funding. Consumer direction may exist in varying degrees and may span many services, but it is not yet common in services for the elderly.

Consumer-directed care is contrasted with agency-directed home health care and managed community-based long-term care. Typical features of these programs are care delivered through a provider agency (vendor) by caregivers who are supervised by medical professionals, case management to coordinate services, service decisions based on the judgments of the agency's case manager or professional staff, and public regulation of providers to ensure quality.

In a full-fledged consumer-directed program, the recipient is a consumer—not a client or patient—and can:

- Decide what is needed to assist with personal care (including devices, home modification, or one or more assistants to help)
- Select, hire, and train the helpers (including family members, if desired)
- Negotiate the work schedule and tasks with the helpers
- Pay the helpers with cash or vouchers from the program
- Fire the helpers if necessary (National Institute on Consumer-Directed Long-Term Services, 1996)

The consumer has the option of using a case manager or service coordinator and agency-directed care for some or all of the assistance, the right to designate a surrogate to make choices if the consumer cannot do so, and the ability to change the approach over time (e.g., start with agency-directed services and move toward self-direction as experience is gained).

In the United States, advocates for people with disabilities were the first to call for consumer direction of services. For the elderly, adoption has been slower. The service systems for the aged are generally more paternalistic and emphasize protection and the need for professional and medical oversight. There has often been an assumption of client or patient dependence, even by many elderly consumers themselves.

Increased attention to issues of autonomy, choice, and control; policy and legislative gains by people with disabilities; and the success of consumer-directed program options in other nations have heightened the interest in consumer direction for the elderly (Tilly, 1999). In several U.S. studies, this model of service delivery increased consumers' autonomy, improved their satisfaction with services and quality of life, improved their self-management of health, and decreased the cost of service provision. These outcomes assume major importance in the face of the growing need for assistance by an expanding aging population—a population with experience "having it their way" (Doty, Benjamin, Matthias, & Franke, 1999).

Commonly voiced fears about consumer direction for the elderly are that older consumers will be abused, exploited, or neglected without agency-provided supervision; their health will deteriorate if their assistants are not trained home health aides; they will be unable to train their assistants or fire them, if need be; the hiring of family members will substitute paid help for unpaid care that was already being provided; consumers will spend the money inappropriately; consumer direction will generate increased liability for the state; and the program will generate negative publicity for public officials. In large-scale studies, none of these negative outcomes is more frequent under consumer-directed care than with agency-directed models (Doty et al., 1999).

Surveys and demonstration programs are beginning to identify which elderly might choose consumer direction. Among elderly home-care recipients who participated in telephone surveys, 30% to 50% were interested in a consumer-directed cash option versus traditional agency-based services. Interest was strongest among those with longer experience receiving home care, those with prior experience hiring help, those dissatisfied with current service, those willing to take on more responsibility for managing their own care, those with informal supports, and ethnic minorities. The degree of desired control varies, from those willing to manage everything to those who are happy with agency-directed care but would like some control over the time of day services are received and the types of tasks performed. Most respondents would prefer to use a fiscal intermediary to do bookkeeping tasks and would want training in employer skills (Simon-Rusinowitz et al., 1997).

Elements of Consumer-Directed Practice

Philosophy, setting, and policies largely determine a program's degree of consumer directedness. The goals of consumer-directed services (as seen in the disability community) are broader than staying out of a nursing home or merely performing activities of daily living on a regular basis. Rather, the goal is to empower the person with the disability to live a full life in the community—to work (if desired),

carry on family life, attend functions and use facilities in the community, get further education, or pursue interests (Simon-Rusinowitz & Hofland, 1993).

In consumer-directed PAS programs, the help provided is called personal assistance or attendant care and is broader than personal care. PAS may include help with some paramedical activities, such as taking medications and managing bowel and bladder programs, or help with activities outside the home, including transportation and communication. Generally, personal assistance involves persons or devices that help individuals with disabilities perform the everyday tasks they would perform by themselves if they were not disabled. Ideally, family and friends can be paid helpers (Simon-Rusinowitz & Hofland, 1993).

Not all consumers desire consumer direction. They may differ in the choices they want to make and how much control they wish to exercise. Consumer-driven care must therefore be characterized by flexibility and respect for the expressed needs and desires of consumers. It must begin with full consumer participation in any assessment; the consumer's viewpoint is the basis for goal setting and service planning.

Consumer-directed care should include options, and extensive sharing of information about the options is essential. Consumers need the same information a case manager would use to make decisions, including the risks involved, the budget available, limits on services, differences between approaches, responsibilities of other parties, quality of providers, and avenues for complaints and appeals. A professional's advice should be available as well. Information about options should be provided in various formats, not just verbally from a case manager.

In consumer-directed service models, supportive services are needed to help everyone who wants to self-direct develop the necessary skills. Such services include information and referral; training in finding, interviewing, selecting, hiring, supervising, evaluating, and firing helpers; assistance in developing a backup plan for when the attendant is unexpectedly absent; a bookkeeping or fiscal intermediary service to do the financial tasks, or training in these tasks (issuing paychecks, withholding and paying relevant taxes, and meeting the requirements of labor laws and regulations); and peer support. Some programs provide a registry of screened and available helpers, assistance in checking the backgrounds of job applicants, help in obtaining assistive equipment and devices, and training in self-advocacy.

Program Supports for Consumer-Directed Practice

A program wishing to support or enhance consumer-directed practice should have a well-articulated philosophy of consumer control; consumer participation in planning and policy making and on boards and task forces; and a Consumer Bill of Rights describing all parties' rights and responsibilities. Documentation of consumer preferences is an essential element of the assessment and service plan. Quality assurance practices should include consumer input and measurement of responsiveness to consumer preferences. Case managers' training should help identify attitudes and biases that may be barriers to consumer direction. The program should have an ethics committee with consumer representation to help case managers examine the problems of supporting consumer choice, and there should be oversight to ensure that consumer choice is not being used as an excuse to neglect or abandon consumers who make unwise or unpopular decisions.

The Future of Consumer-Directed Care

Service systems for the elderly appear to be undergoing a paradigm shift. The appeal of consumer-directed services, including its simplicity, cost-effectiveness, and support for autonomy, seems likely to ensure its expansion. The Cash and Counseling Demonstration and Evaluation, a multiyear study initiated in 1998 in Arkansas, Florida, and New Jersey, will provide information about program features and implementation and help shape future policy. Cosponsored by the Robert Wood Johnson Foundation and the U.S. Department of Health and Human Services, Office of the Assistant Secretary for Planning and Evaluation, the study is comparing the regular Medicaid personal assistance program to a model that offers a cash allowance and information services to consumers, who then purchase the assistance they need as they see fit. Other consumer-directed initiatives in the United States and abroad are being studied for their applicability to aging services (Simon-Rusinowitz, 1999). Results of these studies can guide the organization of care so that it supports each consumer in assuming the amount of control he or she desires over the assistance needed.

BARBARA W. SCHNEIDER
KEVIN J. MAHONEY
LORI SIMON-RUSINOWITZ

Internet Key Words
Autonomy
Consumer Direction
Personal Assistance Services

Internet Resources
Cash and Counseling Demonstration
*http://www.inform.umd.edu/HLHP/AGING/
    CCDemo/index.htm*

Consumer Choice News
*http://www.ncoa.org/consumerdirect/pubs/
    index.htm*

National Council on Aging initiative on consumer-
    directed care
*http://www.ncoa.org/consumerdirect/
    consumer_direct.htm*

CONTINUING CARE RETIREMENT COMMUNITY

The recent proliferation of continuing care retirement communities (CCRCs) makes definitional precision virtually impossible at the present time. However, the following description provides a useful working definition. It is adapted from the American Association of Homes and Services for the Aging, principal CCRC trade association. (In 1994 the American Association of Homes for the Aging [AAHA] changed its name to American Association of Homes and Services for the Aging [AAHSA]. Earlier publications bear the name AAHA, but to avoid confusion, they will also be referenced as AAHSA.)

A CCRC is an organization that provides a full range of living arrangements, residential services, and health care to residents usually 65 and over, as their needs change over time. This continuum includes housing, where most residents live independently and receive services such as meals, transport, maintenance, and housekeeping; additional support services for those who require assistance with activities of daily living; and health care—both ambulatory and institutional—for the temporarily ill or those who require long-term skilled nursing care.

The majority of CCRCs combine all these benefits and services under one master resident contract, usually paid for by a lump-sum entry fee plus monthly maintenance fees and usually self-insured. Depending on the contract, entrance fees may be nonrefundable, partially refundable, or fully refundable (AAHSA & Ernst & Young, 1989).

Overview of Industry

The "universe" of potential CCRCs—entry-fee, rental, and equity-payment plans—may be as large as 1,000 (AAHSA & Ernst & Young, 1993; for estimates prior to 1984, see Winklevoss & Powell, 1984). Entry-fee plans approximating the above description are estimated at 700–800. (Unless otherwise indicated, the statistical data in this article are based on AAHSA's 1991 survey of entry-fee CCRCs [AAHSA & Ernst & Young, 1993]). With an average of 340 residents each, this would mean approximately 250,000 residents in total, less than 1% of the over-65 population.

The average age of existing CCRCs, located in 36 states, is about 23 years. Pennsylvania, California, Florida, and Virginia are in the lead. The Philadelphia/Delaware Valley area has been called the "CCRC Capital of the World," with some 45 communities, partly because of Quaker and other religious leadership and partly because of high public acceptance of the concept.

Rooted in 19th-century charitable homes for the aging, the vast majority (AAHSA estimates 98%) are not-for-profit. However, several large for-profit organizations, including the Marriott hotel chain, have entered the field in recent years.

CCRCs differ widely with respect to philosophy, costs, fees, services, methods of financing capital costs, and so on. (For brief profiles of over 500 individual CCRCs, see Walters [1994]). A major distinction relates to the extent of long-term-care (LTC) insurance coverage for residents needing assisted living or skilled nursing care. About 43% of CCRCs cover most LTC without additional charge beyond the entry and monthly fees (extensive agreement). Twenty-nine percent cover some but not all LTC (modified agreement). Twenty-eight percent require residents to pay for LTC as used (fee for service). Although the extensive contracts are more expensive, they appear to be increasing in number relative to fee-for-service plans.

This distinction is important not only to individual residents but to public policy. Under extensive and, to a lesser extent, modified contracts, the potential cost of institutional LTC is shifted from the individual and the public sector to the CCRC. Underwriting the residents' LTC risk most often protects them against the potential of "spending down" to Medicaid eligibility.

The "Seamless Continuum of Health Care"

For most CCRC residents (average age 81) the assurance of good health care is the major factor in the decision to move to a CCRC. In a 1990 survey of over 900 individuals on one community's waiting list, 91% indicated this as the most important reason for moving (Pennswood Village, 1990).

The term "seamless continuum of care" has received considerable publicity from the American

Hospital Association (Jack & Paone, 1994) and other professional organizations. This has long been the central feature of CCRC health care philosophy. Moreover, most communities have in place the physical and organizational structure to permit the philosophy to be realized. Almost all provide emergency care, assisted living/personal care, skilled nursing, physical therapy, and social services. Most also provide ambulatory care, and a growing minority add home care for minor conditions, prescription drugs, and hospice-type terminal care. All these elements are usually tied together by a form of managed care facilitated by one master contract. Combined with Medicare parts A and B and, as required by most CCRCs, some form of Medigap insurance, the resident is assured nearly complete coverage for all essential health care needs and costs. Even in fee-for-service communities, skilled nursing is usually available on the premises under supervision of the same management.

Thus, in contrast to most other retirement arrangements, the CCRC resident can expect to move from independent living to assisted living to whatever form of health care is needed in the final illness without ever having to leave the community, except for temporary hospitalization for acute illness or severe mental illness. Even hospitalization is usually minimized. At Pennswood Village, for example, of 139 deaths in 1989–1993, only 8% took place in a hospital (Pennswood Village, N. Spears, Executive Director, unpublished data, Feb. 23, 1994). This was by choice. Ninety-nine percent of the residents of this community have signed living wills requesting no heroic measures.

Costs and Affordability

Fees vary widely—by size of apartment, type of resident contract, age of community, geographic location, and so on. The lowest entry fee reported for 1990 was $1,047 (studio apartment, limits on health care covered); the highest, $625,000 (two bedrooms, full health care coverage). For a one-bedroom unit, in 1990 the average low *entry* fee with limited health care coverage was $34,352; the

average high fee with full health care coverage, $95,152. Average annual increases in entry fees ranged from a low of 3.57% in 1992 to a high of 4.84% in 1990.

Average 1990 *monthly* fees for a one-bedroom unit ranged from a low of $694 ($8,328 a year) with limited health care to $1,299 ($15,588 a year) with full health care. Average annual increases in monthly fees ranged from 5.01% (1988) to 6.27% (1991). Because the largest component of costs in most CCRCs is health care, the latter figures may be compared with the consumer price index (CPI) medical component, which, over the same 5-year period, varied from a high of 9.6% (1990) to a low of 6.6% (1992).

As to affordability, comparing the 1990 fees with average 1989 income and net-worth figures for Americans 65 and over (Taeuber, 1992), it appears that at least 25% could have afforded a CCRC at that time, opposed to less than 1% actually in residence. Obviously, there are also nonfinancial barriers. Still, it is clear that the majority of elders could not afford a CCRC under present conditions.

This is an issue for general public policy. Policy makers need to consider whether the CCRC is a desirable model for helping to solve the LTC dilemma in this country, just as it was decided a quarter-century ago that the health maintenance organization (HMO) was a desirable model for general acute care. If the answer is positive, ways could and should be found to make the CCRC more widely available. There are numerous options, such as encouraging or even subsidizing lower-income Americans by partial prepayment of fees through some form of LTC insurance.

Financial Risks

Responsibility for the broad array of CCRC benefits, especially the lifetime guarantee of skilled nursing care if needed, is clearly a weighty one. Moreover, there is evidence that CCRC residents live significantly longer than other Americans. At Pennswood Village, for example, the average age at death in 1993 was 88.3 years (Pennswood Village, N. Spears, Executive Director, unpublished data, Feb. 23, 1994). In the past, some communities with more goodwill than actuarial and managerial skill were unable to meet these responsibilities and failed.

In recent years, however, the bankruptcy rate has declined to almost zero—less than 0.5% a year according to one study (Conover & Sloan, 1994). This rate may be compared to that of HMOs, reported to vary from 0.3% to 0.1%, and is comparable to that for health insurance in general.

This improvement may be attributed to a combination of factors: increasing sophistication on the part of the investment community, better educated and more sophisticated residents (Kytle, 1994), improved actuarial techniques, improved management, the work of the Continuing Care Accreditation Commission, and broader state regulation. Thirty-seven states now regulate the financial aspects of CCRCs; all regulate their health care facilities.

Regulation, however, can be a two-edged sword. The strict financial and other requirements that have contributed to financial stability and consumer protection have also limited investment. The high capital costs and other difficulties associated with starting a new CCRC are limiting the availability of CCRCs in some parts of the country.

What Future for the CCRC?

The only certainty is uncertainty. The constellation of problems facing all LTC in the United States—demographic, political, legal, and ethical—plus the special challenge of administering the complex life-care contract, especially the health care component, suggest that the model itself could be overwhelmed and that the CCRC industry could end up as a handful of little "Camelots" for a fortunate 1% to 2% of the elderly population.

It is possible, however, that the CCRC is an idea whose time has come. Like prepaid group practice in the late 1960s, before the name was changed to HMO and it received the blessing of White House and Congress, the CCRC could be on the threshold of a major expansion. The impressive growth of the past 25 years suggests considerable momentum that, if accelerated over the next few decades, could reach perhaps 5% to 7% of the elderly by the year 2000 and up to 10% by 2020 (Somers & Spears, 1992). For those over age 75, the principal target population, the proportions could reach 15% in 2000 and 25% in 2020.

Hope of reaching any such targets, however, depends not just on projections of elderly income and buying power but on public and private leadership in forging the essential national consensus with respect to LTC financing and on the industry's ability to function within such a consensus and to develop the necessary managerial skills, especially in the health care area. In any case, the CCRC experience—with its emphasis on sheltered housing, resident independence, prevention, managed health care, and insured LTC—should be helpful, even to those who prefer other models.

ANNE R. SOMERS

See Also
American Association of Homes and Services for the Aging
Assisted Living
Home Health Care
Continuum of Care
Health Maintenance
Long-Term-Care Financing: International Perspective

Internet Key Words
Continuing Care Retirement Community
Long-Term Care Insurance
Sheltered Housing

Internet Resources
The Continuing Care Accreditation Commission
http://www.ccaconline.org/

Selecting a Continuing Care Retirement Community, American Association of Homes and Services for the Aged
http://www.aahsa.org/public/ccrc.htm

CONTINUUM OF CARE

A continuum of care (COC) is an integrated system of care composed of services and integrating mechanisms that manages clients through a comprehensive array of health, mental health, and social services over time and across care settings (Evashwick & Weiss, 1987). Both *services* and *integrating* mechanisms must coexist for a system to qualify as a COC. At its core, the services provided to clients must be comprehensive, encompassing health, mental health, and social services delivered in different settings that span all levels of intensity, including acute inpatient, extended care, ambulatory, and home care. In more mature COCs, community outreach programs, wellness promotion, and housing extend these core services to broaden the continuum's comprehensiveness. To be effective, the services need to be patient centered, adaptable to the specific and often quickly changing needs of the client. The goal of a COC system is to provide effective, high quality care that will help clients maximize their functional independence and well-being in an increasingly market-driven, resource-limited health care environment.

The availability of comprehensive services is not sufficient to qualify as a continuum of care. Equally important are infrastructure integrating mechanisms that encompass and bind the services together and enable care coordination through all settings and levels. These mechanisms include case management, clinical/best practice pathways, information/communication systems, and financing.

Conceptual Basis

The continuum of care concept is relatively new and much of the literature has been anecdotal, descriptive, and dominated by an economic orientation (Hall & Oskvig, 1998). Although similar, the term "continuum of care" is not the same as "continuity of care," which in general practice has traditionally meant care from a single care provider, usually the physician, spanning an extended period of time and through multiple episodes of illness.

When a patient became ill, the patient would be admitted to the hospital under the physician's care and stay (often for weeks) until ready to return home. Because the range and level of patient care sites were limited, a single care provider could feasibly deliver this type of care. Reimbursement was also relatively straightforward, determined by a fee-for-service system where physician and other service providers were paid for services rendered. This pattern resulted, however, in the creation of a supply-driven system with few cost management incentives.

The ability of single providers to deliver this type of continuity has dramatically declined in recent years. Disruption of this traditional "cottage industry" approach, and the development of new systems, can be attributed to cost-containment pressures. To curb the projected exponential growth in health care expenditures, a fundamental shift occurred in the early 1980s away from the traditional fee-for-service model towards a prospective payment system (PPS) model. This approach resulted in health care dollars becoming, for the first time, a limited fixed resource, giving purchasers and providers of health care compelling motivations to deliver care at the most appropriate and cost-effective site.

An obvious outcome of the influence of prospective payment was a dramatic reduction in acute hospital utilization, historically the most resource intensive health care service setting, and the growing use of alternative care sites that are theoretically more cost effective. The continuum of care concept arose within this context. The growing number of clients requiring multifaceted, often complex, ongoing care, speaks to the need for a comprehensive, integrated model that is theoretically better positioned than the traditional model to provide this type of care. An integrated service model may also have market and financial advantages that are less prevalent with services provided on an individual basis.

The continuum of care concept is a management systems answer to the traditional, single provider continuity-of-care approach. As impersonal as the systems approach may seem, the human component still remains the single most important aspect of the COC for clients and care providers. An exemplary COC system uses advanced integrating information mechanisms to keep the primary physician informed and involved as the patient is transferred from one level of care or facility to another. Equally important is the system's ability to seamlessly transfer or access patient specific information anywhere along the COC.

Beneficiaries of the Continuum of Care

The main beneficiaries of the continuum of care approach are patients with long-term, chronically active illness/disability; or acute, relatively short-term but complex illnesses.

The first group of clients includes individuals who have ongoing, complex medical problems or functional disabilities, who are unable to care for themselves, and who cannot depend solely on their informal support systems. Although chronically disabled younger adults and the mentally ill are included in COC systems, the most significant population is the frail elderly, the heaviest users of multidisciplinary services whose multiple needs are often not adequately addressed by most fee-for-service systems. Frail seniors often require formal orchestration of an array of services that change over time as their needs change from acute, to acute rehabilitation, to subacute, to home care or long-term nursing home care. The exponential growth of the elderly is one of the primary demographic forces driving health care systems to develop an integrated continuum of care model.

Previously healthy younger or older patients with any major acute illness such as stroke, myocardial infarction, or hip fracture represent the second group most likely to benefit from a continuum of care approach where improved coordination of multiple services could occur. The difference between this acute care group and the former is that because recovery of independent function is anticipated, the need for integrated care is typically episodic and shorter term. Once the services and integrating mechanisms of a continuum are in place, however,

any patient who requires multiple service components of the continuum are likely to benefit from its integrated infrastructure, organization, and operation.

Care Coordination: The Key to Integration

The continuum of care model also brings new responsibilities and challenges to care providers. Increased utilization of alternative care sites, varying services, and multiple care provider involvement even under the same COC organization set the stage for chaos and fragmentation of care. Patient-centered care that includes integrated referrals, tracking, and care continuity does not occur naturally or without effort. *Case management*, also known as care coordination, service management, or service coordination, has been adopted by COC systems as one way to address these challenges. In addition to facilitating access by clients to the various services offered by the COC organization, the case manager's role includes client identification, access to services, assessment, care planning, monitoring, and follow-up.

The COC model requires a relentless commitment to communication among its care providers as patient care is increasingly delivered via a team approach. Without optimal communication between care providers, the COC system will fail to prevent fragmented care, and may even damage the patient-physician relationship as the role of the primary physician is marginalized.

Case Study

Mrs. Smith, an elderly woman with worsening osteoarthritis, undergoes an elective total knee replacement (TKR) at the COC hospital. On postoperative day 3 after an uneventful surgery, Mrs. Smith is transferred to a COC skilled nursing facility (SNF) near her home for rehabilitative and skilled nursing care. At the SNF she develops a postoperative anemia and significant knee swelling but improves clinically and physically to her discharge home on postop day 15. Six weeks postoperatively,

Mrs. Smith is using a cane and walking independently into her orthopedist's office for follow-up. In sum, she had an uneventful surgery, timely treatment in the SNF, and a smooth return to her home, home care nurse, and the care of her PCP. Nothing extraordinary; everything went smoothly.

Behind the scenes and transparent to the patient, however, was an integrated COC system. The process began with a specific orthopedic clinical care pathway developed by the COC hospital team in collaboration with the COC rehab SNF specialist, the rehab nursing home, and the COC home care program. Prior to surgery the patient was identified as needing a short-term rehab SNF admission for immediate postoperative skilled rehab services. The COC rehab SNF closest to the patient's home was identified as the SNF of choice and notified of the pending admission so that preparations could be made for a smooth and timely transfer.

The orthopedic surgeon and the postacute care specialist had previously reviewed and approved postoperative TKR care plans and shared common expectations concerning the timing, duration, and intensity of therapy required, and the projected improvement. The transfer of care from the acute to COC postacute facility occurred with all of the clinical information needed to manage the patient's care as well as the clinical pathway developed by the hospital and postacute care teams.

The complications of anemia and knee swelling that occurred at the SNF were managed by frequent communication between the rehab specialist, the orthopedic surgeon, and the patient's primary physician and eliminated the need for hospital readmission. As the patient approached SNF discharge, the COC case manager arranged for the COC home care agency to assume Mrs. Smith's nursing, rehab, and functional care needs in the community. The case manager ascertained that Mrs. Smith's adult children understood her need for short-term assistance with heavier tasks, such as grocery shopping, until she regained functional independence. On the day of SNF discharge, arrangements were made for the COC durable medical equipment supplier to provide Mrs. Smith with the needed equipment at home.

The transfer of care from the COC SNF specialist to the primary care physician (PCP) happened with a voice mail message to the PCP and a standardized, rehab summary note faxed to the PCP's office outlining issues that needed immediate attention such as anticoagulation. On postop day 17, the final rehab discharge summary was e-mailed to the patient's PCP and orthopedist, thereby closing the physician continuum of care loop.

At home, the COC home care agency arranged for a homemaker twice a week to help Mrs. Smith with personal care and light housekeeping. Communication that occurred between the COC visiting nurse and SNF specialist helped to clarify questions about her rehabilitation discharge instructions. Because the communication loop had been completed among physicians, the PCP assumed responsibility for any subsequent clinical issues that arose at home, such as dizziness or pain management. During her involvement with home care, the COC case manager maintained weekly contact with Mrs. Smith to inquire about her well being. The COC case manager determined that Mrs. Smith did not require additional or extended services. If these services were needed, discussions about the type and length would have occurred among the COC home care team, the patient, and the COC case manager. The COC home care therapist advanced Mrs. Smith to a cane as she continued to improve physically and was completely discharged from home care one month postop.

All services were arranged under the umbrella of the COC organization orchestrated through a single COC case manager. A high level of communication and teamwork occurred among care providers to enhance the common patient record shared by all COC care providers. An integrated financial billing system enabled the COC to maintain a comprehensive record of all services received by Mrs. Smith through her total episode of care. Although the above example has mainly focused on the communication between care providers, the value of proactively including the patient and family in care planning through each component of the COC cannot be underestimated. The critical junctures for patient and family involvement usually revolve around the pre and posttransition points, such as transfer from acute hospital care to rehab to home care. In the end, systematic but individually tailored care delivered in an integrated manner helped to restore Mrs. Smith's independence.

Conclusion

The growing number of older adults, especially those who are frail, coupled with today's resource-limited, competitive health care environment offer two compelling reasons for the emergence of integrated continuum of care systems. Although data are needed to demonstrate that integrated COC systems can deliver cost-effective high quality care, there is no indication that the current trend towards more COCs will be reversed. Many health care organizations, such as managed care and hospital care systems, have restructured themselves to offer comprehensive core COC services. The challenge of integrating mechanism development and refinement, however, still lies ahead and includes clinical data standardization and management, transfacility clinical guidelines or pathways, case management systems, strengthening collaborative professional relationships and communication, client and family involvement in care, and alignment of financial risks and rewards (Boult & Pacala, 1999).

CHARLES T. PU

See Also
Best Practices in Geriatric Care
Intergenerational Care

CONTRACTURES

Because activity and mobility are vital to the total health of the elderly, musculoskeletal problems that limit functional capacity have confounding effects. Physical changes thought to be associated with normal aging are all too often due to inactivity. Aging predisposes the elderly to development of contractures as a result of gradual but progressive loss

of muscle bulk and subsequent formation of fibrous tissue. The phenomenon is precipitated by muscle inactivity, producing the so-called disuse syndrome that follows prolonged (more than three days) bed rest. Immobilization results in a 3% loss of original muscle strength each day for the first seven days; thereafter, the process plateaus. Inactivity of any kind leads to muscle atrophy and replacement of muscle bulk with noncontractile tissue. Elderly patients and their caregivers often underestimate the degree of damage that can come from immobility. When rest is recommended for an older person with functional limitations, this exacerbates disuse and initiates a downward spiral of decline.

Joint contracture is a result of fibrosis of the skeletal muscle tissue attached to bone. The fibrotic tissue causes muscle shortening and fixed resistance to passive stretch of the affected muscle. Dupuytren's contracture—the most studied and discussed contracture in medical literature because it is seen across all adult ages, including 10% to 15% of the elderly—is caused by fibroblastic proliferation in the fine structure of the palmar fascia, resulting in finger deformity. Of unknown etiology, it is more common in diabetics. The condition is painless and may be related to repetitive microtrauma. A contracture can be the result of joint ankylosis and muscle shortening acting in concert or independently. Contractures are generally not painful unless the limbs are moved beyond their range limit.

Contractures are a major health concern for the elderly and are commonly the result of immobilization and disuse during hospitalization. Several chronic neuromuscular and costomuscular conditions common in the elderly predispose them to the development of muscle and joint contractures, including Parkinson's disease, osteo- or rheumatoid arthritis, and Alzheimer's disease. Spasticity and muscular hypertonia associated with neuromuscular conditions often precipitate muscle immobility and contracture. In some patients, however, contractures potentiate spasticity (Halas & Bell, 1990). Thus, the pathophysiological mechanism creates a feedback cycle by which contractures and spasticity augment each other.

Muscular weakness and loss of dexterity can also lead to the development of contractures. For example, muscle contractures are observed as early as two months after a cerebrovascular accident. Contracture of the shoulder joint, or frozen shoulder, is often seen in poorly rehabilitated stroke patients. Adhesive capsulitis is also responsible for post-stroke arthropathies and resultant contractures of the ankles and hips.

Alzheimer's disease poses the greatest risk for contracture development. A landmark study demonstrated that despite late-developing motor function disturbances, nearly a quarter of the patients studied had contractures in early or middle stages of the disease. More than three-quarters of Alzheimer's patients who had lost the ability to walk had contractures. At the end stage of Alzheimer's disease, it was exceptionally rare to find a patient without contractures of hips, knees, elbows, shoulders, and wrists. It is common in nursing homes to find residents with advanced dementia in permanent and disfiguring fetal positions. The federal Minimum Data Set is a potential source of data for identifying the association between Alzheimer's disease and contracture formation and the related progression of each.

A study in Great Britain showed that among 222 residents, 121 (55%) had at least one joint contracture. The study demonstrated statistical significance between lower limb contracture and the frequency of ambulation, and it found that pain may be a significant contributing factor to contracture formation. Pain is likely underreported and underassessed in nursing homes, particularly for those who are cognitively or communication impaired.

Hip-flexion contracture—ambulating with the upper torso tipped forward—is more likely the result of prolonged or restrained sitting than an age-related change. Heel cord contractures often result from the effect of gravity on an unsupported foot while in bed or a lounge-type chair. A high frequency of gait and postural abnormalities and hip-flexion contractures is observed among arthritic, ambulatory elderly patients who have adapted to their painful hips and spines; therefore, they do not report their discomfort to the clinician. Often, costomuscular pain is accepted as an almost normal consequence of aging by both patients and their physicians. However, pain universally leads to the limitation of motion. As a painful disease progresses, the sites and intensity of joint contractures also increase.

Complications from contractures range from aesthetically repugnant and psychologically disturbing body disfigurement to forced immobilization, increased dependence, and predisposition to pressure ulcers over affected bony prominences. End-stage osteoarthritis of the hip results in limitation in external rotation, abduction, and flexion movement of the hip. This motion limitation invariably changes posture and gait and predisposes the patient to immobility and further progression of the hip contracture. If this same patient has osteoarthritic limitations and distortions in other body parts—for example, hands, fingers, spine—the complications related to quality of life, risk factors, and health care needs are further compounded.

Limitations in function caused by contractures can be avoided, except when a resident's intrinsic medical condition makes the negative outcome difficult or impossible to avoid. Use of restraints (an extrinsic factor) should be eliminated or drastically reduced, unless no other course of action can meet the resident's safety needs.

Most contractures appear to be preventable and, if developed and detected in early stages, reversible; yet there is no scientific information that corroborates this theory. Understanding the pathophysiological mechanism of contracture development makes the hypothesis plausible and calls for aggressive avoidance of prolonged immobility of elderly patients and appropriate pain assessment and management.

Contracture risk assessment and avoidance must be an integral part of the community-based intake or institutional admission process. The use of processes to assist in the avoidance of contractures must become as widespread as nutrition assessment.

The 1987 Omnibus Reconciliation Act (OBRA 87) has a regulation that requires facilities to ensure that residents who enter nursing homes without limited range of motion do not experience reduction in range unless the patient's condition demonstrates that such a reduction is unavoidable. The regulation further requires that a patient with limited range of motion receive appropriate treatment and services to increase range or prevent further decrease. In terms of regulatory oversight, some contractures are avoidable; as such, a contracture is a negative outcome of care that can be measured and used to judge the standard of care provided in an institution.

Effective preventive and treatment interventions for patients who have or are predisposed to contractures rely on collaboration among medical, rehabilitation, and nursing services. Medical oversight is needed for pain management and surgical consultation with regard to some forms of flexion contracture, such as those of the knees and hands. Restorative or maintenance rehabilitation therapy includes transcutaneous electric stimulation, traction that capitalizes on drawing and pulling forces, braces, orthotics, and splinting.

The natural history of contracture is sometimes associated with learned dependence that is unknowingly "taught" by caregivers, who anticipate and expedite the functional activities of daily living for the care recipient. In turn, volitional mobility is discouraged, disuse is advanced, and contracture formation is potentiated.

VALERY PORTNOI
PHYLLIS RAMZEL

See Also
Balance
Gait Assessment Instruments
Gait Disturbances
Multidimensional Functional Assessment: Overview
Physical Therapy Services
Restraints

Internet Key Words
Contracture
Movement
Range of Motion

COPING WITH CHRONIC ILLNESS

Chronic illness is a common accompaniment of late middle and old age. Approximately 85% of people 65 years and over suffer from at least one chronic illness, the two most common diseases being arthritis and hypertension. Evidence suggests that older adults generally adjust better psychologically to the diagnosis of a chronic disease than do younger adults (Ganz, Lee, Sim, Polinsky, & Schag, 1992), presumably because of the recognition that advanced age brings an increased risk for illness and disability. For younger adults, serious illness and associated disability are particularly stressful because they are perceived as "off-time" events (i.e., not developmentally normative).

Nevertheless, for both younger and older adults, the diagnosis of a chronic illness is likely to be a distressing event that elicits a range of psychological reactions (Miller, 1992). Most common is a loss of self-esteem as usual roles become difficult to carry out and assistance from others must be accepted. Patients may feel that they are less worthy of the love of others due to their new illness-related roles, particularly dependency. They may fear that others will withdraw or abandon them, and they also feel guilty about the demands their illness places on others. Patients may experience guilt for having become ill, especially if they have been negligent about their health care. They may fear the loss of control over bodily functions and the accompanying shame and humiliation. Anger may also occur, although this is a less common reaction among older adults than younger ones, for whom the diagnosis is perceived as more unjust.

Whether older adults cope with stressors, including physical illness, differently from younger adults has received considerable research attention. Few differences have been found, but older adults appear to engage in less help seeking. It is unclear whether the few coping differences are a function of changes in the way elders cope or of the kinds of stressors elders confront, which in turn evoke different coping strategies.

Coping Strategies

Attempts to master illness-related stressors (as with most kinds of stressors) typically include both problem-focused and emotion-focused coping strategies. Problem-focused coping involves action directed at removing or circumventing the stressor or gathering resources to confront it, such as seeking information and eliciting social support. In contrast, emotion-focused coping involves attempts to reduce or eliminate the emotional distress associated with, or cued by, the stressor (Lazarus, 1990), such as positive reappraisal, minimization, distancing, and accepting responsibility (Lazarus & Folkman, 1984). Emotion-focused coping tends to be used when the situation is not alterable and thus must be tolerated or endured. Emotion-focused strategies may be particularly important and effective for chronically ill people, because although they may have little control over the course of the illness (Taylor, Helgeson, Reed, & Skokan, 1991), they may be able to control their emotional reactions to it.

Cognitive emotion-focused strategies commonly used to cope with the distress of chronic illness include "doing well" and "being healthy," funding positive meaning in the experience, engaging in downward comparisons, and normalizing the plight.

Redefining "Doing Well" and "Being Healthy": Some patients with chronic illnesses are able to feel less ill or debilitated by redefining what it means to be "doing well" and "being healthy." For example, they may now construe "doing well" as staying involved in the activities they value most or continuing daily life much as before the diagnosis. Others may judge themselves healthy if their medical conditions are stable and do not need to see their physicians between scheduled visits. Some patients substitute a spiritual definition of well-being for a physical one. Still others, who believe that physical and mental states are closely tied, are able to think of themselves as healthy by maintaining a positive mental state. In general, these strategies allow patients to experience a subjective sense of physical well-being despite the objective reality of living with a chronic or serious illness.

Engaging in Downward Social Comparisons: Elders may compare their own situations to those of other individuals with the same disease

whose conditions seems poorer, either medically or psychologically. Such downward social comparisons help maintain a sense of emotional well-being. Patients come to feel less victimized by the illness. Self-reminders that "it could be worse" are emotionally reassuring and allow patients to view personal circumstances as less threatening.

Normalizing Their Plight: By normalizing and universalizing suffering, patients can feel less separated from (healthy) others, thus diminishing the sense of difference and victimization. Patients can reason that everyone has a cross to bear, and nobody escapes suffering in life.

Finding Positive Meaning in the Experience: The ability to find positive or constructive meaning in a negative life event may be an important condition for subsequent psychological adjustment (Tait & Silver, 1989). In doing so, patients are apparently able to feel less victimized, and the event is rendered more acceptable. Thus, patients may claim that they feel better now and are "healthier" than before their diagnosis due to the adoption of positive lifestyle responses. Others may report that the supportive response of family and friends has enabled them to realize how much they are loved.

Behavioral and Cognitive Adaptive Tasks

By definition, prospects for recovery are limited in chronic illness. Successful adjustment to these conditions typically requires the mastery of a number of adaptive tasks (Miller, 1992).

Modifying Daily Routines: The chronically ill frequently must change their daily routines to accommodate symptoms (e.g., fatigue, pain) or comply with treatment regimens. Patients often reorganize activities to better conserve limited energy or minimize disruption of valued activities.

Mastering the Information and Skills Required for Self-Care: Chronic illness, due to its protracted nature, often requires considerable patient participation in activities aimed at preserving health and functioning and adhering to treatment demands. Patients must master considerable information to carry out these activities or rely on family and friends to assist them in self-care activities.

Adhering to Treatment Regimens: The treatment regimens for different chronic illnesses vary in their complexity and psychological demands. Medication and dietary nonadherence is common among older as well as younger patients with chronic diseases. Although such behavior may appear self-destructive irrational to health professionals, research on patients' beliefs about illnesses and medications typically reveal that such behavior is rational or at least understandable from the patient's perspective. The associated behavior patterns are often attempts to reassert control or preserve quality of life by avoiding treatment side effects. At other times, however, these behaviors can be a form of denial regarding the illness or its severity.

Coping with Uncertainty: Chronic illnesses often have an erratic course or trajectory. Patients live with considerable uncertainty about whether their conditions will remain stable or worsen. The efficacy of treatment regimens may be uncertain, or there may be considerable variability in the benefits realized. Uncertainty is inherently stressful and difficult to cope with, and it is hard to mount an appropriate coping response under such conditions.

Maintaining a Sense of Control: Illnesses can undermine patients' sense of control or mastery over health and life. Although a sense of control is viewed as a fundamental human need, there has been some debate whether this sense diminishes in late life. Common patient strategies for regaining a sense of control include becoming knowledgeable about the illness, effectively carrying out self-care activities, and adopting alternative therapies as an adjunct to traditional medical care.

Preserving Self-Esteem: Latent negative self-images are often activated by illness. Patients may feel vulnerable, helpless, incapacitated, and dependent on others. Depression associated with chronic illness is common among the elderly and is characterized by lowered self-esteem. Patients may attempt to preserve or restore self-esteem through strategies such as downward comparisons or mastery of illness-related information and self-

care activities. Strategies that enhance patients' sense of control over their illness can enhance self-esteem.

Renegotiating Social Relationships: Being ill is a social state and a physical condition. A diagnosis changes not only the patient's self-concept and behavior but also the reactions of family members and others. As such, illnesses have both positive and negative personal and interpersonal consequences. Estranged family members may become closer and more supportive. Alternatively, the physical caregiving burden and emotional impact of the disease may cause family members to resent the added responsibilities and at the same time experience guilt about such feelings.

Several circumstances unique to the elderly may compromise their ability to cope with chronic illness. The typically attenuated social networks of older adults may limit the availability of informal practical and emotional support. Limited financial resources can restrict access to formal assistance. Older adults may also be less motivated than younger adults to preserve or improve their health status. They may hold fatalistic attitudes toward disability, assuming it to be an unavoidable concomitant of the aging process.

KAROLYN SIEGEL
STANLEY J. ANDERMAN

See Also
Chronic Illness

Internet Key Words
Adaptive Tasks
Chronic Illness
Emotion-Focused Coping
Problem-Focused Coping
Psychosocial Adjustment

Internet Resources
Coping with Illness
*http://www.baltimorepsych.com/
    PhysicianLinks.htm*

Eldercare
http://www/elderweb.com/

GeroWeb
http://www.iog.wayne.edu/IOGlinks.html

Health Management for Older Adults
*http://www.medinfo.ufl.edu/cme/hmoa/
hmanage.html*

CORONARY ARTERY DISEASE

Despite a recent decline in mortality due to cardio-vascular disease, heart disease remains the most common cause of death in older persons. In fact, the majority of cardiac deaths in persons older than 65 are attributable to ischemic heart disease (Wei, 1999).

Angina Pectoris

Clinical Presentation

Similar to young patients, the majority of older patients suffering from clinical coronary artery disease present with angina pectoris. However, certain nonspecific presentations of myocardial ischemia are more prevalent among elderly patients. These include syncope or sweating with exertion, coughing with emotional stress, palpitations upon effort, and episodes of confusion. Because typical cardiac symptoms may be absent, it is imperative to maintain a high level of suspicion for coronary artery disease when evaluating older patients. The vague and often unrecognizable presentation of myocardial ischemia in older persons may contribute to delays in seeking medical attention and in the establishment of a diagnosis and treatment.

Diagnosis

The diagnosis and management of angina pectoris are based primarily on the patient's clinical history. A careful history should include the characteristics of the discomfort, the duration of the attacks, the presence of associated symptoms, and the circum-stances that precipitate or relieve the discomfort. The presence of cardiac risk factors such as hypertension, hyperlipidemia, diabetes mellitus, cigarette smoking, and family history should be elicited and may heighten suspicion for coronary artery disease. A broad differential diagnosis should be maintained when evaluating older patients with chest discomfort, as these patients may have a wide range of potential medical conditions. Nonanginal causes of chest discomfort that should be considered include esophageal disease, chest wall pain, anxiety, pericarditis, pulmonary embolism, aortic disease, cervical disease, and intra-abdominal events.

A physical examination and laboratory tests may reveal the presence of peripheral vascular disease or evidence of hypertension, hyperlipidemia, or diabetes mellitus, but they are often nonspecific. A resting electrocardiogram may show nonspecific ST-T wave changes, or it may be normal. Exercise testing can help establish a diagnosis of coronary artery disease in patients who are able to exercise and reach a target heart rate, but it is often less useful in older patients, who may have abnormal baseline electrocardiograms or are unable to exercise because of comorbidities or deconditioning. Exercise testing may be combined with scintigraphic studies or echocardiography to localize ischemic disease. For those older patients who are unable to exercise adequately, pharmacological stress testing can confirm the presence of myocardial ischemia.

Treatment

The management of angina pectoris may include a variety of medical, surgical, and nonpharmacological therapies. The goals of therapy are to improve the patient's symptoms while maintaining a desired level of activity and to reduce cardiovascular risk factors.

Basic pharmacological options for the management of angina pectoris in older persons are the same as for younger patients. Altered drug metabolism, changes in drug excretion, and a propensity for adverse drug effects in the elderly necessitate a careful approach. Initial drug doses should be low, with gradual titration. Close follow-up should be maintained, with an emphasis on monitoring for effectiveness and adverse reactions. Beta-blockers are effective antianginals and antihypertensives that have been shown to reduce mortality in patients with coronary artery disease. Nitrates can be effective antianginals and may be particularly useful in the presence of heart failure or conduction abnormalities. The degree of nitrate tolerance can be limited by incorporating a minimal 8- to 10-hour nitrate-free period every 24 hours in the regimen. Calcium channel blockers may also be useful anti-ischemic agents. Unless contraindicated, low-dose aspirin should be prescribed for patients with angina.

Antianginal therapy should be accompanied by cardiovascular risk factor modification. Hypertension and hyperlipidemia should be appropriately managed. Smoking cessation, a low-fat diet, and a carefully planned exercise regimen should be recommended.

A care plan should be formulated, with special attention to the older patient's emotional needs and support systems. Depression and social isolation should be identified and managed appropriately. In addition, proper patient education regarding the management of symptoms can help alleviate stress and anxiety.

Revascularization

Intervention in older patients with coronary artery disease is increasing as more of them maintain independent, active lives. A growing number of patients undergoing percutaneous transluminal coronary angioplasty and coronary bypass surgery are now over age 65. Although elderly patients may experience higher mortality and complication rates with revascularization procedures, their outcomes can be excellent. Therefore, chronological age should not be a contraindication. Interventional therapy should be considered in appropriate older patients who may derive significant benefit.

Acute Myocardial Infarction

Clinical Presentation

The clinical presentation of acute myocardial infarction in the elderly may vary widely, and the classic symptoms of chest pain and dyspnea may be absent. Atypical presentations observed in older persons include confusion, agitation, syncope, weakness, stroke, vertigo, cough, and abdominal pain. It is important for health care providers to gain familiarity with the broad spectrum of nonspecific symptoms that may represent myocardial infarction. In addition, older persons may have more silent or nonclinically detected myocardial infarctions (Tresch, 1998).

Diagnosis

The goals of the history and physical examination are to make an early presumptive diagnosis, identify early complications, and guide further evaluation and management. A careful history, including the nature and duration of symptoms, should be obtained. The presence of antecedent coronary artery disease and cardiovascular risk factors should be determined. Findings of the physical exam in older patients may be nonspecific, as in younger patients.

The electrocardiogram may show ischemic changes, but it is often nonspecific in the elderly, in whom there may be an increased incidence of smaller, non-Q wave infarcts. The serum cardiac enzyme profile may demonstrate age-associated differences. Older patients may exhibit less total creatine kinase release, and the diagnosis of myocardial infarction may be complicated by the finding of normal total creatine kinase with an elevated myocardial fraction. Cardiac-specific troponin T and troponin I are new markers for acute myocardial infarction that may have improved sensitivity and specificity. Echocardiographic imaging may be useful for assessing left ventricular function and for diagnosing valvular disease or complications such as papillary muscle infarction or tear or ventricular wall rupture.

Older patients suffering from myocardial infarction have higher rates of cardiac complications, including congestive heart failure, arrhythmias, cardiogenic shock, and ventricular and papillary muscle rupture. The elderly are also at risk for noncardiac complications, including delirium and pressure ulcers. In addition, older patients tend to experience higher in-hospital mortality rates and longer lengths of stay.

Management

The management of acute myocardial infarction in older persons should be prompt and aggressive when indicated. Although advanced age is not a contraindication to the use of recanalization procedures, physicians should be cognizant of the increased risks associated with these procedures in older patients.

Thrombolytic therapy has been reported to significantly reduce mortality in older patients presenting early during the course of a myocardial infarction. Although older patients may experience higher complication rates with thrombolysis, the relative benefit of therapy may actually be greater because of the higher mortality and morbidity in older compared with younger patients who do not receive thrombolytic therapy. Several studies have demonstrated that elderly patients derive significant benefit from thrombolytic therapy (Rich, 1998).

Primary percutaneous transluminal coronary angioplasty is an alternative to thrombolysis for the reestablishment of coronary perfusion. This approach is associated with excellent outcomes for older persons when performed in a timely fashion in specialized centers.

Antiplatelet therapy in the form of low-dose aspirin improves survival in older patients and should be initiated and continued. Both early and long-term use of beta-blockers have been associated with significant mortality reduction for older myocardial infarction patients. The use of angiotensin-converting enzyme inhibitors improves survival and cardiac function in the post–myocardial infarction period. Nitrates have been reported to be beneficial for older patients during myocardial infarction and may be particularly useful in the setting of congestive heart failure or hypertension. The use of adjunctive anticoagulation may be beneficial in certain subsets of elderly patients with acute coronary syndromes in whom there are no contraindications.

Ambulation should begin early during the recovery period, with gradual increases in the level of activity. Prolonged bed rest and inactivity may place patients at risk for the development of venous thrombosis, pulmonary embolism, deconditioning, and skin breakdown. A slow and progressive exercise regimen, including adequate warm-up and cool-down periods, should be implemented.

A multidisciplinary approach is required when caring for older post–myocardial infarction patients. Cardiac risk factors should be identified and managed aggressively. The social and psychological aspects of elderly patient care should be emphasized, as they have an important impact on recovery and subsequent rehabilitation. The presence of depression, social isolation, a lack of support systems, and an inadequate home environment should be identified and addressed optimally.

JEANNE Y. WEI
BRENDA ROSENZWEIG
CATHERINE MORENCY

See Also
Heartburn
Heart Failure Management: Congestive and
    Chronic Heart Failure
Rehabilitation

Internet Key Words
Angina Pectoris
Coronary Artery Disease
Myocardial Infarction
Percutaneous Transluminal Coronary Angioplasty
Revascularization

Internet Key Words
American Heart Association
http://www.americanheart.org

National Heart, Lung, and Blood Institute
http://www.nhlbi.nih.gov

COUGH

Cough is an important defense mechanism that clears the airway of excess secretions and foreign materials. Cough can indicate underlying illness and is one of the most common symptoms that prompts patients to seek medical attention (Widdicombe, 1999).

Cough may be initiated voluntarily or involuntarily. Physiologically, cough is a reflex that has an afferent limb, including the trigeminal, glossopharyngeal, superior laryngeal, and vagus nerves; a center in the medulla; and an efferent limb, including the recurrent laryngeal and spinal nerves. The irritant receptors are located in the larynx, trachea, large bronchi, nose, external auditory canal, pleural surfaces, and diaphragm. Stimulation of these sites may result in cough. It has four phases: a deep inspiration, followed by glottic closure; diaphragmatic relaxation; muscle contraction, accompanied by a markedly increased intrathoracic pressure; and sudden opening of the glottis, with an explosive release of pressure (Irwin et al., 1998). Narrowing of the airway caused by a large transpulmonary pressure gradient leads to a high linear velocity. Such flows generally eliminate mucus and foreign matter.

Causes

Any disorder that irritates, inflames, infiltrates, constricts, or compresses airways may cause coughing, as can bronchospasm. Cough is easily classified as acute or chronic; chronic cough is usually defined as one persisting longer than three weeks.

The most common cause of cough in elders and young people is acute upper respiratory infection. An acute cough may be mild or severe and is usually associated with other symptoms, such as runny nose, sore throat, headache, malaise, fever, and chills. The history and physical findings of rhinitis and pharyngitis, with a normal chest exam, support the diagnosis of acute viral upper respiratory infection. In most patients, the cough resolves within two to three weeks. A chest x-ray is usually not necessary as an initial evaluation in healthy younger persons, unless abnormalities are found on the chest exam. A cough resulting from viral infection, atypical pneumonia, and environmental pollutants is usually nonproductive; a cough due to bacterial bronchitis and pneumonia often produces purulent sputum.

An acute cough associated with wheezing suggests airway obstruction, possibly from asthma or an exacerbation of chronic obstructive pulmonary disease (Irwin & Curley, 1991). It may also suggest pulmonary edema in patients with a history of congestive heart failure (CHF). Acute-onset cough with stridor in a patient with dementia suggests foreign body obstruction. Acute cough with dyspnea in a patient with a swallowing disorder or a patient receiving tube feeding often indicates aspiration.

The most common cause of chronic cough is tobacco smoke. Chronic cough occurs in up to 75% of smokers and is often ignored or minimized. Up to 30% of smokers develop chronic bronchitis, which is defined as chronic productive cough for at least three months a year for at least two consecutive years (Philp, 1997). A changing pattern of cough in a heavy smoker warrants an evaluation for malignancy. X-ray screening for lung cancer in smokers is still controversial.

In nonsmokers, postnasal drip resulting from allergic rhinitis and sinusitis is the most common cause of chronic cough. Nasal discharge, sneezing, frequent throat clearing, and foul smell or taste often suggest postnasal drip. Asthma is a common cause of chronic cough, especially in younger patients. In some asthmatic patients, cough may be the only presenting symptom. Another common cause is gastroesophageal reflux disease (GERD). Cough often occurs at night or is worse when lying down, and it may be accompanied by heartburn or "sour" belches (Irwin et al., 1998). A dry cough resulting from airway hypersensitivity following acute upper respiratory infection may persist for six to eight weeks.

Various medications are also important causes of chronic cough in elders. Angiotensin-converting enzyme inhibitors cause a cough that is nonproductive, has a gradual onset, and resolves after discontinuing the medication. Beta-blockers may cause cough by inducing bronchospasm. Elderly patients' eyedrops should be reviewed, since ophthalmic beta-blocker preparations may produce bronchospasm. Sometimes, inhaled drugs such as bronchodilators, steroids, and cromolyn sodium can cause a dry cough as well.

In elders, cough caused by CHF must be considered, because the prevalence of CHF is very high in this age group. Lung cancer is always a concern, especially when a cough is accompanied by weight loss. Tuberculosis should also be in the differential diagnosis. When any of these conditions is suspected, a chest x-ray is necessary.

If the initial evaluation is unrevealing, less common conditions require investigation: bronchiectasis, interstitial lung disease, pulmonary embolism, tracheal compression (secondary to aortic aneurysm, enlarged thyroid, or mediastinal tumor), and remote occupational exposure (e.g., asbestos, silicon). Psychogenic cough is rare in elders.

Evaluation

A detailed history provides valuable clues for diagnosis. Information should include (1) characteristics of the cough: acute or chronic, productive or nonproductive, character of the sputum, changes with time or position; (2) associated symptoms: fever, chills, wheezing, dyspnea, weight loss, night sweats; (3) allergy, tobacco use, and environmental exposure; (4) symptoms of postnasal drip or GERD; and (5) a complete medication list, including eyedrops.

The physical examination should focus on the ears, nose, throat, and lungs. Mucus, erythema, and "cobblestoning" in the posterior oropharynx suggest postnasal drip as the cause. Chest examination may yield evidence of underlying asthma, pneumonia, or CHF. A localized wheeze may indicate a bronchogenic tumor or foreign body. Newly developed clubbing raises the suspicion of lung cancer.

A chest x-ray is always indicated for a chronic cough in an older person. It may reveal an infiltrate, a mass, hilar adenopathy, or CHF. A sinus computed tomography (CT) scan may help confirm the diagnosis of sinusitis. A high-resolution CT scan is indicated if bronchiectasis, interstitial lung disease, or a mediastinal mass is suspected. Pulmonary function tests (PFTs) are helpful for the diagnosis of both obstructive and restrictive lung diseases. If asthma is strongly suspected and PFTs are normal, bronchoprovocation testing may be useful. When cough persists after empirical treatment and routine evaluation is unrevealing, more invasive testing may be indicated. Bronchoscopy is important for the diagnosis of endobronchial tumor, foreign body, and granulomatous disease. GERD may need confirmation by gastrointestinal endoscopy or overnight monitoring of hydrogen ion concentration (pH).

Management

In general, searching for its cause is more important than suppressing the cough. A productive cough usually should not be suppressed unless it is severe enough to disrupt sleep or produce marked discomfort. Treatment can be definitive or nonspecific (Lawler, 1998).

When a specific cause is identified, the condition can be definitively treated. Smoking cessation is the most important therapy for chronic bronchitis. Postnasal drip from allergic rhinitis can be treated with antihistamines and inhaled steroids. Sinusitis can be treated with antibiotics; GERD, by diet modification, elevation of the head of the bed, and various medications, including antacids, H_2 blockers, proton pump inhibitor, and prokinetic agents (Irwin & Curley, 1991). Asthma can be treated by eliminating allergens and initiating therapy with bronchodilators and steroids. More than one condition can be present at the same time (e.g., asthma and postnasal drip, asthma and GERD).

Nonspecific or symptomatic therapy should be considered when the cause is not treatable or the cough produces marked discomfort and performs no useful function (e.g., coughs secondary to tracheal compression). Among hundreds of cough and decongestant preparations, dextromethorphan is the most effective; it can raise the threshold of the cough center. Others that are effective include diphenhydramine, caramiphen, and benzonatate. If these medications are ineffective, codeine can be tried at a dosage of 15 to 30 mg every three to six hours. The combination of codeine and dextromethorphan is widely used. Other nonspecific modalities are hydration, expectorants, and mucolytic drugs, but their effectiveness needs further study. Clinicians should recognize the potential for delirium and other side effects from these medications.

NANCY XIAOSHUANG YIN

See Also

Chronic Obstructive Pulmonary Disease
Confusion/Delirium: Risk, Diagnosis, Assessment, and Interventions
Heartburn
Heart Failure Management: Congestive and Chronic Heart Failure

Internet Key Words

Asthma
Chronic Bronchitis
Cough
Gastroesophageal Reflux Disease (GERD)
Postnasal Drip
Tobacco Smoke

Internet Resource

American Academy of Family Physicians
http://www.aafp.org/healthinfo

CREATIVITY

Creativity is a powerful source of growth that is vital throughout life as we continually create and re-create ourselves. It is an attitude, an activity, and a philosophy about growing older that knows no age boundaries. For the aging individual, creativity may reflect a response to the uncertainties, losses, and challenges of existence. Creativity can continue until death; it is not necessarily tied to chronological age but rather to the process of self-actualizing one's creative potential (Simonton, 1998).

An obstacle in the discussion of creativity and aging, particularly in the conduct of research, has been establishing its definition. Rollo May's definition that creativity is "the process of bringing something new into being" (1975, p. 39) is both simple and encompassing. The domains of creativity include not only the creative arts but also social creativity and what gerontologist Gene Cohen (2000) calls the distinction between creativity with a big "C" versus creativity with a little "c." Creativity with a big "C" refers to the more sweeping accomplishments that can change a community or a society, whereas creativity with a small "c" refers to those accomplishments that can change a family's or individual's life course.

The concept of the "Ulyssean adult" (McLeish, 1976) views creativity as a learning process by older adults who continue to explore and adventure, or by adults who express creativity for the first time in later life. The Ulyssean life is possible for a greater number of older adults than ever before because many have secure retirement incomes and increased leisure time.

Creativity and creative expression can be stimulated and nurtured in older adults through their involvement in the creative arts: music, dance, theater, writing, and visual arts. This includes the folk or traditional arts, which are anchored in and expressive of shared ways of life, ethnic heritage, and religion. Examples of creative activities include Ukrainian egg-decorating traditions, American Indian beading and basketry, and old-time fiddling. The arts offer possibilities for increased self-esteem, socialization, learning, integration, mastery, joy, and self-discovery. Since prehistoric times, the creative arts have been used as a tool for healing. Babylonians, Greeks, East Indians, and ancient Oriental civilizations used visualization to heal. Music and dance in hunter-gatherer cultures freed "boiling energy"—healing energy for the Kalahari Bushman, a part of the culture's medicine.

The inspiration provided by the creative arts can also increase life satisfaction and quality of life for the well to the frail elderly. A theater and writing project called Timeslips conducts sessions with people who have Alzheimer's and related dementia at adult day centers. A visual image such as a cowboy playing a banjo with a horse leaning over his shoulder, three smiling women playing the accordion, or a small girl standing next to a huge elephant is presented to the group. Participants are asked questions based on the image, and their answers are built into stories that give a glimpse into the experience of living with dementia. At one day center, a person with dementia who never spoke more than two words at a time sang a solo of "Beautiful Blue Eyes" to a stunned audience after participating in the workshop.

This creative process not only provides opportunities for self-expression but also provides staff and caregivers with new vehicles to reach those with dementia.

Principles of Creativity

Self-expression is a basic need throughout the life span that affects overall health, joy, and well-being. Jung (1971) regarded imagination and creativity as healing forces, whereby deep-seated feelings could be symbolically represented and released through the creative act. The creative arts are an opportunity for self-expression, achievement, and reengagement amid losses, voids, and uncertainty.

Clinicians must probe to find out about an older adult's interests, past work, or hobbies. A "creativity assessment" can include these questions: What makes you feel most alive? What projects have given you the most pleasure? What skills do you have that you would like to pass on? What are your sources of imagination? Are there creative issues in your life that are troubling you now? How would you like to express yourself creatively? Engaging an elder in painting, writing a poem about turning 80, working on a pottery wheel, joining a

discussion, moving to the beat in a dance class, or expressing sorrow when listening to a musical piece is a way out of isolation.

The creative arts can help stimulate and compensate for sensory loss—the "thinning of life"—through one-on-one or group activities, working with each sense separately, or as a total sensory experience. There is a natural pathway from sense memory to life review that transforms sensory-inspired stories into reminiscence and art. The University without Walls is a telephone conference call program for home-bound older adults. Taught by volunteers, the classes include Short Stories, Poetry, Play Reading, Dramatic Literature, The Artist's Way, Insights into Opera, Film Studies, Women Artists, and 20th Century Art. Staff members connect participants to the conference call. Classes generally last 50 minutes and meet for 4 to 12 weeks, during which time they provide opportunities for stimulating conversation, discussion, friendship, and lifelong learning.

Creativity may be used as both a strategy for reducing loss and a tool for problem solving. In his later-life poetry, William Carlos Williams wrote of "an old age that adds as it takes away." Loss can be a catalyst for creative expression. Matisse had diminishing vision and suffered from severe intestinal disorders; he created from his wheelchair. Monet continued painting into his 80s following two cataract operations.

The Connecticut Hospice Program offers an exceptional arts program for older adults that specifically addresses loss through the creative arts. The arts program is as indispensable as nursing or dietary service. Art activities for hospice patients range from bedside art to evening concerts, from home-care arts to ongoing exhibits in the main gallery. Hospice-employed artists are oriented to the hospice program, are directly involved with patients and families at the bedside, and are core members of the caregiving team. The hospice program was the first to offer a model for arts in hospice care, resulting in the inclusion of arts in the Connecticut State Health Code—the first time arts were integrated by law into a health care program. All art program activities are recorded in patient charts and discussed as part of the patient assessment.

Many older adults are the keepers of cultures and traditions. Their lives are their life stories—rich natural resources of experience and wisdom. Elders Share the Arts has created Generating Community, a model intergenerational program that brings together older adults in nursing homes, community centers, and senior centers with youths age 5 to 18. For example, teenagers from the Dominican Republic were trained to explore turning points in life. They interviewed older adults about their work histories and how the elders felt about their jobs when they were younger. This process provided the teenagers with role models for solving problems and making decisions in their own lives. One parent credited the program for her child's more respectful behavior at home.

The exciting new area of creative rituals can be used to help mark significant events in a lifetime. These rituals acknowledge celebrations, losses, and transitions that occur as people age, such as losing a driver's license, getting a walker, losing a limb, transferring property, and entering a nursing home. Age markers and celebrations or rites of passage have been created for people turning 50, 60, 70, 80, 90, or 100. The basic structure of ritual is its theme, acknowledging the turning point, marking the event, naming the losses, telling the stories, developing strategies for compensation, and sharing with a community. Elements might include changing the environment through lights, smell, and sound and using props, music, masks, and food.

A ritual about losing a driver's license might involve bringing a group together to acknowledge the loss of the ability to drive and identify alternatives to compensate for that loss. During a driver's license ritual, participants can share stories about cars they have had, trips they have taken, and other significant "car moments." The grief and fear surrounding loss of a driver's license can be discussed, and the group can develop creative solutions to compensate for the loss of mobility, such as friends giving IOUs for car trips to doctors and grocery stores and money for taxicabs.

Clinicians should strive to recognize older adults for the creative individuals they are and bring art and creativity to their practices.

ANDREA SHERMAN

See Also
Activities
Intergenerational Care
Life Review

Internet Key Words
Age Markers
Keepers of Cultures and Traditions
Life Review and Reminiscence
Rituals
Self-Expression
Sensory Loss

Internet Resources
Americans for the Arts
http://www.artsusa.org/

Elders Share the Arts
*http://www.artswire.org/Community/highperf/hp/
     hpstories/ESTA.html*

National Endowment for the Arts
*http://arts.endow.gov/partner/Accessiblity/
     Monograph/OlderContents.html*

The Poetry of Aging
*http://www.gen.umn.edu/faculty_staff/yahnke/
     poetry/poetry6.htm*

University without Walls
*http://www.dorotusa.org/seniors/
     university_walls.htm*

CRIME VICTIMIZATION

Prevalence and Epidemiology

Among all demographic groups stratified by age, older adults have the lowest prevalence of crime victimization. The National Crime Victimization Survey (NCVS), a nationally representative sample of household residents in the United States conducted by the Bureau of Justice Statistics (BJS), found an annual incidence rate of 7.2 violent crime victimizations per 1,000 individuals over the age of 65 (Bureau of Justice Statistics, 1994). In contrast, the highest prevalence of violent victimization was estimated for individuals between the ages of 16 and 19 with an annual incidence rate of 102.5 per 1,000, representing over a tenfold increase in risk over their older counterparts.

Two points regarding the NCVS data merit comment. First, although older adults had the lowest risk of crime victimization among all demographic groups, their rate of victimization exceeded rates of elder abuse (calculated as 2.6 per thousand per year in the most commonly cited study [Pillemer and Finkelhor, 1988]) by over sevenfold. Second, a more detailed analysis of NCVS data by gender and crime type conducted by the authors reveals very disturbing trends (Bachman et al., 1998), which are described below.

Crime Victimization Among Older Adults: A Closer Look

Several aspects of victimization in older adults are alarming. In the NCVS data, older victims of robbery, particularly women, were more likely than younger victims to sustain injuries, and injured elderly victims of violent crime were more likely than younger injured victims to suffer a serious medical injury that required hospital or other medical care. This may reflect an underlying loss of physiologic reserve known to accompany normal aging, which may place the older person at greater risk for significant trauma. For example, given equivalent assaultive force to an extremity, an older woman with osteoporosis is more likely to sustain a fracture than her younger counterpart.

Other important differences in older crime victims point to contextual characteristics of the crime. For example, older victims of crime were more likely to face multiple offenders as well as offenders with guns. Younger homicide victims were more likely to be killed by known offenders, while older adults were just as likely to be killed by a stranger as by someone they know (Bachman, 1992; Bachman & Saltzman, 1995).

This pattern of victim/offender relationship is also present in nonfatal violence against the elderly. For example, while women under the age of 65 were more vulnerable to violence perpetrated by non-strangers compared to strangers, women over the age of 65 were just as likely to be victimized by a stranger as by an intimate or another offender known to the victim. And finally, regarding location of victimization, elderly violent crime victims were more likely to be assaulted at or near their home compared to younger victims (Bachman et al., 1998).

The Impact of Crime on Health in Older Adults

Virtually no data on crime victimization and subsequent specific health outcomes are available, yet nearly every clinician involved in the care of an older person can recall the patient/client who suffered a loss of physical and or psychosocial well being after crime victimization. Many of these individuals were ostensibly well compensated in all spheres (medical, functional, and psychosocial) after experiencing a crime, but victimization set in motion an inexorable spiral that ultimately resulted in loss of independence. For those people who are not well compensated, this spiral is only worse. These cases are striking not only for their trajectory, but also for how sometimes "minor" victimization can insidiously erode quality of life.

The notion that a single event might set into motion a progressive spiral of decline in many domains for an older adult has a basis in aging theory. Normal aging is accompanied by a loss of physiologic reserve in various systems that need not lead to phenotypic decline (Shock, 1985). Rather, when the organism is taxed through stress, illness, or other factors, this loss of physiologic reserve is unmasked. Clinicians skilled in geriatric medicine encounter the effects of this discquilibrium on a daily basis: the ostensibly high functioning older adult who develops new incontinence with simple pneumonia; the patient with mild cognitive impairment who develops alarming confusion with the addition of a medication; or the compensated older person who declines medically after bereavement.

Crime victimization may be such a precipitating event in the life of an older person. It is a stressful experience that may impact on physical health, mental health, and functional independence.

The Role of Multidisciplinary Providers in the Care of Victimized Older Person

Given the paucity of research in this area, what is the clinician's role in the care of an older person who is a victim of crime? Is it simply to treat lacerations and abrasions, or should health care professionals play a more aggressive role in these situations? Or is crime simply not within the purview of the clinician?

We favor the more aggressive stance. A fundamental tenet of gerontology is that medical and social problems conspire to threaten the independence of the older person. Whether such a functional spiral is provoked by an acute medical illness such as shingles, or an acute social problem such as robbery, is ultimately moot. The outcome of such provocation is undeniably medical, undeniably quality-of-life depleting, and undeniably costly when independence is lost.

Besides caring skillfully for acute illnesses, clinicians should recognize that the older person who experiences crime is vulnerable in a great many ways. Elders with chronic diseases that have been well controlled may decompensate for physiologic reasons or because psychological factors related to crime may result in noncompliance or self-neglect. Although data are lacking, older crime victims are probably at risk for psychic distress in a great many ways, ranging from major depressive disorders to post-traumatic stress syndrome. These should be screened for and aggressively treated if identified. Support groups for victims of crime may be useful in this regard, although the issues for older crime

victims may be somewhat different from those for younger ones, and the authors are unaware of support groups geared specifically to the older individual. Social isolation because of the fear of recurrent crime victimization is also a concern. Specific inquiries should be made regarding the size and quality of social interactions, e.g., maintenance of previously cherished hobbies and activities after crime.

Clinicians should also not be lulled into complacency because victimization is perceived as "minor." For many older adults, an episode of victimization, such as burglary, seems trivial because little was stolen, there was no contact with the perpetrator, and insurance was available to replace stolen items. Yet, despite an initial complacent response, a dramatic decline ensued.

Economic Predation of Older Adults

The subject of financial exploitation of older adults by intimates or "con artists" deserves special mention because its prevalence is believed to be increasing. Additionally, the high prevalence of cognitive impairment, diagnosed or not, among older people makes them especially vulnerable. Clinicians may become aware of exploitation when an older person or a family members describes it frankly to them, or other aspects of the clinical presentation make it clear something is amiss in the finances of the older person (such as not purchasing medicines that were previously affordable).

Here the clinical evaluation involves an assessment of cognitive status, and may include the clinician recommending guardianship with respect to finances if decision making capacity is impaired. Consultation with a neuropsychologist should be considered. Susceptibility to economic predation in the context of dementia itself may be evidence of impaired decision making capacity if, for example, the older person chooses to pursue sweepstakes entries rather than purchase insulin.

Appropriate legal authorities (e.g., local law enforcement or district attorneys) should be involved where appropriate. Educating older citizens on how to identify culprits and resist their overtures is a role played by many senior centers and elder advocacy agencies.

MARK LACHS
RONET BACHMAN

See Also
Adult Protective Services
Elder Mistreatment: Overview
Elder Neglect

Internet Resource
Bureau of Justice Statistics
http://www.ojp.usdoj.gov/bjs/

CULTURAL ASSESSMENT

The clinical merit of understanding psychosocial factors is well established in geriatric care, where clinicians address complex physiological, psychological, and social issues in treating older persons with multiple chronic diseases and functional impairments. Health care providers and patients do not need to share the same cultural, ethnic, or social values, but addressing these factors establishes a therapeutic relationship, determines the best approaches to care, and improves patient satisfaction and adherence to treatment regimens. Culturally sensitive health care can improve health care delivery to all older persons.

Culture encompasses such a broad range of beliefs, behaviors, and definitions that the notion of conducting an assessment seems daunting. As a symbolic construct through which social life is patterned, acted, and perceived, culture can be observed as enactment and outcome of behaviors in a changing environment and throughout the life span. There is a dynamic, adaptive quality to this "cognitive map"; cultures are not static. Texts describing cultural patterns should be used with caution. Broad, heterogeneous categories—such as Asian/Pacific Islander, which represents more than 25 ethnic groups, some from the same countries of origin—do not meaningfully reflect intragroup or individual variation in ethnicity and life experience.

The cultural patterns referenced here describe a range of beliefs. As in any clinical situation, the focus should be on the individual; stereotyping should be avoided.

Providers should be aware of the cultural and historical experiences of their older patients when determining their health care expectations for cure, treatment, palliative care, or reassurance about particular conditions. Since practitioners and clients may not share the same perception of illness, their expectations for health care may also vary. Older rural African Americans, as well as older Filipinos and older Mexican Americans, may conceive of health as an attribute of personal spirituality. African Americans may believe that prayer and faith are more significant than preventive health measures. Filipinos and Mexican Americans may perceive poor health as punishment or as the result of malevolent witchcraft. These perceptions suggest that illness should be tolerated until its impact on function is too severe for informal care. Coping skills developed over a lifetime in response to racism or other inequalities may hinder evaluation and diagnosis. These include reluctance to divulge information, especially about alternative therapies or dissatisfaction with Western biomedicine. Whereas geriatricians typically seek information about social support for patients, clients may not envision themselves as the center of a family or community support network. Allocentrism—emphasizing the importance of the group over the individual—is a common cultural value shared by many older Hispanics and Asians. The notion of autonomy in decision making is not universal.

Culture-bound syndromes have received attention, although their significance is not well understood. Cultural norms define illness and its expression. For example, among Western Europeans, there is a higher frequency of stomach ailments among Germans, liver ailments among French, and headaches among English. Hispanics might describe shortness of breath as fatigue (*fatiga*); back pain as a kidney pain (*dolor de los riñones*). Asian Pacific Islanders might complain of a "weak kidney" to indicate sexual dysfunction, because the kidney is believed to be the site of libido. Native

Americans might describe stress in their family support system as the patient having a "bad heart," indicating the lack of harmony with caregivers. Whereas patients experience illness as a cluster of symptoms that affect functioning in the social context of their daily lives, physicians tend to redefine symptoms as disease, devoid of social context. Recognition of the culturally mediate experience of poor health is a significant element in a culturally sensitive patient interview.

Conducting Cultural Assessment and Reducing Access Barriers

Health care providers often do not have the time or the training to conduct a comprehensive assessment during a single encounter. As with other diagnostic tools, cultural assessment is linked to the level of care and professional domain. Physicians use cultural assessment to inform their evaluation of symptoms, choice of screening instruments, discussion and selection of treatment options, care plans, advance directives, and placement options (Espino, 1995). Nurses and other allied health professionals use cultural assessment to implement care plans, assess health status and pain, respond to personal-care issues, and provide appropriate emotional and spiritual support to patients and families (Lipson, Dibble, & Minarik, 1996). Psychosocial assessment of patients in the context of their families and communities is a core social work skill. Care planning is based on an assessment of beliefs about disease, efficacy of treatment, and potential for rehabilitation, as well as the impact of disease on quality of life.

The basic elements of cultural assessment, asked or observed, include personal and medical history, health practices and preferences, information needs, and communication styles. Personal history includes place of birth, length of residence in the United States if foreign born, economic status, major support systems, ethnic affiliation and strength of that association, and religious beliefs and importance of those beliefs to daily life. Clinicians should assess the type, depth, and complexity

of information a patient wishes to be told, and by whom. Dietary preferences, prescriptions, proscriptions, and lifestyle changes should be noted with respect to their potential conflict with cultural values. Differing values about the appropriateness of informal home care or institutional care for specific conditions (e.g., cognitive impairments, incontinence, advanced age) may have to be explored. Communication styles include primary and secondary language, speaking and reading levels, print and oral traditions, and nonverbal expression. Patients' descriptions of symptoms using culture-bound references may need to be interpreted, even if the provider and patient speak the same language. When translators are used, clinicians should speak in short phrases and use simple, nontechnical language. To ensure the accuracy of the translation, the interpreter should report the patient's words exactly; accuracy can be checked by asking the patient to repeat the information or instructions and by monitoring nonverbal communication (e.g., facial expression, body language). When a family member is used as a translator, the purpose of the session should be discussed beforehand with the translator to ascertain his or her comfort level with sensitive topics (e.g., anatomical function, especially across gender and generations; bad news). In some cases, a more appropriate relative or a professional interpreter should be found.

A culturally sensitive medical history and examination do not differ from any thorough examination. The art of patient interviewing may diverge from the structured medical model if patients voice multiple complaints or divulge important symptoms only at the conclusion of the interview. Past medical history should be thoroughly reviewed and discussed. Foreign-born patients may have been exposed to treatment strategies that are not familiar to U.S. physicians, and patients may not have access to the same information about diseases in their countries of origin. Medications that are well controlled in the United States may be sold over the counter or prescribed with few safeguards in other countries. Histories of drug allergies may be more complex to elicit from foreign-born elders. Clinicians may need to schedule several visits before a trusting relationship is established.

Effective communication is always key to good patient care. Cross-cultural barriers to access often occur unintentionally. At a first meeting, clinicians may habitually introduce themselves, shake hands firmly, and promptly determine the reason for the patient encounter. In some cultures, this businesslike attitude would be offensive. Native Americans, for example, would likely prefer a light touch to an aggressive handshake. In other cultures, traditional (i.e., indigenous or folk) medicine incorporates the healing arts of counseling and talk therapy. Patients from diverse cultures relate that their traditional doctors "really know them as a person"; they feel distrustful of an abrupt and impersonal approach (Garrett, 1990; Pourat, Lubben, Wallace, & Moon, 1999; Rhoades, 1990).

Patients' expressions of respect may be misinterpreted by providers. Avoidance of direct eye contact is a common form of respect shown by Native Americans, Mexican Americans, and African Americans and should not be interpreted as furtiveness or untrustworthiness. Giving respect and feeling distrust may overlap, as when an African American patient avoids direct disagreement with a doctor's recommendations. Silence and failure to report adverse reactions or unsatisfactory responses to treatment may simply be a way to respectfully avoid direct confrontations, or it may be the patient's way to shield the clinician from the humiliation associated with treatment failure. Silence may indicate respect, acknowledgment of the discourse, or an opportunity to carefully weigh a response. It does not necessarily indicate discomfort or anger and should be an expected element in pacing an interaction. The clinician may carefully schedule patient care by the clock, but many cultures do not share a similar orientation to time. This may be due to practical difficulties of arriving at a destination at an exact time or to the irrelevance of exact timing to most activities. Clinical questions about a symptom's occurrence, intensity, and effect on the patient and his or her social life may initiate a discussion about cultural values. In addition to avoiding harm in the physician-patient relationship, learning about cultural norms can be an enjoyable aesthetic experience for providers.

Simple forms of etiquette are effective. Most cultures afford respect to elders as well as to health care practitioners. Treating clients with respect may be indicated by appropriate terms of address, such as Mrs. Brown rather than a first name or a term of endearment. Personalized relationships are important to many cultures and are established through noncommercial transactions. Clients may offer food or other token gifts to reduce the formal barrier and create a more personal relationship; refusal would be treated with suspicion. Establishing this type of relationship may also be accomplished by the provider offering a personal disclosure, such as initiating a conversation about a mutual interest (which may entail responding to personal questions), to reduce communication barriers. Clinicians should avoid any temptation to relate to clients of different cultures by using terms that are not in their own vocabulary, such as speaking Black English to African American elders or using putative honorifics that may actually be offensive, such as "chief" to a Native American. Clinicians may adjust their medical and anatomical vocabulary to the patient's education and language, but they need to check that the information provided is understood.

Advance Directives and End-of-Life Care

In a multicultural society, concern for justice brings ethical issues to the fore. Several studies in the United States identified significant differences among ethnic and cultural groups on the completion of advance directives and end-of-life decision making strategies (Blackhall et al., 1995, 1999; Caralis, Davis, Wright, & Marcial, 1993; Hornung et al., 1998; Klessig, 1992). For example, African Americans tend to want all possible life-sustaining treatments; they distrust advance directives and see them as authorizing neglect or inferior care based on racial and socioeconomic factors. Korean Americans may voice a personal wish for a natural death (i.e., no life-prolonging technology) but expect their children to insist on all possible life-saving measures.

To set the context for ethical decision making, clinicians need to ascertain whether patients are reluctant or responsive to discussing end-of-life care. In addition, clinicians should attempt to understand beliefs about death, spiritual issues associated with dying, the nature of the social support system, and attitudes of patients and their families toward the health care system. In many cultures, the direct, frank, structured discussions of death implied in advance directives and end-of-life care planning are considered harmful to the patient's well-being, insinuate hopelessness, increase suffering, and hasten the inevitable outcome. Among Native Americans, for example, the issue is best addressed indirectly, talking about others who have died (using a referent term rather than the personal name) to elicit responses about what would constitute a "good death." Clinicians who can address these issues over time are more likely to reach an understanding of their patients' wishes.

Concepts of autonomy vary and imply different norms in the disclosure of information and decision making. Ethiopian and Persian immigrants believe that bad news should be conveyed to the patient by a family member or close friend, not by a health care provider. In these circumstances, doctors confront the dilemmas of concealment of information, truthfulness in diagnosis or prognosis, and protection of patient confidentiality. Some physicians manage the problem by asking patients how much information they want to know, who else should be informed, and who they want to make decisions with or for them. Another strategy is to encourage patients to ask questions over several visits in order to absorb information.

The role of decision maker varies with cultural norms. Daughters might be the first choice among Hispanics and African Americans, sons among Asians, and spouses among Anglo-Europeans (Hornung et al., 1998). In general, Native American cultures strongly support autonomous decision making, and children are unlikely to interpose their wishes. However, if an elder is without capacity, a family spokesperson would likely emerge to represent the elder's authentic wishes. Clinicians should avoid directing information to, and expecting decisions from, the best-educated family member; this person may not necessarily be culturally empowered with decision-making authority.

B. JOSEA KRAMER

See Also
Advance Directives
Aging African Americans
American Indian Elders
Asian and Pacific Islander American Elders
Communication Issues for Practitioners
Gay and Lesbian Elderly
Hispanic and Latino Elders
Homelessness
Immigrant Elders
Over-the-Counter Drugs and Self-Medication
Palliative Care
Patient-Provider Relationships
Rural Elders

Internet Key Words
Access
Advance Directive
Communication
Cultural Sensitivity
Culture
Diversity
End-of-Life Planning
Ethnicity

CURE VS CARE

It appears strange in a post-modern age to be discussing care and cure in adversarial terms. Cure (the usual aim of medical treatment) is usually so dominant that care (the usual aim of nursing) can only be valued and flourish when pitted successfully against cure. Theoretically, people have a right both to the opportunity to strive for cure and to be cared for at the same time. Indeed, experienced health care professionals from all disciplines believe that the two are inseparable. There is little doubt, however, that more resources are committed to the delivery of curative modalities of health care than to carative modalities, including physical resources, personnel, and research. Although a large share of the health dollar is spent on treatment for elderly people—cardiac surgery is now predominantly performed on people over the age of seventy—caring for people when they insidiously lose independence in their later years is left largely to the persons themselves or their relatives or to untrained caregivers.

Modern medicine has exploited the hopes of all people for diminished mortality. While we have been successful in delaying the moment of death, the record to date on achieving cure is not good. The reality is that there is no cure for old age, and death is inevitable eventually. The cost of prolonging life can be high for the person and society in terms of quality of life and the cost of treatments. Many elderly people live with chronic illness and concomitant frailty; most attempts at curing disease result in a compromised state of health for the sufferer. Perhaps *treatment* should be substituted for the word *cure*—treatment that can be given by medical doctors, physiotherapists, dietitians, social workers, nurses, and so on—for all the symptoms of disease that can plague people as the years mount up. If medicine cannot cure disease, then perhaps a more reasonable expectation is for the elderly to be as comfortable and as able as possible within the boundaries of existing science and limited resources for health care. On the whole, this means that treatment is aimed at controlling and improving physical symptoms within the context of each person's social and psychological situation.

Illness haunts elderly people particularly as they reach the eighties and nineties. Old age, for most people, is accompanied by increasingly frequent episodes of illness that can herald frailty, the factor that makes people look and feel old. Older people turn to medicine and expect treatments in order to live to their maximum potential.

Chronically ill elderly people place great store in the medical profession and its powers to treat their illnesses. They refer to medical opinions they have been given, the expertise of different doctors they have encountered, relish the doctor's time and attention, and remember and recount things that doctors had said to them verbatim (FitzGerald, 1999). In general, chronically ill people acknowledge members of the medical profession as the experts (although this respect can wear thin on occasions with the frustration of unsatisfactory outcomes), and they pin their hopes for improvements, or even cure, upon them. What the chronically ill are most appreciative of, however, is doctors who show them humane concern and care in conjunction with medical expertise.

Although it is not quite as clear-cut as presented here, people of any age generally appreciate two things in health care professionals. These are professional acumen and humane qualities. They describe acumen as being able to make a correct diagnosis and improve or contain the course of their disease process. They describe humane qualities in terms of being made to "feel better" or "really cared for." This care is manifest in the time given to others, the way they are listened to, and the way their particular needs are attended to.

Traditionally, nurses have been associated with care rather than cure. However, as roles among health care professionals become blurred, doctors cannot rely on nurses or other health professionals to assure the caring side of treatment. Nurses have a range of therapies to offer within their caring paradigm. This range of therapies is expanding each year as nurses both extend and expand their practice. The nurse practitioner is well positioned to offer elderly people precisely the combination of cure and care that they require.

Nursing has devoted a great deal of time to identifying and testing theories and work that apply nursing therapy within a fundamental caring context. For example, Lydia Hall's (1969) work at the Loeb Centre in New York put into practice her ideas regarding nursing as core, care, and cure. Pearson et al. (1992) demonstrated the effectiveness of nursing therapy within a caring model. Elderly people admitted to units had a primary need for nursing therapy. These patients were compared with a control group nursed in a typically cure oriented system. Results showed that the experimental group had a higher quality of care, higher levels of independence on discharge, slightly longer hospital stay, but lower cost per day. In an ethnographic study, Ersser (1997) explored the therapeutic activity of nursing and concluded that the core elements of nursing were the presentation of the nurse, relating to the patient, and the tasks and procedures undertaken in the course of nursing.

Neuman's (1995) health care systems model is but one example of a theoretical model of nursing that has been developed carefully over three decades. This work combines therapy and care in relation to the specific needs of individual human beings. Suitable for all health care professions, it provides a useful framework for finding the elusive balance of cure and care in each encounter or professional relationship with a client. Neuman's emphasis is on prevention, health education, wellness, and the management of ill health.

As the health professions continue to develop their roles as therapists to the elderly, it will be interesting to see whether they adopt a medical model of short consultation and prescription, or whether they retain and refine skills of caring for and treating people (i.e., curing) at the same time. Their choice will make a world of difference to the elderly.

MARY FITZGERALD

See Also
Advanced Practice Nursing

CYSTOCELE
See
Uterine, Cystocele/Rectocele, and Rectal Prolapse

D

DAYTIME SLEEPINESS

Excessive daytime sleepiness (EDS) is a common symptom in the elderly. Distinct from fatigue, which is difficulty sustaining a high level of performance, EDS is the inability to maintain alertness, with characteristic hypersomnolence. Causes of EDS include age-related changes in chronobiology, sleep disorders, other medical and psychological disorders, medications, environmental factors, and altered social patterns. It is unclear how much of the change in sleep patterns experienced by older adults is due to normal physiological alterations, pathological events, sleep disorders, or poor sleep hygiene. Complicating the isolation of the cause of EDS is the perception that daytime sleepiness is acceptable in older adults. This perception may prevent elders from seeking medical attention or receiving medical care for daytime sleepiness. The effects of sleep disruption on daily functioning and the resultant caregiver burden are common motivations for the institutionalization of older individuals (National Commission of Sleep Disorders Research, 1994).

Consequences of EDS

Approximately 56,000 car crashes per year are attributed to falling asleep at the wheel (Knipling & Wang, 1994). Contributing to this are the neurobehavioral deficits, such as decreased reaction time and sustained attention, associated with EDS. Daytime sleepiness impairs memory, cognitive processing, affect, and functional status. Many of these symptoms are similar to those found in depression, delirium, and dementia. Thus, when contemplating these diagnoses, sleep-related causes should be considered first.

Causes of EDS

Obstructive Sleep Apnea

Obstructive sleep apnea (OSA) is a condition involving intermittent pharyngeal obstruction that produces cessation of respiratory airflow (for at least 10 seconds) and often oxygen desaturation. Arousal after the apneic event restores upper airway patency, and breathing and airflow resume. By definition, OSA is diagnosed when these events occur at a rate greater than five per hour of sleep, accompanied by snoring, gasping, daytime sleepiness, and impaired daytime functioning. However, it is not uncommon for patients with severe symptoms to experience multiple awakenings in one night, severely fragmenting sleep and reducing deep (stage III or IV) sleep and rapid eye movement (REM) sleep, which are necessary for healthy mental and physical functioning. Among individuals over age 65, as many as 24% have sleep apnea, predominantly the obstructive type (Ancoli-Israel, Kripke, & Mason, 1987). Sleep apnea is both an age-related and an age-dependent condition, with an overlap in both distributions in the 60 to 70 age range (Bliwise, King, & Harris, 1994). In particular, OSA symptoms, including nocturnal witnessed apneas, snoring, snorting, and gasping, have been identified in 24% of independently living elderly (>65 years old), 33% of aged acute-care inpatients, and 42% of nursing home residents (Ancoli-Israel et al., 1987). Other easily observable signs of OSA are morning headache, dry mouth from mouth breathing during sleep, and nocturia. Although some attribute the nocturia to benign prostatic hypertrophy, OSA causes polyuria in both men and women (Umlauf et al., 1999).

Treatment depends on the contributing pathology and on patient preference and includes nocturnal nasal continuous positive airway pressure (CPAP), surgical procedures (palatoplasty to reduce airway encroachment), oral appliances, and weight reduction when obesity is a contributing factor. Nasal CPAP is currently the most effective treatment for OSA, producing improvements in neurobehavioral performance, daytime sleepiness, snoring, and quality of life.

Insomnia

Recent studies comparing younger and older adults found that the elderly (>65 years) are approximately 1.5 times more likely to have sleep difficulties. In addition, women are more likely to report insomnia-like symptoms than men. Insomnia includes complaints of delayed sleep onset, premature waking after sleep onset, and very early arousal that results in shortened total sleep time. Although it is unclear whether insomnia is an organic, psychological, pharmacological, chronobiological, or behavioral problem, it is associated with cardiovascular, respiratory, gastrointestinal, renal, and musculoskeletal disorders. Primary insomnia patients have reduced natural killer cell activity, suggesting impaired immune function.

Insomnia may be transient or chronic, and the perception of sleep duration may not correspond to objective assessment. The frequent awakenings suggestive of insomnia may be a conditioned arousal response due to environmental or behavioral cues. Anxiety associated with emotional conflict, stress, recent loss, feelings of insecurity, or change in living arrangements can also produce insomnia. General anxiety and the conditioned arousal response at sleep onset associated with insomnia may prompt more frequent use of hypnotic medication, a common treatment for insomnia. Although hypnotics may temporarily relieve symptoms, they also affect sleep architecture and can cause a deterioration in the quality of sleep. Thus, a cycle of dependency and abuse can occur. For this reason, as well as the lack of evidence regarding

their safety and efficacy in the elderly, pharmacological interventions may not be the most appropriate response. Behavioral interventions might be considered initially.

The cause and duration of insomnia should be the primary determinants of the treatment selected. For example, insomnia associated with a psychological origin, such as depression or anxiety, is best treated with antidepressants or anxiolytics. If the insomnia is of recent onset, short-term benzodiazepines are best for transient symptoms. When insomnia is "learned," and this maladaptation interferes with the initiation of sleep, behavioral interventions are most appropriate. Such interventions include a review of good sleep hygiene (see below) and biofeedback. The effect of institutional routines on the sleep of elders should be evaluated. Early bedtimes, continence checks, noise, and light levels can contribute to behavioral insomnia.

Restless Leg Syndrome and Periodic Leg Movements

Two neuromuscular dysfunctions that can cause EDS in the elderly are restless leg syndrome (RLS) and periodic leg movements (National Commission of Sleep Disorders Research, 1994). RLS is characterized by an almost irresistible urge to move the limbs, usually associated with disagreeable presleep leg sensations. These sensations often interfere with initiating and maintaining sleep, resulting in daytime hypersomnolence. As a secondary condition, RLS may be caused by iron-deficiency anemia, uremia, neurological lesions, diabetes, Parkinson's disease, rheumatoid arthritis, and certain drugs (e.g., tricyclics, selective serotonin reuptake inhibitors, lithium, dopamine blockers, xanthines). Periodic leg movements, also known as nocturnal myoclonus, are flexions of the leg and foot that may disrupt sleep. Although the cause and associated mechanism of this chronic condition are not well defined, these movements have been linked to metabolic, vascular, and neurological causes. Pharmacological interventions for these two disorders, including benzodiazepines, levodopa, and mild opiates, have been used, but their efficacy for long-term treatment has not been extensively evaluated.

Assessment of EDS

The Stanford Sleepiness Scale (Dement & Carskadon, 1982) and the Epworth Sleepiness Scale (Johns, 1992) are brief, valid, reliable measures used to clinically assess EDS. The likelihood of having OSA can be determined by the Mulivariable Apnea Prediction Index (Maislin et al., 1995). The Functional Outcomes of Sleep Questionnaire (Weaver et al., 1997) can be employed to assess the impact of daytime hypersomnolence on functional status, and the Pittsburgh Sleep Quality Scale (Buysse, Reynolds, Monk, Berman, & Kupfer, 1989) is beneficial in the assessment of insomnia. Most important in the assessment of EDS is evaluation of the patient's knowledge and performance of sleep hygiene.

Sleep Hygiene Measures

Poor sleep habits can significantly contribute to sleep problems. An assessment of sleep hygiene reviews the intake of substances that can adversely affect sleep by causing sleep disruption and a reduction in deep and REM sleep, such as alcohol, nicotine, and caffeine. A variety of strategies can be used to alter a patient's sleep patterns and promote sleep initiation and maintenance. The bed should be used only for sleeping (or intimacy or sex). Consistent and rest-promoting bedtime routines should be developed, such as going to bed at the same time each evening. Exercise should be avoided three to four hours before bedtime, because there is a natural drop in body temperature that normally precedes sleep. Caffeine and nicotine should be avoided after 12 noon, and evening intake of alcoholic beverages should not exceed three drinks. Large meals and excitement should be limited before bedtime. Upon awakening, patients should get out of bed, no matter what time it is. If

they awake during the night, they should avoid looking at the clock; frequent time checks heighten anxiety and make sleep onset more difficult. Patients should also avoid frequent naps during the day longer than 10 to 15 minutes' duration. They should be advised to sleep in a cool, quiet environment.

MARY GRACE UMLAUF
TERRI E. WEAVER

See Also
Confusion/Delirium: Risk, Diagnosis, Assessment, and Interventions

Internet Key Words
Daytime Sleepiness
Insomnia
Neurobehavioral Deficits
Obstructive Sleep Apnea
Restless Leg Syndrome
Sleep

Internet Resources
American Academy of Sleep Medicine
http://www.asda.org

National Center on Sleep Disorders Research
http://www.nhlbi.nih.gov/health/public/sleep/index.htm (for the general public)
http://www.nhlbi.nih.gov/health/prof/sleep/index.htm (for health care professionals)

New Abstracts and Papers in Sleep (NAPS)
http://www.websciences.org/bibliosleep/NAPS/

Sleep Net.Com
http://www.sleepnet.com/

DEATH ANXIETY

Philosophers in the modern era have almost unanimously assumed that the human encounter with death is marked by angst, dread, uncertainty, and fear—in a word, anxiety. At the same time, advances in life-extending technology, diet, and general lifestyle (at least in industrialized countries) have improved the longevity of most people, effectively associating death with old age in the popular imagination. As a result of these two converging trends, one might assume that the elderly, who are statistically "closest" to death, would experience considerable death anxiety. However, the general trend across studies points to a decrease in death anxiety with advancing age (Bengtson, Cuellar, & Ragan, 1977; Fortner & Neimeyer, 1999; Fortner, Neimeyer, & Rybarczeck, 2000). This does not mean that all older people have uniformly low levels of death anxiety, but that as a group, they have lower levels of death anxiety than middle-aged people. In later adulthood, death anxiety tends to stabilize and does not appear to continue decreasing with age. Well-designed surveys have demonstrated that age is a relatively good predictor of fear of death, accounting for more of the variation than other important demographic and social variables, such as education, income, and ethnicity. Future researchers need to pay closer attention to factors specific to older adults, such as perceived nearness of death, quality of life, subjective passing of time, and achievement of the developmental tasks of late adulthood, to gain a clearer picture of the psychological transitions in late life that may affect attitudes toward death.

Research has shown that the gender difference in death anxiety evident in younger adults—with females reporting more fear than men—is not present in older adults. This finding is consistent with research showing that older adults are less differentiated by gender and exhibit a more androgynous gender identity. Ethnicity has been associated with greater death awareness, with African Americans and Hispanics reporting greater familiarity with death and greater exposure to violence than whites. However, studies have been mixed regarding whether these ethnic differences are associated with greater *anxiety* about death among various subgroups.

Because deteriorating health and diminished income necessitate changes in residence for many elderly people, researchers have studied whether living arrangements (e.g., in the community versus in an institution) have an impact on death attitudes (Fry, 1990; Neimeyer & Van Brunt, 1995). Some evidence suggests that nursing home residents report greater death fears than do those who live more independently, but this may be confounded with the association between death anxiety on the one hand and poor health and diminished life satisfaction on the other (Goebel & Boeck, 1987; Neimeyer, 1994). Nonetheless, the findings that the institutionalized elderly are more likely to encounter death, think of it often, have less control over their lives, and suffer deterioration in the quality of life make further study of the death concerns of this vulnerable group a priority in future research.

Many studies in this area correlate death anxiety with another single variable, such as physical health, psychological status, and religiosity (Tomer, 1992). A systematic review of this literature indicates that greater physical and emotional problems predict higher levels of death anxiety in older adults, although more work is needed to clarify the specific medical and psychological characteristics responsible for these general trends. In younger cohorts, people who are more religious generally report lower levels of death anxiety. This relationship is less evident in older adults, although some studies suggest that religious orthodoxy and belief are associated with greater death acceptance, whereas simple church attendance and involvement in religious activities are unrelated to death attitudes.

Perhaps a more enlightened approach would be to focus on individual personality traits, coping styles, and competencies and how they interact with environmental conditions to accentuate or ameliorate fears about death and dying. For example, ego integrity or life satisfaction of elderly respondents has been found to interact with their place of residence: Institutionalized elderly with low ego integrity are especially vulnerable to heightened concerns about their mortality. Other work has concentrated on the particular coping skills used by older persons (e.g., prayer, reminiscence) to deal with specific aspects of death anxiety (e.g., helplessness, questions about the afterlife, the pain of dying).

Sophisticated studies that examine discrete death fears as a function of styles of coping with developmental transitions would contribute to our information on the subject.

Unfortunately, our understanding of the nature and predictors of heightened death fear among older individuals has been hampered by several factors, both theoretical and methodological. Theoretically, researchers have concentrated on easily measured demographic characteristics and measures of physical and mental illness, rather than on the potential resources of older adults (e.g., coping, family support), which could yield a more optimistic view of their ability to face death with equanimity, acceptance, or even affirmation. Methodologically, investigators have relied too heavily on unvalidated, idiosyncratic death anxiety scales that treat death attitudes as a single, unidimensional trait rather than a complex construct with many aspects (e.g., fear of pain associated with dying, anxiety about loss of control, apprehension about punishment in the afterlife). However, valid and reliable multidimensional measures of death anxiety and acceptance are now available and are beginning to add clarity and richness to studies. Likewise, investigators have recently begun to ground their studies in more comprehensive psychological, sociological, and developmental theories, which could give coherence and direction to future research. As we continue to clarify the environmental and personal determinants of death anxiety, we will be in a better position to design educational, counseling, and policy interventions to promote a more humane encounter with death and loss at all points in the life span.

ROBERT A. NEIMEYER
BARRY FORTNER

See Also
Anxiety and Panic Disorders
Coping with Chronic Illness
Spirituality
Suicide

Internet Key Words
Death Anxiety
Thanatology

DECONDITIONING PREVENTION

Physical deconditioning in the elderly is a common physiological phenomenon and can severely limit an individual's ability to perform activities of daily living. General deconditioning typically results from prolonged illness, bed rest, or sedentary lifestyle and is commonly associated with decreased aerobic capacity, muscle atrophy, weakness, and loss of coordination and balance. The association between deconditioning and coordination and balance deficits is of particular concern, as falls are a serious threat to the independence and health of the elderly population.

Some of the physiological changes associated with deconditioning include the following (Kisner & Colby, 1990):

- Decreased work capacity resulting from decreased oxygen uptake and cardiac output
- Decreased circulating blood volume, which can result in tachycardia, orthostatic hypotension, dizziness, and syncope
- Decreased plasma and circulating red blood cells, which can decrease the oxygen carrying capacity of blood
- Decreased lean body mass related to decreased muscle size and strength
- Increased prevalence of osteoporosis and an increased likelihood of fractures upon falling

One of the primary objectives of an exercise intervention is to improve physiological and functional indices to promote independence and quality of life. Through regular exercise and conditioning, the negative aspects of cardiovascular and neuromuscular deconditioning can be reversed and functional status increased (Buchner, Cress, de Lateur, Esselman, & Margherita, 1997; Shumway-Cook, Gruber, Baldwin, & Liao, 1997; Wolfson et al., 1996). Several randomized, controlled studies indicate that a regular exercise program can be beneficial for the elderly (Campbell et al., 1997; MacRae, Feltner, & Reinsch, 1994). In particular, increased balance, improved ambulatory status, and decreased incidence of falls have been documented with various exercise interventions. These studies advocate a variety of approaches, including muscle strengthening, cardiovascular endurance, and balance training.

Exercise Prescription for the Elderly

When developing an exercise program, four basic factors should be considered: (1) mode or type of exercise, (2) frequency of participation, (3) duration of each exercise period, and (4) intensity of the exercise bout. Adequate warm-up and cool-down periods, as well as appropriate stretching, should be incorporated in the exercise regimen. Before beginning any exercise program, individuals should be screened and cleared by their physicians for cardiac risk factors, musculoskeletal limitations, and starting exercise levels.

Resistance Training

Decreased muscular strength has been implicated as a risk factor for falls and is a common cause of mobility problems in the elderly. Previously reported resistance programs to improve muscular strength and endurance in this population have focused on the major muscle groups of the upper and lower extremities. Strengthening exercises should address the major muscle groups of the lower extremity, including the calf, knee extensors, and hip extensors. Increased strength in these muscle groups is associated with improved gait parameters (e.g., velocity and stride length) in the elderly (Judge, Smyers, & Wolfson, 1992). Resistance exercises also should address the arms, shoulders, and back, as these muscle groups are important for upper extremity function and postural stability.

Resistance can be accomplished through a variety of methods, including rubber tubing (e.g., Theraband), dumbbells, ankle or wrist weights, or specialized machines. Participants should begin by performing 10 repetitions of each exercise (one exercise per muscle group). The magnitude of resistance should be such that fatigue occurs following the last repetition. Seventy percent of an individu-

al's one-repetition maximum is a common method for determining the appropriate resistance. Exercises should be progressive over time, such that 3 sets of 10 repetitions can be performed. Each set should be followed by a 30-second rest period, with a 1- to 2-minute rest period between exercises.

It is generally recommended that at least two strengthening sessions per week (preferably three) be performed. Exercises should be conducted slowly and with supervision to ensure proper technique and safety. Home programs should be encouraged once the exercises have been mastered.

Cardiovascular/Aerobic Training

Endurance training in the elderly has been associated with increases in walking ability and balance indices (Buchner et al., 1997). The most common types of cardiovascular exercise are walking programs, exercise machinery (e.g., stationary bicycles), and specialized aerobic movement classes. A typical endurance session lasts one hour and should consist of a 15-minute warm-up, a 35- to 40-minute exercise session, and a 5- to 10-minute cooldown period.

In accordance with the guidelines of the American College of Sports Medicine, the training intensity should be between 60% and 75% of the age-adjusted maximum heart rate (calculated by subtracting the participant's age from 220). The target heart rate should be gradually increased from 50% to 75% of the maximum heart rate over the course of three months (Buchner et al., 1997). Heart rate should be monitored during and after exercise to ensure the proper training intensity.

Balance Training

Specific balance exercises have been shown to increase mobility and postural stability in the elderly. The goal of such exercises is to improve vestibular function and proprioception. Both active and static exercises have been proposed in the literature. Active exercises include tasks such as a tandem heel-toe gait and walking along a straight line; static balance exercises include standing on one leg and maintaining various postures. The level of difficulty of static exercises can be increased by having individuals perform them with their eyes closed or while standing on a foam block. In addition, tasks that employ a narrow base of support or weight shifting are appropriate.

Recently, tai chi has been studied in several randomized, controlled trials and has been shown to be effective in improving postural sway (Shih, 1997) and in delaying the onset of falls in the elderly (Wolf, Barnhart, Ellison, & Coogler, 1997). Tai chi exercises involve full body movements that facilitate muscular control of movement, flexibility, and improved "body awareness." Interventions of approximately four months' duration have been effective in improving balance and mobility indices.

Although different forms of exercise have been shown to improve functional status in the elderly, the optimal exercise regimen for improving physical fitness and balance in the elderly has yet to be defined. Exercise programs should be individualized, based on muscle strength, cardiovascular function, and balance and gait parameters. Emphasis should be placed on the most significant deficits and the goals of the participant. A multidimensional exercise program should be implemented to prevent the deleterious effects of deconditioning.

CHRISTOPHER M. POWERS

See Also
Balance
Exercise and the Cardiovascular Response
Falls Prevention
Iatrogenesis

Internet Key Words
Balance
Deconditioning
Elderly
Exercise
Mobility
Strength

Internet Resources
American Council on Exercise
http://www.acefitness.org

American Heart Association
http://www.americanheart.org

50-Plus Fitness Association
http://www.50plus.org

DEHYDRATION

Dehydration, a decrease in total body water, is the most common fluid and electrolyte disturbance among elderly with acute illness. Although the morbidity of hospitalized elderly with dehydration is seven times more likely than in age-matched cohorts without dehydration (Davis & Minaker, 1999), older persons residing in nursing homes and in the community are also at risk. Older people have increased sensitivity to dehydration as a result of the normal physiological changes of aging. In normal aging, fat is increased and lean body mass is decreased, which corresponds with decreased total body water (TBW) in the elderly. Total body water is 60% of body weight in young adults, compared with 50% in older men and 45% in older women. The decrease in total body water after the fifth decade is due to decreased extracellular fluid (ECF) preceding age 60, and to decreased intracellular fluid (ICF) in the sixth to ninth decades of life (Lye, 1992). Other normal age-related changes that place the elderly at risk for dehydration include increased thirst threshold, the kidneys' reduced ability to conserve water and concentrate urine, increased antidiuretic hormone (ADH) secretion, and impaired renal sodium conservation. Dehydration in the elderly can also be iatrogenic, as a result of failure to adjust the dosage of diuretics. "Obligate" diuresis, in addition to being associated with diuretics and other pharmacological agents, is associated with glycosuria, hypercalciuria (secondary to malignancy and hyperparathyroidism), intravenous mannitol, and x-ray contrast agents. Normal fluid intake in an older person is approximately 1200 to 1500 mL, assuming no cardiovascular or renal conditions that limit fluid intake. Normal output is computed at the rate of 30 mL per hour, or approximately 800 mL urine output in 24 hours.

Etiology

Body fluids are divided into intracellular and extra-cellular fluids, which include interstitial fluid, plasma, and transcellular fluid. A shifting of the fluids can result in dehydration. Dehydration can be categorized as isotonic, hypertonic, or hypotonic, based on the quantity of salt (sodium) loss in relation to water loss. Isotonic dehydration occurs with a loss of ECF without any change in ICF; sodium and water are lost in equal amounts, with normal serum sodium levels (135 to 145 mEq/L) and normal serum osmolarity (280 to 295 mOsm/L). Hypertonic dehydration is characterized by greater water loss than sodium loss, with high serum sodium levels (>145 mEq/L) and high serum osmolarity (>295 mOsm/L). Hypernatremia, an electrolyte disorder, is classically associated with this type of dehydration. Hypotonic dehydration occurs when more sodium than water is lost, with low serum sodium levels (<135 mEq/L) and low serum osmolarity (<280 mOsm/L). ECF is the lowest with hypotonic dehydration.

Dehydration is caused by excess loss of water, failure to recognize the need for increased water, and/or a problem with water ingestion. There are many clinical conditions that increase fluid requirements in the elderly and, if untreated, can cause dehydration. The mnemonic *States of Increased Fluid Requirements in Old Men And Girls* is one way to remember these clinical conditions. *Self-imposed* fluid restriction to avoid incontinence can be problematic in the elderly. Fluid loss through the *integumentary* system is a result of burns, a hot and dry environment, hemorrhage, open fistulas, or wounds (including decubitus ulcers). Decreased *functional ability* in the elderly can limit access to fluids due to immobility, physical restraints, diminished vision, and a change in cognition or sensorium such that the need for water is not recognized. Fluid losses through the *renal system* include conditions that cause polyuria, in addition to dialysis-related fluid losses. *Other clinical states* that increase fluid requirements include fever, hyperventilation, hypotension, and third-spacing of fluids. Many *medications* increase fluid requirements in the elderly, es-

pecially diuretics. *Anabolic* steroids and *alcohol abuse* also alter fluid status. Lastly, losses through the *gastrointestinal tract* associated with emesis, diarrhea, constipation, and nasogastric suctioning increase fluid requirements in the elderly. Enteral feedings without adequate free water supplementation is another cause of dehydration in this age group.

The most common cause of hypernatremia is an absolute or relative decrease in body water (Lye, 1992). This can be caused by multiple factors—often three or more acting simultaneously—including febrile illness, frailty, surgery, nutritional supplementation, intravenous compounds, poorly controlled diabetes mellitus, diarrhea, gastrointestinal bleeding, diuretics, and dialysis-related fluid loss. More subtle causes are respiratory infections, steroid therapy, and laxative abuse. Pyrexia associated with acute infection increases insensible water loss from diaphoresis, tachypnea, and increased cellular catabolism. Diabetes insipidus is relatively uncommon in the elderly and is more often nephrogenic than pituitary in origin. Hypoaldosteronism in the context of Addison's disease and the hyporeninemic state associated with normal aging are associated with increased fluid loss, as is the inadequate vasopressin response to hyperosmolarity seen in persons with Alzheimer's disease. Alcohol abuse (ethanol) and phenytoin (Dilantin) suppress vasopressin release, with reduced tubular reabsorption of water.

Diagnosis

Impending dehydration might be signaled by dark yellow or greenish brown urine and urine output less than 800 mL per day, which warrants intensive monitoring and further assessment. Although an early sign of dehydration might be weight loss, imminent or fulminating dehydration can present with altered mental status, agitation or lethargy, lightheadedness, confusion, and syncope. On physical examination, there is decreased skin turgor, dry and pale mucous membranes, longitudinal tongue furrows, speech difficulties, sunken eyes, tachycardia, and orthostatic hypotension. However, research suggests that orthostatic hypotension and decreased skin turgor are not indicators of dehydration (Mentes et al., 1998; Weinberg, Minaker, & The Council on Scientific Affairs, 1995). Symptoms of hypernatremia in the early stages may be nonspecific, with or without ECF loss. Weakness and lethargy are often misinterpreted and attributed to other conditions prevalent in the elderly (Lye, 1992). Measures of urine osmolarity, or specific gravity, and volume can confirm the presence of hypernatremia. Normal specific gravity is ±1.5. Increased hematocrit, blood urea nitrogen (BUN), and serum creatinine indicate significant volume depletion. There may be little to no increase in urine osmolality because the aging kidney has a diminished ability to concentrate urine; serum sodium may be high, low, or unchanged, depending on the underlying cause of volume depletion. However, output volume will be low. Evidence of pituitary diabetes insipidus is decreased urine volume and increased osmolality; there is no change in nephrogenic diabetes insipidus. Hyperglycemia with subsequent glycosuria and osmotic diuresis, hypercalcemia, and hypokalemia must be ruled out as mechanisms of impaired renal water-conserving ability.

Treatment

The severity of fluid volume depletion is estimated by evaluating blood pressure, orthostasis, skin turgor, and urine output. Comparison of the patient's current weight and premorbid (baseline) weight can help determine the degree of fluid loss. Fluid replacement can be oral, intravenous, enteral, or a combination. Three liters of isotonic fluid can be safely administered intravenously at two sites over 24 hours at 60 mL per hour. An estimate of the volume of free water needed to restore body fluid tonicity to normal can be calculated as:

$$\text{Water deficit} = \frac{60\% \times \text{current body weight (kg)}}{(P_{Na}/140) - 1}$$

where P_{Na} is current plasma sodium concentration (Beck, 1999).

Hypertonic dehydration with hypernatremia is treated with rapid infusion of isotonic saline; if the patient is hemodynamically stable, half the fluid deficit should be replaced within 24 hours, and the remaining volume should be replaced during the next 48 to 72 hours (Davis & Minaker, 1999). To reduce serum osmolality, 5% Dextrose in 0.45 Saline is recommended. Serum electrolytes, particularly sodium, potassium, BUN, and creatinine; central venous flow; blood pressure; skin turgor; heart rate; and urine output should be monitored. It is important to note that mental status changes associated with too rapid rehydration and cerebral edema can last two weeks or longer.

Special Considerations for Tube-Fed Patients

If older persons do not receive free water in addition to the enteral feeding formula, they can be at increased risk for dehydration. The caloric density of enteral formulas range from 1 to 2 kcal/mL. Most formulas contain between 71% and 85% water per liter, but more calorically dense formulas (≥ 1.5 kcal/mL) contain less water (Keeth, 1996). To prevent dehydration from occurring in a tube-fed older patient, additional free water should be administered throughout the day when flushing the tube or giving medications. The amount of additional free water is calculated as:

Additional free water intake =
Total water needs
(about 25 to 30 mL/kg body weight/day) −
Fluid intake from feedings
(the percentage of water in the formula ×
the volume of feeding delivered)

Older persons receiving enteral tube feeding are also at risk for hyponatremia and/or hypotonic dehydration, since most enteral formulas have a low sodium content. In this case, it may be necessary to add table salt to feedings (Keeth, 1996).

Environmental Considerations

Older persons living in the community should be monitored during the summer months, especially when heat advisories are in effect. Severe hot and humid weather can cause dehydration, in addition to hyperthermia. Social services can be instrumental in assessing living conditions and providing assistance in securing fans or air conditioners. Older persons must be encouraged to increase their fluid intake during warm weather and refrain from going outdoors when the sun is strongest.

Prevention

The minimum fluid intake for older adults is 1,500 mL/day. For individuals identified as at risk for dehydration, calculation of daily water intake (as opposed to fluid intake) is essential. Calculation of total water intake, based on fluid and food ingested, is recommended, because the elderly have a high percentage of water intake from food (Gaspar, 1999). The standard water intake formula that provides the most individualization is: (1,600 mL × m^2 of body surface area) × 0.75. Multiplication by .75 considers the loss of total body water by the elderly.

Suggestions for enhancing water intake include providing fluids and foods high in water content throughout the day; considering an individual's previous intake pattern and individual preferences; educating the patient, staff, and informal caregivers regarding the need for water intake; assessing the patient's individual water intake goal; and recommending fluids and foods that are high in water content.

ROSE ANN DIMARIA-GHALILI
AND THE EDITORIAL STAFF OF THE
ENCYCLOPEDIA OF ELDER CARE

Internet Key Words
Dehydration
Fluid Replacement
Fluid Volume Deficit

Hypernatremia
Water Intake
Water Intake Standard

DELIRIUM
See
Confusion/Delirium: Risk, Diagnosis, Assessment, and Interventions

DEMENTIA: NONPHARMACOLOGICAL THERAPY

Older adults with Alzheimer's disease suffer progressive cognitive, behavioral, and affective losses. Alzheimer's disease is the most common type of irreversible, degenerative dementia, accounting for approximately 60% of dementia in older adults. Research to date has focused primarily on the biomedical aspects of the disease, such as pathology and pharmacological interventions for cognitive symptoms. Yet the behavioral symptoms are some of the most difficult manifestations for caregivers (both family and formal) and occur throughout the disease trajectory. Treatment of behavioral symptoms remains primarily pharmacological, with antipsychotics the most common agents utilized. Although pharmacological agents have demonstrated some efficacy for specific symptoms of dementia, such as psychosis, the adverse side effects of these drugs force clinicians to constantly weigh the risk-benefit ratio in prescribing for an older confused client. Research efforts are attempting to identify behavioral strategies to augment pharmacological management. However, with few exceptions, the studies of nonpharmacological behavioral interventions have been anecdotal and have not included rigorous, systematic research designs.

Behavioral Symptoms

Behavioral symptoms of Alzheimer's disease are problematic for older adults with dementia, as well as for the family and formal caregivers, such as nursing staff. Symptoms include physical and verbal aggression or screaming, resistance to care, wandering, difficulties with personal hygiene and care, and apathy or withdrawal. Family caregivers report that behavioral problems are a major source of stress and a major reason for the decision to institutionalize adults with Alzheimer's Disease. Behavioral symptoms result from the four main cognitive effects of Alzheimer's disease: amnesia (loss of memory), aphasia (impairment of communication), agnosia (inability to comprehend or recognize objects), and apraxia (inability to use objects properly). Successful nonpharmacological behavioral interventions are structured to compensate for cognitive deficits and maximize the individual's residual capabilities. Agitation as a behavioral symptom is a commonly cited reason for prescribing antipsychotics in confused older adults. Yet the term "agitation" is nonspecific, and there is no accepted definition among clinicians. Usually the person described as agitated exhibits specific behavioral manifestations, such as physical aggression or screaming. Specifying the types of agitated behaviors is crucial for assessment, treatment, and evaluation of interventions. Cohen-Mansfield (1999) is among those researchers who have developed tools to specify behavioral symptoms for systematic and objective measurement.

Physical aggression, including hitting, kicking, slapping, and biting, can pose a serious threat to confused older adults and to their caregivers. Physically aggressive behaviors often occur when caregivers are attempting to provide personal care. Catastrophic reactions are the most severe overreactions resulting in physical or verbal aggression. Antipsychotics such as Haldol and Risperdal have demonstrated some effectiveness in treating aggressive behaviors. Anticonvulsants such as Tegretol have also been used, with many new agents being tested. The adverse side effects of these drugs—increased confusion, extrapyramidal symptoms, and falls—make judicious use essential. Physical restraints often increase aggression and are not associated with a decrease in falls (Strumpf, Robinson, Wagner, & Evans, 1998).

Researcher, clinicians, and caregivers have reported several successful nonpharmacological approaches to physical aggression (Mace & Rabins, 1981; Teri et al., 1992). Decreasing environmental stimuli (noise, activity) to a more tolerable level may decrease frustration and avoid increasing confusion. Approaching the person from the front and maintaining eye contact before touching him or her may facilitate communication and help decrease misperceptions that may trigger aggression. Persons with Alzheimer's disease may have comorbid sensory losses, and the appropriate prostheses (hearing aides, glasses) should be available to optimize communication and interaction with the environment. Reality orientation (correcting misperceptions) may be useful in the early stages of Alzheimer's disease, but as the person's cognitive deficits increase, this approach may cause frustration.

Validation therapy attempts to uncover the message hidden in the confused person's disrupted communication. The method uses a primarily nonconfrontational style, allowing the client to set the tone and direction of the interaction (Feil, 1993). In any encounter in which a person becomes physically aggressive, it is important to remain calm and approach the person in a nonthreatening, nonjudgmental manner. Gentle distraction or redirection to a preferred activity may diffuse the situation and allow the client an opportunity to regain control.

Verbal aggression or screaming, another common behavioral symptom, is treated primarily with pharmacological agents. With treatment for pain, fewer episodes of aggressive or screaming behavior occur, avoiding the need for physical restraints (Cohen-Mansfield, Marx, & Rosenthal 1990). One-to-one focused activities may be helpful, but these strategies do not eliminate this troublesome symptom.

Communication is a vital component of behavioral treatment strategies. The communication abilities of confused adults vary widely, based on the degree of receptive and expressive aphasia they experience. Sensory impairments (visual and auditory) further complicate the ability to communicate. Several verbal and nonverbal techniques are recommended to enhance communication and facilitate use of the client's residual abilities: speak clearly, using short, simple sentences; allow ample time for the person to respond; when repeating a question, do not rephrase it but repeat the original statement; maintain eye contact and use appropriate nonverbal gestures; attempt to understand the client's verbal communication (as fragmented as it may be); and attend to nonverbal gestures or posturing. Caregivers who are familiar with the person are often more sensitive to subtle verbal and nonverbal communications.

Wandering is another behavioral symptom that can pose a serious threat to the safety of confused older adults. A number of strategies have been attempted, including using structured activities, camouflaging exits, or using visual barriers (black horizontal or crosshatched lines on floor) to deter wandering through an exit. Wandering gardens and indoor environments have been designed that are aesthetically pleasing and provide a safe area to wander. Another safety device is a sensor worn by the client that alerts the staff when he or she is approaching an unsafe area. Some sensor devices lock an exit door as the client approaches. Physical restraints, including geri-chairs, are often used to provide safety but are associated with increased agitation and deconditioning. Pharmacological agents are not effective against wandering and are associated with movement disorders and falls.

Socially inappropriate behaviors may result from misperceptions, impulsivity, or disinhibition. Behaviors such as disrobing in public might signal the need to toilet. Confused adults might act inappropriately, believing staff or unrelated persons to be their family members or even their spouses. A person with Alzheimer's disease might take items that belong to others. Managing these behaviors usually requires that family and staff understand that these actions are based on misconceptions or the person's inability to manipulate environmental demands appropriately. A nonconfrontational approach, redirection, or diversion to preferred activities is usually helpful in dealing with these behaviors. Persons who exhibit sexual behaviors, such as masturbation or fondling, may be removed to a more private area. Institutions and families need to work with Alzheimer's clients, whenever possible, to plan strategies for overt sexual behaviors. In addition, staff and families need to be aware of their own cultural biases regarding overt sexual behavior.

Functional Symptoms

Functional deficits include deficits in dressing, grooming, and toileting. Cued or prompted toileting is a common strategy for incontinence in the early stages of Alzheimer's disease, whereby a caregiver takes the person to the bathroom on a timed schedule based on the individual's pattern of urinating and defecating. Beck et al. (1997) demonstrated that training nursing staff to segment dressing tasks maximized confused adults' independence without significantly increasing staff time. Behavioral interventions should be individualized to maximize remaining abilities and compensate for tasks that the client can no longer manage.

Nutritional problems also prevail throughout the course of the illness, often beginning with difficulty shopping or cooking and progressing to swallowing disorders in the latter stages of the disease. Frequent small meals are recommended for confused adults with poor appetites or for those who cannot tolerate sitting for an entire meal. Finger foods are another nutritional strategy, especially as the client loses the ability to use eating utensils.

Dementia Special Care Units

Many long-term-care facilities have opened dementia special care units (SCUs) (Maas & Buckwalter, 1991). Although these units vary in the types of services they provide, they typically include specialized staff training, environmental safety features, and dementia-specific programming. Efforts are made to include the family on the interdisciplinary team and to offer support to family members in the form of individual or group services. However, due to a lack of standardized criteria and rigorous research methodologies, the efficacy of SCUs has not yet been established.

Family Caregivers

Family caregivers are the most important component in treating those with Alzheimer's disease. Most older adults with dementia live in the community, and the vast majority of them are cared for at home by their families. Caring for these older adults is often physically and emotionally challenging. Several organizations, including the Alzheimer's Association, offer information and services to family caregivers. Mace and Rabins (1981) offer numerous strategies for family members caring for confused adults in the community and in most health care settings. For example, persons who tend to wander at a particular time of day (e.g., early evening) can be distracted by an alternative activity, such as late tea and cookies. Bathing is another difficult activity that can be made more pleasant by innovative strategies such as partial baths, sponge bathing, or scented towels.

Affective Symptoms

Apathy and withdrawal may signal depression in clients with Alzheimer's disease, as well as being symptoms of dementia. Evaluation of depression is more difficult due to the person's limited ability to self-report symptoms. Caregivers are often relied on to provide input about depressive symptoms, but the reliability of this input is not known (Teri et al., 1992). Confused older adults with any mood symptoms should be evaluated by a mental health specialist or geriatric health care professionals. In addition to pharmacological treatments, many nonpharmacological treatments, such as individual and group psychotherapy and activity therapies, may be used to treat depression in confused adults (Sadavoy, Lazarus, Jarvik, & Grossberg, 1996).

MARY SHELKEY

See Also
Behavioral Symptoms in Patients with Dementia
Caregiver Burden
Caregiver Burnout
Dementia: Pharmacological Therapy

Dementia: Special Care Units
Family Care for Elders with Dementia
Reality Orientation
Signage
Validation Therapy

Internet Key Words
Behavioral Strategies
Dementia
Elderly
Nonpharmacologic Strategies

Internet Resources
Alzheimer's Association
http://www.alz.org

Alzheimer's Disease Education and Referral Center
http://www.alzheimers.org

National Family Caregivers Association
http://www.nfcacares.org

DEMENTIA: OVERVIEW

Dementia is diagnosed when deterioration of memory and other cognitive skills leads to decline in the ability to perform social, role, or occupational activities (American Psychiatric Association [APA]), 1994). The prevalence of dementia increases significantly with age; it occurs in 5% to 8% of individuals over age 65, 15% to 20% of those over age 75, and 25% to 50% of those over age 85. Alzheimer's disease is the most common dementia, accounting for 50% to 75% of the total. Vascular dementia is the next most common form, although Lewy body dementia may be more prevalent than previously realized (APA, 1997).

Diagnosis

Dementia is a clinical syndrome of symptoms with many different causes and varying courses. Some dementias, such as certain cases of vascular dementia, have a more stable projection; others, such as Alzheimer's disease, cause more progressive dete-

rioration. Formal criteria for the diagnosis of dementia include a significant decline in function,
considerable impairment in memory, and decline
in at least one other area of cognitive functioning.
Areas of cognitive deterioration include aphasia
(language impairment), apraxia (impairment of motor activities in the absence of motor deficits), agnosia (inability to recognize previously meaningful
stimuli in the absence of sensory difficulties), and
executive functioning (difficulty planning, sequencing, organizing, or engaging in abstract
thinking).

The differential diagnosis of dementia requires
a detailed history and physical and neurological
examinations. Obtaining information regarding the
patient's history through caregivers is essential, as
many patients with memory problems have a difficult time providing pertinent information. The
physical exam evaluates for illnesses that can contribute to cognitive impairment. For example,
screens of metabolic function are performed to rule
out hyper- or hypothyroidism, vitamin B_{12} deficiency, medication effects, and other metabolic disorders. Hearing and vision should also be evaluated
as possible contributing factors. The neurological
exam includes assessment of mental status and the
detection of focal lesions or other organic brain
problems. Neuropsychological testing and brain
imaging may also provide helpful information.
Among the difficulties in diagnosing dementia is
the need to rule out delirium or acute confusional
states. Older adults are particularly susceptible to
acute confusional states because of the higher incidence of systemic illnesses in this population and
their greater vulnerability to the adverse effects of
polypharmacy (APA, 1997). Memory impairment
is present in both delirium and dementia. A sudden
onset of symptoms, impairment in attention, and
fluctuating symptoms suggest delirium. In addition,
the impairments seen in dementia persist, whereas
those seen in delirium typically resolve once the
underlying medical issue is addressed. Although
dementia cannot be diagnosed in the presence of
delirium, it is possible for delirium to be present
in a person with dementia (APA, 1997).

Depression in older adults can cause symptoms similar to dementia (dementia syndrome of
depression or pseudodementia). As with delirium
and dementia, depression and dementia may coexist. Older adults experiencing a major depressive
disorder frequently complain of memory problems
and report difficulties thinking and concentrating
(APA, 1994). On mental status screens, depressed
individuals may have poor attention and think that
they are "failing," although their performance is
usually better than they expect. In dementia syndrome of depression, affective symptoms often appear prior to the cognitive symptoms. With treatment, the cognitive problems often resolve. However, nearly 50% of patients with dementia syndrome of depression go on to develop irreversible
dementia within five years (APA, 1997).

Medications

A wide variety of psychoactive medications can be
used to restore cognitive abilities, prevent further
decline, and increase functional status. These medications include cholinesterase inhibitors (tacrine
and donepezil); alpha tocopherol (vitamin E); selegiline (deprenyl), approved for Parkinson's disease
but studied and used in demented populations; and
ergoloid mesylates (Hydergine), approved for nonspecific cognitive decline. Other medications proposed for the treatment of cognitive decline include
nonsteroidal anti-inflammatories, estrogen supplementation, melatonin, botanical agents (e.g., ginkgo
biloba), and chelating agents. Additional agents are
currently being tested and are available through
clinical trials (APA, 1997; Small et al., 1997).

Behavioral Interventions

Behavioral interventions for hallucinations, delusions, agitation, aggressive behavior, disinhibition,
anxiety, apathy, and sleep disturbances are often
helpful. For example, adequate stimulation during
the day may ameliorate a sleep disorder. If behavioral strategies are insufficient, psychotropic medications can be useful adjuncts in managing behaviors and treating psychiatric symptoms. A patient
with dementia requires supervision according to the

type and severity of specific cognitive limitations. A patient with significant cognitive impairment may not be safe alone at home; he or she might improperly administer medications, be unable to cope with a household emergency, or use the stove, power tools, or other equipment inappropriately.

Risks

Demented individuals are at particular risk for elder abuse or neglect because of their limited ability to protest and the added demands and emotional strain on caregivers. Any concern, especially one raised by the patient, must be thoroughly evaluated. However, corroborating evidence (e.g., from physical examination) should be sought to distinguish delusions, hallucinations, and misinterpretations from actual abuse.

Patients with dementia may be at risk due to wandering. Referrals to the Safe Return Program of the Alzheimer's Association (1-800-621-0379) or similar options provided by local police departments or other organizations are appropriate for many families.

The patient and family must be counseled about driving and other activities that put other people at risk. In California and some other states, physicians are required to report a diagnosis of dementia to the Health Department, which then informs the Department of Motor Vehicles.

Costs

Dementia exacts a major financial toll. The combined direct and indirect costs of medical and long-term care, home care, and lost caregiver productivity approach $100 billion annually for Alzheimer's disease alone. Medicare, Medicaid, and private insurance cover a portion of this expense, but the balance is borne by families who care for demented patients (Small et al., 1997).

Caregivers

Providing care for a loved one with dementia can be stressful; as many as 50% of family caregivers suffer from depression (Small et al., 1997). Caregivers can become isolated, fearful of physical harm, and physically and emotionally overwhelmed by the demented patient's needs. A substantial literature reinforces the value of support groups, especially those combining information with emotional support. The Alzheimer's Association is an excellent resource for support groups.

Caregiver education in managing and living with a person with dementia includes keeping requests and demands relatively simple and avoiding overly complex tasks that might lead to frustration; avoiding confrontation and deferring requests if the patient becomes angered; remaining calm, firm, and supportive if the patient becomes upset; being consistent and avoiding unnecessary change; providing frequent reminders, explanations, and orientation clues; recognizing declines in capacity and adjusting expectations appropriately; and bringing sudden declines in function and the emergence of new symptoms to professional attention (APA, 1997).

Respite care is often beneficial, as it temporarily relieves the caregiver from the responsibilities of caring for a demented individual and can provide additional positive stimulation for the person with dementia. Respite care may last for a few hours a day or for weeks to months (depending on the locus of respite care) and may be provided through companions, home health aides, visiting nurses, day-care programs, and brief nursing home stays or other temporary overnight care.

Legal counsel for patients and families should be recommended early in the course of the illness, while the patient is sufficiently competent to participate in decision making, including testamentary wills and advance directives. When family caregivers are no longer able to care for the demented patient at home, they will benefit from both logistical and emotional support in placing the patient in an assisted living facility or a nursing home.

STACY SCHANTZ WILKINS
STACY S. AMANO

See Also
Cognition Instruments
Cognitive Changes in Aging

Internet Key Words
Caregivers
Dementia
Memory

Internet Resources
Alzheimer's Association
http://www.alz.org/

American Psychiatric Association Online Clinical Resources
http://www.psych.org/clin_res/pg_dementia.html

Internet Mental Health (Dementia)
http://www.mentalhealth.com/dis/p20-or05.html

DEMENTIA: PHARMACOLOGICAL THERAPY

Pharmacological therapies for cognitive impairment are gaining more attention because of the increased prevalence of dementia among the elderly. Due to the multifaceted impact of these diseases on the sufferer's life, multiple pharmacotherapeutic approaches are often used, including antidepressants, anxiolytics, sedatives, and cognitive enhancers.

The focus here is primarily on therapies for Alzheimer's disease, because it is the most common form of dementia in the elderly. It is traditionally classified as a primary neurodegenerative disorder, as there is no obvious antecedent pathology leading to neuronal damage. The pathological sequence of events in Alzheimer's disease is not yet understood in detail, but it is generally thought that increases in levels of the small peptide (amyloid or β-amyloid [Aβ]) in the aging brain (through increased production, reduced clearance, or both) leads to neuronal dysfunction and damage.

Vascular dementia is the second most common cause of dementia, and there is increasing evidence that Alzheimer's disease and vascular dementia exacerbate each other. Vascular dementia is a clinical diagnosis of dementia resulting from a large array of pathologies that cause disturbances in cerebral blood flow, including infarction or rupture (hemorrhage) of large, medium, or small blood vessels. For instance, atherosclerotic cerebrovascular disease, which can occur in large and medium-size arteries supplying the brain, shares the same genetic and dietary risk factors of peripheral atherosclerotic disease. In general, therapies for these disorders aim at reducing risk factors, such as controlling blood pressure, maintaining normal limits of serum cholesterol and triglycerides, reversing or controlling arrhythmias, managing diabetes, and stopping smoking.

Etiology of Alzheimer's Disease and Therapeutic Targets

Our understanding of Alzheimer's disease was greatly enhanced by genetic findings in the 1990s. Mutations in the β-amyloid precursor protein (βAPP) gene, which segregate with early-onset familial Alzheimer's, increase either the total amount of Aβ (one of the primary metabolites of βAPP) or the relative amount of the longer form of $A\beta_{1-42}$ over $A\beta_{1-40}$. Although mutations in the other early-onset familial Alzheimer's genes, the presenilins, have multiple effects in cultured cells and genetically engineered worms, flies, and mice, their main relevance to Alzheimer's patients appears to be their impact on βAPP processing. It has been proposed that presenilin 1 is the γ-secretase responsible for the intramembranous cleavage of βAPP to Aβ (Wolfe et al., 1999).

In summary, the early-onset genetic errors in Alzheimer's disease make the accumulation of Aβ central to the disease process, at least in early-onset cases. By extrapolation, late-onset Alzheimer's disease is assumed to be due to an excess production of Aβ, reduced clearance of Aβ, or both. Although no current medications are aimed at lowering Aβ production, the recent discovery of the β-secretase enzymes ensures that there will be extensive work in this area. At this time, therapies aimed at altering βAPP metabolism are in their infancy, as are other possible therapeutic approaches, such as vaccination against Alzheimer's disease.

Alzheimer's disease progresses unequally with respect to neuronal populations. Cholinergic neurons seem particularly vulnerable, and therapies aimed at boosting failing cholinergic transmission are increasingly available. The greatest risk for Alzheimer's disease is advancing age, but the mechanisms associated with aging that allow the disease to occur are poorly understood. Some of these mechanisms are disease specific, such as changes in βAPP metabolism, and others are associated with changes in the microenvironment of aging neurons. For instance, the well-known ability of an isoform of ApoE (E4) to confer risk for early expression of disease may be an example of a repair system that fails unequally (due to genetic variation) with age. ApoE is important during periods of repair after injury. Similarly, changes in the central nervous system's ability to handle oxidative stress with age may both exacerbate the disease process and be aggravated by it. The exact role of inflammation in Alzheimer's disease has yet to be fully elucidated, but it is clear that certain anti-inflammatory therapies are beneficial. Similarly, therapies that nonspecifically support aging neurons (including therapies that help maintain Ca^{2+} homeostasis), such as trophic factors and neuroprotective factors, are likely to be helpful for an array of cognitive disorders caused by age-related changes. Estrogen, for example, may act to protect, restore, and maintain neurons in the stressful microenvironment of the aging brain.

Cholinergic Therapies

Abnormalities in the cholinergic system are consistently identified in Alzheimer's disease, as well as in Parkinson's disease and Lewy body disease. Systematic biochemical investigation of Alzheimer's patients' brains shows substantial deficits of cholinergic transmission in the hippocampus and cerebral cortex. Consideration of the role of acetylcholine in learning and memory and its potential neurotrophic effects, as well as the association between the muscarinic receptor and the processing of βAPP and the production of Aβ, has given rise to cholinergic therapies. Cholinergic agents include cholinesterase inhibitors, muscarinic and nicotinic receptor agonists, and direct modifiers of acetylcholine release.

Acetylcholinesterase inhibitors include tacrine, donepezil, physostigmine, rivastigmine, galantamine, ENA 713, and metrifonate. Tacrine improves cognitive performance and certain secondary psychiatric symptoms and significantly delays nursing home placement. However, hepatotoxicity and gastrointestinal side effects occur in a number of patients. Donepezil (Aricept), recently approved for use in mild to moderate Alzheimer's disease, improves cognitive performance and patient quality of life over short-term follow-up periods and appears to be less toxic and better tolerated than tacrine. Interestingly, patients with Lewy body disease respond well to acetylcholinesterase inhibitors. However, they must be administered cautiously, especially in patients with bronchospastic pulmonary disease, bradydysrhythmia, decreased cardiac contractility, or peptic ulcer disease, due to the risk of adverse side effects (Francis, Palmer, Snape, & Wilcock, 1999; Wheatley & Smith, 1998). Experimental compounds such as xanomeline, milameline, ABT-418, GTS-21, SB202026, and AF1025B are muscarinic or nicotinic agonists. Other experimental compounds such as montirelin, FKS-508, and T-588 are modifiers of acetylcholine release; however, these compounds are in the early stages of clinical development, and prediction of their eventual success is difficult (Francis et al., 1999).

Antioxidant Therapies

Evidence for the role of oxidative stress in neuronal degeneration comes from several areas. For example, the enzyme glutamine synthatase is particularly sensitive to mixed function oxidation; regional losses have been demonstrated in patients with Alzheimer's disease. Recent findings indicate that the neurotoxicity of Alzheimer's Aβ peptide involves mechanisms that result in the generation of hydrogen peroxide, reactive oxygen species, and lipid peroxidation; the alteration of mitochondrial function; and some loss of calcium homeostasis. Furthermore, it has been shown that ApoE isoforms confer different susceptibility to oxidative stress. E2 has the greatest antioxidant properties, followed by E3 and E4—the reverse order in which they confer risk for expression of Alzheimer's disease. Such observations have resulted in the use of antioxidant therapies for the treatment of dementia; these therapies include monoamine oxidase inhibitors (MAOIs), vitamin E, vitamin C, and calcium channel blockers. Selegiline (a relatively selective MAO-B inhibitor) is believed to act as a free radical scavenger and modulator of monoaminergic neurotransmission and is marketed for the treatment of Parkinson's disease. Although there is conflicting evidence regarding the effectiveness of selegiline on cognitive deterioration, it has been shown to have efficacy in the treatment of mood and behavioral concomitants of dementia. Milacemide and lazabemide are two other MAO-B inhibitors. Milacemide is ineffective in Alzheimer's patients, and lazabemide is under investigation. Patients with Alzheimer's or vascular dementia treated with ginkgo biloba, a compound with putative antioxidant properties that is also under investigation, show only modest improvements in cognition. Other antioxidants such as vitamin E, vitamin C, melatonin, and idebenone are currently under investigation and may ultimately prove beneficial (Felician & Sandson, 1999).

Estrogen Replacement Therapy

Neuronal cell culture investigations demonstrate that estrogen protects against oxidative stress, excitotoxins, and Aβ-induced toxicity. A lower risk of Alzheimer's disease or related dementias was found among postmenopausal women on estrogen replacement therapy (ERT) in comparison to nonusers. A history of longer estrogen use, higher estrogen dose, and earlier age of menarche is associated with a lower risk of Alzheimer's disease. Several studies indicate that women on ERT score significantly higher on cognitive testing at baseline than nonusers, and verbal memory performance improves slightly over time. The effect of estrogen on cognition is independent of age, education, ethnicity, and ApoE genotype. However, additional information is needed to better assess the initial success of ERT (Felician & Sandson, 1999).

Anti-Inflammatory Therapy

Inflammatory dysfunction has been identified in the brains of Alzheimer's patients, indicating that inflammatory and immune mechanisms accompany the neurodegenerative processes. For instance, increased cytokines such as interleukin-1 and interleukin-6, which promote the production of βAPP, are observed in the serum, cortical plaques, and neurons of Alzheimer's patients. In addition, reactive microglia are present in neuritic plaques. It was recently shown that Aβ induces a proinflammatory response in microglia that is evidenced by increased leukotriene B$_4$ release. It has also been demonstrated that dipyridamole, a selective cyclic guanosine monophosphate (cGMP) phosphodiesterase inhibitor, as well as compounds that increase cGMP levels, prevent Aβ-induced microglial inflammation. Propentofylline has also been shown to be a nonselective cyclic adenosine monophosphate (cAMP)–cGMP phosphodiesterase inhibitor that increases the intracellular concentration of cGMP as well as cAMP and results in inhibition of potentially neurotoxic functions while maintaining beneficial ones (Paris et al., 1999). Pharmacological interventions aimed at reducing microglial-mediated inflammation by inhibiting cGMP/phosphodiesterase or other means of elevating cyclic nucleotides may be beneficial in the treatment of Alzheimer's disease.

The potential role of anti-inflammatory agents in the treatment of Alzheimer's disease was first suggested by the discovery of an unexpectedly low prevalence of the disease among patients with rheumatoid arthritis treated with nonsteroidal anti-inflammatory drugs (NSAIDs). Case-control studies demonstrated a reduced risk of Alzheimer's disease in subjects receiving NSAIDs. A slower rate of decline on several cognitive measures is also observed in Alzheimer's patients using NSAIDs. In addition, studies with indomethacin have shown improved cognitive function in Alzheimer's patients. However, drug-related side effects such as gastrointestinal intolerance, renal toxicity, and adverse cognitive effects such as sleep disturbance and delirium may limit their use. Many of these side effects may be related to inhibition of cyclooxygenase-1 (COX-1), which is expressed in various organs and tissues and is a mediator of diverse homeostatic functions. Selective inhibition of cyclooxygenase-2 (COX-2), which is expressed in the neocortex and particularly in limbic structures such as the hippocampus, may result in reduced drug-related side effects (Felician & Sandson, 1999). Further investigations of NSAIDs, selective inhibitors of COX-2, and corticosteroids are ongoing, and more information is required before recommendations regarding their use can be made.

Other Therapy

Clinical trials of other compounds with diverse therapeutic values include substances that facilitate the use of fatty acids, such as acetyl-L-carnitine, and calcium channel blockers, such as nimodipine. Piracetam, oiracetam, pramiracetam, aniracetam, CI933, and BMY 21502 may protect the central nervous system from potential damage due to hypoxia and also enhance microcirculation (Wheatley & Smith, 1998). Memantine, an NMDA receptor (a glutamate receptor subtype) antagonist is under investigation for its involvement in memory function.

Immunization with Aβ in transgenic animal models for Alzheimer's disease is showing significant reduction of Aβ plaque formation, neuritic dystrophy, and astrogliosis, but further investiga-

tion is needed to determine the potential applicability to human subjects. Investigational compounds such as JTP-4819 and Y-29794, inhibitors of prolyl endopeptidase, one of which cleaves Aβ from βAPP, are being investigated (Kato, Fukunari, Sakai, & Nakajima, 1997; Wheatley & Smith, 1998).

The eventual success of any pharmacological interaction is dependent on early identification of individuals at risk for dementia. Persons with mild cognitive impairment or other subtle cognitive deficits associated with prodromal dementia should be targeted to delay the age of disease onset and decrease the rate of progression (Petersen, Smith, Waring, Ivnik, Tangalos, & Kokmen, 1999). Because many of these disorders, particularly Alzheimer's, have their origins many years before even mild impairment becomes noticeable, highly effective therapeutic and prophylactic treatments require the identification of preclinical biomarkers of disease.

DAN RICHARDS
LAILA ABDULLAH
MICHAEL MULLAN

See Also
Dementia: Nonpharmacological Therapy

Internet Key Words
Alzheimer's Disease
Beta-Amyloid
Cholinergic
Ginkgo Biloba
Inflammatory
Oxidative

Internet Resources
Alzheimer's Association
http://www.alzheimersassociation.org/

Alzheimer's Research Forum
http://www.alzforum.org/

National Center for Biotechnology Information
http://www.ncbi.nlm.nih.gov/

Online Mendelian Inheritance in Man
http://www.ncbi.nlm.nih.gov/omim/

DEMENTIA: SPECIAL CARE UNITS

Nursing home special care units (SCUs) are a proposed solution to the problems presented by persons with dementia. In addition to needing assistance with basic activities of daily living (e.g., bathing, grooming, toileting), elders with dementia are often anxious and agitated, frequently pace continuously, and are prone to elopement and combative behaviors. These care problems are difficult for staff to handle, resulting in increased stress and burnout. Family members, who usually find it difficult to institutionalize their loved ones, typically experience guilt and loss, as well as the difficulties of releasing control over relatives' care to staff. The SCU strategy is designed to reduce the anxiety and agitation of persons with dementia, which in turn is expected to decrease staff and family caregiver stress. After more than a decade of rapid growth in the number of SCUs, about 12% of U.S. nursing homes now have some kind of specialty unit for persons with dementia. Some assisted living facilities have developed SCUs and programs for persons with dementia.

Although most gerontology specialists agree that SCUs are useful for managing persons with dementia, a number of issues regarding the benefits to residents, staff, and family members are unresolved. Some disagreement remains between proponents of segregation and nonsegregation, advocates of reduced and increased stimuli, those who believe that rehabilitation strategies are useful and those who contend that environmental modification alone is most effective, and those who reject or embrace the medical model. The lack of resolution on these issues is seen in the absence of standardization of SCUs and in the mixed and inconclusive research findings regarding their effectiveness.

There is increasing consensus that five features make an SCU special: (1) specific admission and discharge criteria, (2) staff selection and training according to accepted standards of care for persons with dementia, (3) activity programming designed for persons with cognitive impairment, (4) family programming and involvement in the care of their relatives with dementia, and (5) segregated and modified physical and social environments that provide reduced but appropriate sensory stimuli (Maas, Swanson, Specht, & Buckwalter, 1994).

Several theoretical frameworks guide programs and interventions for persons with dementia in nursing homes. Underlying the person-environment fit (P-EF), person-environment interaction (P-E), and progressively lowered stress threshold (PLST) models is the notion that the environment must be modified to enhance the function of persons with dementia and assist caregivers in coping with their behaviors. The PLST model is most often used to develop SCUs and to evaluate the effects on residents, families, and staff.

Medical, psychosocial, and rehabilitation models are also prominent. In the medical model, nursing homes, resemble hospitals more than homes, with an emphasis on the medical aspects of care rather than on behavioral approaches, function, daily living, and quality of life. Few purely medical treatments address more than the symptoms of dementia. In contrast, the psychosocial model emphasizes the whole person in context and is more appropriate for addressing behavioral problems associated with irreversible dementia. The rehabilitation model focuses on the maintenance and use of functional abilities to prevent excess disability.

Criteria for admission, discharge, and resident selection are critical policies for an SCU. The diagnosis of dementia should be made after a comprehensive psychoneurological examination. The Alzheimer's Association has published guidelines that increase the reliability and validity of diagnosis. Preadmission assessment should include a description of the resident's behaviors; family and resident preadmission preferences, values, beliefs, and interests; family support, understanding, and acceptance; and cognitive, physical, and social function within the preadmission milieu. Preadmission assessment should address advance directives and patient and family preferences for end-of-life care. Admission criteria should focus on appropriate placement to prevent avoidable relocation of the resident.

SCUs are often coordinated by registered nurses (RNs), but they may also be directed by individuals trained in social science or gerontology who are not RNs. The key concern is not who coordinates the SCU program but that interdisciplinary care is provided and that both the social and the health needs of residents are addressed (Buettner, 1998). Nursing care should be managed by professional nurses with adequate assistance by staff who are trained in the care of persons with dementia. Common ratios of total nursing staff to residents are 1:4 to 1:6. In most SCUs, the staffing level is about the same as in general nursing home care units. Perhaps most lacking is a sufficient number of RNs who specialize in the care of persons with dementia. The ratio of specially trained RNs to residents on SCUs should be no less than 1:15 so that residents receive assessments and interventions by RNs who can also provide appropriate oversight, training, and role modeling for other staff members. A social worker, physician, activities therapist, dietitian, and physical therapist should be members of the care team. Special training for all staff should address the specific needs of institutionalized persons with dementia and include education about dementia, its causes, the common resulting behaviors, the effects of these behaviors on caregivers (staff and family), and the appropriate principles and techniques for providing care. Classroom and clinical experiences are recommended so that staff can discuss and practice best-care approaches.

A philosophy of care that is based primarily on rehabilitative and psychosocial models results in programming that provides (1) interventions and activities appropriate for the cognitive and functional abilities of residents; (2) interventions and activities relevant to the interests, strengths, values, heritages, and remaining abilities of residents; and (3) individual and group activities to promote socialization, exercise, reminiscing, and sensory enjoyment. Through such activities, residents are occupied and purposefully engaged, preventing boredom and agitation, which can result in pacing, rummaging, and other disruptive and socially inaccessible behaviors.

Nursing care, based on the needs of the individual, follows a consistent routine, emphasizing flexibility and unconditional positive regard. Care is provided when the resident can accept it. Waiting five minutes is often enough for the resident to be able to cooperate and be less fearful. Psychoactive medications are used selectively to control symptoms—mainly anxiety, agitation, and depression—that interfere with the resident's responses (which may be out of proportion to stimuli), function, and comfort. Catastrophic reactions, responses that are out of proportion to stimuli, are managed by altering stressors in the environment to fit the abilities of the resident. Examples are reducing the number of people, decreasing noise, or dimming the lights in the environment and using distraction to lessen fear and agitation. An important aspect of care is to maintain resident safety without unduly restricting activity and behavior. Although the medical model should receive less emphasis, nursing staff must be vigilant in monitoring residents' health status for the existence or exacerbation of physical or mental illness and to identify adverse effects from medications or treatments. Because persons with dementia often cannot interpret or communicate pain and discomfort, monitoring for behavioral clues is especially important. Awareness of conditions that are likely to cause discomfort or pain should be maintained to guide assessment and intervention.

SCUs should have a program of support for families, starting with the first contact for admission. A thorough orientation to the physical environment and to policies that govern resident care should be provided. Participation of families in the care of their relatives and decision making for them should be actively supported, and institutional barriers to participation removed. Recommended strategies for family involvement include a "buddy system," a family-nurse liaison, and peer support groups (Maas, Reed, Swanson, & Specht, in press).

Modification of the physical environment should include reduction of overwhelming and disturbing stimuli, provision for safe wandering, environmental cues to support memory, and visual, auditory, and other sensory stimulation. The environment should promote resident function and safety, create a homelike atmosphere that enhances family visitation and involvement, and provide a pleasant, functional workplace for staff. Each component of the environment should be evaluated to consider the message it sends to residents and their families. Most residents tend to interact in a more socially acceptable manner in a homelike, "normal" environment. Finally, care must be taken to avoid residents' sensory deprivation and boredom. Too little stimulation can be as bad as too much, causing feelings and behaviors that are uncomfortable for residents and difficult for staff to manage (Peppard, 1991).

The results of systematic evaluation of the effects of SCUs on resident, family, and staff outcomes are mixed. Consistently reported positive outcomes for SCU residents are reduced agitation, less catastrophic and disruptive behavior, decreased use of physical and chemical restraints, and increased socially accessible interactions and participation in activities (Bellelli, Frisoni, Bianchetti, & Boffelli, 1998; Swanson, Maas, & Buckwalter, 1993; Work Group on Research and Evaluation of Special Care Units, 1996). Although studies of staff effects are few, staff describe their work as being less stressful in SCUs. SCU staff are also more knowledgeable about the care of persons with dementia than are staff working in integrated units. Despite these benefits, staff turnover in SCUs is comparable to turnover in general nursing home units. In the small number of published studies, family members report satisfaction with care in SCUs; positive effects include continued involvement in the care of their relatives following institutionalization.

MERIDEAN L. MAAS
JANET P. SPECHT

See Also
Dementia: Nonpharmacological Therapy
Depression in Dementia
Nursing Home Admission
Reality Orientation
Validation Therapy

Internet Resources
Alzheimer's Association
http://www.alz.org

Alzheimer's Disease Centers, National Institutes of Health
http://www.alzheimers.org/adcdir.html

Dementia Special Care Units
http://www.alz.org/us/rtrlspec.htm

Library: Special Care Units
http://www.alz-nova.org/lbscu.htm

Status of Specialty Care in U.S. Nursing Homes
http://www.projhope.org/CHA/research/devprogabs.htm

DEMOGRAPHY OF AGING

The demographic facts of aging explain the emergence of gerontology as a major discipline. Sociology, economics, psychology, and geography are other disciplines that now devote considerable time to studying the aspects of aging. The fundamental need to understand the process of population aging and how it affects all segments of society has led to the development of a specialized field of study termed the demography of aging. The demography of aging deals with the dynamic processes of aging, as manifested by a population with a rising proportion of older persons. It also examines the aging of birth cohorts (groups of persons born within a defined period) throughout their life course. The dynamic process of aging reflects how a total population structure is transformed over time and how cohorts of persons reaching old age may modify the characteristics of the overall aged population. Therefore, the demography of aging draws theoretical insights and methodological approaches from general demography to explain the evidence of populations that are aging and to forecast future developments.

The idea of aging as a set of lifelong processes implies that the aging of children or of the workforce is also a legitimate focus for the demography of aging. For example, in demography and other disciplines, aging aspects of the baby boom generation (cohorts of persons born between 1946 and

1966) have been the subject of a growing body of literature.

The demography of aging focuses on the current demographic profile of the older population, as well as changes in its numbers, proportionate size, composition, and territorial distribution. Moreover, the field addresses the determinants and consequences of changes in attributes of the older population. A substantial body of literature has emerged concerning a wide range of consequences of population aging. Much of this literature is found in disciplines other than demography.

What Is Oldness?

Two decades ago, Siegel (1980) wrote: "The demography of aging brings demographers to focus holistically on a population group, the elderly, and a demographic process, aging" (p. 345). A focus on the "elderly" rests partly on the notion that there is a meaningful concept of oldness and that it is possible to define a useful measure of when a person, or a cohort, should be called "old." However, the delineation of such a threshold is controversial (Bordelais, 1994).

Concurrently, there is growing pressure on government leaders and organizations to make more effective use of the economic potential of healthy older persons over the age of 65. These developments may help shift the focus of the demography of aging away from the group called "the elderly" and toward aging as a dynamic, lifelong process for cohorts and for whole populations.

Key Contributors to the Development of the Demography of Aging

Initial stimulus for the creation of the discipline came from scholars such as Jean Bourgeois-Pichat at L'Institut National pour l'Etude Démographique in Paris. Their ideas gained international acclaim through Bourgeois-Pichat's work at the United Nations Population Division in the mid-1950s (United Nations, 1956). Aging has now become a prominent focus of work at numerous centers for demographic

training and research, with notable sponsorship by the U.S. National Institute on Aging.

United Nations agencies have been key sources of literature on the demography of aging (Stolnitz, 1992; United Nations, 1992), along with various journals that publish articles in the field. The Population Activities Unit of the Economic Commission for Europe is leading a new set of national and comparative studies on population aging. In the 1980s, the Committee for International Cooperation in National Research in Demography sponsored the preparation and publication of 19 national studies on aspects of population aging. International reviews of population aging and of the older population have also received important stimulus from the International Programs Center at the U.S. Bureau of the Census, supported in part by the U.S. National Institute on Aging (Kinsella & Taeuber, 1993).

Measurement and Analysis of Population Aging

There is a hierarchical linkage among individual, cohort, and population aging. Although alternative statistical measures of aging are possible, cohort aging is commonly measured in terms of changes in the average age of the members of a cohort. Population aging is usually measured by changes in the proportion of designated "older persons" in a population. The latter measure is not simply the growth rate or changes in the absolute size of the designated older population.

The pace of population aging depends on the aging processes and relative sizes of sequences of cohorts passing from birth to extinction (Uhlenberg & Riley, 1996). The impact of population aging on society depends on the pattern of succession among aging cohorts possessing different characteristics (Stone, 1999).

A declining birth rate contributes to population aging by creating a faster growth rate among older persons than among youth. Declining death rates among the elderly also contribute significantly to population aging. These facts support classification of population aging processes in terms of the extent to which they involve aging at the base of the population age pyramid (associated with current fertility rate changes) or aging at the apex of the pyramid. Aging at the apex gradually develops when large birth cohorts from the past reach older ages and are further swelled by reduced mortality at older ages. The latter trends have been particularly important in accounting for recent increases in both the number and the proportion of elderly in developed countries.

Declining birth rates during the depression years of the 1930s in the United States and Europe sparked concern over potential depopulation and population aging. These same concerns were expressed in the 1990s concerning a number of countries in Europe. Lesthaeghe (1999) noted that low fertility due to delayed and completed childbearing is continuing in many European countries. Thus, "hyperaging" seems inevitable for many countries, such as Italy, where one-third of the population is expected to be 65 and over by 2030.

A mathematical model of demographic changes was formulated, and the stable population model enabled demographers to examine the long-term effects of different levels of fertility and mortality on age structure. Many articles on this topic can be found in the literature (Stolnitz, 1992; United Nations, 1992). In addition, there is a large body of literature concerning the consequences of population aging (see Stolnitz, 1992).

Changing Characteristics Among the Older Population

The continuing aging of the population caused by previous levels of high fertility, the fall in the birthrate since the early 1960s, and recent marked declines in mortality at older ages has been accompanied by changes in the characteristics of the aged population. The term "population metabolism" has been used to describe how the aged subpopulation is modified over time by new entrants, selective decrements through differential death rates, and changes in status during later life (Myers, 1985).

The aged population is subject, therefore, to high levels of turnover. A high percentage of the population 65 years of age and older on any given date will die within a decade and be replaced by new entrants to that population. This rapid succession of cohorts contributes to a high level of diversity.

In most developed countries, the oldest old (85 years of age and over) constitute an increasing proportion of the aged population (Suzman, Willis, & Manton, 1992). This aging trend will continue until the post–World War II baby boom cohorts reach old age, beginning in the second decade of the 21st century.

Another important attribute of the older populations of most countries is the sex disparity. There are many more older women than older men, because of the differential mortality risks that favor women. Among a wide range of social and economic variables, gender disparity is a characteristic feature of the aged population.

Several other population attributes have been the focus of attention. These include family structure among older persons (Myers & Eggers, 1996), living arrangements of older persons, patterns of paid and unpaid work among older persons, the geographical redistribution (migration) of the older population (Bean, Myers, Angel, & Galle, 1994), and disease prevalence, disability, and subjective well-being (Manton, Corder, & Stallard, 1997).

The demography of aging provides theoretical and methodological perspectives on various aspects of population aging. The field has been enriched not only by contributions from general demography but also by gerontological research and modeling work by scholars concerned with economic issues. A growing number of countries have research under way and publications in preparation on the demographic aspects of the aging phenomenon. Of particular note are analyses of census data, surveys dealing with the aged population, and projections of disaggregated characteristics of the aged population. Among these efforts is the accumulation of data from longitudinal surveys in the United States, Canada, and other countries (both developed and developing). New efforts are under way to rework historical data so as to reveal patterns of cohort differentiation along a variety of dimensions of population attributes (Stone, 1999). The knowledge generated by these activities will provide a sounder basis for policy decisions in aging societies.

LEROY O. STONE

DENTISTRY
See
Geriatric Dentistry

DENTURES

In Western societies, teeth symbolize youth, potency, strength, and virility. Aging is associated with tooth loss. About 70% of people who have lost their teeth express regret, and 60% consider dentures a handicap. Many studies report that 25% to 30% of complete denture wearers have problems with their dentures, especially the mandibular (lower) denture. In fact, a person may have a technically perfect denture but be unable to tolerate it (Ettinger & Jakobsen, 1997; Moltzer, van der Meulen, & Verheij, 1996; van Waas, 1990).

Most natural teeth are lost due to two chronic, infective diseases: caries (decay) and periodontal (gum) disease. In general, both these diseases are preventable if the oral cavity is kept clean by the regular removal of plaque.

Types of Dentures

Dentures are prosthetic replacements for natural teeth and can be divided into four groups:

1. Complete removable dentures
2. Removable partial dentures
3. Fixed partial dentures
4. Fixed detachable dentures

A complete denture replaces the chewing surface of all the teeth in an arch. It sits on the mucosa or may have some support from the remaining natural teeth, which have been cut down to 1 1/2 to 2 mm above the gingival margin. These tooth-supported prostheses are called overdentures. Alternatively, dentures can be supported by implants.

A removable partial denture replaces some missing teeth in an arch and is held in place by metal clasps or by special attachments that fit into or onto some of the remaining teeth.

A fixed partial denture, or bridge, replaces some missing teeth and is supported with a crown or cap on either side of the missing teeth. Bridges are usually cemented in place and are not removable.

A fixed detachable denture is held in place by screws that go into either implants or natural teeth.

Wearing and Removing Dentures

Every removable prosthesis moves while it is in use and is potentially traumatic. The mouth needs six to eight hours of rest from a denture each day. The best time to remove a denture is during sleep, when the production of saliva required for lubrication and retention is at its lowest, and parafunctional movements such as clenching and grinding are at their highest. A patient being treated for a temporomandibular joint problem may be advised by the dentist not to remove the prosthesis during sleep.

Denture Hygiene and Care

A denture, like teeth, gets covered with plaque and needs to be cleaned regularly—ideally, after meals. If a denture is maintained in the mouth continuously, the plaque can be colonized by commensal organisms. If the denture is not removed and adequately cleaned, organisms such as *Candida albicans* (a fungus) can proliferate and release toxins, resulting in a hypersensitivity reaction of the oral mucosa that looks like contact dermatitis (Berg, Silness, & Sorheim, 1987). Persons who are frail, have xerostomia (dry mouth), are immunocompromised, or have had radiation therapy of the head and neck are at greater risk for candidiasis. It has been shown by Stafford, Arendorf, and Huggett (1986) that if dentures are removed at night and allowed to air dry, the organisms on the denture surface do not proliferate as quickly. However, the dentures need to be hydrated by soaking them in water for several minutes before they are put back in the mouth.

When teeth are extracted, the bone that was produced during eruption of the teeth—the residual bone—begins to resorb. The rate of resorption for the anterior maxilla has been measured at about 0.1 mm per year; for the mandible, the rate is four times greater (0.4 mm) (Tallgren, 1999). This resorption, plus normal wear, results in dentures having a finite life span. The average complete denture needs to be relined or replaced every five to seven years. The life span of a partial denture varies, depending on the amount of tooth support it has. Fixed partial dentures should last at least 10 years, because they are usually made of metal, porcelain, or a combination of the two.

The abutment teeth (the teeth supporting the denture) for removable or fixed partial dentures are at higher risk of plaque accumulation, which can result in root surface caries or periodontal disease. Therefore, both dentures and the natural teeth must be cleaned. Fluoride rinses may be helpful for at-risk patients. For fixed partial dentures and implants, special interproximal brushes and superfloss must be used to clean under and around the bridge (pontic) portion of the prosthesis, or the tissues will become inflamed. A daily mouth rinse with 0.12% chlorhexidine gluconate may be helpful for patients who have difficulty cleaning. The best way to clean dentures is to use a denture brush and, if the patient can afford it, a small ultrasonic cleaner. Toothpaste should not be used to clean dentures, as it is too abrasive and will damage the surface of the dentures. A mild dishwashing detergent works well; commercial soaks are also helpful.

Denture-Induced Oral Disease

Oral lesions associated with the wearing of removable prostheses can be due to microbial colonization of dental plaque, traumatic irritation by the denture, or an allergic response to denture materials.

Microbial Colonization of Dental Plaque

Plaque that is not removed can become colonized by *Streptococcus mutans* and lactobacilli, causing caries on the surface of the remaining teeth. The plaque can also be colonized by a wide variety of aerobic and anaerobic organisms, which can result in bone loss around the teeth, called periodontal disease. If *Candida albicans* colonizes the plaque, especially the palatal tissue surface, a denture stomatitis may result. Plaque accumulation is exacerbated by an elder's progressive loss of normal dexterity, poor eyesight, decrease in salivary flow associated with the use of drugs (e.g., anticholinergics), and diseases (e.g., diabetes, depression, Parkinson's disease) (Ettinger, 1999).

Traumatic Irritation

Traumatic irritation can result in an ulcer due to the changing fit of a denture caused by tissue resorption over time. The ability of oral mucosa to resist this mechanical irritation can be diminished by diabetes, nutritional deficiencies, or xerostomia. Denture wearing can also result in irritation hyperplasia, a chronic inflammatory tissue reaction that leads to edema and tissue overgrowth.

If residual bone is overloaded, or if the tissues over it are chronically inflamed, the bone will resorb and be replaced with fibrous tissue. This commonly occurs in edentulous areas in the anterior maxilla and in tuberosity regions, resulting in so-called flabby tissue, which decreases support for the denture.

Although the main risk factors for oral cancer are the use of tobacco products and alcoholic beverages, there is epidemiological evidence that age, inadequate diet, poor oral hygiene, fractured teeth, and wearing of dentures may be contributing risk factors to the 30,000 new cases diagnosed each year. Therefore, all persons over the age of 45 should have their oral cavities evaluated at least yearly for any changes, such as white, red, or mixed lesions, which may be a sign of oral cancer.

Allergic Responses

A small percentage of the population is allergic to the acrylic resins used in complete and partial dentures. However, a much larger population is allergic to the nickel used in some temporary crowns and in the cast framework of removable partial dentures.

Patient Acceptance of Dentures

The older the patient, the more likely he or she is to have problems wearing complete dentures. Most problems are associated with the mandibular (lower) arch (Ettinger, 1993). Laird and McLaughlin (1989) stated that although the technical aspects of denture fabrication are important, it is critical to evaluate the patient's motivation and adaptive ability to wear dentures. In other words, a dentist can make a technically correct denture, but the patient may not be able to wear it if the tissues are unable to tolerate it or the patient does not have the necessary neurological skills. Another significant factor in the patient's adaptive ability is effective verbal and nonverbal communication between the dentist and the patient.

It has been shown in several studies that systemic disease and multidrug regimens increase the number of visits required to fit dentures. However, some patients cannot accept complete dentures either physically or emotionally; for them, implant-supported dentures may be a solution.

RONALD L. ETTINGER

See Also
Geriatric Dentistry
Oral Health
Oral Health Assessment
Xerostomia

DEPRESSION IN DEMENTIA

Two competing stereotypes often cloud a perspective on depression in the elderly. The first is that the advent of any mental disorder in advanced age is an indication of an underlying dementing process; the second, that old age is characteristically a time of losses and reactive depression. The former stereotype suggests that depression in old age is inconsequential; the latter that depression is normal. Both views can lead to a failure to recognize and treat depression among the elderly. Clinically significant depression in the elderly is an eminently remediable condition, but if untreated, it may become refractory and chronic, result in a shortened life expectancy due to suicide and other mechanisms, or cause social dysfunction and disturbing behaviors.

Clinical Features

Post (1992) gives a succinct and clear description of depressive disorder in old age. The sufferer may report or evince sadness, emptiness, and detachment; anxiety and panic states may occur, with or without euphoria and excitement as an associated, cyclic aspect of the affective state. Speech is slowed and diminished, or repetitive and importuning if anxiety is a dominant symptom. Self-esteem decreases, and the patient loses interest in usual activities. The patient may be convinced that he or she is wicked and has sinned or that his or her bodily contents are impaired and objectionable. In all but the mildest episodes, sleep, appetite, body weight, and other vital functions may be disordered. In the most severely psychotic depressives, delusions of physical ill health, poverty, guilt, and self-deprecation may be expressed. Bizarre hypochondriacal and nihilistic delusions and pseudohallucinations may occur. Some patients present as mute. In some cases, paranoid symptoms may be conjoined with an empty or hostile affect superficially resembling paraphrenia. Neurotic depressives retain insight into their depressive symptoms and frequently exhibit phobias; anxiety is often more obvious than the underlying depression. The anxiety may be communicated as a feeling of restlessness or fluttering in the abdomen.

The clinical diagnosis of depression is made using diagnostic criteria that are set forth in the Diagnostic and Statistical Manual for Mental Disorders (DSM-IV) (American Psychiatric Association, 1994). Although the criteria for major depression and its subtypes are clearly described in the DSM-IV, the tool does not adequately describe the unique manifestation of depression in the elderly.

Specific inclusion criteria for the disorder must be met before a clinical diagnosis of major depression can be made. The essential feature of the disorder is either depressed mood or loss of interest or pleasure in all, or almost all, activities for at least two weeks (American Psychiatric Association, 1994). The symptoms must represent a change from previous functioning and must be relatively persistent, occurring for most of the day, nearly every day, during at least a two-week period. Some of the somatic symptoms listed above may be present for other reasons in an older person. The clinician should assess these symptoms in the context of the person's overall health and cognitive function.

The diagnosis of major depression is made only if it can be established that an underlying organic disorder is not present and that the symptoms are not a normal reaction to the loss of a loved one (i.e., uncomplicated bereavement). Additionally, the diagnosis is not made if the disturbance is superimposed on schizophrenia, schizophreniform disorder, delusional disorder, or psychotic disorder or if the criteria for schizoaffective disorder are met. These criteria relate to the importance of ruling out other psychiatric and organic illnesses that may produce depressive symptoms. However, coexisting organic and affective disorders, such as depression and dementia, are common, and the presence of an organic illness does not rule out depression (NIH Consensus Development Panel on Depression in Late Life, 1992).

Estimates of the frequency of depression and, more broadly, affective disorder depend in large part on the definition and criteria used for diagnosis and on the population under study. Prevalence ratios of clinically significant depression have been estimated to be 10% to 15% among the general elderly population (NIH Consensus Development Panel on Depression in Late Life, 1992), 30% to 40% among elderly psychiatric patients in short-term treatment settings, and 12% to 43% among elderly residents of long-term-care facilities (Abrams & Alexopoulos, 1994; Rovner et al., 1991). Clinically significant depression refers to depressive symptoms of sufficient severity to warrant clinical intervention.

Depressive symptoms of varying magnitude occur in 15% to 30% of patients with Alzheimers' disease (Ballard, Cassidy, Bannister, & Mohan, 1993). Among patients with a diagnosis of dementia, approximately 30% also meet DSM criteria of major depression in some studies (Larson et al., 1984); more conservative estimates are in the range of 4% to 6%, with "case levels" of depression approaching 30% (Lobo, Saz, Marcos, Dia, & Camara, 1995).

The evaluation of depressive signs and symptoms in patients with Alzheimers' disease is complex, because the symptoms and signs of depression overlap with those of dementia (Storandt & VandonBos, 1994). In spite of these evaluation and assessment complexities, it is commonly accepted that depression is often concurrent in dementia (Rovner et al., 1991).

Assessment

Abrams and Alexopoulos (1994) identify four major obstacles to the assessment of depression in dementia: the overlap in clinical manifestations of depression and dementia, the inability of demented patients to provide accurate information about their moods and inner lives, the narrow range of depressive symptoms addressed by instruments designed for severely demented subjects, and the transient nature of depressive symptoms in cognitively impaired individuals.

High rates of concurrent depression and dementia have led to attempts to develop tools to measure depression in older people with reversible or irreversible dementia (Alexopoulos, Abrams, Young, & Shamoian, 1988; Kafonek et al., 1989; Toner, Teresi, Gurland, & Tirumalasetti, 1999). Most of these measures are useful primarily among mildly to moderately demented subjects who can communicate their basic needs. Some measures include informant reports of the presence of depressive symptoms, although studies have found this method to be less valid than direct assessment. Self-report scales of depression have been used with both demented and nondemented subjects but have a number of limitations that contraindicate their use with physically or mentally frail elderly, and particularly demented persons (Toner, Gurland, & Teresi, 1988). Observation scales are limited by the fact that adequate validity and reliability estimates are not yet available, the instruments focus on a narrow range of observable depressive symptoms, and they often exclude items that can be assessed only through direct interaction with the patient.

In comparing various measures of depression in demented patients, Abrams and Alexopoulos (1994) suggest that instruments are needed to augment existing observational and informant methods for detecting depressive symptoms among severely, moderately, and mildly demented patients with impaired communication. A measure that uses direct interview techniques and observation is needed for all levels of dementia. Recent work by Toner et al. (1999) combines direct interview techniques using standard questions about symptoms of depression with observations of affect, or "feeling tone." This approach has shown promise in preliminary studies with older subjects who have moderate to severe communication deficits.

Difficult-to-assess populations such as the cognitively impaired need a multisource approach for valid assessment. An assessment methodology comprising observational, informant, and direct assessment measures is recommended. The Cornell Scale for Depression in Dementia (Alexopoulos et al., 1988) is a useful clinical instrument with an informant focus. The Dementia Mood Assessment Scale (Sunderland et al., 1988) is a useful tool with a focus on clinical observation. The Feeling Tone Questionnaire (Toner et al., 1999) augments these

tools by combining an informant focus and clinical observation methods and adding behaviorally anchored ratings of affect. It was specifically designed for use with communication-impaired demented patients, uses standardized questions with simple wording, and can be used by nonclinically trained staff. Finally, because the Feeling Tone Questionnaire requires only 5 to 10 minutes to administer, it can be repeated on several occasions over a prescribed period to capture fluctuating aspects of mood disorder or to obtain an average estimate of depressive symptoms.

JOHN A. TONER

See Also
Dementia: Overview
Depression Measurement Instruments

DEPRESSION MEASUREMENT INSTRUMENTS

Depression is a highly prevalent but underrecognized and undertreated mental health problem in community-dwelling, medically ill, and institutionalized older adults. Contrary to popular belief, depression is not a natural part of aging. With prompt and appropriate treatment, depression is often reversible. However, untreated depression may result in the onset of physical, cognitive, and social impairment, as well as delayed recovery from medical illness and surgery, increased health care use, and suicide. Clinicians can reduce the negative effects of depression by early recognition, intervention, and referral. Although depression measurement instruments are not a substitute for individualized assessment or diagnostic interview by a mental health professional, these instruments are useful screening tools in a variety of clinical settings, especially when baseline measurements can be compared with subsequent scores.

Self-Report Instruments

The Geriatric Depression Scale (GDS) (Yesavage et al., 1983) is a brief self-report questionnaire that has been tested and used extensively in community, acute, and long-term care settings with a variety of older populations and seems to be the best available assessment approach. The patient is asked to respond yes or no to 30 questions about how he or she felt within the past week. A score of 0 to 10 is normal, 11 to 20 indicates mild depression, and 21 to 30 indicates severe depression. The GDS may be used with the healthy or the medically ill, and with mildly to moderately cognitively impaired older adults. However, research to establish the lower limits of cognitive status for which the tool is reliable is limited. Using a cutoff score of 11, the GDS has 92% sensitivity and 89% specificity when evaluated against diagnostic criteria for depression. The validity and reliability of the tool have been supported in clinical practice and research. The GDS has several advantages over other screening instruments for depression: short completion time, limited number of potentially confounding somatic symptoms, high positive correlation with other depression rating scales, easily comprehended binary response options, and validation in a wide variety of older populations. A shorter, 15-item version of the GDS has proved useful in clinical settings where brevity is important to enhance acceptability and patient cooperation (Sheikh & Yesavage, 1986). A score of 6 or above indicates clinically significant depressive symptoms warranting attention. In general, the shorter version has not performed as well as the longer version in psychometric evaluation (Ingram, 1996). When patients cannot complete the questionnaire by hand, the GDS can be administered by an interviewer without changing its psychometric properties.

The Beck Depression Inventory (BDI) (Beck, Ward, Mendelson, Mock, & Erbaugh, 1961) is a 21-item self-report scale used in primary care research and clinical practice with a variety of clinical and community samples of older adults. The Likert-type scale response may be confusing to older adults. Scores range from 0 to 64; a score of 10 indicates mild depression, 16 mild to moderate depression, 20 moderate to severe depression, and 30 or greater severe depression. A cutoff score of 16 is useful in primary care settings and carries a 50% chance of correctly identifying major depression.

Seven of the 21 items (33%) are somatic features of depression that may be easily confused with symptoms caused by medical illness in older adults and may contribute to false-positive scores. A more recent version, the BDI-II (Beck, Steer, & Brown, 1996), replaces items that dealt with symptoms of weight loss, changes in body image, and somatic preoccupation. Sleep and appetite items were revised to assess both increases and decreases in sleep and appetite. The new edition has better clinical sensitivity and higher reliability than the original BDI. An advantage of the BDI is that it can be used cross-culturally.

The Center for Epidemiological Studies Depression Scale (CES-D) (Radloff, 1977) is a 20-item self-report scale with a Likert-type scoring format used extensively in epidemiological research with clinical and community samples of older adults. It has high internal consistency and reliability, acceptable test-retest stability, and good construct validity in these populations. Scores range from 0 to 60, with a cutoff score of 16 or greater distinguishing those with a high probability of major or clinical depression. The instrument has four subscales: general physical well-being, psychomotor retardation, depressed affect, and interpersonal functioning. Older patients may find the Likert scoring format difficult to comprehend. When patients cannot complete the scale by hand, the CES-D can be administered by a clinician without major changes in sensitivity and specificity. The CES-D is reliable for use with African Americans.

The Zung Self-Rating Depression Scale (SDS) (Zung, 1965) is a 20-item scale widely used in depression research. Although it has been used with older adults, it has been validated primarily with younger patients. Scores range from 0 to 80; a score of 50 or greater is indicative of clinical or major depression, and a score of 20 or less indicates the absence of clinically significant depression. Somatic and behavioral items can account for up to 50% of the total score and may result in false-positive scores in patients with medical comorbidities. Studies with acutely medically ill older adults found that an adjusted cutoff score of 60 yielded 87% specificity for clinical or major depression. The SDS can be completed in a short time, but the graded response format may be confusing for some older adults.

Interviewer-Administered Instruments

The Hamilton Rating Scale for Depression (HamD) (Hamilton, 1960) is the gold standard among outcome measures for treatment studies of depression in psychiatric, medical, and geriatric populations. It is most often used with patients already diagnosed as suffering from a depressive illness. The maximum possible score is 52, with a score of 30 or greater indicating severe depression. Somatic and behavioral items account for at least 50% of the scale. The scale requires clinicians to make medical judgments regarding the cause of depressive symptoms and places less emphasis on symptoms judged to be caused by medical illness. Despite concerns about the ambiguity of somatic symptoms in older adults, the HamD is reliable and valid with geriatric populations.

The Cornell Scale for Depression in Dementia (CS) (Alexopoulos, Abrams, Young, & Shamoian, 1988) is a 19-item clinician-administered scale developed specifically to measure depression in older adults with and without dementia. It is the only depression rating scale validated in both populations. The CS relies on information derived from interviews with caregivers as well as direct observations and interviews with patients. The CS is administered in two steps, preferably by the same clinician. The clinician interviews the patient's caregiver (preferably one with knowledge of the patient over time) on each of the items and then briefly interviews the patient. Discrepancies between the clinician's observations and the caregiver's report warrant a reinterview with the caregiver to clarify the reason for the difference. The CS is then scored on the basis of the clinician's final judgment.

A series of studies (Alexopoulos, Abrams, & Shamoian, 1988; Alexopoulos, Abrams, Young, & Shamoian, 1988) evaluated the CS in depressed patients and controls, in both cognitively intact and demented patients and found that the scale had high internal and interrater reliability in both groups. Several investigators have confirmed that the scale has a high level of concurrent validity relative to other measures of depression in patients with dementia. The CS items were designed to be scored primarily on behavioral observation rather than an elaborate interview with the patient. The severity of each item is rated according to three explicitly defined grades: 0, 1 (mild), and 2 (severe), with a possible range of 0 to 38. The time frame used in the evaluation is the previous week to the previous month, depending on the items assessed. The higher the score, the greater the severity of depression. A score of 13 or greater is indicative of major depression. An etiological approach to counting symptoms of depression is recommended, in that clinicians must judge whether symptoms are due to medical illness. CS items are scored positively or counted toward depression, if they are not associated with medical illness.

The self-report GDS (30 or 15 item) appears to offer the best depression measurement for older adults. For those with mild to moderate dementia, self-ratings of depression using the GDS are reliable and valid. Of the interviewer-administered instruments, the HamD is the most well known, but interrater reliability issues are of some concern. The number of somatic items on the HamD could contribute to false-positive scores in older adults with medical comorbidities. The HamD appears to be most useful when measuring treatment outcomes in patients already diagnosed with a depressive illness. The CS is the only depression instrument validated in both nondemented and demented older adults and has reasonable reliability and validity, even in those with moderate to severe dementia. Further study of the CS is needed to achieve acceptable psychometric and diagnostic performance.

LENORE H. KURLOWICZ

See Also
Depression in Dementia
Hispanic and Latino Elders

Internet Key Words
Assessment
Depression

DIABETES: MANAGEMENT

The clinical presentation and acute complications of diabetes include a variety of symptoms, most commonly polyuria, polydipsia, polyphagia, weight loss, and fatigue. In the elderly, however, these symptoms may be confused with normal age-related changes or symptoms of other medical disorders. Polyuria may be attributed to the use of diuretics or urinary incontinence. Weight loss may be associated with changes in appetite caused by medications and gastrointestinal disturbances. Whereas some individuals are asymptomatic and are diagnosed during routine medical examinations, others may have fully developed acute or chronic complications at the time of diagnosis—for example, when a patient presents in the emergency department with severe hyperglycemia or is having a diagnostic workup for pain and burning in the extremities. Thus, in the elderly, the diagnosis of diabetes must be considered in the workup for a variety of specific and nonspecific symptoms.

Management Goals

The two major goals in the management of diabetes, irrespective of age, are prevention of the acute metabolic derangements of diabetes and prevention of the chronic complications of diabetes. These goals are best achieved through a combination of therapies, such as diet, exercise, and medications, aimed at normalizing of blood glucose levels: fasting or preprandial, 80 to 120 mg; postprandial, less than 180 mg; bedtime, 100 to 140 mg. Long-term glucose control is measured by the glycosylated hemoglobin level, which reflects average glucose readings over the previous three months. The recommendation for glucose control, as established by the American Diabetes Association (ADA), is a glycohemoglobin less than 7%, regardless of age (ADA, 1998; Buse & Hroscikoski, 1998).

Treatment

Diet: Nutrition therapy—an essential element in the management of any patient with diabetes, regardless of age—includes maintenance of near-normal blood glucose levels, normalization of serum lipids, attainment and maintenance of a reasonable body weight, and promotion of overall health (ADA, 1998). Dietary recommendations include the following: caloric intake sufficient to attain and maintain a reasonable body weight; *protein* intake 10% to 20% of total daily caloric intake; *fat* intake approximately 30% of total intake for individuals with normal weight and lipids, less than 30% for overweight or obese individuals or those with elevated low-density lipoproteins, and 40% or less for individuals with elevated triglycerides unresponsive to fat restriction and weight loss; and *carbohydrate* intake calculated as the difference between protein and fat intake. Nutritive sweeteners, such as sorbitol, mannitol, and fructose, are considered carbohydrates. Noncaloric sweeteners, such as aspartame and saccharine, have been approved by the Food and Drug Administration and are safe to consume. The nutritional guidelines of the ADA (1998) stress individualization of the diet in accordance with blood glucose and lipid goals and emphasize the importance of weight loss.

Many elderly adults, however, may be undernourished due to a variety of physiological, psychological, social, and economic factors such as changes in smell, taste, and thirst; side effects of medications; inability to shop for food or prepare meals; cognitive impairment, depression, isolation, and loneliness; and inadequate income to purchase food (Gilden, 1999). For older adults who are unable to meet their nutritional needs through a regular meal plan (Holler & Pastors, 1997), it may be necessary to modify their usual food intake by changing the nutrient content or density, modify food consistency, use medical nutritional supplements, and consider enteral and parenteral nutritional support. Malnourished elders may require increased calories and upward adjustment of oral medications or insulin to maintain normal blood glucose levels.

Exercise: Routine exercise is an essential component of the treatment plan. It has been shown to enhance weight loss and improve glycemic control, insulin sensitivity, plasma lipids, quality of life, and psychological well-being in individuals with diabetes (ADA, 1994). Adding an exercise program to the diabetes treatment plan may decrease the need for insulin or oral medications, but the benefits of exercise must be weighed against the risks and presence of complications in elderly adults with type 2 diabetes. Most older adults can undertake walking programs and should be encouraged to contact local senior citizen centers for physical activity programs. Adults with limited mobility or those confined to chairs can participate in arm or chair exercises. Older adults without significant complications can participate in jogging and swimming programs. Swimming, biking, and water aerobics may be most appropriate for individuals with neuropathy to avoid injury to the feet. Patients with retinopathy may need to avoid exercises such as jogging that may cause retinal bleeding.

Oral Medications: Oral diabetes medications are generally recommended when diet and exercise have failed to achieve optimal glucose control in patients with type 2 diabetes. With the current knowledge regarding the relationship between hyperglycemia and chronic complications, many patients are beginning oral diabetes medications at diagnosis, in combination with diet and exercise. There are currently five classes of oral diabetes medications available in the United States: sulfonylureas, meglitinides, biguanides, thiazolidinediones, and alpha-glucosidase inhibitors (ADA, 1998). Individuals with type 1 diabetes are *not* candidates for oral medications.

Sulfonylureas stimulate insulin secretion from the pancreatic β-cell. The most frequent side effect of sulfonylureas is hypoglycemia (DeFronzo, 1999).

Meglitinides stimulate insulin secretion from the pancreatic β-cell more rapidly than sulfonylureas (ADA, 1998). Meglitinides are taken before meals to reduce postprandial hyperglycemia.

Biguanides decrease hepatic gluconeogenesis and glycogenolysis and increase insulin sensitivity (DeFronzo, 1999). Gastrointestinal problems, including abdominal discomfort, are the most common side effects of metformin, the only biguanide currently available in the United States (DeFronzo, 1999). An infrequent side effect of metformin is

lactic acidosis. Metformin is therefore contraindicated in any condition that causes hypoperfusion, such as severe hepatic, renal, and cardiopulmonary disease.

Thiazolidinediones increase insulin sensitivity (DeFronzo, 1999). Because use of thiazolidinediones has been associated with hepatic dysfunction, the manufacturers provide specific guidelines for liver function testing in patients using these drugs.

Alpha-glucosidase inhibitors interfere with the ability of enzymes in the small intestinal brush border to break down oligosaccharides and disaccharides into monosaccharides, thus retarding glucose entry into the systemic circulation (DeFronzo, 1999). Alpha-glucosidase enzyme inhibitors are associated with a number of gastrointestinal side effects, including bloating, abdominal discomfort, diarrhea, and flatulence. These side effects can usually be minimized by slow titration of the daily dosage. Alpha-glucosidase inhibitors are contraindicated in individuals with inflammatory bowel disease, cirrhosis, or plasma creatinine greater than 2 mg/dL.

Insulin: Insulin is indicated as initial therapy in the following situations (DeFronzo, 1999): (1) patients who have type 2 diabetes, a markedly elevated fasting plasma glucose levels (>280 to 300 mg/dL), and *ketonuria* or *ketonemia*; (2) symptomatic patients who have markedly elevated fasting plasma glucose levels (>280 to 300 mg/dL) (these patients may remain on oral agents or be switched to oral agents after six to eight weeks of good glycemic control); and (3) patients with type 2 diabetes who, after consultation with their health care providers, wish to receive insulin as initial therapy. There are a number of insulin preparations currently available in the United States. Long/intermediate-acting insulin preparations can be combined with short/ultrashort-acting insulin preparations to provide sufficient daily insulin coverage for optimal glycemic control (ADA, 1998). An example of such a combination is a split/mixed twice-a-day insulin regimen. In such a regimen, 70% of the daily insulin dose (~0.5 to 1.0 units/g/kg) is administered before breakfast, and 30% is administered before dinner. The prebreakfast and predinner insulin doses are divided into 70% intermediate-acting and 30% short/ultrashort-acting insulin.

Patients and their family members must be taught to draw up and administer insulin correctly. The loss of fine motor skills and visual impairments associated with both diabetes and aging may necessitate the use of adaptive devices, such as magnifiers and insulin dose counters. A list of such devices is available from the National Federation of the Blind. Elderly patients may need the assistance of home health nurses and family members to draw and administer insulin. Premixed insulin preparations such as 70/30 (70% intermediate insulin/30% regular insulin) and pen devices may be helpful.

Side Effects—Hypoglycemia: Biguanides, thiazolidinediones, and alpha-glucosidase enzyme inhibitors generally do not cause hypoglycemia when used alone. However, these medications may cause hypoglycemia when combined with sulfonylureas or insulin. The intensity of the insulin regimen must be balanced between glycemic control and the risk posed by hypoglycemia for a particular patient. For example, the risk of hip fracture from falling in an 80-year-old woman with severe osteoporosis may outweigh the benefits of excellent glucose control. In this case, blood glucose may be maintained at a slightly higher level. An elderly patient with diabetes is more vulnerable to hypoglycemia if he or she reduces overall caloric intake, skips meals, and exercises more intensely than usual. Hypoglycemia may be potentiated when gastrointestinal symptoms, such as those associated with alpha-glucosidase inhibitors and biguanides, are present. If hypoglycemia occurs when alpha-glucosidase inhibitors and sulfonylureas or insulin are given simultaneously, glucose tablets or gels provide the fastest recovery from hypoglycemia. In elderly subjects, the symptoms of hypoglycemia can be confused with cognitive dysfunction. Therefore, patients and their families must understand how to prevent, recognize, and treat hypoglycemia.

Blood Glucose Monitoring

Self–blood glucose monitoring (SBGM) meters that measure capillary blood glucose levels provide immediate feedback, allow individuals to determine patterns of hyperglycemia and hypoglycemia, and facilitate appropriate decisions about insulin doses. Using SBGM may be a problem for elders with diminished visual acuity and fine motor skills. For these individuals, a meter that has easily readable results and requires the least technical skill is recommended. The most appropriate times to monitor blood glucose levels are before breakfast, lunch, dinner, and bedtime snack. These times should be modified, based on each individual's medication regimen.

Medicare and private insurance coverage for home glucose monitoring more than once a day may be unavailable. Clinicians should try to impress upon patients the importance of tight glucose control and work with patients with limited financial resources. Financial support for equipment and materials may be available from community programs or other resources.

LAURETTA QUINN

See Also
Diabetes: Overview

Internet Key Words
Exercise
Insulin
Nutrition
Oral Diabetes Medications
Self–Blood Glucose Monitoring
Type 2 Diabetes

Internet Resources
American Association of Diabetes Educators
http://www.aadenet.org

American Diabetes Association
http://www.diabetes.org

American Dietetic Association
http://www.eatright.org

National Center on Physical Activity and Disability
http://www.uic.edu/orgs/ncpad

National Federation of the Blind
http://www.nfb.org/diabetes.htm

Parke-Davis Pharmaceuticals
http://www.parke-davis.com

Smith-Kline Beecham
http://www.avandia.com

Takeda-American Pharmaceuticals
http://www.takedaamerica.com

DIABETES: OVERVIEW

There are approximately 15.6 million individuals in the United States with diabetes, of whom 5.4 million remain undiagnosed. Diabetes affects approximately 10.7% of the U.S. population 65 to 74 years old and 10.1% of the population over 75 years of age. The prevalence of diabetes is highest among the elderly minority populations: non-Hispanic whites (12.6%), non-Hispanic blacks (17.5%), and Mexican-Americans (21.7%) (Harris et al., 1998). Approximately 18% of all nursing home residents 55 years of age and older are diagnosed with diabetes (National Diabetes Data Group, 1995). In addition, persons over 55 with diabetes are twice as likely as those without diabetes to reside in a nursing facility (Harris et al., 1998). The distribution of diabetes in the elderly is associated with multiple personal, social, and economic costs.

Diabetes-Related Complications in the Elderly

The acute complications of diabetes include diabetic ketoacidosis (DKA), hyperosmolar hyperglycemic nonketotic syndrome (HHNS), and hypoglycemia. Chronic complications include retinopathy, neuropathy, nephropathy, and cardiovascular, cerebrovascular, and peripheral vascular disease. The interplay between these complications and age-related comorbidities contributes to increased mortality in elderly patients with diabetes. The Diabetes Control and Complications Trial (DCCT) and the United Kingdom Prospective Diabetes Study (UKPDS) demonstrated that sustained hyperglycemia is associated with the development of chronic complications. In addition, the UKPDS demon-

strated that the severity of type 2 diabetes increases with longer duration of the disease. Diabetes in the elderly is a clinically complex disorder and is no longer considered a "mild" disease. The maintenance of normoglycemia in elderly patients with diabetes can ameliorate the untoward effects of hyperglycemia and decrease the occurrence and progression of diabetes-related complications.

Classification of Diabetes and Other Forms of Glucose Intolerance

In 1997, the American Diabetes Association (ADA) introduced new diagnostic and classification criteria for diabetes and other forms of glucose intolerance: Type 1 diabetes, Type 2 diabetes, other specific types of diabetes, and gestational diabetes. Categories of glucose intolerance include impaired fasting glucose (IFG) and impaired glucose tolerance (IGT) (ADA, 1998b).

Type 1 Diabetes: Type 1 diabetes affects less than 10% of the elderly diabetic population (Harris et al., 1998) and is being diagnosed with greater frequency in adults. Evidence suggests that an increasing number of adults over 35 years of age are developing a slowly evolving type 1 diabetes, known as latent autoimmune diabetes in adults (LADA). The primary physiological defect in the development of type 1 diabetes is the destruction of the pancreatic β-cell, resulting in an absolute deficiency of insulin secretion. In most individuals with type 1 diabetes, this insulin deficiency results from an autoimmune destruction of the pancreatic β-cell. At the time of diagnosis, 90% to 95% of individuals with type 1 diabetes have circulating antibodies directed against the pancreatic β-cell. In a small number of patients, however, the cause of this pancreatic β-cell dysfunction is unknown. The inability of the pancreas to secrete insulin, regardless of the cause, results in classic symptoms of type 1 diabetes, including polyuria, polydipsia, weight loss, electrolyte imbalances, and diabetic ketoacidosis.

Type 2 Diabetes: Type 2 diabetes affects more than 90% of the elderly diabetic population and increases in prevalence with age-related alterations in insulin sensitivity and secretion, altered glucose metabolism, dietary changes, obesity, and decreased physical activity. Type 2 diabetes is characterized by decreased liver, muscle, and adipose sensitivity to insulin and a defect in pancreatic β-cell insulin secretion. The development of type 2 diabetes follows a typical course. There is an initial period of hyperinsulinemia in which the pancreatic β-cell is able to overcome resistance and maintain normal glucose tolerance. This is followed by a period of postprandial hyperglycemia and increased insulin resistance because hyperinsulinemia is insufficient to maintain normal postprandial glucose tolerance. In the final stage, fasting hyperglycemia is present due to increased insulin resistance, unrestrained hepatic glucose production, and the toxic effects of hyperglycemia on the β-cell. At this time, the patient usually develops clinical symptoms of type 2 diabetes ranging from polyuria to hyperglycemic nonketotic syndrome, a life-threatening state characterized by severe dehydration, increased serum osmolality, and hyperglycemia.

Other Specific Types of Diabetes: Other specific types of diabetes affect less than 3% of the diabetic population. However, some of the secondary causes of diabetes in this category are more likely to occur in aging populations, such as pancreatic disease, hormonal disease, and medications that cause insulin resistance or decreased insulin secretion (e.g., glucocorticoids).

Impaired Fasting Glucose and Impaired Glucose Tolerance: IFG and IGT are terms used to describe individuals whose plasma glucose levels are higher than normal but are not diagnostic for diabetes. Nevertheless, this classification is a major risk factor for the development of both diabetes and cardiovascular disease. IFG is diagnosed as a fasting plasma glucose between 110 and 126 mg/dL. IGT is diagnosed as a two-hour oral glucose tolerance test (OGTT) plasma glucose between 140 and 200 mg/dL.

Risk Factors for the Development of Diabetes

Risk factors in the development of diabetes include ethnicity (Native Americans, Hispanics, African Americans, and Asian Americans), age over 45 years, having a first-degree relative with diabetes, obesity, other medical disorders (e.g., hypertension, dyslipidemias), and history of glucose intolerance.

Complications

Diabetic Ketoacidosis: DKA is a life-threatening condition in which severe abnormalities in protein, fat, and lipid metabolism occur as a result of an absolute or relative deficiency in insulin secretion. DKA usually occurs in patients with type 1 diabetes but may occur in patients with type 2 diabetes during times of severe stress, such as trauma, infection, myocardial infarction, or surgery. Mortality from DKA increases with advancing age. DKA is characterized by hyperglycemia due to increased glucose production and decreased glucose utilization, dehydration related to an osmotic diuresis, and metabolic acidosis related to increased production and decreased utilization of acetoacetic acid and 3-β-hydroxybutyric acid.

Hyperosmolar Hyperglycemic Nonketotic Syndrome: HHNS usually presents in middle-aged to older individuals with type 2 diabetes or IGT in whom physiological stress results in increased hyperglycemia, severe dehydration, and increased serum osmolality. Elderly patients who cannot compensate for fluid losses induced by hyperosmolar hyperglycemia (e.g., stroke patients who cannot swallow or articulate their need for fluid) are particularly vulnerable to HHNS. Often the patient is unaware of any impairment in glucose tolerance. HHNS differs from DKA in that there is no metabolic acidosis caused by an accumulation of serum ketone bodies. Precipitating factors include medications that cause glucose intolerance, such as glucocorticoids; therapeutic procedures, such as peritoneal dialysis; chronic disease, such as renal failure; and acute situations, such as infection. HHNS is characterized by severe dehydration and serum hyperosmolality. Patients may exhibit neurological manifestations due to intercerebral dehydration and renal insufficiency or failure due to the profound dehydration and hyperosmolality. Mortality rates for HHNS range from 10% to 50% and are usually attributed to underlying causes of or precipitating factors for HHNS.

Hypoglycemia: Hypoglycemia is caused by an imbalance that occurs when glucose utilization exceeds glucose production. The low blood glucose level (usually <60 mg/dL) can cause a variety of adrenergic and neuroglycopenic symptoms. Precipitating factors in the development of hypoglycemia include excess exogenous insulin, excess oral hypoglycemic medications, and a decrease in food intake and increase in physical activity in patients using oral hypoglycemic medications. Several abnormalities in the counterregulatory feedback symptoms of type 1 diabetes can result in frequent hypoglycemia. In type 1 diabetes, glucagon secretion becomes deficient two to five years after diagnosis. With prolonged duration of the disease, the epinephrine response is also impaired due to subclinical autonomic neuropathy. Thus, patients with long-standing type 1 diabetes have difficulty both recognizing hypoglycemic symptoms and recovering from hypoglycemia. In elderly patients, the symptoms of hypoglycemia can be mistaken for changes in cognitive function or symptoms of coexisting diseases. Therefore, elderly patients with diabetes treated with insulin or oral sulfonylureas are at risk for increased morbidity and mortality from hypoglycemic episodes.

Chronic Complications: The DCCT and UKPDS demonstrated that chronic hyperglycemia mediates the occurrence and progression of microvascular complications (retinopathy, neuropathy, and nephropathy) and is also a major contributor to the development of macrovascular complications (cardiovascular, cerebrovascular, and peripheral vascular disease). Although the exact physiological mechanisms by which hyperglycemia mediates these complications are unclear, there are four general theories regarding their pathogenesis: the polyol pathway, the protein kinase C pathway, the glycosylation pathway, and the free radical pathway (Mooradian & Thurman, 1999). Regardless of the physiological cause, hyperglycemia is the primary contributor to the development of diabetes-related complications, but hypertension and lipid abnormalities are also major contributors, especially to the development of macrovascular disease.

LAURETTA QUINN

See Also
Diabetes: Management

Internet Key Words
Diabetes Classification
Diabetes Complications
Diabetes Mellitus
Diabetes Risk Factors
Diabetic Ketoacidosis
Glucose Intolerance
Hyperglycemia
Hyperosmolar Hyperglycemic Nonketotic Syndrome
Hypoglycemia
Nephropathy
Neuropathy
Retinopathy

Internet Resources
American Association of Diabetes Educators
http://www.aadenet.org

American Diabetes Association
http://www.diabetes.org

DISCHARGE PLANNING

Every day, elders transition between health care settings, most often between hospital and home. Some continue to receive health services through skilled home visits, inpatient rehabilitation, or admission to nursing homes. Discharge planning assists patients and their caregivers in making these transitions, while supporting the continuity of care and achieving positive discharge outcomes.

Discharge planning is a process to identify patients in need and develop plans of care to be continued after discharge from formal health care or transfer from one type of formal care to another. Most literature on discharge planning describes it as occurring during hospitalization, but discharge planning should occur in any health care setting to coordinate care at the next level. Several regulatory agencies, including the Joint Commission on Accreditation of Healthcare Organizations, and federal laws (such as the Balanced Budget Act of 1997) mandate that all patients be screened for discharge planning needs.

The discharge planning process usually consists of six steps: (1) screening for discharge planning needs, (2) assessing individual health needs, (3) making decisions regarding the referral modality, (4) choosing a vendor, (5) implementing the plan, and (6) evaluating the plan (Pottoff, Kane, & Franco, 1995).

Screening for Needs

Patients over age 65 account for 40% of all hospital stays and 49% of all days of care in hospitals (National Center for Health Statistics, 1998). The average length of stay has decreased dramatically from 13.7 days in 1990 to 5.2 days in 1996 (National Center for Health Statistics, 1998). As a result, patients are often discharged before the full effect of treatment is evident, before the patient and family fully understand the illness or the treatment plan, and before the patient can assume self-care (Potthoff et al., 1995). This is especially true for elderly patients with complex medical conditions and social situations who are discharged to the informal care of family.

Screening identifies patient characteristics that increase the likelihood of services being required after discharge. The criteria most often used in screening include advanced age (older than 75), living alone, impaired mental status, multiple chronic conditions, dependence in activities of daily living, need for skilled nursing care or rehabilitation, repeat admissions in the past six months, or toileting problems (Potthoff et al., 1995). Despite the importance of accurate identification of high-risk patients, these criteria are often used inconsistently. Therefore, little is known about how to identify those patients most likely to benefit from targeted discharge planning.

Patients may be screened by a *direct service model*, whereby a specialist in discharge planning assesses, refers, and coordinates the discharge, or by a *consultation model*, whereby a nurse, physician, or interdisciplinary team work together to identify patients in need. A weakness of the direct service model is that discharge planners may not know the patients as well as staff nurses or physicians do; discharge planning then becomes a separate task instead of an integrated part of patient care (Dash, Zarle, O'Donnell, & Vince-Whitman, 1996). In contrast, the consultation model relies heavily on the skills of the staff nurse and physician and can vary widely in accountability and effectiveness (Potthoff et al., 1995). Regardless of which model is used, the use of standardized screening criteria to identify patients in need of discharge planning is fundamental.

Assessing Needs

Once a patient is identified as needing a discharge plan, a comprehensive assessment is completed to determine the magnitude of his or her needs. Common assessment areas include the patient's medical diagnosis and prognosis; number and types of comorbid conditions; functional, mental, and emotional status; social support; living arrangements; financial and insurance status; events and responses during the current treatment episode; current symptoms and need for skilled care; level of knowledge and capacity for self-care; mobility and home-bound status; age; educational level; numbers and types of medications; history of nonadherence to therapy; prior resource utilization, such as home care, rehospitalization, and nursing home admission; and patient and caregiver preferences. The assessment determines the next steps of the discharge planning process: deciding which modality is appropriate for the patient and which vendor will best meet the patient's needs and preferences.

There are no standardized assessment instruments for determining a patient's needs or for weighing these factors in decision making. The process is vulnerable to subjectivity and incongruence; there is great variation in the amount and type of information collected (Anderson & Helms, 1994) and in the influence of the various factors used in decision making. Further research is needed to identify which assessment characteristics are important in determining the need for continuation of health care services and to examine the effectiveness of discharge decision making in relation to patient outcomes.

Choosing a Care Modality

Once collected, the assessment information must be translated into a need for services and the options for care. Post-hospital options include admission to a nursing home or rehabilitation center, use of a skilled nursing home-care agency, or discharge to informal family caregivers. The available options, and the implications of each one, must be discussed with the patient and family before the decision-making process. This phase of the process is difficult, because no empirical data are available on the risks and benefits of various options, and it is difficult to foresee conflicts that might arise from the differing value structures of the family, patient, and health care providers (Potthoff et al., 1995). For instance, privacy concerns may overshadow the need for skilled home care. The negative stigma or costs of nursing home admission may, in the mind of the family, outweigh the patient's need for 24-hour supervision. Discharge planners are challenged to develop creative alternatives or strategies to accommodate patient and family preferences. They must be knowledgeable about what the community has to offer and the admission requirements of each discharge option. Further studies must (1) determine how to choose the best discharge option; (2) increase our understanding of and facilitate the decision-making process for patients, caregivers, and health care providers; and (3) link patient characteristics with processes of care and outcomes.

Choosing a Vendor

Regardless of the discharge destination, Medicare regulations require that patients be given a choice of vendors. In order to fully inform patients and caregivers, discharge planners must be knowledgeable about the types of services offered by various vendors; their philosophy, location, costs, admission criteria, and policies; and the quality of their care. Once supplied with this information, the patient and family ultimately make the choice. The choices given to patients may be limited by the number of facilities and agencies available locally or the bias of the discharge planner in a vertically integrated health care system, where there are financial incentives to refer patients to one's own system. Areas for improvement in this step of the process include the development of databases on the attributes of facilities, software that would match patient preferences with vendor attributes, and report cards on quality of care.

Implementing the Discharge Plan

Implementation of the discharge plan is a critical juncture in continuity of care. At this point, the discharge planner must contact the receiving agency, obtain acceptance of the patient, transfer relevant patient information, and supervise the arrival of the patient or the onset of services. Challenges associated with this step include difficulty contacting the appropriate personnel at the receiving agency and the paperwork requirements of different agencies. Also, there are few feedback loops between health care systems to ensure that the services ordered by the discharge planner are received by the patient. In this step, communication is paramount to continuity.

The goal of implementation is to transfer and receive information that is concise, complete, accurate, and valued. It is also a critical step for patients discharged to home without formal services. A comprehensive discharge plan, based on the assessment of individual needs, is essential to adequately prepare patients and caregivers to assume responsibility for care. Discharged patients commonly receive instructions in the management of medications, diet, and activity; follow-up care such as signs and symptoms to report and how to care for wounds; or when to schedule follow-up office visits.

For patients referred for continuing services, discharge planners should consult personnel from the receiving agency to determine their information needs. Communication between agencies may be improved by standardizing the amount and type of information exchanged (Anderson & Helms, 1994). Standardization would decrease the number of follow-up phone calls required to track missing information and the time and effort wasted in reassessing patient needs. If the discharge planning process is followed as outlined above, the discharge planner's's comprehensive assessment can be used by the receiving agency, rather than repeating the assessment. Electronic transfer of patient data would facilitate this process. Research is needed to determine the types and amounts of information necessary, to build the infrastructure to support the information transmission, and to develop the appropriate communication feedback loops.

Evaluating the Discharge Plan

This last step in the discharge planning process is the most neglected. Few discharge planners conduct follow-up evaluations of the outcomes of their decisions. Because of the lack of feedback loops, most have no way of knowing whether the services they arranged were ever provided, and if they were, whether they were successful and satisfactory. A system is needed to gather and analyze information on patients at set intervals after discharge. Of particular importance is patient and caregiver satisfaction, the number of rehospitalizations, unmet patient needs, and other adverse events. Again, information technology will be of great assistance. Computerized reminders can prompt discharge planners to make follow-up calls to patients or agencies to assess patient outcomes. Discharge planners can use information system feedback to improve their discharge planning by tracking patient outcomes in relation to patient characteristics and processes of care.

When properly executed, the discharge planning process allows the successful transition of patient care from one level to another. Efforts to make the process more efficient and effective share two common themes: standardization and communication. The ultimate goal to provide accurate, efficient, and effective discharge planning depends on accurately identifying patients in need, making the proper decisions regarding their disposition at discharge, and maintaining communication in a full feedback loop to ensure continuity of care.

KATHRYN H. BOWLES

Internet Resources
Administration on Aging
http://www.aoa.dhhs.gov

American Association for Continuity of Care
http://www.continuityofcare.com

Discharge Planning for the Elderly: An Educational Program for Nurses (book) by K. Dash et al. (1996)
http://www.springerpub.com

Joint Commission on Accreditation of Health Organizations
http://www.jcaho.org

DIZZINESS

Dizziness is the fourth most common complaint of geriatric patients and the most common complaint of persons age 85 and older (Sloane, Blazer, & George, 1989). As a presenting problem in primary care, it increases in frequency with patient age. Thirty percent of people have had dizziness by age 65; by age 80, 66% of women and 33% men have had some form of dizziness.

The term "dizziness" applies to various subjective sensory experiences. Patients use this term to refer to lightheadedness, faintness, disequilibrium, vertigo, blurred vision, and giddiness. Therefore, the clinician must first determine what the patient means by the term. Dizziness can result from abnormalities of any system related to postural control, including the cerebral cortex, basal ganglia, brain stem, cerebellum, vestibular portion of the inner ear and eighth nerve, proprioceptive nerve endings in the neck or lower extremities and their associated peripheral nerves, skeletal muscle, autonomic nervous system, and cardiovascular system. Dizziness rarely represents a life-threatening condition, but older persons with persistent dizziness often limit their activities because of fear of provoking the dizziness, fear of falling, physical deconditioning, or depression secondary to the dizziness.

Types of Dizziness

Dizziness is classically divided into four types: (1) vertigo—a sense of spinning or movement, usually due to problems in the vestibular system; (2) presyncope—a feeling of being about to faint, due to reduced blood flow to the cerebral cortex; (3) imbalance—a sensation of unsteadiness in the lower extremities, usually due to neurological disease; and (4) other—vague sensations often due to anxiety or depression. Unfortunately, fewer than half of older persons can be categorized according to these types, so this method of evaluating dizziness in older persons is not very helpful. Instead, a detailed history and physical examination are usually needed to create a differential diagnosis.

The sudden onset of dizziness in older persons generally has a single cause, such as an acute illness or a new medication. Chronic dizziness generally has multiple causes, although one factor is usually most important. Physicians who work with older persons should search for treatable causes and contributing factors, including physical deconditioning, anxiety or depression, decreased vision, and medication side effects.

Benign positional vertigo is one of the most common causes of dizziness in elderly persons. Patients usually have a spinning sensation that accompanies changing the position of the head. The sensation is caused by small, dense, calcific particles (otoliths) from the saccule or utricle of the inner ear that break loose and migrate into the posterior semicircular canal. Once positioned in the canal or in its receptor, the particles amplify rotational

movements in the plane of the canal. Thus, whenever the patient moves in the plane of the posterior semicircular canal, a short burst of intense vertigo is experienced. With time, the particles are either absorbed or scarred down, so that symptoms abate.

Clinically, benign positional vertigo is characterized by episodes of intense vertigo lasting less than a minute. Rolling over in bed, getting in and out of bed, and bending over and straightening up commonly provoke these attacks. Typically, patients have vertigo attacks with even slight rotatory movements for a few hours; sometimes these attacks may last a few days, with a mild lightheadedness noted most of the time. Symptoms improve quite rapidly and are typically gone in days to weeks.

A simple clinical test can help diagnose this condition. The patient is positioned supine on the examination table with his or her head over the edge and tilted 30 degrees backward. The patient's head is moved in either direction to align the ear parallel to the floor (the Hallpike-Barany maneuver). This produces rotatory nystagmus (to-and-fro movement of the eyeballs) and reproduces the patient's vertigo. The symptoms often diminish with repeated testing.

Medication is generally not helpful for benign positional vertigo, but physical exercises appear to help many patients and should be recommended. Determine the exact maneuver that most stimulates the vertigo, and instruct the patient to repeat the maneuver several times a day. The exercises should be performed about every three hours, repeating them enough times during each session to fatigue the vertigo response (usually three to five repetitions). After being completely symptom free for several days, patients can stop the exercises (Herdman, 1990).

Labyrinthitis (inflammation of the inner ear) is characterized by abrupt onset of severe vertigo and sparing of other neurological functions. The most common cause is viral. Meclizine and promethazine are helpful; low-dose benzodiazepines may provide some relief but must be used with caution. The condition is self-limited and usually begins to resolve in 24 to 36 hours.

Meniere's disease is a triad of recurrent vertigo, tinnitus, and hearing loss. At first, hearing loss is noted only during vertigo attacks; later, a fixed low-frequency loss can be demonstrated. Attacks of dizziness typically last between 2 and 12 hours. Ear pain and blackouts are not associated with Meniere's disease. Approximately 80% of patients respond favorably to medical management, which includes a low-salt diet, diuretics, and meclizine. Surgical treatment is reserved for patients with severe symptoms who do not respond to medical management.

Presyncopal dizziness is distinctive and is caused by diminished cerebral oxygenation. It leads to a feeling that one is about to pass out. Patients describe it as the desire to sit or lie down; as the room getting dark, like a shade is being pulled over their eyes; or simply as intense lightheadedness. Nausea and weakness often accompany the dizziness. Patients who actually lose consciousness are said to experience syncope. Common causes of presyncope and syncope are the transient conditions that cause postural hypotension. Cardiac arrhythmias, myocardial infarction, aortic stenosis, anemia, excessive diuresis, diabetes mellitus, and adrenal insufficiency are other common causes for the presyncopal type of dizziness.

Chronic dysequilibrium in older persons, especially in those age 85 and older, is commonly due to cerebral ischemia or infarction, often involving small vessels. Persons with this condition generally report a sudden or stepwise onset, have gait abnormalities on physical examination, and demonstrate subcortical white-matter lesions on magnetic resonance imaging of the head.

Anxiety and depression are the most common cause of chronic, continual dizziness in younger populations. In older persons with chronic dizziness, psychiatric dysfunction is quite common as well but is rarely the primary cause. Treatment of these secondary psychiatric conditions can, however, reduce disability and improve function.

A host of other diagnoses can produce continual dizziness. Cerebellar atrophy, which may be idiopathic or secondary to degenerative conditions such as alcoholism, leads to a continual feeling of dysequilibrium. Middle ear disease or sinusitis can produce vertigo or more vague sensations of continual dizziness. Bilateral vestibular hypofunction is another cause of continual dizziness, often the result of aminoglycoside toxicity. Other causes include sarcoidosis, carcinomatous meningitis, syphilis, and bilateral acoustic neuromas.

AHMED A. MIRZA

Internet Key Words
Benign Positional Vertigo
Dizziness
Meniere's Disease
Tinnitus

Internet Resources
Coping with Dizziness
http://www.conciliocreative.com/dizzy/

Mayo Clinic Vestibular Rehabilitation Program
http://www.mayo.edu/vest-rehab/

DRIVING

Older drivers, especially women and those over the age of 75, are involved in a disproportionately high number of motor vehicle accidents. Vehicular accidents are the major cause of accidental death in individuals between the ages of 65 and 74, and the second major cause in those over 75 (Dobbs, Heller, & Schopflocher, 1998). Older drivers have more multiple-vehicle accidents, daytime accidents, and accidents at intersections. They are also more likely to commit right-of-way and traffic signal violations (Williams & Graham, 1995). Accidents not resulting in death usually involve injury. As the population continues to age, there will be a sizable increase in the number of older drivers, many of whom will be unsafe. By the year 2030, it is projected that the number of traffic fatalities among elders will triple, making driving safety a priority for health professionals, policy makers, family members, and the public (Burkhardt, Berger, Creedon, & McGavock, 1998).

Age alone is a poor predictor of motor vehicle accidents. The functional impairments associated with normal aging and the presence of specific illnesses and their treatments are better indicators of driving safety in elders. Physical and/or mental changes may be the first sign that an older adult no longer has the necessary skills to drive safely.

Because vision is the key sensory function related to driving (Williams & Graham, 1995), older adults should be assessed for alterations in vision, such as narrowing of the visual field, decreased sensitivity to light, increased sensitivity to glare, and reduced night vision. The presence of cataracts, glaucoma, and macular degeneration should also be determined and treated.

Changes in posture, loss of coordination, reduced muscle strength, and the presence of foot abnormalities can adversely affect an older adult's ability to drive. Arthritis, cardiovascular disease, diabetes, stroke, and Parkinson's disease may also have a negative impact on the skills needed to drive safely. Elders should be assessed and treated for these changes and conditions. Depending on the condition affecting driving ability, physical therapy may be useful in an effort to enhance safety.

It is important to identify cognitively impaired elders early, because they are involved in more motor vehicle accidents than unimpaired elders are. Older adults should be carefully assessed for symptoms indicative of dementia, such as deficits in memory, language, orientation, and visual-spatial function. Health professionals can use the Traffic Sign Naming Test (Carr, LaBarge, Dunnigan, & Storandt, 1998) to assess the driving ability of older adults. The test is short and reliable and accurately identifies individuals with mild to moderate dementia who are unlikely to be safe drivers.

The use or misuse of alcohol by older adults is another factor that should be considered when determining driving safety. Certain prescription medications, such as central nervous system depressants, nonsteroidal anti-inflammatories, and benzodiazepines, used alone or in conjunction with alcohol may have a negative impact on driving ability (Hemmelgarn, Suissa, Huang, Boivin, & Pinard, 1997; Reuben, Silliman, & Traines, 1988).

Health professionals and family members need to recognize that although many older adults believe that they are safe drivers, they alter their driving habits when their functional abilities decline. Many avoid hazardous weather and heavy congestion, restrict their driving to daylight hours, and take fewer trips of shorter duration. However, these actions may not significantly reduce the older driver's accident risk.

Although the effectiveness of driver retraining has not been well established, it is an option that can be suggested when safety is a concern. Programs designed specifically for older adults focus on driving regulations, traffic signs, and risk reduction.

Health professionals should know whether state laws require them to report unsafe drivers to licensing authorities. In states where reporting is not mandated, professional responsibility for the safety of older adults, other drivers, and pedestrians suggests the importance of voluntary reporting.

Counseling unsafe older drivers to restrict their driving or to forfeit their licenses is an essential but often difficult task for health professionals. It may be complicated by well-intentioned family members who have made subtle attempts to convince the older adult to stop driving. The health professional should acknowledge the role that driving has played in the individual's independence and self-esteem. Older adults need to be given the opportunity to discuss the impact that restricted or lost driving privileges may have on their lives. These include lack of access to essential services, such as shopping and health care; social isolation and loneliness; limited recreational opportunities; increased risk of falls due to the need to walk under potentially dangerous conditions, such as ice and snow; loss of income if still employed; and diminished quality of life. The possibility of early or forced entry into assisted living or nursing home facilities is also a reality that warrants discussion.

Although alternative forms of public transportation should be discussed, elders may find them inconvenient, expensive, inaccessible, or unreliable. Efforts should be made to reduce isolation and help nondriving elders remain engaged in social activities. Friends, family members, and church groups can be queried about their willingness to assist with transportation, particularly when public transportation is unavailable. However, it is important to remember that although family and friends may promise assistance, such commitments are not always honored. Broken promises will add to the frustration and isolation already experienced by the nondriving elder.

The events leading to a reduction in driving or the loss of a license and their consequences occur over time. They are not necessarily sequential, and not all older adults experience them. Because many individuals drive safely well into old age, health care providers should not assume that all elders are unsafe drivers.

JULIE E. JOHNSON

Internet Key Words

Alternate Transportation
Driver Retraining
Driving Safety
Motor Vehicle Accidents
Older Drivers

Internet Resource

Administration on Aging
http://www.aoa.dhhs.gov/research/drivers.html

DYSPNEA (SHORTNESS OF BREATH)

The word "dyspnea" originates from the Greek "dys," meaning labored or difficult, and "pnoia," meaning breath. Clinically, dyspnea, shortness of breath, and breathlessness are often used interchangeably. The sensation of dyspnea is subjective. It encompasses the perception of labored breathing and discomfort and the individual's reaction to it (Carrieri-Kohlmann, Lindsey, & West, 1993). In elders, increased workload of breathing is a normal age-related change that can increase by as much as 20% at age 60 as compared with age 20. Some persons with cardiac or respiratory disease may have a blunted perception of dyspnea, possibly be-

cause of physiological adaptation over time. A longitudinal study of elders with chronic obstructive pulmonary disease (COPD) found that individual ratings of dyspnea were not directly linked to changes in lung impairment (Lareau, Meek, Press, Anholm, & Roos, 1999).

The precise incidence of dyspnea is unknown, but it is a frequently reported symptom of pulmonary pathology, heart disease, neuromuscular disease affecting respiratory muscles, obesity, and anxiety. Dyspnea is among the most common reasons for emergency department visits and has been associated with an increase in the likelihood of hospitalization for patients with COPD or heart failure.

A thorough interview to evaluate dyspnea severity is an important first step in assessment. Dyspneic persons often complain of difficulty moving air, inability to get enough air, chest tightness, poor exercise tolerance, aggravated breathing discomfort during social stress, and, in some cases, relief by using alcohol and sedatives. Mouth breathing, puffing, use of the accessory muscles of breathing, and inability to finish a sentence without pausing to breathe are frequent signs of true dyspnea from air hunger. Painful dyspnea suggests other causes, such as pleural or pulmonary inflammation, traumatized or inflamed intercostal muscles or thoracic cage, or even subdiaphragmatic inflammation, as seen in acute cholecystitis (Shepard & Geraci, 1999).

Asking the person about the amount of exertion that causes dyspnea can indicate its severity. However, some people adjust their physical activities in order not to experience dyspnea; hence, this type of inquiry may not be helpful. Severity of dyspnea on exertion (DOE) can be gauged more objectively by the distance an individual can walk in 6 to 12 minutes on a treadmill. There are also several scales that assess dyspnea. The visual analogue scale for elderly persons with COPD provides a quick and reliable measure of dyspnea. The Pulmonary Functional Status and Dyspnea Questionnaire (Lareau, Carrieri-Kohlmann, Janson-Bjerklie, & Roos, 1994) is another reliable scale that measures dyspnea intensity and changes in functional ability in elders.

Pulmonary function tests (PFTs), exercise testing, and arterial blood gas sampling or pulse oximetry all provide information for differential diagnosis. Spirometry is essential in evaluating dyspnea, because self-reporting of dyspnea does not always correlate with objective PFTs. A chest x-ray is not diagnostic but can provide information about heart size, lung parenchyma, pulmonary vasculature, pleural space, and diaphragm position. Wheezing usually indicates narrowing of the bronchi, as in asthma. Crackles indicate areas of atelectasis from underventilation or mucus plugging. Rhonchi may be heard in persons with increased sputum production. Inflammation in the pleura or pericardium may cause a friction rub.

Dyspnea may occur at different times of the day or be associated with a position or phase of the respiratory cycle. Inspiratory dyspnea is associated with upper airway obstruction, such as aspiration or epiglottitis. Expiratory dyspnea occurs with obstruction of smaller bronchioles, as in asthma. Paroxysmal nocturnal dyspnea, usually associated with heart failure, is the sudden onset of difficult breathing and coughing when sleeping in the recumbent position. It occurs one to two hours after lying down and is relieved on assuming an upright position. Orthopnea is inability to breathe lying down; to obtain relief, several pillows are usually needed to elevate the upper body. Orthopnea is very common in end-stage COPD but also occurs in other pulmonary diseases and left-sided heart failure.

One approach to the treatment of dyspnea is to help individuals self-regulate their breathing. Accurately assessing their perception of sensations and symptoms is an important component of biofeedback. Therefore, dyspnea rating scales such as the Borg Scale can be used not only to classify the severity of dyspnea for purposes of monitoring but also to desensitize high dyspnea responders (Mahler, 1999). With repeated use of dyspnea scales, elders who panic and aggravate their condition can learn to control their responses more effectively.

Dyspnea calls for an interdisciplinary approach. Nurses, physicians, respiratory therapists, social workers, and occupational and physical therapists can assist individuals by offering interventions and self-help strategies. Among these are smoking cessation, pulmonary rehabilitation, avoidance of infection and environmental stressors, disease management, increased adherence, minimization of social isolation, and coping with uncertainty.

Smoking cessation, even after the age of 60, halts the progressive decline in pulmonary function (Higgins et al., 1993). Health care providers should ask about smoking history, attempts to quit, and interest in trying again. Strategies should include use of the nicotine patch, oral medication, and behavioral interventions, with an appreciation of the difficulties associated with long-standing nicotine addiction. Interventions should consider all current medications.

Although pursed-lip breathing is a natural response to reduce the severity of dyspnea, some individuals must be taught this technique. The external resistance to expiration through pursed lips increases airway pressure and prevents airway collapse. Other techniques include paced breathing, diaphragmatic breathing, use of an inspiratory muscle training device, and relaxation to decrease anxiety. Upper body movements such as hair combing are especially demanding, but elders can learn energy-conservation techniques through such programs as the Better Breathers Club offered by the American Lung Association.

Dyspnea can also be a sign of pulmonary infection. Elders' decreased T-lymphocyte function contributes to an altered immune response and increased susceptibility to infection. The elderly should receive annual flu immunizations.

When lung infection occurs, dyspnea can be exacerbated by increased bronchial secretions. Teaching a person to cough productively can improve the clearance of pulmonary secretions. Taking a deep breath and huffing several times mobilizes secretions and induces a strong spontaneous cough. When necessary, aerosolized bronchodilators and nasotracheal suctioning may relieve dyspnea. Elders, particularly those with pulmonary disease, and their caregivers need to observe the color changes of sputum. Whitish sputum that becomes yellow or green or increasingly purulent or foul smelling should be reported at once to the appropriate health care provider. Often, however, the presentation of elders with lung infections is atypical. The classic triad of cough, fever, and pleuritic pain may be absent. Symptoms are even further masked in elders receiving steroid therapy for another condition. Treatment for a lung infection must be swift, especially in elders who show signs of impending septicemia.

Extreme weather often worsens dyspnea. Hot and humid weather causes a higher concentration of airborne irritants and pollutants. Cold and dry air may contribute to bronchoconstriction. Covering the mouth with a scarf during cold weather can help minimize dyspnea. Remaining indoors may reduce breathing discomfort. Symptomatic relief has been reported by directing a fan at the face or sitting before an open window. The air movement is thought to stimulate facial receptors that alter the perception of breathlessness (Mahler, 1999). However, since many elders spend most of their time indoors, it is important to be aware of indoor pollutants that can trigger dyspnea, such as cooking fumes, wood fires, perfume, cleaning agents, dust mites, and dry air.

Obesity and the resulting increased intra-abdominal pressure on the diaphragm and lung can make a dyspneic episode worse. Added pounds increase the workload of the cardiopulmonary system. Diet modifications and graduated exercise are encouraged, keeping in mind age-related changes, financial status, cultural factors, and comorbid conditions such as joint disease, dementia, and sensory problems. Exercise benefits dyspneic individuals, especially those with chronic respiratory diseases. Those who exercise show improved endurance and a subjective reduction in breathlessness.

Fatigue is often a major accompaniment of dyspnea. Elders and their caregivers need to pace periods of activity and rest. Paying attention to the location of elevators and rest rooms and obtaining a handicapped parking permit are helpful ways to decrease fatigue as well as dyspnea. For some chronic conditions, such as renal failure, anemia is a complicating factor that can lead to dyspnea and fatigue. This type of anemia can be treated with synthetic erythropoietin injections.

Adequate management of dyspnea requires that the elder appropriately self-manage the underlying cardiac, pulmonary, or neurological condition. Adherence with medication therapy is key. This might include diuretics, calcium channel blockers, antihypertensives, vasodilators, bronchodilators, or opiates. Elders with complex regimens may need assistance. Two ways to enhance adherence are to provide pillboxes with premeasured medication and to simplify the dose schedule. For elders with impaired vision or cognition, color-coding their medication containers can be helpful. The pharmacist should be asked to use non-childproof containers for elders with arthritis or weakness. Keep in mind that despite lifestyle changes and adherence to complicated regimens, dyspnea is exacerbated by worsening pathophysiology. When this occurs, it is crucial not to engage in "victim blaming."

Elders who live alone need telephones with speed dialing to reach established contacts in an emergency: 911, a nurse, family member, friend, or neighbor. Elders who are incapable of dialing need to have devices that allow them to press a button on a pendant to secure help. Community-dwelling elders with chronic illnesses who minimize the significance of dyspnea and have a poor social support system are at grave risk for exacerbation and hospitalization. These elders need referral to a home health agency for an assessment of disease management and the need for home health services. A socially isolated elder can benefit from community outreach programs or "friendly visitor" church programs. When dyspnea compromises activities of daily living, social isolation can be intensified as the elder tries to conserve strength by limiting activity and remaining house bound. Low-flow oxygen therapy, especially for DOE, is useful to maintain normal arterial-oxygen saturation. However, some elders are reluctant to go out in public with portable oxygen and need encouragement to do so.

Uncertainty about when and where dyspnea may occur can lead to panic, fear, anxiety, worry, and depression. Caregivers should be aware of these potential reactions, help elders express their concerns, and refer them to social services or psychiatry when appropriate. Adequate treatment and reasonable control of dyspnea can improve the quality of life for elders who must deal with it as part of their everyday lives.

ELIZABETH F. MCGANN
CHRISTINE FITZGERALD

See Also
Chronic Obstructive Pulmonary Disease
Heartburn
Immunization
Medication Adherence
Social Isolation

Internet Key Words
Breathlessness
Dyspnea on Exertion
Shortness of Breath

Internet Resources
American Association for Respiratory Care
http://www.aarc.org

American Heart Association
http://americanheart.gov

American Lung Association
http://www.lungusa.org

National Heart, Lung, and Blood Institute
http://www.nhlbi.nih.gov

DYSURIA

Dysuria is defined as painful urination and is caused mainly by inflammation. The pain is not felt over the suprapubic area, but rather at the urethral meatus. Usually, pain is accompanied by urgency and frequency of urination. Dysuria is a nonspecific symptom that can occur in different pathological conditions. Possible causes of dysuria include urinary tract infection, urolithiasis (distal ureteral lithiasis), urethral stricture, prostatic hyperplasia, carcinoma in situ of the bladder, interstitial cystitis, pros-

tatitis, and bladder overactivity. There are no exact data on the incidence of dysuria in these conditions, but the symptom is usually present. Symptomatic treatment without evaluating the underlying disease may lead to postponed diagnosis of bladder cancer or other diseases. Urinary incontinence and dysuria have a significant impact on the quality of life of elders (Koyama et al., 1998).

Evaluation of dysuria starts with the patient's history, with particular attention to a history of hematuria, chronic urinary tract infection, use of indwelling catheters, urolithiasis, or urinary incontinence. Concurrent symptoms such as fevers and chills suggest an infectious cause. Painless hematuria may be caused by carcinoma in situ of the bladder or chronic infection. Dysuria might be the only symptom of chronic urinary tract infection in the elderly. In males, other irritative symptoms such as urgency, frequency, and nocturia, together with obstructive symptoms such as hesitancy, intermittency, and decreased urinary stream, may suggest benign prostatic hyperplasia. In females, these symptoms may suggest vaginal and vulvar atrophy with scarring of the external meatus due to menopausal changes.

Clinical examination should include a thorough examination of the external genitalia, a rectal examination in males to evaluate the prostate, and a vaginal examination in females. The minimal technical investigation should include urine microscopy, urine cytology, urine culture, and residual urine measurement; this will prove or rule out the presence of infection, hematuria, and voiding dysfunction. Because of the lack of specificity and sensitivity of urine cytology, a cystoscopic examination and possible biopsy might be indicated when carcinoma in situ has to be ruled out. Kidney and bladder ultrasound or x-ray will exclude or prove lithiasis. The presence of postvoid residual urine must be assessed, because this can be a factor in recurrent infections. Abnormal urine cytology warrants cystoscopy, biopsy, and evaluation of the upper urinary tract. If no underlying disease can be detected, symptomatic therapy may be indicated. This can consist of increased diuresis, low-dose anticholinergics or antispasmodics, and pelvic floor rehabilitation.

The treatment of dysuria is related to the cause. Low-potency estrogens given topically or orally are an effective treatment for urogenital symptoms in postmenopausal women (Milisom, 1996). If dysuria is persistent or if meatal scarring is present, urethral dilatation or urethrotomy may be indicated. In males, prostatic hyperplasia can be addressed medically or surgically.

DIRK DE RIDDER

See Also

Benign Prostatic Hyperplasia

Urinary Tract Infection

Internet Resources

Madison Nursing

http://www.son.edu/

MCP Hahnemann University

http://www.auhs.edu/

E

EATING AND FEEDING BEHAVIORS

The ability of older adults to ingest adequate amounts of nutrients is not affected by age. However, for some disabled older adults, the ability to ingest food may be hampered by the inability to transport food to the mouth, recognize food, perform the voluntary and nonvoluntary stages of swallowing, or remember to eat or stop eating (Amella, 1999b). This special group of individuals must be assisted at meals, or they will starve, choke, or eat inadequate or inappropriate amounts of food. Therefore, when an older individual has an unplanned weight loss or gain or a change in a nutritional health parameter, many health issues must be explored, including feeding and mealtime behaviors.

An older person with food intake problems must be examined for visual problems—cannot see food or utensils; memory problems or dementia—does not recognize food, does not remember having eaten, does not recall how to use utensils, or forgets the voluntary components of swallowing, such as keeping the mouth closed, chewing food, forming a bolus, and moving food into the pharynx; coordination problems due to neuromuscular conditions—cannot coordinate the act of bringing food to the mouth, cannot coordinate the voluntary phases of swallowing, or has impaired neurological pathways used to coordinate nonvoluntary phases of swallowing; mechanical or obstruction problems—lacks the structures needed to feed self and swallow food, has barriers to the completion of successful ingestion, or has poor dentition; psychogenic problems—has depression or a condition with psychotic components, causing an unwillingness to eat or overeating; and deconditioning—is frail and has loss of muscle mass (sarcopenia), producing severe fatigue that inhibits the ability to self-feed or produces an ineffective swallow.

Weight loss is a multifactorial process that may be organic or nonorganic in origin. Chronic illnesses that predispose an older adult to weight loss include cancer, infection, endocrine disorders, and organ failure. Oral health is a strong predictor of nutritional health. Older adults should receive preventive dental services as well as prompt treatment for all oral problems. Dysphagia, the inability to swallow food, is an important but frequently overlooked factor that causes unwillingness or inability to ingest foods. A study in two nursing homes found that dysphagia was present in a majority of residents who were not eating well but could not report symptoms, such as persons with aphasia or dementia (Kayser-Jones & Pengilly, 1999). Another study of 189 nursing home residents reported that 44% developed aspiration pneumonia related to dysphagia (Langmore, Terpenning, & Schork, 1998). Apraxia of swallowing, the inability to put the steps of swallowing together conceptually despite an intact motor system, is often present in persons with dementia (LeClerc & Wells, 1998). Medications may influence eating or the amount consumed. Digoxin, benzodiapezines, opiates, serotonin reuptake inhibitors, laxatives, thyroxine, corticosteroids, chemotherapeutic agents, antihistamines, anticholinergics, and antibiotics are associated with anorexia, increased metabolism, or disturbed functioning of the gastrointestinal tract (Verdery, 1998).

Cultural traditions or religious prescriptions regarding the healing qualities of food may result in food avoidance to keep from violating taboos or risking of illness. For example, Hindus, Sikhs, and some Muslims practice Ayurvedic medicine, in which foods are "hot" and "cold" (Rajwani, 1996). Hot foods such as fish, eggs, yogurt, and honey are used to treat "cold" diseases such as arthritis, respiratory illnesses, and gastrointestinal problems. In these belief systems, ingestion of incompatible foods (hot and cold) may produce toxins; hence, certain combinations, such as meat and milk, are avoided.

The context and presentation of meals may have a powerful influence on the willingness to eat.

Mechanically altered foods are sometimes unpalatable and people placed on these diets are often relegated to eating them forever. However, they may be able to resume former diets with modification or rehabilitation. It may be possible to move a patient from tube feeding to oral feeding or from a puree diet to a mechanically soft diet, thus preserving the aesthetics of meals. The ability to focus on the process of eating may be hindered by an overly stimulating or distracting environment, such as a communal dining room or one where the TV is playing loudly. Caregivers unable to attend to the individualized needs of older patients may resort to an approach that is not based on each person's tempo, food preferences, or needs.

Mealtime is more than a time to ingest food. In every culture, the sharing of food has ritualized social and even religious connotations. The relationship among people who share meals has profound significance, regardless of where the meal occurs. Thus, whether in a nursing home dining room or the kitchen of a home, when one person helps another complete a meal, a fundamental social interaction occurs. The quality of that interaction may influence the meal and even the amount of food consumed. Amella (1999b) demonstrated that when 53 nursing home caregivers were able to share control in their relationships with 53 residents with late-stage dementia, the residents became more social, positively influencing the amount of food consumed at breakfast. Discrete interactional patterns of behavior that initiated, sustained, or extinguished eating were found among 11 nursing home residents with dementia and their caregivers (VanOrt & Phillips, 1992). Institutional barriers to mealtime interaction such as understaffing, inadequate staff education, and a general lack of individualized care influence behaviors that occur at meals (Kayser-Jones, 1996; Kayser-Jones & Schell, 1997).

Failure to thrive (FTT), a syndrome marked by progressive weight loss, declining function, and decreased muscle mass, must be considered in frail older adults. FTT has contextual and interpersonal components that influence the individual's ability to eat or be fed. Newbern and Krowchuk (1994) suggest a model of a failed human-environmental interaction as part of a holistic approach to FTT.

This model suggests that when assessing an individual with weight loss, the professional consider not only physiological components of wasting but also decline in cognitive function, signs of depression, an inability to give of oneself, and an inability to find meaning in life or attach to others. Problems in maintaining connectedness because of interpersonal loss, dependency, loneliness, and feelings of worthlessness may combine with inadequate food intake or a significant health insult to provoke nutritional compromise leading to FTT syndrome.

ELAINE JENSEN AMELLA

See Also
Cultural Assessment
Deconditioning Prevention
Meals On Wheels
Nutrition Assessment

Internet Key Words
Anorexia
Aspiration Pneumonia
Culture
Dysphagia
Failure to Thrive
Feeding
Medications
Oral Health

Internet Resources
Elderly Nutrition Programs
http://www.aoa.dhhs.gov/nutrition

Food and Nutrition Information Center from the National Agricultural Library
http://www.nal.usda.gov/fnic

General wellness sites
http://www.elderweb.com/body/wellness.htm

ELDER MISTREATMENT: OVERVIEW

The term "elder mistreatment" describes a group of behaviors that cause harm or injury to older adults. Actions usually included in this category are abuse, neglect, and financial exploitation. Many professionals also include self-neglect as a type of elder mistreatment. Elder mistreatment can take place in the home, referred to as domestic elder mistreatment, or in institutional settings such as hospitals, long-term-care facilities, assisted living facilities, and nursing homes, referred to as institutional elder mistreatment (Fulmer & Paveza, 1998; Pritchard, 1996; Quinn & Tomita, 1997).

The exact number of older adults affected by elder mistreatment is unknown. The National Incidence Study on Elder Abuse (National Center on Elder Abuse, 1998) suggests that almost 500,000 older adults are the victims of some type of elder mistreatment each year. Other studies suggest that the number may be closer to 2.5 million (Pillemer & Finkelhor, 1988), and even that number may be low when the various types of financial exploitation and mistreatment in institutional settings are taken into account (Paveza & Hughes-Harrison, 1997; VandeWeerd & Paveza, 1998). Regardless of the actual number, it is clear that many vulnerable older adults are the victims of behaviors intended to hurt them physically, emotionally, or both.

The reporting of elder mistreatment is an important aspect of elder care in the United States. All 50 states have laws addressing this issue. In many states, doctors, nurses, social workers, other types of health care and home-care workers, and law-enforcement officers who suspect that an older person is being mistreated must report this to the appropriate state agency. For these persons, there is often a penalty for failing to report suspected abuse. Penalties range from loss of the right to practice professionally in that state to criminal prosecution, with a fine and/or imprisonment if convicted. In a few states, mandated reporting has been extended to all adults, meaning that any adult who suspects elder mistreatment is expected to report it to the responsible agency.

In many but not all states, the name of the person reporting the suspected mistreatment is confidential. In addition, all states have included a Good Samaritan clause in their legislation. This protects the person reporting the suspected mistreatment from a civil lawsuit should the investigating agency decide that no mistreatment has occurred, as long as the reporting person can show that there was no malicious intent.

Definitions of the types of elder mistreatment vary from state to state and across countries, but general concepts apply. *Elder abuse*, for example, is usually defined as behavior meant to deliberately cause physical pain and harm and usually includes such actions as hitting, punching, kicking, and biting, as well as threatening the person with a weapon or using a weapon. *Sexual abuse* of older adults involves unwanted sexual conduct and may occur in the older person's home or in institutional settings. *Elder neglect* is generally defined as failure to provide for the adequate care of an older adult in terms of housing, food, clothing, personal hygiene, medical care and medications, and other things needed to maintain health and well-being. Neglect can be divided into self-neglect, which occurs when older adults fail to provide these essentials for themselves; active neglect, when caregivers deliberately withhold food, clothing, money, and so forth; and passive neglect, when caregivers do not deliberately withhold necessities but may be unaware of what is needed. *Emotional abuse* is generally described as any physical or verbal action with the deliberate intention of causing emotional, psychological, or mental pain or distress. *Financial exploitation* is the willful obtaining of an older person's money or assets by someone with no legal right to that money for his or her own use or benefit. This can include fraud, embezzlement, or undue influence.

The individuals protected by elder mistreatment legislation vary across the country. In some states, the only requirement for protection under the law is achievement of a certain age, often 60 or 65. In other states, the person must be a certain age and also must be considered "vulnerable"— meaning that the person has some physical, cognitive, or mental impairment. Other states use a combination of these two concepts. The legal restrictions can sometimes frustrate attempts to report suspected elder mistreatment, because in the judgment of the investigator, the alleged victim may not meet the requirements for being a vulnerable adult.

The locations where elder mistreatment occurs are as varied as the places where older adults live.

Much of the early work on elder mistreatment focused on those who were mistreated in the community and lived independently or with family. This focus was appropriate, since most older adults, healthy or frail, lived in the community. Interest in elder mistreatment in institutional settings has increased with the realization that older adults residing in these facilities are among the most vulnerable (Glendenning, 1999). For those in caring roles, it is important to recognize that statements of fear or concern about being taken advantage of may be more than a paranoid or delusional response; they may be indications that the older adult is being mistreated.

Providing a general profile of mistreaters is difficult and is somewhat less accurate than previously thought. With that caution in mind, it is still possible to make some general comments about persons who are likely to mistreat older adults. Men are more likely than women to engage in elder mistreatment when all forms of mistreatment are taken into account (Quinn & Tomita, 1997; Reis & Nahmiash, 1997). Some data indicate that mistreaters often have problems with substance abuse or other emotional or psychological problems that contribute to their propensity to mistreat older adults (Anetzberger, 1987; Paveza & Hughes-Harrison, 1997; Pillemer & Wolf, 1986). The age of the person who commits the mistreatment varies, depending on the relationship between the person who is mistreated and the mistreater. In the case of spouses, the age difference may be only a year or two; with other family members, friends, or strangers, the age span may be 10 or more years. Mistreatment can occur between spouses, adult children and parents, grandchildren and grandparents, other relatives, friends, and complete strangers (Anetzberger, 1987; Biggs, Phillipson, & Kingston, 1995; National Center on Elder Abuse, 1998; Paveza & Hughes-Harrison, 1997; Pillemer & Finkelhor, 1988; Pillemer & Wolf, 1986; Quinn & Tomita, 1997).

Similarly, a profile of the mistreated older adult is difficult to provide and varies with the type of elder mistreatment. Some general trends do emerge, however. Women are more likely to be mistreated than men, and the age of mistreated persons may range from the early 60s to well into the 80s, though the age range is somewhat higher for institutional abuse (Aitken & Griffin, 1996; Anetzberger, 1987; Biggs et al., 1995; Glendenning, 1999; National Center on Elder Abuse, 1998; Pillemer & Wolf, 1986; Quinn & Tomita, 1997). Finally, much of the research suggests that the person who is mistreated may have some physical frailty that makes him or her vulnerable (Anetzberger, 1987; Biggs et al., 1995; National Center on Elder Abuse, 1998; Quinn & Tomita, 1997).

Although an understanding of who is mistreated and who mistreats is useful, in the end, what matters most is what we can do about it and how we can help those who have been mistreated. Intervention programs generally focus on providing both the mistreated and the mistreater with medical and social services (Biggs et al., 1995; Quinn & Tomita, 1997). However, recent discussions have recommended that intervention take the form of law-enforcement involvement and criminal prosecution, suggesting that this may be the only way to protect vulnerable older adults. The most critical intervention that individual caregivers or older adults can engage in is making that initial report of suspected mistreatment.

GREGORY J. PAVEZA

See Also
Elder Neglect
Financial Abuse
Money Management

Internet Resources
National Aging Information Center
http://www.aoa.dhhs.gov/naic

National Center on Elder Abuse
http://www.gwjapan.com/NCEA/

ELDER NEGLECT

It is estimated that in the United States alone, 700,000 to 1.2 million older adults suffer from elder mistreatment each year (Pillemer & Finkelhor, 1988). Physical abuse, physical neglect, psychological abuse, financial exploitation, and violation of rights are all examples of elder mistreatment. Neglect, the most prevalent form of elder mistreatment, accounted for 58.5% of all cases in 1994 (Fulmer & Paveza, 1998).

A common definition of neglect is "the failure of a caregiver to provide goods or services that are necessary for optimal function or to avoid harm" (Swagerty, Takahashi, & Evans, 1999, p. 2408). Although neglect is the most common form of elder mistreatment, it is the most underreported and least understood. Because of the lack of understanding of what causes neglect and what its symptoms are, various legal standards are used in the states to determine what elder neglect is. In some states, such as Massachusetts, an instance of neglect must transpire at the hands of another; in others, such as Connecticut, self-neglect is a reportable subcategory of elder neglect.

Two types of neglect, psychological and physical, are often noted. Psychological neglect is the failure to provide dependent elderly individuals with social stimulation, whereas physical neglect is the failure to provide the goods and services necessary for optimal function (Aravanius et al., 1992). Forms of psychological abuse include verbal harassment or intimidation, threats, treating the elder person like an infant, and isolating the elder from family, friends, or activities. Forms of physical neglect include withholding health care; failing to provide eyeglasses, hearing aids, or other physical aids; and failing to provide safety precautions (Aravanius et al., 1992). Physical neglect is often easier to identify than psychological neglect; however, both forms are difficult to diagnose and may be misdiagnosed.

Neglect has been found to be connected to certain risk factors that significantly increase the rate of occurrence. Studies have found that age is a major factor in elder neglect. Persons older than 75 years of age have a significantly increased chance of becoming neglected, and as the age of the patient increases, the chance of neglect becomes even greater (Lachs, Berkman, Fulmer, & Horo-

witz, 1994; Lachs, Williams, O'Brien, Hurst, & Horowitz, 1997). In addition, minorities and elders with small social networks have an increased chance of becoming neglected (Lachs et al., 1994). Conversely, studies have disproved the notion that frail, female, and cognitively or functionally impaired individuals are more at risk.

When examining elderly patients, health care professionals should assess for neglect. The accompanying table presents indicators of neglect that can be found during the physical examination (Fulmer & Ashley, 1989). Although clinical indicators are available, misdiagnosis and underreporting are still quite high. Both ageism, which creates stereotypical prejudices, and cognitive impairment contribute to underreporting and misdiagnosis.

Elder neglect, although often thought of as being associated with nursing homes, is just as prevalent in the community setting. Neglect occurring in the community can stem from self-neglect, caregiver neglect, or family neglect. In a study of 2,812 community-dwelling older adults followed over a nine-year period, the most common perpetrators were the adult children of the elder (45%), followed by spouses (26%) (Lachs et al., 1997).

Intervention in cases in which neglect is suspected must be handled carefully. The first step is to

Indicators of Elder Neglect

Inadequate/inappropriate clothing
Poor hygiene
Poor nutrition
Poor skin integrity
Contractures
Decubiti
Dehydration
Impaction
Malnutrition
Urine burns/excoriation
Duplication of similar medications
Unusual doses of medications
Dehydration > 15%
Failure of caregiver to respond to warning of obvious
    disease
Repetition of admissions due to probable failure of
    health care surveillance

obtain a history and perform a physical examination away from the caregiver or suspected abuser, as the patient may be intimidated and not respond directly to questions (Aravanius et al., 1992). Then, if neglect is still suspected or has been substantiated, the state's adult protective services department should be notified. A multidisciplinary team consisting of geriatricians, social workers, case management nurses, and representatives from legal, financial, and adult protective services can be very effective in solving the problem (Wolf & Pillemer, 1994).

Overall, neglect of elders is a serious and prevalent form of elder mistreatment that can occur in any setting. Risk factors and indicators must be closely watched by clinicians and other caregivers, and intervention must be performed expediently to avoid further suffering.

TERRY T. FULMER
GREGORY J. PAVEZA
ABRAHAM A. BRODY

See Also
Adult Protective Services
Elder Mistreatment: Overview

Internet Resource
National Center on Elder Abuse
http://www.gwjapan.com/NCEA

EMPLOYMENT

The Older Labor Force

Employment in old age remains the exception among both men and women in the United States. As of 1999, only 4 million of the more than 32 million noninstitutionalized persons age 65 and older, or 12.3% of the total, were working or looking for work (U.S. Bureau of Labor Statistics, 2000). Not surprisingly, older men are more likely than older women to be in the labor force—16.9% versus 8.9% in 1999. Somewhat higher proportions

have *some* paid work experience over the course of a year, but for the large majority of older men and women, participation in the formal labor force is an activity of the past.

Few labor force developments of the postwar era have been as pronounced as the workforce withdrawal of men age 65 and older, 45.8% of whom were employed in 1950. Eligibility for retired worker benefits under Social Security as early as age 62 and, for many, private pension benefits at even younger ages have contributed to a decline in labor force participation among younger workers as well. Men between the ages of 55 and 64, for example, had a labor participation rate of 67.9% in 1999, down from 86.9% in 1950.

The picture is demonstrably different for middle-aged women, millions of whom have accompanied their younger counterparts into the labor force over the past five decades. In 1999, the labor force participation of 55- to 64-year-old women stood at 51.5%, up sharply from 1950's 27%. Conversely, there has been relatively little change among women age 65 and older, whose attachment to the labor force has historically been weak; in 1999, only 8.9% were employed, compared with 9.7% in 1950.

The steady march of middle-aged women into the labor force has not been enough to offset the withdrawal of men; as a result, there are relatively fewer middle-aged and older workers today than in 1950. For example, nearly 43% of the 55-plus population were working or looking for work in 1950; by 1999, that was the case for only 31.8%. Nonetheless, these divergent trends have markedly altered the gender composition of the older workforce (which henceforth refers to persons age 55 or older). Women currently constitute just over 44% of that older labor force, in contrast to about 23% in 1950.

By the mid-1980s, the decline in labor force participation by older men had begun to taper off. Although it is premature to conclude that this development heralds a *reversal* of early retirement trends—participation rates have both risen and fallen slightly in ensuing years—the trend toward ever-earlier retirement seems to have come to an end.

Unemployment

If they are in the labor force, virtually all older persons have jobs. Unemployment rates tend to fall with age, in part because access to retirement benefits gives many older workers the option of leaving the workforce if they lose their jobs. Workers who withdraw from the labor force are not counted among the unemployed, even if they would prefer to be working. Persons 55 and older had an unemployment rate of 2.8% in 1999, versus 4.4% for those under the age of 55.

Lower unemployment rates obscure the formidable barriers older persons face if they decide to undertake a job search. A number of factors, not the least of which is age discrimination, contribute to the difficulties older job seekers experience. One consequence is that older job seekers are more likely than younger ones to find themselves among the long-term unemployed. In 1999, one-third (34%) of all older unemployed workers were out of work for at least 15 weeks; the comparable figure for the under-55 unemployed was 24%. Nonetheless, this figure represents a decline from earlier in the decade. Older workers, like their younger counterparts, have benefited from the robust job growth of recent years. Even so, the average duration of unemployment remains substantially longer for older job losers (18.2 weeks versus 12.9 weeks in 1999). In addition, older displaced workers who manage to become reemployed are more likely than their younger counterparts to experience a sharp drop in earnings (Couch, 1998; U.S. Congressional Budget Office, 1993).

Costs and Benefits of Older Workers

Rare is the survey that fails to reveal exceptionally positive attitudes on the part of employers toward older workers, who receive high marks when it comes to loyalty, dependability, trustworthiness, good work habits, and the like (AARP, 2000; Barth, McNaught, & Rizzi, 1993; Rhine, 1984). Yet those same employers are notably less effusive when it comes to bottom-line attributes such as flexibility and technological know-how, and these are the attributes that employers want in their workers.

Concerns about costs also undermine the positions of older persons in the workforce. Older workers may be more expensive than younger workers because, in general, wages rise with tenure, and tenure increases with age; the per-person cost of certain benefits, particularly health insurance and pension contributions, also rises with age (Clark, 1994). According to Clark (1994, p. 1), "many employers seem to believe that older workers are more costly relative to their value than younger workers," but the key words are "relative to their value." If higher costs are associated with greater productivity or other positive returns to the employer, those costs may be justified.

Meta-analyses of the literature on the relationship between age and job performance reveal an extremely weak relationship between age and performance. In fact, Sterns and McDaniel (1994) concluded that, if anything, performance improves slightly with age, regardless of whether supervisory ratings or more objective measures are used to assess performance, but the relationship is tenuous.

Interest in Employment

Whether and to what extent the nearly 38 million older persons who are not in the labor force could be enticed back are uncertain. Worker surveys and public opinion polls over the years have revealed considerable interest in postretirement employment on the part of preretirees. For example, 80% of baby boomers contend that they expect to work in retirement (AARP, 1998). Similar percentages have been reported for other age groups by the National Institute on Aging (1993) and the Employee Benefit Research Institute (Yakoboski & Dickemper, 1997). Many workers—perhaps one-third to one-half—retire gradually by moving into what might be called postcareer or bridge employment before full retirement (Quinn, 1999), but older nonworkers express little enthusiasm for paid employment. As of 1999, fewer than 800,000 men and women age 55 and older who were not in the labor force—or barely 2%—wished they were working. An even smaller proportion technically qualifies as discouraged workers; that is, they are not bothering to look for work because they do not think they could find it (U.S. Bureau of Labor Statistics, 2000).

One explanation for such disinterest might be the unpleasantness of the job hunt; in addition, the paucity of attractive part-time employment options serves as a deterrent. An expansion of employment opportunities for older workers, especially in the form of good part-time jobs, might generate greater enthusiasm for postretirement employment. Formal phased retirement programs that allow workers to ease into retirement by reducing their work hours in the jobs they have seem to hold promise for retaining workers beyond normal retirement age. In the United States, however, such programs are rare outside of higher education.

Job Characteristics

Three-fourths of all older workers are employed full-time, although interest and involvement in part-time work increase with age. Regardless of age, the large majority of men and women who work part-time do so by choice. In 1999, only 4% of older part-time workers were employed part-time because they could not find full-time work.

Older workers can be found in virtually every industry and occupation. Although older men are disproportionately represented in agriculture, the industry and occupational distributions of older and younger workers are, on the whole, quite comparable. The service industries claim the greatest share of both older and younger workers—somewhat less than 40% of each age group.

Gender differences in occupation and industry are more pronounced than age differences. Half of older women workers, but less than one-third of older men, can be found in service industries; agriculture, construction, and manufacturing claim a greater share of men. Older women are also heavily represented in traditionally female occupations, such as clerical and administrative support.

Again, regardless of age, self-employed workers are in the minority; nonetheless, older workers are substantially more likely than younger ones to work for themselves. As of 1999, some 15% of workers age 55 and older and 22% of all 65-plus workers were self-employed; this was the case for less than 8% of the under-55 workforce.

Public Policy and Older Workers

Despite the aging of the U.S. labor force, policies to promote older worker employment or to facilitate the transition to retirement are uncommon. The United States lacks a national older worker employment policy or, for that matter, a national retirement policy.

To the extent that older workers have been the focal point of public policy in recent years, it has been in terms of how a longer work life might alleviate the soaring public costs of supporting an aging population. Government efforts are perhaps best reflected in two provisions of the 1983 amendments to the Social Security Act: One increased Social Security benefits for each year that workers delay collecting between the ages of 65 and 69, and the other gradually increased the age of eligibility for full Social Security benefits, starting for workers turning 62 in the year 2000. By 2027, full Social Security benefits will not be payable until age 67; workers will still be able to collect benefits at 62, but the reduction for early retirement will be greater than it is now. As policy makers begin to grapple with the long-range solvency of Social Security funds, proposals to raise the full-benefit eligibility age beyond 67 are being raised. Few workers enjoy the prospect of being *required* to work longer for full benefits.

In early 2000, Congress voted to eliminate the Social Security earnings cap for beneficiaries between the ages of 65 and 69, who lost $1 in benefits for every $3 they earned above a set limit ($17,000 in 2000). Both the House of Representatives and the Senate were unanimous in their vote to repeal the cap, even though the change would benefit relatively few older persons. Most Social Security beneficiaries are out of the labor force, so they have no earnings to worry about, and the limit did not apply to beneficiaries age 70 or older. The Congressional Budget Office has estimated that only about 625,000 workers (out of 9.3 million men and women in the affected age group) would see higher Social Security benefits in 2000 as a result of the repeal, which was not extended to younger beneficiaries.

The government has also taken steps to eliminate discrimination against older workers, thereby making it easier for people to keep working. Though it persists, age-based discrimination against workers and job seekers age 40 and older is illegal under the provisions of the Age Discrimination in Employment Act (ADEA) of 1967 and its subsequent amendments. Most occupations saw the end of mandatory retirement with the ADEA amendments of 1986; however, there is little evidence that mandatory retirement was responsible for much of the labor force withdrawal of the past several decades.

To date, federally funded training programs—in particular, those under the Job Training Partnership Act (JTPA) of 1982—that might benefit older workers have, for the most part, been restricted to the economically disadvantaged. The JTPA was not designed to meet the training and retraining needs of the large majority of older workers who are not economically disadvantaged but who might be at risk of obsolescence and vulnerable to unemployment. The job training system was fundamentally overhauled in 1998 with the passage of the Workforce Investment Act. It remains to be seen how well this legislation will serve older workers, who are hardly mentioned and are not singled out as an underserved population.

Funded under Title V of the Older Americans Act, the Senior Community Service Employment Program provides minimum-wage employment to low-income elderly, many of whom are women or minorities. This relatively small program ($444 million in fiscal year 1999) assists a group of very disadvantaged job seekers.

Private-Sector Policies

Since most workers are employed in the private sector, it makes sense to look there for efforts to hire, train, and retain older workers. However, there is scant evidence that employers are actively responding to the aging of the workforce with programs or policies that might keep older workers employed or employable, despite substantial labor and skills shortages (AARP, 2000; Barth et al., 1993).

Into the Future

According to the most recent projections by the Bureau of Labor Statistics (BLS), by 2008, the labor force participation rate of persons age 55 and older will rise to 36.8% (Fullerton, 1999a). Longer projections are speculative at best, especially given rapid technological changes, a global economy, and the growth of markets and labor pools in the developing world. Still, the BLS estimates a participation rate for the 55 and older population of 38.4% in 2015, but this will fall to 33.6% in 2025, when baby boomers will all be between the ages of 61 and 79 (Fullerton, 1999a).

There is reason to believe that participation rates could be higher than indicated in these projections. For one thing, baby boomers themselves, who are marching inexorably toward old age, expect and even want to work in retirement. If they decide to prolong their work lives, boomers may be a force for employers to reckon with. In some respects, the retirement prospects of baby boomers are less promising than those of today's retirees: There is little hope for any significant improvement in Social Security benefits; the growth of defined-contribution pension plans at the expense of defined-benefit plans has shifted much of the responsibility for retirement income from employers to workers; and many workers are not saving adequately for their retirement years. Employers might face greater employment demands from middle-aged workers forced to postpone retirement beyond what is currently the norm. Furthermore, persistent labor shortages might cause employers—many of which now rehire retirees—to place greater emphasis on hiring, retaining, and retraining older workers.

Even if the BLS projections prove to be accurate and participation rates rise only modestly, the older workforce will grow more rapidly than other age groups, and the number of older persons in the workforce will increase. BLS projections point to an additional 7.5 million labor force participants age 55 and older in 2008, when older persons will constitute more than 16% of the labor force, up from just 12.7% in 1999. Most of this increase (73%) will be due to the rise in the number of "younger" older workers—age 55 to 64. Over this period, the median age of the labor force will rise as well, from 38.7 years in 1998 to 40.7 years in 2008—a "record level," according to the BLS (Fullerton, 1999b, p. 19). Employers, policy makers, and others who worry about the impact of an aging workforce on economic growth and competitiveness can take comfort in the fact that the median age of the labor force in 2008 is expected to be only slightly above what it was in the early 1960s, when the economy was anything but sluggish.

SARA E. RIX

Internet Key Words
Job Performance
Labor Force
Retirement
Unemployment

ENVIRONMENTAL MODIFICATIONS

The physical and social environments are essential components of the human behavioral system. They can be resources to enrich everyday life as well as constraining influences on certain types of behavior. People generally learn to seek or create resource-rich environments for themselves (environmental proactivity). They also learn the limits beyond which environments become difficult to shape, thereby adapting their behavior to a variety of constraints (environmental docility). To the extent that an older person is healthy, vigorous, and cognitively intact, his or her environmental needs are no different from those of other people. Many residential features are traditionally designed for aesthetic rather than human functional purposes.

Recent advances toward the goal of universal design have sensitized us to how we can make environments more usable for special-needs groups. And success in designing for special needs has often led to the realization that such improvements may be beneficial to everyone.

Given that universal design is the general goal, poor physical or mental health may be an occasion for user-specific environmental designs for older people. Routes to improvement follow some basic principles of sensitization: Look for ways to modify an environment, and use ordinary common sense to devise useful modifications. Good clinical judgment is necessary to determine when to call in an environmental specialist.

To modify a residential environment in a therapeutic manner, the characteristics of the users must be considered. Frequently, the most salient characteristics are the deficits for which the environment is expected to compensate. All too often, however, concentration on the deficits leads to ignoring the environmental aspects that provide enrichment, growth, and purpose to everyday life. A conceptual approach that gives equal consideration to both deficit-reducing and experience-enhancing environmental features invokes a short checklist of human needs: safety, function, cognition, comfort, order, autonomy, privacy, stimulation, affiliation, individuality, and spirituality. Because these needs are universal, they apply to a person at any level of personal competence and with any type of personal preference. Some examples illustrate how modest modifications to enhance the satisfaction of such needs can be introduced by a nurse, social worker, other care staff, or family member.

The Community Household: Safety and accessibility are likely to have environmental barriers. In-home service workers should look for hazards such as rugs that could cause one to trip, furniture blocking important household routes, inadequate lighting, or light distribution that casts shadows in the wrong place. For cognitively impaired people, important objects or areas of the house can be made more visible, such as by keeping a bathroom door open so that the toilet is visible or putting bright-colored tape on a call bell. For the house bound or those with limited mobility, the behavioral world shrinks. A broad visual sweep through a window may be a substitute for actually traversing the neighborhood. Flat surfaces on either side of a chair, located with an optimal view, increase the feeling of competence—a "control center." Even beyond such basic needs, the environmental professional can provide environmental arrangements, stimulating pictures, personalized furnishings, or furniture placement that is conducive to social behavior. Many elders may be proficient in arranging their own environments, so the tone in achieving such changes should be one of collaboration rather than imposition. Overall, the topic of household design for the 85% or more of older people who live in their own homes has been relatively neglected. A discussion particularly suited to people with dementia can be found in Gitlin (1999).

The Nursing Home: For many reasons, institutional environments tend to lack warm or homelike décor. Regulations enacted during the 1990s drastically reduced gross risks to health and safety in nursing homes. In the process, however, cleanliness and safety considerations have overridden other human needs, so that visual, auditory, and textural facets have developed an "institutional" feel that is antithetical to the fulfillment of many experience-enhancing needs.

It is possible to counteract this institutionalism through relatively inexpensive interventions that may be initiated by staff from a variety of backgrounds. Many people have an instinctive feel for design—the latent architect in all of us. Given a motivated leader and genuine administrative support, a multidisciplinary therapeutic environment committee can upgrade the quality of the nursing home environment and influence environment-relevant policies. Chair placement, for example, is almost always concentrated in high-traffic areas such as building entrances or the nurses' station; this can be augmented with single seats for those who like to observe in a solitary fashion. Staff at all levels enjoy the task of enriching the décor of their care areas with pictures or fabrics, individualizing room designations, or enhancing institution-provided artifacts or personal possessions in concert with the resident and family.

When planning a new facility within an established institution, it is incumbent on sponsors, administrators, and designers to use the accumulated technology of environment-behavior researchers to relate the purposes of the new facility to the structural plans. Less frequent but equally important is the active utilization of caregiving staff in this planning process. Unfortunately, administrators often have to be pressured to involve the staff in planning and design, overlooking the fact that participation in the process is a potent staff motivator.

The professional literature has not done full justice to small-scale bootstrapping efforts to upgrade the quality of institutional environments. Nonetheless, the principles of deficit-reducing and experience-enhancing design in new construction are equally applicable to more modest attempts to enhance already existing institutional environments (Brawley, 1997; Calkins, 1988).

Other Settings: Many current housing options for the elderly are based on assisted living, whose design initiatives are similar to those for nursing homes. Although the impetus for user-oriented design originated in housing for relatively independent elders, activity in recent decades has been in housing for higher-income older people. Lack of interest in new publicly subsidized housing is paralleled by government neglect of the million older units designated for elders. These tenants are aging in place, as is their housing. There is a great need for client-centered input on how to accommodate this housing to the needs of the people who are growing older and sometimes more frail.

Acute-care hospitals and outpatient or community-supportive facilities are also appropriate targets for environmental modification. Because these settings serve people for only limited portions of their lives, their design is less critical for users' overall well-being. Nonetheless, the characteristics of the external structure—the appearance of the waiting room, the richness and "legibility" of decor, and the social density of seating patterns—need to be considered in terms of the ultimate goals of the service.

Universal Design and Special-Needs Design: Innovations in designing for special-needs users can lead to environmental improvements for everyone. Grab bars in tubs and showers clearly benefit everyone. Adults purchasing a home for a lifetime will age in place more comfortably if dimensions such as door widths and bathroom space are designed with aging in mind. Sometimes the limits of universal design are reached when a clearly prosthetic feature, such as a stair lift or protection against falling out of bed, is required. Even in these instances, however, user-specific design can be made more aesthetic and acceptable by mainstreaming design principles.

Expertise in Environmental Design: Although this discussion has stressed modifications that can be made by sensitized care professionals who are not experts in environmental design, there are specialists in this area who can produce optimal prosthetic and therapeutic environments. Occupational and physical therapists are trained in analyzing person-environment fit and in recommending modifications that can affect quality of life. Larger communities are likely to have agencies that specialize in home modification and rehabilitation. Such services may be accessed through area agencies for the aging or community development agencies. For major projects, there is no substitute for a pace-setting designer who can consult on major renovations or new facility design.

Every health care professional and every family caregiver can alter residential situations to compensate for deficits or augment environmental benefits. The nursing profession in particular has an important role. Nurses can apply clinical knowledge to determine what environmental changes may enhance the satisfaction of universal needs. Nurses are frequently in a position to organize a multidisciplinary effort to upgrade the needs-satisfying requirements of households (visiting nurses), congregate facilities, and community services. Adding instruction in environmental enhancement to the training of all caregivers could significantly raise the quality of life of older people.

M. POWELL LAWTON

See Also
Signage

Internet Key Words
Behavioral Design
Environment
Household Design
Modifications
Nursing Home Design
Prosthetic Design

ERGONOMICS

The aging of the population presents a number of challenges for researchers, policy makers, and designers of products, jobs, and environments. Although most older people live active and relatively healthy lives, increased age is associated with a decline in functional capacity and a propensity to disability through accidents and diseases. Approximately 30 million Americans over 65 years old are disabled in one or more aspects of self-care. Of those over 75, approximately 40% have multiple chronic illnesses, including dementia. The elderly use more health care services and incur higher health care costs than younger people (Kane, Ouslander, & Abrass, 1999). As the elderly population increases and people live longer, more people will require help with aspects of daily living or disease management. The likelihood of developing a chronic disease or a disabling condition approximately doubles each year beyond age 65. These estimates point to a clear need for strategies to help healthy older people remain productive and independent and to help those who are frail or disabled receive care and support so that they can live in the community for as long as possible.

Ergonomics—the study of human beings and their interactions with products and environments in the performance of tasks and activities—can enhance the independence and quality of life of older people. The objectives of ergonomics are to improve the fit between people and the designed environment so that performance, safety, comfort, and user satisfaction are maximized. Ergonomics implies a systems approach in which age-related changes in capabilities, tendencies, and preferences are translated into guidelines for design within the contexts of tasks and environments (Rogers, 1997).

The ergonomic focus on user-centered design can address the problems of older adults and help them retain and enjoy independence in their later years. One area that could benefit tremendously from the application of ergonomics is health care. Ergonomic principles and methods can improve the lives of elders in areas such as medication adherence, health care delivery, warnings and instructions, home safety, and the design of assistive devices.

Medication Adherence

Medication nonadherence, or failure to take medications as prescribed, is a common problem among older adults and is a significant predictor of hospital admissions. Generally, the problem involves failing to take a prescribed medication, taking the incorrect dosage, taking medications at improper times or in the wrong combinations, or failing to comply with special instructions such as dietary restrictions. Noncompliance is particularly problematic for older people because of the high medication use among this age group and their greater susceptibility to side effects and drug interactions.

The problem of medication nonadherence is complex and may be attributable to numerous factors, such as an individual's perceptions or beliefs (e.g., the person does not believe that he or she is ill or that the medication is effective), cognitive problems (difficulty comprehending or remembering medication instructions), and ineffective strategies to enhance medication compliance. Many of these problems can be ameliorated by ergonomic solutions, such as products and devices that improve the organization of medications, including calendars, electronic pill dispensers, or compartmentalized containers that are congruent with a medication schedule. These devices help offset memory problems associated with comprehending and integrating medication schedules. Voice mail or beepers might also be employed to help individuals remember to take medications. Automated tele-

phone messaging is effective in improving medication adherence and appointment attendance and in monitoring community-dwelling elderly and chronically ill patients. Other solutions include improvements in medication packaging or labeling. Educating individuals about the nature of their illness and the importance of medication in illness management is another potential solution. It is important to understand the cause of nonadherence so that the appropriate intervention can be selected (see Park & Jones, 1997).

Health Care Delivery

Ergonomic applications can also improve health care delivery for those who are frail or have restricted mobility. Ergonomics may facilitate access to health care information and services and enhance the ability of health care professionals to deliver care. For example, computers and information technologies, such as e-mail and the Internet, can help older people access information about a particular illness, medication, diet, or exercise program. The Internet can also help older people communicate with health care providers or other people with similar problems. Several studies have shown that on-line support groups are beneficial for this population. This type of technology may be particularly beneficial for hearing impaired or aphasic individuals. However, for these applications to be successful, the technology must be relatively simple to use, readily available, and affordable, and adequate training must be provided.

Technology can be used by health care providers to communicate with older patients; remind them of appointments and home health care regimens, such as dietary schedules; or check on a patient's general health status. With telemedicine, physicians can directly assess patients and measure blood pressure, gait, and cognitive status. The term "telemedicine" refers to a wide range of technologies, from simple telephone connections to live two-way video and audio transmissions (interactive television). For example, with telemedicine, physicians can measure vital signs and ask hypertensive patients about disease manifestations and drug side effects. The cost of the technology varies with the sophistication of the system.

Although studies indicate that older people are receptive to using new technologies, they often encounter difficulty because of poor training or failure on the part of designers to consider their needs and preferences. For example, when using computers, older people often have difficulty locating screen targets or reading screen characters; these types of problems can often be corrected by increasing font size or reducing screen clutter. User testing and user-centered design are critical to the success of technical systems. Guidelines for human-computer interaction suggest how computers and other forms of technology can be useful to and usable by older adults (Czaja, 1997). Examples of these guidelines include highlighting important screen information, maximizing the size of icons, avoiding complex command languages, and providing on-line aid.

Assistive Devices and Labeling

Data indicate that older people spend much of their time performing basic living activities and spend a great deal of time at home. Many have difficulty performing home tasks, and the rate of home accidents is high in this population. Thus, there is a critical need to employ strategies that allow older people to live safely and comfortably at home, especially those with impairments. Often this requires some type of home modification or the use of an assistive device such as a walker or grab bar.

Task analysis can identify problems in homes or in the performance of home tasks. There are also guidelines for the design of home environments, products, and assistive devices. It is important to recognize the interaction among personal characteristics, environmental conditions, and device characteristics when selecting devices. Also, given the increasing emphasis on self-care and the growing number of medical devices that will be used at home, knowledge of the aging process is necessary to guide the design and selection of these devices.

Ergonomic input is also important in the design and content of warning labels and instructions in response to age-related changes in sensory and perceptual systems and cognition. Use of ergonomic guidelines can make it easier for older people to perceive and comprehend warning information. These guidelines include using simple and concise language, minimizing the amount of irrelevant information, and including pictures or graphic examples.

Ergonomics is concerned with enhancing the fit among people, tasks, environments, and products. The application of ergonomics to issues associated with aging can improve the health, safety, and quality of life of older people. The basic premise of ergonomics is that improvements in performance result from user-centered design and a fundamental understanding of user capabilities, needs, and preferences. According to this premise, improving the health and quality of life of older people requires that knowledge of aging be applied to the design of products and environments.

SARA J. CZAJA

See Also
Medication Adherence
Support Groups

Internet Key Words
Design
Ergonomics
Health Care
Medication Adherence
Safety

Internet Resources
ErgoWeb, Inc.
http://www.ergoweb.com

NIOSH Database
http://www.cdc.gov/niosh/nioshtic.html

ETHICS CONSULTATION

Clinical ethics consultation is a service provided by an individual or group to help health care professionals, patients, and families identify and resolve ethical conflicts and problems that arise in the care of patients. The practice of offering clinical ethics consultations began informally in the 1970s and over the past 30 years has become an established part of the clinical services of many health care institutions. The increased importance of ethics consultation can be traced to at least two sources: the rapid growth of medical technology, which has presented patients, families, and health care providers with new and difficult ethical choices; and the rise of the patient's rights movement and the correlative attack on medical paternalism. Clinical ethics consultation was also given a major impetus by the Joint Commission for the Accreditation of Health Care Organizations, which in 1992 mandated that all accredited health care institutions have a "mechanism" for dealing with disputes concerning end-of-life care.

Modes of Clinical Ethics Consultation

Clinical ethics consultations can be conducted in various ways: by an ethics committee, by a small team (possibly a subgroup of the ethics committee), or by individual consultants. Consultations by committee are often difficult to organize in a timely fashion and may become bureaucratic and depersonalized, but they can provide multiple perspectives that reveal relevant aspects of a case that might otherwise be overlooked. Consultations by individuals—typically clinicians, lawyers, or philosophers specializing in bioethics—are generally more flexible and personal and can be arranged more expeditiously. Many ethics consultative services require that individual consultants report to an ethics committee or consultation group, either for retrospective review of cases or for help with ongoing cases. This supervision provides peer review and quality assurance and is a way of holding consultants accountable for their activities. Consultations by small teams occupy a middle ground between these two approaches.

Goals of Clinical Ethics Consultation

The main goal of clinical ethics consultation is to improve the quality of care by providing a mechanism for the identification, analysis, and resolution of ethical problems and conflicts that arise in the clinical setting. Other important goals include facilitating institutional efforts at quality improvement by identifying common sources of ethical problems and helping health care providers handle ethical problems by providing education in clinical bioethics (Thomasma, 1991).

Roles of the Clinical Ethics Consultant

The role of the clinical ethics consultant has been described variously as professional colleague, educator, facilitator of moral reflection, mediator of moral conflict, and patient advocate. Associated with each description is a particular set of skills and competencies (American Society for Bioethics & Humanities, 1998; La Puma & Schiedermayer, 1991). Some view the consultant primarily as a patient advocate responsible for protecting the patient's rights and interests; others conceive of the consultant as an impartial mediator whose goal is to forge consensus among the involved parties, all of whose rights, interests, and responsibilities are acknowledged. A limitation of the latter view is that even if consensus is achieved, it may fall outside acceptable moral boundaries. An *ethics* consultant cannot be a pure mediator or facilitator but must strive for morally principled consensus. Ethics consultants, however, should not be considered moral policemen; their authority is qualitatively different.

As an educator, the consultant is conceived to have a degree of moral expertise that his or her professional colleagues lack. This expertise does not consist in knowing "the right answer" but in being able to uncover value conflicts, articulate different moral positions on issues, and apply moral reasoning and ethical theory to the issue at hand.

Ethics consultants possess certain skills and knowledge that can help parties clarify options and resolve conflicts. Properly understood, their role is to offer ethical advice and guidance to patients, families, and health professionals, not to make decisions or to override the views of others. A criticism that has been leveled against the very idea of ethics consultation is that it promotes abdication of one's decision-making responsibility and the surrender of one's judgment to so-called ethics experts. However, this criticism misconstrues the nature of the authority that ethics consultants possess (Agich, 1995).

Issues Addressed by Clinical Ethics Consultants

Clinical ethics consultants may be called on by patients, families, or health care professionals to address a wide range of clinical ethical issues. Issues appropriately addressed by ethics consultants include:

- Confidentiality and privacy
- Decisional capacity
- Informed consent and truth telling
- Surrogate decision making
- Withdrawal or withholding of life-sustaining treatment
- The shift from curative to palliative care
- Allocation of scarce medical resources
- Conflicts among health care providers

The attending physician retains decision-making responsibility and authority with respect to his or her patients. As such, the physician should be informed that an ethics consultation has been requested and the source of the request (patient, family member, or other member of the health care team).

Competencies of the Clinical Ethics Consultant

The core competencies of ethics consultants can be divided into two categories: skills and knowledge. Core skills include the ability to:

- Identify the value conflict or problem in the clinical situation
- Listen attentively, respectfully, and supportively to the involved parties
- Elicit the interests and moral concerns of all involved parties
- Promote effective communication among the involved parties
- Articulate care options and their consequences
- Work toward moral consensus

Core knowledge includes competency in:

- Terms used in the diagnosis, treatment, and prognosis of common medical problems
- Key bioethical concepts, principles, and theories
- Techniques of moral reasoning
- Main ethical positions on important clinical issues
- Applicable health law
- The organization, corporate structure, and culture of the institution

Evaluation of Clinical Ethics Consultation

Evaluation of clinical ethics consultation is a matter of considerable interest and debate among those in the field. Meaningful evaluation of the efficacy of ethics consultation is particularly important in light of efforts to control health care costs by eliminating unnecessary and unprofitable services.

Evaluation of ethics consultation must keep in mind the goals of that consultation. Both the process of consultation and its outcomes need to be evaluated. Process is evaluated by asking whether the consultation was conducted in a timely fashion, whether all interested parties were included in the consultation, and whether participants were satisfied with the quality of communication. Outcomes are evaluated by asking such questions as whether a principled ethical resolution of the problem was achieved, whether the participants were satisfied with the outcome, whether the consultation altered the plan of care, and whether the consultant's services were frequently used. Chart reviews, questionnaires, and interviews are some useful evaluation techniques.

JEFFREY BLUSTEIN

Internet Key Words
Ethics Consultation/Consultant
Knowledge Competencies
Mediation
Moral Authority
Moral Conflict
Morally Principled Consensus
Skill Competencies

Internet Resources
Bioethics Consultation Service, Montefiore Medical Center, Department of Epidemiology and Social Medicine
http://www.bioethicsmontefiore.org/prof/clinical/desm/progserv/bioethic/index.html

Core Competencies for Health Care Ethics Consultation, American Society for Bioethics and the Humanities
http://www.asbh.org/papers/

EUTHANASIA

"Euthanasia" comes from the Greek and means "a gentle and easy death" or "the means of bringing about a gentle and easy death." Most ancient Greek and Roman practitioners, Socrates, Plato, and Stoic philosophers from Zeno to Seneca supported physician-induced death of the sick and suffering (Vanderpool, 1995). In contrast to these dominant Greco-Roman traditions, the Hippocratic oath required physicians to swear "neither to give a deadly drug to anybody if asked for it, nor . . . [to] make a suggestion to this effect" (Edelstein, 1989, p. 6). The oath, which continues to exert a towering influence in Western medicine, reflects the Pythagorean conviction that human beings are owned by God or gods and should abide by a divine determination of life's completion (Carrick, 1985).

In contemporary times, discussion of euthanasia has increasingly dealt with the action of inducing a gentle and easy death. Thus, ethical debates about the permissibility of physician involvement in euthanasia concern the question: Are physicians ethically permitted to act to end a patient's life? This question should be distinguished from other ethical questions that may arise at the end of life. For example, as the term "euthanasia" is commonly used today, it does not apply to refraining from using or continuing life-sustaining treatments (passive euthanasia), nor does it apply to providing patients with the means necessary to end their own lives (assisted suicide). Many who defend the permissibility of *physician-assisted suicide* do not support *physician-assisted euthanasia*. Advocates of physician-assisted suicide approve of letting physicians prescribe medications that patients may use to end their own lives; they do not necessarily approve of letting physicians actually administer lethal injections, for example.

Ethical Perspectives

Contemporary ethical arguments supporting euthanasia are often based on *compassion* for the suffering of a terminally ill and imminently dying patient. These arguments purport to show that physician aid in dying is ethically permissible when the patient's condition is associated with severe and unrelenting suffering that is not the result of inadequate pain control or comfort care.

Alternatively, arguments defending euthanasia make reference to the ethical principle of *autonomy*, which requires one to respect the informed choices of competent patients. Under this approach, physician involvement in euthanasia is ethically limited to situations in which competent patients make informed, repeated requests for aid in dying.

Critics of euthanasia charge that both compassion-based and autonomy-based arguments are inadequate. Arguments invoking compassion are faulted on the ground that there is no principled basis for limiting euthanasia to competent patients who choose it. After all, many suffering patients are not competent. Therefore, if the ethical basis for providing aid in dying is compassion, then aid in dying should logically be extended to incompetent persons.

Arguments relying on the principle of autonomy are also criticized for failing to offer a principled basis for appropriately limiting euthanasia. Autonomy-based arguments do not require limiting euthanasia to patients who experience severe and unrelenting suffering, but would presumably allow euthanasia to be applied to healthy people who wished to die. Critics of autonomy-based arguments also doubt that patients' requests to die reflect autonomous choices. Instead, such requests may be made because of inadequate palliative and comfort care, continued use of invasive and futile interventions, and failure to diagnose and treat other underlying causes of the request, such as depression (Emanuel, 1999). In such cases, complying with a patient's request for assistance in dying is not supported by a principle of respect for patient autonomy.

Both autonomy-based and compassion-based arguments are vulnerable to the further objection that there is no principled basis for restricting euthanasia to persons who are imminently dying. After all, the prospect of suffering for a long time is arguably worse than the prospect of suffering briefly. Likewise, the principle of respect for autonomy presumably applies to all competent individuals, irrespective of whether they are about to die.

Legal Perspectives

Just as the ethical status of euthanasia is controversial, the legal status of both euthanasia and assisted suicide has been the subject of intense debate in the United States. At present, assisted suicide is a criminal act in 29 states; however, the constitutionality of these statutes has been challenged in Michigan and Washington. Defenders of physician-assisted death have placed initiatives on the ballots in several western states to decriminalize euthanasia or assisted suicide, but these initiatives have not yet been successful.

Euthanasia is illegal in all other nations, although it is a common misconception that euthanasia is legal in the Netherlands. Article 293 of the Dutch penal code clearly states that anyone "who takes another person's life even at his explicit and serious request will be punished by imprisonment of at the most 12 years or a fine of the fifth category." Although assisted death is formally a crime in the Netherlands, the Dutch penal code allows for exceptions. For example, persons who assist with death will not be punished if the act is impelled by an "overwhelming power" or by a sudden conflict of duties or interests in a situation in which a choice must be made. This exception to punishment has been invoked over the years by Dutch courts, establishing a legal precedent for excluding certain categories of cases from criminal prosecution.

Euthanasia and the Elderly

Although debates about euthanasia apply to persons of all ages, they bear special relevance to elderly persons. This is because death is nearer in old age, and aging individuals may be more likely than younger persons to think about death and the dying process. Perhaps the aging of the population that is occurring in most developed nations will lead societies to focus greater attention on how to ensure humane care at the end of life. The question of whether euthanasia represents humane medical care for dying patients will continue to be discussed.

NANCY S. JECKER

See Also

Autonomy
Depression in Dementia
Depression Measurement Instruments
Palliative Care
Suicide

Internet Key Words

Death and Dying
Euthanasia
Patient Autonomy
Physician Assisted Suicide

Internet Resources

Bioethics for clinicians: Euthanasia and assisted suicide, Canadian Medical Association Journal
http://www.cma.ca/cmaj/vol-156/issue-10/ 1405.htm

Doctor-Assisted Suicide—a guide to WEB Sites and the Literature, Longwood College Library
http://web.lwc.edu/administrative/library/suic.htm

EXERCISE AND THE CARDIOVASCULAR RESPONSE

Elderly patients' response to physical activity can differ significantly from that of younger adults. These differences represent a confluence of factors that include the impact of aging on the cardiac, vascular, and musculoskeletal systems; comorbid conditions, including cardiovascular, metabolic, neurological, and musculoskeletal diseases that become more prevalent as a population ages; and the progressive adoption of an inactive lifestyle. Understanding the role that each plays in the cardiovascular response to exercise is essential for appropriate assessment and development of activity programs aimed at enhancing an elderly patient's functional status.

Cardiovascular response to exercise in young and middle-aged adults is an integrated response aimed at providing more oxygen-rich blood to the working muscles to facilitate the production of high-energy phosphates needed to meet the increased demand associated with more frequent and intense muscular contractions. The cardiac response is an exercise-related increase in cardiac output, mediated by a significant increase in heart rate (near 300%) and a 25% to 40% increase in stroke volume (the amount of blood ejected with each beat of the heart). The increase in stroke volume is mediated primarily by enhancement of the ejection fraction, with little or no increase in end-diastolic volume. Although selected individuals demonstrate remarkable exercise capacities well into their eighth and ninth decades, there are age-related alterations in cardiovascular physiology and

anatomy that occur independent of comorbid conditions and lifestyle.

Resting heart rates are not significantly changed in the elderly; the most "visible" age-related alteration in cardiac exercise physiology is a linear decrease in maximal achievable heart rate. This change is independent of daily activity, obesity, and many medical conditions. An age-related decrease in the activity of sympathetic receptors in the sinoatrial node is the basis of this physiological alteration. A variety of formulas can predict maximal heart rate, and all adjust only for age (e.g., 220 − age).

A second important age-related change in the cardiovascular system is loss of elastic components and increase in fibrous components in the arterial walls. This results in increased stiffness of these structures, with a concomitant increase in systolic blood pressure and workload of the left ventricle during systole. As a consequence of increased left ventricular systolic workload, the left ventricle wall hypertrophies, and wall stiffness during diastolic filling increases. This occurs even in normotensive elderly patients but is much more significant in those with hypertension (including isolated systolic hypertension) and diabetes and is seen to a greater extent in populations with higher sodium intakes. As a consequence of increased diastolic stiffness of the left ventricle, there is an increase in filling pressure (usually standardized as end-diastolic pressure) and in left atrial size (about 20%).

The cardiac output response to exercise in the elderly is important because it determines maximum exercise tolerance. Early studies demonstrated a significant fall in mean maximum exercise achieved cardiac outputs, cardiac ejection fractions, and exercise tolerance in an elderly population. Subsequent studies demonstrated that these findings were seriously confounded by a population bias. The investigators had used "residential" nursing home populations rather than community-living populations, so their sample contained a high proportion of patients with diagnosed or occult coronary heart disease. When the data were reevaluated to exclude patients with positive exercise electrocardiograms or gated nuclear angiocardiograms, there was no significant decrease in resting ejection fractions; exercise ejection fraction increased, but to a lesser extent than in a younger population (Port, Cobb, Coleman, & Jones, 1980). Another study found that cardiac output with exercise decreased in comparison to a younger population, but to a much smaller degree than had previously been assumed (Ades, Waldman, McCann, & Weaver, 1992). The reduced exercise heart rates during submaximal and peak exercise decreased cardiac output (Cardiac output = Heart rate × Stroke volume), but the slower rate permitted a longer diastolic filling time, with an increase in end-diastolic volume and stroke volume. Thus, there was a modest fall in peak cardiac output achieved and an important alteration in the ratio of heart rate and stroke volume contribution to exercise cardiac output.

Cardiovascular disease is the most significant comorbid condition associated with an altered response to exercise. Studies confirm that 30% of elders have significant cardiovascular disease; autopsy studies indicate that an additional 30% had undetected heart disease (for 60% total disease prevalence). This "majority" subset of the elderly population has reduced ejection fraction, stroke volume, and cardiac output during exercise and significantly reduced exercise tolerance compared with elderly patients without cardiovascular disease. Of interest is the somewhat protective effect of the physiology of aging against cardiac pathology. Reduced exercise heart rates result in lower myocardial oxygen requirements during exercise, estimated by: Change in heart rate × Systolic blood pressure. This relative reduction results in a lower likelihood of ischemia in the presence of coronary artery stenosis.

The negative effects of arthritis and musculoskeletal disease (two of the most common clinical comorbidities among the elderly) on exercise tolerance, and the associated increase in myocardial oxygen requirement and provocation of ischemia with clinical angina, are significant. These clinical conditions increase the physical work and pain associated with any given activity and result in increased cardiac output and myocardial oxygen demand. Exercise tolerance is thus limited, and exercise activities are more likely to provoke ischemia and cardiovascular symptoms.

Elders' adoption of an inactive lifestyle is, despite some notable exceptions, widespread. To some degree, this is a consequence of comorbid conditions that make activity uncomfortable, but to a significant degree, it is a societal norm. As such, many healthy elderly people are sedentary and begin to experience detraining (deconditioning), with a consequent loss of muscle mass and reduction in strength and endurance capacity. The negative feedback effect results in diminished exercise tolerance, activities become less comfortable, and the person tends to become even more sedentary. In addition to a loss of functional capacity, the sedentary elderly person is at greater risk for falls, depression, and obesity. Numerous studies demonstrate that exercise training can reverse much of the loss of strength, endurance, and function and can significantly improve the health profile of an elderly person.

The physician must integrate the age-related alterations in the physiology of exercise, the increasing likelihood of overt or occult coronary disease, and the impact of other comorbid conditions to arrive at an effective and safe exercise program. In appropriate patients, this will increase functional capacity and quality of life (Ades et al., 1992). An exercise electrocardiogram (with nuclear or echocardiographic imaging in appropriate patients) is an important first step. Cardiovascular disease must be detected, assessed, and clinically addressed before an activity program can be safely prescribed. Although structured programs (a given exercise performed at a specific frequency, duration, and intensity) are useful in younger populations and some elderly patients, many elders participate in exercise and improve their functional capacity to a greater extent by engaging in a customary activity (e.g., walking) at a tolerable and perceived exertion for a reasonable amount of time, or by performing a series of shorter-duration activities. Safety (injury is a major complication of activity programs for the elderly), socialization (group or paired activities are recommended), and convenience (a major factor in compliance) are as important as the determination of optimal exercise intensity and modalities. The challenge is to increase daily activity levels to prevent or reverse functional loss and impaired clinical well-being associated with inactivity. The ideal program is interdisciplinary and incorporates behavior strategy to enhance long-term adherence to a regular exercise regimen.

RICHARD STEIN

See Also
Deconditioning Prevention
Falls Prevention
Fractures
Heart Failure Management: Congestive and
    Chronic Heart Failure

Internet Key Words
Exercise
Exercise Physiology
Functional Capacity
Heart Disease

EYE CARE PROVIDERS

The need for eye care services increases significantly with aging. Virtually all persons over 65 years of age require some type of corrective lenses to improve their visual performance at a distance, near, or both, and the prevalence of ocular disease rises substantially. Half of all seniors between 65 and 74 years of age have cataracts. Cataract surgery is the most frequent surgical procedure performed under the Medicare program, which pays for 1 million of the 1.3 million cataract procedures performed annually.

Studies have shown that the aged are underserved with regard to eye care, despite being at higher risk for sight-threatening eye conditions. Blindness and visual impairment are highly prevalent among the elderly, especially among nursing home residents. A significant proportion of their visual impairment can be remedied by refractive correction, medical eye treatment, and cataract surgery. Much of the loss can be treated or prevented with appropriate eye care (Tielsch, Javitt, Coleman, & Sommer, 1995).

Eye care services in the United States are provided largely by two groups: ophthalmologists and optometrists. Eye care shares this characteristic with only a few other health care fields, in that two separate and independent professions are trained to provide it. During the past decade, the very nature of eye care providers has changed extensively and radically. Here, the similarities and differences in the education, training, and licensure of optometrists and ophthalmologists are discussed, and the reimbursement for eye care services to the elderly is reviewed.

Ophthalmologists

Ophthalmologists are medical doctors who specialize in the medical and surgical care of the eyes. Ophthalmology is one of 23 medical specialties certified by the American Board of Medical Specialties. All ophthalmologists complete four years of education in a medical school or college of osteopathic medicine, a one-year internship, and three years of postgraduate medical and clinical training in ophthalmology and must pass written and oral examinations to practice as licensed ophthalmologists. Many ophthalmologists take a postresidency fellowship in a subspecialty area such as retina, cornea, and neuro-ophthalmology. Since 1992, board certification is limited to 10 years and must be renewed. In 1997, there were 132 approved ophthalmology hospital residency programs in the United States; 1,497 physicians were enrolled in three-year residency programs. Approximately 500 board-certified ophthalmologists enter the field each year.

More than two-thirds of all ophthalmology residents are male. Approximately 15,000 board-certified ophthalmologists practice nationwide, most in or near large cities within large medical centers, and more than 90% in office-based settings. Forty percent of ophthalmologists specialize in the cornea, retina, cataracts, and glaucoma. Substantial excesses in subspecialist ophthalmologists are likely to develop, given current training levels (Lee, Jackson, & Relles, 1998).

Ophthalmologists are the only practitioners trained to perform major ocular surgery. Ophthalmologists also treat ocular diseases and conduct basic vision examinations, including refractions. Increased patient enrollment in managed care, along with significant cuts in reimbursement by Medicare and other third-party payers, has driven many ophthalmologists to seek other sources of revenue, such as the dispensing of eyewear. Interest in laser surgery to correct nearsightedness, farsightedness, and astigmatism has stimulated an entirely new market for many ophthalmologists.

Optometrists

Optometrists, or doctors of optometry (O.D.), are independent health care professionals and are the major providers of primary eye care in America. Optometry practice is specifically defined by each state; thus, the scope of practice and licensure requirements vary from state to state. All states authorize optometrists to use prescribed drugs to treat eye infections, allergies, and inflammations, and in 44 states, they may prescribe medication to treat glaucoma (Rosenthal & Soroka, 1998). Licensed optometrists must take specified courses, pass written examinations, and demonstrate clinical aptitude in order to have their licenses extended to use and prescribe pharmaceutical agents. These educational requirements are included in the current optometric curricula and licensing examinations, so new licensees automatically meet state requirements. Vision services developed primarily by optometrists benefit a significant number of elderly Americans with residual vision.

Optometric education consists of a minimum of three years of undergraduate study, followed by four years of professional training in a doctoral program at one of the 16 schools and colleges of optometry in the United States. All states except Nevada, New Mexico, and Oklahoma require optometric graduates to pass national boards administered by the National Board of Examiners in Op-

Ratios of Optometrists and Ophthalmologists by Region

Region	Optometrists per 100,000 Population	Ophthalmologists per 100,000 Population	Ratio of Optometrists to Ophthalmologists
Northeast	12.2	8.2	1.5
Midwest	13.1	5.7	2.3
South	10.6	5.8	1.8
West	13.1	6.0	2.2
U.S. total	12.0	6.3	1.9

From: Shoenman J. A. and Gardner E. N. (1998). Results of the First National Census of Optometrists: Final Report Prepared for the American Optometric Association. Bethesda, MD: The Project HOPE Center for Health Affairs.

tometry, and most states require graduates to pass state-administered practical examinations. Optometric postgraduate clinical residency programs are available in ocular disease, geriatric care, vision rehabilitation, contact lens fitting, and pediatric care. All states have continuing education requirements for relicensure.

Historically, optometry was a profession dominated by white males. However, a growing interest in optometry by women and minorities, especially Asians, has contributed to a dramatic change in the demographic profile of optometrists. As of 1998, 33,045 optometrists were in active practice (Schoenman & Gardner, 1998), an average of more than 12 optometrists for every 100,000 people. Optometrists are in greater supply in the Midwest and West, and almost without exception, all states have more optometrists than ophthalmologists.

Two-thirds of all optometrists practice privately, 20% work for optical chains, and the rest are employed by HMOs, hospital clinics, ophthalmological practices, and the military (Soroka, 1997).

Coverage and Reimbursement for Vision Services

Despite the overwhelming need for vision care among the elderly, Medicare coverage for eye examinations and eyeglasses is limited. Routine eye examinations and refraction are not covered, whether provided by an optometrist or an ophthalmologist. Medicare does not cover eye examinations for prescribing, fitting, or changing eyeglasses or contact lenses for refractive errors. Eye examinations are reimbursable only for patients with complaints or symptoms of an eye disease or injury. Because of the high prevalence of ocular disease among the elderly, medical eye care rendered by optometrists and ophthalmologists is a covered benefit under Medicare. Similarly, surgical eye care is covered. A major gap in coverage exists in the area of rehabilitative services. Reimbursement for low vision aids and appliances is denied. Eyeglasses are not covered under Medicare, with the sole exception of those patients who undergo cataract surgery. However, for them, coverage is limited to a single pair of corrective eyeglasses per surgery per lifetime.

Two-thirds of all state Medicaid programs cover routine eye examinations and eyeglasses annually or biennially, even though it is considered an optional benefit. HMOs frequently offer additional benefits to Medicare patients at little or no extra cost, and eye care coverage is one of the more popular additions. Although few HMOs offer eyeglasses as a cost-free benefit, a number do offer ophthalmic materials with moderate co-payments or discounts (Soroka, 1997). Unlike most other care, which requires patients to visit a gatekeeper or primary care provider initially, most managed care plans waive the referral requirement and allow direct access to an eye care provider (optometrist or opthamalogist). Some plans, however, may require patients to consult an optometrist before seeking ophthalmological care.

Mort Soroka

See Also
Cataracts
Glaucoma
Vision Changes and Care

Internet Key Words
Eye Care Providers
Eye Care Services
Ophthalmology
Optometry

Internet Resources
American Academy of Ophthalmology
http://www.eyenet.org/

American Optometric Association
http://www.aoanet.org/

National Eye Institute
http://www.nei.nih.gov/

F

FALLS PREVENTION

Falls are a critical health care problem for older Americans, surpassing motor vehicle accidents as the leading cause of death from trauma for those 65 years and older (National Center for Health Statistics, 1997). A fall may result in prolonged bed rest, soft tissue changes, fractures, incontinence, immobility, discomfort, dependency, delirium, and depression. Even when falls do not result in physical injury, fear of falling again may result in social isolation and limited activity. Because falls have a major impact on quality of life and health care costs, it is critical that physicians and nurses caring for elders identify those at risk and develop strategies for prevention.

A fall is defined as an unintentional or unplanned event in which the person comes to rest on the ground or floor, with or without subsequent injury. A near-fall describes a situation in which a person is caught or intentionally eased to the floor by another.

Etiology

Falls can be attributed to physiological or accidental causes. Risk factors are therefore either intrinsic or extrinsic.

Intrinsic risk factors are patient characteristics or diagnoses that can be measured to predict the likelihood of falling. Although age alone is not a risk factor, a number of age-related changes contribute to falls, including loss of visual acuity, decreased reaction time and bladder capacity, diminished balance and muscle strength, and orthostatic hypotension.

Identification of intrinsic risk factors begins with a general health history and a physical and functional assessment. Physical assessment includes screening for visual deficits, presence of foot problems, balance and gait disorders, improper use of assistive devices, and presence of orthostatic hypotension. A research-based assessment tool, such as Hendrich's Fall Risk Assessment Tool, is a consistent, quick format that can identify risk with an acceptable level of sensitivity and specificity (Hendrich, Nyhuis, Kippenbrock, & Soja, 1995). Risk factors included in this tool are recent history of falls, altered elimination, confusion/disorientation, depression, dizziness/vertigo, and impaired mobility/generalized weakness. The patient's level of risk increases with the number of risk factors present.

Extrinsic risk factors are external or environmental factors that can contribute to falls. Only a small percentage of falls in the hospital are accidental, whereas falls in the home are usually multifactorial, with environmental hazards contributing to the majority of all falls. Accidental falls are related to factors that impair vision, such as poor lighting or shiny floor surfaces; impair walking, such as poor-fitting shoes or uneven floor surfaces; cause tripping, such as throw rugs, cords, and clutter; cause slipping, such as wet floors and bathtubs and showers without mats; and challenge balance, such as shelves too high or too low or lack of grab bars in bathrooms or on stairways.

Prevention

Reduce Environmental Hazards: Seemingly insignificant environmental hazards can easily be overlooked, yet they can be devastating to frail elders with cognitive or sensory decline. Adequate lighting, a clear path, raised toilet seats, and locked wheels on wheelchairs are important precautions.

Compensate for Functional Limitations: Clinicians can assist elders to compensate for functional limitations by adjusting medications to minimize adverse drug reactions, control pain, correct sensory deficits, and provide exercises to improve strength and balance. Treating common conditions such as cardiac or pulmonary disease improves energy and endurance. A hospitalized elder often has a rapid functional decline, which increases the risk for falls and injury. Certain therapies such as mood-altering drugs, restraints, or Foley catheters can contribute to functional decline and may lead to falls. Health care personnel must be proactive and provide regular ambulation and toileting, reorientation, and rapid discontinuation of unnecessary tubes or lines.

Educate: Not all patients are able to participate in fall prevention education; thus, determining the patient's ability to learn is critical. Confusion, pain, anxiety, sleep disturbances, poor hearing, and impaired vision may prevent patients from directly participating in fall prevention education. Families should be involved when the patient cannot participate. Written materials are recommended to reinforce teaching. Topics may include rising slowly to prevent dizziness, the importance of mobility, the correct use of adaptive devices, and ways to call for help.

Communicate: Once a fall risk has been identified, the patient, family, and health care personnel along the continuum of care must be informed. Clear, standardized documentation of risk factors and both universal and risk-specific interventions should occur at the time of admission to the system, upon transfer from one setting to another, and when the patient's condition changes.

Individualize Interventions: A standard fall prevention plan must identify and attempt to minimize all major factors contributing to falls. For example, patients who routinely experience increased confusion and aggression in the early evening (sundowning) may be more likely to fall at this time. Keeping the room brightly lit or asking a family member to visit at that time of day may be helpful. In contrast, patients with Parkinson's disease may be at highest risk in the early morning, when the effect of their medication has worn off. Additional assistance may be needed at these times. A program of scheduled toileting may help prevent falls in certain situations. For example, confused patients with altered mobility may try to go to the toilet unaided. Diuretics, excessive fluid, and caffeine or alcohol intake often result in an urgency to void, causing the patient to rush to the toilet. Rescheduling of diuretics and avoiding excessive fluids before bed can reduce this risk.

Electronic warning devices (bed and chair alarms) are designed to alert caregivers that the patient is getting up. The efficacy of an alarm system depends on the technology and the response time of the caregiver. Also, the availability of this intervention is directly related to cost containment, since it may not be reimbursed. Most insurance plans, including Medicare and Medicaid, do not cover alarm devices or lifelines. Managed care insurers may make exceptions if there is a clear indication that the use of these devices will prevent serious negative outcomes related to falls.

Falls are a geriatric syndrome with multiple contributing factors that vary among individuals but result in a final common pathway. Fall prevention should be an essential component of the interdisciplinary health assessment of elders in all health care settings.

BARBARA CORRIGAN
MAURA BRENNAN

See Also
"Alert" Safety Systems
Assistive Devices
Balance
Environmental Modifications

Internet Key Words
Extrinsic Risk Factors
Falls
Interventions
Intrinsic Risk Factors

FAMILY CARE FOR ELDERS WITH DEMENTIA

Although many issues are important in understanding the needs of family caregivers of patients with dementia, three are key: (1) the importance of early diagnosis and prognosis, (2) common problems associated with memory loss that families find difficult to cope with, and (3) the family's emotional needs throughout the caregiving experience.

Early Diagnosis and Prognosis: With early diagnosis, it may be possible to begin treatment that can slow the cognitive decline while there is still minimal impairment. Most important for family members is that early diagnosis allows the patient and family time to plan for future needs, such as executing a power of attorney and appointing a health care representative. Family members often fear that after the diagnosis, the physician will abandon them. It is essential that physicians convey to the family that they will continue to be involved with the patient and family and that management issues will be reviewed as they arise (Hendrie, Unverzagt, & Austrom, 1997). Practitioners should also educate the patient and the family regarding disease progression and prognosis, provide support, and monitor judgment and safety issues so that the patient can remain independent or community dwelling as long as possible (Richards & Hendrie, 1999; Schultz, 2000).

Common Problems Associated with Memory Loss: Unlike other illnesses, dementia has the additional problem of memory loss. Most family caregivers have difficulty providing care simply because they do not understand what is happening to the patient in the early stages of the disease, nor do they know how to respond appropriately to changes in the patient's behavior. The accompanying table describes some of the most common problems associated with memory loss and how caregivers should respond.

One of the most important things for family caregivers to remember is that a patient diagnosed with dementia does not behave in these ways intentionally. These behaviors are manifestations of a brain disorder, and caregivers should not take anything the patient says or does personally. This can help avoid conflict, anger, and subsequent feelings of guilt. Patients cannot be held responsible for their behaviors, but all behavior has a purpose. It is up to the caregiver to look for that underlying purpose. For example, a patient may be agitated and

Common Problems Associated with Memory Loss

Problem	Common Response	Correct Response
Is unaware of memory loss or denies it	Patient should remember. Why won't he face it?	Patient cannot remember that he cannot remember. He is not doing this intentionally.
Memory fluctuates from day to day	Patient is not trying. She remembers only what she wants to remember.	Some fluctuation in memory is normal. Take advantage of the "good" days.
Asks repetitive questions	Patient is doing this to annoy me. I have answered him 10 times already. He can control this.	Patient cannot remember asking. Patient no longer knows how to ask for attention.
Makes accusations (e.g., stealing)	Patient is crazy. No one is stealing her possessions.	This is a way for the patient to deal with the insecurity caused by not being able to remember.
Won't bathe; becomes agitated and violent about it	Patient knows that it is important to shower every day.	Patient cannot remember all the steps necessary to shower or gets confused in the bathroom. It is embarrassing for him to ask for help. It is not critical to shower every day.
Insists on driving, although it is obvious that she has trouble behind the wheel	Ignore it and hope that nothing bad happens. Rationally try to explain that driving is dangerous.	Enlist professional help, for example, a physician, lawyer, or insurance agent.
Lowered inhibitions	Patient should be able to control himself. He knows he should be dressed to go outside.	This is a symptom of the disease. The patient cannot help it. He is not doing it to embarrass me.

wander around the house because he has forgotten where the bathroom is and he needs to use it. Or a patient may constantly disrobe because she is too hot. The caregiver should not blame the patient for these behaviors but should remain calm, try to figure out what is causing the behavior, and redirect the patient while protecting his or her dignity.

The Family's Emotional Response to Dementia: A patient with dementia may need care for many years. Successful caregiving is based on understanding the caregiver's emotional response to the disease, to the patient, and to the patient's behaviors, which change over time. Families must endure an ongoing grief process as they strive to cope with the demands of caregiving while watching the psychological death of their loved one and the death of that individual's personality—that quality or assemblage of qualities that makes a person what he or she is. Unfortunately, many caregiving families fail to realize that *grief* is an appropriate response when caring for a patient with dementia (Austrom & Hendrie, 1990).

Denial is a common response when confronted with emotionally difficult information, such as the diagnosis of dementia. Although early denial may lessen the emotional impact of the diagnosis, continued denial is counterproductive. It may foster unrealistic expectations about the patient's capabilities and interferes with appropriate planning for the future. Clinicians should recognize that denial and disbelief are common when caregivers first learn about the diagnosis. Families will need a second opportunity to review the information with the clinician.

Anger is commonly experienced by caregiving families that must provide long-term care. Sometimes this anger is directed at the patient. Often, families are angry at the government or the health care system. The cost of long-term care can be devastating to middle-income families, shattering their plans for retirement. Anger may also be directed at other family members for not understanding the toll that caregiving takes and for criticizing their efforts. Family members who do not live with the patient and have not had to provide constant care may not appreciate the extent of the demands placed on the caregiver and may offer suggestions about how to provide better care. Conflicts among family members are not unusual, and relationships are further strained when they cannot agree on the patient's care. Old resentments may resurface and interfere with sensible problem solving. The decision whether to institutionalize the patient often exacerbates these family conflicts and associated guilt (Austrom & Hendrie, 1990).

The emotions of anger and *guilt* are often intertwined. Family members may experience guilt for many reasons: not being attentive enough to the patient before the illness, unresolved past conflicts, or making decisions that the patient objects to. Some feelings of guilt may be a normal reaction to feelings of anger or wishes that the demented patient would die. When the patient finally dies, bereavement reactions are often mixed with relief that it is finally over and with guilt for having wished that it would end (Austrom & Hendrie, 1990). It is important that caregivers know that such mixed emotions are both understandable and common.

When dealing with patients with dementia, family caregivers face the progressive deterioration of the patient's higher mental functions, the behavioral problems associated with the disease, the financial burden, the eventual institutionalization of the patient, and the grief associated with the loss of the patient as he or she had once been. Caregivers must take the necessary time to deal with their own emotions so that they can continue to function effectively as caregivers. This massive burden is rarely recognized by health care professionals, who should provide appropriate long-term support for these families.

MARY GUERRIERO AUSTROM
HUGH C. HENDRIE

See Also
Caregiver Burden
Caregiver Burnout
Dementia: Overview

FAMILY CARE FOR FRAIL ELDERS

Families are the main providers of medically related and supportive care for elders in the United States. Over the past two decades, the number of families involved in elder care has increased dramatically because of the expanding elder population. In 1996, nearly one in four American households contained at least one person who had provided unpaid assistance to a relative or friend age 50 or older (National Alliance for Caregiving and the American Association for Retired Persons, 1997). The dependence on family care has also increased due to the health care system's efforts to reduce length of hospital stays through early discharge.

Typically, families take on care responsibilities with little or no preparation. Despite the health care system's increasing reliance on families, research has not addressed families' capacity to respond to these expectations or examined variations in the quality of family care.

Provision of home care to frail elders is satisfying, meaningful, and rewarding for family caregivers; however, it has also been linked to caregiver stress and role strain, as well as to negative health outcomes, including depression and prolonged grief. Families vary in their responses to providing care to frail elders. Among families that experience difficulty in providing care, the problem areas vary considerably (Archbold & Stewart, 1996).

Interventions for Elders and Their Family Caregivers

Research-based interventions for improving family care are in the early stages of development.

Psychoeducational interventions are effective in increasing the caregiver's problem-solving skills, feelings of competence, and social skills; they also can reduce caregiver stress and moderate negative responses to family care problems, such as depression and anger. These interventions may be delivered in group, family, or individual formats. Group interventions produce small effect sizes; individual interventions produce moderate effect sizes.

Respite care is a method of relieving the caregiver from the burden and strain of care activities. It can be provided through adult day care or in-home respite. In some cases, the elder can be admitted to a hospital or long-term-care facility overnight or longer. Research suggests that respite can be beneficial to families (Archbold & Stewart, 1996, 1999).

Comprehensive home health interventions can improve family care through use of an interdisciplinary team. In studies evaluating comprehensive home health interventions, in-home services were usually provided for 6 to 12 months. The services included nursing, social work, and medical assessments; skilled nursing care; care planning; 24-hour telephone support; individualized teaching; support groups; and respite care (Knight, Lutzky, & Macofsky-Urban, 1993). Patients who received services had significantly fewer hospitalizations, nursing home admissions, and outpatient visits than controls, and caregivers reported greater satisfaction with care.

Telephone advice lines and monitoring have been used successfully in a variety of health care arenas, including hospice and care for frail elders. Families can receive timely responses to their health and family care questions from a nurse familiar with their situation. Telephone advice has been effective in supporting persons with many illnesses, including diabetes and cancer.

Support groups are by far the most accessible intervention for caregivers in the United States, especially caregivers for persons with Alzheimer's disease. The groups vary considerably in purpose and nature, and rigorous evaluation of their effects is difficult, because people join them when they are ready, thus creating difficulties in randomizing to experimental and control conditions. Psychoeducational groups have a more systematic educational focus than do support groups. Providers can recommend support groups to family caregivers who express a need to interact with people who share a common experience. This recommendation should be made with the understanding that the beneficial effects of such support groups have yet to be demonstrated empirically.

Day-to-Day Care Strategies

Most family caregivers want to do a good job. Thus, family care interventions that focus attention on how care is delivered by the family caregiver may be particularly helpful to elders and acceptable to families.

Clinicians in many settings have contact with elders and their family caregivers during health transitions—times when caregivers may feel unprepared for new aspects of the family care role. Although nurses may have the most contact with elders and families during transitions, all clinicians should assess the family's skill level and provide the support needed for caregivers to learn and practice care activities in a safe environment. It is not uncommon for the family caregiver to be a frail elder. Thus, providing instruction in a manner that attends to normal age-related changes in vision, hearing, processing, and learning time is important. In addition to directing interventions toward specific family care activities, clinicians can help families work out systems of care by creating routines that attend to the needs of the older person and other family members.

Family care is not stable. Elders' health and functional abilities change over time, as do the health and functional abilities of family caregivers. Likewise, the formal and informal supports available to families change over time. An effective family care support system may need refinements as aspects of family care change. Research on a wide array of family care situations suggests that when caregivers experience transitions, they may be faced with issues for which they are unprepared. During such transitions, caregivers may be especially open to interventions, especially nursing interventions.

In light of the changes in health care delivery that have cast families in the role of primary provider of health and supportive care for elders, clinicians must consider families as partners and develop innovative ways in which to collaborate in this endeavor.

Systematically Assess Family Care: Current Medicare guidelines focus on the ill individual. In contrast, systematic assessment of the *family care situation* is needed to identify family strengths and family care issues known to be associated with poor outcomes. Issues identified through assessment should include the family caregiver's skill in specific care activities, unpredictability in the care routine, low mutuality (the quality of the relationship between the caregiver and the elder), lack of resources to support care, caregiver strain, and caregiver and elder health problems.

Focus Interventions on the Family Rather than on the Individual Patient: Managing an illness in the home setting requires the cooperative efforts of family members. Such an approach enhances the health status and satisfaction of both the elder and the family caregiver.

Work with Families to Solve Family Care Issues: Resolution of family care issues is facilitated by families and clinicians working together to identify issues and generate solutions. Success is more likely when families contribute to the solution, rather than having strategies imposed on them by health care providers.

Recognize that Complex Family Care Issues May Require Multiple Intervention Strategies: Working in partnership, family members and the clinician can implement and evaluate multiple strategies that are tailored to the specific family situation until the issue is resolved. Interventions are tailored based on elder, family, and cultural characteristics and on family preferences.

Set up Systems to Detect Problematic Transitions: Transitions occur for many reasons, including illness or death of the elder or caregiver, a change in the family care environment, or a change in the level of support available to the family. Methods to detect and respond to transitions in family care include telephone access to a provider who is familiar with the situation and periodic monitoring of families for changes in circumstances.

The American health care system is based on the needs of individuals. As a result, there is an inherent structural problem in assisting families who provide care to frail elders. Except in rare situations, reimbursement, charting, and interventions focus on only one member of the family unit.

The management of chronic illness and frailty is, however, inherently a family issue.

PATRICIA G. ARCHBOLD
BARBARA J. STEWART

See Also
Caregiver Burden
Caregiver Burnout
Respite Care

Internet Key Words
Caregiving
Family Care
Home Health Interventions
Psycho-Educational Interventions
Respite Care
Support Groups

Internet Resources
Administration on Aging
http://www.aoa.dhhs.gov

American Association of Retired Persons
http://aarp.org

National Alliance for Caregiving
http://caregiving.org

FATIGUE

Fatigue is usually defined as an abnormal state of physical or mental exhaustion. Although it may be related to excessive physical or mental activity, fatigue may also signify the presence of a physical or mental illness. Fatigue can significantly diminish one's quality of life by interfering with daily activities. Many conditions that result in fatigue are treatable once the correct diagnosis is made.

Although fatigue is a relatively common symptom in the elderly, it often escapes the physician's attention. Patients usually either ignore their symptoms or accept them as an inevitable part of getting old. Often the caregiver rather than the patient reports this complaint. In many cases, it is appropriate for the physician to inquire about it.

Patients and caregivers use various terms to describe fatigue. For example, they may complain of being weak, tired, exhausted, or unable to concentrate. Careful observation of behavioral patterns is sometimes the only clue to the presence of fatigue in moderately and severely demented patients.

Assessment

History

Careful history taking by the physician is paramount for discovering the cause of fatigue. A complete history includes the patient's description of fatigue, its onset and duration, progression over time, and provoking and alleviating factors. Past medical history and habits such as diet, sleep, exercise, and caffeine, alcohol, and nicotine use must be elicited, as should a list of all prescription and over-the-counter medications.

It is also important to inquire about the symptoms associated with fatigue. For instance, fatigue accompanied by shortness of breath, especially on exertion, may indicate underlying coronary artery disease, asthma, or intrinsic lung disease. Weight loss may be a sign of malignancy or depression. The presence of fever may be the only sign of tuberculosis, endocarditis, temporal arteritis, and polymyalgia rheumatica. Fatigued patients with daytime somnolence and a history of snoring may suffer from obstructive sleep apnea. Temporal arteritis can present with symptoms of jaw claudication, temporal or generalized headache, and extreme weakness, to the extent that the patient is unable to get out of bed.

In a review of systems, it is also appropriate to inquire about cough, inability to get up from a sitting position or comb one's hair, black stools, joint pain and swelling, anxiety, and vegetative signs such as depressed mood, poor appetite, and insomnia. If necessary, a geriatric depression scale may be used to detect hidden depression. Mood, anxiety, and somatization disorders are common in patients with chronic fatigue but are often undiagnosed. They often coexist with medical disorders.

Psychiatric evaluation is indicated when these conditions are suspected or when no diagnosis can be made after the medical evaluation is completed.

Physical Examination

A complete physical exam focuses on the presence or absence of fever, rash, pallor, goiter, elevated or decreased heart rate, rales, lymphadenopathy, hepatosplenomegaly, and neurological deficits. Muscle bulk, tone, and strength should be tested to rule out a neuromuscular disease. Temporal artery palpation and a joint exam are helpful in detecting rheumatological conditions. Sometimes the general appearance of the patient, such as poor grooming, psychomotor agitation, or depressed affect, suggests the existence of a psychiatric disorder.

Laboratory Tests

Routine screening tests include complete blood count, electrolytes, renal and hepatic function, glucose, total protein, thyroid function, iron studies, erythrocyte sedimentation rate, urinalysis, and stool guaiac in almost all patients. Random extensive laboratory testing was not found to be useful in the diagnosis of fatigue. In individual cases, additional tests may be indicated, such as electrocardiography, Holter monitoring, chest x-ray, purified protein derivative testing, sleep studies (polysomnography), electroencephalography, serum and urine protein electrophoresis, Lyme titers, and antinuclear and antithyroid antibodies.

Causes

Fatigue is divided into acute and chronic types, depending on the rate of onset. Although such a framework may be helpful, some conditions are hard to classify because they have features of both (e.g., depression). Sudden onset of fatigue in a previously healthy individual may suggest an acute illness, such as myocardial infarction (especially in persons with diabetes, who can present without chest pain but with only extreme weakness or syncope), anemia caused by acute blood loss or hemolysis, infections with little or no fever, stroke without neurological deficits, adverse drug reactions, and depression.

The causes of chronic fatigue in elderly individuals are numerous and include inflammatory conditions (systemic lupus erythematosus, rheumatoid and osteoarthritides, polymyalgia rheumatica, temporal arteritis and other vasculitides, inflammatory bowel disease, dermatomyositis, polymyositis); infections (tuberculosis, hepatitis, endocarditis, Lyme disease, infectious mononucleosis); malignancy (colon cancer, breast cancer, lung cancer, renal cell cancer, leukemia, lymphoma, multiple myeloma); endocrinopathies (hyperthyroidism, hypothyroidism, pituitary insufficiency, adrenal insufficiency, uncontrolled diabetes); cardiovascular diseases (congestive heart failure, valvular heart disease); neurological diseases (Parkinson's disease, narcolepsy, amyotrophic lateral sclerosis, myasthenia gravis, Eaton-Lambert syndrome); anemia of various causes (chronic occult blood loss, folate and vitamin B_{12} deficiency); hypoxia (chronic obstructive pulmonary disease, interstitial pulmonary fibrosis, obstructive sleep apnea); psychiatric disorders (depression, dysthymia, anxiety, somatization, alcoholism); medications (corticosteroids, diuretics, some nonsteroidal anti-inflammatory drugs, centrally acting antihypertensives, many beta-blockers, anticonvulsants, some antidepressants, antihistamines, sedatives, tranquilizers, digitalis toxicity); alcohol; chronic pain (may also cause insomnia); chronic fatigue syndrome (presents with low-grade fever, myalgias, painful shotty lymphadenopathy, depressed mood, difficulty concentrating); disuse and immobility (general inactivity, prolonged bed rest, deconditioning); lifestyle issues (inadequate rest, inadequate sleep, excessive exercise, boredom); and environmental causes (noise, high ambient temperature, uncomfortable furniture, exposure to heavy metals or carbon monoxide).

Management

Management of patients with fatigue depends on the underlying cause. Once medical disorders are treated, the fatigue should diminish or disappear. Depression, anxiety disorders, and somatization should be treated with medications or psychotherapy. It is important to realize that medical illness can produce depression, both of which need to be treated. Medications that are suspected of causing fatigue should be discontinued, if possible. Sometimes decreasing the dose is sufficient to cause subjective improvement. It is also important to establish rapport with patients and their caregivers, since treatment is often prolonged.

After the cause of fatigue is understood and treatment of the underlying condition is instituted, additional strategies may be employed. For instance, getting enough sleep (usually at least eight hours a night and napping during the day) and enough rest can reduce symptoms of fatigue. Rest does not mean complete inactivity, which can lead to deconditioning, but rather scheduling quiet times when fatigue peaks (usually midday) and taking breaks during or between tasks, before fatigue sets in. Health care professionals should encourage patients with chronic fatigue to remain active by incorporating increased activity in the daily routine (e.g., walking instead of taking a bus or driving, walking upstairs instead of taking the elevator, keeping a garden).

A structured exercise program can increase the muscle strength of even very frail individuals. Exercises using the whole body can maintain strength, tone, and flexibility (e.g., swimming, outdoor walking, yoga, tai chi chuan). Exercise routines should include 5 to 10 minutes for warm-up and cool-down and stretching exercises. A gradual increase in exercise intensity and maintenance of good posture have physiological and psychological benefits. Physical therapists can help design individual exercise programs.

Patients with cardiovascular and pulmonary disease who undertake exercise programs require medical supervision. Exercise should not be prescribed to patients with uncompensated congestive heart failure, complex ventricular arrhythmias, unstable angina pectoris, hemodynamically significant aortic stenosis, large aortic aneurysms, and uncontrolled diabetes mellitus. In addition, eating properly (avoiding heavy meals, alcohol, and caffeine at night) can help alleviate fatigue.

In some cases, asking family or friends for assistance or hiring a home health aide can help a patient manage daily tasks. Energy-conservation techniques should also be practiced—for instance, sitting when taking a shower, brushing teeth, or shaving; working with the arms below shoulder level; eating slowly and consuming smaller and more frequent meals; pushing rather than pulling objects; putting the full foot on steps while climbing stairs; walking uphill slowly; pacing activities throughout the day, alternating with breaks; breaking tasks up into smaller steps; and limiting unnecessary tasks.

Periodic medical reassessment is important to identify any underlying process that presents itself and to address new problems that may arise.

YAKOV IOFEL

See Also
Anemia
Caloric Intake
Daytime Sleepiness
Deconditioning Prevention
Protein Energy Undernutrition
Weakness

Internet Key Words
Energy Conservation Techniques
Exercise
Fatigue

Internet Resources
Chronic Fatigue Syndrome Home Page, U.S. Centers for Disease Control and Prevention
http://www.cdc.gov/ncidod/diseases/cfs/ cfshome.htm

Fatigue Management, University of Washington Orthopaedics
http://www.orthop.washington.edu

FECAL INCONTINENCE

Fecal incontinence, the involuntary loss of feces, has ramifications for social and personal hygiene.

It is a leading cause nursing home placement. Fecal or urinary incontinence, or both, affects 30% to 50% of nursing home residents. Dementia and immobility are the major factors associated with fecal incontinence, despite intact anorectal function. Weakness of the anal sphincter muscle, impaired rectal sensation, pudendal neuropathy, fecal impaction, or a combination of these factors may also predispose to fecal incontinence. Management should be tailored toward remedying the underlying deficits. The treatment approach depends on the patient's ambulatory status, place of residence (institution or community), and whether the problem is predominantly social or hygienic.

Pathophysiology

Fecal incontinence can be associated with a weak or damaged internal or external anal sphincter, in combination with weakness of the pelvic floor muscles. The loss of endovascular cushions, impaired anorectal sensation, poor rectal compliance (high intrarectal pressure associated with a small volume of stool), compromised accommodation (e.g., due to inflammatory bowel disease, radiation enteritis, or rectal surgery), or neuropathy affecting the pudendal or sacral nerves or the spinal or central nervous system may be causative factors. In some patients, incomplete evacuation of stool, large stool volume, liquid stool, and the irritant effect of bile salts in the rectum may account for fecal incontinence. Fecal incontinence in an elderly institutionalized person is often due to impaction of stool in the distal colon. Liquid stool leaks around the fecal mass. Failure to perceive the arrival of stool in the rectum may produce either severe urgency to defecate or leakage of small amounts of stool, particularly if toileting assistance is not immediate.

Management of Ambulatory Patients with Fecal Incontinence

Drug Therapy

If incontinence is associated with diarrhea, specific treatment of the underlying problem, such as inflammatory bowel disease or bile salt malabsorp-

Common Causes of Fecal Incontinence

Traumatic
  Obstetrical
  Postsurgical
  Sexual
  Accidental

Neurological
  Diabetes mellitus
  Pudendal neuropathy
  Cauda equina lesions
  Cerebrovascular injury
  Multiple sclerosis
  Polyneuropathies
  Dementia
  Mental retardation

Inflammatory
  Ulcerative colitis/Crohn's disease
  Radiation proctitis
  Ischemic colitis

Miscellaneous
  Idiopathic bile salt malabsorption
  Lactose intolerance
  Secretory diarrhea
  Laxative abuse
  Irritable bowel syndrome
  Fecal impaction with overflow

From Rao, S. S. C. (1999). Fecal incontinence. *Clinical Perspectives in Gastroenterology, 5*(2), 277–288.

tion, will yield better control of incontinence. Agents such as loperamide (Imodium) or diphenoxylate (Lomotil) will slow transit and solidify stool and can reduce symptoms. Excessive use of these drugs, however, may cause constipation and fecal impaction. If the stools are hard and difficult to expel, a regimen of controlled evacuation with suppositories or enemas at regular intervals may be necessary.

Biofeedback Therapy

Neuromuscular conditioning using biofeedback techniques can improve fecal incontinence in two-thirds of selected patients, although no controlled trial has been performed. Visual, verbal, or audio feedback techniques may be used to improve anal sphincter muscle strength, rectal sensation, and rectoanal coordination (Rao, 1999). In a patient with a weak anal sphincter muscle, but who can still generate an increase in sphincter pressure with squeezing, repeated volitional squeezing of the anal sphincter muscle can improve muscle strength. Improvement of rectal sensation is achieved by repeatedly distending a balloon in the rectum (to produce the sensation of stooling) and successively lowering the balloon volume when the patient consistently perceives a particular level of distention. Rectoanal coordination training consists of distending a balloon in the rectum and asking the subject to squeeze the external sphincter. Over time, these functions of the voluntary sphincter muscle improve.

Some patients also experience fecal seepage—involuntary leakage of small amounts of stool following an otherwise normal defecation. Most of these patients have normal sphincter function and normal pudendal nerves but have impaired rectal sensation and dyssynergic defecation (inappropriate elevation of anal sphincter pressure during defecation). Biofeedback therapy directed toward improving rectal sensation and timing of anal sphincter relaxation during defecation can ameliorate this symptom.

Surgical Treatment

Fecal incontinence associated with rectal prolapse, rectovaginal fistula, or neurological problems such as spinal cord injury may be amenable to surgical repair. But even experienced colorectal surgeons will misdiagnose up to one-fifth of patients presenting with fecal incontinence if the assessment is based solely on history and physical examination (Keating, Stewart, Eyers, Warner, & Bokey, 1997). Anorectal manometry provides comprehensive information regarding the integrity of the anal sphincter and the intactness of rectal sensation.

Prior to surgery, an anal ultrasound study should be done to assess the integrity of the external and internal anal sphincter muscles. By precisely localizing the damage, more accurate reconstructive surgery can be performed.

Pudendal nerve terminal motor latency measures the neuromuscular integrity of the terminal portion of the pudendal nerve and the anal sphincter muscle. Prolonged nerve latency suggests pudendal neuropathy. This test separates neuropathy from rectal wall disorders and provides an explanation for internal sphincter defects. Reconstructive surgery may not be successful in patients with moderate to severe pudendal neuropathy.

Surgical repair of the external sphincter with sphincteroplasty is associated with substantial improvement in continence in about 75% of patients one year after surgery (Engel, Kamm, Sultan, Bartram, & Nicholls, 1994; Wexner, Marchetti, & Jagelman, 1991). The long-term results are uncertain. The demonstration of atrophy of the external anal sphincter by magnetic resonance imaging is associated with significantly poorer outcome (Briel et al., 1999). Except for colostomy, none of the other surgical interventions can guarantee total continence. It is important that the patient is made aware of this limitation to minimize disappointment and avoid a malpractice suit.

Postanal repair to restore the anorectal angle by plication of the pelvic floor muscles behind the sphincter is associated with complete continence in 41% of patients (Pinho et al., 1992) and may help patients with normal sphincters but mild pudendal neuropathy. For patients with extensive sphincter disruption, or those who failed to benefit from prior surgery, surgical construction of a neo-sphincter is an option. Most commonly, the gracilis muscle is transposed from the inner thigh and wrapped around the anus to form a new sphincter. A small electrical stimulator is implanted to provide the stimulus for continuous tonic contraction. The stimulator can be switched off to allow defecation. Seventy-three to 83% of patients with a neo-sphincter are reported to be continent two years after surgery (Baeten et al., 1995; Seccia, Menconi, Balestri, & Cavina, 1994). A colostomy may be required for some patients with severe and refractory fecal incontinence.

Management of Fecal Incontinence in Institutionalized Individuals

Fecal incontinence associated with fecal impaction requires disimpaction followed by tap-water enemas to cleanse the colon. The patient should then be placed on a fiber-restricted diet, regular laxatives, and a program of enemas administered two or three times a week to prevent impaction and soiling.

Loperamide may be useful when chronic diarrhea is present. When there is impaired reservoir capacity or neurogenic dysfunction, a treatment program of normal fiber intake together with regular defecation can reduce stool volume and help minimize the frequency of incontinence. Persistent soiling can be managed with a combination of loperamide to solidify the stool and enemas two or three times a week to empty the stool and prevent impaction. Fecal incontinence associated with excessive laxative use (often prescribed to prevent constipation and fecal impaction) can be managed by reducing the dose of the laxative.

Studies indicate that incontinent patients are less likely to receive toilet assistance from nursing staff (Burgio, Jones, & Engel, 1988). Nevertheless, it is prudent to place patients with impaired cognitive function or mobility on a regular toileting schedule. Whether this practice reduces the frequency of fecal incontinence and decreases the associated skin complications and urinary tract infections has not been examined.

Management of the Social and Hygienic Aspects of Fecal Incontinence

Fecal incontinence is often associated with loss of self-esteem and self-confidence, leading to disruption of relationships with family and friends and impairment of social and occupational activities. For elderly patients, personnel who are experienced in the management of fecal incontinence can be reassuring. Prompt changing of soiled pads or clothes, storage of soiled material in airtight deodorized containers, and appropriate hygienic measures can manage the odor associated with fecal incontinence. Perineal washes can disguise the smell of feces. Foods that can cause malodorous discharge vary from patient to patient, and limiting their consumption is prudent. For bedridden or un-

conscious patients with severe diarrhea, a fecal collection pouch, rectal tube, or anal plug device may be useful.

Immediate cleansing of the perianal skin and changing of undergarments and clothes are of paramount importance. Moist tissue paper (e.g., baby wipes) that is not abrasive is preferable to dry toilet paper. Barrier creams such as Calmoseptine may be useful in preventing skin excoriations. Perianal fungal infection should be treated with topical antifungal agents (Mycelex). When skin breakdown occurs, aggressive steps should be taken to divert the fecal stream and change the patient's position frequently. Scheduled toileting with a commode at the bedside or a bedpan, and supportive measures to improve the general well-being and nutrition of the patient, may all be effective.

Fecal incontinence is a challenging problem, particularly in the elderly. In addition to local problems such as weak sphincter muscles, impaired rectal sensation, and nerve damage, there are global issues such as impaired cognition, poor mobility, comorbid illness, and psychosocial effects. The use of rational therapies that include biofeedback; scheduled toileting; an appropriate regimen of diet, laxatives, and enemas; and supportive care can improve the quality of life in many patients with fecal incontinence.

FELIX W. LEUNG
SATISH S. C. RAO

See Also
Bowel Function
Urinary Incontinence Assessment
Urinary Incontinence Treatment
Uterine, Cystocele/Restocele, and Rectal Prolapse

FINANCIAL ABUSE

Hundreds of thousands of elderly individuals across the country are affected by elder abuse. Financial abuse—the illegal or improper use of an elder's financial assets for profit or gain by another—is one of the fastest growing and least understood forms of abuse. The definition of financial, or fiduciary, abuse varies among states, and evaluating whether financial abuse has occurred is a complex and often subjective determination.

The number of elder abuse incidents reported to adult protective services (APS) increased 150% between 1986 and 1996. The National Elder Abuse Incidence Study (Administration on Aging, 1999a) found that only one of six incidents was reported. Financial abuse represented 30.2% of the reported incidents. Many victims of this hidden crime suffer in silence. Factors contributing to underreporting include cognitive incapacity, fear of retaliation or loss of independence, shame, humiliation, and death.

Financial abuse occurs in all socioeconomic, ethnic, and racial groups. Most reported incidents involve white women age 75 and older. Financial abuse can be perpetrated by anyone, including a family member, caregiver, friend, acquaintance, neighbor, landlord, contractor, professional fiduciary, or scam artist. Financial abusers may also neglect, emotionally abuse, or physically abuse their victims. Risk factors for abuse include physical and mental impairments that create dependency on others, social isolation, and loss of a spouse. Some abusers have an alcohol, drug, or gambling addiction or are dependent on the elder for money or housing.

Loss of an elder's financial assets can result in increased morbidity and mortality. Cardiovascular disorders and heart attacks, decreased immune system functioning, malnutrition and dehydration, and depression can result. Elder mistreatment is associated with a 3.1 times greater risk of dying, after adjusting for other factors associated with increased mortality in older adults (Lachs, Williams, O'Brien, Pillemer, & Charlson, 1998).

Every state has elder abuse, APS, or at least mental health legislation that authorizes the state to protect and provide services to vulnerable, incapacitated, or disabled adults. Most common among the elder abuse statutes are provisions for confidential mandatory or permissive reporting by health care and social services providers.

Indicators of Possible Financial Abuse

Indicators of possible financial abuse may include bank activity that is uncharacteristic of the elder or inconsistent with the elder's abilities (e.g., the ATM card is being used, but the elder is home-bound; frequent ATM use after a long period of occasional use or inactivity). Suspicious activity on credit card accounts, such as purchases of unnecessary major household items or inappropriate items; documents with altered or suspicious signatures; or documents (e.g., pension checks, stocks, and government payments) missing altogether can signal financial abuse. A power of attorney executed by a confused elder, changes in an elder's property titles or other documents—particularly if the elder is confused—and documents ceding title or ownership to new acquaintances should be assessed for validity. Personal property may be missing, or the elder may lack amenities even though he or she can afford them. Failure to receive prepaid services or receipt of an eviction notice or action to disconnect utilities may indicate inappropriate outside financial influence. An elder may appear unkempt or the residence may be in disarray, despite arrangements for personal and domiciliary care. If an elder's mail is redirected to another address or the elder is unaware of or does not understand recently completed financial transactions, clinicians should consider the possibility of financial abuse.

As with any type of elder abuse or mistreatment, behavioral indicators include, fear, withdrawl, depression, hopelessness, hesitation to talk openly, confusion or disorientation, ambivalent statements unrelated to mental dysfunction, and untreated health problems. The elder may relate improbable stories about potential financial abuse, and these should be investigated even if the elder has a history of aggrandizing and manipulating supposed injuries, personal or otherwise.

Caregivers (family or others) may also provide clues about financial abuse, such as a new acquaintance, particularly someone who takes up residence with the elder. Abusive caregivers may be reluctant to participate or cooperate with service providers in planning for care and may socially isolate the elder or always be present when the elder has visitors, even when the elder does not need assistance.

Health Care Providers' Role in Preventing, Detecting, and Reporting Financial Abuse

Health care providers are in a unique position to prevent and detect elder financial abuse. When a person is diagnosed with a dementing disorder, such as Alzheimer's disease, the patient and family should be warned about the potential for financial abuse and be given practical suggestions for protective interventions (e.g., having a cosigner on bank accounts, contacting the bank about options for protecting the vulnerable senior's assets, consulting an elder law attorney). Individuals in the early stages of dementia are often able to live and function independently. However, one of the first areas adversely affected is the individual's ability to manage his or her finances. Persons who have been independently and competently managing their financial affairs (e.g., writing checks, paying bills, making decisions about investments and charitable donations) may begin to make questionable decisions that raise concern among family members. Unfortunate examples abound: the 82-year-old man who sends $50,000 in cash to a telephone solicitor who told him that it would guarantee his winning the $5 million Canadian lottery; the 75-year-old woman who gives $250 each week to the "nice young man" who takes out her trash.

Transfer of financial authority from an impaired elder to a family member, caregiver, or other person must be done cautiously. It can protect the elder from abuse or, in the hands of the abuser, can be a "license to steal." Some abusers take the time to gain their victims' trust in order to gain control of their assets.

Interactions between the elder and caregiver should be observed. The caregiver should be invited to leave the room to provide an opportunity for the elder to speak privately with the clinician. During this time, the elder should be tactfully asked if he or she is being taken advantage of in any way. The elder is unlikely to bring it up spontaneously. The clinician may ask questions such as: Does the elder have adequate financial resources for basic needs? Are these needs being met? If not, why? Has the elder recently signed any documents without understanding their meaning? These questions may encourage the elder to discuss concerns about his or her living situation, including the potential of financial abuse.

When financial abuse is suspected, the clinician should document the patient's cognitive and functional status, relevant direct quotes, and cues and clues that led to the clinician's suspicions. It is not the health care provider's responsibility to confirm that abuse has occurred; rather, the responsibility is to report a reasonable suspicion to the appropriate investigative entity (e.g., the long-term-care ombudsman program for abuse involving nursing home residents, APS offices, or local law enforcement).

Capacity and Undue Influence

Elders have the right to self-determination and control of their affairs until they delegate responsibility or a court grants responsibility to someone else. All adults have presumptive mental competence unless and until proved otherwise. Hence, an elder has the right to make what may seem to be a poor decision.

Two questions raised in many cases of possible elder financial abuse are: Did the elder have the mental capacity to make a competent decision? Was the elder inappropriately manipulated and therefore unduly influenced?

Although states have different definitions of legal capacity (Griso, 1994), most statutes refer to deficits in certain functional abilities pertaining to decision making and judgment. Two types of capacity that often come into question in financial abuse cases are testamentary capacity to make or change a will and the capacity to enter into a contract, usually defined as understanding the nature and consequences of a transaction. Medical and mental health professionals assist the court, law enforcement, and APS by evaluating individuals with questionable mental capacity.

Undue influence occurs when a dominant person uses his or her role and power to exploit the trust, dependency, and fear of another and to deceptively gain control over the decision making of another. Susceptibility to undue influence may be affected by an elder's mental capacity, environment, and medical and psychological factors.

A court must determine whether an individual consented to a business transaction, purchase, or other action. To exercise consent, the individual must act freely and voluntarily, must not be influenced by threats or force, and must have sufficient mental capacity to make competent choices about whether to do something proposed by another. In the absence of consent, transactions (e.g., wills, contracts, gifts, powers of attorney, joint tenancies) may be rescinded by a court order.

Protecting Against Financial Abuse

Financial education that includes financial literacy, proactive health care, and legal and financial planning (including planning for the possibility of incapacity) should be encouraged by clinicians as a means of preventing financial abuse.

Some states have fiduciary abuse specialist teams (FASTs). Based on the Los Angeles County model, a FAST is an interdisciplinary team that provides ombudsmen, other caseworkers, and expert consultants to help victims recover or prevent further loss of their assets. Volunteer professional consultants include law-enforcement officers, prosecutors, private and public-interest attorneys, retired judges, public and private conservators or guardians, health and mental health service providers, care managers, bankers, securities and real estate brokers, and financial and estate planners.

Assessment of the financial management needs of the elderly can determine what services are necessary and appropriate. Informal measures can range from requiring an elder's attendant to provide receipts for purchases to more complex services involving attorneys. Interventions include,

but are not limited to, arranging for direct-deposit banking, establishing a joint tenancy account, hiring a money management service, executing a power of attorney or durable power of attorney, establishing a trust, or filing for probate conservatorship or guardianship. The least restrictive alternatives for protecting an elder's assets should be used.

Susan I. Bernatz
Susan J. Aziz
Laura Mosqueda

See Also
Elder Mistreatment: Overview
Money Management

Internet Key Words
Elder Abuse
Elder Financial Abuse
Fiduciary Abuse
Financial Abuse
Financial Exploitation
Undue Influence

Internet Resources
Commission on Legal Problems of the Elderly
http://www.abanet.org/elderly/

Elderweb Online Eldercare Sourcebook
http://www.elderweb.com/

National Administration on Aging
http://www.aoa.gov

National Aging Information Center
http://www.aoa.dhhs.gov/naic

National Center on Elder Abuse
http://www.gwjapan.com/NCEA

National Committee for the Prevention of Elder Abuse
http://www.preventelderabuse.org

FOOT PROBLEMS

The management of foot problems in older patients begins with a comprehensive assessment, continual surveillance, and a team approach to primary care. Key elements include demographics and social status history; present illnesses; past medical history and systems review; current medications; dermatological, onychial, orthopedic (biomechanical and pathomechanical), vascular, and neurological evaluations; stratification of risk for neurological, vascular, mechanical, or pressure keratosis; ulcer classification, if present; findings of risk; and a management plan (Helfand, 1999). Factors contributing to the development of foot problems in the elderly include the aging process itself, presence of diseases, degree of ambulation, activity limitation, multiple medications, and injury.

Onychial and Other Toenail Problems

Onychia is inflammation of the posterior toenail wall and bed. It is usually precipitated by local trauma or pressure or is a complication of systemic diseases such as diabetes mellitus and peripheral arterial insufficiency. Onychia is usually an early sign of a developing local infection. Mild erythema, swelling, and pain are the most prevalent findings. Lamb's wool, tube foam, or shoe modification should be considered to reduce pressure on the toe and nail. If the onychia is not treated early, paronychia may develop, with significant infection and abscess of the posterior nail wall. The infection progresses proximally; deeper structures become involved, leading to necrosis, gangrene, and the risk of amputation. Management includes pressure reduction, drainage, culture and sensitivity testing, radiographs and scans as appropriate to detect early bone changes, saline compresses, and appropriate systemic antibiotics.

Deformities of the toenail are the result of repeated microtrauma, degenerative changes, or disease. For example, the continual rubbing of the toenail against the shoe is sufficient to produce change. When debridement is not done periodically, the nail structure hypertrophies, continues to thicken (onychauxis), and becomes deformed. Onychogryphosis, or "ram's horn nail," is usually com-

plicated by fungal infection. The resultant disability and pain can prevent the patient from wearing shoes. The exaggerated curvature may cause the nail to penetrate the skin, leading to infection and ulceration. Traumatic avulsion of the nail is more frequent with this condition. Treatment consists of local debridement, followed by mild keratolytics and emollients as long as the deformity is observed. However, patients with sensory loss may not complain of pain and discomfort (Birrer, Dellacorte, & Grisafi, 1988).

The most common nonbacterial infection of the toenails is onychomycosis, a chronic and communicable disease. In superficial infections, changes appear on the superior surface of the toenail and generally do not invade the deeper structures. In more serious manifestations, the nail bed and nail plate are infected, and there is usually some degree of onycholysis (freeing of the nail from the distal edge). *Candida* is most common in patients with some form of chronic mucocutaneous manifestation. Mycotic onychia; autoavulsion; subungual hemorrhage; a foul, musty odor; and degeneration of the nail plate are common findings. Once the matrix of the nail is involved, hypertrophy and deformity occur, which might not be reversible. Multiple drug use and vascular impairment in older patients may limit systemic management. Initial management includes a topical fungicide solution to permit penetration and, if not contraindicated, systemic medication.

Ingrown toenails in the elderly are usually the end result of deformity, onychodysplasia, and improper attempts at self-care. When the nail penetrates the skin, an abscess and infection result. If this is not recognized early, the development of periungual granulation tissue complicates treatment. In the early stage, a segment of the nail can easily be removed to establish drainage; saline compresses are used, and antibiotics are prescribed as indicated. When granulation tissue is present, excision, fulguration, desiccation, or the use of caustics and astringents may be needed to reduce the granulation tissue. In all cases, removal of the penetrating portion of the nail is primary.

Hyperkeratosis

Common among the elderly are the many forms of hyperkeratotic lesions, such as the tyloma (callus) and the heloma (corn), including hard, soft, vascular, neurofibrous, seed, and subungual types. Intractable keratoma, eccrine poroma, porokeratosis, and verruca must be differentiated from these keratotic lesions, although each may present initially as a hyperkeratotic area. The biomechanical and pathomechanical factors that create these problems are associated with compressive, tensile, or shearing stress. Soft tissue loss associated with aging and atrophy of the plantar fat pad increases pain and limits ambulation. Contractures, gait changes, deformities, incompatibility between foot type and shoe last, and arthritis are additional factors that need consideration. Many factors, including skin tone and elasticity, predispose to the development of keratotic lesions (Helfand & Jessett, 1998).

Management is directed toward the functional and activity needs of the patients. Treatment includes debridement, padding, weight dispersion and diffusion, emollients, shoe modifications and last changes, orthoses, and surgical revision. Keratotic lesions can become primary irritants and produce local avascularity, which can precipitate ulceration. Pressure ulcers in the foot usually begin with subkeratotic hemorrhage. If debrided and managed properly, the ulcers usually heal, but they may recur unless adequate measures are taken to reduce pressure on the localized areas of ulceration. Despite all measures, the problem may persist due to residual deformity and systemic diseases, such as diabetes mellitus.

Dermatological Problems

A common problem in older patients is dryness of the skin, or xerosis, due in part to decreased hydration and lubrication, which is part of the normal aging process. Keratin dysfunction also may be evident. Fissures that develop with associated stress are at risk for ulceration. Initial management includes the use of an emollient and a mild keratolytic. A plastic or Styrofoam heel cup can minimize trauma to the heel (Helfand, 1993). Pruritis is also a common complaint and is usually more severe in cold weather.

Treatment of hyperhidrosis and bromhidrosis depends on the cause—local or systemic. If local, hydrogen peroxide, isopropyl alcohol, and astringents may be used topically to control the excessive perspiration and odor. Neomycin powder helps control the odor by reducing the bacterial decomposition of perspiration. Footwear and stocking modifications should be considered. In colder climates, dampness can predispose a patient to the vasospastic effects of cold temperature.

Contact dermatitis may be a reaction to chemicals used in shoe construction, footwear fabrics, or stockings. Clinical findings are limited and usually consist of bilateral skin lesions. Management includes removal of the primary irritant: mild, wet dressings; and low-dose topical steroids (e.g., 0.5% hydrocortisone).

Stasis dermatitis is a result of venous insufficiency and chronic ulceration and is more common in patients with dependent edema. Management consists of leg elevation; tepid, wet dressings; topical steroids; antibiotics as indicated; and supportive measures to manage the venous disease. Pyodermas and superficial bacterial infections should be managed locally.

Tinea pedis in elderly patients is often an extension of onychomycosis and is more common in warm weather. Poor foot hygiene and the inability to see their feet may motivate patients to seek care only when the condition becomes clinically significant. A wide variety of topical medications can control this condition. Solutions, gels, or creams (water washable or miscible) should be used when the patient is unable to easily remove an ointment base.

Other common dermatological manifestations in the elderly are those associated with atopic dermatitis, nummular eczema, neurodermatitis, and psoriasis.

Simple or hemorrhagic bullae are related to shoe trauma and friction or to systemic diseases such as diabetes mellitus. Hemorrhagic bullae related to diabetes mellitus are usually early ulcerative indicators. Management is directed toward eliminating pressure, with the use of supportive dressings, shoe modifications, protection, and drainage when appropriate. Gait changes in the elderly can magnify many foot-shoe incompatibilities and result in local foot lesions.

The management of ulcerations in the elderly depends on the causes and complications of the problem. General principles of care include supportive measures to reduce trauma and pressure to the ulcerated area, such as dressings, orthoses, shoe modifications, and special shoes. Therapeutic shoes are an entitlement under Medicare and should be considered for diabetic patients. The prevention and control of infection and the maintenance of a clean, healthy base to permit healing are essential. The debridement of keratosis is essential to prevent roofing of the ulcer. Physical modalities and measures such as low-voltage therapy (contractile currents) and exercises can improve the local vascular supply to the ulcer and help to establish a clean base. Atrophy of soft tissue and the residuals of arthritis also provide a focus for the development of ulcerations. Ulcerations associated with systemic disease are usually related to neuropathic change and vascular insufficiency, as with diabetes mellitus. Management focuses on identifying the underlying cause, instituting local supportive measures, treating the related systemic disease, minimizing the potential for osteomyelitis, and maintaining the ambulatory status of the patient for as long as possible.

Older patients who do not use footwear at home expose themselves to the hazards of foreign bodies and foot injury. For example, animal hairs may appear as keratotic plugs and require debridement or excision to relieve pain.

The overall care of geriatric patients should reflect a reasonable approach that reduces foot pain and improves the patient's functional capacity. The ability to remain ambulatory is directly related to foot health, and foot problems are common in the elderly. Practitioners must think comprehensively and recognize that a team approach is an essential part of management, because the quality of life of elders depends largely on their ability to remain mentally alert and ambulatory.

Biomechanical and Pathomechanical Problems

Biomechanical and pathomechanical abnormalities of the feet create pain and functional problems in gait and balance and make it difficult to obtain proper footwear. Conditions include hallux valgus, hallux varus (splay foot), hallux flexus, digitus flexus (hammer toe), digitus quintus varus, overlapping toes, underriding toes, prolapsed metatarsals, pes cavus, pes planus, pronation, hallux limitus, and hallux rigidus. Treatment consists of both nonsurgical and surgical interventions. Age should not be the determining factor in considering surgery. Consideration must also be given to the patient's ability to adapt to change in relation to ambulation; to have an anatomically corrected joint and a patient who cannot ambulate without pain defeats the purpose of treatment.

These abnormalities can produce inflammatory changes such as periarthritis, bursitis, myositis, synovitis, neuritis, tendinitis, sesamoiditis, plantar myofasciitis, plantar fasciitis, calcaneal spurs, periostitis, tenosynovitis, atrophy of the plantar fat pad, metatarsal prolapse, metatarsalgia, anterior imbalance, Haugland's deformity, entrapment syndrome, and neuroma. Conservative interventions include shoe last changes, shoe modifications, orthoses, digital braces, physical medicine, exercises, and mild analgesics for pain.

Fractures of the foot and toes may be the result of direct trauma or stress related to bone loss. Most uncomplicated and closed fractures that are in a good position can be managed with the use of a surgical shoe and supportive dressings, as long as the joints distally and proximally are immobilized. Silicone molds can be used for digital fractures.

Shoe modifications for the elderly include mild calcaneal wedges to limit motion and alter gait; metatarsal bars to transfer weight, Thomas heels to increase calcaneal support, long shoe counters to increase midfoot support and control foot direction, heel flares to add stability, shank fillers or wedges to produce a total weight-bearing surface, steel plates to restrict motion, and rocker bars to prevent flexion and extension. Additional internal modifications include longitudinal arch pads, wedges, bars, lifts, and tongue or bite pads.

Rheumatoid changes cause early-morning stiffness, pain, fibrosis, ankylosis, contracture, deformity, impairment, and the reduction of ambulation. Management includes aspirin, nonsteroidal anti-inflammatory drugs, local steroid injections, physical medicine and assistive modalities, shoe last changes, shoe modifications, and orthoses for weight diffusion or dispersion and for support and stabilization.

Peripheral Arterial, Sensory, and Diabetic Problems

Common foot complaints associated with peripheral arterial, sensory, and diabetic changes include fatigue, resting pain, coldness, burning, color changes, tingling, numbness, diminished hair growth, thickening toenails, ulcerations, a history of phlebitis, cramps, edema, claudication, and repeated foot infections. Primary physical findings include diminished or absent pedal pulses, with similar changes in the entire extremity, depending on the location and degree of occlusion. Hypertensive patients may demonstrate pulsations that are a false reflection of the vascular supply. Color changes include rubor and/or cyanosis. The temperature of the foot is usually cool. Vasospastic changes are more pronounced in colder climates. The skin is usually dry, with pronounced atrophy of the skin and soft tissues. Superficial infections are common, and with extension of the infection, pain becomes more significant. Neurological assessment in relation to foot and associated changes should include Achilles reflex, vibratory sensation, sharp and dull response, superficial plantar response (Babinski), paresthesia, burning, joint position, and response to monofilament neurosensory risk stratification.

Older diabetic patients present special problems in relation to foot health. It is estimated that 50% to 75% of all amputations in patients with diabetes can be prevented by early intervention when pathology is noted, improved health education, and periodic evaluation before the onset of significant symptoms and pathology (Helfand, 1995). Elderly diabetic patients with neuropathy present with insensitive feet that generally exhibit some degree of paresthesia, sensory impairment to pain and temperature, motor weakness, diminished or lost Achilles and patellar reflexes, decreased vibratory sense, loss of proprioception, xerotic changes, anhidrosis, neurotrophic arthropathy, atrophy, neurotrophic ulcers, and possibly a marked difference in size between the two feet. There is a greater incidence of infection, necrosis, and gangrene. Vascular impairment is characterized by pallor, absent or decreased posterior tibial and dorsalis pedis pulse, dependent rubor, decreased venous filling time, skin coolness, trophic changes, numbness, tingling, cramps, and pain. Loss of the plantar metatarsal fat pad predisposes to ulceration in relation to the existing bony deformities of the foot.

Hyperkeratotic lesions form as space replacements and provide a focus for ulceration due to increased pressure on the soft tissues and an associated localized avascularity from direct pressure and counterpressure. Tendon contractures and claw toes (hammer toes) are common. A warm foot with pulsations in an elderly diabetic patient with neuropathy is not uncommon. When ulceration is present, the base is usually covered by keratosis that retards and frequently prevents healing. Necrosis is related to infection, with eventual occlusion and gangrene. Footdrop and loss of position sense are usually present. Pretibial lesions are indicative of this change, as well as microvascular infarction. Arthropathy gives rise to deformity, altered gait patterns, and a higher risk for ulceration and limb loss.

Radiographic findings in the feet of elderly diabetic patients usually include thin trabecular patterns, decalcification, joint position changes, osteophytic formation, osteolysis, deformities, and osteoporosis.

Management begins with a reduction of local trauma by the use of orthotics, shoe modifications, and specialized footwear; efforts to maximize weight diffusion and weight dispersion; vasomodifiers; exercise; and local debridement of the ulcerative site. Asymptomatic elderly patients with diabetes should be evaluated at least twice a year to prevent and manage foot problems.

ARTHUR E. HELFAND

See Also
Diabetes: Management
Gait Disturbances
Pressure Ulcer Prevention and Treatment
Pruritis (Itching)

Internet Key Words
Orthopedics
Podiatry

Internet Resource
General foot disorders; American Podiatric Medical Association
http://www.apma.org/foot.html

FRACTURES

Due to decreased bone mass, increased bone fragility, and increased likelihood of falls, the number of fractures in the elderly increases exponentially with age. Much has been learned about osteoporosis over the last several decades, and there are now pharmacological interventions that will slow if not reverse the bone loss associated with aging. Yet falls continue to occur in this patient population, resulting in fractures. Fall prevention programs have been effective in reducing the number of falls, but falls and thus fractures have not been eliminated.

The types of fractures that occur in the elderly are distinct, and they are generally the result of low-energy falls on osteoporotic bone. Fractures in older people generally occur in metaphyseal trabecular bone, whereas fractures in younger people generally occur in diaphyseal cortical bone as the result of high-energy trauma. In the upper extremity, these fractures are found in the distal radius and proximal humerus and around the elbow. In the axial skeleton, vertebral compression fractures are common, and in the lower extremities, hip fractures (proximal femur) and tibial plateau fractures are common. The prevalence of these fractures differs, as does their treatment, morbidity, nursing requirements, and outcome. The treatment of these patients requires an interdisciplinary approach.

Upper Extremity Fractures

Distal Radius

One of the most common upper extremity fracture in the elderly is the distal radius fracture, or Colles' fracture. These fractures commonly occur as a result of falling onto an outstretched hand. There is usually a marked deformity of the wrist, accompanied by significant swelling. Initially, these fractures are treated with reduction of the bone fragments and placement of a plaster splint. If the swelling is not excessive, a short arm cast may be applied immediately after reduction. If reduction of the fracture fragments is not adequate based on the radiographic findings, the patient's age, and physiological function, operative reduction and stabilization of the fracture may be required. Operative stabilization can be achieved with internal or external fixation or both. In the elderly, however, operative intervention is relatively uncommon, and patients rarely require hospitalization for an isolated distal radius fracture. Thus, the majority of these fractures are managed in the outpatient setting, with initial emphasis on elevation of the affected extremity.

At about 10 to 14 days after fracture reduction, the fracture is usually checked radiographically. Although loss of reduction is possible, necessitating operative intervention, the majority of these fractures remain adequately reduced. The splint is removed, and a short arm cast is applied. The extremity remains casted for an additional 4 to 5 weeks.

When adequate healing has occurred—determined radiograpically as well as clinically—the cast is removed. A removable wrist splint is applied, and occupational therapy is begun to regain motion and strength of the wrist. Prior to this, the only therapy is range-of-motion exercises of the fingers and elbow. Although some residual deformity is common after these injuries, the majority of these fractures heal well, with minimal functional deficit.

Elbow

Fractures of the elbow are usually the result of falls directly onto the elbow. Either the olecranon or the distal supracondylar humerus is fractured. Many of these fractures are displaced and require immediate operative intervention with open reduction and internal fixation. Restoration of function depends of the quality of the reduction and the strength of the fixation. If the joint surfaces have been reduced and there is good fixation of metal to bone, a good functional outcome can be expected. Frequently, however, fixation of the screws and plates to osteoporotic bone is less than optimal, and the elbow must be immobilized longer than usual, resulting in stiffness and a compromised functional outcome. If there is significant comminution of the fracture, so that the joint cannot be restored, total elbow replacement may be necessary.

Proximal Humerus

After distal radius fractures, fractures of the proximal humerus are the second most common fracture of the upper extremity. These fractures also usually occur as a result of falling onto an outstretched hand. Fortunately, the majority of these fractures require little intervention by the orthopedic surgeon and can be treated by placing the arm in a sling to immobilize the shoulder. The duration of immobilization is variable, but with nondisplaced or minimally displaced fractures, only a few weeks of stringent immobilization is necessary. During this early postfracture period, the patient remains quite uncomfortable and often can sleep only in an upright position. Early pendulum exercises can usually be initiated, followed by more formal occupational therapy with the use of overhead pulleys. Generally, these fractures heal well, but restoration of a normal range of motion is problematic, especially regaining active motion of the shoulder with overhead activity. This limitation in motion is caused not by the bony injury but rather by the injury to the rotator cuff of the shoulder.

More comminuted or displaced fractures of the proximal humerus require operative intervention by either stabilizing the fracture fragments with metal hardware or replacing the humeral head with a hemiarthroplasty. The advantage of fixing the fracture operatively is that early mobilization of the shoulder can be instituted, improving the functional outcome. Although replacement of the humeral head with a prosthesis seems like a good solution for the treatment of severally comminuted osteoporotic fractures, it is not, because the rotator cuff has also been severely injured. Thus, the functional outcome of these patients is quite poor, with few patients achieving active motion of the shoulder above that level.

Vertebral Compression Fractures

Although not considered a significant clinical problem from an orthopedic standpoint, since they rarely require surgical intervention, vertebral compression fractures cause significant morbidity. These fractures may be the result of a fall, but they can also be the result of merely lifting a heavy object. Frequently, the patient does not seek medical attention, and in fact, little can be done for these patients other than prescribing analgesics and early mobilization. If the pain is severe or if it persists, the use of a supporting brace can be beneficial. Only if there is severe deformity and significant neurological compromise is operative treatment indicated. If surgery is performed, it is technically difficult to achieve good fixation in osteoporotic bone. Recently, the use of bone cement to augment the compressed vertebral body has been advocated, but this treatment remains experimental.

Pelvic and Acetabular Fractures

Whereas complex, unstable pelvic fractures are generally the result of high-energy trauma in younger patients, the elderly commonly sustain nondisplaced or minimally displaced fractures of the pubis rami. These fractures never require operative treatment, but they can be debilitating injuries that severely limit the patient's ability to ambulate. Despite the pain associated with these injuries, patients should be mobilized as quickly as possible, with weight bearing as tolerated on the affected side. Healing of these fractures is often prolonged, with symptoms generally lasting several months.

Acetabular fractures are rare in the elderly, but when they occur, treatment can be difficult. Minimally displaced fractures can be treated with non–weight bearing on the affected extremity; however, the treatment of displaced fractures remains controversial. It is not clear whether it is better to fix these displaced fractures operatively or to allow them to heal in a malreduced position and perform a delayed total hip arthroplasty. The decision depends on many factors, including the nature of the fracture and the physiological status of the patient.

Lower Extremity Fractures

Ankle

Fractures about the ankle are relatively uncommon in the elderly, occurring more commonly in the younger, more active geriatric population. They are often the result of a twisting injury, and fracture of the distal fibula and tibia may be associated with dislocation of the ankle joint. Operative management is usually necessary, and fixation of the fragments can be difficult due to the comminution and poor-quality bone. Non–weight bearing on the affected ankle is required postoperatively. Stiffness of the ankle joint can be avoided with early range-of-motion exercises. Once healing has occurred, balance and proprioception training should be implemented.

Tibial Plateau

Fractures of the proximal tibia are the result of direct blows to the knee, usually sustained during a fall. Most often, the lateral tibial plateau is involved, and the fracture extends into the articular surface. In young patients, absolute anatomical reduction of these fractures is mandated to avoid degenerative arthritis, but in the elderly, minimally displaced fractures can often be treated conservatively with a hinged knee brace and non–weight bearing on the affected leg. Since most of the load through the knee when walking or standing is through the medial compartment, small articular displacements of the lateral tibial plateau are relatively well tolerated.

However, if there is a large articular step-off or instability of the knee, operative treatment is necessary. As with most geriatric fractures, operative treatment can be technically challenging, since obtaining fixation of the screws in poor-quality bone is difficult. In addition, a bone graft is often needed, and the bone from the patient's iliac crest is often of poor quality as well. Thus, synthetic bone grafting materials must often be used. Failure of the operative fixation is not uncommon, with some collapse and angular deformity of the knee. As with tibial plateau fractures treated nonoperatively, even those fractures with significant deformity do reasonably well clinically. If pain in either group remains or becomes debilitating, a total knee replacement can be performed.

Hip

The morbidity and mortality associated with hip fractures are significant problems for the elderly. The treatment of the orthopedic injury and the subsequent care and rehabilitation of the patient place a heavy burden on the health care system. The majority of hip fractures occur in elderly patients with multiple concomitant medical problems. As the elderly population increases, so will the number of hip fractures and the cost of treating them. It is projected that by the year 2040, the number of hip fractures in the United States will double, to more than 500,000. Since the majority of patients sustaining hip fractures require hospitalization, the projected cost of their treatment and rehabilitation will exceed $16 billion.

Hip fractures are the result of a fall, often directly on the hip region itself. There are two distinct types of hip fractures: femoral neck fractures and intertrochanteric femur fractures. Surgical management with early mobilization has become the treatment of choice for most hip fracture patients regardless of the type. However, the surgical management of the two types of hip fractures differs significantly because of the anatomy in this region.

Intertrochanteric fractures occur through a highly vascular trabecular bone bed and, as a result, will heal if the bone fragments are reduced and fixed appropriately. This is done operatively under radiographic control on a special fracture table, and the bone fragments are fixed with a large lag screw and a side plate on the lateral femur. Timing of the surgery is not critical, but the patient's medical problems must be addressed before proceeding. Postoperatively, early mobilization is paramount, and the patient is taken out of bed on the first postoperative day. Ideally, most of these fractures should be treated with minimal weight bearing on the affected side, but elderly patients often bear weight on the extremity when the pain is tolerable. This may lead to some shortening of the extremity and deformity of the fracture, but healing of the fracture is rarely compromised. In fact, in patients whose medical condition is deemed too poor for surgery, early mobilization to a chair can be instituted, and the fracture will heal, albeit with shortening and some deformity. Rehabilitation of patients with intertrochanteric hip fractures tends to be a little slower than for those with femoral neck fractures. The reason for this is unclear.

Femoral neck fractures occur through the highly osteoporotic cortical bone of the femoral neck. There is little if any trabecular bone in this region. In addition, the blood supply to this region and thus to the femoral head is extremely tenuous. Thus, the treatment for these fractures is different from that for intertrochanteric fractures. Nondis-

placed or minimally displaced fractures can be treated with fixation of the head to the neck in situ with cannulated screws under radiographic control. Even in these relatively minor fractures, there is a 10% chance that the blood supply to the femoral head will be significantly compromised, resulting in osteonecrosis or death of the bone of the femoral head. Postoperatively, patients should put only minimal weight on the affected side, but patients often bear weight as tolerated. There is no risk of dislocation.

Although displaced fractures of the femoral head can be reduced and fixed with cannulated screws, just like nondisplaced fractures, the incidence of osteonecrosis is high (30% to 40%). The standard of care in the United States is to replace the head with a prosthesis, known as a hemiarthroplasty. Although a total hip replacement may seem more appropriate, the dislocation rate has historically been high if the socket is replaced at the same time. Dislocation of a hemiarthroplasty is rare, but it is possible; therefore, patients who have had a hemiarthroplasty should adhere to total hip precautions to prevent the likelihood of dislocation. Thus, patients should avoid excessive flexion and internal rotation of the hip. While in bed, a pillow should be placed between the legs. Alternatively, a knee immobilizer can be placed on the knee, since it is difficult to dislocate a hip if the knee is prevented from flexing. In addition, patients should avoid sleeping on the affected side and avoid low chairs and toilet seats. Since most of these prostheses are cemented into the femur, patients can safely place all their weight on the affected extremity.

Although the orthopedic surgeon may feel that the most important part of hip fracture treatment is the operative intervention, the functional outcome of the patient is equally if not more dependent on good nursing and rehabilitation. Aggressive and long-term physical therapy has proved beneficial in restoring ambulatory function, but despite good postoperative care, patients on average lose a level of function after sustaining a hip fracture. The majority of patients do, however, return to the residences they occupied before their falls. Achievement of this goal requires an interdisciplinary team working to optimize the patient's return to premorbid function.

Prevention of fractures in the elderly, by either improving bone quality or preventing falls, clearly remains an appropriate goal for health care workers caring for the elderly, but fractures will still occur. Thus, understanding the types of fractures the elderly sustain, how they are treated, and their outcomes is important for anyone caring for geriatric patients.

MATHIAS BOSTROM

See Also
Environmental Modifications
Falls Prevention
Hip Fractures
Nutritional Assessment
Osteoporosis

Internet Key Words
Distal Radius Fractures
Femoral Neck Fractures
Fractures
Hip Fractures
Intertrochanteric Fractures
Rehabilitation

Internet Resources
American Academy of Orthopaedic Surgeons, Fractures
http://www.aaos.org/wordhtml/pat_educ/ fracture.htm

Falls and Hip Fractures Among Older Adults, Center for Disease Control and Prevention
http://www.cdc.gov/ncipc/factsheets/falls.htm

FUNCTIONAL ASSESSMENT
See
Multidimensional Functional Assessment: Overview

FUTURE OF CARE

Global aging is the most critical long-term issue facing the human race. Only the deterioration of the environment is as novel a threat to the quality of life of human beings. Wars, epidemics, famines, and natural disasters have happened and will happen again in the future. A world filled with more older people and fewer younger ones will create opportunities for positive individual growth and cultural evolution, although the challenges will be daunting. How we choose to care for the elderly in the future will affect not only our future selves but also our children and their children yet to come.

Planning for the future is a talent that is well developed phylogenetically in human beings, but it will become more important in the future. The evolution of health care systems and our general approach to politics and other activities will be a balance between preserving what has worked in the past and developing new concepts and behaviors that will work in the future (Postrel, 1999). The very concept of health and disease will continue to change and influence how we establish both informal and formal systems for providing care (Jobst, Shostak, & Whitehouse, 1999; Whitehouse, Maurer, & Ballenger, 2000). Broad conceptions of health as biopsychosocial and even spiritual well-being suggest that we should look at similarly broad concepts such as quality of life as the desired outcome of health care. Lessons of the past and present from our own and from culturally different health care systems need to be considered carefully, as fundamental changes are occurring in our conceptions of health, disease, and health care systems. The popularity of complementary and alternative medicine (CAM) suggests that ill persons are seeking nontraditional health concepts and practices. Although not enough CAM is evidence-based (much of allopathic medicine is not evidence-based either), the concepts of personal control, environmental and community health, and holism are attractive to many. Human beings' remarkable flexibility in terms of conceptualizing health and protecting it in the past should serve as a positive model as we face the challenges ahead.

Biomedical ethics should be an important focal point for social discussion, particularly if this field broadens its scope beyond the philosophical analysis of the implications of medical technology to explore the relationships between health and other values (Potter, 1971). Further research on health care systems is an important aspect of preparing for the future. However, the creation of volumes of data to devise new, expensive technologies is not the only kind of new knowledge we need. Attending to the integration of what we already know and the development of processes for making wise social decisions, often with incomplete information, will be important goals for scholarship and practice.

General Social Issues

In considering the future of care, one must ask the question: Who will we be caring for in the future? The simple answer is that we will be caring for more older people who have chronic diseases. We can also expect to be caring for them with sophisticated and powerful technologies provided by the revolution in molecular biology. The promise that we will be able to understand and manipulate our genetic makeup and to cure diseases such as Alzheimer's disease seems exaggerated and more likely to occur in the distant future, if at all.

Along with the aging of the population, a second major, long-term global trend is the deterioration of the environment (Orr, 1994). Environmental factors will likely cause more health problems for the elderly and others through the poor quality or inadequate quantity of water, air, or food. Public health will become even more important than in the past. The science of ecology—conceptually superordinate but politically weaker than molecular biology in medicine—should receive more emphasis because it is critical to our health improvement efforts.

How many resources will society allocate to the future care of the elderly? Questions about how much informal care is and should be provided by families and others and how much formal care should be provided by the health care system will continue to be answered in different ways in different cultures. The political will to support formal health care systems for the elderly will be challenged by the need to support other initiatives, such as improvement of the environment and the health and education of children. Societies may be chal-

lenged to ensure that the resources needed to care for older individuals do not outweigh the societal commitment to nurture younger generations. Intergenerational conflicts over resources can become a political issue at a time of resource scarcity; the total amount of resources committed to caring for the elderly and the young will depend on the economic well-being of the country (Friedland & Summer, 1999). Yet economic development depends, in part, on the educational level of children. If environmental deterioration continues, resources to support the economy will diminish. New approaches to modifying current conceptions of capitalism, including attention to natural resources in the economic equation—natural capitalism, for example—will be needed (Orr, 1994).

Information systems will be an important part of the world's future in relation to the economy and education. It remains to be seen whether the revolution in the distribution of knowledge and the associated shifts in social and economic power will benefit individuals within and among different societies. Information systems have the potential to diminish the digital divide between the haves and have nots (Goodman, 1998). Many of these general social issues, such as the state of the economy and the environment, the political issues around resource allocation, and the development of information systems, will directly and indirectly affect health care systems.

Health System Issues

The major forces driving health care system change will continue to be population aging and chronic diseases. Dementia is not only a prototypical chronic disease but also one of the most common. It adds to the complexity of providing care, in that affected individuals have an impaired decision-making capacity, which means that other people must make health care and other decisions for them (Whitehouse et al., 2000).

To address the growing number of older individuals with dementia and other chronic diseases, better integration of acute and chronic care systems is needed. The National Chronic Care Consortium is an umbrella organization supporting new models of coordinated care. One of the fundamental difficulties with health care systems in the United States and other developed countries is the emphasis on the treatment of acute illnesses in hospitals and doctors' offices. Most of the investment in the future of health care has come from government and industry for the development of new, more effective, and more expensive technologies. The future will require the development of biological approaches to health care problems, as well as psychosocial innovations. The creation of long-term-care systems, particularly those that are community based, will need to be better integrated financially and technologically to address increasingly diverse health care needs.

Evidence to justify social investment in medical and surgical treatments should be sought, and the search should receive sufficient budgetary support. Biological and psychosocial interventions should be challenged to demonstrate through pharmacoeconomic impact studies, such as cost utility analysis, that they improve the quality of life and meet people's health care needs. However, our expectations should not be unreasonably high in geriatric patients who suffer from multiple chronic conditions. Therapeutic goals may be less clear than in single-disease situations. Efforts should continue to ensure that results of outcome studies are incorporated into practice. End-of-life care is a component of the health care system that needs further development. We need to be sure that such care reflects the values of individuals who are dying. Our desire to prevent death must be balanced by our concern that quality of life be preserved to the end.

It is said that the best way to predict the future is to create it. Therefore, it is essential that all health care professionals, particularly those working in geriatrics, educate their patients and communities about the challenges ahead. If we continue to focus on biologically dominated and technologically oriented efforts to cure the chronic diseases of the elderly, out of proportion to the efforts needed to improve the health care system overall, the quality of life of older individuals may suffer. Aging is a worldwide phenomenon. We need to share the creation of health care systems that will serve current and future generations of older individuals, as well as the younger people who will care for their elders and who will eventually need such care themselves.

PETER J. WHITEHOUSE

Internet Key Words
Aging
Care Systems
Future: Ethical Issues
Social Issues

Internet Resources
American Association of Retired Persons
http://www.aarp.org

American Geriatric Society
http://www.american.geriatrics.org

Gerontological Society of America
http://www.geron.org

National Chronic Care Consortium
http://www.nccconline.org

National Institute of Aging
http://www.nih.gov/nia

G

GAIT ASSESSMENT INSTRUMENTS

Evaluation of balance or movement disorders requires a thorough history and comprehensive physical examination. In particular, the patient history should focus on use of alcohol, benzodiazepines, neuroleptic agents, antihypertensives, and vasodilators. A complete physical examination, with emphasis on the musculoskeletal and peripheral and central nervous systems, as well as direct observation of gait, is often required to make a diagnosis. The physical examination must include careful inspection of the lower limbs, including the feet, and evaluation of proprioception (position sense), vibratory sense, and tendon reflexes. Muscles should be evaluated for evidence of atrophy and weakness. The exam should also include an evaluation of cognitive function and mood. When the diagnosis is not apparent, focused laboratory, electrodiagnostic, and imaging studies may be helpful.

Gait abnormalities associated with focal muscle weakness produce a characteristic pattern of movement during the gait cycle. A weak ankle dorsiflexor can be observed on physical examination, but it can often be heard as well, as the foot "slaps" to the ground at heel strike. More pronounced weakness in this same muscle group produces a distinct steppage gait. Similarly, impaired knee extension, secondary to weak quadriceps muscles, can manifest as a "back knee" or genu recurvatum, followed by a "lurching" forward of the trunk to keep the line of gravity in front of the knee to prevent buckling. Structural abnormalities, such as joint ankylosis and leg length discrepancies, can be easily identified during the musculoskeletal exam.

Several neurological diseases likewise produce characteristic abnormalities. A flexed-forward, flat-footed, festinating gait coupled with neurological findings of rigidity, bradykinesia, masked facies, and resting tremor is classic for Parkinson's disease. Less specifically, a frontal gait abnormality suggests central nervous system pathology. A sensory ataxic gait characterized by a wide-based, high-stepping or stamping walk, in association with loss of vibration and position sense, loss of reflexes, and a positive Romberg's sign, is characteristic of a gait disorder secondary to peripheral neuropathy.

In subjective gait analysis, clinicians must consider the phase of gait (weight acceptance, single limb support, or limb advancement), as well as the components of gait and the body parts used, such as heel strike, toe off, terminal extension of the knee, hip flexion and extension, and automatic movements. Each deficit is rated separately on a scale of 0 (normal gait) to 10 (severely abnormal).

The Tinetti Gait and Balance Measure (Tinetti, 1986) is a valid and reliable test of balance and gait abnormalities and is easily done in any setting. Balance maneuvers include nine positions and position changes that stress stability. Assessment includes rating the performance of eight activities in a serial fashion using simple criteria. The better a subject's performance, the higher the score. The maximum score for balance is 16 and for gait is 12, yielding a maximum mobility score of 28.

The "Get up and Go" Test (Mathias, Nayak, & Isaacs, 1986), performed in less than two minutes, examines an individual's ability to stand up from a chair, walk 10 feet, turn, walk back, and sit down again. Chair transfers (without the aid of armrests) are a reliable test of quadriceps and gluteal muscle strength and balance. During the walking and turning maneuvers, body posture, upper extremity movements, gait initiation, step length and height, step continuity and symmetry, width of base of support, walking velocity, deviation of path, and degree of sway or unsteadiness are observed and documented. Performance is rated on a scale of 1 to 5. No evidence of risk of falling is rated 1, moderately abnormal is rated 2, mildly abnormal is rated 3, very slightly abnormal is rated 4, and severely abnormal with risk of falling is rated 5.

The patient's ability to maintain balance over bases of support should also be evaluated: parallel, semitandem, tandem, and single-leg stance, progressing from the least difficult to the most difficult measures of balance. A positive Romberg's sign is indicative of altered proprioceptive and vestibular function, and a sternal nudge test (estimation of postural competence) should be performed.

Gait analysis laboratories can quantify movement through the gait cycle. The data collected can be used to guide treatment strategies. Vestibular and balance laboratories can measure postural sway and righting reflexes. Computerized posturography may be helpful in determining the relative contribution of visual, vestibular, and proprioceptive abnormalities to postural instability (Maki, Holliday, & Topper, 1994) and explain why the patient responds in a particular way to postural challenges.

Tinetti Balance and Gait Evaluation

Balance: Subject is seated in a hard, armless chair. The following maneuvers are tested:

- Sitting balance
    0 = leans or slides in chair
    1 = steady, safe
- Arise
    0 = unable without help
    1 = able but uses arms to help
    2 = able without use of arms
- Attempts to arise
    0 = unable without help
    1 = able but requires more than one attempt
    2 = able to arise with one attempt
- Immediate standing balance (first 5 seconds)
    0 = unsteady (stagger, moves feet, marked trunk sway)
    1 = steady, but uses walker or cane or grabs other object for support
    2 = steady without walker or cane or other support
- Standing balance
    0 = unsteady

    1 = steady, but wide stance (medial heels >4 inches apart) or uses cane or walker or other support
    2 = narrow stance without support
- Nudge (subject at maximum position with feet as close together as possible; examiner pushes lightly)
    0 = begins to fall
    1 = staggers, grabs, but catches self
    2 = steady
- Eyes closed at maximum position
    0 = unsteady
    1 = steady
- Turn 360 degrees
    0 = discontinuous steps
    1 = continuous steps
    0 = unsteady (grabs, staggers)
    1 = steady
- Sit down
    0 = unsafe (misjudges distance, falls into chair)
    1 = uses arm or not a smooth motion
    2 = safe, smooth motion
    ____/16 Balance Score

Gait: Subject stands with examiner, walks down a hallway or across the room, first at his or her usual pace, then back at a rapid but safe pace (using usual walking aids, such as a cane or walker).

- Initiation of gait (immediately after told to go)
    0 = any hesitancy or multiple attempts to start
    1 = no hesitancy
- Step length and height (right foot swing)
    0 = does not pass left foot with step
    1 = passes left foot
    0 = right foot does not clear floor completely with step
    1 = right foot completely clears floor
- Step length and height (left foot swing)
    0 = does not pass right foot with step
    1 = passes right foot
    0 = left foot does not clear floor completely with step
    1 = left foot completely clears floor
- Step symmetry
    0 = right and left step lengths not equal (esti-

mate)

1 = right and left step lengths appear equal

- Step continuity

    0 = stopping or discontinuity between steps

    1 = steps appear continuous

- Path (estimated in relation to floor tiles, 12 inches wide; observe excursion of one foot over about 10 feet).

    0 = marked deviation

    1 = mild to moderate deviation or uses a walking aid

    2 = straight without a walking aid

- Trunk

    0 = marked sway or uses a walking aid

    1 = no sway, but flexion of knees or back or spreads arms out while walking

    2 = no sway, no flexion, no use of arms, and no walking aid

- Walk stance

    0 = heels apart

    1 = heels almost touching while walking

    /12   Gait Score

    /28   Total Mobility Score (Balance + Gait)

BARBARA RESNICK
MICHAEL CORCORAN

See Also

Gait Disturbances

Internet Key Words

Ambulation

Balance

Functional Status

Gait Analysis

Geriatric Assessment

Internet Resources

Neurological examination, University of Florida

*http://www.medinof.ufl.edu/year1/bcs/clist/
neuro.htm*

Practical Functional Assessment in Older Persons, Mayo Clinic of Rodchester, Geriatric Medicine Division

http://www.mayo.edu/geriatrics-rst PFA.html

GAIT DISTURBANCES

Changes in gait commonly accompany aging and can be indicative of health and biological age. Although up to 85% of adults age 65 to 69 report no difficulty in walking, only 66% of individuals between 80 and 84 years of age and 51% of those over 85 years of age report no difficulty (U.S. Department of Health and Human Services, 1996). Of noninstitutionalized older adults with difficulty walking, 8% to 19% require the assistance of another person or special equipment.

Unfortunately, many older adults and health care providers accept gait disorders and decreased mobility as a normal age-related changes. Clinicians working with older adults must aggressively evaluate older patients for these problems, identify potentially treatable conditions, and provide the patient and family with appropriate interventions to compensate for changes.

Normal Age-Related Changes in Gait

Gait speed declines 0.2% per year up to age 63, after which it declines 1.6% per year (Dobbs, Charlett, & Bowes, 1993). Other characteristics of gait that change with aging include decline in step length, stride length, and ankle range of motion; decreased vertical and increased horizontal head excursions; decreased spinal rotation; decreased arm swing; increased length of double-support phase of walking; and reduced propulsive force generalized at the push-off phase.

The combination of decreased sensory input, slowed motor responses, and musculoskeletal limitations leads to increased unsteadiness or postural sway under both static and dynamic conditions. Older adults compensate for changes by using sensory input to augment proprioceptive loss.

Pathological Gait Disorders

Several common pathological gait disorders in older adults (see the accompanying table) can occur singly or in combination. Abnormal patterns of

Gait Disorders

Type of Gait	Description
Frontal lobe gait	Wide base of support Slightly flexed posture Small, shuffling, hesitant steps Poor initiation of gait; slipping clutch syndrome Turns by pivoting both feet in a small circle Cannot control changes in base of support
Sensory ataxic gait	Wide-based stance; foot-stamping walk High step/stamping walk Heel touches first, then foot stamps Visual input used to ambulate Positive Romberg's sign
Cerebellar ataxic gait	Wide-based stance Small, irregular, unsteady steps Drunken veering and lurching Impaired trunk control Difficulty with tandem gait En bloc turning
Spastic gait	Swings affected leg slowly in outward arc; circumduction of the leg Legs trace a semicircle when walking Feet scrape the ground Scissoring occurs Short steps Narrow base
Spastic paraparesis	Legs move slowly and stiffly Short, labored steps with decreased hip and knee movement (bilateral circumduction) Toes scrape the ground Scissoring occurs Short steps Narrow base
Steppage gait	Feet are lifted high off the ground to prevent scraping toes Toes hit first, then heels Head is down to observe foot placement
Peripheral vestibular imbalance	Unsteady gait

Gait Disorders (continued)

Type of Gait	Description
Antalgic and gonalgic gait	Reluctant to put weight on the joint Heel strike avoided on affected foot Push-off avoided Decreased stance and swing phases of gait Decreased walking velocity Knee and foot flexed Decreased hip and knee extension Limp due to leg length discrepancy
Podalgic gait	Pain with ambulation Toe contact occurs for three-quarters of the gait cycle
Dementia-related gait	Decreased walking speed Decreased step length Increased double-support time Increased step-to-step variability Increased postural sway Flexed posture Apraxic gait
Festinating gait	Symmetrical rapid shuffling of feet Trunk bent forward; hips and knees flexed Difficulty stepping
Parkinsonian gait	Festination Marche à petits pas; short, flat-footed shuffles Delayed gait initiation Body moves forward before feet Freezing Wide stance En bloc turning Loss of postural control Retropulsion; falls back in one piece like a log Propulsion
Waddling gait	Lateral trunk movement away from the foot, with exaggerated rotation of the pelvis and rolling of the hips Difficulty with stairs and chair rise
Vestibular ataxic gait	Broad based, with frequent sidestepping Drift toward the side of vestibular impairment Unsteady
Cautious gait	Flexed posture Decreased stride length Decreased walking speed Low center of gravity Wide base Short steps En bloc turning

movement may occur because of spasticity, weakness, or deformity, or the movement may compensate for other problems, such as dizziness or pain. Alterations in gait can also be compromised by cardiovascular, arthritic, and orthopedic disorders but are probably most commonly influenced by neurological disorders.

Neurological Problems

Sixty-six percent of all strokes result in chronic neurological deficits that impair gait and balance (Alexander, 1996a). Impairment depends on the size, location, and nature of the lesion. A stroke may result in hemiplegia, hemiparesis, or paraparesis that potentiates loss of muscle strength and proprioceptive input on the affected side. In hemiplegia, the affected leg is often stiff, slightly flexed at the hip, and extended at the knee, and the foot is plantar flexed. The affected arm is maintained in a position of flexion at the elbow. Weakness—for example, of the pretibial muscles—causes a footdrop that is most pronounced during the swing phase of gait. A "steppage" gait, with increased proximal lower extremity flexion, is adapted to help clear the toes as the leg advances through swing; toe strike, as opposed to heel strike, occurs at the beginning of stance. In cases of partial weakness, there may be enough strength to dorsiflex the ankle and provide toe clearance during swing, but there will be an audible "slap" of the foot against the ground after heel strike.

Sequelae of stroke may include spasticity and alterations in muscle tone that can produce extensor and flexor synergies, often referred to as spastic hemiplegia. The extension synergy pattern includes hip and knee extension, internal rotation of the hip and plantar flexion, and inversion of the foot (equinovarus). The leg is functionally lengthened, requiring circumduction of the limb to clear the toes. Step length is shorter, stance time is longer, and the normal fluid pattern of gait is lost. Flexor synergy presents as a pattern of hip and knee flexion and ankle dorsiflexion, making ambulation difficult if not impossible. Adductor spasticity is sometimes

seen and can result in a scissoring gait in which the affected extremity is pulled toward the midline.

Parkinson's disease affects 1% to 2% of individuals older than 60 years of age. A deficiency of dopamine, which functions as a neurotransmitter in the striatum and other related brain centers, results in the physical changes, cognitive dysfunction, and depressive symptoms that characterize Parkinson's disease. Typically, the gait pattern associated with Parkinson's disease is flexed forward, flat-footed, and festinating (hastening).

Peripheral neuropathy is a serious neurological complication of many common problems affecting older adults. Diseases of the three components of the peripheral nervous system—the nerve cell body (neuron), the axon, and the myelin sheath enveloping the axon—are described as neuronopathies, axonopathies, and myelinopathies, respectively. Peripheral neuropathies are classified as focal or multifocal and may influence a single nerve (mononeuropathy) or multiple nerves (mononeuropathy multiplex). When small myelineated and unmyelineated fibers are involved, there is decreased pain and temperature sensation. When large myelineated fibers are involved, there is areflexia and decreased vibration and position sense. For these individuals, the gait is ataxic and wide based, and steps are high and stamping. Impaired joint proprioception and abnormal sensory feedback can result in forceful knee extension with a sometimes audible heel tap at heel strike, followed by a foot "stamp."

The degenerative changes of cervical spondylosis commonly result in cord compression in the elderly. Increased pressure on the spinal cord and nerve roots within the cord influences lower extremity function. Gait is typically spastic and has elements of hip adduction, resulting in scissoring, plantar flexion manifested by reduced toe clearance (toes scraping the ground), and shortened, stiff-legged steps. Vitamin B_{12} deficiency can result in cord degeneration and myelopathy, in addition to a peripheral neuropathy.

Mild dementia can be associated with a nonspecific "cautious" gait characterized by a widened base of support, shortened stride length, flexed posture, and slow gait speed; movement is "en bloc."

More severe dementia typically presents as a frontal lobe gait with poor gait initiation, small shuffling steps, and impaired equilibrium.

Orthopedic Problems

Structural abnormalities that can affect gait include limb length discrepancy, joint ankylosis, contractures, and a variety of arthritic and foot conditions. A leg length difference of more than 1.5 inches causes a person to walk on the forefoot of the shorter leg to functionally increase its length. Hip hiking or circumduction occurs on the swing side to compensate. Painful or antalgic joints generally cause a shortened stance time on the affected side in an effort to minimize painful weight bearing. The altered gait pattern in hip pain resembles a compensated Trendelenburg gait, with the shoulder of the affected side dipped laterally during stance. Arthritis and other painful conditions of the feet can cause a shuffling, flat-footed gait, with decreased heel strike and little or no rollover.

Medications

A variety of medications can influence gait. The major drug groups include sedating psychotropic medications such as benzodiazepines, tricyclic antidepressants, phenothiazines, anticonvulsants, salicylates, and antivertigo agents. Any medication that causes orthostatic hypotension can impair balance and thereby alter gait.

Idiopathic Senescent Gait Disorder

In 18% of those in the community with gait disorders, no specific disease-related cause can be identified (Bloem, Haan, & Lagaay, 1992). These individuals are believed to have an idiopathic senescent gait disorder with a gait pattern that is broad based with small steps, diminished arm swing, stooped posture, flexion of the hips and knees, uncertainty and stiffness in turning, occasional difficulty initiating steps, and a tendency to fall. Only when clinical and laboratory examinations fail to reveal any specific cause can the diagnosis of an idiopathic senescent gait disorder be made.

Treatment of Gait Disturbances

A rehabilitation team approach to the treatment of gait abnormalities is commonly employed. Goals of treatment include appropriate pharmacological intervention; improved functional mobility, strength, and endurance; prevention of deformity; and development of a safe, energy-efficient gait pattern. Some gait abnormalities can be substantially improved with medications, physical therapy, or both.

Orthotics and assistive devices can greatly improve gait; they provide additional sensory input and can supplement muscle activity. The handle of an assistive device should be near the level of the greater trochanter. A cane is typically held on the side opposite the affected lower extremity; the cane and affected leg are advanced simultaneously.

Regular exercise, such as walking for 20 minutes at least three times a week or performing resistive exercises that focus on strengthening the lower extremities, can also improve gait and balance (Chandler, 1996). All older adults should be encouraged to exercise, as this will prevent further decline in gait and improve gait speed and safety.

BARBARA RESNICK

See Also
Balance
Falls Prevention
Foot Problems
Gait Assessment Instruments

Internet Key Words
Ambulation
Balance
Functional Performance
Gait

Internet Resource
Neurological examination, University of Florida
http://www.medinof.ufl.edu/year1/bcs/clist/ neuro.htm

GASTROINTESTINAL PHYSIOLOGY

As in the younger population, gastrointestinal (GI) complaints lead to a significant percentage of primary care provider and hospital visits in the elderly population. Age-related changes in the gastrointestinal tract may make "common" diseases such as peptic ulcer or diverticulitis more severe. Gastrointestinal emergencies, such as hemorrhage, may be poorly tolerated by the elderly patient who has less physiologic reserve than a younger patient with the same lesion.

Physiologic Changes in the Gastrointestinal Tract

Esophagus: Barium studies indicate that pharyngeoesophageal function appears to change with age involving pharyngeal hypotonicity, incomplete opening of the cricopharyngeus muscle, pooling of barium in the pyriform sinuses, or aspiration of barium. Diminished amplitude of esophageal peristaltic contractions, delayed esophageal emptying, and incomplete lower esophageal sphincter relaxation make gastroesophageal reflux disease (GERD) of concern in the elderly.

Stomach: The incidence of chronic gastritis and gastric atrophy increases with age. Gastric motility and gastric emptying diminish progressively with age. Atrophy of the gastric mucosa occurs, leading to achlorhydria (Altman, 1990). The common use of both prescription and over-the-counter nonsteroidal anti-inflammatory agents (NSAIDs) by the elderly increases the risk of ulcers.

Small Intestine: Poor nutrition in the elderly is not related solely to diminished intake but also to altered intestinal function. Increased frequency and severity of gastrointestinal infections due to altered intestinal immune function is seen with aging. Carbohydrate malabsorption has been documented using the standard d-xylose test. Fat and protein absorption have not been well studied. Small intestinal motility appears to remain intact with aging.

Colon: Constipation is possibly the most frequent digestive complaint of the elderly. The contri-

bution to constipation of altered colonic motility remains unclear. In addition to inactivity, numerous medications, such as anticholinergics and narcotic analgesics are known to affect colonic motility. Anorectal function is also impaired. Studies have demonstrated decreased rectal elasticity and capacity and significant diminution in internal anal sphincter function with aging in both men and women.

Liver: Liver size and hepatic blood flow both decrease progressively with age. Hepatic parenchymal fibrosis is observed, but this fibrosis appears to have no functional significance. Geriatric patients have two to three times the frequency of untoward drug effects compared with the general population. Several classes of drugs are metabolized by the liver; as this metabolism slows with aging, elevated serum drug levels may be noted with "standard" dosing. Examples of such drugs include seizure medications such as phenytoin or carbamazepime, antidepressants such as fluoxetine, and the anticoagulant warfarin. Observations of pharmacokinetic changes with aging are confounded by comorbidity in elderly patients.

Hiatus Hernia

A hiatus hernia (HH) is a prolapse of a portion of the stomach through the diaphragmatic esophageal hiatus up into the chest cavity. While the presence of a HH may predispose to GERD (gastroesphageal reflux disease), not all patients with HH will actually have acid reflux.

The most common type of HH is a sliding or axial hernia (Type I), accounting for more than 95% of cases. In a paraesophageal hernia (Type 2), accounting for up to 5% of cases, the esophagogastric junction is located in the normal intraabdominal location; a portion of the stomach with the peritoneum is pulled into the thoracic cavity through the phrenoesophageal ligament.

One of the most common abnormalities of the gastrointestinal tract seen in the western world, HH is found in approximately 50% of individuals over age 50, more commonly in women than men. The higher prevalence in the western world may be related to a relatively low fiber diet that may result in increased intraabdominal pressure during defecation.

Many patients are either asymptomatic or may have vague symptoms, including epigastric or substernal pain, postprandial fullness, substernal fullness, nausea, and retching. Bleeding, overt or occult, as a complication of HH has been reported, and iron deficiency anemia can develop.

Diagnosis of HH can be made with radiologic, endoscopic, and manometric examinations. A barium contrast study of the upper gastrointestinal tract is probably the most common method of diagnosis. Radiographic criteria for the diagnosis include herniation of at least 2 cm of the gastric cardia above the hiatus. Most endoscopists consider a proximal translocation of the esophagogastric junction >2 cm above the diaphragmatic hiatus.

Gastroesophageal reflux disease (GERD) is the most common clinical manifestation of a hiatal hernia. Standard behavior modification and pharmacotherapy are effective in treating the vast majority of these patients. Patients are typically treated with either an H2 blocker (ranitidine, famotidine, nizatidine, cimetidine) or a proton-pump inhibitor (omeprazole, lansoprazole, rabeprazole). Surgical intervention is reserved for only 5% to 10% of symptomatic individuals who do not respond to nonsurgical measures, i.e., patients with refractory symptoms and complications including esophageal strictures, ulcerations, and pulmonary complications. Recurrent hemorrhage in a large hiatal hernia is also an indication for surgery.

Endoscopic and radiographic studies suggest that 50% to 94% of patients with GERD have a type I HH (Mittal, 1997). In a subject without HH, each swallow results in clearance of acid from the esophagus. Patients with HH have delayed acid clearance; each swallow results in an initial reflux followed by the clearance. Recent studies indicate that patients with HH have more stress reflux than patients without HH. Patients with a low lower esophageal sphincter (LES) pressure and large hiatal hernia are most susceptible to stress reflux.

Diverticular Disease

Diverticular disease is limited almost entirely to the middle-aged and elderly populations. The prevalence of all manifestations of the disease shows a striking positive correlation with advancing age. Diverticulosis refers to the presence of one or more diverticula. Diverticulitis is an inflammatory condition that involves one or more colonic diverticula and is almost always symptomatic.

The presence of colonic diverticula increases to approximately 30% over age 50, 50% over age 70, and as much as 66% in octogenarians. More than three fourths of all diverticulosis occurs exclusively in the sigmoid colon; the sigmoid is involved in over 90% of all cases. Most (80% to 85%) patients are asymptomatic. In patients who present with symptoms, about 75% present with painful diverticular disease; 25% present with diverticulitis or bleeding.

The predominant symptom in painful diverticular disease is colicky or steady left lower quadrant abdominal pain, usually exacerbated after meals, and improved by the passage of flatus or by a bowel movement. Physical examination is often unremarkable.

Diverticulitis

Inflammation of one or more diverticula and pericolic tissues is the most common complication of diverticulosis. The incidence of diverticulitis increases with the duration of the pre-existing diverticulosis and with the age of the patient. The site of inflammation and perforation is almost always the sigmoid colon. The pain of acute diverticulitis is characteristically severe, persistent, abrupt in onset, and may be accompanied by anorexia, nausea, vomiting, and fever. Physical examination may reveal localized tenderness in the left lower quadrant. Twenty-five percent of patients with diverticulitis have occult rectal bleeding. Laboratory examination typically reveals leukocytosis.

The classical presentation of acute diverticulitis as described may be greatly altered in the elderly.

The symptoms and signs are much less prominent. A high index of suspicion is thus required to diagnose diverticulitis in many elderly patients with acute abdominal or pelvic complaints. The total clinical picture may be muted in elderly patients even when the disease is severe.

Treatment of acute diverticulitis in the elderly patient requires hospitalization. Conservative medical therapy is the first line of treatment and consists of broad-spectrum antibiotics with coverage against gram-positive and gram-negative organisms, a clear liquid diet or complete bowel rest (NPO), as well as IV hydration. With a conservative medical regimen, at least 75% of patients will respond. It is prudent to obtain surgical consultation on admission to the hospital.

Clinical response to medical treatment is usually apparent in 3 to 10 days. In those who respond favorably to medical management, the diet can be advanced from clear liquids to a house diet over a matter of days. Signs of worsening inflammation or lack of response to treatment should be considered indications for surgery to excise the inflamed segment of the colon. A two-stage procedure is necessary in most elderly patients. The first stage involves resection of the colonic segment containing the inflamed diverticulum with a diverting colostomy proximal to the resection. The second stage is takedown of colostomy and anastamosis for intestinal continuity.

Diverticular Bleeding

Bleeding from diverticula is the most common cause of major gastrointestinal tract hemorrhage in the elderly. About 10% to 25% of patients with known diverticular disease will have bleeding per rectum at some point. In contrast to diverticulitis, which almost always occurs in a single diverticulum in the left colon, two thirds of cases of diverticular bleeding occur in the right colon. Classically, the presentation is of sudden, mild, lower abdominal discomfort, rectal urgency, and the subsequent passage of large amounts of maroon or melenic stool. About 80% of patients will stop bleeding

spontaneously (Swaim & Wilson, 1999). The recurrence rate of 10% to 25% increases with each further attack of diverticular bleeding. Localization of bleeding can be made with a 99mTc-tagged red blood cell scan or mesenteric arteriography. Colonoscopy is of limited utility when massive bleeding is occurring because of the difficulty in visualizing the colonic mucosa but can be attempted after a rapid colonic lavage with an oral PEG solution (such as Golytely). In the elderly patient who may be unable to take a large volume lavage orally, the lavage can be administered via nasogastric (NG) tube.

The treatment of diverticular bleeding should begin with conservative medical management (Ferzoco et al., 1998). The patient should be hospitalized, with bed and bowel rest, blood transfusion, and correction of coagulopathy when present. Persistent active bleeding defined by angiography may be treated with intraarterial vasopressin or local embolization. Surgical resection of diverticula-containing areas of the colon must be considered on an elective basis in patients with recurrent hemorrhages. A partial colectomy is performed if the site of bleeding is well established. When the site of bleeding is uncertain, a subtotal colectomy is indicated. In the elderly patient, the risk of surgery, especially if emergent, is considerably greater than in younger patients and thus the decision to undertake surgery must take into account the patient's overall status.

Ischemic Disease of the Bowel

Ischemic bowel diseases are a heterogeneous group of disorders usually seen in elderly individuals. They represent ischemic damage to different portions of the bowel and, therefore, produce a variety of clinical syndromes and outcomes. The most common of these disorders, colonic ischemia or ischemic colitis, has a favorable prognosis.

Ischemic Colitis

Ischemia of the colon is the most common vascular disorder of the intestines in the elderly. More than

90% of individuals with colonic ischemia are over age 60 and have evidence of widespread atherosclerosis. In most cases, no specific cause for ischemia is identified. Factors that may predispose the colon to ischemic injury are an inherently low blood flow and an additional decline in perfusion associated with functional motor activity. Constipation in elderly patients may further exacerbate colonic circulatory inadequacy secondary to the effects of straining at stool on systemic arterial and venous pressure.

Ischemic colitis usually presents with sudden, mild, crampy, left lower quadrant abdominal pain followed by bloody diarrhea or bright red blood per rectum within 24 hours. Usually bleeding is not massive; severe hemorrhage suggests another diagnosis. On physical examination, mild to moderate abdominal tenderness may be elicited over the involved segment of the bowel. The splenic flexure, descending, and sigmoid colon are the most common sites of ischemic injury. If signs of peritoneal irritation persist, this indicates potentially catastrophic transmural necrosis.

Thumbprints, or pseudotumors, which disappear on subsequent plain films of the abdomen, are the major radiographic criteria for the diagnosis of ischemia. Thumbprints represent submucosal and mucosal hemorrhages and are present only in the acute stage of ischemic colitis. Once colonic ischemia is suspected and the patient has no signs of peritonitis, colonoscopy or sigmoidoscopy should be performed within 48 hours of the onset of symptoms. In situations where invasive workup would be poorly tolerated, a CT scan can be performed and will demonstrate diffuse thickening of the colon suggestive of colitis.

Treatment of acute ischemic colitis is based on early diagnosis and continued monitoring. Once gangrene or perforation is ruled out, the patient is treated non-interventionally with close observation. Parenteral fluids are administered and the bowel is placed at rest (the patient is made NPO). Broad-spectrum antibiotics reduce the length and severity of bowel damage. Optimization of cardiac function and withdrawal of medications that cause vasoconstriction, including digitalis and vasoconstrictors,

are important. Colonic distension should be decompressed with a rectal tube, and serial plain films of the abdomen should be obtained. Continued monitoring of hemoglobin, white blood cells, and electrolytes are performed. If clinical deterioration is suggested by increasing abdominal tenderness, fever, signs of peritonitis or paralytic ileus, or if the patient experiences diarrhea or bleeding for more than 2 weeks, surgical resection should be considered. The extent of resection should be guided by the distribution of disease as revealed by preoperative studies rather than by the appearance of the serosal surface or the colon at the time of operation because the mucosal injury may be extensive despite a normal appearance of the serosa.

In more than 50% of patients, colonic ischemia is self-limited, and in most patients, ischemic colitis is a solitary event. Only 5% of patients experience recurrent episodes. Generally, the symptoms resolve in 24 to 48 hours and the colon heals in 2 weeks. In severe disease, complete healing may require up to 6 months. In less than 50% of patients with ischemic colitis, irreversible damage occurs, including gangrene and perforation, segmental ulcerating colitis, and the development of colonic stricture. Hemodynamic support encompassing the avoidance of hypotension through adequate hydration and the elimination of unnecessary antihypertensive medications is the main thrust of care.

Patients with few or no symptoms but who have endoscopic evidence of persistent disease should have a follow-up colonoscopy approximately 6 weeks after initial detection to determine whether the colon is healing. Ischemic strictures without symptoms should be observed, but resection is required for those causing obstruction. Patients with persistent symptoms for more than 2 weeks are at high risk for colonic perforation, and early resection is indicated. As with other forms of vascular disease, lifestyle modification, such as smoking cessation, are strongly encouraged.

GI Bleed

GI bleeding emergencies in the elderly require rapid diagnosis and aggressive intervention because of the wide range of etiologies of bleeding and the significant potential for adverse outcomes. Comorbid conditions and decreased physiologic reserve make persons 65 and older particularly vulnerable to the adverse consequences of acute blood loss. Thus gastrointestinal bleeding in the elderly is a significant clinical challenge. In an acute GI bleed, the primary initial clinical concern should be achieving and maintaining hemodynamic stability.

Common variables implicated in GI bleeding include normal aging, cancer, chronic use of nonsteroidal anti-inflammatory agents (NSAIDs) or other medications, vascular abnormalities, inflammatory bowel disease, GI carcinomas, diverticulosis, peptic ulcers, and hemorrhoids. Cirrhosis arising from chronic alcohol abuse or chronic viral hepatitis can result in bleeding esophageal or gastric varices. Depending on the location in the GI tract of the bleed, a patient may vomit bright red or coffee ground-like material, or pass tarry, malodorous stools, mahogany colored stools, or bright red blood per rectum. Long-term bleeding may result in symptoms of anemia. Persons who have sustained substantial blood volume loss more than 40% of their normal total are likely to exhibit tachycardia, low blood pressure, reduced urine output, and pallor. These patients need immediate volume replacement.

Acute Upper GI Bleeding

As much as 45% of all cases of acute upper GI bleeding occur in patients age 60 and older. Upper GI bleeding results in death among patients age 65 and older four to ten times more often than in younger patients. Older patients are more likely to present with more proximal, larger gastric ulcers that are prone to perforate and slower to heal. Older patients are more likely to rebleed acutely from ulcers and at least one half of patients over age 70 with peptic ulcer experience complications. The three major etiologies are peptic ulcer disease, gastritis, and gastric and esophageal varices. Older patients are more likely to have NSAID-related bleeding lesions, which are probably due to age-

related gastric mucosa susceptibility to NSAID damage rather than to more prevalent NSAID use. The intake of prescription and over-the-counter NSAIDs in older persons is as high as 64% and this population usually underreports NSAID use. In an older patient, emesis of any material exhibiting the texture of coffee grounds or containing blood requires immediate evaluation. Melena in any patient is considered to be a sign of upper GI bleeding until proven otherwise. Concurrent vascular disease increases the risk of cardiac ischemia when GI bleeding is brisk. Initial lab studies are vital in guiding intervention, as renal failure, thrombocytopenia, or medications may compromise the coagulation system in older patients. GI bleeding in patients taking warfarin or ticlopidine can be more rapidly life-threatening than a thrombotic event.

Risk factors for mortality include concurrent NSAID use, patients requiring more than 5 units of packed red blood cell (PRBC) transfusion, peptic ulcer as a cause of bleeding, and comorbidities including renal, cardiac, pulmonary, or liver disease.

Treatment: Placement of a nasogastric (NG) tube is warranted for patients with obvious upper GI bleeding or in whom such bleeding is suspected because it helps ascertain briskness of bleeding, clears the stomach so that endoscopic view is not obscured, and also reduces the likelihood of aspiration. Serious upper GI bleed is not excluded by NG aspirate that is free of blood or coffee grounds material. In 10% to 15% of cases, NG lavage is clear, even in the presence of active bleeding (usually duodenal ulcer bleeding).

Immediate Steps: These include placement of large-bore IV access at one or more sites, transfer to an intensive or step down monitored setting, stat labs (CBC, creatinine, PT, PTT, blood crossmatching, 12 lead EKG to rule out ischemia), nasogastric tube placement, and early upper endoscopy. Emergency upper endoscopy discloses the source of bleeding in up to 95% of patients and offers therapeutic control of bleeding in addition to diagnostic localization of a bleeding lesion. Hemostasis is achieved through different modalities, including electrocautery of a bleeding ulcer or angiodysplasia, sclerotherapy with epinephrine to provide local vasoconstriction or, in the case of esophageal varices, band ligation. Hemostasis through invasive endoscopic therapy is achieved in 90% of patients.

Follow-Up: Outpatient care of patients after upper GI bleeding involves maintenance antisecretory therapy, preferably with a proton pump inhibitor. Patients with gastric ulcer bleeding should be reexamined 6 to 8 weeks after the initial bleeding episode to document healing because of the malignant potential of gastric ulcers. Patients who have bled from esophageal varices need portal antihypertensive therapy and repeat endoscopic banding.

Acute Lower GI Bleeding

The most common cause of major or acute lower GI hemorrhage is bleeding colonic diverticula (Vernava et al., 1997). The most common causes of chronic lower GI bleeding are hemorrhoids or colonic neoplasms. Most patients experiencing acute lower GI bleeding will exhibit bright red rectal bleeding. Rectal bleeding in the older patient is likely due to ischemic colitis, diverticular disease, colon cancer, and arteriovenous malformations. Prior abdominopelvic irradiation can lead to bleeding from radiation colitis years later. Approximately 10% to 15% of all cases of rectal bleeding are attributable to a cause that is proximal to the ligament of Treitz. Benign anorectal disease and post polypectomy bleed also account for lower GI bleeding. Daily fecal blood loss required to cause guaiac-positive brown stool is approximately 20 cc. Normal daily fecal blood loss is approximately 2 cc. Fecal blood loss required to cause melena is at least 50 cc.

Ischemic Colitis: Ischemic colitis is more common in patients with atherosclerotic vascular disease, where it is frequently heralded by crampy lower abdominal pain that resolves spontaneously before painless hematochezia (frank bright red blood per rectum) begins.

Diverticular Disease: Nearly one half of patients with diverticular disease experience episodic painless minor hematochezia; approximately 5%

of these patients will experience hemodynamically significant lower GI bleeding that requires blood transfusion. Bleeding stops spontaneously in about 80% of patients who develop significant lower GI bleeding. Approximately 20% with massive diverticular bleeding will have recurrent hemorrhage. The precise trigger of the bleeding is unclear.

Other Causes: Bleeding from colonic vascular ectasias (angiodysplasias) is highly variable in severity and manifestations. Hematochezia resulting from colon cancer is often accompanied by weight loss, change in bowel habits, change in stool caliber, and anemia. Lower GI bleeding in the presence of active abdominal pain places inflammatory bowel disease in the differential diagnosis. Evaluation of material passed via the patient's rectum (i.e., dark blood mixed with stool, stool coated with red blood, melena) may help determine the site of bleeding, rapidity, and the quantity of blood involved.

Treatment: Endoscopic evaluation can occasionally be useful in identifying the source of bleeding. However, it is often difficult to localize the single "culprit" bleeding diverticulum in a colon that is filled with fresh blood and multiple large diverticulae. If a bleeding diverticulum is localized at the time of colonoscopy, then injection around the area with epinephrine to cause vasoconstriction has been shown to have some success in controlling bleeding. Bleeding colonic (or gastric) angiodysplasias are usually easily controlled during colonoscopy (or endoscopy) through the use of electrocautery applied directly through the colonoscope (or endoscope).

Nuclear scanning using radionuclide-tagged red blood cells can detect sources of colonic bleeding at rates as low as 0.1 cc blood per minute. This modality, which is completely noninvasive and does not cause nephrotoxicity, can be quite helpful as a confirmatory study prior to surgical resection of the putative bleeding site.

Selective mesenteric arteriography can detect hemorrhage at a rate of 0.5 to 1 mL/minute and can provide hemostatic therapy via intraarterial infusion of vasopressin or embolization of the bleeding arterial vessel with gel-foam or microscopic coils. The morbidity of arteriography is considerably less than that of urgent hemicolectomy or subtotal colectomy. However, patients with renal impairment typically cannot undergo studies with contrast dye, and elderly patients, because of their lower glomerular filtration rate, may be at greater risk for renal complications from arteriography.

Older Patients

The elderly are commonly affected by many of the same gastrointestinal disorders that involve younger patients. However, diminished physiologic reserve and polypharmacy often make management of these diseases more difficult in the older population. The approach to these diseases is, in many cases, through interventional procedures. This approach must be tempered by the individual patient's overall performance status, the patient's cardiopulmonary status, other comorbidities, and finally, in the case of life threatening situations—such as hemorrhage—the patient's end-of-life wishes.

K. R. BYJU
ROGER D. MITTY

See Also
Bowel Function
Fecal Incontinence

Internet Resource
American Gastroenterology Association
http://www.gastro.org

GAY AND LESBIAN ELDERS

Multidiverse ethnic, racial, and religious communities are widely accepted as contributing members of society. These communities have well-grounded familial and social roots, allowing supportive outreach to promote socialization among both young and old. Another community that is not as well known consists of older lesbians and gay men, who, as members of a youth-oriented heterosexual society, have the distinction of double minority status based on their age and sexual orientation (Deevey, 1990).

Sexuality is often neglected or hesitantly approached in discussions with the elderly. Reluctance to obtain the sexual history of an older client can result in an incomplete physical, mental, and psychosocial record. Having lived through decades of discrimination, the aged gay person may choose not to disclose his or her sexual orientation, fearing ostracism or violence. This apprehension and silence stem from the oppression of the past, when homosexuality was classified as a mental disorder by the medical and psychiatric professions and same-sex relations were prosecuted as criminal acts (Reid, 1995). A 65-year-old gay man or lesbian was a child when the American Psychiatric Association declared homosexuality a psychiatric illness in 1942. It was not until 1973 that the American Psychiatric Association declassified homosexuality as a category of mental illness; the American Psychological Association shortly followed suit (Singer & Deschamps, 1994). Consensual same-sex sexual relations were felonious in every state until 1961 when Illinois became the first state to eliminate such laws (Wolfson & Mower, 1994). Within a few years, the lesbian and gay civil rights movement was unofficially founded. The patrons of the Stonewall, a gay bar in New York City's Greenwich Village, along with community residents, fought back against police, who raided the bar on the night of June 27, 1968. Prior to the "Stonewall Riot," lesbians and gay men were regularly arrested when socializing in bars under anticontact statutes, which prohibited two men or two women from socializing with the intent to engage in consensual sexual relations. Eighteen states still have antisodomy laws that prohibit lesbians and gay men from consensual intimate expression.

Demographics are difficult to obtain on lesbians and gay men 65 years of age and older. Definitive numbers continue to elude researchers for a variety of reasons. First is the complexity of defining the term homosexual: Does an individual have to engage in same-sex sexual activity to be identified as lesbian or gay, or can it be how one defines oneself regardless of behavior? The social stigma attached to being lesbian or gay is the second obstacle in ascertaining accurate figures, often leading to gross underreporting in many studies, especially those involving face-to-face interviews with researchers (Singer & Deschamps, 1994). The most commonly cited figure for the lesbian and gay population in the United States is 10%. This number is derived from the 1948 landmark study by Alfred Kinsey and his colleagues, which found that 10% of the male participants self-identified as predominantly gay; in a subsequent study, 6% of women self-identified as lesbian. Applying statistics derived from the Kinsey studies and the American Association of Retired Persons (AARP) 1996 census of the entire aged U.S. population (not distinguished by sexual orientation), the estimated number of lesbians and gay men 65 years and older is 2.6 to 3.4 million persons (AARP, 1997; Singer & Deschamps, 1994).

The aging lesbian and gay community is diverse. Like their heterosexual counterparts, gay and lesbian individuals are concerned about physical and mental age-related changes, access to affordable health care, social isolation, and housing and legal issues. However, access safe, sensitive, and nondiscriminatory services is a major issue for gays and lesbians, but not for heterosexuals (Deevey, 1990; Quam & Whitford, 1992; Reid, 1995).

Physical and Mental Health

Older lesbians and gay men are affected by the same physiological and mental changes as the aging heterosexual community. Although limited empirical data exist, population-based studies speculate that potential health concerns are an increased risk of cancer among lesbians and sexually transmitted diseases among gay men (Bachus, 1998; Deevey, 1990). Lesbian health problems include decreased health prevention (e.g., breast self-examinations), increased consumption of alcohol and cigarettes, obesity, sedentary lifestyle, and skepticism toward traditional health care and promotion (Deevey,

1990; Deneberg, 1994). Lacking quantitative research, the Institute of Medicine has urged the federal government to support studies to identify conditions for which lesbians are at an increased risk, hoping to confirm or dispel long-held myths regarding their health (Thompson, 1999).

Older gay men can become infected with the human immunodeficiency virus (HIV) and other sexually transmitted diseases. After the death of a life partner, an older gay man can find himself dating again. Unable to meet other older gay men because of ageism within the gay community, he may turn to a younger male sex worker to satisfy physical and emotional needs. Without knowledge of safer sexual practices, particularly condom use, the risk of contracting a sexually transmitted disease increases. Safer sex counseling must be provided as part of health promotion and maintenance to both men and women.

Successful aging of the gay community postulates the achievement of a positive lesbian or gay identity through adaptation and adjustment. Friend (1991) proposes two distinct and dramatically opposite outcomes in terms of identity formation: (1) positive identity through self-acceptance, or (2) negative self-image acquiescing to societal homophobia. Internalized homophobia or denial may result in self-neglect, such that an older gay man or lesbian refuses to accept needed assistance. Health care providers should obtain a thorough mental health history, including assessment for depression, life satisfaction, and alcohol intake, because of the community's increased incidence of depression and alcohol abuse (Deevey, 1990; Reid, 1995).

The Health Care System

Lesbians and gay men, regardless of their age, are reluctant to seek health care from a system that is often discriminatory and unsympathetic. Members of the gay community continually negotiate their daily lives in an atmosphere that can range from prejudicial to supportive. They are acutely aware of a provider's intonations or actions that suggest danger or safety. Hence, they underutilize orthodox medical treatments and screenings and rely instead on holistic interventions and intracommunity support (Deevey, 1990; Reid, 1995; Stevens, 1994). Homophobic attitudes or comments can be traumatic to an older gay person, who may be reluctant to disclose her or his sexual orientation because of past experiences with antigay violence, medical provider discrimination, religious condemnation, or family and peer rejection (Deevey, 1990).

Cost is also an important factor to assess, because older lesbians and gay men lack the marriage rights and spousal benefits afforded heterosexual couples. Medical coverage is commonly provided through Medicare for those individuals who can afford to purchase additional (gap) coverage. The recent establishment of community health and outreach organizations in major metropolitan communities such as New York City and Boston allow older women and men—who may or may not self-identify as lesbian or gay—access to sympathetic, affordable, community-organized health care.

Social Services

A common concern of the gay community is social isolation (Quam & Whitford, 1992; Reid, 1995). An aging gay person can find herself or himself alone, afraid to participate in mainstream geriatric organizations and senior centers because of potential hostility from heterosexuals who do not accept or understand their sexual orientation. Senior Action in a Gay Environment (SAGE) was founded in 1978 by a group of older gay men and lesbians in New York City to address this gap in the gay community's needs. SAGE provides multiple services and outreach, including case management and social services, a drop-in center, the friendly visitors program, an AIDS and the elderly program, the older lesbian project, and a support group for caregivers of Alzheimer's-afflicted individuals.

Housing

When older lesbians and gay men were asked who would provide care if they became sick or disabled,

more than 70% indicated one of three choices: a nonrelative, a health care worker, or no one at all (Cross, 1999). With limited access to traditional family members as a resource for caregiving, the gay community has begun to explore other housing options through the development of lesbian and gay retirement residences. National lesbian and gay organizations have also proposed an initiative to educate existing senior residential centers and long-term-care facilities about the gay community. Retirement communities and nursing homes are often unaware of lesbian or gay relationships or may choose to ignore their existence. Older lesbians and gay men who live openly, disclosing their sexual orientation to family and friends, express concern that they may have to hide their identity when they relocate to a senior retirement or assisted living community (Bragg, 1999).

Legal Issues

Only a few states legally recognize same-sex domestic relationships (Reid, 1995). If an individual is hospitalized, his or her partner may be denied visitation or the right to make medical decisions. A surviving partner might be evicted from their apartment or home. Durable powers of attorney and advance directives must be discussed with all clients. It is imperative that the aging gay community, particularly those in committed relationships, designate a health care proxy, because state regulations may give precedence to blood relatives over nonblood relations.

Strategies

Health care professionals can meet the needs of the aging gay community by providing a safe and nurturing environment in which clients can disclose the information that they deem appropriate. Providers should listen attentively for subtle responses that can indirectly reveal a client's sexual orientation during an interview. Deevey (1990) suggests that if a practitioner suspects that an older client is gay, providing general support is often more effective than seeking disclosure. Older lesbians and gay men can refer to their sexual orientation by indirect verbal cues, such as "people like us," or they may not disclose their life partners of many years, referring instead to their roommates or dear friends (Deevey, 1990; Reid, 1995). Others refuse to label themselves as lesbian or gay and may be offended when reference is made to their sexuality (Deevey, 1990). By asking open-ended questions, such as Who is most important to you? rather than Are you married, widowed, or divorced? health care providers go beyond traditional definitions of family and explore a client's own perception of available social support (Deevey, 1990).

Practitioners need to examine their own values and beliefs about sexuality. Only when health care team members are educated and comfortable about their own perceptions of sexuality will open communication be safe and forthcoming, allowing sensitive, nondiscriminatory treatment. Practitioners must treat all patients with compassion and acceptance and not allow personal values and beliefs to interfere with providing the most comprehensive and nonjudgmental services. Aging lesbians and gays are an overlooked and underexplored group. Further study is needed to obtain an accurate, comprehensive portrait of the community.

JOSEPH B. NARUS
MICHAEL CALLEN

See Also

Advance Directives
Cancer Treatment
Cultural Assessment
Human Immunodeficiency Virus (HIV) and AIDS
Patient-Provider Relationships

Internet Key Words

Gay
Homosexual
Lesbian
Sexual Orientation

Internet Resources

Gay and Lesbian Association of Retired Persons, Inc.
http://gaylesbianretiring.org

Golden Threads (older lesbians)
http://members.aol.com/goldentred

Lesbian and Gay Aging Issues Network
http://www.asagin.org/lgain.html

Pride Senior Network
http://www.pridesenior.org

Prime Time Worldwide (older gay men)
http://primetimer.com

GENETIC FACTORS IN ALZHEIMER'S DISEASE

Although Alzheimer's disease (AD) was characterized and named in the first decade of the 20th century, and the familial nature of the disease was documented over 50 years ago (Sjogren, Sjogren, & Lindgren, 1952), it is only recently, in the era of molecular genetics, that we have been able to identify specific genes that are responsible for the heritability of this devastating neurodegenerative disease. It is reasonable to suppose that the most effective therapeutic intervention will target the disease *process* rather than the disease *consequences*, and that the earlier such treatment can be introduced, the more effective it will be. Autopsy findings in AD-affected brains reflect the consequences of perhaps two decades of disease, and although these findings may implicate proteins in the disease process, they do not necessarily reveal pathological mechanisms. Through the identification of genetic causative or risk factors for AD, the disease can be investigated at the molecular level, and pathological cellular mechanisms that may be ongoing long before the clinical presentation of AD can be studied.

In multiply affected families in which the disease has an early age of onset (<60 years), the inheritance pattern is that of autosomal dominance, suggesting that single genes are responsible. At such an early age, the clinical picture is less likely to be confused by other dementia diagnoses. In addition, family members who are genetically predisposed will likely live long enough to exhibit AD; therefore, there is clear transmission of the disease within the family. DNA samples from family members allow comparison between affected and unaffected individuals and identification of which genetic material is specific to early-onset AD. Early-onset AD families have been invaluable in Alzheimer's research, enabling the discovery of three genes in which mutations can cause early-onset AD. It was reasoned that the protein products of these genes must play a role in the disease process, and that the impact of the disease-causing mutations on the normal function of the protein would uncover potentially pathogenic processes.

One of the neuropathological hallmarks of AD is deposition of aggregates of the β-amyloid peptide (Aβ) in the brain, as noted by Alzheimer in 1907. Aβ is derived from enzymatic cleavage of the much larger transmembrane β-amyloid precursor protein (βAPP, encoded on chromosome 21). Whether or not Aβ deposition is a cause or a consequence of the disease is still unknown. The largest isoform of βAPP ($βAPP_{770}$) is 770 amino acids in length; the major βAPP isoforms exhibit two common processing pathways. The first, the "nonamyloidogenic pathways," involves cleavage between amino acid codon 687 and 688 (the so-called α-secretase site), precluding the formation of Aβ, whereas in the "amyloidogenic pathway," βAPP is cleaved at either end of Aβ (between 671 and 672 and between 711 and 715, the β- and γ-secretase sites, respectively). Cleavage of βAPP at the α and γ sites produces p3, which is the C-terminal fragment of Aβ. In 1991, a mutation in βAPP was discovered as the first genetic error causing early-onset familial AD (Goate et al., 1991, Online Mendelian Inheritance in MAN [OMIM], p. 104760). This mutation, which results in an amino acid change of valine to isoleucine, occurs close to the C-terminal of the Aβ peptide fragment at βAPP codon 717 and predicts age of disease onset at approximately 55 years. Subsequently, two other mutations to the same amino acid were identified, changing the predicted valine to glycine (OMIM, p. 104760) or phenylalanine (OMIM, p. 104760). Following the identification of these mutations near the γ-secretase cleavage site, a double mutation causing early-onset AD was discovered at codon 670/671 (lysine→methionine,

asparagine→leucine; the so-called Swedish mutation, because it was identified in a Swedish family) at the β-secretase cleavage site (Mullan et al., 1992 OMIM, p. 104760). Additional mutations at the γ-secretase site (codon 715 and 716 valine→methionine and isoleucine→valine, respectively) have also been identified that segregate with early-onset AD.

The position of these mutations predicts their role in cleavage of Aβ from the βAPP protein, and their effects in cultured cells prove this to be the case. Mutations at codon 715 and 717 result in an increase in the ratio of longer to shorter length Aβ (increased $Aβ_{1-42}$: $Aβ_{1-40}$), whereas the mutation at codon 670/671 produces an approximately five-fold increase in total Aβ (OMIM, p. 104760). Thus, it appears that an increased amount of total Aβ or a relative increase in the amount of the longer, more amyloidogenic (aggregable) Aβ is detrimental.

Additional mutations *within* the Aβ sequence that are associated with AD or amyloidogenic conditions appear to directly affect the α-secretase cleavage of βAPP. At codon 693, a glutamate→glutamine change causes hereditary cerebral hemorrhage with amyloidosis–Dutch type (HCHWA-D), whereas an alanine→glycine substitution at codon 692 can cause either HCHWA-D or AD (OMIM, p. 104760). The 693 mutation increases production of Aβ and of p3 fragments, abnormally beginning with codon 689 or 690. The 692 mutation increases the amount of Aβ relative to p3 and also increases the amount of abnormal N-terminal truncated Aβ fragments (OMIM, p. 104760; Watson, Selkoe, & Teplow, 1999).

Although the βAPP mutations typically result in an age of onset in the 50s, they do not account for all early-onset AD cases. Genetic linkage studies in early-onset AD families that did not contain βAPP mutations resulted in the localization of a disease-causing gene to chromosome 14 (OMIM, p. 104311), and in 1995, the Presenilin-1 gene (first known as S182) was cloned (OMIM, p. 104311). Typically, these PS-1 mutations cause an age of onset in the 40s, and unlike the βAPP mutations, they occur throughout the length of the protein. Protein database searches quickly identified the homologous PS-2 gene on chromosome 1 (OMIM, p. 104311; OMIM, p. 104300), within which a couple of mutations have also been found to cause AD, though with a wider range of age of onset than either the βAPP or the PS-1 mutations. The PS proteins are predicted to have six to nine transmembrane domains and are cleaved in the large extracellular loop to form two fragments, whose association appears to be necessary for their normal function. PS mutations influence βAPP processing in a similar manner to the βAPP codon 717 mutations, by increasing the relative production of the longer, more amyloidogenic Aβ fragments (OMIM, p. 104760; OMIM, p. 104311). Therefore, mutations in all three early-onset AD genes modify Aβ production.

Mutations in βAPP or the PS proteins can be readily identified in the laboratory using a DNA sample extracted from blood or saliva. Direct amplification of specific gene fragments, using the polymerase chain reaction, can be followed either by DNA sequencing or by cleavage of the fragment with enzymes that are sensitive to the specific mutations. However, as these mutations are extremely rare and account for only 1% of all Alzheimer's cases, it is meaningful to screen for these mutations only in families exhibiting early-onset AD. Such genetic testing was introduced in Europe shortly after the discovery of the first βAPP mutations, but other than in research settings, the identification of those at risk is at the discretion of the individuals concerned and warrants many ethical and disclosure considerations. In terms of risk for disease, these mutations are "causative" as opposed to "risk factors," in that possession of one of these mutations virtually guarantees AD. Therefore, the value to the general population of identifying these rare mutations is that they highlight the central role of Aβ and βAPP processing in the disease and promote research into the normal and pathological production and function of Aβ.

Although the normal function of Aβ remains unknown, cellular mechanisms through which Aβ is produced and mechanisms through which Aβ may trigger potentially damaging effects present targets for therapeutic intervention. In terms of Aβ

production, the secretase enzymes are targets, since the inhibition of β- and/or γ-secretase cleavage, or the stimulation of α-secretase cleavage, should reduce Aβ production. The β-secretase enzyme was recently identified by looking for the Aβ-producing effects of different gene-expressing clones on cells containing the 670/671 mutation, which were thus primed for enhanced β-secretase cleavage (OMIM, p. 604252). Inhibition of this β-site APP-cleaving enzyme (BACE, chromosome 11) is one way to reduce the deleterious effects of Aβ, but it remains to be seen whether the resulting increased nonamyloidogenic βAPP processing will have damaging consequences. The searches for γ-secretase and for a role for the PS-1 protein have now been combined, with the discovery that PS-1 either *is* γ-secretase or at least is necessary for the enzyme's function. These suggestions come from the observations that in animals or cells deficient in PS-1, βAPP C-terminal fragments accumulate that have N-terminals beginning at the α- or β-secretase cleavage sites, but there is no evidence for γ-secretase cleavage (OMIM, p. 104311). PS-1 also may play a role in development and neuronal differentiation, as it is homologous to developmental proteins in other species, and its absence leads to embryological developmental problems in mice (OMIM, p. 104311).

The cytotoxic effects of Aβ are well reported in the literature. It has been proposed that these occur through the disruption of calcium homeostasis or the balance of reactive oxygen species, or through direct initiation of apoptosis (Iversen, Mortishire-Smith, Pollack, & Shearman, 1995; Mattson & Goodman, 1995; Mattson, Partin, & Begley, 1998). However, the ubiquitous expression of Aβ, plus the fact that it has been highly conserved throughout evolution, suggests that it has a normal physiological role that, in response to a stimulus or its overproduction, becomes pathological. Physiological mechanisms for Aβ have also been examined at the molecular level, revealing involvement in vasoactive, inflammatory, and immune response mechanisms that could, if disrupted, lead to cerebrovascular and neuronal damage (Tan et al, 1999; Thomas, Thomas, McLendon, Sutton, & Mullan, 1996). Whether these mechanisms, identified in vitro, are relevant to the disease process requires confirmation in vivo. The early-onset genetic discoveries enabled the production of valid animal models of AD in which potential disease mechanisms can be investigated at different ages, and therapeutic compounds tested for efficacy.

Human AD-causing mutations introduced into such mice (see below) are transmitted in an autosomally dominant fashion, as in humans; thus, with each litter of mice, both transgenic and non-transgenic control littermates are produced. Transgenic mouse models carrying the $\beta APP670/671$ Swedish mutation, the $\beta APP717Val \rightarrow Phe$ mutation, and certain PS-1 mutations (e.g., methionine-146-leucine or leucine-286-valine) are most frequently cited. The Swedish mice display behavioral and pathological features of AD, developing cognitive and behavioral problems at around 8 to 9 months and from 12 months onward displaying classic amyloid pathology (OMIM, p. 104760). The $\beta APPVal \rightarrow Phe$ mice have a great deal of amyloid pathology from an early age and show some evidence of behavioral problems (OMIM, p. 104760), whereas the PS-1 animals demonstrate an increase in the $A\beta_{1-42}:A\beta_{1-40}$ ratio (OMIM, p. 104311). Manipulation of the genetically engineered mice makes it possible to determine environmental and biochemical factors that enhance or suppress expression of the disease; by breeding them with other genetically modified mice, the role of additional factors related to the mechanism of AD can be investigated.

The biggest risk factor for AD is aging, and in humans, even in the presence of one of the βAPP or PS mutations, 40 or more years elapse before there are any clinical signs of a disease in process. Something in the aging process, possibly hormonal, triggers the effects of these mutations. Susceptibility to damaging or debilitating effects of the aging process is likely to be genetically determined. One way to consider the genetic factors in AD may be that, once triggered, the βAPP and PS mutations actively assist the aging and degenerating process (resulting in an earlier age of onset), as opposed to factors that more passively contribute risk by modifying susceptibility to damage (hence a later age of onset).

The only confirmed risk factor for late-onset AD is the apolipoprotein E (apoE) gene on chromosome 19, which encodes a protein involved in cholesterol transport and is also a risk factor for coronary artery disease. Of the three common forms of the APOE gene ($\epsilon 2$, $\epsilon 3$, and $\epsilon 4$), the $\epsilon 4$ form is usually present at a frequency of about 15%, but in a late-onset AD population, this can be as high as 40% (OMIM, p. 104300 and 107741). Genetic variation at the APOE locus does not cause AD but increases the risk for an earlier age of onset; although gender and ethnicity are modifying variables, APOE is a major risk factor for both genders in all ethnic groups studied. Epidemiological studies have shown that 55% of APOE $\epsilon 4/\epsilon 4$ homozygotes develop AD by age 80, whereas only 27% of $\epsilon 3/\epsilon 4$ heterozygotes and 9% of non-$\epsilon 4$ carriers develop AD by age 85. Comparing against non-$\epsilon 4$ carriers, the odds ratio for AD was 3.2 to 3.7 for $\epsilon 3/\epsilon 4$ individuals and 14.9 to 30.1 for $\epsilon 4/\epsilon 4$ homozygotes (Farrer et al., 1997). The different isoforms of the apoE protein have been shown to respond differently to insult (Miyata & Smith, 1996; Nathan et al., 1994). It has been suggested, therefore, that the mode of action of the apoE protein in AD is through altered repair mechanisms in the brain in response to damage. ApoE has been suggested to be involved in plaque formation (perhaps through $A\beta$ clearance mechanisms), because APOE knockout mice crossed with the $\beta APPVal \rightarrow Phe$ "Games" mouse develop no amyloid deposits in animals up to 22 months of age (OMIM, p. 104760). Interestingly, expression of human APOE $\epsilon 3$ and $\epsilon 4$ isoforms in these crossed mice *also* has been shown to reduce $A\beta$ deposition, suggesting that these human APOE isoforms may decrease $A\beta$ aggregation or increase $A\beta$ clearance relative to an environment in which mouse APOE is or is not present (OMIM, p. 107741).

Many other genes have been investigated for their possible association with AD but have failed to be universally replicated like APOE. These genes typically have protein products that interact with or relate to $A\beta$ or apoE or are candidates based on pathological observations of AD brains. Examples include the low density lipoprotein receptor-related protein (LRP) or the very low density lipoprotein receptor (VLDL-R), both of which are receptors for APOE; α-2-macroglobulin, an acute phase protein that binds to the LRP receptor and may be involved in Aβ clearance; α-1-antichymotrypsin, which is associated with Aβ in plaque; and cathepsin D, an aspartyl protease that influences βAPP processing and was once thought to be the β-secretase. The gene for Cystatin C, a systein protease inhibitor, has recently been demonstrated to be associated with risk for AD in individuals of over 80 years of age (Crawford et al., 2000). Increased awareness of the role of the cerebrovasculature in AD has led to the investigation and confirmation of vascular risk factors for AD (hypertension, diabetes, atrial fibrillation) (Skoog, Kalaria, & Breteler, 1999). Consequently, genes encoding proteins with vascular involvement may also be associated with AD (e.g., the genes for angiotensin-converting enzyme or the nitric oxide synthase genes).

Although all the genes that determine normal neuronal function are involved in AD, some are specifically related to the AD process. Confirmation that variation within a particular gene confers risk for AD proposes that the encoded protein is a participant in the neurodegenerative AD mechanisms. It also raises the question of its possible involvement in Aβ or PS-triggered pathways. As more genes are identified that are involved in the AD process, the picture becomes larger and more complex, but at the same time, more possibilities for therapeutic intervention are presented.

FIONA CRAWFORD
MIKE MULLAN

See Also
Dementia: Overview

Internet Resources
Online Mendelian Inheritance in Man (OMIM)
*http://www3.ncbi.nlm.nih.gov/Omim/
    searchomim.html*

GENETIC TESTING

Research advances in molecular genetics have had a profound impact on modern medicine. The "genetic revolution" has spurred the development of novel diagnostic and treatment methods and will continue to expand our medical capabilities. Genetic testing is one area of clinical practice in which genetic knowledge has particular relevance for health care providers. Genetic testing raises complex ethical, legal, and psychological issues for both patients and clinicians.

Genetic testing refers to the analysis of human genetic material for heritable disorders. Purely genetic, or mendelian, disorders are very rare, but genetic factors often influence disease expression in more common disorders (e.g., diabetes, heart disease, stroke). Genetic testing may be done for disorders with specific genetic causes that can be reliably identified; methods include examination of DNA or chromosomes, biochemical assays, and identification of disease markers by linkage testing. Sometimes multiple methodologies—and testing of multiple family members—may be required to yield useful results.

Some uses of genetic testing (e.g., prenatal testing and newborn screening) are unlikely to pertain to older adults, but others are more relevant. For example, symptomatic adults may be tested for diagnostic purposes. Asymptomatic individuals at risk for autosomal dominant disorders (e.g., Huntington's disease) may be assessed for the presence of disease-inducing genes. Susceptibility testing is emerging for complex diseases such as breast cancer and Alzheimer's disease, for which identified genes are significant risk factors if not actual causes. A final option of relevance to older adults is DNA banking, which involves freezing genetic material for future use by the patient or family members.

Before genetic testing occurs, a genetics consultation may be appropriate. Genetics consultations usually take place in a specialty clinic and are conducted by a genetic practitioner (e.g., medical geneticist, genetics counselor). Consultations are used to refine the assessment and management of genetic conditions, determine and communicate specific genetic risk estimates, and arrange for psychosocial support. Genetic testing may be ordered as a result of such consultations or directly by a physician; patients are then usually referred to genetics laboratories housed within teaching hospitals or at independent commercial sites. For some disorders, testing may be available only in research settings. Given the ethical, legal, and psychosocial complexities of genetic testing, extensive informed consent procedures and pre- and post-test counseling and education are appropriate.

Psychological Issues

More than most medical procedures, genetic testing raises important psychological issues for patients and their families. Deciding whether to be tested can be a difficult process, and ambivalence often characterizes patients' decision making. Older adults may view testing as a means of informing health care decisions, monitoring future treatment developments, or reducing uncertainty about their (and their children's) risk. Reasons to decline testing include a lack of treatment options for the disorder in question, possible imprecision of test results, and fear of the psychological burden that would be imposed by a positive result.

Health care professionals should also be aware of psychological issues in patients' response to testing (Marteau & Croyle, 1998). Test results, whether positive or negative, can cause distress in the months immediately after testing. However, this distress is usually transient if patients are provided proper post-test counseling. If test results match pretest expectations, even positive results for severe disorders are usually not overwhelming. Still, extreme distress (e.g., thoughts of suicide) is possible, as are more subtle reactions such as survivor guilt. If a positive result comes out of the blue—as happens in some genetic screening programs—the psychological effects can be extensive.

Patients and their families may have misconceptions about genetic testing. For example, risk assessment involves an understanding of statistical probability that may elude many test candidates. Even if one has a good command of the relevant facts, the emotionally charged nature of genetic testing may lead to irrational responses and interpretations (Wexler, 1992). Furthermore, test results may have ripple effects within families, as genetic information is salient not only to the patient but also to immediate blood relatives (including older

adults). For example, predictive testing of middle-aged adults may involve assessment of their parents, who may not want to know their own risk status. Thus, an appreciation for the complex individual and family dynamics surrounding genetic testing is important for clinicians.

Ethical, Legal, and Social Issues

Genetic testing raises concerns about potential abuses. Genetic research is receiving increased attention in domains as diverse as medicine, law, insurance, biotechnology, and popular culture. Clinicians should be aware of the potential for genetic discrimination in the larger social system. Indeed, genetic information that is inappropriately accessed or interpreted by insurers or employers may threaten health insurance eligibility for individuals at risk for high future expenses. Legal safeguards against discrimination may exist under the Americans with Disabilities Act (ADA) and other federal and state laws. However, the ADA applies only to employers (not insurers), and states vary greatly with regard to legal protections against the misuse of genetic testing results. Ethics of care regarding informed consent for procedures and confidentiality of medical information are of particular importance.

Applications for Later-Onset Disorders

Cancer

Genetic testing for cancer susceptibility is evolving rapidly. Genes have been found that either cause or increase the risk for various forms of colorectal, breast, ovarian, and prostate cancers (Lashley, 1998). For example, researchers have identified genetic mutations that confer 70% to 80% risk for hereditary nonpolyposis colorectal cancer (HNCC), a disorder that accounts for approximately 8% of all colorectal cancers. Another genetic mutation of note is BRCA1, which elevates the risk for certain types of breast, ovarian, and prostate cancers. Both BRCA1 and BRCA2 are believed to confer significant lifetime risk for breast cancer (estimates are as high as 85%) and to account for 5% to 10%

of the 180,000 new cases diagnosed each year. Mutation detection is commercially available for both BRCA1 and HNCC.

The American Society of Clinical Oncology (1996) has provided useful clinical recommendations regarding genetic testing; these guidelines emphasize the importance of providing education and counseling to patients and offering access to surveillance, prevention, and treatment services. The guidelines also describe differences among testing programs with regard to clinical utility and appropriateness.

Alzheimer's Disease

Causes of Alzheimer's disease (AD), a progressive neurodegenerative disorder, are not fully known, but recent research has provided some clues. Some early-onset forms of AD have purely genetic causes—three specific mutations have been identified to date—but these types are quite rare. Causes of the more prevalent later-onset (after age 65) forms are more complex, but genetics also appear to play a role. For example, the apolipoprotein E (APOε) locus on chromosome 19 influences AD expression, depending on which of three alleles (ε2, ε3, or ε4) one possesses; the ε4 allele significantly increases the risk, especially when appearing homozygously. Yet despite its association with AD, ε4 is neither necessary nor sufficient to cause the disorder. Indeed, other genes may be involved in late-onset AD, interacting with biochemical factors such as the proteins tau- and amyloid-beta.

Presymptomatic testing for autosomal dominant, early-onset AD is already done assessing for the presence of the aforementioned genetic mutations. Its clinical utility is limited, however, given test cost, lack of treatment options, and the extreme rarity of these AD subtypes. Susceptibility testing for later-onset AD (e.g., APOE genotyping) is not yet recommended, as its precision and accuracy are questionable. Medical experts have discouraged the introduction of AD susceptibility testing for other reasons as well, including incomplete knowledge of the factors that modify APOE–associated AD

risk, the competing risk of mortality from other causes, and the psychosocial, legal, and financial challenges of predictive testing for complex diseases (Relkin & National Institute on Aging, 1996).

Commercial genetic testing for AD is now available to physicians. Such testing is recommended solely for diagnostic—not predictive—purposes, but its availability may spur requests from asymptomatic individuals. Relatives of AD patients often wonder about their own genetic risk, and continuing advances in research and media coverage of AD will only heighten such curiosity. Health care practitioners may need to educate older patients and their families about the proper uses of genetic testing for the disorder. Currently, such testing is likely to be appropriate only as a diagnostic adjunct.

Genetic testing is performed primarily in younger patients. However, as genetic links to later-onset disorders continue to be discovered, care of the elderly will increasingly involve consideration of such testing for both older patients and their blood relatives. Given the rapid pace of change in this area, practitioners need to regularly reeducate themselves on both the uses and the limitations of genetic testing.

J. SCOTT ROBERTS

See Also
Genetic Factors in Alzheimer's Disease

Internet Key Words
Alzheimer's Disease
Genetic Counseling
Genetic Testing
Predictive Testing

Internet Resources
Gene Tests
http://www.genetests.com

National Coalition for Health Professional Education in Genetics
http://www.nchpeg.org

National Human Genome Research Institute
http://www.nhgri.nih.gov

GERIATRIC CARE

See

Best Practices in Geriatric Care

GERIATRIC DENTISTRY

Geriatric dentistry is defined as the provision of oral health care to adults 65 years and older, but the field lacks clear boundaries. Geriatric dentistry can best be described as the provision of oral health care for adults with chronic, debilitating physical or cognitive ailments and the associated medications and psychosocial concerns. Geriatric dentistry focuses on persons with complex health histories and physical disabilities whose needs may be better met by clinicians with skills and training beyond that received in a traditional dental curriculum.

Geriatric Dental Practitioners

Community-based dental personnel provide the majority of geriatric dental services to independently living older adults who have little if any need for assistance in their activities of daily living. In addition to their community-based offices, dentists may provide care in nontraditional settings, such as patients' homes and long-term-care facilities. University- and hospital-based dentists, who also teach, conduct research, and provide leadership and consultation for the practicing community, may treat more clinically challenging older patients.

Geriatric Dental Education

Geriatric dental education includes didactic and clinical topics related to older adults. This information is pertinent, since older adults differ from younger adults in the patterns and prevalence rates of oral disease, the signs and symptoms of disease, the amount and types of dental treatment needed, and access to appropriate oral health care services.

Education in geriatrics is a relatively recent addition to the dental curriculum and is not pro-vided in all dental schools. The American Association of Dental School curriculum guidelines were developed in 1982 and revised in 1989 to include geriatric content. Although these guidelines suggest topics for inclusion in curricula, dental schools are not required to address each area. In 1994, most U.S. dental schools (83%) had a specific course or series of courses in geriatric dentistry, and 54% had a clinical component. Most programs identified the primary barriers to expansion as finances, limited curricular time, and limited faculty expertise.

The education and training of geriatric dental practitioners must address the normal physiological and psychosocial changes associated with aging and the impact of these changes on the delivery of dental services. Age-related changes in the oral cavity and cognitive changes associated with aging and disease must be addressed, as well as appropriate and effective preventive, restorative, and rehabilitative approaches to clinical care. With additional training and clinical experience in geriatrics, along with sensitivity to the needs of impaired older adults, dental practitioners will be better suited to treat the growing number of frail and functionally dependent elders.

Postdoctoral training opportunities for practicing clinicians, clinical faculty, and geriatric academicians are limited. Geriatric dental postdoctoral programs were developed to address clinicians' limited exposure to the frail and functionally dependent elderly. For graduating dentists, General Practice Residency Programs and Advanced Education in General Dentistry Programs are now addressing geriatric clinical care issues. Treatment of older adults is emphasized in these training programs so as to improve general dentists' skills in diagnosis, treatment planning, and medical management of geriatric patients. These dentists will have an important role in treating oral cancer, soft tissue lesions, salivary gland dysfunction, taste, smell, and swallowing disorders and be more involved in speech and language pathology, smoking cessation and prevention. These specialized training opportunities are key to teaching oral health professionals how to work effectively with members of the interdisciplinary geriatric health team.

Current Workforce

Clinicians with added competency in geriatric dentistry and geriatric dental academicians are not being developed in adequate numbers to meet current and projected workforce needs. During the past 20 years, approximately 110 dentists have completed advanced geriatric dentistry fellowships or master's-level training programs (National Institute on Aging, 1987). Most programs are through or associated with a dental school. Currently, there are about six or seven programs, although the number changes with the availability of federal funding. As geriatric dentistry is not an American Dental Association–recognized specialty, the number of personnel with advanced training and clinical experience in geriatric dentistry is unknown.

A key priority of the 1987 Department of Health and Human Services report to Congress was the need for programs to train geriatric dental care providers (National Institute on Aging, 1987). It was noted that existing programs would not generate the required number of trained dental personnel needed in the future. By the year 2020, 2,000 geriatric dental academics and 8,000 to 12,000 dental providers with training in geriatric dentistry will be needed. These projections were based on several assumptions, including:

1. Approximately 40% of older adults fall into the special-needs patient group.
2. Approximately 50% of these patients will seek professional care.
3. In addition to caring for functionally dependent patients, dentists with advanced training in geriatric dentistry will serve an average of about 1,000 independent older adults, with an anticipated utilization rate of 50%.

Nontraditional Settings

Long-term-care residents and home-bound individuals face significant barriers to obtaining health care, including dental care. This group may be served at dental facilities on the premises or through the use of portable equipment or mobile dental vans.

However, for a variety of reasons, most dentists do not provide care to patients in their homes or in long-term-care facilities, and few nursing homes have equipped dental offices on the premises.

Although still limited, the availability of portable or mobile dental equipment for use in private homes, senior housing, and long-term-care facilities has been growing. Little practical information is available to dental professionals before committing to the purchase of portable equipment or contracting with a facility to provide necessary services. And using portable equipment requires physical effort and time. Governmental programs have fallen short of enacting provisions to support oral health care for home-bound elderly. Most dental providers are inadequately trained in geriatrics and the delivery of care in nontraditional dental care settings, and educational opportunities to address this concern are few.

Financing the Costs of Geriatric Oral Health Care

Unlike medical care, dental care is financed primarily by the private sector. Medicare, designed to pay for hospital and physician services, covers only medically necessary dental services, such as the treatment of oral and maxillofacial conditions that are a direct result of or have a direct impact on an underlying medical condition or its resulting therapy. In the 1990s, less than 5% of elders' dental expenditures was paid by Medicare. State-based Medicaid programs and managed care programs vary with regard to services covered and level of reimbursement. In general, private health insurance coverage has few and limited dental benefits.

Nursing Home Provision of Oral Care

As a result of the nursing home reform provisions of the Omnibus Budget and Reconciliation Act (OBRA) of 1987, all nursing homes receiving Medicaid and Medicare reimbursement are required to assist residents in obtaining routine and 24-hour emergency dental care; provide routine and emergency dental services to meet the needs of each resident (from an outside resource, if necessary); assist the resident in making appointments and obtaining transportation to and from the dentist's office; and promptly refer residents with lost or damaged dentures to a dentist. A nursing home may charge Medicare but not Medicaid beneficiaries an additional amount for routine and emergency dental services. Medicaid beneficiaries are to receive emergency and routine dental services (to the extent covered under the state plan). Replacement of appliances (dental and others) lost in a long-term-care facility is generally the responsibility of the facility.

Dentistry at the beginning of the 21st century uses far different techniques and procedures from a decade or two ago. As a result of advances in medical science, technology, and health care, not only are people living longer but they are receiving care for disorders that were once debilitating if not fatal. With good preventive and routine oral health care, tooth extractions for periodontal disease or dental caries are no longer required. With proper oral health care, dentition can be maintained for a lifetime.

JANET A. YELLOWITZ

See Also
Dentures
Oral Health
Oral Health Assessment

Internet Key Words
Dental Care
Geriatric Dental Education
Oral Health

Internet Resources
American Dental Association
http://www.ada.com

Consumer Health Information, New Wellness
*http://www.netwellness.org/healthtopics/
dentalseniors*

Elders' Link to Dental Education and Research, University of British Columbia
*http://www.dentistry.ubc.ca/Elder/NONFRAME/
index3.htm*

Federation of Special Care Organizations in Dentistry
http://www.bsgm.edu:80/dentistry/foscod

Geriatric Dental Assessment Case from Wisconsin Geriatric Education Center
http://www.marquette.edu/wgec/dental/

MEGA Multidisciplinary Education in Geriatrics and Aging
*http://cpmcnet.columbia.edu/dept/dental/
Dental_Educational_Software/
Gerontology_and_Geriatric Dentistry/introduction.html*

GERIATRIC EVALUATION AND MANAGEMENT UNITS

Geriatric evaluation and management (GEM) units are interdisciplinary systems of care designed to improve the outcomes and quality of life of older patients. The GEM unit couples comprehensive geriatric assessment with specific treatment plans, individualized to the patient's needs. GEM units can be located in multiple settings, including hospitals, rehabilitation units, nursing homes, and outpatient clinics. Some hospital-based GEM units focus on a specific diagnosis, such as stroke.

Distinguishing Features and Principles of GEM Units

Although the structure and approach of GEM units are highly variable, most GEM units, regardless of setting, share a set of core principles and features that distinguish GEM care from traditional medical care. First, GEM units view the patient in a much wider context. Whereas traditional medical care is heavily focused on specific diseases, GEM units focus more on functional status than specific diagnoses, with the goal of maintaining and improving function. Further, GEM units often view the patient's needs in a much larger context, focusing not just on medical needs but also on psychosocial needs, caregiver's needs and capabilities, and the unique opportunities and constraints in the patient's environment. This wide focus is critical to the success of GEM units, because outcomes in older patients are often better explained by psychosocial characteristics and family support than by disease severity (Covinsky et al., 1998).

Second, unlike traditional physician-directed medical care, GEM units are interdisciplinary and team directed. In addition to a physician, GEM units usually include a geriatric nurse (often an advanced practice nurse) and a social worker. Many units also include mental health professionals, nutritionists, and physical and occupational therapists. Empowering all members of the multidisciplinary team makes it possible for the patient to benefit from their expertise and insights, which is important if optimal outcomes are to be achieved.

Components of GEM Interventions

The specific components of GEM interventions differ somewhat, depending on the setting and the goal of the GEM unit. However, there are many similarities across units. Most GEM protocols begin with a specific targeting strategy directed at identifying patients who are most likely to benefit from a comprehensive assessment. The general goal is to identify patients at high risk for adverse outcomes such as functional deterioration and nursing home placement, while at the same time conserving health care resources by not targeting patients who are so ill or dependent that they are unlikely to benefit.

Examples of high-risk patients targeted by GEM units include those who are dependent in one or more basic or instrumental activities of daily living, those with cognitive impairment or depressive symptoms, the recently hospitalized, and patients who live alone or with a stressed caregiver. Targeting is often difficult, however, because the science of risk stratification in geriatrics is still young. In fact, it has been difficult to identify any substantial subgroups among hospitalized elders whose risk of functional deterioration is low enough to make intervention unjustified (Landefeld, Palmer, Kresevic, Fortinsky, & Kowal, 1995).

The second component of most GEM protocols is a comprehensive assessment focused on identifying existing remediable problems and identifying focus areas for prevention. These assessments generally include comprehensive evaluations of functional status. Screens for depressive symptoms and cognitive status are also administered, making use of readily available instruments such as the Geriatric Depression Scale and the Mini-Mental Status Examination.

Patients are also usually screened for common geriatric syndromes such as falls, incontinence, sensory impairments, and mobility impairments. Comprehensive evaluation of medication regimens is generally included. In the past, medication reviews focused on polypharmacy concerns. However, it has recently been recognized that the withholding of appropriate therapies, such as beta-blockers in older post–myocardial infarction patients, is also a serious problem (Covinsky et al., 1998). Comprehensive evaluations usually include an assessment of caregiving needs and caregiver stress. Since many issues relevant to older patients are missed in standard medical evaluations, GEM units frequently identify previously unrecognized problems in need of management.

The third component of most GEM evaluations is the development of specific treatment recommendations. The results of the evaluation are reviewed in a team conference, and specific preventive and restorative recommendations are made, based on input and discussion among all team members. For example, a patient found to be at high risk of falling after screening with the Tinetti Balance Scale may be referred for physical therapy or may receive a recommendation for a benzodiazapine taper. The caregiver of a patient with dementia may be referred to a support group or for training in the nonpharmacological management of problem behaviors. Treatment recommendations are often presented during a conference that includes the patient and family members.

The execution of these recommendations varies considerably among GEM units. Some GEM units play a purely consultative role and make recommendations to a primary caregiver or a hospital team. In other cases, GEM units assume primary responsibility for the patient and have the capacity to write specific orders. For example, in some hospitals, the GEM team assumes all responsibility for the care of the patient. This is often the case in disease-specific units, such as stroke units. In some outpatient GEM units, the GEM team assumes overall responsibility for the patient for a certain length of time.

Effectiveness of GEM Units

Most studies of the effect of GEM units on the process of care have shown significant improvements, such as decreased use of restraints and psychotropic medicines. These studies reveal that one of the key barriers to GEM unit effectiveness is a lack of compliance with recommendations by either the patient or the primary physician. Although not all studies have shown positive effects on outcomes, the weight of the evidence suggests that GEM units improve functional outcomes (Landefield et al., 1995). A meta-analysis demonstrated that patients treated in GEM units have lower rates of mortality, nursing home placement, functional deterioration, and hospital readmission (Stuck, Siu, Wieland, Adams, & Rubenstein, 1993). In addition, there is some evidence that GEM units may improve these outcomes without increasing total health care costs. GEM units associated with the best outcomes are those that have control over implementation of team recommendations and those that include provisions for long-term follow-up.

KENNETH E. COVINSKY

See Also
Best Practices in Geriatric Care
Gait Assessment Instruments
Geriatric Interdisciplinary Health Care Teams
Hospital-Based Services

GERIATRIC INTERDISCIPLINARY HEALTH CARE TEAMS

Elders often endure multiple illnesses, as well as cognitive loss, leading to functional impairments and social isolation. Frequently, each medical condition requires a specific treatment or medication regimen that is difficult to adhere to—remembering to take it at the right time and in the appropriate way. Additionally, prescriptions and over-the-counter medications may interact, causing negative side effects. Thus, elderly patients with a number of chronic or acute conditions benefit when they receive coordinated care from several disciplines working together. However, these professionals must be able to create an integrated plan and coordinate the patient's care. When multiple specialists administer uncoordinated care, services can be at odds with one another, and patients can suffer. The "layering" of orders, one on another, is multidisciplinary care. This is in sharp contrast to interdisciplinary care—coordinated care developed by skilled clinicians who jointly agree on a care plan and work together to implement that plan.

Not all geriatric patients require the care of an interdisciplinary team, and even patients who benefit from teams do not require a team review at every visit. The needs of the patient should drive the number and type of providers delivering care. Generally, patients with complex physical, emotional, social, and economic issues need medical, nursing, and social work services to develop or alter a care plan. Studies of the cost-effectiveness of teams demonstrate that patients benefit from an initial comprehensive geriatric assessment (Rubenstein et al., 1984) and from intermittent team review as the patient's condition changes or as family needs warrant. Transitions between levels of care are important points where patients can benefit from interdisciplinary care, as Naylor et al. (1994) demonstrated in their studies on the effectiveness of nurse practitioners providing home care to elders after hospital discharge for coronary conditions. A frequently cited benefit of teams is improved communication among providers, resulting in better patient outcomes (Baggs, Ryan, Phelps, Richeson, & Johnson, 1992). Among the successful collaborative practice models cited by Siegler and Whitney (1994) were nurses and physicians working together to monitor patients at home who had chronic and unstable medical conditions such as congestive heart failure, hypertension, or diabetes.

In 1995, the John A. Hartford Foundation created a $12 million geriatric interdisciplinary team training (GITT) program to encourage the development of team training models for advanced practice nurses, medical residents, and master's-level social workers. The program cited three forces converging to accelerate the need for interdisciplinary teams to manage patients: (1) the unprecedented growth of older populations whose complex conditions require the skills of several disciplines; (2) the shift of services from inpatient to ambulatory and community-based care; and (3) the growth in managed care financing, which demands that health care professionals become more efficient in delivering care (Hyer, 1998).

Teams imply energy, coordination, and synergy. Baldwin and Tsukuda assert that in interdisciplinary health care teams, professionals "merge different, but complementary, skills or viewpoints in the service of the patient and in the solution of his or her health problem(s)" (1984, p. 84). The focus on the patient requires that team members reach consensus on the goals of care, the priorities of treatment, and the evaluation of the plan's success. Sometimes a full and heated discussion of the patient's needs is required to recognize the trade-offs inherent in the plan of care. Effective health care teamwork requires agreement, and agreement requires keen interpersonal skills so that all members contribute to and feel accountable for the plan.

The team can offer a forum for a full discussion of patient and family needs. Frequently, different clinicians bring not only different knowledge bases to the team meeting but also different perspectives on the patient and the home environment. This group discussion ensures that all perspectives are taken into account and that the plan has a good chance of being accepted and adhered to by the patient and family. It is advisable that the patient and critical family members also be considered members of the team and, if possible, participate in team meetings.

Stages of Team Development

Becoming a team requires work. Teams generally progress through four stages of development that can vary in duration from days to weeks. The stages, labeled by Tuckman (1965), are as follows.

Stage 1—Forming: During this period, members are tentative and want to learn about one another and the reasons they became members of the team. Effective tasks in this stage include icebreakers to help members get to know one another and discussions about the purpose and goals of the health care team.

Stage 2—Storming: In this phase, team roles, tasks, and rules are worked out through conflict. Teams need to develop and implement criteria for decision making, leadership, and protocols for the team process. Only through discussions—frequently including disagreements about assumptions and values—can the criteria for the group process by established.

Stage 3—Norming: During this phase, rules for operating in the group are agreed on. Teams now have a sense of cohesion and membership. Members know what to expect at meetings and share responsibility for the team process.

Stage 4—Performing: In this final stage, the roles, rules, and team structure are so well honed that the team is clearly more than the sum of its members. The group needs to monitor its progress to avoid "group-think," but the close working relationship and excellent communication among members allow the team to provide well-coordinated care for patients.

Running an Effective Team Meeting

It is imperative that the group use its time well. Few clinicians value meetings, and even fewer know how to run an effective meeting. As with other aspects of teamwork, team members have to create the structure for effective meetings and develop members' skills. Although each group can adapt the process to satisfy its own needs, effective team meetings share some common elements:

- An agenda that specifies the purpose or expected outcome of the meeting
- Suggested time frames to accomplish the tasks

- A meeting record summarizing the group's decisions, including specific action plans that identify who is responsible for what activities by a designated date
- A group written or discussed evaluation of the meeting process (essential to improve the team process)
- Identified meeting roles for members, including the leader or facilitator, timekeeper, and recorder (ideally, leadership roles rotate so that all members get experience)
- An external facilitator, especially during team formation, during periods of intense decision making and conflict, and when a team appears to lose focus

Physical Environment: The meeting space should be comfortable and have adequate light, equipment, and seating. The location should be convenient for all team members. Members should be seated around a table so that everyone can be seen and heard; each clinician should feel equally empowered and encouraged to contribute to the group.

Meeting Ground Rules: The team must establish its own rules or norms. Examples include: attendance policy, promptness, types of permitted interruptions (e.g., patient emergencies only), confidentiality, protocol for patient presentations, breaks, premeeting preparation and completion of work between meetings, and rules on side conversations and digressions from the topic. Agreement about such standards usually prevents misunderstandings and improves team members' behavior.

Teaching Team Skills: The GITT program reinforces the importance of having students participate on well-functioning teams. Interdisciplinary teams allow students to witness professionals working together to solve patient problems, share expertise, create a plan of care, and resolve conflicts. Students see the rigor involved in productive meetings and can observe the interpersonal dynamics of teams. The teaching of team skills should focus on the attitudes, knowledge, and skills team members must learn (Counsell, Kennedy, Szwabo, Wadsworth, & Wohlgemuth, 1999; Siegler et al., 1998).

Attitudes: How much respect is there for the roles of all health professionals in the care of older adults and their families or caregivers? Are members willing to collaborate with all health care professionals? Is there an appreciation for the interdisciplinary team approach, especially for patients with functional and psychosocial disabilities?

Knowledge: Team members need to learn about the skills, education, and training of their teammates. In GITT programs, trainees shadow members from other disciplines or interview team members about education and license requirements. Team members also benefit from learning about formal and informal community support services and how to access services such as home health care, hospice, mental health services, care management, telephone reassurance, visitors, companions, homemakers, chore services, meal programs, transportation services, senior centers, adult day care, respite care, and local area agencies on aging.

Skills: Effective teams improve patient care and meet the patient's needs. By identifying the myriad problems and issues involved in elder care, the team is better able to prioritize concerns and galvanize treatments and services. Focusing on the patient and family is a key skill. Other communication skills that are important include active listening, succinct presentations, summary of discussions, testing for agreements, focused questions and probes to clarify issues, and willingness to sacrifice autonomy to group consensus.

KATHRYN HYER

See Also
Best Practices in Geriatric Care
Geriatric Evaluation and Management Units
Hospital-Based Services

GERIATRICIAN

A geriatrician is a physician who specializes in the care of the elderly. It is said that the term "geriatrics" was first used to denote a medical specialty in the early twentieth century by Ignaz Leo Nascher, an Austrian physician affiliated with Mount Sinai Hospital and the Jewish Home and Hospital for

Aged in New York. The specialty of geriatrics was created in a policy context by the National Health Service (NHS) in the United Kingdom after World War II (Brocklehurst, 1997). The founders of the NHS recognized the growing population of older people and created a number of positions for specialists in geriatrics who would function as consultants in the major teaching hospitals. These individuals wrote the major textbooks and were the leaders in the field in its early decades. The United States recognized the distinct body of knowledge emerging from aging research and the special approach needed to care for elderly patients, and a specialty was created in 1988. In that year, the American Board of Internal Medicine and the American Board of Family Practice collaborated (for the first time) in an examination to establish a certificate of additional qualifications in geriatric medicine but not a full-fledged subspecialty (Institute of Medicine, 1993). The first three times the exam was administered—1988, 1990, and 1992—eligibility included accredited fellowship training in geriatrics; a physician could also be "grandfathered" by attestation that the majority of his or her practice consisted of elderly patients. In the early years of certification, the majority of individuals taking the exams were grandfathered. In 1991, there were 6,784 board-certified geriatricians in the United States out of a total physician population of 684,414 (Alliance for Aging Research, 1996). The number of board-certified geriatricians practicing in 1999 was about 7,000. This is less than half the number needed, according to Rand Corporation studies. The Alliance for Aging Research projects that by the year 2030, 36,000 geriatricians will be needed to care for the older population, but only 10,000 geriatricians are expected to actually be in practice (Alliance for Aging Research, 1996).

The Institute of Medicine Committee on Leadership for Academic Geriatric Medicine (1987) reported that approximately 100 physicians complete fellowships in geriatric medicine or geriatric psychiatry each year, or about 3% of all physicians taking fellowship training in internal medicine. Most of the geriatrics programs for medical students and residents are voluntary; less than half the eligible residents take the geriatrics elective. Of the 1992 medical school graduating class, only 2.9% took an elective in geriatrics, a decline from 3.5% in the class of 1988 (Institute of Medicine, 1993). Overall, 62% of residents in family practice, 26% in internal medicine, 17% in physical medicine, 48% in psychiatry, and 3% in neurology took either a required or elective rotation in geriatrics (Institute of Medicine, 1993). Furthermore, out of 98,000 positions supported by Medicare in 1998, only 324 medical residency and fellowship positions were in geriatric medicine and geriatric psychiatry (U.S. Senate Special Committee on Aging, 1998).

The attributes of a geriatrician include expertise not only in the medical diseases common in the elderly, such as Alzheimer's disease, diabetes, coronary artery disease, congestive heart failure, and cancer, but more importantly in the non-disease-oriented syndromes such as confusional states, urinary incontinence, frailty, gait disorders, and postural instability. The geriatrician is trained to focus on declines in function, looking for subtle medical or psychiatric conditions that could be reversible; in the case of progressive, chronic illness, the geriatrician focuses on anticipatory management to try to maintain function as long as possible. Trained in the context of interdisciplinary care and often best situated working as part of an interdisciplinary team, a geriatrician must master the complexities of multiple interacting illnesses and polypharmacy, as well as have a broad expertise that includes aspects of psychiatry, neurology, dermatology, urology, orthopedics, gynecology, and the social and psychological dimensions of aging. Geriatricians are likely to be working with families as well as with individuals, and they care for patients in all types of settings, including hospitals, nursing homes, assisted living facilities, and the community.

CHRISTINE K. CASSEL

Internet Key Words
Academic Geriatric Medicine
Geriatricians
Geriatrics Training

History of Geriatrics Medicine
Interdisciplinary Care

Internet Resources
American Board of Family Practice
http://www.abfp.org

American Board of Internal Medicine
http://www.abim.org

GERONTOLOGICAL SOCIAL WORK

Social work is increasingly important as the population ages and the array of services for the elderly grows. Social workers not only provide direct services to elders and their families in the form of psychosocial assessment and treatment; they also serve as referral links to the complex network of services.

The field of social work began in the late 19th century. With roots in helping the poor, professional social work has expanded into important roles in mental health, health, and social services for individuals of all ages and income levels. Social workers provide direct treatment and consultation as well as develop, administer, and evaluate programs and services at the individual, family, organizational, and community levels.

Education

Social workers with bachelor's degrees (BSW) are prepared for generalist practice, whereas those with master's degrees (MSW) are prepared for advanced and often specialized practice. Both levels of social workers are employed by a variety of organizations working with older adults and their families. In most states, individuals with BSWs and MSWs can obtain licenses to practice (e.g., LGSW, LCSW, LCSW-C). State licensing boards regulate the social work profession to ensure that clients are served by social workers with proper training and common ethical standards. In order for a social worker to

be independently reimbursed by most insurance plans, an MSW, two years of supervised social work practice, and clinical licensure are required in most states.

There is no formal license, specialization, or certification to denote that a social worker is qualified to work in the aging field. These practitioners may have studied, specialized in, or had an internship in gerontology as part of their social work education. Field placements (one for a BSW; two for an MSW) are required during social work degree programs, and many are in settings that serve older individuals.

Knowledge Base

Social workers in the field of aging must be knowledgeable about normal age related biological and psychological changes, the family dynamics of aging, ways to help clients age successfully, the actions of commonly prescribed medications for older adults, the usual course of and treatment for medical problems common to older adults, housing and long-term-care alternatives, and the myriad policies and services that affect older adults financially and socially. They must be able to communicate all of the above to clients and families as necessary; assess older adults for depression, dementia, or other mental illness; and refer clients and their families to appropriate community resources. Several national organizations actively promote the education and postgraduate continuing education of social workers in gerontology and geriatrics, including the Association for Gerontology Education in Social Work, the Council on Social Work Education, and the National Association of Social Workers.

Theoretical Perspectives

The most commonly taught human behavior theories are highly appropriate for understanding older persons and their families. *Person-environment fit* is a social work perspective that holds that how an individual behaves and interacts is not just a matter of what occurs within the individual but is a dynamic and evolving state of the "fit" between the individual and his or her environment—family, housing, work, neighborhood, and community. The *competence model* of aging further suggests that age-related changes may lead to "environmental press," whereby the usual demands of the environment becoming overwhelming (Lawton, 1989).

Gerontological social workers have a developmental, life-course approach to human behavior that suggests that growth and change occur throughout the life cycle in a relatively predictable sequence. Each of Erikson's (1963) eight stages of life contains a stage-specific psychosocial task, culminating with *integrity versus despair* in old age. The aging individual needs to look back on his or her life and make peace with what has gone before, while adapting to changes that come with aging and preparing to accept mortality. The idea that older persons are still "working" psychologically and socially, rather than remaining stagnant, dovetails with social work's belief in the autonomy and basic worth of every individual, regardless of his or her age or stage of life.

Social role theory is also important in work with older adults. This theory states that individuals' sense of identity derives from their roles in society. As persons age, they inevitably lose some of their social roles as their children become adults, they retire from careers, or they become widowed. More subtle changes occur as well. Society's standard of sexual attractiveness, physical strength, and a youth-oriented culture challenge role identity for some aging individuals. Social workers use role theory to assess the way older adults perceive their current level of activity and involvement, and whether ageism is preventing the individual from taking on new roles to replace those that are diminished, changed, or lost (Delon & Wenston, 1989).

The concept of *productive aging* (David & Patterson, 1997) recognizes the enormous diversity and abilities of older adults and denounces mandatory retirement or other age-discriminatory policies. *Quality of life* is another key concept of gerontolog-

ical social work practice and is particularly relevant in long-term care and care for the dying. Recognition of *diversity* among the older population is an additional emphasis of gerontological social workers.

Practice Settings

Social workers offer professional support in health, mental health, and social services for older persons and their families and caregivers in both the public and private sectors.

Health: Elders requiring inpatient health care for acute or chronic problems usually see social workers as part of the interdisciplinary treatment team. In hospitals and rehabilitation facilities, the social worker provides psychosocial assessment, family support, discharge planning, and coordination of services. In long-term-care facilities and continuing care retirement communities, social workers provide psychosocial assessments for new residents, help them adjust to the new environment, program social and support activities, and provide ongoing support for families and residents. Usually it is social workers who assist patients or residents in completing advance directives for health care such as health proxy forms or do-not-resuscitate orders. Social workers help patients with Medicaid applications and other financial and in-kind services, and social workers typically participate on the institutional ethics committee in health and long-term-care facilities. In all health settings, the social worker is often the first line of communication with the family and helps "translate" medical jargon and technical information into language families can understand.

Mental Health: Many mental health services for older adults are offered in group or individual treatment modalities by nonspecialized clinics or medical plans. Social workers serving elders need to assess and intervene in situations of bereavement and other adjustments to loss and be able to differentiate these from major depression, anxiety disorders, and early-stage dementia. Cognitive-behavioral or psychodynamic methods are effective in treating depression in older adults (Thompson, Gallagher, & Breckenridge, 1987); reminiscence or life review (Butler, 1963) is often a major component. Supportive psychotherapy and adjuvant psychotropic medications can be the treatment of choice, with social workers assuming the case management and counseling role.

Geriatric outreach or assessment teams that work with older adults and their families are another model for mental health assessment and treatment in many communities. Not uncommonly, social workers make home visits to increase the accessibility of their services as well as to assess the safety and appropriateness of the individual's living situation. Outpatient and counseling services are covered by Medicare and in some states by Medicaid for appropriately licensed social workers.

Social Services: The Older Americans Act of 1965 marked the beginning of an expansion of age-specific programs and services such as senior centers, nutrition programs, homemaker and home health services, and adult day care (Hooyman & Kiyak, 1999). Administered by local area agencies on aging, these programs employ many social workers as directors, clinicians, and case managers who both provide direct treatment and coordinate other services. Adult protective services units in every state are charged with investigating and protecting vulnerable adults from abuse or neglect. Their staff members, primarily social workers, go to the homes of referred clients to assess the level of danger and the need for outside placement and other services, providing intervention and follow-up.

Advocacy and Social Policy: Social workers actively participate in formulating aging policies through public and private administrative positions and legislative activities such as lobbying for the maintenance of age-based entitlements and programs, such as Social Security. "The geriatric social worker is often able to assume the role of an objective advocate. The aged (especially those who are

ill) may be economically, socially, or psychologically vulnerable. They have little power or connections to power" (Kart, 1997, p. 529). Social work has a long-standing ethos and a commitment to helping vulnerable populations and individuals obtain adequate and justly distributed services.

KATHRYN BETTS ADAMS
CONSTANCE SATLZ CORLEY

See Also
Adult Protective Services
Aging Agencies: Federal Level
Aging Agencies: State Level

Internet Key Words
Advocacy
Assessment
Lobbying
Outreach
Psychotherapy

Internet Resources
Association for Gerontology Education in Social Work
http://www.agesocialwork.org

Council of Social Work Education, SAGE-SW Project: Strengthening Aging and Gerontology Education for Social Work
http://www.cswe.org/sage/htm.

National Association of Social Workers
http://www.naswdc.org.

GERONTOLOGICAL SOCIETY OF AMERICA

The Gerontological Society of America (GSA) is dedicated to promoting the scientific study of aging. It encourages exchanges among researchers and practitioners and fosters the use of gerontological research in forming public policy. It is a leader in the advancement of knowledge, the generation of new ideas and the translation of research findings into practice.

Mohammed Reyazuddin

Internet Resource
Gerontological Society of America
http://www.geron.org

GLAUCOMA

Glaucoma is a disease of the eye in which the optic nerve or nerve fibers associated with the nerve are damaged, often as a consequence of higher than normal intraocular pressure and often associated with characteristic losses in visual field (peripheral or "side" vision).

The most common type of glaucoma is Primary Open Angle Glaucoma (POAG) which accounts for approximately 80% of all cases. The "angle" refers to the anatomic space through which an eye fluid termed "aqueous" passes. The aqueous is produced behind the iris (the tissue responsible for eye color) at a relatively constant rate and circulates from the back to the front of the iris through the pupil (the black circle in the middle of the iris). It is filtered out through the angle, which is also known as the anterior chamber angle. Intraocular pressure (pressure within the eye) is a result of the balance between aqueous production and resistance to outflow. Most cases of glaucoma are due to the inability of the eye to efficiently remove aqueous through the anterior chamber angle. As a result, the intraocular pressure increases. In POAG, the angle, as the name implies, is open and there no visible obstruction to the outflow of aqueous. Despite this "normal" appearance, the disease is present, and the exact cause remains incompletely understood. The word "glaucoma" comes from *glaucoses* and was described by Hippocrates as a known affliction of the eyes. It is derived from *glauco* which means *gray and opaque* and probably referred to several conditions of the eye that were not, at the time of Hippocrates, differentiated from what we now know to be glaucoma.

Other forms of glaucoma include secondary open angle type, which affects 3% of the population; angle-closure of all types, which is found in 5%; and patients who are suspected to have glaucoma but who do not meet all of the diagnostic criteria, which includes 11% of the population.

Diagnosis

POAG most frequently occurs in individuals over the age of 40 and increases in prevalence with age. Risk factors based on population-based studies include older age, ethnic origin (patients of African heritage are 3 to 4 times more likely to develop glaucoma than their Caucasian counterparts), family history, intraocular pressure, and possibly male gender and history of diabetes mellitus, although new research is now questioning diabetes mellitus as a risk factor.

Despite the fact that POAG exists in only approximately 2% to 3% of the world population, it is still one of the leading causes of blindness and visual impairment. The probability of blindness or visual impairment in both eyes resulting from glaucoma is less than 10%; however, patients often have other potentially visually disabling conditions associated with age (e.g., cataracts or macular degeneration). Unlike cataracts—which can cause hazy vision or macular degeneration, which reduces central vision—glaucoma compromises peripheral (side) vision. This phenomenon is often imperceptible to patients because their central vision is usually intact until later in the disease process. In addition, the most common forms of the disease are painless. Because of this silent onset, a large percentage of people with glaucoma are undiagnosed and likely never aware of their condition until it is quite advanced unless they have periodic eye care and examinations.

Glaucoma screenings, a relatively common practice in some communities, are often little more than a check of the intraocular eye pressure (unfortunately mislabeled "the glaucoma test"). Although high eye pressure is indeed a risk factor for the disease, between 35% and 45% of bona fide glau-

Commonly Used Glaucoma Medications

Classification	Trade Names	Cap Color	Side Effect
Beta blockers	Timoptic, Betimol*, Beta-gan, Ocupress*, Betoptic, OptiPranolol*	0.5% yellow 0.25% blue *white cap white	Bronchospasm Slows heart rate Depression, impotence
Prostaglandin analogs	Xalatan	white	Eye inflammation
Alpha adrenergic agonists	Alphagan, Iopidine	purple, white	Eye allergic reaction Dry mouth, fatigue
Carbonic anhydrase inhibitors	Trusopt*, Azopt**, Co-sopt***	orange ***yellow+orange	Bitter taste, *stings **Stings less

***A combination of 5% beta blocker and carbonic anhydrase inhibitor.

coma patients in a population will be missed by a pressure reading alone. There is no one test to diagnose glaucoma. Information on risk factors is aggregated, the eye pressure and angle are checked, visual field (side vision) testing is conducted, and most importantly, the integrity of the optic nerve and nerve fibers is assessed by looking through a dilated pupil (using eye drops) to view the nerve and surrounding tissues. Glaucoma crisis or "acute angle closure glaucoma" is a condition in which the anterior chamber angle is anatomically, pharmacologically, or pathogenically closed shut, allowing little or no drainage of the aqueous from the eye. As a consequence, the eye may become painful, red, hazy, and very firm to the touch. This condition is well known because of its unique clinical presentation, however it is relatively uncommon.

Treatment

After the diagnosis of glaucoma is established, there are several ophthalmic medications used for treatment, all of which have potentially significant side effects. The classifications of medications commonly used in glaucoma, their side effects, dosages, and "color-codes" are summarized in the accompanying table. Patients often do not recall the name of the medication they are using but frequently can remember the color of the cap.

Many drugs used for other conditions are capable of increasing intraocular pressure. Systemic cor-ticosteroids (including inhalers), which may be used in rheumatologic diseases and asthma, can raise intraocular pressure in some patients. Antihistamines and some antipsychotic medications are capable of narrowing the anterior chamber angle. This could be problematic for patients at risk for angle closure glaucoma.

Glaucoma is a chronic disease. As such, patients need counseling so that they understand that, with rare exceptions, this disease will persist throughout their life-span, they will be taking medications, and/or they may need to undergo laser or surgical procedures to increase the efficiency of the drainage system of the eye or to surgically create a new, alternative drainage system. In the early phases of treatment, several different medications alone or in combination may be used in an effort to find a treatment that is appropriate for the patient. Throughout the course of treatment, medication dose or type may need to be adjusted. Most glaucoma patients are followed by their optometrist and/or ophthalmologist at least three to four times per year. Patients with diabetes are generally seen annually unless there exists some degree of treatable diabetic retinopathy. Opticians do not treat or diagnosis glaucoma.

Patient compliance (adherence) is a serious problem in the management of glaucoma. Factors influencing compliance include the patient's understanding of the nature of the condition, the patient's ability to self medicate (taking eye drops can be a challenge), the silent nature of the disease (there is no pain and vision is not affected until very late in the disease process), the chronic nature of the disease, insurance/managed care/cost issues with regard to purchasing medication, and access to continuing care (most postoperative costs are covered by Medicare).

If progression continues despite the use of medications, surgical options are considered to halt further loss. As with any surgical procedure, there is a level of risk associated with the procedure and postoperative course. Cataract formation occurs in up to 10% of cases, and potentially sight-threatening infections occur in 1% to 2% of all cases. The postoperative course generally runs about 30 days, however infections are capable of occurring much later (even beyond one year).

Even under the best of circumstances (proper diagnosis and treatment and a patient who follows the treatment regimen) the disease can progress and cause visual impairment. Disease progression typically occurs very slowly, over the course of months. In general, the more advanced the disease, the more rapid the progression and challenge to manage it. As peripheral vision worsens, patients may experience a compromised awareness of objects that appear to either side of them. They may complain of discomfort in crowded or busy environments (e.g., walking on a busy street) where objects and people appear to suddenly loom into view. Many patients lose the ability to orient themselves and/or navigate through unfamiliar environments, and their quality of life can become seriously impaired.

Counseling, techniques to regain orientation, mobility training, and a low vision examination are important at this time. Continued education regarding the course and nature of glaucoma and addressing concerns about loss of independence associated with vision problems are among the key goals of counseling. Orientation and mobility training seek to maximize the patient's ability to move independently, safely, and purposefully in the environment and may or may not include the use of a cane. In addition to a standard eye exam, the low vision examination includes a comprehensive assessment of the degree of usable vision; the pa-

tient's visual goals, objectives, and expectations; and the assessment of devices to augment usable vision.

A typical low vision assessment of a patient with visual impairment as a result of glaucoma includes specialized visual field testing; an evaluation of the patient's current ability to use remaining vision (how efficiently patients scan with their eyes to assess their environment); and how able the patient is to use assistive devices such as prisms (used somewhat like a side-view mirror on a car to assess objects coming from the side) or telescopic devices to compress the image of the environment into the smaller field of view.

Using eye drops can be a challenging task. One way to ensure the drop is instilled correctly is to have the patient gently pull the skin of the lower eyelid between the thumb and index finger to create a "pocket" for the drop. While looking up, the patient then gently squeezes the bottle so that a drop falls into this pocket. Care must be taken to wash hands and not to touch the tip of the bottle with any surfaces, including the eyelids and lashes. Patients should be told to instill one drop at a time and to gently close the eye following instillation. The patient may also be instructed to press on the opening of their tear duct (the inside corner near the bridge of the nose where the upper and lower lids meet). This helps to keep the medication from being flushed out of the eye and into the tear drainage system. The patient should maintain this position for at least 5 minutes per drop so that absorption into the ocular tissues is enhanced.

When working with elderly patients, it is very important to be cognizant of their need for regular eye care. Patients cannot afford to wait until they experience symptoms before seeking care. If an elderly patient has not had a comprehensive eye examination in the past year, it is important to make the appropriate timely referral. It is also important to ask specifically if the patient is currently using or has ever used eye drops. If the patient is a known glaucoma patient, questions should be asked regarding compliance and follow-up with their eye care provider. If the patient is visually impaired as a result of glaucoma, the patient should be addressed face to face in an effort to optimize the speaker being in the patient's line of sight.

MITCHELL DUL

See Also
Assistive Devices
Low Vision
Vision Changes and Care
Vision Safety

Internet Resources

American Academy of Optometry
http://www.aaopt.org/

American Optometric Association
http://www.aoanet.org/

The Glaucoma Foundation
http://www.glaucoma-foundation.org

National Eye Health Education Program
http://www.nei.nih.gov

GRANDPARENTS AS FAMILY CAREGIVERS

An extensive literature exists on family caregiving, much of it focused on the physical and mental health consequences of caregiving for spousal and filial caregivers. In the early 1990s, gerontologists identified another relatively high-risk population of family caregivers—grandparents who are rearing their grandchildren. In 1997, 2.4 million families, or 7% of the nation's families with children under age 18, were maintained by grandparents and other relatives, an increase of 19% since 1990. Grandparent-headed households with neither parent present rose only 6% between 1970 and 1990 but nearly tripled from 0.5 million in 1990 to 1.5 million in 1997 (Bryson & Casper, 1998).

The lifetime incidence of surrogate parenting for at least six months is now estimated at 10.9% for American grandparents; 56% provide care for more than 3 years, and 20% provide care for more than 10 years (Fuller-Thomson, Minkler, & Driver,

1997). Societal factors contributing to the rapid growth of this caregiver group include changing family demographics (e.g., rising rates of divorce, teen pregnancy, single parenthood); interrelated social health problems, notably substance abuse (implicated in over 80% of cases), the HIV/AIDS epidemic, and increased incarceration of women of childbearing age; and child welfare policies that incorporate kin into traditional foster care systems (Burnette, 1997).

Given these contextual factors, it is not surprising that the financial strain and physical and mental health difficulties commonly associated with family caregiving are worse for grandparents than for filial or spousal caregivers. These problems tend to span multiple generations and often predate the onset of care in grandparent-headed households (Fuller-Thomson et al., 1997; Strawbridge, Wallhagen, Shema, & Kaplan, 1997). It is also important to note that custodial grandparents are disproportionately low-income, urban, ethnic minority grandmothers. Nationally, 77% of grandparent caregivers are women; their average age is 59, and median annual household income is $22,176. Just over 60% are white, 29% are African American, and 10% are Hispanic. Relative care has strong cultural and class components; African Americans are three times (12%) and Hispanics nearly twice (8%) as likely as non-Hispanic white grandparents (4%) to assume a custodial role (Fuller-Thomson et al., 1997). Seventy-five percent of children in grandparent-headed households live in metropolitan areas, up from 59% in the mid-1980s (Harden, Clark, & Maguire, 1997).

The main issues facing ethnic minority grandparent caregivers are economic pressures and physical and psychological problems; these appear to be pressing issues for grandparent caregivers more generally (Pruchno, 1999). Nevertheless, grandparents consistently report benefits associated with caregiving, including improved sense of mastery and control, satisfaction, and devotion and commitment to the children in their care. It is important to recognize and utilize these and other strengths that grandparents bring to the custodial role.

Physical health is a critical factor in grandparents' well-being and in their ability to meet the demands of child rearing. Nationwide, 35% of grandparent caregivers rate their health as fair or poor, over 50% are limited in at least one activity of daily living, and grandparents who are caregivers are more likely than those who are not to be limited in four out of five activities of daily living (Minkler & Fuller-Thomson, 1999). Caregiving often results in chronic fatigue and exhaustion and may affect the development or exacerbation of existing health problems such as chronic conditions, somatic symptoms, and harmful health behaviors such as smoking, alcohol abuse, and poor nutrition (Burton, 1992). Like other types of family caregivers, grandparents tend to minimize their health problems and neglect their own needs in order to focus on the needs of their grandchildren.

Mental health difficulties that affect custodial grandparents' quality of life and role enactment include depressive affect, heightened anxiety, role restriction, isolation, and family conflict. One-quarter of custodial grandparents in the 1992–1994 National Survey of Family and Households reported significant levels of depressive symptoms on the Center for Epidemiological Studies Depression Scale—a rate nearly twice that of noncaregiver counterparts (Minkler et al., 1997). Grandmothers are at greater risk for depression than grandfathers (Scinovacz, DeViney, & Atkinson, 1999). Local studies with urban African American (Burton, 1992) and Latino (Burnette, 1999a) grandparents suggest that these rates may be even higher for ethnic minority grandparents in inner cities.

Current and cumulative health and economic deficits contribute to mental health problems, and role-related strains increase these risks. Primary parenting is an "off-time" role for older adults. Grandparents who are engaged in their own age-related developmental tasks are often unprepared for this level of responsibility, especially given the contextual challenges that accompany the role. Many are mourning the loss of an adult child; shame and stigma surrounding this loss may complicate their grief. Constant concerns about grandchildren's present and future safety and well-being produce emotional turmoil. Caregiving disrupts current routines and future plans, restricts social and leisure activities, and produces conflict in balancing the needs of grandchildren with those of other family members (Burnette, 1999b; Burton, 1992). Child-rearing expenditures and hidden costs of lifestyle changes add to the financial strain. Work roles and financial stability may be threatened, for example, through absenteeism, reduced schedules, and untimely resignation or retirement (Burnette, 1999a; Jendrek, 1993; Minkler & Roe, 1993).

Multidimensional assessment of grandparent caregivers' strengths and service needs should include established sources of risk (e.g., physical and mental health, role performance), financial stability, and resilience. Other important assessment criteria are social and spiritual supports, other caregiving responsibilities (e.g., older relatives, friends, or neighbors), understanding of laws and regulations that govern child placement, and environmental safety and security.

Support groups are the most widespread method for addressing the educational and support needs of grandparent caregivers. The American Association of Retired Persons' Grandparent Caregiver Information Center lists more than 500 such groups nationwide. The leadership and purpose of the groups vary considerably; research is needed to develop appropriate curricula and test their effectiveness. Other significant venues are churches, senior centers, schools, and family health clinics.

The intimate interdependence of older and younger family members in grandparent-headed households, often in the actual or functional absence of members of the linking generation, suggests the need for more multidisciplinary, community-based approaches to assessment and intervention. Such strategies require providers to stretch the traditional funding boundaries of age- and problem-based services that span a vast, uncoordinated system of aging, child welfare, family, health, and income maintenance services with variable eligibility criteria and entry points. Targeting and coordinating these services in local communities could improve the quality of services and reduce access barriers such as poor health, depression, lack of transportation or child care, unfamiliarity with the service delivery system, and cultural and language differences.

Visual tools such as genograms or ecomaps may be helpful in assessing family structure and interpersonal relationships and in understanding the special needs of members of younger generations. For example, children may require help with emotional and developmental problems associated with parental loss, prenatal substance abuse or HIV exposure, or abuse and neglect (Hayslip, Shore, Henderson, & Lambert, 1998). Parents, when present, may need drug rehabilitation, family counseling, or help with chronic health conditions. They should also be involved, when possible, in the long-term custody planning for their children.

Grandparents who are rearing their grandchildren help to preserve crucial kinship ties and provide an inestimable social and economic service to society. Their numbers continue to rise steadily; the 2000 census contained a new set of questions on this family configuration. Tracking this trend will improve efforts to meet the needs of grandparent caregivers and their families. Meanwhile, tailoring ethnically sensitive resources and services to the physical health, mental health, and economic needs of these at-risk families will help improve the quality of their individual and family lives.

DENISE BURNETTE

See Also
Intergenerational Care

Internet Key Words
Family Caregiving
Grandparent Caregiving
Intergenerational Relations

Internet Resources
AARP Grandparent Information Center
*http://www.aarp.org/getans/consumer/
    grandparents.html*

Generations United
http://www.gu.org

Grandparent Caregiver Law Center, Institute on Law and Rights of Older Adults, Brookdale Center on Aging of Hunter College
http://www.hunter.cuny.edu/~aging

Grandparent caregivers: A national guide
*http://www.prisonactivist.org/lspc/ngcm/
    cover.html*

Grandparents Raising Grandchildren: An Educational Web Site
http://www.uwex.edu/ces/gprg/gprg.html

Grandparents Raising Grandchildren by B. Hayslip, Jr., and R. S. Goldberg-Glen (2000)
http://www.springerpub.com

GROUP PSYCHOTHERAPY

Group psychotherapy is an effective and valuable approach to the many mental and emotional challenges of late life. Elderly persons may reject the concept of psychotherapy on an individual basis but benefit from the supportive interaction among group members (Burnside & Schmidt, 1994). Clinical and empirical evidence shows that group psychotherapy is an effective treatment modality, particularly for specific symptoms of depression and life stressors such as loss and bereavement (Leszcz, 1987). The applications of group therapy are abundant and offer motivated clinicians a rich venue for serving the needs of older adults. As with all forms of psychotherapy, individual acceptance, engagement, and response will vary. However, the social support, stimulation, and shared experience of a group setting are often of particular benefit to the elderly.

Group therapy is a flexible model that can be utilized in a variety of settings, including senior and community centers, outpatients clinics, mental health centers, nursing homes, and hospitals.

Indications

A group may focus on specific symptoms or themes, including depression, anxiety, bereavement, loneliness, boredom, loss, and issues of aging. Another focus may be specific illnesses, such as Parkinson's disease, arthritis, or chronic pain, or issues of coping with disease states. The group may be identified by the therapeutic modality chosen, including reminiscence or life review, cognitive-behavioral models, psychodynamic therapy, and support groups for caregivers. Groups may focus on specific activities with the goal of task completion. This type of group can be of particular value for those with cognitive loss and dementia, who benefit from the sense of accomplishment conveyed by the group.

The Therapeutic Process

Preparation and screening before any group therapy are essential to a successful outcome (Yalom, 1995). Elderly patients in group therapy should be screened for the presence of acute illness, cognitive loss, and willingness to participate. Heterogeneous groups are more likely to fail if the population includes those with dementia, severe physical or sensory disabilities, or language barriers. Homogeneous groups are more likely to become cohesive, successful, and therapeutic. A structured group format with well-defined goals is most likely to be successful with older adults. Patients who are experiencing an acute crisis or who are suicidal and require immediate treatment are not appropriate candidates for group treatment (Scheidlinger, 1994). Cognitive loss, sensory deficits, personality disorders, and language differences can often be accommodated in specialized group settings (Grant & Casey, 1995).

The purpose and goals of the group therapy should be clearly stated and addressed at the start. The group leader must take an active role in facilitating communication and interaction among group members. The goals of individual group members must be explored and addressed in order to avoid conflict. The therapist must actively encourage interaction and feedback among group members. Often the therapeutic group is a new entity to older adults, who may be uncertain and reluctant. The basic principles of group therapy apply; older adults require support, recognition, and acceptance in order to develop a therapeutic relationship.

Group members will exhibit a wide range of differences in self-esteem and in their tendency to be either passive or aggressive in response to the group setting. Jealousy, envy, and resentment toward other group members and the facilitator are common and need to be addressed through gentle support and reflection. The concept of mutual support and aid among group members must be developed over time. If time and staffing permit, a co-leader is recommended whenever possible. This allows for an objective view by another clinician who can serve as a source of support and interpretation. Ongoing professional supervision and training are important to maintain efficacy and quality.

Applications

Reminiscence therapy, also called life review, is a means by which elderly persons can evaluate and reintegrate their lives. Reminiscence can greatly increase life satisfaction when an individual is able to acquire a sense of self-esteem, accomplishment, and completion (Butler, 1963). With this type of therapy, there is a risk that some individuals, particularly those who are depressed or withdrawn, may experience guilt or morbid preoccupation with mistakes, disappointments, or losses from the past. Because of its individual focus, reminiscence by itself does not increase social skills.

Caregivers of patients with dementia almost universally suffer from burden, anxiety, and often depression. Support groups for caregivers may reduce depression, alleviate anxiety, improve problem solving, and delay the need for nursing home placement. Such support groups do not improve the cognitive status or abilities of patients with dementia; they increase the coping skills of those providing care.

Group therapy for older adults with dementia can be an effective means of maintaining interpersonal skills and promoting a sense of self. Modeling and positive reinforcement by the group leader are vital to success. Psychodynamic and formal cognitive-behavioral techniques such as interpretation and homework assignments are not appropriate for a cognitively impaired population. Groups directed toward persons with mild and newly diagnosed dementia in the early stages have been very useful (Burnside & Schmidt, 1994).

Group therapy in nursing homes is often used as a means of social support, to increase activity level, and to maintain self-esteem and dignity. Consideration must be given to the cognitive status of group members, with adjustments made to accommodate memory impairments and sensory losses. Approaches that focus on maintaining individual identity and self-worth in the context of institutional life have been very productive (Burnside & Schmidt, 1994).

Reimbursement

Group psychotherapy is reimbursable under Medicare Part B, Medicaid, and private insurance plans from eligible providers. Current Procedural Terminology (CPT) codes 90853 and 90857 are applicable to group psychotherapy. Codes 90846 and 90847 apply specifically to family therapy, with 90849 used for multiple-family groups. Group therapy offered in hospital settings may be billed using inpatient codes. Appropriate documentation for each member in attendance must be provided. As patients are often seen individually before placement in a group, visits are reimbursable using codes appropriate to the level of service provided. Many managed care plans find group therapy financially attractive for their members, but the practitioner must follow the procedures for authorization, treatment plan documentation, and coverage limitations under the specific plan. Unlike Medicare and Medicaid, managed care plans are likely to limit the number of sessions to a contracted maximum. These issues need to be addressed both individually and with the group.

Group psychotherapy is an effective means of dealing with the vulnerabilities that accompany the aging process. Older adults may be more likely to engage in group rather than individual therapy, because it is often perceived as less intense and less threatening. The role of the therapist is one of active listening, reflecting and balancing the needs of the individual with the growth of the group as a whole. Group size may range from as few as three participants to as many as ten. However, if cognitive impairment is present, size should be limited to six to eight members (Burnside & Schmidt, 1994).

MELINDA S. LANTZ

See Also
Life Review
Mental Health Services

Internet Key Words
Group Process
Group Psychotherapy
Medical Psychoanalysis
Psychotherapy for the Aged
Reminiscence Groups
Therapy Group Themes

Internet Resources
American Group Psychotherapy Association
http://www.groupsinc.org

Center for the Study of Group Processes
http://www.uiowa.edu/~grpproc/index

International Society for the Study of Personal Relationships
http://www.isspr.org

GUARDIANSHIP AND CONSERVATORSHIP

One of the most difficult situations for a family member or health care provider to face is an older adult who has become unable to make day-to-day decisions. Aside from the anguish this causes all parties, the issue quickly becomes one of providing sufficient but not intrusive assistance, once the adult's ability to make financial and/or medical decisions is compromised. Guardianship is the most drastic intervention, as it represents the partial or total removal of the person's civil rights and appoints a surrogate to make decisions on his or her behalf.

Guardianship, called conservatorship in some states, is usually divided into two types: personal and estate (or property). The terminology differs from state to state, but here the term guardianship refers to guardianship of both the person (the ward) and the property.

Nationally, it is estimated that 500,000 to 1 million adults have guardians (Associated Press, 1987). Although this is a significant number of people, research on guardianship has been difficult to conduct because it is legislated at the state level but practiced at the local level. Each presiding judge has the power to observe state rules or to ignore them. In areas with relatively large populations, exceptions to state laws or regulations are often made through administrative orders. In less populated, more rural areas, even these orders are sometimes dispensed with. However, many states strive to impose guardianship within the context of the principle of "least restrictive alternative."

All jurisdictions are empowered to appoint a guardian for any adult who is determined to lack the ability to make his or her own decisions, but the standards for that determination differ. In most states, a court investigator or an examining committee makes a formal assessment. These assessments are based on the adult's functional and cognitive capacity and often include such instruments as the Folstein Mini-Mental State Examination or the Short Portable Mental Status Questionnaire. If a person is judged to lack the capacity to make personal or financial decisions and has not already designated a surrogate through an advance directive, a guardian will be appointed for the person and/or the estate, as the circumstances warrant.

There is increasing recognition that older adults may be able to make some decisions while lacking the capacity to make others. For example, older adults may not be able to make medical treatment decisions but may be quite capable of determining other aspects of their personal lives, such as where they will live. To provide an option that can be tailored to the individual's abilities, most states allow for limited guardianship—removing the right to make only certain decisions. However, the limited guardianship option is rarely used, be-cause many judges and lawyers believe that the older adult's condition will continue to deteriorate, eventually requiring full (plenary) guardianship (Keith & Wacker, 1994; Wilber & Reynolds, 1995).

Initiating Guardianship Proceedings

An older adult with impaired capacity often comes to the attention of a family member or a neighbor, who notifies the police or adult protective services. The precipitating factor can include such things as possible elder mistreatment, undue influence, inability to perform activities of daily living or pay bills, and the like. If the older adult is in a long-term-care facility, the trigger may be deterioration in cognitive functioning rather than ability to live independently. In these cases, the need for medical decision making often prompts the petition for guardianship. For simplicity's sake, here the referring person is assumed to be a family member.

Once a guardian is believed necessary, a petition is filed with the probate court. The petition provides information to the court on the person's physical and mental condition and lists the next of kin. The petition may include a request to appoint a guardian, but in some states (e.g., Florida), a separate application is necessary. In many states, an attorney must file the petition.

Once a petition is filed, the probate court will hold a hearing to rule on whether to appoint a guardian and who that guardian will be. The adult in question, the next of kin, and any attorneys involved receive notice of the hearing and have the right to be present and to contest the guardianship.

Appointing a Guardian

Most often the family member closest to the adult in question, either emotionally or geographically, is appointed the guardian. In many cases, however, adult children do not live nearby or do not get along with one another. Appointing an adult child as guardian in such circumstances can become a logistical or emotional problem.

Many states provide for the appointment of a public guardian if there is no one else to act in this role. For older adults with sufficient funds, a bank trust department or attorney can be hired to act as guardian. These services tend to be expensive and are clearly not appropriate for everyone. In states that do not have public guardians, a professional guardian can be appointed; this professional can be an individual, a company, or a social services agency.

Appropriate guardianship varies with the circumstances of the person who needs protection. In some states, an older adult can file a preneed guardian statement, informing the court which individual the person chooses as guardian should one ever be needed. Otherwise, two factors are paramount: the nature of the relationship and potential conflict of interest. In the former case, the issue of emotional closeness and trust must be weighed against family dynamics, ability to manage finances, and proximity (see Wilber & Reynolds, 1996, for a discussion of relationships and financial exploitation). In the latter case, the person or entity appointed as guardian should not also be providing medical or social services (see Kapp, 1992, for a thorough discussion of this type of conflict).

Responsibilities of a Guardian

Guardians of the person must make decisions regarding all kinds of health care, as well as place of residence and aspects of the person's social life. Their duties include (but are not limited to) the following:

- Complying with all filing requirements (the probate court will inform individuals of the specifics in each state)
- Complying with the rules of the jurisdiction for procedures such as changing the residence of a ward, charging a guardian's fee, or any other action that requires court approval (this varies by state and county)
- Determining where the ward will live and under what circumstances (e.g., at home or in an as-

sisted living facility, nursing home, or other placement; alone or with others; receiving or not receiving at-home or community services)
- Consenting on behalf of the ward to medical, surgical, or behavioral treatments recommended by the ward's medical providers
- Terminating life support

Most courts require an annual filing describing any actions the guardian has taken to protect or improve the ward's life and what the guardian plans to do for the ward in the coming year.

Guardians of the property have similar filing requirements. They are charged with receiving all assets and filing an inventory of the ward's property as soon as possible. Many states put a time limit on the filing of an inventory (e.g., 30 or 60 days), but an amended inventory can be filed if subsequent property is found. The guardian must then file a periodic accounting of all income received, bills paid, and property bought and sold on behalf of the ward. In addition to these duties, the guardian of the property has to insure the assets of the estate; pay the ward's living expenses; invest the assets according to the "prudent person" rule; employ attorneys, investment advisers, and other professionals as needed; buy and sell property, including real estate; and pay burial and funeral expenses (some states consider this the duty of the executor, not the guardian). Many of these actions require the prior approval of the court.

Even when a guardian has been appointed, there is no substitute for the vigilance of a caring family member or professional. Even when the guardian rarely interacts with the patient, the guardian can be an advocate for the person. For instance, a health care professional who identifies a patient as overmedicated but lacks the authority to change clinicians or suggest a specialist referral could involve the guardian in the plan of care. By combining the health care professional's knowledge of the patient and the advocacy role of the guardian, the best care for the patient can be ensured.

The role of the guardian of the property is critical and highlights the importance of having a caring professional, family member, or friend who is alert to the situation. Although the guardian is responsible to the court and to the ward for the care provided and usually has all-encompassing power over the ward, all it takes to replace a guardian is a petition to the court and a hearing.

Fortunately, most guardians are diligent, caring, and careful people who treat their wards well. Under the right circumstances, having a guardian for an impaired older adult can be an enormous relief to the family, to providers, and to the ward.

SANDRA L. REYNOLDS

See Also
Advance Directives
Elder Mistreatment: Overview

Elder Neglect
Financial Abuse
Money Management

Internet Resources
Facts about Law and the Elderly, American Bar Association
http://www.abanet.org/media/factbooks/eldtoc.html

Guardians and Trustees, Sarasota, FLA
http://www.sarasota-guardians.com

Law and the Elderly: Guardianship
http://www.uslaw.com/library/article/ABAElderlyGuardianship.html

H

HEADACHE

Headache is not usually associated with significant underlying pathology. Occasionally, however, underlying disease is undetected because headache is such a common complaint, reportedly as high as 17% in a population-based cohort of elderly (Cook, Evans, & Funkestein, 1989). Despite its benign nature, headache often has a significant impact on quality of life and productivity.

Although headache in older and younger adults is similar in most respects, there are some notable differences. Most importantly, some causes are unique to the elderly, such as temporal arteritis. Signs and symptoms of the underlying cause may also be different, such as the absence of fever and meningismus in meningitis. Thus, it is important to take the complaint of headache seriously and begin the evaluation with the idea that the headache could be secondary to an underlying disease. The most important tools for evaluating headache in geriatric patients are the history and physical and neurological examinations. Subsequent diagnostic testing should be guided by the information gained during the clinical evaluation. Careful clinical assessment may obviate the necessity of imaging, which is overutilized.

Secondary Headache

During the initial evaluation, the physician must differentiate underlying pathology causing a symptomatic headache (i.e., secondary headache) from the more benign primary headache, in which the headache itself is the primary problem. Underlying causes should be considered before diagnosing the more benign primary headache (e.g., tension-type headache).

Giant cell arteritis is a granulomatous vasculitis involving large arteries, including the carotid, vertebral, and temporal arteries. It classically presents with scalp pain; temporal headache with tender, enlarged temporal arteries; and jaw claudication. It may occur in association with polymyalgia rheumatica, with a history of several months of malaise, weight loss, generalized weakness, myalgias, and low-grade fever. Diagnosis is based on clinical suspicion, supported by an elevated erythrocyte sedimentation rate, and must be confirmed by temporal artery biopsy. Expeditious treatment with corticosteroids is imperative to prevent compromise of vision.

Central nervous system pathology that causes headache includes intracranial hemorrhage, mass lesions, and meningitis. It is important to keep in mind that in the elderly, subdural hematoma can be present in the absence of focal neurological signs or a history of identifiable trauma. Signs and symptoms suggestive of secondary headache that require further investigation include new headache; significant change in preexisting headache, such as severity, frequency, characteristics, or location; any focal neurological signs or symptoms; nausea and vomiting without a history of migraine; fever; meningismus; prior history of malignancy; sudden, severe, explosive headache; personality change or drowsiness; progressively worsening headache; and seizures.

Causes outside the nervous system that should be kept in mind during the initial evaluation include dental disease with referred pain, sinusitis, and glaucoma.

Primary Headache

Primary headache is categorized into definable entities that suggest prognosis and can facilitate appropriate treatment.

Cervicogenic headache is pain originating primarily from disease in the neck. Pain due to disease in the upper cervical regions can be referred to all parts of the head, especially the occipital region, but also the frontal, temporal, parietal, and even orbital regions. Neck pain may not be a prominent complaint in some patients; direct questioning is necessary to elicit these symptoms and a history of neck trauma. Neck crepitus and cervical paraspinal tenderness are usually present. Results of cervical spine imaging studies must be interpreted carefully, taking into consideration the clinical picture. Patients with significant cervical-related head pain can have normal radiographic studies, whereas asymptomatic patients can have significant cervical spondylosis.

In patients with headache associated with cervical spine dysfunction, treatment involves explaining the cause of the pain, prescribing physiotherapy, and educating the patient about ergonomics, such as using a firm pillow or a cervical collar. The patient should be taught how to avoid holding the neck in one position for long periods, especially in positions in which there is excessive neck flexion, extension, or torsion. Pain can be initiated or exacerbated by long-distance driving or prolonged time spent typing or using a computer. Analgesic therapy includes nonsteroidal anti-inflammatory drugs (NSAIDs) and muscle relaxants. Treatment of coexisting depression or sleep disorders can also relieve pain.

Tension-type headache is a common diagnosis. Criteria established by the Headache Classification Committee of the International Headache Society (1988) include symptoms such as mild to moderate bilateral, nonpulsating, pressing or tightening pain that does not inhibit activity. There is no associated nausea or vomiting and no aggravation of symptoms with routine physical activity. Secondary headache must be excluded to make this diagnosis. Treatment should be multidimensional, including pharmacological and psychological interventions. If treatment with medications is warranted, a dual approach consisting of abortive and possibly prophylactic therapy should be used. With abortive therapy, medication is to be taken as needed at the onset of the headache. Treatment should begin with the least toxic medications, such as acetaminophen or aspirin. If necessary, combination preparations with caffeine can be used, or trials with other NSAIDS (e.g., ibuprofen, naproxen) can be at-

tempted. Failure to respond to one type of NSAID should not immediately prompt treatment with an entirely different class of drugs. Patients who fail to respond to one NSAID may have a good response to another. A careful history for gastric distress, ulcers, or inflammation must be obtained, and the patient and family should be warned about serious side effects that can occur (e.g., gastrointestinal bleeding, ulcers, kidney disease), especially when these drugs are used long term. Overuse of analgesics can lead to rebound headaches and conversion of episodic headaches to a chronic form.

If abortive therapy is used by the patient more than a few times a week, prophylactic therapy must be seriously considered. Tricyclic antidepressants (TCAs) can be used prophylactically and can control pain in patients with or without concomitant depression. Although amitriptyline is the most well-established drug in the TCA class for treating headaches, the secondary amines, such as desipramine and nortriptyline, have fewer sedative and anticholinergic side effects and are effective in treating headaches. Common side effects are constipation, urinary retention, orthostatic hypotension, and electrocardiogram changes—all of which should be carefully monitored. Valproic acid can also be used, but the patient must be carefully monitored for tremor, thrombocytopenia, and liver enzyme elevations. Propranolol should be used cautiously, and patients should be monitored for some of the more common side effects, including fatigue, depression, memory loss, impotence, loss of libido, bradycardia, and hypotension. As part of multidimensional treatment, the patient can be referred to a comprehensive pain management center with expertise in several areas, including neurology, anesthesia, rehabilitation medicine, psychology, and occupational and physical therapy. These centers may also offer alternative treatments not otherwise available to patients.

New onset of migraine is rare in the geriatric population. The suggestion of this diagnosis should prompt an evaluation to rule out secondary causes. Patients with a history of migraine who continue to have headaches through their geriatric years should be treated appropriately.

Trigeminal neuralgia causes severe pain and has an increased incidence in the elderly. It is characterized by paroxysms of high-intensity stabbing pain lasting several seconds, with intervening periods of relief. Usually triggered by any minimal stimulation of the face, episodes of pain usually last one to two hours and may occur over weeks to months, followed by a long period without painful episodes. The pain may be described as "electric," is unilateral, and is usually in the distribution of the second or third division of the trigeminal nerve (i.e., over the cheekbone and jaw). Medication for neuropathic-type pain, such as carbamazepine, is usually the first line of treatment. Phenytoin and baclofen are also effective. Capsaicin local cream should be tried, because there is no risk of systemic toxicity. If multiple drug regimens fail, referral for surgery should be considered.

Treatment Principles

Headaches can be treated with medication, but drugs should be only a small part of the overall approach. Situational, social, and psychological triggers should be addressed. Once a secondary headache is ruled out, the patient needs reassurance that there is no serious underlying pathology causing the headache. This alone may provide symptomatic improvement by relieving anxiety, which can exacerbate pain. A "headache diary" can help improve symptoms by giving the patient a sense of control and might identify triggers. All patients should be carefully assessed for common comorbid disorders—depression, anxiety, and sleep disorders—that can exacerbate headache symptoms. Appropriate treatment can significantly improve symptoms and potentially obviate the need for analgesics.

Stressful life events can be associated with the onset or worsening of headaches; addressing these psychological factors can help alleviate symptoms. Headaches exacerbated by stress might respond to relaxation techniques and stress management. The so-called letdown headache sometimes occurs when a period of stress comes to an end.

Ingested substances that can trigger headache include alcohol, nitrites in hot dogs and processed meats, and monosodium glutamate (e.g., in Chinese food). Carbon monoxide can induce headaches, as can environmental changes (e.g., bright lights; odors; changes in humidity, air pressure, or temperature), hormonal changes, and emotional stress. The patient should be educated about these potential triggers and avoid them if possible. Other triggers may be delayed or missed meals, fatigue, exercise, and sexual activity. Changes in sleep patterns, such as insomnia or excessive sleeping, may contribute to symptoms of pain.

Iatrogenic causes of headache should also be considered, given that the elderly are at risk for polypharmacy. Medications commonly related to new onset or worsening of existing headache are nitroglycerine preparations, hormone replacement therapy, sympathomimetics, cyclosporine, gemfibrozil, and many other medications. If medication-induced headache is suspected, tapering and discontinuation of the agents are warranted.

Although headache is a benign condition, it is associated with significant morbidity that is underestimated by the medical community. Sufferers of chronic headache disorders can have a lower level of function than patients with other chronic medical illnesses (Solomon, Skobieranda, & Graggs, 1993). With appropriate treatment, quality of life can improve significantly, not only for the patient but also for the family or caregivers.

CARY BUCKNER

Internet Resources
American Council for Headache Education
http://www.achenet.org

Mayo Clinic Health Oasis
http://www.mayohealth.org/mayo/9901/htm/headache.htm

National Headache Foundation
http://www.headaches.org

HEALTH MAINTENANCE

A major goal of geriatric medicine is to promote and achieve maximal functional and independence for older adults. The unprecedented increase in the number of individuals older than 65 years of age is perhaps the best testament to the health promotion and disease prevention practices of the past few decades. Following advances in treatment modalities in the 20th century has come a new awareness of the need to promote healthier lifestyles, screen for disease, and control chronic illness. The use of health maintenance and disease prevention to enhance the aging process and reduce disability comes under a variety of labels, including morbidity compression, usual versus successful aging, and preventive gerontology. For clinicians, the term "preventive gerontology"—the study of those elements of lifestyle, environmental control, and health care management that will maximize longevity of the highest possible quality—may best identify those aspects of care that should be considered for older adults (Hazzard, 1999).

Aspects of primary prevention include vaccinations, maintenance of optimal function (avoidance of sensory deficits, falls and immobility, polypharmacy, depression and substance abuse, urinary incontinence, dementia, elder abuse and neglect), healthy lifestyle (exercise, nutrition, dental care, nonsmoking, and socialization and social supports), and chemprophylaxis (aspirin and vitamins). Aspects of secondary prevention include screening laboratory evaluations, disease screening (cardiovascular disease, hyperlipidemia, hypertension, diabetes, and osteoporosis), and cancer screening (breast, colon, prostate, cervical, and other cancers).

Despite the trend toward promoting health and wellness, a significant portion of the elderly community has not been reached. Most available literature on prevention and screening implicitly includes older people, but appropriate upper age limits are often unresolved (Goldberg & Chavin, 1997). Those few studies that do explicitly identify older adults indicate that most elders are not receiving the recommended preventive health procedures or are not receiving them at the proper intervals. A large study of self-reported health utilization found that over half of the elderly patients had never had a mammogram or fecal occult blood testing (Choa, 1987).

Practitioners' neglect of health promotion and disease prevention may stem, in part, from the significant disagreement about what routine screening should include, the age at which screening should commence, and when it should be discontinued (Goldberg & Chavin, 1997). The age recommendations may have to be reevaluated in light of the fact that persons 65 years old are likely to live an additional 15 years and remain independent and active until age 75 (Black, 1990). Because many elders remain active into their eighth decades and beyond, it seems reasonable to extend the age at which we stop screening individuals for diseases such as breast and colon cancer. However, the U.S. Preventive Services Task Force (USPSTF), the Canadian Task Force (CTF) on Periodic Health Examination, and the American Cancer Society (ACS) disagree about what constitute reasonable, safe, and cost-effective practices for cancer screening. Additional research is needed with the young-old, the middle-old, and the old-old before it can be said with certainty which practices extend life and promote wellness. However, research clearly suggests a variety of activities that should be included in a comprehensive program of health promotion and disease prevention for older adults. As with any recommendations, interventions should be based on each patient's individual health profile, life expectancy, health beliefs, and goals for care.

Before embarking on a screening program, certain questions should be addressed by the practitioner (Scheitel, 1996).

1. Is the target disease an important clinical problem? Does the burden of disability warrant early detection?
2. Is the natural history of the target disease understood? Does the disease have a latent or early symptomatic period?
3. Is the screening diagnostic strategy effective? Is the accuracy of testing (sensitivity and specificity) established?
4. Is the test acceptable to patients with low discomfort and risk?
5. If the result of the screening test is positive, will patients accept subsequent diagnostic evaluation?
6. Does a known treatment exist for the target disease? Is the treatment available and effective?

Does the treatment have associated risks, such as adverse drug effects or adverse outcomes of surgical intervention?

7. Is the cost of testing balanced by the benefit of treatment?

Based on the answers, the practitioner and the older patient can determine whether screening is reasonable. Patients should be involved with decision making, because they must understand their liability for potential costs, be able to follow instructions to prepare for the test, and accept the possibility of false-positive and false-negative results.

Primary Prevention

Health Promotion

Vaccinations: Many infectious diseases are preventable by vaccinations that have varying degrees of efficacy. For older adults, the most important vaccinations include influenza, pneumococcal pneumonia, and tetanus-diphtheria. Medicare Part B and private insurance companies cover some vaccinations, including pneumococcal pneumonia and influenza vaccinations.

Maintenance of Optimal Function

Geriatricians and gerontologists have identified certain syndromes that limit elders' ability to live life to the fullest. Screening for these geriatric syndromes and correcting problems when they are identified can greatly enhance the quality of life for an older person.

Sensory Deficits: Screening for auditory and visual deficits should be done for all older adults. Visual disturbances resulting from cataracts, glaucoma, or macular degeneration can result in frustration, depression, falls, and loss of independence. Correcting significant hearing loss can restore an individual's interest in the world and greatly improve his or her quality of life.

Falls and Immobility: Falls are the leading cause of fatal injuries in persons over 75 (Carethers, 1992); most are due to environmental hazards and disorders of balance or gait. Patients should be asked about falls, and efforts should be made to avoid further falls. Home visits and physical therapy evaluations can identify the causes of falls in many cases. Podiatry evaluation as well as careful inspection of footwear can often improve mobility.

Polypharmacy: Perhaps the most difficult problem in older adults with complex medical conditions is minimizing the number of drugs prescribed. Drug interactions and adverse drug reactions are common and can be serious and debilitating. Differences in pharmacodynamics and pharmacokinetics related to age can lead to toxicities, resulting in falls and disorientation. Every effort should be made to review the patient's medication list at each encounter and to remove all unnecessary medications.

Depression and Substance Abuse: Depression is common among the elderly, especially those who are home-bound and isolated. Practitioners should screen for this disorder, as depression in the elderly often presents atypically and may manifest as a lack of initiative or somatization. Many older adults are reluctant to reveal feelings of depression and self-medicate with alcohol or other agents. Providers should explore these issues with all patients.

Urinary Incontinence: The incidence of urinary incontinence increases with age and can lead to social isolation, infection, and pressure ulcers. Urinary incontinence is often the compelling factor in nursing home placement decisions. Patients should be asked about incontinence, and an evaluation undertaken when indicated.

Dementia: An elderly patient with unexplained behavioral changes should be screened for dementia. Any patient being evaluated for dementia should also be screened for depression, as this can cause a pseudodementia. Being able to plan for the future and put affairs in order while cognitive function allows are arguments for the early detection of dementia.

Elder Abuse and Neglect: Often underrecognized and underreported, elder abuse and neglect must always be kept in mind. The elder's living arrangement can sometimes clue the practitioner to abuse and neglect situations.

Healthy Lifestyle

Exercise: The benefits of exercise throughout the life cycle are clear. Inactivity is associated with obesity, diabetes, hypertension, and cardiovascular disease. Exercise in any age group has beneficial effects on strength, mobility, balance, bone density, blood pressure, weight, cognitive function, and socialization (Carethers, 1992). Improving strength and balance can reduce falls among the elderly. Activities should be tailored to each patient's functional and cardiorespiratory status. Recommended activities for more mobile elders include walking, swimming, stationary bicycling, gardening, and dancing. For those with gait instability or physical limitations, range-of-motion exercises and isometric activities, such as using elastic bands, may prove beneficial. Exercises that promote socialization and provide group support may promote better adherence to an exercise program.

Nutrition: Maintaining proper nutrition can be difficult for older individuals; shopping and cooking can be burdensome. Some elders rely on community-based services for their meals. Malnutrition and dehydration are common among the home-bound elderly. Poor dentition, swallowing difficulties, loss of taste, medication side effects, and depression can all contribute to poor intake. Weight should be considered a vital sign and checked at each visit. A diet history that includes fiber and fluids for good bowel habits may be needed to assess for adequate intake. Vitamin supplementation should be considered if patients are unable to ingest adequate amounts of food or if the diet is poorly balanced. Special attention should be given to calcium intake in both men and women.

Dental Care: A careful dental examination at least yearly is indicated to screen for periodontal disease. Fitting patients for dentures and assessing their ability to chew and swallow safely are important considerations.

Smoking: The deleterious effects of smoking on the cardiovascular system are clear. When counseled by a physician, about 6% of patients stop smoking. When attempts at smoking cessation are supported in a group setting with careful follow-up and pharmacological interventions, up to 25% of patients stop smoking (Carethers, 1992).

Socialization and Support System: Socialization and support networks are integral components of wellness. Patients who are isolated have an increased incidence of depression, anxiety, and substance abuse. Recent studies suggest that social interaction on a regular basis slows cognitive decline. Older patients should be encouraged to have meaningful personal contacts on a regular basis.

Chemoprophylaxis

Aspirin: Studies of the efficacy of low-dose aspirin for the prevention of myocardial infarction have produced data that are not entirely clear. Low-dose aspirin to reduce the risk of heart attack can be discussed with patients as a reasonable option (Scheitel, 1996).

Vitamins: There are no specific recommendations for vitamin supplementation, but the elderly can be at risk for nutritional deficits for many reasons, including home-boundness, limited finances, cognitive impairment, dental issues, and some biological changes in the gastrointestinal system that affect absorption. The home-bound, in particular, may have reduced sun exposure and require vitamin D supplementation. Certain antioxidants such as vitamin E and vitamin C have been suggested to slow the progression of macular degeneration and reduce cancer risks (Gleich, 1995). The approach should be dictated by the patient's medical history and dietary intake assessment. Patients have to be cautioned to avoid excess intake of fat-soluble vitamins (K, A, D, and E).

Secondary Prevention

Laboratory Screening

To date, no studies support the use of routine laboratory testing in asymptomatic elderly patients. Despite the absence of recommendations, most practitioners caring for elderly patients perform some basic laboratory tests on a regular basis. Some practitioners suggest that screening for thyroid disease in elderly women should be part of routine health maintenance, as the risk of thyroid disease increases with age, especially in women over 50 (Wolf-Klein, 1989). Laboratory evaluation should generally be minimal and done only for individual risk factors or symptoms.

Disease Screening

Cardiovascular Disease: Cardiac disease can present atypically in older individuals, particularly in elderly woman. Risk factors for cardiovascular disease that should be assessed in patients presenting for primary care are hypertension, diabetes, hyperlipidemia, and smoking.

Hyperlipidemia: Hypercholesterolemia is estimated to increase deaths related to coronary artery disease by 30% in adults 65 to 75 years of age (Scheitel, 1996). Data collected since the 1970s indicate that reducing the cholesterol level can significantly reduce cardiovascular and cerebrovascular events in patients 35 to 59 years old. Research examining the long-term benefits of reducing cholesterol in the elderly is lacking. Current recommendations favor cholesterol screening for patients with significant coronary artery disease risk factors and for those with previous coronary artery or cerebrovascular events.

Hypertension: Hypertension is easily screened within the community. Several studies of the benefits of hypertension screening in all age groups demonstrate particular benefit for older individuals and clearly show reduced risk of stroke and heart disease (up to 30% in some studies) when blood pressure readings are kept below 160/90. Blood pressure monitoring is indicated at every visit to a primary care provider's office, or at least annually. Some studies question the usefulness of blood pressure control after age 80 and raise the question of iatrogenesis from antihypertensive medications (Gleich, 1995). Further research is needed on blood pressure screening with advancing age.

Diabetes: Diabetes mellitus is a major risk factor for cardiovascular disease and is the sixth leading cause of mortality in the United States for patients older than 75 years (Scheitel, 1996). There is no definitive data on disease-related outcomes with respect to diabetes in the elderly. In all age groups, patients with hyperglycemia or diabetes have increased morbidity and mortality. Strict control of blood glucose levels can reduce the complications associated with diabetes. The risk of hypoglycemic events that can occur with tight control has to be compared with the risk of end-organ damage seen in diabetics with poorly controlled glucose levels. There are no official recommendations for screening asymptomatic older individuals, except for those with strong family histories or multiple comorbid illnesses.

Osteoporosis: The devastating effect of osteoporotic fractures in older women is well understood by geriatricians. Hip fracture is associated with significant morbidity and mortality and often leads to institutionalization. For this reason, screening for osteoporosis risk factors in asymptomatic older adults is extremely important. Although postmenopausal women are commonly screened, men over age 75 are also at risk. All postmenopausal women should be encouraged to get adequate calcium and vitamin D and engage in weight-bearing exercise when possible. Patients who have already sustained fractures and those with abnormal bone density tests should receive additional agents such as estrogen, bisphosphonates, or selective estrogen receptor modulators (SERMs) to prevent further bone loss.

Cancer Screening

Breast Cancer: The incidence of breast cancer increases with age; early detection can reduce morbidity and mortality. In 1991, 77% of breast cancer deaths were in women older than 55 years; 31% were in women older than 75 years (Scheitel, 1996). Most clinicians recommend yearly mammograms starting at age 40 until age 65 to 70 years. However, since the mean life expectancy for a 75-year-old woman is 12 years, cessation of screening at age 70 may be premature. Some clinicians recommend yearly or biennial mammogram screening of all women over 40 who have a life expectancy of 6 years or longer. Sensitivity of the self-exam as a screening tool for breast cancer is in the 20% to 30% range and should be recommended as an adjunct to mammography or to those patients refusing mammograms.

Colon Cancer: Colon cancer ranks second among all cancer mortalities and accounts for 54,900 deaths yearly. Recommendations for colon cancer screening remain unclear. Sigmoidoscopy entails some discomfort and risk to the patient. Fecal occult blood testing has been shown to decrease mortality but has a high false-positive rate and a sensitivity of about 26%. Two recent case-controlled studies found that screening sigmoidoscopy decreased colorectal cancer mortality by 60% to 95% (Scheitel, 1996). The American College of Physicians and ACS recommend sigmoidoscopy every 3 to 5 years for all patients over age 50, as well as yearly fecal occult blood testing. The CTF and USPSTF recommend a more risk-associated approach and favor annual fecal occult blood testing in those with a family history of colorectal cancers or those with a personal history of endometrial, ovarian, or breast cancer.

Prostate Cancer: The incidence of prostate cancer increases with age. More than 80% of all prostate cancer cases occur in men older than 64 years (Scheitel, 1996). Most prostate cancers have an indolent course and do not present clinically. Current screening methods include the digital rectal examination (with a positive predictive value [PPV] of 31%), the transrectal ultrasound (PPV up to 41%), and the prostate-specific antigen (PPV of 35%), but they are not accurate enough for use as screening tests. The morbidity and mortality associated with follow-up procedures generated by false-positive results need careful consideration. The USPSTF does not endorse any of these three methods for prostate cancer screening. The ACS supports digital rectal examination yearly.

Cervical Cancer: There is no consensus among experts regarding the usefulness of cervical cancer screening in elderly women. The USPSTF and CTF recommend discontinuing Pap smears at age 65 and 70, respectively assuming previously negative results. The ACS, National Cancer Institute, and others recommend yearly Pap smears throughout the patient's life (Scheitel, 1996). Most experts agree that a patient who has not had adequate screening before age 65 should be screened at least once.

Other Cancers: At present, there are no recommendations for routine screening for lung or ovarian cancer, despite the high mortality associated with both. Asymptomatic patients with either strong family histories or environmental risk factors should be considered for more aggressive screening techniques. Screening for oral cancer or skin cancer can be easily incorporated into the physical exam in those patients at risk. New screening techniques to detect lung cancer are being investigated, but none is recommended at present.

Although recommendations for health screening and health promotion are unclear in many cases, there is no question that practitioners and patients are taking a more active interest in health-promoting behaviors. The art of good medical care is to listen to the patient's agenda, work within the parameters of his or her lifestyle, and tailor recommendations by the experts to each person's individual medical issues and health beliefs.

VERONICA LoFASO

See Also

Gait Assessment Instruments
Glaucoma
Heart Failure Management: Congestive and
    Chronic Heart Failure
Hypertension
Immunization
Low Vision
Nutritional Assessment
Obesity
Oral Health
Osteoporosis
Polypharmacy: Drug-Drug Interactions
Polypharmacy Management
Protein Energy Undernutrition
Sensory Change/Loss: Smell and Taste
Sexual Health
Social Supports (Formal and Informal)
Urinary Incontinence Assessment
Urinary Incontinence Treatment

Internet Key Words
Aspirin
Breast Cancer
Cervical Cancer
Colon Cancer
Dementia
Dental Care
Depression
Diabetes
Disease Prevention
Elder Abuse
Exercise
Falls
Functional Status
Health Promotion
Hyperlipidemia
Hypertension
Immobility
Influenza
Neglect
Nutrition
Osteoporosis
Pneumonia
Polypharmacy
Prostate Cancer
Sensory Deficits

Smoking
Social Supports
Substance Abuse
Tetanus
Urinary Incontinence
Vitamins

HEALTH PROMOTION SCREENING

Health promotion covers a broad range of health care services meant to prevent or limit future disease and disability. Many different organizations gather strong evidence in support of (or against) the use of various preventive measures. Evidence of decreased or increased effectiveness of preventive measures for older people is often lacking because intervention studies are not often conducted with that population.

The focus here is on blood pressure, serum cholesterol and glucose, and urinalysis screening tests in older persons. The target conditions these tests are meant to identify, the usefulness of the particular screening test, and general recommendations regarding these tests are summarized.

Blood Pressure Screening

The prevalence of hypertension increases with age. The upper limit of normal adult blood pressure is 140/90 mm Hg; higher levels are associated with an increased risk of cardiovascular disease. Isolated systolic hypertension is a systolic blood pressure equal to or greater than 140 mm Hg, with a diastolic blood pressure of less than 90 mm Hg. High-normal blood pressure is systolic blood pressure in the range of 130 to 139 mm Hg and/or diastolic blood pressure in the range of 85 to 90 mm Hg.

Diastolic blood pressure rises progressively with age until age 50 or 60, then drops off slightly. However, systolic blood pressure continues to rise throughout life, peaking late in life. The prevalence of isolated systolic hypertension increases from 5% at age 60 to 24% at age 80. Hypertension is more common and has a more serious prognosis among

African Americans than among whites. Several studies report that healthy older patients (men and women) with either systolic and diastolic hypertension or isolated systolic hypertension can be treated with relatively modest doses of antihypertensive medications with a substantial decrease in strokes and stroke-related mortality, coronary artery disease mortality, and mortality overall (Messerli & Grodzicki, 1996). The beneficial effects of lowering blood pressure in persons over age 80 has not been conclusively demonstrated. Some clinicians recommend that rapid lowering of blood pressure in hypertensive older patients (> 65) should be avoided (Messerli & Grodzicki, 1996).

All major prevention organizations recommend blood pressure screening in adults, but specific recommendations for the frequency vary or are left to the discretion of the clinician. The U.S. Preventive Services Task Force (1996) does not suggest a specific interval for screening. The American College of Physicians recommends blood pressure measurement every one to two years, with more frequent monitoring of individuals with high-normal blood pressure and of those at high risk for hypertension (e.g., African Americans, obese individuals, those with first-degree relatives with hypertension, and those with a personal history of hypertension) (U.S. Department of Health and Human Services, 1998). Similarly, the National High Blood Pressure Education Program recommends blood pressure measurements for adults every two years and at each patient visit, if possible. Patients found to have high-normal blood pressure should have their blood pressure rechecked within one year. As a general rule, blood pressure should be estimated by the average of at least two readings—more if the two readings differ by more than 5 mm Hg. Most organizations do not suggest an age cutoff to stop blood pressure screening, but the Canadian Task Force on the Periodic Health Examination recommends blood pressure screening only up to age 84 (U.S. Department of Health and Human Services, 1998).

Unfortunately, many hypertensive patients are not treated effectively because of failure to identify the hypertension and institute treatment and failure of asymptomatic hypertensive patients to adhere to the therapeutic regimen.

Serum Cholesterol

Elevated cholesterol levels increase the risk of coronary artery disease. Total cholesterol levels gradually increase with age; women over 55 have higher cholesterol levels than men of the same age. The total cholesterol level is not a reliable predictor of the relative risk of coronary artery disease, but because the number of coronary events increases with age, the risk "attributable" to the increase in cholesterol actually increases with age. Reduction of cholesterol levels in intervention trials decreases the incidence of and the morbidity and mortality associated with coronary artery disease (American College of Physicians, 1996). Evidence supports the reduction of elevated cholesterol levels in individuals with known coronary artery disease (i.e., for secondary prevention) (American College of Physicians, 1996). Hypercholesterolemia screening is indicated in people with a history of myocardial infarction, angina pectoris, evidence of coronary disease, or other evidence of vascular disease, such as stroke or claudication. Subgroup analyses in secondary prevention trials generally demonstrate that the risk reduction with treatment of hypercholesterolemia in patients over 65 years of age is no different from that in younger patients (La Rosa, 1994). Most prevention organizations do not suggest an age cutoff for cholesterol screening. The recommended lipid analysis usually includes measurement of high-density lipoprotein (HDL) and low-density lipoprotein (LDL) levels to refine the patient's risk classification, help guide the choice of therapy, and monitor the response to treatment.

Evidence is not as strong for the reduction of elevated cholesterol levels for primary prevention of coronary artery disease (i.e., in asymptomatic individuals without known coronary or related disease). Studies indicate that cholesterol reduction in men prevents coronary disease–related deaths, but there is less evidence for women. The presence of other cardiac risk factors increases the likelihood of benefiting from treatment of hypercholesterolemia. The American College of Physicians (1996) does not recommend screening for primary prevention in men and women 75 years of age and older and reports that there is insufficient evidence to either recommend or discourage screening for primary prevention in men and women 65 to 75 years of age. The U.S. Preventive Services Task Force and the Canadian Task Force on the Periodic Health Examination do not recommend cholesterol screening in asymptomatic people aged 65 and older. The recommended screening test is the total cholesterol level, with repeat testing periodically (e.g., every 5 years) in persons whose measured values are near the treatment threshold.

Plasma Glucose

Diabetes mellitus is diagnosed in about one-fifth of people over the age of 65 years; approximately half of all diabetics are over 60 years of age. The prevalence of diabetes varies markedly in different racial and ethnic groups: white Americans, 6%; African Americans, 10%; Mexican Americans, 13%; Japanese Americans, 14%. Most older diabetics have non-insulin-dependent diabetes mellitus (NIDDM). The long-term complications of diabetes—retinopathy, nephropathy, neuropathy, cardiovascular disease, and peripheral vascular disease—can be reduced with control of plasma glucose levels, although the effectiveness of this prevention is not as well established in older people.

The fasting blood glucose level is the main test for the screening and diagnosis of diabetes, with a value of 126 mg/dL or greater meeting the criterion for diabetes. Other criteria that establish the diagnosis include the classic symptoms of diabetes, such as polydipsia, polyuria, polyphagia, and weight loss, accompanied by a random glucose level greater than 200 mg/dL or a two-hour plasma glucose level of 200 mg/dL or greater during an oral glucose tolerance test. Organizations vary in their recommendations for or against diabetes screening. The American College of Physicians does not recommend screening in healthy, asymptomatic individuals, and the U.S. Preventive Services Task Force reports that there is insufficient evidence to recommend for or against screening. However, both groups suggest that it may be prudent to screen adults with risk factors such as increased age, obesity, family history, and high-risk

ethnic groups. The American Diabetes Association recommends diabetes screening every three years in all adults age 45 and older.

Urinalysis and Renal Function

Urinalysis by dipstick is a quick and inexpensive test. Conditions commonly identified when screening asymptomatic individuals include occult bacteriuria, hematuria, and proteinuria. Although asymptomatic bacteriuria is common in older women, treatment does not improve long-term health (U.S. Preventive Services Task Force, 1996). Urine dipsticks are relatively accurate in detecting occult hematuria, which can indicate urinary tract malignancies. However, many benign conditions can cause hematuria. The positive heme dipstick test has a very low positive predictive value for malignancy. Proteinuria is usually caused by conditions that are benign or untreatable. The American College of Physicians does not recommend routine urinalysis screening. Both the Canadian Task Force on the Periodic Health Examination and the U.S. Preventive Services Task Force report insufficient evidence either for or against screening for asymptomatic bacteriuria in older ambulatory women.

Routine screening for renal function (e.g., blood urea nitrogen or serum creatinine) is generally not recommended in healthy, asymptomatic older people. Despite reduced glomerular filtration rate with increased age, the serum creatinine concentration generally does not change. A modestly increased serum creatinine concentration can indicate marked impairment in glomerular filtration rate in older people. However, many older people may have chronic conditions that can affect renal function, so there may be other reasons to test renal function.

CATHY ALESSI

See Also
Health Maintenance

Internet Key Words
Diabetes Mellitus
Hypercholesterolemia

Hypertension
Prevention
Urinalysis

Internet Resources
National Diabetes Information Clearinghouse
http://www.niddk.nih.gov/DiabetesDocs.htm

National Heart, Lung, and Blood Institute Information Center (for information on hypertension and cholesterol screening)
http://www.nhlbi.nih.gov/nhlbi/nhlbi.html

National Institute of Diabetes and Digestive and Kidney Diseases' National Kidney and Urologic Diseases Information Clearinghouse
http://www.niddk.nih.gov/UrologicDocs.htm

HEARING AIDS

Hearing aids are the most common nonmedical treatment for sensorineural hearing loss. Factors in hearing aid selection include the type and amount of hearing loss, the patient's willingness to try a hearing aid and work through the adjustment period, ability to adapt to change, lifestyle, family support, and manual dexterity.

Young, middle-aged, and independent older people who complain of hearing difficulties that interfere with work and social interactions are typically highly motivated to improve their hearing and are excellent hearing aid candidates. Elders who are partially or totally dependent on others for daily care are less likely to be successful long-term users.

It is important to explain and discuss with the patient realistic expectations. One common misconception is that a hearing aid restores normal hearing just as eyeglasses restore normal vision. In general, an aid improves hearing by approximately 50%. The goal is not to restore normal hearing but to improve communicative ability and quality of life.

There are five main types of hearing aids, and despite differences in size and style, all have the same basic components: microphone, amplifier, receiver, and power supply. The microphone changes sound waves into electrical signals that pass through the amplifier and are made louder. The receiver takes the electrical energy and converts it back to sound waves that are delivered to the ear canal through the hearing aid shell. A battery supplies the power that runs the hearing aid.

Behind-the-Ear (BTE) Hearing Aid

The BTE hearing aid fits behind the ear and is connected by a small plastic tube to a soft, custom-made plastic ear mold that fits inside the outer ear. The battery, amplifier, and receiver are all inside the case that fits behind the ear. The microphone is located at the top of the hearing aid near the ear hook. The ear hook curves around the top of the ear and attaches to a small piece of plastic tubing extending from the ear mold. Sound travels through the ear mold into the ear. This type of hearing aid is suitable for patients with severe hearing loss and for those with poor dexterity. Historically, it was the hearing aid of choice. The advantage of a BTE device is that it is suitable for all ages and for any degree of hearing loss, from mild to profound. The microphone is at ear level, which simulates natural sound reception. Disadvantages of a BTE device are that it is bothersome for eyeglass wearers, and a poorly fitting ear mold may cause acoustic feedback. To prevent this, the ear mold may need to be remade periodically to ensure a good acoustic seal.

In-the-Ear (ITE) Hearing Aid

In this one-piece hearing aid, the microphone, receiver, and amplifier all fit completely in the outer ear. Each aid is custom made from hard plastic. The ITE is suitable for patients with mild to severe hearing loss. Many patients prefer this model because it is easy to handle and fits comfortably. The ITE is also cosmetically appealing due to its smaller size, and microphone placement simulates natural sound reception. However, the small size makes the volume control and battery door difficult to use. Ear wax may damage the device, and there is an increased chance of feedback due to the proximity of the microphone and receiver. Acoustic feedback, such as whistling, occurs when the microphone is close to a loudspeaker. There are two types of acoustic feedback. One is produced internally from the hearing aid and indicates that the device needs repair; the second is external feedback produced by leakage of amplified sound out of the ear canal and back into the microphone. Feedback that occurs when the hearing aid is being inserted or removed is common and does not necessarily signal the need for action. BTE hearing aids have a clear advantage over the smaller ITE or in-the-canal (ITC) aids, because feedback is less likely to occur.

In-the-Canal (ITC) Hearing Aid

This custom-made aid is smaller than the ITE and fits into the auditory canal. Because the microphone is closer to the tympanic membrane, the signal has a shorter distance to travel and provides a clearer, more natural amplified sound. As with the ITE, the advantages of the ITC are simulation of natural sound reception and small size; disadvantages are increased chance of feedback and difficulty operating the battery door and volume control. The device is easily damaged by ear wax or ear drainage. Therefore, an ITC is not suitable for people with chronic ear conditions such as impacted cerumen. This device is generally not recommended for children because of the difficulty maintaining a good acoustic seal in an ear that is growing.

Completely in the Canal (CIC) Hearing Aid

The CIC is the smallest hearing aid, and it is essentially hidden in the canal. It is suitable for patients with mild to moderate hearing loss. It is cosmetically appealing and provides the best natural sound. Conventional user-operated volume control is replaced by options such as magnet control and remote control; a removal string attached to the CIC assists with extraction. Reduction in the occlusion effect (i.e., the phenomenon that causes the hearing aid user's own voice to sound loud, with an annoying echolike quality) is also a significant advantage. There are several notable disadvantages

to the CIC. It has a short battery life (10 to 14 days), requires good manual dexterity, and has the highest incidence of feedback. Hearing aids such as the ITC and CIC do not cause infection but are contraindicated in patients with chronic ear conditions, aural drainage, dermatological infections, cholesteatoma, and otitis media.

Body-Style Hearing Aid

The body-style hearing aid has a microphone that is worn in a chest harness or pocket or attached to a belt. A wire connects the aid to a receiver on the ear mold, which fits inside the ear. People with difficulty keeping a hearing aid on the ear or who cannot manipulate small controls may benefit from this style hearing aid. Individuals with profound hearing loss requiring powerful amplification generally must use this style. The disadvantages of the body-style hearing aid are that it is cumbersome due to its size and wires and is not cosmetically appealing. If it is worn under clothing, clothes rubbing against the microphone can cause noise. If it is worn on the chest, the microphone and controls may be damaged by food or liquid spills.

Hearing Aid Circuits

Advances in hearing aid technology are occurring rapidly. Programmable computer circuits can individualize the device, so that certain frequencies can be selectively amplified to closely match a person's hearing loss. There are three basic types of electronics used in hearing aids.

Analog/Adjustable: An analog circuit converts sound (pressure waves) into a voltage waveform. After amplification and filtering, the electrical signal is reconverted to sound by a receiver. An audiologist determines the specifications for the aid, which is then built by a laboratory. The audiologist can make limited adjustments via small screws on the aid. This type of aid is the least complex and the most affordable option.

Analog/Programmable: Although still utilizing analog technology, the programmable aids have additional circuitry that enables adjustments to be made digitally via a computer. Some have remote controls that allow the wearer to change the program according to the particular listening environment.

Digital/Programmable: Digital hearing aids use a computer microchip to digitally process sound; this produces the highest-quality sound for the user. Adjustments can be made with a computer, allowing the audiologist to individualize the aid. These aids are typically the most sophisticated and expensive option.

Advanced-technology hearing aid circuitries employ a recent scientific development in amplification electronics, known as nonlinear or compression amplification. The result is a hearing aid with the ability to limit the level of incoming sound volume. This delivers a more natural loudness throughout the person's entire listening range; sounds do not get too loud or too quiet, as can occur with traditional hearing aids. Individuals with mild to moderately-severe hearing loss, those with a high-frequency loss, and those with reduced tolerance to sound derive great benefit from these circuits.

Some hearing aids allow several settings to be programmed into the same hearing aid to adjust for changes in environmental noise. For example, there might be a program for quiet conversation or music and another for noisy situations when communication is difficult.

Given the wide variety of choices for hearing aid users, there is also a wide range of prices—from $500 to $3,000. Medicare pays for audiology evaluations, but hearing aids are not covered. In many states, Medicaid pays for hearing aids. Some states require that consumers receive a free 30-day trial period to evaluate a hearing aid. Health insurance usually does not cover the cost of hearing aids (but some plans may reimburse 5% to 10% of the cost).

In sum, hearing devices have evolved into sophisticated individualized hearing computers. Most hearing-impaired individuals can be helped with a hearing aid that can provide a better quality of life.

MILLA KAREV
STEVE N. BARTZ

See Also
Hearing Impairment

Internet Key Words
Hearing Aids
Hearing Aid Circuits
Hearing Aid Selection
Hearing Aid Technology
Hearing Aid Use in Older Adults

Internet Resource
American Academy of Audiology
http://www.Audiology.org

HEARING IMPAIRMENT

Hearing impairment is one of the most common chronic health conditions affecting older adults. Among noninstitutionalized elderly, hearing loss is the third most common condition after hypertension and arthritis. Approximately one-third of the U.S. population over the age of 65 has some degree of hearing loss. In nursing homes, 70% of residents have significant hearing loss (Warshaw & Moqeet, 1998). However, despite its prevalence, hearing impairment and its impact on quality of life often go unnoticed. Hearing impairment has been associated with social isolation, depression, and decreased cognitive functioning in the elderly.

Hearing depends on the complete function of the auditory pathway, from the external auditory canal to the central nervous system. The hearing apparatus—external ear, middle ear, inner ear, and eighth cranial nerve—comprises the conductive and sensorineural components of hearing. The external ear, containing the auricular concha and external auditory canal, directs sound medially to the tympanic membrane. When sound waves reach the tympanic membrane, the vibration sets into motion the three middle ear ossicles: the malleus, incus, and stapes. The middle ear transmits acoustic energy from the air to the fluid of the inner ear, where the cochlea converts the sound from mechanical to electrical energy. This electrical energy, or nerve impulse, is transmitted via the eighth cranial nerve to the brain stem and central nervous system (Shohet & Bent, 1998).

There are several age-related physiological changes within the hearing apparatus. The external auditory canal is affected by atrophy of the cerumen glands, resulting in drier cerumen and increased risk of impaction in the elderly. Tympanosclerosis, or atrophic and sclerotic changes of the tympanic membrane caused by past middle ear infections, is not uncommon. Otosclerosis, fixation of the stapes footplate to the oval window, causes conductive hearing loss in up to 10% of the elderly population. Aging changes that may occur in the inner ear include atrophy of the basal end of the cochlea (the area responsible for high-frequency sounds), loss of hair cells, and loss of neurons in the auditory centers of the cortex and brain stem (Warshaw & Moqeet, 1998).

Types of Hearing Loss

There are four types of hearing loss: conductive, sensorineural, mixed, and central auditory processing disorder (CAPD). *Conductive hearing loss* is an impairment in the mechanisms by which sound reaches the inner ear. It is characterized by disorders that impede normal transmission of sound waves though the external canal, tympanic membrane, or middle ear. Thus, there is a reduction in air-conducted but not in bone-conducted sounds. Conditions that frequently result in conductive hearing loss include impacted cerumen, tympanic membrane perforation, otitis media, and discontinuity or fixation of the middle ear ossicles (e.g., otosclerosis) (Shohet & Bent, 1998).

Sensorineural hearing loss occurs when the cochlea or auditory nerve pathway is not functioning properly. It is characterized by equal reduction in air and bone conduction. This type of hearing loss can be either congenital or acquired. The cochlea is the most common site of damage secondary to hair cell damage or ganglion cell loss (Shohet & Bent, 1998).

The most common cause of sensorineural hearing loss is presbycusis, which affects one-third of the population over 75 years of age. Presbycusis is a slow, progressive hearing loss that is most pronounced at high frequencies of 2,000 Hz and above. High-frequency hearing loss affects the ability to discriminate speech sounds in environments with background noise. Persons with presbycusis usually know when they are being spoken to, but they may not always understand what is said. Distinction among higher-frequency consonant sounds such as *f*, *s*, *th*, *h*, and *sh* is difficult. The cause of presbycusis remains unclear. Studies have attempted to link the effects of metabolism, arteriosclerosis, smoking, noise exposure, genetics, diet, and stress. Presbycusis remains a diagnosis of exclusion; other causes of bilateral, progressive sensorineural hearing loss must be ruled out before the diagnosis can be made (Warshaw & Moqeet, 1998).

Mixed hearing loss is the combination of conductive and sensorineural hearing loss. Air and bone conduction are both reduced, but loss of air conduction is greater.

A *central auditory processing disorder* is a deterioration of auditory perceptual abilities even when peripheral hearing is only mildly impaired. An example of CAPD is the reduced ability to understand speech in the presence of background noise, or the inability to understand distorted or rapid speech. A CAPD is often mistaken for cognitive decline.

Diagnosis of Hearing Loss

The first step in identifying hearing loss is obtaining information about the onset of the hearing loss. Whether the onset was unilateral, bilateral, fluctuating, progressive, sudden, or insidious is important. A complaint of adult-onset unilateral hearing loss raises suspicion of a tumor, whereas bilateral gradual onset of hearing loss in persons over the age of 60 is associated with presbycusis. Meniere's disease is associated with tinnitus, episodic dizziness, and fluctuating hearing loss (Shohet & Bent, 1998).

It is necessary to inquire about predisposing factors such as head trauma, exposure to noise, involvement of a neoplastic process, or family history of hearing loss. Medications must be carefully reviewed, since commonly used drugs such as the aminoglycoside antibiotics, loop diuretics, salicylates, antineoplastic agents such as cisplatin, and oral or parenteral erythromycin have ototoxic properties (Nadol, 1993).

On physical examination, any disfigurement of the ear architecture, including the auricle and external auditory canal, should be noted. The tympanic membrane should be closely observed for foreign body, impacted cerumen, perforation, tympanosclerosis, effusion, or infection. A complete head and neck, cranial nerve, and, if indicated, neurological examination should be performed.

The Hearing Handicap Inventory for the Elderly Person, a 10-item self-administered questionnaire, asks patients about emotional and social problems associated with impaired hearing (Warshaw & Moqeet, 1998). Screening techniques include simple tuning-fork tests, such as the Rinne and Weber tests. The Rinne test is performed by placing the stem of a vibrating tuning fork on the mastoid process (bone conduction) and then suspending the fork adjacent to the ear canal (air conduction). The patient is asked to determine in which position the sound is louder. Normally, air conduction is greater than bone conduction, so the sound would be louder with the tuning fork placed in front of the ear. The Weber test is performed by placing a vibrating tuning fork on the midline of the forehead and asking the patient which ear perceives the sound. In conductive hearing loss, the sound is louder in the affected ear; in sensorineural hearing loss, the sound is louder in the unaffected ear (Shohet & Bent, 1998).

An audioscope (a portable hand-held audiometer combined with an otoscope) also can be used as a screening tool. An ear speculum is placed in the external auditory canal, forming a tight seal. A sequence of tones is produced by pushing a button on the audioscope; the patient indicates whether a tone was heard by raising a finger.

If hearing impairment is suspected after an office-based assessment, referral to an audiologist for formal assessment is recommended. Hearing sensitivity is measured in units of sound known as decibels (dB). The higher the intensity of sound (in

dBs) required for a person to hear, the poorer the hearing. An average hearing threshold of 30 dB or less in the speech frequencies of 500, 1,000, and 2,000 Hertz (one cycle per second) is satisfactory for routine listening needs. Amplification may be indicated, however, if psychosocial or work-related factors demand better hearing (Warshaw & Moqeet 1998).

Pure tone audiograms address the type of hearing loss (sensorineural, conductive, or mixed), degree of loss (unilateral or bilateral), and at what frequencies the loss occurs. Speech discrimination, tested for each ear, identifies the degree to which a person understands spoken words (word intelligibility).

A tympanogram, often part of the routine audiological test, provides a measure of tympanic membrane mobility. This analysis can assist in the diagnosis of tympanic membrane perforation, effusion, ossicular fixation, and other causes of conductive hearing loss (Shohet & Bent, 1998).

Measurement of the auditory evoked brain stem response is useful in older patients who may be unable to respond appropriately to routine testing. This electrophysiological test requires only that the patient be cooperative and quiet during the procedure (Nadol, 1993).

Computed tomography (CT) without contrast and magnetic resonance imaging (MRI) with gadolinium contrast are the radiological tests of choice for hearing loss. CT is particularly useful for identifying bony lesions of the temporal bone and mastoid process. MRI is the gold standard in the diagnosis of retrocochlear lesions such as acoustic neuroma and is often used when results of the auditory brain stem response are abnormal (Shohet & Bent, 1998).

When a patient complains of sudden or unilateral hearing loss, a draining ear, or signs of conductive hearing loss; is dizzy; or has a significant reduction in hearing loss and speech discrimination, referral to an otolaryngologist is recommended.

Management of Hearing Loss

A simple procedure such as cerumen extraction may be all that is needed to restore adequate hearing. Antibiotic therapy can be used in infectious processes such as tertiary syphilis and otitis media. Corticosteroids can be used to treat immune-related or viral hearing loss. Surgical procedures are quite successful in most conductive hearing losses and in the removal of acoustic neuromas.

A hearing aid should be considered after a complete otological and audiological evaluation has confirmed bilateral sensorineural and medically untreatable hearing loss. Hearing aids are the most common treatment option for patients with sensorineural hearing loss. The success of auditory rehabilitation depends on a person's auditory and physical capabilities, including manual dexterity and mobility, level of social activity, motivation, and adaptability.

Although hearing aids play an integral part in hearing rehabilitation, other interventions, including speech reading training, listening training, and lip reading instruction, play a role in helping the elderly overcome hearing loss. Many hearing-impaired individuals require assistive listening devices in certain surroundings. For example, several inexpensive amplifiers are available for use on the television or telephone. Other services include closed-captioned entertainment, vibrating alarm clocks, fire alarms that flash or vibrate the bed, accessory headsets for television or radio, and telephone and doorbell signaling devices (Shohet & Bent, 1998).

CYNTHIA HOLZER

See Also
Hearing Aids

Internet Key Words
Auditory Pathway
Hearing Aid
Hearing Impairment
Hearing Loss
Presbycusis

Internet Resources
American Academy of Otolaryngology—Head and
   Neck Surgery
http://www.entnet.org/patient

Better Hearing Institute
http://www.betterhearing.org

Self-Help for Hard-of-Hearing People
http://www.shhh.org

HEARTBURN

Heartburn is one of the most common symptoms of gastroesophageal reflux disease (GERD). Patients with GERD exhibit typical symptoms such as heartburn, regurgitation, belching, and water brash. In other patients, atypical manifestations such as chest pain, asthma, cough, hoarseness, laryngitis, globus sensation, and recurrent pneumonia predominate. Complications in the form of esophageal stricture or cancer may develop.

The cause of GERD is multifactorial and may include (1) decreased esophageal clearance due to decreased peristalsis or saliva, (2) a dysfunctional lower esophageal sphincter (LES) that fails to prevent reflux of gastric contents, (3) delayed gastric emptying and distention, (4) hiatal hernia, and (5) gastric acid hypersecretion. Medications used by the elderly, such as beta antagonists and nitrates, may also decrease LES pressure, thereby promoting reflux.

Noncardiac chest pain is a common atypical manifestation of GERD. The patient typically senses a substernal pressure, with or without accompanying shortness of breath or radiation of pain, which may last from a few minutes to several hours. Approximately 50% of patients who have chest pain and normal cardiac angiography are shown to have gastroesophageal reflux by 24-hour esophageal pH monitoring. Because the prevalence of both GERD and cardiac disease increases with age, it can be difficult to differentiate the two based on clinical symptoms alone. Both GERD and cardiac disease may cause chest pain upon exertion. In fact, exertion may exacerbate GERD in some patients. One study showed that among 11 patients who underwent 24-hour esophageal pH monitoring and had high esophageal 24-hour pH scores, 9 had gastroesophageal reflux associated with chest pain on an exercise stress test (Schofield et al., 1987). In contrast, the relationship of noncardiac chest pain to esophageal motility disorders such as nutcracker esophagus, diffuse esophageal spasm, or hypertensive LES is not as well established. Because of its association with significant morbidity and mortality, a concerted effort should be made to exclude coronary artery disease as a diagnosis.

Diagnosis

Testing to diagnose GERD as a cause of noncardiac chest pain is indicated if empirical therapy fails to produce significant improvement in symptoms. With advancing age, there should be a lower threshold for endoscopy, since the probability of esophageal or gastric cancer is increased. Numerous methods are available to test for GERD. Barium esophagography is preferred for patients with dysphagia. However, the sensitivity of the test is low, and it is not suitable for mild or moderate symptoms of GERD. Its use in patients with chest pain is limited.

The Bernstein test was designed to determine whether chest pain is caused by acid reflux. It involves infusion of 0.1 N hydrogen chloride into the distal esophagus using a saline infusion as a control. The test is positive if the acid infusion reproduces the patient's symptoms and the saline infusion does not cause symptoms. Its sensitivity in patients with noncardiac chest pain is low, but the specificity is about 90%.

Endoscopy allows direct examination and cytology or biopsy of the esophageal mucosa. It is the best method for excluding esophagitis and Barrett's esophagus, which is associated with the development of esophageal adenocarcinoma. Endoscopy is indicated if there are associated "alarm symptoms" such as anemia, dysphagia, or weight loss.

Therapeutic response can be assessed in a patient with noncardiac chest pain who has been given omeprazole orally (40 mg in the morning and 20 mg in the evening) for 7 days. The sensitivity of the omeprazole test for diagnosing acid reflux is 78%, and the specificity is 85%. This test has been demonstrated to be cost-effective in overall management.

Although a low LES resting pressure is associated with GERD, only a minority of patients have an LES pressure less than 10 mm Hg. Esophageal manometry is most helpful in diagnosing conditions other than GERD, for example, when an esophageal motility disorder such as achalasia is suspected. However, the finding of nutcracker esophagus or diffuse esophageal spasm as a cause of chest pain is not common. Nevertheless, manometry plays an essential role in assessing esophageal function before antireflux surgery.

Twenty-four-hour pH monitoring of the esophagus involves inserting a thin pH probe in the patient's esophagus 5 cm above the LES. Data can be collected over 24 hours to calculate the total number of times the esophageal pH is less than 4.0. In addition, reflux events can be correlated with episodes of chest pain or heartburn. The sensitivity of the test is about 85%, and the specificity is 90%.

Management

The treatment goals in GERD are to alleviate symptoms, prevent complications, decrease the number of reflux episodes, heal erosive esophagitis, and maintain remission.

Lifestyle modification remains a key component in overall management. The patient should be instructed to elevate the head of the bed by putting 6 to 8 Styrofoam wedges under the mattress. Dietary modification, such as reducing the intake of fatty foods, chocolate, and excessive alcohol, should be initiated. Certain acidic beverages such as colas and orange juice should also be avoided. The patient should not assume a recumbent position after meals, and meals should not be ingested just before bedrest. Weight reduction, the avoidance of tight-fitting garments, and smoking cessation are also helpful.

Acid-Suppressive Medications

Many clinical trials have established the safety and efficacy of the four H_2 receptor antagonists (cimetidine, ranitidine, famotidine, nizatidine) and three proton pump inhibitors (omeprazole, lansoprazole, rabeprazole) approved for use in the United States. Gastric carcinoids or carcinomas stemming from the chronic use of a proton pump inhibitor have not been reported in humans. Proton pump inhibitors are more effective than H_2 receptor antagonists in healing higher-grade esophagitis (Boyce, 1997). A recent meta-analysis found that proton pump therapy leads to more rapid healing of esophagitis than therapy with H_2 receptor antagonists, sucralfate, or placebo. Relief of heartburn was also more effective with proton pump therapy than with H_2 receptor antagonists—77% versus 47% (Chiba, DeGara, Wilkinson, & Hunt, 1997).

Acid suppressive therapy is dose dependent. Increasing the dose increases the proportion of time that the intragastric pH is above 4.0, therefore resulting in more effective healing. A reasonable starting dose for mild or moderate GERD is cimetidine 400 mg twice a day or ranitidine 150 mg twice a day. An initial dose for more severe symptoms of GERD or erosive esophagitis is omeprazole 20 mg twice a day or lansoprazole 15 mg twice a day. Therapy should be continued for at least eight weeks for typical symptoms of GERD. For atypical symptoms, such as chest pain, the duration of therapy should be a minimum of three months.

Prokinetic Medications

Prokinetic drugs (bethanechol, metoclopramide, cisapride) can promote peristalsis, increase acid clearance, and increase LES pressure. Bethanechol has limited side effects and is no longer widely used. Metoclopramide may cause tardive dyskinesia in some cases. Cisapride is a newer prokinetic agent, but there have been reports of cardiac arrhythmias associated with its use. Cisapride should not be used in combination with macrolide antibiotics (erythromycin, clarithromycin) or with drugs in the imidazole class (ketoconazole, fluconazole, itraconazole, or metronidazole).

Maintenance Therapy

In the absence of maintenance acid suppression therapy, esophagitis will likely recur. Clinical trials have shown that acid-suppressive agents' ability to maintain remission depends on the dose, the class of medication, and the number of medications used. For example, combination therapy with omeprazole 20 mg daily and cisapride 10 mg three times a day is more effective than either agent alone or with ranitidine plus cisapride (Vigneri et al., 1995). For monotherapy, omeprazole was more effective than the other agents. Lansoprazole 30 mg daily has been shown to be more effective than 15 mg daily.

Surgery

Antireflux surgery is an alternative reserved for patients who are refractory to medical management. In selected patients, Nissen fundoplication can result in long-term control of heartburn and chest pain.

Helicobacter Pylori

H. pylori as a cause of GERD has not been firmly established. Moreover, recent evidence suggests that *H. pylori* eradication may increase the risk for GERD. A rational approach is to individualize treatment for each patient after a thorough discussion of the risks, benefits, and consequences.

For patients with typical symptoms not responsive to lifestyle modification or antacid therapy, a trial of empirical therapy with H_2 receptor antagonists or proton pump inhibitors is indicated. If there is no response, endoscopy should be performed to rule out Barrett's esophagus and esophagitis. Patients diagnosed with mild esophagitis can be treated with an H_2 blocker an H_2 blocker plus a prokinetic agent, or a single proton pump inhibitor. Patients found to have more severe esophagitis can undergo proton pump therapy or surgery. Patients with noncardiac chest pain unresponsive to empirical proton pump therapy (omeprazole test) should be tested by 24-hour pH monitoring of the esophagus. If acid reflux is confirmed as the cause of chest pain, the patient should continue on medical therapy for at least three months or be considered for surgery.

Treatment of heartburn and noncardiac chest pain is challenging and requires an understanding of the pathophysiology of GERD. The choice of therapy must be based on efficacy, cost, safety, age, and patient participation.

IMMANUEL K. HO

See Also

Chest Pain: Noncardiac Causes

Gastrointestinal Physiology

Internet Key Words

Chest Pain

Esophagitis

Gastroesophageal Reflux

Heartburn

Internet Resources

American College of Gastroenterology

*http://www.acg.gi.org/patientinfo/
    frame_gerd.html*

American Gastroenterological Association

http://www.gastro.org/heartburn.html

HEART FAILURE MANAGEMENT: CONGESTIVE AND CHRONIC HEART FAILURE

More than 3 million Americans have congestive heart failure (CHF), and approximately half a million new cases develop each year. The incidence of this condition rises markedly above age 65, and its prevalence approaches 10% in individuals over age 80. CHF is the most common discharge diagnosis for elderly persons in the United States. CHF shortens life expectancy, resulting in five-year mortality rates of 50%, and causes substantial morbidity and reduced quality of life. The total treatment costs for heart failure, including drugs, physician costs, and nursing home stays, are estimated to be over $17 billion annually. Current therapy for heart failure can improve survival (and quality of life), and decrease symptoms and morbidity, including hospi-

talizations (American College of Cardiology/ American Heart Association Task Force on Practice Guidelines, 1995). Coronary artery disease is the most common cause of CHF, followed by hypertension. Other causes are metabolic problems such as thyroid disease or vitamin deficiencies, valvular problems (most commonly aortic stenosis and mitral regurgitation), and primary myocardial disorders such as hypertropic cardiomyopathy, amyloidosis, and sarcoidosis. Sometimes no cause can be found, in which case the condition is deemed idiopathic.

Diagnosis

The practitioner must use a combination of historical, physical examination, and imaging data to establish the underlying cardiac disease, the precipitating factors, and the severity. Precipitating causes such as arrhythmias (most commonly atrial fibrillation) and unstable angina must be identified and treated immediately, as they may be life-threatening. Acute infections often precipitate CHF in older adults. Valvular and pericardial causes may be identified by the physical examination and history. The focus here is on heart failure due to dysfunction of the myocardium.

History and Physical Examination

CHF causes cardiorespiratory and nonspecific symptoms. Dyspnea on exertion is probably the most sensitive symptom among patients with heart failure. The classic symptoms of orthopnea and paroxysmal nocturnal dyspnea have a sensitivity of less than 50%. The elderly, particularly those with dementia, may first report nonspecific symptoms such as worsening cognitive function, generalized weakness and tiredness, fatigue, and anorexia. Edema in association with dyspnea may indicate heart failure, but if found alone, it is more often caused by venous insufficiency.

In patients with suspicious symptoms, an elevated jugular venous pressure, hepatojugular reflux, laterally displaced apical pulse, and third heart sound may indicate heart failure. The presence of rales with this complex of signs and symptoms is indicative of heart failure. Wheezing, especially in someone with no prior history of lung disease, should alert one to the possible presence of heart failure. In some patients with mild to moderate CHF, none of these signs of heart failure is apparent. In these cases, the practitioner must rely on cardiac imaging studies to establish the diagnosis.

Diagnostic Studies

Diagnostic studies provide information about the possible precipitating, complicating, or primary causes of heart failure. The following studies should be obtained in all patients: chest radiograph, electrocardiogram (ECG), complete blood count, electrolytes, albumin, urinalysis, thyroid function tests, and imaging with echocardiogram or radionuclides. In symptomatic patients, cardiomegaly seen on a chest radiograph is highly suggestive of heart failure, especially when accompanied by evidence of pulmonary vascular congestion. However, a normal chest x-ray does not rule out CHF. Although there is no specific ECG finding for heart failure, the tracing may reveal possible precipitating factors such as acute ischemia, prior myocardial infarction, arrhythmias, left ventricular (LV) hypertrophy, and conduction abnormalities. Thyroid studies, including thyroxine and thyroid-stimulating hormone levels, should be obtained in all elderly patients with heart failure with or without atrial fibrillation, because heart failure may be the initial manifestation of hyper- or hypothyroidism in patients over age 65.

Because clinical symptoms and signs do not reliably distinguish CHF due to systolic versus diastolic dysfunction, it is essential to measure LV function in patients with suspected heart failure. Noninvasive studies such as radionuclide scanning or echocardiography can reliably distinguish between systolic and diastolic dysfunction. With echocardiography, the presence of vascular and pericardial disease can also be determined. The distinction between diastolic and systolic dysfunction is particularly relevant in the elderly population

because of the adverse consequences of using inappropriate medications.

Management

Pharmacological Management

CHF is a chronic disorder that requires pharmacological and nonpharmacological management. The choice of medication is based on the pathophysiology underlying the CHF—systolic versus diastolic dysfunction—and any comorbidities and contraindications. Most of the reliable data on drug efficacy in CHF come from studies of patients with systolic dysfunction. These studies have established multiple drug regimens of three, four, or sometimes five drugs—some combination of an angiotensin-converting enzyme (ACE) inhibitor, diuretic, beta-blocker, and digoxin—as the standard treatment for heart failure due to systolic dysfunction. Evidence related to the optimal treatment for CHF due to diastolic dysfunction is unavailable. However, on the basis of small studies, beta-blockers and calcium channel blockers are generally recommended. Patients with diastolic heart failure are particularly sensitive to overdiuresis.

Diuretics are indicated in patients with evidence of volume overload—peripheral edema and pulmonary congestion—whether due to systolic or diastolic dysfunction. Persons with evidence of mild heart failure can be managed with thiazide diuretics. Those with moderate or severe heart failure require loop diuretics. The dosage used is severity- and patient-dependent. The diuretic dose should be minimized in persons with urinary urgency or incontinence. Patients managed with thiazide and loop diuretics need to be monitored for possible volume depletion, hypokalemia, hyponatremia, hypomagnesemia, and azotemia. An old potassium-sparing diuretic, spironolactone, has recently been shown to benefit patients with heart failure when given in combination with an ACE inhibitor, a loop diuretic, and other medications; however, the combination of and ACE inhibitor and spironolactone can produce substantial negative side effects.

In the absence of any contraindications, ACE inhibitors should be started in all patients with heart failure due to LV systolic dysfunction. Several studies provide strong evidence that in symptomatic patients with an LV ejection fraction below 35% to 40%, ACE inhibitors reduce mortality and morbidity and improve quality of life. Asymptomatic patients with low ejection fractions due to coronary artery disease have also been shown to have a prolonged survival. Although the largest randomized clinical trials have employed captopril or enalapril, the reduction in mortality in patients with heart failure due to systolic dysfunction is probably a class effect. When using any of these drugs in the elderly, the practitioner should start with a low dose equivalent to 2.5 to 5 mg of enalapril. If tolerated, the dose can be titrated up to the equivalent of 20 mg of enalapril per day.

Contraindications to ACE inhibitors include serum potassium greater than 5.5 mEq/L, history of adverse drug reaction to these agents, symptomatic hypotension, and serum creatinine greater than 3.0 mg/dL or creatinine clearance less than 30 mL/minute. To avoid hypotension, patients should be euvolemic when initiating therapy with an ACE inhibitor. ACE inhibitors should be used with caution in patients with renal insufficiency. However, chronic renal insufficiency is not an absolute contraindication. Potassium supplements and potassium-sparing diuretics should be discontinued before starting ACE inhibitors. Cough is a common complaint in patients on this medication. As cough is also a symptom of heart failure, these patients should be evaluated for possible pulmonary vascular congestion. Angioedema has been reported in some patients and, when it involves the oropharynx, is an absolute contraindication to continuing ACE inhibitors.

Vasodilators such as isosorbide dinitrate and hydralazine are appropriate alternatives for afterload reduction in patients who cannot tolerate ACE inhibitors. This combination has been shown to decrease mortality from 19% to 12% at one year and from 47% to 36% at three years in one trial when compared with placebo. Side effects, including palpitations, headache, and nasal congestion, are somewhat more likely with these agents. Anti-ischemic vasodilating agents are also indicated when transient episodes of ischemia are suspected.

An *angiotensin II receptor antagonist* should be prescribed to patients with heart failure due to systolic dysfunction who cannot tolerate ACE inhibitors because of cough, rash, or altered taste sensation. The other major adverse effects of ACE inhibitors are also seen with angiotensin inhibitors. The ELITE study showed that an angiotensin inhibitor, losartan, was superior to an ACE inhibitor in reducing total mortality. Hospitalization rates for CHF were similar for both drugs. Longer comparisons are necessary to determine whether the preliminary findings of the ELITE study are justified and whether the benefits of losartan can be considered a class effect.

There is growing evidence that modulation of the sympathetic nervous system by *beta receptor blockade* can favorably affect patients with heart failure. The Metoprolol in Dilated Cardiomyopathy Trial found that in symptomatic patients with low LV ejection fractions, the administration of metoprololol slowed clinical deterioration and improved symptoms and cardiac function. Similar benefits plus an improvement in survival have been found with carvedilol, a new beta-blocker that also has alpha-blocker activity and antioxidant effects. Carvedilol is recommended in persons with class II and III heart failure who are taking ACE inhibitors, diuretics, and digoxin. Because of side effects—hypotension, dizziness, and worsening heart failure—carvedilol must be prescribed only by practitioners experienced in its use and limitations.

Although *digoxin* is routinely used in patients with heart failure with atrial fibrillation and a rapid ventricular response, its use remains controversial in those with sinus rhythm. Evidence suggests that digoxin improves functional status and can prevent clinical deterioration in patients with LV systolic dysfunction, but it does not prolong survival. Based on these findings, it is beneficial to start digoxin in patients with severe heart failure due to LV systolic dysfunction who remain symptomatic despite treatment with ACE inhibitors and diuretics. When initiating treatment with digoxin, lower dosages should be used in elderly patients, those with renal insufficiency, and those with baseline conduction abnormalities. Practitioners should be aware of noncardiac signs of digoxin toxicity, including confusion, nausea, visual disturbances, and anorexia. The clinician should be aware that digoxin levels may be raised by medications such as quinidine, verapamil, amiodarone, antibiotics, and anticholinergic agents. Therefore, digoxin levels should be checked approximately one week after starting any of these medications. Digoxin's narrow therapeutic window means that it must be used with care in older patients. More frequent monitoring than in younger patients is warranted.

The negative inotropic effects of *calcium channel blockers* and their undesirable activation of the renin-angiotensin and sympathetic nervous systems remain a major concern in patients with heart failure. In several studies, the degree of clinical deterioration was greater in patients treated with nifedipine or diltiazem than in those receiving placebo or isosorbide dinitrate. However, no deleterious effect was observed in other reports of diltiazem or in reports of amlodipine use in the elderly.

Nonpharmacological Management

Because CHF is a chronic disorder, patient and family education about diet and lifestyle changes is crucial to successful long-term management. In addition to avoiding smoking and limiting salt intake to about 2 grams of sodium per day, exercise is beneficial. Exercise training does not improve cardiac function, but it can increase peak cardiac output, leading to a reduction in symptoms such as dyspnea and fatigue. As part of overall management, home-based interventions involving nurses and other health professionals have decreased the rate of unplanned hospital readmissions and have improved mortality.

Heart failure patients over age 70 have a readmission rate as high as 57% within 90 days of discharge (Tresch, 1997). Factors associated with readmission include inadequate follow-up, poor social situations, noncompliance with a low-salt diet,

and noncompliance with medications. It is essential that these issues be addressed as soon as heart failure is diagnosed. Patients should be taught to recognize the symptoms of worsening heart failure, particularly dyspnea on exertion and weight gain, and should be advised to notify the practitioner if these symptoms arise.

The practitioner should arrange for follow-up contact by phone or in person within one week of instituting new medications or after discharge from the hospital. During this contact, the proper use of medications and compliance with diet should be addressed. Because weight gain is a crucial objective sign of fluid overload, patients should be instructed to weigh themselves almost daily and to report any weight gain of three to five pounds (Stewart, Vandenbroek, Pearson, & Horowitz, 1999). Laboratory studies including electrolytes, blood urea nitrogen, and creatinine should be checked during this visit, and medications adjusted as needed. Repeat imaging studies of the heart are of little value in monitoring the progress of heart failure once its pathophysiology has been characterized.

JERRY C. JOHNSON

See Also
Chest Pain: Noncardiac Causes
Coronary Artery Disease
Cough
Dyspnea (Shortness of Breath)
Heartburn

Internet Key Words
Angiotensin Type II Inhibitor
Beta Receptor Blockade
Calcium Channel Blockers
Congestive Heart Failure
Digoxin
Diuretics
Heart Disease
Vasodilators

Internet Resources
American Heart Association
http://www.americanheart.org/

National Heart, Lung and Blood Institute
http://www.nhlbi.nih.gov/

HIP FRACTURES

Hip fracture is epidemic among the elderly. In the United States, over 250,000 hip fractures occur yearly (Perez, 1994), and hip fracture patients account for half of all hospital days for fractures of all types (Griffin et al., 1990). Hip fracture is significantly related to mortality, with a concentrated effect in the first six months after fracture (Wolinsky, Fitzgerald, & Stump, 1997). There is an immediate increase in mortality following hip fracture in medically ill and functionally impaired patients; among those with no comorbidities and few impairments, there is a gradual increase in mortality that continues for five years after fracture (Magaziner et al., 1997). Hip fracture also significantly increases the likelihood of subsequent hospitalization and has been shown to increase the number of subsequent hip fractures, the number of hospital days by 21%, and total charges by 16% (Wolinsky et al., 1997).

Hip fracture results from two important age-related changes. The first is a well-described age-related loss of postural stability, which leads to an increased incidence of falls (Alexander, 1996). An equally important factor is the decrease in bone density that results from both a linear decrease in bone mass with age and trabecular bone loss and diminished bone strength related to postmenopausal osteoporosis (Raisz, 1997). The incidence of hip fracture doubles every five years after the age of 50 (Lyon & Nevins, 1989). By age 65, it is estimated that 50% of American women have a bone mineral content below the fracture threshold (Perez, 1994). This figure rises to 100% by age 85, as senile osteoporosis causes cortical bone loss, further augmenting menopause-related trabecular bone loss (Black et al., 1990). Women have two to four times as many hip fractures as men (Perez, 1994), and one-third of all women who survive to age 90 have suffered a hip fracture. Prevalence rates vary with ethnicity, with most studies revealing a higher prevalence of hip fracture among white females than among African American females (Karagas et al., 1996).

Types of Fractures

Hip fractures are generally divided into three types: femoral neck, intertrochanteric, and subtrochanteric. Femoral neck and intertrochanteric fractures account for 97% of hip fractures in the aged; subtrochanteric fractures account for 3% of proximal femoral fractures, and their occurrence often raises the suspicion of pathological fracture due to metastatic disease as opposed to osteoporotic or traumatic fracture.

Femoral neck fractures are commonly subtyped based on the Garden classification system (Stein & Felsenthal, 1994). In the Garden system, a type I fracture is an incomplete or impacted fracture of the femoral neck with no displacement of the medical trabeculae. A type II fracture is a complete fracture of the femoral neck with no displacement of the medical trabeculae. A type III fracture is a complete fracture of the femoral neck with vans angulation and displacement of the medial trabeculae, and a type IV fracture is a complete fracture of the femoral neck with total displacement of the fragments. It is important to note that types I and II are nondisplaced fractures, which do well with pinning, and types III and IV are displaced fractures, which are unstable and require more extensive open reduction and internal fixation or arthroplasty (Perez, 1994).

Because the femoral neck and head are contained entirely within the joint capsule, they have no periosteum. The arterial supply to the femoral head is disrupted if fracture fragments are displaced, making nonunion and avascular necrosis relatively common complications of open reduction and fixation. Many surgeons routinely use arthroplasty for all displaced femoral neck fractures to avoid these complications.

Intertrochanteric fractures are usually comminuted and significantly osteoporotic (Raisz, 1997), making it difficult to achieve good anatomical reduction of fragments, resulting in an unstable fixation (Ochs, 1990). As the vascular supply to the femoral head and neck is usually not compromised, avascular necrosis and nonunion are uncommon complications. However, patients with intertrochanteric fractures often suffer delayed weight bearing, poorer functional outcomes, and higher mortality than do patients with femoral neck fractures.

Management

Orthopedic management consists largely of evaluating the type of fracture and degree of instability in order to select the most appropriate surgical intervention: pinning, placement of a nail, fixation with a sliding nail or compression screw, hemiarthroplasty, or total hip arthroplasty. However, careful perioperative medical management and use of an interdisciplinary team are essential to optimize outcome. Common issues in hip fracture management include timing of surgical intervention; use of invasive perioperative hemodynamic monitoring; anticoagulation with low-molecular-weight heparin or low-intensity warfarin for prophylaxis against deep venous thrombosis and pulmonary embolus; diagnosis and treatment of perioperative delirium; early mobilization following surgery; and individualized, intensive rehabilitation aimed at maximal functional outcome.

Outcomes

A comprehensive rehabilitative approach to older patients with hip fracture has been widely implemented. A majority of patients are now discharged within a week after surgery to short-term rehabilitation programs that focus on comprehensive rehabilitation, including coordination of medical, nursing, and physical therapy management. Some patients are able to go directly home after orthopedic surgery and continue with therapy at home. Prognosis after a hip fracture is adversely affected by increased age, a high prefracture dependency level,

dementia, postsurgical delirium, depression, a poor social network, and any subsequent rehospitalization or major fall (Magaziner et al., 1990).

Mortality within one year after hip fracture is approximately 20% to 25%. Risk factors for increased mortality include age greater than 85, dementia, presence of malignancy, and occurrence of postoperative pneumonia or deep wound infection (Wood et al., 1992).

Ability to walk is frequently limited after hip fracture. Given that some individuals were not ambulatory before the fracture and that 33% to 43% required a walking aid, only 25% to 50% of individuals recover to their prefracture ambulation status (Stein & Felsenthal, 1994). However, many individuals experience improved function for up to a year (Magaziner et al., 1990). Unfortunately, some patients never walk again. The risk of becoming nonambulatory is greatest for the oldest patients, although encouraging results have been reported using aggressive physical therapy programs (Perez, 1994).

Prevention

Hip fracture prevention has centered around improving gait stability via guided exercise programs (Gregg, Cauley, Seeley, Ensrud, & Bauer, 1998) and diagnosis and treatment of osteoporosis (Raisz, 1997). Public and physician awareness of osteoporosis has increased sharply, resulting in widespread use of dietary calcium supplementation and lifestyle modifications to reduce risk factors for osteoporosis. Additional osteoporosis treatment modalities include nasal spray calcitonin, alendronate (an aminobisphosphonate that inhibits osteoclast activity), and estrogen replacement therapy in various forms (most notably raloxifene, the first available estrogen receptor modulator). Further investigational agents continue to be developed.

LIDIA POUSADA

See Also
Falls Prevention
Fractures

Gait Disturbances
Osteoporosis

HISPANIC AND LATINO ELDERS

Hispanics, also referred to as Latinos, are a diverse population, differing greatly in their social, economic, and cultural characteristics. Understanding this diversity is central to meeting aging U.S. Hispanics' health care needs.

Although Hispanics share a common language, the groups that constitute this population vary considerably with regard to diet, educational background, and religious beliefs and have numerous linguistic and cultural differences. Mexican Americans, Cubans, Puerto Ricans, and an increasing number of individuals from Central American countries (e.g., Honduras, El Salvador, Guatemala, and Costa Rica) and from South America constitute this second largest and fastest growing ethnic minority in the United States. Studies such as the Hispanic EPESE (Established Populations for Epidemiologic Studies of the Elderly), NHANES III (Third National Health and Nutrition Examination Survey), and HHANES (Hispanic Health and Nutrition Examination Study) provide some of the most comprehensive information regarding Hispanics to date.

Obvious issues regarding minority health care include lower educational levels, lack of health insurance, barriers to health care access, and high poverty levels. One of every five Hispanic elders has completed less than five years of formal education (Council on Scientific Affairs, 1991). Additionally, lack of education contributes to Hispanic families' being more than two and a half times as likely as non-Hispanic whites to live below the poverty level. Less education predisposes individuals toward higher unemployment and lower-wage jobs. One-third of Hispanic elders have neither private health insurance nor coverage through Medicare or Medicaid (Council on Scientific Affairs, 1991). Twice as many Hispanic elders report using a hospital emergency department as a source of primary care as compared with non-Hispanic whites.

Due to the multiple barriers affecting minorities, health care providers are encouraged to focus on the day-to-day management issues of ethnic elderly. These include ethnic-specific medical problems, caregiving and translation issues, American customs of medical care, and folk and nontraditional forms of medical care.

Ethnic-Specific Medical Problems

Mexican American elders have a greater risk of dying from diabetes mellitus type 2 and renal failure than their non-Hispanic white counterparts (Espino, Parra, & Kriehbiel, 1994). The HHANES study revealed that 26% of Puerto Ricans and 24% of Mexican Americans have diabetes mellitus (Bassford, 1995); of these, 40% were receiving no treatment. In the NHANES III, similar results were found, with the prevalence of diabetes mellitus two times higher for Mexican Americans than for non-Hispanic whites (Harris, Eastman, & Cowie, 1999). Self-monitoring of blood sugar was less common in Mexican American elders, and poor glycemic control was more common. The combination of high prevalence and low treatment rates places Mexican American elders at higher risk for complications such as end-stage renal disease.

Special outreach activities and educational campaigns are needed in Latino communities to teach the importance of medical screening to Hispanics who are monolingual Spanish or prefer to speak only Spanish. Additionally, establishing effective protocols for referring individuals to screening centers would benefit Hispanics who commonly receive health care from large public hospitals and rarely experience continuity of care.

Hypertension, especially untreated hypertension, is alarmingly high in Hispanic populations. Despite a reduction in tobacco use among Hispanics over the past decade, only 25% of hypertensive subjects had good blood pressure control (Stroup-Benham, Markides, Espino, & Goodwin, 1999). Nontreatment and undertreatment of hypertension in Hispanics remain a significant public health concern. An insufficient understanding of the consequences of hypertension may diminish the degree of hypertension control among Hispanics.

Symptoms of acute myocardial infarction also differ in Hispanics. Hispanics are more likely than non-Hispanic whites to report chest pain, upper back pain, and palpitations as symptoms of myocardial infarction and are less likely to report arm or jaw pain (Meshack, Goff, Chan, & Ramsey, 1998). Hispanic women are more likely to report fatigue, dyspnea, dizziness, upper back pain, palpitations, and cough and are less likely to report chest pain than are non-Hispanic whites females (Meshack et al., 1998).

Smoking and obesity are major contributors to hypertension in Hispanic populations (Stroup-Benham et al., 1999). The overall increase in smoking rates among Hispanics accounts for nearly a doubling of lung cancer rates over the last decade. Among older Mexican Americans, however, there has been a decrease over time in the percentage of smokers, from 27% to 14%, and a decrease in mean systolic blood pressure (Stroup-Benham et al., 1999). Heightened awareness and improved education and treatment strategies for smoking cessation may be contributing to this decrease.

Hispanics also have a higher incidence of cancer of the pancreas, liver, and gallbladder; cancer of the uterus, cervix, and stomach occur at twice the rate compared with non-Hispanic whites (Bassford, 1995). The incidence of breast cancer in the United States is lower among Hispanic women than among non-Hispanic women. Unfortunately, however, Hispanic women are more likely to be diagnosed at a later stage than non-Hispanic women (Bentley, Delfino, & Taylor, 1998). According to HHANES data, approximately 15% of Hispanic women have never had a Pap smear or have not had one within the past five years. Improved access to screening services alone would have a tremendous impact on the health of the U.S. Hispanic population. According to HHANES, more than 45% of Mexican Americans and more than 30% of Puerto Ricans and Cuban Americans reported not having had a complete medical examination in over five years

(Bassford, 1995). Culturally specific education materials and increased vigilance by primary care providers may help improve cancer screening services.

Screening for psychological impairments in the Hispanic elderly often calls for revising commonly used threshold scores. A translated and culturally adjusted version of the Folstein Mini Mental Status Examination overestimates the degree of cognitive impairment in the Hispanic population (Espino & Lewis, 1998). Even with a revised threshold score, a Spanish translation of the Geriatric Depression Scale was less sensitive than the English version (Espino, Bedolla, Perez, & Baker, 1996), thereby underdetecting depression in this population. Increased focus on and attention to somatic complaints such as headache, stomach pain, constipation, and weakness may help primary care providers recognize psychological distress among Hispanic patients. Anecdotal reports suggest that Hispanic patients are also especially likely to report side effects from antidepressant medications, possibly secondary to receptor hypersensitivity.

Caregiving Issues

Functional disability appears to be higher among community-dwelling Hispanic elders than among non-Hispanic whites (Rudkin, Markides, & Espino, 1997). Elder Mexican American nursing home residents are more functionally impaired than their non-Hispanic white counterparts (Espino & Burge, 1989). Both these studies indicate that Hispanic elders may be staying in the community longer, until higher levels of functional impairment are reached. Although not unique to the Latin culture, an extended family support system may lessen caregiver burden and lengthen community residency. Primary care providers need to recognize that individuals within the family may be responsible for an elder's care. Among Hispanics, this role is often left to the oldest daughter, granddaughter, or daughter-in-law. Dealing with multiple family members or even close friends may be frustrating to health providers at times. However, excluding these individuals may impair efforts to reach the health goals set for the patient.

Translation Issues

In addition to cultural barriers, many Hispanic elders face language barriers. Older adults who do not speak proficient English face complex obstacles when using the health care system. Spanish language instructions can be difficult to comprehend, because many elder Hispanics may be illiterate in Spanish as well, or written materials in Spanish may not be targeted to the literacy level of the aged Hispanic population. Primary care providers must be cautious of oral translation, since the translators—often relatives of the elder—may have varying levels of fluency in Spanish. These relatives may edit or filter important information necessary for a complete patient evaluation; in addition, lack of interpreter-patient linguistic equivalency may lead to inaccurate paraphrasing. Imposition of the interpreter's beliefs or self-perception into the interaction also influences the information gathered. Finally, cultural perceptions, such as the stigmatization of mental illness, may result in poor communication. Effective communication requires that clinicians providing care to Hispanic elders be conversant in Spanish or have trained medical translators or bilingual staff.

An increasing challenge for primary care providers involves discussing health screening recommendations. Commonly, elder Hispanics rely solely on the "expertise" of their providers to make recommendations that they trust or assume will be in their own best interests. This expectation may lead to some confusion as health providers attempt to use a shared decision-making approach. Providers may wish to include other individuals who function as the primary support for an elderly person when discussing preventive health options and advance directives.

Folk Medicine and Traditional Beliefs

The frequency with which individuals use the services of unlicensed healers, herbalists, and spiritual healers probably varies among the Hispanic American subgroups. Examples of these alternative services include the use of Curanderos in the Mexican American communities, Santeria among Cuban Americans, and Espiritismo among Puerto Ricans. In the HHANES study, 4.2% of subjects reported consulting a Curandero, herbalist, or other folk medicine practitioner within the prior 12 months (Higginbotham, Trevino, & Ray, 1990). Health providers should inquire about the use of alternative healers (i.e., *verbas medicinals*). Additionally, practitioners should ascertain where medications are purchased, because many Mexican Americans who live in the Southwest cross the border into Mexico, where medications can be bought without a prescription. A common belief among some Hispanic elders is that health is dependent on a "hot-cold" balance. For example, to cure a "cold" disease, a "hot" treatment is necessary, and vice versa. The majority of treatments involve the use of natural medications, such as herbal teas or soups.

The key to the care of the Hispanic American population is the education of health care providers to the cultural diversity of the populations they serve. Elder Hispanic Americans are the fastest growing segment of the population and share many common cultural links. Many barriers exist to providing optimal health care; however, understanding culturally specific characteristics can enhance the quality of care provided and ensure a mutually healthy relationship.

KURT P. MERKELZ

Internet Key Words
Access to Care
Healthcare
Hispanics
Latinos
Mexican American
Minorities and Aging

Internet Resources
Latino Health Profile
*http://www.latinomed.com/resources/
    latino_profile.html*

Pan American Health Organization
http://www.paho.org/

South Texas Border Initiative Homepage
http://www.uthscsa.edu/stbi/

HOME HEALTH CARE

Before the advent of Medicare home health benefits in 1965, approximately 1,400 agencies provided family-focused health promotion and sickness care services (National Association for Home Care [NAHC], 1999a). Since then, the number of agencies, staff, clients, and visits escalated dramatically in response to the introduction of diagnosis-related groups (DRGs) in the early 1980s, reduced length of hospital stays, passage of national legislation that made it easier to establish new agencies, wider acceptance of home care, and an aging population. Often, the new agencies were hospital based or were proprietary, community-based, for-profit agencies.

By 1996, more than 20,000 Medicare-certified and noncertified home health and hospice agencies offered services to 8 million clients (NAHC, 1999a). This dramatic growth represented a 1,000% increase in the number of agencies from 1965 to 1996 and resulted in competition among agencies, a business and leadership orientation, and other philosophical changes. Agencies diversified their programs, services, and staff to meet the needs of clients, families, and referral sources. They added high-technology procedures, 24-hour care, hospice programs, pharmacy services, durable medical equipment, and telehealth programs. Additional physical therapists, occupational therapists, speech pathologists, nutritionists, and social workers were employed.

The average number of home health visits (all disciplines) per client grew from 33 in 1990 to 76 in 1998. Regulations and legislation, including the Balanced Budget Act of 1997, dramatically restricted the type of clients that could receive services and the number and type of Medicare-reimbursed visits that clients could receive. Between 1997 and 1999, approximately 3,000 home health agencies closed for financial reasons. Although the 1999 amendments to the Balanced Budget Act of 1997 are expected to moderate the trend, reimbursement continues to have a greater influence on many practice decisions than does professional judgment (*Caring* [special Home Care issue], 1999; *Home Health Line* [special PPS Proposal issue], 1999; Martin, 2000; NAHC, 1999a). Medicare and Medicaid regulations and reimbursement limitations often determine whether clients are admitted to services, the frequency and length of visits, which disciplines provide services, and the time of discharge.

Persons 65 years of age and older constitute 65% to 70% of all home health clients. Circulatory system disease accounts for about 25% of referrals; neoplasms and endocrine diseases, especially diabetes, are common primary medical diagnoses. Approximately 200,000, or 13% to 20%, of all currently practicing registered nurses are employed in home health and other community-focused services. More than 40,000 licensed practical nurses, 10,000 physical therapists, 8,000 social workers, 4,000 occupational therapists, and 3,000 speech pathologists are similarly employed (Martin, 2000; NAHC, 1999a).

Home Health Care Practice

Some core principles of home health practice have remained the same since the first agencies delivered nursing care at home; other aspects of practice have changed radically over time (Modly el al., 1997). Home-care agencies employ nurses and other clinicians with diverse education and experience. New nurses are often expert acute-care clinicians who need orientation to community-focused practice, community resources, and the power of the client, who may be an individual, family, group, or community. Effective home health practitioners need a wide variety of technical and interpersonal skills. They need to have common sense and be able to work as partners with diverse clients (e.g., diversity in age, medical and nursing diagnoses, race, religion, culture, income, and values) (Brickner et al., 1997; Rosswurm, 1998).

Agencies use various staffing patterns to cover their geographic areas and the needs of their clients.

Some staff members work independently, and others are members or leaders of multidisciplinary teams. Team leaders are often referred to as case managers. When staff work in a hospital or nursing home, colleagues, technology, supplies, and references are nearby. When clinicians make a visit to a home, clinic, or other site, they are usually alone. Thus, it is essential that clinicians be prepared for diverse challenges and be able to communicate effectively with health care, social services, and educational colleagues within the agency and within other community agencies. In response to the challenges of communication, distance, and outcome management, many agencies have implemented standardized vocabularies and automated clinical information systems (*On-line Journal of Nursing Informatics*, 1999).

When agencies provide diverse services to community residents, practice is more generalized. However, even those agencies that offer diverse services usually employ staff members who have specialized skills related to target client groups, diseases, or treatments, such as cardiovascular, diabetes, infusion therapy, hospice, parent-child, and AIDS specialties. A small number of agencies employ staff who have clinical specialty or advanced nursing practice credentials, such as geriatric nurse practitioners.

Description and Measurement of Practice

Home health clinicians and agencies have always evaluated their practice. Recent developments, including automated clinical informations systems and the Internet, as well as standardized vocabularies, reimbursement, managed care, and quality improvement techniques, have provoked greater interest in this analysis (Martin, 2000).

Problem-Solving Process

The six steps of the nursing process offer a useful strategy for home health nurses and other clinicians to describe and measure their practice, especially when these steps are combined with standardized vocabularies and automated clinical information systems. The steps are assessment, problem identification/diagnosis, plans, interventions, evaluation, and outcome management. One research-based method, the Omaha System, illustrates application of the process in home health practice.

Assessment and Problem Identification/Diagnosis: When home health clients are referred for service, assessment and diagnostic information should accompany the referral. However, as part of the admission process, a focused and comprehensive assessment needs to be completed that addresses environmental, psychosocial, physiological, and health-related issues. In 1999, the Health Care Financing Administration (HCFA) mandated that the staff of Medicare-certified home-care agencies complete a 95-item Outcome and Assessment Information Set (OASIS) when they admit new clients; children and pregnant women are excluded. Other assessment tools are mandated for interim periods and discharge. Implementation is scheduled late in 2000 (*Home Health Line*, 1999; Martin, 2000).

The assessment should involve a team approach, with members of various disciplines communicating and contributing relevant data. A case manager is responsible for obtaining additional information from the referral source and communicating pertinent information to the client's home health care physician and insurance or health plan case manager.

After reviewing referral data and physicians' orders, nurses identify the home health client's problems, for example, income, circulation, nutrition (*On-line Journal of Nursing Informatics*, 1999). Nurses then prioritize the problems in partnership with clients and families—a crucial step in determining which problems they will actively work on.

Plans and Interventions: Just as clients' assessments and diagnoses need to be focused, so do care plans and interventions. Interventions used frequently by the home health multidisciplinary team include health teaching, guidance, and counseling; treatments and procedures; case management; and surveillance (*On-line Journal of Nursing*

Informatics, 1999). Nurses may instruct clients in the use of a glucometer, give injections, collaborate with physical and occupational therapists about clients' post-stroke needs, refer clients to Meals on Wheels or registered dietitians, or monitor environmental modifications to decrease the risk of falling.

The average duration of home health care services is two weeks to two months but is decreasing. Thus, care plans and interventions should be selected for the client's priority problems. Ideally, the client, family, and health care team members share the goal of self-care or achieving the greatest degree of independence possible. Clinicians need to anticipate whether other types of services will be required after discharge from home health care services.

Evaluation and Outcome Management: Simple, valid, and reliable instruments are needed to evaluate home health care. Many managed care companies require home health agencies to provide quantitative data that describe the clients they serve, the types of services they provide, the effects of those services on clients, and the costs of services. The HCFA's OASIS data can also be used to evaluate home-care effectiveness. Agencies that have automated clinical information systems are attempting to analyze the data they collect. All agencies must submit required data to Medicare, but they might never receive analytical feedback about their performance from the HCFA.

Client records are a major source of clinical data for an outcome management program. Clinicians need to ask and document the following: What works best? Has the client improved? By how much? From what perspective? One approach is to use the Omaha System's Problem Rating Scale of Outcomes, a five-point Likert-type scale designed for use with specific client problems. It includes three separate numeric subscales for knowledge, behavior, and status (*On-line Journal of Nursing Informatics,* 1999). Ratings offer cues to help clinicians select the most appropriate interventions for the client and provide a baseline for tracking knowledge, behavior, and status data throughout the duration of service. Data collection is repeated at established intervals and at discharge. The clinical data

produced by the ratings can be analyzed in conjunction with staffing, length of service, cost, and other statistical data.

Related Research

Research offers an important strategy for identifying and disseminating the benefits of home health services. From 1992 to 1999, the Community Nurse Organization Demonstration Project was funded by the HCFA to explore the effects of a capitated payment model on the delivery of home health services (*On-line Journal of Nursing Informatics,* 1999). Although findings have not yet been published, the HCFA is incorporating the information in the proposed home health reimbursement regulations.

A series of trials funded by the National Institute of Nursing Research and National Institutes of Health was designed to investigate the benefits of discharge planning and home follow-up provided by advanced practice nurses (Naylor et al., 1999). One clinical trial, conducted between 1992 and 1996, focused on elders at high risk for poor discharge outcomes. Findings indicated that patients in the intervention group, cared for by advanced practice nurses, were less likely to be readmitted to the hospital and were hospitalized for fewer days than were patients receiving usual care. Interestingly, the intervention group achieved these positive health outcomes at half the Medicare costs incurred by the control group.

KAREN S. MARTIN

Internet Key Words
Clients
Communication
Diversity
Home Health Care
Medicare-Certified Agencies
Nursing Process
Partnership
Standardized Vocabularies

Internet Resources

Health Care Financing Administration
http://www.hcfa.gov/medicare

National Association for Home Care
http://www.vnaa.org

National Association for Home Care
http://www.nahc.org

On-line Journal of Nursing Infomatics (Vol. 3, No. 1, Winter, 1999)
http://cac.psu.edu/~dxm12/ojni.html

HOMELESSNESS

A key problem in studying and understanding the magnitude of elder homelessness in the United States is the lack of a clear definition of "older homeless" individuals. Some researchers, noting that homeless individuals look and behave 10 to 20 years older than their actual ages and have significant physical and psychological problems, define an older homeless person as any homeless individual over age 50. With this criterion in mind, it has been estimated that there are between 60,000 and 400,000 older homeless individuals in the United States today, with the number of homeless people over age 50 projected to double by 2030. It is also predicted that those over 50 in 2030 will include more women and more racial and ethnic groups (Burt, 1996). With the forecast of shifting demographics among the future elder homeless, nurses, social workers, physicians, and others caring for homeless populations need to be prepared to address their increasingly complex health and social needs.

A second problem in the identification and care of homeless elders relates to the definition of "homelessness." Increased homelessness among elderly persons is largely the result of the unavailability of affordable housing and the prevalence of poverty among certain segments of the aging populace. Homeless individuals generally lack a fixed and regular place of residence and may sleep on the street. However, many more reside in supervised public or privately operated shelters designed to provide temporary accommodations. Welfare hotels, congregate shelters, and transitional housing fall into this category.

Men, particularly white men, outnumber women four to one among the aging homeless population (Bissonette & Hijjazi, 1994). Given this male predominance, it is not surprising that most studies of homeless elders' physical or mental health needs have focused on small convenience samples of homeless elderly men. No specific studies address age- or gender-specific health and social concerns among elderly homeless populations. Studying homeless women may be difficult, because women do not use public health care facilities as readily as men do. When they do seek health services, it is often for prenatal care or care for dependent children. Older homeless women are therefore less likely to make contact with the health care system and more likely to be overlooked in clinical research and practice.

Numerous studies have established that older homeless persons suffer substantially more physical illness than younger homeless adults and nonhomeless elders. One of the most comprehensive examinations of homeless persons' health to date, using objective as well as self-reported measures of physical health, was undertaken as part of a study of 521 homeless men in two Los Angeles beach communities. In this study, homeless men over age 50 suffered more chronic illness and functional disabilities than younger individuals and demonstrated patterns of illness similar to those of adults 65 and older in the general population (Gelberg, Linn, & Mayer-Oakes, 1990). Their illnesses, often untreated, led to higher mortality than among their nonhomeless peers.

Many of the health problems experienced by older homeless persons are exacerbated by chronic alcoholism and drug addiction. Although alcohol misuse and prescription drug dependence are more typical among older adults, the prevalence of heroin, crack cocaine, and other substance abuse is also on the rise among aging baby boomers. Once unheard of in homeless communities because of the prohibitive cost, cocaine in its smokable form, crack, has become widely available and popular among homeless populations (Jencks, 1994). The shift in geriatric substance abuse creates new challenges for health care professionals. However, illnesses often associated with homelessness and addiction, such as HIV/AIDS, have only recently been recognized among older adults (Emlet, 1997; Tordoff, 1996).

There is also compelling evidence that homeless elders have higher overall rates of mental illness than the general older population. It has been estimated that 10% to 15% of homeless elders have a serious or chronic mental illness, including alcohol and drug abuse and dependency, anxiety disorders, depression, and schizophrenia. Yet few studies address the mental health needs or assess the possibility of alcohol-related dementia among older homeless persons. Limited research examines drug misuse among older persons and its long-term effects on physical, psychosocial, and spiritual health.

Housing programs alone cannot solve the problems of the homeless elderly. Although stable housing is crucial to the well-being of older homeless persons, recent intervention strategies have been directed toward the development of community-based interdisciplinary programs whose focus is on housing and case management. Thus, there is growing recognition that older homeless men and women have special needs that span the gamut from health factors to safety issues.

As we begin the 21st century and watch the inevitable rise in the number of homeless elders in American society, health care providers need to conduct rigorous scientific research that measures the complex parameters of elder homelessness. Outcome data must be gathered to provide an accurate description of homeless elders and their health care needs, as well as any root problems that may be contributing to prolonged homelessness. Only then can programs be designed to address the specific needs of older homeless persons, policies developed to respond to this growing social problem, and adequate resources allocated to implement both practice and policy.

BARBARA BRUSH

Internet Key Words
Aging and Homelessness

Homeless Elders
Older Homeless Adults

Internet Resource
Health Care Issues for Elderly Homeless People, Health Care for the Homeless Information Center
http://www.prainc.com/hch/bibliographies/ elderly.htm

HOMOSEXUAL ELDERS
See
Gay and Lesbian Elders

HOSPICE

Hospice is a concept of care that offers palliative rather than curative treatment to patients in the final stages of terminal illness. When the prognosis is poor, the focus of care should shift from an aggressive pursuit of cure to the relief of suffering, so that the patient's final days can be spent with dignity and bring a sense of closure for both the patient and the family. According to the World Health Organization (WHO), the hospice philosophy affirms life and regards dying as a normal process. It does not prolong life, and it does not hasten death. It provides relief from pain and other symptoms and integrates the psychological and spiritual aspects of patient care. Hospice offers a support system to help patients live as actively as possible until death and to help the family cope with the patient's illness and their own bereavement (Saunders & Kastenbaum, 1997; Volicer & Hurley, 1998).

Eligibility

In the past, most hospice patients were people with end-stage cancer (Stuart, 1995). However, people in the final stages of heart, lung, and liver disease; dementia; neurological disorders such as amyotrophic lateral sclerosis; and AIDS greatly benefit from hospice care. Patients are usually referred to hospice by their primary care physicians. Referrals are also accepted from patients themselves, other health care professionals, family members, caregivers, social workers, and clergy.

For a patient to be eligible for the Medicare-reimbursed hospice benefit, a physician must certify that to the best of his or her knowledge the patient has a life expectancy of six months or less. This requirement is often responsible for a delay in hospice referral. Because of an uncertain prognosis, physicians may be reluctant to estimate life expectancy, especially with regard to nononcological diseases. The National Hospice Organization (NHO) has established guidelines to help physicians and hospices determine prognosis in noncancer diseases (Stuart, 1995). The criteria are based on decline in functional status, independence in activities of daily living, nutritional status, and symptoms pertaining to specific diagnoses.

Hospice Services

At the time of admission, the hospice obtains the necessary medical information from the referring physician. The hospice medical director must concur with the diagnosis of a terminal illness. Most hospices welcome the opportunity to include primary care physicians on the hospice interdisciplinary teams, which establishes the plan of care. The team usually consists of a physician, hospice nurses, social workers, clergy, home health aides, and volunteers; other participants might include a dietitian, physical therapist, occupational therapist, speech and swallowing therapist, and bereavement counselor. The patient and family are also important members of the team and are encouraged to participate in the decision making and in formulating a plan of care oriented to the patient's comfort. After the plan of care is determined and agreed on by all members of the team, medical orders are placed regarding necessary medications, treatments, frequency of services by the team, and durable medical equipment.

A nursing evaluation is done within the first 24 hours and covers not only the patient's physical status but also his or her psychosocial and spiritual needs. Although nurses work under a physician's supervision, the specially designed system of protocols and "as-needed" orders allows them greater independence in symptom management.

The social worker assists the patient and family with advance directives, management of resources, and insurance, as well as providing emotional and spiritual support. Any unresolved spiritual issues are referred to clergy and spiritual counselors, who also provide support in bereavement counseling.

Hospice care can be provided in a variety of settings, most frequently in the patient's home, but also in acute-care settings (hospitals), long-term-care institutions (assisted living facilities, nursing homes), and hospice residential facilities. The location of hospice care is determined by the individual needs of the patient and the family. Regardless of the setting, which may change during the course of a terminal illness, care is always provided by the same interdisciplinary team. In addition to making frequent visits, team members are available by telephone 24 hours a day. Care can be provided at different levels of intensity, ranging from brief daily visits to continuous home care, depending on the patient's needs.

Most hospices also provide respite care for family caregivers, many of whom are frail elders themselves. To relieve some of the burden, the patient can be temporarily transferred for up to five days to a skilled nursing facility, inpatient hospice, hospital, or nursing home.

Medical care of a hospice patient focuses on aggressive pain and symptom control. Pain and symptom control is achieved through standing orders for around-the-clock analgesia rather than as-needed doses. Additional analgesia is given for breakthrough pain. The WHO concept of an "analgesic ladder" is utilized (National Hospice Organization, 1998). Patients are maintained on acetaminophen or nonsteroidal anti-inflammatory medication until pain breaks through. Low-strength opioids are then tried, and only if the maximal dose fails to control the pain is the regimen switched to stronger opioids.

Common side effects of opioid use are constipation, drowsiness, and nausea. Constipation should be aggressively treated with laxatives. Tolerance to the drowsiness usually develops within a matter of days. Nausea may be suppressed with haloperidol or antiemetics. When opioids are properly prescribed, the chance of addiction, tolerance, and respiratory depression is minimal. Elderly patients require remarkably low doses of analgesics, sometimes only half the dose given to younger patients. Adjuvant drugs, including tricyclic antidepressants, anticonvulsants, antispasmodics, and anxiolytics, are often used to decrease the dose of opioids.

Other common symptoms such as anorexia, dyspnea, confusion, incontinence, or skin breakdown can be a source of great discomfort to dying patients. The hospice team is specially trained in identifying and addressing these problems.

Reimbursement

Medicare Part A, Medicaid, and most private insurance policies cover hospice care. As noted earlier, Medicare requires the primary care physician and hospice medical director to certify that the patient's life expectancy is six months or less. Certification must be renewed after 90 days, followed by another 90-day period, and then for an unlimited number of 60-day periods. Medicare beneficiaries are required to sign a statement electing hospice care, thus waiving traditional Medicare benefits covering curative treatment for the terminal disease. Patients can be hospitalized or undergo surgery if it will palliate their symptoms. A do-not-resuscitate (DNR) order is optional; however, many patients elect not to be resuscitated. The Medicare benefit covers medications, durable medical equipment, treatments related to the terminal diagnosis, nursing visits, home health aide visits, social work, and bereavement care. The Medicare hospice benefit does not cover aggressive curative treatment of the terminal illness, services that are not part of the palliative plan of care or are not preapproved by the interdisciplinary hospice team, or services that

duplicate hospice care and are provided by a facility that does not have a contract with the hospice. However, continued follow-up with the primary care physician is strongly encouraged in an effort to create a sense of security, nonabandonment, and continuity of care. Hospice care can be revoked at any time for any reason. After a patient signs a written consent indicating his or her wish to revoke hospice care, regular Medicare benefits will be reinstated.

Improving Hospice

Open and Honest Communication: One factor in effective end-of-life care is keeping patients and their family informed about the disease and its progression and giving everyone involved ample opportunity to ask questions and bring up unresolved issues. Open and honest communication results in realistic goals and a plan of care that best suits the needs and wishes of the patient and family.

Early Referrals: The patient and family should be given enough time to absorb and accept the idea of hospice. From the hospice team's standpoint, early referral gives the team more time to understand and anticipate the patient's needs.

Nonabandonment: The value of primary care physicians and their involvement in the care of terminally ill patients cannot be overestimated. Although physicians may feel that they have nothing more to offer in terms of cure, their continued presence assures both patients and families that they are not alone and ultimately results in high-quality end-of-life care (Byock, 1997).

ANNA LAMNARI

Internet Key Words
Caregiver Burden
Death
Hospice
Pain: Chronic
Palliative Care
Quality of Life

Internet Resources

American Academy of Hospice and Palliative Medicine

http://www.ahp.org

Center to Improve Care of the Dying

http://www.hospice-cares.com

Hospice Care on the International Scene by D. C. Saunders and R. Kastenbaum (1997)

http://www.springerpub.com

National Hospice Organization

http://www.nho.org

HOSPITAL-BASED SERVICES

In nonfederal hospitals, patients 65 years of age and older account for 37% of all discharges and 47% of inpatient days of care. Rates of hospitalization are more than twice as high for patients age 85 years and older compared with patients age 65 to 74 years, although costs of care are highest for 65- to 74-year-old users (Palmer, 2000). Projections for the 21st century indicate that patients age 75 years and over will represent nearly half of the hospitalized population, and their costs of care will more than double. This trend is a major health care issue, because older patients have longer hospitalizations, higher mortality rates, and higher rates of nursing home placement. Furthermore, for elderly patients, hospitalization is associated with functional decline or physical disability that occurs during the course of treatment for an acute medical illness (Palmer, Counsell, & Landefeld, 1998). Functional decline during hospitalization occurs more often in patients who are older than age 75 years, cognitively impaired, and dependent at baseline in two or more instrumental activities of daily living. Recent clinical trials report that at least some of the poor functional outcomes of hospitalization can be attenuated through improved processes and systems of care. Several promising models of hospital care have evolved in recent years and are potentially adaptable by most hospitals in the United States.

Most of the successful models have been nurse driven, often employing an advanced practice nurse. Interdisciplinary models are not uncommon and frequently include geriatricians, nurse specialists, physical and occupational therapists, speech therapists, social workers, and dietitians. The usual objectives are to improve functional outcomes, reduce hospital lengths of stay, prevent nursing home admissions, and prevent rehospitalization. Combining the best features of these services and models of acute care points to the possibility of improved hospital care of elderly patients in the future (Palmer, 2000).

Acute Care of Elders (ACE) Units

The purpose of an ACE unit is to prevent functional decline that results from processes of care (polypharmacy, imposed bed rest, inadequate medical standards of care, uncoordinated care) and patient characteristics (depression, cognitive impairment, physical disability) that often go undetected or untreated (Palmer et al., 1998). Although the size and objectives of ACE units may vary considerably, they have in common four main components: a prepared environment, patient-centered care, interdisciplinary team rounds and discharge planning, and medical care review. The physical environment is designed to foster independent self-care. A brief functional assessment conducted by the patient's nurse is linked to simple measures to help the patient maintain or regain independent self-care. The interdisciplinary team meets daily to review plans for patient care and discharge. An ACE unit may be either a discrete nursing unit (all beds) or a "virtual unit" with a mixed population of elderly and nonelderly patients. ACE units have shown the potential to reduce the incidence of functional decline in older patients, the length of hospital stay, and the risk of nursing home admission after hospital discharge (Palmer et al., 1998). They are also excellent sites for training health professionals in the care of older patients and conducting quality-assurance studies. To succeed, an ACE unit requires a major commitment of administrative energy and resources, especially when financial incentives (e.g., capitated managed care or bundled prospective payment systems) favor reducing total costs of care for an acute illness. The ACE model of care is transportable to other settings.

Geriatric Care Program

A gerontological clinical nurse specialist working with trained resource nurses—registered nurses with additional training in geriatrics—focuses nursing care on patients at high risk for functional decline. The intervention includes identification and monitoring of frail older patients, twice-weekly rounds of a multidisciplinary geriatric care team, and a nursing-centered educational program. In a clinical trial, when patients were matched on a number of target conditions and risks for functional decline at baseline, the intervention resulted in a significant beneficial effect, with a reduction in functional decline (Palmer et al., 1998). The model is attractive for hospitals that employ advanced practice nurses and have a high-risk (frail) patient population.

Elder Care Program

Patients at risk for incident delirium are identified shortly after hospital admission, using the Confusion Assessment Method (Inouye et al., 1999). An array of protocols targeted at specific risk factors (e.g., sensory deprivation, cognitive impairment) serve to optimize cognitive function (reorientation, therapeutic activities), prevent sleep deprivation (relaxation, noise reduction), avoid immobility (ambulation, exercises), improve vision (visual aids, illumination), improve hearing (hearing devices), and treat dehydration (volume repletion). A clinical trial of intervention protocols targeted at risk factors for delirium resulted in a 40% reduction in the incidence of delirium but had no significant effect on the severity of delirium or on recurrence rates. The overall costs of the program may limit its generalizability, but components of the intervention are inexpensive and could easily be incorporated into standard medical and nursing care in most hospitals.

Case Management

Hospital-based case management most often employs advanced practice nurses to identify vulnerable elderly patients and coordinate discharge planning with physicians and other health professionals (Palmer et al., 1998). In one clinical trail, an advanced practice nurse–centered discharge planning and home-care intervention for at-risk hospitalized elders reduced readmissions, lengthened the time between discharge and readmission, and decreased the costs of providing health care (Naylor et al., 1999).

Stroke Units

Meta-analyses of stroke unit studies indicate reduced rates of patient mortality, institutionalization, and functional dependency in comparison with general medical wards. An interdisciplinary team comprising the patient's physician, social worker, physical and occupational therapists, speech therapist, and neurologist uses a systematic approach to enhance patient recovery after admission and begin diagnostic and rehabilitative services (Palmer et al., 1998). Some stroke units include a geriatrician on the team.

Postacute Services

Closely aligned to the acute-care hospital, postacute services include subacute units, geriatric evaluation and management (GEM) units, and inpatient geriatric rehabilitation hospitals. Subacute units are skilled nursing facilities (SNFs) that provide short-term, goal-oriented, and intensive rehabilitative or skilled nursing services. Patients are often transferred to subacute units for continuing care after their clinical status has been stable for at least 72 hours in the hospital. Some patients, such as those with cellulitis or heart failure, can be admitted directly to the SNF subacute unit from the community under Medicare managed care plans. A study found that nearly one-half of 120 patients admitted for heart failure to an acute-care facility were considered at low risk and could have been managed in a subacute setting with significant cost savings (Palmer, 2000).

In Veterans Affairs hospitals, elderly patients who remain functionally impaired following admission to an acute-care hospital may be eligible for transfer to an inpatient GEM unit. Studies indicate that patients receiving the GEM intervention—comprehensive geriatric assessment, multidisciplinary care, and outpatient continuity of care after discharge—are less likely to spend any time in a nursing home following discharge (Palmer et al., 1998). Similar results are found when comprehensive assessment and multidisciplinary care are provided in a community rehabilitation hospital to elderly patients who are recovering from acute medical or surgical illnesses and are at risk for nursing home placement.

Observation Units (Clinical Decision Units)

Observation units or holding areas in emergency departments allow elderly patients to be monitored or evaluated for less than 24 hours before being admitted to the hospital or discharged home. Studies of patients with conditions such as chest pain and asthma found the use of observation units a feasible alternative to hospitalization. However, controlled trials involving elderly patients are rare, and there is no consensus regarding the indications for admitting and discharging patients from these units (Palmer, 2000).

Future Directions

As financial incentives increase to reduce the costs of care and improve quality, the acute-care hospital will evolve into an integrated health system in which the most cost-effective services will be applied to targeted patients in various subsystems of care (Palmer, 2000). The current patchwork of largely uncoordinated subsystems of care will be coordinated across specialty areas in a system designed to contain costs and optimize clinical outcomes. The subsystems of geriatric care in the future medical center will include acute-care (e.g., ACE, stroke) units, subacute or GEM units, and observation units. Additional services will include case management by advanced practice nurses for vulnerable elderly patients, an elder life program on medical and surgical units, and interdisciplinary assessment and coordination of discharge planning throughout the hospital. These services will be complemented by closely aligned services, including home care, palliative care, and day hospitals, and improved medical information systems (Palmer, 2000). The acute hospitalization of elderly patients will be shorter, reserved for those who cannot be cared for in a less intensive environment, and justified when invasive diagnostic or therapeutic procedures are required.

ROBERT M. PALMER

See Also
Case Management
Clinical Pathways
Discharge Planning
Geriatric Evaluation and Management Units
Home Health Care

Internet Key Words
ACE Units
Case Management
Geriatric Care
Observation Units
Postacute Services

Internet Resources
National Center for Health Statistics
http://www.cdc.gov/nchs

National Chronic Care Consortium
http://www.ncccresourcecenter.org

HUMAN IMMUNODEFICIENCY VIRUS (HIV) AND AIDS

HIV and the Acquired Immunodeficiency Syndrome (AIDS) have replaced syphilis as the "great imitator," especially in older adults. HIV and AIDS have so many signs and symptoms that they are often confused with other illnesses or conditions, making them difficult to diagnose unless suspected in the first place. Because of the low index of suspicion in adults over 50 years old, the infection is often misdiagnosed as bone marrow disease (myelodysplasia), dementia (especially Alzheimer's Disease), or occult malignancies. Many practitioners, including geriatricians and emergency department physicians, do not include HIV infection in the differential diagnosis of an older adult who presents with weight loss, leukopenia, severe fatigue, or failure-to-thrive. There is some inherent ageism in this philosophy since it is based on the assumption that mature adults do not enjoy sex and do not remain sexually active in their later years. This philosophy also stems from the high burden of chronic illness in older adults that could account for signs and symptoms that are actually associated with HIV infection.

The patient may complicate the issue by not self-identifying HIV risk factors such as homosexuality or intravenous drug use. Older adults may also be unaware of their own HIV risk factors or those of their sexual partners since virtually no HIV education has been directed towards mature adults. Changing behavior is always challenging, especially when individuals are unaware they need to do so: Why would older adults begin to practice "safer sex" when they never had to do so when they were younger and perhaps more sexually active? No data exist to suggest that risky behaviors are abandoned in the fifth or sixth decade. It is not surprising then that older adults have adopted few, if any, protective behaviors against HIV infection. Older adults with known risks for HIV rarely use condoms during intercourse and are rarely tested for HIV antibodies (Gordon & Thompson, 1995). Some risk factors, such as transfusion-related HIV, may actually increase in older adults.

These and other complications conspire to make HIV disease in the older adult a silent epidemic. Physician reluctance to discuss sex and sexual behaviors with seniors adds to the problem of diagnosis and management. Many healthcare providers are surprisingly unaware of the prevalence of HIV disease in elders (El-Sadr & Gettler, 1995). Discussion of the severity of HIV disease in adults over 50 can raise the level of suspicion for the diagnosis and can improve the clinical care of those with recognized disease.

Epidemiology

The discussion of HIV disease rarely includes the impact on older adults even though they represent a consistent percentage of cases. Of the 688,200 cumulative cases of AIDS, including 410,800 deaths, 10.5%, or 72,161 cases were reported in adults over 50 years old (Centers for Disease Control and Prevention, 1998); 5.6% or 38,453 cases occurred in adults over the age of 55. Since the beginning of the AIDS epidemic in the early 1980s, older adults consistently account for 7% to 10% of cases diagnosed annually. What these numbers do not include are the AIDS cases that were misdiagnosed and never included in the statistics. Thus, the statistics underrepresent the true prevalence of the disease in older adults.

Regional statistics suggest higher percentages of elders with AIDS in epicenters of the disease, especially in New York City, Los Angeles, and San Francisco (Centers for Disease Control and Prevention, 1999b). During the 1990s, AIDS cases in New York City increased 15% in adults over 50 and 11% in adults 60 years or older. These rate increases are markedly higher than the national average of 3% to 5% for those age groups.

The risk factors for HIV disease are identical for older and younger adults. Whereas the "young old" tend to be exposed via unprotected sexual intercourse or intravenous drug use, the "old old" tend to be infected through tainted blood products (Ship et al., 1991). It is estimated that 29,000 individuals aged 13 to 65 years old received HIV-tainted blood products before 1985. Adults over 50 receive 70% of blood transfusions, hence it is much more likely that HIV/AIDS from blood products would occur in older adults. Most cases of AIDS in adults over 70 years old are related to receipt of HIV-infected blood (Ship et al., 1991). Although testing standards are markedly improved since the mid-1980s, HIV-infected individuals who have not yet seroconverted will not be detected. Given the fact that it takes from 6 weeks to 6 months to develop HIV antibodies in a newly infected individual, a recently infected individual's blood could be missed by current methods of detection.

Confounding factors in the epidemiology of HIV disease in older adults may be due to refusal to self-identify one's sexual preference or drug use in a generation that still suffers under past stigmas. It is estimated that 90% of older adults with an acknowledged risk for HIV have never been tested for antibodies. If they are tested, older adults are much more likely to have an unstated risk for infection. Older adults are also more likely to be diagnosed in the final stages of AIDS when they are moribund and unable to provide risk information. They also are more likely to die within the first month of diagnosis than other age groups, a finding that may be explained by the fact that older adults are often diagnosed at a more advanced stage of the disease.

Immune system changes in older adults may make it easier to become infected with HIV or for viral infection to progress to AIDS. Changes in vaginal mucosa from atrophic vaginitis may reduce the effectiveness of this normal protective barrier. Normal alterations in CD4/CD8 ratios with age result in less suppressor activity, making viral replication easier. Infections with other retroviruses, such as HTLV (Human T cell leukemia virus), and/ or changes in tumor necrosis factor (TNF) levels may allow latent HIV to become activated. The burden of comorbid illness and changes in the aging immune system can directly affect the progression of HIV disease in the older adult. Decreases in mucosal immunity and secretory antibody production make opportunistic encapsulated bacterial infections more likely. Comorbid illnesses such as diabetes or renal insufficiency directly affect the immune system and make HIV progression more likely.

HIV Treatment

HIV/AIDS drug regimens change frequently, and updated treatment recommendations from the Inter-national AIDS Society (Carpenter et al., 1998) and British HIV Association (Gazzard & Moyle, 1998) should be consulted. Age alone is never a barrier to effective HIV treatment but increases the challenges faced by the practitioner. Physiological changes due to the aging process, along with medication use for comorbid illness, increase the likelihood of side effects and drug-drug interactions from HIV therapy. This necessitates a higher level of awareness on the part of clinicians treating HIV in older persons to intervene early when such interactions are suspected.

The first and most important concern is the dosing of medications. Anti-retrovirals, prophylactic agents, and other HIV therapies all have side effects and interactions that are potentially worse in an older individual. Doses suggested by manufacturers listed in the *Physician's Desk Reference* (PDR) are based on the treatment needs of a 70 kg young man and do not consider the age-related physiologic changes in lean tissue/body fat ratios, liver/kidney function, protein status, and hydration status. The doses also do not take into account the possible drug-drug interactions that may occur when two medications are competing for the same protein binding site or cytochrome P450 system for metabolism.

Given the high burden of chronic disease in older adults, it would not be surprising to have a senior being treated for hypertension, coronary artery disease, hyperlipidemia, diabetes, post-menopausal hormone therapy, or prostate disease. Adding to this mix the powerful cocktails used to treat HIV disease and the antibiotic regimens used for *Pneumocystis carinii* pneumonia (PCP), *Mycobacterium avium* complex (MAC), or cytomegalovirus (CMV) prophylaxis could make for disastrous combinations.

Unfortunately, there is little information available about HIV treatment optimization in mature adults. Therefore, it is very important to treat HIV positive elders cautiously and with the help of HIV and/or infectious disease (ID) specialists. The general rule of geriatric pharmacology, "start low and go slow," is applicable. Attention must be paid to comorbid illnesses and their treatment regimens so as to avoid potentially lethal interactions. Also, frequent office visits and phone communication are imperative to monitor for adverse drug effects early.

Preventive medicine in older adults with HIV infection should include yearly influenza vaccine and receipt of the pneumococcal pneumonia vaccine. Regular Pap smears are important since it has been found that cervical cancer spreads more rapidly when the individual is concomitantly infected with HIV. Rigorous blood sugar control in the older diabetic patient is important since elevated blood sugar is associated with altered white blood cell function and poor wound healing. Regular ophthalmologic evaluations are necessary not only to screen for macular degeneration and glaucoma but also to rule out CMV retinitis. Finally, proper nutrition and good sleep habits are beneficial for all seniors but especially for those with immune dysfunction.

Care of the older HIV patient requires a multidisciplinary team including physicians, nurses and nurse practitioners, social workers, rehabilitation therapists, nutritionists, and spiritual counselors. Social workers can provide information and contacts to programs and entitlements such as the EPIC and ADAP drug programs available in New York state. Diminished functional status (activities of daily living) can cause significant depression and anxiety; neither of these mental illnesses should be "expected" or "assumed" in the geriatric population. Overall, care of the elderly patient with HIV/AIDS is more challenging than that of the younger patient. Older HIV patients need a holistic approach to care that takes into consideration the mental, physical, and psychosocial health of the individual as well as the functional and physical limitations of aging.

DENIS KEOHANE

See Also
Coping with Chronic Illness

Internet Key Words
Blood Transfusions
Comorbid Illness
HIV/AIDS

Homosexuality
Intravenous Drug Use

Internet Resources
Centers for Disease Control and Prevention, HIV/
   AIDS Surveillance Report
http://www.cdc.gov/hiv/stats/hasrlink.htm

Department of Health and Human Services HIV/
   AIDS Services
http://www.hrsa.dhhs.gov/hab/default.htm

National Institute of Health HIV/AIDS Information Services
http://sis.nlm.nih.gov/aids/index.html

HYPERTENSION

Hypertension, particularly systolic hypertension, is a common condition with serious ramifications in older persons. Aging is associated with progressive elevation of systolic arterial pressure, which is partially related to structural changes in vessels that result in increased vascular stiffness and reduced vascular compliance. Isolated systolic hypertension, generally defined as a systolic blood pressure of 140 mm Hg or greater with a concurrent diastolic blood pressure less than 90 mm Hg, is the most common form of hypertension in the elderly. Elevated systolic blood pressure in older persons is associated with an increased risk of myocardial infarction, stroke, and cardiovascular mortality. Although elevation of either systolic or diastolic blood pressure is associated with increased morbidity and mortality, the systolic blood pressure is a better predictor of cardiovascular complications in older persons (Applegate, 1999).

An important advance in the understanding of hypertension is the accumulation of strong evidence that older patients derive significant benefit from antihypertensive therapy. Several recent trials involving older persons have demonstrated that antihypertensive therapy significantly reduces the risk of cardiovascular morbidity and mortality in elderly patients with systolic hypertension. The value of therapy has been demonstrated for people beyond

80 years of age. Antihypertensive therapy may exert its greatest impact when applied to older persons, because the proportion of individuals who are hypertensive increases with age, and elderly individuals are at greater risk for cardiovascular events.

Diagnosis

As is the case for younger individuals, older persons frequently present with asymptomatic hypertension detected by routine screening measures. Patients may also present with symptoms related to the cardiovascular or cerebrovascular complications of hypertension. Occasionally, patients present with symptoms related to severe and accelerated hypertension, such as headache, visual disturbances, nausea, and confusion. The goals of evaluation are to identify the need for urgent therapy, recognize possible secondary causes of hypertension, and determine the presence of target organ disease and additional cardiovascular risk factors.

A thorough medical history should be obtained, including symptoms related to acute blood pressure elevation and to the vascular complications of hypertension. Additional cardiovascular risk factors should be identified, and prior medication trials and associated side effects should be reviewed. Use of over-the-counter or prescription medications that may contribute to hypertension, such as nonsteroidal anti-inflammatory drugs, oral sympathomimetics, and steroids, should be ascertained. In addition, an alcohol, diet, and exercise history should be obtained.

Blood pressure measurement should be performed with a well-calibrated sphygmomanometer and a cuff of appropriate size. The blood pressure should be measured in both arms, after at least five minutes of rest, and in both the supine and standing positions, as orthostatic hypotension is frequently observed in older persons. The standing blood pressure is generally used to guide therapy. Unless the blood pressure elevation is severe or associated with complications, the diagnosis of hypertension should be based on the average of at least three measurements obtained on separate occasions. The

physical examination should also include height and weight measurements, a funduscopic exam, a cardiovascular exam, and evaluation for the presence of carotid or abdominal bruits. The appropriate routine laboratory tests for older persons with hypertension are the same as for younger individuals and include measures of renal function, potassium level, fasting glucose, uric acid, complete blood count, urinalysis, and electrocardiogram (National High Blood Pressure Education Program Working Group, 1994). A lipid profile may also be helpful as an indicator of cardiovascular risk. Additional investigations may be indicated when the history, physical examination, or initial laboratory results suggest an identifiable cause for hypertension.

Treatment

Nonpharmacological interventions are an important component of antihypertensive therapy. Modification of lifestyle factors that may contribute to hypertension and that may be helpful in lowering blood pressure include moderate sodium and alcohol restriction, weight loss for overweight individuals, and regular exercise. Other lifestyle modifications to reduce cardiovascular risk, such as smoking cessation and a low-fat diet, should be recommended.

A thiazide diuretic is the preferred initial agent in older persons with isolated systolic hypertension. Several studies have demonstrated that thiazide diuretics significantly reduce the incidence of coronary artery disease, cerebrovascular disease, and cardiovascular mortality in older patients with systolic hypertension. Thiazide diuretics are an effective, inexpensive, and generally well-tolerated class of drugs that may have the added benefit of decreasing urinary calcium excretion. The initial dose should be low, and the potassium level should be carefully monitored. A potassium-sparing diuretic, such as triamterene, may be added for hypokalemia. Long-acting dihydropyridine calcium channel blockers may reduce cardiovascular complications in older persons with systolic hypertension and can be considered a second-line alternative to diuretics (Joint National Committee on Prevention, Detection, Evaluation, and Treatment of High Blood Pressure, 1997).

General recommendations for antihypertensive therapy may be altered in the presence of comorbid medical conditions, which may make particular drugs more or less appropriate. It is often useful to select an antihypertensive drug that will provide additional benefits based on other coexisting medical conditions. For example, angiotensin-converting enzyme (ACE) inhibitors are the drugs of choice in patients with diabetic nephropathy and systolic heart failure. Beta-blockers may be used in patients with angina or prior myordial infarction. Alpha-blockers are useful in elderly men with prostatism.

The goals of drug therapy are to maintain the systolic blood pressure below 140 mm Hg and the diastolic blood pressure below 90 mm Hg. Treatment to lower levels may be useful in certain subsets of elderly patients, as is the case for younger patients. The starting drug dosage should generally be lower in older patients, who often demonstrate altered drug metabolism and excretion. Drug titration should be slow and gradual in older persons, as they are at significant risk for adverse reactions. Patients should be monitored closely for side effects such as dizziness, postural hypotension, falls, confusion, and depression. Drugs that may exacerbate postural hypotension or adversely affect cognition should be used with caution. Long-acting formulations with once-daily dosing are generally preferred.

Patient education is an important component of therapy that can help alleviate anxiety and improve compliance. The medical regimen should be carefully reviewed with older persons on a regular basis to check that they are taking their medications as directed and do not require assistance. Psychosocial and environmental factors that may influence care should be identified and addressed. Family members and caregivers may play an important role in terms of providing support, monitoring for side effects, and ensuring that medication is being taken correctly.

JEANNE Y. WEI
BRENDA ROSENZWEIG
CATHERINE MORENCY

See Also

Heart Failure Management: Congestive and Chronic Heart Failure

Internet Key Words

Antihypertensive Therapy
Cardiovascular Risk Factors
Lifestyle Modifications
Postural Hypotension
Systolic Hypertension

Internet Resources

American Heart Association
http://www.americanheart.org.

National Heart, Lung, and Blood Institute
http://www.nhlbi.nih.gov

I

IATROGENESIS

Iatrogenic illness is any illness that results from a diagnostic procedure or therapeutic intervention and is not a natural consequence of the patient's disease. A broader definition includes illness resulting from environmental events (e.g., falls), underdiagnosis (Gorbien et al., 1992), undertreatment, or negligence. As a rule, an iatrogenic illness is one that results from medication, diagnostic or therapeutic procedures, nosocomial infections, or environmental hazards. Older patients are predisposed to iatrogenic illness because of reduced homeostatic reserves, high levels of comorbid illnesses, and polypharmacy, which increases the probability of adverse drug events.

Most studies of iatrogenesis have been conducted in acute-care hospitals. One study of hospitalized patients age 65 years and older found that 58% suffered at least one iatrogenic complication when hospitalized for 15 or more days. Therapeutic interventions accounted for 44% of complications; diagnostic procedures and errors of omission accounted for approximately 10% each (Riedinger & Robbins, 1998). The Harvard Medical Malpractice Study reported that patients over age 65 were twice as likely to sustain injury during hospitalization as younger patients and were more likely to suffer from the effects of hospital negligence (Riedinger & Robbins, 1998). Most of the events were considered preventable. The high rates of iatrogenesis, medical error, and negligence reported in hospitalized patients led the Institute of Medicine to release a report advocating dramatic, systemwide changes to reduce these rates (Kohn, Corrigan, & Donaldson, 1999).

Adverse Drug Events

Adverse drug events in hospitalized patients are associated with significantly prolonged lengths of stay, higher costs of care, and increased risk of death. Polypharmacy, inappropriate drug prescribing, and prescribing errors are common preventable causes of iatrogenic illness. Virtually all classes of medications can cause adverse events, but antibiotics and cardiovascular drugs are most commonly implicated in studies of hospitalized patients. The prevention of adverse drug events in elderly hospitalized patients can include a variety of strategies:

- Review all medications taken by the patient before hospitalization, and assess the patient's previous compliance with prescriptions.
- Avoid unnecessary polypharmacy by using drugs that treat more than one condition (e.g., calcium channel blockers for hypertension and angina pectoris) whenever practical.
- Recognize the occurrence of an adverse drug event (e.g., wheezing resulting from a beta-blocker) before treating the symptoms with a second drug (e.g., a bronchodilator).
- Avoid, if possible, drugs that inhibit or induce cytochrome P450 hepatic metabolism or are highly bound to albumin.
- Consult a pharmacist, pharmacologist, on-line pharmacology program, or text source when additional drug information is needed.
- "Start low and go slow" to titrate the dose of a medication whose maintenance dose has not been established for elderly patients.
- Use lower than usual adult maintenance doses of medications that are excreted renally (e.g., aminoglycosides).
- When feasible, prescribe chronic medications once or twice daily to enhance patient compliance and reduce caregiver burden at home.
- Consider an adverse drug effect when patients develop new or unexplained medical problems.

Diagnostic and Therapeutic Procedures

Common medical complications of diagnostic studies include contrast-associated renal failure, anemia from extensive venipunctures, and urinary tract infections from bladder catheterizations. The patient's hydration should be maintained before and after a diagnostic study is performed with intravenous radiocontrast dyes. Contrast should be avoided if patients have baseline renal insufficiency

Functional decline, delirium, pressure ulcers, and trauma (e.g., falls) may occur because of inadequate processes of care or poor environmental conditions. Pneumothorax and hematomas from arterial lines placed during surgical or invasive procedures have been reported in hospitalized patients and should be considered carefully before performing the procedure. Complications of immobility, including physical disability and pressure ulcers, are potentially preventable through greater attention to patient mobility, exercise, avoidance of physical restraints, and interdisciplinary care. Patients should be assessed for risk of pressure ulcers, with attention to nutritional repletion, mobility, and skin lubrication to reduce incidence.

Standardized protocols can improve surgical outcomes. Attention to blood pressure regulation, oxygen supplementation, and the avoidance of anticholinergic agents in the perioperative period appear to reduce the incidence of postoperative delirium. Anticoagulation with either low-molecular-weight or unfractionated heparins or the use of leg compression devices reduces the incidence of venous thrombosis in elderly patients undergoing elective or emergent hip or knee surgery (Saint & Matthay, 1998). Histamine-2 receptor antagonists or oral sulcralfate reduces the risk of upper gastrointestinal bleeding in critically ill surgical or medical patients with respiratory failure or coagulopathy (Saint & Matthay, 1998).

Nosocomial Infections

Nosocomial, or hospital-acquired, infections are common causes of iatrogenic illness in both hospitalized and institutionalized elderly patients; they increase the length of stay and the costs of hospitalization and increase overall morbidity and mortality (Riedinger & Robbins, 1998). Common nosocomial infections are those of the urinary tract, those re-

lated to intravascular catheters, and pneumonia. Skin (e.g., methicillin-resistant *Staphylococcus aureus*), oropharyngeal (e.g., *Candida* species), and gastrointestinal (e.g., *Clostridium difficile* colitis) infections are other significant nosocomial infections that are potentially preventable (Riedinger & Robbins, 1998).

Bacterial resistance to usual antibiotic therapy results from interaction among microorganisms, patients, antibiotics, and infection control practices. Factors that contribute to antibiotic resistance are frequent prescribing of antibiotics, intrainstitutional transmission of resistant bacteria by cross-colonization of patients due to poor handwashing practices of health care workers, and transfer of colonized patients between institutions (Struelens, 1998). Resistance to usual antibiotics is now common with infections caused by *Streptococcus pneumoniae*, *Staphylococcus aureus*, and *Enterococcus* species (e.g., vancomycin-resistant enterococci [VRE]).

VRE are likely transmitted from patient to patient by the unwashed hands of health care workers or contaminated medical equipment or environmental surfaces (e.g., mattresses, bed rails, telephones, blood pressure cuffs, doorknobs). Infection and antibiotic control procedures, including restriction of vancomycin use, better selection of empirical antibiotics, education of hospital personnel, early detection and reporting of vancomycin resistance, isolation of colonized patients, and appropriate cleansing of the environment, may prevent the spread of VRE in health care settings.

Resistant urinary tract infections are common in elderly patients following prolonged indwelling urinary catheterization. The incidence of infection increases by approximately 5% per day, although the rate of bacteremia (urosepsis) is lower. Catheter use should be limited to patients with incurable urinary retention who cannot be kept clean and dry with standard nursing measures. Patients who are critically ill, in whom precise measurement of urine output is important, are candidates for temporary catheter placement. However, urinary incontinence protocols that include indications for appropriate catheter use and removal can help ensure that catheters do not remain in place longer than necessary. Prophylactic antimicrobial therapies and routine catheter replacement are not recommended.

Nosocomial pneumonia results from colonization of the upper respiratory and gastrointestinal tracts and occurs most often in critically ill patients who are ventilator dependent. Factors promoting nosocomial pneumonia include gastric aspiration, spread of pathogens on the hands of medical and nursing personnel, fecal-oral spread of pathogens, and cross-contamination from other patients. Infections with gram-negative bacteria, notably *Pseudomonas* species, are common. Prevention includes proper cleaning of respiratory equipment at timely intervals, consistent handwashing between patient contacts, cleaning of mechanical equipment (e.g., stethoscopes) between patient contacts, and maintaining the patient in a semiupright position to minimize the risk of gastric aspiration.

Intravascular infections, most often related to central intravenous or intra-arterial lines, are associated with duration and site of catheter use. To reduce the risk of infection, the patient's skin should be disinfected with chlorhexidine gluconate before catheterization, triple-lumen catheters impregnated with antimicrobial agents should be used when a multilumen catheter is necessary, and the line should be removed when there is clinical evidence of infection from the catheter (Saint & Matthay, 1998).

Environmental Hazards

Aspects of the physical environment should be modified to reduce the potential for an iatrogenic event while enhancing the patient's independent functioning in activities of daily living and mobility. Desirable features of the environment include clutter-free corridors equipped with handrails, carpets in hallways and rooms, diffuse lighting, quiet rooms at night, appropriate signage, and grab bars in bathrooms. Physical barriers to patient mobility, including restraints, should be minimized. Guidelines for alternatives to physical and chemical restraints should be employed.

ROBERT M. PALMER

See Also
Environmental Modifications

Infection Transmission in Institutions: Pneumococcal Pneumonia
Over-the-Counter Drugs and Self-Medication
Polypharmacy: Drug-Drug Interactions
Polypharmacy Management
Pressure Ulcer Prevention and Treatment
Pressure Ulcer Risk Assessment
Restraints
Signage
Urinary Tract Infection

Internet Key Words
Adverse Drug Events
Environmental Hazards
Iatrogenic Illness
Nosocomial Infections
Therapeutic Procedures

Internet Resources
CDC Guidelines on Prevention of Nosocomial Pneumonia
http://www.cdc.gov/ncidod/hip/Guide/overview

Institute of Medicine
http://www4.nas.edu/IOM/IOMHome.nsf

National Guideline Clearinghouse
http://www.guideline.gov

IMMIGRANT ELDERS

In recent years, the world has been subjected to violent ethnic wars for autonomy and secession. The United Nations estimates that catastrophic events such as war have forced one in every 120 people in the world to flee their homes and become refugees or displaced persons. Of those, 40% to 60% have experienced some form of torture or severe trauma (Carrington & Procter, 1995). Although some victims begin to experience mental health problems shortly after such trauma, there can be a latency period of 40 years or more after a return to normal life. Some people have early symptoms that persist for many years; for others, the traumatic stress remains latent until reactivated by the process of aging (Woods, 1999).

The constellation of aging and trauma is becoming more complicated as the number of aged people is increasing worldwide. A recent World Health Organization (WHO) statement on aging and health revealed that while the world population is growing at an annual rate of 1.7%, the population over age 65 is increasing by 2% annually. Assuming that migration rates remain approximately constant, the number of older adults who migrate is likely to increase. Some immigration is necessary due to political upheavals; other immigration results from the search for economic opportunities.

Some immigrants are highly skilled, but those who are unskilled or low-skilled often need additional health services due to the dangerous professions (farm and factory work) available to them. The American Farmworkers Association emphasizes that the frequency or intensity of health problems is greater for farmworkers due to the hazards of farming. The lack of safe drinking water can contribute to dehydration and heatstroke; in the absence of toilet facilities, there is a tendency toward urinary retention, which predisposes to urinary tract infection (American Farmworkers, 1995).

Immigrant workers generally do not earn enough to pay for health care, and the elderly almost never have family supports that can provide health insurance for them (American Farmworkers, 1995). They may lack transportation to get to a clinic or may be afraid of losing wages or even their jobs if they take time off to seek health care. Hence, an older person's ability to be assertive in the context of health and help-seeking behavior is constrained.

Illegal immigrant workers who can afford to pay for health services may be constrained by fears that health care workers will report them to immigration authorities and they will be deported, or they may believe that services are unavailable to them as illegal immigrants. The U.S. Public Health Service funds some migrant health centers to provide care to elderly farmworkers, but there are not nearly enough to meet the need. Due to the different rationales for immigration, targeting services to the population is difficult. Although the most visible programs are for migrants with fewer work skills, some issues and characteristics apply broadly to the category of migrant elders. Health care professionals should be aware of these issues and attend to them as part of the standard clinical assessment.

Immigrant elderly also appear to be at risk for social isolation. The WHO Statement on Aging and Health (1999) reports that traditional intergenerational relationships are rapidly disappearing. Urbanization results in social dislocation of young and old, without adequate social support measures in place. Elders are less likely to have attended school or to have been employed and thus are less likely to interact with members of the host nation (Levkoff, MacArthur, & Bucknall, 1995). Younger migrants may lack the material resources to offer any significant support to the older generation. Families may not be able to provide transportation, fees, or other financial support needed for involvement in planned leisure activities. In addition, elderly migrants' social interactions may be restricted if they are not living in areas with people of similar ethnic backgrounds (Ikels, 1986). Lack of a critical mass of both city-dwelling and rural elderly immigrants means that they are less likely to join social clubs outside their own cultures, as they are not inclined to participate in activities with strangers (Henrard, 1996; Ikels, 1986). In the case of African and Korean communities in Australian and North American cities, low social interaction with others means little awareness about early symptoms that might pressage serious mental and physical health problems (Moon & Pearl, 1991; Ssali, 1999).

Clinical assessment and helping strategies for immigrant elderly must incorporate personal reflection, therapeutic sensitivity, compassion, and understanding. The Queensland Health Information Network guidelines for effective communication with immigrant elderly suggest that health professionals should ascertain the person's preferred language and whether an interpreter is necessary. Lack of English proficiency should not be assumed to be the result of poor language attainment; it could be associated with dysphasia due to a current or previous stroke or other neuromuscular disease. Care should be individualized to the person's customs, beliefs, and practices regarding health, illness, and death, and patients should be asked what they believe is the cause of their problem. The clinician should specifically encourage older patients to talk about any issues, needs, or problems they may be experiencing in the hospital or community setting. Health care professionals should allow each patient to decide on the level of family involvement and what additional networks may be available for informal support, such as religious groups or friends.

Assessment of these issues should be linked to and used to improve clinical care situations. Cultural sensitivity should frame how the older person is questioned. Simply asking a question can be an opportunity for the development of a trusting and effective therapeutic relationship. To achieve these aims and ideals, health care professionals must also identify their own prejudices and biases and what is suggested by them. The following questions can guide this reflective process:

- What are my own feelings toward refugees and migrants? Do I indulge those who are distressed or noncommunicative? Do I fear or dislike them? Do they unsettle me? Am I ambivalent toward them? If so, why?
- How are my ideas, thoughts, and feelings about working with people who speak a different language manifested during clinical practice?
- How do media and popular opinion shape my views?
- To what extent do I encourage and allow patients and their families to make decisions about their own care?

Perhaps the most difficult aspect of health care practice with immigrants is that they access the health care system at the very point of their distress. Nevertheless, from migration to resettlement, health care professionals can assist immigrant elders make sense of a sometimes hostile world, better understand the reasons behind their needs, develop culturally sensitive problem-solving abilities, and appreciate factors that can help in stressful situations. With an informed knowledge base of background issues in the wider world, including the political, clinicians, immigrant elders, and their families can work together to target strategies and support programs with the goal of maximizing ongoing coping and health choices.

NICHOLAS G. PROCTER

See Also
Cultural Assessment
Social Supports (Formal and Informal)

Internet Key Words
Aging
Elderly Health
Immigrant Health
Isolation
Social Support

Internet Resources
American Farmworkers Homepage (1999)
http://ncfh.org/aboutncfh/aboutncfh.htm

Queensland Health Information Network (1998), Checklists for Cultural Assessment
http://www.health.qld.gov.au/hssb/cultdiv/check/home.htm

World Health Organization on Aging and Health (1999)
http://www.who.int/aging/scope.html

IMMUNIZATION

Prevention of infectious disease is an essential component of primary health care of older adults. Normal age-related changes in the immune system increase elders' risk for infection. Immunization for influenza and pneumococcal pneumonia is strongly encouraged, particularly for elders living in institutions. The accompanying table summarizes recommendations regarding indications and timing for immunizations.

Patient Involvement and Consent: The purpose of immunization should be explained, and the patient allowed to accept or reject it. For older adults living in the community, signed consent is occasionally obtained before immunization, but oral consent is acceptable. Nursing home residents with decisional capacity should participate in the choice whether to be immunized. For those unable to make their own decisions, consent should be

Recommendations for Immunizations in Older Adults

Vaccination Type	Indications	Frequency
Influenza	All older adults	Annually, just before the start of influenza season
Pneumococcal pneumonia	All older adults	Once, unless at high risk or vaccinated before age 65
Hepatitis B	Older adults at high risk: IV drug users, homosexuals, hemodialysis patients, those receiving blood transfusions	Doses at 0, 1, and 6 months
Tetanus/ diphtheria	All older adults who have never been immunized	Doses at 0, 2, and 8–14 months
Measles, Mumps, Rubella	Older adults who lack evidence of immunity and are at significant risk of exposure	Two doses
Hepatitis A	Older adults at high risk: IV drug users, sexually active homosexuals, persons living or traveling in areas with high rates of infection	Doses at 0 and 6–12 months
Varicella	Those who have not previously had chicken pox	Two doses; the second 1–2 months after the first

obtained from the person holding the elder's health care power of attorney.

Risk Factors: The most common reasons older adults give for inadequate immunization are lack of knowledge about its importance, not being advised to be immunized by their health care pro-

viders, reimbursement issues, and fear of side effects (Resnick, 1998). Education about benefits versus risks is essential to help them understand the advisability of vaccinations. It should be explained that vaccines are safe but that one may have mild systemic reactions such as weakness, fever, transient local pain, or reddened skin. As with any drug, there is the rare possibility of a serious anaphylactic reaction. Influenza and pneumonia vaccinations should not be administered to individuals who are allergic to any component of the vaccine or to anyone with an acute infection.

Reimbursement: Medicare reimburses providers for the cost of influenza and pneumococcal vaccines. Beneficiaries with Part B coverage can receive the vaccines without co-payment or a deductible.

Influenza Vaccine: Approximately 10,000 deaths occur each year because of influenza, with 80% to 90% occurring in the older population. The vaccine is 70% to 80% effective in preventing influenza in the general population, but only 30% to 40% effective in preventing clinical illness among elders (Centers for Disease Control and Prevention [CDC], 1999a). However, the influenza vaccine reduces the severity of symptoms and is 80% effective in preventing death in elders (National Institute on Aging, 2000). Influenza vaccines are reformulated annually and are trivalent (contain inactivated viruses from three strains: two type A strains and one type B strain). Ideally, the vaccine is given just before the flu season starts. The immune response is delayed for three to four weeks, and antibody titers decline rapidly.

Pneumococcal Vaccine: Pneumococcal disease is a common cause of hospitalization and death in persons age 65 and older. The polysaccharide pneumococcal vaccine contains materials from the 23 types of pneumococcal bacteria that cause 88% of pneumococcal infections and is effective against 80% to 90% of the known strains of bacteria causing infection (CDC, 1999a). The need for subsequent doses of pneumococcal vaccine is unclear; additional data are required. The CDC's Web page advises that elders 65 years or older at the time of the first vaccination do not need a second vaccination; elders vaccinated before age 65 and who received the vaccine five or more years ago should be revaccinated. Results from one epidemiological study suggest that the vaccination may provide protection up to nine years after the initial dose (Butler et al., 1993). Data are not sufficient at this time to ascertain the safety of administering the vaccine three or more times. Antibody response to vaccination may be diminished or absent in immunocompromised patients, those with chronic renal disease on dialysis, and persons with AIDS. Thus, those who are older than 75, have severe chronic disease, or are immunocompromised may need a second vaccination five years after the first. Pneumococcal vaccine can be given at any time during the year. Side effects occur in 50% of recipients but are generally limited to pain, erythema, and swelling at the injection site lasting no longer than 48 hours; fever and myalgia are rare.

Tetanus and Diphtheria: Approximately 84% of elders lack antibodies against diphtheria (CDC, 1999a). The risk of mortality from tetanus or diphtheria is greater than 30%. For older adults who have never been immunized, the tetanus-diphtheria vaccination is indicated. The optimal interval for booster vaccination has not been established. The standard regimen is a booster every 10 years, but persons born in the United States who received the childhood series of vaccination can have intervals of 15 to 35 years. Neurological reactions and severe hypersensitivity after a previous dose are contraindications for repeat vaccinations.

Hepatitis B: Vaccination is recommended for high-risk patients as described in the table. Older adults, because of the declining efficiency of their immune systems, may not achieve as high a level of antibodies as younger persons.

Hepatitis A: Preexposure immunization for hepatitis A is recommended for individuals at risk of exposure: those who live in or travel to countries where the rate of hepatitis A virus is high, homosexual persons, intravenous drug users, and individuals with liver disease. At this time, data are insufficient with regard to periodic booster immunization.

Varicella: Many adults with no history of chicken pox have immunity to varicella; less than 10% of adults in the United States are susceptible to the disease. However, if it is contracted, the disease is more severe than in children. Most clinicians obtain titers and vaccinate if there is an inadequate antibody titer of varicella.

Measles (Rubeola), Mumps, and Rubella: For those over 65, antibody titer and immunization are recommended if there is a significant risk of exposure, for example, grandparents raising young children. Measles and mumps vaccination is recommended for those born after 1956.

BARBARA RESNICK

See Also
Morbidity Compression

Internet Key Words
Flu Vaccines
Hepatitis
Immunizations
Pneumococcal Vaccines
Tetanus/Diphtheria

Internet Resources
American Geriatrics Society
*http://www.americangeriatrics.org/
    immsched.html*

Centers for Disease Control and Prevention
http://www.cdc.gov/epo/mmwr/mmwr.html

Successful Strategies in Adult Immunization, Centers for Disease Control and Prevention
*http://aepo-xdv-www.epo.cdc.gov/wonder/
    prevguid/p0000239/entire.htm*

INCONTINENCE
See
Fecal Incontinence
Urinary Incontinence Assessment
Urinary Incontinence Treatment

INFECTION TRANSMISSION IN INSTITUTIONS: PNEUMOCOCCAL PNEUMONIA

Prevention of infections in nursing homes and other types of chronic care facilities is a critical component of providing state-of-the-art, compassionate care. An overall preventive strategy involves both minimizing host factors that predispose to infection, such as decubitus ulcers, Foley catheters, and aspiration of oral secretions and food, and preventing transmission of pathogenic bacteria. Although the focus here is on the prevention of penicillin-resistant *Streptococcus pneumoniae*, the principles apply to the prevention of many different types of infection.

Each year in the United States, pneumococcal disease accounts for approximately 50,000 cases of bacteremia, 3,000 cases of meningitis, 100,000 to 135,000 hospitalizations, and 7 million cases of otitis media. The overall incidence of pneumococcal bacteremia, estimated at 15 to 30 cases per 100,000 population, is higher for persons age 65 years or older (50 to 83 cases per 100,000 population). At least 40,000 patients die from complications of pneumococcal disease each year.

Children younger than 2 years of age and adults 65 years of age or older are at highest risk for infection. Also, persons of any age with certain underlying medical conditions (e.g., congestive heart disease, diabetes, emphysema, liver disease, sickle cell disease, or HIV) and those living in special environments such as chronic care facilities are at increased risk for developing pneumococcal infection. Nearly 31 million people over age 65 were at high risk for pneumococcal disease in 1985. Yet, as of 1993, only 28% had received the pneumoccal vaccine, compared with annual influenza vaccination rates of 52%.

A major challenge in skilled nursing facilities is the potential emergence of penicillin-resistant pneumococci. The first penicillin-resistant pneumococci were reported in 1967. By 1978, a significant outbreak of multiresistant pneumococci isolate was reported from a children's hospital in South Africa (Jacobs, 1999). Currently, the Centers for Disease Control and Prevention (CDC) reports that as many as 35% of U.S. pneumococcal isolates are penicillin resistant. The molecular mechanism of penicillin resistance in the pneumococci is the uptake of penicillin-binding protein (PBP) genes from less susceptible streptococcal species (*Streptococcus mitis*, for example) and integrating pieces of those genes into the pneumococcal PBP genes. The resulting PBPs have a lower binding affinity for penicillin and other β-lactam antibiotics but are able to synthesize cell walls in an apparently normal fashion. Hence, clinicians must use higher and higher doses of penicillin to kill the pneumococci, but with the potential for significant adverse effects (e.g., seizures).

Penicillin-resistant pneumococci are of concern in the elderly for several reasons. First, pneumococci still account for up to 40% of all significant respiratory isolates that cause pneumonia in the elderly. These organisms are at least 4 times more frequent than *Legionella* species and probably equal the frequency of gram-negative pneumonias. Second, these strains are frequently resistant to many of the oral cephalosporins, the macrolides (erythromycin, clarithromycin, azithromycin), trimethoprim-sulfamethoxazole, and a variety of other agents used to treat pneumonia.

Recent data suggest that infection with penicillin-resistant pneumococci may be associated with a higher mortality than infections caused by susceptible strains. A 10-year prospective study of 504 adults with culture-proven pneumococcal pneumonia found that 29% had penicillin-resistant isolates and 6% had cephalosporin-resistant isolates. The mortality rate was 38% in patients with penicillin-resistant isolates and 24% for those with penicillin-susceptible isolates. Although comorbidity and age were the main factors associated with mortality, the high mortality rate associated with penicillin resistance is a troubling statistic. Hence, the Infectious Disease Society of America recommends that vancomycin or the new antipneumococcal fluoroquinolones be used for the treatment of pneumococcal infections that have penicillin or ceftriaxone minimum inhibitory concentrations greater than 2 μg/mL. Several potent agents with enhanced activity against gram-positive organisms are in development, but their utility for the treatment of serious infections caused by penicillin-resistant pneumococci remains to be determined.

Pneumococcal pneumonia outbreaks in nursing homes are not uncommon. In 1997 in Maryland,

14 cases of pneumococcal pneumonia among 120 residents were reported; the attack rate was approximately 11.7%. During a major pneumococcal pneumonia outbreak in skilled nursing facilities in Oklahoma, 13% (11 of 84) of the residents developed pneumonia (7 of 11 with highly resistant serotype 23F), and three died. Among the 13% that developed pneumonia, a highly resistant strain of pneumococcus, type 23F, was isolated in 64%. Seventeen of the 73 residents who were not infected were colonized with type 23F pneumococcus, suggesting substantial nosocomial transmission of resistant pneumococci. Therefore, if resistant pneumococci are isolated from pulmonary infections within chronic-care facilities, respiratory isolation measures are advisable.

Only 4% of nursing home residents in the Oklahoma outbreak were immunized with the 23-valent polysaccharide vaccine. The low rate of vaccination may be attributable to the many physicians who are unconvinced of the vaccine's efficacy in the elderly. Although conclusions from several studies are conflicting, the efficacy of the vaccine is estimated to be about 58% in the 65- to 74-year-old age group and about 13% in the over-85 age group. Despite this relatively modest efficacy, the significant morbidity and mortality associated with serious pneumococcal disease and the increasing rates of resistance argue strongly for routine vaccination of the elderly on admission to skilled nursing facilities and revaccination every four to six years.

Pneumococcal vaccines with greater immunogenicity are needed to provide protection in the elderly. Conjugate vaccines, in which a protein that elicits a durable immunogenic response is attached to the pneumococcal polysaccharides, are currently being studied. One promising compound couples seven pneumococcal polysaccharides (types 4, 6B, 9V, 14, 18C, 19F, and 23F) to diphtheria CRM_{197} protein. This vaccine has demonstrated efficacy in the prevention of bacteremia and meningitis in children. However, the ability of this or similar vaccines to prevent infections in the elderly remains to be studied. It is hoped that conjugate vaccines will prove to be more effective than the presently available polysaccharide vaccine and that this will inspire more widespread use among the institutionalized elderly.

ROBERT A. BONOMO
LOUIS B. RICE

See Also
Immunization

Internet Key Words
Immunization
Infection

Internet Resource
Centers for Disease Control and Prevention: Prevention of Pneumococcal Disease
*http://www.cdc.gov/epo/mmwr/preview/
mmwrhtml/00047135.htm*

INFORMATION TECHNOLOGY

Interactive technology is radically altering the way individuals communicate, interact, access information, and make health care decisions. Nowhere is this change more profound than in the older population. With 95% of health care providers now accessing the Internet and 68% of Internet users accessing health care information (Internet Healthcare Coalition [IHC], 1999), this communication link is proving to be a lifeline to the elderly community. Two aspects of telecommunication are especially relevant to the older population: remote monitoring for health care and access to health information on the World Wide Web.

Remote Monitoring for Health Care

The use of electronic information and communication technologies to provide and support health care when distance separates participants has gained prominence because of the accessibility and flexibility of technology. Success in remote monitoring of specific populations has significant implications for improved health care outcomes. Recent applications of telehealth include remote monitoring of elders with congestive heart failure (CHF) (Heidenreich, Ruggerio, & Massie, 1999) and monitoring of elders (mean age 73) with multiple illnesses, including chronic obstructive pulmonary disease, diabetes, Parkinson's disease, depression, and emphysema (Lindberg, 1997). Improved compliance with medication and treatment regimens and improved functional status were reported in both studies. The CHF study reported decreased resource utilization and subsequent cost savings associated with reduced hospital admission rates and lengths of stay in the monitored group. Anecdotal data in the Lindberg (1997) study noted improved quality of life and active involvement in care among the monitored elderly. Other studies of remote monitoring reported decreased incidence of depression and isolation among the home-bound elderly. In addition to improving lifestyle and functional outcomes, Web-based remote monitoring applications can have customized features such as e-mail, health reminders, diaries, and health information. The over-70 population has demonstrated a remarkable facility to adapt and use computer-based communications and interactions.

Interactive Health Communication and the Elderly

As a result of the rapid technological advances and ease of use of the Internet, the World Wide Web will become a major source of education and health care transactions in the next century. A fast-growing segment of Internet users consists of health care consumers. "Healthcare Professionals and Consumers Online: Fixing the Broken Link," a panel discussion presented at the 1999 IHC conference, discussed consumers' access to their electronic health care records. Issues that need to be addressed include "consumerizing" health care terminology, marketing health care products on the Internet, the power of networked support groups, and older adults on the Internet.

The over-50 age group is the fastest-growing segment of the population to use the Internet (IHC,

1999). There are approximately 1,500 Web sites devoted to elders and specific issues related to this population. SeniorNet has taught 100,000 seniors about computers; it is estimated that 40% of seniors are self-taught, and 30% of seniors own personal computers. Preliminary studies indicate that older adults new to computers will use them successfully for peer support, to obtain information, and to seek professional advice (Science Panel on Interactive Communication and Health, 1999). As a result, individuals using the Internet feel empowered, are more informed, and are better equipped to participate in decision making about their health care.

Provider and Consumer Issues

The Internet can be a powerful tool for health care providers in communicating with patients and improving health care outcomes. Web-based technology can integrate knowledge sources, clinician expertise, and patient-based preferences (Carty, 2000). How health care professionals will incorporate the medium into their practice and use it as a major source of communication and interaction raises many issues, such as response turnaround time, type of transaction (e.g., prescription, appointment, advice), authentication and confidentiality of communications, and the incorporation of e-mail as part of the medical record. The on-line *Ferguson Report* (Ferguson, 1999) maintains that the Internet will continue to assume a prominent role in the communication of, access to, and dispensing of health care advice and information. Health care providers are advised to "jump" into the Net and cultivate smart patients. A scientific panel (Science Panel on Interactive Communication and Health, 1999) recommended four strategic initiatives:

1. Strengthen evaluation and quality of interactive health communication.
2. Improve basic knowledge and understanding of interactive health communication.
3. Enhance capacity of stakeholders to develop and use interactive health communication (includes consumers).

4. Improve access to interactive health communication for all populations.

The proliferation of Internet-based health care networks also challenges providers as the expansion of electronic health tools and e-commerce enables the public to monitor their health status, report health data, access disease prevention information, and purchase health care products. The professional education systems will have to change to produce and support professional practice in a dynamic, digital society. The fastest-growing segments of the population (those over 50) are enthusiastic about using electronic communication in creative and beneficial ways to improve the quality of their life. Professionals also need to demonstrate competence in incorporating technology into their care protocols and patient interactions.

BARBARA CARTY

Internet Key Words
Healthcare
Telecommunication
Telehealth

Internet Resources
American Association of Retired Persons
http://www.aarp.org/

Assisted Living Information
http://www.alfa.org/alcis.htm

Association on Aging
http://www.aoa.gov

Foundation on Aging
http://www.spry.org

Government Resources
http://www.seniors.gov/

Health Finder Resource
http://www.healthfinder.gov/

Home Health Care
http://www.lifemasters.net

Mental Health and Aging
http://www.mhaging.org/

National Center on Assisted Living
http://www.ncal.org/

National Council on Aging
http://www.ncoa.org

National Family Caregivers Association
http://www.nfcacares.org/

Ten Top Senior Web Sites
http://www.toplinks.com/family/seniors.vote

INFORMATION TRANSFER

Care sites for elderly patients have shifted dramatically in the past decade. Prospective-payment mechanisms have forced hospitals to discharge elderly patients more rapidly, with concomitant reductions in patient teaching and increased referrals to continuing care (Lee, Wasson, Anderson, Stone, & Gittings, 1998). Such patients may experience adverse outcomes (Foreman, Theis, & Anderson, 1993), especially rehospitalization during the early postdischarge period (Anderson, Helms, Hanson, & DeVilder, 1999). Ensuring continuity of care when patients are transferred to skilled care units, extended care facilities, and home health care agencies poses a significant challenge for providers at both the discharging and the receiving facility. Communication about patient care needs, goals, and interests is at the core of such transitions but is hampered by communication breakdowns, gaps, and inaccuracies (Anderson & Helms, 2000). An overview of common patient, professional, and organizational barriers to information transfer across health care settings will assist clinicians in improving patient care communication.

Hospitalization is traumatic for elderly patients and their families. Assimilating the implications of hospitalization requires time, a scarce commodity in today's cost-conscious health system. Thus, planning and coordinating aftercare arrangements are often hurried, leaving patients and family members feeling confused and powerless and as if they have had little input into the process. Patient or family resistance to moving to another care site is not uncommon, and resentment can be demonstrated by poor interpersonal communication with providers or even the refusal of a continuing care referral.

The assessment, planning, and coordination of care for elderly patients across the care continuum must begin early in an episode of illness. Hospital discharge planning and case management are mechanisms that can ensure continuity of care for elderly patients (Anderson & Tredway, 1999), allay patient and family concerns about care in transitional sites, facilitate a smooth transition for the elderly patient and family, and ensure that necessary information is transmitted to providers at the new site.

Information transfer may be impeded by professionals' failure to recognize one another's information needs (Knies, 1996). For example, the perspective of nurses in acute-care settings is often not congruent with that of nurses in extended-care facilities or home-care agencies. Hospital-based nurses deal with the acute aspects of illness, while their counterparts in other settings address more chronic, long-term issues. Such intraprofessional discrepancies may influence communication. When providers from many disciplines participate in the care of a patient, problems of overlapping roles and role conflict can confound communication and the information transfer process (Anderson & Helms, 1993).

Improved communication requires understanding the dynamics of information transfer between health care providers at several levels, including (1) interactions between health care personnel, (2) choices about how to structure communication and the communication process, (3) choices about content or the type of information to send, and (4) the organizational relationships and linkages between the different providers caring for a patient.

Clinicians must recognize practice norm variations in different employment settings and be tolerant of the range of collegial perspectives. Gamesmanship and territoriality, whether intradisciplinary or between individuals of different professional backgrounds, can be minimized, if not avoided completely. For example, nurse-to-nurse communication results in better information transfer across settings, as opposed to multidisciplinary or other communication approaches (Anderson & Helms, 1994).

Accountability for coordination and information transfer is essential to avoid the trap that "when information transfer is everyone's job, it is no one's job." Additionally, the use of liaison personnel, or nurses employed by a transitional care agency who visit patients and families in the hospital and gather the necessary patient data, facilitates information transfer from hospital to home care and from hospital to nursing home (Anderson & Helms, 1993).

The method and format of the referral are important factors in the integrity and usefulness of the information transferred. More data are transferred when standardized referral forms are used and when the information is transmitted in a written format, as opposed to a telephone call from the referring to the receiving agency. Professionals who prepare continuing care referrals should use standardized forms, complete all items, and furnish any additional contextual or environmental information that may enhance care by the next provider.

Providers at the new care site must receive complete and accurate patient data, with sufficient lead time to plan for the delivery of effective care. Frequently, arrangements must be made for additional equipment to be delivered, or ancillary service providers such as physical therapists must be contacted. Continuing care providers should receive necessary information well in advance of the patient's discharge from the hospital (Anderson & Helms, 1998).

Referral information is consistently biased toward patient background and medical information. Nursing care and psychosocial information is much less likely to be included in continuing care referrals (Anderson & Helms, 1998). Generally, specific patient data such as demographics and diagnostic codes are more readily communicated than the more complex, discursive information such as nursing care regimens, social status, and cognitive functioning.

More information is not necessarily better for providers at distant care sites. Yet complex content is critical for the provision of uninterrupted, individualized care for elderly patients with multiple problems and limited capacity. Standardized referral forms in a checklist or flowsheet format can be readily completed by the referring provider; content that does not lend itself to such formats, such as nursing care plans, family and environmental considerations, and discharge summaries, is often omitted (Anderson & Helms, 1994). Thus, providers must take the time to communicate less structured patient care data, which requires commitment of time and professional expertise. Institutions that regularly interact should consider the development and adoption of a standardized nursing language to enable improved communication about patients.

Clinicians seeking to improve information transfer in their organizations should investigate the capabilities of the information management system in use. Small hospitals, particularly in rural areas, usually have modest electronic information management systems, but they also have simpler channels and fewer personnel through which patient-related information must pass before it is sent to the aftercare agency. Thus, potential interferences with or distortions of the data may be reduced. Additionally, patients served by small hospitals tend to be better known by staff, increasing the opportunity for higher-quality referral data to be collected and transmitted. In designing hospital discharge planning or case management programs, providers should attempt to minimize the number of phases or personnel patient-related data must pass through and encourage individualized referral preparation.

Linkages between hospitals and community agencies also affect information transfer. Hospitals and home-care agencies with organizational ties transmit more and better data when elderly patients are referred for further services. Freestanding home-care agencies are less likely to receive the same amount and type of patient information. Attendance at patient care planning conferences, the use of liaison professionals, and nurse-to-nurse communication are strongly recommended (Anderson & Helms, 1993).

Despite the fundamental problems of information transfer, providers at distant care settings can implement practices to ensure that necessary patient data are sent appropriately. Attention to some relatively simple guidelines about the nature of patient care communication can facilitate information transfer between sites and thus assist in the provision of continuity of care for elderly patients.

MARY ANN ANDERSON
LELIA B. HELMS

See Also
Communication Issues for Practitioners

Internet Key Words
Communication
Continuity of Patient Care
Patient Referral for Continuing Care

INJURY AND TRAUMA

Traumatic injury is a primary public health problem in the United States. Trauma occurs in all age groups with varying degrees of severity. As people live longer and lead more active lives, the elderly population will be sustaining more injuries. Injuries sustained by people age 65 or older lead to complications or death more often than do injuries in younger patients (Champion et al., 1989).

Factors that influence the care needs of injured geriatric patients are preexisting disease, diminished physiological capacity, and occult underlying disorders that may adversely affect the ability to heal. It is important to consider the likely causes and effects in injured geriatric patients and to learn as much about their underlying health problems as possible in order to provide the most appropriate care (Knies, 1996). For example, when presented with a confused person after a fall, ask: Did the fall cause a head injury that caused the confusion, or did the patient have a stroke that caused the fall? The stress of trauma can bring to the forefront signs and symptoms of a disease process that had previously been silent or ignored.

Falls

The most common mechanism of injury in the elderly is falling. The three most prevalent causes of falls in the elderly are generalized weakness, environmental hazards, and orthostatic hypotension (Baraff, Della Penna, Williams, & Sanders, 1997; Tideiksaar, 1997). Causes of loss of consciousness in the elderly include heart rhythm disturbances; insufficient blood supply to the heart muscle, causing myocardial ischemia or infarction; stroke; seizure; anemia; and blood loss. Mentation can be altered significantly by alcohol and by prescription medication, which can lead to falls or other significant injuries in the elderly.

Elderly persons who have fallen exhibit signs that may be due to the mechanical fall or may have caused the fall in the first place. Changes in orientation may indicate trauma to the central nervous system. Age-related changes of the brain result in cerebral atrophy and increased venous fragility. Subdural hematoma should be considered in an elderly patient presenting with changes in mental status, headache, disturbances in ambulation, or nonfocal neurological findings. Spontaneous intracranial bleeding, stroke, or other cerebrovascular event may have caused the fall. Other findings in patients older than 65 years with acute changes in mental status may include cardiac dysrhythmia, acute myocardial injury, or metabolic disorders.

Musculoskeletal injuries also occur in older patients who fall. Hip fractures and other extremity fractures are common, primarily due to osteoporosis. A fall may be the initial presentation of pathological fractures due to primary or metastatic tumors. Spinal fractures may cause significant injury to the spinal cord and peripheral nerves. Clinical findings may be subtle, such as constipation or incontinence, or they may be obvious, such as paralysis. Recent controlled trials have demonstrated that focused intervention can prevent falls. Devices that reduce the amount of energy delivered to the body can decrease the chance of injury in "frequent fallers." Hip pads and helmets are examples of this kind of approach (Rivara, Grossman, & Cummings, 1997).

Motor Vehicle Accidents

Motor vehicle accidents are the second most common mechanism of injury and account for 28% of all injuries in those 65 years and older. Of note, motor vehicle crashes are the most common type of fatal event in the elderly through age 80. Air bags and proper use of seat belts can reduce injuries from crashes (Rivara, Grossman, & Cummings, 1997). In addition, clinicians must recognize cognitive, sensory, or physical impairments that may lead to a dangerous diminution of driving skills.

Pedestrian injuries are the third most common mechanism of injury, representing 10% of all injuries for persons age 65 years and older. Interventions targeting pedestrian and driver behaviors have not been successful (Rivara, Grossman, & Cummings, 1997). Factors contributing to auto-pedestrian injuries in the elderly include impaired visual acuity and decreased peripheral vision, postural kyphosis, progressive deafness, and altered gait and strength due to arthritis, muscle atrophy, and chronic neurological disease.

Other Trauma

Stabbing and gunshot wounds represent slightly over 8% of injuries in the elderly. Burns are the cause of approximately 8% of all elderly trauma deaths. Merely having a gun in the house increases the risk of suicide 4.8-fold (Rivara, Grossman, & Cummings, 1997). Stab wounds, as in all penetrating injuries, can be deceiving on the surface. Bacteria and other foreign substances carried to the deep tissues may lead to infection. Patients can present with a spectrum of problems, from local cellulitis to systemic illness. Complicating factors in burn injuries include dehydration from insensible water loss, scarring, joint contractures, and infection. Physical abuse of dependent older adults requires special consideration.

Diagnosis

The diagnostic workup of all trauma patients should include questions about the circumstances surrounding the injury, such as the height of the fall, the speed of the vehicle, associated loss of consciousness, and agility and mobility before and after the event. It should be determined whether the patient lives at home or in an extended-care facility. It is important to ask alert patients whether they can recall instances of rough handling or being struck. For dependent individuals, the caregivers should be asked what happened, to determine whether the stories match.

Using the mechanism of injury and the physical examination as a guide, the following guidelines should be considered in ordering imaging studies:

- *Abdominal ultrasound.* Abdominal ultrasound shows fluid in the abdomen and major organ disruption. In a questionable case of intra-abdominal blood level, peritoneal lavage should be done. A significant drawback to abdominal ultrasound is that the procedure and hence the findings are operator dependent.
- *Computed tomography (CT).* CT of the head is indicated when the patient has had a change of consciousness, seizure, or weakness. A subdural or epidural hematoma would be suspected. Intracranial injury, detected by head CT, is a common cause of death in elderly trauma patients. Abdominal CT offers the same "view" as abdominal ultrasound, but it is not operator dependent.
- *Vascular imaging.* This study may be performed by color-flow Doppler ultrasound, magnetic resonance angiography, or conventional catheter angiography. It is important to note that catheter angiography can also be used for treatment; for example, catheter embolization can occlude a bleeding vessel. A relatively minor blood vessel injury may be life-threatening in an individual with limited physiological reserve. Appropriate studies may detect significant vascular injuries.
- *Other studies.* Knowledge of the mechanism of injury involved will guide the diagnostic workup. Other studies should be obtained based on clinical manifestations and suspicion of injury.

Management

Aggressive intervention in the elderly is warranted regardless of age, unless the patient is known to have a preexisting terminal disease or severe injuries with a low probability of survival. Guidelines for trauma management have been established by the American College of Surgeons Committee on Trauma.

Treatment of geriatric multiple trauma victims should follow standard trauma guidelines, including primary survey to assess critical injuries requiring immediate intervention, stabilization, and secondary survey for more complete patient assessment. Understanding the mechanism of a vehicular injury, such as the condition of the vehicle, and the status of others involved in the collision may help guide treatment recommendations.

Airway assessment is the first priority. An elderly patient may have upper airway obstruction due to relaxed musculature of the oropharynx or displaced dentures impairing airflow. As soon as a patent airway is established, a search for blood loss, even in the face of normal vital signs, must be initiated. In older patients, commonly used parameters to assess hemodynamic status can be misleading; therefore, continuous monitoring of vital signs, cardiac activity, and oxygen saturation is essential.

Hypotension-related bleeding usually requires fluid replacement, which must be used with caution in elderly patients, especially those with underlying cardiovascular disease. Judicious use of intravenous crystalloid fluid is recommended—preferably, Ringer's lactate—to avoid hyperchloremia in older patients with impaired renal capacity. After 1 to 2 liters of crystalloid infusion, blood (packed red blood cells) may be infused. Elderly patients may need platelets and fresh-frozen plasma sooner than younger patients, especially if underlying liver dysfunction or abnormal coagulation is present.

An appreciation of the fragility and limited physiological reserves of geriatric patients is essential for emergency medical service providers, the receiving emergency department personnel, and the trauma center team.

JO E. LINDER

See Also
Burns and Other Safety Issues
Driving

Internet Key Words
Injury
Trauma

Internet Resources
Administration on Aging
http://www.aoa.dhhs.gov/

Advanced Trauma Life Support Course
http://www.facs.org/meeting_events/
  atlsatls.regions.html

Iowa Injury Prevention Research
http://www.pmeh.uiowa.edu/iprc

INSTITUTIONALIZATION

Institutionalization refers to an individual's entry into a long-term care setting and the subsequent adjustment to life in this setting (Krauss & Altman, 1998; Rhodes & Krauss, 1999). It has long been recognized that there is a stress of relocation often associated with institutionalization (often called transfer trauma). Relocation stress results not only from the relocation itself, but also from the stresses of anticipation, disruption, and loss of control (Renardy, 1995). Relocation to a long-term care facility is associated with an increase in mortality (Thorson & Davis, 2000). The relationship between mortality of residents entering a nursing home and relocation stress is not necessarily causal, given that people are more likely to enter a nursing home during a period of decline in their physical and mental health.

Recognition of the physical and emotional stresses associated with institutionalization has led to both formal and informal attempts to improve the medical and psychosocial health of nursing home residents. The Omnibus Budget Reconciliation Act (OBRA) of 1987 emphasized that nursing homes are responsible for assuring that residents reach their maximal level of function and well being in an environment that respects their dignity and autonomy. In response, hospitals and nursing homes have moved to reduce the use of chemical and physical restraints.

In addition, federal regulations mandate a Resident Bill of Rights (Eldercare Online, 2000) that emphasizes preservation of patient dignity, autonomy, and choices in the institutional setting. For instance, residents have the right to choose their personal physician and obtain full information, in advance, and participate in planning and making any changes in their care and treatment. Residents also have the rights to participate in social, religious, and community activities as they choose; have privacy in medical treatment, accommodation, personal visits, written and telephone conversations, and meetings of resident and family groups; as well as a variety of other rights.

Although the goals of nursing home care are to promote optimal function and quality of life within a comfortable and safe setting, these goals are not always met. The experience of institutionalization for the elderly patient may be fraught with losses. Privacy, personal possessions, access to resources, autonomy, and control are lost to a degree that varies with the organization and structure of the nursing home. Society's negative attitude about nursing home placement only accentuates the new resident's feelings of loss and failure.

Loss of privacy is significant; one must live intimately with strangers. Even in settings where private rooms are possible, meals are usually communal. Communal bathing facilities and bathing practices are particularly onerous for residents and families. Often, nursing home staff enter residents' rooms without knocking. The social amenities associated with privacy in one's home are often not properly observed in the nursing home setting. Indeed, as a quality indicator during the nursing home surveys, nursing home staff are observed to see if they knock on the door before entering a resident's room.

Loss of personal material possessions can be traumatic; possessions are symbolic and can have meaningful memories. Many nursing homes encourage residents to decorate their rooms with their own furniture, pictures, and mementos. By providing a familiar and comfortable environment, the nursing home can foster a sense of belonging for the resident. Yet, two-bed units limit the ability of residents to retain personal belongings.

Balancing residents' needs for control of one's daily life with the privacy needs of residents is a consistent theme in nursing homes. For example, determining when visitors are welcome poses a dilemma in meeting residents' personal preferences, maintaining privacy of other residents during personal care, and organizational constraints. On admission, residents often lose access to their prior support systems and previous informational resources, for example a well-stocked library, telephone access, and, for some, the Internet. On the other hand, unlimited telephone access may be disruptive to a roommate.

To empower residents, nursing homes have a resident council where residents meet regularly for discussion of concerns related to living in the facility. The resident council was mandated by OBRA to ensure residents the right to voice their opinion and participate in the workings of the nursing home. A council member may take minutes and a social worker usually facilitates the meetings. Issues are then shared with fellow residents and staff, if necessary.

Nursing home residents live in an environment where the dominant focus is on meeting nursing and medical care needs. The "personhood" of the resident may be secondary. Residents often have concerns regarding their personal environment, specifically their rooms, closets, and space to store clothing and personal possessions. They may also fear fellow residents who pose a threat to them (e.g., those who wander or are combative). Measures can be taken to protect residents by increasing supervision of residents known to wander.

Several strategies have been successfully applied to create a program of individualized care and normalize the environment in nursing homes. Most strategies seek to improve communication and interaction between residents and staff. Providing the resident's primary aides, case managers, nurse, and physician with the resident's life history, values, and preferences, such as spiritual needs, have been shown to increase resident choice and individualization in the plan of care (Kane et al., 1999a).

Improving individualized care may be possible by increasing the personal control of the residents and decreasing the control that nursing staff may exert. This could be achieved by allowing residents to decide on their own meal-time and/or bathing schedule as well as decorating their own personal living space. To respond sensitively to an older person's psychosocial and medical needs, all disciplines must communicate regularly to comprehensively discuss individual residents. Establishing high expectations for residents can motivate residents to establish and achieve personal goals (Resnick, 1999).

Social engagement has been shown to improve quality of life and prolong the life span of nursing home residents (Bassuk et al., 1999). Creativity is needed to develop programs where residents are encouraged to make friends and establish meaningful social networks and activities within their new "home." Some nursing homes have successfully developed intergenerational collaborations between children and older residents through "friendly visiting programs" and joint theater productions. Others have fostered continued family involvement with residents after admission through programs that prepare families to visit residents or that address the guilt, conflict, and lack of socioemotional support that families often feel when a loved one is institutionalized (Gaugler et al., 1999).

Some nursing homes have successfully used facility wide initiatives to achieve individualized resident care. The Eden Alternative (*www.edenalt. com*) uses pets as a vehicle to transform a nursing home to an environment that nurtures social and spiritual growth for nursing home residents and staff. The Nursing Home Pioneers (Rochester, NY) is a movement that brings together nursing homes that seek to support and create a culture of dignity for the person who lives in a nursing home. Homes that subscribe to the Pioneer approach seek to promote resident creativity, growth, potential, and strengths.

Sheila FitzSimmons-Scheurer
Michele G. Greene
Ronald D. Adelman

Internet Key Words
Institutionalization
Long-term Care
Nursing Homes

Internet Resources
Eden Alternative
http://www.edenalt.com

Eldercare Online
http://www.ec-online.net

INTEGUMENTARY SYSTEM

Gray hair and thin, wrinkled, sagging skin are perhaps the most common features associated with aging. Most changes in the skin are due to a combination of senescence (intrinsic changes) and the action of external factors, mainly ultraviolet light (extrinsic changes, photoaging, or actinically induced changes). Health professionals must have a clear understanding of these factors to ensure the appropriate diagnosis and management of skin problems in the elderly. Exposed areas such as the face, neck, back of the hands, and forearms are the most readily inspected, and clinicians can easily fall into the trap of assuming that actinically induced "premature" aging changes are normal. Approximately 90% of the cosmetic changes seen in the skin are due to actinic injury. Indeed, changes due to photoaging—coarse and rough skin surface, wrinkling, mottling, and sagging—are probably synonymous with aging in the public mind. However, studies of areas commonly protected from the sun, such as the buttocks and breasts, have helped differentiate chronological changes in the skin from those caused by photoaging. Knowledge of intrinsic and extrinsic changes will help health professionals manage the skin care needs of the elderly and assist in the early identification of abnormal changes, particularly neoplastic problems.

Skin

Loss of elasticity and wrinkling, furrowing, pigmentary changes such as liver spots (solar lentigines), telangiectasia, and increased susceptibility to neoplastic changes are common features of aging skin. These changes are due to the interaction of intrinsic aging (including the action of hormones) and extrinsic aging (photoaging) on the epidermis, dermis, subcutaneous blood vessels, and skin appendages.

Actinic, sun-induced injury accounts for much of the damage seen in aging. Light-exposed areas show epidermal atrophy with dyskeratosis and reductions in melanocytes and Langerhans' cells. There is degeneration of dermal collagen bundles (photosclerosis) and increased elastosis. The blood vessels of the microcirculation exhibit thickening of the vessel walls. The clinical appearance of the skin is characterized by a thick, leathery texture, with marked loss of elasticity and wrinkling, hyperpigmentation, erythema, and inflammation with vascular ectasis and hemorrhages.

Intrinsic aging is associated with a number of skin changes, including dryness, decreased thickness of both dermal and epidermal layers, fragmentation of elastic fibres, and increased fragility. Flattening of the epidermo-dermal junction and some disruption of epidermal architecture are consistent features of chronological aging that may predispose to shearing-type injuries and blistering. Decreased melanocytes and Langerhans' cells result in uneven tanning and increases the risk of actinic injury. The dermis thins with age and becomes more acellular and avascular; dermal collagen bundles become more closely packed, with less space between collagen fibers; there may be an increase in tensile strength but also some thinning of collagen fibers; and changes occur in the composition of the extracellular matrix. With progressive degeneration of elastic fibers and reduced numbers of fibroblasts and mast cells, aging skin tends to wrinkle and sag. Areas protected from light do not show the same vascular dilation as exposed skin does, but there is some loss of vertical capillaries and dilation of lymphatic vessels.

The combination of intrinsic and extrinsic aging changes results in the characteristic appearance of aged skin. Common problems include xerosis (dry or rough skin), increased fragility, slower healing when injured, and increased susceptibility to bruising.

A decrease in both the number and the function of sweat glands leads to reduced sweat production. Sebaceous glands increase in size, but sebum production is reduced. On exposed areas such as the hands, the enlarged glands may present as yellowish, ring-shaped lesions that can be very similar in appearance to early basal cell carcinoma. Biopsy may be necessary to exclude cancerous changes. Changes in skin glands may also be linked to an increased tendency to develop comedones and cysts.

Epidermal 7-dehydrocholesterol and the release of vitamin D into the blood may be decreased by 75% in the aged. Ultraviolet B radiation converts 7-dehydrocholesterol in the skin into provitamin D (MacLaughlin & Holick, 1985).

Skin immune function is also reduced in aging. The number of epidermal Langerhans' cells is reduced by 20% to 50%, and actinic injury may further reduce the number in exposed areas. The percentage of T cells is reduced, and there is a marked reduction in the production of epidermal cytokines, which help stimulate the response of T cells in the skin. Reduced skin immune response may be a factor in increased susceptibility to chronic skin infections and skin cancers. The elderly seem to be particularly prone to viral and fungal skin infections such as dermatophytosis, herpes zoster, and bullous pemphigoid.

Benign and malignant skin neoplasms are more common in the elderly. Seborrheic keratosis is an atypical benign neoplasm that causes well-defined, raised lesions on the skin of the face, neck, chest, or upper back. The tan to black, warty lesions are covered with a greasy crust that is often itchy and may be scratched off to reveal a raw, pulpy base. Skin tags and cherry angiomas are other common benign growths. The elderly may suffer from the usual range of skin cancers, such as basal cell carcinoma and squamous cell carcinoma, but are also at risk for more unusual malignancies such as angiosarcoma of the face and scalp, Merkel cell tumors, and acantholytic squamous cell carcinoma.

The reduced skin immune response may result in fewer problems such as contact dermatitis but much slower resolution of such problems when they do occur. Patch-test reactions in the elderly are likely to develop much more slowly and are muted in comparison to patch-testing in younger patients. Richey, Richey, and Fenske (1988) suggest following up patch tests at three weeks in the elderly.

Decreased skin sensation and an increased pain threshold are often associated with aging. Free nerve endings do not appear to change with aging, but pacinian and Meissener's corpuscles, structures responsible for pressure and superficial tactile perception, show some functional and histological degeneration. This loss of sensory acuity may cause increased susceptibility to burns and mechanical trauma and reduced control of fine movement or maneuvers.

Hair

Aging is marked by a decline in hair follicle activity, with a marked reduction in the number of hair follicles per unit area, reduced growth duration, diminished shaft diameter, and greater time between growth cycles. More hair follicles are in the resting stage of the growth cycle; the hairs adhere poorly to the follicular unit and tend to fall out after normal shampooing or brushing. On the scalp, many terminal hair follicle regress and produce barely visible vellus-type hairs, such that individual hair length is reduced, hairs are much finer, and overall hair quality is diminished. Gray or white hair results from a loss of melanocytes in the hair bulb. Paradoxically, a number of previously vellus-type hairs develop into thick, prominent terminal hairs, most commonly on the ear, nose, upper lip, and chin. These thick, gray, unsightly hairs are difficult to remove and can embarrass both sexes.

Nails

Aging is associated with a range of changes in the color, contour, and thickness of the nails. The elderly are also prone to a range of nail disorders, or onychodystrophies. Aging nails are often dull, lacking in luster, opaque in appearance, and yellow to gray in color. The crescent-shaped lunula may disappear. The nail develops an increased transverse convexity; transverse superficial furrows (onychorrhexis) and deeper ridges develop. The rate of nail growth is reduced, but nails may show abnormal thickening or thinning. Histology demonstrates increased keratinocyte size, with thickening of the dermal bed blood vessels and degeneration of dermal bed elastic tissue. Increased formation of longitudinal ridges and a more marked separation of the horizontal lamellae of the distal nail plate, coupled with a rough surface and irregular distal edge, make the nails more brittle and easily broken. Brittle nails may develop as a consequence of repetitive cycles of dehydration and rehydration or excessive use of dehydrating agents such as nail varnish or cuticle removers (Wallis, Bowen, & Guin, 1991). Brittle nails can be treated by lukewarm water soaks for 10 to 20 minutes at bedtime, followed by application of a moisturizing agent, preferably under an occlusive cover (Cohen & Scher, 1992). Nail varnish may help retain moisture if applied and removed at weekly intervals. Oral biotin may be beneficial in the treatment of brittle nails (Colombo, Gerber, & Bronhofer, 1990). Chronic trauma and infections are common causes of onychodystrophy. Specific disorders include onychauxis (localized hypertrophy of the nail plate), onychoclavus (hyperkeratosis of the nail area, akin to a subungual corn), onychocryptosis (ingrown toenail), and onychogryphosis (localized or diffuse hyperkeratosis of the lateral or proximal skinfolds). Small hemorrhages and hematomas can form spontaneously under the nails; the great toe is a common site for subungual exostosis (abnormal, benign formation of bone).

COLIN TORRANCE

See Also
Foot Problems
Skin Issues: Bruises and Discoloration
Skin Tears

Internet Key Words
Extrinsic Aging
Hair
Intrinsic Aging
Nails
Photoaging
Skin

Internet Resource
Integumentary System, Georgia Institute of Technology
http://www.anatomy.gatech.edu/aging/skin/ index.htm

INTERGENERATIONAL CARE

Historically, the extended family provided the structure and source for the care and support of all of its members, regardless of age. Generations of grandparents, parents, and children, living under one roof or at least close by, provided economic, educational, and health support, and a common framework for sharing values, history, religions, and cultural traditions. Child-rearing responsibility and caring for frail family members was shared according to the family's ethnic and cultural traditions.

With an increasingly mobile and industrial urban-based society, family members from different generations are less likely to have the kind of day-to-day contact that traditionally occurred. Activities like child-rearing, caring for frail elders, and relaying religious and cultural values are often shifted to people outside the family or to institutions such as day-care centers, schools, and nursing care facilities.

Intergenerational programs provide opportunities to recapture the richness of contact and support available in an age-integrated society. Individuals, families, and communities can once again enjoy the benefits of living and learning in a caring community where people of all ages, diverse backgrounds, and experiences can interact. Ideally, intergenerational programs are structured so that all age groups benefit. In reality, the programs vary widely by the services they provide and are presumed effective because they meet numerous needs of individuals and communities. However, research is needed to determine their overall cost-effectiveness.

Definition of Intergenerational Programs

Generations United, one of the earliest programs, defined intergenerational programming as "the purposeful bringing together of different generations in ongoing mutually beneficial planned activities designed to achieve specified program goals. These activities or programs increase cooperation, interaction and exchange between people of different generations" (Generations United, 2000). Talents and resources are shared across the generations; supportive relationships benefit both old and young and the communities in which they live. Many intergenerational programs tend to be bipolar, with young and old together. Some programs envision participants of all ages—young children, adolescents, and middle-aged and older adults. The age of the participants depends on the specific mission of the program. What defines a particular generation and differentiates it from earlier and later ones is left open.

History

Intergenerational programming began with the Foster Grandparent Program funded by the federal government in the 1960s, followed by the Retired Senior Volunteer Program (RSVP) in 1969. The University of Pittsburgh established Generations Together in 1979; Temple University established the Center for Intergenerational Learning one year later. Both these university-based centers began by developing intergenerational programs, education, training, and research. They continue to serve as clearinghouses for intergenerational resources (McGuire & Hawkins, 1998).

Examples

Intergenerational programs can be a single program or a combination of mentoring programs in which older persons mentor or tutor young people, college students visit frail elders in their homes and provide various types of assistance, or adolescents and older adults are partnered for joint community service. They can also include foster grandparenting, day care for young children along with care for frail elders, and training programs for older adults as child-care providers on intergenerational teams.

Kinship care describes the situation in which grandparents or other relatives are raising children of impaired parents. At present, more than 4 million grandparents or other kin are raising children of impaired and nonimpaired parents, many of whom are taking on this commitment alone, on fixed incomes, and with limited access to the resources provided to parents or foster parents.

Family literacy programs are another type of intergenerational programming. Parents and other adults, along with their children, learn how to improve reading and writing skills by participating in a variety of shared projects. Research indicates that both children and adults learn better when working together rather than in groups of only children or only adults (National Center for Family Literacy, 1994). Parents' success in these programs gives them access to other opportunities, such as employment and continuing education (Come & Fredericks, 1995).

Some intergenerational programs focus specifically on health care issues, even though many programs include health education as part of their activities. For example, an urban program trained an intergenerational group of community residents, including school-aged children, to serve as health educators about cancer prevention, detection, and treatment in an African American neighborhood (Lowe, Barg, Norman, & McCorkle, 1997). The Teen-Age Mothers-Grandmothers program worked with teenage mothers-to-be and their mothers during the pregnancy. Findings suggest that teens whose mothers participated in the program were significantly less likely to drop out of school and had significantly better self-esteem (Roye & Balk, 1996).

Lessons from Successful Programs

Strong and sustainable intergenerational programs conduct the following core activities: conduct a needs assessment; establish measurable goals and objectives; develop a monitoring and evaluation plan; support collaborative efforts; create a realistic program design and budget; construct a plan for recruitment, selection, and matching of participants; train staff and participants; coordinate and supervise; recognize and support participants and staff; and provide regular, planned opportunities for reflection. Reflection refers to the process of key staff sharing their thoughts on what went well in the program and what elements could be improved.

Barriers and pitfalls in developing successful intergenerational programs include failure to involve representatives of the intended participant groups early in the planning process, failure to select staff and volunteers from the diverse cultural and ethnic groups represented in the target population, neglect of the principles of effective communication and teamwork (proximity alone is not sufficient to create effective interaction between participants of different ages), failure to maintain a balance between rewards and costs to volunteers, and forgetting to apply appreciation and celebration liberally.

Future of Intergenerational Programs

Intergenerational programs will be affected by changes in the health care system and in society's determination of the appropriate balance of responsibility—between formal and informal caregivers—for the health care of its elderly citizens. Better integration of acute- and chronic-care models is needed to provide a seamless continuum of care for individuals as they move between different phases of illness and care needs.

Intergenerational work and biopsychosocial models of care will likely increase the importance of geriatric interdisciplinary teams. Just as teams often include various generations of health care providers, they will be interacting with various generations of family involved in the care of one of its members.

The future is also likely to raise ethical issues about the allocation of resources. The opportunity for discussion about important public policy choices should occur in an intergenerational context. To improve the quality of the health care system for all, intergenerational health programs can be a locus for critical discussions to prevent or ameliorate socially harmful intergenerational conflicts around issues of social equity.

NANCY S. WADSWORTH
PETER J. WHITEHOUSE

See Also
Grandparents as Family Caregivers

Internet Key Words
Family Caregiving
Intergenerational Care

Internet Resources
Administration on Aging Intergenerational Programs, Projects, and Training
http://www.aoa.dhhs.gov/aoa/webres/ intergen.htm

Generations Together
http://www.pitt.edu/~gti/

Generations United
http://www.gu.org/

INTERNATIONAL PSYCHOGERIATRIC ASSOCIATION

The International Psychogeriatric Association (IPA) is recognized as the world's leading multidisciplinary, nonprofit organization providing health care professionals and scientists with current information about behavioral and biological geriatric mental health. The IPA is also committed to fostering education, facilitating the advancement of research, and promoting international consensus and understanding in psychogeriatric issues through its innovative educational programs, regional initiatives, scientific congresses, regional meetings, peer-reviewed quarterly journal (*International Psy-*

chogeriatrics, published by Springer Publishing Company, New York), special-focus supplements, quarterly newsletter (*IPA Bulletin*), and Web site.

IPA's goals are to:

- Promote awareness of issues related to mental health of the elderly, including diagnosis and assessment, treatment, and rehabilitation.
- Provide an international forum for the exchange of information by professionals in all relevant disciplines on matters pertaining to the mental health of the elderly.
- Encourage the development of educational resources in basic and applied research in the field of psychogeriatics.
- Support the development of services for maximizing the potential of elderly persons in the community and in institutions.
- Support the role of families and professional caregivers.
- Encourage affiliation to IPA by related organizations.

With over 1,500 members from approximately 70 countries, IPA is a veritable *Who's Who* of the world's preeminent professionals and scientists interested in psychogeriatrics. Members include psychiatrists, internists, geriatricians, general practitioners, primary care providers, family physicians, neurologists, nurses, pharmacologists, psychologists, social workers, and other specialists. Support is provided for a limited number of physicians in developing nations through IPA's Sponsored Member Program.

The IPA Research Awards in Psychogeriatrics, presented every two years at IPA's International Congress, are given to the three best original research papers submitted in the field of psychogeriatrics. These awards have clearly encouraged the pursuit of relevant and important research efforts and have earned a respected level of prominence in the worldwide scientific community. Winning papers are published in *International Psychogeriatrics*.

IPA's scientific meetings are educational vehicles with tremendous global impact. They bring together opinion leaders from around the world to hear and present research papers, often prior to publication. Through plenary sessions, symposia, debates, and poster presentations, IPA meetings cover a vast array of topics, including anxiety, depression, dementia (Alzheimer's, Lewy body, and other types), behavioral and psychological symptoms of dementia, delirium, cognitive and noncognitive impairment, suicide, schizophrenia, transcultural issues, models of service delivery, caregiver issues, and others (not all topics are addressed at all meetings).

DOTTIE ZOLLER

Internet Resource
International Psychogeriatric Association
http://www.ipa-online.org

INTERVIEWING

Highly developed interviewing skills are needed for effective and empathic communication between health providers and older patients. Social science and medical research has shown that the quality of communication between health providers and patients significantly influences patient satisfaction, adherence to medication regimens, recommendations for changes in lifestyle and diagnostic tests, recall of information, anxiety, utilization of future health services, health status, and well-being. Communication also influences the provider's ability to diagnose and treat, his or her satisfaction with the encounter, and the risk of malpractice suits.

Barriers to quality communication between health providers and older patients include time pressures on providers (especially in managed care systems), inadequate education and training of health professionals, and ageism that inhibits accurate diagnosis and treatment. In most care settings, health care professionals are under enormous pressure to complete important and complex technical tasks within a limited time frame. Physicians', nurses', and other professionals' workloads make it difficult to satisfy the communication needs of older patients and their families. Financial incentives for "talk" between providers and patients are often lacking. The focus is on the most expeditiously achieved medical outcome of cure. The "personhood" of the patient or the patient's psychosocial domain is often either trivialized or ignored.

Difficult-to-talk-about subjects are rarely presented in medical or professional school curricula. For instance, novice health care professionals are rarely taught how to sensitively and effectively raise issues such as advance directives, sexual dysfunction, urinary incontinence, cognitive impairment, depression, bereavement, and abuse. These intimate subjects are often left unexplored by providers who do not know how to initiate discussions or do not have the time it would take to deal with them adequately. Although health professionals may not have the time or the skill to undertake psychosocial assessment and intervention, they can learn to detect issues and refer patients to professionals who can explore these issues (Greene & Adelman, 1996).

One way to transcend an ageist perspective is to conduct a life history of the older patient to learn of the patient's past and present circumstances, values, and goals. A life history can explore an individual's identity, dispel ageist stereotypes, and enable the health care professional to provide effective and sensitive care. When a comprehensive understanding of the geriatric patient is unrealized, the patient and provider lose an unprecedented opportunity to collaborate in the healing process (Adelman, Greene, & Ory, 2000).

Interviewing older patients requires many of the same skills required for interviewing younger adults. Practitioners must be attentive listeners and adept at identifying the patient's agenda for the visit. Interviews should be "patient centered" and focus on the patient's major concerns. Open-ended questions to facilitate the patient-centered interview may be inadequate. They may set the stage for an effective, empathic encounter, but practitioners must also carefully attend to patients' responses and listen to what is said as well as what is unsaid. Questioning is just one component of good inter-

viewing skills. Providing sufficient information to the patient in language that is free of technical jargon is also key. Research has shown that patients are often dissatisfied with both the quantity and the quality of information transmitted in the medical encounter. Health care providers may overestimate the amount of time they spend in information giving and underestimate patients' desire for information (Waitzkin, 1984).

It is important to realize that interactions between physicians and older patients may be significantly different from interactions with younger patients. The old-old (individuals 85 years and older) grew up in an age of deep respect for the authority of the physician and may not desire as participatory a role in the medical encounter as younger patients. The health care professional must determine the patient's wishes for participation in decision making and must support patients with specific verbal and nonverbal messages. Research indicates that overall physician responsiveness (i.e., the quality of questioning, informing, and support) is better with younger patients than with older patients; there is less concordance on the major goals and topics of the visit between physicians and older patients and somewhat less joint decision making. Physicians are less likely to be egalitarian, patient, respectful, and engaged and to demonstrate therapeutic optimism with older patients than with younger patients, and older patients are less assertive than younger patients (Greene, Adelman, & Rizzo, 1996).

When interviewing older patients, attention must be paid to factors that influence communication: the presence of multiple chronic conditions, sensory deficits, cognitive limitations, and the presence of an accompanying individual during the medical encounter. Inherent in the older patient's often extensive history (due to longevity) is the occurrence of problems and issues over time as well the presence of current multiple, chronic problems. Attending to present-day multiple problems within the context of the patient's life requires time—a limited commodity in today's health care system. Thus, health care professionals must be skilled at focusing the interview and getting to the immediate issues at hand while achieving and maintaining the relationship's interpersonal aspects. There does not have to be a distinction or choice between providing effective care and maintaining a warm interpersonal relationship.

Sensory deficits in older patients (i.e., problems with hearing and vision) are likely to influence communication in medical encounters. The incidence of hearing loss increases each decade, so that as many as 70% of 70-year-old individuals may have hearing problems. To facilitate communication, health care professionals can try several approaches, including identifying the patient's specific needs in this area, reducing background noise in the office, speaking at a slightly louder level (but avoiding shouting), establishing good visual contact, rephrasing rather than repeating misunderstood phrases, pausing at the end of a topic to allow for questions, and amplifying with a microphone and headset (Cobbs, Duthie, & Murphy, 1999).

Vision loss may also affect provider-patient interactions. Individuals over 65 are more likely than younger individuals to experience a decrease in visual acuity, contrast sensitivity, and visual fields. Sitting close to the older patient and providing environmental supports, such as improved illumination in the office, can facilitate communication with patients with vision problems (Cobbs, Duthie, & Murphy, 1999).

Although the incidence of dementia increases with age, it should not be assumed that all old people have a cognitive impairment. Inappropriately stereotyping patients with any cognitive dysfunction as being incompetent and incapable of participating in their care must be avoided (Adelman, Greene, & Ory, 2000). Each patient must be individually assessed. Given the different communication and language impairments over the course of dementia, health care providers must become adept at identifying the needs of the patient at the particular stage of the illness (Orange & Ryan, 2000).

Older patients are often accompanied to the medical visit by a third party (e.g., spouse, adult child, hired professional caregiver), who may significantly affect communication between the health care professional and the patient. One study found that in three-person encounters, older patients were often referred to as "he" or "she" (making the patient an outsider to the interaction). Moreover, when comparing two- and three-person interactions, it was found that although the content of physician's talk was no different, older patients were less responsive and assertive, and there was less shared laughter and joint decision making in triadic visits (Greene, Majerovitz, Adelman, & Rizzo, 1994). Current research in a primary care setting by the authors has shown that older patients are often accompanied by two or sometimes three other individuals who participate in the medical interview. Thus, the health care professional must also become adept at facilitating small group discussions. However, the health care professional should spend some time alone with the older patient to give him or her the opportunity to express concerns and problems that may be highly personal.

The provision of quality medical care to older patients requires sensitive communication skills on the part of health care professionals and a health care system that supports a patient-centered approach. Recognition of the interdependence of medical cure and interpersonal care is essential for the practice of effective and compassionate medical care.

MICHELE G. GREENE
RONALD D. ADELMAN

See Also

Communication Issues for Practitioners
Hearing Impairment
Patient-Provider Relationships
Vision Changes and Care

Internet Key Words

Health Care Provider-Older Patient Communication
Health Care Provider-Older Patient Relationship
Interviewing Skills

J

JOINT COMMISSION ON ACCREDITATION OF HEALTHCARE ORGANIZATIONS

The Joint Commission on Accreditation of Healthcare Organizations works to improve the safety and quality of care provided to the public through the provision of health care accreditation and related services that support performance improvement in health care organizations. Founded in 1951, the Joint Commission evaluates and accredits nearly 20,000 health care organizations and programs in the United States, including hospitals, health care networks, managed care organizations, and health care organizations that provide home care, long-term care, behavioral health care, laboratory services, and ambulatory care services. The Joint Commission introduced assisted living accreditation services in 2000. The Joint Commission is an independent, not-for-profit organization.

DONNA LARKIN

Internet Resource

Joint Commission on Accreditation of Healthcare Organizations
http://www.jcaho.org

JOINT REPLACEMENT: LOWER EXTREMITY

Total joint replacement surgery is a mechanical solution for joints damaged by osteoarthritis, rheumatoid arthritis, and traumatic injuries such as falls. These morbidities are often associated with older adults and cause joint dysfunction, resulting in extreme pain during normal movement and even at rest (Harris & Sledge, 1990). Often, the pain causes the patient to avoid using the joint, resulting in a weakening of the muscles surrounding the joint, which leads to further pain and deterioration of the joint. Joint replacement may be an option to maintain mobility and function and reduce the pain of these diseased or fractured joints.

Joints are formed at the intersection of two bones, which are separated by thick collagenous tissue known as cartilage. Problems occur when the cartilage breaks down due to disease or the integrity of the joint is lost due to a fracture. During joint replacement surgery, the damaged joint is removed and replaced by an artificial device designed to function as a normal joint.

In the elderly, total joint replacements are performed for fractures or severe arthritic conditions. Indications for surgery include severe pain that restricts activities of daily living and is unrelieved by interventions such as pharmacotherapy, the use of assistive devices, or weight reduction. The extent of bone damage can be evaluated through x-rays to determine the appropriate level of treatment (Haberman, 1986). Since joint replacement surgery has little impact on the physical demands required for the joint to function (Harris & Sledge, 1990), patients 55 and older are considered good candidates. Possible contraindications for emergency surgery after trauma may include poor medical condition, significant peripheral vascular disease, neuropathy affecting the joint, or an active infection (Quinet & Winters, 1992).

Total hip replacement and total knee replacement are the most common joint replacements in the elderly. The hip is a ball and socket joint consisting of the top of the femur, the femoral head, and the cup-shaped bone in the pelvis called the acetabulum. Depending on the extent of bone damage, the hip implant can replace the head of the femur, the acetabular cup, or both. The device to replace the head of the femur, referred to as the hip stem, consists of a metal sphere attached to a long stem. Both the spheres and the stems are available in various sizes, depending on the patient's anatomy. The acetabular component consists of a metal hemisphere with a polyethylene liner. The polyethylene surface is an important part of the device because it is the primary weight-bearing surface of the joint.

Hip replacement surgery involves removing the top of the femur, including the femoral head. Surgical instruments are then used to enlarge both the acetabulum and the medullary canal of the femur. The hip stem and prosthetic acetabular cup are inserted into the enlarged sections of the bones and are fixed with either cement or a porous coating on the devices. The reamed-out portion of the bone varies, depending on which fixation method is used. The joint replacement is then reengaged as a ball and socket joint.

The knee joint acts as a hinge between the femur and the tibia. In normal function, the movement of the knee joint can be very complex, combining flexion, rotation, and medial or lateral motion. As the knee flexes, the femur rolls back on the tibia, increasing the potential flexion of the knee. This posterior movement of the femur also causes lengthening of the quadriceps muscle. It is necessary to duplicate this complex motion in the artificial joint to ensure effective muscle contraction and maximum angular motion of the joint (Harris & Sledge, 1990).

In general, the knee implant is used as a surface replacement on the ends of the femur and tibia. The artificial joint consists of a femoral component, a tibial component, and an optional patella replacement. The contours of these components mimic the shapes of the anatomical structures in the normal functioning knee joint.

The surgical technique for total knee replacement is to replace the damaged surfaces of the joints with the prosthetic knee components. First, the bone is cut at distinct angles to prepare flat surfaces in the damaged bone. These surfaces are orientated to provide proper alignment of the implants. The implants are then fitted over the surfaces and fixed to the bone by either bone cement or a porous coating on the devices. Replacement of the patella and the posterior cruciate ligament, which is the

strongest ligament in the joint, is an option (Harris & Sledge, 1990).

Bone cement is an acrylic plastic that is inserted into the reamed-out portions of the bone. The implant is then placed in the cement and held in the proper orientation until the cement hardens. Press fit, the other fixation method, involves a porous coating applied to the device when it is manufactured. With this method, the section of the bone that is reamed out is approximately the same size as the device itself. The implant is then pressed into the opening and may be secured with pegs or screws. Over time, the porous coating is thought to promote bony ingrowth into the surface of the device and therefore strengthen the stability of the implant. For older adults, cemented implants are more commonly used because they provide a stronger replacement in the short term. Coated implants are generally used in younger or more active patients.

Rehabilitation after joint replacement begins immediately following surgery. A successful outcome requires both replacement of the joint and strengthening of the surrounding muscles. Therefore, strengthening exercises are an important component of rehabilitation. Initially, the rehabilitation program includes exercises specific to the joint that was replaced, with only partial weight bearing using crutches or a walker. The immediate goals of rehabilitation are to bear full weight using an assistive device and to function independently in activities of daily living. Gait training gradually continues with a walker, crutches, or cane until the patient is able to walk without these assistive devices.

The benefits of total joint replacement surgery are pain relief and restoration of joint function. Joint replacement allows patients to carry out many daily activities that were previously restricted by pain. Although these artificial devices improve the functioning of the joint, they do not provide the same range of motion as a normal, healthy joint. The patient can expect to be able to walk, sit, climb stairs, put on socks or shoes, and enter a car, but not perform activities involving repetitive impact. Factors that may affect the outcome of rehabilitation include age, weight, level of activity, and other limited functions in the patient (Quinet & Winters, 1992).

There are relatively few complications from total joint replacement surgery, considering the magnitude of the operation. Loosening may occur in one or more of the components, causing pain in the joint. Generally this happens with cemented implants, due to degradation of the bone cement. Excessive wear of the device can cause fragments of bone cement to break off and travel through the body via the circulatory system. Other complications may include nerve damage from the surgery, dislocation of the joint, bone fracture, infection of the joint, and general infection. As with any major surgery, there are potential systemic complications, including infection, blood clot, and pulmonary embolism. Depending on the level of failure and complication, revision surgery may be required.

Total joint replacements of the hip and knee are among the most successful orthopedic procedures performed in the elderly. The patient can expect the implant to last for 10 years or more and provide pain-free functioning that would not otherwise be possible. Advances in the field of orthopedics are being made to increase the functional level of artificial joints and to extend their longevity.

MARSHA AARON
ELLEN FLAHERTY

See Also
Fractures
Hip Fractures
Nutritional Assessment
Osteoarthritis
Osteoporosis
Prosthetics: Lower Extremity
Rheumatoid Arthritis

Internet Key Words
Degenerative Joint Disease
Osteoarthritis
Rehabilitation
Rheumatoid Arthritis
Total Hip Replacement
Total Knee Replacement

Internet Resource
atOrthopedics.com
http://atorthopedics.com/orthopedics/

L

LATINO ELDERS

See

Hispanic and Latino Elders

LEISURE PROGRAMS

When adults retire, achieving meaningful leisure roles is imperative if they are to maintain a high quality of life. Although independent elders regularly lead active leisure lives through their own initiative, other elders, especially those in institutional or assisted living settings, may benefit from planned leisure activities. For such individuals, an ideal leisure program, developed through a collaborative and cross-disciplinary process, enables them to attain and maintain optimal levels of fulfillment and wellness.

Leisure programs cannot simply follow tradition (e.g., what staff decides is best, what has always been done, or what is available in a facility) or perpetuate stereotypical activities (e.g., cards, bingo, board games, and arts and crafts). Rather, they should address the increasingly unique needs of an older population that, according to demographic trends, will have a higher socioeconomic status and greater racial and ethnic diversity. Furthermore, leisure programs will be required in a greater variety of settings, including traditional government-sponsored recreation departments, senior centers, retirement communities, long-term-care facilities, hospice, assisted living, home health care, and senior day care (Teaff, 1990).

In designing a leisure program, practitioners need to be aware that the meaning of leisure varies considerably for each participant. Thus, one activity may have more than one meaning (e.g., some people may play bingo for competitive reasons, while others play purely for social reasons). In contrast, more than one activity may share the same meaning (e.g., people may read, listen to the radio and cas-sette tapes, and watch television exclusively for spiritual reasons). Additionally, meanings may vary within the context (e.g., listening to music may mean one thing in church and another thing at home). Program design has to consider the unique needs of each participant.

Practitioners should ask the same questions researchers have asked when studying the meaning of leisure: What prompts the participant to choose the activity? What does the activity do for the participant? What needs are satisfied through participation? In meeting clients' needs, practitioners must adopt a holistic approach, focusing on the health of the whole person (Van Andel & Heintzman, 1996). Traditional leisure strategies focused primarily on functional activities aimed toward physical health; an approach that includes the meaning of the activity to the participant forces the practitioner to include activities directed toward social, psychological, and spiritual health as well.

Participation in leisure activities meets various needs for older adults (Lawton, 1993), including continuity, service, education, competition, social interaction, solitude, and finding meaning and purpose in life (Roelofs, 1999). Continuity addresses participation in activities that have been of interest over time, such as reading, doing handiwork, or watching sports on television. A need for service usually involves volunteering in activities that benefit others or contribute to the greater good in some way, such as stuffing envelopes, delivering meals, or working at a homeless shelter. The need for education deals with engaging in activities that increase knowledge or stimulate thinking (reading, going to lectures, taking adult education courses). Competition tests one's abilities and challenges oneself to develop personal skills (playing any type of game). Most people have a need for social interaction that consists of participating in activities that provide companionship, such as joining group activities or visiting family or friends. And although they often seek social interaction and stimulation, people also need solitude, opportunities to do things alone for self-expression, for example, painting, gardening, or engaging in other individual hobbies. Activities such as praying, meditating, and attending religious services promote inner peace and satisfaction, which are important requisites for finding meaning and purpose in life. In fact, satisfaction of spiritual needs may be the most valuable benefit of well-designed leisure activities, since some research suggests that many physical and emotional health outcomes may be the product of a chain of relationships that starts with activities such as prayer, meditation, mindful awareness, and group support (Hawks, Hull, Thalman, & Richins, 1995).

Leisure experiences contribute to a sense of wholeness and life satisfaction largely because of the psychospiritual nature of the experience. People often discover what it means to be truly human, truly alive. One theory suggests that optimal leisure experiences contribute to a state of being in which participants become so absorbed in the activity that nothing else seems to matter (Csikszentmihalyi, 1990). This "flow" theory suggests that the greatest satisfaction comes from activities that are intrinsically rewarding and moderately challenging, provide immediate feedback, and allow for a sense of control. Therefore, practitioners are encouraged to craft opportunities for meaningful leisure experiences that will help individuals find meaning and purpose in their lives, develop a healthy appreciation of who they are as unique persons, promote nurturing and empowering relationships with others, facilitate a sense of control by making choices, instill a sense of hope for the future, and provide opportunities for successful experiences (Nicosia, 1994).

The goals of leisure can be achieved most effectively if practitioners use a lifestyle approach to program planning rather than the traditional activity-centered model. Lifestyle programming asks the participant, "What would you do today if you were fully functional and able to make your own choices?" With this model, activity options are more varied because they reflect the needs and interests of the participants rather than those of the staff or the facility. The daily activity schedule is more open to the spontaneous interests of a few participants rather than a structured, scheduled calendar of events. Activity environments should be homelike, with plants, animals, and areas for engaging in special interests and hobbies. These settings might include play spaces for children and learning centers for adults of all ages, as well as beauty shops, coffee shops, libraries, greenhouses, and worship areas. Participants should be able to join their friends, families, and visitors in a wide variety of leisure activities.

Moving from a structured activity-based program to a person-centered, lifestyle model requires comprehensive lifestyle assessments to identify those life experiences and interests that have brought meaning and purpose to participants. Included in an assessment should be favorite foods and recipes, entertainment preferences (music, reading materials, movies, television), religious traditions and customs, hobbies, friends and social contacts, and work interests. The more older adults participate in an assessment, the greater their sense of control and choice regarding future leisure activities.

Practitioners should be aware of the meaning of leisure as a guide for designing person-centered lifestyle programs that assist clients in meeting their needs not only for physical health but also for social, psychological, and spiritual health. Practitioners educated in the leisure disciplines may be in the best position to serve as facilitators of interdisciplinary teams that develop these programs.

Implementing person-centered lifestyle programs for older adults empowers them to achieve meaningful leisure roles and experience a high quality of life throughout their retirement.

LOIS ROELOFS
GLEN VAN ANDEL

See Also
Active Life Expectancy
Activities
Adult Education

Creativity
Retirement

Internet Key Words
Activities
Leisure Participation
Leisure Programs
Person-Centered Lifestyles

Internet Resources
American Alliance for Health, Physical Education,
    Recreation and Dance
http://www.aahperd.org

National Recreation and Parks Association
http://www.active parks.org

Therapeutic Recreation
http://www.recreationtherapy.com

LIFE EVENTS

Occasionally, a concept or theory is so irresistible
that it takes the scientific community by storm.
Introduction of the concept of life events to the
social and biomedical sciences had precisely that
effect, and it remains a major topic in aging research
today. Life events are identifiable, discrete changes
in life patterns that create stress and can lead to the
onset of illness or the exacerbation of preexisting
illness. Life events research traces its roots to the
pioneering biomedical research of Holmes and
Rahe (1967) and the classic sociological study of
stress known as the Midtown Manhattan Study
(Langer & Michael, 1963). Subsequently, thousands of studies examined the impact of life events
on physical and mental health and the factors that
mediate and moderate the effects of life events on
health outcomes.

Why was the concept of life events so compelling? Numerous other variables are equally powerful predictors of physical and mental illness. Primarily, life events offered a potential social risk
factor compatible with epidemiological theories of
illness onset, theories that emphasize the role of
environmental agents on the health of human be-

ings. Although there is virtual consensus that social
environments play a powerful role in health and
illness, isolating relevant parameters of social environments and documenting their effects have been
challenging. Life events are especially attractive
because they represent a social risk factor rivaling
physical risk factors in terms of being objective
(i.e., occurrence of the event can be verified), potentially quantifiable, and occurring before illness onset (thus clarifying causal order).

Interest in life events research did not originate
in aging research, but many gerontologists have
examined the effects of life events on health in later
life. The life-events perspective is compatible with
the crisis orientation that characterized early research on aging—an orientation that focused on the
losses that are common in later life. Many of those
losses (e.g., widowhood, economic problems) can
be viewed as life events. Research has shown that
older adults actually experience fewer life events
than do young adults. Interest in the impact of
life events on late-life health continues because of
evidence that (1) the life events experienced by
older adults are likely to involve major losses, especially bereavement, and (2) resources for coping
with stress typically decrease in later life. Thus,
older adults may be more vulnerable than younger
adults to the adaptive challenges posed by life
events.

Evolution of the Life-Events Perspective

Initially, research emphasized only one element of
life events: the degree to which they disrupt established behavior patterns. Early research explicitly
stated that subjective perceptions of stress are unim-
portant; it is the degree of change that calibrates
the stressfulness of life events. Investigators using
a more interactionist perspective demonstrated con-
vincingly, however, that change itself is not the
"active ingredient" in the link between life events
and illness. Rather, perceptions of stress are crucial
for understanding the effects of life events on
health. Consequently, measures of life events are
now routinely restricted to events that are perceived

as negative or stressful by the individuals who experience them. The importance of experiential measures of life events is further documented by evidence that there is no life event that is uniformly described as positive or negative—stress is in the eye of the beholder.

In early research, life events were considered synonymous with social stress. It is now clear, however, that they are only one category of stressful experiences. Also important are chronic stressors, which are ongoing stressful experiences that do not represent sudden changes in behavior patterns but persist over long periods (e.g., chronic poverty, chronic marital conflict, long-term commitments to providing care for an impaired relative). Thus, social stress is a broad concept, with life events representing only one important area of inquiry.

Major Issues in Life-Events Research

Research on life events covers a broad range of issues. Three major research areas of special relevance to late life are reviewed here.

Mediators and Moderators

Without question, the major focus of life-events research has been understanding the effects of life events on health outcomes. It was observed early on that life events have variable outcomes; although they are statistically significant predictors of illness, most individuals who experience life events do not become ill. The central research question then became: Under what conditions do life events lead to negative health outcomes? More than two decades of sophisticated research has focused on the causal pathways between stress and illness and, more recently, the antecedents of social stress (see Krause, 1999 for a recent review of these issues and late-life health).

It is now clearly documented that whether stressful life events harm health is a function of both the strength of the life event (e.g., death of a loved one poses a greater threat than retirement) and the resources available to the individual for responding to the stress. Two major types of social resources are especially powerful in offsetting the effects of stress: economic resources and social support. The value of economic resources is straightforward: many stressful situations can be remedied or at least diminished by adequate financial resources (e.g., one can live comfortably through involuntary job loss, the economic consequences of widowhood are minimized). Social support, which refers to the tangible and intangible forms of assistance provided by family and friends, is a broader resource. A number of typologies have been suggested for understanding the multiple functions of social support. One useful typology consists of three types of social support: (1) instrumental support, which refers to the provision of tangible assistance (e.g., provision of transportation or personal care); (2) informational support, which refers to the provision of essential information about relevant resources external to the support system (e.g., information about community services); and (3) emotional support, which refers to the comfort, self-validation, and companionship offered by intimate others. Major works concerning the offsetting effects of resources include those by Holahan and Moos (1991), Krause (1986), and Ensel (1991).

Although both economic resources and social support have been shown to mediate the effects of stress on health outcomes, other research suggests that the relationship is statistically interactive rather than mediating. This issue is commonly referred to as the stress-buffering hypothesis (Lin, Woelfel, & Light, 1985). Its proponents suggest that social support protects health only under conditions of stress, rather than having a more general protective effect that exists independent of stress. Research provides empirical confirmation for both the stress-buffering hypothesis and the alternative main-effects hypothesis. Either way, however, social support plays a vital role in reducing the likelihood that stress will have negative effects on health.

There also is convincing evidence that psychosocial resources such as self-esteem and a sense of mastery mediate some of the effects of life events on health (Murrell, Meeks, & Walker, 1991; Pearlin, 1989). These relationships are complex.

On the one hand, high levels of self-esteem and mastery offset some of the potentially harmful effects of life events on health outcomes. On the other hand, stressful life events can erode self-esteem and mastery, increasing an individual's vulnerability to stress-related health problems. One of the important functions of effective social support is bolstering individuals' psychosocial resources so that they can better meet the challenges of stressful life events.

Life events (and stressors more generally) also have been investigated as mediators of the effects of prior risk factors on health. A number of investigators, especially Aneshensel (1992) and Pearlin (1989), have suggested that the different roles and locations in social structures that individuals occupy may affect their amount of exposure to life events and thus ultimately affect health. The primary factors examined as potential antecedents of exposure to life events have been standard demographic variables (age, sex, race) and socioeconomic status. This is a complex research literature, but the general pattern of findings suggests that life events mediate some but not all of the effects of demographic characteristics and socioeconomic status on health. The unmediated effects of these variables have served as the foundation for theoretical speculation that specific subgroups of the population may be differentially vulnerable to life events and other stressors.

Developmental Perspectives on Life Events

Although the volume of research is small, psychologists have begun to articulate the links between psychological development and life events. Early examples of this perspective can be seen in the work of "biographical" psychologists, who enjoyed considerable attention in the mid-20th century (Buhler & Massarik, 1968). More recently, efforts to understand psychological growth, subjective well-being, and positive mental health have spurred additional research on the developmental consequences of life events. The major foci of this research are the ways in which life events often trigger developmental changes (e.g., in personality and

self-concept) and how the effects of life events can differ, depending on the age or developmental stage of the individuals who experience them. Ryff's work in this area has been especially impressive (Ryff & Dunn, 1985; Ryff & Essex, 1992; Showers & Ryff, 1997). In addition, a variety of other research ranging from investigation of the differential effects of parental divorce, depending on the age of the child; the benefits and rewards of caring for an impaired older adult; and the rewards of late-life marriage all speak to the issue of changes in personality that can result from life events. Unlike research from the stress and illness perspective, this research suggests that life events can have positive as well as negative effects.

Life Events in the Life-Course Perspective

Sociological studies of the life course also pay substantial attention to life events, although the term "life-course transitions" is used more frequently in that research tradition. One important element of life-course perspectives is an emphasis on the sequences of transitions that create long-term trajectories or pathways. Life-course studies have focused primarily on the timing and sequencing of events or transitions and the broad range of outcomes related to them (e.g., socioeconomic achievement, marital stability) (see George, 1993 for a review of research in this tradition). Most life-course studies have focused on sequences of transitions in early adulthood. An important contribution to aging research, however, is the body of work by O'Rand and Henretta (1999), demonstrating that economic status in late life is, to a substantial degree, a result of work and family decisions made earlier in life.

After more than 30 years, the antecedents and consequences of life events continue to engage the energies of social, behavioral, and biomedical scientists. The core of this research has been the links between stress and illness. More recently, life events have proved to be important for understanding psychological development throughout adulthood and the ways that status and well-being in late life arc, in large part, the result of events experienced decades earlier.

LINDA K. GEORGE

See Also
Coping with Chronic Illness
Life Review
Validation Therapy

LIFE EXTENSION

Life extension can be evaluated only by statistical analyses of populations. Although life expectancy from birth (estimated average length of life) is widely employed in such evaluations of human populations, median length of life is commonly used in experimental studies with animal models, wherein life extension is inferred if a manipulation increases the median length of life. However, many investigators use the term "life extension" only for manipulations that increase the maximum length of life (the length of life of the last member of a birth cohort to die). Indeed, maximum length of life was long considered a reliable index of the rate of aging. Based on this belief, when a manipulation increased the maximum length of life, it was thought to do so by slowing the rate of aging. In recent years, this concept has been challenged by the recognition that the maximum length of life is influenced by population size, tending to increase as the population size increases (Gavrilov & Gavrilova, 1990). Moreover, even when comparing populations of similar sizes, factors other than the rate of aging have been shown to influence the maximum length of life (Promislow, 1993). In studies with animal models, these problems can be partially overcome by utilizing the age of the longest-lived survivors of a cohort (e.g., the age of the 100th percentile survivors) to provide an index of maximum length of life (Yu, 1995). In the total human population, the oldest age reached by only 1 in 100 million people is believed to be a reasonable index of maximum length of human life.

Experimental Findings with Model Organisms

The first study to show that maximum length of life can be extended was that of Loeb and Northrop (1917), who found that among populations of fruit flies (*Drosophila melanogaster*) living in temperatures of 10° to 30°C, those living at lower temperatures have a longer maximum length of life than those living at higher temperatures. This finding has also been obtained with other poikilothermic species such as nematodes and fish, but not with homeothermic species such as mammals and birds. This increase in maximum length of life is thought to be due to the fact that lowering the environmental temperature of a poikilotherm decreases its metabolic rate (McCarter, 1995). However, there is disagreement whether reduction in metabolic rate plays an important role in life extension.

Decreasing the dietary energy intake, a manipulation called "caloric restriction" or "dietary restriction," is the only experimental manipulation that has consistently been found to extend the maximum length of life of a mammalian species. This was first described in a study of rats by McCay, Crowell, and Maynard (1935); since then, caloric restriction has been found to extend life in a variety of rat and mouse strains, as well as hamsters (Masoro, 1996). It is not known whether caloric restriction causes life extension in other mammalian species, since the relevant research has not been done; compared with rodent species, their larger size and longer life span make such research extremely costly. However, such studies are now under way on nonhuman primates, although it will be many years before the results will be available (Roth, Ingram, & Lane, 1999). In addition to increasing the maximum length of life of rodent species, caloric restriction has had this action in a variety of poikilothermic species, including fish (Finch, 1990).

Studies on the effect of exercise on life extension in rats have a long and controversial history. It now appears that there are two reasons for the conflicting findings: the use of rats suffering chronic infections, and stressful procedures used to make the rats exercise. In studies free of these problems, exercise programs were found to extend the median length of life but not the maximum length of life of rat populations (Holloszy & Kohrt, 1995).

Genetic manipulations have extended length of life in both mammalian and invertebrate species. Selective breeding has resulted in long-lived strains of fruit flies (Arking, Buck, Berrios, Dwyer, & Baker, 1991) and nematodes (Johnson, 1990). Single gene mutations have been found to increase the length of life of nematodes (Lithgow & Kirkwood, 1996) and yeast (Kennedy, Austriaco, Zhang, & Guarente, 1995). In addition, single gene mutations have extended the length of life of mice (Miller, 1999). Transgenic fruit flies that overexpress both superoxide dismutase and catalase also have an extended length of life (Orr & Sohal, 1994).

Many pharmacological agents have been used in attempts to extend the length of life in animal models. Since oxidative stress is thought to be, at least in part, a cause of aging, antioxidants have been much studied in this regard. Naturally occurring antioxidants such as vitamin E, vitamin C, and beta-carotene, as well as synthetic antioxidants such as butylated hydroxytoluene, have been administered either singly or in mixtures to rats and mice. None of these agents has had a significant effect on either the median or the maximum length of life in rat and mice populations (Yu, 1995). The pharmacological agent that has shown the most promise as a life-extending agent is deprenyl. Knoll (1988) reported that deprenyl markedly increased the length of life when administered to old rats. Since then, other investigators have also found that deprenyl extends the life of rats, mice, and hamsters, but the magnitude of this effect was much less in the more recent studies than that reported by Knoll. The administration of dehydroepiandrosterone (DHEA) has been claimed to extend the life of rodents and to have other antiaging actions (Bellino, Daynes, Hornsby, Lavrin, & Nestler, 1995). However, it is unclear whether these actions are due to the fact that DHEA reduces food intake; that is, it may be causing caloric restriction rather than having a direct life-prolonging action. Although it has been claimed that melatonin added to the drinking water extends the life of rats and mice, these findings must be viewed with caution, since only small numbers of animals were studied and food intake was not carefully monitored (Reiter, 1995).

Information on Humans

Probably the most remarkable biological phenomenon of the 20th century is the marked increase in human life expectancy from birth, increasing in the United States from about 47 years for both sexes in 1890 to 73 years for men and 78 years for women in 1990. Public health measures, such as sanitary engineering, have certainly played a major role in this increase. Other factors include medical advances, nutrition, protection from damaging environments (e.g., air conditioning, central heating), and possibly others that have yet to be identified. Much of the increase relates to the prevention of premature death at young ages. However, in the last half of the century, significant increases in life expectancy from age 60 also occurred (Vaupel, 1997). It is difficult to know with certainty whether the maximum length of human life has been extended during the 20th century, but the increasing fraction of centenarians in the world population makes it likely that there has been such an extension.

Health food stores and pharmacies sell many food additives and other agents that purportedly extend life. None of these has been proved to do so. However, there is one intervention that does increase life expectancy in humans: habitual exercise (Lee, Paffenbarger, & Hennekens, 1997). Whether the maximum length of life is also affected by such exercise has yet to be addressed.

EDWARD J. MASORO

See Also
Morbidity Compression

Internet Key Words
Caloric Restriction
Deprenyl
Exercise

Life Expectancy
Maximum Length of Life
Median Length of Life

LIFE REVIEW

The tendency of older persons toward self-reflection and reminiscence used to be considered a sign of loss of recent memory and therefore of aging-related pathology. However, Butler (1963) postulated that reminiscence in the aged was part of a normal life review process brought about by the realization of approaching dissolution and death. It is characterized by the progressive return to consciousness of past experiences and particularly the resurgence of unresolved conflicts for reexamination and reintegration. If the reintegration is successful, such reminiscence can give new significance and meaning to life and prepare the person for death by mitigating fear and anxiety.

This evaluative process is believed to occur universally in all persons in the final years of their lives, although they may not be totally aware of it and may in part defend themselves against realizing its presence. It is spontaneous, unselective, and seen in other age groups as well (adolescence, middle age), especially when individuals are confronted by death or a major crisis, but the intensity and emphasis on putting one's life in order are most striking in old age. In late life, people have a particularly vivid imagination and memory for the past and can recall with sudden and remarkable clarity early life events. They often experience a renewed ability to free-associate and to bring up material from the unconscious. Individuals realize that their own personal myth of invulnerability and immortality can no longer be maintained. All this results in a reassessment of life, which, depending on the individual, may bring depression, acceptance, or satisfaction.

The life review can occur in a mild form through mild nostalgia, mild regret, a tendency to reminisce or tell stories, and the like. Often the life story will be told to anyone who will listen. At other times it is conducted in monologue in private and is not meant to be overheard. It is in many ways similar to the psychotherapeutic situation in which a person is reviewing his or her life in order to understand present circumstances (Haight & Webster, 1995).

As part of the life review, one may experience a sense of regret that is increasingly painful. In severe forms, it can lead to anxiety, guilt, despair, and depression. And in extreme cases, if a person is unable to resolve problems or accept them, terror, panic, and suicide can result. The most tragic life review is one in which a person decides that his or her life was a total waste.

Some of the positive results of a life review can be the righting of old wrongs, making up with enemies, coming to accept one's mortality, and gaining a sense of serenity, pride in accomplishment, and a feeling of having done one's best. Life review gives people an opportunity to decide what to do with the time left to them and work out emotional and material legacies. People become ready to die, although they are in no hurry. Possibly the qualities of serenity, philosophical development, and wisdom observable in some older people reflect a state of resolution of their life conflicts. This is usually accompanied by a lively capacity to live in the present, including the direct enjoyment of elemental pleasures such as nature, children, forms, colors, warmth, love, and humor. Some become more capable of mutuality, with a comfortable acceptance of the life cycle, the universe, and the generations. Creative works may result, such as memoirs, art, and music. People may put together family albums and scrapbooks and study their genealogies.

One of the great difficulties for younger persons (including mental health personnel) is to listen thoughtfully to the reminiscences of older people. Nostalgia is viewed as dysfunctional behavior, representative of living in the past, a preoccupation with self, as well as being boring, meaningless, and time consuming. Yet, as a natural healing process, it represents one of the underlying human capacities on which all psychotherapy depends. The life review is a necessary and healthy process and should be recognized in daily life as well as used in the mental health care of older people (Lewis & Butler, 1974). Life review therapy and reminiscence therapy are employed, for example, in nursing homes (Burnside, 1988). Many people are taping their lives for their families.

There have been a number of studies related to life review and life review therapy, and there has been growing interest in related topics, such as storytelling, oral history, guided autobiography (Birren & Deutchman, 1991), and narrative or experiential gerontology (Haight & Webster, 1995). The International Society of Reminiscence and Life Review was established in 1995.

ROBERT N. BUTLER

See Also
Creativity
Depression in Dementia
Spirituality
Validation Therapy

LONG-TERM-CARE FINANCING: INTERNATIONAL PERSPECTIVE

Two opposing views characterize how long-term care (LTC) should be financed. One is the residual, safety-net model, which places the primary responsibility with the individual or family. Under this model, the state assumes a role in financing LTC only when the family cannot provide support and when the individual has exhausted all assets. The other is the entitlement model, in which the state provides for LTC as a basic human right, regardless of the availability of family support and assets. Among the developed countries, the United States has taken the former view, while the Scandinavian countries have taken the latter. Other countries lie somewhere in between; however, many countries are refocusing their policies toward the entitlement model. Germany in 1995 and Japan in 2000 initiated public LTC insurance (Ikegami, 1997); the United Kingdom's Royal Commission on Long Term Care

recommended that LTC should be provided free to everyone who needs it.

The reason for this shift lies in demographic and social changes. The "oldest old" population (age 80 and older) and the number of "elderly households" have grown dramatically, and this trend will accelerate in the future. However, the number of female relatives, who traditionally provided the bulk of informal care, has been decreasing. At the same time, the proportion of women of working age who are in paid employment has been increasing (Royal Commission on Long Term Care, 1999). The magnitude of these changes is likely to outpace any decreases in burden that may result from the compression of morbidity and the improvement in functional status of the elderly. Moreover, there is growing awareness that the existing health care and social services systems are inadequate and inefficient for providing LTC.

In the United States, expanding private LTC insurance has been advocated as a way to reconcile these two opposing models of public versus private responsibility. However, the problems associated with private health insurance become magnified in LTC. Those in greatest need, the elderly, are the least likely to be able to afford it. Therefore, people would have to enroll when they are comparatively young and at low risk. And premiums would have to be paid until they become old and at high risk. However, should they stop paying, they forfeit all rights to benefits. For the insurers, there are the problems of moral hazard, adverse selection, and the difficulty of predicting overall costs (Organisation for Economic Cooperation and Development [OECD], 1996). Because of these reasons, Germany rejected the idea of expanding private LTC insurance.

Public funding for LTC can take two forms: taxes or social insurance. Since social insurance takes the form of mandatory payroll deductions (as in Medicare), there may appear to be little difference between the two. However, social insurance contributions are reserved solely for the provision of specific services. Therefore, they tend to be more fiscally stable and more sheltered from the general economic situation. They are also less progressive than taxes (i.e., the contribution ratio is fixed, regardless of income level). Moreover, these differences in funding may lead to real differences in the provision of LTC. Under a tax-based system, the LTC services tend to be targeted toward those having less informal support or lower incomes. Care managers employed by the local government decide on the services for each individual, and providers are usually given a monopoly in the community in which they operate. In contrast, under a social insurance system, benefit levels are determined by explicit eligibility criteria based on functional or cognitive performance. Thus, they are set irrespective of the amount of informal support or income, though premiums or co-payments may be waived for those with low incomes.

Whether a country opts for a tax or social insurance model is largely determined by the existing administrative structure. Countries such as Denmark, Sweden, Norway, and the United Kingdom, which finance their health care through taxes, also finance LTC this way. In contrast, Germany, Austria, and Japan rely on social insurance for health care and LTC. However, it should be noted that the accessibility of LTC services is likely to be a function of the absolute amount of resources available. Sweden, with public spending for LTC amounting to 2.7% of gross domestic product, and Denmark with 2.24%, have much higher levels of LTC provision than Germany, which spends only 0.82% (OECD, 1999). The high spending in Scandinavian countries has been justified on the grounds of normalization and improving the quality of life. Thus, these aspects need to be considered when evaluating the justification for providing community services, not just the prevention of nursing home admissions. If the public sector does not pay for an adequate level of LTC, its costs will be borne in the private sector by the daughters or spouses who become exhausted providing care, or by the rapid spend-down of a lifetime of savings.

For publicly financed LTC programs to obtain popular support, they must be perceived to be fair (Campbell & Ikegami, 1999). In a tax-based system, there needs to be implicit trust in the competence and integrity of the care managers to decide

who is going to get what amount of benefits. As individual needs must be balanced with the constraints of the total budget, this is a difficult task. In a social insurance system, where the amount of benefits is determined by explicit eligibility criteria, the way in which the classification program is designed and administered is important. For example, the integrity and competence of the people who decide on eligibility must be assured, the reliability of the assessment procedure should be high (so that similar cases produce similar results), and the categories must be logically delineated and understandable. In both systems, there must be mechanisms for monitoring the eligibility process and for appeals of adverse decisions. A scientific basis for evaluating the eligibility status is still in the process of being developed. The methodology developed for case-mix grouping in nursing homes, such as resource utilization groups, could be utilized.

NAOKI IKEGAMI

See Also
Future of Care
Long-Term-Care Policy
Medicaid
Medicare
Medicare Managed Care
Pensions

Internet Key Words
Care Managers
Eligibility
Informal Care
Normalization
Social Insurance

Internet Resource
U.K. Royal Commission Report
http://www.open.gov.uk/royal-commission-elderly/index.html

LONG-TERM-CARE POLICY

Public policy on long-term care is extremely important to people with disabilities and to providers of nursing home care and home and community-based services. First, long-term care is heavily dependent on public financing from Medicaid, Medicare, the Department of Veterans Affairs, and state-funded programs (Braden et al., 1998). Small policy changes may have a big impact on providers and consumers. Second, although it should not be overstated, some people with disabilities are vulnerable to exploitation and poor care. As a result, the federal and state governments bear a special responsibility to ensure adequate quality of care, especially from publicly funded providers.

Public vs. Private Financing

A key public policy issue is the extent to which long-term care ought to be financed through the private or the public sector. During the late 1980s and early 1990s, there were numerous proposals for new social insurance programs (Rivlin & Wiener, 1988; Wiener, Illston, & Hanley, 1994). However, after the collapse of the Clinton health plan in 1994, most financing proposals have promoted private long-term-care insurance rather than public programs.

Perhaps the most important barrier to the expansion of private long-term-care insurance is that premiums are expensive. Most studies estimate that only 10% to 20% of the older population can afford good-quality private long-term-care-insurance policies (Crown, Capitman, & Leutz, 1992; Rivlin & Wiener, 1988; Wiener, Illston, & Hanley, 1994).

Given the limitations of the current market, public subsidies to promote purchase are frequently proposed, and some have been enacted (Wiener & Stevenson, 1998). One approach is to allow employers to deduct their contributions for group private long-term care insurance as a business expense. A second strategy is to provide tax deductions or credits to individuals who purchase private insurance. Tax incentives for employers and individuals were part of the Health Insurance Portability and Accountability Act of 1996, and additional incentives have been actively debated in Congress

in recent years. A final strategy is to allow purchasers of state-approved private long-term-care insurance to be eligible for Medicaid while retaining more in assets than is normally allowed, an approach being tried in New York, Connecticut, Indiana, and California (Meiners, 1998).

Opponents of public subsidies for private long-term-care insurance emphasize that the initiatives tend to benefit upper- and upper-middle-income older people (Wiener, Illston, & Hanley, 1994). Even with subsidies, most people will not have private coverage (Rivlin & Wiener, 1988; Wiener, Illston, & Hanley, 1994). Moreover, these initiatives will help mostly people who have already purchased insurance, making the cost per additional person insured very large. Finally, the high overhead costs of private long-term-care insurance mean that it is an inefficient source of funding.

Balance Between Institutional and Home and Community-Based Care

A second key public policy issue concerns the balance between nursing home care and home and community-based services. Many policy makers would like to spend more on home care and less on institutional care (Wiener & Stevenson, 1998). First, most disabled older people live in the community but receive only unpaid, informal care from family and friends, often imposing a great burden on these caregivers (Liu, Manton, & Aragon, 1998). Second, most public long-term-care expenditures are for institutional rather than noninstitutional services. Only 14% of Medicaid long-term-care expenditures for older people in 1997 was for home and community-based services (Urban Institute, 2000). Third, policy makers believe that expansion of home care will result in a less expensive long-term-care system and reduce the growth in public expenditures. However, a large, rigorous (though rather old) research literature strongly suggests that expanding home care is more likely to increase rather than decrease total long-term-care costs (Weissert & Henrick, 1994; Wiener & Hanley, 1992). Expenditures rise because large increases in home care more than offset relatively small reductions in nursing home use. More recent research is more optimistic about the possibility of cost savings, but these studies have less rigorous research designs (Greene, Ondrich, & Laditka, 1998).

One of the main barriers to expanding coverage of home and community-based services, especially in the context of open-ended entitlement programs such as Medicare and Medicaid, is the fear that utilization will increase uncontrollably. This potential increase in use is called the "woodwork effect," referring to the notion that people would "come out of the woodwork" to obtain services.

Quality of Care in Nursing Homes

A third policy issue concerns poor quality of care and ineffective regulation of nursing homes, an issue that dates back to the mid-1970s (New York State Moreland Act Commission on Nursing Homes and Residential Facilities, 1975; U.S. Senate Special Committee on Aging, 1974; Wiener, 1981). The Omnibus Budget Reconciliation Act of 1987 (OBRA 87) raised quality-of-care standards for facilities that participate in Medicare and Medicaid and strengthened federal and state oversight. Following the implementation of OBRA 87, several studies found an improvement in quality of care in nursing facilities, especially related to the use of physical and chemical restraints, prevalence of dehydration and stasis ulcers, and use of catheters (Hawes et al., 1997; Mosely, 1996; Phillips et al., 1996, 1997).

Despite these improvements, there is substantial evidence of continuing poor-quality care in nursing facilities and problematic oversight. In a series of studies, the U.S. General Accounting Office (1997; 1998; 1999a–d) found serious problems. One-fourth of nursing facilities nationwide had serious deficiencies that caused actual harm to residents or placed them at risk of death or serious injury, with about 40% of these homes having repeated serious deficiencies. In addition, even when serious deficiencies were identified, enforcement policies were not effective in ensuring that deficiencies were corrected.

In recent years, federal and state regulators have stepped up their inspection and enforcement of nursing home quality standards, which has been applauded by consumer groups. However, many nursing home providers believe that the federal standards do not adequately measure quality and that nursing facility staff are demoralized as a result of the negative approach of regulators.

Role of Medicare in Long-Term Care

A fourth issue is the extent to which Medicare should finance long-term-care services. Medicare covers services provided by skilled nursing facilities, home health agencies, rehabilitation hospitals, hospice agencies, various therapy services, and durable medical equipment. Historically, however, Medicare covered only very narrow, medically oriented postacute care, mostly as an alternative to hospital care, and policy emphasized that Medicare did not cover long-term care. In 1986, skilled nursing facility expenditures and home health expenditures accounted for only 3% of Medicare spending, but rose to 15% by 1996 (Liu et al., 1998).

Medicare spending for postacute-care services began to increase rapidly in 1989, when the Health Care Financing Administration made what were thought to be slight liberalizations of the coverage rules for home health and skilled nursing facility care in response to lawsuits challenging the agency's restrictive coverage interpretations. Expenditure growth was partly the result of providing services to more people and a greater supply of providers. Home health expenditures increased largely because beneficiaries received many more visits. As a result of the coverage change, Medicare began providing home health services to medically fragile and long-term-care populations that it previously did not serve (Leon, Neuman, & Parente, 1997). In addition, there have been reports of fraud and abuse involving Medicare home health services, which may have increased expenditures (U.S. General Accounting Office, 1997).

In order to curb the rate of expenditure growth, the Balanced Budget Act (BBA) of 1997 made major changes to Medicare skilled nursing facility and home health reimbursement methods that are highly controversial and have resulted in large savings. The BBA established a case-mix-adjusted prospective payment for nursing facilities that bundles nursing, therapy, and capital payments into a single per diem amount. This change has strongly constrained the use of therapies, which was a major cause of the expenditure increases. Some observers argue that the methodology does not adequately account for the costs of nontherapy ancillaries, such as prescription drugs (Medicare Payment Advisory Commission, 1998). Several large nursing facility chains have experienced severe financial problems in the wake of BBA, and some have filed for bankruptcy protection, although it is not clear to what extent their financial problems are attributable to the Medicare payment methodology (U.S. General Accounting Office, 1999c).

The BBA has also dramatically changed reimbursement for home health agencies. Ultimately, home health agencies will be reimbursed using a case-mix-adjusted prospective payment system, but until then, agency reimbursement is set through an "interim payment system." While this methodology has many cost constraints, the most important one is a per beneficiary ceiling on payments, which makes long-stay patients less financially attractive. Thus, the reimbursement changes will likely result in a reduction in the use of Medicare home health services by beneficiaries needing long-term care. Although the impact on access is unclear, these changes have had a major impact on the home health industry, causing many agencies to close or reduce services (U.S. General Accounting Office, 1999d).

Long-term care is an important policy issue that affects billions of dollars in public expenditures and millions of people with disabilities. Over the next 30 years, long-term care will rise on the national political agenda as the baby boomers become elderly. However, the near-term aging of the parents of the baby boomers and their need for nursing home and home care may raise the policy profile of this issue sooner rather than later.

JOSHUA M. WIENER

See Also

Demography of Aging
Long-Term-Care Financing: International Perspective
Medicaid
Medicare
Pensions

LOW VISION

Vision impairment refers to altered ability to see normally due to an anatomical or physiological change in the visual system. Examples of vision impairments include decreased central vision, loss of detail, and color blindness. Low vision refers to vision loss that cannot be corrected medically, surgically, or with conventional lenses and that causes difficulty in carrying out daily tasks. Symptoms of low vision may include decreased visual acuity, cloudy vision, constricted visual fields, sensitivity to glare, abnormal color perception, and difficulty seeing in low-contrast conditions (Tomsak, 1997). Individuals with low vision are not legally blind, and although they may experience substantial visual loss, they have some useful vision.

Leading causes of low vision among the elderly include macular degeneration, diabetic retinopathy, complicated cataracts, glaucoma, optic atrophy, retinitis pigmentosa, and corneal dystrophy. Low vision is increasingly prevalent because eye disorders appear with advanced age and with conditions associated with aging. Over 75% of visually impaired individuals in the United States are 65 years or older (Faye, 1999).

Individuals with low vision cannot see the world in the same way they previously did. Although a change in vision can radically alter one's ability to care for oneself and navigate the environment, the disability is not readily apparent to others. Although the person looks and feels the same, his or her functioning has decreased.

Social isolation may ensue when individuals cannot see facial details to identify people or pursue usual pleasurable activities that require eyesight, such as playing cards, attending theater, or viewing movies. The inability to drive due to visual loss may further compound social isolation and reduce social support.

Frustration, depression, grief, fear of blindness, and concerns about driving, loss of former pleasures, loss of independence, and loss of self-esteem are commonly experienced by those with low vision. Clinically, depending on severity, these reactions may lead to diagnoses of adjustment disorder or major depression. Individuals who demonstrate apathy, extreme withdrawal, or unwillingness to engage in activities may need treatment for depression. Also, particular care needs to be paid to those individuals with low vision who already have some degree of cognitive or hearing impairment.

When speaking with a person who has low vision, the speaker should give a self-introduction and stand close to the person. Instructions should be explained verbally, rather than with hand gestures. If the environment has little contrast, the person should be oriented to the room and objects within it (Seidman, 1995). If written instructions are provided, print should be large and in high contrast. The size of the print depends on the individual; the letters need to be large enough to meet the person's needs. When low vision has been affirmed, a hearing assessment should be performed to rule out (or treat) a dual sensory loss.

Vision rehabilitation is the logical, but frequently overlooked, referral that should follow a determination that low vision exists. A vision rehabilitation program addresses visual impairments by assessing visual loss, prescribing and educating the person in the use of low-vision aids, and teaching the person about vision-enhancing and vision-substituting skills and potential changes in the environment. Although adaptive devices cannot make the vision normal, they may improve the ability to conduct activities of daily living by making it easier to see objects.

Low-vision devices include convex lens aids, such as spectacles and magnifiers (from 1× to 60×); telescopic systems; adaptive devices to assist in daily activities requiring reading or close work, such as large-print watches, clocks, timers, or tele-phones and magnifiers for insulin syringes; tints and coatings (light or medium gray lenses to reduce light intensity; amber or yellow lenses to improve contrast); and electronic reading systems that use a closed-circuit television reading machine (Faye, 1999).

Patient support groups are helpful in the adjustment to low vision. Suggestions may be offered by professional caregivers, but recommendations originating from others with low vision are more readily accepted. Creative strategies reported by low-vision individuals include greater reliance on hearing to recognize others, reliance on memory for telephone numbers, ordering foods in public that do not require cutting, and recording recipes and other instructions on audiocassette tapes (Moore, 1999). Unfortunately, dependence on others to provide transportation may be a barrier for some who wish to attend support groups.

Principles of environmental modification applicable to low vision include simplifying the layout to promote order and predictability, increasing contrast sensitivity as needed, and attending to lighting levels. Passageways should be free of both large and small obstacles, and everyday items should be stored in designated places to aid in retrieval. Contrasting colors for walls, floor, and furniture promote visual orientation. Keys can be marked with large letters to indicate use; house or apartment entrances should be adequately lit. Edges of steps should be clearly marked with high-contrast strips (Seidman, 1995). Directional or instructional signs should use large, high-contrast letters and receive adequate lighting. Many people with low vision see better in sunlight. Good lighting improves the contrast of written materials. A sheet of yellow acetate may be used to increase contrast of written materials, if necessary. For reading, a 60-watt bulb in a lamp with a shade that focuses the light on the page is recommended. The page may need to be held close to the eyes, but this is not harmful (Brown, 1997).

In summary, individuals with low vision must learn to live with an irreversible process that poses challenges in conducting daily activities. Once the determination is made that no additional treatment will improve vision, the ophthalmologist or other health care provider should ensure that a referral is made for visual rehabilitation. Caregivers and family members can modify the environment to increase contrast and support the person in the adjustment to visual limitations.

ELAINE SOUDER

See Also
Cataracts
Eye Care Providers
Glaucoma
Vision Changes and Care
Vision Safety

Internet Key Words
Color Contrast
Macular Degeneration

Optical Low Vision Devices
Vision Rehabilitation

Internet Resources
American Academy of Ophthalmology
http://www.eyenet.org

American Foundation for the Blind
http://www.afb.org

American Optometric Association
http://www.aoanet.org

The Lighthouse
http://www.lighthouse.org

Macular Degeneration Foundation Education, Inc.
http://www.eyesight.org

National Eye Institute
http://www.nei.nih.gov

M

MANAGED CARE

See

Medicare Managed Care

MEALS ON WHEELS

Any discussion about providing comprehensive care to today's and tomorrow's elderly would be incomplete without a serious look at the role of elderly nutrition programs (ENPs) in enhancing independence and quality of life. There is sufficient evidence to demonstrate that ENPs improve the nutritional health of the individuals who participate in them. Improvement of nutritional status, in turn, can postpone institutionalization for individuals and thus reduce some of the national costs associated with Medicare and other health care programs.

The term often used to refer to ENPs is a generic one—Meals On Wheels. However, there is more to ENPs than the name Meals On Wheels implies. Besides home-delivered meal programs (Meals On Wheels), ENPs provide congregate meal programs—those carried out in facilities such as senior centers, where seniors assemble and partake of meals together. Although these two types of nutrition programs have similar purposes, they differ in terms of the populations served. Eligibility to receive services is the same: individuals age 60 or older may participate. Because demand far outstrips available services, by law, programs are targeted to those in greatest economic and social need.

The principal law governing the operation and practices of ENPs and providing the largest source of federal funding is the Older Americans Act (OAA) (Ponza, Ohls, & Millen, 1996). As mandated by OAA, the primary objective of both home-delivered and congregate meal programs is to furnish hot (standardized as 140 degrees or higher), nutritious meals to needy seniors at least five days per week. Each meal must meet the minimum standard of furnishing at least one-third of the recommended dietary allowance (RDA) of key nutrients. Most meals actually exceed this RDA minimum, approximating 40% to 50% of the daily requirement, and the meals are typically "nutrient dense," that is, the ratio of nutrients to calories is high. As a result, the daily intake of key nutrients is greater for program participants than it is for similar individuals who do not participate in the program. Increasingly, programs are incorporating special diets, to the extent that budgets allow. Most programs offer low-sodium meals, some offer kosher meals (depending on the population served), and some offer other ethnic meals. Most try to provide special meals at holidays. In addition, some programs provide emergency and shelf-stable meals, so that participants will have food in the event of a blizzard or other emergency that would prevent regular meal delivery. OAA ENPs are prohibited from charging fees, but they may request voluntary contributions. No one can be denied services for inability or unwillingness to pay. Some private programs do charge clients fees. The cost of providing a senior citizen Meals On Wheels for one year is roughly equivalent to the cost of one hospital day for a Medicare patient.

Nutrition screening and education are also provided in over half of all ENPs, and more than a third include nutrition assessment and counseling. Depending on the program's size and capacity, nutrition screening and counseling may be available to both the elder and the caregiver. All programs funded through the OAA use a simple 10-question checklist, "Determine Your Nutritional Health," that was developed by the Nutrition Screening Initiative (NSI). The NSI also developed two other screening and assessment tools (the Level I and Level II Screens). Programs are not required to use these additional tools, and few programs have the ability to use the Level II Screen, which includes basic lab work (serum albumin and serum choles- terol). Non–federally funded programs may or may not use such screens, depending on their resources and program structures. Some states may require more comprehensive assessments beyond the NSI checklist. Use of screening tools is expected to expand in the future as the understanding of their importance increases.

ENPs provide participants with more than just a meal. Although it is accomplished in different ways and to a different degree in congregate sites and in participants' homes, socialization—or at least the reduction of social isolation—is a critical benefit of all ENPs. Those participating in senior meal programs of both types have more social con- tact than similarly situated nonparticipants. This is true despite the fact that, compared with the general elderly population, individuals who participate in ENPs are more than twice as likely to reside alone.

Other than setting, another difference between program types is the demographic characteristics of the program participants. In both cases, program participants are older, predominantly female, more likely to be minority, and poorer than the overall eligible population. Home-delivered meal recipi- ents, on average, are older, poorer, and frailer than their counterparts in congregate programs. Nearly 59% of home-delivery participants have three or more chronic medical conditions, compared with 41% of the congregate population. Additionally, the majority of home-bound participants are unable to perform one or more activities of daily living or instrumental activities of daily living (Ponza, Ohls, & Millen, 1996).

The health status of these populations can be directly improved through nutrition or nutrition- related interventions. Weight is one example. "Nu- tritional status has been shown to affect the age- related rate of functional decline for many organs and to be a determinant of changes in body compo- sition associated with aging, such as loss of bone and lean body mass. Furthermore, diet and nutrition have been related to the etiology of many chronic diseases affecting elderly people, such as osteopo- rosis, atherosclerosis, diabetes, hypertension, and certain forms of cancer" (Ponza, Ohls, & Millen,

1996, p. 21). Body mass index (BMI) can be tied to nutritional status and is affected by nutritional interventions. BMI outside the normal range places individuals at risk for a number of chronic condi- tions. Almost two-thirds of all ENP participants are at increased risk of health and nutritional problems because they are either over- or underweight.

One of the principal assets of ENPs is that they have evolved throughout the years, responding to changing needs and incorporating scientific and technological advances. As ENPs have expanded their services and their reach into communities, the benefits have become more profound. The degree to which ENPs can contribute to improving the health of America's seniors relies on public sup- port—in the form of federal, state, and local fund- ing; financial contributions from individuals, the corporate sector, and foundations; and the invest- ment of time and personal resources of the volun- teers who prepare, serve, and deliver meals.

According to the Administration on Aging, approximately 4,000 ENPs provide services in the United States today. The Eldercare Locator, acces- sible toll-free from anywhere in the United States, can assist seniors and their caregivers in locating a senior meal program in the community (1-800- 677-1116).

ENID A. BORDEN
MARGARET B. INGRAHAM

See Also

Nutritional Assessment
Senior Centers

Internet Key Words

Administration on Aging
Elderly Nutrition Program (ENP)
Nutritional Intervention
Nutrition Assessment and Screening
Older Americans Act (OAA)
Recommended Dietary Allowances (RDA)

Internet Resources

Administration on Aging
http://www.aoa.org

Meals on Wheels Association of America
http://www.mealsonwheelsassn.org

MEASUREMENT

Contemporary health care professionals practice in an era that emphasizes quality, and measurable outcomes must include the use of assessment tools, scales, and indices. Health care providers are expected to prove their effectiveness in clinical, administrative, and financial areas. This has stimulated the development of standardized tools for assessment and evaluation purposes. Interest in and utilization of these tools are increasingly dependent on the strength of empirical (i.e., observable) evidence that has objective, scientific credibility to assess, identify, or determine the extent of a phenomenon, event, or concept. The process of using empirically sound scales, tools, and indices to examine a phenomenon, event, or concept is called *measurement.*

Measurement is commonly understood as the process of assigning numerical values to concepts, phenomena, or events. Numerical values are objective (as opposed to subjective) in nature and thereby enhance the reliability and validity of the information: every event or phenomenon with an assigned number is similar to others with the same assigned number. For the clinician, the availability of measurement or assessment tools is invaluable, because the instruments have been developed and evaluated to determine their level of creditability (reliability) and soundness (validity).

Determining the reliability and validity of a particular instrument is extremely important. *Reliability* is defined as the extent to which the instrument yields the same results on repeated measures. It is concerned with consistency, accuracy, and precision; if the same person is tested several times, the same score will result. *Validity* refers to whether a measurement instrument actually measures what it is supposed to measure. Reliability and validity are complex concepts to measure in and of themselves; sophisticated statistical methods are routinely used to establish the reliability and validity of assessment tools and other forms of measurement.

There are several general guidelines to follow when considering which tools to use in clinical practice. Tools that have been used repeatedly and with different groups are usually reliable. In many instances, the creator of a scale or index provides a "score" or statistical coefficient to summarize the level of confidence of the measure. Many different statistical techniques are used for these purposes. As a general rule, the higher the score, the more confidence one should have in the measure (e.g., .83 reflects strong confidence, .53 moderate confidence; typically, a score under .50 is unacceptable).

Types of Measurement

Scale: One of the most familiar forms of measurement in health care is the scale, which provides an easy method for measuring the magnitude of a phenomenon or event. Examples include a patient's pain assessment using a scale of 0 to 10 or a customer satisfaction survey that uses a Likert scale: 1 = very satisfied, 2 = satisfied, 3 = not satisfied, 4 = very unsatisfied.

Checklist: Another common measurement tool is the checklist approach, which can be used when collecting information on types of procedures performed. Numbers may be assigned to represent steps in the procedure, such as: 1 = inserted IV, 2 = drew blood, 3 = removed Foley catheter, and so forth. These numerical values make it possible to aggregate or group the responses in a meaningful way. It is then possible to *assess* similarities and differences between responses or groups of responses. These assessments are conducted through scoring data and the performance of statistical analyses.

There is an important difference, however, between the numerical values associated with, for example, a pain scale, and the numerical values assigned to steps in a procedure. With regard to the checklist of procedures, the assignment of numbers is arbitrary. There is no intrinsic value to the 1, 2, or 3. This measurement uses *nominal* data, data that are purely for categorization or classification purposes. In contrast, the numbers used in the pain scale have intrinsic value; a score of 1 means less pain than 2, and 0 means no pain. A pain score of 8 is understood as twice as painful as a score of 4. The pain scale is an example of ratio measurement—rankings on a scale with equal intervals between numbers and an absolute zero. When investigating an instrument for possible use, clinicians should be aware of whether data are scale or nominal, because such differences may have implications for the tool's ease of use and interpretation for clinical staff at all levels.

Ordinal and Interval: Ordinal measures show the relative ranking of people, objects, or events. The classic example of ordinal data is the class ranking of students. One student is ranked higher than another, but the actual difference between the students may differ more widely than the ranking. Interval data also provide a means of ranking, but the zero point is not arbitrary. The most familiar use of interval measurement is the Fahrenheit thermometer.

Interpretation of Numerical Values

The primary benefit of measuring people, phenomena, and events using numerical values is the ability to quantify the results using statistical methods. Hence, it is important to distinguish between the different types of measures and data, because the statistical options used are related to the type of measurement used.

The most basic statistical methods used to interpret numerical data are called measures of central tendency, and they describe the "middle" or the "average" of the group of responses. The *mean* (score) is the arithmetic average of all the numerical responses. The *median* is the middle score, or the number at which 50% of the responses are above it and 50% are below it. The *mode* is the most frequent score or response. It is possible to have more than one mode.

Selecting an Assessment Tool in Geriatrics

The availability of standardized assessment tools permits a practitioner to assess, evaluate, and measure where a particular patient fits on any number of clinical, administrative, and educational variables. Once a patient has been assessed, the practitioner can proceed with appropriate interventions. There are many good measures in geriatrics. However, before selecting a particular measure, it is necessary to evaluate its appropriateness for one's specific needs.

First, the practitioner must decide what exactly he or she wants to measure and determine who or what is being measured. For example, is it the needs of a patient with a particular diagnosis, the evaluation of a program, or staff satisfaction? If a type of person is being assessed, is it a specific diagnosis, age group, or patients in a particular setting? In addition, the clinician must decide how much detail he or she wants or needs from a measurement tool. It is important to balance the pros and cons of using more detailed instruments, which are more time-consuming, or using shorter measures that may provide less detail but are more efficient.

When selecting an appropriate measure, clinicians should get as much background information as possible about the development and application of the instruments. This information is often available in the published literature. When reviewing the written material, it is important to examine the size of the group on which the instrument was tested and the type of population on which the instrument was used. Ideally, the clinician should select a measure that has been used extensively with the same population he or she plans to assess. In short, if the clinician plans to assess nursing home residents, it would be best to use a measure developed for this population rather than selecting one for home care or acute care.

Evaluating the validity and reliability of a measurement tool can be difficult. It is useful to know how many tests were performed on the instrument to determine its reliability and validity. It is also helpful to know how many different users have tested the instrument and how large the groups of respondents were.

Finally, the user should consider the method for scoring and interpreting the results when selecting a tool. Difficult scoring methods and complicated statistical interpretations can be too time-consuming for a busy clinician. In some cases, copy-written instruments may require a fee for scoring and interpreting the results.

The utility and benefit of standardized measurement tools in health care are obvious to any clinician, administrator, educator, or researcher. In a health care environment that emphasizes clinical standards and quality outcomes, in addition to regulatory compliance and fiscal responsibility, reliable and valid measures, indices, scales, and surveys are critical to providing quality health care. Whether for clinical decision making, quality performance and improvement activities, or research purposes, there is little doubt that all health care providers will be involved in measurement activities in their daily practices.

PERI ROSENFELD

See Also
Best Practices in Geriatric Care

Internet Key Words
Measures of Central Tendency
Reliability
Statistical Methods
Validity

MEASURING PHYSICAL FUNCTION

Physicians, nurse practitioners, and rehabilitation therapists administer tests to assess elderly clients' physical abilities and to determine their fall risk and activities of daily living (ADL) and mobility limitations. This information can help determine the need for intervention, which may include referral to a physical or occupational therapist, use of a home health aide, or referral to social services for community-based resources. Physical and occupational therapists also perform observational analyses of the quality of movement and its impact on physical function.

There are three ways to assess physical function: self-report, interviewer-administered questionnaire, and direct observation. Self-report measures have been used since the 1940s, but their validity, reliability, and sensitivity may be limited due to unclear definitions of the activity being assessed, assumptions about the client's maximal capacity, and the impact of poor cognitive function on the test results (Guralnik, Branch, Cummings, & Curb, 1989).

In contrast, observation of physical performance has increased face validity as the client performs the task being assessed. Many physical performance tests have documented reliability and sensitivity characteristics. A number of physical performance tests demonstrate adequate psychometric properties. Most of the tests were developed for community-based clients. The Berg Balance Scale was developed specifically for frail elders and those in rehabilitation programs. Research indicates that some test results can be used to predict ADL decline or fall risk (Gill, Williams, & Tinetti, 1995). The disadvantages of physical performance tests include the time needed to test performance and, in some instances, special equipment or training required for test completion. If the test is performed in an office setting, the performance may not represent the client's abilities in the home setting.

A comparison of these measurement options concluded that a composite measure of all three types may provide the most accurate information regarding an individual's physical functioning (Reuben, Valle, Hays, & Siu, 1995).

Performance-Oriented Mobility Assessment

Dr. Mary Tinetti was one of the pioneers who recognized that the standard neuromuscular evaluation does not adequately identify mobility problems (Tinetti & Ginter, 1988). She also noted that the disease-oriented approach may fail to recognize the multifactorial nature of impaired mobility and falls, problems common to the geriatric population. The Performance-Oriented Assessment of Mobility (POMA), also referred to as the Tinetti Balance and Gait Scale (Tinetti, 1986), requires little space and equipment and no specialized training. It evaluates positional changes and gait characteristics used during normal daily activities.

The balance section of the Tinetti Scale uses an ordinal scale to rate the client's ability to hold positions or perform positional changes. It has eight areas—sitting balance, arising from a chair, immediate and prolonged standing balance, withstanding a nudge on the sternum, balance with eyes closed, turning balance, and sitting down—for a total of 16 possible points. The gait section of the scale observes gait in a serial manner and assesses eight components of gait—initiation, step height and length, step continuity, symmetry, path deviation, trunk sway, walking stance, and turning while walking—for a maximum score of 12 points. An earlier version designed for use with an ambulatory institutionalized population led to a second version for community-dwelling individuals. Referred to as the POMA II, it includes an additional five balance tasks: neck turning while standing, back extension, reaching up, bending over to pick up an object, and standing on one leg.

As an objective measure of fall risk, clinicians use a combined score on the balance and gait scales, such that a score of less than 19 indicates a high risk of falls, 19 to 23 indicates an increased risk of falls, and greater than 23 is indicative of a low fall risk. Tinetti encouraged test administrators to use the test results and observations to identify possible restorative, preventive, or adaptive measures and provided a table listing possible fall etiologies and preventive measures (Tinetti, 1986).

EPESE Short Physical Performance Battery

As part of the Established Populations for the Epidemiological Study of the Elderly (EPESE), a test battery was designed to assess lower extremity performance in community-dwelling individuals (Guralnik et al., 1994). The research study was performed by trained individuals, but the test can be administered without specific training if the instructions are followed closely. Longitudinal data analysis demonstrates that the summary score is a strong predictor of death and nursing home admission and, in those with no disability at baseline, is a strong predictor of ADL disability and mobility disability (Guralnik et al., 1995).

Three primary areas are tested: standing balance, measured walk, and chair stands. Each area has a maximum of 4 points, for a total test score of 2 to 12 points. The standing balance score is determined by assessing side-by-side stance, semi-tandem stance, and tandem stance. Each position is held for 10 seconds. The score is determined based on the time held in each position and the ability to achieve and maintain the three positions. Measured walk time is measured by having the individual walk at a normal pace for eight feet. The scores are defined by time: 1 = more than 5.7 seconds, 2 = 4.1 to 5.6 seconds, 3 = 3.2 to 4.0 seconds, 4 = less than 3.1 seconds. Time is also used to score the individual's ability to perform five repeated chair stands from an armless, 18-inch-high chair: 1 = more than 16.7 seconds, 2 = 13.7 to 16.6 seconds, 3 = 11.2 to 13.6 seconds, 4 = less than 11.1 seconds. The timed component allows the test administrator to record declines over time and structure intervention accordingly.

Physical Performance Test

The Physical Performance Test (PPT), developed by Reuben and Siu (1990), is used to identify ADL and instrumental ADL deficits in community-dwelling older adults. It requires minimal equipment and no specific training beyond the instructions accompanying the test and can be administered in 10 minutes, depending on the client's level of ability. There are two versions, one with seven items (write a sentence, simulated eating, lift a book to a shelf, put on and remove a jacket, pick up a penny from the floor, turn 360 degrees, walk 50 feet) and one with nine items (addition of climbing one flight of stairs and climbing stairs up to four flights).

Scoring is based on time for all scores with two exceptions: turning 360 degrees (scored for continuity and stability) and climbing multiple flights of stairs (scored based on the number of flights). As with the EPESE battery, the timed component can be used to demonstrate decline over time and thus guide interventions, or it can be used to show response to an intervention, such as rehabilitation services. Research has not determined the ability of the tool to predict decline over time, but use of the PPT in a home health setting demonstrated that a score of about 17 indicated that patients were no longer home-bound (Crews, Brown, & Norton, 1997).

Berg Balance Scale

The Berg Balance Scale monitors patient status over time and can evaluate the effectiveness of rehabilitation interventions (Berg, Wood-Dauphinee, Williams, & Gayton, 1989). Other researchers have used it to predict the probability of falls for community-dwelling older adults (Shumway-Cook, Baldwin, Polissar, & Gruber, 1997). Scoring is based on an independence-to-dependence continuum. There are 14 test positions (examples include sitting without back support but feet supported, transfers, standing unassisted with variations, turning neck, turning 360 degrees, reaching forward, picking up an object) with a variety of timed, distance, or supervision requirements for each position. Little equipment is needed, and no formalized training is necessary.

The Berg Balance Scale demonstrates excellent sensitivity to change and is often used to measure the response to rehabilitation interventions. However, it can also be used to predict the probability of falls and therefore can help guide referrals for intervention.

There are also tests available for specific patient populations, such as those with arthritis,

stroke, and Parkinson's disease. The choice of which test to use depends on the client's abilities; for example, the PPT is a higher-level test of physical performance than the Berg Balance Scale or the POMA. Observation of physical function should be accompanied by queries regarding the client's performance of activities in the home environment and in the community.

Physical performance tests should be part of routine medical office visits or part of the examination process of physical or occupational therapists. The tests could also be administered by a nurse practitioner or physician's assistant during routine visits. The tests provide an objective measure of the patient's performance to guide the utilization of further health care resources.

ANNE COFFMAN

See Also
Activities of Daily Living
Falls Prevention
Gait Assessment Instruments
Measurement
Multidimensional Functional Assessment: Instruments
Occupational Therapy Assessment and Evaluation
Physical Therapy Services

Internet Key Words
ADL Deficits
Fall Risk
Functional Assessment
Physical Performance

Internet Resources
Aging and Functional Status
http://www.fhcrc.org/science/phs/cvdeab/chpt19.html

Functional Status
http://gwis.circ.gwu.edu/~cicd/toolkit/function.htm

Women's Health and Aging Study
http://www.nih.gov/nia/edb/whasbook/tablcont.htm

MEDICAID

Medicaid was enacted in 1965, under Title XIX of the Social Security Act, to assist states in paying for the health care of the very poor. By setting minimum standards, Medicaid was designed to give states flexibility in their programs, but also to ensure that some specific groups of people would be assisted and that some core services would be provided across the country. Eligibility is based on both a categorical test and a financial needs test. One must usually be aged, blind, disabled, or a member of a single-parent family with dependent children. People who meet the categorical requirements, however, must also have income and assets below specified levels. Eligibility has historical ties to the Aid to Families with Dependent Children (AFDC) and the Supplemental Security Insurance (SSI) cash assistance programs. People whose income is too high but who face relatively high medical expenses may become eligible for Medicaid by "spending down." Recipients of adoption or foster care assistance and pregnant women and children whose family income falls below 133% of the federal poverty level are also eligible.

States have a tremendous amount of latitude in determining the administrative structure, what services they will pay for, how much and by what method they will pay providers, how they calculate income and assets, and in which optional categories benefits will be covered. Consequently, each of the 50 states, the District of Columbia, and the five American territories have different Medicaid programs. Someone who is eligible in one state may not be eligible in another state. States receive federal matching funds for every dollar spent on Medicaid services. The precise federal match, or participation rate, is inversely related to the state's fiscal capacity (based primarily on per capita income), ranging from no more than 83% for the poorest states to no less than 50% for the richest states.

In 1997, 40.6 million people—more than one in seven—were enrolled in Medicaid. Children accounted for the largest proportion—52%—of enrollees. Adults in families accounted for the next largest proportion (21%), followed by nonelderly blind and disabled (17%) and the elderly (10%).

Total Medicaid spending in 1997 reached $161.2 billion. The federal government financed approximately 56%, and states financed the rest. Some 17% of Medicaid payments were for children, and 11% for adults in families. Payments to the nonelderly disabled accounted for 42% of Medicaid spending, and those to the elderly some 30%.

Children and nondisabled adults constitute the largest number of beneficiaries (73%) but consume the smallest share of the expenditures. For example, Medicaid expenditures averaged about $1,157 per child in 1997. Costs for low-income elderly and disabled are significantly higher, owing to the fact that both the elderly and the nonelderly disabled on Medicaid often need substantial health care services. Per capita costs for older beneficiaries in 1997 averaged $10,803.

Mandatory health services provided under Medicaid include inpatient and outpatient hospital care, physician services, laboratory and x-ray services, primary and preventive care, nursing facility and home health care, and other medically necessary services. Medicaid also covers health and health-related services that are not covered by Medicare. In 1995, Medicaid provided supplemental health coverage for nearly 6 million Medicare beneficiaries. Although this accounted for only 17% of all Medicare enrollees, it constituted 35% of the $53 billion in total Medicaid expenditures made on behalf of older people (Waid, 1998). For certain low-income Medicare beneficiaries, Medicaid pays the Medicare Part B premium (which covers physician services), as well as the Medicare deductibles and co-payments.

Medicaid is a major source of coverage for long-term care, covering about half of all nursing home costs (Health Insurance Association of America, 1999). About one-third of all nursing home costs are paid out-of-pocket by individuals and their families; Medicare pays some 12%, and private long-term-care insurance pays even less. For many of these beneficiaries, Medicaid eligibility was obtained after the costs of health care and long-term care exhausted their resources.

Medicaid was originally tied to eligibility for other public assistance programs such as AFDC and SSI. However, beginning in 1984, a series of expansions in Medicaid coverage reflected a significant shift in the philosophical underpinnings of the program. In 1984, changes were made to provide coverage to pregnant women before their receipt of AFDC. This was a critical change to ensure the coverage of prenatal care. In 1989, there were dramatic expansions in Medicaid eligibility for coverage of pregnant women and children that dropped the link to other public assistance programs entirely.

In 1988, changes were made that enabled Medicaid to assist low-income Medicare beneficiaries with limited resources to pay their Medicare deductibles, co-payments, and premiums. These persons do not, however, automatically receive Medicaid coverage for services not covered by Medicare. For elderly individuals with income below 100% of the federal poverty level, Medicaid pays Medicare (Part A and B) premiums, deductibles, and co-insurance; for those with income between 100% and 120% of the poverty level, Medicaid pays the Medicare supplementary medical insurance (Part B) premium. Individuals who were receiving Medicare due to a disability but lost entitlement to Medicare benefits because they returned to work may purchase Part A of Medicare. Individuals whose income is below 200% of the federal poverty level and who are not eligible for Medicaid benefits may qualify to have Medicaid pay their monthly Medicare Part A premiums.

In 1990, federal law expanded the coverage for poor children and pregnant women. Pregnant women and infants whose income is below 133% of poverty (at a minimum) are covered, and states have the option of including pregnant women and infants with family income between 133% and 185% of poverty. Children from age 1 to 6 whose family income is below 133% of poverty are also eligible for coverage. Eligibility for children age 7 to 19 whose family income is less than 100% of poverty is being phased in through 2001. Between 1988 and 1991, 1.8 million more children became eligible for Medicaid because of these expansions (Evens & Friedland, 1994). By 1996, there were more than 16.7 million children under age 21 eligible for Medicaid, accounting for over 46% of the entire Medicaid population.

From 1985 to 1988, Medicaid spending grew at an average annual rate of 10%. Between 1989 and 1992, Medicaid expenditures accelerated dramatically, averaging 18% (The Kaiser Commission on the Future of Medicaid, 1993). In part, this reflects the expansions to poor children and poor elderly. Between 1992 and 1993, expenditures increased 11%, and from 1993 to 1996, 6.9% per year. More recently, however, growth rates have decreased. Although the low growth rates of 3.3% in 1996 and 3.9% in 1997 in overall Medicaid spending might not continue, a return to double-digit rates of increase is not currently expected.

Although the majority of Medicaid services have traditionally been delivered on a fee-for-service basis, managed care is playing an increasing role in the delivery of Medicaid services. Managed care is designed to decrease costs by eliminating inappropriate and unnecessary services and by relying more heavily on primary care and coordination of care. Between 1991 and 1994, the number of Medicaid beneficiaries enrolled in managed care plans nearly tripled from 2.7 million to 7.8 million (The Kaiser Commission on Medicaid and the Uninsured, 1999). Some 16.6 million were enrolled in 1998—more than a sixfold increase since 1991. With the exception of Arkansas and Wyoming, all states are pursuing some managed care initiative. Over half (54%) of all Medicaid beneficiaries in 1997, predominantly poor children and their parents, received services through managed care plans.

States are currently looking at different ways to restructure their Medicaid programs as a way to control state spending for Medicaid and other health programs while expanding coverage to the uninsured. With federal approval, states can propose demonstration programs that would be exempt from federal requirements; they would still be able to receive federal matching funds as long as the program promoted Medicaid objectives and the effect on the federal government was budget neutral.

In addition, many states have proposed extending coverage to low-income individuals who are ineligible for Medicaid because they fail to meet income or asset requirements for eligibility in programs such as AFDC and SSI (McCloskey &

Holahan, 1995). In fact, many states provide Medicaid coverage for other groups, including infants up to age one and pregnant women whose family income is below 185% of the federal poverty level; children under age 21 who meet their state's 1996 AFDC income and resource requirements; and disabled, institutionalized, or medically needy persons below specified income and resource limits.

The Program of All-Inclusive Care for the Elderly (PACE) is a unique capitated managed care benefit for the frail elderly that features a comprehensive medical and social services delivery system financed by integrated Medicare and Medicaid dollars. The Balanced Budget Act (BBA) of 1997 established PACE as a permanent entity within the Medicare program and as a state option under the Medicaid program. Prior to the BBA, PACE existed as a federal demonstration waiver. States that operated PACE under federal authority were allowed to extend their demonstration waivers. As of July 1999, 13 states had approved demonstration sites, and 8 states had elected to provide PACE services to Medicaid beneficiaries as a Medicaid option. PACE allows people age 55 and older who meet their state's eligibility criteria for nursing home care to receive all Medicare and Medicaid services in an adult day health center, at home, or in an inpatient facility. A capitated financing system allows providers to deliver all the services needed by PACE enrollees, rather than only those that are reimbursable under the Medicare and Medicaid fee-for-service systems.

The BBA authorized the creation of the State Child Health Insurance Program, also known as Title XXI, which enables states to initiate and expand health care to uninsured, low-income children. States may create a new health insurance coverage program that defines the amount, duration, and scope of benefits, expand eligibility for children under the state's current Medicaid program, or both. Each state must submit a plan and obtain approval from the Health Care Financing Administration. As of May 1999, 14 states (or territories) had created new programs, 27 had expanded their current Medicaid programs, and 12 had combination plans.

ROBERT B. FRIEDLAND

See Also

Long-Term-Care Financing: International Perspective
Long-Term-Care Policy
Medicare Managed Care
Program of All-Inclusive Care for the Elderly (PACE)
Pensions

Internet Key Words

Aid to Families with Dependent Children (AFDC)
Social Security Act
Title XIX

Internet Resources

Health Care Financing Administration PACE homepage
http://www.hcfa.gov/medicare/pace/pacehmpg.htm

HCFA 2082 Report (abstract)
http://www.hcfa.gov/medicaid/m2082.htm

MEDICARE

Medicare—Title XVIII of the Social Security Act, enacted in 1965—is a federal program of health insurance for persons age 65 or older who are eligible for Social Security benefits. The program was extended, in 1972, to people under age 65 who were entitled to federal disability benefits for at least two years and to certain individuals with end-stage renal disease. The program was enacted in response to a growing awareness of the need to help older persons obtain and pay for necessary medical care. In the early 1960s, only about half of older Americans had any health insurance (compared with 75% of those under 65). Those seeking to purchase private coverage were often denied on the basis of age or preexisting conditions or found private insurance unaffordable (Moon, 1993).

Medicare is a national program (administered locally by private insurance companies) that pays for medical care provided by private health care providers. Eligibility is based on prior employment in a job covered by Social Security and, unlike Medicaid, is not subject to any tests of financial need. Today, the Medicare program covers approximately 39 million Americans. Most beneficiaries are elderly; about 10% are under the age of 65 (Health Care Financing Administration, 1998).

Medicare expenditures for 1997 totaled $207 billion. Of this amount, about 43% went toward inpatient hospital care; another 20% for physician services; 8% for outpatient hospital services; and 17% for skilled nursing care, postacute home health care, medical equipment, and other professional services; the remaining 12% paid managed care plans to provide Medicare-covered services. In addition, Medicare has experienced rapid growth in payments for home health services and hospital outpatient departments; these payments increased from 3% of Medicare expenditures in 1980 to 17% in 1997. Administrative costs of the program were about 2% (Health Care Financing Administration, 1998).

Medicare is the fastest growing program in the federal budget, growing at the rate of nearly 16% each year during its first 25 years (Moon, 1993). Medicare payments account for 52% of health care spending by the nation's elderly, private sources pay 9%, out-of-pocket payments account for 20%, Medicaid pays 13%, and other public sources pay 5%. Given the nature of medical expenses, it should not be surprising that Medicare spending is concentrated on a relatively small percentage of beneficiaries. Five percent of beneficiaries account for 45% of the expenditures; 10% account for 63% of expenditures; and 23% of beneficiaries account for 85% of expenditures (Health Care Financing Administration, 1998).

Traditional Medicare is really two programs: hospital insurance (known as Part A), and supplementary medical insurance (known as Part B). Medicare Part A is financed by current workers and their employers through a payroll tax (1.45%). This participation is compulsory, but the benefit is automatic Part A coverage at age 65 (or younger for the disabled and those with end-stage renal disease). Part B is not financed the same way and is a voluntary program. One-quarter of the premium is paid for by the beneficiary, and the rest is financed through general tax revenues. In 2000, the monthly premium for beneficiaries of Part B was $45.50. Most older persons (over 96%) are covered by both Parts A and B.

In 1997, Congress passed legislation that added the Medicare Plus Choice (M+C) program (known as Medicare Part C). M+C offers beneficiaries an array of new managed care and other health plan choices that are more similar to those available in the private sector. The Medicare program pays M+C plans a fixed monthly amount for each enrollee to cover the same basic benefits package as offered under the traditional program. Enrollees in M+C plans pay the monthly Medicare Part B premium and additional out-of-pocket costs, depending on which plan they choose.

Part A of Medicare pays for all "reasonable" inpatient hospital care for the first 60 days less a deductible ($776 in 2000) for each benefit period (period associated with one acute illness). For days 61 to 90 in a hospital, a beneficiary must pay a co-payment of $194 per day. For stays beyond 90 days in a benefit period, an insured person may elect to draw on a 60-day lifetime reserve, which requires a co-payment of $388 per day.

Following a hospitalization of at least three days, Part A also pays for up to 100 days of skilled nursing care or skilled rehabilitation services if such care associated with recuperation is warranted. The first 20 days in the skilled nursing facility are free from cost sharing, but for days 21 to 100, a daily co-payment ($97 in 2000) is required. In addition, medically necessary home health care is covered, as is hospice care for terminally ill Medicare beneficiaries. There are no co-payments or deductibles for qualified home health care or hospice care.

Part B of Medicare pays 80% of physicians' "reasonable charges" for most covered services in excess of an annual $100 deductible (50% for most outpatient mental health services). Covered services include medically necessary physician services, laboratory and other diagnostic tests, x-ray and other radiation therapy, outpatient services at a hospital or comprehensive outpatient rehabilitation facility, rural health clinic services, home dialysis supplies and equipment, prostheses (other than dental), physical and speech therapy, and ambulance services. Biennial mammography screening coverage (one of the few preventive services offered by Medicare) was added in 1990, and in 1993, vaccines for influenza were covered. In 1998, glucose monitoring for diabetic beneficiaries, diabetes education, and bone mass measurement were added to Medicare-covered services. For those with only Part B coverage, medically necessary home health visits are also covered.

Under Part C of Medicare, beneficiaries, have the option of enrolling in health maintenance organizations (HMOs), preferred provider organizations (PPOs), provider-sponsored organizations (PSOs), private fee-for-service plans, or medical savings account plans (MSAs). In 1998, approximately 18% of Medicare beneficiaries were enrolled in managed care plans (U.S. General Accounting Office, 1999). These plans must provide all the services that Medicare covers. Enrollment is growing rapidly; from 1993 to 1997, growth in Medicare risk enrollment was over 27%, while enrollment growth among all HMO options was slightly over 13% (Health Care Financing Administration, 1998).

Although Medicare provides significant coverage to older and disabled beneficiaries, it does not pay for all health care costs. On average, about one-half of the elders' total health care expenses is estimated to be paid by Medicare (Health Care Financing Administration, 1998). Estimates from 1999 suggest that, on average, people age 65 and older spent $2,149 out-of-pocket on health care (excluding the out-of-pocket costs of nursing home care and home care), or about 19% of average per capita income (American Association of Retired Persons, 1999).

The largest source of out-of-pocket expenses is for services not covered—most notably, outpatient prescription drugs. But the co-payments and deductibles, particularly for multiple hospital admissions, can be considerable. Other relatively common services not covered by Medicare include routine physical examinations, nonsurgical dental services, hearing aids, and eyeglasses. Other less common but very expensive services that are not

covered include custodial care in a nursing home, stays in nursing facilities beyond 100 days, and private duty nurses.

The most common sources of out-of-pocket expenses are the Medicare Part B premiums and deductibles and co-payments for covered services. Also, some physicians do not accept Medicare "assignment" of their charges—that is, they do not accept the Medicare payment as full payment. Although there are now limits on the amount by which the physician's bill can exceed the Medicare payment, beneficiaries are liable for the difference.

Efforts to control Medicare costs have taken many forms, including professional review organizations to protect against unnecessary hospitalizations and surgical procedures. Legislation enacted in 1983 created a prospective payment system, which set limits on hospital reimbursement for inpatient procedures by establishing a set amount to be paid for each diagnosis-related group. The justification for this new system was to encourage hospitals to manage the care of patients within a fixed budget.

More recently, Medicare revamped the method by which physicians are paid, phasing in a national fee schedule that began on January 1, 1992, and was completed by 1997. The fee schedule is based on efforts to systematically estimate the relative value of specific medical services based on the resources used to provide those services. Before this, Medicare paid physicians based on the charges that were submitted relative to the average charge in the community.

The impetus for the M+C program was the hope that competition among plans would hold down Medicare expenses. This aspect of Medicare is too new to evaluate, but so far, the competition and subsequent marketplace efficiencies have not yet emerged.

Medicare is the largest single source of health insurance in the United States, serving more people than the citizenry of most industrialized nations. The service area is as vast and diverse as the nation, and the population served is marked by differences in expectations, shaped by a lifetime of different experiences and cultural backgrounds. The program is made even more complex by the fact that there are categorical designations of services and distinct groups of entitlement. In response to legislated changes, court challenges, and the dramatic changes in the diagnosis and treatment of diseases and injuries, Medicare has undergone and is undergoing constant change. Many recent reform proposals, such as the addition of a prescription drug benefit, suggest that there will be additional changes in the future. Despite all the changes, the program has ensured access to health care to some of the nation's frailest and most medically vulnerable citizens. The ensuing financial protection, which is pooled broadly across society, is critical for individual patients and their families.

ROBERT B. FRIEDLAND

See Also

Long-Term-Care Financing: International Perspective

Long-Term-Care Policy

Medicaid

Medicare Managed Care

Internet Key Words

Social Security Act

Title XVIII

Internet Resource

Health Care Financing Administration Medicare Homepage

http://www.hcfa.gov/medicare/medicare.htm

MEDICARE MANAGED CARE

Medicare History and Benefits

The enactment of Medicare in 1965 provided government-supported health insurance and benefits for most elderly. To participate, an individual must be eligible to receive Social Security or railroad retirement benefits. The number of elderly eligible for Medicare grew from just over 19 million in 1968 to almost 34 million in 1998, representing over 95% of those 65 years and older.

Medicare Part A provides insurance for inpatient hospital stays, qualifying skilled nursing facility care, skilled home health care, and hospice care. Part B helps pay the costs of physician services, outpatient hospital services, medical equipment and supplies, and other health services and supplies. All eligible adults over 65 are entitled to Part A, but individuals must enroll and pay a monthly premium ($45.50 in 2000) for Part B coverage.

Benefits are generous but restrictive. Medicare does not cover prescription medications, hearing aids, dental care, and most home and community-based long-term care. Most beneficiaries purchase various supplemental insurance policies to cover some of these "gaps" in coverage and to help with co-payments.

Supplemental policies and the cost of medication have become unaffordable for many older adults. In addition, the out-of-pocket expenses associated with Medicare benefits have been increasing due to rising supplemental policy costs. Many elderly left traditional Medicare and joined Medicare managed care organizations (MMCOs) because of these rising out-of-pocket costs and the expanded benefits that most MMCOs offer.

Medicare Managed Care

At the outset, Medicare did not alter the financial incentives and relationships among physicians and other providers; thus, opportunities for cost control were limited. As a result, unanticipated demand for services and other factors combined to place unsustainable financial pressure on the Medicare Trust Fund (Iglehart, 1999).

In 1982, Congress passed the Tax Equity and Fiscal Responsibility Act (TEFRA) in an attempt to rein in escalating health care costs. TEFRA provided incentives for health maintenance organizations (HMOs) to enroll Medicare beneficiaries on an "at-risk" basis. Under this arrangement, participating HMOs (called MMCOs) receive monthly fixed capitation for their Medicare enrollees with which they must fund all their health care expenses. Prior to capitation, HMOs received funding based on complex formulas linked to their members' utilization of services. Under capitation, the incentive is to carefully manage utilization in order to keep spending within the allocation. Capitation was actuarially set at 95% of the actuarially adjusted average per capita cost (AAPCC) of caring for a Medicare enrollee in the local Medicare market. Health plans assume full financial risk for their Medicare members and in return receive fixed monthly payments from the Health Care Financing Administration (HCFA). Plans are responsible for all regulatory and administrative expenses and must provide all Medicare-covered services. They cannot request additional funds from HCFA if expenses exceed revenues.

The growth of TEFRA MMCOs began slowly, but in the last few years of the 1990s, growth was rapid. Medicare managed care (MMC) is attractive to beneficiaries because of the virtual absence of paperwork, elimination of the need for costly supplemental Medigap insurance, low deductibles and co-payments, and a rich array of extra benefits, including various levels of prescription medication coverage, dental care, customized health club memberships, transportation, and eyeglasses. The trade-off, however, is that beneficiaries lose unrestricted choice of providers; they must use only plan physicians and other providers in all but urgent and emergency situations.

MMC has been very attractive to health plans primarily in metropolitan areas, where monthly capitations are highest and there is an oversupply of providers, including hospitals and primary care and specialty physicians. Intense competition has kept beneficiary costs such as premiums and co-payments down in many areas of the country. Conversely, low reimbursement and difficult recruitment from among the smaller pool of available providers have made MMC less attractive in rural areas. Rural MMCOs have discontinued service in many areas of the country. Of the more than 9 million rural Medicare beneficiaries, 73% live in counties with no MMC, and 17% live in rural counties with only one MMC plan. This situation is viewed as an inequity by Medicare beneficiaries.

Scrutiny of MMCOs' success focuses on overpayment to the plans, the draw-off of health care dollars for marketing, reductions in access to care, withdrawal of benefits such as prescription coverage, and enrollment biases toward low-risk seniors that would favorably affect a plan's financial health.

Medicare Plus Choice

The Balanced Budget Act (BBA) of 1997 expanded the number of provider types that can assume primary risk by creating the Medicare Plus Choice (M+C) program. M+C plans arose from traditional HMOs to include provider-sponsored organizations (PSOs) and preferred provider organizations (PPOs). In addition, certain beneficiaries may be able to choose another type of coordinated care plan, the Religious Fraternal Benefit Society plan. Under BBA regulations, M+C organizations are required to assess the health status of all new members within 90 days of enrollment and must have ways to identify and manage those with medically complex chronic conditions.

Impact of MMC on Patient Care

It was expected that integrated financing with capitation would control costs and lead to improved care for older people, particularly with regard to the use of skilled nursing care, population screening and risk stratification, chronic disease management, primary care, and preventive services (Wagner, 1996). Evidence of improvement has been slow to emerge and is uneven. To date, the promise of improved care for the elderly has exceeded performance realities. Some data suggest that the elderly and the chronically ill do not fare as well in MMC compared with traditional fee-for-service Medicare (Ware, Bayliss, Rogers, Kosinski, & Tarlov, 1996).

Several MMCO programs, however, appear to facilitate higher-quality skilled nursing facility care models by using nurse practitioners (Reuben et al., 1999), screening and identifying at-risk seniors (Boult et al., 1993; Pacala, Boult, Reed, & Aliberti, 1997), running cooperative care clinics (Scott,

Gade, McKenzie, & Venohr, 1998), coordinating care and disease management programs, and forging partnerships with community service agencies and providers. There is growing evidence that MMC beneficiaries have greater access to preventive services such as immunizations. A controversial study reported that not-for profit managed care plans performed better than for-profit plans in all measures, including significantly higher rates of prescribing beneficial medication after heart attack, performing annual diabetic eye examinations, and doing mammography (Himelstein, Woolhandler, Hellander, & Wolf, 1999).

HCFA as Purchaser

HCFA oversees and regulates traditional Medicare and M+C. As its role changed from payer to purchaser, regulations surrounding M+C have become stricter. HCFA gathers information about MMCOs for quality measurement and to provide beneficiaries with information on benefits, premiums and co-payments, and disenrollment statistics.

HCFA is collaborating with the National Committee for Quality Assurance (NCQA) to create a report card ("Medicare Compare") on Medicare-specific measures. NCQA is a private, nonprofit organization that assesses and reports on the quality of managed care plans. Measures include mammography rates, prescription of beta-blockers after heart attack, percentage of primary care and specialty physicians with board certification, and percentage of diabetics receiving retinal exams.

HCFA is also surveying samples of MMCO beneficiaries to determine how well physicians communicate with them, how easy it is to see specialists, and the overall perceived quality of their health plans. Much of this information is available on the Internet. Most MMC beneficiaries are satisfied, but 25% say that they would not encourage a friend to join their plan.

Social HMOs and PACE

Two comprehensive variations of MMC are the social health maintenance organization (SHMO) and the Program of All-Inclusive Care for the Elderly (PACE).

Congress authorized the SHMO demonstration in 1984 to determine whether investing in long-term-care benefits for Medicare HMO enrollees could save money by coordinating care and providing services that might prevent more costly medical complications. Additional benefits provided by SHMOs include personal care aides, homemakers, medical transportation, adult day health care, respite care, and case management in a community setting. Services have monthly dollar limitations and co-payments, and recipients of expanded services must meet eligibility thresholds. A SHMO uses funds from Medicare risk-adjusted capitation, co-payments, premiums, and Medicaid to pay for the extra services. The BBA significantly increased the number of beneficiaries a SHMO could enroll and authorized the transition of the programs from demonstrations to permanent status.

PACE programs provide a broad array of health and social services to enrollees who are both nursing home and Medicaid eligible. PACE programs blend Medicare and Medicaid dollars with the goal of maintaining members at home rather than placing them in nursing homes. Services are similar to those offered by SHMOs but are more intensive and do not have dollar limits. One significant difference between the two programs is that PACE members must attend adult day health care centers several times a week; they are the hub of care and provide a number of social and health services. If a PACE member enters a nursing care facility, the PACE site or organization pays for it. PACE has permanent status pursuant to the BBA of 1997.

Cost Savings Under MMC

MMCOs have held down their costs, which has allowed them to enhance benefits to members. The largest portion of savings comes from negotiating favorable contracts with providers (purchasing on the margin), managing hospital utilization, and requiring preauthorization (preapproval) for specialty referrals and expensive diagnostic testing.

Areas with the greatest oversupply of providers are advantageous to MMCOs, as lower reimbursement is the norm. But lower reimbursement has created enmity and distrust among traditional providers and between providers and health plans.

The Future of MMC

Uncertainty about the future of MMC focuses on the phasing out and replacement of the AAPCC by a new reimbursement strategy constructed from utilization and diagnostic data from across continuum of care. HCFA's stated intent is to reward a higher reimbursement to those MMCOs serving members with a higher disease burden.

As a result of this change and concerns about lower reimbursement, some MMC plans are withdrawing from certain markets, including areas of California, Florida, Arizona, Louisiana, Connecticut, New York, Colorado, and many other states. They are scaling back their marketing to Medicare beneficiaries, increasing beneficiaries' co-payments, becoming more restrictive in their drug benefits, or introducing monthly premiums.

RICHARD D. DELLA PENNA
MONA ROSENTHAL

See Also
Medicaid
Medicare
Program of All-Inclusive Care of the Elderly
(PACE)

Internet Key Words
Adjusted Average Per Capita Cost
Balanced Budget Amendment
Capitation
Health Plan Employer Data and Information Set
Health Maintenance Organization
Medicare Compare
Medicare Managed Care
Medicare Managed Care Organization
Medicare Plus Choice
Medicare Risk
National Committee for Quality Assurance

Nurse Practitioners
Part A Medicare
Part B Medicare
Preferred Provider Organization
Provider Service Organization
Program of All Inclusive Care of the Elderly
Social Health Maintenance Organization
Tax Equity and Fiscal Responsibility Act

Internet Resources
Medicare Compare
http://www.medicare.gov/comparison/default.asp

National Committee for Quality Assurance
http://www.ncqa.org/Pages/Main/index.htm

PACE
*http://managedcare.hhs.gov/
    program_descriptions/medicare/pace.htm*

SHMO
*http://managedcare.hhs.gov/
    program_descriptions/medicare/
    social_hmos.htm*

MEDICATION ADHERENCE

Increasing age is associated with a higher prevalence of chronic diseases that necessitate long-term or lifelong medication intake. Patients older than 65 years are the largest consumers of medications, receiving nearly half of all prescribed medicines. Most of these people are on at least one prescription drug regimen. Correct medication taking is a prerequisite for successful management of chronic illnesses. Nonadherence (noncompliance)—that is, patients not taking medications as prescribed—may result in increased morbidity, mortality, and health care costs and a decreased quality of life (De Geest, von Renteln-Kruse, Steeman, Degraeve, & Abraham, 1998). Patients may not take medications as prescribed for a variety of reasons, including financial inability to pay for medications, health beliefs about medications, and complicated medication regimens.

Assessing Adherence with Medication Regimens

There is no gold standard to evaluate adherence with a medication regimen. Direct methods include assay of medication, medication by-products, or tracers (e.g., digoxin) in bodily substances and observation of medication administration. The reliability of assay for adherence assessment is dependent on the half-life of the substance under scrutiny. Assay does not provide information about actual intake and regularity of medication taking. Observation allows evaluation of complex medication behaviors (e.g., insulin injection, use of inhalers or nebulizers) and can also be done to assess whether older patients are still capable of independent medication management despite visual or physical impairments. Formerly, observation was feasible only during a clinical encounter between patient and health care provider, but the use of electronic communication methods such as videophone or "Net-Meeting" has extended the scope of the traditional clinical setting.

Indirect measurement methods are self-report, collateral report, prescription refills, pill count, and electronic event monitoring. Self-reports and collateral reports by significant others or clinicians underestimate medication adherence. Memory distortion is an important limiting factor for reliable self-report, even in those patients with normal cognitive functioning, who accurately remember medication intake only over the past three to five days. Moreover, no information about medication-taking dynamics is provided by collateral or self-report methods. Tracking prescription refills assesses whether patients have redeemed prescriptions, but this is only a rough measure of medication-taking dynamics. Pill count is another indirect method that compares the number of pills remaining at the time of a clinical encounter to the number of pills prescribed over the period under study. This method does not prove that the patient took the pills, nor does it give any information concerning regularity of medication taking. Electronic event monitoring uses a pill bottle that is fitted with a cap that contains a microelectronic circuit. The date and time of each bottle opening and closing are recorded as a presumptive dose. Recorded data can be downloaded to a computer that lists and graphically depicts individual medication-taking dynamics. Although indirect, electronic event monitoring has superior sensitivity compared with other direct and indirect methods and allows assessment of nonadherence continually and in a multidimensional manner (De Geest et al., 1998).

Risk Factors for Nonadherence in the Elderly

Knowledge of determinants of nonadherence is helpful to identify patients at risk for not taking their medications correctly. Moreover, modifiable determinants provide a basis for developing preventive and restorative adherence interventions. Older age per se is not associated with a higher risk for medication nonadherence, but a number of processes associated with aging may negatively influence patients' ability to manage their medications independently and correctly. More specifically, functional and sensory capabilities and cognitive functioning may be more impaired in geriatric patients. For instance, it can be more difficult for the elderly to handle childproof caps, blister packages, or nebulizers or to swallow large pills. Labels may be misread, and colors of pills may not be recognized. Forgetfulness is a determinant of medication nonadherence in all age groups. Cognitive impairment, a problem that increases with age, compromises patients' ability to manage their medications, necessitating the assistance of significant others or clinicians to guarantee correct medication intake. The fact that many elderly live alone and are relatively socially isolated deprives them of necessary social support and places them at risk for depression, both known risk factors for medication nonadherence (De Geest et al., 1998; Park et al., 1999).

Knowledge about the medication regimen does not guarantee adherence. Patients have to be aware of the intended effect of the medication, how to administer it, possible side effects, and other relevant aspects. A significant proportion of the older population has inadequate or marginal functional health literacy, making it difficult to process the health information and instructions given to them. Patients' perceptions about their illness and treatment, as well as their symptom experience, influence adherence. Inappropriate or incongruent illness representations can jeopardize patients' correct medication intake. For instance, patients may perceive medications as "poisonous" and thus stop taking them, or they may wrongly think that a disease such as hypertension is an acute disease that does not require long-term medication taking (De Geest et al., 1998; Park et al., 1999). Patients with personality disorders are prone to more behavioral and management problems, resulting in a higher rate of nonadherence.

Treatment-related factors such as duration, complexity, and cost of medication regimens can also negatively affect adherence. Physiological effects of aging make older patients more prone to the occurrence of adverse events. Medication side effects and, more specifically, the perceived symptom experience of these side effects may inhibit correct medication intake. For instance, beta-blockers can induce fatigue, impotence, and sleeplessness—symptoms that may negatively influence adherence (De Geest et al., 1998).

Strategies to Enhance Adherence with Medication Regimens

Achieving adherence with medication regimens is a major challenge. Strategies to prevent nonadherence or to enhance adherence begin with an assessment of the patient's ability to independently manage a medication regimen. Functional and sensory abilities, cognitive functioning, literacy, knowledge, motivation, illness representations, sources of social support, and financial status should be carefully evaluated during a standard geriatric clinical assessment (De Geest et al., 1998).

Patients need to be educated about the different aspects of their therapeutic regimens. This can be achieved using traditional educational tools, but Internet teaching applications or interactive computer programs with touch screens are valuable and cost-effective teaching interventions.

In addition to educational interventions, behavioral strategies should be included in an adherence intervention program. This can be achieved by tailoring the regimen to patients' lifestyles, teaching patients self-monitoring strategies (e.g., medication booklet), suggesting the use of medication aids (e.g., medication box), and cueing (e.g., leaving containers in a particular location) (De Geest et al., 1998). Evaluation of coping strategies may also be indicated, as it is known that avoidance coping is associated with a higher risk for nonadherent behavior.

Social support from significant others and family members is also helpful. Support can include preparing the patient's medication, reminding the patient to take the medicine, refilling prescriptions, and helping the patient decide to contact a health care worker if a problem arises. Clinicians should also assess the patient's ability to pay for needed medications. A variety of local, state, and federal programs may be available to assist with medication costs. A videophone or telephone is an effective tool to help patients living alone manage their medication schedules correctly (Bennett, 1999; De Geest et al., 1998; Fulmer et al., 1999).

Nonadherence with medication regimens is an important behavioral problem that can be ameliorated with adequate assessment, screening for risk factors, and the implementation of preventive and restorative interventions. Electronic event monitoring as a future monitoring and intervention tool for patients with chronic illness at high risk for nonadherence (e.g., heart failure) is promising (De Geest et al., 1998; Fulmer et al., 1999).

SABINA DE GEEST

See Also
Assistive Devices
Medicaid
Medicare
Over-the-Counter Drugs and Self-Medication
Polypharmacy Management
Social Supports (Formal and Informal)

MEMORY IMPAIRMENT
See
Life Events

MENTAL CAPACITY ASSESSMENT

Assessment of mental capacity should be a routine part of medical care, but many health care professionals lack the training to do this accurately and well. The law recognizes different types and degrees of mental capacity. Health care professionals have traditionally been asked to assess three types: the capacity to make medical treatment decisions, to provide basic care for oneself, and to handle finances (Blum, 1999). Controversial capacity assessments include determining a person's ability to drive a car and deciding whether someone has the capacity to end life-sustaining treatment.

Informed Consent for Medical Treatment

The provision of informed consent regarding health care decisions depends on three elements: information disclosure by the physician, voluntary participation by the patient, and competence of the patient.

Information Disclosure: The first step of information disclosure is to convey the information in readily understandable language. This may require the use of a translator and multilingual forms. Educational and cultural factors must be considered, as well as emotional and cognitive states. Once these issues are addressed, it is the health care professional's obligation to present the following information: the nature of the medical problem, the proposed remedy and reasonable alternatives, the attendant risks and benefits of the treatment, and the option of no treatment. Relevant and significant information, from the patient's perspective, must be disclosed. However, a physician can limit disclosure to the therapeutic issues, thereby omitting information concerning the patient's nonmedical interests, such as business affairs. The ethical goal is to encourage patient autonomy and allow knowledgeable judgment regarding treatment options.

Voluntary Participation by the Patient: People must be allowed the freedom to accept or reject treatment without being subject to undue influence. Coercion by family, friends, or even the health care team compromises an individual's voluntary participation, as can psychiatric symptoms such as delusions, depression, or dementia.

Competency: Competency is a legal term that means to be "duly qualified: answering all requirements; having sufficient capacity, ability, or authority; possessing the requisite physical, mental, natural or legal qualifications" (Black, 1991). Only a court can decide whether a given person is competent to perform a particular act.

There are many forms of competency. For example, in the U.S. legal code, an adult is presumed to have *general competence* unless a special court proceeding finds otherwise. As a result, and unless otherwise specified, adults are allowed to vote, own property, and choose lifestyles without interference from others. *Specific competence* refers to specific skills required for a particular act. Common examples include the competence to drive a car, perform medical procedures, or manage funds. Making informed decisions regarding health care is a type specific competence.

A third type of competence is *consequence-dependence competency*. Similar to "the ends justify the means" thinking, this is commonly seen in medical settings when a patient's competency is questioned if he or she disagrees with recommended treatment. Conversely, if a patient does not object to a recommended treatment, his or her competence is rarely questioned. In the first case, pressuring someone to change a decision undermines his or her autonomy and the therapeutic relationship. In the second case, compliance by the patient is confused with choice and ignores the important possibility that the patient may not be making a knowledgeable, informed, and rational decision. In both cases, and even though the medical team hopes to achieve whatever it considers to be a good outcome, discounting a patient's understanding or reasoning is legally and ethically incorrect. When evaluating the ability to make treatment decisions, the process is more important than the outcome.

When a court finds a person to be incompetent, a judge appoints a surrogate decision maker. There are two categories of surrogates: guardians and conservators. In most states, a guardian has the legal authority to make decisions directly affecting the incompetent person's life, such as providing consent for medical treatment or placement into a nursing care facility. A conservator's legal responsibility is limited to making decisions about the person's assets and property. Some states have combined the two functions, so that only one designation is used; others reserve the term guardian for surrogate decision makers for children.

Mental Capacity Assessment

Mental capacity assessment refers to the analysis of a person's cognitive and emotional status as it relates to specific observable or legal behaviors. Three approaches to assessing mental capacity are used, singly or in combination:

1. Functional (or behavioral)
2. Medical
3. Philosophical

Functional Approach: The functional approach focuses on the appropriateness of a person's actions and statements. If a person behaves in an odd, unusual, or abnormal manner, concerns may exist about whether the person is cognitively impaired. *Cognition* here refers to a combination of conscious awareness, ability to maintain attention, memory, and information processing.

Once a concern exists, the next step is to categorize *how* the behavior is inappropriate. Twelve mental functions are necessary to make knowledgeable, reasoned, and rational decisions. Describing these categories, and the associated abnormal behaviors, helps determine the extent and severity of a person's decision-making problem. The model presented here is used by health care professionals, social services agencies, law enforcement, and legal professionals throughout much of the United States. It is useful for assessing a person's capacity to consent to treatment (White, 1994) and to handle financial or business matters (Blum, 1998).

The first and most basic function is the ability to *express desires.* This may be done in any form—speech, writing, pointing, moving, and so forth. Next, a person must be able to *respond to the environment.* This refers to the ability to recognize significant changes in one's surroundings, such as the presence or absence of people or items of desire. These two functions are the most basic and are impaired only in extreme situations (delirium, advanced dementia, coma, medication effects).

The next major function is the ability to *focus attention.* One of the most important and fundamental aspects of the brain is the ability to sift through the multitude of stimuli that constantly bombard the senses and prioritize them. If this ability is impaired or overwhelmed, the affected person becomes confused, anxious, and unable to learn or assess new information. In order to make knowledgeable decisions, the person must be able to assess whether new information is pertinent. This ability may be impaired if the patient suffers from delirium, dementia, head trauma, psychosis, or mood or anxiety disorders, or it may be a side effect of medication.

The next major function to be assessed is *memory.* Once presented with pertinent information, the patient must remember the information long enough to make a decision based on it.

The patient must also know his or her *options,* the *roles and responsibilities of all pertinent parties,* and the likely *consequences* of each option. Essential related functions are the abilities to *use abstract concepts* and *think strategically* (described in more detail in the section on executive functions). Also, the person must be *free from delusions.* A *delusion* is an unwavering false belief that is not culturally appropriate and is maintained despite contrary evidence. Common delusions found in dementia patients include paranoia and persecution, abandonment, infidelity, the belief that a spouse or caregiver is an impostor, and the belief that the person's house is not his or her real home. A person with a delusion may still have adequate capacity if the delusion does not affect the decision at hand. For example, a patient may be able to consent to medical treatment yet believe that his spouse is an impostor, if this belief has no bearing on the decision to accept or reject treatment.

Once a person is able to learn and remember pertinent information and knows the major options, likely consequences, and pertinent parties, he or she needs to make a decision. This involves two functions: *choosing an option* and *maintaining* (or not excessively changing) *the decision.* Several medical conditions may impair these abilities, including (but not limited to) medication effects, major depressive disorder, anxiety disorders, schizophrenia, and dementia.

Many times, an elder (or disabled adult) attempts to hide cognitive problems by asking questions, using humor, or minimizing contact. Therefore, the evaluator should consider the elder's comments and behavior in multiple settings, if such information is available. One helpful strategy is to ask the person to summarize the pertinent information regarding treatment or other issues and to explain the reasons for the final decision. If the elder cooperates and is able to do this, the evaluator can quickly assess whether there is significant impairment.

One benefit of the functional model is that the approach is logical and appeals to common sense. It does not require specialized expertise to evaluate the information. Drawbacks of the model are that it requires review of observations by both interested and objective parties (more time and therefore greater initial expense, especially if the evaluation is to be used for legal purposes). Also, the model does not provide information regarding causation, treatment, or likely prognosis. Medical and psychiatric evaluation is necessary to answer these questions.

Medical Approach: Ideally, a behavioral assessment should be done before a medical, that is, psychiatric or psychological, assessment. The medical approach relies on signs and symptoms of disease to determine capacity. The presence of symptoms is equated with cognitive impairment and, therefore, incapacity. Typically, this model requires evaluation of attention, orientation, memory, thought processes, executive and other cognitive functions, hallucinations, delusions, obsessions, and mood and anxiety disorders. Because situations exist in which a person has symptoms yet retains decisional capacity, the medical approach may be inadequate and incomplete if the findings are not correlated to actual behavior. Furthermore, medical and neuropsychological evaluation requires specialized knowledge, skills, and experience that is not always readily available (Blum & Eth, 2000).

The benefit of the medical approach is that record review may be limited to medical records; hence, it is less time-consuming and less expensive than evaluations that include behavioral observations outside a medical environment. Also, this approach may provide information regarding causation, treatment, or likely prognosis. There are two major drawbacks of this approach. First, many medical personnel do not have the knowledge, skills, or experience to correctly assess cognitive impairment. Therefore, when reviewing conclusions from experts in fields other than mental health, it is necessary to confirm that the correct assessments were performed. Second, the presence of signs and symptoms must be compared to the person's observable behaviors for the evaluation to be valid.

Philosophical Approach: The philosophical approach uses vague and imprecise concepts to define mental capacity. The classic formulation is that a person has adequate capacity if he or she is able to express his or her desires, understands and appreciates the situation, and is rational. Although this approach uses accepted legal language, definitions of terms vary, and there is no consensus regarding their evaluation, making this approach difficult to use.

Executive Functions

The set of mental abilities needed to plan, organize, and carry out actions—executive functions—is used to shift focus, analyze tasks, form concepts and strategies, and evaluate one's own behavior. The most important but often unrecognized or overlooked capability is the ability to think strategically. When impaired, a person cannot readily assess the consequences of his or her decisions. Subtle signs of dysfunction are usually dismissed as trouble concentrating, mild memory loss, eccentricity, or personal rigidity. Although such conditions and personality styles exist, new onset or exacerbation of these behaviors suggests early impairment of executive functions and requires psychiatric investigation.

The executive functions are critical to attention, memory, and learning. Impairment of executive functions is often found in elders who have psychiatric disorders, such as dementia, delirium, severe depression, and mania. Impairment may occur before there are more noticeable neurological problems.

Behavioral problems include (1) impaired organization of material to be learned; (2) poor recall of recent and remote information; (3) automatic imitation of the gestures and actions of others; (4) automatic and inappropriate use of objects in the environment; and (5) unnecessarily repeating words, phrases, or actions. Other common co-occurring behaviors include irritability, tactlessness, impulsivity, undue familiarity, absence of social constraint, fatuous and inappropriate euphoria, rapid mood fluctuations, mania, and obsessions and compulsions. Symptoms of profound apathy, monosyllabic responses, and indifference to pain occur less often.

Neuropsychological tests of executive functions may be used to clarify the nature and extent of cognitive deficits detected on a mental status examination. Frequently administered tests of executive functions include the Wisconsin Card Sorting Test, the Stroop Color Word Test, the Rey-Osterrieth Complex Figure Copying Test, and the Trails Making Tests of the Halstead-Reitan Neuropsychological Battery. However, these tests are limited by the considerable variability in performance and the lack of normative values across the geriatric age span. In addition, there is only a moderate relationship between test results and behavior in the real world. The evaluator must therefore draw on many different sources of data when assessing someone's functional capacities.

Tests of Mental Capacity

There is no single test of mental capacity. The evaluator needs to know what task—the specific capacity—is to be assessed. Assessing the capacity to make medical treatment decisions is different from assessing the ability to make financial decisions. The evaluator then uses the functional and medical models described earlier. Formal psychological testing, such as tests of gross cognitive functions (e.g., the Mini-Mental Status Examination), memory (e.g., the Rey-Osterrieth Complex Figure Copying Test), or executive functions, may also be used, but only as part of an overall assessment.

Many areas of capacity assessment do not have associated standardized tests. Examples include assessing the capacity to participate in medical research, complete advance directives, drive, or marry. For these situations, practitioners are advised to perform thorough functional and medical assessments, using psychiatric or psychological tests as adjunctive tools only. When providing conclusions regarding these types of capacity assessments, practitioners must emphasize the limitations of their evaluations and conclusions.

The Mini-Mental Status Examination (MMSE) (Folstein, Folstein, & McHugh, 1975) is often inappropriately used to assess mental capacity. The MMSE is a crude screening tool of cognitive abilities and has numerous limitations. It is not a diagnostic test, nor can it assess the impact of an emotional condition. Scores on the MMSE do not necessarily correlate with real-world function. People with high scores may be unable to function without supervision, and people with low scores may be able to function independently. For example, if the test suggests that there is significant impairment, such as dementia, the MMSE cannot demonstrate if it affects the person's mental capacity to make financial transactions or designate a health care proxy. Scores on the MMSE are affected by age, education, and physical illness. Older people and those with less formal education tend to score lower, though they may be functionally unimpaired. Ill people tend to score lower, even if the medical symptoms are minimal or absent. Scores on the MMSE are affected by all of the following transitory conditions: alcohol and illicit drug use, medications, sleep deprivation, low blood sugar, low blood pressure, headache, or pain anywhere in the body.

If an elder's capacity to make medical decisions; participate in health care discussions; provide self-care; execute wills, trusts, or contracts; or engage in any other financial transaction is questioned, the MMSE should be used only as part of an overall evaluation. Relying solely on the MMSE during such evaluations is a misuse of the test.

The MacCAT-T is a reliable and easy-to-use test of mental capacity to make medical decisions. Based on the philosophical model of mental capacity assessment, the MacArthur Competence Assessment Tool for Treatment (MacCAT-T) evaluates the patient's understanding, appreciation, reasoning, and communication related to his or her medical condition by asking a series of questions about the condition and treatment decisions. The test requires minimal preparation time and can be performed at the bedside. Initial evaluation of the MacCAT-T compared healthy people living in the general community to those hospitalized for depression, schizophrenia, or ischemic heart disease. Significant impairments were found for a minority of people in all groups. People in the schizophrenia

and depression groups displayed poorer understanding of treatment disclosures, poorer reasoning in decision making regarding treatment, and more impaired appreciation of the ramifications of their illness or potential treatment. These problems were more pronounced among people with schizophrenia (Appelbaum & Grisso, 1995).

BENNETT BLUM

See Also
Adult Protective Services
Advance Directives
Autonomy
Cognitive Screening Tests: The Mini-Mental Status Exam
Depression in Dementia
Elder Neglect
Financial Abuse
Guardianship and Conservatorship

MENTAL HEALTH SERVICES

The prevalence of most mental illnesses, including substance abuse, is lower among older adults than among younger adults. However, research also indicates that rates for depression and cognitive disorders are higher for older adults (George, 1992).

Older adults use many mental health services. Traditional services include those provided through community-based agencies and through total and partial hospitalization. Many of these programs treat mental illness through medication and counseling techniques such as psychoanalysis, group therapy, and cognitive and behavioral therapies. Elders with more severe, persistent mental illness may use case management, in which a caseworker coordinates services within a network of providers.

Community watch systems have been developed whereby workers who have regular contact with older adults are trained to look for potential problems. For instance, mail carriers are asked to watch for uncollected mail or other signs that a resident may have a problem. If a problem is suspected, the carrier contacts a participating state or community agency, which checks on the resident.

This system supports elders who are living alone and may have difficulties that affect their daily functioning. Many communities use telephone contact programs through which a worker maintains contact with elders known to be at risk for mental health or other problems. Many senior centers offer day-care programs for those with cognitive impairments. These programs offer such services as recreation, personal care, and nutrition management. Finally, many health clinics and senior and drop-in centers screen for mental illness. To prevent problems, these sites often provide educational workshops to inform elders of risk factors associated with mental illness. If mental health problems are identified, older adults must be referred to a mental health professional.

Several residential programs serve elders with mental illness. In addition to services provided at inpatient psychiatric and state hospital settings, older adults living in skilled nursing facilities can receive mental health treatment as part of their care plan. Many Alzheimer's disease units or campuses have been developed for patient care as the disease progresses. Thus, patients can be admitted during the early stages of the disease and be moved to specialized units on the campus as the disease becomes more debilitating.

Professionals and Programs

Physicians, psychiatrists, psychologists, social workers, and psychiatric nurse practitioners are some of the professionals who offer geriatric mental health care. Professionals in these disciplines may choose to obtain specialized training in gerontology. Services provided by these professionals include referral, advocacy, psychotherapy, medication, and service coordination and management. Also, many master's-level programs in gerontology have been developed. Graduates of these courses provide mostly management services and are generally involved in developing and administering mental health service programs for older adults rather than providing direct mental health services.

Medicare and Medicaid are the two programs that provide coverage for the majority of elders needing mental health services. Medicare, the main health insurance program for people 65 and over, covers a limited amount of mental health services. Part A (hospital insurance) of Medicare pays 80% of the costs for inpatient psychiatric care, up to a lifetime limit of 190 days. Part B (medical insurance) of Medicare pays 50% of outpatient mental health services (Health Care Financing Administration, 1999). However, older adults must pay a monthly premium for Part B coverage. Also, Medicare currently does not cover prescription costs, including medications used to treat mental illness. Several supplemental insurance programs (e.g., Medigap) provide varying coverage for mental health services not covered by the basic Medicare program. These supplemental programs also require elders to pay monthly premiums.

Medicaid, the health insurance program for low-income individuals, pays for nursing home care if older adults cannot afford these services. Medicaid also pays for a variety of mental health services, including medication costs, but the type of services covered, the amount paid by Medicaid, and eligibility for services vary from state to state. Medicare and Medicaid programs have moved toward a managed care model of providing services in many communities. In these instances, the older adult must obtain a referral from his or her primary care physician to receive mental health services. If the patient does not participate in a managed care program, he or she needs to seek a provider who accepts Medicare or Medicaid payments.

Many older adults rely on private insurance for their health care. Most health insurance companies offer limited mental health services, and some cover the costs of medications. Older adults also can elect to obtain mental health services through a fee-for-service agreement. Finally, many community mental health centers offer services based on a sliding fee scale for elders who have low incomes, fixed budgets, or both.

Older adults can complete advance directives for mental health care. These directives enable them to specify the type and scope of treatment they would desire if they were unable to articulate their wishes.

Gaps in Availability and Coverage

A major concern about elders with mental illness is their low rate of service use. Many do not seek services from mental health professionals because of the stigma attached to receiving such services. Consequently, many elders visit their primary care physicians complaining of symptoms that may be caused by mental illness. Unfortunately, many health care providers are not trained to recognize symptoms of mental illness in older adults or to differentiate between mental and other health problems. Health care providers holding stereotypical views of aging may assume that symptoms of mental illness are part of normal aging. For this reason, elders with mental illness may be misdiagnosed or not referred to mental health specialists. Further, there is a lack of communication and collaboration between professionals in health care settings and community mental health centers, so many seniors are not receiving comprehensive care. When referred for mental health problems, they find few agencies specializing in treating older adults. Currently, many agencies focus their dollars on programs for children and young adults with chronic mental illness. Moreover, many agencies lack outreach services, and many older adults who are ill, frail, disabled, or home-bound or who lack transportation cannot access services. Programs that offer outreach services, such as home health care and delivered meals, do not offer mental health services. Finally, many of the mental health services that are available are cost prohibitive to elders on fixed incomes.

Although Medicare and Medicaid cover some mental health services, these programs are designed to focus on health and long-term care. If older adults use Medicare Part B for mental health services, the cost can be prohibitive, because they must pay 50%. And if older adults suffer from severe or persistent mental illness, a limit of 190 days for hospitalization is usually not sufficient for adequate treatment. Purchasing supplemental insurance is not an option for many low-income seniors.

Elders can choose to participate in managed care plans, which often provide some mental health benefits. However, managed care programs are designed for physically functional individuals who have acute health problems. Also, the emphasis in managed care is on cost containment. Thus, managed care systems tend to restrict access to specialized psychiatric care or to limit the type and scope of treatment individuals can receive. Further, many mental health providers outside the managed care system cannot survive financially, which limits elders' choice of providers.

Recent legislation addressed mental health service coverage among health insurance companies, but gaps in coverage still exist. The Mental Health Parity Act requires that all group health plans, and insurance coverage offered in connection with group plans, place dollar limits on mental health benefits equal to those placed on medical and surgical benefits. However, this act does not require health plans to provide mental health services. It also does not mandate guidelines on cost sharing, types of mental health services available, or amount of mental health services one can receive (Department of Health and Human Services, 1999).

ANISSA T. ROGERS
ROBERT W. DUFF

See Also
Cognitive Changes in Aging
Medicaid
Medicare
Psychotropic Medications

Internet Key Words
Community Mental Health Centers
Medicaid
Medicare
Mental Illness

Internet Resources
Administration on Aging
http://www.aoa.dhhs.gov/

American Association of Retired Persons
http://www.aarp.org/

Health Care Financing Administration
http://www.hcfa.gov/

National Institute on Aging
http://www.nih.gov/

MINI-MENTAL STATUS EXAM
See
Cognitive Screening Tests: The Mini-Mental
  Status Exam

MINI NUTRITIONAL ASSESSMENT
See
Nutritional Assessment

MONEY MANAGEMENT

The ability to access cash and pay bills when due
is necessary for elders to remain living independently in the community. While seemingly uncomplicated, the ability to carry out these essential and
routine tasks of daily money management (DMM),
or finding another person to assist with them, can
be the difference between living independently in
the community and living in an institutional facility.

Case Study: Mrs. Oliver, 87 years old, lived
in Penn South for 37 years. The Penn South Program for Seniors (PSPS), a program providing support services at a New York City middle-income
housing cooperative that had become a naturally
occurring retirement community, learned from the
building management that Mrs. Oliver had not been
paying her rent. Mrs. Oliver had been an active
PSPS volunteer until a year ago, when her eyesight
began to fail and her husband died. When the program's social worker went to Mrs. Oliver's home,
she found an overwhelmed woman whose finances
were in complete disarray. Mrs. Oliver had been
asking various neighbors to write her checks for
her. She did not know what had or had not been
paid, nor did she know her bank balance. Mrs.
Oliver told the social worker that her husband had
always paid the bills. Her children lived far away,
and she did not want to admit to them that she

could not handle her affairs. She feared that they
would want to place her in a nursing home or insist
that she move in with them, but Mrs. Oliver did
not want to give up her home and her independence.

Mrs. Oliver's situation is an example of the
predicament that 5% to 10% of people over age 65
and 24% of those over age 85 find themselves
in; they can no longer manage the tasks of DMM
(Wilber & Buturain, 1993). These tasks include
such things as accessing cash for routine purchases,
paying bills, and filling out medical insurance
forms, as well as keeping track of one's own money.
Over one-third of the seniors that the PSPS serves
need some form of money management assistance.

Factors contributing to a senior's inability to
handle day-to-day finances include visual impairment, physical frailty, or emotional illness. An unsafe neighborhood can prevent a person from going
to the bank. Arthritis and other medical conditions
may limit the ability to write. Memory problems,
varying from forgetfulness and confusion to dementia, often hinder the ability to follow through on
tasks. Those unable to manage their own money
must be assisted if they are to remain living in the
community safely.

The majority of seniors who need help with
DMM rely on family or friends. However, for
many, like Mrs. Oliver, families are not accessible,
or the senior may have outlived the individuals he
or she would have entrusted with personal affairs.
Instead, seniors in need of DMM assistance may
seek help from financial, legal, and social work
professionals. A new type of practitioner, the daily
money manager, may also be available. But finding
help in money management is not easy; people do
not know where to look, and there is no consistent
referral source nationwide. Who provides DMM
assistance is determined by considerations such as
the type of resources available in the community,
the accessibility of the resources, the person's capacity, the complexity of the situation, and the person's ability and willingness to pay for assistance.
DMM assistance does not take the place of professionals from other fields. A good DMM provider
refers clients as needed to other professionals and
is one part of a long-term-care team that helps
seniors remain living at home.

Good DMM, no matter who provides it, begins with a thorough assessment of the client, including the identification and evaluation of existing helpers and an analysis of available financial resources, current expenditures, and the overall situation. The daily money manager must distinguish between the client's inability to handle money tasks and his or her inability to make decisions about money. If there are questions about the client's cognitive capacity, a psychiatric assessment is necessary to determine whether the client is an appropriate candidate for DMM. If capacity is found to be limited, a referral to adult protective services may be necessary to seek guardianship (Amerman & Schneider, 1995).

Tasks of DMM may include the following:

- Sifting through letters and bills to create a filing system.
- Helping with medical insurance claims, and advocating for proper payment on claims.
- Assisting with paying bills, including straightening out any incorrect bills and over- and underpayments, writing checks for the senior to sign, or arranging for a bank to pay bills automatically.
- Helping to prepare bank deposits, transfer money between accounts, arrange for direct deposit of regularly received checks, and balance checkbooks.
- Setting up a method for the senior to get cash for daily living expenses.
- Determining budgets and tracking expenditures.
- Filling out forms.
- Helping with entitlement applications and advocacy.

At PSPS, most clients need additional social services. Money management problems are often early signs that a person needs other support, such as help accessing medical care, home care, and transportation. Also, clients often need to talk to someone about their fears and loss of independence.

Although social work agencies have long recognized the need to provide DMM, the scope and availability of such services vary greatly around the country. Most agencies have chosen not to extend their care to include power of attorney, representative payee, or guardianship services because of the labor-intensive costs of these tools and because there is no regular funding stream that subsidizes that level of care. There are also fears of liability. Few free or low-cost programs provide a full range of DMM.

Supportive service programs in naturally occurring retirement communities, such as PSPS, are good sites for DMM programs. Being on-site, the program makes efficient use of staff and interns. For those clients who can no longer do their own banking, PSPS helps them access their own money. PSPS banks money sent by families or guardians, distributes it to home-care workers for daily living expenses, and monitors the use of it. Such activities help eliminate the potential for abuse.

A money manager should employ the least restrictive interventions available to help the client manage his or her own affairs. If there is a trusted family member or friend, establishing a durable power of attorney may prevent the need for a guardian in the future if the senior loses capacity. The powers can be broad or limited by the client. It should be noted that a power of attorney can be abused, because there is no legal oversight. For clients who have lost the capacity to handle their finances, the Social Security Administration (and some pensions) may be petitioned, with a doctor's supporting evidence, to assign a third party as the client's representative payee, who receives and uses the senior's pension for the senior's benefit. A guardian can also be named by the state court to care for a client's financial affairs. Guardianship is the most restrictive measure, as it takes away the client's rights, but it also provides the most formal oversight of the client's affairs.

All successful relationships are built on trust, with clear communication about needs, goals, tasks, and responsibilities. It is particularly important that providers of DMM establish written policies and contracts that delineate the roles and responsibilities of the daily money manager. Clear records should be kept about all transactions. Carefully following written policies and procedures helps protect against potential lawsuits. Experts in the area of

DMM report that lawsuits are rare. However, a program or practitioner should be either bonded or insured.

As seniors age and find it harder to handle their finances and access their own funds, they risk financial abuse and institutionalization. DMM is becoming an essential component of any successful long-term-care system whose goal is to help seniors live safely in the community.

KAREN BASSUK

See Also
Activities of Daily Living
Financial Abuse

Internet Resources
American Association of Daily Money Managers
http://www.aadmm.com

Berlin, L. (1997). "Daily money management? A new kind of service"
http://www.bizmonthly.com/news1997/February/berlin.html

Daily Money Management Services
http://www.aarp.org/confacts/money/dailymon.html

MORBIDITY COMPRESSION

The health of seniors is an increasingly important health and socioeconomic issue as the population ages in developed nations. In the absence of a theoretical paradigm for health education and prevention, the health promotion community has been criticized for mistaking association for causality or for promoting a world of disabled and demented individuals.

The compression of morbidity paradigm envisions a reduction in overall morbidity (and health care costs) by narrowing the period of morbidity between the onset of disability and death (Fries, 1980). The healthy life is seen as a life that is vigorous and vital until shortly before its natural end. Intuitively, the concept of delaying the onset of disability through the prevention of disease and the reduction of health risks seems natural. However, in the early and middle years of the 20th century, movement was away from this ideal, with a steady increase in the portion of life spent ill or infirm. Acute illnesses had given way to chronic diseases with longer periods of disability and morbidity. This phenomenon has been termed "the failure of success."

It was suggested that as people took better care of themselves and lived longer, they would live into those later years in which disability is greatest and would experience an increase in overall lifetime disability. Critics feared that good behavioral health habits would lead to an epidemic of Alzheimer's disease and a huge population of enfeebled, demented elders who would pose an immense strain on medical resources. Thus, the direct test of compression (or extension) of morbidity depends on the effects, studied prospectively and longitudinally, of reduced health risks on cumulative lifetime disability and on mortality. Will age-specific disability decline more rapidly than age-specific mortality, or vice versa?

New and emerging data suggest that early fears were unfounded. First, life expectancy from advanced ages has plateaued rather than increasing markedly, as predicted. In the United States, the life expectancy of women from age 65 increased only 0.6 year from 1980 to 1997. From age 85, female life expectancy in the United States has remained constant at 6.4 years.

Second, recent longitudinal data document the ability to greatly postpone the onset of disability with age. Researchers at the Stanford University School of Medicine have, since 1985, studied the effects of long-distance running and other vigorous exercise on patient outcomes in 537 members of a runners club, compared with 423 age-matched community controls; participants in both groups are at least 50 years old. The study was designed as a test of the compression of morbidity hypothesis. Appropriate controls for self-selection bias included longitudinal study; x-rays of hands, knees, and hips; intention-to-treat analyses; and statistical adjustment for other variables. Disability levels are assessed yearly, allowing the area under the disability curve to be assessed and approximating cumulative lifetime disability. When a person is said to be disabled, it means that he or she has difficulty performing one or more activities of daily living. Runners, exercising vigorously for an average of 280 minutes per week, delayed the onset of disability by about 10 years compared with controls. Among both male and female runners, disability increased at a rate only one-third that of the controls, after adjusting for age, initial disability, educational level, smoking behavior, body mass index, history of arthritis, and the presence of comorbid disease. As these subjects moved from age 58 toward age 70, the differences in physical function between the exercising and the control populations actually increased rather than decreased. The lifetime disability rate in exercisers is only one-third to one-half that in sedentary individuals.

In a University of Pennsylvania alumni study, 1,741 university attendees were surveyed in 1939 and 1940; again in 1962, at an average age of 43; and then annually since 1986. This unique data set contains over 50 years of longitudinal data, including demographics, health care utilization, health risks, disability, and mortality. Health risk strata were developed for persons at high, moderate, and low risk, based on the three risk factors of smoking, body mass index, and lack of exercise. Cumulative disability from 1986 (at an average age of 67) to 1994 (at an average age of 75) or until death served as a surrogate for lifetime disability. Persons with high health risks in 1962 or in 1986 had approximately twice the cumulative disability than those with low health risks. Results were consistent across survivors, deceased subjects, males, females, and those without disability in 1986 and over the last one and two years of observation. Deceased low-risk subjects had only one-half the disability of high-risk subjects in their last one and two years of life. High-risk subjects, despite having increased mortality, had greatly increased lifetime disability. Onset of disability was postponed by approximately 7.75 years in the low-risk stratum as compared with the high-risk stratum. The 100% reduction in disability rates was balanced against

only a 50% reduction in mortality rates, documenting compression of morbidity (Vita, Terry, Hubert, & Fries, 1998).

Recent major studies by other groups confirm these findings. Daviglus and colleagues (1998) showed substantial decreases in Medicare costs for those with few risk factors in midlife. Freedman and Martin (1998) showed significant age-specific functional improvement in seniors over a seven-year period. Reed and colleagues (1998) related healthy aging to prospectively determined health risks, with results closely similar to those of the Stanford University group.

Compression of morbidity is readily demonstrable in those who exercise vigorously compared with those who do not, in those with low behavioral health risks versus those with high risks, and in those with high educational attainment compared with low. Health risk behaviors as determined in midlife and late adulthood strongly predict subsequent lifetime disability. Both cumulative morbidity and morbidity at the end of life are decreased in those with good health habits. Morbidity is postponed and compressed into fewer years in those with fewer health risks.

Randomized controlled trials demonstrate that health improvement and risk reduction programs can reduce risks, improve health status, and reduce the need and demand for medical services. The Bank of America study of 4,700 retirees randomized into a health improvement program or a control group reduced costs by 20% and improved health indices by 10% to 20% (Fries, Block, Harrington, Richardson, & Beck, 1993). The California Public Employees Retirement System study of 57,000 seniors yielded similar results (Fries, Harrington, Edmund, Kent, & Richardson, 1994). Chronic disease self-management programs in arthritis and in Parkinson's disease documented the effectiveness of health improvement interventions in persons with chronic illness. Self-management programs in healthy seniors have improved health and reduced costs (Fries, Koop, Sokolov, & Beadle, 1998).

Three components support a national health policy to improve senior health. First is the conceptual base, represented by the compression of morbidity paradigm. Second is the epidemiological data associating behaviors with health outcomes, comparing effects on morbidity with those on mortality, and providing proof of concept. Third is the randomized controlled trials, now available, that prove that effective behavioral interventions can decrease senior morbidity and medical care costs (Fries, Koop, Sokolov, & Beadle, 1998).

The paradigm of a long, healthy life with a relatively rapid terminal decline represents an attainable ideal. Health policies must be directed at modifying those health risks that precede and cause morbidity if this ideal is to be approached for a population.

JAMES F. FRIES

See Also
Life Extension

Internet Key Words
Health Promotion
Healthy Aging
Morbidity Compression

Internet Resource
Health Project
http://healthproject.stanford.edu

MULTIDIMENSIONAL FUNCTIONAL ASSESSMENT: OVERVIEW

Increase in the number and proportion of elderly persons in the population, as well as recognition that their problems are even more likely than those of younger persons to be interrelated, has resulted in concern for comprehensive multidimensional functional assessment. Such assessments are intended not only to provide an overall view of personal status, identifying those areas where status is adequate as well as those where decrement is present, but also to provide a rational basis, within an integrated framework, for allocating services and assessing service impact. Emphasis is on functional

status rather than on diagnosis, because the level of personal independence is the relevant issue, for which diagnosis is not necessarily informative.

Historically, such evaluations have been used in surveys of the elderly and when it was considered important to obtain comprehensive, detailed information. As such, time constraints were somewhat relaxed. With increased recognition of the value of a multidimensional approach, however, there have been attempts to develop brief measures for use in the regular outpatient setting and as intake screens by service agencies.

A review of the major English-language multidimensional assessments indicates an important core of agreement regarding seven dimensions that should be examined: (1) activities of daily living (ADL), (2) physical health, (3) mental health, (4) social resources, (5) economic resources, (6) environmental matters, and (7) level of strain on the caregiver. The first three dimensions relate to personal functioning; the other four dimensions reflect both broader societal concerns and external conditions influencing continued community residence.

Each of these seven core dimensions is itself multidimensional. They vary regarding their subdimensional content and the population for which they are intended. For example, inclusion of basic ADL and instrumental ADL may vary, depending on whether the focus is on community-dwelling elderly, the institutionalized, or both. Physical health assessments are typically based on a combination of topics selected from self-assessments of health, presence of physical symptoms, diagnosed illnesses and conditions and the extent to which these interfere with usual activities, prescribed medications, level of activity and measures of incapacity, and use of medical services. Similarly, mental health assessments may include brief screens of cognitive functioning, depression, and psychiatric disorder; personal assessment of emotional well-being; and indicators of the quality of mental health functioning. Assessment of social resources generally includes information on the extent and adequacy of contact with family and friends, the anticipated availability of help from these sources, and participation in social activities. The underlying concern in the area of economic resources is to determine whether income is adequate. Sources of income may be reviewed to promote an accurate response and identify sources for which the individual may be eligible.

Because environmental matters (e.g., physical accessibility, social milieu) can affect the feasibility of independent living, some multidimensional assessments inquire about the structural properties of the dwelling, access to services, and local ambiance. Special measures such as the Multiphasic Environmental Assessment Procedure (MEAP) (Moos & Lemke, 1984), whose primary focus is the residential environment and its livability, have been developed.

Help from the family may determine the feasibility of home living, but such help often places considerable stress on the caregiver. As a result, assessment of caregiver strain and service needs has become a concern (Gwyther & Strulowitz, 1998; Ory, Hoffman, Yee, Tennstedt & Schulz, 1999). Such measures, however, are not yet routinely included in multidimensional assessments. Also absent, but potentially important because of its apparent impact on health status, is information on religiousness or spirituality.

Many multidimensional assessment measures have been developed, but validity and reliability have been determined for only a few. Primary among these are the Comprehensive Assessment and Referral Evaluation (CARE) (Gurland et al., 1977–78), Iowa Self-Assessment Inventory (ISAI) (Morris et al., 1990), Multilevel Assessment Inventory (MAI) (Lawton, Moss, Fulcomer, & Kleban, 1982), Older Americans Resources and Service (OARS) (Fillenbaum, 1988; Maddox, 1979), and Functional Assessment Inventory (FAI) (Cairl, Pfeiffer, Keller, Burke, & Samis, 1983), which is a modification of OARS. With the exception of the ISAI, they are typically administered by trained nonprofessionals, although OARS can also be self-administered.

CARE was designed to be a clinical instrument to examine medical, psychiatric, and social problems of elderly persons in the community. Considerably refined since its introduction, CARE now

Overview of Content, Samples, and Administration of Selected U.S. Multidimensional Functional Assessment Questionnaires

	CARE[a]	MAI	OARS	SOS[b]
Activities of daily living				
Instrumental	++	++	++	++
Basic	++	++	++	++
Physical health				
Self-assessment	+	+	+	+
Symptoms	++	0	0	0
Diagnosis, medications	+	+	++	+
Medical services	+	+	+	+
Level of activity	+	++	++	0
Social impairment (bed days)	++	++	++	0
Mental health				
Cognitive functioning	++	++	++	+
Symptoms/ diagnosis	++	+	+	0
Self-assessment	+	+	+	+
Excellence of functioning	+	0	+	0
Social resources				
Contacts with family, friends	++	++	++	+
Availability of help	++	+	++	0
Economic				
Income, source	+	+	++	+
Environmental				
Environmental matters	++	++	0	+
Caregiver				
Level of strain	++	0	0	0
Other areas examined				
Nutrition	++	+	+	0
Time use	0	++	0	0
Services	+	+	++	0
Summary scores	++	++	++	+

0 = absent or essentially so, or not applicable.
+ = present, minimal; or type.
++ = present, adequate.
[a]Information applies to the original CARE instrument. For recent modified versions, consult references.
[b]Service Outcome Screen, now called Pathfinder Profile.

affords users the unique opportunity of selecting from a variety of brief, statistically derived scales to tailor a purpose-specific questionnaire (see Gurland, Golden, Teresi, & Challop, 1984). The MAI is based on Lawton's (1972) conceptual model of the well-being of older people. It is a "nested" instrument, permitting the extraction of medium- and short-length forms when time does not permit administration of the preferred full-length questionnaire. The OARS questionnaire operationalizes a program evaluation and resource allocation model. It places equal emphasis on determination of service use and assessment of all levels of functional status. Accordingly, it facilitates assessment of service impact. OARS data from various studies are archived, permitting OARS users to compare their data with those of others. The comparability of CARE, ISAI, MAI, and OARS in assessing functional status remains to be determined. They represent, however, a set of valid and reliable instruments of proven value.

Brief and therefore less comprehensive multidimensional assessments include the 36-item Medical Outcome Study Short-Form Health Survey (SF-36) (Ware, 1996; Ware & Sherbourne, 1992), the 17-item Duke Health Profile (Parkerson, 1998; Parkerson, Broadhead, & Tse, 1990), and the Functional Status Questionnaire (FSQ) (Jette, Davis, & Cleary, 1986). Administration time is 5 to 10 minutes for the SF-36, less than 5 minutes for the Duke Profile, and 15 minutes for the FSQ. The last two measures are specifically designed for primary care use. Each assesses physical and mental health, ADL, and social functioning. In addition, the FSQ is concerned with work performance and sexual relationships.

A measure originally called the Service Outcome Screen (SOS) and recently renamed the Pathfinder Profile (Maddox & Bratesman, 1997) is unique, in that it links agency requirements for comprehensive information with a rapid multidimensional review of client status. In particular, it identifies the urgency with which intervention is needed. Like the FSQ, which is designed to print out a patient profile, the Pathfinder Profile provides printouts that describe patient status and fulfill mandated agency reporting requirements.

There has been extensive discussion of and agreement on the need for multidimensional functional assessments for general use in geriatric evaluation (Rubenstein & Rubenstein, 1998). Compared with multidimensional functional assessment, geriatric assessment has a stronger focus on identifying undiagnosed conditions and instituting relevant intervention. When targeted to those patients most likely to benefit (those neither too well nor too sick), declines in recurrent hospitalization, length of stay, and mortality have been found, and a cost saving generally occurs (Rubenstein & Rubenstein, 1998). To date, no uniform measure has been developed, but there is interest in this area.

Current attempts to reduce health care costs may be linked to increasing attempts to develop brief comprehensive functional assessments for use in primary care, so that problems can be identified at an earlier, more treatable, stage. How far it is possible to do that remains to be determined.

GERDA G. FILLENBAUM

See Also
Multidimensional Functional Assessment: Instruments

Internet Key Words
ADL
Disability
Geriatric Evaluation
IADL
Multidimensional Assessment
Multidimensional Environmental Assessment Procedure
Quality of Life

MULTIDIMENSIONAL FUNCTIONAL ASSESSMENT: INSTRUMENTS

Multidimensional function assessment was developed to facilitate comprehensive assessment and effective *resource allocation* among older people. A variety of methods and tools are available for those purposes: Older Americans Resources and

Services (OARS), Multilevel Assessment Inventory (MAI), Minimum Data Set (MDS), and SPICES (an acronym for sleep disorders, problems eating or feeding, incontinence, confusion, evidence of falls, and skin breakdown).

OARS

The assessment form for OARS is the OARS Multidimensional Functional Assessment Questionnaire, or OMFAQ. The OMFAQ has been updated to reflect changes in the aging population and has been used in hundreds of research studies and geriatric clinical settings across the country. Normative data sets on community elderly are based on data from Durham, North Carolina (minimum age 65, N = 998), Virginia (minimum age 65, N = 1,530), and Cleveland, Ohio (minimum age 65, N = 1,609). The Cleveland study includes longitudinal data.

The OMFAQ is best understood in the context of the OARS model, which consists of three elements: (1) population classification according to functional status, (2) classification of service utilization, and (3) development of a transition matrix of services and functional states.

The OMFAQ has two parts. Part A assesses overall personal functioning in five dimensions: social, economic, mental health, physical health, and self-care capacity. The social resource dimension consists of 14 items, including marital status, living arrangement, social supports, and interaction. The economic resource dimension has 16 items about employment, income, and insurance status. The mental health dimension has 6 items, including life satisfaction and depression. The physical health dimension has 19 items, including inpatient days, nursing home days and medication in the last six months, current illness, other physical problems, and supportive devices. The self-care capacity dimension has 15 items consisting of both instrumental activities of daily living (IADL) and physical activities of daily living (PADL) sections.

The theoretical basis and practical relevance of the IADL section of the OMFAQ were based on the earlier works of Lawton and Brody (1969). IADL consist of seven items: telephone use, travel,

shopping, meal preparation, housework, taking own medicine, and handling personal finances. Unlike the original Lawton and Brody scale, all items are asked of both men and women. The assessment of performance level is the same for all items: each is evaluated according to whether the activity in question can be performed unaided, whether some help is needed, or whether the activity cannot be performed at all. The PADL section of the OMFAQ, developed from the Katz ADL (described later), consists of seven items: eating, dressing, grooming, walking, getting in and out of bed, bathing, and toileting. Part A includes a final summary rating that has been developed for each dimension. Possible scores range from excellent functioning (1) to totally impaired (6). These summary ratings provide a profile of individual functioning, highlighting particular areas of functioning that require further attention.

Part B of the OMFAQ assesses the extent of utilization in each of 24 generically defined service areas, including transportation, social and recreational services, employment services, educational services, mental health services, nursing care, physical therapy, continuous supervision, relocation and placement services, homemaker services, and systematic multidimensional evaluation. Service providers can be both formal and informal. For each service area, the elderly person is asked about actual use in the past six months and self-perception of the need for such services.

A trained interviewer who has at least a high school education typically fills in the OMFAQ, but the items can also be self-administered. By administering parts A and B to the same subjects at periodic intervals, the data can be used to develop a transition matrix that can assess the impact of generically defined service packages on a target population (Fillenbaum, 1988).

MAI

The MAI was developed at the Philadelphia Geriatric Center in 1982. It is based on a conceptual model of the well-being of older people defined by Lawton and colleagues (1982), who described human competence as a variety of tasks. The lowest task level, life maintenance, is followed by more complex levels of functional health, perception-cognition, physical self-maintenance, instrumental self-maintenance, effectance (activity emanating from motivation to explore), and social behavior. The MAI systematically assesses human well-being in the domains of physical health (self-rated health, health behavior, health conditions), cognition (mental status, cognitive symptoms), ADL (PADL, IADL), time use, social interaction (friends, family), personal adjustment (morale, psychiatric symptoms), and perceived environment (housing quality, neighborhood quality, personal security).

MDS

The MDS differs from other multidimensional functional assessment instruments, in that its purpose is to provide a scientific basis for planning and delivering care. The MDS was developed by a Health Care Financing Administration contract, and its use is mandated in all nursing homes certified to participate in Medicare and Medicaid programs. The information collected primarily by nursing staff includes cognitive pattern, communication problems, vision problems, mood and behavior, psychosocial well-being, physical functioning, continence, disease, health conditions, oral and nutritional status, skin condition, activity pursuit patterns, medications, special treatments, and discharge potential (Morris, Murphy, & Nonemaker, 1995).

The MDS's value in clinical and research settings has been internationally recognized. It has been translated and validated in more than 15 countries. A home-care version of the MDS, the MDS-HC, includes two-thirds of the nursing home MDS items and has additional items in social functioning, informal support, IADL, and environment. Some of the nursing home MDS items were deleted to cover the different and complex needs in a home-care setting.

SPICES

The SPICES is a very short assessment tool for the elderly in a variety of settings (Fulmer, 1991).

Although easy to remember, the SPICES should not be used as a replacement for a complete nursing assessment.

Physical Function Instruments

Instruments for physical function include the Katz ADL and the Barthel index. The Katz ADL is the most widely used in a variety of settings and includes bathing, dressing, eating, transferring, continence, and grooming. The Barthel index has been particularly useful in rehabilitation settings to monitor improvement over time. It includes bathing, grooming, continence, stair climbing, and the ability to propel a wheelchair.

YUKARI YAMADA
NAOKI IKEGAMI

See Also

Activities of Daily Living
Geriatric Evaluation and Management Units
Geriatric Interdisciplinary Health Care Teams
Measuring Physical Function
Multidimensional Functional Assessment: Overview
Occupational Therapy Assessment and Evaluation

Internet Key Words
Functional Assessment

Internet Resources
Duke University Center for the Study of Aging and Human Development University
http://www.geri.duke.edu/index.html

John A. Hartford Foundation Institute for Geriatric Nursing
http://www.nyu.edu/education/nursing/hartford.institute.

MUSCLE CRAMPS

Painful and prolonged involuntary muscle contractions, or cramps, are common among older individuals. More than 70% of elderly people (Man-Son-Hing & Wells, 1995) and between 35% and 95% of all adults have experienced muscle cramps at least once in their lives (Leclerc & Landry, 1996). This prevalence might be underestimated, since muscle cramps are most likely reported to the physician only if they are experienced frequently (Oboler, Prochazaka, & Meyer, 1991).

Modern neurophysiological research strongly suggests that muscle cramps are caused by excitation of spinal motoneurons, mediated by changes in presynaptic input (Jansen, Lecluse, & Verbeek, 1999). Careful attention to signs and symptoms should differentiate between true muscle cramps and contracture, tetany, and dystonia (McGee, 1990).

True muscle cramps include ordinary or idiopathic cramps of benign origin, as well as muscle cramps associated with specific conditions.

True muscle cramps are related to electrical motoneuron hyperactivity that produces sustained muscle spasms. They often are preceded by repetitive contractions of isolated motor units, clinically visible by muscle twitches or fasciculations.

A *contracture* is an electrically silent, involuntary muscle contraction related to missing muscle relaxation. Unlike ordinary cramps, contractures do

Etiologies of True Muscle Cramps

1. Ordinary (idiopathic) cramps
2. Muscle cramps associated with specific conditions
   - Altered fluid and electrolyte levels
     Hypoglycemia
     Severe hyponatremia with salt depletion
   - Hemodialysis
   - Drug-induced cramps
     Nifedipine
     Beta agonists
     Clofibrate
     Penicillamin
     Alcohol
   - Lower motoneuron disease
     Amyotrophyic lateral sclerosis
     Polyneuropathies
     Recovered poliomyelitis
     Peripheral nerve injury
     Nerve root compression

not occur at rest but occur during exercise and in association with metabolic myopathies or thyroid disorders.

Tetany is a clinical syndrome of generalized nerve hyperexcitability producing muscle spasms, sensory hyperactivity, and paresthesias. It is usually caused by hypocalcemia, respiratory alkalosis, hypokalemia, or hypomagnesemia.

Dystonia is characterized by sustained contraction of agonist and antagonist muscles. It may manifest as occupational cramps in patients who have spent years mastering fine hand motor control for a specific task (e.g., pianist, typist), or it may be induced by antipsychotic medications.

Ordinary (Idiopathic) Cramps

Ordinary cramps, the most common form of muscle cramps, are localized, involuntary, visible, and usually painful contractions of skeletal muscle. They occur preferentially and sporadically at night in the lower extremities (calf muscles) and usually last a few seconds to a few minutes. Ordinary cramps can be familial (Jacobsen, Rosenberg, Huttenlocher, & Spire, 1986) and are more frequent in individuals with well-developed muscles (Muir, Davidson, Percy-Robb, Walsh, & Passmore, 1970) or cirrhosis (Abrams, Concato, & Fallon, 1996). The cramp begins with a voluntary contraction starting at a muscle's shortest position. This may explain the susceptibility to night cramps, when, because of ankle plantar flexion (i.e., under heavy bedcovers), the calf muscles are most shortened and therefore in their most vulnerable position.

The most effective treatment for an acute muscle cramp is stretching the affected muscle (Leclerc & Landry, 1996). Calf muscle cramps are relieved by placing the foot firmly in dorsiflexion, either by hand or by standing. The calves can be stretched by leaning forward against a wall from a distance of two feet, keeping the feet flat and heels on the floor. Simple mechanical prevention of benign nocturnal leg cramps includes stretching the calf muscles several times a day, using light bedclothes or a footboard during sleep, or lying prone with the feet dangling over the edge of the bed. Pharmacological prophylaxis of ordinary muscle cramps should be considered only if these simple mechanical measures fail and the patient's quality of life is affected significantly.

Quinine sulfate, commonly used at higher doses for malaria treatment, is effective but controversial for cramp prophylaxis. In a meta-analysis including 107 patients from six trials, quinine reduced the number of nocturnal leg cramps in elderly people by 27%, compared with placebo (Man-Son-Hing & Wells, 1995). The U.S. Food and Drug Administration bans quinine-based over-the-counter preparations for nocturnal leg cramps due to its potentially severe side effects. The typical dose for cramp prevention is 200 to 300 mg, taken regularly at bedtime as opposed to on an as-needed basis; four weeks of treatment may be necessary to show a beneficial effect (Man-Son-Hing & Wells, 1995). Drug interactions and serious side effects such as quinine-induced thrombocytopenia, potentially fatal hypersensitivity reactions, visual toxicity, cinchonism (nausea, vomiting, tinnitus, and deafness), and ventricular arrhythmia require close patient monitoring and careful consideration of risks versus benefits. Patients need to be educated realistically and advised of the potential efficacy and side effects of quinine (Leclerc & Landry, 1996). Other proposed pharmacological treatments include calcium blockers, vitamin E, carbamazepine, diphenhydramine, phenytoin, methocarbamol, and riboflavin. To date, no randomized controlled trials proving the efficacy of such pharmacological treatments have been reported.

Muscle Cramps Associated with Specific Conditions

Fluid and electrolyte disorders due to decreased fluid intake, drug side effects, or hypoglycemia can cause muscle cramps in older adults. *Hyponatremic cramps*, however, occur only when hyponatremia is accompanied by salt depletion (McGee, 1990), as induced by severe diarrhea, a salt-free diet, or hemodialysis treatment. Hyponatremic cramps are cured by fluid and saline substitution. No cramps occur in patients with hyponatremia and normal or expanded total body sodium, such as in the syndrome of inappropriate secretion of antidiuretic hormone or uremia. Commonly, prescribed drugs such as diuretics, selective serotonin reuptake inhibitors, or carbamazepine can induce hyponatremia in the elderly and have to be considered as a potential cause when cramps occur. Muscle cramps related to hypoglycemia are relieved by glucose substitution.

Hemodialysis leg cramps are a common complication during dialysis treatment. Cramps can occur toward the end of a dialysis session or under a high ultrafiltration rate, when large amounts of fluid must be drawn. Dialysis cramps are immediately relieved by intravenous injection of hypertonic dextrose or hypertonic saline and can be prevented by using a high-sodium dialysate.

Drug-induced cramps are associated with nifedipine, beta agonists, clofibrate, penicillamine, and alcohol. The cramps usually reverse after drug cessation.

Lower motoneuron diseases such as amyotrophic lateral sclerosis, polyneuropathies, recovered poliomyelitis, peripheral nerve injury, and nerve root compression are associated with muscle weakness, muscle hypotrophy, and other signs of muscle denervation. These cramps are particularly difficult to treat; the prognosis is closely linked to the underlying disease.

RETO W. KRESSIG
JEAN-PIERRE MICHEL

Internet Key Words
Hemodialysis Cramps
Leg Cramps
Muscle Cramps
Quinine Sulfate
Stretching
Tetany

Internet Resources
Mayo Foundation for Medical Education and Research
http://www.mayohealth.org

University of Iowa Health Care: The Virtual Hospital
http://www.vh.org

NATIONAL ASIAN PACIFIC CENTER ON AGING

The National Asian Pacific Center on Aging is a nonprofit advocacy organization that promotes the well-being of Asian and Pacific Islander American older adults. The organization operates employment programs for older workers and conducts policy and research projects among 10 geographic multiethnic and multilingual Asian and Pacific Islander American elder communities. A particular focus is the development of useful, accessible, and accurate information on health care access. Materials on the Medicare, Medicaid, and Dual Eligible State Buy-in programs and affordable health care for low-income elders and other information are available on its multilingual Web site: www.napca.org. The center can also be contacted at its offices: National Asian Pacific Center on Aging, 1511 Third Avenue, Suite 914, Seattle, WA 98101-1626, telephone 206-624-1221, fax 206-624-1023.

DONNA L. YEE

NATIONAL COUNCIL ON THE AGING

The National Council on the Aging (NCOA) is the nation's first association of organizations and professionals dedicated to promoting the dignity, self-determination, well-being, and contributions of older persons. NCOA members include senior centers, adult day facilities, congregate meals sites, faith congregations, area agencies on aging, and other community service organizations.

Founded in 1950, NCOA helps community organizations enhance the lives of older adults, turns creative ideas into programs and services that help older people in hundreds of communities, and is a national voice and powerful advocate for public policies, societal attitudes, and business practices that promote vital aging. NCOA was instrumental in the development of Foster Grandparents, Meals On Wheels, Family Friends, and dozens of other innovative programs for older adults.

NCOA helps community service organizations by providing leadership, technical assistance, tools, and training to community organizations. NCOA conducts research and demonstration projects on the impact of promising innovations and supports the adaptation of proven innovations throughout the nation. NCOA also partners with thousands of community organizations to help older persons by sponsoring community service jobs for more than 8,400 low-income seniors each year in 19 states.

See NCOA's Web site at www.ncoa.org for more information, or write to National Council on the Aging, 409 Third Street NW, Suite 200, Washington, DC 20024.

MICHAEL REINEMER
KANIKA P. MODY

NATIONAL INDIAN COUNCIL ON AGING

Since it was founded by the National Tribal Chairmen's Association in 1976, the National Indian Council on Aging (NICOA) has served as the nation's foremost nonprofit advocate for American Indian and Alaskan Native elders. For two decades, NICOA has provided leadership and effective advocacy for Indian aging issues. The organization has been actively involved in public policy and research efforts on federal, state, and local levels.

Board of Directors: The organization is governed by a 13-member board of directors—all Indian elders—representing each of the federal Bureau of Indian Affairs regions, plus a representative of the National Association of Title VI Grantees, who represents the interests of 229 reservation-based senior programs. Board members, who serve four-year terms, are elected by tribal elders from their respective regions. The organization's mission—"to bring about improved, comprehensive services to American Indian and Alaska Native elders"—is based on a five-point action plan formulated from recommendations at the 1976 and 1978 national conferences on Indian aging.

Publications and Advocacy: NICOA's publications on a variety of Indian aging issues have been widely distributed and cited. *The NICOA Report: Health and Long-Term Care for Indian Elders,* published in 1996, remains the most comprehensive attempt to document the health and functional status of American Indian elders. The organization continues to advocate on the basis of its foremost policy statement, the *National Indian Aging Agenda for the Future,* which was last updated in 1994.

NICOA is a recognized authority on issues of demographics, quality of life, and public policy issues pertaining to Indian elders. The organization has presented expert testimony before subcommittees of the U.S. Congress on many occasions and has been actively involved in multiple reauthorizations of the Older Americans Act (OAA). Both the OAA and the Indian Health Care Improvement Act contain significant NICOA recommendations.

Indian Health Data Projects: Since the establishment in 1996 of its Indian Health Data Bureau, NICOA has established a series of interrelated, ongoing grants with the Administration on Aging, Administration for Native Americans, and Indian Health Service (IHS). Focusing on the provision of current, accurate health information to tribal, IHS, and urban Indian health care providers, NICOA has acquired all Medicare and Medicaid data for Indians. This data, when compared with IHS patient data, the U.S. census, and many other national and state databases, is providing a depth and range of analysis never before available in Indian country (or possibly anywhere else).

NICOA's Indian Health Data Bureau team offers one of the nation's foremost Geographic Information System (GIS) mapping capabilities. NICOA's pioneer work with GIS mapping for health care providers led to a new contract with the $150 million IHS diabetes program. That project is resulting in the first automated GIS reporting and analysis of clinic and hospital data.

NICOA, as one of 10 national contractors, also administers a $6 million Department of Labor program that provides Senior Community Service Employment Program (SCSEP) job training and placement for more than 800 elders in 14 states. The organization maintains its national headquarters in Albuquerque, New Mexico, and smaller SCSEP offices in Phoenix, Arizona, Lawton and Oklahoma City, Oklahoma, Southfield, Michigan, and several other states.

DAVID BALDRIDGE

NATIONAL LONG TERM CARE OMBUDSMAN PROGRAM

Nursing home residents and their families and friends often face situations that they would like to discuss with someone who could help them understand and resolve the issues. The long-term-care ombudsman is a trained advocate to protect the health, safety, welfare, and rights of residents of nursing homes, board and care facilities, and assisted living facilities, many of whom are elderly, frail, and isolated from their communities. Ombudsmen work to promote residents' rights, quality of care, and quality of life and to facilitate change at local, state, and national levels to improve care. Thousands of volunteers and paid staff regularly visit long-term-care facilities to monitor care, observe staff-resident interactions and the physical environment, and voice the concerns of residents who lack family or are otherwise unable to speak and advocate for themselves.

The word "ombudsman" derives from the Swedish language and connotes a public official appointed to investigate citizens' complaints against local or national government agencies that may be infringing on individuals' rights. Ombudsmen are also described as liaisons, supporters, and friends who can provide information and guidance.

The Long Term Care Ombudsman Program (LTCOP) first appeared in 1972 as a five-state demonstration program in response to widespread reports of poor quality in nursing homes. The LTCOP operates in all 50 states, Washington D.C., and Puerto Rico. Under the aegis of the Older Americans Act (OAA) since 1978, the LTCOP is funded under Title VII (Vulnerable Elder Rights Protection Activities), with additional funding from state and local governments and community agencies. The Office of the State LTCOP is most often housed in the state unit on aging. Each state designates a long-term-care ombudsman who is responsible for developing a statewide program to identify, investigate, and submit complaints on behalf of residents. Local or regional ombudsmen are typically located in area agencies on aging; however, many are sponsored by private programs and legal services.

Ombudsmen are not regulators or surveyors. Their role is to identify and resolve problems on behalf of residents. Their responsibilities, described in Title VII of the OAA, include representing residents' interests before government agencies and seeking administrative, legal, and other means to redress those interests; consumer education and facilitation of public comment on laws, policies, and actions; technical support for new or ongoing resident and family councils; and analysis and comment on recommended changes in laws and regulations applicable to long-term-care residents.

Ombudsmen receive training on residents' rights and how these rights should be respected by long-term-care facilities. Additional training can include problem-solving and conflict-resolution skills, communication skills, the administrative and clinical structure and process of nursing home care, and interviewing skills.

In 1998, more than 900 paid ombudsmen and 7,000 certified volunteer ombudsmen visited facilities and addressed 201,053 complaints. These complaints were about meals; lost items; resident care (32% of all issues), including personal care, restraints, and rehabilitation; and residents' rights (31%), such as personal rights, admissions and evic-

tions, and abuse and neglect. More than 70% of complaints are resolved to the satisfaction of the resident or complainant. If a problem is not resolved, the ombudsman can suggest alternatives, such as filing a complaint with the state's regulatory agency.

Ombudsmen meet with facility staff and residents to educate them about residents' rights. Ombudsmen can attend resident and family councils in the facility and encourage residents and families to do so, also.

A resident or family member's name cannot be used in the follow-up of a complaint, unless permitted by the resident or person acting on the resident's behalf. Ombudsman records are also confidential and cannot be read by facility staff.

Ombudsmen encourage residents and families to use the care planning process to ensure that residents receive individualized care. Care planning sessions are held quarterly or when changes in the resident's condition warrant and should include facility staff (including nursing assistants), family, and the resident. The ombudsman can suggest how to make these sessions more productive and how to use the sessions to address problems. At the resident's request, the ombudsman may attend the care planning session to advocate for and assist the resident. In addition, ombudsmen work to address systemic issues to improve quality of care.

Families use the LTCOP for assistance in finding facilities for their loved ones. Among other things, families are advised to examine the most recent survey (inspection) results, which must be posted in a visible, accessible location in every nursing home and can also be found on the Nursing Home Compare Web site. Families are encouraged to visit facilities at different times of the day to observe the care being given and to ask specific questions about staff availability and training, activities, and meals. Many ombudsman programs have directories of facilities in the area that are made available to prospective residents and families. In 1998, over 200,000 queries were answered, including explanations of the differences between nursing homes and assisted living facilities and information about community alternatives such as adult day care and home care.

Program effectiveness varies from program to program, often due to the limited number of ombudsmen. There is a limited ombudsman presence in board and care facilities. The Institute of Medicine's evaluation of the LTCOP (1995) reported that the program performed a vital public service and had improved long-term care by its system-wide advocacy and educational efforts and its communication with state and federal policy makers, regulatory agencies, provider associations, and other consumer interest groups. Quality-of-care and quality-of-life improvements supported and enhanced by the LTCOP include increased personal needs allowances, protection from involuntary discharge and room change, reduced use of psychoactive medications and physical restraints, improved building and fire safety codes, and advance directives education and utilization (Institute of Medicine, 1995). In Oregon, the LTCOP opened nursing homes to public scrutiny, resolved specific care issues, promoted and protected residents' rights, and improved the quality of care (Nelson, 1995). Additional research is needed on the association between ombudsman efforts at the individual and system levels and specific quality-of-care and quality-of-life outcomes.

The Ombudsman Resource Center, housed at the National Citizens' Coalition for Nursing Home Reform in Washington, D.C., provides technical assistance, training, and referrals. To find an ombudsman, contact the center at 202-332-2275, or visit its Web site.

ALICE H. HEDT

Internet Key Words
Care Plans
Long-Term Care
Nursing Home
Ombudsmen
Resident Rights

Internet Resources
Consumer Coalition on Assisted Living
http://www.ccal.org

Eldercare Locator
http://www.aoa.dhhs.gov/AOA/dir/91.html

National Citizens' Coalition for Nursing Home Reform
http://www.nccnhr.org

Nursing Home Compare, Health Care Financing Administration
http://www.medicare.gov/nursing/home.asp

NATURALLY OCCURRING RETIREMENT COMMUNITIES

Naturally occurring retirement communities (NORCs) are housing developments, buildings, or neighborhoods that were not planned or designed for older people but have evolved to house primarily people 60 years of age or older (Hunt & Gunter-Hunt, 1985; UJA, 1992). A 1992 survey by the American Association of Retired Persons found that 27% of all older Americans live in NORCs, compared with 6% in planned senior housing or retirement communities. In 1992, 86% of older Americans wanted to stay where they were and not move as they aged, compared with 78% in the late 1980s (Lanspery & Callahan, 1994). "Since NORCs were not designed for the elderly, they usually lack the health and social services as well as the physical amenities that a growing share of older residents need to maintain satisfying and independent lives" (UJA, 1992, p. 48). Literature on NORCs includes how to define and identify a NORC, descriptions of NORCs that currently exist, how NORCs are funded and staffed, and how to create a NORC. However, little literature exists on the delivery of interdisciplinary health care services for older adults living in NORCs.

NORCs vary by physical size, population characteristics and size, and reasons for existence. Populations age differently: accumulation occurs when older people are left behind in an area that more mobile residents are leaving; recomposition occurs when older people are drawn to areas that other residents are leaving; congregation occurs when migrants of all ages are drawn to an area, but older people are drawn more than younger people (Lanspery & Callahan, 1994). NORCs have been defined by how populations age. Recomposition and congregation NORCs have been subdivided into three types. In rural areas, for example, amenity NORCs typically attract younger, healthier, better educated people seeking a lifestyle rich in rural amenities such as lakes and forests; convenience NORCs typically attract people who are moving from a rural area to a nearby rural community offering a more convenient lifestyle, often because of declining health, retirement, or widowhood; and bifocal NORCs tend to attract people similar to those in amenity NORCs but who wish to remain close to family and friends, like those who move to convenience NORCs (Lanspery & Callahan, 1994).

The success of NORC programs is dependent on a responsible lead agency, organization, or group that is recognized in the geographical area and experienced in leadership, entrepreneur efforts, data collection and research, and cooperation with other agencies, organizations, or groups (Lanspery & Callahan, 1994). Hunt and Ross (1990) found three main attributes important in attracting older people to apartment complexes that became NORCs: (1) location, with proximity to shopping, service facilities, and friends and family; (2) management, in terms of a well-maintained complex and the ability to maintain a stream of referrals; and (3) design, to promote independent living. Staffing of supportive service programs in NORCs varies but generally includes social workers, secretaries, case managers, volunteers, and occasionally nurses. Hiring of professional staff is often dependent on outside funding. NORCs generally have no specific sources of funding and may depend on insurance, government payments, block grants, and out-of-pocket monies. Supportive services are financed through government funds, NORC owners, user fees, refinanced mortgages, charitable contributions, and insurance plans (Schwartz, 1991).

NORCs are located in rural and urban areas throughout the United States. There are 10 NORCs in the New York City area alone, mainly due to the dense elderly population. These NORCs have service programs in place, but few provide the interdisciplinary primary care services needed to deal with the frailty that comes with increasing age and elders' preference for conveniently located health care services. For example, Penn South Houses in New York City is a 2,820-unit co-op with 6,200 residents, over 70% of whom are elderly (Schwartz, 1991). It has the largest share of elderly people of any NORC in New York City. The Penn South Program for Seniors (PSPS) was the first NORC supportive services program in the United States, providing basic health and social services for older adults in the Penn South community since 1986. Residents in the program vary in terms of health and functional status, social problems, and needs. PSPS staff, as well as residents of the Penn South NORC themselves, began seeking out primary care services in the local area to assist in coordination of care. Many residents had not seen a primary care provider in many years or were physically unable to get to their primary care providers and thus had unmet medical, social, and psychological needs requiring assessment, management, treatment, and prevention.

To meet the health care needs of Penn South residents, Saint Vincents Hospital and Medical Center of New York's Section of Geriatric Medicine developed a model for the provision of care to clients in a NORC. A satellite outpatient geriatric medical practice, Saint Vincents Senior Health at Penn South (Senior Health), opened in June 1995 in the Penn South NORC. Many of the referrals received from the registered nurses or social workers at the PSPS are for primary care, because many NORC residents lack primary care providers. Senior Health is staffed by an interdisciplinary team of geriatricians, geriatric fellows, geropsychiatrists and geropsychiatric fellows, social worker, gerontological nurse practitioner, registered nurse, licensed practical nurse, and secretary. An intake interview to assess each client's needs may be conducted in person or by telephone by a nurse or social worker if the resident or family requests it before the first visit. Patients with dementia, urinary incontinence, complex teaching needs, or noncomplex medical diagnoses are usually assigned to the nurse practitioner; more complex medical cases are assigned to one of the physicians, all of whom have admitting privileges at Saint Vincents Hospital Medical Center. Senior Health staff provide both primary care and consultative services for residents with their own primary care providers. The most frequent use of consultative services is for geropsychiatric evaluation, treatment, and management of depression, anxiety, and dementia, especially when combined with psychotic features.

Interdisciplinary team members assess and manage each client's physical, social, psychological, functional, nutritional, financial, and spiritual needs. Health maintenance and disease prevention needs are continually monitored, as well as medication issues such as polypharmacy and drug-drug interactions. Additionally, health teaching is given to patients, caregivers, companions, home health aides, home attendants, family members, neighbors, or other health care professionals as needed. No particular health education topic predominates; rather, topics are individualized to patients' needs. All those involved in the care of the patient are considered integral members of the health care team. The cost to the patient is the same as for a visit to any health care provider. Most Penn South residents have Medicare or Medicaid, which pays for their health care visits.

Senior Health appeals to many Penn South residents because it is on the premises and Saint Vincents Hospital is the receiving tertiary care hospital for 911 and emergency calls. Many Penn South residents chose Senior Health for primary care either because they did not have primary care providers or because they wanted primary providers close to home. The focus of all care is on health promotion, disease prevention, and the treatment and management of current medical, psychiatric, functional, nutritional, social, and financial problems. Care also includes the identification and management of geriatric syndromes, such as urinary incontinence, falls, malnutrition, and immobility. Senior Health and PSPS staff work collaboratively; referrals are made to and from each program. Providing on-site primary care services minimizes hospitalization and institutionalization and assists NORC staff in keeping residents in their homes and communities.

Although literature exists about NORCs in both rural and urban areas in the United States, it is unclear whether NORCs exist in other countries and, if so, how they are defined, staffed, and funded. It would be interesting to compare and contrast housing developments, specifically NORCs, in the United States and other countries. These are areas for further research. One can find information about NORCs from local area agencies on aging; local housing authorities; the American Association of Retired Persons; various local and federal foundations that support social, health, and general service programs, such as the UJA–Federation of Jewish Philanthropies in New York and the Robert Wood Johnson Foundation; and the U.S. Census Bureau (Lanspery & Callahan, 1994).

SHERRY A. GREENBERG

Internet Key Words
Housing and Aged
Naturally Occurring Retirement Communities
Retirement Communities

NURSING HOME ADMISSION

In Western industrialized countries, admission to a nursing home is a significant event in the lives of older people. The number of people requiring nursing home admission is expected to rise as the population ages and as the number of frail aged and those suffering with dementia increases. Placement of a relative in a nursing home is generally frowned upon by society; home care is viewed as best, and nursing home care as a last resort (Smallegan, 1981). The decision to place a relative in a nursing home often occurs after a long and stressful period of home care. For many older people with declining health, nursing homes are equipped to offer better care than the home environment. To qualify for nursing home admission, a person must require continuous nursing care or be chronically ill but not so sick as to require hospital care. It is of some concern, therefore, that cultural perceptions undermine the critical social value of nursing homes.

An extensive body of literature describes the prevalence of caregiver stress in the community; a growing amount of literature is specifically concerned with caregiver stress as it relates to nursing homes. Increasingly, caregivers are themselves elderly. Several studies indicate that caring has different meanings and results in different responses in elderly caregivers, especially spouses, as compared with younger ones. Wenger (1990) suggests that for a spouse, caring is associated with intimacy, companionship, and reciprocity. When relationships are seen to be enduring and rewarding, attitudes toward care are more positive. Elder people were found to be less inclined to complain or to seek help and more likely to sustain a caring role despite the severe incapacity of the care recipient and greater physical stress for the caregiver (Wenger, 1990). In cases in which the caregiver was younger, such as an adult child, caring intruded more into the caregiver's primary relationships and was viewed as more stressful, isolating, and constraining.

Relocation and Its Effect on Caregivers

It is generally assumed that nursing home admission, by eliminating the need for 24-hour care, would result in reduced caregiver stress. Evidence to date does not support this expectation. Most studies suggest that the transition and postplacement period is a major life stressor associated with guilt, anger, despair, resentment, and general psychological distress.

The transition from home to nursing home can be made less traumatic by promoting an understanding of relocation as just another phase in the normal life cycle of the family. Nursing home entry can then be reconceptualized as a process rather than as a single life event. Participation in simple rituals that say good-bye to old contexts and assist in the acceptance of the new living environment can prepare families for the relocation.

Nursing Home Staff and Relatives

Nursing home staff have to be made aware of relatives' needs so that they can work with relatives to reduce the stress of nursing home admission. The admission process usually focuses entirely on the needs of the resident, to the extent that nursing home staff largely ignore the difficulties and stress that relatives experience. Caregivers often become the "hidden clients" of nursing homes (Duncan & Morgan 1994).

Relatives' guilt and grief are often manifested as anger directed at nursing home staff. Other attitudinal barriers can obstruct good communication between relatives and nursing home staff. One such barrier is the commonly held view that families cease to be interested in their relatives once they are placed in nursing home care. Yet studies consistently show that, in most instances, relatives continue to care for the nursing home resident, and although their role as caregiver is transformed, many relatives emphasize the importance of their contribution in the nursing home context. Preadmission meetings can mark the beginning of an ongoing relationship not only with the resident but also with the family and can help alleviate such communication and perception barriers.

In general, relatives' involvement with nursing home residents can have positive outcomes for both relatives and residents. When caring is divided into technical and nontechnical tasks—with relatives being relegated the nontechnical tasks—studies show that relatives' competence in delivering nontechnical care is generally high and that this is an acceptable approach for nursing home administrators who are concerned about avoiding litigation. Other studies suggest that a more useful approach is to explore the purpose of tasks and emphasize collaboration between staff and relatives rather than task division (Thompson, 1990).

Understanding the Needs of Caregivers

An extensive study of the nursing home admission process from the perspective of relatives and significant others by Pearson and colleagues (1998) observed that nursing home admission generates complex reactions and that society's general view of nursing homes shapes these reactions. Their data showed that relatives and significant others experienced a range of responses and emotions, including feeling bewildered, left out, guilty, or alone, or they experienced a sense of panic, failure, sorrow, powerlessness, or remorse. Intervention in the early stages of admission can improve the outcomes for relatives, significant others, and the new resident.

Pearson and colleagues (1998) also examined societal views and "rules" about nursing home admission identified in public discourses from diverse sources. Themes identified in the professional literature include the idea that "home is best" and that society has a filial responsibility to provide care for the elderly; women are generally the primary caregivers; caring stress must be reduced to prevent premature nursing home admission; and nursing home entry results in guilt and continued stress for caregivers. Themes identified in the general media include the following: the costs of a graying society must be curbed; filial responsibility is expected and must be encouraged by government support; those who can afford to pay their own way should be required to do so; the family home is highly valued and should not be taken from old people to pay for care; nursing homes are to be feared and avoided; and the frail aged must be cared for "at home." Themes in government publications include that the cost of aging must be curbed, managed, and privatized. Although the research identified few differences in the dominant rules among the three texts, the voices of informal caregivers, the recipients of care, and professional caregivers were missing from the dialogue.

Studies suggest that nursing home admission, in many instances, shifts rather than eliminates the terms of caring and the reasons for continued stress. Categories of inquiry used to analyze caregiver stress are in most instances predetermined and therefore tend to prescribe and impose a rigid and fixed understanding of caregiver roles. Nursing home admission requires a support program for relatives and significant others if the transition from home care to nursing home care is to be successful.

ALAN PEARSON

See Also
Caregiver Burden
Caregiving Relationships

Nursing Homes
Relocation Stress

NURSING HOMES

A nursing home (NH) may be certified as a Medicare skilled nursing facility, a Medicaid nursing facility, or both, or it may be licensed as an NH by the state health department or some other state or federal agency. The NH must provide on-site supervision by a licensed nurse (registered nurse [RN] or licensed practical nurse [LPN]) 24 hours a day, 7 days a week (Rhoades, Potter, & Krauss, 1998). Overall goals are to maintain or improve physical and mental function, eliminate or reduce pain and discomfort, provide social involvement and recreational activities in a safe environment, reduce unnecessary hospitalizations and emergency room use, and provide a dignified death. Nursing homes are caring for an older, frailer population and are devoting more of their resources to the care and treatment needs of special populations (Spillman, Krauss, & Altman, 1997).

Facility Characteristics

In 1996 there were slightly more than 16,800 NHs with 1.8 million certified or licensed beds (average size: 104 beds). Seventy-five percent of NHs have fewer than 125 beds, but the 25% with more than 125 beds account for almost half of all NH beds. There are more residents in NHs that have independent living or personal care units in the Midwest than in any other region. Sixty-six percent of homes are for-profit, and most are located in the South; 26% are not-for-profit; the remaining homes are government owned. More than two-thirds of for-profit NHs, compared with less than one-third of not-for-profit NHs, are group or chain affiliated.

Despite a 17% reduction in bed supply between 1991 and 1996 and a growing elderly population, occupancy rates are dropping. Home care and assisted living are increasingly viable options to meet chronic long-term-care needs and posthospital short-term subacute or skilled nursing needs. Nevertheless, at least 40% of elders are expected to spend some time in an NH at some point.

Risk Factors for Admission and Services Provided

Approximately 1.56 million people (slightly less than 6% of the elderly cohort) are in NHs on any given day. Risk factors for admission include advanced age, medical diagnosis, living alone, loss of self-care ability, impaired mental status, white race, lack of informal supports, poverty, hospital admission, confinement to bed, and female gender. Mandated preadmission screening prevents homes from admitting or retaining persons who are mentally ill, developmentally disabled, or mentally retarded unless the individual needs the kind of skilled nursing services the home provides. NHs must provide dental, podiatric, and medical specialty consultation services. Some have fully equipped dental and x-ray suites, laboratory facilities, and pharmacies. Virtually all provide rehabilitative services, but the intensity of service—skilled or maintenance—varies with the home's program and Medicare participation.

More than 3,000 NHs have special care units (SCUs) that are formally defined. Constituting almost 7% (120,400) of all NH beds, SCUs are more common in not-for-profit and chain NHs and in larger homes. Principles of SCU operation include specific admission and discharge criteria, specially trained staff, special programs, resident and family education, mechanisms for evaluating care, and a distinctly identifiable area. SCUs are for care related to ventilator dependence, traumatic brain injury, oncology, pressure sores, AIDS, skilled rehabilitation (short-term stay), subacute care (medically unstable patients), Alzheimer's disease or dementia care, and so forth. Facilities with a large Medicaid population tend not to have SCUs, and those with many Medicare residents tend not to have Alzheimer's SCUs.

Resident Characteristics

Most NH residents are white (88.7%), female (72%), and married (17%). Only 8.9% are African

American, although this number is increasing. The most frequent diagnosis is dementia, followed by heart disease and hypertension. Thirty-six percent of residents are hearing impaired; 39% are visually impaired. At least 60% have some kind of communication problem.

Approximately 30% of residents are age 75 to 84; nearly half are 85 or older. Fewer than 10% are younger than 65. More residents under 65 are in hospital-based nursing homes (HBNH) and government (e.g., Veterans Affairs) facilities than in not-for-profit or for-profit facilities. In this group, the most frequent diagnoses are seizure disorder, hypertension, stroke, diabetes mellitus, and dementia.

The number of residents needing assistance with activities of daily living (ADL) increased from 72% in 1987 to 83% in 1996 (Krauss & Altman, 1997). Only 3% require no ADL assistance and are not demented. Two-thirds of all residents need assistance with mobility, and almost 80% need toileting assistance (transfer, reminders, hygiene). Slightly less than 40% are incontinent of bowel and bladder.

Nearly 50% of all residents (and more than half of those over 85) have dementia alone or in combination with other mental disorders. Twenty percent have at least one symptom of depression, particularly those age 65 to 84 years. More than half have short- and long-term memory loss and are disoriented. Behavior problems are less common than cognitive impairment. Fifteen percent of residents exhibit inappropriate behavior: making harsh sounds, sexual acting out, disrobing, smearing food or feces. Men are more likely to resist care and be verbally and physically abusive.

Length of Stay and Discharge Planning

More residents were in NHs for less than six months in 1996 than in 1987. The average length of stay (LOS) in HBNHs is 14.7 days, as compared with 32 days in Medicare skilled nursing facilities. The average LOS for long-term residents is 2.5 years. Predicting discharge is especially important for those with long-term-care insurance that limits coverage or trades fewer NH-covered days for more home-care-covered days or that protects the resident's income and assets with "up-front" coverage of the costs of NH care.

Staffing

Two-thirds of NH staff are full-time employees. Homes must have an RN on duty for at least eight consecutive hours a day and licensed personnel (LPNs) to provide care around the clock, seven days a week, unless the home has been granted a waiver (due to a shortage of qualified personnel and when resident health and safety are not in jeopardy). On average, current staffing per resident day is RNs, 0.5 hour; LPNs, 0.8 hours; nurse assistants, 2 hours. Total nursing care hours per resident day in 1995 was 3.3 hours, an increase from 3.1 in 1991 (Harrington, Carrillo, Thollaug, & Summers, 1996). Having more full-time-equivalent RNs is positively related to resident functionality (Braun, 1991) and negatively related to restraint use (Zinn, 1993).

All homes must have a full-time licensed administrator and a full- or part-time medical director and in-service instructor. The number of social workers, activity therapists, and nutritionists vary with bed size. Twenty percent of NHs have no physical, occupational, or speech language therapists on staff or under contract. Less than 1% have a geriatric nurse practitioner or physician assistant. This is changing, however, as homes recognize the improved care and cost savings associated with geriatric nurse practitioners (Shaughnessy, Kramer, Hittle, & Steiner, 1995).

Costs and Reimbursement

Federal and state governments spent approximately $80 billion in 1996 for NH care. About 62% of residents are dually eligible (Medicare and Medicaid) beneficiaries; about 30% have Medicare only. The Medicare component of the NH program remains essentially restricted to posthospital skilled

nursing, rehabilitation, or both. The percentage of Medicare-covered residents grew from 5% (cost: $2 billion) in 1990 to 13% (cost: $9.6 billion) in 1996. Program growth is attributed to the 1987 Omnibus Budget Reconciliation Act (OBRA) mandate that residents be restored to their maximal physical and mental potential. OBRA increased the number of residents who received skilled rehabilitation and reduced hospital LOS, with a concomitant increase in short-stay subacute care. The Medicare daily rate is generally higher than Medicaid reimbursement, but it can be less than the private pay rate.

More than half of all NH Medicaid beneficiaries are institutionalized all year. Medicaid coverage grew from 47% of residents in 1989 to 69% in 1996 (cost: $40.8 billion). Private pay accounted for 44% of NH revenue in 1985, but only 28% in 1996; private insurance, 3% to 4%; public or charity, 3%. Expenses are higher in the Northeast, where the share paid by Medicaid is the highest in the nation.

NHs are under growing pressure to increase revenue and reduce costs. As of mid-1998, Medicare skilled nursing facilities are no longer reimbursed on a cost-based system but on a prospective payment system that classifies residents to reflect the resource use of different patient types. The payment rate covers all costs of furnishing covered skilled nursing services (i.e., routine, ancillary, and capital-related costs). Seventy percent of Medicare-covered NH stays in 1996 were provided to residents receiving skilled rehabilitation (Health Care Financing Review, 1998). Medicare skilled nursing facilities received, on average, $182 per patient day, or $5,866 per admission. In contrast, HBNHs received $330 per day, or $4,939 per admission.

At least 17 states are using some kind of case-mix reimbursement system for Medicaid that classifies residents into homogeneous resource utilization groups and links reimbursement to residents' needs. It is more equitable for skilled nursing facilities with different caseloads. As such, case-mix has the potential to improve access for heavy-care patients, increase facility efficiency, and avoid excessive payment for light-care cases. It acts as a disincentive to admit light-care residents. In the absence of a rigorous quality-of-care survey system, however, case-mix can also create perverse incentives to increase residents' functional dependence and morbidity.

Capitated Medicare managed care contracts with NHs vary. Risk can be shared between the managed care company and the NH, or full risk can lie with the managed care company. Rates are negotiated for certain diagnostic categories or skilled nursing treatments using, for some arrangements, a case-mix pricing model. Managed care's influence is seen in NHs' positioning themselves for competition by providing subacute-type services.

Nursing Home Reform

An Institute of Medicine (IOM) study in 1986 found that the quality of life and care in most NHs was poor. IOM recommendations were enacted in sweeping reforms contained in OBRA, 87 which included resident rights, minimum nurse staffing, an interdisciplinary team approach for evaluation and care planning, measurable goals, creation of a uniform resident assessment instrument (RAI) focused on functionality and self-care, development of a minimum data set (MDS) for comprehensive care planning and outcome norms, justification and monitoring of physical and chemical restraints, mandatory training and certification for all nurse assistants, and a revised quality-of-care survey that was resident-centered.

NHs have unannounced surveys every 9 to 15 months conducted by the state health department. There can also be "look-behind" surveys by federal examiners. Accreditation by the Joint Commission on Accreditation of Healthcare Organizations is optional for NHs, but HBNHs and homes seeking managed care contracts or affiliations must be accredited. The 10 most frequently cited deficiencies in 1997 were food sanitation, comprehensive assessment, comprehensive care planning, hazard-free environment, pressure sores, physical restraints, highest practicable care, housekeeping, dignity, and accident prevention.

The Future

The Northeast and West lost a significant number of beds in the late 1990s. Overall, the number of disabled elderly has diminished, but this may be an artifact of the definition of disability used for program eligibility. Some disabled elders who were in NHs 10 years ago are now meeting their needs elsewhere. (For example, Medicare home health care beneficiaries increased from 5.3% in 1985 to 9.6% in 1995.) It is predicted that the number of NHs will continue to diminish slightly but that the number of beds will rise to 2.2 million by 2020. Half a million more beds will be needed by 2010. NH residents currently in Medicare skilled nursing facilities were in hospitals 10 years ago.

Fifty-one percent of all NH residents have some form of advance directive, including do-not-resuscitate orders; 4% have do-not-hospitalize orders. Of residents with dementia, 14% have directives. The steady increase in advance directives that state a person's refusal of hospitalization and other life-sustaining interventions will likely result in fewer hospitalizations and more "planned deaths" in NHs.

Assisted living, expected to capture as much as 50% of light-care NH residents, is a new option for those with physical as well as cognitive disabilities and is a new market for private-pay clients. Consolidation and integration of care networks (e.g., adding an assisted living or personal care unit) is expected to continue. Case management, associated with managed care, is likely to increase and influence discharge planning. The growing presence of geriatric nurse practitioners in NHs will allow homes to care safely for more medically complex and functionally dependent residents.

ETHEL L. MITTY

See Also
Advance Directives
Assisted Living
Dementia: Special Care Units

Internet Key Words
Advance Directives
Managed Care

Medicaid
Medicare
Minimum Data Set
Nursing Homes
Special Care Units

Internet Resources

American Association of Homes and Services for
    Aging
http://www.aahsa.org

American Health Care Association
http://www.ahca.org

NUTRITIONAL ASSESSMENT

Standard, comprehensive geriatric assessment of overall health status in the elderly employs simple, rapid, inexpensive, and internationally validated scales for rating cognitive function, functional status, walking, balance, and socioeconomic status. Resulting corrective interventions help to lower mortality, improve quality of life, and save health care costs. The standard workup should also incorporate a nutritional status rating, since malnutrition is a prognostic factor closely related to mortality and morbidity and its prevalence, which is relatively low in free-living elderly (5% to 10%), rises considerably (30% to 60%) in hospitalized or institutionalized elderly.

Conventional malnutrition assessment using anthropometrics, dietary recall, and laboratory investigation is too long and expensive as a first-line strategy. A number of simple, rapid tests for detecting or diagnosing malnutrition in the elderly have been developed and in some cases validated. Nutrition screening involves identifying the characteristics associated with nutritional problems in the general population. The aim is to detect the subjects at risk from malnutrition, identify the causes, and guide corrective action. If malnourishment is suspected, these tests are supplemented by conventional nutritional assessment before planning treatment. These standard tests are also used in the overall workup for diagnosing malnutrition states,

in combination with the medical and nutritional history, medical treatments, clinical examination, anthropometric measures, and laboratory results.

The Mini Nutritional Assessment

The Mini Nutritional Assessment (MNA) (Guigoz, Vellas, & Garry, 1997) is an 18-item nutritional questionnaire for the elderly developed and validated jointly by the Center for Internal Medicine and Clinical Gerontology of Toulouse (France), the Clinical Nutrition Program at the University of New Mexico (United States), and the Nestlé Research Center in Lausanne (Switzerland). It was validated in a total of 600 elderly of varying health status (free-living, at-risk, and institutionalized) after first being developed in Toulouse in 1991 and then further validated in 1993. It was also tested in Albuquerque, New Mexico, in free-living elderly sharing a cultural context different from that in Toulouse and who had been participating in a longitudinal study started over 15 years previously. Since their nutritional status was fully documented, they were the ideal population for validating the test. The MNA was compared with a gold standard defined as the opinion of two nutritional physicians with access to the results of a full nutritional assessment comprising laboratory results, anthropometrics, body weight, dietary recall, and complete medical records. Good correlations were found with plasma albumin and prealbumin.

The test items are anthropometrics (calf and arm circumference) as a measure of fat and muscle mass; body mass index (weight [kg] / height [m]^2); number of drugs (consumption of more than three drugs daily can cause anorexia); acute disease in the previous three months; bedsores; mobility; appetite; eating habits—number of meals and daily consumption of protein, vegetables, fruits, and liquids; and subjective health (a good reflection of health status in the elderly). In cognitively impaired subjects, the test requires help from the family or health care personnel. Subjects are classified into three levels on the basis of scores from 0 to 30: 24 or higher, satisfactory nutritional status; 17 to 23.5, risk of malnutrition; less than 17, protein energy malnutrition. The MNA is easy to administer, patient-friendly, inexpensive (no laboratory investigations are required), very sensitive (96%), highly specific (98%), and reproducible.

Recent reanalysis of the MNA data collected in Toulouse and in Mataro, Spain, identified six items strongly correlated with conventional (physician) nutritional assessment. These were used to redesign the MNA into a validated questionnaire for use in healthy elderly. It still contains 18 items but is now administered in two stages (Rubenstein, 1998). Stage 1 is a screening questionnaire using the six strongly correlated items. It takes three minutes to administer, versus 10 minutes for the overall questionnaire. The maximum score is 14, and a score of 12 or higher indicates satisfactory nutritional status, with no requirement to proceed to stage 2 (assessment). A screening score of 11 or less is an indication to proceed to assessment stage 2, which consists of the remainder of the MNA; in this case, the screening and assessment scores are totaled: total scores of 17 to 23.5 indicate a risk of malnutrition, and scores less than 17 indicate protein energy malnutrition. The screening stage of this new form of the MNA (MNA-NF) can be viewed as a preliminary nutritional assessment, reserving the full MNA to confirm the diagnosis and above all to guide nutritional intervention. The full MNA is best used with at-risk or ill elderly with a high likelihood of malnutrition.

Malnutrition Screening in Free-Living Elderly: Nutritional status should be monitored even in healthy elderly to detect and prevent deficits and their consequences. Primary prevention of this kind can be performed by using the MNA-NF, which identifies those elderly at risk of malnutrition, and giving them the necessary information for improving their diets. Other studies using the MNA have confirmed that few free-living elderly are malnourished (0.5%) but that the percentage increases (5.7%) when they become institutionalized. Malnutrition may therefore be a predictive factor for institutionalization. The MNA-NF could be used as a screening tool for malnutrition risk by general practitioners during routine health checks of elderly

persons. In the New Mexico study, MNA scores correlated with overall health status: very healthy elderly all scored greater than 24, the malnutrition risk threshold (27.6 ± 1.82). Scores of 27 to 30 could be markers of successful aging.

Malnutrition Screening in Frail or Institutionalized Elderly: The MNA can be used for nutritional assessment of frail or institutionalized elderly. A study using the MNA showed that half the elderly in institutions are at risk of malnutrition. The MNA is highly sensitive (96%) in evaluating nutritional status in frail elderly (Guigoz, Vellas, & Garry, 1997).

Nutritional Assessment in Hospitalized Elderly: MNA scores correlate with duration and cost of stay. A study in 166 inpatients older than 70 years of age showed that admission MNA scores of less than 17 correlated with hospital stays an average of 10 days longer than for those with MNA scores greater than 17 (Quadri et al., 1998). MNA scores also correlate with functional status during hospitalization, risk of bedsores, and transfer to an institution (Cohendy, 1998). MNA scores correlate well with Mini-Mental Status Exam scores, indicating that declining cognitive function can be responsible for declining nutritional status. The MNA is also useful for assessing nutritional status in non-hospitalized elderly, such as those with Alzheimer's disease, who are at particular risk. Scores correlate with the weight loss often found in Alzheimer's disease. We have developed an education program that trains Alzheimer's caregivers to monitor their relatives' nutritional status using the MNA.

Management and Nutritional Intervention Guidelines Using the MNA: The MNA was specifically designed to guide nutritional intervention by identifying the risk factors requiring correction. An MNA score of 24 or higher indicates satisfactory nutritional status. Reweigh the patient three times monthly, and offer basic balanced diet advice.

An MNA score of 17 to 23.5 indicates malnutrition risk, with a good prognosis given early intervention. Analyze the MNA results to identify the reasons for the low score—that is, examine those items where points were lost. For example, if the patient is taking more than three drugs daily (item

H), determine with the physician whether the number can be reduced. If the patient eats only two proper meals a day or is anorexic (items J and A), determine whether he or she has an eating disorder. If the family cannot provide meals, arrange for Meals On Wheels or domestic help. If the patient does not eat the requisite foods (fruits, vegetables, dairy products, meat) or take enough fluid (items K–M), advise about varying the diet and offer domestic help or Meals On Wheels. For skin lesions (item I), offer advice on high-calorie and high-protein supplements. When possible, perform a detailed dietary interview. Follow up with a repeat MNA in three months.

An MNA score less than 17 indicates protein energy malnutrition. Analyze the score as described earlier, and perform a dietary interview. Investigate for other causes of malnutrition, such as depression, decreased functional status, or cognitive decline. Tailor advice to the MNA results and the dietary interview—for example, fractionate eating (eat more frequent meals and snacks with less substance but greater nutritional density)—and address the underlying causes. Repeat the MNA in three months.

We recently performed an MNA-assisted interventional study in retirement home elderly. Those scoring higher than 24 were not supplemented. Those scoring 17 to 23.5 were randomized into two groups with and without supplementation (300 to 500 kcal daily using varied high-protein nutrient mixes). Those scoring less than 17 were all supplemented. Supplementation was well accepted, as approximately 400 kcal daily was ingested for two months. Most of the supplemented elderly increased their body weight and improved their MNA scores at the end of the study. The study shows the utility of the MNA in nutritional assessment and follow-up (Lauque, Arnaud-Battandier, Mansourian, & Guigoz, 1999).

Nutritional evaluation tools allow the early detection of malnutrition and should be integrated into the standard gerontological workup. They help in planning preventive action and averting incipient malnutrition by rapid and appropriate nutritional intervention. In particular, the MNA not only as-

sesses nutritional status but also guides nutritional intervention. It is thus an effective reference tool for diagnosing protein energy malnutrition in the elderly (Morley, Miller, Perry, Guigoz, & Vellas, 1997).

BRUNO C. VELLAS
SYLVIE LAUQUE

See Also
Body Composition
Caloric Intake
Eating and Feeding Behaviors
Measurement
Protein Energy Undernutrition

Internet Key Words
Malnutrition
Nutritional Evaluation

Internet Resources
American Academy of Family Physicians
http://www.aafp.org/

Determine Your Health Questionaire
*http://www.oznet.ksu.edu/___library/fntr2/
samplers/gt332e.htm*

Healthy People 2000
http://odphp.osophs.dhhs.gov/pubs/hp2000/

Nutrition Screening Initiative
http://www.aafp.org/nsi/

NUTRITIONISTS

A registered dietitian (RD) is the most common nutrition professional and is a designated member of the health care team responsible for nutrition care as defined by the Health Care Financing Administration's Joint Commission on Accreditation of Health Care Organizations. In 1997, there were 57,243 members of the American Dietetic Association (ADA), the primary organization for dietetic professionals (Bryk & Soto, 1999).

Academic and Clinical Preparation of the RD

The Commission on Dietetic Registration (CDR) of the ADA confers the RD credential on an individual who meets specific academic and clinical requirements and passes a national registration examination.

An RD must have a minimum of a bachelor's degree from a university or college in the United States that is regionally accredited. The university's course work must be approved by the Commission on Accreditation for Dietetics Education (CADE) of the ADA. In 1999–2000, there were 234 academic programs approved by CADE in the United States and Puerto Rico. Academic requirements include course work in physiology, anatomy, biochemistry, the psychosocial sciences, management, and nutrition and food science. Courses must include information about assessment of nutritional status of the elderly; the age-related effects on metabolism, nutrition needs, and food choices; adaptive feeding techniques and alternative feeding modalities; nutrition counseling; and the effects of socioeconomic, cultural, and psychological factors on food and nutrition behavior. Courses are also required in economics, organizational management, large-volume feeding, and food service management.

In addition to academic education, a minimum of 900 hours of supervised clinical practice in a CADE-approved or -accredited practice program must be completed. There are 289 of these programs in the United States and Puerto Rico. Experiences are planned to develop basic skills in nutritional assessment and management of food and nutritional needs for people across the life span. Supervised practice programs may include food and nutrition experiences in acute and ambulatory care settings, skilled nursing facilities, home-care programs, congregate feeding and home-delivered meal programs for the elderly, and other community programs. Following successful completion of the supervised clinical practice, the person is eligible to take the national registration examination and obtain the RD certification. In addition to national registration, 27 states have licensure statutes.

Other Nutrition Providers

A certified nutrition specialist (CNS) has a graduate degree in nutrition, has completed either 1,000 hours of supervised practice or 4,000 hours of unsupervised practice, and has passed a certification examination. Requirements for academic and clinical preparation for this credential are not as specific, nor are the requirements for clinical preparation as rigorous, as those for a dietitian.

A dietetic technician, registered (DTR) provides support to the dietitian in all health care settings. The DTR credential is conferred by the CDR on a person who has successfully completed an associate of science degree or a bachelor's degree in dietetics and has specific clinical experience in a program accredited by CADE. The DTR works under the supervision of the RD and may provide the following services: screening for nutrition risk, intervention for patients with less complex nutrition problems, and preventive nutrition services.

A certified dietary manager (CDM) most commonly works in a skilled nursing or long-term-care facility, under the supervision of an RD. In the absence of an RD, the CDM directs food and nutrition services. The CDM is trained in a certificate program, usually in a community college.

Advanced-Level Practice

In addition to establishing and enforcing standards for entry-level dietetic education, the CDR oversees the continuing professional education (CPE) of RDs. RDs must complete 75 hours of CPE every five years. This continuing education is reviewed and approved by committees in each state and by the CDR. Beginning in 2001, requirements for continuing education will also include a periodic assessment of learning needs and a plan to update needed knowledge and skills. It is expected that this change in the continuing education requirements will stimulate dietitians to seek CPE that meets their specific professional goals. For example, a dietitian working with the geriatric population would need to show evidence that CPE addresses concepts that are important in geriatric nutrition.

Although almost 50% of the members of the ADA have master's or doctoral degrees, there is no specific advanced-level degree or credential in geriatric nutrition. There is, however, a dietetic practice group for gerontological nutritionists in ADA that has developed standards of practice for nutritionists working with older people (Shoaf, Bishirjian, & Schenkler, 1999). The standards relate to the provision of services, application of research, communication and application of knowledge, utilization and management of resources, and maintenance of competence in the area of geriatric nutrition. A newsletter that focuses on food and nutrition needs of the elderly is published by the practice group. The ADA has been an active advocate for the role of nutrition in maintaining health and good functional status in older people. Examples of position papers and educational resources can be found on the ADA Web site.

Role of Nutritionist in Health Care

The ADA maintains a database of its members (RDs and DTRs). Responses from 38,000 members (95% RD, 5% DTR) indicated that 35% worked in acute-care hospitals, 10% in ambulatory settings, 11% in extended care facilities, 11% in community and public health programs (including congregate feeding and home-delivered meals for the elderly), 12% as consultants or in private practice, and 2% in home-care programs (Bryk & Soto, 1999).

Older people are more likely to have chronic conditions and functional impairments that interfere with the maintenance of good nutritional status. Lack of attention to dietary intake and poor nutritional status can have a negative impact on many chronic diseases and contribute to declining health. There is consistent evidence that interventions for malnutrition or as therapy for chronic disease, particularly cardiovascular diseases, obesity, diabetes, renal failure, and osteoporosis, have positive effects on outcome and are cost-effective (Sheils, Rubin, & Stapleton, 1999).

The geriatric population is considered to be at high nutritional risk in most health care settings.

The dietitian's role is to assess this nutritional risk, work with the health care team to plan interventions, and evaluate the outcome of care. Interventions may include individual or group counseling to prevent disease or reduce the effects and progression of disease, addressing food insecurity, recommending and providing modified diets, supplementing energy and nutrient intake, and recommending and monitoring nutrition support (enteral and parenteral). In some settings, particularly home care, the dietitian may develop nutrition education materials and educate other health professionals, such as nurses, and informal caregivers who provide direct care to patients.

There is little direct reimbursement by third-party payers for nutrition services at the present time. In the acute-care setting and in skilled nursing facilities, dietitians' services are part of the daily patient costs, which also include other basic services such as nursing and food service. In ambulatory care and home care, dietitians' services are part of the administrative overhead. The one exception to reimbursement is the new Medicare reimbursement for diabetes self-management, which requires that an RD be part of the teaching team.

CAROL PORTER

Internet Key Words
American Dietetic Association
Certified Dietary Manager
Certified Nutrition Specialist
Commission on Accreditation/Approval for Dietetics Education
Commission on Dietetic Registration
Dietetic Technician, Registered
Gerontological Nutritionist
Nutritionist
Registered Dietitian

Internet Resources
American Dietetic Association
http://www.eatright.org

Gerontological Nutritionists Practice Group
http://trc.ucdavis.edu/gerinutr

O

OBESITY

Obesity, defined as a body mass index (BMI) of more than 27, is highly prevalent among all age groups in the United States. BMI is calculated by dividing weight (in kilograms) by height (in meters) squared; it correlates well with obesity. Approximately 40% of men and women between the ages of 60 and 79 are overweight. After age 80, 18% of men and 26% of women are overweight, irrespective of race and ethnic group. Obesity tends to be more prevalent among older women than older men.

Men and women tend to gain weight up to age 50 to 60, after which both tend to lose weight. The magnitude of weight gain decreases with aging, and while weight loss increases. A 2-kg (4-lb) weight increase per decade of life between ages 30 and 55 is a normal age-related change and is thought to be due to a decline in energy expenditure as a result of two phenomena:

- Decline in activity level due to physical limitations or to the misperception that a sedentary lifestyle is normal with aging.
- Decline in the basal metabolic rate (responsible for about 70% of energy expenditure) secondary to a decline in lean body mass.

The role of genetic predisposition to obesity is not fully understood. The hormone leptin, produced by adipose cells, appears to exercise a negative feedback on the hypothalamus, leading to suppressed appetite. Leptin deficiency can lead to early obesity due to loss of this critical feedback (Bray, 1999; Montague et al., 1997).

Until 1995, the presence of overweight in older persons was considered acceptable and sometimes desirable. This position was largely based on conclusions from the National Health and Nutrition Examination Survey (NHANES), suggesting a significant increase in mortality with a low BMI, compared with a small or no increase in mortality with a high BMI. Recent studies, adjusted for smoking and the presence of other illnesses masking a low BMI, reveal a 2:1 increase in mortality with a BMI above 28.5 (Stevens et al., 1998). Regional fat distribution (i.e., waist-to-hip ratio), rather than BMI, is increasingly linked to poor health outcomes. Fat within the abdomen is more metabolically active than fat in the thighs and buttocks and results in an increased concentration of fatty acids in the portal system, potentially harmful lipids, and insulin changes. Longitudinal studies show an association between increased abdominal upper body fat and increased overall mortality (Higgins, Kamel, Garrison, Pinsky, & Stokes, 1988). This concept of "central obesity" is likely to gain credence as an important indicator of health.

Obesity is associated with the development of several diseases, including hypertension, coronary artery disease, hyperlipidemia, diabetes mellitus, obstructive sleep apnea, and several cancers. It is also associated with psychological disorders related to perception of body image, including depression, anorexia nervosa, and bulimia. Obesity is associated with functional decline, especially mobility, among older persons and with a substantial burden of illness. Overweight (BMI ≥ 25) is a risk factor for clinically relevant nosocomial infections in surgical patients (Choban, Heckler, Burge, & Flancbaum, 1995) and for wound infection and incisional hernia after midline abdominal surgery (Israelsson & Jonsson, 1997). The economic impact of obesity on health care is significant. For example, the estimated cost associated with treatment and health outcomes of the current overweight (BMI ≥ 25) population of middle-aged American women during the next 25 years is $16 billion, compared with $5.89 billion for women with a BMI of 23 to 24.9 (Gorsky, Pamuk, Williamson, Shaffer, & Kaplan, 1996; Wolf & Colditz, 1996).

Management

There is no gold standard for how aggressively to pursue weight loss in overweight or obese older persons. It is, however, recommended in the presence of comorbidities such as high blood pressure, diabetes, and osteoarthritis.

The most effective weight-loss programs are multidisciplinary and comprehensive and combine behavior modification (lifestyle change), nutritional education and diet, exercise, medication (where appropriate), and long-term maintenance support. An initial medical evaluation before designing a weight-loss program should exclude medical causes of weight gain, such as hypothyroidism, and false obesity in cases of fluid overload and edema. Continued medical and nutritional supervision and intervention may be necessary as weight-loss goals are met, including readjusting antihypertensives and diabetic medications. Some patients may require cardiopulmonary and musculoskeletal evaluation before engaging in exercise as part of the weight-loss program.

Goals should be realistic and aiming at accomplishing a "lighter" weight rather than a "light" weight. The lighter weight goal should be set individually to ameliorate health risks and medical problems and normalize functional capacity. A reduction of 10% of body weight, and then maintaining it, is a reasonable initial goal. Weight loss should not exceed 1 lb per week.

Changing the lifestyle of obese older persons can be challenging. The focus is alteration of daily habits, specifically eating patterns and attitudes. Participation of family members, particularly the caregiver, is critical to the success of any program. Small changes in lifestyle can have enormous benefits in terms of function, health, and quality of life even without large weight losses.

The use of a very low energy diet (less than 1000 kcal/day for women and 1200 kcal/day for men) or a low-protein diet is strongly discouraged. Possible complications include excessive fluid and electrolyte shifts and quicker but short-lived weight loss. A dietary guideline for obese older persons should focus on adopting a healthful diet that mod-erately reduces energy (1200 to 1500 kcal for women; 1500 to 1800 kcal for men), is low in fat, and is high fiber. Maintaining adequate intakes of protein (1 g/kg/day), fluids, and other nutrients is essential. Alcoholic beverage provide "empty" energy, meaning calories without nutrients. Reasonable alcohol intake should not exceed two beers or glasses of wine per day for men and half that for women.

Exercise programs must be individualized for age, diseases, and disability. "Burning" calories can be accomplished by aerobic exercise. Maintaining a higher basal metabolic rate (to increase the calories burned at rest) requires increasing the muscle mass, which is accomplished by strengthening exercises. Ideally, 30-minute workouts five to six times a week are recommended. Routine activities such as mall walking, water aerobics, dancing, playing with grandchildren, and gardening should be encouraged. Strength and resistance exercises can be done using resistance machines, elastic bands, or simple weight-lifting devices.

Appetite suppressants are generally discouraged but may be necessary in extreme situations. Since these medications are serotonin reuptake inhibitors, concomitant use of antidepressants of the same group should be discouraged. Sibutramine is commonly used, and it results in weight loss of up to 10%, followed by maintenance of lost weight for up to two years if treatment is continued. Side effects include a small rise in blood pressure (2 to 3 mm Hg). This drug replaced the older serotonin reuptake inhibitors dexfenfluramine and the combination of fenfluramine and phentermine (fen/phen), which were associated with valvular problems and pulmonary hypertension (Jick et al., 1998; Kahn et al., 1998). A new class of medications, the so-called fat blockers such as orlistat, decreases caloric intake by inhibiting intestinal lipase, thus suppressing fat absorption. Side effects such as oily stools and spotting, fecal urgency, and occasional fecal incontinence limit their utility. New promising drugs that act as peripheral satiety factors are being developed (cholecystokinin agonists), and trials of recombinant human leptin are under way.

Long-term maintenance is the final component of the weight-loss program. Simply reducing serving sizes and eliminating high-fat snacks may be sufficient if patients are able to increase their physical activity. Modest reductions in caloric intake (e.g., 1,200 to 1,500 kcal for women and 1,500 to 1,800 kcal for men) are recommended. To help patients continue a healthy eating pattern, office appointments can be scheduled at gradually lengthening intervals, starting with monthly visits. As an alternative, patients may regularly attend meetings of support groups such as Weight Watchers or Overeaters Anonymous.

M. Louay Omran

John E. Morley

See Also

Body Composition

Caloric Intake

Exercise and the Cardiovascular Response

Nutritional Assessment

Internet Key Words

Calories

Exercise

Fat

Nutrition

Obesity

Overweight

Internet Resources

International Journal of Obesity

http://www.stockton-press.co.uk/ijo/index.html

Journal of Obesity Research

http://www.naaso.org/obres

Overeaters Anonymous

http://www.overeatersanonymous.org

Shape Up America

http://shapeup.org

Take Off Pounds Sensibly (TOPS)

http://www.tops.org

Weight Control Institute of Diabetes and Digestive and Kidney Diseases

http://www.niddk.nih.gov

OCCUPATIONAL THERAPISTS

Occupational therapy centers on the concept of occupation—the everyday, meaningful activities that make up the daily lives of all individuals. For older individuals, occupation is particularly important, because it is through occupation that the elderly engage in meaningful and productive pursuits. Occupation facilitates the process of finding or maintaining meaningfulness in later years and allows older adults to maintain dignity and position in their environment while pursuing role development and goal achievement. Recognizing the importance of meaningfulness in daily activities significantly increases the likelihood that the aging process will be a positive experience. This framework of occupational therapy addresses the quality of life of elderly persons as manifested through daily activities, tasks, and expectations.

By definition, "occupational therapy is the art and science of helping people do the day-to-day activities that are important to them despite impairment, disability, or handicap" (Niestadt & Crepeau, 1998, p. 5). It views the individual through a lens centered on life performance. For a therapist, performance of life tasks is a determinant of health, so the goal of therapy is to prevent or minimize performance dysfunction.

Occupational therapists embrace the concept of *adaptation* as a means of intervening with older adults. Adaptation refers to any adjustment or change of behavior in response to the challenges or demands of aging (Christiansen, 1991). Successful aging includes the ability to adapt to events such as changes in physical appearance, decreasing physical strength and declining health, retirement and reduction in income, deaths of family members and friends, one's own impending death, and the fear of engaging in new leisure activities. The therapist helps elderly individuals assume adaptive behaviors—those that "result in an improvement in the fit between the individual and the environment" (Christiansen, 1991, p. 32). If performance deficits are identified, occupational therapy facilitates the individual's ability to adapt and thereby resolve the deficit. When successful adaptation occurs, the older adult experiences aging as successful performance in new and existing situations.

It is generally assumed that problems in older adulthood are the result of disease or disability. Rather than focusing on illness, however, the occupational therapist addresses the person's ability or inability to perform or participate in life's tasks and strives for older adults to remain engaged in life. Age should not force a person to forsake earlier interests. Both the therapist and the older person can be aware of limitations that require participation to take on different forms, but participation can still be achieved and maintained (American Occupational Therapy Association, 1994).

Performance areas are broad categories of human activity that are typically part of daily life: activities of daily living, work and productive activities, and leisure activities. The therapist addresses these performance areas when determining abilities and deficits. It is especially important that older adults strive for balance among the performance areas. As interests change, abilities decline, and patterns of rest and relaxation vary, older individuals may not recognize the magnitude of change that has occurred in one or another performance area. They may be focused on one area, with major neglect of others. Once the therapist determines the patterns of attention and neglect, it is possible to identify the reasons for those patterns and move toward balance among all the performance areas.

Activities of daily living, also known as self-care or self-maintenance tasks, are those tasks that individuals perform routinely: grooming, bathing, toileting, dressing, feeding, socializing, communicating. These skills are basic, but the ability to successfully engage in them changes over the life span. Some older adults may be inclined to give up and let others meet their self-care needs. Others may feel that they are a burden to family and friends and avoid seeking the help they need. To maximize the elderly client's self-care abilities, the therapist, together with the client, should identify those self-care activities that can be accomplished independently, as well as the appropriate level of assistance that may be required for other tasks.

Work and productive activity may have emphasized "paid work" during early adulthood, but it shifts to a focus on productive activity for elderly individuals. An area frequently neglected by other disciplines, productivity builds self-confidence and self-reliance, which contribute to a healthy aging experience. To determine the types of productive activity in which an aging person may engage, the therapist assesses mental and physical function, work history, previous interests and activities, social and family relationships, and current goals and directions. For older adults, productive activity may take the form of a retired elementary school teacher who becomes an at-home tutor to neighborhood children or an aging drama teacher who joins a community program that includes theater outings and discussion groups. Occupational therapists believe that opportunities to participate in society can significantly increase overall health and quality of life for older adults.

Leisure also undergoes change throughout adulthood and aging. Through leisure, elderly individuals can have positive physical and mental experiences, and they are more likely to increase or develop social relationships and networks. Yet, after a lifetime of paid work or home responsibilities, older adults may have difficulty engaging in leisure pursuits. They may feel guilty about relaxation without a work mandate or lack knowledge about how to engage in leisure activities. The therapist, using information from assessment of current status and performance history, works with the client to determine potential leisure activities that can be developed or maintained.

Performance components—sensorimotor, cognitive, and psychosocial—are the elements of performance that occupational therapists assess and for which they provide interventions to improve performance. These fundamental human abilities, in varying degrees and in differing combinations, are required for successful engagement in performance areas. For example, an elderly individual recovering from a cerebrovascular accident may wish to live in a community setting that requires competence in activities of daily living and productive activities. Thus, specific performance components such as muscle tone, gross motor coordination, postural control, and self-management need to be addressed and interventions designed that facilitate the achievement of that outcome. Or an older individual may begin to avoid leisure activities involving friends who regularly assemble for a card game. The therapist's assessment indicates that the individual is experiencing cognitive deficits that interfere with his ability to sequence the cards correctly. His embarrassment leads him to avoid not only the card games but also his friends. In this scenario, the therapist helps the individual relearn methods of card playing that compensate for the cognitive deficit and allow him to gradually reenter the game. In anticipation of further deficits, the therapist also helps the client adopt new roles, such as scorekeeper, that he will be prepared to assume if the deficit reoccurs and interferes with his playing ability on a long-term basis. The therapist recognizes the importance of both productivity and leisure pursuits as part of the aging process. This, coupled with an understanding of cognitive processes and their contribution to task accomplishment, allows the therapist to facilitate the client's involvement in the personally meaningful occupation of playing cards.

Performance context is the third dimension of occupational therapy—factors that influence an individual's engagement in desired or required performance areas. The performance context is both *temporal* and *environmental.* For example, a decision for an older adult to move to a supervised community may be partly a function of age but may also relate to environmental contexts. If the individual has experienced episodes of falling, a supervised setting with deliberately reduced environmental obstacles may be a better choice than attempting to circumvent the risks of the home environment.

Maximal functioning in the performance areas is the primary concern of occupational therapists. In order to achieve identified goals, performance components and performance contexts must interact with the performance areas, and all three must be considered relative to how they contribute to the ability of older individuals to participate in their environments. The ultimate goal of occupational therapy is to help the elderly remain within the mainstream of human activity. The focus is on quality of life, with an expected outcome of increased performance and participation through routine engagement in meaningful occupation.

A referral for services is appropriate when there is a reasonable expectation that occupational therapy will benefit the client. Depending on state licensure laws, a client may self-refer to occupational therapy. Referrals may also emanate from physicians, other health professionals, teachers, caregivers, employers, insurance companies, or other agencies that recognize the need for occupational therapy interventions. Historically, occupational therapists have worked primarily in hospitals and other health care institutions. Over the past decade, therapists have begun to work in a variety of other settings, including private practice arenas, home health agencies, community programs, adult day-care centers, and industrial sites.

In planning therapeutic interventions, occupational therapy practitioners join colleagues in other disciplines such as medicine, speech-language pathology, physical therapy, social work, psychology, nursing, education, community planning, administration, and industry. Occupational therapists work with other professionals to provide comprehensive services to individuals, groups, families, and others who are directly or indirectly addressing the physical, social, emotional, or developmental challenges of aging. Depending on the setting, whether it is community or hospital based, the occupational therapist may be accountable to the team leader, the referral source, or the reimbursement source for the services provided.

Like other health professionals, occupational therapists are responsible for documenting their services. This usually occurs through written client evaluations, progress reports, staff notes, daily treatment notes, Medicare reports, and discharge notices. Accurate records are essential for evaluating progress, reporting to physicians and caretakers, and billing for services. Occupational therapists are reimbursed through individual fees for service as well as through third-party payers, including Medicare and private insurance companies. The therapist is frequently challenged to articulate the validity of occupational therapy in such a way that the reimbursement agency recognizes the importance of providing the necessary reimbursement.

Occupational therapy can at times be confused with other therapies, such as physical or recreational therapy. Although all three disciplines focus on the quality of life of older adults, it is important to remember that there are distinct differences. Occupational therapy emphasizes performance and the ability of clients, especially older clients, to participate in occupations that are meaningful to their lives.

CYNTHIA HUGHES HARRIS

See Also
Occupational Therapy Assessment and Evaluation

Internet Resource
American Occupational Therapy Association
http://www.aota.org

OCCUPATIONAL THERAPY ASSESSMENT AND EVALUATION

Occupational therapists have a distinct role in determining a client's functional diagnosis, including the cause of dysfunction and dependency. An essential component of the evaluation process, the functional diagnosis ascertains the individual's capacity to perform required and valued activities of everyday living (self-care, work, and leisure). The overarching goal of occupational therapy (OT) is to enable clients to gain a sense of efficacy from being able to engage with satisfaction in their life roles and associated tasks. Occupational functioning and occupational performance are the terms most widely used to conceptualize this goal (Neuhaus & Miller, 1995). The Commission on Practice (1995) of the American Occupational Therapy Association clarified its use of two other terms—assessment and evaluation—to be more consistent with other health professionals organizations and external standard-setting organizations. Evaluation is the process of obtaining and interpreting data necessary for intervention. This activity includes planning for

and documenting the evaluation process and results. Clinical assessment is the use of specific tools or instruments to measure domains of function in the evaluation process.

The occupational therapist's contribution to the plan of care is to generate hypotheses that clarify the client's performance strengths and deficits. Identifying the interrelationship of performance components (physical, cognitive, affective, social, environmental) can determine where and how to intervene to maximize performance in the basic activities of daily living (BADL) and instrumental activities of daily living (IADL). The strengths and deficits in the performance components of function affect patients' motivation, skills, and habits in the context of their physical and social environment, all of which influence treatment planning and goal setting. While observing needed and desired tasks of everyday living, occupational therapists work with patients to remediate or compensate for performance difficulties. "Repeat assessments of task performance are valuable for *monitoring* progress and the effects of multi-disciplinary interventions on function. For example, occupational therapists will alert physicians if they suspect that deterioration of task performance signals the need for adjusting the dosage of a medication causing adverse side effects" (Miller & Bear-Lehman, 2000, p. 288).

The following case study illustrates the nature of the evaluation performed by occupational therapists. A summary of the case is presented, followed by an initial interview and examination that is responsive to the client's clinical complaints. The initial interview and examination focus on several domains of function simultaneously in order to understand the underlying causes of functional decline and dependency. As the therapist gathers information, she or he forms hypotheses about the cause of the dysfunction in order to establish appropriate treatment plans and goals.

Case Study: Mr. Frank

This 73-year-old male client sustained a left cerebrovascular accident with a residual right hemiparesis three weeks ago. He has a history (five years ago) of one myocardial infarction. Transfer from a neurology service to a skilled nursing facility and referral for occupational therapy occurred within the last week.

Functional Status

Mr. Frank walks independently without assistive devices, demonstrating a slow, unsteady gait and a stooped posture. He is unshaven, and his hair is uncombed. His right dominant affected upper extremity has one-quarter to one-half active range of motion throughout the limb, and he is presently unable to perform bilateral motor tasks effectively. He needs minimal to moderate physical assistance with most self-care activities.

Clinical Complaints

Mr. Frank states that he has no appetite and that he knows that OT can't help him. "I built my own house, and look at me," he states. He complains of right shoulder pain and was observed pounding his fist when he unsuccessfully tried to use the telephone. "I couldn't even remember my own number, and, to make things worse, I fell out of bed last night," he exclaimed.

Social History

Mr. Frank is married to a 64-year-old woman. They have no children. They have stated that their greatest pleasure is taking weekend trips to the home in the country that Mr. Frank built for them 12 years ago. He retired five years ago from a successful contracting business. Until this recent cerebrovascular accident, Mr. Frank did some consulting for his previous business, met retired friends for lunch and golf, and spent a lot of his time making and fixing things for his home and his relatives.

Initial Occupational Therapy Interview and Examination of Mr. Frank

After an initial conversation about Mr. Frank's feelings and concerns, Mr. Frank reveals that he has pain in his right shoulder. The OT assessment reveals that the right shoulder is painful because of a subluxation (a partial dislocation). The therapist then demonstrates to Mr. Frank how to reduce pain while walking and sitting. As part of the cognitive assessment, the therapist asks Mr. Frank to sit and walk using the new methods demonstrated. Can he retain the instructions? Do the instructions need to be broken down into one or two steps? How often do the instructions need to be repeated before Mr. Frank can follow through with the treatment regimen? Prognosis for recovery will be based in part on his cognition. *(A referral to physical therapy for ambulation training with a cane is essential.)*

Next, the therapist, with input from Mr. Frank, assesses the occupations, or daily tasks related to work, play, and leisure, that will both increase his functional abilities and be most important to him. He will participate in his rehabilitation only if he finds BADL and IADL meaningful. Selecting activities that are a priority for Mr. Frank, with or without the use of adaptive devices, will reduce his depression, increase the strength in his affected arm, and reduce his shoulder pain. *(Specific dynamic or static positioning will be recommended for each activity.)*

The therapist also discusses with Mr. Frank how he fell out of bed and asks him how he thinks it happened. A BADL assessment that focuses on transfers in and out of the hospital bed is a priority. Is the bed higher than the one he has at home? A new environment increases fall risk and must be assessed immediately. Motivation to accept OT for transfer training should increase if Mr. Frank can be helped to recognize the difference it could make to his safety and therefore his well-being. Mr. Frank's awareness or lack thereof of safety issues may indicate cognitive perceptual-sensory deficits or clinical depression. Further assessments to differentiate these conditions, if not already conducted by members of the team, are critical before effective treatment can be planned and implemented. Correct transfer techniques will be demonstrated to nursing staff when the cause of the problem is determined. The occupational therapist also considers whether antihypertensive medication may have caused postural hypotension or whether sedatives or other medications may be contributing to confusion or dizziness—all risk factors for falls. After hearing Mr. Frank's response and reviewing his medical chart, a consultation with the nursing staff or the physician may be initiated.

Conversation with Mr. Frank also reveals that he was upset about not being able to remember his telephone number. The Assessment of Motor and Process Skills (AMPS), an IADL assessment performed by occupational therapists trained in its use, starts with the client's priority (in this case, use of the telephone). The AMPS reflects a person's ability to use the underlying motor (movement) skills and process (organization to complete a task) skills effectively during full performance of tasks (Neuhaus & Miller, 1995). The occupational therapist also determines whether any members of the team have conducted other cognitive or psychological assessments. If conducting these assessments is not the role of other team members in this setting, the occupational therapist can administer the Folstein Mental Status Examination and the Geriatric Depression Scale (Gallo, Fulmer, Pavezo, & Reichel, 2000). The differential diagnosis between a cognitive impairment and depressive pseudodementia is a challenge and is the responsibility of the team. The client's strong expressions of distress and unkempt appearance and the location of the vascular lesion may signal depression as a comorbid condition.

After an evaluation of Mr. Frank's immediate needs, the occupational therapist discusses future needs and plans so as to target OT toward reducing performance deficits in the activities of everyday living and enhancing the abilities and roles that are important, specifically, to Mr. Frank.

Summary

Occupational therapists frame function in occupation, or the ability to perform the daily tasks related to work, play, and leisure. Functional assessment instruments in OT, standardized and nonstandardized, address multiple domains of function to determine causes of limitations and present or potential strengths. They include history taking and ascertaining what is currently most important to clients, as well as their needs and hopes for the future. Narrative interviews, self-reports, and observation of performance (e.g., self-care, work, leisure) in simulated or natural environments, alone and with significant others, are critical components of the evaluation process. Periodic reassessments are conducted to determine whether treatment and goals need to be modified or have been attained.

Referral to OT is indicated when older adults are demonstrating early functional decline and are at risk for increased morbidity and dependency, or after they have sustained an illness or injury that impedes function. An OT evaluation is beneficial to older adults, as well as their caregivers, in determining how much assistance in BADL and IADL is required and how much improvement is possible. Through remediation or compensatory strategies, occupational therapists assist older adults to remain as independent as possible.

PATRICIA MILLER

See Also
Activities of Daily Living
Deconditioning Prevention
Measuring Physical Function
Multidimensional Functional Assessment: Instruments
Multidimensional Functional Assessment: Overview
Rehabilitation

Internet Resource
American Occupational Therapy Association
http://www.aota.org

OLDER AMERICANS ACT

In 1965, the U.S. Congress enacted proposals that established three major federal programs to address the needs of older Americans. Signed into law by President Lyndon B. Johnson, these laws created the nation's Medicare and Medicaid programs and the programs funded under the Older Americans Act (OAA).

The federally administered Medicare health insurance program provides most older Americans with access to and coverage for medical care delivered in hospitals, clinics, private physicians' offices, and, to a limited degree, the home. The state-administered federal Medicaid program covers the costs of institutional long-term care and limited home and community-based care for nursing home–eligible persons with limited incomes and assets. Both Medicare and Medicaid are entitlement programs.

In sharp contrast, the OAA was given a broader agenda but has received more modest appropriations from Congress for nonentitlement formula grants to states. The act authorizes and funds Grants for State and Community Programs on Aging (Title III), specifically Supportive Services and Senior Centers (Title IIIB). It also authorizes and funds Congregate Nutrition Services (Title IIIC1) and Home Delivered Nutrition Services (Title IIIC2); Training, Research, and Discretionary Projects and Programs (Title IV); Grants for Native Americans (Title VI); and Allotments for Vulnerable Elder Rights Protection Activities (Title VII). These programs are administered through a partnership of the federal Administration on Aging, state units on aging, and Area Agencies on Aging.

The OAA provides a broad set of 10 core national values and objectives that articulate a vision for America's's elderly population. The act accommodates the multidimensional needs of America's diverse aging population, irrespective of socioeconomic status, but calls for specific attention to persons who are in greatest economic and social need, with particular attention to low-income minority individuals. It establishes the Administration on Aging as an operating division within the U.S. Department of Health and Human Services and also establishes an Aging Network of state and Area Agencies on Aging. The latter have as designated responsibilities the assessment of needs at the state and local levels and, in response to these, the "development and implementation of [new or improved] programs and comprehensive and coordinated systems to serve older individuals." Together, the Aging Network leverages available funding to stimulate greater investments of state, local, and private dollars to meet the identified needs of older persons.

Aims of the Older Americans Act

The 10 objectives detailed in Title I of the OAA specify requisites for an adequate quality of life for older persons. They include a recognition of the need for an "adequate income in retirement in accordance with the American standard of living," "the best possible physical and mental health," and the importance of "a comprehensive array of community-based, long-term care services . . . to appropriately sustain older people . . . [and] support . . . family members and other persons providing voluntary care." Title I also calls for opportunities for employment; meaningful civic, cultural, educational, training, and recreational activities; and other desirable opportunities and conditions.

Some have criticized the aims of the OAA as being too broad and too multidimensional in scope. Others have objected to the act's singular focus on older persons. For still others, the act articulates a set of aspirations and commitments that serve as a societal benchmark for a segment of the population that might otherwise be overlooked.

Because explicit implementation programs are not identified, the Administration on Aging and the Aging Network have considerable latitude in their advocacy for and pursuit of implementation activities. Since enactment of the OAA in 1965, a number of initiatives have been launched by the Administration on Aging and the Aging Network. The most recent examples are the proposed National Family Caregiver Support Program and National Life Course Planning Program, which were introduced during the first session of the 106th Congress as part of the Clinton administration's OAA reauthorization proposal.

Structure and Roles of the Aging Network

The Aging Network, established by the OAA and led by the Administration on Aging, includes 57 state units on aging and approximately 650 area agencies on aging—all administrative units that were authorized by the 1973 amendments. More than 220 tribal organizations representing some 300 Native American tribes and about 27,000 service providers are also part of the Aging Network. In addition, more than 40 national organizations represent service provider, advocacy, minority elderly American, retiree labor, and other aging groups. Many of these organizations receive OAA funds through Titles III, IV, V, VI, and VII.

When last reauthorized in 1992, the OAA was amended to elevate the commissioner of the Administration on Aging to assistant secretary for aging in the Office of the Secretary.

The responsibilities borne by state and area agencies on aging have expanded over the last two decades, with one-third of all state units administering programs for disabled adults, including Medicaid waiver and large state-funded home and community-based service programs, as well as OAA programs and services. Such an evolution in the role of state units is in keeping with the OAA's aim of systems development.

Reauthorization and Funding of the Older Americans Act

Although OAA programs and services continue to receive support through Congressional appropriations, the Act was last reauthorized in 1992 for a three-year period. Reauthorization proposals were considered in 1997 and in 1999 during the 105th and 106th Congresses, respectively. During the 105th Congress, a number of issues were central to the reauthorization discussion. Minority targeting, cost sharing, and the allocation of Title V (Community Employment Service for Older Americans) funds to national sponsors and to the states were among the more thorny issues, all of which continued to dominate the discussion in the 106th Congress. However, many believed that the Title V funding issue was the most critical one, with strong differences often falling along partisan lines (Kirchoff, 1999).

The reauthorization proposal introduced by the Clinton administration during the first session of the 106th Congress included several new program proposals. For example, the Administration on Aging proposed to amend Title IIID, In-Home Services for Frail Older Individuals, by renaming it the National Family Caregiver Support Program and authorizing states to provide information and assistance, respite, education, support groups, counseling, and limited supplemental services. Additionally, the Administration on Aging proposed to amend Section 741 (Title VIIA, Chapter 5), Outreach, Counseling, and Assistance Program, to establish a National Lifecourse Planning Program.

To enable states to modernize service delivery systems, the Administration on Aging proposed yet another initiative through its reauthorization proposal. It called for the amendment of Title III, Section 307(a), to establish an option that would enable states to use the greater of 4% or $300,000, in the instance of the 50 states, the district of Columbia, and the Commonwealth of Puerto Rico, or $50,000, in the case of Guam, the Virgin Islands, American Samoa, and the Commonwealth of the Northern Mariana Islands, of nutrition and supportive services funding to support the development, testing, and implementation of innovative, cost-effective methods of OAA service delivery to older persons and their families. It also proposed to redesignate Title IV, Training, Research, and Discretionary Projects and Programs to be State and Local Innovations and Programs of National Significance. Finally, in relation to Title V, the administration proposed to maintain the central features of the program. It suggested that the Senior Community Service Employment Program be enhanced by increasing accountability, improving customer service, improving the linkages to the one-stop delivery system developed under the Workforce Investment Act of 1998, and pursuing "more equitable allocation of resources among states" (Shalala & Herman, letter to Congress, March 31, 1999).

From 1995 through 1999, the Administration on Aging requested and received small budget increases. In fiscal year 2000, the administration proposed an unprecedented $165 million increase, with

$125 million for the proposed caregiver support program. Also proposed were a $35 million increase for home-delivered meals, an additional $4 million to focus on the elimination of health disparities among minority elders, $1.035 million more for program administration, and an additional $210,000 for Operation Restore Trust. The National Family Caregiver Support Program received broad bipartisan support, was embraced by the Democratic minority in the Senate, and was a Clinton administration priority. At the close of the first session of the 106th Congress, monies for the proposed caregiver program were initially placed in the budget and were then removed. The total appropriation for the Administration on Aging was thus $52.4 million—the single largest increase in the agency's budget in the last decade of the 20th century.

The Future

Demographic projections for 21st-century America suggest that ours will be a nation with more older adults than younger persons. Between 2011 and 2030, the baby boom generation, comprising some 76 million individuals, will become older Americans. Thus, as many policy makers have observed, the aging of the population will be one of the major challenges facing them in the immediate future and for the long term.

The Aging network had demonstrated through everyday practice and through its addressing of emerging concerns that it is interested in evolving its service delivery models by using new technologies and evidence-based interventions, in response to the diverse needs of differing cohorts of older Americans. The future of OAA services and programs will depend on how proactively and creatively the executive and legislative branches of federal, state, and local governments, including the Administration on Aging and the governmental members of the Aging Network, utilize the flexibility inherent in the law. It will depend on whether policy makers use the act's transcendent goals and aspirations to deal with issues that affect the quality

of life of older persons and their families. If such an approach is embraced in the future, the Administration on Aging and the OAA will be able to flourish and respond effectively to the needs of our nation's growing elderly population.

JEANETTE C. TAKAMURA

The views expressed are solely those of the author and do not reflect an official position of the federal government, the U.S. Department of Health and Human Services, or the Administration on Aging.

Internet Resource
Administration on Aging
http://www.aoa.gov

ORAL HEALTH

Dentists currently provide care for more older Americans than at any time in history. The 1989 National Health Interview Survey reported a 30% increase since 1983 in the number of patients over the age of 65 seen by dentists. This trend is accompanied by the fact that more older patients have their own teeth, and many elders have received dental care throughout their lifetimes.

Dental Restorations

Most of the teeth of older persons have dental restorations, many of which exhibit signs of mechanical breakdown associated with length of service or recurrent disease. As teeth are re-restored, the restorations become larger.

After a number of re-restorations, restorations within the tooth are no longer possible; prosthetic crowns or other extensive restorations are necessary. A growing number of sophisticated single- and multitooth crowns and bridges are placed in elderly patients (Meskin, Dillenberg, & Heft, 1990). The teeth of aging individuals are different from those of younger persons. As teeth age, they are subject to various chewing forces (attrition), abrasion (from a toothbrush), and the placement of multiple restorations. Over time, the nerve chambers in the teeth become narrower and shorter. This results in teeth that are more brittle and therefore more likely to fracture. In addition, these teeth do not respond to stimuli in the same way as the teeth of younger persons. It is common for patients in their late 60s to undergo re-restoration of teeth without local anesthesia. Conversely, the lack of sensitivity on the part of aging teeth can be hazardous. Not uncommonly, older patients present for dental care with large areas of dental caries and no symptoms. This fact alone justifies routine dental care, including radiographic surveys, even in the absence of symptoms.

Periodontium

In healthy aging patients with proper oral hygiene, the periodontium appears normal. Periodontal degeneration is not associated with aging. In the absence of appropriate oral hygiene or in the presence of systemic disease, older patients can suffer a full range of periodontal diseases that are manifest by a variety of inflammatory and degenerative states. Recent studies challenge the long-held belief that loss of dentition is caused by periodontal disease. Periodontal disease in older patients is a process of limited, local, active disease rather than a generalized, chronic disease state; it progresses slowly and is treatable. Treatment is conservative and generally limited to minimally invasive procedures for most patients.

Oral Hygiene

The single most important factor relating to retention and health of the dentition centers on oral hygiene. This premise applies to patients who are aging with relatively good health or to those with acute or chronic disease. With aging, it becomes more difficult to maintain proper oral hygiene because of multiple interacting factors, including the number and status of dental restorations, anatomical relationships between adjacent teeth that have been altered due to failing restorations or loss of bony

height and recession of the gingiva, fixed and removable prostheses, root surfaces exposed by disease or as a result of surgical repair of gingival disease processes, impaired visual acuity, compromised manual dexterity, restricted range of motion, and pharmacologically induced xerostomia.

Proper technique for oral hygiene begins with selecting a toothbrush with soft bristles and educating the person about why this is recommended. Patients often believe that stiffer bristles are more effective at debriding the surfaces of the teeth when, in fact, medium- and hard-bristle brushes lack the flexibility needed to efficiently access the normal curvature of teeth surfaces.

Fluoride

The essential factor in toothpaste selection is ensuring that the toothpaste contains fluoride. Considered essential for pediatric patients, fluoride can have a significant positive impact on the oral health of aging patients as well. Patients who are healthy and free from acute or chronic dental disease are well served by over-the-counter fluoride-containing toothpastes. Those with chronic diseases or xerostomia will benefit from the increased fluoride levels contained in prescription-strength toothpaste.

Three fluoride formulations are available by prescription: acidulated phosphofluoride, stannous fluoride, and neutral sodium fluoride. Of the three, pH-neutral sodium fluoride is generally the preferred formulation for fluoride delivery in older patients. Xerostomic patients do not have adequate character, rate, and flow of saliva to keep the oral mucosal tissues moist. Acidulated phosphofluoride contains phosphoric acid as a mechanism of etching tooth enamel to facilitate the uptake of the fluoride ion. It can be very irritating to dry mucosa and is not recommended for patients with xerostomia.

Stannous fluoride is an effective fluoride vehicle with some unwanted side effects. Tin in the stannous formulation causes extrinsic staining on the crowns and exposed root surfaces of teeth and has a metallic taste that some patients find unpleasant. Studies indicate that compliance with home-care instructions for oral care decreases if the patient finds the taste of the toothpaste unpleasant.

Prescription-strength fluoride rinses are available, but there are some problems with their use by elders. They are designed for topical delivery of fluoride and should not be swallowed. Patients who are cognitively impaired due to stroke or progressive neurological disorders may have difficulty rinsing and expectorating the fluoride rinse. Prescribing a topical fluoride rinse increases the number of steps in the oral hygiene routine, whereas substituting a prescription-strength toothpaste does not. Use of prescription-strength fluoride toothpaste increases the benefit of fluoride and probably achieves greater compliance than does use of a rinse.

Dental Floss

Flossing daily is the only effective technique for removing bacterial plaque, the primary causative agent in periodontal (gum) disease, between adjacent teeth. Typically, patients floss their teeth using both hands to hold the piece of floss. Those who are unable to hold the floss by the fingers of both hands can use any commercially available floss aid. These devices hold the dental floss so that the patient can use one hand to guide the floss between adjacent teeth. Floss aids are particularly helpful for patients who have restricted range of motion in the shoulder, elbow, or hand; patients with arthritis; and those with early cognitive impairments. Disposable floss aids eliminate the need to thread the floss onto the aid and are easy to use.

Saliva

Saliva, essential for oral health, has four main functions: it is a buffering agent, has antimicrobial properties, aids in remineralization, and is a lubricant. Since xerostomia is a common side effect of many prescription medications taken by the elderly, elders warrant careful evaluation for this oral problem. Practitioners should assess patients who complain of dry mouth and observe for sparse saliva

that is excessively mucinous or completely absent. Initially, health care providers can attempt to provide palliative relief with commercially available synthetic saliva. Patients' attempts to relieve the symptoms of xerostomia by sucking on candy provides short-term relief but often causes rampant dental caries.

Healthy persons who are aging can expect to retain their teeth with a minimal commitment to daily oral hygiene. Those persons with acute or chronic illness can look forward to a similar outcome, assuming that the dentist develops with them a daily oral hygiene regimen that is appropriate with regard to diagnosis, prognosis, and pharmacological management and that the patient is able and willing to carry out the practices.

ALAN STARK

See Also
Dentures
Geriatric Dentistry
Oral Health Assessment
Xerostomia

Internet Key Words
Dental Restoration
Dentistry
Fluoride
Oral Health
Oral Hygiene
Peridontal Disease

Internet Resources
American Society of Geriatric Dentistry
*http://www.bgsm.edu:80/dentistry/foscod/
    asgd.htm*

Elders Link to Dental Education (University of British Columbia)
*http://www.dentistry.ubc.ca/Elder/NONFRAME/
    index3.htm*

Geriatric Dentistry Referral
http://www.dentalsite.com/dentists/ger.html

Oral Health and Aging (University of Maryland at Baltimore)

http://cpmcnet.columbia.edu:80/dept/dental/
  Dental_Educational_Software/
  Gerontology_and_Geriatric_Dentistry/
  Oral_Health/oral_health_indx.html

ORAL HEALTH ASSESSMENT

Despite increasing awareness of the need for dental care, many older people have poor oral health, inadequate oral hygiene, and carious teeth; are edentulous; and lack or have poorly fitting dentures. Oral health problems often affect an individual's well-being, self-esteem, and quality of life (Kayser-Jones, Bird, Paul, Long, & Schell, 1995). Maintaining oral health in old age is important, because untreated dental problems can cause pain and discomfort that may interfere with eating and swallowing, resulting in inadequate nutritional intake. Oral infection may be a minor problem for a healthy younger person, but it can cause a serious systemic infection in an older frail person.

Oral carcinoma is a concern among elders. The American Cancer Society estimated that nearly 22,000 new cases of oral cancer would be diagnosed in the United States in 1999, and 50% of oral cancer cases reported between 1985 and 1996 were among people 65 years of age and older. Early detection and treatment are the most important factors in reducing oral cancer morbidity and mortality (Shiboski, Shiboski & Silverman, in press).

Although edentulism (loss of all teeth) has declined since the 1970s, approximately 41% of people 65 years and older are edentulous (Burt, 1992). As edentulism decreased, the occurrence of caries (tooth decay) in old age increased. Today, many people reach adulthood with most of their teeth free of caries, but people can and do develop caries in old age. Although dental health is improving, lower socioeconomic status, geographic region, cultural factors, and education influence oral health care (Ettinger, 1997). Among older Americans, 51.3% of those with fewer than 8 years of education are edentulous, compared with 28.6% of those with 12 or more years of education (Burt, 1992).

The oral health status of nursing home residents is a continuing concern. Numerous studies document the high prevalence of oral disease among nursing home residents and the need for evaluation and treatment (Altieri, Vogler, Goldblatt, & Katz, 1993; Warren, Kambhu, & Hand, 1994). Barriers to obtaining oral health care in nursing homes include poor Medicaid coverage for adult dental services and no Medicare coverage for dental care. Few older people have private dental insurance, and due to multiple functional and cognitive impairments, access to dental care is limited for many residents.

Federal regulations require that an initial oral health assessment be completed by a dentist within 14 days of admission to a nursing home. Federal regulations also state that nursing facilities must provide routine and emergency dental services to meet the needs of residents; must, if necessary, assist residents in making dental appointments and arranging transportation to and from the dentist's office; and must promptly refer residents with lost or damaged dentures to a dentist. States may also have oral health care regulations, and how care is provided may vary from state to state. In California, for example, many nursing homes do not have dentists on staff; dental services are provided by contract, usually during monthly visits. If emergency dental care is needed, it is provided only if the nurse, resident, or family requests care.

Kayser-Jones Brief Oral Health Status Examination

Nursing staff (registered nurses [RNs], licensed vocational nurses [LVNs], and certified nursing assistants [CNAs]) are responsible for providing oral hygiene and are therefore in a position to identify oral health problems. However, most nursing staff have limited or no preparation in assessing oral health.

The Kayser-Jones Brief Oral Health Status Examination (BOHSE) can be used by nursing staff to assess the oral health status of residents. It was developed based on recommendations from the

Resident's Name _____   Date _____

Examiner's Name _____   **TOTAL SCORE** _____

CATEGORY	MEASUREMENT	0	1	2
LYMPH NODES	Observe and feel nodes	No enlargement	Enlarged, not tender	Enlarged and tender*
LIPS	Observe, feel tissue and ask resident, family or staff (e.g. primary caregiver)	Smooth, pink, moist	Dry, chapped, or red at corners*	White or red patch, bleeding or ulcer for 2 weeks*
TONGUE	Observe, feel tissue and ask resident, family or staff (e.g. primary caregiver)	Normal roughness, pink and moist	Coated, smooth, patchy, severely fissured or some redness	Red, smooth, white or red patch; ulcer for 2 weeks*
TISSUE INSIDE CHEEK, FLOOR AND ROOF OF MOUTH	Observe, feel tissue and ask resident, family or staff (e.g. primary caregiver)	Pink and moist	Dry, shiny, rough red, or swollen*	White or red patch, bleeding, hardness; ulcer for 2 weeks*
GUMS BETWEEN TEETH AND/OR UNDER ARTIFICIAL TEETH	Gently press gums with tip of tongue blade	Pink, small indentations; firm, smooth and pink under artificial teeth	Redness at border around 1-6 teeth; one red area or sore spot under artificial teeth*	Swollen or bleeding gums, redness at border around 7 or more teeth, loose teeth; generalized redness or sores under artificial teeth*
SALIVA (EFFECT ON TISSUE)	Touch tongue blade to center of tongue and floor of mouth	Tissues moist, saliva free flowing and watery	Tissues dry and sticky	Tissues parched and red, no saliva*
CONDITION OF NATURAL TEETH	Observe and count number of decayed or broken teeth	No decayed or broken teeth/roots	1-3 decayed or broken teeth/roots*	4 or more decayed or broken teeth/roots; fewer than 4 teeth in either jaw*
CONDITION OF ARTIFICIAL TEETH	Observe and ask patient, family or staff (e.g. primary caregiver)	Unbroken teeth, worn most of the time	1 broken/missing tooth, or worn for eating or cosmetics only	More than 1 broken or missing tooth, or either denture missing or never worn*
PAIRS OF TEETH IN CHEWING POSITION (NATURAL OR ARTIFICIAL)	Observe and count pairs of teeth in chewing position	12 or more pairs of teeth in chewing position	8-11 pairs of teeth in chewing position	0 7 pairs of teeth in chewing position*
ORAL CLEANLINESS	Observe appearance of teeth or dentures	Clean, no food particles/ tartar in the mouth or on artificial teeth	Food particles/tartar in one or two places in the mouth or on artificial teeth	Food particles/tartar in most places in the mouth or on artificial teeth

Upper dentures labeled: Yes ☐   No ☐   None ☐   Lower dentures labeled: Yes ☐   No ☐   None ☐   *Underlined = refer to dentist immediately
Is your mouth comfortable? Yes ☐   No ☐   If No, explain:

_____

Additional comments: _____

_____

—kayser ones reure

FIGURE 1   Kayser-Jones Brief Oral Health Status Examination. ©1995, Regents of the University of California, San Francisco. All rights reserved. Used with permission. The author acknowledges the assistance of William F. Bird, D.D.S., Dr.P.H., in the development of this instrument.

American Dental Association, a review of available oral assessment guides, and consultation with dental school faculty. Ten items reflect the status of oral health and function (lymph nodes; lips; tongue; tissue inside cheek, floor, and roof of mouth; gums between teeth or under artificial teeth; saliva; condition of natural teeth; condition of artificial teeth; pairs of teeth in chewing position; and oral cleanliness). Each item has three descriptors and is rated on a three-point scale (0, 1, 2)—0 indicating the healthy end, and 2 the unhealthy end of the scale (see the accompanying figure). The final score is the sum of the scores from the 10 categories and can range from 0 (very healthy) to 20 (very unhealthy).

In the initial test of the BOHSE instrument, the assessment scores compiled by the nursing staff were compared with those of a dentist to determine whether the staff could identify and evaluate oral health problems as well as the dentist. Findings disclosed that RNs and LVNs had scores very similar to those of the dentist. CNAs had the lowest correlation scores with the dentist but still did remarkably well.

Staff who knew the residents well were more successful than the dentist in completing examinations of cognitively impaired residents. When examining residents who refused to be examined by the dentist, the nursing staff spoke to them kindly, waited patiently, and gently persuaded them to cooperate.

After the BOHSE assessment, several residents were referred for immediate dental care for conditions discovered during data collection, including an infected root tip, a moderately severe case of lichen planus (an inflammatory condition of the mucous membranes), a gum abscess, gingivitis, several cases of candidiasis, and poorly fitting or broken dentures. The residents suffering these conditions had not complained to the staff; these problems were discovered only through their participation in the research project (Kayser-Jones et al., 1995).

Before using the BOHSE, nursing staff should receive four to six hours of in-service education from a professional dentist. School of dentistry faculty, dentists in private practice, or dentists who contract to provide dental services to nursing homes can educate nursing staff. Slides should be used to illustrate healthy and pathological conditions, emphasizing how each item (i.e., lips, tongue, tissue) is scored. The dentist-educator then examines a resident, demonstrating use of the instrument. Subsequently, the nursing staff should examine the same resident under the dentist's supervision. This procedure should be repeated until the nursing staff produce scores very similar to those of the dentist.

Advantages of Oral Health Examinations by Nursing Staff

Since the nurses are well acquainted with the residents and familiar with their habits, behaviors, likes, and dislikes, they are more likely than a visiting dentist to be successful in examining residents. A resident who refuses to be examined one day can be examined the following day. Nursing staff can examine residents on a regular basis, which is especially important for people with cognitive impairment, who may be unable to report pain or discomfort. Moreover, if there is a change in a resident's behavior, such as refusal to eat, the nurse can do an examination to rule out oral disease. Involving the nursing staff leads to earlier recognition of problems and may prevent systemic infections or other situations such as weight loss. Teaching nursing staff to do an oral health assessment increases staff awareness of the importance of dental health and oral hygiene.

It must be emphasized, however, that the BOHSE is for screening purposes only. It is not a diagnostic tool, and it does not replace the need for periodic examination by a professional dentist.

Given the high level of unmet dental needs in nursing homes and the relationship between oral health status and physical health and well-being, nurses, dentists, and physicians should collaborate in addressing this important problem. The BOHSE can also be used in day-care centers and when providing home care to older people.

JEANIE KAYSER-JONES

See Also
Oral Health

Internet Key Words
Dental Caries
Edentulousness
Nursing Homes

Internet Resources
American Dental Association
http://www.ada.org/

International Association for Dental Research
http://www.iadr.com/start.html

National Institute of Dental Research
http://www.nidr.gov/

Oral Health America
http://www.oralhealthamerica.org/home.html

OSTEOARTHRITIS

Osteoarthritis (OA), the most common rheumatic disease, disproportionately affects older populations. Prevalence figures vary, but it is estimated that 12% of the U.S. population is affected. It may be found radiographically in almost 75% of those over age 70. However, studies indicate that early radiographic disease may begin even in the fifth and sixth decades. As a chronic condition, OA can potentiate loss of function and independence and is a leading cause of total joint replacement of the knee and hip.

The most commonly affected sites include the hand, knee, hip, and spinal facet joints. Though joints may have radiographic evidence of disease, they may not be symptomatic. This is particularly true of the proximal interphalangeal (PIP) and distal interphalangeal (DIP) joints of the hand. These joints may develop bony, hard swellings on the superolateral aspect known as Bouchard's (PIP) and Heberden's (DIP) nodes.

The discrepancy between radiographic OA and symptoms has been noted in many studies (Hochberg, Lawrence, Everett, & Coroni-Huntley, 1989). Women tend to be more symptomatic; the correlation between the radiographic grade of OA and pain is strongest for the knee and the hip, although some patients with Kellergran grades 3 and 4 OA are asymptomatic. The DIP joint tends to be the least symptomatic, regardless of sex.

Pathophysiology

OA is a disease in which hyaline articular cartilage, the lining tissue of all synovial joints, fails to perform. Long regarded as a "wear and tear" consequence of aging, OA is now construed as a disequilibrium between synthetic and destructive elements with a failure to maintain cartilage homeostasis.

Recent advances related to the pathophysiology of OA provide not only a fuller understanding of the complexity of the disease process but also insights into potential therapeutic interventions. Matrix Metalloprotease (MMPs) enzymes play a key role in cartilage degradation. The MMPs consist of three main groups: collagenase, gelatinase, and stromelysin. Tissue inhibitors of Matrix Metalloproteases (TIMPs) balance the activity of the MMPs. In OA, the equation tips toward the MMPs and cartilage degradation. Interleukins, especially IL-1, play a key role not only in triggering MMP synthesis and nitric oxide production but also in inhibiting cartilage repair. Nitric oxide levels are increased in the OA joint and also affect MMP levels.

Risk Factors

Risk factors for OA include age, gender, heredity, obesity, occupation, trauma, and hypermobility. Other risk factors identified in the Framingham osteoarthritis study include low vitamin D and C levels (McAlindon et al., 1996). It appears that 25-hydroxyvitamin D levels are associated with progression but not incidence of disease of the knee. Moreover, quadriceps muscle weakness has been implicated as promoting the development of knee OA. This finding clearly has important implications for therapeutic intervention. Presently, dosing guidelines for vitamins C and D in patients with OA do not exist. However, doses within the newest recommended guidelines would be prudent.

Advancing age is one of the strongest risk factors for OA. Prior to age 50, men have a higher prevalence of OA, but after age 50, women have a higher incidence and prevalence. Generalized OA, a variant that involves either nodal or non-nodal hand OA in combination with other large-joint OA, appears to have a genetic component and is seen more commonly in women.

Obesity is correlated with the development of both knee and hand OA but not hip disease. In the elderly, the combination of obesity and heavy physical activity for at least three hours a day increases the risk of developing OA of the knee 13-fold (McAlindon, Wilson, Aliabadi, Weissman, & Felson, 1999). More importantly, weight loss is associated with reduced risk of developing symptomatic knee OA. Whether weight loss can improve symptoms in those who already have the disease has not been tested, but weight reduction appears to be prudent.

In 1986, the American College of Rheumatology published criteria to define and diagnose OA of the knee (Altman, Asch, & Bloch, 1986). Subsequently, similar models were developed for OA of the hip and hand. The criteria are based on a combination of signs, symptoms, and simple laboratory procedures arranged in a traditional inclusion criteria model or the superior decision tree analysis. It should be emphasized that not all joint pain is articular and that nonarticular conditions such as bursitis and tendinitis need to be ruled out.

Management

Management of OA requires a multifaceted approach using a combination of nonpharmocological and pharmacological interventions. The goals of treatment are to eliminate pain and inflammation, prevent degradation of the cartilage lining, and in so doing, maintain or improve the quality of life. With current therapeutic strategies, pain and inflammation can be controlled, but there is no approved pharmacological or proven nutritional supplement that will prevent or repair cartilage damage.

Nonpharmacological Treatment

The American College of Rheumatology published treatment guidelines for knee and hip OA that emphasize the importance of nonpharmacological interventions (Hochberg et al., 1995a, 1995b). Education about the disease process should include cautionary advice about the numerous claims made by manufacturers of nutritional supplements; however, the role of physical therapy (PT) and occupational therapy (OT) cannot be overemphasized. Effective PT includes localized modalities such as heat, ultrasound, and ice, along with a graduated exercise regimen to improve joint range of motion and motor strength. Hydrotherapy, a low-impact protocol, is often very effective in individuals with OA of the knee and hip.

Physical therapists can fit patients with supportive devices and orthotics that reduce pain and improve function. Those individuals with unicompartmental tibiofemoral arthritis may benefit from a lightweight "unloading" knee brace. In addition, the use of a vasoelastic insert or a wedged insole may be helpful for those with medial compartment OA of the knee. A properly used cane can also provide effective "unloading" of an involved knee or hip. The cane length should be equal to the distance from the bottom of the shoe heel to the wrist crease when the arm is held at the side. For the cane to be effective, it must be used on the side contralateral to the involved leg. Mobility in this manner often takes practice but can be taught effectively by a physical therapist.

The choice between heat and ice is somewhat arbitrary. Ice is an effective analgesic and anti-inflammatory in the acute traumatic situation, but it is not necessarily better than moist heat in the chronic setting of OA. Both moist heat and ice raise the pain threshold and reduce spasm. Moist heat should be applied cautiously and for no more than 20 minutes at a time. These modalities can be used on an as-needed basis three to four times daily. Parrafin baths are effective means of providing heat to symptomatic DIPs and PIP joints.

Berman et al. (1999) published a randomized, controlled trial evaluating the efficacy of acupuncture as an adjunctive therapy for elderly patients with OA of the knee who remained on their standard care of anti-inflammatory and analgesic medications. The study showed a 40% improvement in Western Ontario and McMaster University Osteoarthritis Index scores over baseline and supports the notion that acupuncture is a safe and effective addition to the armamentarium. However, the analysis lasted only 12 weeks, and there was some loss of effectiveness 4 weeks following the final acupuncture treatment.

Pharmacological Treatment

Rubifacients have been used for many years by patients seeking relief, but only recently has scientific evidence supported the use of one topical treatment, capsaicin. Capsaicin depletes substance P by enhancing its release from unmyelinated C nerve fibers. Various strengths are available. Lower concentrations (.025%) need to be applied four times daily, whereas higher concentrations (.075%) can be applied twice daily. The most common side effects are local erythema, burning, and stinging, which tend to diminish with repeated use. Handwashing after application is strongly urged, since capsaicin can cause marked eye irritation if the individual happens to rub his or her eye.

Glucosamine and chondroitin sulfate, two over-the-counter nutritional supplements, have become very popular. To date, there is no definitive study of glucosamine and chondroitin supplements, but some studies suggest efficacy of these agents in OA of the knee and hip when compared with traditional nonsteroidal anti-inflammatory drugs (NSAIDs). However, many of these studies have methodological problems. Consumers need to be aware that there is no guarantee of the reliability of these supplements.

NSAIDs have been the cornerstone of OA treatment. These drugs provide both analgesic and anti-inflammatory activity. However, because appreciable inflammation is often absent, and considering the risk of gastropathy, analgesics frequently suffice. Acetaminophen in doses up to 4 g daily has been proved efficacious, although in a survey of OA patients, 67% of patients preferred an NSAID to acetaminophen (Wolfe, Zhao, & Lane, 2000).

Despite the estimated 70 million prescriptions written annually, NSAIDs can have serious side effects. NSAID-associated mortality was reported to be equivalent to the mortality figures for AIDS (Singh & Triadafilopiolus, 1999). Risk factors for NSAID gastropathy include age greater than 60, history of ulcer, concomitant use of corticosteroids, concomitant use of warfarin, and the use of high-dose and combination NSAIDs. Age-related risk is independent and linear. Nevertheless, some patients do not respond to analgesic use and present with signs of inflammation. Before the release of COX-2–selective NSAIDs, some gastrointestinal (GI) risk reduction could be achieved with the use of nonacetylated salicylates.

Other attempts to reduce GI risk center on the use of an NSAID in combination with misoprostol H_2 blockers and proton pump inhibitors. There is strong evidence for the efficacy of misoprostol in doses of 200 µg four times daily. The proton pump inhibitor omeprazole 20 mg daily was also shown to reduce gastric ulcer disease. The H_2 blockers are less effective in protecting the gastric mucosa. High-dose famotidine 40 mg twice daily shows some benefit, but only misoprostol has been approved for the prevention of NSAID-induced ulcers. When taken four times a day, misoprostol can cause diarrhea; this can be ameliorated by dosing two or three times daily, but there is some loss of efficacy.

The new COX-2–selective NSAIDs celecoxib and rofecoxib, both approved for the treatment of OA, offer efficacy and an improved GI safety profile and do not interfere with platelet function. Older patients whose symptoms are not relieved with acetaminophen or who have other risk factors are candidates for a COX-2–selective NSAID.

Intra-articular Corticosteroids and Surgery

The judicious use of intra-articular corticosteroid injections (no more frequent than three to four times a year) can be very effective, especially in those patients with a single painful joint. Commonly used preparations include triamcinolone acetonide and methylprednisolone at a dose of 40 mg, often mixed with 1 to 2 mL of 2% lidocaine for a large joint like the knee. There is no role for systemic corticosteroids in the treatment of OA.

Two forms of hyaluronic acid supplementation are available for intra-articular knee injection. Hyalgan and Synvisc differ in their molecular weights and dosing frequency. Hyalgan is given as a series of five intra-articular injections over four weeks, whereas Synvisc is a series of three injections over two weeks. Patients with advanced OA of the knee (Kellergran grade 4) are less likely to respond to injection treatment, whereas those with milder disease radiographically may respond for as long as six months. Patients need to be advised that benefits do not start immediately after receiving the injections; it may take 8 to 10 weeks before maximal benefits are realized. Knee injections are very well tolerated, and the nominal pain of the injection can be reduced by using a topical spray such as ethyl chloride.

For some patients, surgical options such as arthroplasty, arthroscopy, and osteotomy may be reasonable choices. However, some individuals with concomitant medical illnesses, cognitive disability that makes them unable to follow postoperative treatment plans, or marked muscle weakness may not be appropriate candidates. Indications for total joint replacement are refractory pain that is not controlled with a medical and PT regimen and significant loss of independence accompanied by decreasing quality of life.

There are no remitative agents available for OA. Areas of active investigation include the development of specific cytokine inhibitor agents that would promote repair of damaged cartilage. The National Institutes of Health is currently sponsoring a study on doxycycline, an MMP inhibitor, in bilateral knee OA progression. Ongoing longitudinal and epidemiological studies are likely to expand avenues for prevention and treatment.

STUART GREEN

See Also
Fractures
Joint Replacement: Lower Extremity
Occupational Therapy Assessment and Evaluation
Physical Therapy Services

Internet Key Words
Non-Pharmacologic
NSAIDS
Osteoarthritis
Pharmacologic
Treatment

Internet Resources
American College of Rheumatology
http://www.rheumatology.org

Arthritis Foundation
http://www.arthritis.org

Johns Hopkins Arthritis Center
http://www.hopkins-arthritis.som.jhmi.edu

National Institute of Arthritis and Musculoskeletal and Skin Diseases, National Institutes of Health
http://www.nih.gov/niams/

OSTEOPOROSIS

Metabolic bone disease produces diffusely decreased bone density and diminished bone strength, predisposing to fractures. Osteopenia, or abnormally decreased bone density, results from four conditions that involve different pathological processes: osteoporosis, osteomalacia, hyperparathy-roidism, and myeloma. Osteoporosis, the most common metabolic bone disease, is characterized by low bone mass and microarchitectural deterioration of bone tissue, leading to enhanced bone fragility and a consequent increase in fracture risk (Babbitt, 1994). It must be differentiated from osteomalacia, defined as an increase in unmineralized matrix independent of the amount of bone mass. Osteoporosis affects 15 to 20 million U.S. citizens and causes approximately 1.5 million fractures annually (Babbitt, 1994). The personal and economic costs of severe osteoporotic fractures are high, and these fractures cause substantial morbidity and decline in quality of life. Ten billion dollars is spent annually on treatment.

In most cases, osteoporosis is primary, and no other disease causing bone loss can be diagnosed. Only 5% of all cases of osteoporosis are secondary to other disease processes, including gastrointestinal disorders (malabsorption disorders, hepatic disease), neoplasm (multiple myeloma), and various endocrinopathies (hyperthyroidism, hyperparathyriodism, hypogonadism, diabetes mellitus). Primary osteoporosis is further divided into postmenopausal (type 1) and senile (type 2). Type 1 is more common and is seen in the 50- to 75-year age group, with a female-to-male ratio of 6:1, involving primarily an accelerated rate of bone loss; usual fracture sites include the vertebrae and radii. Type 2 occurs after age 70, and the female-to-male ratio is 2:1; bone loss is both trabecular and cortical, and fracture sites include the hips and vertebrae. Factors associated with the risk of osteoporosis include age, female sex, white race, family history, low bone mass, prior fractures after age 50, smoking, amenorrhea, sedentary lifestyle, hyperthyroidism, corticosteroids, anticonvulsant therapy, psychotropic drugs, and alcoholism.

Clinical Presentation and Diagnosis

Osteoporosis is usually asymptomatic until fractures occur. Back pain and deformity of the spine are the most frequent symptoms of vertebral body fractures. Height loss is a cardinal sign of vertebral osteoporosis. Compression fractures are often multiple and most commonly occur between T11 and L2. Wrist, hip, and pelvic fractures may be the first manifestation of osteoporosis. Generalized skeletal pain is uncommon, and between fractures, most patients are pain free.

Laboratory findings are usually normal. X-rays may show osteopenia, but 30% of bone mineral density must be lost before osteopenia is apparent radiographically. Bone densitometry is a screening tool for osteopenia in high-risk individuals and facilitates assessment of response to therapy. Several noninvasive techniques exist that readily detect bone mass, varying in accuracy, precision, radiation dose, convenience, and cost. Currently available techniques include single-photon absorptiometry (SPA), dual-photon absorptiometry (DPA), quantitative computed tomography (CT), and dual-energy x-ray absorptiometry (DEXA). For several reasons, DEXA is the preferred diagnostic method. DEXA measures the current trabecular and cortical bone mineral content of the hip and spine. Results are provided using the "age-matched" (Z-score) or the "young-normal" (T-score). The T-score is given in units of standard deviations (SDs) compared with the bones of a healthy 30-year-old. According to the criteria established by the World Health Organization, a T-score of −1.0 to −2.5 SDs below the mean constitutes osteopenia; a T-score below −2.5 SDs is defined as osteoporosis.

Prevention and Therapy

The goal in the general population is to improve the peak bone mass attained in adolescence and to promote specific behaviors to maintain bone mass. Such goals include altering lifestyles with respect to general nutrition and calcium intake, maximizing physical activity, exposure to sunlight, and reducing or eliminating risk factors that are detrimental to bone. The goals for patients currently at greatest risk for developing osteoporosis are early identification and early intervention, prevention of further bone loss, and reduction of the number of falls and fractures (Gueldner, Burke, & Wright, in press).

Three modalities—calcium, exercise, and estrogen—have received the greatest attention in the prevention of menopause- and age-related bone loss. Calcitonin, pharmacological doses of vitamin D, bisphosphonates, sodium fluoride, and other treatment strategies are usually reserved for the treatment of established disease.

Calcium

Calcium has an essential role in the development and maintenance of a healthy skeleton. It is well known that many individuals lack adequate calcium intake. It has been shown that increased dietary calcium can increase peak bone density in adolescents and young adults, as well as slow age-related loss in older individuals (Piper, Galsworthy, & Bockman, 1995). The 1994 consensus development conference on optimal calcium intake recommended increases in the recommended dietary allowance (RDA) of calcium in most age groups (Table 1).

The preferred source of calcium is food (Table 2). However, if dietary sources provide only 600 mg of calcium per day, 900 mg should be added in supplement. Since the aim is total calcium intake of 1,000 to 1,500 mg per day. Calcium supplements should be taken with meals and no more than 500 to 600 mg at a time.

Vitamin D

Vitamin D deficiency is common in older individuals. Thus, an intake of 400 to 800 IU of vitamin D is recommended.

TABLE 1   Recommended Total Daily Calcium Intake (mg)

Age Range	Men	Women
11–24	1,200–1,500	1,200–1,500
25–50	800	1,000
51–60	1,000	1,500 (1,000 if taking estrogen)
Over 65	1,500	1,500 (1,000 if taking estrogen)

TABLE 2   Calcium Content Food Sources

Food*	Calcium Content per Serving (mg)
Skim Milk	302
Yogurt, plain, low-fat	250–400
Ice cream	176
Fruit juice with added calcium	300
Sardines (with bones) (1/2 cup)	375
Tofu (1/2 cup)	150
Broccoli, fresh, cooked	136
Carrots	100
Cottage cheese, low-fat (1/2 cup)	78

*Based on 1 cup serving, unless otherwise noted.

Exercise

Immobility and disuse (bed rest, weightlessness, casted limbs, paralysis) may lead to accelerated bone loss and osteoporosis. Exercise is markedly beneficial for the musculoskeletal system, increasing muscle strength, flexibility, and balance. A regimen of moderate weight-bearing exercise is reasonable, such as walking for 30 to 45 minutes three to five times per week. Other forms of exercise must be determined on an individual basis.

Estrogen

Estrogen replacement can protect postmenopausal women against osteoporosis, decreasing the incidence of vertebral and hip fractures by 50% in some studies (Schussheim & Siris, 1998). In addition, estrogen replacement therapy (ERT) improves cardiolipid profile, enhances mental function, protects dentition, and improves urinary physiology. ERT appears to be most effective in the first few years of menopause but also has some benefit in women with established osteoporosis.

Unopposed estrogen increases the risk of endometrial cancer; adding progestin brings the risk down to that of the general population. Increased risk of breast cancer from menopausal estrogen replacement is small but significant and should be weighed carefully by patient and physician.

Bisphosphonates

Bisphosphonates inhibit osteoclast-induced bone resorption, thus decreasing bone turnover and shifting the balance toward bone formation. Alendronate (Fosamax), one 10-mg tablet daily, effectively increases bone density and reduces the fracture rate by about 50% (Piper, Galsworthy, & Bockman, 1995). It should be taken in the morning on an empty stomach with a glass of water. The patient should then wait 30 minutes, in an upright position, before eating or drinking anything else, even coffee or juice (Schussheim & Siris, 1998).

Alendronate can cause esophagitis. It is contraindicated for patients with esophageal strictures or achalasia and must be used with caution in those with a history of hiatal hernia, dysphagia, gastritis, or peptic ulcer disease.

Calcitonin

Calcitonin also increases bone density. However, bone density increments with calcitonin are not nearly as marked as with estrogen or bisphosphonates. No reliable fracture prevention data are available for calcitonin, but it is reported to have significant analgesic properties, making it useful for fracture pain. A nasal spray (Miacalcin) is available; the usual dose is 200 IU per day (a single spray), all in one nostril, alternating nostrils daily. Nasal dryness and irritation are the most common complaints.

Raloxifene

Raloxifene (Evista) is a selective estrogen receptor modulator. Its bone anabolic effects are only about 50% as effective as estrogen. However, this level is sufficient for the prevention of osteoporosis. The chief advantage of raloxifene is that it does not cause any stimulation of the ovary, uterus, or breast. There are no data on the protective effect of raloxifene for heart attack or stroke or its effect on cognition. The most common side effects are mild leg cramps and a minor increase in the incidence of hot flashes.

AHMED MIRZA

Internet Resources
Doctors' Guide to Osteoporosis Information
http://www.psigroup.com/osteoporosis.htm

National Osteoporosis Foundation
http://www.nof.org/

Osteoporosis Society of Canada
http://www.osteoporosis.ca/

OVER-THE-COUNTER DRUGS AND SELF-MEDICATION

Age-related physiological and psychosocial changes, increasing prevalence of chronic disease, and the likelihood of multiple pathologies predispose older people to experience problems with medications. Medication noncompliance (nonadherence), polypharmacy, and the use and abuse of over-the-counter (OTC) or nonprescription drugs are common. The elderly are the major users and abusers of both OTC and prescription drugs (Roe, 1984). Market surveys suggest that although older Americans spend less on OTC drugs than on prescription drugs, the elderly account for 40% to 50% of all OTC drug purchases, a pattern similar to that in 14 European and North American countries. A study of the elderly in an Australian community found that 18% of men and 25% of women were currently using three or more types of prescription drugs. More women than men (44% vs. 29%) used two or more types of OTC drugs (Simons, Tett, & Simmons, 1992). Fifty-six percent of men and 76% of women using multiple prescribed medications also used multiple OTC drugs (Simons et al., 1992), indicating the depth of the problem of self-medication.

Patients and even health professionals may think of OTC drugs as safe and of little pharmacological significance. All drugs, including OTC medications, carry a level of risk that is less easily characterized in the elderly. Polypharmacy greatly increases the occurrence of adverse reactions and interactions. In addition to "traditional" OTC drugs, self-medication may include the use of herbal medicines and complementary therapies. There is scant research regarding possible interactions among prescription, OTC, and herbal medications.

In general, an OTC drug is purchased without a prescription. However, the range of OTC drugs available varies among countries. For example, in the United States, the nonsteroidal anti-inflammatory drug naproxen was switched from prescription only to OTC in 1994, but it still requires a prescription in the United Kingdom. OTC drugs can also be differentiated by those that are freely available—for example, in supermarkets—and those that can be provided only by licensed pharmacies. In some countries, certain OTC drugs may be dispensed only when a pharmacist is present.

A number of preparations once available by prescription only are now available OTC. The H_2-receptor antagonist cimetidine is now available in the United Kingdom without prescription to adults, provided the pack does not contain more than a two-week supply. Yet side effects of this drug include diarrhea, other gastrointestinal disturbances, headache, dizziness, tiredness, and rash, and some reports indicate that the drug can mask the symptoms of gastric cancer in those of middle age or older. Less common side effects of H_2-receptor antagonists noted in the elderly include confusion, depression, and hallucinations. Potentially hazardous interactions of cimetidine reported in the British National Formulary include increased plasma concentration of a number of antiarrhythmics, warfarin and nicoumalone enhancement, inhibited metabolism (i.e., increased concentration) of antiepileptics (phenytoin, carbamazepine, valproate), possible increases in cyclosporin level, and increased plasma theophylline level. Cimetidine also interacts with some tricyclic antidepressants, antidiabetics, antipsychotics, benzodiazepines, and beta-blockers. A study of Australian veterans identified cimetidine use in combination with benzodiazepines, tricyclic antidepressants, theophylline, and carbamazepine (Parkes & Cooper, 1997). Although the benefits usually outweigh the risks, the total clinical and pharmacological situation needs to be carefully considered. OTC drugs have to be treated with the same caution and respect as prescription-only medications.

U.S. Food and Drug Administration (FDA) OTC drug categories include allergy treatments, analgesia and antipyretic products, antimicrobials, bronchodilators, dermatological products, emetics, hematinics, laxatives, sedatives, stimulants, vitamin-mineral supplements, and weight-loss aids. The OTC drugs used most commonly by the elderly tend to be drugs for the treatment of pain and fever, coughs, colds, or allergy; insomnia; heartburn and acid reflux; constipation; diarrhea; and nausea and vomiting.

These drugs tend to be used to treat symptoms. However, they may mask a more serious pathology. Although OTC drugs are usually sold with clear warnings on the labels about the need to consult a medical practitioner if symptoms persist, the patient or family may equate controlling symptoms with treating the underlying disease. Late presentation for diagnosis or reduced compliance with prescribed medications can result. To illustrate some of the issues surrounding OTC drugs and polypharmacy in the elderly, delirium and adverse effects of antacids are considered in more detail.

Delirium: Delirium can be present on hospital admission in 22% of elderly patients and develops in as many as 31% of inpatients (Flaherty, 1998). The contribution of prescription drugs to confusion and delirium in the elderly is well recognized, but the role of OTC medications is less appreciated. OTC drugs may contribute to the development of delirium through their direct actions or by interaction with other drugs. Flaherty (1998) illustrated the risk of OTC medications in the case of an elderly man with early dementia, coronary artery disease, peripheral vascular disease, and benign prostatic hypertrophy: The patient takes an OTC cold medication to treat sinus congestion; the preparation contains an alpha-adrenergic agonist that increases bladder sphincter tone, resulting in urinary retention that causes agitation and confusion; admitted to an acute-care hospital, the patient

might well be placed on an antipsychotic that increases confusion.

OTC analgesics such as the salicylates, ibuprofen, or paracetamol (acetaminophen) can cause delirium in the elderly as a result of chronic use or taking a large dose (Grigor, Spitz, & Furst, 1987; Steele & Morton, 1986). OTC medications for insomnia can also cause problems, because many of these products contain sedating antihistamines such as diphenhydramine or doxylamine, alone or in combination with an analgesic. In some countries, preparations containing hyoscine (scopolamine) are available as OTC sleep aids, as well as for motion sickness. Some herbal medicines may contain atropine and hyoscine. Antihistamines cause confusion at high doses after the first dose in susceptible elderly patients (Tejera, Saravay, & Goldman, 1994).

Perhaps the area of greatest concern is the proliferation of combination preparations for coughs, colds, and allergies. These preparations may contain many different ingredients, including antihistamines, sympathomimetics, analgesics, and expectorants. Mental status changes after the use of such preparations, including those designed for nasal inhalation, have been reported (Brown, Golden, & Evans, 1990; Snow, Logan, & Hollender, 1980). A study of the psychiatric side effects of phenylpropanolamine in 37 cases found more reactions due to the use of OTC preparations than to use of the same drug obtained by prescription (Lake, Mason, & Quirk, 1988). Numerous authors have suggested that every drug has the potential to cause delirium in the elderly. The active ingredients of an OTC need consideration when reviewing a patient's medications.

Antacids: Drug-induced malnutrition is an underappreciated aspect of OTC medications. Authors such as Roe (1984) suggest that excessive consumption of OTC drugs such as antacids and laxatives is the most significant cause of drug-induced malnutrition. All antacids can interact with other drugs, including antidepressants, antibiotics, cardiac glycosides, antipsychotics, and antiepileptics. Antacids may impair absorption; some attack the enteric coatings on pills, exposing the drug to gastric acids.

Antacids may be prescribed, but more commonly they are used as self-medications to cope with nonspecific gastrointestinal symptoms such as indigestion, gas, flatulence, bloating, acid reflux, gastrointestinal discomfort, and heartburn. Antacids are used to treat epigastric symptoms of gastrointestinal or cardiovascular origin. Dyspepsia typically presents as heartburn or food-induced discomfort (indigestion), is often of uncertain origin, and may cause little concern. However, underlying causes include hiatus hernia, peptic or esophageal ulcer, esophageal sclerosis, alcoholic gastritis, angina, congestive heart failure, and dyspnea due to emphysema (Roe, 1984).

Antacids are often aluminum or magnesium compounds that may be taken in liquid or chewable tablet form; relief lasts three to four hours. Aluminum-based antacids have a tendency to cause constipation; magnesium-based products tend to have laxative effects. Aluminum and magnesium compounds are contraindicated in patients with impaired renal function. Aluminum can be absorbed from the gastrointestinal tract; evidence suggests greater retention in renal impairment. Magnesium overload is reported in patients with renal failure using magnesium-based antacids. Hypophosphatemia is a possible consequence of excessive use of both aluminum and magnesium hydroxide antacids—a problem that may be underreported due to its insidious development and the tendency to ascribe symptoms to other aging-related changes (Roe, 1984).

Folate malabsorption is another possible adverse effect of antacid use in the elderly, although the evidence is not strong. An interaction between excessive intake of milk and antacids, the milk-alkali syndrome, has also been described. Elderly patients, particularly those with renal impairment, are at risk. Hypercalcemia occurs without accompanying hypercalciuria or hyperphosphaturia. General symptoms include nausea, vomiting, anorexia, headache, and weakness. Calcium deposition in the cornea can cause band keratitis. Sodium bicarbonate used in excess can cause alkalosis and sodium overload, a particular risk with preexisting heart disease.

Although the news media overstate the case that some OTC drugs are killers, it is clear that OTC drugs are not innocuous; they have the same potential to cause harm as prescribed drugs. The problem of polypharmacy and reducing the negative effects of drugs in the elderly requires a full assessment of both prescribed and OTC drugs. Clinician review and careful education of the patient and family are necessary to ensure the correct use of all drugs.

COLIN TORRANCE

See Also
Cough
Heartburn
Medication Adherence
Polypharmacy: Drug-Drug Interactions

P

PAIN ASSESSMENT INSTRUMENTS

Assessment of pain for older adults is an essential part of pain management. When clinicians fail to ask about specific pain symptoms or pain effects, older clients are placed at risk for unidentified, misdiagnosed, and undertreated pain. Pain management is most effective when pain assessment identifies the underlying cause of pain. It is also necessary to distinguish acute from chronic pain in order to plan care accordingly. Because there are no objective biological markers of pain, the patient's self-report is the most reliable method by which to gather information regarding pain.

There are many standardized tools for objective pain assessment in older adults, but these instruments are not always used by clinicians. In their research on pain assessment tools used with the elderly, Herr and Mobily (1993, p. 39) stated, "although pain is a multidimensional concept, subjective intensity is probably the component most often measured in both clinical practice and in treatment-outcome research." In a comparison of nurses' and patients' ratings of pain, Weiner and colleagues (1995) found that patients' ratings were routinely higher. These findings underscore the need to use an objective pain assessment instrument for initial and ongoing pain assessment.

The most frequently used pain assessment measure is a numeric rating scale in which the client is asked to choose a position on a scale of 1 (very little pain) to 10 (the worst pain imaginable). However, the scale is difficult for some older adults because of its abstract design (Weiner, Ladd, Pieper, & Keefe, 1995). Furthermore, the scale has not been found to be reliable in a cognitively impaired population (Ferrell, Ferrell, & Rivera, 1995).

Given factors such as sensory deficits and cognitive impairments, simple questions and tools that can be easily used are most effective with an older population. The Visual Analog Scale (VAS) is a straight, horizontal 100-mm line anchored with "no pain" on the left and "worst possible pain" or "pain as bad as it could possibly be" on the right (Carr, Jacox, & Chapman, 1992). The patient is asked to indicate where the pain is on the scale. The "faces scale" depicts facial expressions on a scale of 0 to 6, with 0 a smile and 6 a crying grimace. Studies of VAS for pain assessment in an elderly population have demonstrated reliability and validity.

The McGill Pain Questionnaire (MPQ) (Melzack, 1975; Melzack & Katz, 1992) is a widely used instrument consisting of 78 words categorized into 20 groups, a drawing of the body, and a Present Pain Intensity (PPI) subscale (a 6-point ordinal scale). Among older adults, the PPI can be used apart from the entire MPQ. The tool is effective with both cognitively intact and cognitively impaired older adults (Ferrell, Ferrell, & Rivera, 1995). However, Herr and Mobily (1993) found that older adults with visual and hearing impairments had difficulty understanding this questionnaire and became tired during the assessment. The Verbal Descriptor Scale (VDS) is a variation of the PPI that uses simple language to describe pain. Herr and Mobily (1993) reported that the VDS had a low failure rate and was highly correlated with other pain measures.

Ethnic, cultural, and spiritual factors play an important role in a patient's perception of and response to pain. Cultural variations exist that must be explored with the patient and family. A descriptive, correlational study of 411 Mexican American and non-Hispanic white subjects age 65 to 74 revealed that the MPQ is a valid tool for pain assessment in an older, community-based, multicultural sample (Escalante, Lichtenstein, White, Rios, & Hazuda, 1995).

A descriptive study of 1,193 subjects, age 67 to 99, that compared the SF-36 health survey with two osteoarthritis indices two to seven years after knee replacement, supports the inclusion of both a

Advantages and Disadvantages of Pain Assessment Instruments

Assessment Instrument	Advantages	Disadvantages
Numeric Rating Scale (NRS)	No instrument necessary. Older adult is simply asked to rate pain on an abstract scale of 1 to 10	The abstract design of the scale makes it difficult for some older adults to understand, especially the cognitively impaired.
Visual Analogue Scale (VAS)	Easy to develop and administer to most older adults, including those who are cognitively impaired or who have language, hearing, and speech deficits. Good reliability and validity.	The inconvenience of administering these paper-and-pencil scales may preclude use in fast-paced clinical environments.
McGill Pain Questionnaire (MPQ)	The PPI and VDS subscales are effective apart from the entire MPQ. The tool is reliable with cognitively intact, cognitively impaired, and community-based multicultural older adults.	Lengthy instrument that requires completion of paper-and-pencil scale. Older adults with visual and hearing impairments have difficulty understanding the entire questionnaire and become tired during the assessment.
SF-36 Health Survey	Effective way to measure the effect of pain on activities of daily living	Lengthy paper-and-pencil scale that has not demonstrated effectiveness in the cognitively impaired.
Prompting Intensity Pain Electronic Recorder (PIPER)	Used effectively in combination with VAS to measure recent and current pain. Highly valuable in fully understanding older adults' pain experience when used by itself and with other instruments.	Requires an electronic recording device that is not readily available in most clinical settings. Most effective when used with another instrument.
Discomfort with Dementia of the Alzheimer's Type (DS-DAT)	Appropriate for assessing pain in Alzheimer's patients.	The inconvenience of administering this specialized paper-and-pencil scale may preclude its use in diverse clinical settings. Most appropriately used in specialized settings targeting cognitively impaired older adults.

generic and a disease-specific health-related quality of life measure to assess patient pain outcomes fully (Bombardier et al., 1995). Cognitively intact subjects older than 65 years with back pain (n = 39) were administered both self-report and observational instruments to assess pain, pain behavior, and disability. The data revealed that pain behavior observation, especially during activities of daily living, is a more sensitive and valid way of assessing pain behavior in older adults than is self-report of pain (Weiner, Pieper, McConnell, Martinez, & Keefe, 1996).

Lewis and colleagues (1995) conducted three studies to assess the reliability and validity of the Prompting Intensity of Pain Electronic Recorder (PIPER) in combination with the VAS and other instruments to assess recent and current pain. Findings indicated that frequent measures of pain should

be taken to fully understand patients' pain experience and that the PIPER is a valid and reliable instrument that is easily used by the elderly, either by itself or with other instruments.

Ferrell and colleagues (1995) found that cognitive impairment is a substantial barrier to objective pain assessment, yet patients were able to make most of their needs known in a qualitative if not quantitative way. In a descriptive study of 46 cognitively impaired older adults, subjects were assessed on admission to the hospital for pain, cognitive impairment, and functional status using several instruments. The researchers concluded that current standards of practice that rely on patient self-report of discomfort are inadequate for confused older adults (Miller et al., 1996). When working with acutely confused elderly patients, clinicians need to anticipate the likelihood of discomfort as they assist with activities of daily living and intervene appropriately (Miller et al., 1996).

Alternatives to objective pain assessment in cognitively impaired older adults were evaluated in three studies conducted to develop and validate the Discomfort with Dementia of the Alzheimer's Type (DS-DAT) pain assessment tool for use with an Alzheimer's population. The study resulted in the conclusion that the nine-item DS-DAT tool is an appropriate instrument for assessing pain in Alzheimer's patients (Hurley, Volicer, Hanrahan, Houde, & Volicer, 1992).

No single measure of pain is applicable to all older adults. Herr and Mobily (1993) recommend ascertaining an elderly client's best match with a pain measurement tool rather than using one tool exclusively. A consensus panel of the Agency for Health Care Policy and Research recommends that pain assessment instruments be chosen with regard to the patient's age and physical, emotional, and cognitive condition, as well as available time and knowledge of the clinician administering the instrument.

MEREDITH WALLACE
ELLEN FLAHERTY

See Also

Over-the-Counter Drugs and Self-Medication
Pain: Acute
Pain: Chronic
Polypharmacy Management

PAIN: ACUTE

"Pain is most poorly managed in those most defenseless against it—the young and the old" (Liebskind & Melzack, 1988, p. 131). It is estimated that 70% to 80% of elders have experienced pain at one time or another (Herr & Mobily, 1992). Approximately 50% of community-dwelling elders experience serious pain problems. For the most part, studies of pain among elders have focused primarily on the prevalence of chronic pain, rather than identifying the frequency and severity of acute pain.

Therefore, it is not known whether chronic pain is in fact the most common problem for older adults. Chronic health problems in the later years often cause older adults to report discomfort on a daily basis, which can involve both acute pain (characterized by sudden onset, obvious pathology, and usually less than three months' duration) and persistent chronic pain of longer duration. Osteoarthritis, the most common long-term painful disorder in older adults, is characterized by both chronic and acute types of pain. Acute pain syndromes in elders include postoperative pain, traumatic fractures of the hips or long bones, compression fractures from osteoporosis, initial herpes zoster–related pain and diabetic neuropathy, exacerbations of rheumatoid arthritis, and acute chest pain.

Age-related changes affect the perception of and response to acute pain and complicate its treatment. Women with osteoporosis can fracture a rib with a vigorous sneeze or cough and present with sudden point tenderness. Slowed neurological function with age can alter the nociceptive or sensing component of pain, leading to delayed discomfort, particularly in the case of trauma. It is not uncommon for elderly persons to fall or otherwise injure themselves but fail to report pain until hours or even days later. Cognitive problems such as dementia or depression can compromise a patient's ability to express pain. Cardiac conditions, which are common during the later years, may present with atypical chest, jaw, or arm pain. Chronic diseases that compromise circulation, such as atherosclerosis and diabetes, result in peripheral vascular disease that can cause acute ischemic pain. Likewise, the accompanying neuropathy from chronic diabetes often presents as acute pain. A decline in function in the gastrointestinal and genitourinary systems may cause atypical responses to medications designed to treat acute pain. Overall, elders are more likely to have an adverse response to analgesics, particularly those affecting liver function and the gastrointestinal and genitourinary systems. In addition, elders are susceptible to side effects from opioids, which are often used to treat moderate to severe acute pain.

The misconception by health care professionals that older patients do not require as much or as frequent pain medication contributes to inadequate dosing of analgesics, despite observation of the disabling effects of pain. Patients' own values about taking medications and fears that opioid use will lead to addiction may make them reluctant to report pain and take medication, resulting in needless suffering.

The pain experience comprises a variety of physiological, emotional, and behavioral responses. Autonomic responses such as tachycardia, elevated or depressed blood pressure, diaphoresis, and pupillary changes can be the body's first line of defense to indicate tissue injury (Dellasega & Keiser, 1997). Anxiety may also be a symptom associated with acute pain. Pain can have profound effects on maintaining self-care activities and the responsibilities of independent living; it also disrupts sleep patterns, because it can interfere with the ability to rest.

Differentiating acute from chronic pain in older adults using existing scales is a major challenge. The Agency for Health Care Policy and Research (AHCPR) clinical care guidelines for acute pain refer to operative or medical procedures and trauma as the etiological factors in pain (Acute Pain Management Guideline Panel, 1992), but for many older adults, acute flare-ups of a chronic painful condition related to disease states can cause acute pain. Most pain measures are generic; that is, they fail to distinguish sources and patterns of acute and chronic pain. Several items from the Brief Pain Inventory (BPI), advocated by the AHCPR, ask the respondent to report levels of present, average, least, and worst pain, but elderly patients must sometimes be asked about these ratings in relationship to multiple sources of pain. The Pain History Questionnaire currently undergoing psychometric testing by Dellasega, Simons, and Weaver (in preparation) is a series of specific questions that ask the respondent to describe the history of the pain so that acute forms can be separated from chronic pain and the coexistence of both types can be detected. Using pain scales with patients who cannot clearly communicate due to cognitive or sensory impairments can be difficult if not impossible. Although pictorial representations of faces can be used to elicit reports of pain from cognitively impaired

elders, this type of scale does not measure physiological and behavioral manifestations (changes in heart and respiratory rate, anxiety, agitation, and groaning) that are more likely to indicate acute pain (Dial, 1999). A pain flow sheet offers a way to monitor and document behavioral deviations from baseline status in patients who cannot communicate, using such clues as facial grimacing, splinting or guarding of painful areas, and a reluctance to move about or a noticeable decline in the ability to tolerate activity. However, many instruments for the evaluation of acute pain have not been tested extensively with older patients; clinicians often rely on clinical judgment.

In the absence of an acceptable scale or instrument that can measure characteristics of acute pain and acute pain superimposed on or coexisting with chronic pain, clinicians need to interview and observe patients to evaluate their comfort levels. Patients should be asked to provide as detailed a history of their pain as possible and the strategies they have used to manage pain (including nonprescription medications and complementary or alternative therapies). The effectiveness of each strategy and the individual's daily routine for managing acute episodes of pain should be evaluated. The onset of pain should be differentiated from preexisting chronic pain by the duration, pattern, and precipitating factors. Among hospitalized older patients, 45% of the old old (80 years and older) complained of pain, with 19% experiencing moderate to severe pain (Desbiens, Mueller-Rizner, Connors, Hamel, & Wenger, 1997). Acute postoperative pain is among the most common sources of acute pain. It is well established that aggressive analgesic therapy following surgery leads to better postoperative outcomes, but special considerations must be given to care of the elderly. Postsurgical elders are more prone to delirium, which can be aggravated by acute pain or interfere with the ability to accurately report discomfort. Those most at risk for pain had orthopedic problems (hip and other fractures) as opposed to other medical diagnoses. Pain that was evident during hospitalization was likely to persist following discharge, making it clear that plans are needed for effective pain control when patients return home or to a long-term-care facility.

Standard analgesics used to treat acute pain include acetaminophen, nonsteroidal anti-inflammatory drugs (NSAIDs), nonopioid and opioid combinations (acetaminophen plus codeine [Tylenol 2, 3, 4], acetaminophen plus hydrocodone [Vicodin, Lortab, Lorcet], acetaminophen plus oxycodone [Percocet, Roxicet]), and lower doses of other opioids (oxycodone, morphine) alone for mild to moderate pain. Hepatotoxicity can occur with doses of acetaminophen in excess of 4,000 mg per day; gastric bleeding, renal impairment, and platelet dysfunction are associated with the use of some NSAIDs. Adverse effects from opioid analgesics may include increased constipation, altered mentation that can lead to falls, and an increased risk for respiratory depression. Information on usual starting doses, dosing guidelines, and adverse effects of analgesics can be found in the Acute Pain Management Guideline Panel (1992) and the American Geriatrics Society Panel on Chronic Pain in Older Persons (1998).

Patient-controlled analgesia (PCA), a mainstay for the treatment of postoperative pain, is safe and efficacious for elders (Egbert et al., 1990). Because elders are more likely to experience higher analgesic peaks, longer duration of action, and increased side effects, selection of an appropriate opioid and appropriate dosing parameters, such as the PCA self-administered demand dose and continuous or basal rate background infusion, is critical, especially for those with impaired renal function. Morphine, the opioid of choice for PCA in both the young and the old, is superior to meperidine (Demerol) in relieving pain. However, the incidence and severity of side effects such as nausea, mood disturbances, and unusual dreams appear to be similar for both drugs. Meperidine should be avoided in elders, as it possesses an active metabolite, normeperidine, which can accumulate with repeated doses. Morphine also has active metabolites that can reach toxic levels if they accumulate, but this is rarely a problem if PCA is administered with usual doses and a short duration of administration. The initial starting dose of morphine for opioid-naïve patients is 0.5 to 1 mg per hour, with a demand PCA or self-administered dose of 1 mg every 10 to 15 minutes. For the oldest old, it is recommended that intermittent demand dosing alone be used until the response to therapy can be evaluated. Other opioids such as hydromorphone (Dilaudid) in doses of 0.2 mg per hour, with a 0.2-mg demand dose every 10 to 15 minutes, and fentanyl (Sublimaze), which should be dosed according to recommendations by a pain expert (e.g., anesthesiologist), are effective alternatives for patients with renal impairment, unmanageable nausea and vomiting, or other adverse effects from morphine.

Older patients may require verbal and written information prior to surgery to emphasize the principles of PCA therapy, the need to self-medicate before the pain worsens, and the importance of reporting unrelieved pain. In general, older patients are more reluctant to access demand doses because they are concerned about the effects of the medication or afraid of becoming addicted. Elder patients' concerns about using opioid analgesics should be addressed both pre- and postoperatively.

Epidural analgesia is associated with improved pain control and respiratory function and a lower incidence of chronic pain problems following surgery (e.g., postamputation phantom limb pain) when compared with other types of analgesia (Wulf, 1998), especially for a patient with a thoracic or abdominal incision; there is earlier recovery of gastrointestinal function and decreased risk of thromboemboli. A combination of an opioid (e.g., morphine or fentanyl) and a local anesthetic (e.g., bupivacaine [Marcaine] or ropivacaine) is routinely administered by continuous infusion, although either may be administered alone. Fentanyl, because of less rostral (vertical) spread to higher levels of the central nervous system, is preferred over morphine to reduce the risk of respiratory depression, especially for patients with preexisting pulmonary disease. Local anesthetics such as bupivacaine that affect sensory and motor neurons are associated with a greater incidence of orthostatic hypotension and lower motor weakness than are anesthetics selective for sensory nerves, such as ropivacaine. Intermittent bolus injections of epidural morphine or a longer-acting local anesthetic, for example, lidocaine, can be administered.

Important considerations with epidural therapy include an increased risk of respiratory depression if systemic opioids are administered with epidural opioids; altered cognitive status, particularly with epidural local anesthetics; and urinary retention, especially in males with preexisting bladder problems and benign prostatic hypertrophy. Patients receiving epidural therapy require hourly respiratory rates for the first 24 hours, especially if morphine was administered or if patients have preexisting pulmonary conditions. Lower concentrations of a local anesthetic (e.g., solutions of 0.05% [0.5 mg/mL] or 0.0625% [0.625 mg/mL]) such as bupivacaine can be administered to minimize the incidence and severity of orthostatic hypotension and lower motor weakness. Patients receiving a local anesthetic need assistance getting out of bed or ambulating, especially for the first 24 hours. In addition, frequent repositioning is necessary to prevent pressure ulcers and to maintain circulation, as patients may experience a decrease in sensation. Cognitive status should be assessed, especially if patients received epidural bolus injections of a local anesthetic, because accumulative effects can lead to toxic serum levels.

Elderly patients and their caregivers are often poorly prepared to deal with posthospital needs. Pain is one of the most frequent problems experienced after leaving the hospital and results in additional telephone calls and outpatient visits for help with pain relief. Teaching a complex medication regimen prior to discharge is difficult, given the pressure for shorter hospitalizations. Nevertheless, realistic goals for pain control should be established, and patients should be aware that acute and chronic pain can coexist. Discharge planning should include preparation for the degree of pain, which may not steadily decline but may worsen with increased activity; an effective analgesic regimen that can be tolerated, along with specific dosing guidelines; instructions for managing side effects such as constipation from opioids; reassurance that the patient will not become addicted with continued use of opioid therapy; and alternative methods of pain control, such as heat, massage, relaxation, proper alignment of body parts, and scheduling of alternating periods of rest and activity.

Home health nursing visits should be ordered to monitor pain levels. Less than a quarter of all elderly hospital patients receive home-care services after discharge (Dansky, Dellasega, Shellenberger, & Russo, 1996). Older persons who have not been hospitalized typically receive no formal home care unless it is reimbursed by private or personal funds. If resources are available to help an older person with activities of daily living, a home health aide might reduce physical exertion that can exacerbate pain. Scheduling activities and structuring the environment to place minimal demands on an elder with acute pain may help conserve energy and promote recovery.

ROSEMARY POLOMANO
CHERYL DELLASEGA

Internet Key Words
Acute Pain
Epidural Analgesia
Nonsteroidal Anti-inflammatory Drugs (NSAIDs)
Opioid Analgesics
Patient-Controlled Analgesia (PCA)

Internet Resources
American Pain Society
http://www.ampainsoc.org

American Society for Pain Management Nurses
http://www.ajn.org/people/nsorgs/aspmn

International Association for the Study of Pain
http://www.halcyon.com/iasp

Purdue Pharma L.P.
http://www.partnersagainstpain.com

Resource Center for State Cancer Pain Initiatives
http://www.wisc.edu/molpharm/wcpi

Roxane Pain Institute Laboratories
http://www.pain.roxane.com

PAIN: CHRONIC

Chronic pain is one of the most pervasive yet undertreated problems in older adults. Complaints vary, but it is estimated that about half of all community-dwelling older adults and about two-thirds of all nursing home residents have problems with pain. Physiological changes of aging, sensory deficits, cognitive impairment, underreporting, and under-treating affect all aspects of this complex phenomenon. Research on the assessment and management of chronic pain is complicated due to the factors cited above (Ferrell & Ferrell, 1996).

Unrelieved pain results in depression, decreased socialization, sleep disturbances, impaired ambulation, and increased health care utilization and costs. The detection and management of pain must include routine pain assessment and reassessment, careful use of analgesic drugs, and nonpharmacological interventions such as physical therapy and nontraditional approaches.

Presentation

Several attributes of aging may affect the clinical presentation of pain in older people. The mechanisms that send pain impulses to the brain are presumably subject to age-related physiological changes. Psychological factors include the belief among both elders and practitioners that pain is a normal aspect of aging, thus inhibiting reporting and treatment (Farrell, Gibson, & Helme, 1996).

A multidisciplinary approach is needed for this complex phenomenon. Pain involves dimensions of functionality, social well-being and relationships, and spiritual wellness related to suffering and the meaning of pain (Ferrell, 1996). These interrelated phenomena have a particular impact on the elderly both in the community and in institutional settings.

A study of noninstitutionalized rural elderly 65 years and older (n = 4,592) found that 86% (n = 2,477) had reported pain of some type in the past year; 59% (n = 1,827) had reported multiple pain complaints (Mobily, Herr, Clark, & Wallace, 1994). Those over 85 years old had less pain than 65- to 84-year-olds. Joint pain was the most prevalent site, followed by night leg pain, back pain, and leg pain while walking.

Of 97 nursing home residents with an average age of 89, 71% reported pain, 34% constantly, and 66% intermittently. Back pain was the most frequent complaint (40%), followed by arthritis of appendicular joints (24%), previous fracture sites (14%), and neuropathies (11%) (Ferrell, Ferrell, & Osterweil, 1990).

Barriers to pain assessment among cognitively impaired older adults was illustrated in a descriptive study of 46 residents, ages 73 to 95, on two medical units of one hospital. Subjects were assessed on admission for pain, cognitive impairment, and functional status using several instruments. The researchers concluded that current standards of practice that relied on patient self-report of discomfort were inadequate for confused older adults. Cognitive impairment was a substantial barrier to objective pain assessment: "these patients were not comatose or incapable of feeling pain. They were able to make most of their needs known in a qualitative but not always quantitative way" (Ferrell, Ferrell, & Rivera, 1995, p. 197).

Assessment

Older clients are at risk for unidentified, misdiagnosed, and undertreated pain. Pain management is most effective when pain assessment—including specific pain symptoms or pain effects—identifies the underlying cause of pain. It is necessary to distinguish acute from chronic pain to plan care accordingly. Because there are no objective biological markers of pain, self-report is the most effective method for information gathering regarding pain.

The Visual Analog Scale (Ferrell, Ferrell, & Rivera, 1995) is the most useful pain scale and can be successfully administered to all older adults.

Management

Pain management is complicated by adverse drug reactions and analgesic sensitivity. Safe, effective analgesic therapy requires in-depth knowledge of age-related changes in pharmacokinetics and pharmacodynamics (Popp & Portenoy, 1996). A reduction in the therapeutic index of most drugs prescribed for older adults further complicates treatment, as there is often little leeway between favorable effects and adverse effects. The old adage "start low and go slow" is strongly recommended.

For the most effective pain management, medications should be given on a regular basis, with extra doses added during treatments and activities likely to result in pain. There is scant research about the benefit of each administration route: by mouth; intravenous bolus, continuous medication, or self-controlled pump; intramuscular; and subcutaneous.

Nonpharmacological pain management includes relaxation, guided imagery, transelectrical nerve stimulation, and other interventions such as humor, massage therapy, therapeutic touch, and acupuncture. Additional solutions, including the use of computers to create artwork and to stay connected to the outside world through the Internet, need to be explored. Regular exercise "with praise" was effective in reducing pain in older adults in one study (Miller & LeLieuvre, 1982).

Research is needed on patients' perceptions of their pain management and how pain affects function, mood, general state of well-being, and quality of life. Enabling older adults to provide data regarding their perceptions of the barriers to pain management can provide practitioners with important data. Investigation into nurses' attitudes with regard to pain management in nursing homes can provide insight into potential barriers and guide educational in-service programs. From a regulatory perspective, pain management is a patient's right as well as a key factor in quality of life.

Overcaution when prescribing analgesics, out of concern about side effects such as increased confusion and frequent falls, can lead to the neglect of pain management and can be harmful to elderly patients. Research focusing on the refinement and development of reliable and valid assessment tools, the safe use of analgesic drugs, and nonpharmacological interventions is also needed.

Evidence about the importance of nonpharmacological interventions to manage pain suggests additional studies are needed, such as, the uses of distraction, humor, massage therapy, therapeutic touch, acupuncture, and imagery. Creative solutions like using computers to create artwork and connect to the world through the internet need to be explored.

ELLEN FLAHERTY

See Also
Over-the-Counter Drugs and Self-Medication
Pain Assessment Instruments
Pain: Acute
Polypharmacy Management

Internet Key Words
Joint Pain
Opioid Drugs
Osteoarthritis
Visual Analog Scale

Internet Resources
The Management of Chronic Pain in Older Persons, American Geriatrics Society Consensus Guidelines, available from the National Guideline Clearing Househttp
http://www.guideline.gov/index.asp

OncoLink: Pain Management
http://www.oncolink.upenn.edu/specialty/pain/

PALLIATIVE AND END-OF-LIFE CARE

The rapidly aging population presents unprecedented challenges to our health care system. Although an increasing number of elderly citizens continue to enjoy good health and high quality of life, providing compassionate, high-quality care for patients with life-limiting and terminal illnesses continues to be an elusive goal. Unmet needs in palliative and end-of-life care are even more evident in the elderly population. This is particularly important for rural, minority, and poor elderly populations, for whom access to care is frequently problematic and undertreatment of pain and other symptoms is more common (Lo, 1995).

Numerous barriers exist to providing excellent palliative and end-of-life care that could be overcome by enhanced educational opportunities for health care professionals and members of the community. Health care providers sometimes have unfounded concerns about causing drug addiction in patients treated with opioid analgesics. Elderly patients may have reduced expectations for pain control and frequently interpret pain as an inevitable consequence of aging. Death is often interpreted as failure by care providers and may result in dying patients' experiencing a sense of abandonment. Dying in America is often characterized by dying in pain and dying alone.

America has also been referred to as a "death-denying society." Physicians often have difficulty deciding when cure is no longer possible and when intensive, high-quality comfort care should be provided instead. Patients and their families sometimes have unrealistic expectations regarding medicine's ability to cure patients with severe life-limiting disease.

The SUPPORT study (1995) demonstrated a failure to respond to critical needs of patients at the end of life. These included pain management and do-not-resuscitate (DNR) implementation, among others.

Principles of Palliative Care

Palliative care (from the Latin, *pallium*, meaning cloak) refers to the comprehensive care provided to patients and their families when cure is no longer possible. Palliative care affirms life, regards death as a normal process, and neither hastens nor postpones death. It emphasizes all-inclusive care that addresses the physical, emotional, social, and spiritual needs of patients and families. Palliative care models may incorporate disease-specific therapies as well as more general supportive and comfort care approaches (Post & Dubler, 1997).

Although palliative medicine is a rapidly emerging specialty, most palliative care is provided by primary care physicians and nurse practitioners. Hospice is a valuable service providing all-inclu-

sive palliative care, but it remains underutilized in the United States. Barriers to hospice utilization include relatively low patient referral rates by physicians and limited patient and family awareness of this invaluable resource. The National Hospice Organization (1998) recently published eligibility criteria for patients with noncancer diagnoses, such as end-stage congestive heart failure, chronic obstructive pulmonary disease, Alzheimer's disease, and other clinically advanced life-limiting conditions.

Goals of End-of-Life Care

The primary goals of providing high-quality end-of-life care have been outlined by the Robert Wood Johnson Foundation: (1) remove barriers to good care at the end of life, (2) develop policies and implement practices that promote excellence in care, (3) create public understanding about end-of-life care choices, (4) solicit and respond to public priorities, and (5) monitor the impact of efforts to improve end-of-life care.

Palliative and end-of-life care issues include the critical goals of providing high-quality management of pain and other symptoms, such as agitation, constipation, dyspnea, and depression. Assisting with the spiritual and psychosocial needs of the terminally ill patient and providing family support are also important goals. Respite care for family caregivers and anticipatory and after-death grief support are import dimensions of comprehensive palliative care. A values history questionnaire can be used to assess the specific preferences of geriatric patients nearing the end of life.

The potential impact of ethnic and cultural diversity on end-of-life decision making should be considered. The language of death and dying may reflect deeply held cultural and religious values (Irish, Lundquist, & Nelson, 1993). Health care providers should be aware of the unique needs of terminally ill patients and their families, which may reflect long-standing and firmly held beliefs and traditions. Bioethical dimensions of end-of-life care address advance care planning as well as possible withholding or withdrawing of medical interventions, including nutrition and hydration.

Ethical Domains of End-of-Life Care

The principles of clinical ethics include autonomy, beneficence, nonmaleficence, justice, and truth-telling. These principles may have a direct impact on end-of-life care. Autonomy refers to the patient's right to choose and may include implementation of an advance directive, such as a living will or durable power of attorney for health care. The risks and benefits of potential therapeutic interventions must be accurately communicated by the clinician to facilitate the patient's informed decision making.

The principle of autonomy may conflict with the principle of justice, which refers to the broader needs of society as a whole. This is relevant in an era of limited health care resources, when individual needs and preferences may be expected to yield to those of the collective society. Providing care when there is no reasonable expectation for clinical recovery or meaningful survival illustrates one source of potential conflict (Gillick, 1996). Beneficence implies an obligation to benefit patients, whereas nonmaleficence obligates the health care provider to try to prevent harm and to minimize risks. Truth-telling requires that honesty form the basis of the patient-provider relationship.

Pain Assessment

Accurate pain assessment of geriatric patients, performed in conjunction with reliable and effective pain intervention, is vital but often challenging. These needs are well documented in long-term-care facilities, where studies indicate a high incidence of untreated or undertreated pain (Keay & Schonwetter, 1998). Although approximately 90% of pain can be managed, studies indicate that between 25% and 40% of nursing home residents may experience untreated or undertreated pain. A recent cross-sectional investigation determined that 26% of 49,971 nursing home residents experienced pain on a daily basis. Approximately 25% of persons with daily pain received no analgesics (Won, Lapane, Gambassi, Bernabei, Mor, & Lipsitz, 1999). Residents who were more than 85 years old, cognitively impaired, male, or a racial minority were at greater risk of not receiving analgesics.

The concept of pain as a fifth vital sign is acknowledged as an effective mechanism to assess pain. This approach may prove useful for geriatric patients, since pain symptoms are often not adequately assessed by health care professionals or clearly communicated by patients. Many elderly patients have reduced expectations for pain relief and may consider unrelieved pain an inevitable consequence of aging and terminal illness. Pain assessment instruments include the utilization of the 0 to 10 Likert scale. A score of 0 indicates no pain, and 10 indicates the worst pain possible.

Frail elderly patients may have difficulty clearly expressing the fact that they are in pain or identifying exactly where the pain is located and its intensity. Cognitively impaired patients often express pain by frowning, grimacing, fidgeting, groaning, or heavy breathing. A pain assessment instrument consisting of a succession of faces, from smiling to frowning, may be more useful than a numerical scale to evaluate pain in patients with dementia.

It is important for health care providers to specifically inquire whether pain is present and to request details concerning its features. This is preferable to expecting geriatric patients to volunteer this information. Physical examination of patients complaining of pain is also necessary.

Pain Management

Although the science of pain management has advanced dramatically over the past 10 years, the SUPPORT study (1995) clearly indicates a major need to improve pain control for dying patients. Approximately 50% of dying noncomatose patients in this four-year study experienced moderate to severe pain during the last days of life. These findings contrast with the fact that adequate pain control is regarded as a patient's right. There is a corresponding responsibility for the physician and nurse practitioner to provide appropriate pain relief.

Information on effective pain interventions is readily available to health care professionals. The World Health Organization's three-step analgesic ladder and the Department of Health and Human Services' publications are excellent examples of valuable educational resources.

Physicians sometimes express misplaced concern about the addiction potential of potent analgesics when used to treat terminally ill patients. Medical-legal concerns and perceived threats from licensing boards may limit their willingness to administer such medications. Physicians are also concerned about causing the premature death of patients while providing necessary pain management (Foley, 1997).

Special Challenges

Physicians may encounter difficulty during the transition from the *cure* mode to the *comfort* mode. This may result in futile care, with no expectation of improving the quality of life or meaningful survival of the patient. Prolonging the dying process may be the unintended result. The importance of communication involving all stakeholders cannot be overemphasized. This facilitates optimal care and helps avoid measures compromising quality of care at the end of life (Byock, 1997).

Educational initiatives for health care professionals and members of the community should be given high priority to help achieve the goal of providing high-quality comfort care for terminally ill patients.

Spiritual and emotional support for patients and family caregivers may be neglected. The recognition that anticipatory grief will be experienced by the patient and by family caregivers is an opportunity for meaningful intervention by health professionals. Support mechanisms can also be developed to assist health care professionals in the grieving process following the death of a patient.

The Future

High-quality end-of-life care for elderly patients and their families will be achieved by the collaboration of physicians, nurses, pharmacists, social workers, clergy, and other professionals. As members of interdisciplinary palliative care teams, they are committed to enhancing the quality of life for terminally ill patients and their families and will play an increasingly important role. Hospice is an effective mechanism that provides this comprehensive care.

The influence of spirituality and the role of the faith community in helping patients experience a peaceful and good death are also the subjects of renewed interest. Increased public dialogue with health care professionals and policy makers will ultimately accomplish much to enhance end-of-life care.

JEROME E. KURENT

See Also
Pain Assessment Instruments

Internet Key Words
Bioethics
End-of-Life Care
Geriatrics
Pain Management
Palliative Care
Quality of Life
Spirituality

Internet Resources
American Academy of Hospice and Palliative Medicine
http://www.aahpm.org

American Geriatrics Society
http://www.americangeriatrics.org

American Pain Society
http://www.ampainsoc.org

Americans for Better Care of the Dying
http://www.abcd-caring.com

Growthhouse
http://www.growthhouse.org

Last Acts
http://www.lastacts.org

Project on Death in America
http://www.soros.org/death

Robert Wood Johnson Foundation
http://www.RWJF.ORG.main.html

PARKINSONISM

Parkinson's disease afflicts the majority of patients with Parkinson's syndrome. The term *idiopathic Parkinson's disease* is often used to stress the elusiveness of the causes of the condition. Parkinsonism is a clinical symptom complex similar but not identical to idiopathic Parkinson's. There are many causes of parkinsonism, including infectious diseases, vascular causes, drug-related disorders, tumor, head trauma, striatonigral degeneration, and other neurological conditions. James Parkinson first described the condition in 1817 as "shaking palsy." The discovery of an association between a neurotoxin, the meperidine analogue 1-methyl-4-phenyl-1, 2, 3, 6-tetrahydropyridine (MPTP), and parkinsonism has greatly advanced our knowledge of the biochemical mechanism of the disease symptoms. This neurotoxin was a contaminant of the illicit drug supply in northern California and produced persistent parkinsonism in young drug abusers. MPTP-induced parkinsonism in animals can be used to study the disease.

Epidemiology

Approximately 1% of those over 60 years of age in the United States are affected by idiopathic Parkinson's disease. Incidence increases with age but peaks at age 75 years. Population-based studies suggest that the prevalence of parkinsonism continues to increase in the oldest age groups and is associated with a twofold increase in the risk of death (Bennett et al., 1996).

Pathogenesis

Parkinson's disease and parkinsonism secondary to other causes, such as medication or cerebrovascular trauma, do not differ in the pathogenesis of symptom production. Extrapyramidal dopamine deficiency is common to all types of parkinsonism. However, the formation of an intracellular inclusion body in the brain—the Lewy body—is observed only with Parkinson's disease. One popular theory is that the accumulation of free radicals secondary to oxidative stress results in selective nigral cell destruction (Lang & Lozano, 1998).

Clinical Manifestations

The clinical triad of tremor, muscular rigidity, and bradykinesia characterizes parkinsonism. In Parkinson's disease, resting tremor is commonly the initial symptom, but in 15% to 30% cases, tremor is absent or less prominent as the disease progresses. Absence of tremor should not obscure the diagnosis.

Muscular rigidity is usually readily elicited during passive range of motion. It is often perceived by a patient as stiffness and pain and can be confused with arthritis or bursitis. Passive range of motion elicits smooth resistance ("lead pipe") with or without superimposed ratchetlike jerks ("cogwheel" phenomenon).

Bradykinesia is difficulty initiating movement and delayed or slowed execution of movement. The inability to execute simultaneous or sequential movements is one of the most fundamental motor disturbances in Parkinson's disease and is prominent when patients are turning, walking, and talking at the same time. The use of visual, auditory, and other sensory cues to overcome freezing suggests that patients have an intact motor capacity but have difficulty accessing that capacity. A tendency to fall forward (propulsion) or backward (retropulsion) results from the loss of postural reflexes. Postural instability and falls are the most disabling of all parkinsonian symptoms.

Dementia was not originally recognized as part of Parkinson's disease. In the original description of the disease, Parkinson categorically stated that "the senses and intellect are uninjured." It is currently estimated that 30% of Parkinson's disease patients develop dementia (Bennett et al., 1996). In one study, patients with Parkinson's disease had a four times higher cumulative risk of developing dementia than did a control group (Mayeux, 1993). In advanced and untreated Parkinson's disease, depression occurs in 40% to 60% of patients. The overlap of depressive and extrapyramidal symptoms, such as apathy and psychomotor retardation, makes the reliable diagnosis of depression very difficult in these patients. Superimposed depression clouds the prognosis and course of the disease.

The age of the patient and the age of onset of the disease influence its course. Older patients have a more rapid and malignant course, with more rapid progression of disability, poorer treatment response, and greater likelihood of developing dementia.

Diagnosis

The diagnosis of Parkinson's disease is made on clinical grounds only; caution must be exercised to avoid an erroneous diagnosis. It is important to distinguish secondary parkinsonism from idiopathic Parkinson's disease. The clinical symptom complex of parkinsonism can be triggered by certain medications, notably, psycholeptics. The disease can be a consequence of cerebrovascular injury, so-called vascular parkinsonism. Cases have been described of parkinsonism secondary to head trauma, hydrocephalus, hypothyroid and hypoparathyroid status, brain tumor or paraneoplastic disorders, and infections such as encephalitis, slow virus-associated infections, and human immunodeficiency virus. Secondary parkinsonism can be a component of Alzheimer's disease, Lewy body inclusion dementia, progressive supranuclear palsy, and Shy-Drager syndrome.

Treatment

Treatment can be subdivided into neuroprotective or preventive and symptomatic. The initial hope that the selective monoamine oxidase B inhibitor selegiline would have a neuroprotective effect and therefore delay the onset of disability has not been confirmed by clinical observations (Lang & Lozano, 1998). Levodopa, a precursor to dopamine, has been the cornerstone of symptomatic therapy since the 1970s. Unfortunately, its effectiveness decreases after five to seven years of administration. The common practice of delaying initiation of treatment with levodopa until the latter stage of the disease, out of fear of drug toxicity, was recently challenged in the medical literature (Lang & Lozano, 1998). Levodopa treatment often requires concomitant medications to control the drug's many disturbing symptoms, such as psychiatric complications and dyskinesias.

Neuroablation with modern stereotactic techniques permits targeting of the affected brain tissue with millimeter accuracy. Deep brain stimulation has been introduced as an alternative to neuroablation. Neurosurgery should be reserved only for medically refractory cases (Arle & Alterman, 1999).

Psychosocial Implications and Management

Understanding and managing the psychosocial ramifications of Parkinson's disease are critical, as they can affect the quality of life of both the patient and family members, who may be the primary caregivers as well. Ongoing attention must be given to assessing and reassessing the person's psychosocial situation in the context of his or her changing physical, mental, emotional, spiritual, environmental, and financial status, as well as that of the caregivers. It should be kept in mind that the coping strategies and capacities of older persons with Parkinson's may not be as robust as those of younger affected persons.

Parkinson's disease affects a person's motor ability and emotional and mental health, with the potential to dramatically interrupt, in varying degrees, the person's life and that of his or her partner or family. At the time of initial diagnosis, the risk of depression and dementia should be assessed. This time should also be used as an opportunity for the affected person and caregivers to learn about the disease in all its manifestations. Identifying important quality-of-life issues and planning for the future may ameliorate feelings of helplessness (Koplas et al., 1999). At a more advanced stage of the disease, when the patient's cognitive and physical functions may be seriously impaired, the practitioner should focus on the caregiver's understanding of the disease, involvement in treatment planning, and ability to provide care. Because treatment is more palliative than curative, with the attendant uncertainties as research continues and treatment options are tried, the psychosocial aspects of care become very important.

Individual counseling and support groups for affected persons and caregivers can be an effective component of the total treatment plan, particularly since the prevalence of depression is high (Andersen, 1999). Such services can help patients and caregivers define the quality of life for each person with Parkinson's. Individualized activities should be encouraged, as they form the basis of effective coping strategies to manage life within emerging limitations.

Home-care services are an important consideration in the management and treatment plan, as they allow the patient to remain at home and facilitate the caregiver's capacity to manage care at home. The level and amount of services vary, based on the availability of home-care services in a locality, level of disability and need, insurance coverage, ability to pay for services privately, and capacity of the affected person or family member to direct care. A full home evaluation for physical safety should not be overlooked, especially with the high risk of falls among people with mobility disabilities.

Institutional care may be considered when the disease is advanced and the affected person can no longer care for himself or herself or when patient's needs are so extensive that care cannot be provided at home. Temporary nursing home placement for respite care can provide short-term relief for caretakers.

The complexities and progressive nature of this disease require periodic care plan adjustments. A close, dynamic, interdisciplinary relationship among the patient, caregiver, physician, and other professional and paraprofessional staff will enhance the effective management of this disease and its effect on the patient's quality of life.

VALERY A. PORTNOI
RENEE WARSHOFSKY-ALTHOLZ

See Also
Cognitive Changes in Aging
Depression in Dementia
Eating and Feeding Behaviors
Falls Prevention
Gait Disturbances
Social Isolation

Internet Key Words
Falls
Movement Disorders
Parkinsonism

Internet Resources
American Parkinson Disease Association, Inc.
http://www.apdaparkinson.com

National Parkinson Foundation, Inc.
http://www.parkinson.org

Parkinson's Disease Foundation
http://www.parkinsonsfoundation.org

United Parkinson's Foundation
http://www.pdf.org

PATIENT-PROVIDER RELATIONSHIPS

Provider-patient relationships are built on more than just good habits such as active listening, sitting face-to-face with a patient, and using appropriate titles such as "Mr." and "Ms." These habits are necessary, but they are not sufficient for a successful relationship. A provider-patient relationship is a union of people with inherently unequal knowledge, power, and responsibilities. A provider who fails to address these issues can experience a variety of difficult relationships with patients. These include the "demanding," "noncompliant," or "difficult" patient, or even a lawsuit. Providers who want to avoid these problems and develop successful relationships with their patients should acquire the skills of effective decision making, avoid conflicts of interest, and foster relationships with other providers.

Effective Decision Making

Much of the provider-patient relationship involves decisions such as choosing treatments, selecting a home-care agency, and planning follow-up care. The substance of any decision is information about the benefits and risks of choices. Paternalistic models of decision making argue that the provider should privately review the information and make the choice for the patient. However, most contemporary models of provider-patient relationships advance a model of shared decision making.

Shared decision making incorporates the principle of respect for autonomy into the provider-patient relationship. The provider is obligated to explicitly discuss information and prompt the patient to make a choice (Brock, 1993). The goal of this model is a patient who understands the decision. The challenges are what kinds of information ought to be discussed and the degree of participation the patient wants in decision making. The failure to address these challenges has costs for health care professionals and society. Patients who feel that they did not participate in decisions feel dissatisfied and change providers (Kaplan, Greenfield, Gandek, Rogers, & Ware, 1996), and most malpractice claims are prompted by the perception that the provider did not listen to the patient (Hickson, Clayton, & Entman, 1994).

Shared decision making requires the provider to recognize how much information and decisional control the patient wants. As a general rule, patients want information, but studies of patients with a variety of diseases consistently show that patients' desire for control varies along a continuum from active ("I prefer to make the decision about which treatment I will receive") to passive ("I prefer to leave all decisions regarding treatment to my doctor"). Only 3% to 22% of patients want an active role. Most patients who want either an active or collaborative role ("I prefer that my doctor and I share responsibility for deciding which treatment is best· for me") are younger than 65 and have greater than a high school education (Beaver et al., 1996; Degner et al., 1997; Strull, Lo, & Charles, 1984).

Clinician characteristics can influence the patient's ability to achieve the desired degree of decision-making control. Physicians typically underestimate the amount of information patients want but overestimate the degree that patients want to participate in treatment decisions (Strull, Lo, & Charles, 1984). Consequently, patients often fail to achieve the decisional role they desire. Of a cohort of 1,012 women with breast cancer, 58% desired a more active role than they actually experienced (Degner et al., 1997), and 61% of a cohort of 150 women with breast cancer did not achieve their preferred role (Beaver et al., 1996). Physician characteristics associated with fostering a patient's desired role include training in primary care or interviewing skills, a "low-volume" practice (defined as <70 outpatients/week), and the clinician's satisfaction with his or her level of professional autonomy (Kaplan et al., 1996). These data suggest that characteristics of the provider and the provider's practice influence the quality of the provider-patient relationship.

These data provide useful information for providers who care for elderly patients. A provider is likely to be younger and more educated than the patient. Hence, the provider will likely overestimate the patient's desire for participation in decisions about medical care. The key point in fostering effective decision making is that patients want information about treatments but differ significantly in the amount of control they want in making a treatment decision.

Avoidance of Conflicts of Interest

Relationships with and obligations to institutions, other professionals, and other patients can affect the quality of a provider-patient relationship because they can create conflicts of interest. For there to be a conflict of interest, the commitment or relationship must involve a role with these core features: it is socially designed and chosen by the person, the role serves the good of others, the role involves discretion and judgment as part of its function, and the people who benefit from the person's work trust the person simply because he or she holds the role (Erde, 1996). Conflicts of interest matter because they threaten the provider's ability to supply fair and impartial service to a patient and they undermine the virtue of trust that facilitates relationships between people who differ in knowledge and power. The following cases illustrate potential conflicts of interest.

Case 1:   The son of a nursing home resident gives a gift certificate to the head nurse with a note of thanks for her hard work. Two weeks later, the son requests that the head nurse assign an additional nurse's aide to assist his mother with bathing. The nurse feels used and regrets that she accepted the gift certificate.

Case 2:   A physical therapist evaluates an elderly man with a gait disorder caused by a peripheral neuropathy. The therapist is the principal investigator in a study testing different techniques of gait retraining. This man meets the study's entry criteria. The man explains that his son is burdened by having to bring him to his physical therapy appointments. Participation in the study requires extra visits compared with usual care. The therapist debates whether he should recommend enrollment in the study.

Case 3:   One week after a geriatrician signed a contract to speak at a pharmaceutical company–sponsored conference on hypertension to be held in Hawaii, she receives a box of free samples of the company's latest antihypertensive. While reviewing the package insert, she finds that there are limited data on the drug's effectiveness and safety in elderly patients. Many of her elderly patients

have systolic hypertension and complain about the cost of their medicines. She wonders whether she should cancel the talk, give the talk but refuse the fee, or leave the free medicines in the box. She regrets that she fostered this relationship.

Case 1 illustrates how giving gifts to providers can invoke complex emotions and motivations on the part of the gift giver and receiver (Lyckholm, 1998). The nurse worries that her professional judgment in allocating staff fairly according to available resources and residents' needs is being "bought" by the patient's son. Although some professional codes of ethics recommend the refusal of all gifts, this extreme position fails to recognize that gift giving can represent a socially and culturally appropriate way for people to express affection and thanks. The challenge to a provider is to establish the limits of this expression: Make sure that the gift is modest and goes to those who deserve it, and "close the exchange" by means of a thank-you note.

In case 2, the physical therapist faces the classic dilemma of the physician-scientist (Jonas, 1969). He feels obligated to promote the health and well-being of the patient, but he also wants to conduct research that might produce generalizable knowledge. Among the key protections for human research subjects is informed consent. The physical therapist should candidly discuss the risks, burdens, and benefits of the research project versus the standard care.

In case 3, the physician faces a number of conflicting obligations. The lecture in Hawaii promises recognition, a free vacation, and a generous fee. Her patients will appreciate free medications. But a physician is obligated to assess a drug's merits fairly and objectively. Can she recommend a drug that has limited data on its efficacy in her patients? Should patients know that she is a paid lecturer hired by the company that produces the free samples she gives away? Typical mechanisms to address such conflicts of interest are avoiding the relationship or, when it cannot be avoided, disclosing it to people who may be affected by it. In this case, the physician should inform her hypertensive patients about the risks and benefits of the drug and her relationship with the company that produces it.

These diverse cases suggest that many factors can affect the provider-patient relationship. A useful guideline for addressing conflicts of interest is to recognize that the provider's authority and status rest on the patient's trust that the provider will act in his or her best interests and will avoid or at least disclose any potential relationships that might hinder this goal.

Relationships with Other Providers

Care of an elderly person often requires the continuous delivery of medical, nursing, and social services to a patient with chronic diseases. This diversity of services means that an elderly patient is likely to have relationships with multiple providers. Hence, providers must foster relationships with one another. A nursing home resident relies on nurses, nurses' aides, physicians, and therapists. Quality long-term care requires effective communication and collaboration that bridge the cultures of these disciplines. Health care professionals of different disciplines generally train independently. The skills of interdisciplinary communication and collaboration are acquired only in the later stages of training, during clinical rotations, and are not so much taught as learned by trial and error and by modeling the behavior of mentors.

Patients depend on providers to serve their best interests. A patient-provider relationship is not a partnership of coequals. Society recognizes the importance and the uniqueness of this relationship and seeks to preserve it with laws that uphold its confidentiality and that protect the patient from abandonment and malpractice. Although a provider brings his or her own personality to a relationship, developing effective decision-making skills, addressing conflicts of interest, and fostering relationships with other providers can be the foundation for a successful and ethical relationship with a patient.

JASON H. T. KARLAWISH

PENSIONS

Employer pensions are one of the primary sources of income for many retirees, and pension income is often the determining factor in whether individuals are able to continue their preretirement levels of consumption throughout their retirement years. Expenditures on pensions represent a significant component of labor costs for many firms in the United States. Firms establish pension plans as part of their human resources policies in an effort to attract, retain, motivate, and ultimately retire workers.

The two primary types of pensions are defined benefit and defined contribution plans. Defined benefit plans promise a specified benefit in retirement. Firms must have sufficient funds to pay the promised benefits. In defined contribution plans, firms and employees make periodic contributions to a pension account. The retirement benefit depends on the contributions and the returns on those investments. Some plans, such as cash balance plans, have characteristics of both types (McGill, Brown, Haley, & Schieber, 1996).

The benefit in defined benefit plans can be determined by (1) multiplying a generosity factor times years of service times a salary average or (2) multiplying a dollar amount by the number of years of service. For most workers covered by nonunion, single-employer plans, benefits are based on the first formula; multiemployer and collectively bargained plans tend to adopt the second formula. Most defined benefit plans in the private sector are funded solely by employer contributions, whereas such plans in the public sector typically require employee contributions.

Several types of defined contribution plans exist, including profit sharing, money purchase, and 401(k) plans. These plans can be the only pension offered by a company, or they can be in addition to a defined benefit plan. Many 401(k) plans are funded predominantly by employee contributions, which are often partially matched by employer contributions. In most defined contribution plans, the worker must decide whether to make a contribution, how much to contribute, and how to invest these funds. The trend toward greater use of defined contribution plans raises new issues concerning the future of retirement income in the 21st century. These include whether workers will choose to contribute sufficient funds to support an adequate retirement benefit, whether workers will invest wisely for retirement, and whether workers will withdraw funds prior to retirement.

History

The first private pensions in the United States were established in the late 19th century; however, widespread pension coverage did not occur until the second half of the 20th century. Prior to World War II, the proportion of the labor force covered by employer pensions was only about 15%. Coverage reached 25% of the labor force around 1950 and continued to increase until the mid-1970s, when approximately 50% of the labor force was covered by employer-provided pensions. Since the mid-1970s, the proportion of the labor force participating in pension plans has remained at about the same level, with coverage increasing somewhat for women and declining for men. Public employees have always had higher coverage rates than private workers, and over 90% of public employees are covered by pensions (Employee Benefit Research Institute, 1997).

Historically, most pension participants were enrolled in defined benefit plans. This was especially true among large employers; small firms were more likely to offer defined contribution plans. Since the passage of the Employee Retirement Income Security Act (ERISA) in 1974, there has been a large and continuing trend toward greater use of defined contribution plans, especially 401(k) plans. Changes in government regulation increased the cost of providing defined benefit plans relative to defined contribution plans, and subsequent changes in policy allowed pretax employee contributions to 401(k) plans. In response to these changes, employers have increasingly opted for defined contribution plans. In addition, changes in the industrial structure of the economy and the composition of the labor force have resulted in more workers being covered by defined contribution plans (Clark & McDermed, 1990).

Why Do Pensions Exist?

Pensions exist as a form of compensation because of worker and firm preferences. In general, individuals prefer current cash payments to in-kind benefits or deferred compensation. Differential tax treatment of various forms of compensation alters the relative desirability of certain types of compensation. Income tax policy allows firms with qualified pension plans to treat contributions to pension funds as a current expense, and workers do not report these contributions or the returns on them as current income. Benefits are taxable when they are received during retirement. Pension compensation is never subject to payroll taxes such as those that finance Social Security retirement benefits and Medicare.

This preferential tax treatment means that workers covered by pensions can receive greater total compensation for the same cost to the firm than if all compensation were paid in the form of cash wages. Since the tax advantage is greater for those with higher incomes, coverage is likely to be more desirable for high-wage workers, and benefits for these workers tend to be more generous. High-wage workers are also more interested in obtaining pension coverage, because their income replacement ratio from Social Security is lower than that for low-wage workers.

Employers prefer to offer pensions in an effort to modify worker behavior. Provisions in most pension plans impose a penalty on workers who quit or are fired prior to retirement age (Ippolito, 1999). This penalty means that workers who leave the firm will receive less in future pension benefits than will workers who stay. As a result, companies that provide pension plans may reduce turnover rates, thus reducing hiring and training costs. From the firm's perspective, this means that a dollar allocated to a pension fund may cost the firm less than an extra dollar allocated to cash wages.

Vesting, Portability, and Turnover

Vesting refers to the portion of benefits that a worker would receive if he or she left the firm. For example, workers with zero vesting would not receive any pension benefits if their employment was terminated, whereas workers who are 100% vested would receive the promised benefit at the specified retirement age. ERISA currently requires that firms adopt one of two vesting standards: (1)

100% vesting after five years, or (2) 20% vesting after three years, increasing to 100% vesting after seven years. Virtually all defined benefit plans firms have adopted the first option.

Workers covered by defined benefit plans who leave the firm prior to retirement accumulate lower retirement benefits than do similar workers who remain with the firm until retirement, even with 100% vesting. This penalty for leaving results from benefits being frozen when the worker terminates employment; that is, the benefit at retirement is not increased for inflation or the growth in real wages. Workers typically cannot transfer credit for years of service from one firm to another. This lack of portability implies that workers who move from job to job among firms with identical pension plans will have substantially lower benefits than those workers who remain with a single firm. This penalty tends to reduce turnover and increase tenure in those firms that offer defined benefit pension plans.

Retirement Incentives

Pension plans alter retirement incentives for older workers by changing the value of continued employment. The change in the present value of retirement benefits is an important concept that affects retirement decisions. Before a worker becomes eligible to retire, the value of future benefits increases as the number of years of service and annual salary increase. For participants in defined benefit plans, the increase in the value of pension benefits increases as they approach retirement age. The present value of pension benefits is altered after the worker reaches the age of eligibility for retirement benefits. Continued employment may increase annual benefits, but the worker must forgo current benefits in order to remain on the job. This produces a sharp decline in total compensation, providing an incentive for the worker to retire. Considerable research indicates that participation in a defined benefit plan does alter the timing of retirement (Kotlikoff & Wise, 1990).

Pension provisions can be altered by plan sponsors to increase or decrease these retirement incentives. For example, employers can introduce penalties for early retirement, increases in benefits for delayed retirement, or maximum benefit provisions to influence retirement decisions. During the past two decades, many employers have attempted to reduce the size of their workforce by encouraging early retirement. Most early retirement programs provide for short-term modifications to the pension plan to provide large, temporary incentives for workers to retire. Defined contribution plans are more neutral in their retirement incentives.

ROBERT L. CLARK

See Also
Retirement

PERCUTANEOUS ENDOSCOPIC GASTROSTOMY (PEG) TUBE FEEDING

Gastrostomy tubes are small, pliable feeding tubes that pass through the abdominal wall and into the stomach cavity. Gastrostomy tubes are placed for the purpose of providing nutrition, hydration, and medications when the patient is unable to consume adequate nutrition by mouth (Loser, Wolters, & Folsch, 1998). Although the technique of delivering nutrition into the stomach through a surgical incision in the abdominal wall has been practiced for over 100 years, techniques requiring a less extensive surgical procedure have been available for only about 20 years. The most common of these techniques is called *percutaneous endoscopic gastrostomy* (PEG). These tubes are typically placed by a gastroenterologist with the aid of an endoscope. Gastrostomy tubes may also be placed using radiological techniques, in which case the operator uses a fluoroscope to guide tube placement. Thus, gastrostomy tubes can be placed surgically, endoscopically, or radiologically.

Over 125,000 older adults receive PEG tubes each year in the United States, and this number has been increasing steadily for the past decade. A significant part of this increase is due to the growing population of older adults who have difficulty eating. Approximately 2% of community-dwelling older adults and 40% of older adults residing in nursing homes require assistance with eating. Difficulty eating or swallowing (dysphagia) may be caused by a wide range of conditions, including head and neck cancer, stroke, and dementing disorders such as Alzheimer's disease. However, a patient may also have difficulty eating due to the combination of several illnesses, an acute illness such as pneumonia, or trauma. Still other patients may have a PEG tube placed because they tend to aspirate or choke on their food. It is believed that repeated aspiration of food or saliva may increase the risk of pneumonia (aspiration pneumonia). Patients suspected of aspirating food may be recommended to undergo a swallowing study, technically referred to as a videofluoroscopic deglutition examination (VDE). The VDE may allow the health care team to match specific dysphagia rehabilitation or management techniques to the swallowing defect identified. VDE can also be used to assess the success of these management techniques. VDE and dysphagia management techniques are typically supervised by a speech therapist, who may recommend swallowing techniques instead of a PEG tube for certain patients. Although there are no absolute nutritional parameters for determining the need for a PEG tube, patients who continue to lose weight or have low serum albumin levels despite assistance with eating are often considered for artificial feeding. When the need for artificial feeding is expected to be long term, PEG tubes rather than nasogastric tubes are typically used.

The PEG procedure itself is generally safe, enjoying over a 95% success rate for tube placement (Rabeneck, Wray, & Petersen, 1996). About 50% of PEG tubes are placed during an acute hospitalization, with the remainder placed in outpatient settings. Only about 4% of patients experience a major complication such as infection in the skin, infection in the abdomen, bleeding, or leaking around the outside of the tube (Rabeneck et al., 1996). Unfortunately, the majority of patients experience minor complications or problems such as irritation or pain at the PEG tube site, nausea, vom-

iting, diarrhea, constipation, abdominal pain, or obstruction or clogging of the tube. Long-term follow-up studies generally find that 10% to 20% of patients who receive a PEG tube eventually return to independent feeding.

In many studies, 15% to 20% of patients receiving PEG tubes died within 30 days of the procedure, and up to 50% died within one year (Rabeneck et al., 1996). These deaths are nearly always attributed to the patient's underlying disease rather than to complications related to the PEG tube. However, this high mortality rate and the high frequency of minor complications have raised questions about which patients really benefit from PEG tubes. Recent data suggest that interventions such as tube feeding rarely improve survival or maintain quality of life in patients suffering from dementia (Finucane, Christman, & Travis, 1999; McCann, 1999). Decisions about whether to proceed with PEG tube feeding are intensely personal (Callahan, Haag, Buchanan, & Nisi, 1999). However, it is generally recognized that artificial feeding is a medical intervention, and as such, patients may choose to accept or refuse artificial feeding based on their own wishes.

The PEG tubes consist of three main parts visible outside of the patient's abdomen. Farthest away from the patient is a feeding adapter port, which has a wider lumen than the PEG tube, and allows one to introduce food, water, or medications into the tube. This adapter port connects to the PEG tube that leads directly into the patient's stomach. Just before the PEG tube enters the patient's stomach, there is a retention device near the skin that keeps the entire tube from slipping into the patient's abdomen.

The tube exit site should be cleansed daily with soap and water, and the area should be kept dry. Unless otherwise indicated, a dressing should not be used. Observe the site for redness, irritation, or gastric leakage. Be sure that the retention device is close to the skin. It should be snug but not overly tight against the skin. A good rule of thumb is that the edge of a dime should be able to fit in between the skin and the retention device. During the healing or "maturing" of the tube incision site (approximately two months), it is important that the tube be anchored securely so that there is no undue movement. Otherwise, the tube could migrate into the peritoneal cavity and cause leakage of formula into the abdominal cavity, leading to peritonitis. Signs of peritonitis include fever, vomiting, pain, and abdominal distention. Feeding should be stopped immediately and the physician notified if this is suspected. As the tube incision site heals, the ostomy through which the tube passes becomes lined with skin or scar tissue (this passageway is referred to as a tract). If dislodging of the PEG tube occurs before formation of the tract, the tube will have to be replaced endoscopically.

Approximately 24 hours after placement of the PEG tube, feedings can be initiated using one of the many commercially available formulas. Formula selection is based on the patient's medical condition and nutritional requirements. Prior to initiation of feeding, a dietitian is typically consulted to aid in the choice and quantity of formula and to assess water requirements. The formula is administered by either continuous or bolus feeding. Continuous feeding is given by pump around the clock and started at a slow rate to monitor the patient for tolerance. Over the next several days, the feeding can be increased to the targeted caloric requirement as tolerated by the patient. When the PEG tube is inserted as an outpatient procedure, bolus feedings are usually selected. The total amount of formula required daily is divided over three to five feedings (Loser et al., 1998). A 60-mL syringe is connected to the feeding port, and the formula is poured in slowly. Formula should be given at room temperature. When administrating tube feeding, the head of the bed should be elevated at least 30 degrees and remain elevated for at least two hours after feeding. Monitor the patient closely for signs of nausea, vomiting, and diarrhea. The formula type, rate, and schedule may need to be reevaluated based on the patient's response. Vomiting can lead to aspiration, which may lead to pneumonia. Diarrhea can signify feeding intolerance, but other causes such as fecal impaction, medication side effects, or infection should also be investigated.

The caloric and fluid needs of each patient differ. It is important for patients and caregivers to completely understand the feeding plan. Depending on their condition, some patients may be able to take some oral nourishment. It is important to understand what type of nourishment can safely be consumed by mouth, if any. For instance, some patients may be able to take solid foods without difficulty, but thin liquids may present an aspiration risk. Written instructions should be given to the caregiver concerning the feeding schedule, care of the tube, and common troubleshooting measures. If the patient is living at home, the feeding schedule can be tailored to the needs of the caregiver as well as the patient. All medications given through the tube must be in liquid form or crushed and mixed with warm water; however, it is important to confirm that the medication can be safely given crushed. Inform the pharmacist when a prescription is being filled that the medication is to be administered through a feeding tube. Liquid medications should be provided when possible. If the medication should not be crushed, the pharmacist will notify the physician and may recommend an alternative medication. Medications should never be added to the formula and should never be mixed together.

The tube may have to be replaced as often as once or twice a year, or it may last for several years without replacement. The most common problem related to PEG feeding is clogging of the tube. It is best to prevent clogging by completely flushing the tube with water before and after bolus feeding and medication administration. If the patient is receiving continuous feeding, the tube should be flushed with water at least every four hours. If the tube becomes clogged, use a piston-type syringe and flush the tube with warm water by applying gentle pressure. Sometimes gently "milking" the outside of the tube helps unclog it. The feeding adapter can crack, or the tip may fall off or lose its elasticity, resulting in leakage of the formula. If this occurs, the damaged adapter can be cut off with scissors and a universal adapter can be fitted into the tube. This avoids the need to replace the entire tube. Routine changing of the feeding tube is not usually recommended; however, a tube may eventually deteriorate. The tube may elongate, bubbles may appear on the outside of the tube, or leakage along the tube length may occur. By this time, the tract has probably healed sufficiently, and an experienced person such as a physician or nurse can remove the tube at the bedside and replace it with a gastrostomy tube specifically designed for bedside insertion.

CHRISTOPHER M. CALLAHAN
KATHY M. HAAG

See Also
Eating and Feeding Behaviors
Nutritional Assessment
Swallowing Disorders and Aspiration

Internet Key Words
Dysphagia
Gastrostomy
Percutaneous Endoscopic Gastrostomy Tube
Swallowing Study

PERIPHERAL VASCULAR DISEASE

Atherosclerosis is the most frequent cause of chronic peripheral arterial occlusive disease, the prevalence of which increases significantly with age (Weitz et al., 1996). It is a disease process that involves both large and small arteries. Although peripheral vascular disease (PVD) by definition includes the carotid arteries, this section focuses on disease of the lower extremities.

Approximately 15% of people over age 50 have lower extremity PVD, as diagnosed by noninvasive techniques (Weitz et al., 1996). In the Framingham heart study (Kannel, 1996), the biennial incidence rate of symptomatic PVD for ages 35 to 74 was 7.1 in 1,000 for men and 3.6 in 1,000 for women. The incidence in individuals over 65 was one and a half to three times greater than in the 35 to 64 age group (Kannel, 1996). Arteriosclerosis of limb arteries is associated with coronary and cerebral atherosclerosis. Symptoms of lower extremity arterial insufficiency are local manifestations of a generalized disease process; diagnosis of PVD should help identify patients who are at increased risk of cardiovascular events such as stroke and myocardial infarction (Kannel, 1996). Atherosclerotic disease of the lower extremities causes pain and can lead to limb loss; thus, early identification can not only decrease mortality associated with other diseases but improve quality of life as well.

The risk factors that predispose patients to develop atherosclerotic vascular lesions include cigarette smoking, hypertension, diabetes mellitus, hyperlipidemia, and family history of vascular disease. A thorough history and physical examination can help confirm the diagnosis of PVD and suggest the location of the lesion.

History and Clinical Manifestations

Intermittent claudication is one of the most characteristic symptoms of PVD. It is usually described by the patient as exercise-induced pain in the lower extremity (calf, thigh, or buttock) or as profound fatigue that is relieved by rest within minutes. Symptoms appear distal to the site of occlusive lesions.

Ischemic rest pain develops when the blood supply is severely compromised and is inadequate even at rest. Patients may describe burning or pain that is exacerbated by elevation of lower extremities and relieved by slow walking or by keeping the foot in a dependent position. Gangrene occurs when arterial flow is inadequate to maintain viability of the tissues.

Physical examination of the lower extremities should include careful inspection of both legs, checking for discoloration, nail or skin atrophy, dependent rubor, and ulceration. The clinician should palpate the femoral, posterior tibial, and dorsalis pedis arteries and auscultate the femoral arteries for bruits.

Diagnostic Testing

An absence or decrease in the force of the pedal pulses is an indication for obtaining an ankle-brachial index (ABI), a simple method of identifying the degree of vascular insufficiency. In diabetic patients and those with end-stage renal disease, the ABI can be falsely elevated because of tibial artery calcification. Ischemic wounds and ankle surgery may interfere with pressure measurements.

Normally, distal blood pressures are higher than brachial blood pressures. Acceptable ABIs are 1.0 or greater. For ABIs between 1.0 and 0.9, testing should be repeated every two to three years. If the ABI is less than 0.9 but greater than 0.5, consistent with mild to moderate PVD, it should be repeated within three months (Orchard & Strandness, 1993). If the ABI is still less than 0.9, intensive risk factor modification (see below) should be started, and the ABI should be checked annually. Risk factor modification has also been recommended if ankle systolic pressure is more than 300 mm Hg, if ankle blood pressure is 75 mm Hg or more above arm pressure, or if the ABI is greater than 1.3, as these are signs of peripheral arterial calcification. Patients with an ABI less than 0.5 should undergo angiography to determine vessel anatomy (Orchard & Strandness, 1993).

Another useful test in assessing PVD is sequential limb pressures, which are obtained by placing cuffs at different levels on the lower extremities. A 20 to 30 mm Hg discrepancy between extremities is indicative of occlusion proximal to the cuff in the extremity with decreased pressure. A decrease in pressure of more than 30 mm Hg in two consecutive levels in the same extremity suggests a disease process at the level proximal to the cuff (Jaff & Dorros, 1998).

Pulse volume recordings are another test to identify occlusive lesions. These are plethysmographic tracings that demonstrate changes in blood flow through a lower extremity. Attenuation of the waveform (which is normally a rapid systemic upstroke) is indicative of PVD.

Toe systolic pressure measurement might help assess occlusive disease in diabetic patients, to determine whether calcification of the arteries has extended to digital vessels. An absolute pressure of 30 mm Hg or less is considered insufficient for the healing of ulcers (Orchard & Strandness, 1993).

Duplex ultrasonography is an important method of identifying the degree of stenosis in the arteries by determining the velocity of blood in various vessels. This test is especially useful when choosing the appropriate access to an occlusive lesion amenable to endovascular therapy.

Although studies demonstrate the importance of transcutaneous oxymetry in predicting wound healing, the method has limitations. Results are affected by oxygen concentration in inspired air, hemoglobin saturation, pulmonary function, cardiac output, skin thickness, and presence of inflammation. Each case must be assessed individually before determining the applicability of the test.

Patients referred to vascular laboratories for "noninvasive studies" will have ABI, segmental pressures, and pulse volume recordings. Other studies must be specified by name.

Angiography is an invasive method that outlines the vessel diameter and thus identifies the level and anatomy of the occlusion. Angiography is necessary if the clinician is contemplating surgery or interventional radiographic procedures. Risks of the procedure include allergic reaction to the contrast material, renal failure, and damage to the arteries or limb.

Treatment

Exercise remains the most important conservative therapy for intermittent claudication, together with smoking cessation. Aspirin (or other platelet inhibitors) and cholesterol-lowering agents are also recommended. At present, the U.S. Food and Drug Administration has approved two drugs to treat symptoms of claudication: pentoxifylline (Trental) reduces blood viscosity and improves red cell deformability; cilostazol (Pletal) inhibits platelet aggregation and acts as a vasodilator. Cilostazol is contraindicated in patients with congestive heart failure. These drugs are rarely effective and may not be worth the expense.

Invasive modalities include percutaneous transluminal angioplasty, placement of intraluminal stents, bypass surgery, and amputation. These remain the major options for patients whose pain does not respond to lifestyle modifications and drug therapy and for those with limb-threatening ischemia.

Angioplasty has a good long-term patency rate in the common iliac artery and is further improved by the use of vascular stents (Weitz et al., 1996). Patency rates are poorer in the superficial femoral artery, particularly with longer lesions. In most reported series, the use of intravascular stents below the inguinal ligament has not added significantly to long-term patency. Current research is focusing on methods to prevent restenosis.

Lower extremity bypass surgery is most useful when the limb is threatened by ischemia; success rates approach 85% to 90% (Weitz et al., 1996). The success and complication rates of bypass surgery appear to be better than those of amputation, and surgeons should attempt to salvage limbs in all but the most debilitated, nonambulatory patients (Weitz et al., 1996).

In the future, therapeutic angiogenesis, "the clinical use of growth factors to enhance or promote the development of collateral blood vessels in ischaemic tissue" (Henry, 1999, p. 1536), may become the treatment of choice for patients with PVD. Among the many growth factors currently under study, vascular endothelial growth factor and fibroblast growth factors show the most promise. Maximum benefit may occur through the administration of combinations of growth factors or the genes encoding them.

NATALYA KOZLOVA
EUGENIA L. SIEGLER

Internet Resource

American Heart Association
*http://americanheart.org/Scientific/statements/
1996/1201.html*

PHYSICAL FUNCTION
See
Measuring Physical Function

PHYSICAL THERAPISTS

The Office of the Surgeon General established physical therapy during World War I through the Division of Special Hospitals and Physical Reconstruction. Over 2,000 "reconstruction aides" restored function to patients with poliomyelitis and other disabilities. Conditions that limit function affect one in seven Americans; care costs approach more than $170 billion annually. Physical therapy services are thus an integral component of an interdisciplinary care team dedicated to meeting the needs of elders with functional limitations.

More than 115,000 physical therapists (PTs) currently practice in the United States and provide care to over 1 million clients per day. The minimum educational requirement is a baccalaureate degree in physical therapy from a program accredited by the Commission on Accreditation in Physical Therapy Education (CAPTE) (American Physical Therapy Association [APTA], 1997), but most physical therapy programs offer a master's degree. In 2002, CAPTE will limit its scope to accrediting only those professional programs that award postgraduate degrees. Currently, 173 colleges and universities have educational programs in physical therapy (APTA, 1998). Candidates must pass a state-administrated examination, but if a candidate's score is high enough, some states honor licensure granted by other states (APTA, 1997).

A physical therapist assistant (PTA) assists the PT in procedures and tasks as delegated by a supervising PT (APTA, 1997). PTAs must graduate from an associate's degree program accredited by an agency recognized by the APTA. The PT is directly responsible for the actions of the PTA (APTA, 1997).

The American Board of Physical Therapy Specialties (ABPTS) certifies PTs who have acquired specialized knowledge and have extensive clinical experience in geriatrics through the APTA. Geriatric clinical specialists must have a minimum of 6,000 hours of direct patient care before they can apply to take the geriatric clinical specialist's exam (APTA, 1998). In 1999, 394 geriatric clinical specialists were certified in the United States.

The diversity of settings in which geriatric PTs practice—hospitals, homes, physical therapy offices, rehabilitation facilities, subacute care facilities, nursing homes, hospices, fitness centers, and academic and research centers—reflects the versatility of their skill and knowledge base. PTs specializing in geriatrics take a comprehensive approach when caring for older patients with acute or chronic illnesses that limit function. Physical therapy assessments are not standardized and may vary according to practitioner as well as the locus of care.

Physical therapy, as defined by the state practice acts adopted by the APTA, includes examining patients with impairments, functional limitations, and disabilities in order to determine a diagnosis, prognosis, and intervention; alleviating impairments and functional limitations by designing, implementing, and modifying therapeutic interventions; preventing injury, impairments, functional limitation, and disability, including the promotion and maintenance of fitness, health, and quality of life; and engaging in consultation, education, and research (APTA, 1998).

Major sources of reimbursement for physical therapy in the geriatric population are HMOs, Medicare, and Medicaid. At the time of this writing, Medicare Part B allowed $1,500 a year for outpatient physical therapy. Many managed care companies have caps on the number of PT visits, depending on diagnosis. In many states, patients have direct access to PTs without a physician's referral. The APTA lists those states that allow direct access to physical therapy services on its Web site.

It is projected that by the year 2020, somewhere between 9.7 million and 13.6 million older people will have moderate to severe functional disability. PTs provide care through direct intervention and to individuals who are not necessarily ill but may benefit from professional consultation with a PT.

Physical therapy services for elders focus on rehabilitation, including the treatment of impairments and functional limitations leading to disability. Disability in the older population is usually multicausal, making physical therapy management of geriatric patients a challenge.

The PT examines the geriatric patient and performs a comprehensive evaluation based on the patient's past medical history, current diagnosis, functional impairment, disability, and review of relevant systems. Specific tests are selected and ad-

ministered to obtain baseline data regarding the patient's cardiovascular, neurological, pulmonary, and musculoskeletal systems. Evaluations performed by a PT may include motor function, joint integrity and mobility, range of motion, muscle performance, posture, gait and balance, pain, neuromotor development, sensory integration, aerobic endurance, ventilation, respiration, and circulation evaluations. Data obtained during the evaluation determine the degree to which specific individualized interventions are likely to produce changes in the patient's condition in order to achieve desired outcomes. The patient, PT, and other members of the health care team (e.g., occupational therapist, geriatrician, social worker, geriatric nurse practitioner) determine the goals of treatment.

Therapy services can be either restorative or maintenance. For restorative services, Medicare requires that therapy be provided by a PT or PTA, and progress must be shown in a predictable period of time. The goal of maintenance physical therapy is to maintain the patient's present level of function. There is no standard duration of treatment for particular events such as stroke or hemiparesis. However, the *Guide to Physical Therapist Practice* lists the expected number of visits per episode of care for particular diagnoses (APTA, 1997).

The model of physical therapy is evidence-based practice that uses clinical experience and concomitant external evidence such as research reports, clinical practice guidelines, and quality indicators. By integrating such data into clinical practice, PTs are working to incorporate newly acquired information into higher-quality care for geriatric patients (Levi, 1999).

Sandy B. Ganz

See Also

Physical Therapy Services
Rehabilitation

Internet Key Words

Acute Care Settings
Geriatric Clinical Specialist (GCS)
Nursing Homes

Physical Therapist (PT)
Physical Therapist Assistant
Rehabilitation Centers

Internet Resource
American Physical Therapy Association
http://www.apta.org

PHYSICAL THERAPY SERVICES

The management of geriatric patients with and
without active pathology is a challenge to physical
therapists due to the myriad functional impairments
and limitations commonly seen in the elderly. The
goals of physical therapy are to alleviate pain; prevent the onset and progression of physical impairment, functional limitation, disability, or changes
in physical function and health status that result
from disease or injury; and restore, maintain, and
promote general fitness, health, and quality of life
(Guccione, 1993).

The physical therapist integrates five key elements of management to maximize functional outcomes: examination, evaluation, diagnosis, prognosis, and intervention. These elements guide the
physical therapist toward the desired outcome. The
three-part examination consists of patient history,
systems review, and specific tests and measurements (Guccione, 1993).

History: The patient's health status, both
past and present, is obtained. Specific complaints,
along with health risk factors and concomitant problems that have implications for therapeutic intervention, are identified. Data obtained from a patient
history may include general demographics, social
history, occupation or employment, medications,
living environment, history of current condition,
functional status and activity level, laboratory and
diagnostic tests, and past medical and surgical history. Family history, self-reported health status, and
cognitive and communication ability should also
be included.

Systems Review: Systems review information assists the therapist in formulating a diagnosis,
prognosis, plan of care, and appropriate interven-

tions. The physical therapist may review the cardiopulmonary, integumentary, musculoskeletal, and
neuromuscular systems.

Specific Tests and Measures: After synthesis of all pertinent information gathered from the
history and systems review, the physical therapist
determines which tests and measures are needed to
elicit additional information. A thorough physical
examination should include a functional assessment
of upper extremity function, bed mobility and transfer status, ambulation analysis, strength and range
of motion, posture assessment, and cognitive and
perceptual evaluation, as well as the patient's ability
to perform activities of daily living.

Upper extremity function is measured by the
patient's ability to perform hand-to-mouth activities, finger dexterity, dressing, grooming, and hygiene and the ability to use appropriate assistive
devices for ambulation. Rolling is an important
aspect of bed mobility; patients who are unable to
reposition themselves in bed are at risk of developing pressure sores. If patients are unable to roll
from side to side, a side rail may be used as an
enabler. For home-bound elders who are unable to
turn without a side rail and for whom a hospital
bed is not an option, a child's bed railings may be
used. However, such railings may not be as stable
as railings specifically designed for adults.

Transfer status is the ability to perform movements to and from various heights and surfaces,
such as transferring in and out of bed, on and off
a chair or wheelchair, on and off the toilet, and in
and out of the tub. Adaptive or durable medical
equipment is used to make transferring safe and
easy if the patient is unable to transfer independently. A sliding board may be used for transferring
from bed to wheelchair. A drop-arm commode may
be used if a patient is unable to make the multiple
transfers necessary between the bed and the toilet
in the bathroom. A transfer tub bench may be helpful for transfers in and out of the tub. Use of such
equipment is dependent on the patient's sitting balance ability.

Ambulation status—cadence, velocity, and
step height—is determined through an observational gait analysis. If the patient is ambulating with

an assistive device such as crutches, walker, or cane, the assessment includes evaluating walking technique and determining whether the assistive device is the correct height and used properly (Tinetti, 1986).

Strength is evaluated through manual muscle testing techniques that determine the extent and degree of muscular weakness resulting from disease, injury, or disuse. Therapeutic exercise should be initiated if strength deficits are noted. Range of motion is the ability of a joint to move through a complete arc of motion. Most movements of the extremities are measured in degrees from a specified starting point. Although normative ranges have been established for adults, few data are available on normative ranges for the geriatric population. As a result of the aging process, range of motion decreases; it is important for the therapist to distinguish between normal range of motion and functional range of motion. An appropriate therapeutic exercise program directed at increasing range of motion should be initiated in cases of functional impairment with a concomitant decrease in range of motion (Guccione, 1993).

Posture assessments are routinely performed in conjunction with the musculoskeletal exam to ascertain whether any postural deformities are influencing the patient's functional status. Standing posture is a simple way to evaluate balance. Is the patient standing with or without support from an external device such as a cane? Is the patient's base of support wide or narrow? A wide base of support indicates decreased stability in the upright position.

Included in geriatric physical therapy functional assessment are performance-based measures such at the Tinetti Gait and Balance Assessment (Tinetti, 1986), Functional Reach (Duncan, Weiner, Chandler, & Studenski, 1990), and Timed "Get Up and Go" Test (Podsiadlo & Richardson, 1991). All these evaluative tools provide the therapist with a baseline and ongoing status.

SANDY B. GANZ

See Also
Balance
Falls Prevention
Gait Assessment Instruments

Internet Key Words
Functional Assessment
Performance Based Measures
Physical Therapy
Tests and Measures

Internet Resources
American Physical Therapy Association (APTA)
http://apta.org

APTA Section on Geriatrics
http://geriatricspt.org

PODIATRIC MEDICINE

Podogeriatrics is the branch of podiatric medicine that focuses on the treatment of foot and related disabilities, deformities, and complications occurring in later life. Foot conditions may be local or the result of complications associated with multiple chronic diseases, as well as the result of changes associated with the aging process itself. Given the fact that older persons tend to react to illness, deformity, and disease differently from younger persons, caring for them includes understanding the specific syndromes that older patients experience and being part of a team that manages multiple diseases.

Foot problems are universal in their occurrence. By age 65, almost 90% of the elderly have had one or more foot problems that caused some level of ambulatory dysfunction or functional disability. The factors contributing to the development of foot problems in older patients include changes in gait, duration of hospitalization or bed confinement, prior foot care and management, environmental factors, emotional modifications, current medications and therapeutic programs, past foot conditions or manifestations of systemic disease, increased medication schedules, changes in mental status, increased level of drug sensitivity, increased susceptibility to local infection due to neurovascular impairment, and related systemic diseases. Most older patients and those with chronic diseases such as diabetes mellitus, peripheral arterial insufficiency, and the various forms of arthritis exhibit foot complaints, and many chronic systemic diseases are first manifest by foot symptoms (Helfand, 1999). The ability to walk pain-free is often the catalyst that prevents a sedentary lifestyle in the elderly and the chronically ill. Concern about the prevention and early detection of disease, deformity, and disability, as well as the ability to stratify risk, offers a window of opportunity to significantly reduce the prevalence of foot disability in later life (Helfand & Jessett, 1998). Programs for older patients must focus on prevention—primary, secondary, and tertiary—and offer comprehensive services and access based on patient need, with the goals of providing quality care and maintaining the quality of life.

Education

A doctor of podiatric medicine is licensed to practice in all states and the District of Columbia. Attaining this degree requires four years of academic study at a college of podiatric medicine accredited by the Council on Podiatric Medical Education of the American Podiatric Medical Association, preceded by undergraduate education, including a pre-professional academic degree.

Primary training in clinical geriatrics is completed through the first professional degree and residencies. Additional training is provided by geriatric education centers and through continuing education programs. Future fellowships are being developed to focus on the academic elements to enhance clinical knowledge. A large number of podiatrists are in solo or small group practices, but interdisciplinary and team care is being expanded. Most programs have followed the paradigm of the Department of Veterans Affairs.

The basic curriculum at the seven colleges of podiatric medicine provides an introduction to the basic medical sciences such as anatomy, physiology, biochemistry, microbiology, and pathology. In addition, the basic clinical aspects of podiatric and medical care are covered, including biomechanics, pathomechanics, medicine, gerontology, and clinical podiatric medicine. The third year includes medical areas such as peripheral vascular disease, neurology, dermatology, podiatric surgery, and

public and preventive health. The fourth year is primarily clinical and includes ambulatory clinical care, clerkships, and institutional externships.

The core program prepares new practitioners for their roles as primary providers of foot care and prepares students for educational programs beyond the first professional degree, such as residencies, preceptorships, fellowships, graduate education, and continuing education. Licensure is obtained in the manner prescribed by each state and may include examinations provided by the National Board of Podiatric Medical Examiners or appropriate examinations or endorsements provided by the various state licensing agencies.

Podiatric Medical Management

Immobility can translate to social segregation, lost efficiency, and declining health and can create personality and emotional changes. Foot health is needed by older patients to increase personal comfort, reduce the probability of medical or surgical complications, reduce institutional length of stay, reduce the possibility of hospitalization related to foot infections, and reduce the emotional distress associated with foot discomfort.

Care provided by podiatrists includes services for ambulatory patients, hospitalized patients, and those in long-term-care settings (Evans & Williams, 1992). Generally, care components include the history, physical examination, radiographs, laboratory studies, and other special diagnostic tests, such as those related to biomechanics and vascular analysis.

Neurological, vascular, and other related conditions should be managed primarily, with appropriate consultation as indicated. Debridement, pathomechanical, orthopedic, biomechanical, radiographic, orthotic, and dermatological procedures are employed as elements of total patient management (Hazzard, Blass, Ettinger, Halter, & Ouslander, 1999). Appropriate topical, systemic, and controlled drugs should be utilized based on the diagnosis and therapeutic indications. Conservative as well as surgical management may be carried out as indicated and appropriate. Health education should be a major component of care for older patients.

Reimbursement

Insurance reimbursement for podiatric services is similar to Medicare reimbursement. In 1967, podiatrists were included in the Medicare regulations and considered physicians with respect to the services they are legally authorized to perform by the state. Certain types of treatment or foot care are excluded, whether performed by a doctor of medicine, doctor of osteopathic medicine, or doctor of podiatric medicine (Helfand, 1993). These exclusions are generally defined as routine foot care in the absence of localized illness, injury, or symptoms involving the foot. Services are covered when systemic conditions such as metabolic, neurological, or peripheral vascular disease result in circulatory embarrassment or areas of diminished sensation in the feet or legs and evidence documents that inappropriate care would be hazardous for the patient because of underlying systemic disease. Covered diseases include but are not limited to arteriosclerosis obliterans, arteriosclerosis of extremities, occlusive peripheral arteriosclerosis, peripheral arterial disease, Buerger's disease, chronic venous insufficiency, diabetes mellitus, intractable edema secondary to congestive heart failure, renal insufficiency, kidney disease, hypothyroidism, peripheral neuropathies, multiple sclerosis, and Raynaud's disease.

Pain alone is not an indication for coverage. The patient must be under active care and present with nontraumatic amputation, absent posterior tibial pulse, advanced trophic changes, change in hair growth (decreased or absent), nail changes (thickening), pigmentary changes (discoloration), change in skin texture (thin, shiny) or color (rubor or redness), absent dorsalis pedis pulse, claudication, temperature changes (cold feet), edema, paresthesias, or burning.

The key component is the recognition of the fact that because the vast majority of foot problems in older patients are chronic and related to systemic disease, management must include continuing surveillance, assessment, and care.

ARTHUR E. HELFAND

See Also
Foot Problems

Internet Resource
American Podiatric Medical Association
http://www.apma.org/

POLYPHARMACY: DRUG-DRUG INTERACTIONS

Older people take more prescription and over-the-counter (OTC) medications than younger people. Although those over 65 years old constitute only 12% of the population, they consume 30% of all prescribed drugs. Some medications are absorbed, distributed, metabolized, and excreted (pharmacokinetics) differently in the elderly, and the action of drugs (pharmacodynamics) may be exaggerated or diminished. Of special significance in the geriatric population, different drugs interact with each other either by pharmacokinetic inhibition or induction of drug metabolism or by pharmacodynamic potentiation or antagonism. Knowledge of these pharmacological pathways has a profound impact on the quality of geriatric medical care. Polypharmacy is defined here as the use of five or more chronic medications, although some define it as the long-term simultaneous use of two or more drugs (Hanlon et al., 1997).

Medication mishaps in the elderly occur for numerous reasons. Multiple providers are often unaware of one another's new prescriptions or medication changes, especially after hospitalization. Older patients often have visual and cognitive impairments that lead to errors in self-administration. Patients may be unable to afford their medicines, so they take only some of what is prescribed based on how they are feeling. Functional illiteracy, which is not uncommon among the elderly, makes adherence to a medical regimen difficult. Finally, the average clinician is not knowledgeable about the vast number of possible drug interactions. Computer databases may not reflect all interactions or may show too many, making it difficult to identify the clinically important ones.

Changes in drug metabolism in the healthy elderly are often minimal and not clinically significant. However, the clinical impact of these changes in older people with kidney or liver disease can be considerable. Adverse drug reactions are two to three times more likely to occur in older patients. In general, drug absorption is complete in older persons, although it often occurs at a slower rate. Bioavailability, the fraction of an oral drug reaching the systemic circulation, depends on absorption and first-pass metabolism. Some drugs have increased bioavailability in the elderly (e.g., labetalol, levodopa, nifedipine, omeprazole).

Drug distribution can change due to age-related alterations in body composition. Weight is reduced, percentage of body fat is increased, and total body water and lean mass are decreased. Hydrophilic drugs have a higher concentration, since they are distributed in a smaller volume of body water. Lipophilic drugs have a larger volume of distribution and a longer half-life, since they are distributed in a larger volume of fat. Decreased albumin and other binding proteins may or may not affect the active drug (free) concentration.

Hepatic metabolism varies greatly among individuals based on age, sex, lifestyle, hepatic blood flow, presence of liver disease, and other factors. Although enzymes are usually unchanged by aging, many drugs are metabolized more slowly in older people due to a reduction in hepatic blood flow. Renal excretion of drugs diminishes by 35% to 50% due to decreased glomerular filtration rate (GFR). However, because of decreased muscle mass, measurement of serum creatinine does not reflect GFR, and many formulas to estimate creatinine clearance based on age, weight, and creatinine are inaccurate. Obtaining a 24-hour urine collection for creatinine clearance is the most accurate way to estimate GFR. The approach to medication dose reduction for geriatric patients is similar to that for patients with kidney dysfunction.

These age-related pharmacokinetic changes result in a longer drug half-life, a diminished clearance, and a longer time to reach a steady state. This is reflected in different serum levels for a given dose. Because of age-related pharmacodynamic changes, drugs have a different effect at the same serum level. For example, opioids have a greater analgesic effect, benzodiazepines have a greater sedative effect, and anticoagulants are associated with a higher risk of bleeding. Beta-blockers, in contrast, are less effective in the elderly (Swanger & Burbank, 1995).

Drug-Drug Interactions

When more than one drug is taken, age-related pharmacokinetic and pharmacodynamic considerations may complicate the drug interactions. Some interactions result in less drug being available through the mechanisms of impaired absorption, induced hepatic enzymes, and inhibition of cellular uptake. Impaired absorption can be due to binding by a concurrently administered drug, such as cholestyramine-binding digoxin and thyroxine. Administering these drugs two hours apart helps minimize this interaction.

Certain drugs induce hepatic metabolic enzymes in the cytochrome P-450 system (CYP). It may take weeks for these enzymes to become maximally active. Increased enzyme supply breaks down the active drug and causes less drug delivery. This occurs with drugs such as phenobarbital, rifampin, and phenytoin. Smoking and chronic alcohol use can induce similar effects. Hepatic enzyme induction results in lower levels of warfarin, quinidine, verapamil, cyclosporine, methadone, and many other medications. Inhibition of cellular uptake or binding may produce less drug availability. The interaction of clonidine and tricyclic antidepressants occurs through this mechanism, diminishing efficacy of both drugs.

Interactions that result in more drug availability include inhibition of metabolic enzymes and inhibition of renal excretion. Inhibition of metabolism leads to increased half-life, accumulation of the drug, and potential toxicity. Inhibition, unlike induction, can occur immediately. The recent understanding of the mixed function oxidase system and its isoforms allows prediction of potential interactions. For example, CYP-3A metabolizes cyclosporine, quinidine, lovastatin, warfarin, nifedipine, lidocaine, astemizole, cisapride, erythromycin, methylprednisolone, carbamazepine, and triazolam. Many of these medications are also inhibitors of the same CYP oxidase system. Cyclosporine can reach toxic levels if coadministered with erythromycin. A well-known interaction involving CYP-3A is cisapride with ketoconazole, which can produce a polymorphic ventricular tachycardia. CYP-2D6 is inhibited by quinidine and blocks the conversion of codeine to morphine, which makes codeine less effective (Rizack & Gardner, 1998). Inhibition of renal excretion causes more drug availability. For example, probenecid inhibits the excretion of penicillin. Clinicians can utilize this interaction to prolong the half-life of penicillin.

Pharmacodynamic interactions are those in which the actions of the different drugs affect the same end point. Warfarin and aspirin interact by increasing the likelihood of bleeding through separate pathways. Similarly, warfarin and nonsteroidal anti-inflammatory drugs make gastrointestinal bleeding more likely. NSAIDs also raise blood pressure and may undermine the action of antihypertensive agents.

Managing a Patient on Multiple Medications

With careful attention to the drug regimen, physicians, nurses, pharmacists, and older patients themselves can minimize and avoid serious drug-drug interactions. The health professional must be aware of all medications the patient is taking, including OTCs, vitamins, and herbal remedies. One method is to ask the patient to bring all his or her medications to each office visit ("the brown paper bag"). The practitioner may be surprised by the medications other providers have prescribed, by old prescriptions still being refilled at the pharmacy, and

by the range of OTC drugs the patient is using. One approach to avoid polypharmacy is to provide an older patient with a "medication passport" that lists all medicines he or she is taking. The patient shows this list to subspecialists and to the primary provider at each visit.

Start all new medications at low doses and increase the strength slowly ("start low, go slow"). Limit the number of medications to as few as necessary, and routinely review all drugs. Attempt to withdraw any unnecessary agents and, in some cases, consider nondrug therapy. Be cautious with newly released drugs. Report any adverse reactions, working closely with pharmacists, drug manufacturers, and public health departments. Investigate all complaints, as they may point to drug-drug interactions. Drug toxicity and drug interactions should be part of the differential diagnosis for altered mental status, fatigue, incontinence, gait disorder, and many other geriatric syndromes.

In an institutional setting, there should be a plan to monitor drug treatment programs on a regular basis. In the home setting, use of reminders, pillboxes, and other memory aids can minimize errors and enhance compliance. Attention to the patient's individual needs, with compensation for any specific functional or cognitive impairment, is critical. Family members and personal caregivers should be trained to monitor medication adherence and to report any difficulties.

Computer databases and other drug interaction resources should be used to check for known interactions. If the pharmacy has such a system, it should have a record of all the medications the patient is taking, even those supplied elsewhere.

Avoid giving a medication to counter the effects of another medication—for example, giving antiparkinsonian medication to treat the rigidity caused by antipsychotics or metoclopramide. The best approach is to lower the dose or substitute another agent for the offending one. Food-drug interactions, such as grapefruit juice potentiating buspirone or felodipine, can be checked using the databases and resources.

Polypharmacy in the elderly can be minimized, and drug-drug interactions can be avoided, with judicious use of medications while providing excellent medical care.

BARRIE L. RAIK

See Also
Polypharmacy: Management

Internet Key Words
Adverse Drug Reactions
Drug Interactions
Polypharmacy

Internet Resources
Physician's Desk Reference Online
http://www.pdr.net/index.html

POLYPHARMACY: MANAGEMENT

Polypharmacy, the excessive prescription and self-administration of medications, is an enormous problem among older adults (Stewart & Cooper, 1994). For most of the diseases associated with aging, medications are a mainstay of health care and are often the most cost-effective treatment choice. The number of chronic diseases and the number of medications taken by older adults are highly correlated. However, if more than five long-term medications are self-administered, older patients have increasing difficulty managing their treatment regimens and have poor adherence or compliance (Murray, Darnell, Weinberger, & Martz, 1986). Therein lies the conundrum of appropriately prescribing medications that are absolutely necessary but not overwhelming the patient with too many pills to manage throughout the day. In addition to noncompliance, polypharmacy includes increased risk of adverse effects, drug-drug interactions, and higher costs. Solutions to these problems include limiting the numbers of drugs prescribed by requiring a clear rationale for new prescriptions; increasing the awareness of all drugs patients are taking, including over-the-counter (OTC) and herbal supplements; providing patient education;

and using devices to assist patients with their medication regimens, such as medication cards and special drug packaging techniques.

Physicians and other prescribers should provide a clear rationale and therapeutic objectives for each drug prescribed. Far too often, a medication is prescribed for long-term administration with little attention to whether therapeutic goals are being met. With multiple prescribers, patients accumulate and aimlessly self-administer their medications without good integration or feedback to a therapeutic plan. A key rule should be to carefully withdraw any medications that have no clear rationale, as well as those not achieving the intended treatment goals.

Polypharmacy can be prevented by appropriate communication between patients and providers, awareness of all drugs in a patient's regimen, and careful supervision of drug administration in patients with cognitive impairment (see Weintraub, 1990). Multiple caregivers should be brought into the communication loop. Patients often tell a nurse or a pharmacist how they actually take their medications but tell physicians what they want to hear, namely, that they take their medications exactly as prescribed. Also, subtle drug-induced problems can be vividly apparent to one health professional or close caregiver but missed by another. For example, the patient's spouse or a visiting nurse might readily notice mild drug-associated cognitive impairment that could be missed by a primary care provider who visits infrequently. Enhancing communication among health care professionals requires access to a current, well-kept record of all the drugs the patient is receiving. Advances in electronic medical records and widespread Internet access could eventually provide protected access to a centralized drug record for all.

Physicians need to be aware of all the medications that they and other primary caregivers have prescribed. Although cumbersome, it can be helpful to have patients bring all their medications with them to office visits. In addition to prescription medications, physicians should ask their patients about OTC drugs, herbal medications, and nostrums and concoctions that patients might be taking without realizing their potential pharmacodynamic effects. This potpourri of medicaments can put the patient at risk for adverse reactions, counteract important therapeutic goals, or confound the patient's management of the essential medications being prescribed.

Patients must be educated about all the drugs they receive. Patient drug education is often haphazard and without assurances that patients are receiving important information about their medications. Patients need to know the generic or brand names of their medications, the disease or problem each medication is intended to treat, how to take the medication, whether there are any important food or nutrient interactions, and the commonly encountered adverse effects. This information should be provided verbally and in written form. Prioritizing medications for patients is also important, since they often believe that all their pills are equally important. A stool softener capsule may have the same priority in a patient's mind as a tablet for heart failure.

Patients with complicated medication regimens or cognitive impairment have difficulty remembering all the information about their medications. Such patients benefit from a medication card that contains the needed information with a picture of the pill or an actual pill affixed directly to the card. If patients or caregivers prepare such medication cards, they need to be reminded to modify the cards pursuant to changes in the medication regimen. During in-home visits to patients given medication cards, clinicians sometimes find that the patient is no longer taking the medication, pills are missing, or the patient's grandchildren are playing with the cards.

Patient behaviors and habits, such as hoarding or sharing medications with relatives or neighbors, should be considered (Darnell, Murray, Martz, & Weinberger, 1986). Outdated medications can be ineffective and, in rare instances, toxic. Alcoholism, smoking, and other drug abuse may result in preoccupations that supersede the prescribed regimen or cause memory impairments as to when critical prescription medications were last taken. Substance abuse may augment or nullify the metabolism of prescribed drugs by inducing or inhibiting gastrointestinal and hepatic enzymes.

Even when specially prepared medication cards are available, some patients are unable to self-administer their medications. Such patients may require special packaging of their medications. A small prospective, randomized trial demonstrated that "unit-of-use" packaging increased patient compliance and reduced the variability with which medications were self-administered (Murray, Birt, Manatunga, & Darnell, 1993). All the patient's medications to be administered in the morning were placed in a single "yellow dot" dosing cup; all the evening medications were placed in a "blue dot" cup. Least-square mean compliance and standard error over six months for patients (n = 9) receiving twice-daily unit-of-use packaging was 93 ± 2, compared with 83 ± 2 for patients (n = 10) randomized to conventional pill containers and twice-daily regimens, and 79 ± 2 for patients (n = 12) randomized to variable medication administration regimens and conventional containers. Special packaging techniques such as unit-of-use and compartmentalized containers can be very helpful, but they are labor-intensive and require commitment from the primary caregiver and pharmacist. Such medication packaging programs may not be readily available to patients, and when they are, patients will likely have to pay out of pocket for the service.

Medications are the central intervention for improving the survival and quality of life of older adults. Polypharmacy results because older patients often have multiple chronic diseases and disorders for which medications are the treatment of choice. Patients requiring multiple medications are at increased risk of adverse effects and drug interactions. OTC and herbal medications, as well as abused substances, often add to patients' confusion and can also produce potent pharmacological effects and interactions. When the number of medications exceeds the patient's functional ability to carefully and appropriately self-administer and manage those medications, noncompliance occurs. Communication among providers, patients, and their families, though difficult, is essential for the prevention of polypharmacy. Clear treatment goals for each medication should be apparent to patients and all caregivers. Selected patients may benefit from special medication packaging programs that facilitate medication management and administration. Systems to support safe self-management of medications by patients are needed but have not been forthcoming.

MICHAEL D. MURRAY

See Also
Medication Adherence
Over-the-Counter Drugs and Self-Medication
Polypharmacy: Drug-Drug Interactions

Internet Key Words
Adverse Drug Effects
Drug Therapy
Noncompliance
Polypharmacy

Internet Resources
American Nurses Association's Statement on Poly-
    pharmacy
*http://www.ana.org/readroom/position/drug/
    drpoly.htm*

National Institute on Aging Age Page
*http://www.aoa.dhhs.gov/aoa/pages/agepages/
    medicine.html*

Polypharmacy lecture by George J. Caranasos,
    M.D., University of Florida, Geriatric Educa-
    tion Center
*http://www.medinfo.ufl.edu/cme/hmoa2/poly/
    poly.html*

POVERTY

Although the poverty rate has decreased over the last three decades, poverty is a major concern for the elderly. For purposes of determining the federal poverty level, income is defined as all cash payments received by an individual or family from carnings, government benefits, or any other source. For a family unit of one, the poverty threshold is $8,240; for a family of two, $11,060 (Elderly Nutrition Project, 1999). This poverty line is adjusted upward each year as the cost of supporting a family rises. Longer life, higher costs for medical care, smaller and more distant families, and greater needs for daily living support constitute the poverty scenario for elders. Generally, elders prefer to maintain their dignity by hiding their financial needs rather than asking for help or burdening their caregivers. Even when elders agree to accept financial aid, they need to be guided through the intricacies of the application process. It is predicted that 40% of the elderly in America will experience poverty for at least one year between the ages of 60 and 90 (Rank & Hirschl, 1999).

Certain groups of elders have higher levels of poverty. Age, gender, widowhood, race, and ethnicity impact financial status. The poverty rate of elderly women is nearly twice the level of elderly men. However, Social Security keeps many women out of poverty. Even though women's total Social Security contributions are 38% of total contributions, due to income levels and years of work, women receive 53% of total benefits (Center on Budget and Policy Priorities, 1998). The poverty rate of elderly African Americans and Hispanics is more than twice the level of elderly whites in the United States (U.S. Bureau of the Census, 1997). A lifetime of low wages predisposes minority elderly to poverty. Elderly legal immigrants were severely affected by the passage of the 1996 Welfare Reform Act. Many are no longer eligible for safety-net programs such as supplemental security income (SSI), food stamps, and Medicaid. However, some states have adopted laws that maintain the previous coverage for legal immigrants (Yan, 1999).

Social Security and SSI are a safety net for older adults. Social Security provides at least 50% of the total income for at least 50% of elders in the United States. It is the major financial antipoverty program and successfully reduces the severity or impact of poverty among this group. With the introduction of SSI in the mid-1970s, the face of poverty changed. Elders who are not eligible for Social Security because of a limited work history may now access a minimum level of financial aid through SSI. In addition, SSI recipients receive state medical, prescription, and transportation services.

However, even after receiving federal benefits, 14.7% of women and 8.2% of men remain impoverished. SSI, Social Security, and private pensions kept the elderly poverty rate in the United States at 11.9% in 1997 (Center on Budget and Policy Priorities, 1998). Elders can access a variety of services through local and regional service providers. Agencies use a computer program that includes a comprehensive listing of federal and state programs. When workers input a client's financial, medical, and social needs, the program matches the person to appropriate resources and services.

The elderly are the most underrepresented group receiving food stamps (Elderly Nutrition Project, 1999). Many elders do not know that they are eligible for the program, have difficulty completing the required paperwork, and do not understand how food stamps or food cards are used.

Financial safety net programs have strict eligibility guidelines that leave many individuals ineligible for assistance. These are the near poor, or "tweeners"—those who, because they have some income and assets, are ineligible for many assistance programs. The near-poor income level is 25% above the poverty threshold. The proportion of elderly at this income level (19%) is higher than the percentage of poor in the general population (9%) (Food Research and Action Center, 1997). Widowed middle-class women who lose spousal pensions when their husbands die also find themselves in this financial situation (Villa, Wallace, & Markides, 1997).

Several programs exist to help the near poor. Elders who are "house rich and cash poor" can apply for reverse mortgage programs. These programs allow home owners to receive monthly checks based on the equity in their homes. Banks are familiar with the rules and regulations. Prescription drug plans for low-income elders are available in at least 15 states. Some states also offer to pay Medicare premiums for near-poor elders. The Medicare assistance program pays all or part of the Medicare premium based on an elder's monthly income and total assets. The Medicare office can assist with the application and processing (Health Care Financing Administration, 2000). The Senior Community Service Employment Program

(SCSEP) is a federal workforce initiative that offers job training to individuals age 55 or older who meet low-income guidelines. Supported by the U.S. Department of Labor, the SCSEP provides retraining, employment, and community service opportunities for eligible elders. Salaries are subsidized for work performed in nonprofit agencies.

Securing a place to live is a challenge for many elders due to the declining availability of affordable housing and the loss of medical subsidy programs for the poor and near poor. The Department of Housing and Urban Development Section 8 housing and voucher programs reduce the rate of poverty among the elderly (Shashaty, 1999). Although the number and types of housing subsidies and vouchers are decreasing, opportunities for public housing units are still available. Eligibility requirements for services differ for each authority. The local, county, or state housing authority can be contacted for information. Homeless elders living on the street have difficulty accessing services; they avoid shelters for fear of being victimized by other residents. Homeless elders have untreated medical conditions, suffer more than other homeless persons, and need assistance in negotiating the eligibility processes to obtain services (National Coalition for the Homeless, 1999). Some states provide limited subsidies for assisted living facilities for the poor and near poor. The state department of aging can be contacted for information and assistance regarding assisted living initiatives.

Legal Counsel for the Elderly, an affiliate of the American Association of Retired Persons, offers legal aid for low-income elders regarding landlord-tenant problems, eligibility for benefits, pension disagreements, and employment discrimination. This service is available to nonmembers. The Department of Veterans Affairs offers medical, pharmaceutical, residential, and social supports for veterans and their families. The Gray Panthers, a social action group, and Senior Action in a Gay Environment (SAGE) are also valuable resources.

ROSALIND KOPFSTEIN

See Also
Access to Care
Homelessness

Meals On Wheels
Medicaid
Social Security

Internet Key Words
Housing
Near-Poor
Poverty Rate
Social Security Income

Internet Resources
Administration on Aging
http://www.aoa.dhhs.gov/

American Association of Retired Persons
http://www.aarp.org

Center on Budget and Policy Priorities
http://www.cbpp.org/

Gray Panthers
http://www.graypanthers.org

Senior Action in a Gay Environment
http://www.sageusa.org

PRESSURE ULCER PREVENTION AND TREATMENT

In the past decade, the evaluation, treatment, and prevention of pressure ulcers have changed dramatically. This evolution has taken place for several reasons:

1. Explosive growth in the number and type of wound care products
2. Maturation of the field of pressure ulcer and wound care research
3. Examination of pressure ulcer development as a quality indicator by the Joint Commission on Accreditation of Healthcare Organizations (JCAHO) and the Health Care Financing Administration (HCFA)

Although pressure ulcers are not within the exclusive purview of geriatrics, their multifactorial origins and multidisciplinary treatment make them the quintessential geriatric syndrome. That wound care research comes from nursing, physical therapy, geriatric medicine, nutrition, dermatology, and surgical literature reflects how intractable and how challenging pressure ulcers can be.

Pressure ulcers develop because of the amount of pressure, the duration of pressure, and something particular about the person—his or her "innate tissue tolerance" (Braden & Bergstrom, 1987). It is difficult to capture or measure those intrinsic factors, so a variety of tools have been developed to assess an individual's risk for development of pressure ulcers. Presently, two risk assessment tools, the Braden and Norton scales, are widely used (Bergstrom, Allman, & Carlson, 1992). These scales weigh slightly different factors: The Norton scale includes physical condition, mental condition, activity, mobility, and incontinence; the Braden scale uses sensory perception, moisture, activity, mobility, nutrition, and friction or shear. Low scores indicate that a person is at high risk for developing a pressure ulcer. Other scales are under development. Most institutions use these scales to target patients for early intervention through pressure relief, friction reduction, and nutritional supplementation.

Prevention techniques may seem onerous at first, but in the long run, they are less costly, less time consuming, and less emotionally draining than the lengthy process of treating a pressure ulcer. Both hospital staff and family caregivers must learn the correct techniques for positioning the patient, and health professionals should take the time to observe how the patient is being moved:

- Is a drawsheet or other lifting device being used correctly to minimize friction and shear when the patient is moved?
- Is the patient being repositioned frequently enough? Every two hours is the standard recommendation for most individuals in bed, but if stage I ulcers are appearing, even this schedule may be inadequate.
- Is the patient optimally placed to minimize pressure on bony prominences? For example, when lying on the side, is the patient positioned on the buttocks rather than the hip? Are the heels elevated?

- Do caregivers recognize that ulcers can occur anywhere and in any position? Are they providing adequate pressure relief through cushioning on support surfaces and repositioning every hour when the patient is out of bed? Caregivers may be meticulous in their attention to the patient's skin care needs when he or she is in bed but may leave the patient in a chair for hours at a time, under the false assumption that pressure ulcers (bedsores) will not occur when the patient is sitting up.
- Is the patient capable of initiating any spontaneous movements? If so, caregivers and aides should encourage the patient to reposition himself or herself every 15 minutes or whenever possible.

Pressure-relief devices, such as a foam overlay or air mattresses and cushions, may be appropriate, depending on the setting and the patient's level of risk. For patients who are malnourished, a dietitian or other nutrition specialist can recommend the appropriate fluid and nutrient intake, determine the need for supplementation with vitamins or minerals, and assess adherence; families must work with the health care team to implement the dietary advice. Families and clinicians should determine how well the patient is eating; whether the patient is having difficulty eating because of poor dentition, constipation, dysphagia, dementia, or visual impairment; or whether other illnesses, such as depression, may be affecting the patient's desire for food. With some creativity, patients' nutritional status can improve without having to resort to tube feedings.

The National Pressure Ulcer Advisory Panel (NPUAP, 1998b) recommends the following staging criteria for pressure ulcers:

- Stage I: an observable pressure-related alteration of intact skin as compared with the adjacent or opposite area on the body. Changes may include one or more of the following: skin temperature (warmth or coolness), tissue consistency (firm or boggy), and/or sensation (pain, itching). The ulcer appears as a defined area of persistent redness in lightly pigmented skin; in darker skin, the ulcer may appear with persistent red, blue, or purple hues.
- Stage II: partial-thickness skin loss involving epidermis and/or dermis.
- Stage III: full-thickness skin loss involving damage or necrosis of subcutaneous tissue that may extend down to, but not through, underlying fascia.
- Stage IV: full-thickness skin loss with extensive destruction, tissue necrosis, or damage to muscle, bone, or supporting structures.

Staging is a method of classifying the amount of tissue destroyed. Pressure ulcers can only progress to higher stages, not regress to lower ones. A healing ulcer does not change from a stage III to a stage II; it is always a stage III wound. Tools specific to the healing process can be used to describe changes in size, exudate, and signs of tissue healing. Areas that have healed are always at higher risk for redevelopment of ulcers.

Pressure relief is essential to both prevention and treatment. The techniques and concerns are the same as those for prevention: Once an ulcer has occurred, it is important to avoid any pressure on the area if possible. The clinician should also determine whether family or staff reeducation about pressure relief is necessary.

In addition to adequate turning and positioning, pressure-reducing support surfaces are essential for the management of pressure ulcers. These include foam mattresses; special pads or mattresses of air, water, or gel; air flotation beds; and air fluidized beds. The HCFA divides these surfaces into three separate groups, and each has specific criteria for Medicare Part B reimbursement in the home setting (Wound, Ostomy and Continence Nurses Society, 1998). Costs can vary more than 10-fold. To determine the most reasonable support surface for a patient, clinicians should take into account the location and stage of the ulcers, the effectiveness of pressure reduction, the weight of the mattress or bed, and the noise that motors can make.

As in the case of pressure ulcer prevention, adequate nutrition is a must for treatment. The clinician must take the new ulcer into account when making recommendations about protein and caloric needs: How exudative is the wound, and how much protein loss is occurring? Is the wound of sufficient size to cause extra catabolic stress on the patient? Do antibiotics, pain medications, or limitations in positioning of the patient have any impact on the patient's ability to eat? Is there a need to supplement vitamins?

The key to wound care is the establishment and maintenance of a moist, pink, granulating wound bed. This necessitates removal of all nonvital tissue, including eschar (the brown or black dead tissue that can cover the wound), slough (the yellow or white, often fibrous material that clings to the wound bed), and devitalized connective tissue, which may appear dusky red or frankly necrotic. Debridement can be surgical, mechanical (using irrigation or swabbing with gauze), autolytic (taking advantage of the body's own proteins by using synthetic dressings), or enzymatic (using commercially prepared agents such as collagenase) (Bergstrom, Bennett, & Carlson, 1994).

The most common mistake that wound care novices make is inappropriate wound assessment. Visual inspection alone is inadequate. Darkly pigmented skin can mask stage I ulcers. Soft, brown eschar can also superficially resemble hyperpigmented, normal skin, but eschar feels devitalized, spongy, and cool, unlike healthy tissue. Palpation may be able to detect changes in skin temperature compared with the surrounding tissue. Palpation of stage III and IV wounds also enables the clinician to determine the extent of undermining, lyse adhesions, probe for abscess pockets and tunneling, and feel for exposed bits of bone. The clinician should also assess the odor of the wound; foul-smelling exudate may be the first sign of abscess or necrosis (Maklebust & Sieggreen, 1996).

An extensive description of the products available for wound care is available in the literature (Hess, 1999). However, despite the wide variety of materials available, there are few studies comparing products, and some clinicians develop preferences based on familiarity with a limited number of products rather than research results.

Wounds heal most effectively when moist. To maintain the proper amount of moisture, wounds usually require some form of hydration. The most popular hydrating agents are wound hydrogels, which can be applied directly to shallow wounds or can be rubbed into moistened gauze that is then used to pack the wound. Simple gauze moistened with normal saline alone is cheap and effective, but it works best if changed several times a day and cannot be allowed to dry out; such labor-intensive regimens are rarely feasible unless the patient or family is willing and able to change dressings.

Hydrocolloid dressings work well for stage I and II wounds, and they can often be left on the wound for many days. However, they can damage friable skin, especially if they must be removed daily because of exudate or soilage. Hydrocolloid dressings may also be less effective in diabetic patients.

Highly exudative wounds require hydrophilic dressings, and many are now available. These include sodium chloride–impregnated gauze, foam dressings, alginates, and many of the mixed dressings. These dressings can often be left on for several days, depending on the amount of exudate.

A variety of tools are available to assess wound healing. One such tool is the Pressure Ulcer Scale for Healing (PUSH), launched by the NPUAP (1998a). It provides a scoring system for healing and cues the clinician to reassess the care plan if healing is not occurring appropriately. The clinician generates a score based on ulcer size, exudate, and tissue type and follows the score over time to determine whether the wound is healing appropriately.

Good wound care requires frequent assessment, pressure relief, adequate nutrition, maintenance of cleanliness, and readiness to try a different wound care product if the regimen is not working. Inadequate healing requires reassessment of all factors: Is there sufficient pressure relief? Is the wound truly clean, or is there still dead tissue or bacterial colonization? Is the patient receiving adequate nutrition? Is the wound being dressed properly?

With proper care, most pressure ulcers heal. But given how costly they are and how painful to endure, prevention is paramount.

EUGENIA L. SIEGLER
ELIZABETH A. AYELLO

See Also
Pressure Ulcer Risk Assessment

Internet Resources
American Academy of Wound Management
http://www.aawm.org/

National Pressure Ulcer Advisory Panel
http://www.npuap.org

Wound Care Communications Network
http://www.woundcarenet.com

Wound, Ostomy and Continence Nurses Society
http://www.wocn.org

PRESSURE ULCER RISK ASSESSMENT

Pressure ulcers are a serious health problem in the elderly and occur in all health care settings. Healthy People 2010 created a national indicator related to pressure ulcers. Objective 16 is to reduce the proportion of nursing home residents with a current diagnosis of pressure ulcers. The target is 8 diagnoses per 1,000 residents, reduced from the current prevalence of 16 per 1,000 nursing home residents.

Braden Scale

One of the most widely used risk assessment tools for the early identification of persons at risk for pressure ulcers is the Braden scale (Agency for Health Care Policy and Research, 1992; Braden & Bergstrom, 1987). The scale consists of six factors associated with the development of pressure ulcers: mobility, activity, sensory perception, moisture, nutrition, and friction or shear. The scores for each of the six factors are added to obtain the total risk score. The risk cutoff score for the general population is 16; scores of 16 or lower indicate high risk for developing a pressure ulcer.

The changes inherent in aging skin and the dark pigment in black and Latino patients require that scores be adjusted for these patient populations. A cutoff score of 17 to 18 has been proposed for

older patients and for black and Latino patients (Lyder et al., 1998). The risk assessment score is an adjunct to nursing judgment about when to implement pressure ulcer prevention strategies.

Risk Assessment Frequency

Clinical practice guidelines of the Agency for Health Care Policy and Research (AHCPR, 1992) recommend pressure ulcer risk assessment on admission to a facility and then periodically, based on the clinical setting and changes in the patient's condition. In acute care, reassessment should be done on most patients within 48 hours after admission or after 24 to 48 hours for intensive care unit patients. Reassessment should also be done when major changes occur in the patient's condition. In long-term care, reassessment should be done weekly for the first four weeks after admission and then quarterly. In home care, reassessment should be done at every visit by the nurse, based on the patient's illness severity.

Prevalence and Assessment of Stage I Pressure Ulcers

Several national surveys (Barczak, Barnett, Childs, & Bosley, 1997) support the idea that the largest percentage of pressure ulcers are stage I or II. Color, especially erythema, has been the gold standard used by clinicians to identify stage I pressure ulcers. Patients with darkly pigmented skin have the lowest prevalence of stage I pressure ulcers and a significantly higher prevalence of higher-stage, full-thickness ulcers. Because intact, darkly pigmented skin does not change color (does not blanch) when pressure is applied over a bony prominence, reliance on the classic NPUAP definition of a stage I pressure ulcer as "non-blanchable erythema of intact skin" might account for missed identification of early skin injuries in persons with darkly pigmented skin.

Rather than redness, the new indicator for assessing stage I pressure ulcers in persons with darkly pigmented skin would be darkening of the client's skin tone from the usual skin color, which may present as blue, gray, or purple (Bennett, 1995; Henderson et al., 1997). Adequate light is essential when assessing clients with darkly pigmented skin. Natural or halogen light sources are better than fluorescent lights, which cast a bluish hue and can interfere with detection of stage I pressure ulcers. Clinicians should avoid wearing tinted glasses that alter their ability to make color assessments.

Clinicians also need to include factors other than color, such as the temperature of the skin over bony prominences, to ascertain differences from the surrounding skin. Initially, an area of early skin injury feels warmer. As the capillaries collapse as a result of pressure and the tissue dies, the skin temperature cools. Since pressure ulcers occur most frequently on the sacrum, and secondarily on the heels, clinicians should pay particular attention to these areas.

Tissue consistency may also be an indicator of a stage I pressure ulcer. Clinicians should palpate for a firm or boggy feel. The revised NPUAP 1998 stage I definition also alerts clinicians to include sensation as an indicator. Sensation may present as pain or itching.

As soon as a stage I pressure ulcer is suspected, the patient should be positioned so that he or she is not lying or sitting on the ulcer. Pressure ulcer prevention guidelines as recommended by the AHCPR and NPUAP should be implemented.

Nutritional Assessment

Malnutrition and nutritional deficiencies have been linked with pressure ulcer formation. AHCPR guidelines recommend an abbreviated nutritional assessment at least every three months for patients at risk for malnutrition. Clinically significant malnutrition is serum albumin less than 3.5 g/dL, total lymphocyte count less than 1,800/mm^3, or more than 15% decrease in body weight (AHCPR, 1994). Clients should also be assessed for oral and cutaneous signs of vitamin and mineral deficiencies. For example, extreme transparency of the skin on the hands, cellophane or tissue-paper skin, and purplish blotches on lightly traumatized areas (due to capillary fragility and subepithelial hemorrhage) reflect vitamin C deficiency. Superficial flaking of the epidermis suggests a deficiency of essential fatty acids and, in nonpigmented skin, vitamin A deficiency. Dry, reddened skin around the nose and eyebrows is a sign of zinc deficiency. Adequate nutrition includes appropriate protein, calories, vitamins, minerals, and fluids and is important for clients at risk of developing pressure ulcers, as well as those for whom wound healing is the goal (Ayello, Thomas, & Litchford, 1999).

A comprehensive plan of care must include assessment of the risk for developing pressure ulcers. The Braden scale is a valid and reliable instrument that should be used on admission and periodically thereafter, based on the particular clinical setting and the patient's condition. Indicators for early pressure ulcer injury (stage I) need to be modified for clients with darkly pigmented skin. Nutritional assessment is an important part of risk assessment and for healing of pressure ulcers.

ELIZABETH A. AYELLO

See Also
Integumentary System
Pressure Ulcer Prevention and Treatment
Skin Issues: Bruises and Discoloration
Skin Tears

Internet Resources
American Academy of Wound Management
http://www.aawm.org

Healthy People 2010
http://www.health.gov/healthypeople/

National Pressure Ulcer Advisory Panel
http://www.npuap.org

Wound, Ostomy & Continence Nurses Society
http://www.wocm.org

PRIMARY CARE PRACTICE

The concept of primary care for patients who are beyond their reproductive and working years is relatively new for organized medicine. The blossoming number of healthy, active elders living into their 80s, 90s, and occasionally 100s is a phenomenon of the late 20th century and one that is certain to increase in the next millennium (Ham & Sloane, 1997). Although chronic diseases remain prevalent in later life, successes in treatment have allowed patients with a variety of disorders to reach old age. The quality of life in those later years becomes the overarching consideration for older patients, their caregivers, their providers, and health policy planners. Optimal quality of life throughout the life span is the goal of primary care. With the increase in the older population, the primary care of older persons will occur mostly in traditional primary care offices (family practice, general internal medicine). Patients in special situations (such as those with disabilities in three or more areas of basic function or in one area with behavioral disturbances) may benefit from transfer to a geriatric primary care practice.

Quality of life for most patients is measured in terms of functional capacity: which daily tasks they can accomplish independently, with assistance, or not at all. Functional assessment is an integral part of all geriatric care, in every site of care. Accurate functional assessment leads to treatment plans that are highly satisfying to patients and families.

Ideally, geriatric primary care remains forward looking. For most patients, preservation of autonomy throughout the life span is a crucial part of quality of life (Forciea & Lavizzo-Mourey, 1996). Functional assessment leads to the identification of problems that often can be ameliorated. Prevention of new illnesses can be attempted. Counseling and preparation can lead to successful outcomes in dealing with the expected challenges of this stage of life.

Access to the office is facilitated if it is close to mass transit, since elders' driving skills may become impaired. The office telephone should be answered by a receptionist. If electronic menus are unavoidable, the message should include no more than three options, and "stay on the line for help" should always be included as a choice for those with rotary phones or those who are confused.

Office hours during daylight are ideal, since patients' vision may be limited, they may fear for their physical safety after dark, or travel fees for seniors may be reduced in off-peak hours. Conversely, if a family member needs to accompany the patient, hours in the evening or on weekends are ideal.

Offices need to be wheelchair friendly; most sites have appropriate ramps and elevators for entrance. Offices should have adequate waiting and examination room space to accommodate patients in wheelchairs, and doors must be wide enough to allow wheelchair entry. Special examination tables that are wider and elevating are optimal, allowing patients to transfer from chair to table without risk of falls. If expense precludes having these special tables in all exam rooms, at least one room should be equipped, and older patients should be seen in that room. Step stools with attached handrails make the ascent to and descent from the exam table safer for older patients. Office chairs should have arms and be of sufficient weight to withstand "pushing off" by patients as they attempt to stand.

Since the odds of an older patient having an acute event are high, the practitioner should be available to provide care or offer firm links to care at a variety of sites. Most physicians still offer inpatient hospital care either by themselves or by their practice group. Many hospitals are developing special acute medical-surgical floors specifically designed and staffed for older inpatients. Patients cared for on these floors usually exhibit less functional decline during hospitalization and have shorter lengths of stay. Acute medical events in older patients often require rehabilitation, either in a specialized facility or in a nursing home. A new concept in the United States is the outpatient day hospital for short-term outpatient rehabilitation (Yoshikawa, Cobbs, & Brummel-Smith, 1998). Firm links to a home health care agency for visiting nurses and at-home physical and occupational therapy are essential. Social work services should also be available through these agencies. During the duration of the caring relationship, the provision of house calls by the primary care provider may be required. Similarly, admission to a nursing home may be necessary; the presence of a trusted primary care provider in the facility may ease the pain of this transition for patient and family. Some managed care companies are offering financial incentives for primary care physicians to remain with their patients after transfer to a long-term-care facility.

The diagnosis or treatment of illness in older patients often requires consultation with specialists. In an ideal situation, specialists are available on the same site as the primary care office. Older patients may have difficulty obtaining transportation to another site or become confused with multisite care. Most geriatric primary care practices offer advanced practice nursing, social work, rehabilitation, and psychiatry or neurology care as part of ongoing team care (Hazzard, Blass, Ettinger, Halter, & Ouslander, 1999). The identification of consulting specialists with a special interest in and expertise with older patients is critical to useful consultation.

Older patients require longer visits than are standard for general practice, because they have more active problems, take more medications, and frequently require conversation with a secondary informant (e.g., spouse, adult child, caregiver). Most geriatric primary care offices use a standard of 60 minutes for new patients and 20 to 30 minutes for return visits. When visits of this length are impossible, a variety of strategies can be used: A complete new patient assessment may be developed over several visits, or previsit questionnaires or telephone interviews may be used to collect useful information in the medical history. Some practices use "group visits" of patients with similar problems on a given half day. Education and counseling are facilitated, and the group members can support one another. Geriatric primary care practices have found support for unbillable services such as nursing evaluation and teaching and social work assessment and case management from a variety of sources. Hospitals may help support the practice as a source of patients for inpatient care and specialty referrals. Some practices have generated endowment income, and others have the support of grants for innovative care or teaching.

The initial visit establishes the foundation for ongoing care. It is essential that the patient feel welcome, understood, and secure. Although attention to the office structure, staffing, and interdisciplinary links may facilitate these feelings, the initial encounter with the provider is the keystone of the experience. Scheduling adequate time for the encounter allows the visit to be relaxed for both the patient and the provider. Although the principles of the medical history, physical examination, and medical decision making are preserved in the evaluation of older patients, certain areas deserve special emphasis.

Information gathering may be more complicated in older patients; family or caregivers accompanying the patient may have critical information to provide. One strategy is to invite everyone to the exam room to initiate the visit. The patient is asked if the family member may stay for the history (after assuring the patient that he or she will have private time with the provider) and if the family member may be addressed privately. The previsit telephone call may have prepared the patient and the family for this conversation.

The social history is of paramount importance in older patients. It is recommended that providers ascertain information about the living situation (apartment, house, stairs, bathroom and bedroom location), support systems available (family, neighbors, agencies), and occupational history at the beginning of the history. In the private conversation between provider and patient, questions about possible mistreatment or fear of family or caregivers should always be asked.

A functional assessment should always be done by self-report, family report, or direct observation. A complete medication review should be attempted. Patients should be instructed to bring all medications with them to the visit. Ideally, the patient should demonstrate an ability to open the bottles and recognition of the purpose of the pills and the dosing schedule.

A dialogue about advance directives should begin at the initial visit. Patients should be encouraged to bring in existing documents so that they can be reviewed and kept on file by the medical doctor. Patients without living wills and durable powers of attorney for health should be given information about them for discussion at future visits. Patients should be encouraged to select surrogate decision makers and review their wishes with family members.

The review of systems should pursue common syndromes seen in older patients: incontinence, memory loss, falls, immobility, vision and hearing loss, and dentition changes, to name only a few.

The physical examination may need to be performed over several visits due to time constraints or patient fatigue. Attention to postural blood pressure changes, alterations in the skin, and musculoskeletal and neurological exams is especially important.

Patient instructions should be clear and succinct. Not all problems need to be addressed and solved during the initial visit, although the patient should be reassured that all complaints will eventually receive attention. Written instructions for new medications or exercise may facilitate compliance.

The schedule of return visits should be frequent enough to complete initial evaluations, monitor chronic problems or new therapies, and reassure the patient or family. Education about illnesses or health promotion may be scheduled with a member of the team, if available. Interdisciplinary meetings and plans may be scheduled around return visits.

Primary care visits for older patients should allow the collection of full information regarding their functional and social status and active medical problems. Prevention of unnecessary disability is crucial. Preparation of the patient for the challenges of late life should be ongoing. Multiple medical problems should be managed to allow maximal function and independence throughout the life span.

MARY ANN FORCIEA

See Also

Best Practices in Geriatric Care

Multidimensional Functional Assessment: Overview

Internet Key Words

Ambulatory Geriatric Medicine

Continuum of Care

Primary Care

Internet Resources

American Geriatrics Society
http://www.americangeriatrics.org

Institute on Aging, University of Pennsylvania
http://www.ageweb@mail.med.upenn.edu

PRISON-RESIDING ELDERS

It is hard to create just programs in an unjust society. Nowhere is this maxim easier to support than in the context of elderly persons in prisons and jails. A legal activist founded Project for Older Prisoners in response to the story of one prisoner at Louisiana State Prison at Angola—a 50-year-old drifter with a low IQ who had been sentenced to 30 years in a maximum-security prison for stealing $117 and a cherry pie from a convenience store. At 67, the prisoner was suffering from bleeding ulcers and emphysema. There were no special programs for this sort of person. Indeed, in Angola in 1996, more inmates died in prison than were released on parole. Despite the fact that the popular image of prisoners is young and tough, the demographics of the criminal justice system are challenging that image and changing the reality.

The goal of health care is to diagnose, comfort, and cure. The goal of corrections is to confine and punish. These are mutually incompatible goals, yet they must coexist in the world of correctional health care. It is difficult to gain the trust of young inmates who come for intermittent illness, disease, or trauma. For most of these inmates, however, the interaction with the health service is brief and, given the self-limited nature of most ambulatory care, successful. The task of caring for elderly patients who are increasingly frail and needy and whose illness is likely to lead to death presents far greater obstacles. Yet this is the most rapidly increasing cohort of those being held behind bars.

It is difficult to provide sensitive, high-quality health care in correctional settings. The budgets are never adequate, and the facilities are hard to renovate as notions of quality change. Inmates come to the infirmary for many reasons: to address symptoms, to avoid onerous tasks, to have the opportunity to speak to someone who is not part of the guard or security staff. All these ancillary uses put added stress on inadequate staff.

The graying of the prisons is but the last in a line of events that have marked the consequences of America's policy of incarceration. In 1998, the United States incarcerated more than 1.8 million persons in city and county jails, in state prisons, and in the lockups and prisons of the federal Bureau of Prisons. By the end of that year, state prisons were operating at 13% to 22% above capacity, and federal prisons were operating at 27% above capacity.

Much of this growth in the prison population is the result of tougher sentencing laws, minimum mandatory sentences, and "three strikes and you're out" laws that sentence an inmate to life without parole for even minor charges. The impact on communities of color, which disproportionately account for many of the prison population, is extraordinary. The Sentencing Project in Washington, D.C., estimates that one out of three black men between the ages of 19 and 29 is under the jurisdiction of the criminal justice system and is in jail, in prison, or under the supervision of probation or parole. Overlooked in these statistics are special populations who are likely to need more than the typical young inmate requires. These are elderly inmates sentenced to life in prison or merely caught in the trap of determinate sentences and mandatory minimum sentences so that their aging, and likely their death, will occur in the prison system.

Chronological age of inmates may not be a valid predictor of physiological age; many prisoners tend to exhibit the chronic illnesses of older-aged persons well before their free-world cohorts do. The reasons for this are not obscure. Inmates tend to be poor and did not have easy access to medical care while in the community. Socioeconomically, they have had less access to housing and jobs and the sorts of lifestyles that keep one healthy. While in prison, the constant stress, isolation, loneliness, cigarettes, enforced idleness, heavy food, and lack of real relaxation all take their toll.

The U.S. Department of Justice found that by 2005, inmates 50 and older will make up 16% of the prison population, up from 11% today (Baker, 1999). These inmates will exhibit all the conditions of old age that exist outside of prisons but will do so under harsher conditions and in less sympathetic surroundings. Cancer, heart disease, congestive heart failure, chronic obstructive pulmonary disease, emphysema, Alzheimer's disease, and strokes will occur in this population. But the ability of individual inmates and of the health care providers to manipulate the environment to lessen the debilitating effects of diseases is hampered in a correctional setting.

Many of the effects of chronic disease can be moderated by regular medical support and changes in the physical environment, such as air quality and special diets. None of these effective, low-tech solutions is likely to be available in correctional settings. Creating a less dusty environment helps emphysema; limiting long walks helps the symptoms of heart failure; eating special diets that are rich in carbohydrates may mitigate some of the discomfort of chemotherapy for cancer. Prison buildings are large and hard to clean, and it is difficult to control temperature; living quarters are often far from the dining and recreation areas; kitchens are institutional in the extreme and are not designed, equipped, or inclined to meet the special nutritional needs of any inmate.

The processes and components of life lead to and require interaction with others. Despite the highly mobile nature of society, successful aging generally envisions some relationship with children, grandchildren, and others who have contributed to the fabric of one's life. The sharing, stories, memories, and reconciliation that most older people experience are denied to prisoners.

Finances are also an issue. Inmates in prisons and jails are not eligible for either Medicaid or Medicare. State and federal appropriations must fund all their needed and increasingly expensive medical care. The numbers are substantial. The annual cost of care for the average inmate is $20,000, whereas the cost for an elderly inmate is between $60,000 and $69,000. Most states have not yet begun to analyze and project the need for special facilities and staffing when standard prisons become unable to care for demented and disabled inmates.

In the world outside of correctional institutions, the last decade of the 20th century made substantial progress in accommodating the needs and wants of dying patients and their loved ones. Physicians, nurses, and social workers have enhanced their communication skills, facilitating open and honest discussion of diagnoses and prognoses, even when the choices are difficult and the future is uncertain and not promising. Enhanced communication depends on the implicit assumptions of trust that largely mirror the commitments and concerns in the doctor-patient and provider-patient relationship. The understanding of both patient and family is that the medical care system will use all its skill and expertise to support and extend life, but when that is no longer possible, it will use its talents to control symptoms and ensure a comfortable death. The willingness of medicine to acknowledge its limits has been the basis for discussions of decent end-of-life care. But basic to that dynamic of care is the understanding that all that could be done to extend a patient's life has been done.

The discussion in prisons is very different. Many prisons are the subject of litigation directed at the quality of medical care and whether the care provided meets the constitutional standard. Prisoners have a constitutionally protected right to medical care. The Supreme Court in *Estelle v Gamble* (1976) reasoned that to put prisoners behind bars, where they could not secure their own care, and not to provide care could result in the sort of torture the Eighth Amendment was designed to prohibit. The standard crafted by the Court stated that inmates must be provided with care that is not "deliberately indifferent" to "serious medical needs." This language has been the source of thousands of lawsuits by inmates challenging the adequacy of the care they received and whether the care provided meets the constitutional standard. To date, few cases have raised questions about the care provided to elderly inmates.

Basic to end-of-life care and to caring for those with chronic disease that will lead to death is the notion of trust and the forging of alliances. These sorts of relationships are hard to create in correctional settings. It is difficult to bond with a health care professional when one is accompanied on a trip to the hospital by a prison guard whose task is to ensure that one's shackles are properly fastened. Medical relationships never take place outside the accepted bounds of security policy.

One answer to this dilemma would be the use of compassionate release as a way of placing persons in community settings, programs, and funding streams. This alternative, permitting a prisoner to die at home or in a noncorrectional institution, has little support from local correctional and judicial authorities. Some states and the Federal Bureau of Prisons do not entertain the possibility of compassionate release as a policy option. For those that do, it requires the concerted efforts of health care staff, the district attorney's office, the local parole board, and the community facility that would accept medical oversight. The difficulty of getting these disparate services to work together has made compassionate release a theoretical rather than a real possibility in most areas of the country. In New York City, the authorities wait until inmates are near death to release them, to be sure that they will be unable to commit any more antisocial acts. As a result, most inmates who are released are far too sick to be cared for by family and spend their last days in a long-term-care facility. States are beginning to deal with the issue of a graying prison population. In 1999 Virginia opened a special facility for ailing and aged prisoners and Washington opened a similar facility in 1997 (Baker, 1999).

Adequate care for elderly inmates with chronic health problems must address the following:

- Correctional and medical staff should be taught to regard and treat terminally ill inmates as patients approaching the end of life, not as individuals for whom suffering and dying are yet another appropriate phase of punishment.
- Palliative care protocols should be in place to ensure that the care team can accurately assess the level of physical discomfort and provide an effective response. In this circumstance, an inmate's history of drug use should not disqualify him or her from receiving adequate analgesia and even opioids when required for pain control.

- The prison formulary should stock adequate pharmaceuticals, keeping them secure but available for the effective management of pain and other symptoms.
- Special foods and fluids should be made available on request, and staff should assist those who cannot manage on their own.
- Visiting rules should be relaxed to permit family members, friends, and other loved ones increased access to the prisoner.
- Chaplains and other spiritual advisers, including other inmates, should also be permitted enhanced access to the inmate.

Accommodating the needs of elderly and dying prisoners requires a sea change in the attitudes and approaches of corrections officials. At some point in the process of aging, most prisoners will acquire a chronic illness and lose the ability to harm others. At some point, their designation as the punished will need to make room for the label of the dying. This is a particularly difficult shift in focus, given the structure and management of most correctional institutions. It is the bare minimum that a compassionate society must provide.

NANCY N. DUBLER

See Also
Advance Directives
Autonomy

Internet Key Words
Incarceration

PROGRAM OF ALL-INCLUSIVE CARE FOR THE ELDERLY (PACE)

The Program of All-Inclusive Care for the Elderly (PACE) is a prepaid, capitated managed care program in support of comprehensive primary, long-term, and acute care delivered by not-for-profit or public providers. PACE is intended for frail, disabled elderly patients: Enrollees must be at least 55 years old (although participants' average age is about 82), live in a PACE catchment area, and be state certified as eligible for nursing home care. PACE's goals are to maximize each participant's autonomy and continued community tenure and to provide quality care at lower cost to Medicare, Medicaid, and private payers than would be expected under alternative financing and service arrangements. PACE sites attempt to achieve these goals by providing comprehensive assessment and intervention in community settings (including PACE centers, program-related housing facilities, and participants' homes) and by minimizing inappropriate, avoidable, and expensive use of institutional (hospital and nursing home) care that would have an adverse fiscal impact on the site (Bodenheimer, 1999; Eng, Pedulla, Eleazer, McCann, & Fox, 1997).

For those who are eligible, the basic principles of PACE are (1) enrollment is voluntary, not affected by changes in health status, and continues as long as enrollees desire; (2) the Medicare and Medicaid capitation is considered payment in full (there are no deductibles or co-payments); and (3) the local PACE organization assumes full financial risk for enrollees' care without limitation on service scope, amount, or duration. The financial underpinnings of PACE are complex and evolving. To date, the Medicare "share" of PACE is the monthly capitation rate paid by the Health Care Financing Administration (HCFA) to each local PACE provider. The rate is based on the locally adjusted per capita cost calculated by the HCFA for HMO reimbursement using a uniform frailty-adjustment factor. The payment methodologies are still under development, but PACE providers in the future will continue to receive their capitated Medicare support through such adjustments of rates. Under Medicaid, the monthly capitation is negotiated and annually contracted between the local PACE provider and the state. Generally, states base their payments on their reimbursements for a nursing home–eligible population, including both nursing home residents and community-based long-term-care recipients. Combined Medicare and Medicaid capitation payments currently average approximately $3,200 to $3,400 per participant per month (White, 1998).

Ninety-five percent of enrollees have been dually eligible (i.e., recipients of both Medicare and Medicaid). The National PACE Association—a nonprofit voluntary organization of PACE providers—has identified broadening PACE's accessibility to middle-class enrollees as one of several strategic priorities.

At the core of PACE services is the interdisciplinary team, consisting of professional and paraprofessional staff. Each PACE team includes a primary care physician, nurse, social worker, physical therapist, occupational therapist, recreation therapist or activity coordinator, dietitian, PACE center supervisor, home-care liaison, health workers or aides, and drivers. To integrate care provision, the team assesses participants' needs; develops care plans encompassing all Medicare- and Medicaid-covered services, including institutional, home, community, and end-of-life-care; and directly delivers all or most services. Oversight of progress and delivery of care is facilitated by PACE's emphasis on day-center attendance, but congregate housing options and in-home services have been developed and integrated into care planning and provision. Some PACE organizations allow for limited, team-supervised care provision under contract with independent care providers. Recently, in part to address a common reason for failure to enroll in PACE, some sites have contracted with independent primary care physicians to allow them to continue to follow their patients after PACE enrollment. Under this arrangement, PACE provides the participant and his or her physician with all team services, including care coordination and nurse practitioners or physician assistants. Social and medical services are delivered primarily in PACE's adult day health centers (center-based services must include primary care services, social services, restorative therapies, personal care and supportive services, nutritional counseling, recreational therapy, and meals). These are supplemented by in-home and referral services in accordance with the participant's needs.

The PACE model was developed over many years in San Francisco by On Lok, a not-for-profit, community-based organization formed to address geriatric long-term-care needs in Chinatown. PACE was initiated in the late 1980s as an HCFA demonstration to replicate and evaluate the On Lok model in other locations (Eng et al., 1997). By late 1997, 12 fully capitated programs were operating and providing care in nine states under demonstration (waiver) status. In part because of evaluation results suggesting positive PACE impacts on participant outcomes, participant satisfaction, and Medicare costs (Chatterji, Burstein, Kidder, & White, 1998; White, 1998), the Balanced Budget Act of 1997 made PACE a permanent Medicare provider, allowing states the option of paying PACE under Medicaid.

Subsequently, the PACE model has been fully established by 25 programs in 13 states. As of the effective date of federal regulations (still pending), new programs will operate under provider authority, and existing demonstration programs will transition to provider status. With their proliferation as authorized providers, PACE programs will have to meet several of the same requirements as sites in the demonstration period. For example, each program will comprise a not-for-profit local service organization or partnership capable of providing the broad range of acute and long-term services, operating with risk reserves adequate to ensure fiscal viability in the face of possible losses occasioned by the management of extremely intensive, expensive cases. (The Balanced Budget Act also authorizes a new four-year PACE demonstration for up to 10 for-profit entities.) Regulations are intended to be sufficiently flexible to allow the basic service delivery model to make the best use of the provider organization's assets in meeting the particular enrollee community's needs and preferences.

DARRYL WIELAND

Internet Key Words
Capitated Care
Frail Elderly
Interdisciplinary Team
Long-Term Care
Managed Care

Medicaid
Medicare

Internet Resources

Health Care Finance Administration's PACE Information
*http://www.hcfa.gov/medicaid/pace/
pacehmpg.htm*

National Pace Association
http://www.natlpaceassn.org

PROSTHETICS: LOWER EXTREMITY

A prosthesis is a device designed to replace a missing part of the body or to enhance the performance of a body part. In the elderly, lower limb prosthetics are commonly used for rehabilitation after amputation. Due to an increased incidence of cardiovascular disease, complications of diabetes mellitus, vascular occlusion, and lower limb malignancy, amputation rates increase after age 55 (Cutson & Bongiorni, 1996).

In general, the two types of lower limb amputations are transfemoral and transtibial. Transfemoral amputation, removal of the limb above the knee, may be performed to ensure proper healing of the stump, especially in older patients. The prosthesis for an above-the-knee amputation consists of a quadrilateral total contact socket, thigh section, articulated knee joint, lower extremity shank, and foot. Variations can include hydraulic knee joints and energy storage feet.

Below-the-knee, or transtibial, amputation is more common because it yields a higher chance for successful rehabilitation. Below-the-knee devices consist of a patellar tendon–bearing prosthesis with a solid ankle cushion heel, foot, and lightweight plastic socket with total contact using a liner. Another type of knee prosthesis utilizes a thigh corset and a waist belt (Chan & Tan, 1990; Cutson & Bongiorni, 1996).

Benefits of prosthetic devices may include regaining lower limb function and mobility, restoring symmetry of the limbs, and cosmetic enhancement. A direct correlation exists between use of a prosthesis and increased level of independence in self-care and mobility. Independence in mobility and the ability to perform activities of daily living independently are essential to maintain quality of life in geriatric patients.

The main focus of rehabilitation with a prosthetic device is to restore function using the minimum amount of energy. Goals for rehabilitation and the range of function regained with prostheses vary among patients. Some individuals utilize prosthetic devices exclusively for transfer; others resume full mobility. Because of the high level of energy expenditure needed for prosthetic use, younger patients have a greater chance of regaining full mobility than do older adults.

Rehabilitation with a prosthetic device is not always a realistic intervention. Factors in this decision include the type of amputation and the existence of other comorbidities, such as cardiopulmonary disease. The energy requirement for rehabilitation after a bilateral transfemoral amputation may exceed the patient's capabilities. Sixty percent of transtibial and 20% to 30% of transfemoral amputees are recommended for rehabilitation with a prosthetic device, and approximately 80% to 90% of older amputees are fitted for prostheses (Cutson & Bongiorni, 1996).

The level of amputation, and thus the type of prosthetic device, has a direct effect on the success of rehabilitation. Because the knee joint plays an important role in the biomechanics of ambulation and mobility, it is desirable to preserve the knee during amputation. The anatomical structure of the knee joint is designed for normal loading conditions, which regularly exceed the patient's body weight. An artificial device is less than optimal and requires more energy than a normal functional knee. When the entire knee joint is removed during a transfemoral amputation, more energy is required to use the prosthetic device. The more proximal the prosthesis, the higher the energy demand will be in the prosthetic gait. Therefore, below-the-knee amputation is more desirable in the elderly to increase the chance of a successful rehabilitation.

The early stages of prosthetic rehabilitation are important both physically and emotionally. Rehabilitation immediately after the operation focuses on promotion of stump healing, reduction of edema, and prevention of the complications of bed rest. The psychological impact of losing a limb can be devastating; therefore, early rehabilitation should focus on maximizing the patient's functional independence. Activities early in the rehabilitation program can include transfers; strengthening the lower extremities, including the amputated extremity; increasing upper body strength; and balance and coordination activities. Early ambulation training in healthy individuals with one lower limb intact can begin with crutches or a walker in the immediate postoperative period. Prosthetic rehabilitation in older adults may begin with a walker because of the increased energy expenditure and the balance required to ambulate with crutches. However, the initial use of a walker may increase the difficulty of transitioning to a normal gait pattern with a prosthesis. After wound healing and the removal of sutures or staples, a standard prosthetic fitting can be done.

After fitting, rehabilitation with the prosthetic device focuses on teaching patients an energy-efficient gait. With both below- and above-the-knee amputations, movement with the prosthetic device is different from normal joint movement. Therefore, the patient needs to learn how to move efficiently with these devices. This includes learning good balance, weight shifting, and an energy-efficient gait pattern.

Because a prosthesis cannot fully restore the normal function of the limb, living with a prosthesis requires modifications in the patient's daily lifestyle. Advanced training can include walking around obstacles, through doorways, and on uneven terrain. Patients can eventually learn activities such as sitting on the floor, picking up objects off the ground, walking on ramps, and using escalators. The strength; and balance and coordination of the patient determine the level of advanced activity achieved with the prosthesis (Pandian & Kowalske, 1999).

Many factors affect the outcome of rehabilitation with a prosthetic device, including gender, comorbidities, general health condition, and the type of amputation. In general, male patients tend to rehabilitate better than female patients. Medical conditions that directly influence rehabilitation include cardiovascular disease, arthropathies, central nervous system diseases, diabetes mellitus, and dementia (Chan & Tan, 1990). Complications from amputation surgery such as myocardial infarction, congestive heart failure, cerebrovascular accident, and delirium are common in older patients (Cutson & Bongiorni, 1996). The condition of the stump after surgery can also affect rehabilitation. The presence of comorbidities correlates with the prognosis for rehabilitation. Age, as an exclusive variable, does not appear to have an effect on the overall success of rehabilitation.

Complications from prosthetic devices include stump pain, pressure ulcers, and infections of the stump in the form of skin abscesses. Some stump pain is considered "phantom limb pain," referring to pain in the amputated limb. This perception is real pain caused by intact nerve endings signaling the brain, despite the amputation. Complications associated with amputations may interfere with the use of the prosthesis. Poorly fitting sockets and floppy stumps are reasons given by patients for discontinuing prosthesis use; often the prosthesis is discarded after only several months of training and use (Chan & Tan, 1990).

The psychological adjustment to losing a limb can be very difficult for older amputees. Patients sometimes view the loss of a limb as evidence of their body's deterioration and therefore a symbol of mortality. These psychological effects can negatively impact patients' rehabilitation and further alter their quality of life. Successful use of a prosthetic device is extremely important in restoring the psychological health of elderly patients, even if the prosthesis is used exclusively for cosmetic purposes.

MARSHA AARON
ELLEN FLAHERTY

See Also

Fractures
Gait Assessment Instruments
Rehabilitation

PROTEIN ENERGY UNDERNUTRITION

Weight loss occurs in most people over the age of 70 and appears to be associated with the physiological anorexia of aging. This places the older person at risk of severe weight loss if a concurrent disease develops. The condition of predominant weight loss is known as *marasmus*. Severe protein loss with a decrease in serum albumin is called *kwashiorkor*.

The physiological anorexia of aging is responsible for decreased intake accounting for approximately one-third of calories over the life span. Decreased food intake is associated with diminished sensations of taste and smell, early satiation due to decreased fundic compliance and therefore increased antral stretch of the stomach, increased satiating effect of cholecystokinin, and altered central neurotransmitter drive. Leptin, a protein hormone produced by adipose cells, decreases food intake. In older women, leptin levels decline; in older men, they increase due to the decline in testosterone levels. This increase in leptin in men accounts for their greater decrease in food intake compared with women.

Protein Energy Malnutrition

Most causes of protein energy malnutrition are treatable and are best remembered by the mnemonic MEALS ON WHEELS:

Medications (e.g., digoxin, theophylline, fluoxetine)
Emotions (depression)
Alcohol, elder abuse, anorexia tardive
Late-life paranoia
Swallowing problems

Oral problems
Nosocomial infections (tuberculosis, *Helicobacter pylori, Clostridium difficile*)

Wandering and other dementia-related behaviors
Hyperthyroidism, hypoadrenalism, hypercalcemia
Enteric problems (gluten enteropathy, pancreatic insufficiency)
Eating problems

Low-salt, low-cholesterol diet
Stones (gallstones)

Depression, the most common cause of weight loss in older persons, accounts for 30% of those with protein energy malnutrition. In addition, social problems such as poverty, inability to shop, and inability to prepare meals need to be considered. Cancer accounts for less than 20% of all cases. The effects of protein energy malnutrition include hip fractures; decreased cognition; anemia; decreased immune function, especially a decline in CD4+ T lymphocytes; and increased infection.

Diagnosis

Weight loss of 5% in one month or 10% in three months is pathognomonic of protein energy malnutrition. A body mass index (BMI) of less than 19 is also highly suggestive of undernutrition. Measurements of midarm muscle circumference and skinfold thickness (triceps or subcapsular) can be used to detect loss of protein and fat, respectively. Low albumin (< 3.5 mg/dL) may reflect either protein energy undernutrition or a cytokine-related aging process or disease. Cytokines not only block albumin production but also cause extravasation of albumin from the intravascular space. Other proteins with a shorter half-life than albumin (e.g., prealbumin and retinol binding protein) are also decreased by both cytokines and undernutrition. Transferrin is a poor marker of undernutrition in older persons because of the high prevalence of anemia of chronic disease and iron deficiency. Total lymphocyte count may be helpful in making the diagnosis of undernutrition in some older individuals.

The Mini Nutritional Assessment (Guigoz, Vellas, & Garry, 1996) is an excellent tool for judging whether older persons are malnourished or at risk for malnutrition and does not require any blood tests. In contrast, the Nutritional Screening Index (Omran & Morley, 2000) is a poor tool because of its low sensitivity and specificity.

Management

It can be extremely difficult to get older persons to increase their food intake. The first approach is to discuss food preferences and permit the person to indulge in his or her favorite foods. Therapeutic diets should be discontinued unless the patient has renal failure or severe congestive heart failure. Snacks should be encouraged between meals and before bedtime. Ill-fitting dentures and dental or oral cavity problems should be addressed. Flavor enhancers and salt or salt substitutes may compensate for the reduced sense of taste and smell. Encouraging the person to eat in social situations (e.g., at a senior center) is often helpful. Adequate fluid intake needs to be encouraged because of the risk of dehydration.

Liquid caloric supplements, if used, should be taken between meals (at least two hours before the next meal) and not used as a substitute for meals. Studies suggest that older persons ingest more calories in liquid form than in solid form.

If dysphagia is suspected, a swallowing evaluation should be done and appropriate management instituted, such as a change in food consistency. Assistive eating devices and mealtime feeding assistance may be necessary. Some persons with dementia may have apraxia of swallowing and need to be reminded to swallow.

In all persons with undernutrition, a depression screen and medication review should be done to exclude anorectic drugs. In the absence of a pathological or pharmacological cause, an orexigenic drug such as an anabolic steroid (e.g., testosterone, nandrolone, oxandrolone) and megestrol acetate (Megace) can be tried. Although growth hormone reverses nitrogen loss in catabolic patients, its use in malnourished patients is associated with increased mortality (Takala, Ruckonen, Webster, Nielsen, & Zandstra, 1999). Dronabinol has limited success in increasing weight (Volicer, Stelly, Morris, Mclaughlin, & Volicer, 1997).

If the person is unable to swallow, enteral feeding should be considered. For long-term use, a percutaneous endoscopic gastrostomy tube is recommended. A major advantage of enteral feeding over parenteral feeding is maintenance of gut mucosal integrity and prevention of translocation of bacteria from the gut lumen that is associated with marked cytokine activation. However, recent studies question the efficacy and safety of long-term enteral tube feeding in older persons (Mitchell, Kiely, & Lipsitz, 1998).

For the short term, when the gut cannot be used to feed, peripheral parenteral nutrition is preferred over total parenteral nutrition. Total parenteral nutrition should be reserved for patients in whom no oral intake is expected for at least 7 days. The person should be encouraged to ingest orally whenever possible in combination with peripheral parenteral nutrition. Studies of older patients indicate very low caloric intake during hospitalization. All persons ingesting less than 1,000 calories a day should take a multivitamin, as this level of caloric intake is inadequate to meet the recommended daily requirement for vitamins.

Finally, it needs to be recognized that it is ethically acceptable to decline food intake at the end of life. A person's advance directive with regard to artificial nutrition should be respected.

JOHN E. MORLEY
M. LOUAY OMRAN

See Also
Assistive Devices
Caloric Intake
Dentures
Depression Measurement Instruments
Eating and Feeding Behaviors
Meals On Wheels
Nutritional Assessment
Oral Health
Percutaneous Endoscopic Gastrostomy (PEG) Tube Feeding
Polypharmacy: Drug-Drug Interactions
Sensory Change/Loss: Smell and Taste
Total Parenteral Nutrition

Internet Key Words
Percutaneous Endoscopic Gastrostomy Tube
Total Parenteral Nutrition

PRURITUS (ITCHING)

Age-related pruritus or itching tends to be distributed bilaterally and symmetrically and is commonly experienced on the scalp, back, and legs. Changes in the skin due to the aging process increase both the frequency and the severity of the sensation. The stratum corneum of aged skin has a lower water, sebum, and fatty acid content. The pathophysiology of senescent pruritus is related to factors such as dermal neuropeptide levels, altered stratum corneum barrier function, and altered sensory threshold for subepidermal unmyelinated nerves. Pruritus may be secondary to systematic disease such as renal failure, hyperthyroidism, hepatobiliary disease, and anemia.

Variations in the degree to which pruritus is intermittent or constant and interrupts a person's lifestyle make it difficult to estimate prevalence. Current prevalence estimates are between 30% and 45%, although most elderly people have intermittent symptoms (Gilchrest, 1995).

Many skin diseases cause pruritus. Xerosis (dry skin), the most common cause of pruritus in elderly people, can be caused by frequent bathing, use of strong soaps and detergents, cold weather, low humidity, rough clothing, malnutrition, and use of diuretics and cholesterol-lowering drugs. Atopic dermatitis, a chronic inflammatory skin disease, can begin in late life. The skin may itch before an eruption appears. Skin thickening (lichenification) is typical.

Scabies, caused by the parasite *Sarcoptes scabiei,* can cause intractible itching. The person presents with red papules on the trunk, in genital areas, or at the finger web spaces. Scabies is a common cause of epidemic generalized pruritus in nursing homes. Other pruritic skin conditions include psoriasis, skin infections, drug rash, insect bite, and anogenital (pruritus ani) and vulval pruritus. Severe pruritus may also be due to malignancy and psychogenic causes. Neurodermatitis can arise in anxious or depressed people who scratch themselves, creating a primary skin lesion.

Treatment

Treatment of pruritus due to systemic or skin disease is directed at the underlying cause. Topical antiprurititic agents have limited effectiveness. Calamine lotion cools by evaporation effect. Menthol preparations have a cooling action, while camphor and phenol have a local anesthetic effect. Ultraviolet light has been successfully employed in a range of pruritic disorders. Acupuncture and transcutaneous nerve stimulation are alternative treatments of unknown efficacy.

Pruritic skin is best managed by using liberal amounts of topical emollients that soothe and hydrate the skin. Their occlusive effect retards transepidermal water loss. Emollients are short acting and should therefore be applied regularly and frequently, even after improvement occurs. Simple emollient preparations such as aqueous creams are often as effective as the more complex proprietary formulations. Creams are emulsions of oil and water and are typically well absorbed into the skin. Ointments are greasy preparations, often with a paraffin base; they are normally anhydrous and insoluble in water. They are particularly suitable for chronic, dry lesions. Creams are less greasy and are therefore cosmetically more acceptable than ointments. Greasy medicaments may affect a person's adherence to treatment. Patients should be encouraged to experiment with different emollients. Those containing urea can alleviate itch, reduce xerosis, and enhance the skin's barrier function. Commonly used emollients include the following: aqueous cream, emulsifying ointment, white or yellow soft paraffin (petroleum jelly), Diprobase, zinc cream or ointment, E45, Unguentum, and Aveeno (colloidal oatmeal). Typical emollient bath additives are Aveeno, E45, Balnetar, and Oilatum.

Emollients reduce rapid evaporation of moisture. Ideally, they should be applied four times a day from head to toe using one direction of stroke, following the direction of the hair follicle to prevent the development of folliculitis. Using a spatula to remove the emollient, checking its shelf life, and using gloves for application may prevent cross-infection.

Typically, topical steroids are used for several days to calm inflamed skin and are then replaced with emollients. Steroids are contraindicated for pruritus due to xerosis alone and are no more effective than emollients, which are safer and cheaper. Sedative antihistamines commonly used include hydroxyzine hydrochloride (Atarax), with an initial dosage of 25 mg at night, increasing to 25 mg three to four times daily if necessary. Tricyclic antidepressants may also be prescribed, such as doxepin. Pruritus ani may be relieved using mild astringents such as bismuth subgallate and zinc oxide.

Prevention

A survey of noninstitutionalized elderly suggests that skin problems are common and skin care needs are largely unmet (Beauregard & Gilchrest, 1987). On examination, 85% of older persons were found to have xerosis. Typically, those studied had poor understanding of their skin problems and had sought virtually no professional assistance.

Patient education can help the person understand how to reduce those factors that promote itching and enhance alleviating factors. Self-care activities such as effective diet and skin care regimens may help prevent dry, pruritic skin. Key nutritional requirements include an adequate intake of water, vitamins C and A, minerals such as zinc, quality protein, and fatty acids.

Excessive use of soap should be avoided, because it dissolves sebum. Soap substitutes or non-perfumed superfatted soaps provide an excess of emollient and prevent moisture loss. Hard water often leads to greater use of soap and, in turn, bath oil additives. In a bathtub, pouring water over the nonimmersed areas or use of a continuous spray over all body parts for up to 10 minutes adds moisture to the skin through a soaking action. Warm water is preferable and safer than hot water and minimizes evaporation from the skin. Patting the skin dry with a cotton towel and avoiding rubbing help maintain a natural protective barrier.

Care is needed to avoid unduly frequent bathing. Partial baths or assisted bowl washes may be more suitable. Restrictions on bathing may be imposed by mobility and staffing problems in institutional settings. Caution must be exercised when helping the elderly transfer out of the bath following the use of bath oils. Fit elderly people may wish to consider swimming to help hydrate the skin, although showering afterward to remove chlorine or salt residues is advisable.

Elderly people are at risk for contact dermatitis due to deterioration of the skin's barrier function and its ability to clear chemicals from the dermis. Teaching can raise awareness of common allergens such as soaps, detergents, nickel, rubber (sometimes found in old underclothes), balsams (in many cosmetics), perfumes, plant dyes, lanolin, and esters (medicinal preservatives).

Environmental factors include humidity and temperature of living areas. Dry skin is common when humidity is less than 30%, measured with a hygrometer. Therefore, maintaining a humid environment when bathing is helpful. Central heating or use of an electric blanket may have an excessive drying effect and lead to pruritus. Use of cool compresses or a spray water mister may be helpful, although the effectiveness of such measures is not clear and is likely to be short acting at best. Cotton bed linen and clothing are helpful in achieving adequate skin ventilation and comfort. Winter itch, associated with low temperatures, humidity, and wind, is common. Advice about skin protection is important (Hardy, 1992).

Scratching is a reflex function at the spinal cord level. Although it may provide temporary relief, it can promote further itching. The itch-scratch cycle should be explained and efforts made to break it, using strategies such as diversional therapy.

Pruritus and scratching may be triggered by psychological factors such as stress, anxiety, depression, and loneliness. Because of the intensity of the itch or its continuous nature, pruritus can have a profound effect on concentration, relaxation, sleep, and rest. Insomnia may be a direct consequence of pruritus. The embarrassment of itching and the need to scratch should not be underestimated.

STEVEN J. ERSSER

See Also
Skin Issues: Bruises and Discoloration

Internet Resources
American Academy of Dermatology
http://www.aad.org

New Zealand Dermatological Society
http://www.dermnet.org.nz/

PSYCHIATRIC DIAGNOSIS AND THE *DSM-IV*

In spite of the fact that most of the mental illnesses of late life are treatable and virtually all allow for helpful intervention, older adults account for only 4% of those seen in community mental health centers and less than 2% of those seen in private practice.

Clinical work with older adults has many special characteristics. An integrated, holistic, biopsychosocial approach is necessary to negate societal stereotypes of older age as a time of decline and older people as rigid and unable or unwilling to change. Myths and misconceptions of aging often block the accuracy of assessments and diagnoses (Marino, 1996). Further, older adults may shy away from intervention due to the stigma associated with a lack of understanding.

The current cohort of older adults often views mental health interventions as strategies for "crazy people." Treatment may conflict with their principles of independence and self-reliance in solving their own problems. However, aging baby boomers' experiences with and perceptions of mental health care are probably more positive and interventions may be more acceptable in the future.

Although categorization of mental illness has useful applications for clinical, research, and educational purposes, the emphasis on pathology encourages a focus on weaknesses and limitations rather than on strengths. In turn, this compounds society's, and particularly the elderly's, stigmatization of mental illness. It burdens older persons by emphasizing the negative aspects of functioning at a time of life when concern about deterioration and decline is common. Although an accurate diagnosis is critical, it is important to avoid overpathologizing mental health problems (Turner, 1992). Part of the diagnostic challenge with older people relates to the rapid and sometimes dramatic changes— improvement as well as deterioration—that take place, necessitating frequent reevaluations and rediagnosis.

DSM-IV

The *Diagnostic and Statistical Manual of Mental Disorders* (*DSM*) is an assessment tool used by mental health and medical practitioners to make diagnoses, plan treatments, measure impacts, and predict outcomes. A multiaxial system views the individual as part of broader systems, taking into account family and other relationships, presence of medical comorbidities, and previous levels of functioning. Axis I specifies mental disorders, axis II personality disorders, axis III medical illnesses and conditions, axis IV severity of psychosocial stressors, and axis V overall functioning. The most recently published *DSM* is the *DSM-IV* (American Psychiatric Association, 1994).

The prevalence of mental health problems in older adults is high. Between 15% and 20% suffer from *DSM* axis I illnesses. Approximately 10% of community-dwelling elderly are purported to have Alzheimer's disease. Although only 1% may meet the criteria for major depressive episodes (MDEs), 6% of those with significant medical illness also have MDEs. This percentage increases in the hospital. Furthermore, 15% of nursing home residents also have MDEs. Other older adults experience several symptoms of MDEs but do not meet all the criteria for this diagnosis. Other common axis I illnesses include delirium, psychoses, anxiety disorders, somatoform disorders, and adjustment disorders.

Although *DSM-IV* is a comprehensive effort that establishes a categorical classification that divides mental disorders into types based on criteria with defining features, two important caveats are offered:

1. Clinical judgment is necessary. The diagnostic criteria are guidelines that are to be informed by clinical judgment, not used as a cookbook application. Thus, a certain diagnosis may be given in the absence of sufficient symptoms to meet all the criteria.

2. Each category of mental disorder is not a totally discrete entity that unequivocally differentiates it from other mental disorders or from no mental disorder at all. Each disorder may present with different combinations of symptoms that are influenced by personality, environmental influences, presence of comorbid conditions, and previous history.

Since the publication of *DSM-IV* in 1994, several developments have occurred that will need to be incorporated into future revisions. For example, dementia categorization needs to be expanded. Specifically, Lewy body dementia (LBD) and fronto-temporal dementia (FTD) are histopathologically and clinically distinguishable from other dementias (McKeith, in press; Kertesz, in press). It is increasingly clear that each dementia requires different treatment and management approaches. For example, conventional neuroleptic medication is not recommended for the psychotic symptoms of LBD because of the likelihood of accentuating already present neurological symptoms. Currently marketed cholinesterase inhibitors appear to slow the progression of Alzheimer's disease, but this has not yet been demonstrated in most other dementias.

Over the past several years, increasing clinical and research attention has focused on the behavioral and psychological symptoms of dementia (Finkel & Burns, in press). In *DSM-IV*, dementia of the Alzheimer's type and vascular dementia can include the presence of delirium, delusions, and depressed mood, with the additional specification of behavioral disturbances. However, data suggest that the psychosis of Alzheimer's disease may be a distinct and discrete entity distinguishable from other late-life psychoses (Jeste & Finkel, 2000).

Another emerging common clinical entity is subsyndromal depression. Many older adults suffer from serious and disabling depressive symptoms but with too few symptoms to establish a diagnosis of MDE. Yet for many, antidepressant medications are helpful in resolving or managing the symptoms. In these "nondiagnosable" situations, issues related to family, comorbid medical illnesses, social supports, personality, and strengths are all crucial considerations.

Many geriatric mental health practitioners continue to bemoan the loss of the diagnosis of paraphrenia (late-life schizophrenia) since *DSM-II* days. This term continues to be used in the literature but is diagnosed as psychotic disorder not otherwise specified in *DSM-IV*. Paraphrenia is a late-life psychosis, predominantly of women, characterized by bizarre (typically paranoid and often sexual) delusions as well as hallucinations. It is associated with sensory deficits, premorbid personality disorders, and lower rates of marriage. A person with paraphrenia often involves the police or the legal system in her delusional pattern and poses a difficult challenge to mental health practitioners. A combination of supportive psychotherapy and antipsychotic medication is usually recommended, but patient cooperation and compliance are uncommon. Sometimes, a housekeeper or paraprofessional (under professional supervision) can form a positive relationship that can be helpful.

Age-related cognitive decline (ARCD) is a new diagnosis in *DSM-IV*. ARCD supplants previous terms such as age-associated memory impairment and benign senescent forgetfulness. ARCD refers to an "objectively identified decline in cognitive functioning consequent to the aging process that is within normal limits given the person's age. Individuals with this condition may report problems remembering names or appointments or may experience difficulty in solving complex problems" (Small et al., 1997, p. 684). Recent clinical and research findings suggest that there is a common clinical picture of word-finding difficulties, misplacement of objects, and problems remembering names that does not lead to dementia. However, another clinical condition, minimal cognitive impairment, is a significant risk factor for Alzheimer's disease (Small et al., 1997).

Medical illnesses are most common in the elderly population. Mental disorders and medical illnesses may contribute to each other. For example, MDEs predispose an elder to a wide range of medical illnesses, including cerebrovascular and cardiovascular disease, diabetes mellitus, and Parkinson's disease, whereas each of these conditions is a risk factor for MDEs. Medical illnesses are the paramount cause of delirium; sensory deficits contribute to late-life psychoses. A medical evaluation is a customary and generally necessary part of a comprehensive evaluation of an older adult.

The axis V Global Assessment Scale (GAS) is a seven-point scale measuring social and occupational functioning. It first appeared in *DSM-III* (1980) and is derived from the Health-Sickness Rating Scale (Luborsky, 1962). With the publication of *DSM III-R*, the GAS was replaced by the Global Assessment of Functioning (GAF), a 90-point scale emphasizing social, occupational, educational, and psychological functioning over the past year, as well as at the time of administration. The *DSM-IV* uses a revised version of the GAF, with a 100-point scale. The GAF appears most highly correlated with occupational functioning (Goldman, Skodal, & Lave, 1995).

The GAF is a widely accepted tool, used by the Health Care Financing Administration and Medicare managed care organizations to measure the impact of psychiatric treatment and to predict intervention outcomes. Medicare and other HMOs use the scale to determine reimbursement for treatment.

Although the inclusion of a focus on function is laudable, the GAF causes several concerns. First, it has never been validated as a measurement, seriously compromising additional or smaller research studies. Next, as it applies to older adults, two of the three domains—school and occupational functioning—have little or no relevance to typical older adulthood. Therefore, the clinician is limited to the third domain, social functioning, which also can be problematic for this group. Medical illnesses, loss of loved ones, sensory deficits, and transportation problems can separately or in combination result in decline of social functioning but may not be indicative of concomitant mental illness. Thus, an individual may be psychologically healthy yet score low on the GAF due to other factors. Clearly, future *DSM*s must pay more attention to the functional markers for older adults.

Categorization of mental disorders is necessary and important for clinical, educational, and research purposes. *DSM*s have continued to integrate a greater understanding of these disorders, as well as the biopsychosocial factors that contribute to affect, or modify their course. *DSM-IV* built on previous research, expanded several diagnoses, and added others. With the rapid growth of research, additions and modifications will be required in *DSM-V*. A more meaningful functional scale is needed for older adults. It is critical that our emphasis on symptoms and pathology not impair our understanding of individual strengths and support systems.

Sanford I. Finkel
Robyn L. Golden

See Also

Cognition Instruments

Cognitive Screening Tests: The Mini-Mental Status Exam

Vascular and Lewy Body Dementia

Internet Key Words

Mental Illness

Psychiatry

Psychology

Social Work

Internet Resources

American Psychiatric Association
http://www.psych.org

Electronic Gerontology Related Resources
http://www.cs.umd.edu/users/connie

Internet Mental Health
http://www.mentalhealth.com

Medscape
http://www.medscape.com

National Association of Social Workers
http://www.naswdc.org

PSYCHOLOGICAL/MENTAL STATUS ASSESSMENT

The format for mental status assessment of elderly clients does not differ significantly from that used with younger clients. The health professional takes a psychiatric history, including a history of the presenting complaint and the medical, psychiatric, personal, and family history. The mental status examination should include the usual areas of appearance and behavior, speech, mood and affect, form and content of thought, perception, sensorium and cognition, and insight and judgment. Nevertheless, some considerations specific to the elderly must be considered.

It is generally acknowledged that the elderly population is increasing dramatically in developed countries. This, coupled with the fact that the risk of occurrence or reoccurrence of mental illness increases with age, means that the ability to carry out a careful mental status assessment is an important skill for any health professional working with elderly clients.

While it is true that the risk of mental illness increases with age, mental illness itself should not be seen as a natural and therefore inevitable consequence of the aging process. There is a tendency for health professionals to view a decline in cognitive functioning as natural in an elderly person and to see depression, due to the loss of functional abilities or other losses, as an expected response. Such beliefs may lead the professional to see what she or he expects to see. In such circumstances, the health professional may fail to appreciate the complex interplay of social, biological, and psychological factors involved in the aging process; misdiagnosis and inadequate treatment may follow. The almost inevitable outcome is a poorer quality of life for the elderly client.

It seems that the attitude of the health professional is the starting point of any psychological assessment of an elderly client. How health professionals comport themselves during the encounter with the client and the beliefs they bring to the assessment are as important as the questions asked of the client and the tests undertaken.

Psychological assessment requires an engagement with the client at a much deeper level than that of mere data collection. The psychological assessment is a therapeutic interaction whose goal is to understand the client more fully in order to meet the client's needs. Clients who feel alienated from the assessor or misunderstood are not reliable historians and will not be cooperative in the assessment process. Many authors have written on the process of mental health assessment, including questioning techniques and assessment tools (see Othmer & Othmer, 1994). However, here we deal with the health professional's ways of being with the patient that can inform the assessment process (Walsh, 1999).

Psychological assessment should have as its base the notion of shared humanity. That is, we are all more similar than we are different. The client is seen first and foremost as a fellow human being in the world who can be understood by the clinician through what they have in common—their humanity. In this way, the clinician is less likely to treat the client as a data-filled object to be examined and categorized. Three elements are essential to "being with" the client in shared humanity: understanding, possibility, and care-full concern (Walsh, 1999). "Being with" is more than a physical proximity to the client; it also entails an emotional and psychological proximity.

Being with the client in understanding means taking the point of view that everyone can be understood at some level, even if they are confused, angry, or anxious. Behavior, including delusional and confused behavior, has meaning and can be interpreted. From this perspective, the clinician is more likely to explore issues and behaviors rather than to simply label them as being related to depression, dementia, or the aging process. It also means attending to the client. Understanding can be submerged through an overemphasis on the task-oriented, stepwise process of data collection. We can become intent on asking certain questions or doing certain tests and fail to pick up on the subtle, nonverbal clues in the client's manner or behavior.

As human beings, we are all imbued with possibility. The treatment options and paths open to elderly clients should be based on their uniqueness as human beings and their unique sociocultural circumstances. The possibilities open to a client should not be cut off on the basis of a diagnosis alone or some other apparently constraining factor.

A recognition of possibility also safeguards the clinician from jumping to hasty conclusions and from making decisions based on a paucity of evidence or a proclivity to act on a preconceived notion or prejudice.

In being with the client in understanding and possibility, the clinician is exposing his or her care and concern for the client. This genuineness brings about an engagement that is more likely to yield fruitful results for both the client and the clinician.

The Four Ds

An elderly person can experience the same mental health issues as a younger person in the community. Nevertheless, the complex interplay of biological, psychological, and physical factors may present the clinician with a confusing picture. This is especially true for what is sometimes called the four Ds of psychiatric assessment in the elderly: depression, dementia, delirium, and delusions (Stuart & Laraia, 1998). Failure to differentiate among these can have serious consequences.

The incidence of depression rises with age. Unfortunately, this very treatable illness is often overlooked or misdiagnosed. Depression in the elderly may be a consequence of loss or physical illness, or it may be endogenous. The presentation of depression in the elderly can be quite different from depression occurring in other populations. The client may present with numerous physical complaints for which a cause cannot be found (this may also be a feature of an individual suffering from a high degree of anxiety). Lack of energy and less interest in activities can sometimes be regarded as consequences of the aging process and hence dismissed, but they may be related to an underlying depression. The depressed client may present with a lack of appetite, weight loss, irritability, agitation, preoccupation with the past, and lack of engagement with family and friends. Paranoid ideation and suspiciousness may also be present. An elderly client presenting with such features should also be assessed for suicidal ideation and should undergo a complete physical examination, as many physical illnesses have depressive symptoms as part of their presentation. Some of these disorders include myocardial infarction, Parkinson's disease, various cancers, thyroid disorders, and stroke.

Depression can also be mistaken for dementia. However, the two disorders can coexist. Depression has a rapid onset, usually weeks to months, whereas dementia usually has a gradual course and develops over a period of years. Feelings of sadness, worry, or guilt are present in the depressed person. Guilt is usually absent in dementia, although the client's mood may be labile. Sleep disturbance is more usual in depression—either insomnia (sometimes with early morning waking) or excess sleeping. Impaired drawing abilities and object-naming abilities are usually absent in dementia. Cognition in a depressed person may seem impaired due to difficulty concentrating and apathy. The patient may reply to testing with many "I don't know" answers. In dementia, the cardinal sign is a disturbance of executive functioning (planning, organizing, sequencing, and abstracting). These functions are readily and quickly assessed using the Mini-Mental Status Exam (Othmer & Othmer, 1994).

Delirium caused by an underlying organic disturbance typically has a rapid onset of hours to days and a fluctuating course and is characterized by a disturbed level of consciousness ranging from hyperalert to difficult to arouse. Level of consciousness is usually intact in both dementia and depression (Stuart & Laraia, 1998).

Although content disturbance is a feature of all of the above disorders, the presentation varies. In a patient suffering from delirium, thought content may be characterized as incoherent and confused. Delusions may be present. Thought content in a depressed elderly client may include negative thoughts and thoughts of death, as well as hypochondria and nihilistic delusions. A client with dementia may appear paranoid, disorganized, and delusional. Judgment is poor in all these disorders, but socially inappropriate behavior is more common in clients suffering from dementia (Stuart & Laraia, 1998).

Although the distinguishing features of the four Ds may be helpful, it is paramount that a clinician undertaking a mental status assessment of an elderly client be sensitive to the client's perception of the situation and not ignore the fact that the client is a fellow human being and is situated in a unique sociocultural environment with a unique history as a person. Such a focus on the personhood of the individual ensures that the clinician undertakes an assessment that is sensitive, inquiring, and thoughtful and has the potential to bring about the best outcome.

KENNETH WALSH
MARY KATSIKITIS

See Also

Cognitive Screening Tests: The Mini-Mental Status Exam

Confusion/Delirium: Risk, Diagnosis, Assessment, and Interventions

Dementia: Overview

Depression in Dementia

Depression Measurement Instruments

PSYCHOPATHOLOGY

Psychopathology refers to the study of the symptoms and signs of mental disorder and other abnormal psychological processes. For some purposes, it is useful to view psychopathology as an indicator of an underlying disorder or other abnormal process; this can guide the application of operational diagnoses, selection of appropriate treatment, and research into etiology. For other purposes, psychopathology is regarded as a cause of undesirable outcomes, such as distress or danger to the individual or those around. Outcomes also include impairments of functioning in cognitive faculties, feelings, social behaviors, work and other roles, leisure activities, and the adaptive tasks of daily living. Prevention and relief of undesirable outcomes should motivate the relationship between patient and therapist, drive the management of mental health care, and set goals for research into the determinants of the course of psychopathology.

The distinction between psychopathology and normality is as clear in elders as it is in younger persons. Declines in memory or other cognitive capacities sufficient to cause impairments in the adaptive tasks of daily living signal a high probability of the presence of psychopathology, whereas continued competence suggests normal aging, even if accompanied by complaints of changes in memory (Livingston et al., 1990). Only a small proportion—usually reported as 2% to 4%, with some reports as high as 10%—of those 65 years or older suffer from dementia (Folstein et al., 1991; Livingston et al., 1990). Elders are typically neither sad nor demoralized. No more than 12% to 18% have a degree of persistent and intense depression that warrants clinical attention (Blazer & Williams, 1980; Gurland et al., 1983; Livingston et al., 1990), whereas less than 10% qualify for a formal diagnosis of depressive disorder, and only about 1% to 6% reach criteria for major depression (Johnson, Weissman, & Klerman, 1992; Kennedy, Kelman, & Thomas, 1990). Other causes of chronic psychopathology involving enduring alterations in thinking, loss of insight, reduced emotional responsiveness, and withdrawal are rare, such as conditions related to schizophrenia (Keith, Regier, & Rae, 1991). Long-standing deteriorative processes of schizophrenia may end in extensive loss of communicatory powers and require careful observation to distinguish from the later stages of dementia or the effects of neurological disorders.

If not suffering from psychopathology, elders continue to have a full range of intellectual, emotional, and social skills. There is no consistent relationship between aging and psychopathology. It is true that rates of dementia climb steeply with advancing age (Fratiglioni et al., 1991), but most even extremely old persons—85 years or older—do not show psychopathology of memory and other cognitive functions. Some centenarians are not cognitively impaired. The frequency of other mental disorders that are found in adults of any age may rise or fall with increasing age. Depressive disorders severe enough to interfere with social roles and meet criteria for major depression probably lessen in frequency or plateau in the years after middle age. Somewhat at odds with the preceding remarks on age variation in depression, suicides increase with age in white males, although survived attempts decrease. Schizophrenia or other persistent paranoid disorders show a sharp drop in frequency after age 45 years, though they can occur later (Jeste, 1997). Alcoholism and substance abuse also are less common in older age (Helzer, Burman, & McEvoy, 1991). Anxiety disorders diminish with age (Regier, Narrow, & Rae, 1990), though they remain prevalent enough, around 4% to 10% (Swartz, Landerman, George, Blazer, & Escobar, 1991), to raise public health concerns.

Certain conditions actually improve with age. Aging tends to dampen the intensity of the positive symptoms of schizophrenia, although the negative symptoms may persist (Soni & Mallik, 1993). Perhaps a third of schizophrenics achieve a complete recovery by old age. Late-onset paranoid states are characteristically circumscribed, allowing the sufferer to manage an independent existence, even though the accompanying behaviors may be annoying to others.

Some features of the mode of presentation of psychopathology in elders are not specific to a particular disorder but rather may be true of many disorders. The use of somatic language (i.e., bodily complaints) to describe the symptoms of mental disorder (somatization) is encountered more frequently in older patients (Dworkin, 1990) as a denial or minimization of emotional symptoms. Cognitive impairments may be precipitated by several syndromes of psychopathology, including depression, and a variety of psychotropic medications.

Differences that seem to be linked to age are due to cohort characteristics. The excesses of generations prone to alcohol and drug consumption progress in waves up the age range (Glynn et al., 1985). Illiteracy and poor education appear to increase vulnerability to dementia in old age (Murden, McRae, Kaner, & Bucknam, 1991), with the possibility that successive cohorts with progressively improving educational experience will have lower rates or later onset than currently found for dementia. Cohorts who are sophisticated in articulation of

psychopathology without feeling stigmatized have not yet reached the phase of old age. Some other age differences may be largely determined by attitudes of health providers who, for example, may not encourage free expression of emotional symptoms in older patients or who may dismiss general somatic symptoms, such as lack of energy or appetite, as normal for old age without making the connection to psychopathology.

There is also a characteristic tendency of psychopathological conditions to be concomitant with some other health condition. Depression especially tends to occur in the context of another mental disorder, such as dementia, or with a physical illness or disability. Late-onset schizophrenics, more often than younger cases, are found to have lesions disclosed by neuroimaging. Not only the biological substrate but also the social context undergoes changes with age that can, in turn, affect psychopathology. The features of the social network, especially its quality (George, Blazer, & Hughes, 1989), may alter the frequency of depression in old age. Phobic states may be disguised as real fears of crime or as residual limitations on mobility occasioned by physical disease, a confounding variable that is aggravated by the high association between the late onset of agoraphobias and the advent of physical illness or injury. Because of the resulting mislabeling, opportunities for enhancing activities through treatment of phobias may be lost. Lessened social expectations of elders make failing function associated with psychopathology less likely to be noticed.

A knowledge of psychopathology in advancing age can highlight the predominant normality of elders, and the delineation of disordered conditions that call for treatment. Psychopathology is a complication, not a natural part, of the aging process.

BARRY GURLAND

See Also
Behavioral Symptoms in Patients with Dementia
Crime Victimization
Dementia Overview
Depression in Dementia
Psychiatric Diagnosis and the DSM-IV

Internet Key Words
Depression
Psychopathology

PSYCHOTIC DISORDERS AND MANIA

Treatment options for psychotic disorders and mania among older adults have improved considerably. However, similarities among late-onset schizophrenia, delusional disorder, mania, and the psychoses of depression, dementia, and stroke can make the diagnosis difficult. With skill and perseverance, the practitioner will be able to help the patient preserve relationships and secure needed care, even though insight and judgment may not be fully restored.

The practitioner's priority is an alliance with the patient or, if that is not possible, with the family or third party who is trying to help. The patient's agenda rarely includes correction of false beliefs or perceptions. However, denial and lack of insight do not preclude a working relationship. The practitioner's willingness to understand the patient's perspective may offer welcome relief to persons who have been subjected to ridicule. At the same time, pathologically suspicious patients are wary of intimacy, despite painful loneliness. They need trust, and trust cannot be established merely by good intentions or professional status. Demonstrated reliability, availability, and respect for the patient's individuality can allow sufficient trust to sustain the diagnosis and treatment process. The practitioner need not agree with the delusions or verbally acknowledge the presence of hallucinations but should accept the patient's psychotic perceptions as evidence of a need for assistance. Directive, confrontational efforts to correct the false beliefs of delusions or to categorically deny the reality of hallucinations will only provoke an argument.

It is possible to identify distressing issues with the patient and third party, form a consensus, and share plans to reduce the problem. For example, with a delusional person who is convinced that the landlord is piping poison gas into the apartment via the radiator, it may be sufficient to reply, "I may not be able to fight the landlord for you, but I can see how stressful the situation is for you. Let me give you something to reduce the effects of the stress so that you can manage your [insomnia, nerves, depression] better." Similarly, a husband whose jealous delusions lead to threats against his wife may agree to take medication to avoid loss of control and separation. Focusing on the physical effects of the distress may facilitate diagnostic studies needed to identify reversible contributors to the disturbance, such as impaired vision or hearing, polypharmacy, endocrine or metabolic disorders, undetected central nervous system diseases, or neoplasms.

Changes in the patient's circumstances may bring a long-standing but previously hidden problem to the practitioner's attention. Caring family members or neighbors may have sheltered the person throughout life, compensating for interpersonal deficits that became apparent only when the social network began to fail due to death, disability, or illness. Alternatively, the patient may have managed independently until personal assistance needs, loss of privacy, or invasive medical procedures unsettled a fragile sense of security. The practitioner's foresight and efforts to restore equilibrium are basic to success with all psychotic disorders.

Gaining the patient's permission to share the results of diagnostic procedures and treatment recommendations with other concerned parties sets the groundwork for future collaboration. Providing a clear expectation that "we work as a team" avoids accusations of betrayed confidence. Interdisciplinary teamwork is also critical to clinical efficiency. Particularly when disturbances in judgment or impulse control are expected, initial networking with team members will be time well spent when crises arise. Family and other involved parties need guidance to cope with the patient's difficulties. They should be made aware that complete insight is an unrealistic goal, but greater social compatibility and independence are not. Delusions that do not interfere with care or personal well-being and do not overtly distress others may be ignored. However, some friends or family may take the accusations as a personal affront or feel compelled to argue for their falsity. It can be difficult to be forgiving when one's best efforts result in groundless insults of malicious intent. Yet families are remarkably resourceful and resilient. Family therapy should be directed at understanding strengths and how to shore up the caring relationship. It may be impossible to resolve long-standing conflicts, but mediation can restore equilibrium.

Schizophrenia

The premature mortality and lost personal productivity and autonomy make schizophrenia the most devastating mental illness of adult life. Critical elements of successful treatment are a comprehensive, individualized approach, including medication, family support and education, and aggressive case management. Although 90% of persons with schizophrenia receive antipsychotic medication, only 50% receive the recommended array of psychosocial and rehabilitative services. Failure to add an antidepressant for depressive episodes or to provide for psychosocial treatments (family intervention, therapeutic day programs) is the most frequent inadequacy (Lehman & Sreinwachs, 1998).

Most cases of schizophrenia in males occur in the second and third decades of life. Among women, the illness demonstrates a bimodal age of onset, with a significant second peak occurring in the menopausal years. Late- and early-onset schizophrenia are similar in several ways. Positive symptoms are more prominent in women, and impairments in visual and auditory processing are also common, as is a family history of psychosis and a personal history of adjustment problems in childhood (Jeste, 1997). Sensory impairments, particularly hearing difficulties, that were thought to contribute to the onset of paranoia may be the result of difficulty acquiring and accommodating to glasses and hearing aids as a result of the illness (Rabins et al., 1997). The overall pattern of cognitive impairment is similar in early- and late-onset disease, as are findings of nonspecific white matter

and ventricular abnormalities on brain magnetic resonance imaging (MRI). The course is persistent, and mortality is increased. Late-onset disease is more frequent in women, has fewer severe negative symptoms, and is mostly delusional and paranoid in character. Impairments in learning, abstraction, and cognitive flexibility are not as severe. Symptoms may be reduced with lower doses of typical antipsychotics than would be required in early life. However, the prevalence of tardive dyskinesia approaches 50% within 24 months of typical antipsychotic treatment (Woerner et al., 1998). Newer atypical antipsychotics at low doses are recommended.

Delusional (Paranoid) Disorder

Delusions are false beliefs and inferences that seriously impair social judgment and cannot be interpreted as originating from religious or cultural group norms. Pathological suspiciousness, jealousy, exaggerated self-regard, and erotic obsession are the most frequent manifestations. Most aspects of personality and cognitive performance remain intact. However, failure to pay bills or to attend to physical illness or disability, along with accusations against others for which there is no basis in fact, brings these people to the attention of social services agencies. In most instances, antipsychotic medication substantially restores the person's capacity to manage, but only one in four patients abandons the delusions and gains clear insight into the problem.

Mania

Late-onset mania is often misdiagnosed and is probably more common than reported. The presentation is more complex and less typical than classic bipolar (manic-depressive) illness. Diagnostic certainty is often clouded by the presence of cognitive impairment suggesting dementia. Late-onset mania is often secondary to or closely associated with other medical disorders, most commonly stroke, dementia, or hyperthyroidism. It can also be associated with medications, including antidepressants, steroids, estrogens, and other agents with known central nervous system properties (McDonald & Nemeroff, 1996).

The diagnostic workup identifies indicators, such as structural brain changes or dementia, that assist in prognosis, which is generally less favorable in late-onset mania. A careful history from family may uncover hypomanic episodes that did not seriously impair the individual but in retrospect are clear indicators of early-onset disease. The difficulties associated with contributing conditions age-related vulnerability to medication side effects, and structural brain changes make treatment difficult. Structural brain changes most frequently include subcortical hyperintensities seen on MRI (McDonald, Krishnan, Doraiswamy, & Blazer, 1991).

Late-onset mania is more frequent among men. The manic episode often presents with confusion, disorientation, distractibility, and irritability rather than elevated, positive mood (Young & Klerman, 1992). The clinical interview may be characterized by irrelevant content delivered with an argumentative, intense, yet fluent quality. Patently unrealistic plans concerning finances or travel, exaggerated self-regard, and contentious claims to certainty in the face of evidence to the contrary are also seen. The statements may be plausible but are too improbable to be real. An unsuspecting examiner may be puzzled by the difficulty of the interchange until the diagnosis of mania is considered.

The presence of psychosis, sleep disturbance, and aggressiveness, particularly in a nursing home setting, may suggest dementia or depressive disorder rather than mania. Because mania in late life is genuinely less frequent than depression or dementia and less frequently recognized, these patients are most often treated with antipsychotics, antidepressants, or benzodiazepines, which provide only partial relief.

Pharmacological Treatment

Although seniors who have experienced good results with lithium should not be switched to an alternative, a number of concerns argue against lithium for initial treatment in older adults. Advanced age, absence of family history of bipolar disorder, and mania secondary to another medical condition, particularly stroke or dementia, all predict a poor response to lithium. Because the kidneys clear lithium, the age-related decline in renal function places older adults at increased risk of toxicity. Structural brain changes, which may not be clinically apparent, are associated with a higher risk of lithium intolerance. Interactions with psychotropics and other medications further complicate the use of lithium in older adults.

Although antipsychotics are frequently prescribed for mania, there is a growing consensus that anticonvulsants, called mood stabilizers in this context, are preferable both for acute treatment and for the prevention of recurrence (Sussman, 1998). The anticonvulsant valproic acid is increasingly a first choice for augmentation of antidepressant therapy in the treatment and prevention of mania. The risk of hepatic toxicity is low when a therapeutic blood level is achieved. However, valproic acid inhibits hepatic enzymes that metabolize a variety of medications frequently used by older adults. Patients taking beta-blockers, type 1C antiarrhythmics, benzodiazepines, or anticoagulants may be prescribed valproic acid but should be monitored closely until the dose is stabilized. There is considerable enthusiasm for gabapentin and lamotrigine, recently approved for the treatment of epilepsy, but there are limited data on their use as mood stabilizers in late-life mania, depression, or psychosis. The low rate of drug interaction, protein binding, lack of need to monitor therapeutic levels or liver toxicity, and unique clearance make them particularly attractive (Semenchuk & Labiner, 1997).

Antipsychotics and Movement Disorders

Practitioners may encounter older persons with lengthy histories of antipsychotic administration, most frequently with haloperidol, perphenazine, or thioridazine. These typical antipsychotics tend to induce movement disorders as the dose and duration of administration increase. Although movement disorders also emerge spontaneously with advanced age, they are rarely disabling unless they evolve into Parkinson's disease or drug-induced pseudo-parkinsonism. Although the typical antipsychotics are the most frequent cause of pseudo-parkinsonism, the selective serotonin reuptake inhibitors and the tricyclic antidepressants have also been implicated.

Drug-induced movement disorders are a direct result of dopamine receptor D_2 blockade and have several manifestations. Akathisia is a jittery, restless feeling that may be difficult to distinguish from anxiety but worsens as the dose increases. The onset of drug-induced parkinsonism (extrapyramidal side effects) with typical antipsychotics is dose related. It can be alleviated with anticholinergics, but these agents pose problems for older adults whose cholinergic tone is reduced by age or other medications with anticholinergic properties, such as the tricyclic antidepressants.

Tardive dyskinesia follows longer-term treatment with typical antipsychotics. It may be irreversible even when the medication is stopped or an atypical antipsychotic is substituted. Compared with younger patients, older adults have a three- to fivefold increased risk of developing tardive dyskinesia. Spasticity of the tongue and lips and writhing movements of the trunk can be disfiguring if not disabling.

Movement disorders can be minimized by limiting the choice of antipsychotics to the atypicals: olanzepine (Zyprexa), risperidone (Risperdal), quetiapine (Seroquel). These drugs exhibit less dopamine D_2 receptor antagonism and more serotonin 5-HT_2 receptor antagonism than do typical antipsychotics. Thioridazine is less likely to provoke extrapyramidal side effects and irreversible movement disorders, but it lowers blood pressure and impairs balance as the dose is increased. Movement disorders are frequently seen but are usually not disabling with low doses of haloperidol and perphenazine. Thus, if symptoms are well controlled and the movement disorder does not impair function or appearance, patient and practitioner may prefer not to change to the newer atypical antipsychotics. If cost is a concern and treatment will be short term, the typical antipsychotics are acceptable.

If treatment becomes long term, an atypical should be substituted by cross-tapering. The typical antipsychotic is reduced by 25% every three to four days until only 25% of the original dose remains, at which point the atypical is introduced at the lowest dose. In three to four days, the atypical is increased, and the typical is eliminated. The atypical is increased further, based on the recurrence of symptoms, usually sleep disturbance. For example, to cross-taper from 2 mg haloperidol to olanzepine, the haloperidol is reduced by 0.5 mg every three to four days until the patient is taking 0.5 mg, at which point 2.5 mg olanzepine is introduced. The haloperidol is eliminated three to four days later, and the patient is observed. If sleep becomes disturbed or signs of the psychosis exacerbate, the olanzepine should be increased to 5 mg, with further increases depending on the emergence or persistence of symptoms.

Quetiapine or olanzepine may be useful for the treatment of psychosis in Parkinson's disease. Use of these drugs avoids the worsening bradykinesia seen with typical antipsychotics and with doses of 2 mg or more of risperidone. Clozapine, the first atypical antipsychotic medication, does not induce movement disorders. At doses given to young persons with schizophrenia, it is anticholinergic and induces agranulocytosis in as many as 2% of patients. However, at low doses (6.25 to 50 mg) given to older patients with levodopa-induced psychosis it significantly reduces psychosis, and tremor without impairing cognition (Cummings, 1999). Thus, despite its side effect profile, clozapine at low doses and with white blood cell count monitoring may be beneficial for a small but clearly defined set of older patients.

Psychosis arising out of delirium and in combination with dangerous behavior (agitation, pulling out life supports) requires haloperidol, which remains the drug of choice due to its relative freedom from cardiovascular side effects. There are insufficient data to recommend the atypical antipsychotics for psychotic depression or bipolar disorder. The complex pattern of receptor blockade (dopaminergic, serotonergic, cholinergic) brought about by the atypicals raises questions about their use with antidepressants possessing significant serotonergic or anticholinergic properties.

Some patients will not take oral medications but will accept a long-acting injectable antipsychotic. However, this is no substitute for an ongoing, trusting relationship. The prolonged half-life of haloperidol decanoate means that stabilizing the patient's symptoms with an initial injection of 12.5 to 25 mg and subsequent monthly increases will take time. Long-acting injectable antipsychotics are best used upon discharge from an inpatient stay during which symptoms were controlled with oral medication. Haloperidol decanoate is given by deep intramuscular injection every 28 days at 25 times the daily oral dose that stabilized the psychosis (e.g., a daily oral haloperidol dose of 2 mg equals 50 mg of haloperidol decanoate every 28 days). The total dose should not exceed 200 mg every 28 days.

Electroconvulsive Therapy for Mania or Psychosis

Electroconvulsive therapy (ECT) has a long history in the treatment of mania and psychotic depression and may be indicated for a severely disturbed older psychotic patient when agitation or the threat of violence becomes extreme. It may be particularly useful in cases of medication inefficacy or intolerance, psychotic depression, imminent suicidal risk, or morbid nutritional status. However, advanced age, concurrent antidepressants, and cardiovascular compromise increase the risk of adverse reactions, with cardiovascular complications being the most frequent events. Cognitive impairment associated with ECT includes temporary postictal confusion, transient anterograde or retrograde amnesia, and, less commonly, permanent amnestic syndrome, in which events surrounding the treatment are forgotten. Treatments can be limited to twice weekly and applied unilaterally to the nondominant hemisphere to minimize confusion.

Psychosis and mania impair the patient's capacity to make health decisions and to maintain intimate relations and social connections. A trustworthy, matter-of-fact approach targeted on immediate needs and concrete goals can alleviate distress and ultimately help the person not to act on hallucinations or delusions. Forging a lasting, workable alliance with the patient, family, and other caregivers, despite denial and lack of insight, is one of the more satisfying aspects of clinical practice. Antipsychotics and mood stabilizers can be beneficial once an alliance has been formed.

GARY J. KENNEDY

See Also
Psychiatric Diagnosis and the DSM-IV

Internet Key Words
Bipolar Disorder
Delusional Disorder
Mania
Neuroleptics
Psychosis
Schizophrenia

Internet Resources
National Alliance for the Mentally Ill
http://www.nami.org

National Depressive and Manic-Depressive Association
http://www.ndmda.org/

National Institutes of Mental Health
http://www.nimh.nih.gov/

PSYCHOTROPIC MEDICATIONS

The use of psychotropic medications in long-term-care facilities is the subject of federal regulation, mandatory monitoring, and clinical concern. It is an area in which the prescribing of specific agents is subject to regulatory constraints unique to clinical medicine practiced in nursing homes.

Subtitle C, the Nursing Home Reform Act, of the Omnibus Budget Reconciliation Act of 1987 (OBRA-87) (PL 100-203), included a provision stating that residents of nursing facilities must be "free from the use of physical and chemical restraints." The regulations and subsequent interpretive guidelines issued by the Health Care Financing Administration (HCFA) were considered controversial by many due to their broad scope and were revised and finalized in 1990 (Schorr, Fought, & Wray, 1994).

A chemical restraint is defined as a psychopharmacological drug used for discipline or convenience and not required to treat medical symptoms. The misuse of psychotropic medications as a means of chemical restraint was identified by the HCFA as a form of abuse and neglect of nursing home residents. Hence, the prescribing of psychotropic medications must be clinically indicated, necessary, and appropriate. Specific indications, dose ranges, and documentation of ongoing need are required by the OBRA-87 regulations.

Many aspects of the OBRA-87 regulations reflect basic geriatric medical and psychiatric principles. The requirements for assessment, diagnosis, and documentation of indications for and side effects of psychotropic medications have highlighted the need for, and importance of, psychiatric interventions in nursing homes. Long-term-care facilities and those who practice in these settings should recognize the basic intent of these regulations, that is, to ensure the clinically necessary and judicious use of psychotropic medications (Lasser & Sunderland, 1998).

The terms *antipsychotic* and *psychotropic* are often confused and used incorrectly as synonyms. Antipsychotic medications are one broad category of psychotropic medications that include antipsychotic, antidepressant, anxiolytic, and sedative-hypnotic agents. Antipsychotic medications are subject to the greatest degree of regulation under OBRA-87.

The interpretive guidelines issued by the HCFA focus on antipsychotic medications and stress that these particular agents must not be administered unless necessary to treat a specific condition. Conditions warranting antipsychotic medications are included in the guidelines: schizophrenia, schizoaffective disorder, delusional disorder, mania and depression with psychotic features, acute psychotic episode, brief reactive disorders, atypical psychosis, Tourette's disorder, Huntington's disease, and organic mental syndromes, including dementia with behavioral disturbances that present a danger to the resident or others, disrupt care, or cause the resident distress. However, facilities are also instructed that antipsychotic agents may not be used when certain disruptive behaviors are the only indications. These behaviors include wandering, poor self-care, restlessness, impaired memory, anxiety, depression, insomnia, unsociability, indifference, fidgeting, nervousness, lack of cooperation, and behaviors that do not represent a danger to the resident or others.

The use of psychotropic medications on an as-needed (PRN) basis is strongly discouraged. Physician reevaluation is required if PRN medications are ordered and used more than twice in a seven-day period. The clinician is encouraged to formulate a treatment plan that avoids PRN medication orders.

The prevalence of psychiatric disorders among residents of long-term-care facilities is formidable. Nursing homes have become de facto mental health treatment centers. Tariot and colleague (1993) reported the presence of at least one behavioral or emotional problem in 90% of a sample of nursing home residents, with 50% displaying four or more concurrent symptoms. As nursing homes must address the mental health needs of residents, rates of psychotropic drug prescriptions have become widely used as monitoring tools. Comparative benchmark values are being used as standards by which facilities are measured. These values are available from state health departments, but they vary markedly. Antipsychotic medication prescription rates have been reported as low as 10% and as high as 86% (Lasser & Sunderland, 1998); the national average, which is published by the HCFA, is 30%. However, the quality of the psychiatric and mental health care provided to nursing home residents cannot be adequately or thoroughly measured by the rate of psychotropic drug use. It is important to be aware of the facility's and the practitioner's rates of use in comparison with others as a starting point for performance improvement efforts.

Dementia is associated with a high prevalence of psychiatric symptoms and behavioral disturbances. Agitation, the most commonly reported complication of dementia, refers to symptoms of disruptive behaviors, including inappropriate verbal outbursts, physical aggression, and nondeliberate motor activity. Ninety percent of patients with dementia have at least one behavioral symptom, and 60% have four or more problem behaviors (Tariot et al., 1993). This spectrum of behavioral symptoms includes physical destructiveness, verbal disruption, intrusiveness, impulsivity, and resistance to caregivers. Insomnia and sleep-wake cycle disturbances may develop, and nighttime wandering may compromise the patient's safety. Disrobing and sexually inappropriate behaviors often occur.

A major problem in dealing with disruptive behaviors is the recognition and treatment of depression and depressive features among nursing home residents. Depression in a frail elderly nursing home resident with dementia may present as agitated, disruptive behavior. A screening test such as the Geriatric Depression Scale may help identify symptoms in older nursing home residents (Sheikh & Yesavage, 1986). In patients with more advanced dementia, depression is often not even considered as a cause of behavioral disturbance. The clinician must maintain a high index of suspicion and consider depression in the differential diagnosis of all behavioral disturbances in the elderly. Other contributing factors include pain, hunger, thirst, or generalized discomfort.

Facilities may implement a variety of therapies to care for residents with behavioral problems. Behavioral management is often the key to therapy, as well as an important adjunct to treatment when psychiatric medications are indicated. Environmental and milieu management is extremely important, particularly for behaviors such as wandering, restlessness, and pacing. Therapeutic recreational modalities are vital to daily life and can help address issues such as loneliness, isolation, and boredom, which can lead to a variety of behavioral disturbances. Access to physical, occupational, and speech therapies; correction of sensory deficits; and promotion of independence are important aspects of any behavioral intervention.

The appropriate use of psychotropic medications in long-term-care facilities requires assessment and diagnosis, identification of target symptoms, attention to dosing, and monitoring. Desirable effects as well as negative side effects of the medication must be monitored and documented. Attempts at dose reduction must be made at least every six months for antipsychotic agents and every four months for benzodiazepine-type anxiolytic medications. Weekly nursing documentation for all patients on psychotropic medications is recommended. The duration of therapy must be individualized as part of a care plan that includes alternatives and adjuncts to medication, including behavioral, environmental, and activity therapies. Each category of psychotropic medication is subject to maximum daily dose recommendations. Clinicians are prompted by the regulations to screen for sedation, orthostatic hypotension, extrapyramidal symptoms, and the development of involuntary movements. An annual screening test such as the Abnormal Involuntary Movement Scale (AIMS) is easy to administer and can detect early signs of tardive dyskinesia (Munetz & Benjamin,1988). Evidence shows that more frequent AIMS testing is beneficial.

The pharmacist is an important member of the interdisciplinary team when evaluating the psychotropic medication regimen and documentation. Federal guidelines call for a monthly drug regimen review by a pharmacist for every resident of a long-term-care facility. The pharmacist should review the documentation for medication effectiveness and adverse reactions and screen for any irregularities in drug therapy that may arise in the course of treatment.

The OBRA-87 regulations appear to have had a significant impact on psychotropic medication prescribing practices in long-term-care facilities. There has been an overall reduction in the use of antipsychotic agents, although reports still indicate a rate as high as 42% (Lasser & Sunderland, 1998). Use of antidepressant agents appears to be slowly increasing and may indicate a greater recognition and treatment of depression. Use of PRN psychotropic medications has diminished significantly.

Unfortunately, published studies have not correlated the rate of psychotropic drug use with clinical outcomes (Lasser & Sunderland, 1998).

The use of psychotropic agents in long-term-care facilities remains a controversial subject because of the high prevalence of psychiatric disorders and concerns regarding access to mental health services in nursing homes. Staff must be educated in managing behavior. Psychotropic medications play an important role in the treatment of the mental disorders that are common among the residents of long-term-care facilities. However, the use of these agents must be accompanied by careful clinical judgment, assessment, and monitoring. As with all successful nursing home care, the interdisciplinary team treatment plan is an integral part of therapy.

MELINDA S. LANTZ
MARY SHELKEY

See Also
Behavioral Symptoms in Patients with Dementia
Depression in Dementia

Elder Mistreatment: An Overview
Restraints
Wandering

Internet Key Words
Chemical Restraints
Psychotropic Medications

Internet Resources
American Academy of Neurology
http://www.aan.com/public/practiceguidelines/list

American Journal of Geriatric Psychiatry
http://www.ajgp.psychiatryonline.org/gci/content/full/6/2/S41

American Psychiatric Association
http://www.psych.org/clin_res/pg_dementia

Family Caregiver Alliance
http://www.caregiver.org/factsheets/fs_rlnhC

Health Care Financing Administration
http://www.hcfa.gov

Q

QUALITY-OF-LIFE ASSESSMENT

Improved quality of life is the primary goal for clients receiving long-term care, but despite the best intentions, it can be easily obscured by ongoing pressures associated with scheduling of services and record-keeping requirements. With careful planning, a strategy can be developed to advance a quality-of-life focus in nursing care planning and documentation, clinical case conferences, and administration. Here, we illustrate the techniques using the example of working with home-care agencies through the Matching Home Care Service Time Patterns Project funded by the Fan Fox and Leslie R. Samuels Foundation, New York.

Introducing the Objective

The process begins by introducing the objective of advancing quality of life. A consultant or internal quality-of-life committee can create a partnership between the nursing service and the consultants. Activities are aimed at increasing staff and client communication about quality of life, which can be accomplished by directing staff attention to gathering the information required to match the time patterns of assisted functional activity to the time patterns of client need for such assistance. Other client needs can be addressed by communicating with clients about how they prefer routine services to be delivered. Increasing staff awareness of client-specific quality-of-life information can be accomplished in the course of expanding staff-client communication about quality of life.

Reviewing Current Practices

To determine how quality-of-life issues can be better integrated into care patterns, an outside consultant or internal committee should review patient charts and observe clinical case conferences to determine the nursing service's current practices in incorporating quality-of-life information, introducing systematic information on quality of life into the service setting, training nursing staff in techniques for obtaining quality-of-life information, and collaborating with staff in finding ways to address quality-of-life issues in clinical practice. The staff's role is to endorse the goals of the partnership and to supplement the information-gathering techniques with their current record keeping, care planning, documentation, and case conferences.

Nurses are well suited to be centrally involved in this process because, as professional advocates for the client, they are concerned about client preferences, and they are traditionally responsible for planning care and scheduling services. However, all members of the nursing team (e.g., director, administrator, nurses, social workers) must be informed of the objectives and activities so that changes will be understood and supported.

To demonstrate the service's practices in incorporating information on clients' quality of life, a few charts should be reviewed for a record of client-staff interaction on specific quality-of-life topics, including discussion of the care plan schedule, client's approval of the care plan schedule, client's personal care preferences, and the effect of health-related problems or services on quality of life. References to quality of life can be recognized by explicit descriptions of the client's emotional comfort, physical comfort, levels of mobility and task performance, social activities and personal relationships, awareness of needs, perceptions of current health and expectations for future health, sense of freedom to make decisions and choices, and views on the treatment and services received.

Obtaining New Quality-of-Life Information

A variety of instruments can be used to measure quality-of-life information for application to clinical service. Quality-of-life status can be measured using systematic interview guides such as the Quality of Life 100 Point Scale (QoL-100) questionnaire (Barrett et al., 2000). This visual analog obtains information on 10 domains: emotional comfort, physical comfort, mobility, task performance, social relationships and activities, awareness, health in general, future health, choices, and view of treatment and services. Supplementary scales can be selected from instruments such as the Clinical Comprehensive Assessment and Referral Evaluation (CLIN-CARE) (Gurland et al., 1990). This comprehensive geriatric assessment instrument measures activities of daily living, living conditions, depression, self-perceived health, chronic pain, effort tolerance, critical incidents, fears, stress, satisfaction with services, involvement in decisions, and positive qualities.

Needs and preferences can be measured using the paradigm of the BADL Time-Activity Pattern Schedule (Gurland, 2000), which records the match between the client's current time patterns of assistance in the basic activities of daily living (BADLs), including bathing, dressing, toileting, eating, and transferring, and the patterns for these activities in the period just prior to their needing assistance. The BADL Needs and Preferences Interview (Gurland, 2000) addresses client's general satisfaction and dissatisfaction with the time patterns involved in BADL care assistance and time-related preferences, such as how long a BADL takes and how frequently it occurs. Problems resulting from unmet needs are also reported.

Observing Clinical Case Conferences

Observing clinical case conferences is an opportunity to determine the degree to which quality-of-life issues are considered by the staff when assessing the needs of clients, planning care, or evaluating interventions. The Clinical Case Conference Inventory (Gurland & Barrett, 2000) lists the quality-of-life topic areas and is accompanied by a glossary of terms that also provides examples. As each topic area is addressed by members of the nursing

staff, the conversation can be rated, based on whether the topic was mentioned once (1), followed by a discussion (2), or resulted in a care plan (3).

Sections of the Clinical Case Conference Inventory address quality-of-life status expressed in ordinary language and in the quantitative terms used in the QoL-100 scale. Needs and preferences are expressed through client satisfaction with the timing of BADL assistance and with the relative timing of two or more assisted BADLs. Preferences for change include the time that assistance is given, the duration and frequency of an activity, the pace of an activity, the sequence of steps involved (e.g., to avoid a chill, the client prefers that the bath be run and bathing equipment collected before she is assisted with undressing), and the caregiver's attitude. Schedule issues include reasons for planned care not being received, awareness of contingency plans when planned assistance is not available, and the use of services (e.g., delivered meals) that reduce the need for personal assistance. Recording the number and variety of quality-of-life references made by the nursing staff and the progression to a plan of action can enhance suggestions made later to reinforce the results of the quality-of-life analysis.

Sensitizing the Staff

To emphasize the connection among patterns of service delivery, clients' views on their needs and preferences, and clients' self-perceived quality of life, and to ensure a change in clinical practice, staff must be sensitized. For example, staff can be provided with summaries, presented in bar-graph form, of clients' perceptions of the degree that health-related problems have adversely affected their quality of life in the 10 domains of the QoL-100. Before viewing the bar graph, the nurse marks on a QoL-100 schedule how he or she thinks the client reported his or her quality of life in each domain. Differences may represent a conflict in how the nurse and client view the effect of health problems on quality of life and indicate a need to implement interventions aimed at improving quality of life. The varying lengths of the bars may serve as a guide to prioritizing interventions.

Reviewing Public-Relations Materials

Written materials used by the nursing agency to promote its program should be reviewed to identify the presence of a quality-of-life mission statement and to back up suggestions that will be made later to reinforce the results of the quality-of-life analysis.

Reinforcing the Process

Once the charts of all the clients in the project have been reviewed, the intervention with the nurses completed, the case conferences observed and rated, and the public-relations materials reviewed, the staff should be presented with information on how to advance a quality-of-life focus in charting practices, clinical case conferences, and outside agency communications. For each of these areas, existing practice should be acknowledged, additional changes suggested, and the quality-of-life rationale stated. For example, if there is a lack of quality-of-life focus in the initial intake forms, a space in the client's record could be created stating that the care plan and schedule have been discussed and are agreeable to the client (or a surrogate, if the client is cognitively impaired). This provides documentation that clients of the nursing agency actively participate in the care plan and schedule from the onset, and it implies that clients' needs and preferences are considered.

Clinical case conferences often follow a format of presenting a client or clients and focusing on problems involved with care. The nursing team members involved with the client being discussed and sometimes the entire staff attend these conferences. In addition to the traditional approach, it could be suggested that they use quality-of-life issues (or domains) as a focus or theme. For example, the theme "freedom to make decisions about my life" could be discussed with respect to the following: How much control does the client have over how his or her life is conducted? Have health-related problems affected his or her range of choices related to daily activities, types of treatments received, and services scheduled? What is the importance of this freedom to the individual? How do clients respond to loss of freedom? What can be done to reaffirm in clients the feeling that they have *not* been excluded from choices about their lives? This approach changes the emphasis of the conference and forces a discussion of the effects of health-related problems on quality of life from the clients' perspective. With improved quality of life as the goal, discussion would move more swiftly in the direction of solutions and would encourage greater participation by everyone in attendance.

Regarding the mission statement of the nursing agency, it may need to be changed to directly state that quality of life is the goal of service.

VIRGINIA W. BARRETT
BARRY J. GURLAND

See Also
Autonomy
Elder Mistreatment: Overview
Home Health Care
Pain Assessment Instruments
Patient-Provider Relationships

R

REALITY ORIENTATION

Reality orientation (RO) is the presentation and repetition of specific information to confused and disoriented people. It was developed in 1958, in an attempt to rehabilitate severely disturbed war veterans. By the early 1960s, the first RO programs for people with dementia were described (Folsom, 1966), and two types of RO were differentiated. Twenty-four-hour RO was an active approach, with staff orienting people to reality at all times and in every interaction. Classroom RO was intended as a supplement for the more confused, consisting of sessions in which people were presented with RO material, typically in a classroom-type setting. RO boards, containing information such as the name of the hospital, its location, the time, day, date, month, year, and next meal, were commonly integrated into both approaches. The formal establishment of RO marked the beginnings of a breakthrough in dementia care. Its use appeared logical, as recent memory and orientation are two of the primary deficits in dementia.

RO is based on a number of theoretical models. First, there is evidence that people with dementia can learn. Second, it has been argued that keeping the brain stimulated helps activate and maintain neurons in both Alzheimer's disease and normal aging. Hence, the mental stimulation of RO might reduce the rate of decline in dementia. Third, Kitwood's (1993) theory of malignant social psychology identifies the way in which the environment in many care settings and treatment by staff (including labeling and stigmatization) frequently perpetuate the symptoms of dementia. The original aim of RO was to encourage positive interaction between staff and clients, working directly against dementia stereotypes. Fourth, dementia, and indeed normal aging, are often accompanied by many sensory losses. RO might bring the person more in touch with his or her environment by increasing sensory and environmental stimulation.

Twenty-four-hour RO involves orientation as part of every interaction; for example, "Good morning, Mr. Palmer, my name is Mrs. Greene. It's nice weather for October, isn't it? We will be eating our Sunday roast in an hour, at one o'clock." It is commonly accompanied by environmental cues such as pictures on doors demonstrating the function of the room, color coding, and signposts indicating the direction of facilities. These cues can be used by staff to prompt appropriate behavior and encourage orientation. In classroom RO, groups of approximately four to eight people usually sit in a "classroom," where they are presented with information on a board, which is repeated. Groups are run by one or more facilitator; sessions last for 30 to 60 minutes and can occur daily or only once a week. Activities might involve asking people to name another person in the group or a member of staff, using clocks and calendars, and repeating the day, date, location, season, and any forthcoming events or festivals. Other activities might include preparing food, designing and using maps, creating collages for discussion, gardening, and practicing using money. Discussion can be prompted through newspapers, television, short films, radio, and slide shows. Multisensory stimulation, such as touching objects, smelling, and tasting, can facilitate orientation; for example, summer flowers or autumn leaves can be used. Praise and rewards are integral to the success of the process. Sessions are often enhanced with tea and refreshments, greetings and handshakes, and a pleasant environment (e.g., a living room).

RO waned in popularity around the mid-1980s, largely due to criticism that the approach was sometimes too prescriptive. More recent RO work, or cognitive stimulation, involves outpatients attending twice-weekly stimulation groups (Breuil et al., 1994). Tasks include naming and categorizing objects, developing lists of associated words, and doing arithmetic, such as calculating prices on shopping lists. The significant improvements in cognition that have been demonstrated indicate that this treatment may be more theoretically advanced than previous work. It is based on growing knowledge of the neuropsychology of dementia.

Of equal importance to the content of the RO sessions is the way information is presented. It is best to use short, simple sentences; encourage response and repetition; use past experiences as a bridge to the present; keep conversation specific; encourage humor; and provide a commentary on events (Holden & Woods, 1982). One should never agree with a person who is clearly wrong, but choose a more appropriate response option. This might involve gentle correction (on less sensitive issues), distraction, or validating the emotional content of the person's utterance. Recent research has demonstrated the importance of the method of "teaching." For example, errorless learning has been used to teach face-name associations for those with early dementia. People are taught names through repeated exposure and repetition, but in order to reduce error, they are encouraged to respond only if they are sure that the answer is correct. Results have been promising and suggest that this method facilitates implicit memory, which is relatively well preserved in Alzheimer's disease. Additionally, as errors are minimized, participants have an increased experience of success.

Research has tended to focus on classroom RO. Results are mixed, although many studies demonstrate improvements in verbal orientation, cognition, memory, learning, language, and behavior. In some studies, staff have reported that RO helps them get to know patients better; other staff dislike the approach. A Cochrane systematic review (Spector, Orrell, Davies, & Woods, 1998) was conducted to collate evidence of the effectiveness of classroom RO for dementia. The outcomes from six randomized controlled trials with 125 subjects were combined, and results showed significant improvement in both cognition and behavior after RO. Some of the control groups engaged in social activities, suggesting that results were due to the specific effects of RO rather than merely the nonspecific effects of interaction. This evidence-based review

was valuable in demonstrating how a treatment that is somewhat out of fashion has the potential to significantly improve cognition and behavior.

RO has generated considerable debate over the years. Critics contend that it has been applied in a rigid, impersonal way by uninspired staff who focus on information and instructions rather than on human, interactional processes. Further, it has been suggested that changes are artifactual, having no real impact on people's day-to-day lives, and that reorientation can actually reduce self-esteem. Folsom's (1966) early ideas have perhaps been interpreted by some as a banal "cure" for dementia by using stringent corrective measures. What appears imperative for RO's success is that it be applied sensitively to people who *want* to receive it, that people be treated as individuals, and that the approach be adapted accordingly.

There is some debate about for whom RO is suitable. Clearly, the most likely recipients are people who want to be oriented, such as those at early stages of dementia who are aware of their memory losses and want to maintain their independence. In contrast, those at later stages of the illness may have less interest in factual information and may find the approach distressing. The highly successful cognitive stimulation groups reported by Breuil and colleagues (1994) are for people living in the community, recently diagnosed with dementia, who have *chosen* to attend the groups to slow their cognitive decline, with the aim of being able to function independently for a prolonged period. Some interactions may trigger inappropriate responses, such as when a person asks for his or her (dead) mother. Here, validating the person's feelings by taking the time to talk about his or her mother might be more comforting than simply reminding the person that she is dead. The sensitivity and ability to use flexible approaches depend on good staff-patient relationships and the thought put into every interaction.

The Cochrane review suggested that treatment effects of RO may be difficult to sustain once the program terminates; hence, weekly maintenance groups may be valuable. For maximum effect, RO could become part of a broader rehabilitative program designed for the individual, in which behavior learned in sessions is reinforced through the 24-hour approach. For example, a person's activities of daily living could be assisted by the use of memory aids or diaries, and environmental manipulation could be used to aid self-care and locating the bathroom. As people are more likely to learn covertly, rather than overtly, introducing more personal as opposed to factual information might be more effective. The resources offered by television, media, radio, and so on can be excellent in raising interest and bringing relevance to the RO information offered.

In conclusion, the benefits of RO have been demonstrated by research. RO creates a social environment and has the potential to improve staff-patient relationships through increased interaction and perhaps a greater understanding of each other. Costing little in terms of time and resources, RO requires that staff undergo brief rather than extensive training. Each person must be treated with respect and dignity, with the question being asked: What is most important to this person? Does he or she want to be oriented and more cognitively aware? How does being more oriented affect the person in other ways? The benefits of RO depend on it being applied in a sensitive and flexible manner with willing participants.

AIMEE SPECTOR
MARTIN ORRELL

See Also
Dementia: Overview
Validation Therapy
Vascular and Lewy Body Dementia

Internet Key Words
Alzheimer's Disease
Cognition
Dementia
Memory
Orientation

Internet Resources
Reality Orientation
*http://easyweb.easynet.co.uk/~jollespio/
demen31.htm*

RO Cochrane Review
*http://www.update-software.com/ccweb/cochrane/
revabstr/ab001119.htm*

RECREATION
See
Therapeutic Recreation Specialists and Recreation Therapists

RECTAL PROLAPSE
See
Uterine, Cystocele/Rectocele, and Rectal Prolapse

REHABILITATION

Disability is disproportionately high among the elderly. Appropriate rehabilitation can restore form or function after illness or injury through the use of a variety of interventions directed at the disablement process.

The disablement process is the process by which disability, or limitations in functional ability, occurs. The most recent model of the disablement process disseminated by the World Health Organization (WHO) suggests that disability occurs when impairment of one or more organ systems, for example, muscle weakness, causes difficulty carrying out self-care activities (WHO, 1999). Impairment or disability that causes difficulty participating in society is sometimes called a handicap.

These relationships are thought to be bidirectional, such that difficulty at the activities (disability) level may cause impairment at the organ system level. For example, people who use wheelchairs are at greater risk than others for developing shoulder pain. Whether someone has a disability or experiences a handicap may be influenced by contextual factors such as environmental barriers or lack of social support. For example, someone who uses a wheelchair for mobility may be able to continue employment if the workplace environment is wheelchair accessible. The most recent WHO model also emphasizes the use of neutral terms, such as "difficulty with activities" rather than the negative term "disability."

Interventions

Rehabilitation interventions address the individual's ability to interact with the environment through restoration of full, modified-independent, or assisted function. Some diseases and disease pathology may affect the benefit gained from specific rehabilitation interventions or alter the way the interventions are applied. Rehabilitation interventions include, but are not limited to, exercise prescription, adaptive techniques (modifications in the way an activity is performed), assistive technology (e.g., canes, walkers, wheelchairs), physical modalities (e.g., heat, cold, ultrasound), orthotic devices (braces, splints), and prostheses (artificial limbs).

Exercise can help prevent disability and, even in the presence of pathology, can reduce the extent of disability and counteract the adverse effects of immobility, whether due to systemic illness or trauma. An increasing number of people are discovering the beneficial effects of exercise through voluntary participation in organized community-based programs. However, older adults, particularly those who are medically ill or disabled, may benefit from an evaluation by a rehabilitation specialist, who can design an individualized program that takes into consideration conditions that may affect response to exercise. Exercise programs typically include activities to increase flexibility, muscular strength, aerobic endurance, posture, balance and bone health.

Adaptive techniques involve modifying a task so that it can be performed despite physical limitations. Proficiency with adaptive techniques enables individuals to interact with the environment more favorably—for example, learning to dress independently using one arm following onset of hemiplegia after a stroke. An assistive device, such as a raised toilet seat or a cane, may be recommended to make a task easier and safer. Assistive technology, such

as a power wheelchair, can be costly initially, but it can dramatically increase an individual's ability to maintain independence. Assistive devices can be purchased with or without a prescription and with or without the guidance of a rehabilitation specialist, although a doctor's prescription is usually needed for insurance reimbursement. Health care insurers do not, however, uniformly cover all types of assistive technology; for example, reimbursement may be available for a wheelchair but not for a wheelchair ramp. Rehabilitation specialists can recommend which devices will be most helpful in improving function and facilitating independence, and they may assist with equipment design and purchase (e.g., provide advice about insurance reimbursement). Rehabilitation therapists are skilled in fitting and training patients (and caregivers) to use their new equipment safely and efficiently.

Physical modalities include a variety of interventions that promote healing by decreasing local inflammation, muscle spasm, or pain. Physical modalities include ultrasound, diathermy, transelectronic nerve stimulation (TENS), electrical stimulation, whirlpool, massage, and application of heat or cold. Research data on the efficacy of many physical modalities are limited, but there is ample anecdotal support for the benefits of cold and heat, such as cold compresses following acute joint injury and the application of heat to reduce arthritic pain.

Orthotic and prosthetic devices can be critical elements in restoring function, but the cost can be prohibitive, and insurance does not necessarily cover these devices, sometimes driving patients to forgo their purchase. Reimbursement varies, depending on the specific device and the specific insurance program and the intermediary interpreting the regulations. Physicians and rehabilitation professionals should make every effort to advocate for patients who require orthotic or prosthetic equipment.

Professional Disciplines

There are many different types of rehabilitation professionals who assist with the process. A physi-

Factors that May Affect Rehabilitation Interventions, Prognosis, or Goals

Factor	Potential Effect
Cognitive impairment	Goals may be more limited. Take advantage of skills the patient already has; use interventions that don't require carryover.
Disability has been present for many years	Goals may be more limited and directed to compensatory strategies or treatment of deconditioning.
Motivation is limited	Goals need to be well defined and reached in measurable steps.
Patient had prior rehabilitation for the same problem	Rehabilitation may be limited unless new functional decline has occurred.
Terminal illness	Intervention is directed toward reducing pain and caregiver burden.
Severity of disability	Extremely mild disability may not require intervention. Extremely severe disability may have limited potential for benefit.
Cultural circumstances	Absence of a caregiver, financial limitations, cultural beliefs may preclude use of certain techniques or technologies.
Malnutrition	Unable to build muscle; rehabilitative interventions may be limited.
Delirium or altered level of consciousness	Unable to learn or cooperate; rehabilitation may not be appropriate until resolved.
Hemodynamic instability	May make it unsafe to carry out certain types of exercise.
Occult fracture, bony metastasis	Weight-bearing exercise could worsen fracture or cause fracture; rehabilitation interventions may be limited.
Acute infection (e.g., bladder infection, pneumonia)	May cause confusion, fatigue, or hypotension; rehabilitation may not be appropriate until problem resolved.
Acute skin or joint infection	May cause fatigue, pain, or muscle splinting; rehabilitation may not be appropriate until resolved.
Acute inflammatory disease (e.g., certain rheumatological and neuromuscular conditions)	Resistive exercise may impair recovery; rehabilitative interventions may be limited.
Acute orthopedic conditions	Joint instability may preclude use of certain exercises, and functional goals may be limited.
Medications that may affect rehabilitation treatment: psychotropics, beta-blockers, those with parkinsonian side effects, antihypertensives, anticoagulants	May alter ability to cooperate during rehabilitation or to carry out certain types of exercise or reduce the effectiveness of treatment. Rehabilitation may not be appropriate until problem resolved.

Modified from Hoenig, H., Nusbaum, N., & Brummel-Smith, K. (1997). Geriatric Rehabilitation: State of the Art. *Journal of the American Geriatrics Society,* 45(11), 1371–1381.

atrist is a physician who specializes in rehabilitation. Occupational therapists, physical therapists, and speech-language pathologists are educated in didactic and internship programs, increasingly at the postgraduate level, and certified or licensed to provide rehabilitation services. They provide rehabilitation services in all patient care environments, including community- and home-based settings. Other rehabilitation providers may be involved in the patient's care, depending on the setting and the patient's unique needs, including dietitians, kinesi-otherapists, music therapists, psychologists, recreational therapists, and vocational rehabilitation specialists.

Programs

Rehabilitation programs are provided in many different settings. In the acute-care hospital, rehabilitation may be initiated, but the primary focus of the rehabilitative treatment is facilitating discharge planning and discharge to a less acute level of care or home. Rehabilitation hospitals and subacute-care settings (including some skilled care facilities) provide the most comprehensive rehabilitation, typically using an organized team of rehabilitation providers and daily treatment. Rehabilitation services in long-term-care settings vary considerably from site to site, varying both in the frequency of treatment and in the types of practitioners who provide treatment. This is because the patients in these settings may not need the same frequency and type of therapy as do patients recovering from a recent stroke, amputation, or spinal cord injury, and because of differences in reimbursement for therapy in different settings.

Home health care enables patients to receive treatment from rehabilitation professionals in their homes and permits the therapist to address the patient's functional limitations as they are encountered in his or her own environment. A certain number of home visits are covered by Medicare; HMO reimbursement is variable. However, the frequency of therapy provided in the home is often limited (e.g., three visits a week from a physical therapist) compared with the inpatient setting (e.g., daily or twice-daily physical therapy).

Community-residing patients can utilize outpatient rehabilitation services. In some states, patients can self-refer for outpatient physical therapy; however, insurance companies usually require physician referral for reimbursement of the therapy. Primary care providers may refer their patients to licensed rehabilitation professionals for fitness programs so that the program can be individually designed, with consideration of coexisting pathology. Insurance reimbursement may or may not be available for fitness programs, depending on the patient's diagnosis. It is always advisable to contact the insurance carrier before starting rehabilitation to ensure that reimbursement will be available, as there is a fair amount of variation among third-party reimbursers (e.g., Medicare, private insurance), and regulations about reimbursement for specific conditions or treatments change frequently. For example, reimbursement may be available for a fitness pro-

gram as part of cardiac rehabilitation or to treat deconditioning.

Reimbursement for geriatric rehabilitation comes from four primary sources: Medicare (Part A and Part B supplemental), Medicaid, the Department of Veterans Affairs, and private insurers. Reimbursement may vary considerably, depending on the insurance company or fiscal intermediary (e.g., Medicare).

Treatment supervised by a rehabilitation therapist is generally considered appropriate and usually will be reimbursed, as long as the patient is making progress toward a measurable functional goal. Other rehabilitation goals include determining realistic goals in light of the patient's own goals and condition, educating patients about their potential for recovery, teaching self-care in the presence of an impairment, and advising patients and caregivers on the type of care that may be necessary when transitioning from one environment to another (e.g., skilled nursing facility to home) (Gresham et al., 1995). When function has been restored to an optimal or maximal level and the patient and family or caregivers have safely mastered new skills, the discharge plan is discussed and implemented. Recommendations may be given for independently continuing certain treatment interventions for the purpose of maintaining and promoting optimal function. Community-based programs are, in increasing numbers, able to accommodate special needs of the geriatric population and may be utilized to maintain function after formal rehabilitation has ended. Recreational therapists are fairly common in these programs. Occupational therapists, physical therapists, and physicians are more likely to play a consultative role.

Individuals who may benefit from rehabilitation referral are those with preexisting limitations in functional ability; those with a change in functional ability due to disease, illness, or injury; or those at increased risk for loss of functional abilities. Referral to a rehabilitation professional can be of great benefit when guidance is needed to eliminate, minimize, or modify the contextual factors that, in many instances, magnify or create disabilities. For example, persons who are medically unstable may not be appropriate candidates for rehabilitation services. However, primary care providers must continue to reassess the need for rehabilitation interventions, particularly when the prognosis for recovery is favorable; the detrimental effects of immobility are much easier to prevent than to remediate.

HELEN HOENIG
CORRIE J. ODOM

See Also
Activities of Daily Living
Assistive Devices
Deconditioning Prevention
Exercise and the Cardiovascular Response
Fractures
Gait Disturbances
Multidimensional Function Assessment: Instruments
Occupational Therapy Assessment and Evaluation
Physical Therapy Services

Internet Key Words
Assistive Technology
Disability
Exercise
Mobility

Internet Resources
ABLEDATA
http://www.abledata.com

American Academy of Physical Medicine and Rehabilitation
http://www.aapmr.org

American Occupational Therapy Association
http://www.aota.org

American Physical Therapy Association
http://www.apta.org

American Speech and Hearing Association
http://www.asha.org

American Therapeutic Recreation Association
http://www.atra-tr-org

RELOCATION STRESS

Older adults commonly experience three relocations as they age (Litwak & Longino, 1987). First, they may relocate to a desirable geographic location shortly after retirement. Some years later, they may return to an area to be close to family members and desired and needed medical care. As their health and care needs change, they may relocate to more supportive housing, such as assisted living, certified retirement communities, group homes, or senior apartments that provide various services.

Many older adults must relocate due to declining functional ability or worsening health. They are particularly likely to move if they are cognitively impaired and have limited assistance in their homes (Miller, Longino, Anderson, James, & Worley, 1999). This move is commonly to a long-term-care facility when their needs surpass the availability and capacity of community services and family caregivers. Among those with cognitive impairment, particular sequelae have been associated with relocation to a nursing home: incontinence, difficulty walking, and excessive nighttime activity (Hope, Keene, Gedling, Fairburn, & Jacoby, 1998). African American elders with Alzheimer's disease are most likely to relocate to nursing homes if they are unmarried and have a limited ability to perform activities of daily living (Miller, Prohaska, & Furner, 1999). Rural residents relocate to nursing homes because of limited housing options, inappropriate referrals, and burdened caregivers (Congdon & Magilvy, 1998). Residents of more intense care environments, such as graduated retirement homes or nursing homes, have to relocate one or more times within the facility or to a new facility as their needs change or if the facility changes its mission.

Nearly all professionals working with older adults have had the experience of trying to help an older adult adjust to the process of relocation, in either a professional or a personal capacity. The widespread incidence of relocation stress in varying intensities among older adults should be of concern to health care professionals, who may be well positioned to prevent and help older adults recover from some of the stress of relocation and facilitate adjustment to the new home.

Identifying Relocation Stress

Two questions commonly considered by those helping older adults through the relocation process are (1) who is most vulnerable to the stress associated with relocation? and (2) how is this stress identified? Relocation stress may occur with greater regularity and intensity when an older adult relocates precipitously, with limited choice or input into the decision. Involuntary moves in which there is little perceived improvement in living conditions may be especially stressful. Perhaps the most stressful situation, and ironically one of the most common, is when the relocation occurs in response to a health crisis. In this situation, the older adult may feel rushed into the relocation, have little choice in the location of the new home, and have minimal time to prepare. The decision to relocate is commonly made by the older adult's physician, adult child, or another family member (Johnson, Schwiebert, & Rosenmann, 1994).

Any older adult who relocates, but particularly those who fit the potentially most stressful scenario (precipitous move; little choice, especially due to cognitive impairment; little preparation; and limited perceived improvement in living conditions), may show signs of depression, anxiety, withdrawal, and morbidity (Lander, Brazill, & Ladrigan, 1997). These may be accompanied by weight loss, anorexia, poor nutrition, falls, a decline in self-perceived health, reduced social support, and loss of a sense of coherence (Johnson, in press). Any of these manifestations may occur during the process of relocation decision making, during the actual move, or within three months afterward. The first month following relocation may be the most difficult period, and signs of stress may be most obvious.

Careful, multidisciplinary assessment during these periods is critical to identify signs of relocation stress. Use of a well-tested instrument, such

as the Geriatric Depression Scale (Yesavage et al., 1983) or the Self-Efficacy for Functional Activities Scale (Resnick, 1999), may assist in early identification of relocation stress. However, it is also necessary to assess nutritional status, engagement in social behavior, physical activity level, sleep, and morbidity. These areas are equally important to assess in those who are cognitively impaired.

Preventing Severe Relocation Stress

As with most negative health situations, preventing severe relocation stress is a better approach than trying to minimize it once it is already present. Ensuring the older adult's participation in making the decision to relocate is the earliest preventive measure. This is a multidisciplinary task for those helping the older adult in the situation preceding relocation. Clearly, not all older adults will want to participate, and these persons should not be expected to do so. Participation necessarily assumes some cognitive capacity and includes considering alternatives and exercising choice. Recognition that relocation is needed may not be as immediate for the older adult as for participating family members. Thus, professionals working with families in which relocation is being contemplated need to ascertain the degree of difference in view, if any, between the older adult and family members. This is best done by interviewing them separately. Family members should not pressure the older adult into the decision but instead discuss with him or her the benefits of the new location in terms of how it may improve living conditions. Participation and choice in selecting the new residence may help prevent severe relocation stress.

Beyond choice, careful preparation may minimize the stress of relocation. This also involves a multidisciplinary approach. For example, the case manager or social worker may establish the groundwork by having staff members of the new residence contact the older adult and family to begin to establish rapport and an identification with the new location. The nurse practitioner or primary care physician should conduct a thorough assessment and transmit the results to those responsible for new residents at the new location. This may help ensure that the most appropriate and beneficial services and care are made available to the relocated older adult.

Each of these measures assumes that there is no emergency or acute need for relocation. However, this is commonly not the case. Hospitalized older adults often must relocate to the first available place (e.g., a nursing home) when their inpatient days have elapsed, or to a place that they may have tentatively considered years earlier when they put their name on a waiting list. Alternatively, hospitalized elders might have to relocate to an adult child's or other relative's house with home care. When relocation occurs precipitously, it may seem that choice and participation are not possible. However, even limited participation by the older adult in the decision can be helpful. Deciding which belongings to take to the new residence, what new things are needed, and what will be done with belongings not taken along may help minimize relocation stress, assuming that this is done in a fairly unhurried manner. Decisions about the new residence, such as room color and furniture arrangement, may help prevent severe relocation stress.

Anticipatory planning and preparation involve making visits to the new residence at different times of the day and week, meeting and visiting with residents and staff, having meals there, and viewing the new living space. It also includes asking and receiving answers to questions about such things as policies, special services, programming, and events.

Treating Severe Relocation Stress

Although preventive strategies are the optimal approach in relocating older adults, these may not always be used or effective. Most who relocate experience varying degrees of stress. Such stress may facilitate adaptation, as changes in behavior, attitudes, or both are necessary for older adults to adjust to a new residence. To minimize severe relocation stress, reminiscence therapy, massage,

therapeutic touch, bibliotherapy, active listening, prayer, music therapy, art therapy, social programming, and linking new residents with a "buddy" who is not a newcomer for social support may be helpful. Pharmacological treatment for depression and anxiety may be needed in severe cases. Continuing assessment of nutritional status, self-care, and social behavior is needed to monitor progress. Older adults and their family members should participate in the process of assessment and intervention.

Despite more than three decades of research demonstrating the profound effects of relocation stress on older adults, precipitous, relatively unplanned, and involuntary relocation still occurs regularly. Professionals who work with older adults and their families should advocate for the older adults' choice, participation, and preparation to decrease the incidence of this largely preventable yet potentially damaging situation.

REBECCA A. JOHNSON

Internet Key Words
Institutionalization
Nursing Home Placement
Relocation Stress Syndrome

Internet Resources
Appropriations for rural housing
*http://www.aarp.org/wwstand/testimony/
    1998hm3935.html*

Caregiving-housing options
http://www.aarp.org/caregive/4-house.html

Consumer Information Center
http://www.pueblo.gsa.gov

RESPITE CARE

Respite care refers to short-term supervisory, personal, and nursing care provided to impaired older adults, typically those who cannot be left alone because of physical or mental disabilities. The purpose of respite care is to provide the informal caregivers of impaired older adults with temporary relief or respite from their caregiving responsibilities. Of all the services designed for community-dwelling impaired older adults, respite care is most firmly rooted in recognition of the social, primarily family context within which caregiving occurs.

The need and rationale for respite care services emerged primarily as a result of overwhelming research evidence that caregivers are at substantial risk for psychological distress, clinical depression, social isolation, and perhaps exacerbation of physical illness and financial problems. Nearly two decades of evidence document the high proportion of caregivers who experience one or more of these problems (see Aneshensel et al., 1995; Wright, Clipp, & George, 1993). In addition, 25 years of research on the predictors of institutionalization of impaired older adults clearly indicate that older persons at greatest risk of nursing home placement are those without families and those whose families are no longer willing or able to tolerate the demands of home care. And there is considerable evidence that caregivers' levels of stress and well-being strongly predict institutionalization of the impaired relatives for whom they care (Colerick & George, 1986; Pruchno, Michaels, & Potashnik, 1990). These facts fueled hope that respite care might also delay institutional placements of impaired older adults.

There have been few controlled trials examining the effects of respite care services on caregiver well-being or on the risk of institutionalization of impaired older adults. Evidence to date is less than compelling. Montgomery and Borgatta (1989) reported no differences in the amount of time caregivers spent in their caregiving tasks and only modest decreases in caregiver burden for respite recipients. Lawton, Brody, and Saperstein (1989) found a statistically significant but modest delay (22 days, on average) in institutional placement among patients whose caregivers received respite services. Despite high levels of satisfaction with respite care reported by recipients, respite care had no demonstrable effect on caregiver well-being. George, Fillenbaum, and Burchett (1988) found receipt of respite care services to be associated with decreased time devoted to caregiving, increased participation in social activities, and increased satisfaction with social

participation, but they observed no improvements in caregivers' levels of stress symptoms or physical health. Clients reported high levels of satisfaction with the respite services they received. A common denominator of all three studies was difficulty in recruiting sufficient numbers of caregivers—even when the respite care services were heavily subsidized—to evaluate their effects.

More recent research has focused largely on the factors that make respite care services more or less attractive to caregivers and the characteristics of caregivers who do and do not use respite care services. An example of the former is a study by Kosloski and Montgomery (1993) that examined the effects of perceptions of respite care characteristics and quality on utilization of such services. Caregivers' perceptions of the quality of the service, its convenience, and its usefulness were shown to be significant predictors of respite care use after characteristics of the caregiver and impaired older adult were statistically controlled. In an Australian study, Braithwaite (1998) compared caregivers who did and did not use institutional respite care services. Not surprisingly, she found that respite care use was significantly related to higher task demands of caregivers. More surprising was the importance of the quality of the caregiver–care recipient relationship. Poor-quality, dysfunctional relationships between caregiver and care recipient were the strongest predictor of respite care use, whereas caregivers' appraisals of intimacy were associated with decreased use of respite care.

There is consistent evidence that caregivers who use respite care services are likely to use other community-based services (e.g., help with housekeeping, private sitters) as well (George et al., 1988; Kosloski & Montgomery, 1993). The meaning of this propensity to use service is unclear. It may be that some caregivers are simply "service oriented," a viewpoint supported by evidence that high levels of education, income, and community involvement are related to the use of multiple services to assist in caregiving. Alternatively, use of multiple services may reflect the realities of the caregiving situation, indexing higher caregiving demands and more impaired care recipients. This explanation is also supported by research evidence that caregivers who receive community services are more likely to institutionalize their impaired relatives than are those who do not use such services.

Research on respite care continues to be hampered by conceptual ambiguity. The term *respite care* is defined differently across studies, being variously operationalized as in-home service, a brief period of institutional care, and even a synonym for adult day care. As long as the research base rests on a variety of respite care models, consistent and cumulative findings are unlikely. A sensible approach would be to examine adult day care as a separate type of service and to consistently distinguish between in-home and institutional respite, which undoubtedly differ in terms of their predictors, availability, and cost.

Two final issues merit note. First, so long as respite care services are not covered or only marginally covered by private health insurance, Medicare, and Medicaid, it will be difficult to disentangle the effects of reimbursement on respite care utilization. Second, respite care seems to have evolved to less of a research issue (in terms of its effectiveness, cost-effectiveness, and so forth) and more of a consumer issue. Regardless of the lack of compelling scientific evidence documenting its cost-effectiveness, advocacy groups such as the Alzheimer's Association and the American Association of Retired Persons strongly favor the inclusion of respite care in public health care financing programs. For the foreseeable future, political processes will probably be more important than research results in placing respite care services on the policy agenda for community-based long-term-care services.

LINDA K. GEORGE

See Also
Caregiver Burden
Consumer-Directed Care

RESTRAINTS

The prevalence of physical restraints in nonpsychiatric settings, estimated in 1989 to affect 500,000 elderly persons daily in hospitals and nursing homes, led many to conclude that a restraint crisis existed. High prevalence in the United States was sharply contrasted with what appeared to be lesser use in several Western European countries. The historical antecedents for these differences appeared to be related to American beliefs that had become embedded by the end of the 19th century: restraint use was therapeutically sound, was necessary to control troublesome behavior, and prevented tragic accidents and injuries.

For nearly 100 years, those beliefs were largely unchallenged; debate concerning the efficacy of physical restraint was limited, and alternative interventions were rarely considered. The efforts of advocacy groups and committed clinicians, changes in nursing home regulation and accreditation standards for hospitals, warnings from the Food and Drug Administration (FDA), and research demonstrating successful restraint reduction have forced a complete reexamination of their use. Although prevalence has declined in U.S. nursing homes since 1990, restraint use and the problems associated with it remain a global concern.

A physical restraint is attached to or adjacent to a person's body, cannot be removed easily, and restricts freedom of movement. Bilateral full-length side rails are also considered restraints. It is important to keep in mind that physical restraints are often used in conjunction with psychopharmacological restraints. When these drugs are given for purposes of discipline or convenience and are not required to treat specific medical or psychiatric conditions, they are considered chemical restraints.

Physical restraints are applied in hospitals and nursing homes primarily for three reasons: fall risk, treatment interference, and behavioral symptoms. To date, no scientific evidence has demonstrated the efficacy of restraints in safeguarding patients from injury, protecting treatment devices, or alleviating such behavioral symptoms as wandering or agitation. Several recent studies, in fact, suggest relationships between physical restraints and falls, serious injuries, or worsened cognitive function (Capezuti, Strumpf, Evans, Grisso, & Maislin, 1998).

Nevertheless, most health care professionals and other caregivers see few alternatives to restraint use in some situations. Hospitals and nursing homes often do not have personnel with expertise in aging or with the requisite skills for assessing and treating clinical problems specific to older adults. Fears about legal liability (albeit misplaced), lack of interdisciplinary discussions about decisions to restrain, and staff perceptions about resident behaviors also influence restraint practices. Insufficient staffing levels and the costs of hiring additional employees have long been regarded as obstacles to minimal use of physical restraints; however, data show that caring for nursing home residents without restraints is less costly than caring for residents who are restrained (Phillips, Hawes, & Fries, 1993).

Continued use of physical restraints is paradoxical in view of mounting knowledge about their considerable ability to do harm. Physical restraints reduce functional capacity as the person quickly loses muscle strength, steadiness, and balance when restricted to a bed or chair. Physical restraint can also cause problems of elimination, aspiration pneumonia, circulatory obstruction, cardiac stress, skin abrasions or breakdown, poor appetite and dehydration, and accidental death by asphyxiation. Immobilization of elderly patients by prolonged use of restraints can lead to many serious biochemical and physiological effects as well. Abnormal changes in body chemistry, basal metabolic rate, and blood volume; orthostatic hypotension; contractures; lower extremity edema and pressure ulcers; decreased muscle mass, tone, and strength; bone demineralization; overgrowth of opportunistic organisms; and electroencephalographic changes have been documented (Castle & Mor, 1998; Evans & Strumpf, 1989).

Attempting to restrain a frightened, delirious patient increases his or her panic and fear of danger and can produce angry, belligerent, or combative behavior. Other emotional responses include a sense of abandonment or desolation, loss of control, reduced self-esteem, depression, and withdrawal. A vicious circle occurs when the harms of restraint are combined with the characteristics of persons likely to be restrained—usually those of advanced age who are physically and mentally frail, prone to injury and confusion, and experiencing invasive

treatments. Although the answer to the question "Which comes first, physical restraint or decline in function?" is not entirely clear, the evidence is compelling that prolonged physical restraint contributes to frailty and dysfunction.

Several reports of restraint reduction in nursing homes and one clinical trial show that the prevalence of physical restraints can be significantly reduced without increasing serious injuries or hiring more staff (Evans et al., 1997). This can be accomplished through implementation of a range of alternative approaches to assessment of, prevention of, and response to the behaviors routinely leading to restraint. However, changes in fundamental philosophy and attitudes among institutions and caregivers must occur. In settings where restraints have been reduced, there is a strong emphasis on individualized, person-centered care; normal risk taking; rehabilitation and choice; interdisciplinary team practice; environmental features that support independent, safe functioning; involvement of family and community; and administrative and caregiver sanction and support for change. The presence of professional expertise, particularly medical directors and expert nurses with education and skill in geriatrics, is crucial for sustained cultural change.

Although legislation and other forms of external regulation or control do not change beliefs or entirely alter entrenched practice, the Nursing Home Reform Act, enacted in 1990, helped raise standards in nursing homes. Guidelines for surveyors of long-term-care facilities state that use of restraints must occur within a context of rigorous clinical assessment and comprehensive care planning that weighs risks against any potential benefit and informs the resident or representative of those risks. Interventions for specific problems must be tried and documented; if restraints are used, specific indicators warranting such use must also be documented. Likewise, the guidelines require careful scrutiny for the use of unnecessary drugs, especially long-acting benzodiazepines, hypnotics, anxiolytics, sedatives, certain drugs for sleep induction, barbiturates, antipsychotics, and antidepressants. The FDA, in response to the known risks of physical restraints and reports of restraint-related deaths, mandates that all devices carry a warning label concerning potential hazards.

Following a decade of emphasis on restraint reduction or elimination in nursing homes, clinicians, researchers, and regulators have recently focused attention on these practices in acute-care settings. The Joint Commission on Accreditation of Healthcare Organizations and the Health Care Financing Administration define restraint use as both physical and chemical. Standards mandate that restraints be used only to improve well-being in cases in which less restrictive measures have failed to protect the patient or others from harm. In addition, continual individualized assessment and reevaluation of the patient by clinicians and consultation with the patient's own provider must occur with restraint use. Direct care staff must also be trained in the proper and safe use of restraining devices.

Current approaches to restraint reduction vary along a continuum from promotion of restraint-free care to an attitude of tolerance for restraint use under certain circumstances. To some extent, successful (although incomplete) reduction of physical and chemical restraints in nursing homes underscores the need to achieve the same changes in hospitals, where a disproportionately high incidence of iatrogenesis occurs, much of it exacerbated by the use of physical restraints and adverse reactions to psychoactive drugs. The resulting complications—especially delirium, pressure ulcers, infections, and fall-related injuries—can add dramatically to the cost of care by contributing to further loss of function.

Nonuse of physical restraints and appropriate use of psychoactive drugs are increasingly considered the standard of care for elders in all health care settings. Such a standard will challenge professional caregivers to use comprehensive assessments to make sense of individual behaviors and to employ a range of interventions that enhance physical, psychological, and social function, as well as to acknowledge and affirm the uniqueness and dignity of the older person.

NEVILLE E. STRUMPF
LOIS K. EVANS
MEG BOURBONNIERE

See Also
Behavioral Symptoms in Patients with Dementia
Psychotropic Medications

Internet Resources
American Geriatrics Association, position paper on restraints
*http://www.americangeriatrics.org/
positionpapers/restrain.html*

Health Care Financing Administration
*http://www.hcfa.gov/news/newsltrs/
aug99hw.htm#1*

U.S. Food and Drug Administration
*http://www.fda.gov/opacom/backgrounders/
safeuse.html*

RETIREMENT

The term *retirement* has multiple meanings. On a broad level, it signifies a set of economic and societal practices that help to manage the size of the labor force. The term also designates a social status that has come to be considered a historically new stage of the life course. At the level of the individual, retirement can refer to a changed relationship to the economy, to the temporal event of withdrawal from work, and to a personal transition from one life role to another that entails a process of adaptation.

Various criteria are used to classify individuals as retired—for example, reduced labor force participation, cessation of a career, receipt of pension income, willingness to self-identify as retired, or some combination of these. Most people understand retirement to be an older worker's cessation of full-time employment coupled with reliance on public or private pensions for a significant portion of income.

Retirement can be a crisp, single event, perhaps concluding a career or occupation. Retirement can also occur as a set of transitions—exit and reentry, part-time or intermittent employment—that bridges the adult work role with final withdrawal from the labor force. Recent research has raised awareness of more variable and indeterminate patterns of retirement that have arisen due to changing labor markets and pension incentives.

Retirement as a Societal Practice

Retirement as we know it today—withdrawal from the labor force, coupled with the earned entitlement to pension income—is a relatively recent development in industrial societies (Guillemard & Rein, 1993). As a matter of economic policy, retirement can be seen as a device for drawing older workers out of the labor force and financing partial replacement of their lost wages. The emergence of modern retirement practices to support this policy awaited several historical trends, among them the enhanced productive capacity of industrialized economies, growth in the size of older populations, and the rising power of governments to manage state pension systems (Donahue, Orbach, & Pollak, 1960).

Before 1950, older workers who left the American labor force had to rely largely on personal and family financial resources, as ample or as meager as they might be. Only a small proportion of the labor force was covered by any sort of pension arrangement, usually at the discretion of the employer. Even so, labor force activity by older people was declining in the aggregate—for example, from 67% of men age 65 and older in 1890 to 54% in 1930. This trend occurred as industrialized wage and salary work supplanted the small business and craft occupations in which older people more readily retained positions. Industrialization, however, had mixed consequences. Whereas a new dedication to progress and efficiency threatened the jobs of some older workers, others enjoyed higher wages and accumulated wealth that allowed them to reduce work or contemplate retiring (Haber & Gratton, 1994).

State-sponsored access to a postretirement income was first enunciated in the United States in the Social Security Act of 1935, which established a federal old-age pension program. Multiple motives attended the design of this legislation. Coming in the midst of the Great Depression, such public pensions would reduce unemployment by removing the aged from the labor force and thus create jobs for younger workers. Considerable public sentiment also favored the relief of economic distress among older persons out of work. This public pension system would be financed by a wage and payroll tax; participation was compulsory for workers employed in commerce and industry, the groups initially covered. Benefits levels were linked to previous earnings—high earners getting a larger benefit—but benefits levels were devised to provide only a floor of protection. In theory, personal savings and private pensions would also provide a share of retirement income. The eligibility age of 65, already a conventional pension age, would eventually be popularly established as the expected, legitimate age for retirement.

Old-age pensions under Social Security were expanded over the years. During the 1950s and 1960s, coverage was extended to almost all occupational groups. The option of early retirement at age 62 with actuarially reduced benefits was added. In the early 1970s, a major increase in benefits was granted, and a mechanism was set in place for regularly adjusting benefits for increases in the cost of living. Throughout this period, political parties steadfastly promised the maintenance of Social Security, and efforts were made to sell a positive image of retirement to the public (Graebner, 1980).

The years after World War II saw a concurrent expansion of private or job-specific pensions, whose benefits often dovetailed with Social Security provisions. Pensions were useful to employers as a way of creating employee loyalty and promoting the orderly turnover of personnel. Unions, permitted after 1949 to include pensions as a collective-bargaining issue, saw pensions as a way to reduce unemployment and augment Social Security income. Federal tax provisions encouraged the creation of pension plans in industry and among the self-employed. The 1974 Employee Retirement Income Security Act established federal controls over the administration of private pensions in order to protect covered workers against benefits losses. Individuals were eventually allowed to establish tax-deferred individual retirement accounts as a way of saving for retirement.

U.S. Labor Force Participation Rate (Percent) by Gender and Age, 1950–1998

Year	Men		Women	
	55–64	65+	55–64	65+
1950	86.9	45.8	27.0	9.7
1960	86.8	33.1	37.2	10.8
1970	83.0	26.8	43.0	9.7
1980	72.1	19.0	41.3	8.1
1990	67.7	16.4	45.3	8.7
1998	68.1	16.5	51.2	8.6

Source: Rosenfeld, C., & Brown, S. C. (1979). The labor force status of older workers. *Monthly Labor Review, 102*(11), 13; U.S. Bureau of Labor Statistics. *Employment and Earnings*, various January issues. Washington, DC: U.S. Government Printing Office.

These developments contributed to a correlative decline in the labor force participation of older adults. In 1950, 45.8% of men age 65 and over were in the labor force, compared with only 16.5% in 1998 (with roughly half of this employment consisting of part-time work). Participation among men age 55 to 64 has dropped as well, to 68.1% in 1998. Women also had higher retirement rates over these decades, though this trend is masked by women's historical rise in job holding. Retirement practices have evolved to such an extent that so-called early retirement—prior to age 65—is now the norm. In 1998, 67% of men and 69% of women who were new Social Security beneficiaries opted for their first benefits payments before reaching age 65 (Social Security Administration, 1999).

Retirement has now come to be accepted as a new stage of life and an appropriate role for older people. The provision of public and private pensions, though not complete or adequate in all respects, has transformed retirement from a status of possible dependency and financial deprivation to a period of life that is widely anticipated for its leisure possibilities. The overwhelming majority of individuals now view retirement favorably and demand that retirement income be protected as an entitlement for former membership in the labor force.

The Decision to Retire

Though most older workers eventually retire, there can be great latitude in the timing of this life step. Factors involved in the retirement decision include pension availability, prospects for income security over the long term, opportunities for continued employment, job conditions, workplace norms and administrative rules, personal dispositions with regard to work and retirement, family circumstances, and continued ability to perform on the job. Married persons may make mutual decisions, taking the spouse's employment, pension, and health into account. Among these many considerations, wealth and health are of paramount importance in the decision to retire (Quinn & Burkhauser, 1994).

Few adults can manage to fund retirement from personal savings alone. Thus, pensions are necessary to replace income lost by withdrawal from work. Pension eligibility and age incentives, in turn, bring various societal, organizational, and personal objectives to bear on the individual decision.

Retired worker pensions under Social Security can be claimed at age 65 for full benefits or at age 62 for reduced benefits, after having worked and contributed for at least 10 years. If a Social Security recipient exceeds the maximum allowable earnings ceiling ($10,080 to $17,000 annually in 2000, depending on age), benefits are sharply reduced. This "retirement test," in effect until age 70, discourages extensive employment after retirement.

New policies, however, will favor extended work careers. Beginning in 2003, the age for full Social Security benefits will advance gradually to age 66 and then to age 67 by 2027. Early retirement at age 62 will still be possible, but with a greater reduction in benefits. Delayed retirement will bring a larger benefit. The retirement test, too, will be liberalized.

Employer-provided pensions also guide the timing of retirement. Pension plans have widely varying characteristics and are unevenly distributed among occupational groups. They are more common, and benefits levels more generous, in industries characterized by large, highly organized firms and strong labor unions. Most government workers are covered. In the mid-1990s, two-thirds of all full-time employees participated in one or more retirement plans, a rate of coverage that has remained constant. The mix of plan types, however, was shifting from traditional "defined benefit" plans to a nearly equal proportion of "defined contribution" plans as primary or supplemental pension sources.

In defined benefit plans, employees are promised a steady benefit that can be claimed after meeting an age and length-of-service requirement. Virtually all such plans have early retirement options, and benefits formulas are often structured to encourage retirement at an optimal age beyond which there is little financial advantage to stay on the job. Unlike Social Security, few such pension benefits are regularly adjusted for increases in the cost of living, leaving their projected purchasing power vulnerable to inflation. Under defined contribution plans, the employer and often the employee make contributions of a specified amount to the employee's account. There is no fixed future benefit; retirement income is drawn from the earnings of the account, which has been managed or directed by the employee. This shift from corporate paternalism to employee self-reliance puts greater responsibility, as well as future risk, on the individual. Withdrawals from defined contribution plans, such as the 401(k) type, can typically begin at age 59 1/2 but must begin by age 70 1/2. At this point, the prospective retiree must decide how to annuitize these sums to provide a long-term stream of income.

Given a reduction in the expenses associated with working, retirees need to replace about 65% to 85% of previous earnings to maintain the same standard of living, a goal that eludes the majority of retirees if they rely on employer and Social Security pensions alone (Schulz, 1995). Despite employer pensions and efforts to encourage private saving, Social Security remains the major source of income for two-thirds of white beneficiaries and three-fourths of minority beneficiaries.

Another major factor in retirement decisions is the worker's health and continued ability to perform on the job. Approximately 20% to 30% of all retirements involve ill health or disability as the primary reason for retirement. Access to employer-sponsored health insurance is also a consideration for workers who want to exit their jobs prior to the Medicare eligibility age of 65.

Retirement under mandatory or compulsory age rules is no longer the significant factor that it was in the past. Mandatory retirement at a stated age traditionally was seen as an efficient management tool that could terminate older workers equitably, if impersonally, in advance of probable obsolescence. Labor unions frequently conceded to management the prerogative of mandatory retirement rules in exchange for more extensive pension benefits and the protection of seniority systems. Evidence from the 1960s showed that far more older male workers faced mandatory retirement than were actually terminated involuntarily. Nevertheless, the issue of forced retirement had enough symbolic importance to prompt successive amendments to the Age Discrimination in Employment Act, which raised the permissible mandatory retirement age for jobs in private industry from 65 to 70, and then virtually eliminated mandatory retirement. This legislated prohibition should not obscure the fact that older workers may still face subtle age discrimination, negative perceptions about their productivity, and local pressure to retire (Sterns, Sterns, & Hollis, 1996).

One significant development has been the growth in workers' expressed willingness to retire, to the point that a preference for leisure has become the primary self-reported reason for retirement. Marketers of travel, recreation, and financial services have encouraged strong demand for retirement by celebrating the consumption of leisure as a legitimate lifestyle. Retirement is promoted as a time for release and self-development in advance of the frailty that heralds death (Blaikie, 1999).

In general, workers who have had regular, stable employment and enjoy relative social class advantages control their own retirement decisions. However, some workers leaving the labor force trade a history of disability or chronic unemployment for early retirement. Strained financial circumstances cause other workers to forgo retirement. Finally, professionals and the self-employed, who may have continued employment opportunities and stronger personal investments in their identity as workers, are more likely to stretch out their labor force attachment.

Adaptation to Retirement

Early conceptions about possible retirement adaptation, developed while the retirement role was still gaining popular acceptance, foresaw retirement as a stressful transition and a "crisis" in personal identity. Instead, research reports from longitudinal studies of workers' experiences before and after retirement concluded that there is likely to be continuity of well-being and activities during the transition from work to retirement. This research has disclosed that a more satisfactory retirement experience is had by individuals with better health, greater income, a stronger social network, and an adaptive personality—circumstances that favor adjustment at all stages of life. Serious dissatisfaction with retirement happens when retirement occurs unexpectedly or coincides with another negative life event, such a spouse's illness. Most knowledge about retirement adaptation has been generated by studies of men, but evidence indicates that most conclusions apply to women as well.

Adaptation is enhanced by a process of having anticipated and accepted retirement in advance of the event. Although few older workers have the opportunity to participate in formal planning programs, financial and lifestyle advice about retirement is widely available. Most preparation for retirement is informal and involves some information seeking and anticipatory rehearsal of the retirement role. Although the role of retired persons may never develop the specificity of the work role, there are certain expectations that retirees will remain active, independent, and self-reliant. The vagueness and flexibility of the retirement role is a particular benefit to retirees whose diminished health and financial resources limit their ability to participate.

The temporal event of retirement, symbolized by the last day on the job, may be marked by some ceremony. Although a majority of the public favors the idea of a gradual withdrawal, phased retirement options are not common. The satisfaction with retirement expressed by most retirees may be tinged with ambivalence about the simultaneous freedom and marginality of a status that leaves work behind (Savishinsky, 2000). Overall, the maintenance of personal resources such as income and health has a lot to do with the quality of the retirement experience over the long term.

Contrary to the widely held notion that retirement often has negative consequences for health, epidemiological studies have consistently demonstrated that the event of retirement does not influence the risk of decline in physical or mental health. Indeed, considerable numbers of retirees report that retirement has a beneficial effect on their health. Marital discord, another supposed negative outcome of retirement, is also far less common than is generally thought. Nevertheless, the persistence of such myths about retirement's disruptive effects indicates continuing public wariness about the quality of life in retirement.

Depending on their circumstances, retirees exhibit a wide variety of lifestyles. No particular level of leisure participation or social engagement has been shown to be the sole formula for a satisfactory retirement. Approximately one-quarter of retirees do some part-time work. An enthusiastic minority also step into worklike volunteer activities. Despite the powerful, popular image of Sun Belt retirement, fewer than 5% of retirement-age people migrate across state lines in a five-year period. Retirement is a status largely lived out in one's same community, where there is a continuity of roles, activities, and relationships.

Prospects for the Future

As the American way of retirement evolves, changes will occur primarily to benefit those political and market interests on whose behalf modern retirement practices have evolved (Quadagno & Hardy, 1996). These interests do not all coincide. With an eye to the fiscal solvency of the Social

Security program, policy adjustments since 1983 have aimed at slowing rates of retirement. Despite increased longevity, workers want more than ever to retire, but with flexibility. They also want protection for their retirement benefits and pension rights. Firms will continue to use pension incentives to manipulate retirement timing in response to changing labor markets and the need for older workers. Cultural forces will shape the image of retirement as the financial services industry competes for the savings of workers and marketers compete for sales to retirees. As large birth cohorts from the 1940s and 1950s approach later life, further adjustments to retirement systems, in the form of possible retrenchments or new forms of pension financing, will receive consideration. Despite alarms from some quarters, provision can be made for the retirement of the baby boom generation without incurring social divisiveness (Radner, 1998).

DAVID J. EKERDT

See Also
Pensions
Social Security
Tax Policy

RETIREMENT COMMUNITIES
See
Naturally Occurring Retirement Communities

RHEUMATOID ARTHRITIS

Rheumatoid arthritis is a progressive disabling condition that results in pain, joint damage, and functional loss and is associated with profound economic consequences in terms of lost wages, disability benefits, hospital costs, nursing home costs, home-care costs, and professional and medication expenses. Loss of independence is a fear of many patients with rheumatoid arthritis. The disease has been associated with psychological distress, depres-

sion, and anxiety. A multidisciplinary approach to the treatment of patients with rheumatoid arthritis, as well as early initiation of treatment, may lead to improved quality of life and outcome.

Epidemiology

Among the general adult population, rheumatoid arthritis most commonly occurs in the third to fifth decades, with a prevalence rate in the population of 0.3% to 3%. This disease has been divided into two subsets by several authors: adult-onset rheumatoid arthritis (AORA) and elderly-onset (after age 60) rheumatoid arthritis (EORA), each with its own distinctive characteristics. In most people, disease onset occurs prior to age 60; however, there is a distinct category of patients who develop this disease after the age of 60. New cases of EORA have been reported among 14% to 55% of rheumatoid arthritis cases. In the AORA population, women are more frequently affected than men, with a ratio of 2:1 to 3:1; however, with increasing age, the incidence in men rises, with the ratio becoming 1:1 (Van Schaardenburg & Breedveld, 1994).

Diagnosis

The diagnosis of rheumatoid arthritis in the elderly is made when characteristic symptoms are present. The majority of patients may have an indolent onset of disease, with progressive pain, swelling, and morning stiffness of characteristic joints: proximal interphalangeal joints, metacarpophalangeal joints, wrists, elbows, shoulders, hips, knees, ankles, and metatarsophalangeal joints. In 30% of patients, the onset is acute, with an asymmetrical presentation. A complete history, physical examination, and laboratory evidence of inflammation are necessary to make the diagnosis of rheumatoid arthritis. Radiographs of the hands and other involved joints with evidence of periarticular osteopenia or marginal erosions support the diagnosis. A complete blood count, chemistries, and serologies such as rheumatoid factor, antinuclear antibody, erythrocyte sedimentation rate (ESR), and C-reactive protein (CRP) should also be obtained.

Despite an increasing prevalence of rheumatoid factor positivity in up to 40% of healthy elderly, rheumatoid factor is positive in only 32% to 58% of elderly patients presenting with new-onset disease (Lance & Curran, 1993). This compares with 80% positivity in AORA. Although the onset may be abrupt, several authors describe a milder presentation in elderly seronegative patients, with polymyalgia symptoms and frequent axial (shoulder and hip) involvement. Constitutional symptoms such as weight loss and a functional decline may accompany these symptoms. Patients who are rheumatoid factor positive have been described as having more persistently active disease, greater functional decline, more radiographic erosions, and increased mortality when compared with seronegative patients. Other differences between EORA and AORA include less frequent metatarsophalangeal joint involvement and fewer subcutaneous nodules in patients who are rheumatoid factor negative.

Older patients with longstanding disease may have joint deformities, synovitis, nodules, and radiographic erosions. Laboratory tests such as ESR and CRP are elevated. It is not unusual for a patient with rheumatoid arthritis to have anemia. Often the indices are consistent with anemia of chronic disease; however, a mixed etiology may be present and should be evaluated thoroughly.

Two other disease entities have been described in the elderly and may be difficult to distinguish from seronegative rheumatoid arthritis. These diseases have been described as variants of rheumatoid arthritis or as overlapping with the spectrum of the disease. First, polymyalgia rheumatica needs to be considered when evaluating an elderly patient with an acute presentation of symmetrical shoulder pain and an elevated ESR and CRP. The distinction may be difficult to make, and treatment with corticosteroids is effective for both entities. The other differential diagnosis to consider is remitting seronegative symmetrical synovitis with pitting edema. This is a relatively rare disease described in elderly men, with a typical presentation of acute synovitis involving the wrist, carpal joints, and flexor digitorum tendon sheaths and evidence of pitting edema of the dorsum of the hands. Other upper extremity and lower extremity joints may be involved. Rheumatoid factor is negative, and HLA association is with B27 and B7.

Crystal arthropathies such as polyarticular gout and calcium pyrophosphate deposition disease may present similarly to rheumatoid arthritis. Synovial fluid analysis and radiographs help make the distinction. Elderly patients with new-onset disease may be more difficult to diagnose, especially if they present with polyarticular joint involvement, myalgias, negative serologies, and constitutional symptoms. In these situations, other diagnostic possibilities need to be entertained, such as viral and bacterial infections, connective tissue disorders such as systemic lupus erythematosus and Sjögren's syndrome, metabolic disorders, and malignancy.

Although the differential diagnoses appear straightforward, diagnosis in the elderly may be hindered by difficulty eliciting information from cognitively impaired patients; the presence of multiple coexisting diseases, which may confound history and physical examination; the increasing frequency of positive serologies in elderly patients; and difficulty obtaining laboratory studies and radiographs in nonambulatory patients, who may not have the social and financial support to obtain appropriate medical care.

Treatment

Although the treatment strategy for rheumatoid arthritis in the elderly is similar to that used for younger patients, treatment in this high-risk group is a challenge (Albert & Schwab, in press). The goal of therapy is to relieve pain, diminish disability and joint destruction, and improve quality of life and functional outcome. The problems encountered with increasing age are multiple. The risk of drug toxicity is increased secondary to age-related alterations in pharmacokinetics and pharmacodynamics. Polypharmacy, often found in the elderly, can substantially increase the risk of drug interactions and adverse reactions. Multiple comorbid conditions may alter the metabolism and excretion of prescribed medications, resulting in lower initial doses for most older patients. "Start low and go slow" is the general therapeutic guideline for drug dosages. An outpatient or in-home exercise program should also be prescribed for all patients to improve functional outcomes.

In addition to the concerns already listed, cognitive dysfunction, financial constraints, and limited social support influence treatment decisions. For instance, a practitioner may decide on a milder, less toxic treatment regimen for a home-bound or nursing home patient with multiple comorbid conditions in order to diminish drug toxicity. This is more prudent than a potentially toxic regimen, especially if physician visits are infrequent and blood monitoring is not available in these settings. Often a multidisciplinary team approach helps diminish some of the obstacles encountered by the elderly and by practitioners.

The choice of drug therapy for a particular person depends on the progression of the disease, toxicity of the drug, and the individual's coexisting conditions (Glennas et al., 1997). The current treatment strategy includes the early initiation of disease-modifying agents to retard damage and ultimately diminish disability. The potential toxicity of any drug needs to be weighed against the morbidity caused by the disease. These drugs are equally effective in the elderly as they are in younger persons. Their toxicities, however, may be more profound because of alterations in pharmacokinetics and pharmacodynamics, drug interactions, and increased frequency of coexisting conditions which may alter drug metabolism and excretion.

Simple analgesics such as acetaminophen are ineffective in this disease. Nonsteroidal anti-inflammatory drugs (NSAIDs) are effective in patients with mild to moderate disease but may not be well tolerated in the elderly because of gastrointestinal toxicity, which includes inflammation, ulceration (reported in the small and large bowel), bleeding, and perforation. Other potential toxicities include nephrotoxicity, hepatoxicity, cardiovascular effects of volume overload, central nervous system effects of confusion and cognitive impairment,

and hematological abnormalities such as thrombocytopenia, neutropenia, and hemolytic anemia. Increasing age, history of peptic ulcer disease, concomitant corticosteroid or anticoagulant use, cigarette smoking, alcohol use, and ingestion of multiple NSAIDs have been identified as risk factors for gastrointestinal complications. The new cyclooxygenase-2 inhibitors may provide increased protection in these patients; however, further trials need to be conducted to assess their safety in the elderly.

Hydroxychloroquine seems to be well tolerated in the elderly. Side effects include gastrointestinal discomfort, rash, photosensitivity, and retinal toxicity which is rare. Patients should obtain biannual ophthalmological examinations, especially if they have coexisting ocular disease such as cataracts or macular degeneration. The role of corticosteroids in the treatment of rheumatoid arthritis is extremely controversial because of the potential toxicities of osteoporosis, diabetes, cataracts, glaucoma, anxiety, delirium, and atherosclerosis. Although these agents are effective in reducing inflammation, they should be used with great caution and at the lowest effective dose. Antiresorptive therapy should be started simultaneously to diminish potential bone loss, especially in a population in which osteoporosis is prevalent.

Other second-line agents such as methotrexate and sulfasalazine appear to be well tolerated in the elderly. Methotrexate is often the agent of first choice because of its early onset of action and high efficacy–to–low toxicity ratio. The primary toxicity of methotrexate is hepatic and hematological. Frequent surveillance of liver function and blood count should occur. Starting doses are often low in the elderly (2.5 mg a week) and slowly titrated. The primary toxicity of sulfasalazine is gastrointestinal problems; liver function and hematological abnormalities are rare. Other agents such as cyclosporine, azathioprine, parenteral gold, and oral auranofin, and penicillamine are less popular in this age group. Specific side effects include bone marrow toxicity, nephrotic syndrome with parenteral gold, and skin rash and dysgeusia with penicillamine. Of the newer agents leflunomide, an inhibitor of pyrimidine biosynthesis, has been shown to be effective in reducing disease activity. Side effects include rash, liver function abnormalities, alopecia, and gastrointestinal symptoms.

Tumor necrosis factor (TNF) is a proinflammatory cytokine that contributes to the pathogenesis of rheumatoid arthritis. Agents have been developed to inhibit this cytokine, including etanercept, a recombinant human TNF receptor fusion protein, and infliximab, an anti-TNF alpha antibody. Although recent studies have shown both agents to be well tolerated and effective in reducing disease activity, more data are needed to assess their long-term efficacy, safety, and toxicity in both younger and older patients.

EDNA P. SCHWAB

See Also
Osteoarthritis

Internet Resource
Arthritis Foundation
http://www.arthritis.org

RISK ASSESSMENT AND IDENTIFICATION

Risk assessment allows clinicians to target services and interventions to individuals. This has the potential to increase access to services for individuals who are most likely to benefit, resulting in improved outcomes and efficient use of resources. Certain general medical conditions, markers of physical and cognitive function, and socioeconomic factors place elders at risk for functional decline, institutionalization, or death.

Many of these risk factors are seen in the group of elders frequently labeled as "frail." The exact definition of frailty varies across studies and across experts. We prefer to identify this group of elders as those at highest risk for functional decline or death. This approach recognizes frailty as more than the sum of existing disabilities and underscores the importance of assessing ongoing pathology and adverse environmental conditions that significantly increase the risk of functional decline or death for all elders. Although every elder identified as frail is at risk, not every elder who is at risk can be labeled frail. Risks are not necessarily fixed; an individual's risk may decrease or increase over time as conditions evolve or the environment changes.

Clinicians can use clinical and demographic characteristics to identify elders at increased risk for functional decline or death. However, although the natural history and contribution of many of these risk factors are understood, it is often difficult to isolate which risk factor is most important. Inconsistency across studies reflects widely varying approaches in how risk factors are defined and measured, the selection of the interval for risk exposure, the number and combination of risk factors considered, and variations in examined outcomes. We focus here on risk factors that have immediate or intermediate impact—typically less than five years—as opposed to risk factors with more delayed impact.

Screening and preventive services are frequently targeted to groups of persons at risk for developing disease or decline. Some preventive services, such as immunization, use age as the primary targeting criterion. Some services are targeted to persons with existing morbidity. For example, persons with diabetes mellitus are at increased risk for vision loss and are therefore targeted for periodic eye examinations. Preventive services that address risks specific to a particular condition or syndrome and approaches for evaluating and treating individual risk factors (e.g., nutritional risk assessment, pressure ulcer risk assessment, fall risk assessment) are discussed elsewhere in this text.

Medical conditions and syndromes that are important risk factors for functional decline, institutionalization, and death include alcohol abuse, tobacco use, anxiety disorder, asthma, arthritis, cancer, cerebrovascular disease or stroke, chronic obstructive pulmonary disease, cognitive impairment, cardiovascular disease (e.g., myocardial infarction, heart failure, valvular heart disease, hypertension), depression, diabetes mellitus, falls, hip fracture, malnutrition, inadequate exercise, and sensory (vision or hearing) impairment. Clinical risk assessment and intervention require consideration of the severity of each condition, the interaction of conditions, and the impact on the individual. Poor health practices (e.g., tobacco use, lack of regular exercise) not only contribute to the development of many conditions but also affect condition management and can be independent risk factors for functional decline or death.

On a population basis, a simple count of conditions, or a "weighted summary of conditions," provides overall risk assessment. In addition, symptom summaries, medication class or type, and medication count may serve as measures of risk because they may reflect illness severity, iatrogenic risk, and/or quality of medical care. Summaries of the use of other medical services (e.g., number of physician office visits, number of prior hospitalizations, past use of any institutional services) may identify persons at increased risk for future utilization and decline.

The accompanying table lists markers of functional status associated with increased risk for functional decline, institutionalization, and death. The table lists three general groups of functional status measures: physical function or functional limitation, instrumental activities of daily living (IADL), and basic activities of daily living (BADL). While many of the markers of physical function are self-reported, the direct assessment of functional limitations can be an important and simple part of the geriatric physical examination. Mobility and ability to rise from a chair unaided can be quickly observed and identify high-risk individuals at an early point on the trajectory of decline. IADL and BADL represent tasks important to independent living and self-maintenance. For the individual patient, it is particularly important to note that current functional disability in IADL or BADL does not equate with inevitable decline. Indeed, the majority of persons with IADL or BADL limitations will remain stable over two years, and a significant percentage will evidence a decrease in the number of dependencies on follow-up questioning (Crimmons, Saito, & Reynolds, 1997; Manton, Corder, & Stallard, 1993).

Markers of Functional Status Associated with Increased Risk for Death or Decline

Functional Limitation

Difficulty with physical function, including:
  Doing heavy work around the house
  Walking 1/2 mile
  Climbing one flight of stairs
  Standing for long periods
  Lifting or carrying weights of 10 lbs
  Using hands or fingers
  Pulling or pushing large objects
  Stooping, bending, or kneeling
  Reaching with either or both arms
  Picking up an object from the floor
  Rising from a chair
  Standing on one foot

Instrumental Activities of Daily Living

Difficulty or needing help with:
  Using telephone
  Shopping
  Preparing meals
  Doing laundry
  Maintaining house
  Managing finances
  Taking own medications
  Driving car, using bus or taxi

Basic Activities of Daily Living

Difficulty or needing help with:
  Bathing
  Continence
  Walking across room
  Grooming/hygiene
  Toileting
  Transferring
  Dressing
  Feeding

An individual's global rating of his or her own health as "fair" or "poor" is a strong and consistent risk factor for future functional decline or death. Self-rated health may be so effective because of its ability to capture unmeasured disease severity, unmeasured functional impairment, and difficult-to-measure factors such as self-efficacy or locus of control.

Age is an important independent predictor of functional decline and death, even after accounting for multiple other risk factors. For elders, the interaction of ongoing disease or pathology with organ- or system-level functional reserve predicts both functional decline and mortality. Age continues to describe unmeasured risk, despite advances in measuring function and in understanding the mechanisms of disease and overall aging. One objective of current aging research is better delineation of the reasons that advanced age serves as an independent risk factor for decline.

Other demographic factors and markers of socioeconomic status that identify elders at risk for functional decline, death, institutionalization, social isolation, or abuse include low household income and assets, low educational level, minority ethnic status, and marital status, especially recent loss of a spouse. A significant number of elders have incomes below the poverty level, limiting their access to health care goods and services that require some out-of pocket expenditures. Functional dependency, cognitive impairment, and low level of education place many elders at risk for financial fraud and health care mismanagement. Whereas the overall population over the age of 65 is expected to double over the next half century, the number of ethnic elders (Hispanic, African American, and Asian and Pacific Islander) is expected to increase even more rapidly. Minority elders may be at "double jeopardy"—different disease patterns, beliefs, and practices; physical hardship; lack of formal education; stress; poverty; a history of inadequate access to preventive health care; and inadequate access to culturally appropriate care result from and contribute to patterns of discrimination (Ham & Sloane, 1997).

The quality and extent of social contacts and social support are highly associated with affect, functional status, mortality, and institutionalization. Elders who are recently widowed, retired, or relocated are at greater risk for physical and emotional decline. Elders whose dependency needs exceed the family's ability to provide support are at high risk for elder abuse. Conversely, caregivers who are overly taxed and have a history of mental illness or alcohol or drug abuse are at risk of becoming abusers. It is critical that the "shadows of the elderly"—abuse, alcohol, depression, and dementia—be routinely assessed with tools that address social isolation and caregiver burden (Zarit, Reever, & Bach-Peterson, 1980).

The physical home environment interacts with many risk factors. Elders make up a large proportion of persons involved in home accidents leading to significant injury. Many accidents are preventable. A simple checklist that highlights key home safety measures includes uncluttered and even floors, good lighting, bed height that allows feet to touch the floor, placement of other furnishings to provide support, grab bars by the toilet and in the shower, and toilet and shower seats that allow for easy transfer (Rubenstein, Josephson, & Robbins, 1994).

Formal risk assessment tools vary in the number of items included, the domains covered, and the outcomes assessed. Some are disease specific, some predict utilization, and others predict the need for selected services. The breadth of factors that can place an individual at risk underscores the importance of tailoring interventions based on assessments of individual risk. An interdisciplinary team approach is particularly well suited, efficient, and cost-effective for assessing, identifying, and integrating the multiple factors needed for a thorough risk assessment (Weiland, Kramer, Waite, & Rubinstein, 1996).

DEBRA SALIBA
MARTHA S. WAITE

See Also

Activities of Daily Living
Environmental Modifications
Falls Prevention
Nutritional Assessment
Pressure Ulcer Risk Assessment
Urinary Incontinence Assessment

Internet Key Words
Activities of Daily Living
Functional Status
Home Safety
Self-Rated Health

Internet Resources
American Geriatrics Association
http://www.americangeriatrics.org

Los Angeles Alzheimer's Association
http://www.alzla.org/healthpro.htm

RURAL ELDERS

Over 25% of the U.S. elderly population and half of the world elderly population live in rural areas. The aging process, health status, and kind of health care elders receive vary between rural and urban settings. Cross-cultural studies indicate that rural elders in developed and developing countries have greater problems accessing appropriate, affordable health care than do their urban counterparts. Research, policy decisions, and funding for services are affected by differences in the definition of rural. Rural is often defined dichotomously as urban versus rural or as metropolitan versus nonmetropolitan areas. These definitions focus on the quantitative dimensions of population density and spatial distribution. They fail to capture the qualitative variability among groups and within groups. More recent definitions of rural describe a continuum, with small remote areas at one end and large metropolitan cities at the other (Coward & Krout, 1998). Factors such as distance to emergency services, economics, geography, and occupation are included to capture a clearer picture of rurality.

Cultural orientation and historical development of rural communities also contribute to socioeconomic differences and diversity in health status factors, especially between farming and nonfarming regions. Rural elders in farming regions generally have better health and functioning than rural elders in nonfarming regions, and closer family interactions and support are associated with farming regions. Less than 10% of U.S. rural elders, however, live in farming regions (Coward, Bull, Kukulka, & Galliher, 1994).

Although rural diversity exists, common characteristics and needs of elders living in rural environments are well documented. Economic disadvantage appears to be a common feature of rural elderly life. Lack of employment opportunities causes large migrations of younger adults to urban areas, weakening the family care and support systems. Rural elders have higher rates of poverty, less formal education, poorer housing, greater transportation problems, and more chronic illnesses and functional disabilities than do their urban counterparts (Krout, 1994). Socioeconomic and environmental factors contribute to the limited access to preventive health care and to less healthy lifestyle patterns. Disabilities resulting from cardiac and respiratory diseases, arthritis, and diabetes are common among rural elders. Rural areas also have a growing population of frail elders.

Affordable and acceptable health care and support services for rural elders are critically lacking (Krout, 1994). The economy, geography, and transportation problems of rural areas, as well as the lack of trained health care providers, are major barriers to the development of needed services. Primary care is affected by problems in the recruitment, retention, and distribution of physicians in rural communities. Nurse practitioners and physician assistants, however, are helping to meet the need for primary care services. Gaps are especially apparent in the continuum of care between independent living and institutional care.

Family caregivers provide the bulk of home care for dependent rural elders. They often assume the caregiver role without any preparation, training, or assistance from formal providers. Frail rural elders rely almost exclusively on informal caregivers. Such long-term caregiving results in negative consequences for rural family caregivers, who have more than twice the stress and burden of urban caregivers (Dwyer & Miller, 1990). Elderly women fulfill the traditional role of nurturer and primary caregiver. Because rural elders must travel to more urban areas for hospitalization, transitions from acute-care settings, especially when the hospital is some distance from the rural residence, are made without including rural elders and family caregivers in discharge planning. The only options for unprepared rural elders and family caregivers may be nursing home care or home care without assistance. Rural families may promise relatives that they will never place them in a nursing home, viewing the nursing home option as a failure to care for their own. Geographic distances, transportation problems, and economic factors can delay nursing home placement until the family caregiver's health fails.

Broadly, rural elders tend to be self-reliant and conservative. They have strong ties with family and church and are emotionally attached to their land, which commonly has been home for several family generations. Traditional values and practices thrive in rural communities, especially in communities that are more isolated from outside influences. These values and belief systems, while valuable in maintaining group cohesiveness and shared values related to health and illness, often further isolate rural elders from mainstream health care and limit their willingness to seek formal health services.

Community-based support services such as respite care, senior centers, and meals-on-wheels are limited (Krout, 1994). Adult day-care services in rural areas are predominantly social models of care. Family caregivers of rural elders describe adult day care as costly, difficult to access, and unable to meet the needs of frail elders. Health insurance coverage of rural residents has been traditionally low; long-term health insurance to cover respite services is cost-prohibitive. Nurses often work closely with family caregivers and serve as the liaison between families and other formal health care providers. Access to home-care nursing services, however, has decreased as a result of Medicare reimbursement changes.

Long-term health care and support services are needed for the growing rural population of physically and cognitively dependent elders, especially those over the age of 85 years. Necessary long-term-care services include preventive, therapeutic, rehabilitative, supportive, and maintenance services. Without such services to assist in caregiving, the stress and burden of family caregivers will intensify. As care recipients become more dependent, social interactions of caregivers decrease. Isolation of these vulnerable rural elderly caregivers and care recipients is a major concern because of its negative consequences on health and well-being. Financial assistance for family caregivers from state and national sources is fragmented or nonexistent. If unpaid family caregivers of frail rural elders were no longer able to provide care, health care costs would increase by billions of dollars.

Comprehensive, culturally sensitive systems of care must be developed and implemented. Urban care models cannot simply be transplanted into rural communities. Acceptable services must involve the community. Case management models are needed to reduce the fragmentation of care across health care settings. Recent findings of demonstration projects indicate that primary care of rural elders could be strengthened by collaborative practices including physicians, advanced practice nurses, and social workers and by geriatric education programs for providers. In 1992 the John A. Hartford Foundation initiated several demonstration projects to improve geriatric health care. Two rural projects were the Geriatric Collaborative Care Model at the Carle Clinic in Illinois and the Geriatric Interdisciplinary Team Training project at the University of North Carolina. The outcomes of these projects have been positive, indicating improvement in collaborative efforts and access to quality care. The Community Nursing Organization, a demonstration project funded by the Health Care Financing Administration, is a national coordinated care model for rural elders. This capitated model of nurse-managed health care provides community-based care to healthy elders and to elders at high risk for poor outcomes. Health education and case management by nurses are essential components of the model. Preliminary outcomes of this project reflect high client satisfaction and cost-effective, quality-care outcomes.

In the past decade, several federally supported rural health initiatives have been implemented to address the health care delivery, education, and technological needs of rural areas. Partnerships between rural communities and universities have formed. However, special attention to the care of rural elders and family caregiving needs has been limited. Although these demonstration programs have met some of the immediate health care needs in rural communities, the long-term outcomes remain uncertain. Other health care delivery models for rural elders need to be developed and evaluated. Telecommunications and the potential contributions of other technologies should be creatively explored as resources for addressing the distinctive health care and supportive needs of rural elders. Community volunteers and leaders must be involved in developing and maintaining community-based programs for rural elders.

MARY ANN ROSSWURM

See Also

Caregiver Burden
Caregiver Burnout
Poverty
Transportation

Internet Resources

American Health Care Association
http://www.ahca.org/secure/growth.htm

National Family Caregivers Association
http://www.nfcacares.org

Rural Information Center
http://www.nal.usda.gov/ric/richs/stats.htm

S

SATISFACTION MEASUREMENT

Satisfaction is defined as "fulfillment of a need or want" and "the quality or state of being satisfied." Patient satisfaction is considered a valid indicator of the quality of health care and health care services. To improve quality of care and evaluate health care services provided by interdisciplinary staff, it is necessary to develop valid and reliable instruments to measure patient satisfaction with that aspect of health care services (Merkouris, Ifantopoulos, Lanara, & Lemonidou, 1999).

Few theoretical frameworks have been established for satisfaction, despite the abundance of surveys reported in the health care and general literature. Satisfaction can be conceptualized as a performance evaluation, an affect-based assessment, or an equity-based assessment. A satisfaction measurement instrument should have a theoretical base for the measure's validity. Criteria of satisfaction vary by setting; what is satisfactory in home care may not be satisfying to a nursing home resident, and vice versa. Notions of satisfaction, in some sense similar to those of quality, can be different between providers and patients. A patient may value the quality of the interaction with primary caregivers, whereas the provider may value no adverse sequelae after surgery.

The value of the interdisciplinary team in meeting individual or population health care goals also needs to be reflected in the measurement. The researcher must have a clear idea of what is to be measured (Schommer & Kucukarslan, 1997) and what perceptions of satisfaction the operative definition represents. Overall, the instrument should be patient driven and patient centered.

Research methodology of patient satisfaction can be qualitative or quantitative. Self-reporting questionnaires and telephone surveys are commonly used to assess patient satisfaction. The success of patient satisfaction surveys depends not only on question and answer wording but also on the layout, formatting, and appearance of the questionnaire. Self-report questionnaires returned via mail can avoid the limitations of patients' responses in face-to-face interviews. To produce an optimal response rate, personalized letters for prenotification and follow-up are essential. Interactive technology to administer questionnaires may provide rapid feedback but might inhibit criticism if anonymity cannot be assured (Krowinski & Steiber, 1996).

To assess reliability and validity, the questionnaire should be given to a representative sample of respondents. The patient satisfaction questionnaire should be practical with regard to its length, language, and complexity of response. No single standard measure of patient satisfaction is applicable to all health care situations. Researchers should either use an existing measure with demonstrated reliability and validity or develop a new measure by using a systematic process (Schommer & Kucukarslan, 1997).

Studies of patient feedback indicate that most patients are satisfied with their health care services and that patient loyalty increases if the patient considers the health care experience a very satisfying one (Schommer & Kucukarslan, 1997). Research indicates a multitude of factors that compromise patient satisfaction, including culture, impressions, expectations, perceptions, and feelings toward health care. As countries develop health care systems and incorporate advanced medical technology, notions of satisfaction will change as a reflection of expectations, access, and outcomes. The concept of satisfaction may come to include the quality of the relationship with the primary providers and the ethical behavior embedded in the interaction.

TOSHIKO ABE

See Also
Cultural Assessment
Cure vs Care

Quality-of-Life Assessment
Spirituality

SELF-CARE

Self-care is more than independent performance of activities of daily living (ADL). It encompasses all assessments, decisions, and actions in daily life to obtain and meet basic personal needs, intelligent participation in disease prevention or management, and promotion of personal well-being or development. Self-care helps an elderly person survive safely as long as possible, maintain autonomy and independence, control symptoms and discomfort, and attain and keep some degree of life satisfaction. Performing self-care itself involves decisions and choices, from food selection to personal hygiene. Seeking professional medical assistance and adhering to a medical regimen are examples of excellent self-care.

Well intentioned "overcaring" by family members or health care providers can make a functionally impaired elderly more dependent than necessary. Dependency usually is assessed in terms of ADL. Although this approach certainly has merit, it is conceptually limited with respect to the complexity of human functioning. Assessments based on impaired physical mobility tend to be overemphasized and generalized to all human functions. For example, a wheelchair-dependent person is treated as incapable of decision making. Systematic review of the elderly's self-care ability can prevent unnecessary dependency.

Adequate self-care means performing the correct assessments, making the right decisions, and executing those decisions safely and correctly (Orem, 1985, 1995). Self-care assessment includes assessment of one's own behavior and the environmental conditions that may impose a health risk or cause a decrease in comfort. Self-care decisions assess how conditions can be maintained or changed, if necessary, what will be done, and which course of action to follow. Self-care performance includes preparing oneself, materials, or the environment for self-care procedures; performing them; and ending them.

Adequacy of self-care draws on scientific principles, cultural norms, and personal preferences. For example, restricting fluid intake to 500 mL per 24 hours is based on scientific principles for a patient with certain types of nephropathy; restricting food intake during the day may be based on religious norms; taking a hot shower in the evening may be based on personal preference.

Adequacy in meeting basic personal needs includes sufficient intake of air, fluid, and food; provision of care related to body hygiene and elimination; balance among sleep, rest, and activity and activity diversity; balance between solitude and social contacts; prevention of falls, trauma, infection, intoxication, suffocation, cold, and burns; and social support from a social network (Evers, 1998a).

Intelligent participation in health maintenance includes timely symptom detection of, for example, fever, persistent cough, blood in urine, white stool, or a breast nodule and seeking and securing appropriate medical advice. Adequate health maintenance self-care includes effectively performing, for example, medication intake, intermittent bladder catheterization, blood glucose monitoring, colostomy care, or muscle exercises and monitoring treatment effects such as polyuria, diarrhea, hypoglycemia, or bleeding. Self-care may include necessary changes in lifestyle, for example, diet, smoking reduction or cessation, or increase in social contacts (Evers, 1998).

Comfort and well-being can be enhanced by adequate control of symptoms such as pain, dyspnea, or dry mouth. Well-being can be enhanced by changes in self-image when confronted with an amputation, stoma, device implantation, prosthesis, or even use of a wheelchair. Well-being can be promoted by mitigation of damage caused by loss of body parts, paralysis of extremities, or loss of a partner, children, friends, beloved personal belongings, or a home (Evers, 1998a).

Adequate self-care to promote personal development includes overcoming damage caused by neglect by a partner or children, by personal contempt, physical abuse, or rape. Personal development is enhanced by undisturbed transition to retirement stage, the infirmity of old age, and ultimately dying (Evers, 1998a).

The elderly cannot make adequate self-care assessments or decisions, nor can they put these decisions into action, if they lack the necessary capabilities. Capability for self-care develops in the course of daily living, aided by intellectual curiosity, instruction from and supervision by others, and successful experience in performing self-care. Self-care capabilities encompass attention and vigilance, controlled use of energy, control of body position, knowledge acquisition, reasoning, motivation, decision-making ability, the ability to set priorities, and the integration of self-care into personal and family life (Evers, 1989; Orem, 1985).

Assessment of the adequacy of an elderly person's self-care or self-care capabilities should focus on determination of the presence of, or risk for, self-care problems. In other words, what is the elderly person able to do in this particular situation to adequately supply basic personal needs, intelligently participate in disease prevention or management, and promote personal well-being and development (Evers, 1994)?

Fundamentally, the intent is to assay the needs and patterns of self-care, identify any risk, and strategize how the actual or potential need or normal functional pattern can be restored or an imbalance removed. It is rather like looking at the burdens of care imposed on the elderly person and the consequences if those care needs are unmet.

- What is sufficient intake of air, fluid, and food for this person?
- What are the usual hygiene and excretion patterns and rituals, and how can the usual patterns be restored?
- What is the balance between rest and activity, solitude and social interaction?
- What are the preventable health risks, and how can the elder be taught to detect and control risks?
- What is the social network and informal support system, and how can lack of social support be amended?
- What symptoms or treatment effects must be monitored by the elderly, and how can they be monitored?
- What medical assistance must be secured?

- What medication administration must be carried out by the elderly? If this is not done, who will be responsible?
- What lifestyle change is necessary? What if this doesn't happen?
- What symptoms and discomfort need to be controlled by the elderly? If this doesn't happen, how can control be secured?
- What change in self-image is necessary for this elderly person? What if it doesn't happen? How can changes be enhanced?
- What kind of loss or neglect cause damage, and how can this be mitigated?
- What would be an undisturbed life phase transition for this elderly person, and how can this be enhanced?

Information about the elderly's self-care capabilities can be more specific with the following detailed questions:

- Does the elderly person have the necessary attention and vigilance for correct assessments, the right decisions, and correct performance of technical procedures?
- Can the elderly person acquire the necessary knowledge, and can he or she adequately reason?
- Does the elderly person have sufficient motivation and decision-making capability?
- Does the elderly person have the necessary repertoire of technical or procedural skills?
- Can the elderly person set priorities and integrate new self-care in his or her personal or family life?

Determination of the presence of, or risk for, self-care problems is based on a comparison of the requirements for adequate self-care and the self-care capabilities of the elderly person. Self-care problems can be formulated as absent self-care, unstable self-care, inadequate self-care, and risk for self-care problem (Evers, 1998b).

GEORGE C. M. EVERS

See Also
Activities of Daily Living
Assessment Tools and Devices

Autonomy

Case Management

Multidimensional Functional Assessment: Overview

Internet Resources

Khon Kaen University Faculty of Nursing, Self-Care Deficit Research Webpage

http://www.kku.ac.th/~nu.dean/screseng1.htm

University of Missouri–Columbia, Self-Care Deficit Web Page

http://www.hsc.missouri.edu/son/scdnt/scdnt.html

SENIOR CENTERS

Begun in the early 1940s, senior centers in this country now number between 10,000 and 12,000. The large majority of elderly are aware of the existence and know the location of such centers, and studies indicate that between 15% and 20% of older persons participate in center activities. Many senior centers are multipurpose and provide a wide range of health, social, recreational, and educational services. The Older Americans Act (OAA) targeted senior centers to serve as community focal points for comprehensive service coordination and delivery at the local level. Thus, senior centers not only provide services for older persons but also play important information and referral roles through linkages with a wide variety of other community organizations. Senior centers are often used by other agencies as delivery sites for programs such as congregate meals and health education (Krout, 1990).

Senior centers generally receive funding from a large number of sources, including the OAA, state and local government, fund-raising, and participant contributions. Larger senior centers generally have professional, paid staffs but also rely heavily on older persons as volunteers. Longitudinal data reveal that senior center budgets and programming have increased in the past decade. The mean number of activities and services offered by centers in a national sample of centers studied in the mid-1990s was 11 and 18, respectively (Krout, 1990).

Nine of 10 centers offered information and referral, transportation, and congregate meal services, and 70% offered home-delivered meals. Three-quarters of the centers offered health screening and maintenance, health education, and nutrition education. Information and assistance services (e.g., consumer protection; housing; crime prevention; financial, tax, and legal aid; social security) were offered by around two-thirds of the centers) (Krout, 1990).

The majority of research suggests that senior center users generally have higher levels of health, social interaction, and life satisfaction and lower levels of income than do nonusers (Krout, 1983, 1990). However, throughout their history, senior centers have responded to the needs of frail and well older persons, with some centers developing a greater emphasis on one group or the other (Krout, 1995, 1996). Depending on the senior center and the geographic area in which it is located, considerable diversity is found in who attends a center, the number of programs it offers, and the size of its facility, staff, and resources. Centers in many big cities, as well as suburbs and even some rural areas, are facing challenges and opportunities associated with a growth in the number of elders from diverse ethnic backgrounds. High levels of immigration in the 1990s meant increasing numbers of older adults who do not speak English. Many user populations have "aged in place," and attendance for some center programs, such as congregate meals, has leveled off or declined. Some centers are also having difficulty attracting newly retired older persons in their early 60s, creating problems for funding, volunteering, and center leadership. In response, many senior centers have expanded their focus on nutrition and socialization or recreational services to include wellness and lifelong learning activities.

Despite their prevalence and 60-year history, senior centers have largely been ignored by gerontological researchers. We do not have adequate data on center utilization patterns, management, or program content and evaluation. Longitudinal and trend data, other than counts of congregate meal participants, are lacking. This lack of data has not kept senior center advocates from working to strengthen center management and programming.

The National Council on the Aging recently promulgated standards and guidelines for centers to gain accreditation.

In conclusion, it is clear that senior centers have grown and diversified over the years. Shifts in federal and state spending; evolving priorities in health and social services for the elderly, with a greater focus on cost containment and targeting those at risk and changing demographics and retirement patterns present considerable challenges for senior center programming. Among the strengths of senior centers are their diversity and their ability to serve different segments of the older population in many different ways. Senior centers do many things well with relatively few resources and are certainly capable of improving and expanding existing functions, given the appropriate resources and mission. Although clearly a part of the community-based services system, their role in the long-term-care continuum is still evolving and needs to be better defined. It would be regrettable if the wealth of talent, energy, and dedication found in senior center professionals, volunteers, and participants were not utilized to the fullest in the 21st century.

JOHN A. KROUT

See Also
Adult Day Care
Meals On Wheels
Older Americans Act

Internet Resource
National Institute of Senior Centers
http://ncoa.org/nisc/nisc.htm

SENSORY CHANGE/LOSS: SMELL AND TASTE

Chronic problems with taste and smell are common in the elderly population, and although they are seldom fatal, they have a serious impact on quality of life. Aging affects the sense of smell more than the sense of taste. An estimated 50% of individuals age 65 or older have lost some sense of smell. The U.S. National Health Interview Survey reported that of the 2.7 million adults who have chronic difficulty with their sense of smell, 40% were 65 years of age or older, and of the 1.1 million adults who have chronic problems with taste, 41% were 65 or older (Hoffmann, Ishii, & MacTurk, 1998). Problems with the sense of smell can result from normal aging, environmental exposure, medications, surgical interventions, and illnesses such as Alzheimer's disease (Finkelstein & Schiffman, 1999).

Approximately two-thirds of taste acuity is dependent on smell. Age-related declines in the number and acuity of taste buds and in the production of saliva, which assists in dissolving food, may contribute to a diminished sense of taste. Smoking, poor oral hygiene, dentures that rub the tongue, nasal polyps, sinusitis, radiation therapy to the head and neck, and side effects of medications are associated with taste disorders. The sense of taste can be affected by over 250 medications that are commonly used by community-residing elderly, including furosemide and clindamycin (Finkelstein & Schiffman, 1999).

The senses of smell and taste are important components of a holistic assessment of older adults. An appropriate time to evaluate the senses is during cranial nerve examination. The olfactory function (smell) of the first cranial nerve, can be assessed by having the patient close his or her eyes, occlude one nostril, and sniff through the other nostril to identify an odor (e.g., coffee, cinnamon); this is repeated with the other nostril using a second scent after a brief rest period (Lueckenotte, 1996). The University of Pennsylvania Smell Identification Test, a 40-question, multiple-choice, scratch-and-sniff tool, evaluates the ability to smell by asking the person to identify various odors (Hoffman et al., 1998; Nusbaum, 1999).

Taste function can be evaluated during the assessment of the facial (seventh) and glossopharyngeal (ninth) cranial nerves by application of a salty solution on the anterior third of the tongue, sweet on the tip, bitter on the posterior third, and sour on the middle third (Lueckenotte, 1996). The taste buds used to detect salty and sweet are the ones most affected by aging. The tendency to compensate by adding excessive amounts of sugar and salt to foods (Nusbaum, 1999) can be problematic, if not dangerous, in older individuals with conditions in which excess use is contraindicated (e.g., diabetes, heart failure, hypertension).

Taste and smell are protective mechanisms. Dangerous situations that can result from impairment of these senses include the inability to smell smoke or a gas leak or to detect that food may be spoiled (Lueckenotte, 1996).

Loss of taste and smell can impact the quality of life for older adults (Nusbaum, 1999). Scents that may have been associated with enjoyment or pleasure, such as the smell of a Christmas tree, coffee brewing, and the fragrance of flowers, may no longer be detectable. The aroma of foods may be lost and can contribute to nutritional problems (Nusbaum, 1999). Elders may complain to family and caregivers that their cooking is unappealing; such complaints are common in institutional settings such as nursing homes. Older adults may also change the composition of their diet due to their lack of ability to taste salty and sweet foods, leading to overeating and obesity (Nusbaum, 1999).

Issues related to hygiene and cleanliness of the living environment may result from the inability to notice offensive personal, pet, or housekeeping odors. Overuse of perfume or cologne by individuals with a decreased sense of smell may somewhat mask body odor but may be offensive to others. Patient behavior may be misinterpreted as forgetful or demented if hygiene is poor secondary to a decrease sense of smell.

A variety of strategies can be implemented to stimulate and enhance what remains of the sense of smell and taste. Stimulating the sense of smell is important because the scent of food is an appetite stimulant and activates the sense of taste in the taste buds. Mouth care should be performed prior to each meal, with assistance as needed. The teeth, gums, and tongue should be brushed using a soft toothbrush, or the mouth can be rinsed with a solution of half-strength mint mouthwash and warm water. Lemon and glycerin swabs should be avoided, as they can contribute to drying of the oral mucosa. To promote nutritional intake, foods should be served separately in an attractive manner, varying in color, shape, texture, and temperature. Blended foods are the most difficult to taste; texture should be added to foods whenever possible, if this is not contraindicated by dietary restrictions such as liquid or pureed diets. For example, chunky instead of smooth applesauce could be used for individuals with swallowing difficulty who are not at risk of aspiration. Warm foods tend to be aromatic and should be served with each meal, allowing the aroma to permeate the environment. Sugar and salt substitutes, such as herbs and spices, can be used to enhance the flavor of foods.

Smoke alarms and pilot lights should be installed or, if financially feasible, gas ranges should be replaced with electric ones. Smoke alarms should be tested frequently for adequate functioning. Elderly individuals should be instructed to read expiration dates on perishable foods prior to eating them. Open foods should be dated to avoid spoiling. The sense of smell can be stimulated through the use of pleasant fragrances, such as lotions, colognes, and flowers. A drop of cologne on a lightbulb or a room freshener can give the room a pleasant and familiar odor to stimulate the remaining sense of smell.

Although not usually life threatening, changes in the ability to smell and taste influence the quality of life of older adults. This impact can be observed in the areas of nutrition, safety, hygiene, and overall enjoyment of life. Smell- and taste-enhancing strategies need to be employed that can stimulate remaining levels of functioning.

ANN MARIE SPELLBRING
ELIZABETH E. HILL-WESTMORELAND

Internet Resources
Health Watch: Sensory Changes in Later Life
http://www.working-solutions.com

Living with Sensory Loss
http://www.healthandage.com

Smell and Taste Disorder, National Institutes of Health
http://www.nih.gov/health/chip/nidced/smell/

SEXUAL HEALTH

Human sexuality has biological, affective, motivational, and cognitive aspects. Sex is one of the four primary human drives (the others are avoidance of thirst, hunger, and pain) and is not synonymous with sexual behavior or activity or with sexuality (Woodson, 1997). Sexual behavior or activity involves the genitalia and erogenous zones. Sexuality, however, is a broader term that includes sexual behavior, sex, emotions, attitudes, and relationships (Woodson, 1997). Normal age-related changes associated with sexual health and sexuality include biopsychosocial factors, none of which cause problems with sexual function if disease and adverse drug effects are absent (Meston, 1997). Estimates of the prevalence of sexual dysfunction among the elderly are probably low—because of social, cultural, and ageist stereotypes and clinicians' reluctance to include sexual function in assessment—to the same degree that reports describing seniors' active sex lives are inflated because data come from respondents interested in sexual activity. Conservatively, at least half of all married couples age 60 and over have sexual relations at least monthly; slightly more unmarried men (53%) than women (41%) have daily or weekly sexual relations (Janus & Janus, 1993). Women tend to attribute their less frequent sexual activity to the death of partners. Factors associated with sexual health and activity in older men and women include health status, medications, lifestyle, relationship issues, belief system, self-image, prior sexual activity and enjoyment, knowledge, and emotions. The neuroendocrine system mediates sexual desire, excitement, arousal, climax (orgasm), and recovery. Sight, smell, and touch stimuli affect sexual responsiveness differentially. Loss of privacy in a nursing home or living with younger relatives is also a major barrier to developing and maintaining a satisfying sexual relationship. Having to act furtively is destructive not only to sexual activity but also to a mature relationship between consenting adults.

Clinician reluctance to query patients about their sexual activity is based, in part, on the societal norm that sexuality and sexual behavior are private.

Gender differences and cultural factors among both clinicians and patients related to the discussion of genital anatomy and sexual practices may be factors in the diagnosis and management of sexual dysfunction. Vague terms such as "down there" may be less embarrassing for older adults and probably reflect generational differences regarding talking about sexual activity. Additionally, people frequently differ on what is meant by *sex*. The term is often used interchangeably to refer to a person's gender, to kissing or caressing, and to oral, vaginal, and anal intercourse. Hence, an accurate history is often difficult to obtain if clinicians or patients are unwilling to use the appropriate terms.

A study by Johnson (1996) somewhat supported the stereotype of the differences between men and women on what constitutes a satisfying sexual relationship. Talking, sitting, making themselves attractive, and saying loving words were less important to men than to women. Men more than women valued erotic reading material, erotic movies, sexual daydreams, physical intimacy (including caressing and intercourse), oral sex, and masturbation.

Sexual dysfunction (such as impotence, premature ejaculation, or anorgasmy) or simply the inability to appreciate or want an intimate physical experience is often driven by proscriptive societal, religious, and cultural norms about appropriate sexual behavior for older people and notions of physical attractiveness. Sex education for senior citizens has not been a high-priority health issue; few elderly are knowledgeable about changes in sexual response associated with aging or how to compensate for them. Fear, avoidance, ridicule, and, for some, anxiety and depression might be avoided with information. Additionally, the teaching of safe sexual practices, such as the use of condoms, is crucial to help prevent sexually transmitted diseases and the spread of the human immunodeficiency virus (HIV).

Abrupt loss of a sexual partner or of the ability to have a pleasurable sexual activity, including masturbation, can lead to problematic behavior, particularly in an institutional setting such as a nursing home. Staff have limited knowledge about the sexual practices of older adults and even fewer strategies for intervention into so-called sexually aggressive behavior. Although nursing homes are comfortable with sexual relations between married partners, staff are very uneasy and lack guidance about sexual activity among nonmarried, homosexual, or bisexual residents. The issue is burdened, also, with the complex reactions of the pairing partners' own progeny, concerns about the competency of the partners to understand what they are doing, and mutual willingness (particularly for the female partner).

The Sexual Response Cycle

In general, more time is needed by older sexually active men and women to be sexually aroused, complete intercourse, achieve orgasm, and be re-aroused in comparison to younger sexually active individuals. Testosterone decrease, a normal age-related change in men, reduces the tone of erectile tissue. Changes in collagen and the vascular endothelium may impair erection stiffness or frequency. Erection can take longer to achieve, be less full, and be maintained without ejaculation in comparison to younger men. Force of the ejaculation is decreased; volume of seminal fluid is less; and there are fewer contractions with orgasm, rapid loss of erection, and a longer refractory period. The notion of "male menopause" has not been verified by scientific fact, unlike the female climacteric or menopause, which has distinct characteristics.

Women also experience fewer orgasmic contractions; vasocongestion reduces more rapidly in older than in younger women. Fatty tissue loss in the pelvic area may predispose the clitoris to becoming more easily irritated. Vaginal estrogen cream or water-based lubricants can be applied directly to the vagina to treat such irritation. Diminished libido is more likely related to increased age, dyspareunia, body-image change secondary to breast or gynecological surgery, and psychosocial factors rather than to the physiology of menopause. Hormone replacement alone is not sufficient to restore flagging libido or loss of interest in sexual activity.

On average, intercourse lasts 10 to 15 minutes. The oxygen usage or "cost" approximates walking 2 to 2.5 mph or climbing one to two flights of stairs. The average pulse rate range in most individuals is 90 to 160 (systolic blood pressure can double), and respirations can be as high as high as 60 per minute. Men experience slightly more elevated vital signs than women, which may be related to sexual position (Hazzard, Blass, Ettinger, Halter, & Ouslander, 1999). Arousal without intercourse has a slightly lower impact on vital signs.

Diseases and Diagnostics Associated with Sexual Dysfunction

Various diseases and their treatments can affect sexual health. Hypo- and hyperthyroidism, diabetes mellitus, renal failure, stroke, dementia, depression, hypertension, coronary heart disease, cirrhosis, chronic obstructive pulmonary disease, Parkinson's disease, recurrent cystitis, and chronic infections can impair sexual function. Medications may alter sexual function by interfering with the release of hormones and acting on the autonomic and central nervous systems, interfering with libido, erection, ejaculation, or orgasm (Miller, 1995). Medications that may cause erectile dysfunction (ED) include certain antidepressants such as selective serotonin reuptake inhibitors, monoamine oxidase inhibitors, and tricyclic antipressants; tranquilizers; anticholinergics; phenothiazines; and antihypertensives such as beta-adrenergic blockers (Doerfler, 1999). Cimetidine may cause ED as well as gynecomastia (Doerfler, 1999). Imipramine may reduce sexual desire; methyldopa may inhibit erection by reducing pelvic blood flow. Little is known about medication effects on female sexuality, but it is thought that medications may diminish lubrication, decrease libido, or interfere with achieving orgasm (Miller, 1995).

Postmenopausal estrogen deficiency causes changes in the entire pelvic region, including reduction in the length of the vaginal vault, atrophy of the vaginal epithelium, and reduced amount and acidity of vaginal secretions, all of which predispose to infections and can cause dyspareunia. Urinary incontinence and irritation of the bladder and urethra because of thinning of the vaginal wall may also disincline a woman to have sexual relations.

Rosen (1998) reviewed several physiological methods to assess male sexual function, such as Rigiscan, volumetric circumferential plethysmography, and erectionmeter. Self-report measures examined included the International Index of Erectile Function, Brief Male Sexual Function Inventory, and the Center for Marital and Sexual Health Questionnaire. ED, formerly known as impotence, should not be diagnosed unless it occurs at least 25% of the time in encounters with the same sexual partner (Butler, Lewis, Hoffman, & Whitehead, 1994). It is associated with certain medications (e.g., antihypertensives), surgery (prostatectomy), diseases affecting the autonomic nervous system, circulatory diseases, diabetes, chronic prostatitis, hypercholesterolemia, renal failure, and central nervous system stimulants (National Institutes of Health, 1992). Emotional factors associated with ED include anxiety, depression, alcohol use, and fatigue. Treatment approaches include self-injection of intracavernosal medications that are smooth muscle relaxants (prostaglandin E, papaverine, and phentolamine in a combined low dose), oral ingestion of sildenafil citrate (Viagra) one to two hours before sexual activity, external vacuum devices, implants or prostheses, and revascularization. Viagra should not be taken more than once daily and is contraindicated in patients with cardiovascular disease and those who require nitrates. Vacuum devices are the least invasive; newer models reduce ejaculatory pain associated with earlier devices.

Physical performance is likely to be affected after a stroke. Men can have erectile and ejaculatory difficulties, and women may experience reduced vaginal secretions. Parkinson's disease is associated with loss of sexual desire and other sexual dysfunction. L-dopa improves sexual performance because it elicits a richer sense of well-being and increased mobility. Three months after a myocardial infarction (MI), if the patient can climb two flights of steps without chest pain, sexual activity can be resumed. A stress test is recommended prior to resuming sexual activity. Clinicians suggest that sexual activity should occur in the morning in a comfortable, nonstressful environment and that the post-MI patient should be in the position that requires the least exertion, that is, on the side or on the bottom. One-third of MI patients resume their normal sexual activity (Butler, Lewis, Hoffman, & Whitehead, 1994).

Interventions for and Prevention of Sexual Dysfunction

Lifestyle changes in midlife, such as regular exercise, reduced-fat diet, and smoking cessation, increase the probability of remaining potent and sexually active. Interventions and prevention strategies must, of necessity, be contingent on careful assessment and identification of contributing factors. As with many topics in professional and lay education, knowledge acquisition does not automatically become a blueprint for changed behavior or the desirable outcome, such as attitude change. Clinicians and caregivers may need time to recognize and understand their own feelings about sexual activity, the source of their attitudes, and the range of options for older people.

Those who are socially isolated by virtue of geographic location, finances, loss of partner (through death or institutional placement), or language proficiency may require counseling and community-based support services. Chronic medical conditions, including pain, have to be addressed by the appropriate health care provider. The benefit, burden, and consequences of each treatment option have to be addressed with the individual. Appropriate physical fitness programs for cardiac patients can moderate the physical signs and anxiety associated with sexual activity. Patients need counseling and support to resume sexual activity after an MI, coronary artery bypass surgery, and stroke. Those suffering from arthritis (women more than men) can achieve sexual pleasure with a combination of effective arthritis management strategies and sexual position change. Moist heat before sexual relations can reduce muscle spasms.

Individuals with sexual dysfunction need to be counseled to reduce alcohol consumption and smoking, since tobacco is a vasoconstrictor and may cause ED. All medications should be reviewed and adjusted as needed to treat and prevent sexual dysfunction and ED. Additionally, patients should be assessed for and counseled not to use illicit drugs such as cocaine, amphetamines, barbiturates, and opiates, which may cause sexual dysfunction (Doerfler, 1999).

Clinicians should assess the elderly patient's sexuality, validate and reassure age-related sexuality changes, counsel and educate patients who must cope with altered body image and age effects, and refer patients for special diagnostics and therapy when indicated. For this to happen, clinicians must be sensitive to the culture, constraints, language, and sexual interests of their patients. A therapeutic environment between provider and patient recognizes the embarrassment, for some, associated with sexual topics and language (Meerabeau, 1999). The clinician must balance support of an older person's possible disinterest in sexual activity based on personal standards while dispelling the myths and stereotypes that reduced physical intimacy is a natural consequence of aging.

SHERRY A. GREENBERG

See Also
Atrophic Vaginitis
Gay and Lesbian Elders
Social Isolation
Social Supports
Urinary Tract Infection

Internet Key Words
Dyspareunia
Erectile Dysfunction
Sexuality

Internet Resources
American Association of Sex Educators, Counselors, and Therapists
http://www.aasect.org

Sexual Health Information Center: Sex and Aging
http://www.sexhealth.org/infocenter/SexAging/sexaging.htm

Sexual Health Network
http://www.sexualhealth.com

SIGNAGE

Signs are a way of compensating for an unfamiliar environment and would be unnecessary if all environments were designed to be self-explanatory. It is an irony that some of the most unfamiliar environments are hospitals and large residential institutions and that the people who have to interact with them may have the most difficulty reading or understanding signs.

The ability to navigate independently through an environment enhances autonomy. An environment that is difficult to navigate can make people feel confused, anxious, irritable, or frustrated. They may lose confidence or form a poor image of the institution. Staff may become frustrated by visitors frequently asking them how to find their destination.

Older adults tend to be of smaller stature than younger adults are; some may have a slight forward head tilt. Poor eyesight and dementia, common in elderly people, can make reading and understanding signs difficult. Age-related vision changes include opacities in the central lens (cataracts), opacities in the periphery of the lens, changes in the vitreous (resulting in an increased scattering of light), deterioration in visual acuity (even in the absence of cataracts), yellowing of the lens, and decreased upward gaze in some people. These changes may cause difficulty reading small writing or indistinct lettering, unevenness in the perception of color, and a sensitivity to glare. Sensitivity to glare becomes even more problematic in bright light or in environments with bright surfaces. Glaucoma, if present, can constrict peripheral vision. The effects of dementia that can affect a person's ability to recognize signs are a decreased ability to read, interpret abstract symbols, reason or problem solve, and make a mental map of a building or space and an unreliable memory of recent events.

Signs should be easily seen, easily understood, and attractive. They should be designed primarily for people unfamiliar with the environment, such as visitors or people with poor memory, rather than for staff. The lettering should be clear and simple. Generally, a sans serif font is easier to read for short signs. A combination of upper and lower case is preferable, because the use of all upper-case letters removes the word's "shape" and decreases legibility.

Dark lettering on a light background with minimal use of different colors is easiest to read. Blue can be difficult to differentiate from black. Red on black is difficult for color-blind people. Yet certain colors are now internationally associated with safety. Red indicates prohibition or stop. Yellow indicates caution or risk of danger, such as where infectious or hazardous materials are present. Blue indicates some mandatory action, such as "break glass in case of fire." Green indicates a safe action or safe condition, such as a fire exit (Department of Health and Social Security and the Welsh Office, 1984; Baron, 1987). Conventions in current practice may not carry the same meaning for people with dementia, if their memory is of an earlier time (Calkins, 1988).

The size of a sign is determined by its location and the target population. The minimum recommended letter height for the general population is that capital letters should be 1 inch high for every 30 feet of viewing distance (or 3 cm high for every 10 m viewing distance) (McLendon & Blackistone, 1982). The size should be larger for an elderly population.

Images can enhance comprehension but should be used in addition to words, not as a replacement for them. The image should be realistic rather than abstract. People with dementia find a realistic picture of a toilet more recognizable than the international symbol of a male or female stick figure. Potential images should be tested on the target population (Wilkinson, Henschke, & Handscombe, 1995).

Sign location is best determined by assuming the role of a visitor coming to the building for the first time. Every point throughout the building that requires a decision by the visitor should have a sign. Signs protruding perpendicular to a wall may

be more visible in some situations, but they should not be placed too high.

In a complex environment, a hierarchy of signs can be helpful (MacKenzie & Krusberg, 1996). Directions can be given to a general area, and as a person approaches the desired destination, more specific directions can be given. Ideally, a sign should provide only enough information to allow someone to reach the next decision point. The more information that is provided, the longer it takes to read and the harder it is to remember. Signs intended for staff use can be differentiated from those intended for public use. Signs for staff can be smaller, in a different color, and placed below those for the public.

Lighting should be sufficient but should not create glare. Signs placed in a "puddle" of darkness between two bright areas are harder to see and less likely to be regarded as important. Glare comes not only from inappropriate lighting but also from shiny or polished surfaces.

To the extent possible, language should be in "plain English" and use natural speech, such as would be used when talking to a friend. Depending on the local culture, "toilet" is usually preferable to "rest room," "lavatory," "bathroom," or "powder room." "X-Ray Department" is more easily understood by laypeople than "Department of Medical Imaging" or "Radiology." People understand "ear, nose, and throat" more easily than "otorhinolaryngology." The language should be friendly and positive. For example, "No Parking" can be made more positive by erecting an arrow and sign to the "Visitors' Car Park."

Signs are not just written labels or symbols. Latent clues, such as placing chairs outside a room designed for sitting, are a form of sign that may be more comprehensible that writing. Labeling of some areas, such as toilets, may require a combination of sign clues. Some people may follow a clearly written sign, some may recognize a bright canopy above the door, some may be guided by the door color, and others may need personal guidance. Ensuring that all toilet doors are a particular color, all exit doors are a different color, and all cupboard doors are the same color as the walls are important indicators of their function. A reception area is more recognizable if it is in an open, accessible space, has a counter with someone behind it, and is well lit.

Signs should be considered in the overall design of a building, rather than as an afterthought. The environment should be as self-explanatory as possible and should not rely on the people being able to remember where they are or how they got there (Judd, Marshall, & Phippen, 1998). Signs are necessary when there has been a failure or inability to achieve this. They should avoid unnecessary detail and be simple, attractive, clearly written, well placed, and designed with the first-time visitor in mind.

TIM J. WILKINSON

See Also
Environmental Modification
Primary Care Practice

Internet Key Words
Building Design
Dementia
Signs
Vision

Internet Resources
American/Canadian School and Hospital Maintenance
http://www.facilitymanagement.com/artadasi.htm

National Council on Disability
http://www.ncd.gov/index.html

SKIN ISSUES: BRUISES AND DISCOLORATION

General health, diet, heredity, activity, and exposure influence the rate at which age-related skin changes occur. Changes in the function and appearance of the skin due to aging alone, known as intrinsic aging, include decreased wound healing, decreased elasticity and tensile strength, diminished

ability to respond to injury, decreased mechanical protection and insulation, and diminished ability to thermoregulate (Hazzard, Blass, Ettinger, Halter, & Ouslander, 1999).

Most age-related skin changes are not the result of aging alone but are due to a combination of aging and chronic environmental exposure, primarily sun exposure (Hazzard et al., 1999; Miller, 1999). This process, photoaging, is responsible for wrinkling and yellowing of the skin and thickening of the epidermis on sun-exposed areas. In addition, sebaceous glands enlarge, blood vessels become dilated and tortuous, and skin pigmentation becomes mottled. The physiological consequences of intrinsic aging result in the characteristic features commonly observed in the skin of older adults, such as fragility, tears, discoloration, and bruising. However, discoloration and bruising may be accelerated or exacerbated by the effects of photoaging.

Although it is commonly believed that the epidermis thins with advanced age, research shows flattening of the dermal-epidermal junction due to retraction of the papillae that connect the dermis to the epidermis. The result is a reduction in the surface area of the skin rather than an actual thinning of the epidermis (Gilchrest, 1992) and leads to poor nutrition and adhesion between these two layers and an overall decrease in the resilience of the skin. The combination of these factors results in separation of these layers and the likelihood of skin tears. Older adults are therefore more susceptible to bruises, blistering, and abrasions from mechanical stress or shear-type injuries.

Changes in the pigmentation, or coloration, of the skin are also noticeable with age. Melanocytes in the epidermis show some decline in function; the remaining cells may be unevenly distributed and not functioning normally. As a result, the skin becomes blotchy and unevenly pigmented, with areas of brown, spotty pigmentation frequently occurring on the scalp, neck, face, arms, and hands. These benign macular lesions are termed *liver spots* (solar lentigines) or *senile freckles*. A decrease in melanocytes also reduces tanning, and the ability of the remaining melanocytes to shield the underlying dermis from ultraviolet rays is diminished. Older adults are at increased risk for sun-exposure skin damage, predisposing to both benign and malignant skin changes.

The density, cellularity, and vascularity of the dermis progressively diminish with aging, resulting in loss of elasticity and turgor and less "give" under stress. The characteristic pale, thin, paperlike quality of the skin further contributes to tear-type injuries. Vascular changes in the dermis predispose older adults to petechiae, or minor bruising. The thin-walled, fragile blood vessels lose their connective tissue support. Following minor trauma, petechiae develop due to the fragile nature of the skin and increased capillary fragility. Areas of ecchymosis subsequently develop. These well-defined red-brown macules, termed *senile purpura*, vary in size from a few millimeters to several centimeters (Hazzard et al., 1999; Noble, Greene, Levinson, Modest, & Young, 1996). They occur most commonly on the exposed surfaces of the forearms but can occur elsewhere as well.

Minor bruising is a normal and common finding in older adults because of age-related vascular changes, but it may also be an indicator of a pathological process, such as acute leukemia or Cushing's syndrome. In addition, a number of pharmacological agents can induce purpuric bleeding, such as sulfas, aspirin, nonsteroidal anti-inflammatories, thiazides, procaine penicillin, phenytoin, methyldopa, barbiturates, and coumadin. Bleeding ceases when the drug is withdrawn.

A system of dating bruises by color is helpful to determine the stage of healing of a bruise. The approximate age of a bruise can be categorized as 0 to 2 days, red (swollen and tender); 2 to 5 days, red to blue; 5 to 7 days, green to yellow; 10 to 14 days, brown; and 14 to 28 days, clearing (Noble et al., 1996, p. 1684). The color presentation of a bruise is influenced by many factors, including the amount of force and area of injury, health status of the older adult, condition of the skin, and medications known to induce purpuric bleeding. Major bruising can occur in individuals who have coagulation deficiencies, liver disease, and a warfarin overdose (Hazzard et al., 1996). Often, laboratory studies such as platelet count, bleeding time, prothrom-bin time, or partial thromboplastin time are performed when there is a question about the extent and amount of bruising present.

A thorough history and careful examination of the skin can provide important information about the health status of the older adult and possibly serious problems, such as falls, neglect, or abuse. Multiple bruises in various stages of healing may alert the clinician to problems of physical abuse, alcoholism, or self-neglect (Miller, 1999).

Clinical evaluation is based on a thorough understanding of the normal skin changes associated with aging. This knowledge is essential in order to distinguish changes that may signal the presence of a more serious problem requiring further evaluation.

Maintaining skin integrity and preventing injury are important goals for clinicians working with older adults. Miller (1999) offers some helpful tips on skin care. Strategies to promote healthy skin include adequate amounts of vitamins A and C and fluids in the diet. Humidification and the application of emollient lotions at least twice a day, particularly after bathing, help prevent dryness, which makes the skin more susceptible to tears. Mild soaps such as Dove or Tone should be used when bathing. Daily bathing, however, should be discouraged, as this can further dry the skin. Skin care products that contain alcohol or perfumes should not be used because of their drying effect.

Older adults can avoid sun damage by wearing sun visors, wide-brimmed hats, and long-sleeved cotton shirts while in the sun. Sunscreen with an SPF of at least 15 should be frequently applied, and older adults should be encouraged to avoid sun exposure during the late morning and early afternoon hours.

Frequent changes in position are important for older adults who have activity or mobility limitations. Skin breakdown is more likely to occur in the presence of impaired circulation and external pressure. The use of pressure-relieving appliances is useful in maintaining skin integrity. Proper positioning in bed with the head of the bed elevated and the knees flexed and supported helps prevent shearing of the skin against the bed surface.

Environmental factors such as cluttered rooms; poor lighting; slippery floors; low, soft furniture; and sharp-cornered objects are the cause of many accidental injuries. These potential hazards can be avoided by creating an environment that is safe, comfortable, and stimulating for the older adult. Prevention plays a key role in maintaining skin integrity and reducing potential problems.

BARBARA J. EDLUND

See Also
Elder Mistreatment: Overview
Elder Neglect
Integumentary System
Pressure Ulcer Prevention and Treatment
Pressure Ulcer Risk Assessment
Skin Tears

Internet Key Words
Bruises
Discoloration
Fragile
Photoaging
Skin Tears

Internet Resources
National Institutes of Arthritis and Musculoskeletal and Skin Diseases, National Institutes of Health
http://www.nih.gov/niams/

National Skin Center of Singapore, Common Skin Diseases
http://www.nsc.gov.sg/commskin/skin.html

SKIN TEARS

Skin tears are a common occurrence among the elderly, especially those who are institutionalized. The reported incidence of skin tears in nursing homes approaches one per resident per year (Malone, Rozario, Gavinski, & Goodwin, 1991), for an estimated of 1.5 million skin tears annually, and is probably significantly underreported. Estimated costs of caring for skin tears approaches $10 per day for 10 to 30 days (Mason, 1997), for an estimated annual expense of $4.5 billion.

Although rarely life threatening or disabling, skin tears can be a site of infection and a source of pain. Families and residents are disturbed by the unsightly presence of skin tears, often perceiving them as the result of poor care. Frequent skin tears may also be predictive indicators of imminent health problems, such as delayed healing. Physicians' orders for specific treatment are rarely sought; hence, facility protocols or the personal preferences of caregivers produce wide variations in treatment.

Definition and Classification

A skin tear is a separation of the epidermis from the underlying dermis or connective tissue, thus creating a flap. Tears are jagged or L-shaped in appearance and are most often associated with minor trauma, friction, or shear (friction of the skin against an unforgiving surface).

Skin tears can be classified into three groups. Category I tears are linear; the epidermal flap can be completely placed back over the open wound, yielding no loss of skin coverage. A skin tear is category II when up to 75% of the epidermal flap is lost. A category III tear has lost the entire epidermis; there is no viable flap of tissue to replace on the wound (Payne & Martin, 1990). The most common sites for skin tears are the forearm, hand, dorsum, elbow, arm, and lower leg.

Cause

As skin ages, the epidermis gradually thins, and the rete pegs that meet at the junction of the epidermis and dermis flatten, thus diminishing adhesion of the dermis to the epidermis. These anatomical changes, combined with loss of elasticity and loosening of elastin fibers, allow the layers of skin to be more easily separated. When the skin of an elderly patient is rubbed or sheared, the attachment, already weakened by the aging process, further weakens, and the layers of the skin separate. Bleeding or weeping may occur between the layers, exacerbating the separation. A blister can form and eventually rupture, leaving a flap of separated skin. Skin tears may also occur immediately after trauma or shear.

Institutionalized elders are at a higher risk for skin tears because of their need for transfer assistance, wheelchair use, and repositioning. Poorly padded wheelchair hand brakes and armrests frequently cause skin tears. Many skin tears are not observed at the time they occur. Not uncommonly, the patient discovers the tear, cleans it, and covers it with a tissue. It can be several hours before the tear is brought to someone's attention, by which time the wound surface, appearance, and potential for proper healing have been compromised.

Documentation

A thorough description of the wound includes the precise location, size, and appearance (color, ecchymosis, drainage); whether pain is associated with the wound; and treatment measures pursuant to physician orders or facility protocols. The circumstances of the incident—for example, transfer or lifting—and plan of care to protect the patient from future skin tears should also be noted. Some facilities require an incident report, particularly if the skin tear is considered an injury of unknown origin. It may be advisable to notify the patient's family or significant other; notification can be state mandated, facility policy, or optional. Facilities can elect to include skin tears in their data collection for quality improvement. Continuing assessment and description of the wound should be done at least weekly.

Treatment

The wound should be gently irrigated with normal saline to remove blood and detritus. If the edges of the tear can be approximated relative to the epidermal flap, the flap is replaced on the wound edges, using sterile technique. The edges of the flap are affixed to the surrounding skin with wound closure strips (Steri-strips). If only part of the epidermal flap remains, the edges are approximated and secured. The open wound should be covered with a nonocclusive, moisture-retentive dressing, such as Vaseline gauze or opaque foam dressing (Thomas, Goode, LaMaster, Tennyson, & Parnell, 1999). Wounds with complete loss of epidermal tissue should be similarly covered. Adaptic Nonadhering Dressing (Johnson & Johnson Medical, Inc.) does not stick to the wound and is recommended. The dressing should then be covered with rolled gauze and nylon netting or stockinette for added protection from further trauma. Dressing changes should be performed as often as necessary to keep the area free from exudate and contamination. Steri-strips are not removed until the flap has healed. Pain medications administered on a regular basis may be necessary for the patient to continue with daily activities. Although no allergic reactions to Vaseline or clear adhesive dressings have been reported, allergy to ointments is possible.

Treatments to Avoid

Transparent film dressings appear to provide needed skin protection, but they have critical disadvantages, such as maceration of tissue related to increased heat and moisture and added skin trauma with dressing removal because these dressings stick to the skin. The excessive moisture associated with this type of dressing can lead to accumulation of exudate, promoting separation of epidermis and dermis. If a patient is admitted with a clear occlusive dressing on a skin tear, it is better to leave it in place than risk causing further tears by changing to a better type of dressing.

If an area of liquefied blood appears under the dressing, it can be left alone unless the site appears infected. To remove the dressing, push the skin away from the dressing rather than removing the dressing from the skin. Tape should not be applied to at-risk skin surfaces. Antimicrobials such as Betadine, Hibiclens, and PhisoHex deter the healing rate of wounds and are not recommended.

Prevention

Patients should be bathed no more than three times weekly; emollient soaps appear to be effective in preventing skin tears in patients at risk (Mason, 1997), but they have to be used cautiously, as the patient's body surfaces become sticky and places the patient at risk for falls. The average healing time for a skin tear on the leg is twice as long as for one on the arm. If the skin tear site becomes infected, it is most likely with a staphylococcal or streptococcal organism. The Norton or Braden scale can be helpful in identifying patients at risk for skin tears. Systemic cortisone use can predispose a patient's skin to tearing.

Chairs should have soft armrests, and wheelchair brake handles should be padded. If reasonable and acceptable to patients, wearing long-sleeved garments can protect the arms and elbows. Comprehensive in-service training of staff caregivers and family members involved in transfers and lifting heightens awareness of the dangers in the physical environment and in direct care. Gait belts can be helpful when lifting and moving residents because they reduce the tendency to pull on the forearms.

LISSA CLARK
JOYCE BLACK
STEVEN B. BLACK

See Also
Integumentary System
Pressure Ulcer Prevention and Treatment
Pressure Ulcer Risk Assessment
Skin Issues: Bruises and Discoloration

Internet Key Words
Paper Thin Skin
Skin Tears

Internet Resource
Wound, Ostomy and Continence Nurses Society (WOCN)
http://www.wocn.org/

SMELL
See
Sensory Change/Loss: Smell and Taste

SOCIAL ISOLATION

The term *isolation* has diverse meanings in the gerontological literature. Social isolation generally denotes an absence of social interaction, contacts, and relationships with family, friends, neighbors, and society (Kahana, 1995). Age-related changes such as widowhood, retirement, and family dispersion have created the popular view of the elderly as more isolated than the general population. However, total social isolation, or the absence of social contacts, interaction, and social support networks, is relatively rare among the elderly. Studies indicate that only 4% of the elderly report extreme isolation (Kahana, 1995). More frequent is emotional isolation due to the loss of a partner or close friends or to relocation. Furthermore, some elderly have a lifelong history of withdrawal and may have long-term personality traits that foster a solitary existence. Nonetheless, the documented decrease in social interaction experienced by most older people is involuntary.

The theory of socioemotional selectivity posits that older people narrow their range of social partners and are more likely to choose long-term friends and loved ones for social interaction; having positive emotional experiences is more important than other motives such as information seeking or self-definition (Carstensen, Isaacowitz, & Charles, 1999). With a smaller network, elders can conserve physical and cognitive resources, freeing time and energy for selected social relationships.

Research on the connection between social relations and well-being supports the everyday understanding of the important social, psychological, and behavioral functions of human interaction. Exactly how and why social relationships influence well-being are not understood, but there is growing awareness of the complexity of social support that has a bearing on understanding social isolation. Social support or relationships can be viewed as having three components: quantitative structure and composition of the social network, such as type and amount of social support functions (e.g., emotional support; tangible aid or instrumental support and guidance); and the qualitative perceived adequacy of the support provided (Oxman & Berkman, 1990).

Studies since the 1980s indicate that people with strong social relations have lower morbidity and mortality, including a better possibility of recovery after illness (Rowe & Kahn, 1998). Greater frequency of emotional support from social networks, particularly among those reporting low frequency of instrumental support, had a favorable impact on functional outcomes. Social connectedness has also been linked to emotional well-being, with studies showing that the quantity and quality of relationships are significant predictors of depressive symptoms in patients who are middle-aged and older. Although not tested in research studies, the consistent findings regarding friendship and morale have been attributed to the fact that unlike family relationships, friendships are often built by choice and involve reciprocal exchange of assistance between peers (Hooyman & Kiyak, 1996). Conversely, low levels of social activity and social contacts are associated with poor functional outcomes. Studies of the negative aspects of social relationships underscore that social relationships of a conflictual or negative nature can adversely affect health (Bowling, 1991; Rowe & Kahn, 1998).

Changes associated with aging can result in reduced contact between older adults and family, friends, and associates. Death of a spouse or other family members, hearing and vision limitations, decline in mobility, residential relocation, and loss of social roles are factors that may singly or in combination place an older person at increased risk for social isolation. The tendency is to assume that living alone, being single, or not having a family necessarily implies social isolation; however, studies find that older women in particular are able to receive help and emotional support from both old and new relationships. It is more likely that the

interaction of multiple changes and losses, rather than bereavement alone, places most older adults at greater risk of isolation (Hooyman & Kiyak, 1996).

The relationship between social isolation and elder abuse, neglect, or mistreatment has received a lot of attention. Both self-neglect and abuse by family are most likely to occur when the patient has dementia or late-life depression, conditions that impose relatively high psychological and physical burdens on caregivers. Social isolation also can be a strategy for keeping abuse secret. Elders who are homeless, have a history of chronic mental illness or substance abuse, and live in inner-city neighborhoods are also at high risk for social isolation (Kahana, 1995).

Assessment

Overall social functioning and identification of barriers to interaction with others must be assessed. Health care practitioners from multiple disciplines are critical links in identifying and addressing social isolation in older adults, including attention to possible root causes such as sensory impairment, compromised mobility, and clinical depression. When social isolation is unrecognized, opportunities are missed to improve physical and mental health and prevent abuse. As in all client assessments, a relatively simple initial screening can lead to more complex secondary and tertiary strategies. Paraprofessional case managers and case aides may screen the client to determine the existence or absence of social support. Practitioners with more clinical training can then explore the dynamics of any support system's strengths and weaknesses and pursue necessary social intervention. Among the best known measures of social functioning is the Older Adult Resources and Services (OARS) Social Resource Scale, part of the larger OARS battery (Kane & Kane, 1981). OARS addresses social parameters such as marital status, circle of friends, frequency of contact, presence of a confidant, and possibility and level of social support in the future. Assessors must be aware that clients may overestimate the quality or quantity of relationships for a variety of reasons, including wishful thinking, protection of family members, pride, or feeling that something besides the truth is the desired response.

Prevention and Interventions

Important prevention strategies include education of the general population, but especially older people and family caregivers, about the importance of remaining socially engaged through all avenues, from telephone visitation to active participation in organizations. Modifying the concept of a financial portfolio, Cohen (1995) encourages the development of "social portfolios," a program of meaningful activities and interpersonal relationships that usher individuals into old age. Addressing social isolation also requires social policies and programs that promote social exchange and interaction. In the process of facilitating or providing services, communities need to consider the diversity of their older residents, including differences in race, cultural background, socioeconomics, home settings, and general functioning. Many communities have a comprehensive network of senior centers, opportunities for volunteerism, and recreational activities; however, they may not be user-friendly for elders who are frail, belong to ethnic minorities, or require specialized transportation. The physical design of both private homes and communal environments (community centers and senior housing, for example) can promote or impede communication between friends, participation in recreational activities, or vicarious enjoyment through watching others. Practitioners need to develop individualized "social prescriptions" to ensure that older adults have the physiological ability to contact and communicate with others, other people with whom to interact, opportunity for meaningful interaction, and the motivation or desire to interact. Specific plans may involve introducing effective aids for vision or hearing deficits, treating clinical depression, modifying homes for access, helping with transportation, and securing referrals to self-help or support groups or to community social or educational programs with peers.

In addition to this direct approach, practitioners should avoid reducing an elder's opportunity for social activity and meaning by overemphasizing health and safety issues. Family caregivers of dependent older adults may experience stress and require assistance in planning ways to enrich the lives of elderly relatives and alleviate stress.

Daily Physical and Emotional Care

Although research is limited, participation in adult day-care programs and self-help and support groups appears to be beneficial in preventing isolation of older adults and family caregivers and therefore may shield against elder abuse (U.S. Department of Health and Human Services, 1999). Outcomes of self-help bereavement programs found that participants experienced fewer depressive symptoms, recovered their activities, and developed new relationships more quickly than those without such support.

Older individuals with a lifelong history of isolation and those who have no children or spouse are more likely to depend on the active assistance of social agencies and mental health services. Elders may not always welcome outside assistance, fearing loss of control or denying the need for help. Effective formal outreach programs require carefully trained personnel to overcome this resistance. Cases of social isolation due to self-neglect or abuse require referral to the state-designated programs serving abused and neglected elders.

Technology's role in decreasing social isolation is a recent and as yet unstudied phenomenon. There are several senior and caregiver sites on the World Wide Web; the Internet is being promoted as a means by which both disabled and active older adults can access information and communicate with friends and family. Given the increasing likelihood of disability in later life, helping older adults maintain or establish meaningful social ties and avoid social isolation is critical in helping them maintain their independence and overall well-being.

NANCY L. WILSON

Internet Key Words
Friendship
Social Functioning
Social Prescription
Social Relations
Social Support

Internet Resources
Administration on Aging
http//www.aoa.dhhs.gov/

Eldercare Web
http://www.elderweb.com

National Council on the Aging
http://www.ncoa.org

SOCIAL SECURITY

The national pension system of the United States is officially named Old-Age, Survivors, and Disability Insurance (OASDI). The general term *Social Security* is often used to describe this program, but it has a different meaning to different people in different countries. Included with the term Social Security at times are the medical-expenses benefits of the Medicare program, unemployment insurance, and work-connected injury benefits.

Individual Equity and Social Adequacy

Before discussing the evolution of Social Security, two basic terms must be defined: *individual equity*

and *social adequacy* (Myers, 1993). Individual equity is present when a covered individual receives benefit protection actuarially equivalent to the contributions paid. Thus, individual equity can be satisfied on a "benefit protection" basis, without a requirement that the person always receive benefits at least equal to total contributions, with interest. Conversely, social adequacy means that benefits are paid according to presumed needs, not necessarily in relation to contributions paid.

Occasionally, two other terms are used instead—*insurance* and *welfare*—but they have possibly misleading connotations. Welfare implies means tests administered by social workers or public officials, and insurance implies strict individual equity, which is not the case for either social insurance or private group insurance. However, Social Security is, in fact, insurance in the broad sense of the term because it involves the pooling of risks. Benefits under Social Security are partially, although not completely, related to earnings level and length of coverage but are heavily weighted in favor of lower-paid persons.

In practice, OASDI has always provided benefits that involve a mixture of individual equity and social adequacy. It is sometimes believed that the benefit structure under the original 1935 law was on a completely individual equity basis, but this was not the case. Much larger benefits, from a relative standpoint (in relation to taxes paid), were provided for those retiring in the early years than for those retiring later (just as is usually done in private pension plans through the use of prior-service credits). Also, relatively larger benefits have been provided for low-paid persons than for high-paid ones and for those with dependents.

In the early years of the program, the general tendency in OASDI was to move more in the direction of social adequacy and away from individual equity. There have, however, been instances in which the trend was somewhat in the opposite direction.

OASDI Coverage

The 1935 act in its initial coverage applied only to workers in commerce and industry, representing about 60% of the total workforce. Excluded were groups for which administrative problems or questions of constitutionality were present or who already had pension plans (such as federal civilian and military employees). Coverage of agricultural and domestic workers and the self-employed seemed to pose administrative problems. Coverage of employment in state and local governments was thought to be impossible on a compulsory basis because of constitutional considerations. Coverage of employees of nonprofit organizations of a charitable, educational, or religious nature was not provided because of these organizations' traditional tax-exempt nature.

Legislation in the early 1950s extended coverage to most of the groups previously excluded. Program operations had indicated that the administrative problems of covering farm and domestic workers and self-employed persons could be handled. Coverage for employees of state and local governments and nonprofit organizations was handled by permitting coverage on an elective basis; at present, about 80% of state and local employees are covered. In 1984, coverage of nonprofit employees was made mandatory. Members of the armed forces were covered in 1957 and retained in full their previous pension plan. Permanent federal civilian employees who have their own pension system were not covered in the legislation of the 1950s, but beginning in 1984, all newly hired employees are covered and have a modified, supplemental pension plan. The roughly 300,000 temporary civilian employees were covered in 1951.

The principal groups not covered as of 2000 are federal civilian employees with their own independent retirement system hired before 1984, some state and local government employees who have their own retirement system, and relatively low-paid and short-service workers in farm and domestic employment.

Coverage has always applied throughout the 50 states and the District of Columbia. Coverage was later extended to all other geographical areas under the jurisdiction of the United States, such as Puerto Rico and Guam. In addition, persons working on American ships and airplanes are covered, as are citizens and resident aliens going abroad to work for American employers (or, on an elective basis, for subsidiaries of American employers).

OASDI Benefit Categories

The original act provided monthly benefits only for retired workers age 65 and older. In addition, lump-sum refund benefits were available that "guaranteed" each covered individual that total payments would be at least equal to the employee contributions, plus some allowance for interest. In 1956 for women and in 1961 for men, actuarially reduced benefits were made available at ages 62 to 64.

Auxiliary benefits for retired workers were provided in 1939 for spouses age 65 and older (later lowered to age 62, with reduced amounts payable) and children. At the same time, survivor benefits were provided for widows age 65 and older (later reduced to age 60 or to age 50 if disabled, with reduced amounts), and for children and their mothers or fathers.

The benefit rates for auxiliary beneficiaries (spouses and children) and for survivors are expressed as percentages of the primary amount (that which is received by a retiree at the normal retirement age, currently 65). These rates (before the effect of the maximum family benefit provision, discussed later) are 50% for auxiliary beneficiaries, with reduced amounts for spouses without eligible children who claim benefits at ages 62 to 64; 100% for widows and widowers age 65 or older at initial claim, with reduced amounts for claim at earlier ages (ages 60 to 64 if nondisabled and ages 50 to 59 if disabled); and 75% for eligible child survivors and their widowed parents.

In 1983 the normal retirement age was changed. It will increase gradually from age 65 to 67, beginning in 2003, and reach age 67 in 2027. The early retirement ages will not change, but the reductions for early retirement will, in general, be larger percentages. This change was made to recognize increased longevity and thereby assist in alleviating long-range financing problems.

Benefits are increased for persons who defer retirement beyond the normal retirement age. For those attaining such age before 1990, the increase is 3% per year of delay. For those who reach such age in 1990 and later, the increases will become larger (3.5% for 1990–91 attainments, 4% for 1992–93 attainments, 4.5% for 1994–95 attainments, etc.), until it reaches 8% for those attaining age 66 in 2009.

In the past, unequal treatment occurred between men and women regarding auxiliary and survivor benefits. During some periods, men could receive such benefits only on proof of dependence, whereas dependency was always presumed for women. Also, child survivor protection was not available for female workers on as broad a basis as for male workers. Over the years, because of both legislative changes and court decisions, this unequal treatment has been eliminated.

There has always been an "antiduplication" provision, according to which persons receive only the largest of the several benefits for which they may be eligible in a particular month. For example, if a person is eligible for a benefit as a retired worker on her or his own earnings record, and also for a survivor benefit on the basis of a deceased spouse's earnings record, only the larger of the two amounts is payable. This procedure is followed on grounds of social adequacy. This principle dictates that only one benefit should be payable for a given month, to satisfy the social-responsibility aspects of the program, rather than paying all benefits for which the person is eligible. The latter approach, if followed, would be justified solely on individual-equity grounds.

Even if a person does not receive a benefit on his or her own earnings record because another available benefit is larger, this does not mean that the individual has received nothing for the contributions paid. Actually, such an individual has already received valuable disability and survivor protection in the past. Further, such an individual may have received a retirement benefit based on his or her own earnings record before the spouse retired and claimed benefits, at which time the auxiliary benefit was not available.

Disability insurance (DI) benefits were added in 1956, with the amount being the same as for a retiree at age 65. At first, these were payable only to disabled workers ages 50 to 64. Later, auxiliary benefits for spouses and children, comparable to those available for retired workers, were added. Still later, the age 50 limitation was removed, so that disabled workers could qualify for disability benefits at any age younger than 65.

Eligibility for benefits depends on having the required number of "coverage credits." In 2000, a credit is given for each $780 of earnings, up to a maximum of four per year. For retirement and survivor benefits, no more than 40 credits are required; for disability benefits, generally 20 credits in the last 10 years are required. The amount of earnings required for a credit is adjusted from year to year by the change in average wages.

Level of OASDI Benefits

The level of OASDI benefits has always been influenced by several general principles. The benefit structure has provided weighted amounts, so that those with lower earnings receive *relatively* higher amounts than those with higher earnings (Rejda, 1994).

The floor-of-protection concept has always been present. Specifically, the underlying principle has always been that OASDI benefits should not provide complete retirement income for all persons by being at a level approximating previous after-tax earnings. Rather, it is assumed that most workers can be expected to supplement their OASDI benefits with other forms of income, such as private pensions, individual savings, and home ownership, which produces significant imputed income. However, considerable differences of opinion exist as to the relative height of the floor of protection. In considering the level of OASDI benefits over the years, it is important to consider them relative to earlier earnings and not in absolute dollar amounts. The latter are affected by changes in economic conditions, and the standard of comparison should be a relative one, not an absolute one.

The original act provided benefit amounts that were larger for those with high earnings than for those with low earnings, although not proportionately so. A maximum limit was imposed on the amount of annual earnings creditable toward the computation of benefits. This amount has been increased over the years. Similarly, larger benefits were provided for those with long periods of coverage than for those who would retire in the early years of the act's operation, but again, not proportionately so. The benefit structure was not based entirely on individual equity; considerable elements of social adequacy were present.

In the 1939 act, adopted before monthly benefits were first payable, the design of the benefit structure was also changed by adding auxiliary benefits for spouses and children. Less emphasis was given to increasing the benefit amounts to recognize length of coverage, and the principle of relatively higher benefits for low-paying than for high-paying persons was continued. The 1950 act moved further in this direction by eliminating any recognition of length of coverage in computing benefit amounts for persons with steady, continuous employment histories. This was done by removing the provision for a 1% increment in computing the benefit for each year of coverage.

The formula for the basic benefit, the primary insurance amount (PIA), under the 1950 act and subsequent legislation is of the following general form: A% of the first $X of average lifetime covered earnings, plus B% of the next $Y of average earnings, plus C% of the next $Z of average earnings, and so on (where A, B, and C are the "benefit percentages"—which decrease from A to B, from B to C, etc.—and where $X, $Y, and $Z are termed *bend points*). Thus, the formula is heavily weighted in favor of lower-paid persons, so that the social-adequacy element applies. Average earnings are computed over the total potential working lifetime during which the person could have been covered (after age 21 and before age 62), but with provision for eliminating a few years of low earnings. Generally, five years of low earnings are dropped, and years of good earnings before age 21 or after age 61 can be substituted for lower-income years.

Following the 1950 act, several ad hoc benefit increases were made. The changes in 1954 represented a real increase in the benefit level, but the changes in other years up through the 1965 act merely kept the level up-to-date with changes in prices, as measured by the consumer price index (CPI). The increase in benefits under the 1967 act and the three legislative enactments between 1969 and 1972 made significant real increases in the benefit level. For example, the increases in 1969 to 1972 represented, in the aggregate, a "real" rise of about 23%.

The 1972 act, besides increasing benefits currently being paid by 20%, inaugurated automatic indexing of the benefit level, based on increases in the CPI being applied to the benefits in current payment status and to the benefit percentages in the benefit formula, beginning in 1975. The indexing procedure adopted in computing the initial benefit amounts, which was the same as had been done successfully on an ad hoc basis during the 1950s and 1960s, turned out to be faulty under the changed economic conditions of the 1970s and most likely for subsequent periods.

This faulty procedure has incorrectly been described as "double indexing," because (1) the benefit percentages that were applied to the average earnings were increased by rises in the CPI, and (2) average earnings would be higher as a result of inflation. However, a third factor was present—namely, the bend points in the benefit formula, expressed in dollars, remain unchanged, so that a smaller portion of average earnings would fall in the first, much higher, band below the first bend point. The interaction of these three elements, however, is such that replacement rates (initial benefit as percentage of final earnings) would likely be unstable for future generations of retirees, depending on economic conditions.

Only with certain economic circumstances could stability occur: if wages rise by about 4% per year and prices increase by about 2% per year. However, if both wages and prices increase considerably more than this, with wages not rising much more rapidly than prices (or even with this relationship reversed), then replacement rates would get out of control and increase continuously over future years. This would result in serious cost problems.

Conversely, if wages should rise considerably more than prices, especially if the latter did not increase much, replacement rates would decrease, and the result would be *lower* program costs than anticipated.

Some of the overexpansion of the benefit level resulting from this faulty indexing procedure, and also some of the "real" increases in the level resulting from the 1969 to 1972 legislation, were later eliminated. This was the result of (1) the ad hoc benefit increases in 1974 being less than what they would have been if they had been based on the actual CPI rises, and (2) the benefit formula under the 1977 act being designed to produce levels about 6% lower than would have resulted under previous law for persons newly eligible in 1979 and later.

The 1977 act also provided that, in the computation of average earnings for the benefit calculation, the earnings record was to be indexed. The earnings of each past year are updated approximately to the time of eligibility (i.e., to age 60 for retirement cases) to reflect changes in general earnings levels over the years. Also, a new basic benefit formula was adopted. The benefit percentages will not be increased by the rises in the CPI (as formerly); instead, the bend points will be increased by the rises in average wages. The benefit formula for those attaining age 62 in 2000 (or dying or becoming disabled before age 62 in 2000) is 90% of the first $531 of average indexed monthly earnings (AIME), plus 32% of the next $2,771 of AIME, plus 15% of AIME in excess of $3,202. For example, for an AIME of $2,000 (about what results for a steady average-wage worker), the basic benefit is $948, whereas for an AIME twice as large, it is $1,484, or only 57% higher.

The result of the new computation procedure, as can be mathematically demonstrated, is to produce stability of replacement rates. For persons retiring at the normal retirement age (currently, age 65) or retiring on account of disability, the benefit rate for a steadily employed worker with average earnings at all times will be about 41% of final earnings, regardless of the year of retirement. The corresponding figure for a worker with low earnings (about half of the average) will be about 55%,

whereas the figure for a worker with maximum covered earnings (about 2.4 times the average earnings) will be about 28%.

The 1983 act introduced a financial stabilizing device to be applicable when the OASDI trust funds are at a relatively low level. The indexing of benefits in current payment status will then be based on the CPI increase or the rise in the general wage level, whichever is lower. If the trust funds later build up to a moderately high level, any decrease in the indexing resulting from using wage increases would be restored, prospectively.

Restrictions on OASDI Benefits

Several restrictions apply to the benefits payable. The antiduplication provision results in individuals receiving only the largest of the benefits for which they are eligible each month. Specifically, if a person is eligible for both a benefit based on his or her own earnings record and one based on the spouse's earnings record, the amount payable is equal to the larger of the two.

An overall limit is placed on the benefit available to the family members of a retired, disabled, or deceased worker. In general, full benefits at the percentage rates mentioned previously are payable only when there are no more than two beneficiaries in the family (such as a retired worker and spouse, or a widow and one child) in retirement and survivor cases. In essence, a partial benefit is payable to the third family member if there are three beneficiaries in the family (e.g., a widow and two children), but no benefit recognition is given for family members in excess of three.

The maximum family benefit formula for retirement and survivor cases at attainment of age 62 in 2000 (or death before age 62 in 2000) is 150% of the first $679 of the PIA, plus 272% of the next $301 of PIA, plus 134% of the next $298 of PIA, plus 175% of PIA in excess of $1,278. The net effect of this formula for those with PIAs of $1,278 or more is to provide a maximum of 175% of PIA. For disabled worker beneficiaries, the maximum is lower, generally not providing full benefits when more than two beneficiaries are present (i.e., the maximum generally is 150% of the PIA).

Another restriction on OASDI benefits is the so-called retirement test, more accurately described as the earnings test, because it applies not only to retirees but also to most other types of beneficiaries. However, the test does not apply to retirees age 70 or older. Such a test has always been present on the basis of the philosophy that benefits should be available only when there are no substantial earnings from employment. The general basis of the test is that a certain amount of earnings is allowed without any reduction in the OASDI benefits payable, but that for earnings in excess of such an amount, $1 of benefits is withheld for each $2 of "excess" earnings. These annual exemptions in 2000 were $17,000 (or about 56% of the nationwide average wage) for beneficiaries ages 65 to 69 and $10,080 for other beneficiaries. Such amounts are adjusted upward in future years on the basis of changes in the general wage level. Beginning in 1990, the basis for withholding of benefits in excess of the annual exempt amount is on a $1-for-$3 basis for persons at and above the normal retirement age.

A limitation on the benefit paid to disabled workers applies if they receive disability pensions under other governmental plans. The total disability payments received cannot exceed 80% of recent average earnings before disablement. Should this be the case, the OASDI benefit is reduced accordingly, but the total payable will not be less than the OASDI benefit would be in the absence of the offset provision.

Benefits payable to spouses of insured workers (whether as a wife, husband, widow, or widower) are reduced if the spouse is receiving a pension under a government employee retirement system under which the members were not covered under OASDI on the last day of service of such spouse. Under such circumstances, any OASDI benefit payable to the individual from the earnings record of the spouse is reduced by two-thirds of the amount of the government employee pension.

OASDI benefits are also reduced in the case of workers who receive pensions that are based, in whole or part, on earnings from noncovered employment (in the past or in the future), such as work for federal, state, and local governments and nonprofit organizations. This provision applies only to those who attain age 62 (or are disabled) after 1985.

Income Taxation of OASDI Benefits

Until 1984, OASDI benefits were not subject to federal, state, or local income taxes. After 1984, up to 50% of these benefits is considered taxable income for federal income tax purposes for high-income persons. The income limits are $25,000 per year for single persons and $32,000 for married couples filing joint returns (and zero if separate returns are filed when the couple live together at some time in the year), with only 50% of OASDI benefits being counted as "income" for this purpose. Beginning in 1994, up to 85% of the benefits are considered taxable income for income tax purposes. Unlike many elements of the OASDI program, these threshold amounts are not indexed in the future. Thus, over time, an increasing proportion of beneficiaries will be affected (only about 25% currently). State and local income taxes do not apply to OASDI benefits in almost all states.

Financing OASDI

The payroll taxes and other sources of income—such as interest on investments and, since 1984, the receipts from income taxes on OASDI benefits (but not those based on the increase from 50% to 85% in 1994, which anomalously go to the Hospital Insurance Trust Fund)—are deposited in two trust funds: one for the old-age and survivor benefits and the other for disability benefits. Similarly, the benefits and administrative expenses are paid out of these trust funds, with almost all of the benefit payments going out at the beginning of each month.

Over the years, the basic financing concept of OASDI has been that it should be supported entirely from the scheduled payroll taxes. The only exceptions are certain closed groups for which benefits are financed from general revenues (e.g., persons who became age 72 before the early 1970s and were not "insured"). For a period during the 1940s, the law provided for the possibility of general revenues being injected into OASDI (introduced because the tax rate was frozen during the 1940s), but this was eliminated by the 1950 act. The 1983 act, however, introduced some general revenue financing in indirect manners. By far the major such method is to return the federal income taxes on OASDI benefits to the trust funds.

When the program was initiated in 1935, funding was on what might be called a modified reserve basis, under which a rather sizable fund would be developed. The interest thereon would finance a substantial part of the ultimate costs. Over the years, the emphasis on building up a relatively large fund lessened. In recent years, for all practical purposes, the OASDI system was financed on a pay-as-you-go (or current cost) basis. However, this procedure will not be followed in the future if present law is not changed. Beginning in the early 1990s, trust funds have built up substantial balances and will increasingly do so for several decades, but these will decline and become exhausted by about 2034, according to the intermediate cost estimate.

Another basic principle is that a schedule of increasing tax rates has always been incorporated into the law, extending into the future. This gives the public an indication of likely future cost trends, which is especially important for programs financed on a pay-as-you-go basis.

The payroll tax rate regarding employees has always been divided equally between the employer and employee (with the exception of 1984, when there was a small difference). This has not been dictated by actuarial or economic reasons but rather by what might be called aesthetic logic: that each party involved should share the cost equally. The combined employer-employee tax rate began at 2% in 1937 and continued at this level until 1950. Since then it has gradually increased and was 11.4% in 1986 (with an additional 2.9% for the hospital insurance [HI] portion of Medicare). The rate increased to 12.12% in 1988 and to 12.4% in 1990, with an additional 2.9% for HI. The 12.4% rate is subdivided between the OASI and DI trust funds; the

latter will receive 1.88% in 1994 to 1996, 1.70% in 1997 to 1999, and 1.80% thereafter.

When the self-employed were first covered by the 1950 act, there was a considerable difference of opinion as to the appropriate tax rate to charge. Some argued that the basis should be the combined employer-employee rate, on the grounds that the system should receive the same amount of income regardless of whether the covered worker is an employee or a self-employed person. Others argued that the self-employed should pay only the employee rate, on the grounds that both would receive the same benefit amount for a given earnings record. Under the political compromise reached, the self-employed paid 1.5 times the employee rate. The 1983 act, to raise additional revenues, provided the "logical" basis of the self-employed paying the combined employer-employee rate, but with a credit to reflect the fact that employers can count such taxes as business expenses for income tax purposes. The trust funds were credited with the entire employer-employee tax, with the difference coming from the General Fund of the Treasury. Beginning in 1990, the self-employed paid the full employer-employee tax rate but received an income tax credit against their net profit equal to 50% of such tax.

The OASDI payroll tax has always been applied up to a certain amount of annual earnings, which is called the "earnings base," and benefits are determined only on such amounts. Initially the earnings base was established at $3,000, which covered about 92% of all earnings in covered employment. Beginning in 1951, the earnings base was increased periodically by ad hoc legislation, and until 1972 it covered about 80% (in some years, somewhat less) of the total earnings in covered employment.

The 1972 act and subsequent legislation provided for several significant increases in the earnings base, with automatic adjustment thereof being applicable to 1975 and thereafter. This would have stabilized the proportion of the total covered earnings that would be taxable at about 85%. The 1977 act provided for ad hoc increases in the earnings base for 1979 to 1981 that raised it more than the automatic adjustments would have done. The automatic adjustments apply after 1981, and the proportion of total earnings taxable is now about 85%, with the base in 2000 being $76,200. The HI earnings base was the same as the OASDI one until 1991, when it was made higher (and, for 1994 and after, has been eliminated).

Legislation in 1981 provided for possible interfund borrowing (repayable with appropriate interest) among the OASI, DI, and HI trust funds. The purpose of such borrowing was to make additional financing resources available to the OASI trust fund, whose balance had been declining steadily ever since 1975 and was expected to decrease in the future; by late 1982, it would have problems meeting benefit payments. Such borrowing was permitted only during 1982 and could not exceed amounts sufficient to pay benefits during the first six months of 1983. The 1983 act extended this provision through 1987, with repayment of any loans being required before 1990.

The investments of the trust funds have always been required to be in obligations of the federal government. These obligations can be obtained in any of three ways: purchases on the open market, purchases of obligations available to the public when they are first offered for sale, and purchases of special issues.

Almost all the investments have been in special issues, which are available only to the trust funds and are part of the national debt. The law prescribes the interest rate on special issues, namely, the average market rate on all federal obligations having four or more years to go until maturity. As a result, all newly invested monies that are placed in special issues carry an interest rate close to that available if purchase had been in the open market and consisted of medium- and long-range securities. The special issues are redeemable at par when funds are needed to meet benefit payments. The law does not prescribe the duration until maturity of special issues, but the procedure has been to issue, initially, short-term certificates maturing on the next June 30; then, at that time, long-term securities are issued, with the aim of having the total investment portfolio of special issues being,

as nearly as possible, in 15 equal blocks maturing over the next 15 years.

Potential Problems of OASDI

In the early 1980s, the OASDI system was confronted by significant financing problems over both the short and the long term. The balance in the OASI trust fund was falling rapidly (but not that in the DI trust fund) and would soon have been exhausted if legislative action had not been taken. At the same time, it was estimated that the outgo in the next 75 years (the valuation period) would be significantly higher than the income.

Accordingly, with bipartisan support, legislation was enacted in 1983 to restore the financial integrity of the program according to the best available actuarial cost estimates (Myers, 1992). As to the short run (the remainder of the 1980s), the financing was based on the assumption that economic conditions would be unfavorable. The actual experience in that period was much better than the expectation, and accordingly a much larger fund balance was built up than had been projected.

The 1983 legislation involved many changes so that all groups—present and future beneficiaries, workers, employers, and the general taxpayer—bore some part of the burden of providing additional financial resources. Among the many changes made were the following: (1) permanently delaying by six months each year the cost-of-living adjustments; (2) increasing the payroll tax rates for 1984, 1988, and 1989; (3) increasing the tax rates for the self-employed to the combined employer-employee rate; (4) making up to half of the benefits taxable for high-income beneficiaries; (5) enforcing compulsory coverage of newly hired employees of the federal government and all employees of nonprofit organizations; (6) appropriating monies from the General Fund for payment of the costs arising with respect to certain military service wage credits; and (7) increasing the normal retirement age from 65 to 67, beginning in 2003 and to be phased in over the following 24 years.

Another cash-flow financing crisis in the near future seems unlikely. It would take a severe depression, with high unemployment and prices rising as rapidly as wages (or more so), to cause such a crisis.

Further, the demographic situation in the early 2000s will continue to be relatively favorable. The number of persons reaching retirement age each year will level off (or even decrease somewhat) because the annual numbers of births between 1925 and 1939 were smaller than in the preceding and succeeding years.

Significant financial pressures will appear gradually after 2010, when the post–World War II baby boom enters the retirement age group. This financial strain will be met, to a considerable extent, by the higher tax rates that went in effect after 1989 and by the increase in the normal retirement age.

Although the actuarial estimates that were made in 1983 showed that the OASDI program would be in actuarial balance (i.e., total outgo would be met by total income) over the next 75 years, later estimates showed a less favorable picture. The 1999 estimates projected that the program will be financed only through 2034, or for only 35 years (Board of Trustees, 1999). This results from several factors, including adverse disability experience, assumed longer life expectancy, lower birthrates, and smaller projected increases in real wages.

The foregoing situation does not mean that OASDI is doomed to collapse. Rather, small changes can gradually be made—and even legislated now—that can prevent this. For example, benefit costs can be reduced over the long run by a somewhat more rapid increase in the normal retirement age. Similarly, small increases in the tax rates can be phased in beginning some 30 years hence.

Administration of OASDI

The OASDI program is administered principally by the Social Security Administration, until April 1995, a component of the Department of Health and Human Services, but thereafter an independent agency (originally it was administered by the Social Security Board, an independent agency). The only exceptions are that the Treasury Department collects the payroll taxes, handles the trust funds, and prepares the benefit checks, and that the determinations of disability are made by state agencies, usually the vocational rehabilitation ones.

The Social Security Administration has a central office (in Baltimore, Maryland), six program centers, and about 1,350 district and branch offices. The OASDI trust funds are the general responsibility of the board of trustees, composed of the secretary of the treasury as managing trustee, the secretary of health and human services, the secretary of labor, and two members from the public.

Possible Future Developments of OASDI

It seems unlikely that significant benefit changes will be made in the OASDI program in the next few years, except possibly to resolve the previously mentioned long-range financing problem (Steurle & Bakija, 1994). The program may be left to "settle down" after the significant amendments in 1983 and until public confidence in it is restored.

Considerable support existed in the past for a major revision of OASDI so as to have so-called earnings sharing. Under this approach, the earnings of a married couple would be pooled and divided equally between them. The purpose would be to produce an earnings record for the homemaker in one-salary families that would carry through life, whether or not the marriage continued. Under present law, the spouse who is not in the labor force receives benefits on the earnings record of the other spouse on the basis of legal status. Divorced persons whose marriage lasted at least 10 years also qualify. The earnings-sharing proposal has great merit in theory, but it would be difficult to implement in an equitable manner without a significant increase in the cost of the program. An immediate complete shift to this basis would result in many "losers" (both men and women) as well as "winners." Currently, little support for this proposal is present— possibly because of its complexity and cost.

Currently, some persons who are primarily concerned about the general budget deficit urge that the Social Security benefit outgo should be reduced immediately. They assert that such action will reduce the budget deficit and hold down the growth in the national debt. Actually, such a result is not the case, because if the excess of income over outgo of the OASDI system is increased, this merely means that the trust funds will hold more of the national debt, and the general public will hold less, but the total will be the same.

Specifically, these persons propose that OASDI benefits should be means tested, although in actuality they are really suggesting income testing (i.e., not considering assets at all). The weakness of this approach is that it would likely result in less national savings—a most undesirable outcome—because many people would reduce their savings on the grounds that, by saving, they would lose their OASDI benefits.

In the past few years, many proposals have been put forward to privatize OASDI, in whole or in part. This would consist of reducing the scope of OASDI—both benefit levels and contribution rates—and substituting individual savings accounts backed up by investment in the private sector. However, when it comes to specific details, there are so many different basic approaches, which are not compatible with each other, that it seems unlikely that any action in this direction will occur. Most importantly, this approach would destroy the floor of protection that OASDI provides. For example, great differences in benefit protection could arise, depending on the level of the stock market when the person retires. Then, too, if the accounts are managed by the federal government (as in some proposals), the danger of governmental control over private industry arises.

Currently, there is considerable lack of confidence that the OASDI program will be there some decades hence to pay retirement benefits to the then-aged population. However, it must be remembered that it is a flexible program that can be (and, in the past, was) adjusted from time to time to reflect changing demographic, economic, and political conditions. For this reason, it seems likely that OASDI will be with us forever.

ROBERT J. MYERS

See Also

Medicaid

Medicare

Pensions

Internet Resource

U.S. Social Security Administration

http://www.ssa.gov

SOCIAL SUPPORTS (FORMAL AND INFORMAL)

Extensive evidence indicates that social support systems moderate the deleterious effects of stress on caregivers' health and mental health status (Biegel & Schulz, 1999; Gwyther, 1994). Informal social support is provided to the elderly by family members, friends, neighbors, clergy, or coworkers. Formal support is the broad array of social service, welfare, health, and mental health services.

Informal Support Systems: Strengths and Limitations

Family members provide extensive support to the elderly and represent the elderly's most significant social resource (Hooyman & Gonyea, 1995). Adult children provide substantial amounts of support, with children of frail elders assuming an even larger share of activities necessary to meet the elder's care needs. Noninstitutionalized elderly receive most medically related personal care services from their children.

Friends and neighbors are also significant providers of social support for the elderly because of their physical proximity. They may be the most valuable resource to the elderly in times of emergency. Community groups and associations are important to the well-being of the elderly, even though elders participate to a lesser degree than other age groups. Most participation in voluntary organizations by the elderly is church or synagogue related.

Limitations of the informal support system suggest interventions to strengthen the care of the elderly. A small but significant number of the elderly population is socially and physically isolated, with few if any significant others to turn to for help and assistance (Buckingham, 1994). These elderly are at greater risk for health and mental health problems and are at greater risk for institutionalization (Harel & Biegel, 1995). Formal care providers must assess the needs of this at-risk population group and provide or arrange for the needed formal supports.

A growing body of research on family burden indicates that, for many families, caregiving is an emotional, physical, and, at times, financial burden (Biegel & Schulz, 1999). A significant number of older people are caregivers themselves to an aging parent, spouse, or grandchild. Changing demographics and lifestyle choices have resulted in more older persons with chronic health problems, fewer children and fewer siblings to share the caregiving burden, and decreased availability of female caregivers due to their full-time employment—all of which heightens the need for support groups, education, and respite care.

As the elderly grow older, so do their friends. Thus, illness and death can weaken the friendship network when it is needed the most. Similarly, those elderly living in age-segregated neighborhoods may have few younger neighbors to assist them. Friends and neighbors may be unwilling or unable to provide assistance that requires equipment or expertise (e.g., skilled personal care). Elderly individuals needing specialized care cannot realistically expect—nor should their case managers expect—that this care can be provided either by elderly friends or by neighbors.

For informal community-based support systems to be most effective, linkages are needed among family caregivers, neighbors, friends, clergy, and other helpers. Often, this does not occur due to the fragmentation of helping networks. A variety of interface models for such linkages has been suggested, such as community empowerment, volunteer linking, consultation with natural helpers, and consultation and technical assistance to support groups (Lyons & Zarit, 1999; Naparstek, Biegel, & Spiro, 1982; Noelker & David, 1994).

Barriers to the Use of Formal Services

Studies of service utilization patterns indicate that many older people eligible for service programs do not use them (Pedlar & Smyth, 1999). Service delivery barriers are reported for minority elderly, those who are blind or visually disabled or have mental health problems, and those whose service needs span both the aging network and other more specialized service delivery systems (Biegel, Farkas, & Song, 1997; Urdaneta, Saldana, & Winkler, 1995; Young, French, & Catague, 1995).

Effective interventions to address the needs of the elderly who may be underutilizing formal services must define the barriers and identify their locus or source. The latter is especially important, as strategies to address these barriers must be tailored to their specific locus in order to be most effective. Barriers can be conceptualized as system, agency, community, or individual (Biegel, Johnsen, & Shafran, 1997). System-level barriers include political, economic, and social forces that shape policy and influence service availability and cost, location and availability of transportation, hours of service, auspices of service, provision of information about services to potential referral agents and users of services, and linkages with formal and informal service systems and providers. Agency-level barriers are those that directly affect service delivery, such as skills, attitudes, and behavior of staff and cultural sensitivity. Community-level barriers are attitudes and behaviors by the lay community toward service use—for example, the stigma associated with the use of mental health services. Individual-level barriers refer to personal and family attitudes and behaviors toward the formal services offered, such as lack of knowledge about services in general, not understanding how services can be helpful, and not knowing where to go for help with specific problems or issues; negative attitudes toward formal services and unwillingness to accept help; and role of family and other informal network members in discouraging or preventing service use. Other barriers relate to the health status characteristics of the elderly and may include health problems such as chronic illness or physical mobility problems and activity restrictions.

Strengthening Support Systems of the Elderly

Efforts by human services professionals to strengthen the social support systems of the elderly should begin with psychosocial assessments and a specific focus on the strengths and weakness of the elderly individuals' informal and formal social support systems. It is necessary to determine the kinds of resources being provided to the individual by the informal network and the ways in which this network helps the individual meet daily living and safety needs and enhances quality of life. The kinds of formal assistance provided and the elderly individual's or caregiver's satisfaction with that assistance should also be assessed.

The assessment should also identify the needs that are not currently being met by the elderly individual's informal and formal support networks. This includes daily living needs, as well as the need for love, affection, and companionship. These weaknesses may be due to an insufficient number of individuals in the network, the quality or quantity of their involvement, or their lack of expertise in tasks that involve specific technical skills (e.g., health care needs). Another important area to examine is the stability and potential continuity of an individual's informal network; the elder might be dependent on an individual who may also be elderly, in poor health, or have other equally pressing demands, such as child raising. This kind of situation requires close monitoring and perhaps the establishment of a special category of at-risk clients—whose support systems are adequate at present but have a tenuous or questionable future. In addition, problems experienced by the elderly individual in the use of formal services should be assessed.

In conclusion, human services professionals need to know not only that certain needs and wants are not being met but also, in the elderly person's opinion, why this is so.

DAVID E. BIEGEL

See Also
Aging Agencies: City and County Level

Aging Agencies: State Level
Caregiver Burden
Caregiver Burnout
Caregiving Benefits
Caregiving Relationships
Family Care for Elders with Dementia
Family Care for Frail Elders
Grandparents as Family Caregivers
Naturally Occurring Retirement Communities

Internet Resources
Agenet
http://www.agenet.org

American Association of Retired Persons
http://www.aarp.org

SPEECH THERAPISTS

Speech Therapists (ST), also known as Speech Language Therapists (SLT) and Speech Language Pathologists (SLP), are specialists in the prevention, diagnosis, and treatment of speech, language, voice, swallowing, fluency, and related disorders. SLPs work with people having problems with, for example, making speech sounds, stuttering, understanding and producing language, oral motor function that causes eating and swallowing difficulties, and cognitive communication impairment, such as attention and problem-solving disorders.

SLPs are licensed in 44 states and almost all require a master's degree or equivalent for practice in health care settings. Continuing education units (CEUs) are mandatory in 34 states to retain licensure. Some states require only a license for general practice, some require national certification (Certificate of Clinical Competency) administered by the American Speech-Language-Hearing Association (ASHA), and others require both. Applicants for the Certificate are required to complete a clinical fellowship successfully in addition to required academic and practicum experience and to pass a national exam in speech-language pathology.

Approximately 230 colleges and universities have graduate programs in SLP. In addition to anatomy, physiology, and developmental courses related to speech, language and hearing, speech disorders, acoustics, and psychological components of communication are studied. The clinical fellowship is an important transitional phase between the supervised graduate level practicum and independent delivery of services. Some states require that the clinical fellow register with the licensing board or obtain temporary licensure.

In 1996, there were approximately 87,000 SLPs in the United States, almost half in schools and colleges or universities. Of those not in academic settings, hospitals employ 1 in 10 SLPs; other work settings include physician offices, private practice SLP offices, and home health care agencies. Some SLPs contract to work in schools, hospitals, and nursing homes; others are on staff. Growing demand for SLPs (and audiologists) is associated with increased public awareness of communication disorders and their impact on quality of life, funded mandates for speech therapy, technological advances that improve survival post-stroke and neurological trauma, and access to reimbursement. Given the projected increase of the elderly population, and the age-related changes in speech, language, and hearing that baby boomers will experience, practitioners are experiencing greater workloads associated with increased demand, fewer reimbursed treatment sessions in which to attain specific outcomes, and increased documentation requirements.

SLPs need to be able to explain to a patient why a diagnostic test is indicated and what will be done, effectively communicate test results to patients and other clinicians, diagnose and create an individualized treatment plan projecting total duration of the course of treatment, number and length of sessions, and estimated costs to the patient, if any. Oral and written tests and special instruments diagnose the extent of impairment and track changes. Augmentative alternative communication methods are selected for those with little or no speech capability. SLPs may conduct pure-tone air conduction screening and screening tympanometry. Special recognition or certification for dysphagia evaluation skills identifies practitioners with specialty training in instrumental testing of swallowing. Varying by state, modified barium swallow or the videofluoroscopy test can be performed by an SLP with the x-ray technician in consultation with a radiologist as needed. Education and counseling are critical aspects of speech therapy to help patients and their families cope with stress, frustration, and misunderstanding.

Medicare, Medicaid, and private health insurance usually require licensure for SLP eligibility for reimbursement. SLPs can be reimbursed under Medicare provided there is a physician's order ("certification") for treatment and re-certification of the need for services, including a plan of care every 30 days; home health and hospice re-certification is required every 60 days. Medicare guidelines define "reasonable and necessary" speech-language pathology services, require documentation of functional progress, differentiate skilled from unskilled services, and define covered maintenance programs. Dysphagia services are covered by Medicare if there is documentation that the patient is "motivated," moderately alert, and has some degree of deglutition. Evidence is also needed of patient and caregiver training in feeding and swallowing techniques, and the SLPs assistance in, and analysis of, tests such as videofluoroscopy, upper GI series, or endoscopy.

Speech therapy is an optional Medicaid service and varies by state. Many state Medicaid programs and private insurers use the Medicare Fee Schedule (MFS) rankings to guide reimbursement. Individuals with speech, language, and hearing disorders can be Medicaid-covered for diagnostic, screening, preventive, or corrective services provided by or under the direction and supervision of a speech language pathologist (or audiologist). The patient must have a physician's referral order.

Health plans, insurance policies, and managed care organizations may provide coverage for communication disorders pursuant to illness or accidents but might not cover these disorders if they are developmental or congenital. Reimbursement might be limited to specific provider settings or only to licensed practitioners. A physician's order or prescription for services also might be required. Most health maintenance organizations (HMO) limit services to 60 days or two months. The HMO has the right to decide if the patient will not benefit from short-term treatment and can therefore deny coverage. However, HMOs can extend coverage for as long as the service is deemed medically necessary. Medicare HMO enrollees are entitled to the same amount of coverage as that available to non-HMO Medicare beneficiaries in the same geographic area. SLPs must use billing codes, such as those in the ICD-9 and Physician's Current Procedural Terminology (CPT) when filing claim forms.

Speech-language pathology assistants, also known as communication aides or service extenders, perform tasks only as prescribed, directed and supervised by SLPs. Assistants generally have some training, as prescribed by each state, and a limited scope of responsibility. An SLP assistant cannot be employed without a supervising SLP. Many states limit the number of assistants an SLP may supervise and define the parameters of their activities. The need for such assistants varies by setting; more are needed in acute care than in institutional or residential care. An assistant can conduct speech-language screening, follow a treatment plan and document progress, assist with assessment, and perform other administrative tasks. Assistants are not permitted to perform standardized or nonstandardized diagnostic tests, formal or informal evaluations, interpret test results, counsel a client or family, or write or revise a plan of care.

EDITORIAL STAFF

See Also
Audiology Assessment
Hearing Impairment

Internet Key Words
Audiology
Speech Language Pathologist
Speech Language Therapist
Speech Therapist

Internet Resources
American Speech-Language-Hearing Association
http://www.asha.org

Occupational Outlook Handbook, Bureau of Labor Statistics, Speech Language Pathologists and Audiologists
http://stats.bls.gov/oco/ocos085.htm

SPIRITUALITY

With recent scientific research positively correlating "well-being" with improved health status, the need for health care providers to recognize and address the connection between spirituality and health in their patients has been widely encouraged (Barnum, 1996). Health care professionals caring for elderly patients, in particular, need to promote spiritual caregiving as older individuals grapple with social change (i.e., retirement, loss of friends, family, or partner), chronic or terminal illness, and/or quality-of-life issues.

Over the past decade, gerontologists have attempted to understand more about the degree to which older persons are involved in religious and spiritual practices and the outcomes of such practices on physical and mental health. Although most of the research is descriptive, there appears to be a positive correlation between religion or religiosity and improved health status. One study of chronically ill elders found that individuals who engaged in regular religious activities and/or spiritually-based practices lead more productive and adaptive lives than nonreligious chronically-ill older adults (Ainlay, Singleton, & Swingert, 1992). Another study found a significant positive correlation between spirituality and coping among older adult care recipients and their caregivers (Forbes, 1994). Forbes also noted that older African American men and women, along with older women, tend to be more religious and coped better than older white men and women in general, and men in particular. These findings and others suggest that gender and race may influence older persons' spiritual expression and practice. Regular participation in religious activities among the elderly, such as attending weekly religious services, reading the Bible regularly, and engaging in routine prayer, have also been shown to decrease blood pressure and depression (Koenig, George, Cohen, Hays, Larson, & Blazer, 1998; Koenig, George, & Peterson, 1998).

While many studies of spirituality and aging focus specifically on Judeo-Christian practices, others interpret spirituality more broadly to include religion, love and belonging, morality, and death and dying. Hungelmann, Kenkel-Rossi, Klassen, and Stollenwerk (1996) developed the JAREL Spiritual Well-Being Scale after analyzing 31 in-depth interviews with older subjects about their spiritual well-being. Their work with older adults led them to define spirituality as a multidimensional construct encompassing broad dimensions of relationships with self and others as well as ties to past, present, and future events.

Understanding the link between patient spirituality, time, and relationship building is important when assessing spirituality in older adults. The elderly, like all individuals, have spiritual needs. Whether they are formally affiliated with an organized religion or not, older patients need to express their spiritual concerns and needs within the context of their life histories.

Despite this apparent logic, however, the spiritual needs of older adults are seldom addressed. Caregivers' failure to do a spiritual assessment has been associated with numerous factors, including the ambiguous meaning of spirituality, provider discomfort with spirituality, patient reluctance to share spiritual concerns, and lack of clinical time to devote to patients' spiritual care needs (Brush & Daly, in press). Moreover, few practice models operationalize spiritual assessment and care into clinical practice or measure clinical outcomes of the older person's spiritual well being. Attempts to quantify spirituality through pre-tested instruments measuring various parameters of spirituality have often been flawed or statistically insignificant. Nonetheless, most researchers and clinicians agree that spirituality is important to older adults and suggest that providers include spiritual assessment as part of routine practice. Nurse and social work educators in particular are encouraged to prepare future nurses and social workers to incorporate spirituality into a holistic care paradigm.

First, providers need to show respect for the elder's religious or spiritual articles and practices that symbolize individual faith and values. Certain religiously based dietary restrictions or care philosophies, for example, may not coincide with Western medical thought but should be considered in a patient's overall care plan. Second, health care providers need to provide an atmosphere where expressions of spirituality are accepted and encouraged. Providers must also recognize that not all individuals wish to express their spirituality within a religious framework and should encourage them to express their spirituality through other means, such as sharing life stories or personal perspectives on life meaning. Finally, providers could engage in spiritual practices with patients such as prayer, meditation, or healing modalities in accordance with their own degree of comfort. Providers need to explore their own spiritual perspectives and practices in order to offer adequate spiritual interventions in situations where they are most needed.

For many older people, religious commitment and a sense of spirituality are important aspects of how they age and how they approach the end of life. Future research in spirituality and aging must assess outcomes of spiritual care, explore the meaning of spirituality across diverse groups of elders, and demonstrate clear educational and practice goals for nurses, physicians, social workers, and others caring for older patients. Only then can health care providers understand how older persons, whether community dwelling or institutionally based, experience the religious and spiritual dimensions of later life.

BARBARA BRUSH
EILEEN M. MCGEE

See Also
Subjective Well-Being

Internet Key Words
Religiosity
Spirituality
Spiritual Care
Spiritual Well-Being

Internet Resources
Forum on Religion, Spirituality, & Aging
http://www.asaging.org/forsa.html

National Interfaith Coalition on Aging (National Institute on Aging)
http://www.aoa.dhhs.gov/aoa/dir/174.html

STAFF DEVELOPMENT

To prepare health care professionals for a constantly evolving health care environment, educational strategies for students and professionals must be transformed to reflect the realities of practice. This transformation needs to occur in professional schools of education and in the workplace (American Association of Colleges of Nursing, 1998). When designing educational strategies, important components include generic competencies within the context of the changing health care environment, specific competencies focused on the care of the elderly, and teaching methods for the enhancement of professional learning.

Generic Competencies

The evolution of the health care delivery system into "managed care" proposes to control and constrain costs for payers and to preserve quality for consumers. In response to a major paradigm shift in the methods, location, and participants in health care delivery, health care professionals are redefining their roles and relationships with each other and with the patient. Health care professionals need to have a systems orientation, collaborative skills, and the ability to produce documentation that addresses the information needs of multiple and diverse stakeholders. The education environment also must move beyond the safe and convenient skill/knowledge/task model of education to one that includes decision-making skills and critical thinking.

A systems orientation broadens the perspective of the provider to a view of health care as a system of multiple placement options offering the patient choices for recovery and/or support. Clinical experiences for patients have broadened to include acute, ambulatory, subacute, home, community-based, and long-term care. Consequently, all providers need to learn the goals of health care embedded in each of these settings so as to be able to consider the clinical, sociological, and financial implications for the patient. Knowledge about managed care, "business-like" approaches such as cus-

tomer service, and evidence-based practice have become basic concepts for virtually all health care professional groups. Consequently, the educational implications for understanding and functioning within this multidimensional, interdisciplinary model of health care complicate teaching in many ways.

The effective management of transitions is a relatively new perspective of accountability that relies on a systems orientation. The "hand off" is the critical point of distinction between success and failure for a patient. Learning to appreciate the complexity of transitions is a foundation of expert practice. Planning comprehensive care requires collaboration by health professionals to make the right choices with and for patients and families. Technology is an integral part of health system operations that incorporate clinical, informational, research, professional, and financial aspects of care. Information technology significantly alters the way health care providers and patients interact, communicate, and control outcomes. Informatics education is a relevant staff development investment in that it enhances quality patient care, accountability, critical thinking, research, decision making, research, evidence-based practice, and continuing education/staff development.

Interdisciplinary accountability brings home the point that a group of people with complementary skills can experience more success than an individual alone, particularly when dealing with complex problems. Teams can produce striking achievements with impressive energy, synergy, and efficiency (Pfeiffer, 1998). Thus, it is not unreasonable to expect the professional disciplines engaged in the delivery of health care to work together to produce the best quality outcomes for patients, families, and communities. Collaboration can replace task-based practice with clinical leadership, competition with cooperation, and opinions with knowledge. Health care professionals need to learn how to look for and identify commonalities and shared vision among professionals (Barnum, 1999) to move past traditional notions of turf and ownership of the patient. Documentation is one type of concrete evidence of practice. In a sense, documentation is a "patient story" that structures practice, captures the clinical understanding of the situation, and provides a teaching/learning strategy for improved practice (Benner et al., 1999). Inherent in telling the patient story are decisions about where to begin the story, what to tell, what to leave out, and where to end. Novice professionals are often not prepared to tell "the story," and often feel frustrated with their abilities to translate their practice into succinct, meaningful notes. An in-depth understanding of the clinical, legal, and financial value of the medical record also needs to be addressed in basic and continuing education.

Specific Competencies

The specific competencies necessary for the care of the older adult are responsive to three main themes: complexity, fragility, and vulnerability. It is within this framework that specific competencies can be outlined.

Managing primary health problems while simultaneously controlling a variety of coexisting conditions is a challenge for the novice as well as the experienced professional. Concepts and theories of aging, age-related changes, and implications for care can be discussed in a discrete course or in workshops. "Best practices" findings need broad dissemination that reaches providers and educators, not just researchers and regulators.

Familiarity with risk assessment tools as well as a concentration of health promotion and maintenance efforts in all practice disciplines is key to the quality care and life of the older adult. Common assessment tools include the evaluation of risk for falls and loss of skin integrity, cognition, and need for pain control. Incorporating these tools into staff education is fundamental to "defensive practice" and the promotion of quality of life for the elderly. The vulnerability of the older adult means that the professional caregiver is a "watchful eye" for concerns such as elder mistreatment, insurance scams, financial threats, education overload, bureaucratic confusion, polypharmacy, and so on.

Teaching Methodologies

The shift from the industrial age to the information age suggests the need for a paradigm shift from teacher to learning agent who would in turn create a learning "infostructure" that transcends the four walls of a classroom (Tapscott, 1996). To engage in an educational dialogue rather than listening to a lecture expresses the shift in the educational process. Learning experts report that students retain only 20% of what they hear and even less of what they read on their own. However, more is retained when individuals are involved in their own learning, especially when performance is integrated with real life situations. Educators, both in schools and in institutions of work, can benefit from this revival by integrating small group learning and interactive approaches to education. The "safety" of the classroom may be sacrificed for the experience of "learning while you work." In the end, however, the "bedside" as classroom will prove to be worth the effort.

Strategies for information age education include internship experiences where learning can be safe yet active and participatory; case based learning that brings the realities of care delivery into the learning experience; story-telling about patient care that encourages logical reasoning and reflective judgment; and the integration of skills and assessment data for problem solving and generation of care plans. Team teaching is a powerful example of role-modeling and incentive to actualize interdisciplinary collaboration at the bedside

MARIA L. VEZINA

See Also
Best Practices in Geriatric Care
Communication Issues for Practitioners
Geriatric Interdisciplinary Health Care Teams
Information Transfer

Internet Key Words
Competencies
Nursing Education
Learning
Staff Development

Internet Resources

Learning Bridges
http://www.learningbridges.com

National Nursing Staff Development Organization
http://www.nnsdo.org

STRESS

See
Relocation Stress

STROKE/CEREBRAL VASCULAR ACCIDENT

Definition

A stroke is the abrupt onset of focal or global neurologic symptoms caused by ischemia or hemorrhage within or around the brain resulting from diseases of the cerebral blood vessels and in which neurologic symptoms continue for more than 24 hours. If the deficit lasts less than 24 hours the event is termed a transient ischemic attack (TIA). Despite a significant decline in the fatality rate over the past 20 years, stroke remains the third leading cause of death in the United States after cardiovascular diseases and cancer (Anderson, Kochanek, & Murphy, 1997).

Prevalence and Risk Factors

For those over the age of 65 in the United States, the average annual rate of cerebrovascular disease (stroke or TIA) is 67 per 1,000 persons. This rate climbs to 97 per 1,000 persons for those over 85 years of age (Anderson et al., 1997).

Risk factors for stroke that cannot be modified include ethnicity, family history, age, and gender. After age 55, stroke risk doubles with each decade. Prior to age 75, people of African American descent are twice as likely to suffer a stroke as are people of European origin. Modifiable risk factors include diabetes, hyperlipidemia, smoking, elevated blood pressure, some types of heart disease including atrial fibrillation, carotid artery stenosis, and excessive alcohol use. Transient ischemic attack is associated with a 12% risk of a first stroke in the year following the TIA. A meta-analysis of 33 studies found that smoking was associated with a significant increase in the risk of stroke (relative risk = 1.5, 95% confidence interval 1.4 to 1.6). Atrial fibrillation, which increases in frequency with age, also increases the risk for stroke. It is modifiable in stroke prevention either through anticoagulation or through cardioversion into a stable heart rhythm.

Prognosis

Despite treatment advances, the prognosis following an acute stroke remains grave for many patients. The 30-day mortality rate for ischemic stroke is 10% while the rate for hemorrhagic stroke is as high as 52% (Bamford, Dennis, Sandercock, Burn, & Warlow, 1990). The mortality rate is considerably higher in patients who were functionally dependent prior to the stroke and for older patients. First year recurrence rate after a stroke is 13% and subsequent stroke risk is 4% per year. Five years after a first stroke, 30% of persons will have suffered a second stroke. This is approximately 9 times the risk of the general population, and is independent of age at the time of the first stroke.

Cognitive impairment is a common complication of stroke. In one study, 36% of patients with no prior history of cognitive impairment demonstrated significant cognitive impairment 6 months following a stroke as measured with the graded neurologic scale and the Mattis dementia rating scale (Schmidt, Mechtler, Kinkel, Fazekas, Kinkel, & Freidl, 1993).

Additional complications of stroke include an increased risk of emotional lability and the new onset of seizures. The prevalence of silent myocardial infarction accompanying stroke in patients 75 years of age and older may be as high as 33%. Post-stroke depression is a frequent complication of stroke, occurring in 27% of patients (Paolucci, Antonucci, Pratesi, Traballesi, Grasso, & Lubich,

1999). Strokes affecting the left hemisphere pre-frontal or basal ganglia structures are associated with depressed mood following stroke, as compared with strokes on the right side or other parts of the left brain (Morris, Robinson, Raphael, & Hopwood, 1996).

Screening and Prevention

According to the U.S. Preventive Services Task Force, all persons should be screened for high blood pressure, and all smokers should receive counseling toward smoking cessation. There is insufficient evidence to date to argue for or against screening for carotid artery stenosis in persons who have no symptoms of stroke or TIA. Recent studies demonstrate a significant reduction in stroke risk and regression of carotid artery plaques with the use of certain cholesterol lowering medications (Wardlaw, Warlow, & Counsell, 1997). The relationship between alcohol consumption and stroke risk is not linear. Persons who consume one to two drinks of any alcoholic beverage per day reduce their risk of stroke as compared with nondrinkers. Greater amounts of alcohol, however, are associated with increased stroke risk.

Active treatment of hypertension in the elderly is associated with significant reductions in the risk of stroke (SHEP Cooperative Research Group, 1991). It is estimated that adequate control of hypertension could prevent more than 200,000 strokes per year and smoking cessation could prevent 60,000 strokes per year for a combined savings of 15 billion healthcare dollars per year. Thus, almost 300,000 strokes per year are potentially avoidable. Smokers who cease tobacco use benefit from a reduction in stroke risk that, approximately five years after quitting, approaches the risk level they would have been at had they never smoked.

For patients with a history of transient ischemic attack, subsequent use of drugs that inhibit platelet function, such as aspirin, can decrease the risk of stroke. At least 70% of patients who suffer a transient ischemic attack and have carotid artery stenosis may benefit from carotid endarterectomy.

The risk of stroke due to atrial fibrillation can be reduced to the level of stroke risk in persons without atrial fibrillation with the use of medications such as warfarin.

Treatment

The goals of stroke treatment are to minimize or prevent ischemic brain infarction, minimize the acute complications of stroke, optimize functional recovery, and prevent stroke recurrence.

Specialized inpatient stroke units, as compared with standard care units, have a significant positive impact on long-term mortality reduction, the combined outcomes of death or dependency, and the combined outcomes of death or institutionalization (Stroke Unit Trialists' Collaboration, 1997). An important consideration in early stroke treatment is the use of a post-stroke rehabilitation program. Patients cared for in such programs show greater improvements in activities of daily living and visual-perception function than patients given conventional care. These improvements are more pronounced the earlier that patients are enrolled in these programs following the stroke (Ottenbacher & Jannell, 1993).

Medications used in the treatment of acute ischemic stroke include aspirin, heparin, warfarin, corticosteroids, tissue plasminogen activator, and streptokinase.

Heparin: In one large trial (Wardlaw et al., 1997), heparin did not significantly decrease the death rate as compared with placebo in patients at 14 days or 6 months after stroke. At 6 months the percentage of persons dead or functionally dependent was identical (62.9%). Other studies confirm that heparin therapy in the immediate post-stroke period does not improve mortality. On the other hand, early heparin therapy in patients with presumed or confirmed ischemic stroke does decrease the risk of post-stroke deep vein thrombosis by 80%.

Warfarin: While studies are inconclusive with regard to the benefit of warfarin following stroke, one large meta-analysis of warfarin use in patients with ischemic stroke or transient ischemic attack demonstrated no significant improvement in mortality rate, functional status, or recurrent stroke (Wardlaw et al., 1997). There was however an increase in the risk of fatal intracranial hemorrhage in patients treated with warfarin.

Aspirin: Two recent large trials, CAST (Chinese Acute Stroke Trial) and IST (International Stroke Trial Pilot Study Collaborative Group), demonstrated a significant reduction in mortality for patients with ischemic stroke who were treated with aspirin as compared with placebo. There was also a significant reduction in stroke recurrence in patients treated with aspirin as compared with those given placebos and in the risk of combined endpoint of death or non-fatal stroke in the aspirin-treated group.

Corticosteroids: The use of corticosteroids in acute ischemic stroke has not demonstrated a survival benefit or improvement in functional outcomes.

Thrombolytic Therapy: Streptokinase and Tissue Plasminogen Activator (t-PA): The use of thrombolytic therapy in acute ischemic stroke remains controversial. In some studies, thrombolytic therapy for acute ischemic stroke was associated with a significant excess of early deaths and total deaths (Wardlaw et al., 1997). However, there is also evidence that this therapy may significantly decrease the combined outcome of death or dependence. One large study comparing the use of t-PA with placebo showed that patients treated with t-PA were more likely to have a favorable outcome at 6 and 12 months according to their score in the Barthel index, modified Rankin scale, and Glasgow Outcome scale than those who received placebo (Wardlaw et al., 1997). Data are lacking on head to head studies of t-PA and streptokinase.

JEREMY BOAL

See Also
Activities of Daily Living
Falls Prevention
Fecal Incontinence
Hypertension

Multidimensional Functional Assessment: Overview
Occupational Therapists
Physical Therapists
Speech Therapists
Swallowing Disorders and Aspiration
Urinary Incontinence: Treatment

Internet Key Words
Cerebrovascular Accident
Stroke
Tissue Plasminogen Activator (t-PA)
Transient Ischemic Attack (TIA)

SUBACUTE CARE

The shift toward the provision of "subacute" care is in large part the result of substantial changes in federal reimbursement patterns. The 1965 Medicare legislation entitled a Medicare eligible patient to receive Medicare Skilled Services in a Medicare certified skilled nursing unit or facility for a period of 100 days after at least a three-day qualifying hospital stay. For services to be eligible for coverage, a patient needs ongoing rehabilitation post-orthopedic injury or a cerebrovascular accident; treatments and therapies often associated with hospital based care such as intravenous therapy or enteral nutrition, or observation and monitoring of a medical condition that occurred during the hospital stay. The 1983 implementation of the Prospective Payment System (PPS) for hospitals and the associated Diagnostic Related Groups (DRGs) encouraged hospitals to discharge Medicare beneficiaries from the hospital into alternative levels of care sooner than had occurred in the past, in order to optimize reimbursement through PPS. Hospitals quickly recognized the benefit of having access to Medicare skilled nursing facilities that could follow the patient's plan of care in a "step-down" environment.

Some hospitals with falling occupancy rates seized the opportunity and established a hospital based skilled nursing facility within the walls of the hospital. Thus, hospitals that cared for a Medicare beneficiary in the acute care hospital received a full DRG payment and then discharged the patient to the hospital's Medicare skilled unit on their own campus. The hospital received about $300 a day for the skilled nursing facility stay in addition to the DRG payment. Implementation of DRGs was the most significant change that "morphed" Medicare skilled care into subacute care as it is currently known.

Many nursing home providers began to use the term "subacute" care rather than Medicare skilled care to differentiate the type of care provided to medically complex patients from traditional long-term care. Regulations and reimbursement associated with Medicare skilled care coverage continued to dictate the parameters for subacute care. To date, there are no specific regulations specific for "subacute" care in skilled nursing facilities.

Although reimbursement through the Catastrophic Coverage Act of 1988 was short lived—the legislation was repealed after less than two years—the availability of expanded coverage for skilled nursing services made nursing homes aware of the substantial reimbursement opportunities available through the provision of subacute care. The result was an expansion of subacute care services provided in nursing homes. Continued growth in the managed care market promises continued financial incentives for the provision of subacute care, as managed care companies continue to search for less expensive levels of care for services typically provided in a hospital. Insurers benefit from placing not only Medicare beneficiaries but also younger adults with complex medical problems and/or rehabilitation needs into a less expensive skilled nursing facility.

The professional service hours, including nursing and rehabilitation therapies, are higher in the subacute level of care than in a traditional nursing facility because of the complexity of care and higher resource utilization. Patients in subacute care require rehabilitation post-orthopedic surgery; physical, occupational, and speech therapy post-cerebrovascular accident; complex medical man-

agement of diabetes mellitus, particularly in the presence of comorbid conditions; intravenous therapy, enteral feeding, total parenteral nutrition, chemotherapy, intravenous antibiotics, or other medications; complex wound management, care of recent ostomies, management of stage III and IV pressure sores and ulcers related to peripheral vascular disease; ventilator management, complex respiratory management including tracheostomy care, frequent suctioning, and/or respiratory treatments. Direct nursing care hours per resident per day average between 4.5 and 5.5 direct care hours. The average length of stay in a subacute unit or facility is currently about 15 days according to industry data (Levenson, 1996). Subacute care can be provided in a distinct-part subacute unit or dispersed throughout the facility.

Implementation of the PPS for Medicare skilled care in 1998 reimburses a nursing facility for Medicare skilled/subacute care according to a case mix hierarchy of skilled rehabilitative and nursing needs. A Resource Utilization Group (RUG) construct based on functional and health status, treatments, and therapy needs is drawn from the Minimum Data Set (MDS) instrument that categorizes nursing residents into one of 44 RUG categories of care. Twenty-six of the 44 RUGs represent eligibility for Medicare skilled care reimbursement. On April 10, 2000, as published in the *Federal Register*, the Health Care Financing Administration (HCFA) has proposed changes in PPS through the use of 178 RUG categories (HCFA, 2000). These proposed changes are HCFA's response to the Balanced Budget Refinement Act of 1999 in which Congress directed HCFA to readjust PPS payments in order to be more equitable to subacute providers.

Subacute care will continue to evolve as providers report the clinical data for each resident receiving this level of care to HCFA. The electronic submission of MDS data has been used to create 24 indicators of care that are currently being used to measure quality in subacute units. The 24 quality indicators, including medication utilization, weight loss, and hydration, are used to compare the resident profiles from one nursing facility to all nursing facilities within their state. The quality indicators are used by state surveyors to direct their focus during the annual Medicare/Medicaid recertification survey to areas that indicate substandard care.

MARY T. KNAPP

See Also
Home Health Care
Hospital-Based Services
Medicaid
Medicare
Medicare Managed Care
Nursing Homes

Internet Key Words
Medicare Skilled Care
Prospective Payment System
Subacute Care

Internet Resources
American Health Care Association
http://www.ahca.org

American Hospital Association
http://www.aha.org

Code of Federal Regulations
http://www.access.gpo.gov/nara/cfr

Department of Health and Human Services
http://www.os.dhhs.gov

Health Care Financing Administration
http://www.hcfa.gov

Social Security Online
http://www.ssa.gov

ZA Consulting, LLC
http://www.zaconsulting.com

SUBJECTIVE WELL-BEING

Researchers continue to use the term *subjective well-being* (SWB) as an abstract, superordinate construct encompassing the affective and cognitive reactions of individuals to their life experiences (Diener, Suh, Lucas, & Smith, 1999). This definition underscores that the focus is on the subjective experience instead of the objective conditions of life. Diener et al. (1999) distinguish among pleasant affect, unpleasant affect, and life satisfaction. Affect involves moods and emotions which represent people's on-line evaluations of their current experiences. In contrast, life satisfaction is primarily a cognitive evaluation of the quality of one's experiences over the life course. Models of successful aging often include SWB-related variables because they assess whether people are experiencing life in positive ways.

Measurement of SWB

Measures used to assess the SWB of older adults have been developed by four groups of researchers:

1. Social scientists using surveys to identify aggregate level trends in subjective social indicators.
2. Social scientists investigating how structural and individual variables influence variation in levels of life satisfaction.
3. Gerontologists studying individual differences in happiness among older adults.
4. Life-span developmental psychologists interested in age trends in psychological well-being.

Using items from morale and life satisfaction scales, Shmotkin and Hadari (1996) found support for a second-order factor model of SWB in which the first-order factors were Reconciled Aging, Unstrained Affect, General Contentment, Present Happiness, and Past Self-Fulfillment. Using the items on Ryff's (1989) measure, Ryff and Keyes (1995) confirmed a second-order factor of psychological well-being in which the first-order factors were Autonomy, Environmental Mastery, Personal Growth, Positive Relations with Others, Purpose in Life, and Self-Acceptance.

Correlates of SWB

Ascribed and Achieved Social Statuses: Quantitative research syntheses during the 1980s revealed that variables such as age, gender, race, education, marital status, and occupational status share no more than 3 percent of the variance with SWB (Okun & Stock, 1987). Subsequent research on age trends in SWB reveals a more complex picture. Diener and Suh (1998) concluded from a survey of 60,000 adults from 40 nations that pleasant affect declined with age whereas life satisfaction and unpleasant affect did not. Keyes and Ryff (1999) reported that some dimensions of psychological well-being increase from young adulthood to midlife (e.g., autonomy), some dimensions decrease from midlife to old age (e.g., purpose in life), and other dimensions show no consistent age trends (e.g., positive relations with others).

Material and Social Resources: Several material and social resources that decline with age have been linked to SWB, including health, income, and activity participation (Okun & Stock, 1987). This state of affairs has been called the "paradox" of contentment in old age—that is, how do older people maintain their life satisfaction in the face of threat and loss (Filipp, 1996)? We begin our examination of this paradox by considering the influence of personality on SWB.

Personality: In a meta-analytic review, De-Neve and Cooper (1998) reported that neuroticism ($R^2 = .06$) was an inverse correlate and conscientiousness ($R^2 = .05$) was a positive correlate of life satisfaction. In addition, neuroticism ($R^2 = .06$) was an inverse correlate and extraversion ($R^2 = .07$) and agreeableness ($R^2 = .04$) were positive correlates of happiness. Personality may influence how people appraise life events, perceive social support, and cope with stressful roles.

Self-Concept: Older adults appear to use self-protective processes to maintain their SWB. Self-protective processes include lowering ideal self-assessments, reducing aspirations, envisioning positive future selves, reorganizing the salience of identity hierarchies, and engaging in downward social comparisons (Brandtstädter & Greve, 1994). Heckhausen and Brim (1997), for example, examined social downgrading across adulthood. Adults of all ages rated the problems of age peers to be more serious than their own problems. When individuals were personally affected by problems in a given life domain, the social downgrading of age peers was more pronounced. Furthermore, the social downgrading of age peers when individuals were personally affected by problems in non-normative life domains increased with age.

Motivation and Emotion: Diverse conceptions have been used in efforts to understand the links between motivation and SWB (Filipp, 1996). One approach has been to view motivation through the lens of goals. LaPierre, Bouffard, and Bastin (1997) performed a content analysis on the goals generated by older people in response to a series of sentence stems. Higher life satisfaction ratings were associated with generative goals and personal growth goals, and lower life satisfaction ratings were associated with health preservation goals. Freund and Baltes (1998) demonstrated that optimization—allocating resources to attain desired goals—is positively related to SWB among older adults.

A second approach has been to view motivation through the lens of social motives. Carstensen (1993) theorizes that older people become more selective in their relationships because, with age, the affect regulation function of the self becomes increasingly important. By restricting themselves to smaller but emotional satisfying groups of friends and family members, older people may avoid experiencing negative affects (Gross, Carstensen, Pasupathi, Tsai, Skorpen, & Hsu, 1997). Furthermore, age has been shown to correlate positively with self-reports of emotion regulation abilities (Gross et al., 1997). Consequently, older adults may be less adversely affected by unpleasant emotions than younger adults.

Coping: Research on the relation between coping and SWB has been informed by both life span and stress and coping frameworks. Heckhausen and Schulz (1995) proposed a life span model of coping in which the dimensions were type of control (primary versus secondary) and strategy (selection versus compensation). Selective primary control refers to actions aimed directly at attaining age-appropriate goals, whereas compensatory secondary control refers to strategies aimed at minimizing the inimical effects of failure. Heckhausen (1997) found that older adults disengage from goals that are linked to obsolete roles and tenaciously pursue goals that are associated with maintaining current levels of functioning. She also observed that with age the use of various secondary compensatory control strategies (e.g., rating one's subjective age as lower than one's actual age) increases. Secondary compensatory control strategies may attenuate the effects of negative life events on SWB.

Smider, Essex, and Ryff (1996) employed a stress and coping framework to investigate how appraisal and psychological resources jointly influence adjustment to community relocation. Among older women, the relation between the appraisal of the difficulty of the move and increases in aggravation were buffered by both environmental mastery and autonomy. Thus, older women may use their psychological resources to maintain their SWB when coping with life transitions perceived to be difficult.

Social Support: The relation between social support and SWB is complex. One source of complexity is that not all "supportive" transactions are welcomed by the recipient. Smith and Goodnow (1999) found that age was inversely related to the occurrence of unsolicited support. Furthermore, age did not exert a direct effect on unpleasantness ratings of unsolicited support, because, relative to younger adults, older adults make greater use of active discounting of what unsolicited support providers say and believe that unsolicited support violates the norm of down-playing the weaknesses of the support recipient. Okun and Keith (1998) observed that older adults report fewer negative social exchanges than younger adults.

A second source of complexity vis-a-vis the relation between support and SWB is that older adults are providers as well as recipients of support. In a longitudinal study of intergenerational exchange, Davey and Eggebeen (1998) found that "overbenefitting"—support received exceeds support provided given prior patterns of support received and provided—was associated with lower psychological well-being among older adults. However, in the face of a life transition, older adults had

higher psychological well-being if they (a) received more support than would be predicted on the basis of prior levels of exchange; and (b) provided less support than would be predicted on the basis of prior levels of exchange. Thus the relation between social support and SWB appears to be moderated by several contextual variables.

Religiosity, Volunteering, and Wisdom: Several studies have investigated the relations between informal social roles (church member, volunteer, and wise elder) and SWB. Church attendance is positively related to SWB among older adults (Okun & Stock, 1987). In a meta-analysis of the benefits of volunteering, Wheeler, Gorey, and Greenblatt (1998) observed that the average older volunteer is located at the 70th percentile in the distribution of older non-volunteers with respect to SWB. Ardelt (1997) examined the relation between wisdom and life satisfaction. Wisdom was conceptualized as a dimension of psychosocial development and was assessed using a combination of cognitive, reflective, and affective indicators. The correlation between wisdom and life satisfaction was strong for both older women ($r = .77$) and men ($r = .64$). Opportunities to enact meaningful informal roles may enhance the SWB of older adults.

Conclusion

There is a growing emphasis on investigating SWB in groups varying in culture, ethnicity, gender, and age (Diener & Suh, 1998; Keyes & Ryff, 1999). In addition, there is increased interest in understanding the dynamics of older adults' SWB. Several methodological challenges can be addressed by (a) developing measures that do not rely on self-report; (b) field testing interventions; (c) conducting laboratory experiments; (d) examining data from panel studies; (e) carrying out micro-level analyses; and (f) performing macro-level analyses. For now, we can describe an older person who is high in SWB as: high in extraversion, low in neuroticism, wise, skilled in emotion-regulation, actively working on generative and personal growth goals that are age-appropriate, utilizing self-protective strategies, selective in the choice of social partners, judicious in both the provision and receipt of help from network members, participating in pleasurable activities and informal roles, perceiving his/her health to be good or excellent, and free of major functional limitations and financial concerns.

MORRIS A. OKUN

See Also
Satisfaction Measurement
Spirituality
Volunteerism

SUBSTITUTE DECISION MAKING

Because the decisional capacity of the elderly may be diminished, fluctuating, or lapsed entirely, it is important to anticipate the need for others to make health care decisions for patients who may lack the capacity to decide about medical treatment for themselves. Although in the past elderly patients may have had and even expressed preferences about treatment, they may have lost the ability to participate in planning their care. These responsibilities then fall to substitute decision makers. The medical, legal, and ethical communities have responded to the formerly capacitated with two approaches: advance directives and surrogate decision making.

Through advance directives—living wills and health care proxy appointments—a capable person can articulate wishes, values, and directions so they can be communicated and respected after capacity has lapsed. These legal authorizations are intended to ensure that the voice of the formerly capacitated person will still be heard in the care planning discussion. Amplification of the patient's voice is the responsibility of the health care proxy and the provider who interpret and honor the provisions of the living will.

Health Care Proxies and Surrogates

A health care proxy is a competent adult over 18 years of age who has been selected and legally appointed by a patient to make any and all medical decisions whenever the patient is unable to make those decisions. The advantage of a proxy is that health care decisions are made for the patient by a person who can interact with the care team. That person can then respond to a changing medical situation in light of the patient's known values and preferences, as well as the diagnosis and prognosis. The proxy appointment presupposes a trusting relationship and the willingness of the proxy to use his or her judgment in the patient's interest rather than rigidly following a set of instructions. Through the appointment, the patient is saying, "I believe you know me and understand what I consider important. I trust you to make the decisions that you think I would make if I knew what you will know about my condition."

Surrogate decision making seeks to identify and amplify the voice of the currently incapacitated patient who never executed an advance directive but may have verbally communicated treatment wishes, as well as the patient who never articulated health care preferences of any kind. Absent explicit instructions, it is necessary to search for guidance in what is known about the patient or what will promote his or her best interest. In every state except New York and Missouri, these health care decisions are made by others based on either substituted judgment (when the person's wishes are known or can be inferred) or the best interest standard (when the person did not have or did not articulate treatment preferences). Substituted judgment looks to what the patient would want if able to choose, based on prior statements, behaviors, and patterns of decision making. The best interest standard weighs the benefits and burdens to the patient of therapeutic options as evaluated by the substitute decider.

A health care surrogate may be any competent adult over 18 years of age who, although not specifically chosen or legally appointed by a patient, assumes the responsibility for making health care decisions on behalf of a patient who has lost the ability to do so. A surrogate by state law is a person whose authority to make health care decisions for someone else is based on state statute or case law.

An informal surrogate is a person, usually someone close to the patient, who is asked by the medical team to participate in making treatment decisions because there is no one who has been specifically appointed or legally authorized (Post, Blustein, & Dubler, 1999). As a rule, both medicine and law are more comfortable providing than withholding treatment and the law accords considerable authority to surrogates, especially next of kin, in consenting to treatment. Decisions about limiting treatment are more problematic and depend on the state in which the patient is treated (Sabatino, 1999).

Most substitute decision makers, whether legally appointed proxies or surrogates, are family members who are presumed by tradition and often by law to know patients most intimately and act in their best interest. It is expected in many cultures that, in times of trouble and serious decision making, families are the best source of insight, information, and support. This notion underlies the practice of health care providers, law enforcement officials, teachers, and others who care for people to seek family assistance in making decisions.

How patients select their proxies can have significant implications. Especially within families, this can assume the aura of a contest for affection or a referendum on relationships. It is not uncommon to hear, "I would have to choose my sister. She would be so hurt if I picked someone else," or "My daughter actually knows me better, but if I appoint her my son will think I don't trust him." Everyone is better served if the selection is based on the needs of the patient, the decision-making characteristics and accessibility of the potential proxy, and above all, the strength of their relationship.

Compounding the difficulties faced by proxies and surrogates is the fact that their responsibilities are often triggered at times of medical crisis, intense emotion, and family turmoil. Sadness about impending death, concerns about their own aging, and uneasiness with role reversals are layered on top of old rivalries, lingering resentments, and guilt related to unfulfilled obligations. The unresolved conflicts that characterize family dynamics are heightened in an atmosphere of crisis, illness, and

death. When family members are placed in the position of making decisions—often about life and death—for loved ones, the added stress can be overwhelming.

Physician Guidelines

Because decision making for others carries such significant implications, considerable discussion should precede the lapse of capacity or the onset of illness. Adults of any age and health status should speak with their families, close friends, and care providers about their treatment wishes, thoughts about comfort and function, and notions of an acceptable quality of life. Whenever possible, health care proxies should be appointed and included in physician-patient discussions (Zeleznick, Post, Mulvihill, Jacobs, Burton, & Dubler, 1999).

To promote collaboration between doctors and their patients' proxies and surrogates, the following guidelines were developed and appear in *Making Health Care Decisions for Others: A Guide to Being a Health Care Proxy or Surrogate—A Quick Reference for Physicians* (Montefiore Medical Center, Division of Bioethics, 1999).

1. Determine if the patient has designated a health care proxy and review all documents that provide guidance about the patient's prior wishes. If there is no appointed proxy, determine who will be acting as the patient's surrogate in the event that capacity is lost.
2. Ensure that all relevant information and documents are placed in the patient's medical record.
3. Discuss general treatment preferences with the patient prior to loss of capacity and note preferences in the medical record.
4. Encourage patients to routinely articulate preferences and select a health care proxy prior to an admission or emergency situation.
5. With the patient's consent, include the proxy or surrogate in discussions with the patient.
6. Upon determination that the patient has lost capacity, provide the proxy or surrogate with

the same medical information (diagnosis, prognosis, treatment options, and recommendations) that a capacitated patient would receive in order to make informed decisions. The proxy or surrogate may rely on the patient's previously expressed wishes; knowledge of the patient's values, beliefs, and attitudes; or on his/her own judgment as to what is in the patient's best interest. Remember that the patient trusts the proxy and presumably the surrogate to make the best decision possible under the circumstances.

7. Support the proxy or surrogate in the process of deciding for another, which is often more difficult than deciding for one's self.

8. Alert the proxy or surrogate to specific changes in the patient's medical condition.

9. Apply patient confidentiality standards to information provided to the proxy or surrogate.

10. Avoid burdening the proxy or surrogate with "false" decisions (e.g., when a patient is to be transferred out of the ICU, do not present the issue to the proxy as a choice he or she must make).

11. Explain that a do-not-resuscitate (DNR) order does not mean do not treat. Explore palliative care options when cure is no longer the goal. Assure the proxy or surrogate that the patient will not be abandoned.

12. Advise the proxy or surrogate of institutional resources, such as bioethics consultants, patient advocates, social workers, translators, and spiritual advisors.

13. Use institutional resources to support your relationship with the proxy or surrogate. Request a bioethics consultation or assistance from patient services and social services whenever there is confusion, uncertainty, or conflict about decision making.

Perhaps the most important clinical resource physicians bring to the interaction with substitute decision makers is their support. Decisions about starting, continuing, or terminating treatment, especially life-sustaining measures, can be painful and often paralyzing for those who are asked to act on

behalf of their loved ones. If abandoned to make these difficult choices alone, the proxy or surrogate can feel solely responsible for the outcome. Care providers, especially physicians, have an obligation to become familiar with their patient's wishes and shoulder part of the decision-making burden (Post, Blustein, & Dubler, 1999).

LINDA FARBER POST

See Also
Advance Directives

Internet Key Words
Bioethics
Health Care Proxies
Health Care Surrogates
Living Wills

Internet Resources
American Society of Law, Medicine, and Ethics, Health Law Resources
http://www.aslme.org

Montefiore Medical Center Home page
http://www.bioethicsmontefiore.org

SUICIDE

Despite numerous methodological problems in studies of suicide (Pearson, Caine, Lindesay, Conwell, & Clark, 1999), in western societies suicide seems to occur most frequently in older people and especially among white men over the age of seventy. Comprising only 13 percent of the U.S. population, individuals age 65 and older account for 20 percent of all suicide deaths, with white males being particularly vulnerable. The highest rate is for white men age 85 and older: 65.3 deaths per 100,000 persons in 1996, about 6 times the national U.S. rate of 10.8 per 100,000 (Lambert & Fowler, 1997). In contrast, in younger white populations, suicide is more common among women although the rates of suicide are increasing in nonwhite populations (Chance, Kaslow, Sum merville, & Wood, 1998). The situation in geriatric care is rendered more difficult by the number of "covered up" suicides. For example, an elderly lady refuses to eat, rejects therapeutic assistance, and becomes steadily weaker leading to cardiac decompensation, lung edema, and ultimately death. In such a case, suicide is unlikely to be identified as a causal factor. Seniors who live alone are the most important suicide risk group, but suicide also occurs in the hospital, in long-term care facilities, and in retirement homes. Moreover, seniors tend to use "harder" methods, resulting in fewer suicide attempts but more successful suicides (van Casteren, van der Veken, Tafforeau, & van Oyen, 1993; McIntosh, Santos, Hubbard, & Overholser, 1994; Moscicki, E. K, 1997).

Most suicides take place in the course of depression. The danger is greatest in psychotic depressions. Suicide risk is very high when depression is linked with generalized anxiety or social anxiety (Angst, Angst, & Stassen, 1999; Pearson, Conwell, & Lyness, 1997).

Suicide can also take place outside a typical depression, for example in the course of a psychotic state (Szanto, Prigerson, Aouck, Ehrenprreis, & Reynolds, 1997). Similarly, certain patients suffering from personality disorders are more exposed (Kjellander, Bongar, & King, 1998). In a borderline personality disorder, for example, patients show a persistent fear of being abandoned while being prone to high impulsiveness. For patients with a narcissistic personality disorder, an important loss can also lead to suicide (narcissistic death). For example, an older man for whom physical strength has always been a key component of life may be unable to accept a permanent disability (McIntosh et al., 1994).

Suicide prevention should be a multifaceted social welfare initiative, including social activities in the neighborhood, activity programs about retirement planning, and support groups for those in mourning. A substantial proportion of individuals who attempt suicide have contact with their primary care provider prior to death: 20 percent on the same day, 40 percent within one week, and 70 percent within one month of the suicide (Pearson, Conwell, & Lyness, 1997). Such behavior offers an important avenue for suicide prevention. Professionals and nonprofessionals need appropriate education in the psychosocial aspects, challenges, and tasks of caring for elders. Finally, psychiatry and clinical psychology specialists should be called in for assistance at home, in the community, and in health care retirement communities.

The treatment for suicidal elders usually comprises a combination of psychotherapy and medication. Antidepressive drugs are indicated in the presence of a depressive state with suicidal ideation. Older medications such as tricyclic antidepressants (TCAs) and monoamine oxidase inhibitors (MAOIs) can be difficult to tolerate due to side effects or, in the case of MAOIs, dietary and medication restrictions. Newer medications, such as the selective serotonin reuptake inhibitors (SSRIs), have fewer side effects than the older drugs, making it easier for patients, including older adults, to adhere to treatment. Clinicians need to be aware that the beneficial effects of such medication usually take longer to appear in older patients. In the case of a major depressive disorder with a clear vital inhibition, the suicide risk can paradoxically increase following the initiation of drug therapy, as the negative feelings remain while the inhibitory forces gradually decrease.

Psychotherapy for the elderly has gathered momentum since the 1980s, with evidence that treatment is effective for a number of psychological problems experienced by seniors (Conwell, 1997). For suicidal patients, the combination of family psychotherapy with individual psychotherapy is indicated in order to elicit specific information about patients' competencies and positive behaviors while attempting to neutralize negative attributes. Some misconceptions regarding suicide can act as barriers to an appropriate course of action. It is therefore important to formulate some premises very clearly:

• Some people suppose that a person who talks about suicide will not actually commit suicide. This is a wrong assumption: although one can have the impression that suicide happens unexpectedly, most of these suicides and suicide attempts are preceded by one or other verbal (or non-verbal) signal.

- To talk with the patient about his suicide thoughts is the best diagnosis, prevention, and therapy (see below). It is a misconception that talking about suicide can further push the patient along the path of negative thoughts (see below).
- Most often it is not true that a person who is suicidal definitely wishes to die. Hesitation and ambivalence are clues about the patient's own sense of reality. Furthermore, many suicide attempts are an alarm signal, and are often a last resort to get attention.
- On geriatric departments one can have the wrong idea that the depression (and anxiety) seen in patients is solely the result of their physical pain, so all attention should be concentrated on this aspect. Depression in older adults, however, can also be attributed to a variety of "losses" including physical capacity and social connections. Appropriate suicide prevention should examine all aspects of "loss" in addition to control of physical pain.
- Some people wrongly assume that the suicidal patient does not think about the well-being of his/her partner or child(ren), but conversations with these patients reveal that they are, in fact, concerned about their loved ones: "If I were to disappear, their lives would become much more bearable after a while."

Communication sessions with the suicidal elderly patient and a professional specialist offer the best diagnosis, prevention, and therapy. It is the only way that the suicidal potential of the patient can be properly established. Extra attention is required for spotting and interpreting certain statements (e.g., "Everything is pointless . . . a waste of time and energy"). A variety of aspects must be analyzed: How concrete or well developed is the suicidal plan? Is there still some perspective for the future? Is the patient still capable of controlling his aggression in an adequate manner, given that the risks are higher in the presence of aggressive outbursts? Is there a lot of loss at stake? Does the patient still have a good link with key people from his environment?

Preventive steps on the basis of this dialogue include giving extra attention, sessions with a geri-psychiatrist or geri-psychologist, or admission to a geriatric psychiatry department. Every person in contact with the patient plays an important signaling function in this context. Professionally-led sessions offer the best therapy because they build trust into the relationship with the patient. The specialist makes it known to the patient that he/she is really concerned with the patient's problems, relying on empathic listening skills, and is available in case of need. The patient should be encouraged to ask for further sessions if required. As the supportive, trusting therapeutic mileu forms, the patient should be invited to talk about feelings of guilt and aggression. These feelings should initially be taken as is, i.e., the therapist should not try to discuss them or in some way comment on their relative importance. Sessions should be private and handled with professionalism and sensitivity, so that patients do not feel pressure to reveal their fears and anxieties. Yet, the term "suicide" should not be avoided as this might delude the patient into thinking that the therapist does not want to talk about it or, alternately, finds the subject too threatening.

Attention must be given to the family also. It is not uncommon that family members are shocked at the idea that a parent or partner may be suicidal. Family assistance means that emotional support is given to the family members, and that well-meaning yet counter-productive actions of the family are avoided (e.g., "Father, why don't you go on holiday for a couple of weeks!" or "Go to a party and your spirits will be lifted!"). Should the patient unfortunately still commit suicide, it is imperative to guide the family. Guilt feelings may be very strong and can lead to pathological mourning.

The quality of the relationship between the patient and therapist is of utmost importance in suicide prevention. Establishing rapport and developing a consistent communication path must be an unbroken thread in this critical process.

LUC VAN DE VEN

See Also
Anxiety and Panic Disorders
Depression in Dementia
Euthanasia
Social Isolation

Internet Key Words

Depression

Family

Psychotherapy

Suicide

Internet Resources

Older Adults: Depression and Suicide Facts, National Institute of Mental Health

*http://www.nimh.nih.gov/publicat/
elderlydepsuicide.cfm*

Selected Bibliography on Suicide Research (1997), National Institute of Mental Health

http://www.nimh.nih.gov/research/suibib.htm

SUNDOWN SYNDROME

Sundown syndrome, is also referred to as "sundowning," "twilight transient confusional states," "senile nocturnal delirium," "acute confusional state," and "nocturnal confusional episodes." Sundowning is frequently linked to delirium, and is often described as a nocturnal exacerbation of delirium.

Definition and Description

Sundown syndrome is generally defined as increasing confusion and agitation that occurs near sunset or evening hours (Drake, Drake, & Curwen, 1997). The variety of behaviors associated with this phenomenon include wandering, restlessness, reduced attention span, sleep-wake disturbances, and altered psychomotor behavior. The overall incidence of sundown syndrome is uncertain. Further, little agreement exists about the frequency of disruptive behaviors during specific evening hours. Most studies report data that cover the entire period of wakefulness rather than the specific period of sunset.

A major factor in the definition and description of sundown syndrome is related to the onset of darkness. Although the etiology is unclear, deterioration of the suprachiasmatic nucleus of the hypo-

thalamus, the principle pacemaker of the circadian system, may be the underlying reason for disruptions in the sleep-wake cycle. Studies of dementia patients indicate that deterioration of the hypothalamus is related to the suprachiasmatic nucleus. Other studies found that people with the syndrome are most likely to have an underlying dementia. Sensory deprivation, decreased ambient light, sleep apnea, alterations in metabolism, medications, fatigue, disrupted sleep-wake cycle, and a decreased stress threshold have also been associated with sundown syndrome.

Diagnosis and Treatment

The lack of a consistent definition and the variability of symptoms contribute to the difficulties in assessing, diagnosing, treating, managing, and investigating this syndrome. Diagnosis is generally based on observation of excessive confusion and agitation at a particular time of day.

Treatment of sundown syndrome begins with identification of the underlying causes of the delirium, such as toxicity, infection, metabolic alterations, or medications. Nonpharmacological treatment of sundown syndrome is recommended, with interventions that modify the environment and provide activities during the day. For example, restricting of sleep and increasing activities during the daytime have resulted in decreased wakefulness at night.

Pharmacologic treatment is sometimes used in acute situations. Low dose neuroleptics such as haloperidol (Haldol) or thioridazine hydrochloride (Mellaril) have been administered, however the side effects of these medications, particularly falls and fractures, indicate that these medications should be used judiciously and cautiously. Neuroleptic medications recently on the market such as olanzepine or risperidol in low dosages may have similar therapeutic outcomes with fewer side effects. Other drugs such as melatonin (Brusco et al., 1999) and donepezil (Wengel et al., 1998) have shown effectiveness in decreasing symptoms of sundown syndrome. Some studies support the use of beta-blockers such as propranolol and pindolol for agitated behavior (Burney-Puckett, 1996), however the Food and Drug Administration has not approved these drugs for the treatment of sundown syndrome. Benzodiazepine hypnotics have minimal effects on sundowning behaviors.

Environmental Changes

In a hospital setting, it is important to decrease the noise and confusion surrounding older patients. Constant traffic created by transporting patients, delivery of meal trays, numerous interruptions for laboratory tests, loud talking by staff, and loud televisions on the unit can be minimized if the door to the patient's room is kept closed and a quiet environment is provided. Patients should be approached slowly and quietly, and a sitter or family member who provides consistency and reassurance can be helpful. The television should be turned on only for particular programs that are familiar to patients. To promote normal circadian rhythms, the patient's room should be brightly lit during the day and dark during the night. Family members can be encouraged to sit with the patient to provide a level of calm as well as social stimulation. Activity such as walking during the day should be encouraged, and awakening during the night for vital signs and medications should be limited.

In long-term care settings, a structured environment reduces the onset of sundown syndrome (Hall & Gerdner, 1999). A consistent routine with appropriate environmental cues, such as light and activity during the day and darkness and quiet during the night, may be effective in minimizing agitation and confusion. Some practicing clinicians report that a short rest with soothing music during the afternoon promotes sleep at night (Matteson, unpublished). Others have found that if the resident is agitated at the same time every day, i.e., late afternoon, a stimulant such as coffee given about 30 minutes before the usual event is effective in reducing behavioral symptoms (Matteson, unpublished). Conversely, other clinicians argue that reducing stimulants such as caffeine is effective (Ka-

nowski et al., 1995). Other effective interventions in reducing some symptoms include regular exercise at least three times per week, decreased stimulation, and bright light therapy (two hours, from 7:00 to 9:00 PM). Unfortunately, no treatment approach has demonstrated consistent effectiveness.

MARY ANN MATTESON

See Also
Behavioral Symptoms in Patients with Dementia
Confusion/Delirium: Risk, Diagnosis, Assessment, and Interventions
Wandering

Internet Key Words
Agitation
Confusion
Delirium
Sleep-Wake Disturbances
Sundown Syndrome

Internet Resources
Health-Center: Symptoms: Confusion or Disorientation
http://site.health-center.com/brain/symptoms/ confusion.htm

Virtual Hospital: University of Iowa Family Practice Handbook, third ed: Neurology: Delirium
http://www.vh.org/Providers/ClinRef/ FPHandbook/Chapter14/03-14.html

SUPPORT GROUPS

Support groups are a generic type of interpersonal network that, depending on objectives, initiation, leadership, and composition, may be variously described as self-help, mutual support, or treatment groups. The social and emotional bonding of individuals into networks of persons who perceive a shared fate and affirm mutual responsibilities for one another is a basic process observed in all stable social groups. Kinship groups are prototypic mutual-support networks that are intended to provide timely and appropriate information, practical services, and emotional support when needed. In complex, socially differentiated societies, kinshiplike interpersonal networks appear in large numbers and in great variety, apparently to compensate in part for the attenuation of traditional kinship ties and the limited capacity of kin groups to provide needed support in a timely way.

In psychosocial terms, the essence of support groupings is the reliable availability of interpersonal networks in which participants perceive themselves to be accepted and understood and expect to receive timely information and supportive assistance in mastering problems of everyday life. Specifically, support groups provide (1) models of emotional mastery in responding to potentially traumatic events and circumstances; (2) guidance in cognitive interpretation and response to these events and circumstances; (3) consensual validation of self-esteem or reinvention of the self in the face of significant challenges or loss; and (4) instrumental, palpable, practical help in securing and using resources required to cope. Although research has increasingly documented the extent and variety of mutual support groups worldwide, evidence regarding the outcomes and effectiveness of such groups remains sparse (Lieberman & Snowden, 1993; Litwak, 1985; Schopler & Galinsky, 1993).

Support groups are observed worldwide, but this kind of informal provision of informal care, particularly for aging populations, is especially common in the United States. Similar informal groups have been documented in the United Kingdom, Holland, Poland, Yugoslavia, and Latin America (Maddox, 1984).

Even in the United States, however, the actual number and variety of support groups is a matter of conjecture. In the 1980s, for example, an estimated one half million groups involving millions of persons existed. The Center for Self-Help Research at the University of California, San Francisco, estimated from various surveys in the early 1990s that 7.5 million adults participated in one or another support group designed to assist individuals facing every conceivable problem (Lieberman & Snowden, 1993). Participants were found to be predominantly white, middle class, and middle aged. Lifetime participation in mutual support groups was estimated to involve 3.6% of adult men and 2.2% of adult women. One estimate suggests that in 1990, 3% of adults in the United States had attended a meeting of the more than 27,000 local chapters of Alcoholics Anonymous, a prototypic support group, in the previous year. The Alzheimer's Disease and Related Disorders Association in the 1980s published principles for creating support groups, and by the 1990s reported 221 chapters and over 2,800 support groups in all 50 states.

Support groups have come to be a regular component of therapeutic interventions designed to prevent illness and improve health (Macauley & Katula, 1999). Conceptually, support groupings are a component of the more general phenomena of social networking and social integration that have been established to have positive implications for health and well-being (Unger et al., 1999).

Although support groups appear typically to produce benign or beneficial effects, systematic research is scarce on outcomes that would assist in specifying what kinds of support are beneficial for what individuals under what circumstances. The fluidity of membership in support groups and the complex number of variables that remain uncontrolled make definitive research difficult. Additional research will be required to develop decision rules for professionals regarding whether and how to use support groups to assist clients and patients. Of particular importance is knowledge about the limits of using support groups effectively and how to enhance the complementarity of formal and informal care services for older adults and their families (Litwak, 1985). Further, the timing, not just the provision of emotional support, information, and practical care, may be of importance in the management of care.

GEORGE L. MADDOX

See Also
Alzheimer's Association
Group Psychotherapy

Internet Key Words
Mutual Support
Self Help

SWALLOWING DISORDERS AND ASPIRATION

Swallowing is a complex mechanism that requires the intricate coordination of several cranial nerves and a very large number of muscles of the face, mouth, pharynx, and esophagus. This enables the important physiological task of transporting liquids and firm food (i.e., the bolus) from the mouth into the esophagus while crossing a complicated anatomical region that is not only involved in swallowing but also in respiration and speech. The main causes of swallowing disorders are neuromuscular diseases (e.g., stroke), local structural abnormalities (e.g., Zenker diverticulum), and motility disorders of the upper esophageal sphincter.

Swallowing disorders occur in all age groups and can be produced by a wide variety of pathologies (neurologic and medical problems or structural abnormalities). The resulting impairment may range from very mild to life threatening. It is, however, necessary to distinguish the effect of normal aging from the effects of specific diseases or degenerative changes. Indeed, nondysphagic elderly may have altered function without impairment. With increasing age, several changes can be observed: increased stiffness of the upper esophageal sphincter, progressive lengthening of the duration of the hyoid movement, shortening of the duration of the laryngeal closure, etc. These changes are congruent with a general impression that aging per se does not lead to pathology, but it puts the aging person in a less favorable or at-risk position. As such, aspiration is likely to be the result of pathology and not due to normal aging.

Dysphagia (difficulty in swallowing) is a surprisingly common symptom and one that spans all ages. It is helpful to divide dysphagia into two types: oropharyngeal and esophageal. Dysphagia secondary to a lesion above or proximal to the esophagus is called oropharyngeal dysphagia. This symptom is often characterized as a transfer problem: the patient has trouble transferring food from the mouth into the pharynx and esophagus. Patients with esophageal dysphagia have difficulty transporting food down the esophagus once the bolus has been successfully transferred through the pharynx.

Here the term dysphagia refers only to oropharyngeal dysphagia. In many cases, dysphagia is only one of many symptoms, but it may also occur solitarily. A patient who complains about difficulties swallowing certain foods or liquids may have a swallowing disorder. Coughing usually indicates that liquid or food has entered the airway, but some patients do not cough when they aspirate ("silent aspiration"). A gurgly voice quality after swallowing indicates that food remains in the larynx. The patient's mouth should be examined for residual food after swallowing.

Globus sensation or a feeling of a lump in the throat is usually not related to swallowing and should not be confused with dysphagia. Regurgitation or the return of undigested food or liquid may have different causes; delayed regurgitation of undigested food is suggestive of a Zenker diverticulum. Odynophagia or painful swallowing is most commonly due to acute disorders such as a pharyngitis. Aspiration is defined as food (liquids or solids) entering the airway below the level of the true vocal cords (Logemann, 1986). Aspiration may lead to an aspiration pneumonia. Penetration refers to food that has entered the larynx but remains above the true vocal cords.

Evaluation

Investigation of a swallowing disorder requires a multidisciplinary approach. It should always start with a careful clinical evaluation, preferably done by a speech-language pathologist.

Radiologic imaging is central to evaluating, diagnosing, guiding the management, and assessing the interventions for swallowing disorders (Jones & Donner, 1991). A modified barium swallow/videofluorographic study of swallowing (Logemann, 1983) provides vital information for the management of patients with aspiration. The moment of aspiration in relation to the pharyngeal stage of deglutition (before, during, or after deglutition) (Logemann, 1986) seems to be crucial.

Manometry is only useful in combination with radiology, a correlation best obtained with manofluoroscopy. This technique simultaneously records pressure at different heights, anatomic events, and bolus transit on a single videoscreen along with timing numbers. It may be an important tool in assessing the indication for extramucosal myotomy of the upper esophageal sphincter.

Management

Once a clear insight into the patient's swallowing problem in terms of anatomic or physiologic abnormalities has been obtained, treatment can be considered. This involves a team approach and an individualized treatment plan (Leonard & Kendall, 1997), but there are some general rules when dealing with elders who have swallowing problems:

- No attempt to feed orally should be made unless the patient is fully alert.
- Maintain a calm environment as the patient should not become distracted.
- Sit next to the patient, at the same height, and ensure that the food is placed within his visual field. An upright position is best with the head in the midline.
- Allow sufficient time.
- Provide small quantities.
- Observe the patient and assist when necessary. Self-feeding sometimes improves swallowing.
- Offer another spoon or fork-full of food only when the previous one is swallowed. Do not presume that another spoonful will help move the previous one. Place the food in the mouth centrally or at the best side (i.e., left side if there is a right-sided paresis).
- Be sure that the patient's mouth and teeth (or dentures) are clean before eating; check for any oral residue after the meal.
- Don't encourage the patient to speak during mealtime. The person assisting should not initiate a conversation. Limit the talking to short clear messages such as "open your mouth, chew, etc."
- Ask advice concerning the utensils to use during eating.

Other interventions, however, must be individualized, including compensatory measures, swallow maneuvers, medication, and surgery. Compensatory measures include postural changes. A chin down position is recommended when there is a delay in triggering the pharyngeal phase; when there is a unilateral pharyngeal weakness, the head should be turned to the weaker side.

The patient should receive the food consistency best adapted to his situation. Mixed or pureed food is advisable if there is a chewing problem or in the healing stages postoperatively; liquids are indicated when there is an Upper Esophageal Sphincter (UES) opening problem. In case of tongue and pharyngeal weakness, it may be necessary to alternate liquid and solid. A food thickener can be used in patients with problems swallowing liquids.

Swallowing maneuvers require much more cooperation and understanding from the patient. Among the most frequently used are:

- The Mendelsohn maneuver, which prolongs the UES opening. In this maneuver, the patient is instructed to voluntarily elevate and hold the larynx in an upright position.
- The effortful swallow, which enhances the tongue thrust.
- The supraglottic technique, which teaches the patient to close the true vocal cords before and during the swallow and to clear any residue that may have entered the laryngeal vestibule.

The supraglottic technique is designed for patients with reduced laryngeal closure who are at risk of aspiration. It involves taking a deep breath, holding the breath while swallowing, and coughing immediately after the swallow. The supersupraglottic swallow is very similar to the supraglottic technique with the addition to bear down during breath holding.

In rare cases, medication can improve swallowing, e.g., in patients with myasthenia gravis. Several surgical procedures can improve swallowing disorders. For significant cases of upper esophageal sphincter (UES) dysfunction, an extramucosal myotomy of this sphincter may help correct the problem. A Zenker diverticulum can be treated by a diverticulopexy in combination with extramucosal myotomy; endoscopic treatment is an alternative. Medialization of a paralyzed vocal cord can be performed through an injection or an implant technique. This intervention can be proposed for voice improvement and to avoid aspiration. If oral feeding places the patient at too great a risk, percutaneous endoscopic gastrostomy (PEG) tube feeding should be considered as a temporary solution to allow recuperation and revalidation to take place without the burden of a nasogastric tube.

EDDY DEJAEGER
ANN GOELEVEN

See Also
Percutaneous Endoscopic Gastrostomy (PEG) Tube Feeding

Internet Key Words
Aspiration
Deglutition
Dysphagia
Manofluoroscopy
Modified Barium Swallow
Swallowing Maneuvers

Internet Resources
American Speech-Language-Hearing Association
http://www.asha.org

Dysphagia Resource Center
http://www.dysphagia.com

T

TASTE

See

Sensory Change/Loss: Smell and Taste

TAX POLICY

Taxation is an increasingly important but often overlooked component of federal, state, and local policies for the elderly. The tendency has been to focus on budget outlays or direct expenditures, such as Social Security, Medicare, Medicaid, and the Older Americans Act, and less on tax credits, deductions, and exemptions (known as tax expenditures). Federal-level tax policies, rather than state and local policies, also have commanded center stage. Since the late 1970s and early 1980s, however, greater attention has been paid to the impacts of tax policies at all governmental levels on the aged and on the larger society.

In the 1950s and 1960s, several rationales were advanced for special tax treatment of the elderly. These included higher rates of poverty, being categorized as the "deserving poor," and "payback" for their prior contributions to society. As a result, an additional exemption was enacted for persons aged 65 and older; Social Security benefits, and often other retirement income, were not subject to federal and state income tax. In the name of fairness, elders also benefitted from tax policies that were available to all age groups (e.g., deductions for medical expenses in excess of a certain proportion of income).

Federal Tax Policy

In the 1980s, several factors converged to change the tax status of the elderly, including concerns over the viability of the Social Security system and perceptions that higher-income elders were "greedy geezers" receiving Social Security payments they did not really need. The largest federal tax expenditures ($18 billion) for the retired were the additional age-based exemption and the exclusion of Social Security benefits (Gist, 1988; Zahn & Gold, 1984). Social Security reforms enacted in 1983 subjected benefits to income taxation, with differential (but unindexed) thresholds for singles and couples; the resulting revenue was credited to the Old Age and Survivors Insurance (OASDI) trust fund. The Tax Reform Act of 1986 eliminated the special exemption for the aged but doubled the personal exemption and boosted the standard deduction for all taxpayers including the elderly. More recent tax reforms during the 1990s increased the percentage of benefits subjected to the income tax while also raising the thresholds of Social Security benefits subject to that tax. Despite this action, calls for means-testing Social Security benefits have continued to be heard, although they have subsided some in the face of discussions over privatization of Social Security.

Also reflecting a shift from preferential treatment for older Americans were changes in the one-time exclusion on the capital gains realized from the sale of the primary residence of those aged 55 and over. This policy was recently replaced by a provision covering *all* homeowners, with a capital gains exclusion of $250,000 for single individuals and $500,000 for couples, and no requirement to buy another primary residence within two years. This enables older homeowners, especially those who are "house poor," to access higher levels of equity in their homes for other non-housing expenses, such as long-term care.

The raising of the threshold for estate taxes, first to $600,000 and gradually to $1 million in the new millennium; and dependent care tax credits for employed taxpayers caring for elderly spouses or parents (as well as children) tended to favor the elderly. Some of these federal enactments were subsequently adopted by many of the states, especially those with income tax policies closely tied to federal policy. More recent proposals have included annual tax credits ranging from $1,000 to $3,000 for families caring for the elderly. It is estimated that the U.S. Treasury will forgo $26 billion over ten years, if the higher figure is adopted.

A major tax expenditure (more than $81 billion) is for pensions (employer and self-employed plans such as Keoghs and Individual Retirement Accounts [IRAs]) that permit contributions to be deducted from present tax obligations by the future elderly population. The 1986 tax law restricted the levels of both employer and employee pension contributions. Although the ceilings have increased over time, restrictions are still in effect. The amount of tax-deferred contributions to IRAs for higher-income individuals enrolled in an employer plan also was restricted. More recent enactments have raised the amount that married couples can contribute to IRAs.

These pension-related tax expenditures have given rise to discussions about which income groups and types of employees (public or private) benefit most from these exclusions. This has led to advocacy for taxing the value of pensions, all or in part (see, for example, Munnell, 1992; Salisbury, 1993; Schieber, 1990), and similarly to proposals for taxing the value of employer-provided Child and Disabled Dependent Care credit. A handful of states have enacted deductions or exemptions for taxpayers caring for an older or disabled person, as well as tax credits for energy expenditures.

The property tax is usually perceived as the most onerous for the elderly in comparison with younger households (Reschovsky, 1994). In retirement, the ratio of property wealth tends to rise as income falls, raising the property tax burden. In the 1970s, and again in the 1990s, this problem was exacerbated by residential property values increasing at a faster rate than other property values. This tax burden has often been perceived as the least fair among common taxes (U.S. Advisory Commission on Intergovernmental Relations, 1972–1994) and gave rise to tax revolts in 24 states, especially California and Massachusetts. The property tax generates 75% of county and 50% of municipal tax revenues. Most visible in its financing of public education, it is often the only significant tax paid by the growing middle-class elderly population to support state and local services (Mackey & Carter, 1994).

Every state except Wyoming has either a homestead exemption or "circuit breaker" program of tax relief. Nearly 75% of the states have enacted homestead exemptions and credits; 24 favor seniors by limiting participation to that age group or providing them with greater benefits than are available to the general population. Four states leave this policy up to the discretion of their local governments; 16 states require elderly homeowners to meet income criteria. Recent research shows higher awareness levels and application rates for these programs among the elderly than for other property tax programs (Baer, 1998).

"Circuit breakers" in 32 states prevent property taxes from placing a tax burden overload on taxpayers. Unlike the homestead programs, these state programs are carefully targeted to low- and moderate-income taxpayers and can benefit both homeowners and renters. Most (22 states) limit eligibility to the elderly and, in some cases, disabled homeowners to delay the payment of taxes to a later date. If income guidelines are met, part or all of those taxes can be postponed until the owner's death or the sale of the property. Recent legislation in California allows elders who sell their homes to transfer their existing, generally lower, property tax rates to new in-state locations in eight counties having reciprocity. In addition, a handful of states have enacted property tax freezes and abatements for home repairs, generally restricted to the aged. Elders' knowledge about these programs is considerably less than their familiarity with the homestead program (Baer, 1998).

A final category of state and local taxes is the sales tax, a major source of revenue for nearly all states, counties, and cities. Very few states provide a credit or rebate to elders for part of the sales tax paid. Perhaps the greatest sales tax boon for the aged is the exemption for food and especially prescription drugs, because they purchase more prescriptions than do younger persons. Because they characteristically spend a higher proportion of their income on such nontaxable items, their sales tax burden is lowered (Mackey & Carter, 1994).

Basic Issues Arising from Tax Preferences for the Aged

Basic issues are raised by the use of tax expenditures, not the least of which is that the actual cost of an aging society is not known. The extent of forgone tax revenues is not well documented, especially at the subnational level. Most state legislatures have not yet analyzed how tax benefits for the elderly now affect or will impact their revenue systems (Mackey, 1995).

Another issue is one of intragenerational equity, that is, the distribution of tax forgiveness among different elderly income groups. Many tax policies, especially deductions, are more beneficial to upper-income elderly than to low-income aged. The more extensive taxation of Social Security benefits enacted in the 1980s and 1990s has tended to right that imbalance, but tax-preferred pensions still are less likely to benefit low-income families. Workers in families with incomes below $15,000 get relatively little benefit from pension-related tax incentives (Schieber, 1990). Furthermore, state property tax relief programs generally favor homeowners over renters, the segment of the elderly population most likely to pay excessive costs for housing (Liebig, 1998).

Other issues spring from the relative efficiency of tax laws in promoting the welfare of those elders who need help the most. These include questions of whether an increase in direct spending on low-income elders or using tax credits would be more effective, or whether age is an appropriate factor on which to base tax relief. In addition, little is known about the effectiveness of many of these provisions in increasing the well-being of the aged. For example, we do not know if property tax breaks help older persons keep their homes or cause them to attempt to maintain their homes at considerable financial and personal health risk, or if federal and state dependent care tax credits help elders maintain relative independence, enhance quality of life, and avoid nursing home placement.

Tax policies can be an important mechanism for achieving important policy objectives, such as homeownership. Major questions, however, need to be addressed regarding the impact of tax expenditures on the general welfare of elders, on different subgroups of the aged, and on society as the proportion of the elderly increases in the 21st century.

PHOEBE S. LIEBIG

See Also
Pensions
Retirement
Social Security

THERAPEUTIC RECREATION SPECIALISTS AND RECREATION THERAPISTS

The terms *therapeutic recreation specialist* and *recreation therapist*, used interchangeably, describe bachelor or master's level professionals with academic preparation in leisure and therapeutic recreation theory and practice.

Geriatric service settings that benefit from the expertise of therapeutic recreation specialists include long-term care, assisted living, congregate housing, dementia care, physical rehabilitation, respite care, geriatric day care, gero-psychiatric clinics, senior citizens' activity programs, municipal recreation programs, and intergenerational programs (Avery, 1997). Recreation specialists are knowledgeable about physical, social, cognitive, and psychological problems commonly associated with aging, recognize the critical importance of health promotion and maintenance, and utilize meaningful recreation and leisure as both a means and an end for achieving a reasonable quality of life. Recreational therapists know how to assess behavior, design appropriate adaptations and interventions that fit clients' needs and interests, and evaluate their effectiveness. Modalities common in therapeutic recreation programming for older persons include remotivation, resocialization, reminiscence, expressive arts, movement and music, stress management, assertiveness training, physical exercise, cognitive retraining, reality orientation, sensory programs, behavior management, pet-assisted therapy, horticulture and therapeutic gardening, aquatics, travel, community service, special interest groups (hobbies, collections, etc.), and computer technology activities.

Approximately 38,000 recreational therapists practiced in the United States in 1996. The National Council for Therapeutic Recreation Certification (NCTRC), the credentialing agency, reports that approximately 16,000 individuals are currently "Certified Therapeutic Recreation Specialists" (CTRS). Candidates for testing/certification must have at least a bachelor's degree in therapeutic recreation or recreation with a specialization in therapeutic recreation, specific courses in recreation/leisure theory, therapeutic recreation theory, abnormal psychology, anatomy and physiology, growth and development and other human service disciplines, and an internship with a CTRS. Certification is not always required in clinical settings.

An "activities specialist" or "recreation specialist" works in recreation programs but is not required to have a bachelor's degree in therapeutic recreation or national certification.

Individuals with degrees in related fields can qualify to take the certification exam, but additional course work and experience are required. NCTRC is recognized by the National Commission for Certifying Agencies (NCCA) for compliance with high standards of quality and integrity in the certification and competency assurance process that serves the interests of the public, employers, and certificants. The National Association of Activity Professionals, an industry supported association, focuses on support and training for activities directors who lack a college degree or who have degrees in fields not specific to therapeutic recreation. Its membership is open to CTRS as well. All associations have the same goal: to promote quality recreation programming for long-term care residents.

Nursing homes receiving Medicaid or Medicare funding must offer planned and organized recreation/activity services that address each resident's individual needs and interests. Resident-focused services include leisure assessments, participation in care planning, program depth and breadth, and documentation of residents' progress toward treatment goals. Federal regulations do not require that recreational therapists working in nursing homes be baccalaureate prepared or nationally certified. Nursing home operators/administrators determine who is better prepared to provide recreational services to a particular patient population. State surveyors, using Health Care Finance Administration (HCFA) guidelines, hold the agency accountable for quality of care. All activities personnel complete the Minimum Data Set (MDS), the nationally used interdisciplinary assessment tool in long-term care, but at this time, only Certified Therapeutic Recreation Specialists complete a special section of the MDS (Sec. T) that identifies further, more specific need for therapeutic recreation as treatment. The National Therapeutic Recreation Society and the American Therapeutic Recreation Association encourage employment of CTRSs in long-term care, especially in those agencies that also have subacute and rehabilitation services, or specialized dementia care.

Assisted living, a less costly option for elder care than nursing home care, helps residents remain as active and independent as possible, providing supportive services as needed. Many assisted living agencies also offer dementia care. Proposed industry standards for assisted living recognize the value of structured and organized recreation services based on a well developed "service plan" that identifies, like the treatment plan in long-term care, the needs of each resident. As in long-term care, assisted living facilities are not required to hire a CTRS, but the same benefits to residents must apply and should be an incentive to hire qualified, well prepared professionals in consultant or full-time positions.

The care of dementia patients in special units, day care, or general psychiatric hospitals requires professional skills well suited to therapeutic recreation specialists. Since the activity level of patients in the early and middle stages of Alzheimer's disease remains high, treatment plans should not over-stimulate an easily confused patient but, instead, fully utilize all the residual strengths and cognitive abilities. Therapeutic recreation specialists understand the degenerative nature of dementia and provide structured and creative activities that tap into residents' past interests, keep social connections with family and friends at their optimal level, and monitor cognitive functioning so that new adaptations to the environment can be made.

Recreation employees who are not degreed or specially trained may provide senior programming in community recreation settings, senior centers, and retirement communities. Under the purview of a therapeutic recreation specialist, however, programs operate within a health promotion/disease prevention model. Activities are designed to maintain high levels of fitness, emotional well being, intellectual stimulation, and social interaction. Drawing on research that elders continue to learn and expand their areas of interest, activities that stimulate new learning and maximize the use of existing skills are stressed.

Leisure education, a standard programming technique used by therapeutic recreation specialists, helps program participants understand the value of leisure and recreation and develop or maintain a healthy leisure lifestyle. Education and service oriented activities complement social opportunities and physical activity to create a well-rounded, solid program foundation. Seniors in these settings are encouraged to contribute to the community at large and to their peers in ways that support meaningfulness in later life. Since leisure implies a level of personal freedom, motivation from within, and a desire to deepen personal happiness and life satisfaction, the role of the therapeutic recreation specialist in promoting the quality of life is essential.

CATHY O'KEEFE

See Also
Assisted Living
Leisure Programs
Nursing Homes

Internet Key Words
Assisted Living
Nursing Homes
Recreation Therapy
Therapeutic Recreation

Internet Resources
The National Therapeutic Recreation Society
http://www.NTRSNRPA@aol.com

The American Therapeutic Recreation Society
http://www.atra-tr.org

The National Council for Therapeutic Recreation
    Certification
http://www.nctrc@NCTRC.org

TIME

Time and temporality are central to analysis of aging and life course patterns. As essential as both concepts are to personal development and to the study of aging, they are also elusive: the more precisely scrutinized, the more difficult to pin down. Scholarly consideration of things temporal is often an implicit reflection of chronological time keeping, is paradigm specific, or revolves around time cast in one of two ways. Time is viewed either as an exogenous factor grounded in cosmological evolution and the directionality of an expanding universe, or as an endogenous process based on properties of the system in question, whether that be an organism, individual, or cultural system (Pixten, 1995; Fraser, 1996). Examination of time and temporality in gerontology has focused on (a) time as an unidirectional index of age-related changes, (b) temporal orientation as a function of biological, psychological, or social change, (c) temporality as a consequence of personality factors, social involvements, or awareness of aging, and (d) shifts in sense of futurity as a concomitant of aging itself (Hendricks, in press).

Cross Disciplinary Analyses

Explorations of temporal aspects of aging processes are found in the physical and biological sciences, in psychology, and in an array of social and cultural investigations. In the first case, there is no doubt that life has definite rhythms and periodicities embedded in the very "stuff" of life. Indeed, many facets of our biological clocks seem to oscillate from within and demonstrate diurnal patterns, evidencing 24-hour cycles. At a physiological level, these biological clocks are part and parcel of temporal experience and there are no fewer than two dozen such rhythms emerging from micro- and macro-level physiological processes. Recurrent metabolic cycles of cellular and bodily functions or the light-dark cycle of photoperiod effects provide elemental temporal pacemakers marking sequela or time passage. Circadian rhythms fluctuate on 24-hour cycles emanating from within the organism, but they are also capable of being reset. For example, the daily run of hormonal, neurological, and metabolic functions, like sleep, breathing, hunger, temperature, and even cognitive variations, or monthly events such as estrous cycles, can be offset and recalibrated by external conditions. The well-known "jet lag," characteristic of a disruption in temporal coordination, followed by a recalibrated sleep cycle, is a case in point. Underlying each and posing as possible causal agents are enzyme activation and inhibition, oxidative metabolism in the brain, brain rhythms, ionic diffusion across cell membranes, RNA synthesis, and hypothalamus regulatory activity to name but a few.

To date, a "time organ" wherein time sense originates has not been identified. The presence of diverse biological rhythms suggests emergent harmonic periodicities, perhaps based on fundamental photo- or thermal periods, which, although they may not be dependent on cognition for their existence, do involve cognition and reflection if they have meaning for temporal awareness. Yet even without mentation, biological cycles may be the ontogenesis of temporal reckoning. Because physiological periodicities are thought to help synchronize inherent biotemporality with psychological and social time frames, disruption of these regularities by stress, suppressors, or disease may have far-reaching effects on all components of temporal orientation (Hendricks, forthcoming).

Despite the physiological substrate involved in time's composition, no sense organ exists to help us sense time the way we sense other stimuli. This is not to say, however, that psychological factors are not involved. For 100 years psychologists have sought to study time as other sensory processes are studied without having a definitive physical stimulus involved. There is reasonable consensus that among the psychological facets of time perception, perceptual and cognitive dimensions, as well as internal and external phenomena, are involved. Sense of time is recognized to reflect a number of psychological characteristics, including sensorimotor skills, emotional states, processing speed, memory, and reminiscence, as well as other functions carried out in the cerebral cortex.

Sequential thoughts and perhaps sequential actions are also axiomatic to time reckoning for through them we develop a kind of extended consciousness vital for elementary psychological and social psychological aspects of human life (Damasio, 1999). Temporal awareness is recognized as a developmental property essential for personality, self-concept, sense of well-being, and control. As early as age two, some anticipatory mindset begins to become evident and by age six or seven higher order thinking, consonant with an adult-like sense of temporality, is involved in both cognition and behavior. Not surprising, time is thought to be critical to formal operations and to any type of ordering principles involving extension or proleptic imagery.

Perception and cognition are linked to experience, including time, and dependent in turn on information processing. From a psychological standpoint, changes in temporality thought to be associated with age may be due to delayed processing times associated with stimulus masking, sequential stimulus integration, motivation, schemata for and amount of information already stored, peculiarities of memory function, interaction with others, or slowing of an internal clock. Sense of time is a highly personal characteristic, and the results of time research at the psychological level are anything but consistent, and the concomitants of testing presently preclude definitive conclusions.

Whether time estimation, a comparison of perceived versus clock time, is sufficient to account for widely documented performance differences is a difficult question. Further, whether accuracy in time estimation over the short run has any relationship to longer temporal perspective has not been demonstrated. As valuable as laboratory research is for certain aspects of time perception, the link with lived temporal experience awaits validation.

Temporality and Life Course

One thing is clear, the way we think about time is unique to humans and the ground upon which recurrent patterns in the environment are incorporated into mental models of what life is all about (Friedman, 1990). In addition to internal bodily processing and psychological functioning, time sense is rooted in interaction and membership categories. Time is ingrained in cultural and social systems and in many respects membership groups provide coordination of the various times operative in people's lives, and their patterns of participation are reflected in their awareness of and attitudes about time. Beginning with family life, education, and work, and moving on to involvement in diverse organizations, the mandates of public policy and the mundane aspects of life such as meal functions and weekly activities, life is punctuated by social influences. Among these, age grading, age norms, and socially prescribed transitions provide an inexorable link between individual definitions of temporality, definitions of the life course, individual experience, and recognized societal transitions. There is dynamic but loose coupling between lives and social structures that makes temporal integration, or isochronalism, difficult to maintain but important.

Social correlates constitute another dimension of temporality and definitions of the life course. By and large these social dimensions are held in accord by the synthetic regularity of clock time,

yet asynchrony may occur (Jerrome, 1994). Nonetheless, time is widely thought to be culturally contingent and socially emergent, evolving out of primary activities and the need to impose organizational schemata. The pace of time arises from biological functioning, is couched in the rhythm of life or the structure of the language, but is seldom written in stone. The variability comes from the role contextual factors play in determining individual timelines. The gamut of variables incorporated under the social dimension is broad, ranging from general normative temporal orientations to the impact of immediate and mundane variables such as place of residence and occupational pursuits. To the extent that individuals have differential involvements in social activities, work, or family life, they also are possessed of distinct temporal horizons (Jerrome, 1994). Of course, gender and a host of socially recognized differences are an inescapable part of the equation as well and serve to nuance the relationship. In other words, sense of time, timing, and temporality, their construction and perception, reflect the many varieties and the relativistic nature of temporal worlds that coexist for individuals embedded in a given social system.

Through the mechanism of shared time, the relational quality of inner and outer time is established. One of the more obvious patterns imposed on individuals is the conception of life course, age norms, and the number and breadth of temporal gradations held by a particular society or social group, but there are countless other similar consensual perspectives on temporality. Childhood and old age, as we now perceive them, are largely consequences of a modern industrial temporal orientation. Historical awareness adds yet another facet to temporal perspective. Locating our kind and ourselves in the long-term flow of events, whether through religion or mythology provides an ideational dimension to time. Time viewed as a consequence of history also adds a feeling of linearity to time's passage: the past receding away in one direction while the future rushes at us, like an arrow, seemingly of its own volition, from the other direction.

Time in Aging Research

In gerontology, time is utilized to provide a basic index for behavior, serving as a framework subsuming a number of time-dependent processes; but in and of itself time has little explanatory power. Time is often discussed as analogous to space, the metaphors of simultaneity, direction, and duration being used in both cases. Temporal orientation itself is likely a composite of four linked elements. These elements are biological, personal, social, and historical in nature, with each possessing its own ontological status. At any point, temporality reflects the interweaving of these elements as well as the transitory primacy of one of them. All too frequently there is an overemphasis on chronological time to the exclusion of other dimensions of temporality.

A common assumption is that older individuals are less future oriented than younger people, but those who distinguish cognitive and subjective components of futurity have challenged this idea. What does seem clear is that those persons who have a positive outlook are more future oriented. In a structural equation analysis of five national data sets in Israel, Shmotkin (1991) suggested that although futurity declines and orientation to the past increases with age, neither stands free of a sense of life satisfaction. It may well be the case that a sense of autonomy, control, and futurity are interconnected (Fingerman & Perlmutter, 1995). In contrast, Kulys and Tobin (1980) asserted that individuals with high future concerns are less personally secure than those focused in the present. Another psychological characteristic thought to be associated with foreshortened time perspectives is awareness of finitude (Rappaport, Foster, Bross, & Gilden, 1993). Likely as not, any relationship—and the evidence is not yet complete—is between foreshortened subjective futures and awareness of death. It is important to bear in mind that the latter cannot be operationalized simply by chronological age. Further, retrospection and reminiscence, the converse of futurity, may increase with age but are not as obsessive or as simplistic a component of time orientation as has often been assumed (Hendricks, 1995).

One of the paradoxes of temporal reckoning is that the time of memory or of an anticipated future does not exist in isolation from the present. Time is continuously experienced and interleaved; it can be partitioned only for purposes of analysis. Past, present, and future commingle in the here and now, though the focal length of time perspective may be long or short, depending on momentary concerns. The future is thus created, not necessarily of equal intervals, out of an actor's current agenda, the same as is the case for the past. Both are important components of motivation for those who positively anticipate the coming of tomorrow.

JON HENDRICKS

See Also
Life Events
Life Review

Internet Key Words
Consciousness
Memory
Metabolism
Temporality

TOTAL PARENTERAL NUTRITION

For those elderly whose gastrointestinal systems cannot handle feedings, nutrition can be provided directly into the vascular system by a parenteral route. Total Parenteral Nutrition (TPN), known also as Central Parenteral Nutrition (CPN), can make the difference between healing and nonhealing, recovery and nonrecovery, and, for some, life and death.

Parenteral nutrition, usually administered through a catheter threaded into the subclavian vein to the superior vena cava, allows large amounts of large molecules and hypertonic solutions to infuse into an area of rapid blood flow so that the nutrients are immediately diluted and distributed throughout the body. The parenteral solution thus diluted can be administered for long periods of time without the damage that hyperosmolality can cause smaller blood vessels.

Peripheral Parenteral Nutrition (PPN) is administered via smaller peripheral veins, but is limited to isotonic solutions that provide only limited kcalories (kcal) and protein (5% to 10% dextrose and 3% to 5% amino acids); it is used for short-term nutritional support. Nutrition provided by TPN and PPN is also known as hyperalimentation, as it is given outside the alimentary canal. It can be used alone or given in addition to an eaten, enteral diet.

Components of Parenteral Nutrition Solutions

Parenteral solutions must be carefully calculated daily after careful consideration of lab values in order to provide all required nutrients in the required amounts. Commercially prepared solutions are available as well as individually compounded solutions. The basic components include water, dextrose, amino acids, electrolytes, vitamins, and trace elements. Lipids are also given, either as piggybacked solutions or mixed in with the other nutrients. Intravenous (IV) medications that can be mixed in with TPN solutions include antibiotics, vasopressors, narcotics, diuretics, and insulin.

Energy should be supplied by carbohydrates and fats. The most common parenteral carbohydrate is dextrose monohydrate, which yields 3.4 kcal/g because of its hydrated form. It may be given in concentrations between 5% and 70%. Higher concentrations can cause hyperglycemia, but may be used when fluid is restricted.

Lipid emulsions are used to provide energy and to prevent the development of essential fatty acid deficiency. Commercial lipid emulsions are formulations of safflower oil, soybean oil, or a combination of the two with glycerol for isotonicity and egg phospholipid as an emulsifier. A 10% lipid emulsion yields 1.1 kcal/mL; 20% solutions are also used. Soybean and safflower oils are rich in linoleic acid, large amounts of which can alter prostaglandin function. Thus, large amounts of these lipid additives can cause immunosuppression.

TABLE 1   Parenteral Nutrition Complications

Mechanical Complications	Metabolic Complications
Pneumothorax	Dehydration from
Hemothorax	osmotic diuresis
Hydrothorax	Hyperosmolar, nonke-
Tension pneumothorax	totic, hyperglycemic
Subcutaneous emphysema	coma
Brachial plexus injury	Rebound hypoglycemia
Subclavian vein injury	on sudden cessation
or hematoma	of parenteral nutrition
Central vein thrombo-	Hypomagnesemia
phlebitis	Hypercalcemia and
Arteriovenous fistula	hypocalcemia
Thoracic duct injury	Hyperphosphatemia and
Hydromediastinum	hypophosphatemia
Air embolism	Hyperchloremic
Catheter fragment	metabolic acidosis
embolism	Uremia
Catheter misplacement	Hyperammonemia
Cardiac perforation	Electrolyte imbalance
Endocarditis	Trace mineral
	deficiencies
Infection and Sepsis	Essential fatty acid
Catheter entrance site	deficiency
Contamination during	Hyperlipidemia
insertion	
Long-term catheter	*Gastrointestinal*
placement	*Complications*
Catheter seeding from	Clolestasis
blood borne or distant	Hepatic abnormalities
infection	Gastrointestinal villous
Solution contamination	atrophy

Adapted from Mahan, K., and Escott-Stump, S. (Eds.) (2000). *Krause's food, nutrition, and diet therapy.* Philadelphia: W. B. Saunders. Used with permission.

Therefore, lipid infusions are usually restricted to less than 300 kcalories of total daily calories and should not exceed 2g/kg/day.

Essential and nonessential amino acids, administered in sufficient amounts for protein synthesis, are provided in 10% solution of amino acids that supply 100 grams of protein per liter. Although amino acids provide 4 kcalories per gram of protein for energy, many facilities do not calculate amino acids as kcal intake because of their prescribed use in tissue repair. Some facilities use particular combinations of amino acids for treatment of conditions such as renal failure, liver failure, and trauma, but there is some debate about the efficacy of this approach.

Electrolytes, vitamins, and trace elements are components of and facilitate multiple biochemical reactions. Most patients requiring TPN start out malnourished and suffer from deficiencies or excesses of these substances. Magnesium, phosphate, and potassium requirements in particular increase in severely malnourished patients (Hamaoui & Kodsi, 1997). Multivitamins are added usually on a once a day basis according to recommendations by the American Medical Association Nutrition Advisory Group (1979). The amounts of infused vitamins and minerals are somewhat less than Recommended Dietary Allowances due to immediate parenteral absorption. Vitamin K is given separately in a weekly IM or IV injection.

The maximum volume of TPN fluid administered is usually not greater than three liters per day. Total intake must consider other therapeutic fluids, such as IV antibiotics, chemotherapy, and blood products. This is especially important when fluid overload is significant, as with patients experiencing congestive heart failure, or renal or hepatic failure. TPN is usually infused initially at 42 mL/ h, and then increased in stages until the chosen goal is reached. TPN should not be stopped abruptly due to rebound hypoglycemia. The infusion should be tapered when it is ended, or a 10% dextrose solution hung, if the TPN solution is interrupted. In more stable patients, cyclic TPN can be given at night, so that the patient is free from IV infusion and more readily able to accept enteral feedings during the daytime.

Complications

Although lifesaving, TPN may involve significant complications. It should only be considered if the gastrointestinal system is nonfunctional and all approaches have been considered, such as prokinetic agents and postpyloric tubes, and all other changes in feeding. Enteral feeding, if at all possible, provides more benefit with few risks. Possible compli-

TABLE 2 Recommendations for Monitoring Patients Receiving TPN

Baseline	Routine	As Clinically Indicated
Weight, height, body surface area	*Every 8 hours* Vital Signs Temperature Urine fractionals	*Fluid Disorders* Urine sodium or fractional sodium excretion
Body composition (arm fat and muscle areas, bioelectrical impedence, subjective or functional measures)	*Daily* Weight Fluid intake and output Serum electrolytes, glucose, creatinine, blood urea nitrogen until stable, then twice weekly	Serum osmolality Urine specific gravity *Protein Status* Nitrogen balance, serum prealbumin *Lipid Disorders* Serum triglycerides or lipid clearance test Respiratory quotient
Serum electrolytes, glucose, creatinine, blood urea nitrogen		
Serum magnesium, calcium, phosphorous	*Weekly* Serum magnesium, calcium, phosphorus, albumin Liver function tests Complete blood count Review of actual oral, enteral, and TPN intake	Essential fatty acids (if fat-free TPN is nec.) *Hepatic Encephalopathy* Plasma amino acids *Gastrointestinal Losses* Serum trace elements Stool electrolytes
Serum triglycerides and cholesterol		
Liver function tests		
Serum albumin or pre-albumin		
Energy (estimated or measured), protein, fluid and micronutrient needs	*Long-term TPN* Body composition measures Serum trace elements Vitamins	*Respiratory Compromise* PaCO$_2$ Indirect calorimetry, respiratory quotient *Acid-base Disorders* Blood pH Anion gap

Adapted from Lenssen, P. (1998). Management of total parenteral nutrition. In: Skipper, A. (Ed.), *Dietitian's handbook of enteral and parenteral nutrition*, 2nd ed. Rockville, MD: Aspen. Used with permission.

cations of parenteral nutrition can be classified as mechanical, metabolic, gastrointestinal, and related to infection and sepsis, as is summarized in Table 1 (Mahan & Escott-Stump, 2000).

Monitoring Guidelines: Because of multiple possible complications, close attention must be paid to monitoring the patient on TPN. Although exact protocols may vary, most facilities conduct baseline, routine, and "as indicated" monitoring, which are illustrated in Table 2 (Lenssen, 1998). In addition, the catheter insertion site must be carefully monitored and kept covered with a clear occlusive dressing or changed every 48–72 hours using sterile technique.

DOROTHY G. HERRON
LYNN B. GREENBERG

See Also
Caloric Intake
Protein Energy Undernutrition

Internet Key Words
Hyperalimentation
Peripheral Parenteral Nutrition

Internet Resource
The American Society of Parenteral and Enteral Nutrition (ASPEN)
http//www.clinnutr.org

TRANSPORTATION

A key feature of the way of life in many countries is the high level of personal mobility through using automobiles. With the exception of individuals living in extreme poverty or dwelling in cities, there are few impediments to personal mobility until age-related changes prevent individuals from safely operating a vehicle (Schaie & Pietrucha, 2000). The U.S. Department of Transportation (1997) states that it is in the best interest of the individual to prolong automotive mobility as long as possible, i.e., as long as a person can safely drive. Public transportation is available in major cities, but is often limited in outlying suburban and rural areas.

A person may also be unable to use regular public transportation due to a functional or cognitive limitation. Special senior transportation is available to address the mobility needs of those older adults who can no longer drive or use public transportation.

History and Policies

Due to the rise of the automobile as primary means of personal mobility, the United States is less connected by public transportation today than it was in the late 1920s (Wacker, Roberto, & Piper, 1998). This is true for other industrialized countries only to a much lesser extent. An amendment to the Urban Mass Transportation Act of 1964, passed in 1970, was a first step toward addressing the transportation needs of older adults and persons with disabilities and led to a significant increase of special transportation services (Wacker, Roberto, & Piper, 1998). The Rehabilitation Act of 1973 (Sec. 504) requires transportation systems design such that they can be used effectively by older adults and at half fare during off-peak hours. This led to the creation of paratransit systems—special door-to-door transportation services for older adults.

The Americans with Disabilities Act (ADA) of 1990 mandates wheelchair access to all new public transportation vehicles and that paratransit systems are comparable to fixed route systems in services and fares. The Intermodal Surface Transportation Efficiency Act (ISTEA) of 1991 was the impetus for better coordination of local and regional transportation systems by including them as an essential factor in funding decisions. This Act also permitted transfer of funds between highway and transit programs (Wacker, Roberto, & Piper, 1998).

Transportation Programs and Funding

Non-driving older adults rely on a wide range of transportation providers, including public transit authorities, private for-profit and non-profit organizations, religious groups, and informal support systems like caregivers and friends. Most public transportation services are fixed-route systems. However, transit authorities can fulfill the equal access mandate by operating deviated-fixed-route systems that operate on a fixed route from which the driver may deviate if an eligible rider makes a request.

Another common approach is paratransit or demand-responsive systems that are usually available only to an eligible subgroup of the population, such as older adults. The services are more flexible, bringing a passenger from one specific location to another. Some demand-responsive systems operate on short notice, while others require up to 24-hours advance notice. The systems also differ in the extent of services rendered. Some providers will only stop at the curb, while others will come to the door or even inside a home to pick up a passenger. Incidental transit is transportation provided by human service organizations for their clients, such as van services that transport clients to and from an adult day care center.

A significant amount of capital equipment costs, operational expenses, and administrative costs of public and private non-profit senior transportation providers are covered by the Federal Transit Administration (FTA). A second important source of funds is the Department of Health and Human Services (DHHS) that administers funding provided by the Older Americans Act (OAA), Title XIX of the Social Security Act (1965), and the Community Services Block Grant (Wacker, Roberto, & Piper, 1998). Many state and local agencies receive their transportation funding through these programs.

The costs to the user of senior transportation services vary, depending on the services used. For example, door-to-door transportation services may cost up to twice the regular fare, while off-peak travel using regular public transportation costs half the usual fare. If a client is going to a physician's visit and lacks individual transportation, Medicaid will often cover the costs of transportation to the selected provider. Human service organizations often charge a nominal fee for the use of their transportation services.

Accessing Services

Procedures for assisting clients in accessing transportation services are as varied as the systems in place. Typically, the process begins with assessment of the client's individual needs. Does the client have specific physical or cognitive limitations that need to be addressed? What are the client's reasons for seeking senior transportation services? The practitioner should then educate the client about the available local transportation and senior transportation systems. Which service matches the client's individual transportation needs? Does the client fit the eligibility requirements? Typical eligibility requirements are age-related (e.g., 60 years and older) or certain levels of disability or chronic illness. The client may need assistance gathering the required documentation and completing an application. An eligibility letter from a physician might be required. It is helpful to rehearse with the client how to use the senior transportation system and provide a written list of steps: calling, setting up a pick-up time, using transportation passes, etc. This may be especially important for clients planning to use demand-responsive transportation, as these systems require the client to make a request in advance. Finally, if it appears necessary, the practitioner should arrange for someone to accompany the client the first time that senior transportation is used.

Recommendations

A U.S. Department of Transportation (DOT) report on transportation for an aging society recommended construction of safer highway systems; development of systems that aid in identifying and evaluating when driving becomes problematic or unsafe; performance-aiding technology like collision warning and avoidance systems; and the provision of non-driving senior transportation alternatives (U.S. Department of Transportation, 1997).

MATTHIAS J. NALEPPA

See Also
Americans with Disabilities Act

Internet Key Words
Demand-responsive System
Deviated-fixed-route System

Fixed-route System
Incidental Transit
Paratransit

Internet Resources

The U.S. Department of Transportation website offers a wide range of resources, including information on transportation needs of older adults.
http://www.nhtsa.dot.gov/people

The U.S. Administration on Aging website provides background information about the impact
of increasing longevity on transportation needs
of older adults.
*http://www.aoadhhs.gov/Factsheets/
transportation.html*

TRAUMA
See
Injury and Trauma

TUBE FEEDING
See
Percutaneous Endoscopic Gastrostomy (PEG)
Tube Feeding

TUBERCULOSIS

Epidemiology

Tuberculosis (TB) is the most common infectious
agent causing death. In the late 1980s, the United
States witnessed a dramatic reversal of a decades
long trend towards fewer cases of tuberculosis. Factors that influenced a resurgence of the disease
included trends in immigration and HIV infection,
and a deteriorating public health infrastructure
(Brudney & Dobkin, 1991). These events resulted
in an excess of more than 50,000 TB cases than
would have been predicted by the trends of the
prior 30 years.

Most troubling among the events of the 1990s
was the rapid rise in the levels of multi-drug resistance (MDR) among strains of tuberculosis (Dooley
et al., 1992). MDR strains have been identified in
virtually every state in the country but its prevalence
has been greatest in regions with high endemic
rates of TB and HIV infection. The loss of immune
function associated with HIV infection creates a
greater predisposition to the development of active
tuberculosis at a relatively early stage of HIV infection. This predisposition is a more potent promoter
of TB than any previously identified risk factor.

Persons over age 65 represent a substantial
proportion of active TB cases in this country each
year. In 1998, there were 18,361 cases of TB. Those
age 65 and older accounted for nearly 3,000 cases
(17%). Only the 25–44 year old age group accounted for more cases—a difference of only 200
cases. Those 25–44 year olds have the highest rates
of HIV infection and were disproportionately affected by TB during its resurgence in the 1980s
and 1990s. Case rates for those 65 and older are
higher than any other segment of the population.

Foreign born cases of tuberculosis have increased annually in the United States since the early
1980s. In 1999, 43% of all TB cases occurred
among foreign born persons. The annual incidence
of foreign born TB disease has increased on a percentage basis as native born disease has declined
in the U.S. but the actual numbers of foreign born
cases have increased as well. Elderly foreign born
persons are particularly susceptible to reactivation
TB as they assimilate a new culture and the stresses
entailed in that process.

Tuberculosis in the Elderly

Tuberculous disease among older persons is generally assumed to represent reactivation of latent foci
of infection acquired through exposure to an active
case at some prior time. The relatively constant
occurrence of disease in the elderly population as
opposed to the upsurge that occurred in 25 to 44
year olds supports this hypothesis. Our understanding of latent tuberculosis infection is limited but

many elderly develop active disease years after their initial exposure. Alterations in immune function, such as the development of diabetes mellitus, treatment with immunosuppressive medications, the occurrence of neoplastic disease, and involuntary weight loss, all of which occur in the elderly, predispose to reactivation of latent infection and the progression to active tuberculosis.

Diagnosis

The constitutional and respiratory symptoms of tuberculosis are sufficiently nonspecific to confuse detection. The clinician must maintain a level of suspicion for the diagnosis or it will be overlooked among other possible disease states in the elderly. Tuberculosis is often mistaken for community-acquired pneumonias or malignancies. Most episodes of unprotected exposure to tuberculosis result from exposure to patients admitted with a suspicion of lung cancer who are subsequently diagnosed with tuberculosis. Tuberculosis has also been misdiagnosed as chronic pneumonias such as actinomycosis, histoplasmosis, non-tuberculous mycobacterial infections, nocardia, cryptococcosis, and coccidioidomycosis. Central nervous system tuberculosis can be mistaken for dementia although its progression is much more rapid than typical dementing diseases.

All persons with chronic respiratory complaints of cough (greater than two weeks duration), weight loss, fever, loss of appetite, and night sweats should be considered TB suspects while further evaluation is undertaken. Maintaining an appropriate level of suspicion for clinical tuberculosis is the key to avoiding unnecessary exposures, particularly in an institutional setting.

Clinicians are required to report all suspect or proven cases of tuberculosis to their local health department. The diagnosis of tuberculosis infection still relies on tuberculin skin testing (purified protein derivative–PPD). A positive tuberculin skin test indicates prior exposure to Mycobacterium tuberculosis. The infected state is not well understood. In this state the organism is harbored in the host but does not cause tissue damage and is not contagious. At a later time the organism may reactivate and progress to active tuberculosis disease usually involving the lungs, but in up to 20% of patients an extrapulmonary site will be the only focus of disease. The greatest risk of developing active tuberculosis is during the first one to two years following infection when 5% of patients will develop active TB disease. Another 5% of patients will develop active disease at a later time.

The tools for diagnosing active disease include chest radiography, sputum acid fast bacillus (AFB) smears, and mycobacterial cultures. Culture remains the gold standard for demonstrating active tuberculosis. Current methods of liquid culture media have reduced the time for isolation of Mycobacterium tuberculosis by one to two weeks. Newer technologies based on DNA amplification can predict with a high degree of precision which AFB smear positive specimens will be culture positive for Mycobacterium tuberculosis. DNA amplification of smear negative specimens is less reliable in predicting culture positivity for TB. Decisions regarding hospitalization, isolation, initiation of therapy, and investigation of new contacts should be based on clinical and radiographic grounds rather than DNA amplification. Sputum acid fast smears are positive in the majority of cases, but when TB is suspected at an extrapulmonary site both smears and cultures are positive in less than half the cases due to the smaller burden of organisms at the site.

Seventy to eighty percent of active tuberculosis cases involve the lungs. Adult reactivation disease usually involves the upper lung zones whereas primary disease may present with lower lung field abnormalities on the radiograph as well as mediastinal lymphadenopathy. Many patients are left with scarring at the site of the original radiographic abnormality following the completion of a treatment course.

All adults should have a baseline PPD performed as part of a general health assessment. Subsequent tuberculin testing in adults should be based on occupational requirements or following exposures to contagious cases of TB. Nursing homes residents should be screened for tuberculosis with a tuberculin skin test at the time of admission. Those who have converted from a negative to a positive result should receive preventive therapy without regard to age. Further evaluation of tuberculin reactors requires a chest x-ray and, if abnormal, investigation for active disease through the procurement of sputum or tissue specimens.

Assessment of the full risk of developing active disease is dependent on the presence of underlying conditions that promote progression from infection to primary disease. HIV infected persons are the most susceptible hosts with a 200-fold greater risk for active disease than immunologically intact hosts (Johnson et al., 1998). Conditions that promote development of active TB disease may also inhibit the cutaneous T-cell response necessary to mount a tuberculin reaction. Therefore, clinical suspicion, judgment, and the appearance of the chest radiograph should supersede the tuberculin skin test as a guide to the evaluation of TB suspects.

Vaccination

Many immigrants to the United States were vaccinated in their country of origin with Bacillus-Calmette-Guerin (BCG), a live attenuated strain of M. bovis. BCG is effective in ameliorating some of the more severe forms of tuberculosis, particularly in children. Its benefits include reduction in the incidence of miliary disease and TB meningitis. However, BCG does not reduce the incidence of adult reactivation tuberculosis. The dilemma of interpreting tuberculin skin tests following BCG vaccination is that vaccinees often originate in areas highly endemic for tuberculosis and/or they cannot recall if they were vaccinated. This raises the question of whether the cutaneous response to the PPD is due to a past exposure to TB or to cross reactivity between BCG and PPD.

The results of tuberculin skin testing in BCG vaccinees should be interpreted without regard to BCG status. BCG reactions are typically smaller than those associated with TB infection and the size of the BCG response wanes over time. Since BCG is usually administered in childhood, the response to a tuberculin skin test induced by BCG should have diminished by adulthood.

Adult tuberculin reactions should be interpreted according to Centers for Disease Control and American Thoracic Society guidelines and treated accordingly (American Thoracic Society, 1994). Those persons with 5 millimeter reactions who are household contacts of cases, HIV infected, or have abnormal x-rays should receive treatment for latent infection. Persons with underlying medical conditions such as heart and lung disease, diabetes, and organ failure should receive preventive therapy for 10 mm tuberculin reactions. Those with 15 mm reactions and no risk factors for TB exposure and disease require preventive therapy because a tuberculin reaction of that dimension is unlikely to be due to any factor other than TB infection. Smaller reactions to PPD can be due to exposure to nontuberculous mycobacteria, however preventive therapy is indicated for smaller reactions when the patient is at particular risk for developing active disease (e.g., due to HIV infection or diabetes mellitus). The larger the tuberculin reaction, the more likely it is due to exposure to Mycobacterium tuberculosis rather than BCG or a non-tuberculous infection.

Treatment

Current regimens for the treatment of latent tuberculous infection include isoniazid for 6 or 9 months or the newly recommended regimen of rifampin and pyrazinamide for 2 months. The 2-month regimen is as effective as 12 months of isoniazid in HIV infected patients for the treatment of latent infection (Halsey, 1998) but the 2-month regimen is likely to be associated with a higher incidence of medication side effects.

At least two active drugs must always be administered for the treatment of active TB disease. Patients with TB that is fully susceptible to all of the first line drugs should be treated for six months. Initial therapy for active tuberculosis requires isoniazid, rifampin, pyrazinamide, and ethambutol for the first two months followed by isoniazid and rifampin for an additional four months. The rationale for the polypharmacy therapeutic approach is that the level of isoniazid resistance is sufficiently high in most areas that four drugs are necessary at the outset of treatment to insure that at least two drugs will be active. Preventive therapy and the treatment of active disease can be administered either daily or on an intermittent basis twice or three times weekly. Appropriate doses are indicated in the ATS/CDC treatment guidelines (American Thoracic Society, 1994).

Patients should be monitored during therapy with weekly sputum specimens until the sputum has been cleared of acid fast organisms. Thereafter, monthly sputum specimens should be monitored for continued response to treatment. A follow-up chest x-ray film should be obtained 2–3 months into treatment to document radiographic improvement. At the completion of therapy, an end of therapy sputum culture and chest x-ray film should be obtained. Ninety-eight percent of patients completing a six-month treatment regimen will be cured.

Adverse Drug Reactions and Drug-Drug Interactions

TB treatment is a minefield of adverse drug reactions and drug-drug interactions. Some of these problems are particularly germane to the elderly. Isoniazid and rifampin both induce hepatocellular injury. Although the frequency and severity of isoniazid induced hepatitis is a feared complication it is a relatively infrequent occurrence. Many patients develop some liver enzyme abnormality but it is without clinical consequences. Isoniazid induced hepatitis is age-related and increases significantly after the age of 35. Gout attacks may be induced by pyrazinamide, which blocks uric acid excretion. Ethambutol can precipitate optic neuritis or changes in color vision, but this is rare.

The entire daily dose of all medications should be taken at one time. Better drug absorption is achieved when administering isoniazid and rifampin on an empty stomach, however taking so many pills at one time may lead to the most common adverse drug reaction—gastrointestinal intolerance. Better tolerance may be achieved at the expense of serum drug levels by crushing tablets and capsules and mixing them in juices or soft foods (applesauce, puddings, ice creams). Drug absorption may be delayed and peak concentrations diminished by administration of the medications with food. There are no specific foods that should be avoided while taking the antituberculous medications. If administration with food does not improve GI tolerance, it may be necessary to separate the medications and take each at a different time of day. This approach may diminish the level of adherence to the prescribed regimen, however.

Rifampin, the key agent in the treatment of tuberculosis, is a potent inducer of the Cytochrome p450 system, and as such can play havoc with the serum levels of other drugs metabolized through this pathway. The interactions between rifampin and other drugs are so numerous that drug-drug interactions should be considered whenever new medications are administered to a patient taking rifampin. The serum levels of theophylline, antiarrhythmics, oral contraceptives, antifungals, beta-blockers, corticosteroids, phenytoin, oral hypoglycemics, and warfarin may all be reduced by rifampin. Among persons with HIV infection, rifampin may induce hepatic cytochromes to the extent that the serum levels of the protease inhibitors are reduced. Subtherapeutic levels of the protease inhibitors may promote the emergence of resistant HIV in the patient. In patients taking protease inhibitors, rifabutin, another compound in the rifamycin class of drugs, should be substituted for rifampin. Rifabutin induces the cytochrome system to a lesser degree than rifampin and spares the protease inhibitors.

Implications for Public Health

In the United States, a substantial increase in federal support for local health departments to control TB and programs in which patients are directly observed while taking their medications greatly improved the rate of completed treatment courses and reduced the levels of resistance. U.S. cases of TB are again on the decline and have fallen each year since 1992.

Yet, despite the public health successes of the United States, reports from other countries are sobering. Tuberculosis is rampant in many areas. Eight million new cases of TB are diagnosed every year, mostly in Asia, Africa, and Latin America. Three million people were estimated to have died of tuberculosis in the year 2000, making TB the largest single cause of death from an infectious disease. While the caseload in the U.S. is of manageable proportion and declining, travel from areas of the world with high endemic rates of TB makes the specter of further epidemics and higher levels of drug resistance in this country a plausible realistic concern.

PATRICK J. BRENNAN
AMY MORGAN

See Also
Caloric Intake
Cancer Treatment
Cough
Diabetes: Overview
Health Promotion Screening
Human Immunodeficiency Virus (HIV) and AIDS
Immigrant Elders
Infection Transmission in Institutions: Pneumoccoccal Pneumonia
Protein Energy Undernutrition

Internet Key Words
HIV Infection
Immune Function
Multi-Drug Resistance (MDR)
Mycobacterium Tuberculosis
Neoplastic Disease
Purified Protein Derivatives (PPD)
Respiratory Infections
Tuberculosis (TB)

U

URINARY INCONTINENCE ASSESSMENT

Urinary incontinence (UI) is a common symptom in older adults that is often curable if appropriately identified, assessed, and treated (Kane, Ouslander, & Abrass, 1999; Wyman, 1999). Seventeen to fifty-five percent of older women and 11% to 34% of older men experience UI (Thom, 1998). The physical, psychological, social, and economic impacts of UI are significant. UI is associated with physical problems, such as urinary tract infections, pressure ulcers, rashes, and falls. These problems can be compounded by psychosocial consequences such as embarrassment, social isolation, depression, and diminished quality of life. Caregiver stress increases significantly when the care recipient becomes incontinent. Therefore, it is not surprising that UI is a common precipitant of nursing home admission. Financially, the annual cost of incontinence care has been estimated at $11.2 billion for community-dwelling people of all ages and $5.2 billion for nursing home residents (Hu, 1994).

Identifying Urinary Incontinence

UI is too frequently neglected. Patients often delay or do not seek professional treatment for UI because they are embarrassed to talk about it, erroneously believe that UI is part of the normal aging process, or that it is not treatable. Additionally, health care providers do not typically ask patients if they are incontinent.

Because many people do not seek care for UI, direct care providers from any discipline should question their patients/clients and refer them to their primary care physician or nurse practitioner for UI assessment. Some simple questions to ask might be

- "Are you having any problems with your bladder"

- "Do you have trouble holding your urine?"
- "Do you lose urine when you don't want to?"
- "Do you ever wear a pad or other protective device to collect your urine?"
- "Do you ever lose urine when you cough or laugh, or on the way to the bathroom?"

By asking these questions, care providers can identify incontinent older adults. Referral for assessment and treatment will help prevent patients from unnecessarily suffering the costly distress of UI.

An interdisciplinary approach to UI is ideal because the factors that contribute to UI and its therapies may involve nursing, medicine, psychology, enterostomal therapy, physical and occupational therapies, and others. While treatment and management of urinary incontinence is discussed elsewhere, clinicians should be aware of the essential components of the UI assessment that can aid in diagnosing, treating, and managing the syndrome.

Assessment

The importance of conducting the basic assessment of UI cannot be overstated; *assessment is critical to determining optimal treatment and management of UI.* Assessment is brief, simple, and makes minimal demands on the patient. The basic assessment can be done in an outpatient setting by primary care providers, usually nurse practitioners or physicians, and is a Medicare reimbursable service. Typically, the patient is mailed or given an incontinence monitoring record to complete and a subsequent appointment is scheduled for the UI assessment. If the patient is cognitively impaired, a caregiver is required to participate in the assessment and management plan.

Basic assessment of UI includes a history, physical examination, incontinence monitoring record, stress test, post-void residual volume, and urinalysis (Agency for Health Care Policy and Research [AHCPR], 1996). This process can lead to a diagnosis of the specific type of UI, an indication that additional studies are needed, and implementation of appropriate therapy.

The goal of therapy is to attain maximal dryness and ameliorate symptoms to the patient's satisfaction. It is important to identify the symptom that is most bothersome to the patient and tailor treatment to relieve that symptom. For example, the patient's main concern may be nighttime wetness (though infrequent) that interferes with sleep, while the clinician might feel that patient's more severe daytime incontinence is the more glaring symptom to address. In this case, nocturnal incontinence is the patient's priority and should be the focus of initial treatment.

History: The patient interview should consist of a detailed characterization of the UI and a focused medical, psychologic, neurologic, and genitourinary history. This includes an assessment of medications, functional status, social issues, environmental factors, and other UI risk factors (see DIAPPERS below).

Eliminating risk factors of acute UI is essential to achieving and maintaining continence. The acronym DIAPPERS helps clinicians remember the reversible causes of acute UI: *D*elirium, *I*nfection of the urinary tract, *A*trophic vaginitis or urethritis, *P*harmaceuticals, *P*sychological causes, *E*xcess fluid, *R*estricted mobility, *S*tool Impaction (AHCPR, 1992).

Essential components of the clinical history are: the patient's perception of the most bothersome symptom, the patient's treatment preferences, motivation, and expectations for outcomes of treatment. The patient should be asked about the duration and characteristics (stress, urge, dribbling, etc.) of UI, and previous treatments and their effects. History taking must include frequency of UI, timing, amount of continent voids, and precipitants of incontinence (situational antecedents, cough, certain exercises, surgery, injury, previous pelvic radiation therapy, trauma, new onset of diseases, new medications). Asking the amount and the type of pad, brief, or other protective devices used can help quantify leakage. The patient should be assessed

for other urinary tract symptoms (nocturia, dysuria, hesitancy, interrupted stream, straining, hematuria) and for alteration in bowel habits or sexual function. The relationship between UI and intake of fluid, caffeine, alcohol, or medications that can affect continence status should be determined. Caffeine, alcohol, and diuretics cause polyuria, urgency, and frequency; sedative hypnotics (including alcohol) reduce awareness of the need to void and cause muscle relaxation. Anticholinergics cause sedation, rigidity, and immobility. Alpha- and beta-adrenergic agonists and calcium channel blockers decrease outlet resistance and stress incontinence. Narcotic analgesics can cause urinary retention, fecal impaction, sedation, and delirium.

Besides medical history, it is important to assess for cognitive, functional, psychological, social, and environmental factors that can influence UI and its management, in order to effectively treat the incontinent patient. Determining the patient's ability to ambulate, undress, and position oneself appropriately for voiding, and the ability to comprehend the signal to void or to trigger appropriate toileting behavior, are critical. Clothing should be evaluated for the need to be adapted with Velcro fasteners and elastic waistbands to facilitate speed and ease of undressing to void. Psychological factors should be considered, such as depression that can lead to apathy or lack of motivation to toilet. Social factors such as living arrangement and the availability of caregiver assistance must be assessed. Environmental assessment for factors that can affect continence can be valuable. For example, is the path to the bathroom well lit? Is the bathroom well marked? Is the distance to the bathroom manageable for the patient? Can the bathroom accommodate assistive devices, contain grab bars, and have a raised toilet seat if needed? Are the bed and chairs 16–18 inches high to facilitate ease in rising to stand before walking to the bathroom?

Incontinence Monitoring Record: The incontinence monitoring record provides valuable information to help determine the type of UI, possible cause of UI, and helps establish an appropriate management regimen. For 3–7 days, the patient or caregiver chronicles when voiding occurs, estimated volume (small or large), incontinence episodes (including leakage or dribbling), and associated events such as the presence of "urge," coughing, use of diuretic, requests for toileting assistance, bowel movements and if accompanied by straining or fecal incontinence, and fluid intake. The record can indicate which behavioral interventions would be effective and provides a baseline to evaluate treatment efficacy. A Sample Bladder Record for ambulatory patients is shown in Figure 1 (AHCPR, 1996), and a Sample Incontinence Monitoring Record for patients in the hospital or nursing home is shown in Figure 2 (Ouslander, Urman, & Uman, 1988; AHCPR, 1996). These records can be used or modified to suit the individual patient.

Physical Examination: The physical assessment includes general, abdominal, rectal, genital, and neurological examinations. The general exam focuses on conditions such as edema, neurological abnormalities, mobility, cognition, and manual dexterity related to toileting skills. Abdominal exam is done to identify bladder distention, organomegaly, suprapubic discomfort, masses, and signs of fluid collection or increased abdominal pressure. Rectal exam is done to note perineal sensation, sphincter tone, fecal impaction, rectal mass, or prostate abnormalities. Genital exam is done to identify skin abnormalities, atrophy, inflammation, masses, pelvic organ prolapse, and pelvic muscle tone. Neurological exam is done to detect focal abnormalities that may indicate stroke, multiple sclerosis, or spinal core compression

Stress Test: The patient is instructed to drink 32 oz. of fluid one hour prior to the appointment. If this is difficult for the patient, the clinician can measure the post-void residual (PVR) volume first, then keeping the catheter in place, fill the bladder with saline before performing the stress test. With a full bladder, the individual is asked to cough vigorously while the examiner observes for any urine loss. If immediate leakage of any urine occurs with the cough, then stress incontinence is likely; if leakage is delayed or persists after the cough, the patient probably has uninhibited bladder contractions. If this test is done with the patient lying on the exam table and there is no leakage, the test should be repeated in the standing position.

Post-Void Residual (PVR) Volume: The PVR is the volume of urine remaining in the bladder immediately after voiding. The patient is asked to void while being observed for hesitancy, straining, slow, or interrupted stream indicating obstruction, a contractility problem, or both. The PVR is determined by subsequent catheterization or by pelvic ultrasound. PVRs less than 50 mL are considered normal bladder emptying. Repetitive PVRs of 100–200 mL or higher are signs of inadequate bladder emptying. These patients need referral for more specialized evaluation.

Laboratory Tests: A clean catch urine specimen should be obtained for urinalysis (UA). The UA can detect pyuria and bacteriuria suggesting infection as a possible cause of UI. If infection is

NAME: _____

DATE: _____

INSTRUCTIONS: Place a check in the appropriate column next to the time you urinated in the toilet or when an incontinence episode occurred. Note the reason for the incontinence and describe your liquid intake (for example, coffee, water) and estimate the amount (for example, one cup).

Time interval	Urinated in toilet	Had a small incontinence episode	Had a large incontinence episode	Reason for incontinence episode	Type/amount of liquid intake
6–8 a.m.					
8–10 a.m.					
10–noon					
Noon–2 p.m.					
2–4 p.m.					
4–6 p.m.					
6–8 p.m.					
8–10 p.m.					
10–midnight					
Overnight					

No. of pads used today: _____ No. of episodes: _____

Comments: _____

_____

_____

FIGURE 1   Sample bladder record for ambulatory patients (AHCPR, 1996).

INCONTINENCE MONITORING RECORD

INSTRUCTIONS: EACH TIME THE PATIENT IS CHECKED:
1) Mark *one* of the shapes in the BLADDER section at the hour closest to the time the patient is checked.
2) Make an X in the BOWEL section if the patient has had an incontinent or normal bowel movement.

◻ = Incontinent, small amount   ∅ = Dry   X = Incontinent BOWEL

◻ = Incontinent, large amount   ⊿ = Voided correctly   X = Normal BOWEL

PATIENT NAME _____ ROOM # _____ DATE _____

| | **BLADDER** | | | | **BOWEL** | | | |
	INCONTINENT OF URINE		DRY	VOIDED CORRECTLY	INCONTINENT X	NORMAL X	INITIALS	COMMENTS
12 am	◻	◻	◯	△ cc___				
1	◻	◻	◯	△ cc___				
2	◻	◻	◯	△ cc___				
3	◻	◻	◯	△ cc___				
4	◻	◻	◯	△ cc___				
5	◻	◻	◯	△ cc___				
6	◻	◻	◯	△ cc___				
7	◻	◻	◯	△ cc___				
8	◻	◻	◯	△ cc___				
9	◻	◻	◯	△ cc___				
10	◻	◻	◯	△ cc___				
11	◻	◻	◯	△ cc___				
12 pm	◻	◻	◯	△ cc___				
1	◻	◻	◯	△ cc___				
2	◻	◻	◯	△ cc___				
3	◻	◻	◯	△ cc___				
4	◻	◻	◯	△ cc___				
5	◻	◻	◯	△ cc___				
6	◻	◻	◯	△ cc___				
7	◻	◻	◯	△ cc___				
8	◻	◻	◯	△ cc___				
9	◻	◻	◯	△ cc___				
10	◻	◻	◯	△ cc___				
11	◻	◻	◯	△ cc___				
TOTALS:								

©1984

FIGURE 2 Sample incontinence monitoring record for patients in hospitals or nursing homes (AHCPR, 1996; Ouslander, Urman, & Uman, 1988).

suspected, urine culture and sensitivity should be done to specify bacteria and drug sensitivity. Hematuria without bacteruria suggests cancer, stone, or infection; glycosuria indicates uncontrolled diabetes; and proteinuria indicates renal disease. Blood tests including blood urea nitrogen, creatinine, glucose, and calcium are suggested for patients with diminished renal function or polyuria.

Specialized Tests: Referrals for specialized tests are recommended for patients who fail to respond to treatment, are considering surgical intervention, have hematuria without infection, comorbid conditions such as recurrent urinary tract infection, persistent difficulty with bladder emptying, previous anti-incontinence or radical pelvic surgery, symptomatic genital prolapse, prostate abnormalities suggestive of cancer, abnormal PVR, and neurological conditions. Specialized tests include urodynamic evaluation to determine the anatomical and functional status of the bladder and the urethra, cystoscopy to visualize the bladder and urethra, and imaging tests such as ultrasonic and fluoroscopic studies.

Accurate identification and assessment of incontinent older adults is critical to successfully treating them. The assessment leads to a specific diagnosis and treatment plan, which can cure or improve the patient's UI.

LYNNE MORISHITA

See Also
Atrophic Vaginitis
Fecal Incontinence
Sexual Health
Urinary Incontinence Treatment
Urinary Tract Infection

Internet Key Word
Urinary Incontinence

Internet Resources
American Foundation for Urologic Disease
http://www.afud.org/

International Continence Society

http://www.klinikum.rwth-aachen.de/webpages/urologie/info.html

National Association for Continence
http://www.sfcs.org.sg/continet/resource/res005.html

Wound, Ostomy and Continence Nurses Society
http://www.wocn.org/

URINARY INCONTINENCE TREATMENT

Urinary incontinence, or involuntary loss of urine, occurs in 15% to 35% of elderly persons living at home, and among 40% to 50% of elderly persons living in nursing homes (Fantl et al., 1996; Steeman & Defever, 1998). For elderly women living at home the prevalence rates range between 12% and 49% and are approximately twice as high as for men (7% and 22%) (Steeman & Defever, 1998).

Diagnosis

Urge incontinence is associated with an abrupt and strong desire to void, often related to local irritants such as coffee, infection, tumors, or to neurological disorders such as stroke or multiple sclerosis. Volume loss varies, depending on the amount of urine in the bladder when it contracts. Stress incontinence is leakage of small amounts of urine during activities that increase intra-abdominal pressure, such as coughing and sneezing. It is caused by weakness of the pelvic muscles after childbirth, for example, or by weakness of the urethral sphincter after prostatectomy. Mixed incontinence is a combination of urge and stress incontinence, and is common among elderly persons. Overflow incontinence is a result of overdistention of the bladder due to bladder hypotonia, which may be caused by the adverse effect of certain medications, for example, or may be due to urethral obstruction that can be associated with fecal impaction. It may present as constant dribbling, but urge and stress symptoms may occur as well. Reflex incontinence occurs without warning or sensory awareness, and may be seen in spinal cord injury. Functional incontinence is due to factors outside the urinary tract such as (other) physical, cognitive, and perceptual deficits.

Retention: Retention can be an acute condition or one that develops slowly. Inability to urinate is not uncommon in the elderly and is associated with prostatic hypertrophy in men, uterine prolapse in women, and fecal impaction in both. A patient unable to void has a palpable suprapubic mass with tenderness or discomfort; there may be some dribbling urination. Medications causing retention include amphetamines, opioids, anticholinergic agents, antihistamines, antiparkinson's drugs (e.g., Levodopa), calcium antagonists, ganglion blockers, muscle relaxants, tricyclic antidepressants, estrogen combinations, alpha-adrenergic agonists, and beta-adrenergic blocking agents. Treatment includes urethral dilatation, local application of estrogens, medications, surgery, or permanent placement of a drainage tube (cystotomy or nephrostomy), depending on the cause. Uncorrectable retention is managed with intermittent catheterization.

Although the elderly are more prone to the development of incontinence, it should not be considered a normal part of aging. Basic evaluation of urinary incontinence requires a focused medical, neurological, and genitourinary history and physical exam, direct observation of urine loss (cough stress test), estimation of post-void residual volume, and urinalysis to determine the type, cause, and if it is a permanent or a transient and reversible condition (Fantl et al., 1996). A symptoms history includes documentation of the frequency, volume, timing, precipitants, and circumstances of incontinence; duration of incontinence; daily voiding habits; fluid intake pattern; bowel pattern and history of fecal impaction; medications, mental status, mobility, sensorium, living environment and social factors; previous incontinence management/treatment and results; and most bothersome symptoms to the patient.

Treatment

Behavioral techniques, pharmacotherapy and surgical procedures are the major treatment categories for urinary incontinence in the elderly (Fantl et al., 1996).

Behavioral Techniques: Behavioral techniques entail teaching particular behaviors to enhance bladder or sphincter control. Habit training, prompted voiding, and bladder retraining are based on scheduled toileting with varying degrees of active participation by the older person. Habit training and prompted voiding may yield improvement in mild to moderate incontinence even among cognitively impaired elderly or when a short trial has been successful. Bladder retraining is primarily applicable to urge and mixed incontinence, but has demonstrated its usefulness for stress incontinence. Pelvic muscle exercises are indicated for stress incontinence, but may be effective for urge and mixed incontinence as well. In general, behavioral techniques yield only modest levels of cure of incontinence (<25%), but a marked level of improvement (>50%). The advantage of these techniques is the absence of side effects, but success is dependent on highly motivated, skilled, and adherent patients and caregivers on a long-term basis.

Pharmacotherapy: Pharmacotherapy for urinary incontinence relies on two major groups of medications: one aiming at controlling involuntary bladder contraction (urge and reflex incontinence), such as Oxybutynin, and the other targeted at increasing strength of urethral closure (stress incontinence), such as phenylpropanolamine. This medication should be combined with behavioral techniques. Oxybutynin treatment requires slow upward titration and monitoring for residual urine (Thüroff et al., 1998). Pharmacotherapy is easier to comply with, yet may have considerable side effects. Improvement rates are considerable, yet the cure rates are modest; high cure rates are usually accompanied by high side-effect rates.

Surgical Procedures: Surgical procedures, the most invasive treatment modality, are used to alleviate urethral obstruction, increase urethral resistance, or correct the anatomical position of the urethra. In general, surgery yields high cure rates (>70%), but complications are common. In addition to general risks related to any surgical procedure, complications can be severe and irreversible, or require further invasive treatment. The most common complications are urinary retention among women and stress incontinence among men. Recent less invasive surgical techniques, such as periurethral injections with bulking agents and endoscopic bladder-neck suspension, are promising for treatment of stress incontinence in the elderly (Stanton, 1998).

Incontinence Aids: Incontinence aids, including absorbent, drainage, and occlusive devices should be able to preserve hygienic, mental, and social comfort, and permit the incontinent elder to have as normal a life as possible. Such aids are important for those incontinent elderly for whom total cure is not attainable, or may be needed during treatment as a provisional measure until continence is achieved.

Absorbent Devices: *Absorbent devices* is an umbrella term for a variety of pads with different absorbent capacities that are disposable or reusable (e.g., cloth pads). Caregivers should assist the incontinent person to choose the most suitable absorbent product and regularly review patient satisfaction. In addition to economic and ecologic considerations, the choice should be primarily based on quality-of-life considerations that include availability of the product, ease-of-handling, wearing comfort, maximum absorbency, odor containment, skin protection, and ease of product disposal or laundering (Füsgen et al., 1998).

Drainage Devices: Drainage devices include catheters for transurethral, suprapubic or intermittent use, and external collecting devices. Catheterization is indicated only for patients with release problems that are not amenable to other interventions, such as urinary retention and overflow incontinence. Clean intermittent catheterization, a self-performed technique that can be learned even by very old people, has the advantage of independent control over the bladder, and mitigates the need for bulky pads or an indwelling catheter (Füsgen et al., 1998). Transurethral catheters for longterm use are associated with infection, blockage due to encrustation, and injury to the urethra (Füsgen et al., 1998). A suprapubic catheter is a better alternative for long-term use if intermittent catheterization is not a viable option. External collecting devices such as a condom-catheter may be an alternative

to pads, but require careful adhesion, adequate hygiene, skin protection, and at least once daily change to prevent infections, skin breakdown, and ischemic complications.

Occlusive Devices: Occlusive devices, for the alleviation of stress incontinence, such as a penile clamp or a urethral plug, are used infrequently.

Other Care Strategies

A comprehensive patient-training program on how to use an appliance and regular evaluation is advised for all incontinence devices. Other general care strategies for prevention and management of urinary incontinence rely, first of all, on a positive attitude of all concerned and active involvement of the elderly person and/or the primary caregiver.

An individually tailored program should take into account the elder's history of successful incontinence-management strategies and consider cognitive ability, motivation (history of successful undertakings), mobility, manual dexterity, psychological factors, the environment, and compliance (Füsgen et al., 1998). Caregivers can facilitate continence by adjusting the environment (e.g., good lighting and signage, commodes or urinals within reach, grab bars, raised toilet seats, Velcro closures of clothing) and by enhancing mobility to improve access to the toilet. Equally important is an individualized toileting schedule and prompt response to calls for assistance.

Continence may be further promoted by educating the elderly person or caregivers about fluid and dietary management. Dietary caffeine (e.g., caffeinated coffee, tea, cola beverages, and chocolate) should be restricted or eliminated. Adequate fluid and fiber intake is necessary to prevent urinary infection and fecal impaction that may both cause or exacerbate urinary incontinence. Most fluids should be consumed during day-time and restricted after dinner to minimize night time incontinence. Strategies for management of nocturia for persons with edema of the lower extremities include leg elevation in the afternoon to promote natural diuresis during daytime (Fantl et al., 1992).

In general, care for elderly persons with urinary incontinence should aim toward reducing or eliminating incontinent episodes and at promoting physical, social, and mental comfort. Elders should not have recurrent bladder infections or skin breakdown, become self-isolating, or feel ashamed because of their incontinence.

ELS STEEMAN
MIA DEFEVER

See Also
Bowel Function
Fecal Incontinence
Urinary Incontinence Assessment
Urinary Tract Infection

Internet Key Words
Continence Devices
General Care Strategies
Retention
Treatment
Urinary Incontinence

Internet Resources
International Continence Society
http://www.continet.org.sg

National Association for Continence
http://www.nafc.org

Tower Urology Institute for Continence
http://towerurology.com

URINARY TRACT INFECTION

Urinary tract infection is one of the most common infections in the elderly. In view of increasing life expectancy, understanding the significance and appropriate management of urinary tract infections (UTI) in this expanding elderly population is becoming increasingly important.

The prevalence of UTI in the elderly is much higher than in younger populations. The causes for heightened susceptibility to UTIs include age-related changes such as immune dysregulation and endocrine changes such as estrogen deficiency in the elderly woman and a decrease in the bactericidal activity of prostatic fluid in the elderly man. Elderly patients more frequently have external and internal drainage devices that promote colonization of the urinary tract and, hence, UTIs (Nicolle, 1994). There is also an increased risk of urinary infection due to fecal and urinary incontinence, urethral instrumentation, and catheterization in this population group. Associated diseases such as diabetes, neurologic disorders accompanied by a neurogenic bladder, and prostatic hypertrophy are major contributing factors. In addition, modern medicine's gift of longer life spans to the elderly and patients with metabolic, neoplastic, or immune-deficiency disorders places them at increased risk for infection.

Diagnosis

Most elderly patients have asymptomatic bacteriuria that is frequently persistent or recurrent. Many patients will have documented persistent bacteriuria for years, often associated with pyuria. It has been suggested that the presence or absence of pyuria can be used to discriminate between colonization and infection of the urinary tract and thus identify patients who need treatment. This criterion cannot be used in the bacteriuric elderly, however, where virtually all persons have pyuria. Nevertheless, the absence of pyuria is a good predictor of the absence of bacteriuria. A subgroup of elderly patients have voiding abnormalities that are managed with long-term use of indwelling catheters. The prevalence of bacteriuria among nursing home residents with catheters approaches 100 percent. Long-term indwelling catheter use is also associated with an increased likelihood of invasive infection, secondary to mucosal trauma from the catheter.

UTI can present with irritative symptoms (dysuria, urgency, frequency), alterations in continence status, and acute change in cognitive status. Invasive infections present as pyelonephritis or bacteremia with potential evolution to septic shock. Metastatic infection following bacteremia may occur, leading to infections of prosthetic devices or endocarditis. In the absence of obstruction or associated renal disease, a UTI cannot cause a decline in renal function.

The bacteriologic features of UTI in the elderly differ from those in the young. E. coli still is the most common infecting organism, but its frequency relative to other organisms is reduced in the elderly compared to the younger population (where it is isolated in about 85% of infections). Bacteriuria due to gram-positive bacteria is much more common in elderly men than elderly women. The shift in uropathogens and the high frequency of mixed infections and of antimicrobacterial resistance are due in large part to the high frequency of institutionalization, hospitalization, catheterization, and antimicrobacterial usage in the elderly population.

Treatment

Prospective randomized comparative trials of therapy and no-therapy in institutionalized elderly with asymptomatic bacteriuria have not shown a decrease in morbidity or mortality due to UTI in those receiving antibiotic therapy. Few studies examine treatment of asymptomatic bacteriuria in the non-institutionalized elderly population, but it seems doubtful that measurable benefits would be achieved by a screening and treatment approach in this group.

Urinary tract infections in the presence of structural urinary tract abnormalities or systemic conditions (e.g., diabetes) are clinically significant and should be treated promptly. In addition, urinary infections caused by urea-splitting bacteria (which cause formation of infected stones) may lead in the long run to severe renal damage and should therefore also be treated.

There is no question that symptomatic urinary infection should be treated. The choice of antimicrobial agent is similar to that for a younger population; that is, it should be an antimicrobial to which the infecting organism is sensitive and that is tolerated by the patient. Agents used for first-line therapy are those for which there is the greatest experi-ence, which have been shown to be highly effective, and that are least expensive. Because of the high rate of recurrent infection, the use of post-therapy cultures is questionable. Bear in mind that the elderly population is more susceptible to the toxic and adverse effects of antimicrobial agents and that the safety margin between therapeutic and toxic doses is significantly narrowed.

Antibiotic therapy for an institutionalized patient should be used only if there are clear indications because institutionalization provides the potential for cross-contamination and facilitates spread of organisms among patients (Cefalu & Agcaoili, 1998). Thus, antimicrobial resistance may emerge rapidly in an individual and disseminate throughout the institution, subsequently complicating therapy of infections in general. When it is not certain whether a recent change in symptoms is attributable to urinary infection, it may be appropriate to consider a trial of antibiotic therapy. If there are no alterations in symptoms after a therapeutic course of 7 days, then antibiotic therapy should be stopped.

The optimal duration of therapy for the elderly patient has not been established. Available data suggest that at least 3 to 7 days of therapy should be given to treat urinary infection in elderly women with irritative lower tract symptoms (Schaeffer, 1998). If symptoms suggest an invasive infection or pyelonephritis, a 2-week course of therapy is appropriate. For men, short-course therapy is not effective, so a 2-week therapy should be given as initial treatment.

There are only two indications for the use of long-term therapy. The first is in elderly woman with recurrent irritative lower tract symptoms associated with recurrent urinary tract infections. The second indication is in elderly patients with genitourinary abnormalities that may not always be curable by surgery. If these patients experience recurrent symptomatic infections, long-term suppressive therapy is indicated.

Prevention

Prevention of urinary tract infection in the elderly must be the primary goal. Adequate fluid intake and achieving a good diuresis in this population can be an on-going challenge. Cranberry juice reduces the risk of bacteriuria and pyuria in elderly women (Avorn et al., 1994). Local factors, such as atrophic vaginitis that leads to colonization of the meatal region with coliform bacteria, may be treated with estrogens. Underlying disease, such as benign prostatic hyperplasia with post void residual, should be addressed properly. Antibiotics alone are seldom the right or only treatment for this complex problem.

BEN VAN CLEYENBREUGEL
LUC BAERT
DIRK DE RIDDER

See Also
Dehydration
Urinary Incontinence
Urinary Incontinence Assessment

Internet Key Words
Antibiotics
Bacteriuria
Prevention
Pyuria
Urinary Tract Infection

Internet Resource
U.S. National Institute of Diabetes and Digestive and Kidney Diseases (NIDDK)
http://www.NIDDK.NIH.gov/health/urolog/ (8 nowember 1999).

UTERINE, CYSTOCELE/ RECTOCELE, AND RECTAL PROLAPSE

Pelvic floor relaxation is common in the elder population, and results in cystocele, rectocele, uterine prolapse, and rectal prolapse. These pelvic floor changes are more common in females. Often women cease gynecological care after childbearing years, making it important for health care providers to be familiar with these conditions.

Diagnosis

Cystocle and rectocele occur when the vaginal wall weakens. Patients may complain of pressure, incomplete voiding, or urinary incontinence. During pelvic examination, with the patient bearing down, an anterior vaginal wall bulge will be visualized with cystocele; rectocele appears as bulging of the posterior vaginal wall. Some patients may need to be examined in a standing/squatting position. Cystocele and rectocele can occur with or without uterine prolapse.

Uterine prolapse occurs when the supporting structures no longer maintain anatomical integrity and the uterus descends into the vagina. Risk factors include multiple childbirths, large babies, advancing age, and Caucasian race. Increased intra-abdominal pressure is caused by obesity, abdominal and pelvic tumors, chronic pulmonary diseases, chronic constipation, connective and neurogenic disorders, malignancy, or occupations requiring heavy lifting that place women at risk. Signs and symptoms include pelvic pressure, low back pain, vaginal bulging/pressure, dyspareunia, and difficulty with urination/defecation. Physical exam reveals the cervix or uterus prolapsing toward the introitus or beyond. Examination while coughing and/or bearing down or in a standing position may be necessary for mild prolapses to be observed. Varying degrees of severity of uterine prolapse (Hall et al., 1996) include: (I) prolapse to the ischial spine; (II) prolapse to the introitus; (III) prolapse to just beyond the introitus; and (IV) complete uterine and vaginal inversion involving bladder and bowel (Dambro, 1998). With uterine prolapse, the normal environment of the uterine lining becomes disrupted and tissue becomes dry and irritated. If the irritation worsens, the area is susceptible to abrasion, ulceration, and infection.

Treatment

Treatment is based on age, sexual activity, prolapse severity, and presence of pathology. Conservative treatment includes pelvic muscle exercises, i.e., the

Kegel exercise (Kegel, 1950), hormone therapy, weight loss, management of constipation, pessary placement, and avoidance of activity increasing intra-abdominal pressure.

Teaching patients to perform the pelvic muscle exercise (PME) correctly is challenging. Patients must be motivated to perform 30–80 PMEs a day. Teaching is best done during a pelvic examination; the clinician should ask the patient to contract the pelvic floor muscles ("squeeze the vagina muscles") around the gloved examination finger while assessing for proper technique and strength of muscle contraction. Often patients squeeze the wrong muscles (e.g., abdominal or gluteal). Biofeedback therapy is beneficial for the correct training of pelvic musculature. Vaginal weighted cones also help with correct performance of the pelvic exercise, but have been used mostly with premenopausal populations (U.S. Agency for Health Care Policy and Research, 1996).

Pessaries maintain anatomical position of the uterus mechanically and may be an appropriately conservative treatment option for some women. Pessaries are classified as support (ring) or space filling (cube). While the more popular pessary is the support version, consensus as to which one is more effective is lacking among urogynecologists. Patients who are either unfit for surgery, awaiting surgery, or refuse surgery should be considered candidates for pessary use. Other considerations include hormone status, sexual activity, hysterectomy, pubococcygeal strength, cognitive and functional status, and stage of prolapse. Pessary use has risks. Improper care may lead to vaginal wall ulceration, fistula formation, or bowel herniation. Patients and/or caregivers require instruction on proper pessary care, i.e., daily removal and cleansing with mild soap and water. Some experts suggest that a pessary may remain in place a maximum of 6–8 weeks with well-estrogenized vaginal tissue. Biannual gynecological examinations are recommended for preventive care. Evaluation for pessary replacement should occur at 12–18 month intervals. Maintaining a well-estrogenized vaginal environment either with oral or vaginal estrogen promotes healthy tissue. Long-term care residents with cogni-

tive decline are a high-risk population for pessary use. Health care for this population must be especially vigilant (Cundiff & Addison, 1998; Roberge, McCandish, & Dorfsman, 1999; Ryan, 1999).

Surgical repair for pelvic floor relaxation includes hysterectomy, and cystocele/rectocele repairs. Colpocleisis or vaginal obliteration may be a treatment option for uterine prolapse and cystocele for non-sexually active, frail older women. Other surgical procedures are sacrospinous fixation, iliococcygeus suspension, uterosacral suspension, or abdominal sacral colpopexy. Preoperatively, many tests may be performed, such as defecograpy, fluoroscopy, cystoscopy, or urodynamics. These diagnostic tests aid in confirming diagnosis and may reveal unanticipated anatomical abnormalities to be considered during a surgical procedure. Defecograpy views the anorectum anatomy and fluoroscopy views pelvic floor structures in real time; contrast is used in both to identify anatomical structures. Cystoscopy and urodynamics provide information about the function of the bladder and urethra. Weidner (1998) provides a review of these diagnostic tests for pelvic floor evaluation. Postoperatively, the patient is instructed to avoid lifting anything heavier than ten pounds for a minimum of three months, and afterwards, anything greater than twenty pounds, possibly for life. Weight loss and management of medical conditions (i.e., chronic obstructive pulmonary disease) that increase intra-abdominal pressure aid in the treatment of prolapse. Pelvic muscle exercises may complement surgical outcomes (Cundiff & Addison, 1998).

Rectal prolapse, protrusion of the rectum through the anus, may be partial, involving only the rectal mucosa, or complete prolapse where rectal folds prolapse through the anus. Some rectal prolapses may not be seen on visual examination, requiring defecography for diagnosis. Risk factors for rectal prolapse include myelomeningocele, exostrophy of the bladder, cystic fibrosis, chronic constipation or diarrhea, imperforate anus, multiple sclerosis, stroke/paralysis, and dementia. Signs and symptoms may include anorectal pain or discomfort during defecation, feeling of incomplete evacuation, incontinence, rectal bleeding or discharge. Treatment during the acute phase requires manual reduction of the prolapse and management of diarrhea/constipation. Patient education is focused on avoidance of constipation or diarrhea. Surgical options for recurrent rectal prolapses include transabdominal or transrectal rectopexy (Dambro, 1998; Nagle, 1999).

Health care providers need to be more familiar with pelvic floor dysfunctions as the elder population increases. While not necessarily life threatening, these conditions are important to the quality of life. A generalist can easily teach the conservative measures discussed above while patients await the recommendations of specialists.

ANNEMARIE DOWLING-CASTRONOVO

See Also
Bowel Function
Fecal Incontinence
Social Isolation
Urinary Incontinence: Assessment

Internet Key Words
Cystocele
Kegel's Exercises
Pelvic Muscle Exercises
Pessary
Rectal Prolapse
Rectocele
Uterine Prolapse

Internet Resources
American Society of Colon and Rectal Surgeons
http://www.fascrs.org

FAQ's Woman's Diagnostic Cyber
http://www.wdxcyber.com

Griffith's 5 minute Clinical Consult
http://www.5mcc.com

V

VALIDATION THERAPY

Validation Therapy, developed through clinical practice with Alzheimer-type nursing home residents, is based on a developmental theory that in old age, when controls loosen, disoriented, very old persons need to express buried emotions in order to die in peace. This final life struggle is called "Resolution." Validation techniques are based on the principle that when emotions are suppressed they fester and can become toxic. When emotions are expressed to someone who listens with empathy (Validation), the person is relieved. Validation therapy uses fifteen verbal and nonverbal techniques to communicate with those very old elders diagnosed with an Alzheimer-type dementia and includes a method for forming Validation Groups with time-confused elders. Validation therapy assumes an attitude of respect for old people diagnosed with a dementia.

Validation is an interdisciplinary helping method. The goal of Validation is to improve the quality of life. Nonverbal Validation techniques such as "music" and "mirroring" often restore dignity and well-being for early onset Alzheimer populations. Administrators, nurses, social workers, psychologists, physical and occupational therapists report significant improvement after six months of Validation Therapy with *late onset* Alzheimer populations. Results include decreased staff burnout, increased communication between nursing staff and disoriented residents, decreased agitation, increased family visits, less movement to "Vegetation," fewer tranquilizing medications, and increased communication within Validation Groups (Blumenthal, 1999; de Klerk-Rubin, 1994; Feil, 1993; Fine & Rouse-Bane, 1995; Lewis & Feil, 1996; Sharp, 1999).

Verbal Validation Techniques

Verbal Validation techniques restore well-being to older people in Phase One of Resolution, "the Mal-oriented." These individuals are mostly oriented to time and place, have no history of mental illness, are verbal, but repeat things that are not true in present time and often accuse others, projecting their frustration. The Maloriented use present day people to vent emotions that they never could express to important people in their lives in the past. In response, the Validating caregiver recognizes that the older person must use symbols (people or things in present time that substitute for people from the past) to express emotions. The Validating caregiver must accept the Maloriented where they are, and let them heal by venting to someone who listens with empathy.

A caregiver full of hurt cannot listen with empathy. Step One in Validation means that the caregivers free themselves from their own emotions, so as to enable them to accept the emotions of the older person and feel what they feel. The caregiver must step into the older person's shoes by "Centering." Verbal validation techniques for the Maloriented include:

1. Breathe deeply, inhaling from the nose and exhaling from the "Center" (a spot about three inches below the waist).
2. Avoid "feeling" words. Ask nonthreatening factual questions: Who? What? Where? When? How? Avoid asking "Why."
3. Rephrase, repeating their key word, picking up their tempo.
4. Reflect the look in their eyes.
5. Listen to their verbs. Use their preferred sense. Speak their language.
6. Ask the extreme: "How bad?" "How often?"
7. Reminisce.
8. Help the person find a familiar coping method.

The following is an example of an interaction using verbal validation techniques:

> The nurse examines a 90 year old who is physically not hurting, but always complaining: "My

back hurts. I have a pain in my chest. My neck hurts. I wish I were dead."

The Validating nurse builds trust, and helps this old woman express her psychological pain. The woman is terrified of dying alone. Using the *kinesthetic* sense the nurse asks, "Does it *feel* like a hammer pounding on your head, or is it more like a dull ache?"

The woman responds (loosening her grip on the nurse's arm relieved to be understood), "Yes. Just like a hammer. That's right."

The nurse (asking the *extreme*), "When is the pain the worst?"

The woman responds, "It hurts all the time, but at night, when I'm alone, the pain is horrible."

Re-phrasing and reminiscing, the nurse asks, "When no one is with you, does the pain get worse? Have you ever had this terrible pain before?"

As the 90 year old woman begins to trust, her voice becomes less harsh and shrill, demonstrating that she feels safe. "When my husband died, I had the same pain in my head."

The nurse then attempts to *find a familiar coping method*: "How did you stand it when he died? What did you do?"

The woman responds, "I listened to the Strauss waltzes we danced to. We loved to dance. That's how I got through the night."

The nurse provides an answer, "I can get you some Strauss waltzes. When the pain gets bad, turn on your tape recorder. If you need me, I'll be here."

The woman responds, relieved, "You're a sweet girl. You can go now, honey. I know you have other people to take care of. But you'll come back with the waltzes?"

The nurse keeps her promise and the older woman complains less. She is not cured, but she trusts the nurse and is no longer so afraid to be alone. This Validating communication took five minutes. Validation does not take much clock-time, but it does take energy, focus, and caring. These techniques have been documented in videos (Feil, 1997).

Nonverbal Validation Techniques

Those in Phase 2 of Resolution, "the Time Confused, "are very old people with more physical deterioration. They can no longer tell chronological clock-time. They go by memories, not minutes.

Their emotions spill. They lose social controls. They retreat to the past, partly because they are no longer able to tell present time and partly because they need to restore the past to resolve it before they die. The Validating caregiver accepts the physical deterioration and psychological needs, using both verbal and nonverbal Validation.

Actions involved in nonverbal validation include:

1. Observe the emotion.
2. Say their emotion with the same emotion.
3. Genuinely mirror their movements.
4. Use close, genuine eye contact.
5. Touch the patient using soft movements, for example the "Mother's touch" is a gentle, circular motion on the upper cheek.

The following conversation is an example of an interaction using nonverbal validation techniques:

An 88 year old woman screams, "Get out of my way. I have to see Mother."

The physical therapist, mirroring the woman's anxiety and moving with her responds, "Has something happened to your mother?"

The older woman responds, "Yes. She is sick. She's all alone. I have to help her."

The physical therapist, while gently touching the old woman on the upper cheek, using close eye contact, mirroring her fear, responds, "Are you afraid that you'll lose her?"

The older woman, in response, stops moving, looks close into the physical therapist's eyes, nods her head and cries, "I lost her. She died."

On a deep level of awareness, the old woman knew that her mother was dead. She had buried that knowledge. Now, in old old age, she restores her mother to express her grief. The physical therapist shares her grief. Crying brings relief. Within four minutes, the time-confused woman smiles at the physical therapist, "You're a nice girl. I like you."

Each time the time-confused woman needs to see her mother, she is Validated. After three weeks, she no longer looks for her mother. Her feelings have been expressed and she is relieved.

In Phase 3, "Repetitive Motion," older Alzheimer's patients may have lost speech, but still retain the human needs to express emotions and feel safe and useful. They use movements of lips, tongue, teeth, jaw, and body to express needs despite the inability to communicate those needs in a way that is understandable to caregivers. Nonverbal techniques that can assist in the Validation process with older adults in Phase 3 include music and ambiguity. Music, especially childhood songs with emotional memories, can help those in Repetitive Motion to express emotions. Ambiguous responses including vague pronouns and numbers can provide safe options for patient-caregiver interactions.

The following is an example of an interaction using nonverbal Validation techniques with an individual in Phase 3:

> A 94 year old says, "He titled on the beetlebum."
> The caregiver responds using *ambiguity*, "Did it hurt him?" (*him* is an ambiguous pronoun).
> The 94 year old responds laughingly, "No. We twiddled all the time."
> The older adult and the caregiver laugh together, singing, "A Bicycle Built for Two."
> The 94 year old communicates until she dies.

Using vague pronouns to substitute for unique word combinations in Phase 3 stimulates interaction and prevents withdrawal inward. As a result, the individual's emotions and human needs are met resulting in a death with dignity.

NAOMI FEIL

See Also
Reality Orientation

Internet Key Words
Malorientation
Nonverbal Validation
Repetitive Motion
Resolution
Time-Confusion
Validation Therapy
Verbal Validation

Internet Resources
Knowledge link, Inc.
http://www.knowledgelinc.com

Österreichisches Institut für Validation
http://start.at/validation

Tertianum ZIP, Switzerland
http://zfp.tertianum.ch

Validation Training Institute
http://www.vfvalidation.org

VASCULAR AND LEWY BODY DEMENTIAS

Alzheimer's disease has received the majority of attention from the public and clinical scientists. Yet Lewy body disease is approximately one third of all dementias making it more common than vascular dementia (VaD). Difficult to treat behavioral disturbances attributed to Alzheimer's disease may be the result of unrecognized diffuse Lewy body disease.

At the present time, risk reduction for Alzheimer's disease seems distant. However, our ability to reduce cardiovascular risk factors that complicate if not cause dementia is already at hand. Despite improved survival rates associated with cardiovascular disease, an increase in vascular related dementia is expected even if present cases are over diagnosed. Given the higher prevalence of cardiovascular disease in educationally disadvantaged minorities, this population may also be disproportionately affected by the increasing prevalence of vascular dementia (Lilienfeld & Perl, 1994).

The Comprehensive Approach to Dementia

The comprehensive approach to dementia care seeks to push dementia-related disability to the end of the life span. It consists of six elements:

• accurate diagnosis of the specific dementia and comorbid conditions for treatment and prognosis;

- patient and family education particularly given the transient, misleading improvements in VaD or Lewy body disease;
- caregiver counseling and support;
- pharmacological palliation of cognitive impairment;
- interventions to lessen behavioral disturbances;
- discussion about end-of-life decisions and institutional care.

Diagnostic Procedures

The diagnosis of "probable" Alzheimer's dementia can be accurately made in 90% of cases by history and clinical examination. "Possible" cases may have atypical features but no identifiable alternative diagnosis. Computerized tomography (CT) of the brain, without contrast, should be performed when focal neurological signs are present, when change in mental status is sudden, or when trauma or mass effects are suspected. Magnetic resonance imaging (MRI) may be indicated when VaD is suspected or when the course of illness is stair-step rather than the smooth decline of Alzheimer's disease. However, the white matter changes seen on T2 weighted images are not necessarily indicative of dementia. The use of MRI to confirm the clinical diagnosis also has therapeutic and prognostic significance for patient and family.

Genetics

Early onset Alzheimer's disease is more often familial rather than sporadic. Familial early onset (before 65) Alzheimer's disease has been linked to mutations on chromosomes 1 (presenilin 2), 14 (presenilin 1), and 21, with some forms of late onset linked to chromosome 12.

Individuals inheriting one of the presenilin producing genes will develop Alzheimer's disease at an early age. The apolipoprotein allele E-4 (APOE-4), located on chromosome 19, is associated with the more common late onset Alzheimer's disease but is not associated with markedly elevated risk as with the presenilins. However, the determi-

nation of true risk is complicated for both the familial and sporadic cases of Alzheimer's disease. The age of onset varies, susceptible persons may not survive to the age of onset, and other mediating factors associated with advanced age may modify risk. As a result there is no consensus on genetic screening for Alzheimer's disease (Small et al., 1997). Indeed families who proceed with testing frequently overinterpret the meaning of both positive and negative results despite caveats from the clinical team. The genetics of VaD, beyond that which may be explained by vascular disease, remain elusive. There appears to be no heritable component of Lewy body dementia.

Differential Diagnosis

Alzheimer's disease is the most prevalent dementia and with Pick's disease represents the cortical dementias with primary neuronal degeneration. Huntington's and Parkinson's disease represent the subcortical dementias. Dementia associated with Lewy bodies overlaps with both Parkinson's and Alzheimer's disease in presentation and distribution of pathology.

A finding of widespread amyloid plaques and neurofibrillary tangles from histological examination of the neocortex are the criteria for "definite" Alzheimer's disease. Lewy body dementia is characterized by round neocortical inclusion bodies with neurofibrillary tangles being few to none. Amyloid plaques may be present as in Alzheimer's disease. Lewy bodies are also seen in Parkinson's disease without dementia. The dopamine depletion of Parkinson's disease is due to neuronal death in the nigrostriatum. However, both dopaminergic and cholinergic deficits are seen in Lewy body dementia and may exceed those found in Alzheimer's disease (Luis et al., 1999). The secondary neuronal degeneration of the vascular dementias (multi-infarct, Binswanger's) is due to angiopathic disorders, most commonly ischemic heart disease, arrhythmias, hypertension, and diabetes. However, the pathology of VaD is frequently of a mixed type (cortical and subcortical) with diverse presentations in which the

COGNITIVE DEFICITS ARE PROGRESSING AND INVOLVE MORE THAN MEMORY IMPAIRMENT

Insidious onset with smooth decline and motor function minimally impaired?

Abrupt onset or fluctuating course

YES NO➔

↓

ALZHEIMER'S DEMENTIA

↓

History of stroke or significant ischemic brain injury on CAT scan or MRI?

Marked fluctuation in cognitive impairment, hallucinations prominent, signs of Parkinson's syndrome evident, falls

YES NO➔

↓

VASCULAR DEMENTIA   YES

↓

LEWY BODY DISEASE

FIGURE 1   Algorithm for differential diagnosis of the dementias, capturing 95% of cases.

loss of brain volume, ventricular dilatation, bradykinesia, and the cognitive deficits are difficult to distinguish from Alzheimer's disease.

The algorithm shown in Figure 1 captures 95% of the dementias. Key elements are character of onset and decline, evidence of ischemic brain injury, prominent hallucinations, and Parkinsonian features. Smooth decline following insidious onset is probable Alzheimer's disease. The dementia is likely to be vascular when the onset is abrupt, the course fluctuates, and there is significant evidence of ischemic brain injury. Hemiparesis, gait disorder, and other signs of past stroke also suggest vascular dementia (Bowler et al., 1999). Although perceptual distortions are common in dementia, Lewy body disease is the probable diagnosis when visual or auditory hallucinations are prominent, signs of Parkinson's disease are evident, and the course is characterized by lucid moments alternating with confusion. Paranoid delusions, falls, and depression may also characterize diffuse Lewy body dementia (Luis et al., 1999). The cognitive impairment associated with stroke or acute traumatic brain injury is abrupt, and may predispose the person to dementia but may also improve over the six months following the incident. Similarly, persons with Parkinson's disease may experience bradyphrenia (cognitive slowing) and hallucinations but it is the motor rather than cognitive impairment that is most disabling.

General Treatment Considerations

The diversity in dementia presentation, comorbid conditions, and clinical course highlights the importance of measures to prevent the onset and progression of cardiovascular disease and diabetes. Weight control, exercise, lowering of cholesterol, treatment of diabetes and hypertension, elimination of tobacco use, and minimizing alcohol intake represent good preventive health behaviors at any age. For patients with suspected VaD, optimizing the treatment of cardiovascular disease and the use of 325 mg aspirin daily should be recommended (Nyenhuis & Gorelick, 1998).

Medications to Palliate Cognitive Impairment

Cholinesterase inhibitors improve cholinergic neurotransmission and are FDA approved for mild to moderate Alzheimer's disease. However, because the common dementias may overlap diagnostically, some practitioners offer every patient a cholinesterase inhibitor but will discontinue it if the response is equivocal at three months. Cholinesterase inhibitors may improve cognition, delay decline, lessen the disability in activities of daily living, improve psychological and behavioral disturbances including psychosis, and forestall nursing home admission. Transient side effects of cholinergic enhancement occurring at the initiation of treatment are nausea, diarrhea, sweating, bradycardia, and insomnia.

The cholinesterase inhibitor donepezil (Aricept) is started at 5 mg. Increasing to 10 mg after one month will provide a slightly greater chance of improving cognition. Patients in the earliest stage of dementia when cognitive decline is relatively slow are most likely to benefit. Middle stage patients in whom the slope of decline is more acute exhibit more transient improvement. Benefits may take three months to become evident; after ten months of treatment, most responders will revert to their level of disability when donepezil was started.

Another cholinesterase inhibitor rivastigmine (Exelon) dosed from 6 mg to a maximum of 12 mg twice daily should be gradually titrated up over a month to six weeks to avoid nausea. Galantamine at 6 mg to 12 mg twice daily has similar activity and side effects as donepezil and rivastigmine but also modulates nicotinic receptor activity. It is not clear whether persons not responding to one cholinesterase inhibitor will respond to another. However, those who are intolerant of one agent deserve a trial of the other. Many practitioners will add vitamin E (alpha-tocopherol) to the regimen. When used alone and at 2,000 IU daily, vitamin E reduces the rate of functional decline and delays nursing home placement in moderately impaired persons with dementia. However cognitive performance is not enhanced. Adding extract of *Ginkgo biloba* should be discouraged due to its minimal benefit, cost, and risk of coagulopathy (Cummings, 2000).

Management of Behavioral and Psychological Signs and Symptoms

The number and effectiveness of medications to reverse cognitive impairment remains limited, but a variety of pharmacological and nonpharmacological approaches may counter the disability of behavioral disturbances. Problem behaviors are often an expression of the caregiving context and the caregiver's capacities as well as the patient's disease (Mittleman, Ferris, Shulman, & Steinberg, 1995). The fluctuating course of both Lewy body and vascular dementia make the certainty of benefits and ease of adjustments difficult. Teri's (1997) characterization of the three-point sequence or ABCs of problematic behavior is the central management strategy.

- First identify the "antecedents" or triggering events, such as changes in daily routine or the environment, interpersonal conflict, emotional or physical stressors. The antecedents can then be removed or minimized as a preventive measure.
- Then describe the "behavior" in detail, how often it occurs, when and where it is most likely to

happen, and how long it lasts. Caregivers may need to step back and watch or take notes to provide sufficient detail and to set the baseline for measurement of improvement. This observation period also refines recognition of antecedents and how the problem behavior fits into other aspects of the patient's life.
- Finally, identify the "consequences" of the behavior, how the caregiver or others react to reinforce or deter the activity, and what happens when the activity ceases.

Delusions, Hallucinations, Unwarranted Suspiciousness, Falls

It is important to distinguish persistent false beliefs or perceptions from transitory illusions that result from impairments in vision, hearing, and cortical deficits. If the psychosis does not interfere with care or distress the patient, medication may be unnecessary. However, when patients act on their delusions through seclusiveness, threats, accusations, or assault, antipsychotic medication will be necessary.

Gait disturbance due to apraxia, quadriceps weakness, rigidity, sedatives, and poor vision predispose the dementia patient to falls, soft tissue injury, and fractures (Cohen-Mansfield & Werner, 1997). Physical therapy and change in medications may reduce the risk but not eliminate falls completely. Yet measures to completely eliminate falls such as physical restraints degrade quality of life. This dilemma should be discussed with staff and family to reach a balance between safety and freedom.

Medications to Lessen Behavioral and Psychological Disturbances

A short acting benzodiazepine (lorazepam 0.5 mg oral or intramuscular) can help the patient through procedures such as CT scan or MRI. Low doses of the sedating antidepressant trazodone (25 mg) may be effective for sleep disturbance but hypotension will ensue as the dose is increased. The sedating antipsychotic thioridazine in low doses (10 mg) may also be beneficial. Haloperidol (0.5 mg), thioridazine (10 mg), and risperidone (0.25 mg), available in liquid forms, assist patients with difficulty swallowing or facilitate disguising medication from patients with difficulty accepting pills. The benefits of haloperidol are its relatively low sedative and hypotensive effects. Preferable for longer term treatment, although more expensive, risperidone (Risperdal) is a mildly sedative atypical antipsychotic and is not hypotensive. At low doses (1–2 mg daily) it is superior to placebo for the treatment of suspiciousness and aggressive behavior in dementia. However above 2 mg, extrapyramidal signs begin to appear (Katz et al., 1999). Olanzapine (Zyprexa) at 2.5–10 mg is less likely to induce extrapyramidal effects than resperidone. It may cause somnolence and gait disorder but is rarely hypotensive. It also reduces agitation and aggression. Quetiapine (Seroquel) is more sedative than olanzapine but no more likely to cause extrapyramidal effects. It is also relatively free of interactions with other drugs. At a mean dose of 100 mg (25 mg initially then up to 50 mg bid) it reduces behavioral disturbances, most notably hostility.

Although antipsychotics have been the main treatment for aggression, the antiepileptic valproate (125 mg twice daily to start) is increasingly recognized as an antiaggression agent as well as a mood stabilizer. It is relatively safe, is neither amnestic, arrhythmogenic, nor hypotensive but should be monitored with therapeutic levels. Gabapentin and lamotragine are newer antiepileptic agents with reports of antimanic properties (Semenchuk & Labiner, 1997); however, their use in older patients with dementia is not well documented.

GARY J. KENNEDY

See Also
Advance Directives
Substitute Decision Making

Internet Key Words
Alzheimer's Disease
Lewy Body
Parkinson's Disease
Vascular Dementia

Internet Resources
Alzheimer's Association
http://www.alz.org

Caregiver home page
http://www.alzwell.com

For research updates
http://www.alzforum.org

Regarding Lewy body dementia
http://www.ccc.nottingham.ac.uk

VETERANS AFFAIRS (VA) HEALTH CARE SERVICES FOR ELDERLY VETERANS

The U.S. Department of Veterans Affairs' (VA) mission is to serve men and women honorably discharged from the U.S. military service and their families. Three branches of VA are involved:

1. The Veterans Benefits Administration (VBA) provides compensation and pension benefits to veterans with service-connected, and non-service connected disabilities, and to eligible dependents of deceased veterans.
2. The Veterans Health Administration (VHA) provides comprehensive health care services for eligible veterans.
3. The Veterans Cemetery System provides the final resting place for veterans and eligible family members.

VA is not authorized to provide health care services directly to dependents of veterans, other than in emergency situations, but a number of programs for veterans, such as respite care, indirectly assist the veteran's family caregiver.

Overall, the total veteran population is decreasing and currently stands at approximately 25 million. However, the proportion of veterans 65 and older has been steadily increasing over the past two and a half decades to 9.3 million currently (U.S. National Center for Veterans Statistics and Analysis, 1994). This large elderly population reflects the aging of the World War II and Korean Conflict veteran cohorts.

History of VA

Created in the early 1930s, the VA achieved department-level status in the executive branch of the federal government in the late 1980s. A number of historic events have influenced the evolvement of VHA into the largest, integrated health care system in the U.S. Affiliation agreements with medical schools were initiated by VHA in 1945. These agreements served as a foundation on which VHA has built an academically oriented, integrated health care system.

Major research advances under the auspices of the VHA with application to the general public include the development and testing of prosthetic devices for amputees, lithotripsy to treat renal calculi, development of the first antihypertensive medications, and effective treatment strategies for post-traumatic stress disorder that benefited veterans as well as those who experienced natural and man-made disasters, such as Hurricane Andrew and the Oklahoma City bombing.

VHA currently has affiliation agreements with nearly 1,000 health professional schools, colleges, and university health science centers. As part of these affiliation agreements, health care education is provided by VHA for more than 105,000 trainees on an annual basis. Trainees include resident physicians and fellows, medical students, and nursing and other associated health students.

This three-faceted approach—research, education, and training—in combination with clinical capabilities to address prevalent health problems, was employed by VHA in the early 1970s to strategically plan for the aging of the World War II and Korean War veteran population. The plan included establishment of Geriatric Research, Education and Clinical Centers (GRECCs), Geriatric Medicine Fellowship Programs, and Interdisciplinary Team Training Programs in Geriatrics (ITTP).

Geriatric Research, Education and Clinical Centers (GRECCs): Established in 1975, GRECCs are charged with advancing knowledge of the aging process and geriatric health problems through research; training of medical, dental, and associated health students and VA staff in care of the elderly; and the development and evaluation of improved models of care for elderly patients. At present, there are twenty GRECCs systemwide. Evidence of their effectiveness includes research in the early 1980s at the Sepulveda (CA) GRECC that demonstrated the utility of geriatric evaluation and management units for a targeted group of frail individuals following acute hospital admission; the pioneering work of the Palo Alto (CA) GRECC in development of the understanding of insulin resistance syndromes in the elderly; and the establishment and evaluation of a palliative care unit at the Bedford (MA) GRECC for veterans with late-stage dementia, the first such unit in the U.S. (Goodwin & Morley, 1994).

Geriatric Medicine Fellowship Programs: Established in 1978, VHA became the single largest training site in the U.S. for geriatricians. Between 1978 and 1992, 275 physicians graduated from VHA's 2-year geriatric medicine fellowship program.

Interdisciplinary Team Training Program in Geriatrics (ITTG): Established at 12 VHA sites in 1978, ITTGs have been major training sites and sources for students and VA staff in developing an interdisciplinary approach to care of frail elderly veterans.

Development of Geriatric and Long-Term Care Services in VHA

Nursing home care for elderly and disabled veterans, initiated in the 1960s, includes VA operated, contract community, and state veterans' homes. Domiciliary and community residential care have been even longer standing programs in VHA. Home and community-based services developed more slowly, beginning in the 1970s with home-based primary care. Currently, a spectrum of services is provided directly by VHA or by contract with a community provider. These services include geriatric evaluation and management, home-based primary care, fee basis home care, homemaker/home health aides, VA and contract adult day health care, respite care, hospice, community residential care, domiciliary care, and nursing home care (Cooley, Goodwin-Beck, & Salerno, 1998). Trainees of all health disciplines have planned clinical rotations through the geriatrics and long-term care programs at multiple VA sites.

VA has played an instrumental role in the care of spinal cord injuries, amputations, mental health problems, homelessness, geriatrics, and physical rehabilitation. Special programs for patients with Alzheimer's and related dementias have also been developed at several VHA sites. VA also has a large substance abuse program including inpatient and outpatient treatment programs. These are a component of our overall mental health program for veterans. Effectiveness of several of these care programs in terms of patient outcomes and cost, has been evaluated and published by GRECC researchers and those in VHA's Health Services Research and Development Centers.

The challenge of meeting the health care needs of a large elderly population is ongoing. Transformation of VHA from a hospital-based system to a primary- and ambulatory care-based health system, which began in the mid-1990s within the context of changes in the overall U.S. health care system and competing demands for limited resources, is a continuing challenge to our organization and the stakeholders involved in the process. Planning and budgeting care delivery is currently decentralized to 22 Veterans Integrated Service Networks (VISNs), established in 1995, to facilitate this transformation (Kizer, 1996). Implementation of the Veterans Millennium Health Care and Benefits Act, enacted in November 1999, is another major challenge. One section of this Act mandates the provision of extended care services for eligible veterans as specified in the legislation. Prior to this Act, these services were discretionary.

VA provides care either directly or through contract for veterans enrolled in the VA health care system. We currently provide care for about 3.5 million of the 25 million veterans in the country.

VA only pays for care of a veteran by a private physician or in a non-VA hospital provided through a contract arrangement. VA is currently not authorized to collect Medicaid or Medicare payments for care of veterans in VA.

The direction being taken by VHA and supported by the legislation includes maintaining the current capacity in VA nursing homes while expanding access to home and community-based care of the elderly. The VHA's experience in developing improved models of care for the elderly, evaluating the effectiveness of those models, and training students and staff to acquire additional clinical expertise is a strong foundation for meeting these current and future challenges.

MARSHA E. GOODWIN-BECK

See Also
Geriatrician
Health Maintenance

Internet Key Words
Department of Veterans Affairs (VA)
Extended Care
Geriatric Research, Education and Clinical Centers (GRECCs)
Geriatrics
Veterans Health Administration (VHA)

Internet Resources
Department of Veterans Affairs
http://www.va.gov

Geriatric Research, Education and Clinical Centers (GRECCs)
http://155.100.207/40/greccweb

VISION CHANGES AND CARE

Caring for the older adult's visual system is multifaceted and multidisciplinary. It involves optimizing sight, vision, eye coordination, visual perception, and eye health. One key factor in eye and vision care in the older adult is understanding that there is a difference between sight and vision. Measuring *visual acuity* is synonymous with sight whereas vision refers to how one functions or how one interacts with the environment. Decreased visual functioning ability can lead to other sensory, motor, and emotional disorders that are risk factors for dependency and institutionalization. Sensory loss has a significant impact on how an individual interacts with the environment. The ability to function in the environment with ease and assurance of safety contributes to successful aging (Fangmeier, 1998; Kosnick et al., 1988).

Visual Impairment

Visual impairment ranging from mild to severe is a term used when good visual functioning ability cannot be achieved with conventional glasses or contact lenses. Mild visual impairment is present when everyday activities such as reading the newspaper, reading medication labels, and threading needles are not easily or efficiently accomplished. When activities of daily living are affected because of a drop in best corrected visual acuity of 20/50 or worse, a person is also said to have a mild visual impairment otherwise known as "functional" visual impairment. Mild visual impairments are not necessarily caused by ocular disease. The visual system undergoes various age-related changes that affect visual functioning. Many of the changes are "normal" because they are not caused by pathology, for example, presbyopia denotes decline in near visual acuity because of a decrease in the focusing (accommodating) ability of the lens and invariably leads to the need for reading glasses or bifocal lenses.

Retinal Sensitivity: Distance and near visual acuity may also decline because retinal sensitivity to light decreases as a result of the decrease in the number and function of retinal cells. Consequences of lowered retinal sensitivity include decreased visual acuity, fatigue with reading, and difficulty functioning in dimly illuminated places like restaurants, churches, and subway stations. Patients with decreased retinal sensitivity can be helped by increasing the lighting to a minimum of 75–100 watts.

Incandescent lighting is preferred to fluorescent because the flickering nature of fluorescent lights can result in discomfort. Portable lighting such as mini flashlights may be of significant help with reading or when ambulating in dimly lit situations. Central lighting is better than task lighting. The goal is to have the room uniformly illuminated, as with a torch lamp, thereby decreasing glare and shadows.

Contrast Sensitivity: Contrast sensitivity or the luminance difference between the object and background also declines with age because of decreased retinal sensitivity to light, pupillary miosis that decreases the amount of light into the eye, and age related central nervous system changes. Reduced contrast sensitivity can lead to blurry vision, decreased reading efficiency, and reading fatigue. Increased lighting and enhancing the contrast of the reading material by magnifying the print with eyeglasses or magnifiers, placing yellow filter paper over the poorly contrasted reading material, or using black lettering on a white background can be helpful. Announcements or mealtime specials in senior centers or nursing homes should use black letters on white background or white on black background rather than pastel color chalk or light-colored markers against a white background, as is typically done. Decreased contrast sensitivity can also threaten safety. The older adult may have difficulty ambulating up and down stairs or curbs because of problems in judging the change in heights. Improved contrast can be created by affixing a strip of bright carpet at the base of stairs or orange paint can be used on curbs to cue the change in height.

Useful Field of View: Age related neuronal degeneration in the visual cortex affects the speed of visual information processing. The older adult does not see as fast and thus loses efficiency in interacting with the environment. Reading is slower and safety becomes an issue. Decline in the speed of visual information processing results in reduction in the useful field of view (UFOV), a type of dynamic visual field that evaluates the amount of information acquirable in a single glance. It is a measure of the field of visual attention. Unlike standard visual field testing that looks at the size of the visual field, UFOV involves detecting, localizing, and identifying targets in the presence of distracting information. Daily activities that utilize this visual parameter are spotting faces in a crowd, detecting a single traffic sign in the midst of many, and navigating an intersection when driving or crossing a street. A limited UFOV forces a person to make more fixations in order to scan a visual scene. Clinically, UFOV status is determined by case history or it can be measured with a Visual Attention Analyzer (Ball & Owsley, 1993). When managing patients suspected of having a UFOV deficit, educating the patient about their altered field may result in the elder paying more attention to, or avoiding difficult situations like, driving during rush hour. Vision therapy exercises such as tachistoscopic training, which allows a patient to practice visually localizing a target in the presence of distracting stimuli may be helpful (Sekuler & Ball, 1986).

Dark/Light Adaptation: Abrupt changes in lighting may also be problematic with aging due to a decrease in dark/light adaptation. The rod and cone retinal cell function slows and pupil miosis causes less light to enter the eye. It takes longer for the visual system to adjust to a dimly lit area after having been in bright light, or adjust to a bright area after being in the dark. The classic consequence is difficulty with night driving, maneuvering through tunnels, and entering or exiting dark theaters. To ensure safety, education about the changes—wearing sunglasses outdoors to keep the pupil in a darker adapted state, and maintaining a constant level of lighting indoors whenever possible—is recommended.

Glare: Glare is a common daily problem due to pupil miosis that makes light scatter more. Media opacities from common ocular disease like cataracts will also cause light to scatter more as it enters the eye. Also, the retina becomes less able to process extraneous light. A bright light can create a dazzling sensation that produces discomfort and/or interferes with optimal vision. Consequences of glare include difficulty driving at night, reduced vision, and difficulty reading (especially with glossy paper where more light is reflected). Wearing sunglasses, hats or visors, placing shades/sheers on windows, and antiglare coating on eyeglasses, as well as cleaning windows, windshields, headlights, and dashboards are recommended.

Assessment

So-called "normal" visual changes often go undetected because the older adult does not report difficulty in these areas for various reasons. The changes are considered part of normal aging and it is assumed that nothing can be done. The symptoms may also be vague and therefore overlooked by the clinician, or the elder may not have been questioned about them.

One of the key assessment strategies for screening for mild vision impairment is to ask pertinent, directed questions. The questions provided below can be used to uncover functional vision problems. If the answer to any of the questions is "yes," a full eye exam is recommended (Northwestern University and the University of Calgary, 1990):

1. While Wearing Your Eyeglasses or Contact Lenses: Is there anything that you would like to do but are unable to because of your vision?
2. While Wearing Your Eyeglasses: Does your vision interfere with shopping, cooking, watching television, handling money (writing checks, paying bills), reading the newspaper, using the telephone, using the telephone book, reading the Bible, reading medication labels, or threading a needle?
3. In the past month, have you needed assistance with any of the above listed activities in #2?
4. How many falls have you experienced in the last year?
5. Do you feel that your falls are related to vision difficulties?
6. When did you obtain your current eyeglasses or contact lenses?
7. When wearing your eyeglasses or contact lenses, how would you rate the quality of your vision?
8. Do you have any eye problems? If so, what kind?
9. Are you bothered by glare?

10. Do you have problems adjusting to bright light when coming from a dark place such as exiting a movie theater?
11. Do you have problems seeing in dim light such as in restaurants, houses of worship, movies, on the subway?
12. Does it take a long time for your eyes to adjust to dim light when coming from a bright room such as entering a movie theater?
13. Do you have night vision problems?
14. Do you experience difficulty driving at night? During the day?
15. After reading for 30 minutes or less, does the print become blurry?
16. When reading, does the print become double?
17. Do you skip lines when reading?
18. Do you have good lighting in the home?
19. Do you have problems seeing with your side vision?

Low Vision

Severe visual impairment is described as legal blindness or partial sightedness. Legal blindness is a decline in the best corrected visual acuity of 20/200 or less in the better eye or a decrease in the visual field below 20 degrees. If the best corrected visual acuity in the better eye is 20/70 or worse, or the visual field is 30 degrees or less, then a person is partially sighted.

Legal blindness and partial sightedness are usually caused by an ocular disease. According to the U.S. National Center for Health Statistics (1996), the rate of visual impairment in the noninstitutional population is 35% for those 65 year of age and older versus 13% for those under the age of 65. The American Optometric Association reports that 51.7% of those legally blind, 53.7% of the partially sighted, and 70% of the functionally visually impaired are age 65 and over.

Low vision describes visual impairment whether it is legal blindness, partial sightedness, or functional visual impairment. Eye care practitioners who specialize in caring for patients with low vision will perform a low vision examination and make

recommendations on the appropriate optical or non-optical device to enhance visual functioning.

TANYA L. CARTER

See Also
Cataracts
Glaucoma
Low Vision
Vision Safety

Internet Key Words
Contrast Sensitivity
Eye and Vision Care
Functional Visual Impairment
Glare
Low Vision
Visual Functioning
Visual Impairment

Internet Resources
Aging and Your Eyes
http://www.seniors.tcnet.org/articles/ article06.htm

Resources for Individuals with Visual Impairment, American Academy of Ophthalmology
http://www.eyenet.org/public/pi/resources/ low_vision_resources.html

Visual Care of the Elderly
http://www.optom.demon.co.uk/elderly/ elderly.htm

VISION SAFETY

The prevalence of vision impairment among older adults continues to increase, especially since the definition of visual impairment was broadened from legally blind or partially sighted to include "any other trouble seeing" (Crews, 1994, p. 64). It is estimated that two thirds of persons with low vision are over the age of 65.

Low vision exists when ordinary glasses, contact lenses, medical treatment, and/or surgery are unable to correct sight to the normal range. Legal blindness is defined as 20/200 or less with corrected vision (glasses or contact lenses) or less than 20 degrees of visual field in the better eye. Visual acuity, contrast and glare sensitivity, depth perception, and visual fields decline with age (Miller, 1999; Tideiksaar, 1997). Studies indicate that visually impaired older adults have more functional limitations, health problems, and postural instability that place them at risk for falls than older adults who are not visually impaired (Miller, 1999). Diminished visual acuity can lead to depression caused by threats to, or actual loss of, personal and socioeconomic independence.

The major causes of visual impairment in older adults are cataracts, glaucoma, macular degeneration, and diabetic retinopathy. Some of these conditions may be surgically corrected or prevented from causing additional visual loss with administration of specific medications. For persons with low vision, it is essential to determine to what extent the older adult can interact safely within the environment to enhance remaining vision and provide adaptations for daily living.

Care Strategies for Low Vision

Low vision can affect an older adult's ability to perform activities of daily living such as dressing, using the telephone, shopping for food, preparing meals, taking medications safely, managing money, and driving or arranging for other forms of transportation independently. In addition to affecting basic IADLs, low vision can also impact on leisure activities such as reading, watching TV, and activities requiring near vision such as needlepoint (Miller, 1999).

The plan of care should address environmental adaptations for daily living, safety and security measures, glare and illumination, magnification devices, use of contrasting colors, and community resources. Environmental modifications can aid the older adult in performing everyday activities optimally. Frequently called numbers should be in large bold print near the telephone or programmed into an electronic telephone that has a large face and

contains approximately 10 numbers. Door exits should be identified with contrasting colors. Stove, oven, and microwave dials can be marked with colored tape to create contrast. Batteries need to be replaced regularly in smoke detectors. Keys can be identified with contrasting colors or large labels. Medication container labels should use large print to indicate dosage, administration time, and any specific directions or drug warnings. Bright colored dots can be used to distinguish medication containers for those who have the ability to differentiate colors; otherwise, tactile markers can be used. Daily or weekly pill dispensers and "talking" watches and clocks can help maintain a medication schedule.

Safety and Security Measures

A room by room review for cluttered furniture, loose carpeting, and other pathway hazards can identify environmental risk. Lighting is an aspect of the environment that is most easily adapted to improve visualization and reduce risk of falling (Ryan & Spellbring, 1996). One-hundred watt bulbs increase low lighting levels. Fluorescent lighting provides greater illumination and is more cost effective than incandescent lighting. Blue fluorescent lighting is evenly spread and free of shadows (Tideiksaar, 1997). A night light in bedroom, bathroom, and kitchen will aid eye accommodation in going from dark to light. Sensor operated night lights that turn on as the older adult enters the darkened room are inexpensive. Extra lighting is needed in high-risk locations—bedroom, stairs, halls, and bathroom. The path from the bedside to the bathroom needs extra lighting at night for bathroom use. Three way lamp switches and lighted switchplates increase illumination, and a flashlight at the bedside can aid in safety (Tideiksaar, 1997).

Glare: Decreased tolerance to glare places an older adult at risk for falls (Tideiksaar, 1997). The use of thin draperies or adjustable blinds can reduce the amount of sunlight entering the room and reduce glare. Sunglasses and tinted lenses should be worn when going outdoors to ease accommodation to bright light. The glare of bright lights may impair driving at dusk and at night.

Color Contrasts: Older adults have difficulty distinguishing between blue and green but can better differentiate red, orange, and yellow. Impaired depth perception may make it difficult for the older adult to judge distances and surface levels, which can increase the risk for falling. Contrasting colors should be placed on doorways and stairs to indicate a change in the height or depth of a surface (Tideiksaar, 1997). Toilet seat covers can also be contrasted to wall and floor surfaces (Miller, 1999).

Assistive Devices to Aid Magnification: Magnification for low vision includes optical and nonoptical products that bring objects closer or make them bigger. Pocket magnifiers allow for near vision enhancement such as money identification and putting on makeup. Hand or stand magnifiers can be used for reading assistance, and portable video magnifiers are used for reading and writing. A hand held monocular telescope can be used for spotting distant objects.

ANN MARIE SPELLBRING

See Also
Eye Care Providers
Low Vision

Internet Key Words
Color Contrast
Glare
Illumination
Low Vision
Magnification

Internet Resources
American Academy of Ophthalmology
http://www.eyenet.org

American Optometric Association
http://www.aoanet.org

The Lighthouse National Center for Vision and Aging
http://www.lighthouse.org

National Eye Institute
http://www.nei.nih.gov

VOLUNTEERISM

Volunteer activity tends to be distinguished from economic endeavors in the sense that much of it is bound to return to its producers in the form of moral gratification, ego security, and social recognition rather than income benefits. In the specific case of the elderly, volunteerism is often regarded as an antidote of loneliness and as a means of reaffirming a sense of usefulness and self-respect. It is also claimed that it compensates for the loss of those lifetime roles that defined and were central to a person's identity (Rapoport & Rapoport, 1975). There is not, however, a broad base of consensus when attempting to define volunteerism in operational terms. A maximalist position would not distinguish between formal and informal activities. It, therefore, equates the continuous and systematic tasks performed in a hospital, for instance, with joining community groups for recreational purposes and the occasional help exchanged among neighbors. A more restrictive definition limits volunteerism to programmatic activities under the auspices of a formal service organization. The first of these positions bolsters the optimistic contention that volunteerism constitutes, in fact, a generalized and *sui generis* attribute of the aging lifestyle or experience.

Fischer, Mueller, and Cooper (1991) point out that 52% of the elderly in Minnesota were found to volunteer for formal or charitable organizations. A closer scrutiny of their data reveals, however, that 38%—or two thirds—were actually engaged in church work, including proselytizing, singing in a choir, and ushering. Only 16% volunteered for hospitals, mental health clinics, and orphanages. A definition of volunteering, from this viewpoint, seems to also encompass social participation and organizational affiliation. An earlier study by Babchuck, Peters, Hoyt, and Kayser (1979) already noted that, although voluntaristic activity was limited to a small minority of the elderly, membership in voluntary organizations was far more pervasive and that the rate of affiliation did not actually decline with age. For a critical discussion of issues in estimating volunteerism in late life, see Herzog, Kahn, Morgan, Jackson, and Antonucci (1989).

Earlier studies confirmed that only a minority of the aged do volunteer, and those who do so are middle-class persons who continue a lifestyle pattern acquired earlier in life (Morris, Lambert, & Guberman, 1964; Pffeifer & Davis, 1971). When reviewing a series of studies that cover a quarter-century span, Cnaan and Cwikel (1992) similarly conclude that although formal volunteering has increased since the early 1960s, it reached a plateau and remained static since the mid-1970s. Moreover, only a small percentage of the aged actually volunteer, and there is no evidence that their rate of involvement exceeds that of any other age cohort.

If voluneerism is not a universal occurrence among the elderly, what are its predictors, that is, what is the likely profile of the actual or potential volunteer? Monk and Cyrus (1974) found that individuals interested in volunteering are better educated, tend to be homeowners, and have a sense of rootedness in their communities. They also have extensive knowledge of community resources, are financially secure, have an expressed commitment to their age peers, are in good health, and have developed a self-perception that includes multiple social and recreational networks, as well as a belief that one should help others. Income and education reappear as correlates in other studies (Chambre, 1984; Lemke & Moos, 1989; Ozawa & Morrow-Howell, 1988). Moral imperatives, such as helping others, fulfilling a moral obligation, and a sense of usefulness or remaining productive, were positively related to frequency of volunteering in the Marriot Seniors Volunteerism Study (Okun, 1994). Similar findings were reported by Kuehne and Sears (1993) in a study of volunteers assisting families who have chronically sick and disabled children living at home. At the end of the 9-month term of commitment it appeared that those who persisted and continued in their volunteer role were more educated, had a higher socioeconomic status, and tended to volunteer as well in other community pursuits.

Inquiries about whether men and women differ in the volunteering patterns have produced, however, inconclusive evidence. Nathanson and Eggleton (1993) studied the contrasting motivations to volunteer in a county ombudsman program and concluded that a program's intrinsic characteristics were far more influential in attracting and then retaining elderly volunteers than either gender or age chronological differences. Chambre (1984) noted, in turn, that women exhibit more continuity as volunteers through the life cycle, and Babchuck, Peters, Hoyt, and Kayser (1979), in the previously noted study, observed an affiliative cleavage, with women more likely to belong to church-related and socially expressive groups, and men predominating in work-related associations. Indeed, gender plays a role in shaping the volunteer career of older persons, but, as noted by Fischer, Rapkin, and Rappaport (1991), it cannot be interpreted without the concurrent consideration of a person's developmental, occupational, and employment background. Work experience appears to be more decisive for men than for women, when observing their placement in leadership roles.

What are the preferred volunteer tasks among the elderly? The Harris survey (1979) established a higher participation in health and mental health programs, including work with alcoholics and substance abusers, followed by psychological support, involvement in civic affairs, family youth and children-oriented services, recreational activities, and education. A survey in Calgary, Alberta, established that older volunteers choose to work with members of their own age group (Bramwell, 1993). Earlier, Gustaitis (1980) reported that the aged are particularly effective as advocates for programs that benefit their own cohort, but Monk, Kaye, and Litwin (1984) cautioned that advocacy and ombudsman programs take a substantial toll in the form of high stress, burnout, and turnover rates.

The federal government has spearheaded the creation and sponsorship of many volunteering programs earmarked for older persons. Historically, the largest is the Retired Senior Volunteer Program, launched in 1971 with the intent of recruiting volunteers aged 60 and older, and placing them in public and nonprofit social service programs. Most volunteers carry on an assortment of health, nutrition, education, housing, counseling, and social services functions. Their assignments are designed in a way that paid staff in participating agencies are not displaced. The Foster Grandparent Program was initiated to bring low-income seniors in contact with problem children for companionship, mentoring, and recreation. Placements include institutions for the mentally retarded, physically handicapped, and the emotionally disturbed, plus correctional facilities, hospitals, and schools. The Senior Companion Program assists the frail and vulnerable aged in their own homes, and the Service Corps of Retired Executives encourages retired executives and professionals to apply their expertise as consultants to national and international enterprises and ventures. Other programs worth mentioning are Family Friends, which matches senior volunteers with families who have children with chronic disabilities living at home, and "Grandma Please," a telephone hotline serving latchkey children.

It is often taken for granted that volunteering is "good" for the aged and that it improves their self-esteem and life satisfaction, but empirical studies testing these assumptions remain inconclusive. Hunter and Linn (1980–1981) found that elderly volunteers are more satisfied with life, have a stronger will to live, and report fewer somatic or depressive symptoms than those who do not volunteer, but Carp (1968) reported that increased happiness and richness of social life were found only among elderly engaged in paid rather than volunteer activities. Ward (1979) found that active group participation was indeed associated with higher life satisfaction, but cautioned that this may result because of the better health and higher socioeconomic status of the volunteers. Moreover, helping others is not necessarily associated with life satisfaction but involvement in peer-group activities through leadership and nonleadership roles provides more meaningful feelings of personal fulfillment (Perry, 1983).

It must be born in mind that not all forms of volunteerism are practiced for sheer altruism. Cahn and Rowe (1992) describe a barter system in the form of "time dollars" that volunteers can earn in exchange for services performed. Credits are assigned for each hour-unit of community work and stored in a "time bank." The volunteers can transform those credits at any point when they themselves may be in need of services, such as home care and household chores transportation. Ozawa and Morrow-Howell (1993) reported on a similar respite care program in Missouri that uses a credit system as an incentive for volunteers to participate.

There is no consensus whether elderly volunteers play an essential role for society, whether they really make a difference. There are critics who chastise the aged for not doing more, being too self-involved, and not helping their communities to the fullest. Implicit in this judgment is the accusation that the aged are faring too well and have more resources at their disposal than young people (Dowling, 1993). There are also those who caution that volunteers cannot fill every gap and compensate for the myriad social programs that are being reduced or dismantled for lack of budgetary resources (Cnaan & Cwikel, 1992), and there is a trend originated in the trade union field that is simply opposed to the very idea of volunteering because it is regarded as a form of exploitation that also takes away jobs from wage earners (Hutton, 1981).

Regardless of whether older persons volunteer to find life satisfaction, sociability, or just to fill time, volunteerism is no longer left to the spontaneous whims of casual joiners. Large-scale programs are dedicated to the systematic outreach, recruitment, training, motivation, and retention of older volunteers. These programs seek to legitimize voluntarism as an essential component, perhaps the very core, of aging lifestyle.

ABRAHAM MONK

See Also
Grandparents as Family Caregivers

Internet Key Words
Foster Grandparent Program
Retired Senior Volunteer Program
Voluntarism

W

WANDERING

Delirium is a syndrome that may occur as a result of multiple complex interacting neurotransmitter systems and pathological processes; wandering is just one of its many symptoms. Wandering may occur in any setting and present some of the most difficult care management problems. Acute care staffs often do not know how to intervene or care for wandering patients. Older adults are frequently denied admission to nursing homes because of the disruptions caused by wandering, such as upsetting or frightening other residents by entering other residents' rooms and/or endangering patients themselves. In the community setting, unfortunately, wandering associated with delirium may not be recognized, particularly when elders are isolated or live alone.

Prevalence by Setting

In hospital settings, delirium occurs in 14% to 56% of older patients (Inouye et al., 1999). This enormous range occurs because delirium is not always identified and recognized in older patients and varies according to the setting. Wandering associated with delirium occurs less frequently in hospital settings than in nursing home or community settings because older hospital patients are often immobilized as a result of acute physical illnesses such as stroke or existing chronic illnesses such as arthritis. Wandering occurs in 4% to 11% of nursing homes residents and in 25% to 70% of community residing older adults.

Definition, Description, and Diagnosis of Wandering

A clear, operational definition of wandering does not exist. It has been described as aimless move-

ment, functional purposive behavior, a means of dissipating tension and coping with stress, moving about without a fixed course, aim, or goal, or towards an inappropriate, impossible goal. Behaviors associated with wandering include pacing, agitation, aggressiveness, and incontinence.

The etiology of wandering associated with delirium is commonly multifactorial, so that it is difficult to determine a single cause (Hall & Gerdner, 1999). Etiological factors include mental impairment, confusion, darkened or unfamiliar environment, acute or chronic pain, boredom, stress, tension, anxiety, lack of control, need of exercise, diseases of the central nervous system, and cardiac decompensation.

Given the absence of a clear, operational definition in the research or clinical literature, the diagnosis of wandering associated with delirium is difficult. In general, diagnosis is generally based on observation of excessive pacing or agitation.

Care Strategies

Prior to any treatment or behavioral intervention, it is necessary to assess the possible causes and patterns of wandering. The relationship between specific medications (dose, frequency, and time of administration) and wandering behaviors, specifically agitation, and measures of circulation and oxygenation in relation to wandering behaviors should be explored. Also, infectious processes or other pre-acute conditions may first evidence themselves in restlessness and wandering behavior. Chemical and physical restraints should be avoided if possible, particularly with delirious patients. Safety is a prime concern in the treatment of wandering patients because other patients try to defend themselves or their property, and, as a result, may strike out.

Hall and colleagues (1995) identified five behavior patterns associated with wandering in dementia patients: tactile wanderings, environmentally cued wandering, reminiscent or fantasy wandering, recreational wandering, and agitated purposeful wandering. The first four patterns are generally calm behaviors, whereas the last is associated with agitated behavior. These behavior patterns can provide a basis for environmental modifications or day-to-day interventions. However, agitated purposeful wandering requires a more intensive analysis and focused intervention.

Tactile Wanderings: Patients who demonstrate tactile wanderings tend to use their hands to explore the environment. They may have lost their ability to communicate verbally and appear to have lost their ability to interpret visual environmental cues. To prevent wandering away from a unit, redirection and diversion away from hallways and doors, by walking along and engaging the resident in discussion is often effective. Doorknobs and locks can be modified by hiding them or changing them periodically, and alarms may also be helpful. Alarms modify behavior by negative reinforcement and alert the staff that someone is attempting to leave.

Environmentally Cued Wandering: These wanderers tend to follow cues in the environment. For example, hallways are cues to walking, and chairs are cues to sitting. Patients respond well to nonrestrictive environmental barriers such as disguising doors by covering them with curtains or mirrors, using red STOP signs, or providing diversionary activities such as music, arts and crafts, or games. Well-placed chairs can encourage sitting and resting.

Reminiscent or Fantasy Wandering: These wanderers are determined to go to a place associated with their past, e.g., going to work, to a previous residence, or to see their parents. Confronting people with the reality that these things are no longer possible often results in increased agitation. In some cases, it is helpful to discuss the meaning of the delusion with the patient, such as describing the patients' past work or their parents. In other cases, caregivers may create a pseudo-experiential area in which patients may have some type of activity that mimics the past, such as a desk with papers, a small

garden, or a job helping out on the unit by setting the tables, folding napkins, or assisting with activities.

Recreational Wandering: This type of wandering is usually associated with a need for exercise and activity. The wandering is purposeful and often occurs at the same time every day. Caregivers should take the patients for a walk or engage them in activities in a daily routine. Walks should follow the same route every day, and activities should be structured consistently.

Agitated Purposeful Wandering: Patients who demonstrate agitated behavior and wandering are generally extremely confused, frightened, and panicked. They cannot be reasoned with and may be combative or assault anyone who confronts them. The immediate action is to attempt to diffuse the situation. The patient should not be confronted and should be allowed maximum control of the situation. Assuring safety and security is of utmost importance. It is helpful to divert attention to other things if possible, and to speak in a calm, reassuring voice. Sometimes singing with the patient provides comfort. To prevent further episodes, attempt to determine what precipitated the agitation and wandering. If a person is delirious, the cause could be any one of the etiologies mentioned above. A protected environment with reduced environmental stimuli often benefits dementia patients. It would also be helpful to identify if this behavior occurs at the same time each day.

In the community, it may be advisable to contact the "Safe Return Program" of the Alzheimer's Association for referral to local or community-based programs that assist the wanderer and caregiver in safety issues.

MARY ANN MATTESON

See Also
Alzheimer's Association
Confusion/Delirium: Risk, Diagnosis, Assessment, and Interventions

Internet Key Words
Agitation
Delirium
Safety First
Wandering

Internet Resources

Alzheimer's Association
http://www.alz.org

Health-Center: Symptoms: Confusion or Disorientation
*http://site.health-center.com/brain/symptoms/
confusion.htm*

Virtual Hospital: University of Iowa Family Practice Handbook, third ed: Neurology: Delirium
*http://www.vh.org/Providers/ClinRef/
FPHandbook/Chapter14/03-14.html*

WEAKNESS

Weakness is one of the most common, yet complex symptoms appearing in older adults. It represents one of the primary risk factors associated with falls, fractures, and disability in this population and has far reaching physical, psychosocial, and financial consequences. Despite the prevalence of weakness in older adults, it has received minimal attention in the literature. Current textbooks in the areas of aging and gerontology do not address weakness as an explicit and "unique" symptom. Rather weakness appears primarily in the medical and health-related literature as part of a constellation of other symptoms associated with a wide spectrum of disease processes and as a secondary outcome of disabling conditions that produce alterations in mobility. Defined as a decrement in strength, endurance, or power, weakness is a multifactorial phenomenon characterized by a poorly understood interplay of factors.

Assessment of Weakness

Risk Assessment

Weakness is one of the most challenging symptoms for clinicians to assess, diagnose, and manage. An in-depth history should elicit risk factors predisposing the client to weakness. Risk factors for weak-

ness in the older adult include biologic aging, cumulative of multiple acute and chronic diseases and treatments used for their management, nutritional inadequacies, and/or lifestyle factors.

Age: A client's age is a risk factor for weakness. Increasing age brings structural/physiological changes in the musculoskeletal system that make weakness an inevitable sequela of aging. By 80 years of age, muscle strength may decline by as much as 50% owing to loss of muscle mass from decreases in size and number of muscle fibers and the motor neurons that control function of strength producing skeletal muscles (type II fast twitch) (Miller, 1999). Declines in muscle strength are greatest in muscles of the legs compared to those of the arms and back and account for the predictable pattern of loss in functional ability that occurs with aging. Activities requiring greater strength and endurance such as running errands and doing chores are lost before those requiring less strength such as self-care activities (Daltroy, Logigian, Iversen, & Liang, 1992).

Acute/Chronic Illness: The accumulation of acute and chronic illnesses and their accompanying treatments represent a major risk factor for weakness in older adults. Superimposing illness on the age-related changes in muscle may potentiate existing muscle weakness. Many disease processes contribute to weakness by producing specific effects on skeletal muscle or limiting mobility/activity (Fiatarone & Evans, 1993). Dramatic decreases in weakness may be induced by changes in mobility with a 1% to 3% loss of muscle strength occurring for every day of immobility (Halas & Bell, 1990).

Iatrogenic causes of weakness, especially the medications used in the treatment of illness, increase older client risk. The high incidence of chronic illnesses requiring multiple drug use, coupled with age related changes in the absorption, distribution, metabolism and excretion of medications, put this population at high risk for drug-induced weakness. Drug-induced weakness may be caused directly from adverse drug reactions, drug-drug-alcohol interactions, or overmedication or indirectly from fluid or electrolyte imbalances precipitated by certain medications commonly used in the older population such as diuretics, electrolyte supplements, and/or laxatives. On the rise in elderly patients and often going unrecognized is weakness associated with drug abuse, dependence, or withdrawal and alcohol withdrawal (Chew & Bimbaumer, 1999).

Nutritional Deficiencies: Older adults with chronic nutritional deficiencies are at risk for the development of weakness. Age related physiological and biochemical changes, the presence of chronic illnesses, including dementia and depression, and medications may lower the client's nutritional status and precipitate weakness. Although inadequate intakes of total energy, protein, and micro-nutrients (vitamin D, calcium, magnesium, zinc) have been linked to structural and functional changes in muscle, evidence reveals micro-nutrients to predict muscle strength better than other nutritional indices (Fiatarone & Evans, 1993).

Inactivity: The role of lifestyle has been underscored as a risk factor for weakness in the older adult. Not only current activity but an older person's lifelong patterns of exercise influence muscle strength. Past and present activity patterns characterized by sedentariness increase the older adult's risk of muscle weakness. Often older adults withdraw from activity, a tendency reinforced by the stereotype of aging as a time of decline and frailty. At the very time when age-related changes are contributing to decline, and activity should be increasing, it is on the decrease. Aging accompanied by an increasingly sedentary lifestyle will most always result in muscle atrophy and a corresponding decline in muscle strength. Existing sensory changes and mobility problems, such as gait impairment, a history of falls, and balance problems may also contribute to inactivity and increase client risk of developing weakness.

Symptom Assessment

Consisting of both subjective and objective features, weakness requires careful symptom assessment. Perceived subjectively by clients, weakness is often described in vague and nonspecific terms, such as "feeling weak." Not only is weakness interpreted differently from individual to individual, but it is often used interchangeably with other symptoms such as fatigue, dyspnea, dizziness, and pain. Its lack of specificity makes weakness challenging to assess. Greater specificity can be achieved through basic symptom assessment including the time pattern, quality, location, aggravating and relieving factors, and accompanying symptoms. Eliciting the patient, family, or caregiver's description of the weakness experience should assist in differentiating weakness from other symptoms.

Assessment of the impact of weakness on the client's daily activities needs to be made. In assessing the extent to which weakness has interfered with a client's capacity to perform self-care activities, it is important to determine the client's current self-care capacity compared with what they were able to do prior to the development of weakness. Since those activities requiring greater muscle strength such as running errands, shopping, yard and house work are lost before self-care activities, changes in these activities also need to be assessed. Standardized instruments measuring functional loss of self-care and instrumental activities of daily living are available and may be incorporated into assessment. Determining the type of assistance the client requires to perform daily activities and the impact of the weakness on the client's family or primary caregiver will provide further insight into the nature, extent, and severity of the weakness.

Clients may experience feelings of weakness without objective, measurable loss of strength. Objective evidence of weakness occurs after only one week of immobility and the corresponding cumulative declines in muscle strength (Halas & Bell, 1990). Numerical scales are the simplest objective measures of muscle strength, generally grading client's ability to reach, grip, push, and lift along a continuum ranging from no strength to active movement against gravity and full resistance. If available, more accurate quantitative measures of weakness may be obtained using tensiometry and dynamometry (hand-grip or computerized), which generate the amount of strength exerted during movement. Anthropometric measurements (body weight, skin fold thickness, and limb circumference) and urinary markers of skeletal muscle mass loss (creatinine and 3-methylhistidine), may also provide additional information about the extent of muscle atrophy and changes in lean body mass that result in weakness (Kasper, 1993).

Prevention and Management of Weakness

Prevention

Some causes of weakness are preventable, such as those resulting from lifestyle practices, treatments such as polypharmacy, and fluid and electrolyte imbalances. To slow inevitable declines in muscle strength, older adults need to be encouraged to adopt healthy lifestyles with attention to their nutritional and activity practices. Medication reviews should be conducted regularly with older clients to reduce the risk of drug induced weakness. Regular monitoring of a client's fluid, electrolyte (potassium, calcium, phosphate, magnesium), and nutritional status should also serve to detect problems and correct them before weakness develops.

Management

Where weakness cannot be reversed by treating the cause, management of weakness must be aimed at breaking the self-perpetuating cycle of weakness and preventing further declines in muscle function. Without early intervention strategies, declines in muscle strength will lead to decreased endurance, further limiting activity and mobility, and in turn creating more weakness. Interventions directed at weakness vary according to the nature of the weakness, its cause(s), the expected trajectory of the patient's illness, and the goals of the patient and family (Nail & Winningham, 1995). Management of weakness must be multidisciplinary, bringing together the necessary expertise from physical/occupational therapy, exercise/fitness, pharmacology, nutrition, medicine, nursing, and social work as a particular situation warrants. Management strategies should always include the client and caregiv-

er(s) and what they have found helpful in managing the weakness.

Strength training may be used to alter, even reverse, age-related losses in muscle strength and mass and to improve muscle quality, or the strength per unit of muscle mass. Muscle strength may be enhanced among healthy, community-dwelling older adults and frail, institutionalized elders using a variety of training stimuli (type of training, amount of resistance, duration, etc.) (Fiatarone, O'Neill, Ryan, et al., 1994). Use of progressive, high-intensity dynamic resistance exercise will provide the greatest gains in muscle strength and size but static or low-intensity dynamic resistance exercise is not without its benefits as well.

Strength training/conditioning programs vary depending upon a clients' particular needs, including their past and current activity pattern, functional status, level of mobility, and health history. Clients beginning such a program would need medical clearance before proceeding. Any strength training program, regardless of setting, requires that certain training principles be followed. The goal of strength training is to make muscles work harder than they are accustomed to working but to minimize the risk of injury. For clients who are bedridden or immobilized, static and dynamic bed exercises that promote regular muscle contraction help to maintain or increase muscle strength. Strength training with older adults with more limited functional capacity can be encouraged using exercises that are simple and do not require specialized equipment. Strength training programs using an armchair and the client's own body weight have reportedly made older clients feel stronger and less fearful about falling (Brill, Cornman, Davis, et al., 1999). For healthy older adults two to three strength training sessions per week on non-consecutive days are recommended (Pearle, Mutell, Romanelli, 1997). Exercise sessions generally involve anywhere from one to three sets of exercises, at six to twelve repetitions for each of eight to twelve exercises, with resistance between 60% to 80% of maximum for all major muscle groups. The number of sets, repetitions, and amount of resistance is progressively increased to maximize muscle responsiveness as tolerated by the client. Aerobic exercises that provide a weight-bearing effect, while providing minimal enhancement of muscle strength, should be encouraged in order to preserve or increase lean body mass.

The efficacy of nutritional supplementation in the prevention or reversal of muscle weakness has yet to be established in older adults with mild, chronic undernutrition. Nutritional supplementation does not provide the near-universal benefit provided by resistance training and may be most beneficial in elderly subpopulations with specific nutritional deficiencies (Fiatarone & Evans, 1993). Given that disuse and undernutrition produce similar changes in skeletal muscle and often appear together in the elderly, it would seem reasonable that muscle retraining and nutritional supplementation used together would yield the greatest gains in muscle strength. Yet multinutrient supplementation used in conjunction with resistance training in older adults does not surpass gains in muscle strength obtained through training alone (Fiatarone, O'Neill, Ryan, et al., 1994). Experts in strength training, nevertheless, advocate for nutritional supplementation that provides extra calories, protein, and micronutrients to offset the increased energy expenditure accompanying strength training.

Ensuring safety is important in managing weakness to prevent falls. The use of mobility aids may need to be used with a client experiencing irreversible weakness. Modifications to the home environment may need to be considered in clients with severe weakness.

KATHY L. RUSH

See Also
Deconditioning Prevention
Dyspnea (Shortness of Breath)
Fatigue

Internet Key Words
Assessment of Weakness
Lifestyle
Muscle Strength
Muscle Weakness
Musculoskeletal Changes with Aging
Strength Training

Internet Resources
National Institute of Arthritis and Musculoskeletal and Skin Diseases
http://www.nih.gov/niams/

National Institute of Neurological Disorders and Stroke
http://www.nih.gov/ninds/

NIDRR Rehabilitation Research and Training Center in Neuromuscular Diseases at UC Davis
http://medpmr.ucdavis.edu/

WELL-BEING
See
Subjective Well-Being

X

XEROSTOMIA

Saliva plays a major role in maintaining oral health, and alterations in salivary gland function may compromise oral tissues and functions. Reduction in salivary flow most commonly manifests as symptoms of oral dryness. The subjective complaint of dry mouth is termed xerostomia, while the objective alterations in salivary performance are termed salivary gland dysfunction (Atkinson & Fox, 1992). This semantic distinction is important as not all dry mouth complaints are the result of salivary dysfunction. The term xerostomia should be reserved for the symptoms of oral dryness only and not be used synonymously for reduced salivary function.

Multiple functions of saliva have been detected. Saliva is important for taste, mastication, deglutition, digestion, maintenance of oral and soft tissues, control of microbial populations, voice, and speech articulation. Lubrication of the oral cavity is another important function of saliva. Adequate salivary flow enhances movement of the tongue and lips, which aids in cleansing the oral cavity of food debris and bacteria. Saliva also allows for proper tongue and lip movement necessary for clear articulation.

Dysphagia in the elderly can be caused by various conditions: neurologic, systemic, psychologic, environmental, and oral changes. Patients with salivary gland dysfunction often complain of difficulty swallowing. The oral preparatory phase of swallowing requires mastication and the formation of a food bolus. Efficient mastication and bolus formation are dependent upon a moist, lubricated oral mucosa, an intact dentition and periodontium, and fluid to wet the food. Transport during swallowing requires the lubricatory and wetting properties of salivary secretions. Saliva also plays a role in digestion by helping the upper gastrointestinal tract rinse gastric secretions from the esophageal regions.

Causes of Salivary Gland Dysfunction and Xerostomia

Xerostomia is most commonly associated with diminished salivary gland function. However, the subjective complaint of dry mouth does not always correlate with the objective finding of decreased measured salivary flow rates (Fox, Busch, & Baum, 1987). Nonsalivary gland circumstances, including changes in the patient's cognitive state, psychologic distress, mouth breathing, and sensory alterations in the oral cavity may lead to the perception of dry mouth. Therefore, it is important to determine if salivary gland function is actually decreased using objective measuring techniques.

Medication usage is the most common cause of decreased salivary gland function. Anticholinergic medications such as antihistamines are most likely to cause hypofunction. Multiple classes of medications are associated with xerostomia although the exact mechanism is not understood. Sedatives, antipsychotics, antidepressants, and diuretics are common drugs that induce xerostomia. Medication-induced xerostomia is reversed if the medication is discontinued or a medication can be switched to one that causes less xerostomia.

Aside from medications, there are other iatrogenic causes of salivary gland dysfunction. Head and neck cancer is frequently treated with ionizing radiation (Johnson et al., 1993). Salivary glands are often in the field of radiation and permanent salivary gland destruction occurs. Fibrosis of the muscles of mastication and pharyngeal muscles results in chewing and swallowing difficulties. Other iatrogenic causes include cytotoxic chemotherapy, internal radionuclides, and bone marrow transplantation. Salivary gland surgery may involve removal of the gland or may cause damage to the gland or its innervation. Surgery is often performed for tumor removal, infection, stone removal, or duct stricture.

Systemic disease is also a common cause of salivary gland dysfunction. Sjögren's syndrome

(SS) is a chronic autoimmune disease with lymphocyte-mediated destruction of the salivary glands and other exocrine glands. This autoimmune disease primarily affects postmenopausal females. Other systemic conditions with prominent salivary involvement include cystic fibrosis, Bell's palsy, Diabetes mellitus, amyloidosis, HIV, thyroid disease, malnutrition (dehydration, anorexia), and psychologic factors (affective disorder).

Contrary to common belief, salivary gland function is generally well preserved with age in the healthy elderly. However, xerostomia is a common complaint found in up to 25% of institutionalized older adults and is often caused by systemic disease or its treatments (Schubert & Izutsu, 1987).

Hypersalivation is another common complaint in the geriatric patient population. However, frequently salivary flow rates are normal but motor function has been compromised, leading to decreased swallowing efficiency and the perception of increased salivary flow. Some medications can cause increased saliva production. Patients who wear dentures for the first time can experience a mild transient increase in salivary flow.

Diagnosis

There are several specific questions that help differentiate salivary gland hypofunction from the subjective complaint of dry mouth (Valdez & Fox, 1993). These questions focus on oral activities that require adequate saliva production. Questions helpful in evaluating patients with complaints of dry mouth include:

1. Do you have difficulty swallowing dry foods?
2. Does your mouth feel dry while eating a meal?
3. Do you sip liquids to aid swallowing dry food?
4. Does the amount of saliva in your mouth most of the time seem to be too little, too much or you don't notice?

A positive response to questions 1–3 or the perception of too little saliva is significantly associated with reduced salivary gland function.

Saliva collection determines if the salivary glands are producing within a normal range. Several methods of saliva collection exist and are helpful with diagnosis and appropriate treatment. Dentists with advanced education in oral medicine are trained to perform salivary function evaluations. Salivary flow rates are essential for diagnostic and research purposes. Existing salivary gland function is determined by objective measurement techniques. Salivary flow rates can be determined individually from the major glands or from all the glands (whole saliva).

Whole saliva is the mixed fluid contents of the mouth. The main methods of whole saliva collection include the draining method, spitting method, suction method, and the absorbent (swab) method. The draining and the spitting method are more reliable and reproducible for whole saliva collection. Stimulated whole saliva can also be obtained by having the patient chew on an inert material such as paraffin wax, unflavored gum base, or a rubber band.

Individual gland collection is performed using Carlson-Crittenden collectors. Individual collectors are placed over the Stenson's duct orifices and held in place with gentle suction. Submandibular-sublingual individual gland collection uses a suction device or an alginate held collector called a segregator. As saliva is produced, it flows through tubing and is collected in a pre-weighted vessel.

Stimulated individual gland saliva is obtained by applying 2% citric acid bilaterally to the dorsal surface of the tongue at 30 second intervals. Preweighed tubes are used for individual salivary gland collections. Tubes are weighed again after collection and flow rates are determined in mL per min per gland. Saliva collected for analysis should be collected on ice and frozen until analysis. Normal stimulated flow rate is considered approximately one mL per minute per gland.

The specific mechanisms of most causes of hypofunction are not known. Irradiation results in a reduction in cell number, a decrease in gland size, and fibrosis of the glandular parenchyma. Autoimmune exocrine disease is associated with inflammation and eventual loss of acini. The anticholinergic properties of medications also result in decreased salivary function. The specific mechanisms leading to these alterations have not been fully described. Even though the mechanisms of destruction are different, the result in each circumstances is the same—decreased salivary gland function and compromised oral functions.

Treatment of Salivary Gland Dysfunction and Xerostomia

Treatment for xerostomia is limited and, therefore, preventive measures must be emphasized.

Prevention

Proper shielding and positioning during radiation therapy to the head and neck region protects the salivary glands and other tissues. Radioprotective agents such as amifostine may protect the salivary glands during head and neck radiation therapy and allow less salivary gland destruction.

Dry mouth patients have increased susceptibility to dental caries (American Academy of Oral Medicine, 1999). Applying topical fluoride has been shown to reduce caries and help preserve the dentition. Fluorides are available as rinses and higher concentration gels. The gels can be applied by brush or in custom-made carriers that hold the material against the teeth. The frequency and mode of application must be determined for each patient based on the extent of salivary hypofunction and caries activity. Stannous fluoride gel or neutral sodium fluoride gel is effective for caries prevention. Neutral sodium fluoride is recommended when the taste of stannous fluoride is not tolerated. Neutral sodium is also recommended for patients with multiple ceramic dental restorations. When coupled with increased attention to dental hygiene and frequent professional dental care (every 3–4 months), supplemental fluoride can protect against the rampant dental decay that can accompany salivary dysfunction. However, sometimes even with meticulous oral hygiene, patients may continue to have increased caries.

Patients with salivary gland dysfunction have an increased incidence of salivary gland infections, and preventive measures are helpful. Patients should be encouraged to milk their salivary glands daily by gentle massage of the major glands. Adequate fluid intake and hydration are important for infection prevention. Sucking on sugarless candies or wiping the oral cavity with glycerine swabs will stimulate salivary flow and help prevent mucous plug formation and salivary gland infections. Dry mouth patients also have an increased incidence of oral fungal infections. Due to the increased incidence of caries, sugarless antifungal agents such as nystatin powder or clotimazole troches are recommended.

Treatment

Salivary gland destruction is not reversible regardless of the method of damage. Potential treatments for salivary gland destruction are active areas of research. Currently, treatment for salivary gland dysfunction is limited to symptomatic treatments or systemic sialogogues. Saliva substitutes are available as rinses and gels; however, often patients prefer sipping water. The taste and mechanical stimulation of salivation from sugarless candy and chewing gum provide relief for some patients.

Several systemic sialogogues have been investigated. These agents are only useful for patients with remaining functioning salivary glands. Limited clinical trials have shown anetholetrithione to be effective for mild medication-induced xerostomia. Clinical trials with the mucolytic agent bromhexine have been conducted but with varying results. Pilocarpine hydrochloride is a parasympathetic agonist that increases salivary flow. It is the most widely tested secretogogue and has been shown to be effective in radiation- and Sjögren's syndrome-induced hypofunction. Side effects are common but often tolerated. Among other contraindications, pilocarpine should be used with caution in patients with a history of respiratory difficulty, heart disease, or glaucoma. Recently, another para-sympathetic agonist, cevimeline, was approved by the FDA as a systemic sialogue.

Clinical trials have been conducted investigating autoimmune disease-related xerostomia and treatments with the antirheumatic medication hydroxychloroquine and with other disease modifying antirheumatic agents such as prednisone. Patients with SS are frequently prescribed such systemic medications. Serologic signs of disease activity improve and there are reports of improved salivary gland function. However, the side effects may be severe and the effects on underlying gland pathology have not been demonstrated. These therapies cannot be recommended for treatment of salivary gland dysfunction in primary SS. Nonsteroidal anti-inflammatories have not been shown to reduce dry mouth symptoms or to improve salivary flow rates.

Multicenter phase three clinical trials for alpha interferon lozenges for the treatment of salivary gland dysfunction secondary to autoimmune disease are currently being conducted. Initial studies suggest possible reversal of salivary gland destruction from autoimmune disease. Researchers are currently investigating the potential use of gene therapy as a future treatment modality for patients with radiation- or autoimmune-related salivary gland destruction.

Conclusion

Salivary gland hypofunction is not reversible. However, preventive measures and conservative treatments can avoid or limit mucosal breakdown, infections, and permanent damage to teeth. Symptomatic relief may be obtained with local measures and systemic secretogues in many patients. As clinicians, we should establish clear diagnoses, make certain that patients understand the causes of their dry mouth, and deliver the most efficacious preventive and management techniques available.

The presence of saliva impacts our daily activities. Saliva is required for support of the basic functions of the oral cavity—alimentation and communication. Management of symptoms and increasing saliva output may help patients feel more comfortable and improve the quality of their lives.

MARGARET M. GRISIUS

See Also
Swallowing Disorders and Aspiration

Internet Key Words
Dentition
Salivary Glands
Xerostomia

Internet Resources
Internet Resources for People With Sjögren's Syndrome
http://www.dry.org

National Institute of Dental and Craniofacial Research
http://www.nidcr.nih.gov

National Institutes of Health
http://www.nih.gov

REFERENCES

Abraham, I., Bottrell, M. M., Fulmer, T. T., & Mezey, M. D. (1999). *Geriatric Nursing Protocols for Best Practice*. New York: Springer Publishing Company.

Abrams, G. A., Concato, J., & Fallon, M. B. (1996). Muscle cramps in patients with cirrhosis. *American Journal of Gastroenterology, 91*(7), 1363–1366.

Abrams, R., & Alexopoulos, G. (1994). Assessment of depression in dementia. *Alzheimer's Disease & Associated Disorders, 8*(Suppl. 1), 227–229.

Abrams, R. C., Teresi, J. A., & Butin, D. N. (1992). Depression in nursing home residents. *Clinics in Geriatric Medicine, 2,* 309–322.

Acute Pain Management Guideline Panel, U.S. Department of Health and Human Services. (1992). *Clinical Practice Guidelines Acute Pain: Management of operative or medical procedures or trauma.* Rockville, MD: American Health Care Providers Resource.

Adair, N. (1999). Chronic airflow and respiratory failure. In W. R. Hazzard, J. P. Blass, R. L. Ettinger, J. B. Halter, & J. G. Ouslander (Eds.), *Principles of Geriatric Medicine and Gerontology* (4th ed., Chap. 56, pp. 745–755). New York: McGraw-Hill.

Adams-Price, C. E. (1998). *Creativity and Successful Aging*. New York: Springer Publishing Company.

Adams, W. E. (1995). The incarceration of older criminals: Balancing safety, costs, and humanitarian concerns. *Nova Law Review, 19,* 465–485.

Adelman, R., Greene, M., & Ory, M. (2000). Older patients and their physicians. *Clinics in Geriatric Medicine, 16,* in press.

Ader, R., & Cohen, N. (1991). The influence of conditioning on immune responses. In R. Ader, D. L. Felten, & N. Cohen (Eds.), *Psychoneuroimmunology* (2nd ed., pp. 611–646). San Diego: Academic Press.

Ades, P. A., Waldman, M. L., McCann, W. J., & Weaver, S. O. (1992). Predictors of cardiac rehabilitation participation in older coronary patients. *Archives of Internal Medicine, 152,* 1033–1035.

Administration on Aging. (1997). *1997 National Ombudsman Reporting System Data Tables* [Web Page]. URL http://www.aoa.gov/ltcombudsman/97NORS/default.htm [1997, March 3].

Administration on Aging. (1999). *National Elder Abuse Incidence Study* [Web Page]. URL http://www.aoa.dhhs.gov/abuse/report/default/htm [1999, November 19].

Administration on Aging. (1999). *Profile of Older Americans* [Web Page]. [2000, February 29].

Agency for Health Care Policy and Research. (1992). *Pressure Ulcers in Adults: Prediction and Prevention, Clinical Practice Guideline Number 3* [Web Page]. URL http://www.ahrq.gov/clinic/cpgonline.htm.

Agency for Health Care Policy and Research. (1994). *Treatment of Pressure Ulcers, Clinical Guideline Number 15* [Web Page]. URL http://www.ahrq.gov/clinic/cpgonline.htm.

Agency for Health Care Policy and Research. (1996). *Urinary Incontinence in Adults: Acute and Chronic Management Clinical Practice Guideline Number 2 (1996 Update)* [Web Page]. URL http://www.ahrq.gov/clinic/cpgonline.htm [2000, September 10].

Agency for Healthcare Research and Quality. (1999). *AHRQ Overview* [Web Page]. [2000, February 28].

Agich, G. (1995). Authority in ethics consultation. *Journal of Law, Medicine & Ethics, 23,* 273–283.

Ahronheim, J. C., Moreno, J. D., & Zuckerman, C. (2000). *Ethics in Clinical Practice* (2nd ed.). Gaithersburg, MD: Aspen Publishers.

Ainlay, S., Singleton, R., & Swingert, V. (1992). Aging and religious participation: Reconsidering the effects of health. *Journal for the Scientific Study of Religion, 31*(2), 175–188.

Aitken, L., & Griffin, G. (1996). *Gender Issues in Elder Abuse*. London, England: Sage Publications.

Albert, D. A., & Schwab, E. P. (In press). *Geriatric Rheumatology: Treatment of the Rheumatic Diseases*. Philadelphia: W. B. Saunders Company.

Albert, S. M., & Logsdon, R. G. (Eds.). (2000). *Assessing Quality of Life in Alzheimer's Disease*. New York: Springer Publishing Company.

Alessi, C. A., Stuck, A. E., Aronow, H. U., Yuhas, K. E., Bula, C. J., Madison, R., Gold, M., Segal-Gidan, F., Fanello, R., Rubenstein, L. Z., & Beck, J. C. (1997). The process of care in preventive in-home comprehensive geriatric assessment. *Journal of the American Geriatrics Society, 45*(9), 1044–1050.

Alexander, N. (1996). Gait disorders in older adults. *Journal of the American Geriatrics Society, 44,* 434–451.

Alexander, N. B. (1996). Using technology-based techniques to assess postural control and gait in older adults. *Clinical Geriatric Medicine, 12*(4), 725–744.

Alexopoulos, G. S., Abrams, R. C., & Shamoian, C. A. (1988). Use of the Cornell Scale in non-demented

691

patients. *Journal of the American Geriatrics Society, 36,* 230–235.

Alexopoulos, G. S., Abrams, R. C., Young, R. C., & Shamoian, C. A. (1988). Cornell Scale for depression in dementia. *Biological Psychiatry, 23,* 271–284.

Allan, J., Mayo, K., & Michel, Y. (1993). Body size values of white and black women. *Research in Nursing & Health, 16,* 323.

Allen, K., & Blascovich, J. (1996). The value of service dogs for people with severe ambulatory disabilities: A randomized trial. *JAMA, 275,* 1001–1006.

Alliance for Aging Research. (1996). *Will you still treat me when I'm 65—The national shortage of geriatricians.* Washington, DC: Author.

Almberg, B., Grafstrom, M., & Winblad, B. (1997). Caring for a demented elderly person: Burden and burnout among caregiving relatives. *Journal of Advanced Nursing, 25,* 109–116.

Altieri, J. V., Vogler, J. C., Goldblatt, R., & Katz, R. V. (1993). The dental status of dentate in institutionalized older adults. *Special Care in Dentistry, 13*(2), 66–70.

Altman, D. F. (1990). Changes in gastrointestinal, pancreatic, biliary, and hepatic function with aging. *Gastroenterology Clinics of North America, 19*(2), 227–234.

Alzheimer's Association. (1999). *Physician and Health Care Professionals General Statistics/Demographics* [Web Page]. URL http://www.alz.org/hc/overview/stats.htm [2000, July 27].

Amella, E. J. (1999). Dysphagia: The differential diagnosis in long-term care. *Lippincott's Primary Care Practice, 3*(2), 135–149.

Amella, E. J. (1999). Factors influencing the proportion of food consumed by nursing home residents with dementia. *Journal of the American Geriatrics Society, 47,* 879–885.

American Academy of Oral Medicine. (1999). *Clinician's Guide to Oral Health in Geriatric Patients.* Baltimore, MD: Author.

American Academy of Sleep Medicine. (1999). Sleep-related breathing disorders in adults: Recommendations for syndrome definition and measurement techniques in clinical research. The Report of an American Academy of Sleep Medicine Task Force. *Sleep, 22,* 667–689.

American Association of Colleges of Nursing (AACN). (1998). *The Essentials of Baccalaureate Education for Professional Nursing Practice.* Washington, DC: American Association of Colleges of Nursing.

American Association of Colleges of Nursing (AACN). (1998). *1997–98 Enrollment and graduations in baccalaureate and graduate programs in nursing*

(Publication No. 97-98-1). Washington, DC: Author.

American Association of Homes and Services for the Aging and Ernst and Young. (1989). *Continuing care retirement communities: An industry in action: Analysis and developing trends.* Washington, DC: American Association of Homes and Services for the Aging.

American Association of Homes and Services for the Aging & Ernst and Young. (1993). *Continuing care retirement communities: An industry in action: Analysis and developing trends* (Vol. 1 & 2). Washington, DC: American Association of Homes and Services for the Aging.

American Association of Retired Persons. (2000). *American Business and Older Employees.* Washington, DC: American Association of Retired Persons.

American Association of Retired Persons. (2000). *Boomers Look Toward Retirement.* Washington, DC: American Association of Retired Persons.

American Association of Retired Persons. (1997). *A profile of older Americans: 1997.* Washington, DC: Author.

American College of Cardiology/American Heart Association Task Force on Practice Guidelines. (1995). Guidelines for the evaluation and management of heart failure. Report of the American College of Cardiology/American Heart Association Task Force on Practice Guidelines (Committee on Evaluation and Management of Heart Failure). *Journal of the American College of Cardiology, 26*(5), 1376–1398.

American College of Physicians. (1996). Guidelines for using serum cholesterol, high-density lipoprotein cholesterol, and triglyceride levels as screening tests for preventing coronary heart disease in adults. *Annals of Internal Medicine, 124,* 515–517.

American Diabetes Association. (1994). *The fitness book: For people with diabetes.* Alexandria, VA: American Diabetes Association.

American Diabetes Association. (2000). Insulin administration. *Diabetes Care, 23*(1), S86–S89.

American Diabetes Association. (1998). *Medical management of type 2 diabetes.* Alexandria, VA: American Diabetes Association.

American Diabetes Association. (1998). Report of the expert committee on the diagnosis and the classification of diabetes mellitus. *Diabetes Care, 21*(Suppl 1), S5–S19.

American Dietetic Association. (1998). *Pocket Resource for Nutrition Assessment, CD-HCF Practice Group.* Chicago, IL: Author.

American Farmworkers. (1995). *Homepage* [Web Page]. [1999, November 21].

American Geriatrics Society Panel on Chronic Pain in Older Persons. (1998). The management of chronic pain in older persons. *Journal of the American Geriatrics Society, 46*(5), 635–651.

American Indian Health Care Association. (1993). *Our Voices.* Denver, CO: American Indian Health Care Association.

American Medical Association, Department of Foods and Nutrition. (1979). Multivitamin preparations for parenteral use: A statement by the nutrition advisory group. *Journal of Parenteral & Enteral Nutrition, 3,* 258.

American Occupational Therapy Association. (1994). Uniform terminology for occupational therapy. *American Journal of Occupational Therapy, 48,* 1047–1054.

American Physical Therapy Association. (1998). Report No. E-24. Alexandria, VA: Author.

American Physical Therapy Association. (1998). Report No. PR-23-A. Alexandria, VA: Author.

American Physical Therapy Association. (1997). Guide to Physical Therapist Practice. *Physical Therapy, 77,* 1163–1650.

American Psychiatric Association. (1994). *The Diagnostic and Statistical Manual of Mental Disorders*, 4th ed. (DSM-IV). Washington, DC: Author.

American Psychiatric Association. (1997). Practice guidelines for the treatment of patients with Alzheimer's disease and other dementias of late life. *American Journal of Psychiatry, 154*(5 supplement), 1–39.

American Society for Bioethics and Humanities. (1998). *Core Competencies for Health Care Ethics Consultation.* Glenview, IL: American Society for Bioethics and Humanities.

American Society of Clinical Oncology. (1996). Statement of the American Society of Clinical Oncology: Genetic testing for cancer susceptibility. *Journal of Clinical Oncology, 14,* 1730–1736.

American Thoracic Society. (1994). Treatment of tuberculosis and tuberculosis infection in adults and children. *Monaldi Archives for Chest Disease, 49*(4), 327–345.

Amerman, E., & Schneider, B. (1995). *Clinical protocol for problems with money management. Clinical protocol series for care managers.* Philadelphia, PA: Philadelphia Corporation for Aging.

Ancoli-Israel, S., Kripke, D. F., & Mason, W. (1987). Characteristics of obstructive and central sleep apnea in the elderly: An interim report. *Biological Psychiatry, 22,* 741–750.

Andersen, S. (1999). Patient perspective and self-help. *Neurology, 52*(3), S26–S28.

Anderson, D. C. (1997 July). Aging Behind Bars. *The New York Times,* p. 28.

Anderson, M. A., & Helms, L. B. (1993). An assessment of discharge planning models: Communication in referrals for home care. *Orthopaedic Nursing Journal, 12*(4), 41–49.

Anderson, M. A., & Helms, L. B. (1998). A comparison of continuing care communication. *Image: Journal of Nursing Scholarship, 30*(3), 261–266.

Anderson, M. A., & Helms, L. B. (1994). Quality improvement in discharge planning: An evaluation of factors in communication between health care providers. *Journal of Nursing Care Quality, 8*(2), 62–72.

Anderson, M. A., & Helms, L. B. (2000). Talking about patients: Communication in continuity of care. *Journal of Cardiovascular Nursing.*

Anderson, M. A., Helms, L. B., Hanson, K. S., & DeVilder, N. (in press). Unplanned hospital readmission: A home care perspective. *Nursing Research.*

Anderson, M. A., & Tredway, C. A. (1999). Communication: An outcome of case management. *Nursing Case Management, 4*(3), 104–121.

Anderson, R. N., Kochanek, K. D., Murphy, S. L., & National Center for Health Statistics. (1997). Report of Final Mortality Statistics, 1995. *Monthly Vital Statistics Report, 45*(11), supp 2, table 7.

Aneshensel, C. S. (1992). Social stress: Theory and research. *Annual Review of Sociology, 18,* 18–38.

Aneshensel, C. S., Pearlin, L. I., Mullen, J. T., Zarit, S. H., & Whitlatch, C. J. (1995). *Profile in Caregiving: The Unexpected Career.* San Diego, CA: Academic Press.

Anetzberger, G. J. (1987). *The Etiology of Elder Abuse by Adult Offspring.* Springfield, IL: Charles C. Thomas, Publisher.

Angst, J., Angst, F., & Stassen, H. H. (1999). Suicide risk in patients with major depressive disorder. [Review] [56 refs]. *Journal of Clinical Psychiatry, 60*(Suppl 2), 57–62.

Anonymous. (1997). Expert Committee on the Diagnosis and Classification of Diabetes Mellitus. Report of the Expert Committee on the Diagnosis and Classification of Diabetes Mellitus. *Diabetes Care, 20*(7), 1183–1197.

Anonymous. (1999). *Health Watch—Sensory changes in later life* [Web Page]. URL http://www.workingsolutions.com/lient/newsletter/newsletterarchive/hel_sensechange.html [1999, December 13].

Anonymous. (1999). *On-line Journal of Nursing Informatics* [Web Page]. URL http://cac.psu.edu/~dxm12/ [2000, January 4].

Anonymous. (1982). Review Symposium. *Aging & Society, 2,* 383–392.

Anonymous. (1999). Special Home Care Issue. *Caring, 18*(8), 1–75.

Anonymous. (1999). Special PPS Proposal Edition. *Home Health Line, 24*(41), 1–15.

Antonucci, T. C. (1986). Social support networks: A hierarchical mapping technique. *Generations, 10*(4), 10–12.

Antonucci, T. C., Sherman, A. M., & Vandewater, E. A. (1997). Measures of social support and caregiver burden. *Generations, 21*(1), 48–51.

Appelbaum, P., & Grisso, T. (1995). The MacArthur Competence Study I-III. *Law & Human Behavior, 19*(2), 105–174.

Applebaum, R., & Austin, C. (1990). *Long Term Care Case Management: Design and Evaluation.* New York: Springer Publishing Company.

Applegate, W. B. (1999). Hypertension. In W. R. Hazzard, J. P. Blass, W. H. Ettinger, J. B. Halter, & Ouslander J. G. (Eds.), *Principles of Geriatric Medicine and Gerontology* (4th ed., Vol. 52, pp. 713–720). New York: McGraw-Hill Publishing Company.

Aravanis, S. C., Adelman, R. D., Breckman, R., Fulmer, T., Holder, E., Lachs, M. S., O'Brien, J. G., & Sanders, A. B. (1992). *Diagnostic and Treatment Guidelines on Elder Abuse and Neglect.* Chicago, IL: American Medical Association.

Archbold, P. G., & Stewart, B. J. (1996). The nature of the family caregiving role and nursing interventions for caregiving families. In E. A. Swanson, & T. Tripp-Reimer (Eds.), *Advances in Gerontological Nursing* (pp. 133–156). New York: Springer Publishing Co.

Archbold, P. G., & Stewart, B. J. (1999). Strengthening Family-Based Care for Frail Elders: Lessons from the U.S. and U.K. Background Paper Prepared for the WHO Expert Panel on Home Care. Available from the Author.

Ardelt, M. (1997). Wisdom and life satisfaction in old age. *Journals of Gerontology, 52B,* P15–P27.

Arden, N. K., Nevitt, M. C., Lane, N. E., Gore, L. R., Hochberg, M. C., Scott, J. C., Pressman, A. R., & Cummings, S. R. (1999). Osteoarthritis and risk of falls, rates of bone loss, and osteoporotic fractures. Study of Osteoporotic Fractures Research Group. *Arthritis & Rheumatism, 42*(7), 1378–1385.

Arguelles, T., & Lowenstein, D. A. (1997). Research says si to development of culturally appropriate cognitive assessment tools. *Generations, 21*(1), 30–31.

Ariyoshi, S. (1984). *The Twilight Years.* Japan: Kodansha America.

Arking, R., Buck, S., Berrios, A., Dwyer, S., & Baker, G. T. (1991). Elevated paraquat resistance can be used as a bioassay for longevity in a genetically based long-lived strain of Drosophila. *Developmental Genetics, 12,* 362–370.

Arle, J. E., & Alterman, R. L. (1999). Surgical options in Parkinson's disease. *Medical Clinics of North America, 83*(2), 483–498.

Arno, P. S., Levine, C., & Memmott, M. M. (1999). The economic value of informal caregiving. *Health Affairs, 18*(2), 182–188.

Aronow, H. U., Gold, M., & Beck, J. C. (1999). *In Home Preventive Healthcare Program: Program Manual.* Santa Monica, CA: Center for Healthy Aging.

Assisted Living Federation of America. (1998). *The Assisted Living Industry: An Overview—1998.* Washington, DC: Assisted Living Federation of America.

Associated Press. (1987). *A Special Report: Guardians of the Elderly.* New York: Author.

Association of Retired Persons. (1999). *Out-of-Pocket Spending on Health Care by Medicare Beneficiaries Age 65 and Older: 1999 Projections.* Washington, D. C: Association of Retired Persons Public Policy Institute.

Atchley, R. C. (1993). Continuity theory and the evolution of activity in later adulthood. In J. R. Kelly (Ed.), *Activity and Aging* (pp. 5–16). Newbury Park, CA: Sage.

Atkinson, J. C., & Fox, P. C. (1992). Salivary gland dysfunction. *Clinics in Geriatric Medicine, 8,* 499–511.

Austrom, M. G., & Hendrie, H. C. (1990). Death of the personality: The grief response of the Alzheimer's disease family caregiver. *The American Journal of Alzheimer's Care & Related Disorders & Research,* (March/April), 16–27.

Avery, L. F. (1997). *Activity Programming in Long Term Care.* New York: Springer Publishing Company.

Avorn, J., Monane, M., Gurwitz, J. H., Glyn, R. J., Choodnovsky, I., & Lipzitz, L. A. (1994). Reduction of bacteriuria and pyuria after ingestion of cranberry juice. *JAMA, 271*(10), 751–754.

Ayello, E. A., Thomas, D. R., & Litchford, M. A. (1999). Nutritional aspects of wound healing. *Home Healthcare Nurse, 17*(11), 719–730.

Aziz, S. J. (2000). Los Angeles County Fiduciary Abuse Specialist Team: A model for collaboration. *Journal of Elder Abuse & Neglect, 12*(2).

Babbit, A. M. (1994). Review osteoporosis. *Orthopedics, 17*(10), 935–941.

Babchuck, N., Peters, G. R., Hoyt, D. R., & Kayser, M. A. (1979). The voluntary associations of the aged. *Journals of Gerontology, 34,* 579–587.

Bachman, R. (1992). *Crime Victimization in the City, Suburban, and Rural Areas.* (Report No. 135943). Washington, DC: Bureau of Labor Statistics, U.S. Department of Justice.

Bachman, R., Dillway, H., & Lachs, M. S. (1998). Violence against the elderly: A comparative analysis of robbery and assault across age groups. *Research on Aging, 20*(2), 183–198.

Bachman, R., & Saltzman, L. (1995). *Violence against women: Estimates from the redesigned survey.* (Report No. 154438). Washington, DC: Bureau of Labor Statistics, U.S. Department of Justice.

Bachmann, G. A. (1994). Vulvovaginal complaints. In R. A. Lobo (Ed.), *Treatment of the Post-Menopausal Woman: Basic and Clinical Aspects* (pp. 137–142). New York: Raven Press.

Bachus, M. A. (1998). HIV and the older adult. *Journal of Gerontological Nursing, 24*(11), 41–46.

Baer, D. (1998). *Awareness and popularity of property tax relief programs.* Washington, DC: AARP Public Policy Institute.

Baert, L., Ameye, F., Astrahan, M., & Petrovich, Z. (1993). Transurethral microwave hyperthermia for benign prostatic hyperplasia: The Leuven clinical experience. *Journal of Endourology, 7*(1), 61–69.

Baeton, G. C., Geerdes, B. P., Adang, E. M., Heineman, E., Konsten, J., Engle, G. L., Kester, A. D., Spaans, F., & Soeters, P. B. (1995). Anal dynamic graciloplasty in the treatment of intractable fecal incontinence. *NEJM, 332,* 1600–1606.

Baggs, J. G., Ryan, S. A., Phelps, C. E., Richeson, J. F., & Johnson, J. E. (1992). The association between interdisciplinary collaboration and patient outcome in a medical intensive care unit. *Heart & Lung, 21*(1), 18–24.

Bahadur, G. G., & Sinskey, R. M. (2000). *Manual of cataract surgery.* Boston, MA: Butterworth-Heineman.

Baker, D. P. (1999). *Virginia opens special prison for aging inmates* [Web Page]. URL http://www.dallasnews.com/national/0703nat4prison.htm [2000, August 8].

Baker, M. E. (1997). Service needs, usage, and delivery: A look at the imbalance for African American elderly. *Journal of Poverty, 1*(1), 93–108.

Balducci, L., & Extermann, M. (1997). Cancer chemotherapy in the older person: What the medical oncologist needs to know. *Cancer, 80*(7), 1317–1322.

Balducci, L., Silliman, R. A., & Baekey, P. (1998). Breast Cancer: an oncological perspective—Part I. In L. Balducci, G. Lyman, & W. Ershler (Eds.), *Comprehensive Geriatric Oncology* (pp. 629–660). Harwood Academic Publishers.

Baldwin, D. C. Jr., & Tsukuda, R. A. (1984). Interdisciplinary teams. In C. Cassel, & J. R. Walsh (Eds.), *Geriatric Medicine, Volume II, Fundamentals of Geriatric Medicine* (Chap. 30). New York: Springer-Verlag.

Ball, K., & Owsley, C. (1993). The useful field of view test: A new technique for evaluating age-related declines in visual function. *Journal of the American Optometric Association, 64,* 71–79.

Ballard, C., Cassidy, G., Bannister, C., & Mohan, R. (1993). Prevalence, symptom profile, and aetiology of depression in dementia sufferers. *Journal of Affective Disorders, 29,* 1–6.

Baltes, P. B., & Baltes, M. M. (1990). Selective optimization with compensation. In P. B. Baltes, & M. M. Baltes (Eds.), *Successful Aging: Perspectives from the Behavioral Sciences.* New York: Cambridge University Press.

Bamford, C., Gregson, B., Farrow, G., Buck, D., Dowswell, T., McNamee, P., & Bond, J. (1998). Mental and physical frailty in older people. *Aging & Society, 18,* 317–354.

Bamford, J., Dennis, M., Sandercock, P., Burn, J., & Warlow, C. (1990). The frequency, causes and timing of death within 30 days of a first stroke: The Oxfordshire Community Stroke Project. *Journal of Neurology, Neurosurgery & Psychiatry, 53*(10), 824–829.

Baraff, L. J., Della Penna, R., Williams, N., & Sanders, A. (1997). Practice guideline for the management of falls in community-dwelling elderly persons. [Review] [152 references]. *Annals of Emergency Medicine, 30*(4), 480–492.

Barber, H. R. K. (1994). Gynecological problems. In B. A. Eskin (Ed.), *The Menopause: Comprehensive Management* (3rd ed., pp. 183–210). New York: McGraw-Hill, Inc.

Barczak, C. A., Barnett, R. I., Childs, E. J., & Bosley, L. M. (1997). Fourth national pressure ulcer prevalence survey. *Advances in Wound Care, 10*(4), 18–26.

Barnes, J. F. (1990). *Myofascial release: The search for excellence.* Paoli, PA: Rehabilitation Services, Inc.

Barnes, P. (1992). Poorly perceived asthma. *Thorax, 47,* 408–409.

Barnum, B. (1996). *Spirituality in Nursing: From Traditional to New Age.* New York: Springer Publishing Company.

Barnum, B. (1999). *Teaching Nursing in the Era of Managed Care.* New York: Springer Publishing Company.

Baron, J. H. (1987). Signpost your hospital. *BMJ, 295,* 482–484.

Barrett, V. W., Gurland, B. J., & Chin, J. (2000). The QoL-100: A new instrument for measuring quality of life, preliminary utilities. *Journal of Nursing Measurement, in press.*

Barth, M. C., McNaught, W., & Rizzi, P. (1993). Corporations and the aging workforce. In P. H. Mirvis, *Building the Competitive Workforce.* New York: John Wiley & Sons.

Bassford, T. L. (1995). Health status of Hispanic elders. *Clinics In Geriatric Medicine: Ethnogeriatrics, 11*(1), 25–38.

Bassuk, S. S., Glass, T. A., & Berkman, L. F. (1999). Social disengagement and incident cognitive decline in community-dwelling elderly persons. *Annals of Internal Medicine, 131,* 3.

Bean, F. D., Myers, G. C., Angel, J. L., & Galle, O. R. (1994). Geographic concentration, migration, and population redistribution among the elderly. In L. G. Martin, & S. H. Preston (Eds.), *Demography of aging* (pp. 319–355). Washington, DC: National Academy Press.

Beauchamp, T. L., & Childress, J. F. (1994). *Principles of Biomedical Ethics* (4th ed.). New York: Oxford University Press.

Beauregard, S., & Gilchrest, B. A. (1987). A survey of skin problems and skin care regimes in the elderly. *Archives of Dermatology, 123,* 1638–1643.

Beaver, K., Luker, K. A., Owens, R. G., Leinster, S. J., Degner, L. F., & Sloan, J. A. (1996). Treatment decision making in women newly diagnosed with breast cancer. *Cancer Nursing, 19*(1), 8–19.

Beck, A. T., Steer, R. A., & Brown, G. K. (1996). *Beck Depression Inventory II.* San Antonio, TX: The Psychological Corporation.

Beck, A. T., Ward, C. H., Mendelson, M., Mock, J., & Erbaugh, J. (1961). An inventory for measuring depression. *Archives of General Psychiatry, 4,* 561–571.

Beck, C., Heacock, P., Mercer, S. O., Walls, R. L., Rapp, C. G., & Vogelpohl, T. S. (1997). Improving dressing behavior in cognitively impaired nursing home residents. *Nursing Research, 46*(3), 126–132.

Beck, L. H. (1999). Aging changes in renal function. In W. R. Hazzard, J. P. Blass, W. H. Ettinger, J. B. Halter, & J. G. Ouslander (Eds.), *Principles of Geriatric Medicine and Gerontology* (4th ed., pp. 767–776). New York: McGraw-Hill.

Bellelli, G., Frisoni, G., Bianchetti, A., & Bofelli, S. (1998). Special care units for demented patients: A multi-center study. *The Gerontologist, 38*(4), 456–462.

Bellino, F., et al. (Eds.). (1995). Dehyroepiandrosterone (DHEA) and aging. *Annals of the New York Academy of Sciences, 774,* 1–350.

Bengtson, V. L., Cuellar, J. B., & Ragan, P. K. (1977). Stratum contrasts and similarities in attitudes toward death. *Journal of Gerontology, 32,* 76–88.

Benner, P., Hooper-Kyriakidis, P., & Stannard, D. (1999). *Clinical Wisdom Interventions in Critical Care: A Thinking Action in Approach.* Philadelphia, PA: W. B. Saunders Company.

Bennett, D., Beckett, L. A., Murray, A., Shannon, K., Goetz, C. G., Pilgrim, D. M., & Evans, D. A. (1996). Prevalence of parkinsonian signs and associated mortality in a community population of older people. *NEJM, 334*(2), 71–76.

Bennett, J. (1999). Medication adherence in rheumatoid arthritis patients: Older is wiser. *Journal of the American Geriatrics Society, 47,* 172–183.

Bennett, M. A. (1995). Report of the task force in the implications for darkly pigmented intact skin in the prediction and prevention of pressure ulcers. *Advances in Wound Care, 8*(6), 34–35.

Bentely, J. R., Delfino, R. J., & Taylor, T. H. (1998). Differences in breast cancer stage at diagnosis between non-Hispanic white and Hispanic populations, San Diego County. *Breast Cancer Research Treatment, 50*(1), 1–9.

Berg, D. (1993). Dermatology. In A. H. Goroll, *Primary Care Medicine* (p. 505).

Berg, D., Worzala, K., & Pachner, R. (1998). *Handbook of Primary Care Medicine.* New York: Lippincott Williams & Wilkins Publishers.

Berg, K., Wood-Dauphinee, S., Williams, J. I., & Gayton, D. (1989). Measuring balance in the elderly: Preliminary development of an instrument. *Physiotherapy Canada, 41*(6), 304–311.

Berge, M., Silness, J., & Sorheim, E. (1987). Professional plaque control in the treatment of stomatitis prosthetica. *Gerodontics, 3,* 113–116.

Bergstrom, N., Allman, R. M., & Carlson, E. D. (1992). *Pressure Ulcers in Adults: Prediction and Prevention. Clinical Practice Guideline.* Rockville, MD: U.S. Department of Health and Human Services, Public Health Service, Agency for Health Care Policy and Research.

Bergstrom, N., Bennett, M. A., & Carlson, E. D. (1994). *Treatment of Pressure Ulcers. Clinical Practice Guideline, No. 15.* Rockville, MD: U.S. Department of Health and Human Services, Public Health Service, Agency for Health Care Policy and Research.

Berkman, B., Dobrof, R., Harry, L., & Damon-Rodriguez, J. (1998). Social Work. In S. Klein (Ed.), *A National Agenda for Geriatric Education: White Papers.* New York: Springer.

Berman, B. M., Singh, B. B., Lao, L., Langenberg, P., Li, H., Hadhazy, V., Bareta, J., & Hochberg, M. (1999). A randomized trial of acupuncture as an adjunctive therapy in osteoarthritis of the knee. *Rheumatology (Oxford), 38*(4), 346–354.

Biegel, D., Farkas, K., & Song, L. (1997). Barriers to the use of mental health services by African-American and Hispanic elderly persons. *Journal of Gerontological Social Work, 29*(1), 23–44.

Biegel, D. E., Johnsen, J. A., & Shafran, R. (1997). Overcoming barriers faced by African-American families with a family member with mental illness. *Family Relations, 46*(2), 163–178.

Biegel, D. E., & Schulz, R. (1999). Caregiving and caregiver interventions in aging and mental illness. *Family Relations, 48*(4), 345–354.

Biggs, S., Phillipson, C., & Kingston, P. (1995). *Elder Abuse in Perspective.* Buckingham, England: Open University Press.

Binstock, R. H. (1991). From the great society to the aging society: 25 years of the Older Americans Act. *Generations, 15*(3), 11–18.

Birren, J. E., & Deutchman, D. E. (1991). *Guiding autobiography*. Baltimore: Johns Hopkins University Press.

Birrer, R. B., Dellacorte, M. P., & Grisafi, P. J. (1998). *Common Foot Problems in Primary Care*, 2nd ed. Philadelphia, PA: Henley & Belfus, Inc.

Biship, C. E. (1999). Where are the missing elders? The decline in nursing home use, 1985 and 1995. *Health Affairs, 18*(4), 146–155.

Bissonnette, A., & Hijjazi, K. M. (1994). Elder homelessness: A community perspective. *Nursing Clinics of North America, 29*, 409–416.

Black, D. M., Browner, W. S., & Cauley, J. A. (1990). Appendicular bone density and age predict hip fracture in women: The Study of Osteoporotic Fractures Research Group. *JAMA, 263*, 665–668.

Black, H. C. (1991). *Black's Law Dictionary* (6th ed.). St. Paul, MN: West Publishing Co.

Black, J. (1990). Health promotion and disease prevention in the elderly. *Archives of Internal Medicine, 150*, 389–393.

Blackhall, L. J., Frank, G., Murphy, S. T., Michel, V., Palmer, J. M., & Azen, S. P. (1999). Ethnicity and attitudes toward life sustaining technology. *Social Science & Medicine, 48*(12), 1779–1789.

Blackhall, L. J., Murphy, S. T., Frank, G., Michel, V., & Azen, S. P. (1995). Ethnicity and attitudes toward patient autonomy. *JAMA, 274*(10), 820–825.

Blaikie, A. (1999). *Aging and Popular Culture*. New York: Cambridge University Press.

Blazer, D., & Williams, C. D. (1980). Epidemiology of dysphoria and depression in an elderly population. *American Journal of Psychiatry, 137*, 439–444.

Blinder, M. A. (1998). Anemia and transfusion therapy. In *The Washington Manual of Medical Therapeutics* (29th ed., Chap. 19). New York: Lippincott-Raven.

Bliwise, D. L., King, A. C., & Harris, R. B. (1994). Habitual sleep durations and health in a 50-65 year old population. *Journal of Clinical Epidemiology, 47*, 35–41.

Bloem, B., Haan, J., & Lagaay, A. (1992). Investigation of gait in elderly subjects over 88 years of age. *Journal of Geriatric Psychiatry & Neurology, 5*, 78–84.

Blum, B. (1998). *Mental Capacity Inventory*. Tucson, AZ: Author.

Blum, B., & Eth, S. (2000). Forensic issues in geriatric psychiatry. In B. Sadock, & V. Sadock (Eds.), *Kaplan and Sadock's Comprehensive Textbook of Psychiatry* (7th ed., pp. 3150–3158). Baltimore, MD: Williams and Wilkins.

Blum, B. (1999). *Undue Influence In Elder Financial Abuse—Testimony to the Senate Committee on Commerce, Science and Transportation, Hearing on "Fraud: Targeting America's Seniors."* [Web Page]. URL http://www.senate.gov/~commerce/hearings/hearings.htm [1999, August 4].

Blumenthal, H. (1999). A View of the Aging Disease Relationship From Age 85. *Journals of Gerontology, 54A*(6), B255–259.

Bodenheimer, T. (1999). Long-term care for the frail elderly—the On Lok Model. *NEJM, 341*(17), 1324–1328.

Bombardier, C., Melfi, C. A., Paul, J., Green, R., Hawker, G., Wright, J., & Coyte, P. (1995). Comparison of a generic and a disease-specific measure of pain and physical function after knee replacement surgery. *Medical Care, 33*(4 Supplement), AS131–AS144.

Bonomo, R., & Rice, L. B. (1999). Emerging issues in antibiotic resistant infections in long-term care facilities. *Journals of Gerontology, 54A*(6), B260–267.

Bordelais, P. (1994). Le vieillissement de la population: Question d'actualité ou notion obsolete? *Le Debat, Novembre-Decembre, 82*, 173–192.

Boult, C., David, B., McCaffrey, D., Boult, L., Hernandez, R., & Krulewitch, H. (1993). Screening elders for risk of hospital admission. *Journal of the American Geriatrics Society, 41*(8), 811–817.

Boult, C., & Pacala, J. T. (1999). Integrating healthcare for older populations. *American Journal of Managed Care, 6*, 45–52.

Bowe, J. (1986). *Changing the Rules*. Silver Spring, MD: T. J. Publishers.

Bowler, J. V., Steenhuis, R., & Hachinski, V. (1999). Conceptual background to vascular cognitive impairment. *Alzheimer's Disease & Associated Disorders, 13*(3 Supplement), S21–29.

Bowling, A. (1991). Social support and social networks: Their relationship to the successful and unsuccessful survival of elderly people in the community: An analysis of concepts and a review of the evidence. *Family Practice, 88*, 68–83.

Boyce, H. W. (1997). Therapeutic approaches to healing esophagitis. *American Journal of Gastroenterology, 92*(4 supplement), 22S-27S.

Braden, B., & Bergstrom, N. A. (1987). A conceptual schema for the study of the etiology of pressure sores. *Rehabilitation Nursing, 12*(1), 8–12.

Braden, B. R., Cowan, C. A., Lazenby, H. C., Martin, A. B., McDonnell, P. A., Sensing, A. L., Stiller, J. M., Whittle, L. S., D.C. S., L. A. M., & Stewart, M. W. (1998). National Health Expenditures, 1997. *Health Care Financing Review, 20*(1), 127–132.

Bragg, R. (1999 October). Fearing isolation in old age, gay generation seeks haven. *The New York Times*, pp. 1, 16.

Braithwaite, V. (1998). Institutional respite care: Breaking chores or social bonds? *Gerontologist, 38,* 610–617.

Bramwell, R. D. (1993). Seniors as volunteers and their training. *Journal of Volunteer Administration, 12,* 47–57.

Branch, L. G., Coulam, R. F., & Zimmerman, Y. A. (1995). The PACE evaluation: Initial findings. *The Gerontologist, 35,* 349–359.

Brand, C. A., Jolley, D., Tellus, M., Muirden, K. D., & Wark, J. D. (1999). Risk factors for osteoporosis and fracture in patients attending rheumatology outpatient clinics. *Australian & New Zealand Journal of Medicine, 29*(2), 197–202.

Brandtstader, J., & Greve, W. (1994). The aging self: Stabilizing and protective processes. *Developmental Review, 14,* 52–80.

Braun, B. (1991). The effect of nursing home quality on patient outcomes. *Journal of the American Geriatrics Society, 40,* 811–816.

Braun, K. L., Horwitz, K. J., & Kaku, J. M. (1988). Successful foster caregivers of geriatric patients. *Health & Social Work, 13*(1), 25–34.

Brawley, E. G. (1997). *Designing for Alzheimer's Disease.* New York: Wiley Publishing Co.

Bray, G. (1996). Hazards of obesity. *Endocrinology & Metabolism Clinics of North America, 25*(4), 907–919.

Bray, G. (1992). Pathophysiology of obesity. *American Journal of Clinical Nutrition, 55*(suppl.), 488S–494S.

Breuil, V., De Rotrou, J., Forette, F., Tortrat, D., Ganansia-Ganem, A., Frambourt, A., Moulin, F., & Boller, F. (1994). Cognitive stimulation of patients with dementia: Preliminary results. *International Journal of Geriatric Psychiatry, 9,* 211–217.

Brickner, P. W., Kellog, F. R., Lechich, A. K., Lipsman, R., & Sharer, L. K. (1997). *Geriatric Home Health Care: The Collaboration of Physicians, Nurses and Social Workers.* New York: Springer Publishing Company.

Briel, J. W., Stoker, J., Rociu, J. E., Lameris, J. S., Hop, W. C., & Schouten, W. R. (1999). External anal sphincter atrophy on endoanal magnetic resonance imaging adversely affects continence after sphincteroplasty. *British Journal of Surgery, 86,* 1322–1327.

Brill, P. A., Cornman, C. B., Davis, D. R., Lane, M. J., Mustafa, T., Sanderson, M., & Macera, C. A. (1999). The value of strength training for older adults. *Home Care Provider, 4*(2), 62–66.

Brock, D. W. (1993). The ideal of shared decision making between physicians and patients. *Life and Death: Philosophical Essays in Biomedical Ethics* (pp. 55–79). New York: Cambridge University Press.

Brocklehurst, J. C. (1997). Geriatric medicine in Britain: The growth of a specialty. *Age & Ageing, 26*(Supplement 4), 5–8.

Brown, B. (1997). *The Low Vision Handbook.* Thorofare, NJ: Slack, Inc.

Brown, J. B., Beck, A., Boles, M., & Barrett, P. (1999). Practical methods to increase use of advance medical directives. *Journal of General Internal Medicine, 14*(1), 21–26.

Brown, T. M., Golden, R. N., & Evans, D. L. (1990). Organic affective psychosis associated with the routine use of non-prescription cold preparations. *British Journal of Psychiatry, 156,* 572–575.

Brudney, K., & Dobkin, J. (1991). Resurgent tuberculosis in New York City. *American Review of Respiratory Disease, 144,* 745–749.

Brusco, L. I., Fainstein, I., Marquez, M., & Cardinali, D. P. (1999). Effect of melatonin in selected populations of sleep-disturbed patients. *Biological Signals & Receptors, 8*(1–2), 126–131.

Brush, B. L., & Daly, P. R. (in press). Assessing spirituality in primary care practice: Is there time? *Clinical Excellence for Nurse Practitioners.*

Bryk, J. A., & Soto, T. K. (1999). Report of the 1997 membership database of the American Dietetic Association. *Journal of the American Dietetic Association, 99,* 102–107.

Bryson, K., & Casper, L. M. (1999). *Coresident grandparent and grandchildren.* (Report No. P23–198). Washington, DC: U.S. Bureau of the Census.

Buchner, D. M., Cress, M. E., de Lateur, B. J., Esselman, P. C., & Margherita, A. J. (1997). A comparison of the effects of three types of endurance training on balance and other fall risk factors in older adults. *Aging & Clinical Experimental Research, 9,* 112–119.

Buckingham, R. W. (1994). *When living alone means living at risk: A guide for caregivers and families.* Buffalo, NY: Prometheus Books.

Buettner, L. L. (1998). A team approach to dynamic programming on the special care unit. *Journal of Gerontological Nursing, 24*(1), 23–30.

Buhler, C., & Massarik, F. (Eds.). (1968). *The Course of Human Life.* New York: Springer.

Bureau of Justice Statistics. (1994) *Criminal Victimization in the United States, 1994* [Web Page]. URL http://www.ojp.usdoj.gov/bjs/abstract/cvius94.htm.

Burgio, L. D., Jones, L. T., & Engel, B. T. (1988). Studying incontinence in an urban nursing home. *Journal of Gerontological Nursing, 14,* 40–45.

Burkhardt, J. E., Berger, A. M., Creedon, M., & McGavock, A. T. (1998). *Mobility and Independence: Changes and Challenges for Older Drivers* [Web Page]. URL http://www.aoa.dhhs.gov/research/drivers.html [2000, June 28].

Burnette, D. (1997). Grandparents raising grandchildren in the inner city. *Families in Society, 78*(5), 489–499.

Burnette, D. (1999). Physical and emotional well-being of Latino grandparent caregivers. *American Journal of Orthopsychiatry, 69*(3), 305–318.

Burnette, D. (1999). Social relationships of Latino grandparent caregivers: A role theory perspective. *The Gerontologist, 37*(1), 49–58.

Burney-Puckett, M. (1996). Sundown syndrome: Etiology and management. *Journal of Psychosocial Nursing & Mental Health Services, 34*(5), 40–43.

Burney-Puckett, M. (1988). I. Burnside (Ed.), *Nursing and the Aged: A Self Care Approach.* New York: McGraw-Hill.

Burnside, I., & Schmidt, M. G. (1994). Working with older adults: Group processes and techniques, 3rd ed. Boston: Jones and Bartlett.

Burt, B. B. (1992). Epidemiology of dental disease in the elderly. *Clinics in Geriatric Medicine, 8,* 447–459.

Burt, M. R. (1996). Homelessness: Definitions and Counts. In J. Baumohl (Ed.), *Homeless in America* (pp. 15–23). Phoenix, AZ: Oryx Press.

Burton, L. M. (1992). Black grandparent rearing grandchildren of drug addicted parents: Stressors, outcomes, and social service needs. *The Gerontologist, 32*(6), 744–751.

Buschke, H., Kuslansky, G., Katz, M., Stewart, W. F., Sliwinski, M. J., Eckholdt, H. M., & Lipton, R. B. (1999). Screening for dementia with the Memory Impairment Screen. *Neurology, 52*(2), 231–238.

Buse, J. B., & Hroscikoski, M. (1998). The case for postprandial glucose monitoring in diabetes management. *Journal of Family Practice, 47*(5), S29–S36.

Butler, J. C., Breiman, R. F., Campbell, J. F., Lipman, H. B., Broome, C. V., & Facklam, R. R. (1993). Pneumococcal polysaccharide vaccine efficacy: An evaluation of current recommendations. *JAMA, 270,* 1826–1831.

Butler, R. (1963). The life review: An interpretation of reminiscence in the aged. *Psychiatry, 26,* 65–70.

Butler, R. M., & Lewis, M. (1977). *Love and Sex After 60.* New York: Harper & Row.

Butler, R. N. (1969). Ageism: Another form of bigotry. *The Gerontologist, 9,* 243–246.

Butler, R. N. (1963). The life review: An interpretation of reminiscence in the aged. *Psychiatry, 26,* 65–69.

Butler, R. N., Lewis, M. I., Hoffman, E., & Whitehead, E. D. (1994). Love and sex after 60: How to evaluate and treat the impotent older man. *Geriatrics, 49,* 27–32.

Buysse, D. J., Reynolds, C. F., Monk, T. H., Berman, S. R., & Kupfer, D. J. (1989). The Pittsburgh Sleep Quality Index: A new instrument for psychiatric practice and research. *Psychiatry Research, 28,* 193–213.

Byock, I. (1997). *Dying Well: A Prospect for Growth at the End of Life.* New York: Riverhead Books.

Byock, I. (1998). *Dying Well: Peace and Possibilities at the End of Life.* New York: Riverhead Books.

Bytheway, B. (1995). *Ageism. Rethinking Aging Series.* Buckingham, Great Britain: Open University Press.

Byyny, R. L., & Speroff, L. (1996). *A Clinical Guide for the Care of Older Women: Primary and Preventive Care* (2nd ed.). Baltimore: Williams & Wilkins.

Cahn, E., & Rowe, J. (1992). *Time Dollars: The New Currency that Enables Americans to Turn Their Hidden Resource-Time into Personal Security and Community Renewal.* Emmaus, PA: Rodale Press.

Cairl, R. E., Pfeiffer, E., Keller, D. M., Burke, H., & Samis, H. V. (1983). An evaluation of the reliability and validity of the Functional Assessment Inventory. *Journal of the American Geriatrics Society, 31*(10), 607–612.

Calkins, M. P. (1988). *Designing for Dementia: Planning Environments for the Elderly and the Confused.* Owings Mills, MD: National Health Publishing.

Callahan, C. M., Haag, K. M., Buchanan, N. N., & Nisi, R. (1999). Decision-making for percutaneous endoscopic gastrostomy among older adults in a community setting. *Journal of the American Geriatrics Society, 47*(9), 1105–1109.

Campbell, A. J., Robertson, M. C., Gardner, M. M., Norton, R. N., Tilyard, M. W., & Buchner, D. M. (1997). Randomised controlled trial of a general practice programme of home based exercise to prevent falls in elderly women. *BMJ, 315,* 1065–1069.

Campbell, J. C., & Ikegami, N. (1999). *Long Term Care for Frail Older People—Reaching for the Ideal System.* Tokyo: Springer-Verlag.

Capezuti, E., Strumpf, N., Evans, L. K., Grisso, J. A., & Maislin, G. (1998). The relationship between physical restraint removal and falls and injuries among nursing home residents. *Journals of Gerontology, 53A,* M47–M52.

Capra, F. (1996). *The Web of Life: A New Scientific Understanding of Living Systems.* New York: Doubleday.

Caralis, P. V., Davis, B., Wright, K., & Marcial, E. (1993). The influence of ethnicity and race on attitudes toward advance directives, life-prolonging treatment, and euthanasia. *Journal of Clinical Ethics, 4*(2), 155–165.

Carethers, M. (1992). Health promotion in the elderly. *American Family Physician, 45*(5), 2253–2259.

Carlsen, M. B. (1991). *Creative Aging.* New York: Norton and Company.

Carp, F. M. (1968). Some components of disengagement. *Journal of Gerontology, 23,* 382–386.

Carpenter, C. C., Fischl, M. A., Hammer, S. M., Hirsch, M. S., Jacobsen, D. M., Katzenstein, D. A., Montaner, J. S., Richman, D. D., Saag, M. S., Schooley, R. T., Thompson, M. A., Vella, S., Yeni, P. G., & Volberding, P. A. (1998). Antiretroviral therapy for HIV infection in 1998: Updated recommendations of the International AIDS Society—USA Panel. *JAMA, 280*(1), 78–86.

Carr, D. B., Jacox, A. X., & Chapman, B. (1992). *Acute pain management: Operative or medical procedures or trauma, Clinical Practice Guideline.* (Report No. 92-0032). Rockville, MD: U.S. Department of Health and Human Services, Public Health Agency, Agency for Health Care Policy and Research.

Carr, D. B., LaBarge, E., Dunnigan, K., & Storandt, M. (1998). Differentiating drivers with dementia of the Alzheimer type from healthy older persons with a Traffic Sign Naming test. *Journals of Gerontology, 53A*(2), M135–139.

Carrick, P. (1985). *Medical Ethics in Antiquity: Philosophical Perspectives on Abortion and Euthanasia.* Dordrecht, Netherlands: D. Reidel.

Carrieri-Kohlmann, V., Lindsey, A., & West, C. (1993). *Pathophysiological Phenomenon in Nursing: Human Responses in Illness* (2nd ed.). Philadelphia, PA: W. B. Saunders, Co.

Carrington, G., & Proctor, N. G. (1995). Identifying and responding to the needs of refugees: A global nursing concern. *Holistic Nursing Practice, 9*(2), 9–18.

Carstensen, L. L. (1993). Motivation for social contact across the life span. In J. E. Jacobs (Ed.), *Developmental Perspectives on Motivation* (pp. 209–254). Lincoln, NE: Nebraska University Press.

Carstensen, L. L., Isaacowitz, D. M., & Charles, S. T. (1999). Taking time seriously: A theory of socioemotional selectivity. *American Psychologist, 54*(3), 165–168.

Carter, M., Van Andel, G., & Robb, G. (1997). *Therapeutic Recreation: A Practical Approach.* Prospect Heights, IL: Waveland Press.

Carty, B. (2000). *Nursing Informatics: Education for Practice.* New York: Springer Publishing Company.

Castle, N. G., & Mor, V. (1998). Physical restraints in nursing homes: A review of the literature since the Nursing Home Reform Act of 1987. *Medical Care Research & Review, 53,* 139–170.

Catherall, D. (1999). Treating traumatized families. In C. Figley (Ed.), *Burnout in Families: The Systemic Costs of Caring.* Boca Raton, FL: CRC Press.

Caygill, C. P., Charlett, A., & Hill, M. J. (1998). Relationship between intake of high fiber foods and energy risk of cancer of the large bowel and breast. *European Journal of Cancer Prevention, 7*(Suppl. 2), 11–17.

Cefalu, C. A., & Agcaoli, D. J. (1998). Preventing antibiotic misuse in older patients. *Hospital Medicine, 34*(8), 41–50.

Center for Health Law and Ethics, University of New Mexico School of Law. *Values History Form* [Web Page]. URL http://www.hospicefed.org/hospice_pages/valuform.html [1999, August 16].

Center on Budget and Policy Priorities. (1998). *Social Security and Poverty Among the Elderly: A National and State Perspective.* Washington, DC: Author.

Centers for Disease Control and Prevention. (1999). *HIV/AIDS Surveillance Report: U.S. HIV and AIDS cases reported through December 1999 Year-End edition* [Web Page]. URL http://www.cdc.gov/hiv/stats/hasr1102.htm [2000, January 15].

Centers for Disease Control and Prevention. (1999). *Recommendations of the Advisory Committee on Immunization Practices* [Web Page]. URL http://www.americangeriatrics.org/immsched.html [2000, January 1].

Centers for Disease Control and Prevention, National Center for Health Statistics. (1999). Mortality from Alzheimer's Disease: An Update. *National Vital Statistics Reports, 47*(20), 1–8.

Chambre, S. M. (1984). Is volunteering a substitute for role loss in old age? An empirical test of the activity theory. *Gerontologist, 24,* 292–298.

Champion, H. R., Copes, S. W., Buyer, D., Flanagan, M. E., Bain, L., & Sacco, W. J. (1989). Major trauma in the geriatric patient. *American Journal of Public Health, 79*(9), 1278–1282.

Chan, K. M., & Tan, E. S. (1990). Use of lower limb prosthesis among elderly amputees. *Annals of the Academy of Medicine, Singapore, 19*(6), 811–816.

Chance, S. E., Kaslow, N. J., Summerville, M. B., & Wood, K. (1998). Suicidal behavior in African American individuals: current status and future directions. *Cultural Diversity & Mental Health, 4*(1), 19–37.

Chandler, J. (1996). Understanding the relationship between strength and mobility in frail older persons: A review of the literature. *Topics in Geriatric Rehabilitation, 11,* 20–37.

Chatta, G. S., & Lipschitz, D. A. (1999). Anemia. In *Principles of Geriatric Medicine and Gerontology* (4th ed., Chap. 69, pp. 899–906). New York: McGraw-Hill.

Chatterji, P., Burstein, N. R., Kidder, D., & White, A. J. (1998). *Evaluation of the Program of All-Inclusive Care for the Elderly: The Impact of Pace on Participant Outcomes.* Cambridge, MA: ABT Associates.

Chew, W. M., & Bimbaumer, D. M. (1999). Evaluation of the elderly patient with weakness: An evidenced based approach. *Emergency Medical Clinics of North America, 17*(1), 265–278.

Chiba, H. W., DeGara, C. J., Wilkinson, J. M., & Hunt, R. H. (1997). Speed of healing and symptom relief in grade II to IV gastroesophageal reflux disease: A meta-analysis. *Gastroenterology, 112*(6), 1798–1810.

Chiriboga, D. A., & Pierce, R. C. (1993). Changing contexts of activity. In J. R. Kelly *Activity and Aging* (pp. 42–59). Newbury Park, CA: Sage.

Choa, M. S. (1987). Use of preventive care by the elderly. *Preventive Medicine, 16*, 710–722.

Choban, P. S., Heckler, R., Burge, J. C., & Flancbaum. (1995). Increased incidence of nosocomial infections in obese surgical patients. *American Surgeon, 61*(11), 1001–1005.

Christiansen, C. (1991). Occupational therapy: Intervention for life performance. In C. Christiansen, & C. Baum (Eds.), *Occupational Therapy: Overcoming Human Performance*. Thorofare, NJ: Slack, Inc.

Citro, J., & Hermanson, S. (1999). *Assisted Living in the United States*. Washington, DC: Public Policy Institute.

Clark, R., & McDermed, A. (1990). *The Choice of Pension Plans in a Changing Regulatory Environment*. Washington, DC: The American Enterprise Institute.

Clark, R. L. (1994). Employment Costs and the Older Worker. In S. E. Rix (Ed.), *Older Workers: How Do They Measure Up?* Washington, DC: American Association of Retired Persons.

Closs, J. (1996). Pain and elderly patients: A survey of nurses' knowledge and experiences. *Journal of Advanced Nursing, 23*, 237–242.

Cnaan, R., & Cwikel, J. G. (1992). Elderly volunteers: Assessing their potential as an untapped resource. *Journal of Aging & Social Policy, 4*, 125–147.

Cobbs, E. L., Duthie, E. H., & Murphy, J. B. (Eds.). (1999). Geriatrics Review Syllabus: A Core Curriculum in Geriatric Medicine. 4th ed. Dubuque, Iowa: Kendall/Hunt Publishing Company for the American Geriatrics Society.

Cohen, C. I. (1999). Aging and homelessness. *The Gerontologist, 39*(1), 5–14.

Cohen, G. (2000). *Creative Age: Awakening Human Potential in the Second Half of Life*. New York: Avon Books.

Cohen, G. D. (1995). Mental health promotion in later life: The case for social portfolio. *American Journal of Geriatric Psychiatry, 3*, 277–279.

Cohen-Mansfield, J. (1999). Measurement of inappropriate behavior associated with dementia. *Journal of Gerontological Nursing, 25*(2), 42–51.

Cohen-Mansfield, J., Besansky, J., Watson, V., & Bernhard, L. J. (1994). Underutilization of adult day care:

An exploratory study. *Journal of Gerontological Social Work, 22*, 21–39.

Cohen-Mansfield, J., Marx, M. S., & Rosenthal, A. A. (1990). A description of agitation in the nursing home. *Journals of Gerontology, 44A*, M77–84.

Cohen-Mansfield, J., & Werner, P. (1997). Management of verbally disruptive behaviors in nursing home residents. *Journals of Gerontology, 52A*, M369–M377.

Cohen, P. R., & Scher, R. K. (1992). Geriatric nail disorders: Diagnosis and treatment. *Journal of the American Academy of Dermatology, 26*, 521–531.

Cohendy, R. (1998). The Mini Nutritional Assessment for preoperative nutritional evaluation: a study on 419 elderly surgical patients. In B. Vellas, P. J. Garry, & Y. Guigoz (Eds.), *Mini Nutritional Assessment (MNA): Research and Practice in Elderly* (Vol. 1, pp. 117–122). Philadelphia, PA: Lippincott-Raven.

Cole, T. R., & Winkler, M. G. (Eds.). (1994). *The Oxford Book of Aging*. Oxford, England: Oxford University Press.

Colerick, E. J., & George, L. K. (1986). Predictor of institutionalization among caregivers of patients with Alzheimer's Disease. *Journal of the American Geriatrics Society, 34*, 493–498.

Colombo, V. E., Gerber, F., & Bronhofer, M. (1990). Treatment of brittle fingernails and onychoschizia with biotin: Scanning electron microscopy. *Journal of the American Academy of Dermatology, 23*, 1127–1132.

Come, B., & Fredericks, A. D. (1995). Family literacy in urban schools: Meeting the needs of at-risk children. *Reading Teacher, 48*(7), 566–570.

Commission on Practice, the American Occupational Therapy Association. (1995). New practice changes. *American Journal of Occupational Therapy, 49*(10), 1072–1073.

Congdon, J. G., & Magilvy, J. K. (1998). Rural nursing homes: A housing option for older adults. *Geriatric Nursing, 19*(3), 157–159.

Connecticut Community Care, Inc. (1994). *Guidelines for long term care management practice: A report of the national committee on long term care case management*. Bristol, CT: Author.

Connecticut State Department of Health. (1979). Hospice Care Program. Chapter 368A, 13-D4b, Department of Public Health. Title 19A. Public Health and Well Being.

Connell, B. R., & Wolf, S. L. (1997). Environmental and behavioral circumstances associated with falls at home among healthy elderly individuals. Atlanta FICSIT Group. *Archives of Physical Medicine & Rehabilitation, 78*(2), 179–186.

Conover, C. J., & Sloan, F. A. (unpublished). Bankruptcy risk and state regulation of continuing care retirement communities.

Conrad, K. J., Hanrahan, P., & Hughes, S. L. (1990). Survey of adult day care in the United States: National and regional findings. *Research on Aging, 12,* 21–39.

Conwell, Y. (1997). Management of suicidal behavior in the elderly. *Psychiatric Clinics of North America, 20*(3), 667–683.

Cook, N. R., Evans, D. A., & Funkestein, H. H. (1989). Correlates of headache in a population-based cohort of elderly. *Archives of Neurology, 46,* 1338–1344.

Cooley, S. G., Goodwin-Beck, M. E., & Salerno, J. A. (1998). United States Department of Veterans Affairs health care for aging veterans. In B. I. Villas, J. P. Michel, & L. Z. Rubenstein (Eds.), *Geriatric Programs & Departments Around the World.* New York: Springer Publishing Company.

Costa, P. T., Williams, T. F., & Somerfield, M. (1996). *Early Identification of Alzheimer's Disease and Related Dementias. Clinical Practice Guideline. Quick Reference for Clinicians.* (Report No. 97-0703). Rockville, MD: U.S. Department of Health and Human Services, Public Health Service, Agency for Health Care Policy and Research.

Couch, K. A. (1998). Late Life Job Displacement. *The Gerontologist, 38*(1), 7–17.

Council on Social Work Education. (1999). *SAGE-SW Project: Strengthening Aging and Gerontology Education for Social Work* [Web Page]. URL http://www.cswe.org/sage./htm. [1999, December 21].

Council on Scientific Affairs. (1991). Hispanic Health in the United States. *JAMA, 265*(2), 248–252.

Counsell, S. R., Kennedy, R. D., Szwabo, P., Wadsworth, N. S., & Wohlgemuth, C. (1999). Curriculum recommendations for resident training in geriatrics interdisciplinary team care. *Journal of the American Geriatrics Society, 47*(9), 1145–1148.

Cova, D., Beretta, G., & Balducci, L. (1998). Cancer chemotherapy in the older patient. In L. Balducci, G. H. Lyman, & W. Ershler (Eds.), *Comprehensive Geriatric Oncology* (pp. 429–442). Harwood Academic Publishers.

Covinsky, K. E., Palmer, R. M., Kresevic, D. M., Kahana, E., Fortinsky, R. H., & Landefeld, C. S. (1998). Improving functional outcomes in older patients: Lessons from an acute care for elders unit. *Joint Commission Journal on Quality Improvement, 24,* 63–76.

Coward, R., Bull, C., Kukulka, G., & Galliher, J. (1994). *Health Services for Rural Elders.* New York: Springer Publishing Co.

Coward, R., & Krout, J. (1990). *Aging in Rural Settings.* New York: Springer Publishing Co.

Crawford, F., Freeman, M., Shinka, J., Abdullah, L., Morris, M., Krivian, K., Richards, D., & Duara, R. M. M. (in press). A functional polymorphism in the Cystatin C gene (CST3) is a novel genetic risk factor for late-onset Alzheimer's Disease.

Crews, J. E. (1994). The demographic, social, and conceptual contexts of aging and vision loss. *Journal of the American Optometric Association, 65*(1), 63–68.

Crews, L., Brown, M., & Norton, B. J. (1997). *Effectiveness of the physical performance test for detecting and monitoring changes in functional status in elderly patients receiving home physical therapy services.* St. Louis, MO: Unpublished Master's Thesis, Washington University.

Crimmins, E. M., Saito, Y., & Reynolds, S. L. (1997). Further evidence on recent trends in the prevalence and incidence of disability among older Americans from two sources: The LSOA and the NHIS. *Journals of Gerontology, 52B,* S59–S71.

Cross, P. S. (1999). *Report on assistive housing for elderly gays and lesbians in New York City: Extent of need and the preferences of elderly gays and lesbians.* New York: Brookdale Center on Aging at Hunter College.

Crown, W. H., Capitman, J., & Leutz, W. N. (1992). Economic rationality, the affordability of long-term care insurance and the role of public policy. *Gerontologist, 32,* 478–485.

Csikszentmihalyi, M. (1990). *Flow: The psychology of optimal experience.* New York: Harper & Row.

Cullen, K., & Moran, R. (1992). *Technology and the Elders* (p. 76). Brussels, Belgium: Commission of the European Communities.

Cumming, E. & H. W. (1961). *Growing old: The process of disengagement.* New York: Basic Books.

Cummings, J. L. (2000). Cholinesterase inhibitors: A new class of psychotropic compounds. *American Journal of Psychiatry, 157,* 4–15.

Cummings, J. L. (1999). Managing psychosis in patients with Parkinson's disease. *NEJM, 340,* 801–803.

Cummings, S. R., Rubin, S. M., & Black, D. (1990). The future of hip fractures in the United States: Numbers, costs and potential effects of postmenopausal estrogen. *Clinical Orthopaedics & Related Research, 252,* 63–66.

Cundiff, G. W., & Addison, W. A. (1998). Management of pelvic organ prolapse. *Obstetrics & Gynecology Clinics of North America, 1225*(4), 907–921.

Cutler, S. J., & Danigelis, N. L. (1993). Organized contexts of activity. In J. R. Kelly (Ed.), *Activity and aging* (pp. 146–163). Newbury Park, CA: Sage.

Cutson, T. M., & Bongiorni, D. R. (1996). Rehabilitation of the older lower limb amputee: A brief review. *Journal of the American Geriatrics Society, 44*(11), 1388–1393.

Czaja, S. J. (1997). Computer technology and the older adult. In M. G. Helander, T. K. Landauer, & P. V. Prabhu (Eds.), *Handbook of Human-Computer Interaction* (pp. 797–812). Amsterdam: Elsevier Science B.V.

Daltroy, L. H., Logigian, M., Iversen, M. D., & Liang, M. H. (1992). Does musculoskeletal function deteriorate in a predictable sequence in the elderly? *Arthritis Care & Research, 5*(3), 146–150.

Damasio, A. R. (1999). *The feeling of what happens: Body and emotion in the making of consciousness.* New York: Harcourt Brace & Company.

Dambro, M. R. (1998). *Griffth's 5 Minute Clinical Consult.* Philadelphia, PA: Williams & Wilkins.

Damron-Rodriguez, J., Wallace, S., & Kington, R. (1994). Service utilization and minority elderly: Appropriateness, accessibility, and acceptability. *Journal of Gerontology & Geriatric Education, 15*(1), 45–63.

Dansky, K. H., Dellasega, C., Shellenberger, T., & Russo, P. (1997). After hospitalization: Home health care for elderly persons. *Clinical Nursing Research, 5*(2), 185–198.

Darnell, J. C., Murray, M. D., Martz, B. L., & Weinberger, M. (1986). Medication use by ambulatory elderly. An in-home survey. *Journal of the American Geriatrics Society, 34*(1), 1–4.

Dash, K., Zarle, N. C., O'Donnell, L., & Vince-Whitman, C. (1996). *Discharge Planning for the Elderly: An Educational Program for Nurses.* New York: Springer Publishing Company.

Davey, A., & Eggebeen, D. J. (1998). Patterns of intergenerational exchange and mental health. *Journals of Gerontology, 53B,* P86–95.

David, G., & Patterson, G. (1997). Productive aging: 1995 White House Conference on Aging, challenges for public policy and social work practice. In C. C. Saltz (Ed.), *Social work response to the White House Conference on Aging: From issues to actions* (pp. 9–25). Binghamton, NY: Haworth Press.

Daviglus, M. L., Liu, K., Greenland, P., Dyer, A. R., Garside, J. B., Mannheim, L., Lowe, L. P., Rodin, M., Lubitz, J., & Stamler, J. (1998). Benefits of a favorable cardiovascular risk-factor profile in middle age with respect to Medicare costs. *NEJM, 339,* 1122–1129.

Davis, C. M. (1997). *Complementary Therapies in Rehabilitation: Holistic Approaches for Prevention and Wellness.* Thorofare, NJ: Slack, Inc.

Davis, G. C. (1998). Nursing's role in pain management across the health care continuum. *Nursing Outlook, 46,* 19–23.

Davis, K. M., & Minaker, K. L. (1999). Disorders of fluid balance: Dehydration and hyponatremia. In W.

R. Hazzard, J. P. Blass, W. H. Ettinger, J. B. Halter, & J. G. Ouslander (Eds.), *Principles of Geriatric Medicine and Gerontology* (4th ed., pp. 1429–1441). New York: McGraw-Hill.

Dawson, P., Wells, D. L., & Kline, K. (1993). *Enhancing the Abilities of Persons with Alzheimer's Disease and Related Dementias: A Nursing Perspective.* New York: Springer Publishing Company.

Dawson, S. E. (1992). Navajo uranium workers and the effects of occupational illnesses: A case study. *Human Organization, 51*(54), 389–397.

Dayhoff, N., Suhrheinrich, J., Wigglesworth, J., Topp, R., & Moore, S. (1998). Balance and muscle strength as predictors of frailty among older adults. *Journal of Gerontological Nursing, 24,* 18–27.

De Geest, S., von Renteln-Kruse, W., Steeman, E., Degraeve, S., & Abraham, I. (1998). Adherence issues with the geriatric population. Complexity with aging. *Nursing Clinics of North America, 33*(3), 467–480.

De Klerk-Rubin, V. (1994). How validation is misunderstood. *Journal of Dementia Care, 2*(2), 14–16.

Deevey, S. (1990). Older lesbian women an invisible minority. *Journal of Gerontological Nursing, 16*(5), 35–38.

DeFronzo, R. A. (1999). Pharmacology therapy for type 2 diabetes mellitus. *Annals of Internal Medicine, 131*(4), 281–303.

Degner, L. F., Kristjanson, L. J., Bowman, D., et al. (1997). Information needs and decisional preferences in women with breast cancer. *JAMA, 277,* 1485–1492.

Dellasega, C., & Keiser, C. (1997). Pharmacologic approaches to chronic pain in the older adult. *Nurse Practitioner, 22*(5), 20–35.

Delon, M., & Wenston, S. R. (1989). An integrated theoretical guide to intervention with depressed elderly clients. *Journal of Gerontological Social Work, 14*(3/4), 131–146.

Demallie, D. A., North, C. S., & Smith, E. M. (1997). Psychiatric disorders among the homeless: A comparison of older and younger groups. *The Gerontologist, 37*(1), 61–66.

Dement, W. C., & Carskadon, M. A. (1982). Current perspectives on daytime sleepiness: The issues. *Sleep, 5,* S56–66.

Deneberg, R. (1994). *Report on lesbian health care.* Washington, DC: National Gay and Lesbian Task Force Publications.

DeNeve, K. M., & Cooper, H. (1998). The happy personality: A meta-analysis of 137 personality traits and subjective well-being. *Psychological Bulletin, 124,* 197–229.

Department of Health and Human Services. (1999). *Mental Health Parity Act: Medicare Consumer Informa-*

tion [Web Page]. URL http://www.hcfa.gov/medic-aid/bbamhp.htm [1999, September 15].

Department of Health and Human Services, Health Care Financing Administration. (2000). *Federal Register, 42 CFR 411 and 489 Medicare Program; Prospective Payment System and Consolidated Billing for Skilled Nursing Facilities—Update Proposed Rule.* Washington, DC: Department of Health and Human Services, Health Care Financing Administration.

Department of Health and Social Security and the Welsh Office. (1984). (Report No. 65: Building Components). London: Her Majesty's Stationery Office.

Deprez, R. D., Conelehan, J., & Hart, S. K. (1985). Hypertension prevalence among Penobscot Indians of Maine. *American Journal of Public Health, 75,* 653–654.

Desbiens, N. A., Mueller-Rizner, N., Connors, A. F., Hamel, M. B., & Wenger, N. S. (1997). Pain in the oldest-old during hospitalization and up to one year later. *Journal of the American Geriatrics Society, 45*(10), 1167–1172.

Destefano, F., Conelehan, J., & Wiant, M. K. (1979). Blood pressure survey on the Navaho Reservation. *American Journal of Epidemiology, 109,* 335–345.

Dewar, M. (1996). Soft tissue injury. In M. B. Mengel, & L. P. Scheibert (Eds.), *Ambulatory Medicine: The Primary Care of Families.* Stamford, CT: Appleton & Lange.

Dial, L. (1999). *Conditions of Aging.* Baltimore, MD: Williams & Wilkins.

Diener, E., & Suh, E. M. (1998). Age and subjective well-being: An international analysis. *Annual Review of Gerontology and Geriatrics, 17,* 304–324.

Diener, E., Suh, E. M., Lucas, R. E., & Smith, H. L. (1999). Subjective well-being: Three decades of progress. *Psychological Bulletin, 125,* 276–302.

Dipiro, J. T., Talbert, R. L., Yee, G. C., Matzke, G. R., Wells, B. G., & Posey, L. M. (1999). Pharmacotherapy: A pathophysiologic approach (4th ed.). Connecticut: Appleton & Lange.

Division of Nursing, Health Resources and Services Administration. (1996). *Notes from the national sample survey of registered nurses* [Web Page]. URL http://158.72.83.8/bhpr/dn/survnote.htm [1999, August 15].

Dobbs, A. R., Heller, R. B., & Schopflocher, D. (1998). A comparative approach to identify unsafe older drivers. *Accident Analysis & Prevention, 30*(3), 363–337.

Dobbs, R., Charlett, A., & Bowes, S. (1993). Is this walk normal. *Age & Ageing, 22,* 27–30.

Doerfler, E. (1999). Male erectile dysfunction: A guide for clinician management. *Journal of the American Academy of Nurse Practitioners, 11*(3), 117–123.

Donahue, W., Orbach, H. L., & Pollak, O. (1960). Retirement: The emerging social pattern. In C. Tibbitts (Ed.), *Handbook of social gerontology: Societal aspects of aging* (pp. 330–406). Chicago: University of Chicago Press.

Donahue, W. T. (Ed.). (1955). *Earning Opportunities for Older Workers.* Ann Arbor: University of Michigan Press.

Dooley, S. W., Jarvis, W. R., & Martone, W. J. (1992). Multi-drug resistant tuberculosis. *Annals of Internal Medicine, 117,* 257–259.

Doress, P. B., & Siegal, D. L. (1987). *Ourselves, Growing Older: Women Aging with Knowledge and Power.* New York: Simon & Schuster.

Dossey, L. (1997). The healing power of pets: A look at animal-assisted therapy. *Alternative Therapies in Health & Medicine, 3*(4), 8–16.

Doty, P., Benjamin, A. E., Matthias, R. E., & Franke, T. M. (1999). (Report No. 100-94-0022). Baltimore, MD: Office of Disability, and Aging and Long-Term Care Policy of the Office of the Assistant Secretary for Planning and Evaluation.

Dow, J. A., & Mest, C. G. (1997). Psychosocial interventions for patients with chronic obstructive pulmonary disease. *Home Healthcare Nurse, 15*(6), 414–420.

Dowling, K. (1993). Seniors should still do their fair share. *U.S. Catholic, 8,* 22.

Drake, L., Drake, V., & Curwen, J. (1997). A new account of sundown syndrome. *Nursing Standard, 12*(7), 37–40.

Dubler, N. (1998). The collision of confinement and care: End-of-life care in prisons and jails. *The Journal of Law, Medicine & Ethics, 26,* 149–156.

Duke University Center for the Study of Aging and Human Development. (1978). *Multidimensional Functional Assessment: The OARS Methodology* (2nd ed.). Durham, NC: Duke University.

Duncan, M. T., & Morgan, D. L. (1994). Sharing the caring: Family caregivers' views of their relationships with nursing home staff. *The Gerontologist, 34*(2), 235–244.

Duncan, P. W., Weiner, D. K., Chandler, J., & Studenski, S. (1990). Functional reach: A new clinical measure of balance. *Journals of Gerontology, 45A*(6), M192–197.

Duvall, S., Lens, A., & Werner, E. B. (1999). *Cataract and Glaucoma for Eyecare Paraprofessionals.* Thorofare, NJ: Slack, Inc.

Dworkin, S. F., von Korloff, M., & LeResche, L. (1990). Multiple pains and psychiatric disturbance: An epidemiologic investigation. *Archives of General Psychiatry, 47,* 239–244.

Dwyer, J., & Miller, M. (1990). Determinants of primary caregiver stress and burden: Area of residence and the caregiving networks of frail elders. *Journal of Rural Health, 6,* 162–184.

Dykes, P. C., & Wheeler, K. (1997). *Planning, Implementing, and Evaluating Critical Pathways: A Guide for Health Care Survival Into the 21st Century.* New York: Springer Publishing Company.

Edelstein, L. (1989). [O. Temkin, & C. L. Temkin (Eds.)], *Ancient Medicine: Selected Papers of Ludwig Edelstein.* Baltimore: John Hopkins University Press.

Egbert, A. M., Parks, L. H., et al. (1990). Randomized trail of postoperative patient-controlled analgesia vs. intramuscular narcotics in the frail elderly. *NEJM, 150*(9), 1890–1903.

El-Sadr, W., & Gettler, J. (1995). Unrecognized Human Immunodeficiency Virus infection in the elderly. *Archives of Internal Medicine, 155,* 184–186.

Eldercare Online. (2000). *Rights of nursing home residents* [Web Page]. URL http://www.ec-online.net/knowledge/articles/resrights.html [2000, March 20].

Elderly Nutrition Project. (1999). *Elderly Nutrition Project* [Web Page]. URL http://www.projectmeal.org/faqs6.htm [1999, November 16].

Emanuel, L. L. (1999). Facing requests for physician-assisted suicide. *Journal of the American Medical Association, 280*(7), 643–647.

Emlet, C., Crabtree, J., Condon, V., & Treml, L. (1996). *In Home Assessment of Older Adults: An Interdisciplinary Approach.* Gaithersburg, MD: Aspen Publishers.

Emlet, C. A. (1997). HIV/AIDS in the elderly: A hidden population. *Home Care Provider, 2*(2), 69–75.

Employee Benefit Research Institute. (1997). *EBRI Databook on Employee Benefits.* Washington, DC: Author.

Eng, C., Pedulla, J., Eleazer, G. P., McCann, R., & Fox, N. (1997). Program of All Inclusive Care for the Elderly [PACE]: An innovative model of integrated geriatric care and financing. *Journal of the American Geriatrics Society, 45*(2), 2223–2232.

Engel, A. B., Kamm, M. A., Sultan, A. H., Bartran, C. I., & Nicholls, R. J. (1994). Anterior anal sphincter repair in patients with obstetric trauma. *British Journal of Surgery, 81,* 1231–1234.

Ensel, W. M. (1991). "Important" life events and depression among older adults: The role of psychological and social resources. *Journal of Aging & Health, 3,* 546–566.

Erde, E. L. (1996). Conflicts of interest in medicine: A philosophical and ethical morphology. In R. G. Spece, D. S. Shimm, & A. E. Buchanan (Eds.), *Conflicts of Interest in Clinical Research and Practice* (pp. 12–41). New York: Oxford University Press.

Erikson, E. H. (1963). *Childhood and Society* (2nd ed.). New York: Norton.

Ersser, S. (1997). *Nursing as a Therapeutic Activity: An Ethnography.* Avebury, England: Aldershot Publishing.

Escalente, A., Lichtenstein, M. J., White, K., Riso, N., & Hazuda, H. P. (1995). A method for scoring the pain map of the McGill pain questionnaire for use in epidemiologic studies. *Aging Milano, 7,* 358–366.

Espino, D. (Ed.). (1995). Ethnogeriatrics. *Clinics in Geriatric Medicine, 11*(1).

Espino, D. V., Bedolla, M., Perez, M., & Baker, F. M. (1996). Validation of the Geriatric Depression Scale in an elder Mexican American ambulatory population: A pilot study. *Clinical Geriatrics, 16,* 55–67.

Espino, D. V., & Burge, S. K. (1989). Comparison of aged Mexican-American and non Hispanic White nursing home residents. *Family Medicine, 21*(3), 191–194.

Espino, D. V., & Lewis, R. (1998). Cognitive disorders in older minority populations: Issues of prevalence, diagnosis, and treatment. *American Journal of Geriatric Psychiatry, 6,* 519–525.

Espino D. V., Parra, E. D., & Kriehbiel, R. (1994). Mortality differences between elderly Mexican Americans and non-Hispanic whites in San Antonio, Texas. *Journal of the American Geriatrics Society, 42,* 604–608.

Ettinger, R. L. (1999). Epidemiology of dental caries: A broad view. *Dental Clinics of North America, 43,* 679–694.

Ettinger, R. L. (1993). Managing and treating the atrophic mandible. *Journal of the American Dental Association, 124,* 234–241.

Ettinger, R. L. (1997). The unique oral health needs of an aging population. *Dental Clinics of North America, 41,* 633–649.

Ettinger, R. L., & Jakobsen, J. R. (1997). A comparison of patient satisfaction with dentist's evaluation of overdenture therapy. *Community Dentistry & Oral Epidemiology, 25,* 223–227.

Ettinger, R. L., & Mulligan, N. (1999). The future of dental care for the elderly population. *Elder Care: Journal of the California Dental Association, 27*(9), 687–692.

Evans, J. G., & Williams, T. F. (1992). *Oxford Textbook of Geriatric Medicine.* Oxford, England: Oxford University Press.

Evans, L. K., & Strumpf, N. (1989). Tying down the elderly: A review on physical restraint. *Journal of the American Geriatrics Society, 37,* 67–74.

Evans, L. K., Strumpf, N. E., Allen-Taylor, S. L., Capezuti, E., Maislin, G., & Jacobsen, B. (1997). A clinical trial to reduce restraints in nursing homes. *Journal of the American Geriatrics Society, 45,* 675–681.

Evans, W. J., & Campbell, W. W. (1993). Sarcopenia and age-related changes in body composition and functional capacity. *Journal of Nutrition, 123,* 465–468.

Evans, W. J., & Cyr-Campbell, D. (1997). Nutrition, exercise and healthy aging. *Journal of the American Dietetic Association, 97,* 632–638.

Evashwick, C. J., & Weiss, L. J. (1987). *Managing the Continuum of Care.* Rockville, MD: Aspen Publishers, Inc.

Evens, A., & Friedland, R. B. (1994). *Financing and Delivery of Health Care for Children.* Washington, D.C.: National Academy of Social Insurance.

Evers, G. (1989). *Appraisal of Self Care Agency (ASA) Scale.* Van Gorcum, Assen, Netherlands: Author.

Evers, G. (1998). Die Selbstpflegedefezit-Theorie von Dorothea Orem (The self care deficit theory of Dorothea Orem). In J. Osterbring (Ed.), *Erster Internationale Pflegetheorien Kongress Nürnberg (First International Nursing Theory Conference Nürnberg)* (pp. 104–133). Bern, Switzerland: Hans Huber Verlag.

Evers, G. (1998). *Meten van zelfzorg (Measurement of self care).* Leuven, Belgium: University Press Leuven.

Evers, G. (1994). Orem's self care concepts in nursing practice. In G. Mashaba, & H. Brink (Eds.), *Nursing Education: An International Perspective* (pp. 89–111). Kenwyn, South Africa: Junta & Co.

Fadiman, A. (1997). *The Spirit Catches You and You Fall Down.* New York: Noonday Press.

Fagin, L., Carson, J., DeVilliers, N., Bartlett, H., O'Malley, P., West, M., McElfatrick, S., & Brown, D. (1996). Stress, coping, and burnout in mental health nurses: Findings from three research studies. *International Journal of Social Psychiatry, 42*(2), 102–111.

Fangmeier, R. (1998). *The World Through Their Eyes: Understanding Vision Loss: A Manual for Nursing Home Staff.* New York: The Lighthouse, Inc.

Fantl, J., Newman, D., Colling, J., DeLancey, J., Keeys, C., & Loughery, R. (1996). *Urinary Incontinence in Adults: Acute and Chronic Management: Clinical Practice Guideline Number 2, 1996 Update.* (Report No. 96-0682). Rockville, MD: Department of Health and Human Services, Agency for Health Care Policy and Research.

Farran, C. J., Keane-Hagerty, E., Salloway, S., Kupferer, S., & Wilken, C. S. (1991). Finding meaning: An alternative paradigm for Alzheimer's disease family caregivers. *The Gerontologist, 31,* 483–489.

Farrell, M. J., Gibson, S. J., & Helme, R. J. (1996). Chronic nonmalignant pain in older people. In International Association for the Study of Pain, *Pain in the Elderly: A report of the Task Force on Pain in the Elderly.* Seattle, WA: Author.

Farrer, L. A., Cupples, L. A., Haines, J. L., Hyman, B., Kukull, W. A., Mayeux, R., Myers, R. H., Pericak-Vance, M. A., Risch, N., & van Duijn, C. M. (1997). Effects of age, sex, and ethnicity on the association between apolipoprotein E genotype and Alzheimer disease. A meta-analysis. APOE and Alzheimer Disease Meta Analysis Consortium. *JAMA, 278*(16), 1349–1356.

Fauci, A. S., Braunwald, E., Isselbacher, K. J., Wilson, J. D., Martin, J. B., Kasper, D. L., Hauser, S. L., & Longo, D. L. (1998). *Harrison's Principles of Internal Medicine* (14th ed.). New York: McGraw-Hill.

Faye, E. E. (1999). Low vision. In D. Vaughan, T. Asbury, & P. Riordan-Eva (Eds.), *General Ophthalmology* (pp. 377–383). Stamford, CT: Appleton & Lange.

Federal Register. (1999). Washington, D.C.

Feil, E. R. (1997). *Myrna, the Maloriented.* E. R. Feil (Producer). Cleveland, OH: Validation Training Institute.

Feil, N. (1993). *The Validation Breakthrough: Simple Techniques for Communicating with People with Alzheimer's-Type Dementia.* Baltimore: Health Professions Press.

Feldman, R. H. L., & Fulwood, R. (1999). The three leading causes of death in African Americans: Barriers to reducing excess disparity and to improving health behaviors. *Journal of Health Care for the Poor & Underserved, 10*(1), 45–71.

Felician, O., & Sandson, T. A. (1999). The neurobiology and pharmacotherapy of Alzheimer's disease. *Journal of Neuropsychiatry & Clinical Neurosciences, 11*(1), 19–31.

Felsenthal, G., Garrison, S., & Steinberg, F. (1994). Rehabilitation of the aging and elderly patient. *Assessment and Management of the Disabled Elderly* (pp. 84–85). Baltimore: Williams & Wilkins.

Felton, J. S. (1998). Burnout as a clinical entity—its importance in health care workers. *Occupational Medicine, 48*(4), 237–250.

Fenske, N. A., & Lober, C. W. (1990). Skin changes of aging: Pathological implications. *Geriatrics, 45*(Mar), 27–35.

Ferguson, T. (Ed.). (1999). *Ferguson Report* [Web Page]. URL http://ferguson-report.sparklist.com [2000, March 14].

Ferrell, B. A. (1996). Overview of pain in the elderly. In B. A. Ferrell, & B. R. Ferrell (Eds.), *Pain in the Elderly: A Report of the Task Force on Pain in the Elderly.* Seattle, WA: International Association for the Study of Pain.

Ferrell, B. A. (1995). Pain evaluation and management in the nursing home. *Annals of Internal Medicine, 132*(9), 681–687.

Ferrell, B. A., & Ferrell, B. R. (1996). *Pain in the Elderly: A Report of the Task Force on Pain in the Elderly.* Seattle, WA: International Association for the study of Pain.

Ferrell, B. A., Ferrell, B. R., & Osterweil, D. (1990). Pain in the nursing home. *Journal of the American Geriatrics Society, 38,* 409–414.

Ferrell, B. A., Ferrell, B. R., & Rivera, L. (1995). Pain in cognitively impaired nursing home patients. *Journal of Pain & Symptom Management, 10,* 591–598.

Ferrell, B. A., Wisdom, C., & Wenzl, C. (1989). Quality of life as an outcome variable in management of cancer pain. *Cancer, 63,* 2331–2329.

Ferrucci, L., Cecchi, F., Guralnik, J. M., Giampaoli, S., Noce, C. L., & Salani, B. (1996). Does the clock drawing test predict cognitive decline in older persons independent of the mini-mental state examination? *Journal of the American Geriatrics Society, 44,* 1326–1331.

Ferzoco, L. B., Raptopoulos, V., & Silen, W. (1998). Acute diverticulitis. *NEJM, 338*(21), 1521–1526.

Fiatarone, M. A., & Evans, W. J. (1993). The etiology and reversibility of muscle dysfunction in the aged. *Journals of Gerontology, 48*(Special Issue), 77–83.

Fiatarone, M. A., O'Neill, E. F., Ryan, N. D., Clements, K. M., Solares, G. R., Nelson, M. E., Roberts, S. B., Kehayias, J. J., Lipsitz, L. A., & Evans, W. J. (1994). Exercise training and nutritional supplementation for physical frailty in very elderly people. *NEJM, 330*(25), 1769–1775.

Filipp, S. H. (1996). Motivation and emotion. In J. E. Birren, & K. W. Schaie (Eds.), *Handbook of the Psychology of Aging* (4th ed.). San Diego: Academic Press.

Fillenbaum, G. G., George, L. K., & Blazer, D. G. (1988). Scoring nonresponse on the Mni-Mental State Examination. *Psychological Medicine, 18,* 1021–1025.

Fillenbaum, G. G. (1988). *Multidimensional Functional Assessment of Older Adults: The Duke Older Americans Resources and Services Procedures.* Mahwah, NJ: Lawrence Erlbaum & Associates.

Finch, C. E. (1990). *Longevity, Senescence, and the Genome.* Chicago: University of Chicago Press.

Fine, J., & Rouse-Bane, S. (1995). Using validation techniques to improve communication with cognitively impaired older adults. *Journal of Gerontological Nursing, 21,* 39–45.

Fingerman, K., & Perlmutter, M. (1995). Future time perspective and life events across adulthood. *The Journal of General Psychology, 122,* 95–111.

Fink, S. V., & Picot, S. J. (1995). Decisions for nursing home placement and adjustment in Black and White caregivers. *Journal of Gerontological Nursing, 21*(12), 35–42.

Finkel, S. I., & Burns, A. (in press). Behavioral and Psychological Symptoms of Dementia: Clinical and Research Update.

Finkelstein, J. A., & Schiffman, S. S. (1999). Workshop on taste and smell in the elderly: An overview. *Physiology & Behavior, 66*(2), 173–176.

Finucane, T. E., Christman, C., & Travis, K. (1999). Tube feeding in patients with advanced dementia: A review of the evidence. *Journal of the American Medical Association, 282,* 1365–1370.

Fischer, L. R., Mueller, D. P., & Cooper, P. W. (1991). Older volunteers: A discussion of the Minnesota Study. *Gerontologist, 31,* 183–194.

Fisher, K., Rapkin, B. D., & Rappaport, J. (1991). Gender and work history in the placement and perceptions of elder community volunteers. *Psychology of Women Quarterly, 15,* 261–279.

Fitzgerald, M. (1999). The experience of chronic illness: A phenomenological approach. In R. Nay, & S. Garratt (Ed.), *Nursing, Older People, Issues and Innovations* (pp. 40–64). Sydney, Australia: Maclennan & Petty.

Flaherty, J. H. (1998). Commonly prescribed and over-the-counter medications: Causes of confusion. *Clinics in Geriatric Medicine, 14*(1), 101–127.

Fleming, A. W., Sterling-Scott, R. P., Carabello, G., Imari-Williams, I., Allmond, B., Foster, R. S., Kennedy, F., & Shoemaker, W. C. (1994). Injury and violence in Los Angeles: Impact on access to health care and surgical education. *Archives of Surgery, 127*(4), 671–676.

Foley, K. M. (1997). Competent care for the dying instead of physician assisted suicide. *NEJM, 336,* 54–58.

Folsom, J. C. (1996). Reality Orientation for the Mental Patient. Presented at 122nd annual meeting of the American Psychiatric Association, Atlantic City, NJ.

Folstein, M. F., Folstein, S. E., & McHugh, P. R. (1975). "Mini-mental state": A practical method for grading the cognitive state of patients for the clinician. *Journal of Psychiatric Research, 12*(3), 189–198.

Food Research and Action Center (FRAC). (1997). *1996 Poverty Facts.* Washington, DC: Author.

Forbes, E. J. (1994). Spirituality, aging, and the community-dwelling caregiver and care recipient. *Geriatric Nursing, 15,* 297–302.

Forciea, M. A., & Lavizzo Mourey, R. (1996). *Geriatric Secrets.* Philadelphia, PA: Hanley & Belfus.

Foreman, M. D., Fletcher, K., Mion, L. C., Trygstad, L. J., & the NICHE Faculty. (1999). Assessing cognitive function. In I. Abraham, M. M. Bottrell, T. T. Fulmer, & M. D. Mezey (Eds.), *Geriatric Nursing Protocols for Best Practice* (Chap. 5, pp. 51–61). New York: Springer Publishing Company.

Foreman, M. D., Mion, L. C., Trygstad, L. J., Fletcher, K., & the NICHE Faculty. (1999). Acute confusion/delirium: Strategies for assessing and treating. In I. Abraham, M. M. Bottrell, T. T. Fulmer, & M. D. Mezey (Eds.), *Geriatric Nursing Protocols for Best Practice* (Chap. 6, pp. 63–75). New York: Springer Publishing Company.

Foreman, M. D., Theis, S. L., & Anderson, M. A. (1993). Adverse outcomes of hospitalization in the elderly. *Clinical Nursing Research, 2*(3), 360–370.

Fortner, B. V., & Neimeyer, R. A. (1999). Death anxiety in older adults: A quantitative review. *Death Studies, 23,* 387–411.

Fortner, B. V., Neimeyer, R. A., & Rybarczeck, B. (2000). Correlates of death anxiety in older adults: A comprehensive review. In A. Tomer (Ed.), *Death Attitudes and the Older Adult.* Philadelphia, PA: Taylor and Francis.

Fox, P. C., Busch, K. A., & Baum, B. J. (1987). Subjective reports of xerostomia and objective measures of salivary gland performance. *Journal of the American Dental Association, 115,* 581–584.

Francis, P. T., Palmer, A. M., Snape, M., & Wilcock, G. K. (1999). The cholinergic hypothesis of Alzheimer's disease: A review of progress. *Journal of Neurology, Neurosurgery & Psychiatry, 66*(2), 137–147.

Fraser, J. T. (1996). Time and the origin of life. In J. T. Fraser, & M. P. Soulsby (Eds.), *Dimensions of Time and Life: The study of Time VIII* (pp. 3–17). Madison, CT: International Universities Press, Inc.

Fratiglioni, L., Grug, M., Forsell, Y., et al. (1991). Prevalence of Alzheimer's disease and other dementias in an elderly urban population: Relationship with age, sex and education. *Neurology, 41,* 1886–1892.

Freedman, M., Leach, L., Kaplan, E., Winocur, G., Shulman, K. T., & Delis, D.C. (1994). *Clock Drawing: A Neuropsychological Analysis.* New York: Oxford University Press.

Freedman, M. L. (1998). Hematologic disorders. In W. B. Abrams, & R. Berkow, *The Merck Manual of Geriatrics* (2nd ed., Chap. 70–72). Whitehouse Station, NJ: Merck & Co.

Freedman, V. A., & Martin, L. G. (1998). Understanding trends in functional limitations among older Americans. *American Journal of Public Health, 88*(10), 1457–1462.

Freund, A., & Baltes, P. B. (1998). Selection, optimization, and compensation as strategies of life management: Correlations with subjective indicators of successful aging. *Psychology and Aging, 13,* 531–543.

Friedman, W. (1990). *About Time: Inventing the Fourth Dimension.* Cambridge, MA: MIT Press.

Frieland, R., & Summer, L. (1999). *Demography Is Not Destiny.* Washington, DC: National Academy on Aging Society: A Polity Institute of the Gerontological Society of America.

Friend, R. A. (1991). Older lesbian and gay people: A theory of successful aging. In J. A. Lee (Ed.), *Gay Midlife and Maturity* (pp. 99–118). Binghamton, NY: Haworth Press, Inc.

Fries, J. F. (1980). Aging, natural death, and the compression of morbidity. *NEJM, 303,* 130–136.

Fries, J. F., Block, D. A., Harrington, H., Richardson, N., & Beck, R. (1993). Two-year results of a randomized controlled trial of a health promotion program in a retiree population: The Bank of America Study. *American Journal of Medicine, 94,* 455–462.

Fries, J. F., & Crapo, L. M. (1981). *Vitality and Aging: Implications of the Rectangular Curve.* San Francisco: Freeman.

Fries, J. F., Harrington, H., Edmund, R., Kent, L. A., & Richardson, N. (1994). Randomized controlled trial of cost reductions from a health education program: The California Public Employees' Retirement System (PERS) Study. *American Journal of Health Promotion, 8,* 216–223.

Fries, J. F., Koop, C. E., Sokolov, J., & Beadle, C. E. (1998). Beyond health promotion: Reducing health care costs by reducing need and demand for medical care. *Health Affairs, 17,* 70–84.

Froehlich, T. E., Robinson, J. T., & Inouye, S. K. (1998). Screening for dementia in the outpatient setting: The time and change test. *Journal of the American Geriatrics Society, 46*(12), 1506–1511.

Fry, P. S. (1990). A factor analytic investigation of homebound elderly individuals' concerns about death and dying, and their coping responses. *Journal of Clinical Psychology, 46,* 737–748.

Fuld, P. A. (1978). Differential diagnosis of dementias. In R. Katzman, R. D. Terry, & K. L. Bic (Eds.), *Alzheimer's Disease: Senile Dementias and Related Disorders.* New York: Raven Press.

Fuller-Thomson, E., Minkler, M., & Driver, D. (1997). A profile of grandparents raising grandchildren in the United States. *The Gerontologist, 37*(3), 406–411.

Fullerton, H. N. Jr. (1999). Labor Force Participation: 75 Years of Change, 1950–1998 and 1998–2025. *Monthly Labor Reviews, 122*(12), 3–12.

Fullerton, H. N. Jr. (1999). Labor Force Projections to 2008: Stay Growth and Changing Composition. *Monthly Labor Reviews, 122*(11), 19–32.

Fulmer, T. T. (1991). Grow your own experts in hospital elder care. *Geriatric Nursing, 12*(2), 64–66.

Fulmer, T. T., & Ashley, J. (1989). Clinical indicators of neglect. *Applied Nursing Research, 2*(4), 161–167.

Fulmer, T. T., & Ashley, J. (1986). Neglect: What part of abuse? *Pride Institute Journal for Long Term Home Health Care, 5*(4), 18–24.

Fulmer, T. T., Flaherty, E., & Bottrell, M. (2000). Acute and critical nursing care of the elderly. In T. T. Yoshikawa, & D.C. Norman (Eds.), *Acute Emergencies and Critical Care of the Geriatric Patient*. New York: Lippincott, Williams & Wilkins.

Fulmer, T. T., Hollander Feldman, P., Sook Kim, T., Carty, B., Beers, M., Molina, M., & Putnam, M. (1999). An intervention study to enhance medication compliance in community dwelling elderly. *Journal of Gerontological Nursing, 25*(8), 6–14.

Fulmer, T. T., & Paveza, G. (1998). Neglect in the elderly patient. *Nursing Clinics of North America, 33*(3), 457–466.

Füsgen, I., Bienstein, C., Böhmer, F., Busch, B., Cottenden, A., & Heidler, H. (1998). Interdisciplinary care for urinary incontinence in the elderly. *World Journal of Urology, 16*(Suppl. 1), S62–S70.

Gaeta, T. J., LaPolla, C., & Melendez, E. (1995). AIDS in the Elderly: New York City Vital Statistics. *The Journal of Emergency Medicine, 14*(1), 19–23.

Gallagher, H. (personal communication). (1992).

Gallo, J. J., Fulmer, T., Pavezo, G. J., & Reichel, W. (2000). *Handbook of Geriatric Assessment*. Gaitherburg, MD: Aspen Publishers.

Ganz, P. A., Lee, J. J., Sim, M. S., Polinsky, M. L., & Schag, C. A. (1992). Exploring the influence of multiple variables on the relationship of age to quality of life in women with breast cancer. *Journal of Clinical Epidemiology, 45*, 473–485.

Gardner, R., Oliver-Munoz, S., Fisher, L., & Empting, L. (1981). Mattis Dementia Rating Scale: Internal reliability study using a diffusely impaired population. *Journal of Clinical Neuropsychology, 3*, 271–275.

Garrett, J. T. (1990). *Indian Health: Values, Beliefs and Practices*. (Report No. HRS (P-DV-90-4)). Washington, DC: U.S. Government Printing Office.

Gaspar, P. M. (1999). Water intake of nursing home residents. *Journal of Gerontological Nursing, 25*(4), 23–29.

Gastel, B. (1994). *Working with Your Older Patient: A Clinician's Handbook*. Bethesda, MD: National Institute on Aging, National Institutes of Health.

Gaugler, J. E., Zarit, S. H., & Pearlin, L. I. (1999). Caregiving and institutionalization: Perceptions of family conflict and socioemotional support. *International Aging and Human Development, 49*(1), 1–25.

Gavrilov, L. A., & Gavrilova, N. S. (1990). *The Biology of Life Span: A Quantitative Approach*. Chur, Switzerland: Harwood Academic Publishers.

Gazzard, B., & Moyle, G. (1998). 1998 Revision to the British HIV Association Guidelines for Antiretroviral Treatment of HIV Seropositive Individuals. *Lancet, 9124*, 314–316.

Gelberg, L., Linn, L. S., & Mayer-Oakes, S. A. (1990). Differences in health status between older and younger homeless adults. *Journal of the American Geriatrics Society, 38*, 1220–1229.

Gelfand, D. E. (1999). *The Aging Network: Programs and Services* (5th ed.). New York: Springer Publishing Company.

General Accounting Office. (1999). *Medicare Managed Care: Greater Oversight Needed to Protect Beneficiary Rights*. Washington, DC: U.S. General Accounting Office.

Generations United. (2000). *Generations United Homepage* [Web Page]. URL http://www.gu.org/ [2000, March 12].

Genevay, B. (1997). See me! Hear me! Know who I am! An experience of being assessed. *Generations, 21*(1), 16–18.

George, L. (1997). Choosing among established assessment tools: Scientific demands and practical consideration. *Generations, 21*(1), 32–36.

George, L. (1992). Community and home care for mentally ill older adults. In J. E. Birren, R. B. Sloane, & G. D. Cohen (Eds.), *Handbook of Mental Health and Aging* (2nd ed., pp. 793–813). San Diego, CA: Academic Press.

George, L. K. (1993). Sociological perspectives on life transitions. *Annual Review of Sociology, 19*, 353–373.

George, L. K., Blazer, D. G., Hughes, D. C., et al. (1989). Social support and the outcome of major depression. *British Journal of Psychiatry, 154*, 478–485.

George, L. K., Fillenbaum, G. C., & Burchett, B. M. (1988). *Respite Care: A Strategy for Easing Caregiver Burden*. Durham, NC: Duke University Center for the Study of Aging and Human Development.

Geron, S. (1997). Taking the measure of assessment. *Generations, 21*(1), 5–9.

Gift, A. (1989). Validation of a vertical visual analogue scale as a measure of clinical dyspnea. *Rehabilitation Nursing, 4*, 323–325.

Gilchrest, B. A. (1995). Pruritus in the elderly. *Seminars in Dermatology, 14*(4), 317–319.

Gilchrest, B. A. (1992). Skin diseases in the elderly. In E. Calkins, A. B. Ford, & P. R. Katz (Eds.), *Principles of Geriatrics*. Philadelphia, PA: W. B. Saunders.

Gilden, J. L. (1999). Nutrition and the older diabetic. *Clinics in Geriatric Medicine, 15*(2), 371–390.

Gill, T. M., Williams, C. S., & Tinetti, M. E. (1995). Assessing risk for the onset of functional dependence among older adults: The role of physical performance. *Journal of the American Geriatrics Society, 43*, 603–609.

Gillick, M., Berkman, S., & Cullen, L. (1999). A patient-centered approach to advance medical planning in the nursing home. *Journal of the American Geriatrics Society, 47*(2), 227–230.

Gillum, R. F., Gillum, D. S., & Smith, L. (1984). Cardiovascular risk factors among urban American Indians: Blood pressure, serum lipids, smoking, diabetes, health knowledge, and behavior. *American Heart Journal, 107,* 765–776.

Gist, J. R. (1988). *The effects of state income tax reform.* Washington, DC: AARP Public Policy Institute.

Gitlin, L. N. (1999). Testing home modification interventions. In R. Schulz, G. Maddox, & M. P. Lawton (Eds.), *Annual Review of Gerontology and Geriatrics: Interventions in Research with Older Adults.* New York: Springer Publishing Company.

Gleich, G. (1995). Health maintenance and prevention in the elderly. *Prevention in Practice, 22*(4), 697–710.

Glendenning, F. (1999). Elder abuse and neglect in residential setting: The need for inclusiveness in elder abuse research. *Journal of Elder Abuse & Neglect, 10*(1/2), 1–12.

Glennas, A., Kvien, T. K., Andrup, O., Clarke-Jenssen, O., Karstensen, B., & Brodin, U. (1997). Auranofin is safe and superior to placebo in elderly-onset rheumatoid arthritis. *British Journal of Rheumatology, 36,* 870–877.

Glillick, M. (1996). *Choosing Medical Care in Old Age: What Kind, How Much, When to Stop.* Boston: Harvard University Press.

Glynn, R. J., Bouchard, G. R., LoCastro, J. S., & Laird, N. M. (1985). Aging and generational effects on drinking behaviors in men: Results from the Normative Aging Study. *American Journal of Public Health, 75,* 1413–1419.

Goate, A., Chartier-Harlin, M. C., Mullan, M., Brown, J., Crawford, F., Fidani, L., Giuffra, L., Haynes, A., Irving, N., James, L., et al. (1991). Segregation of a missense mutation in the amyloid precursor protein gene with familial Alzheimer's disease. *Nature, 349*(6311), 704–706.

Goebel, B. L., & Boeck, B. E. (1987). Ego integrity and fear of death: A comparison of institutionalized and independently living older adults. *Death Studies, 11,* 193–204.

Gold, M. (1998). Beyond coverage and supply: Measuring access to healthcare in today's market. The concept of access and managed care. *Health Services Research, 33*(3), 625–632.

Goldberg, T. H., & Chavin, S. I. (1997). Preventive medicine and screening in older adults. *Journal of the American Geriatrics Society, 45*(3), 344–354.

Goldman, H. H., Skodal, A. E., & Lave, T. R. (1995). Revising Axis V for DSM IV: A review of measure of social functioning. *Journal of Child Psychology & Psychiatry & Allied Disciplines, 36*(5), 787–792.

Golledge, J. (1997). Lower-limb arterial disease. *Lancet, 350*(9089), 1459–1465.

Goodman, C. (1989). *Reaching out to a multi-cultural community: Challenges for adult day care centers.* Long Beach, CA: California State University, Department of Social Work.

Goodman, K. (1998). *Ethics, Computing, and Medicine: Informatics and the Transformation of Health Care.* Cambridge: Cambridge University Press.

Goodwin, M. E., & Morley, J. E. (1994). Geriatric research, education, and clinical centers: Their impact in the development of American geriatrics. *Journal of the American Geriatrics Society, 42,* 1012–1019.

Gorbien, M. J., Bishop, J., Beers, M. H., Norman, D., Osterweil, D., & Rubenstein, L. Z. (1992). Iatrogenic illness in hospitalized elderly people. *Journal of the American Geriatrics Society, 40,* 1031–1042.

Gordon, C., Gaitz, C. M., & Scott, J. B. (1976). Leisure and lives: Personal expressivity across the life span. In R. H. Binstock, & E. Shanas (Eds.), *Handbook of aging and the social sciences* (pp. 310–341). New York: Van Nostrand Reinhold.

Gordon, S. M., & Thompson, S. (1995). The changing epidemiology of Human Immunodeficiency Virus infection in older persons. *Journal of the American Geriatrics Society, 43,* 7–9.

Gorsky, R. D., Pamuk, E., Williamson, D. F., Shaffer, P. A., & Kaplan, J. P. (1996). The 25 year health costs of women who remain overweight after 40 years of age. *American Journal of Preventive Medicine, 12,* 388–394.

Gould, E., & Mariano, C. (1999). Sensory changes and communication with older adults. In C. Mariano, E. Gould, M. Mezey, & T. Fulmer (Eds.), *Best Nursing Practices in Care for Older Adults.* New York: The John A. Hartford Foundation For Geriatric Nursing.

Goulding, M. M. (1997). Coping with loss in older years: A personal and professional perspective. In J. Zeig (Editor), *The Evolution of Psychotherapy: The Third Conference.* New York: Brunner/Mazel, Inc.

Graebner, W. (1980). *A History of Retirement.* New Haven, CT: Yale University Press.

Grant, R. W., & Casey, D. A. (1995). Adapting cognitive behavioral therapy for the frail elderly. *International Psychogeriatrics, 7,* 561–571.

Greene, M. G., & Adelman, R. D. (1996). Psychosocial factors in older patients' medical encounters. *Research on Aging, 18,* 84–102.

Greene, M. G., Adelman, R. D., & Rizzo, C. (1996). Problems in communication between physicians and older patients. *Journal of Geriatric Psychiatry, 29,* 13–32.

Greene, M. G., Majerovitz, S. D., Adelman, R. D., & Rizzo, C. (1994). The effects of the presence of a third person on the physician-older patient medical interview. *Journal of the American Geriatrics Society, 42,* 413–419.

Greene, V., Ondrich, J., & Laditka, S. (1998). Can home care services achieve cost savings in long-term care for older people? *Journals of Gerontology, 53B,* S228–238.

Gregg, E. W., Cauley, J. A., Seely, D. G., Ensrud, K. E., & Bauer, D.C. (1998). Physical activity and osteoporotic fracture risk in older women. Study of Osteoporotic Fractures Research Group. *Annals of Internal Medicine, 129*(2), 81–88.

Gresham, G. E., Duncan, P. W., & Stason, W. B. (1995). *Post Stroke Rehabilitation: Assessment, Referral, and Patient Management.* (Report No. 95-0663). Rockville, MD: U.S. Department of Health and Human Services.

Griffin, M. R., Melton, L. J. I., & Ray, W. A. (1990). Incidence of hip fracture in Saskatchewan Canada, 1976-1985. *American Journal of Epidemiology, 131,* 502–509.

Grigor, R. R., Spitz, P. W., & Furst, D. E. (1987). Salicylate toxicity in elderly patients with rheumatoid arthritis. *Journal of Rheumatology, 14,* 60–66.

Griso, T. (1994). Clinical assessment for legal competence of older adults. In M. Storandt, & G. VandenBos (Eds.), *Neuropsychological Assessment of Dementia and Depression in Older Adults* (pp. 119–139). Washington, DC: American Psychological Association.

Gross, J. J., Carstensen, L. L., Pasupathi, M., Tsai, J., Skorpen, C. G., & Hsu, A. Y. C. (1997). Emotion and aging: Experience, expression, and control. *Developmental Psychology, 12*(4), 590–599.

Grossman, M., Mickanin, J., Onishi, K., Hughes, E., D'Esposito, M., Ding, X.-S., Alavi, A., & Reivich, M. (1996). Progressive non-fluent aphasia: Language, cognitive and PET measures contrasted with probable Alzheimer's disease. *Journal of Cognitive Neuroscience, 8,* 135–154.

Grundy, E. (1984). Mortality and morbidity among the old. *BMJ, 288,* 663–664.

Guccione, A. (1993). *Geriatric Physical Therapy.* St. Louis, MO: Mosby Publishing Company.

Gueldner, S. H., Burke, S., & Wright, H. (in press). *Preventing and Managing Osteoporosis.* New York: Springer Publishing Company.

Guigoz, Y., Vellas, B., & Garry, P. (1996). Assessing the nutritional status of the elderly: The mini-nutritional assessment as part of the geriatric evaluation. *Nutrition Reviews, 54*(1, Pt 2), S59–65.

Guillemard, A. M., & Rein, M. (1993). Comparative patterns of retirement: Recent trends in developed societies. *Annual Review of Sociology, 19,* 469–503.

Guralnik, J. M., Branch, L. G., Cummings, S. R., & Curb, J. D. (1989). Physical performance measures in aging research. *Journals of Gerontology, 44A,* M141–146.

Guralnik, J. M., Ferrucci, L., Simonsick, E. M., Salive, M. E., & Wallace, R. B. (1995). Lower extremity function in persons over the age of 70 as a predictor of subsequent disability. *NEJM, 332,* 556–561.

Guralnik, J. M., Simonsick, E. M., Ferrucci, L., Glynn, R. J., Berkman, L. F., Blazer, D. G., Scherr, P. A., & Wallace, R. B. (1994). A short physical performance battery assessing lower extremity function: Association with self-reported disability and prediction of mortality and nursing home admission. *Journals of Gerontology, 49A,* M85–M94.

Gurland, B., Golden, R. R., Teresi, J. A., & Challop, J. (1984). The SHORT-CARE: An instrument for the assessment of depression, dementia, and disability. *Journal of Gerontology, 39*(166–169).

Gurland, B., Kuriansky, J., Sharpe, L., Simon, R., Stiller, P., & Birkett, P. (1977–1978). The Comprehensive Assessment and Referral Evaluation (CARE): Rationale, development, and reliability. *International Journal of Aging & Human Development, 8*(1), 9–42.

Gurland, B. J., & Barrett, V. W. (2000). Clinical Case Conference Inventory. *The Stroud Report, 110.*

Gurland, B. J., Copeland, J., Kuransky, J., et al. (1983). *The Mind and Mood of Aging.* New York: Haworth Press.

Gurland, B. J., Lantiqua, R. A., & Teresi, J. A. (1990). The CLIN-CARE: A technique for comprehensive geriatric assessment. In J. D. Frengley, P. Murray, & M. L. Wykle, *Practicing Rehabilitation with Geriatric Clients* (pp. 78–95, 220–237). New York: Springer Publishing Company.

Gurley, R., Lum, N., Sande, M., Lo, B., & Katz, M. (1996). Persons found in their homes helpless or dead. *NEJM, 334,* 1710–1716.

Gustaitis, R. (1980). Old versus young in Florida: Review of an aging America. *Saturday Review.*

Guttman, R., & Seleski, M. (1999). *The Diagnosis, Management and Treatment of Dementia: A Practical Guide for Primary Care Physicians.* Chicago, IL: American Medical Association.

Gwyther, L. (1994). Service delivery and utilization: Research directions and clinical implications. In E. Light, G. Niederehe, & B. Lebowitz (Eds.), *Stress Effects on Family Caregivers of Alzheimer's Patients: Research and Interventions.* New York: Springer Publishing Company.

Gwyther, L. P., & Strulowitz, S. Y. (1988). Care-giver stress. *Current Opinion in Psychiatry, 11,* 431–434.

Haber, C., & Gratton, B. (1994). *Old Age and the Search for Security: An American Social History.* Bloomington, IN: Indiana University Press.

Haberman, E. (1986). Total Joint Replacement: An Overview. *Seminars in Roentgenology, 11*(1), 7–9.

Haight, B. K., & Webster, J. D. (Eds.). (1995). The art and science of reminiscing: Theory, research, methods, and applications. Washington, DC: Taylor & Francis.

Halas, E. M., & Bell, K. L. (1998). Contracture and other deleterious effects of immobility. In J. A. Delisa, B. M. Gans, & W. L. Bockenek (Eds.), *Rehabilitation Medicine: Principles and Practice* (3rd ed., pp. 1113–1133). Philadelphia, PA: W. B. Saunders.

Hall, A. F., Theofrestous, J. P., Cundiff, G. W., Harris, R. L., Hamilton, L. F., Swifty, S., & Bump, R. C. (1996). Interobserver and intraobserver reliability of the proposed International Continence Society, Society of Gynecologic Surgeons and American Urogynecologic Society pelvic organ prolapse classification. *American Journal of Obstetrics & Gynecology, 175*(6), 1470–1471.

Hall, G., Buckwalter, K., & Stolley, J. (1995). Standardized care plan: Managing Alzheimer's patients at home. *Journal of Gerontological Nursing, 21*(1), 43–44.

Hall, G., & Buckwalter, K. C. (1987). Progressively lowered stress threshold: A conceptual model for care of adults with Alzheimer's Disease. *Archives of Psychiatric Nursing, 1,* 309–312.

Hall, G. R., & Gerdner, L. A. (1999). Managing problem behaviors. In J. K. Stone, J. F. Wyman, & S. A. Salisbury (Eds.), *Clinical Gerontological Nursing: A Guide to Advanced Practice* (2nd ed., pp. 623–643). Philadelphia, PA: W. B. Saunders.

Hall, L. (1969). The Loeb Center for Nursing and Rehabilitation, Montefiore Hospital and Medical Center. *International Journal of Nursing Studies, 6,* 81–95.

Hall, W. J., & Oskvig, R. O. (1988). Transitional Care: Hospital to Home. *Clinics in Geriatric Medicine, 14,* 799–813.

Halsey, N. A., Coberly, J. S., Desormeaux, J., Losikoff, P., Atkinson, J., Moulton, L. H., Contave, M., Johnson, M., Davis, H., Geiter, L., Johnson, E., Huebner, R., Boulos, R., & Chaisson, R. E. (1998). Randomised trial of isoniazid versus rifampicin and pyrazinamide for prevention of tuberculosis in HIV-1 infection. *Lancet, 351*(9105), 786–792.

Ham, R. J., & Sloane, P. D. (1997). *Primary Care Geriatrics: A Case Based Approach* (3rd ed.). St. Louis, MO: Mosby Year-Book.

Hamaoui, E., & Kodsi, R. (1997). Complications of enteral feeding and their prevention. In J. Rombeau & R. Rolandelli (Eds.), *Clinical Nutrition: Enteral and Tube Feeding* (p. 554). Philadelphia: W.B. Saunders.

Hamilton, M. (1960). A rating scale for depression. *Journal of Neurology, Neurosurgery, & Psychiatry, 23,* 56–62.

Hanlon, J. T., Schmader, K. E., Koronkowski, M. J., Weinberger, M., Landsman, P. B., Samsa, G. P., & Lewis, I. K. (1997). Adverse drug events in high risk older outpatients. *Journal of the American Geriatrics Society, 45*(8), 945–948.

Harden, A. W., Clark, R., & Maguire, K. (1997). *Informal and Formal Kinship Care. Report for the Office of the Assistant Secretary for Planning and Evaluation.* (Report No. HHS-100-95-0021). Washington, DC: Department of Health and Human Services.

Hardy, M. A. (1992). Dry skin. In G. M. Bulecheck (Ed.), *Nursing Interventions: Essential Nursing Treatments.* Philadelphia, PA: W. B. Saunders.

Hare, J., Pratt, C., & Nelson, C. (1992). Agreement between patients and their self-selected surrogates on difficult medical decisions. *Archives of Internal Medicine, 152*(5), 1049–1054.

Harel, Z., & Biegel, D. E. (1995). Aging, ethnicity, and mental health services: Social work perspectives on need and use. In D. K. Padgett (Ed.), *Handbook on Ethnicity, Aging and Mental Health* (pp. 217–241). Westport, CT: Greenwood Press.

Harrington, C., Carrillo, M. S., Thollaug, S. C., & Summers, P. R. (1996). *Nursing Facilities, Staffing, Residents and Facility Deficiencies, 1991–95.* San Francisco, CA: University of California at San Francisco School of Nursing.

Harris, L., & Associates. (1979). *The Myth and Reality of Aging in America.* Washington, DC: The National Council on Aging.

Harris, M. I., Eastman, R. C., & Cowie, C. C. (1999). Racial and ethnic differences in glycemic control of adults with type 2 diabetes. *Diabetes Care, 22*(3), 403–408.

Harris, M. I., Flegal, K. M., Cowie, C. C., Eberhardt, M. S., Goldstein, D. E., Little, R. R., Wiedmeyer, H., & Byrd-Holt, D. (1998). Prevalence of diabetes, impaired fasting glucose, and impaired glucose tolerance in U.S. adults. *Diabetes Care, 21*(4), 518–524.

Harris, W., & Sledge, C. (1990). Total Hip and Total Knee (1). *NEJM, 323*(11), 725–731.

Harris, W., & Sledge, C. (1990). Total Hip and Total Knee (2). *NEJM, 323*(12), 801–807.

Hartman, A. (1995). Family Therapy. In National Association of Social Work *Encyclopedia of Social Work.* Washington, DC: NASW Press.

Hatcher, R. A., Trussell, J., Stewart, F., Cates, W., Stewart, G., Guest, F., & Kowal, D. (1998). *Contraceptive Technology* (17th ed.). New York: Ardent Media.

Hausdorff, J., Levy, B., & Wei, J. (1999). The power of ageism on physical function of older persons: Reversibility of age-related gait changes. *Journal of the American Geriatrics Society, 47,* 1346–1349.

Hawes, C., Mor, V., Phillips, C. D., Fries, B. E., Morris, J. D., Steel-Friedlob, E., Greene, A. M., & Nennestiel, M. (1997). The OBRA-87 nursing home regulations and implementation of the Resident Assessment Instrument: Effect on process quality. *Journal of the American Geriatrics Society, 45,* 977–985.

Hawkes, S. R., Hull, M. L., Thalman, R. L., & Richins, P. M. (1995). Review of spiritual health: Definitions, role, and intervention strategies in health promotion. *American Journal of Health Promotion, 9*(5), 371–378.

Hayslip, B., Jr., Shore, R. J., Henderson, C. E., & Lambert, P. L. (1998). Custodial grandparenting and the impact of grandchildren with problems on role satisfaction and role meaning. *Journals of Gerontology, 53B,* S164–S173.

Hazzard, W. R. (1999). Preventive gerontology: A personalized, designer approach to a life of maximum quality and quantity. In W. R. Hazzard, J. P. Blass, W. H. Ettinger, J. B. Halter, & Ouslander J. G. (Eds.), *Principles of Geriatric Medicine and Gerontology* (4th ed., Chap. 17, pp. 239–245). New York: McGraw-Hill.

Hazzard, W. R., Blass, J. P., Ettinger, W. H., Halter, J. B., & Ouslander, J. G. (1999). *Principles of Geriatric Medicine and Gerontology* (4th ed.). New York: McGraw-Hill.

Headache Classification Committee of the International Headache Society. (1988). Classification and diagnositic criteria for headache disorders, cranial neuralgias and facial pain. *Cephalgia, 8*(Suppl 7), 1–96.

Health Care Financing Administration. (1997). *Health Care Financing Review: Medicare and Medicaid Statistical Supplement.* Baltimore, MD: U.S. Department of Helath and Human Services.

Health Care Financing Administration. (1998). *Medicare: A Profile of Medicare Beneficiaries.*

Health Care Financing Administration. (2000). *Medicare & You.* (Report No. HCFA Publication No. 10050). Baltimore, MD: Author.

Health Care Financing Administration. (1999). *Overview of the Medicare Program* [Web Page]. URL http://www.hcfa.gov/medicare/mcarcnsm.htm [1999, November 15].

Health Care Financing Administration. (1999). *Survey Procedures and Interpretive Guidelines for Skilled Nursing Facilities and Intermediate Care Facilities. In State Operations Manual: Provider Certification.* Washington, DC: U.S. Government Printing Office.

Health Insurance Association of America. (1999). *Guide to Long Term Care.* Washington, DC: Author.

Heckhausen, J. (1997). Developmental regulation across adulthood: Primary and secondary control of age-related challenges. *Developmental Psychology, 33,* 176–187.

Heckhausen, J., & Brim, O. G. (1997). Perceived problems for self and others: Self-protection by social downgrading throughout adulthood. *Psychology and Aging, 12,* 610–619.

Heckhausen, J., & Schulz, R. (1995). A life-span theory of control. *Psychological Review, 102,* 284–304.

Heidenreich, P., Ruggerio, C., & Massie, B. (1999). Effect of a home monitoring system on hospitalization and resource use for patients with heart failure. *American Heart Journal, 138*(4), 633–640.

Helfand, A. E. (1995). *Feet First.* Harrisburg, PA: Professional (Pennsylvania) Diabetes Academy.

Helfand, A. E. (Ed.). (1993). The geriatric patient and considerations of aging. *Clinics in Podiatric Medicine & Surgery, 10*(2), 1–250.

Helfand, A. E. (1999). Public health strategies to develop a comprehensive chronic disease and podogeriatric protocol. *National Academies of Practice Forum, 1*(1), 49–57.

Helfand, A. E., & Jessett, D. F. (1998). Foot problems. In J. M. S. Pathy (Ed.), *Principles and and Practice of Geriatric Medicine* (3rd ed.). Edinburgh: John Wiley & Sons.

Helzer, J. E., Burnam, A., & McEvoy, L. T. (1991). Alcohol abuse and dependence. In L. N. Robins, & D. A. Regier (Eds.), *Psychiatric Disorders in America: The Epidemiologic Cachment Area Study* (pp. 81–115). New York: The Free Press.

Hemmelgarn, B., Suissa, S., Huang, A., Boivin. J. F., & Pinard, G. (1997). Benzodiazepine use and the risk of motor vehicle crash in the elderly. *JAMA, 278*(1), 27–31.

Henderson, C. T., Ayello, E. A., Sussman, C., Leiby, D. M., Bennett, M. A., Dungog, E. F., Sprigle, S., & Woodruff, L. (1997). Draft definition of stage I pressure ulcers: Inclusion of persons with darkly pigmented skin. *Advances in Wound Care, 10*(5), 16–19.

Hendrich, N. A., Kippenbrock, T., & Soja, M. E. (1995). Hospital falls: Development of a predictive model for clinical practice. *Applied Nursing Research, 8*(3), 129–139.

Hendricks, J. (forthcoming). It's about time. In S. McFadden, & R. C. Atchley (Eds.), *Aging and the Meaning of Time.* New York: Springer Publishing Company.

Hendricks, J. (Ed.). (1995). The meaning of reminiscence and life review. Amityville, NY: Baywood.

Hendrie, H. C., Unverzagt, F. W., & Austrom, M. G. (1997). The dementing disorders. *Psychiatric Quarterly, 68*(3), 261–279.

Henrard, J. C. (1996). Cultural problems of aging especially regarding gender and intergenerational equity. *Social Science & Medicine, 43*(5), 667–680.

Henry, R. G. (1999). Alzheimer's disease and cognitively impaired elderly: Providing dental care. *Elder Care:*

Journal of the California Dental Association, 27(9), 709–717.

Henry, T. D. (1999). Therapeutic angiogenesis. *BMJ, 318*(7197), 1536–1539.

Hepburn, K., & Reed, R. (1995). Ethical and cultural issues with Native American Elders: End-of-life decision making. *Clinics in Geriatric Medicine, 11,* 97–111.

Herdman, S. J. (1990). Treatment of benign paroxysmal positional vertigo. *Physical Therapy, 70,* 381–388.

Herr, K. A., & Mobily, P. R. (1992). Chronic pain and depression. *Journal of Psychosocial Nursing & Mental Health Services, 30*(9), 7–12.

Herr, K. A., & Mobily, P. R. (1993). Comparison of selected pain assessment tools for use with the elderly. *Applied Nursing Research, 6,* 39–46.

Herzog, A. R., Kahn, R. L., Morgan, J. N., Jackson, J. S., & Antonucci, T. C. (1989). Age differences in productive activities. *Journals of Gerontology, 44B,* S129–S138.

Hess, C. T. (1999). Resources in wound care: 1999 Directory. *Advances in Wound Care, 12*(4), 155–223.

Hickson, G. B., Clayton, E. W., & Entman, S. S. (1994). Obstetrician's prior malpractice experience and patient's satisfaction with care. *JAMA, 272,* 1583–1587.

Higginbotham, J. C., Trevino, F. M., & Ray, L. A. (1990). Utilization of Curanderos by Mexican Americans: Prevalance and Predictors Findings from NHANES 1982–84. *American Journal of Public Health, 80*(suppl), 32–35.

Higgins, M., Enright, P., Kronmal, R., Schenker, M., Anton-Culver, H., & Lyles, M. (1993). Smoking and lung function in elderly men and women. *JAMA, 269,* 2741–2748.

Higgins, M., Kamel, W., Garrison, R., Pinsky, J., & Stokes, 3. (1988). Hazards of Obesity: The Framingham Experience. *Acta Medica Scandinavica, 723,* 23–36.

High, D. M. (1993). Advance directives and the elderly: A study of intervention strategies to increase use. *Gerontologist, 33*(3), 342–349.

Himmelstein, D. U., Woolander, S., Hellander, I., & Wolf, S. M. (1999). Quality of care in investor-owned vs. not-for-profit HMOs. *JAMA, 282*(2), 159–63.

Hochberg, M. C., Altman, R. D., Brandt, K. D., Clark, B. M., Dieppe, P. A., Griffin, M. R., Moskowitz, R. W., & Schnitzer, T. J. (1995). Guidelines for the medical management of osteoarthritis. Part II. Osteoarthritis of the knee. American College of Rheumatology [see comments]. *Arthritis & Rheumatism, 38*(11), 1541–1546.

Hodges, J. R., Patterson, K., Oxbury, S., & Funnell, E. (1992). Semantic dementia: Progressive fluent aphasia with temporal lobe atrophy. *Brain, 115,* 1783–1806.

Hoenig, H., Nusbaum, N., & Brummel-Smith, K. (1997). Geriatric rehabilitation: State of the art. *Journal of the American Geriatrics Society, 45*(11), 1371–1381.

Hoffman, C., Rice, D., & Sung, H. Y. (1996). Persons with chronic conditions: Their prevalence and costs. *JAMA, 276*(18), 1473–1476.

Hoffman, H. J., Ishii, E. K., & MacTurk, R. H. (1998). Age-related changes in the prevalence of smell/taste problems among the United States adult population. Results of the 1994 disability supplement to the National Health Interview Survey. *Annals of the New York Academy of Sciences, 855,* 716–722.

Hoffman, N. B. (1991). Dehydration in the elderly: Insidious and manageable. *Geriatrics, 46*(6), 35–38.

Hoffman, S. B., & Kaplan, M. (1996). *Special Care Programs for People with Dementia.* Baltimore, MD: Health Professions Press.

Hoffman, S. B., & Platt, C. A. (2000). *Comforting the Confused: Strategies for Managing Dementia,* 2nd ed. New York: Springer Publishing Company.

Holahan, C. J., & Moos, R. H. (1991). Life stressors, personal and social resources, and depression: A 4-year structural model. *Journal of Abnormal Psychology, 100,* 31–38.

Holden, U. P., & Woods, R. T. (1982). *Reality Orientation: Psychological Approaches to the Confused Elderly.* London, UK: Churchill-Livingstone.

Holler, H. H., & Pastors, J. G. (1997). *Diabetes Medical Nutrition Therapy.* Alexandria, VA: American Diabetes Association.

Holloszy, J. O., & Kohrt, W. M. (1995). Exercise. In E. J. Masoro (Ed.), *Handbook of Physiology* (pp. 633–666). New York: Oxford University Press.

Holmes, R. B., & Rahe, R. H. (1967). The social readjustment rating scale. *Journal of Psychosomatic Research, 11,* 213–218.

Hooyman, N., & Kiyak, H. A. (1999). *Social Gerontology: A Multidisciplinary Perspective* (5th ed.). Boston: Allyn and Bacon.

Hooyman, N. R., & Gonyea, J. (1995). Feminist perspectives on family care: Policies for gender justice. *Family Caregiver Application Series* (Vol. 6). Newbury Park, CA: Sage Publications.

Hope, T., Keene, J., Gedling, K., Fairburn, C. G., & Jacoby, R. (1998). Predictors of institutionalization for people with dementia living at home with a carer. *International Journal of Geriatric Psychiatry, 13*(10), 682–690.

Hopp, F. P. (1999). Patterns and predictors of formal and informal care among elderly persons living in board and care homes. *The Gerontologist, 39,* 167–176.

Hornung, C. A., Eleazar, G. P., Strothers, H. S. 3rd, Wieland, G. D., Eng, C., McCann, R., & Sapir, M. (1998). Ethnicity and decision-makers in a group of frail older people. *Journal of the American Geriatrics Society, 46*(3), 280–286.

Horowitz, A. M., Goodman, H. S., Yellowitz, J. A., & Nourjah, P. A. (1996). The need for health promotion on oral cancer prevention and early detection. *Journal of Public Health Dentistry, 56,* 319–330.

Hoyl, M. T. et al. (1999). Development and testing of a five-item version of the geriatric depression scale. *Journal of the American Geriatrics Society, 47,* 873–878.

Hu, M., & Woollacott, M. (1994). Multisensory training of standing balance in older adults. *Journals of Gerontology, 49A,* M52–M61.

Hu, T. (1994). Clinical guidelines and cost implications in the case of urinary incontinence. *Geriatric Nephrology & Urology, 4,* 85–91.

Huggins, C. M., & Phillips, C. Y. (1998). Using case management with clinical plans to improve patient outcomes. *Home Healthcare Nurse, 16*(1), 15–20.

Hull, R. (1995). *Hearing in Aging.* San Diego: Singular Publishing.

Hungelmann, J., Kenkel-Rossi, E., Klassen, L., & Stollenwerk, R. (1996). Focus on spiritual well-being: Harmonious interconnectedness of mind-body-spirit-use of the JAREL spiritual well-being scale. *Geriatric Nursing, 17*(6), 262–266.

Hunt, M. E., & Gunter-Hunt, G. (1985). Naturally occurring retirement communities. *Journal of Housing for the Elderly, 3*(3/4), 3–21.

Hunt, M. E., & Ross, L. E. (1990). Naturally occurring retirement communities: A multiattribute examination of desirability factors. *The Gerontologist, 30*(5), 667–674.

Hunt, V. V. (1996). Infinite Mind: The Science of the Human Vibrations of Consciousness, 2nd ed. Malibu, CA: Malibu Publishing Co.

Hunter, K. I., & Linn, M. W. (1980–1981). Psychosocial differences between elderly volunteers and non-volunteers. *International Journal of Aging & Human Development, 12,* 205–213.

Hurley, A. C., Volicer, B. J., Hanrahan, P., Houde, S., & Volicer, L. (1992). Assessment of discomfort in Alzheimer's patients. *Research in Nursing & Health, 15,* 369–377.

Hutton, W. R. (1981). Volunteering: Unaffordable luxury for the elderly. *Generations, 4,* 12–13.

Hyer, K. (1998). The John A. Hartford Foundation Geriatric Interdisciplinary Team Training Program. In E. L. Siegler, K. Hyer, T. Fulmer, & M. Mezey (Eds.), *Geriatric Interdisciplinary Team Training.* New York: Springer Publishing Company.

Ikegami, N. (1997). Public long-term care insurance in Japan. *JAMA, 278*(16), 1310–1314.

Ikels, C. (1986). Older immigrants and natural helpers. *Journal of Cross-Cultural Gerontology, 1*(14), 209–222.

Inglehart, J. (1999). *Medicare and Managed Care: A Primer.* New York: Health Affairs and the California Healthcare Foundation.

Ingram, F. (1996). The short Geriatric Depression Scale: A comparison with the standard form in independent older adults. *Clinical Gerontologist, 16,* 49–56.

Inouye, S. K., Bogardus, S. T. Jr., Charpentier, P. A., Leo-Summers, L., Acampora, D., Holford, T. R., & Cooney, L. M. Jr. (1999). A multicomponent intervention to prevent delirium in hospitalized older patients. *NEJM, 340,* 669–676.

Inouye, S. K., & Charpentier, P. A. (1996). Precipitating factors for delirium in hospitalized elderly persons: Predictive model and interrelationship with baseline vulnerability. *JAMA, 275,* 852–857.

Inouye, S. K., Rushing, J. T., Foreman, M. D., Palmer, R. M., & Pompei, P. (1998). Does delirium contribute to poor hospital outcomes? *Journal of General Internal Medicine, 13,* 234–242.

Institute for Health and Aging, USCF. (1996). *Chronic Care in America: A 21st Century Challenge.* Robert Wood Foundation.

Institute of Medicine. (1995) *Real People Real Problems: An Evaluation of the Long-term Care Ombudsman Programs of the Older Americans Act (Summary)* [Web Page]. URL http://www.nap.edu [2000, March 10].

Institute of Medicine. (1993). *Strengthening Training in Geriatrics for Physicians.* Washington, DC: National Academy Press.

Institute of Medicine Committee on Leadership for Academic Geriatric Medicine. (1987). Academic Geriatrics for the Year 2000. *NEJM, 316,* 1425–1428.

Internet Healthcare Coalition. (1999) *Proceedings: Quality Healthcare Information on the Net '99* [Web Page]. URL http://www.ihealthcoalition.org/community/conf_order.html.

Ippolito, R. (1997). *Pension Plans and Employee Performance: Evidence, Analysis, and Policy.* Chicago: University of Chicago Press.

Irish, D. P., Lundquist, K. F., & Nelson, V. J. (1993). *Ethnic Variations in Dying, Death, and Grief.* Washington D.C.: Taylor and Francis.

Irwin, R. S., Boulet, L. P., Cloutier, M. M., Fuller, R., Gold, P. M., Hoffstein, V., Ing, A. J., McCool, F. D., O'Byrne, P., Poe, R. H., Prakash, U. B., Pratter, M. R., & Rubin, B. K. (1998). Managing cough as a defense mechanism and as a symptom. A consensus panel report of the American College of Chest Physicians. *Chest, 114*(2 Suppl Managing), 133S–181S.

Irwin, R. S., & Curley, F. J. (1991). The treatment of cough: A comprehensive review. *Chest, 99*(6), 1477–1484.

Iso-ahola, S. E., Jackson, E. L., & Dunn, E. (1994). Starting, ceasing, and replacing leisure activities over the human life-span. *Journal of Leisure Research, 26*(3), 227–249.

Israelsson, L. A., & Jonssen, T. (1997). Overweight and healing of midline incisions: The importance of suture technique. *European Journal of Surgery, 163*(3), 175–180.

Iverson, L. L., Mortishire-Smith, R., Pollack, S., & Shearman, M. (1995). The toxicity of in vitro beta-amyloid protein. *Biochemical Journal, 311,* 1–16.

Jack, C. M., & Paone, D. L. (1994). Toward creating a seamless continuum of care: Addressing chronic care needs. *American Hospital Association, Section for Aging and Long-Term Services.*

Jackson, J. S., Antonucci, T. C., & Gibson, R. C. (1995). Ethnic and cultural factors in research on aging and mental health: A life course perspective. In D. K. Padgett (Ed.), *Handbook on Ethnicity, Aging and Mental Health* (pp. 22–46). Westport, CT: Greenwood Press.

Jacobs, M. R. (1999). Drug resistant streptococcus pneumoniae: Rational antibiotic choices. *American Journal of Medicine, 106*(5A—Special Supplement), 19S–25S.

Jacobsen, J. H., Rosenberg, R. S., Huttenlocher, P. R., & Spire, J. (1986). Familial nocturnal cramping. *Sleep, 9*(1), 54–60.

Jaff, M. R., & Dorros, G. (1998). The vascular laboratory: A critical component required for successful management of peripheral arterial occlusive disease. *Journal of Endovascular Surgery, 5*(2), 146–158.

Jansen, P. H. P., Lecluse, R. G. M., & Verbeek, A. L. M. (1999). Past and current understanding of the pathophysiology of muscle cramps: Why treatment of varicose veins does not relieve leg cramps. *Journal of the European Academy of Dermatology and Venereology, 12*(3), 222–229.

Janus, S. S., & Janus, C. L. (1993). *The Janus Report on Sexual Behavior.* New York: John Wiley & Sons.

Jecker, N. S., & Self, D. J. (1991). Medical ethics in the 21st century: Respect for autonomy in care of the elderly patient. *Journal of Critical Care, 6*(1), 46–51.

Jencks, C. (1994). *The Homeless.* Boston: Harvard University Press.

Jendrek, M. P. (1993). Grandparents who parent their grandchildren: Effects on lifestyle. *Journal of Marriage & Family, 55,* 609–621.

Jensen, G. L., & Rogers, J. (1998). Obesity in older persons. *Journal of the American Dietetic Association, 98,* 1308–1311.

Jerrome, D. (1994). Time, changing, and continuity in family life. *Aging & Society, 14,* 1–27.

Jeste, D. V. (1997). Special Issue on schizophrenia, antipsychotics, and aging. *Schizophrenia Research, 27,* 101–267.

Jeste, D. V., & Finkel, S. I. (2000). Psychosis of Alzheimer's Disease and Related Dementias: Diagnostic Criteria for a Distinct Syndrome. *American Journal of Geriatric Psychiatry, 8,* 1.

Jette, A. M., & Davies, A. R. C. P. D. (1986). The Functional Status Questionaire: Reliability and validity when used in primary care. *Journal of Geriatric Internal Medicine, 1,* 143–149.

Jick, H., Vasilakis, C., Weinrauch, L. A., Meier, C. R., Jick, S. S., & Derby, L. E. (1998). A population-based study of appetite suppressant drugs and the risk of cardiac-valve regurgitation. *NEJM, 339,* 719–724.

Jobst, K., Shostak, D., & Whitehouse, P. (1999). Diseases of meaning: Manifestations of health, and metaphos. *The Journal of Alternative & Complementary Medicine: Research on Paradigm, Practice, & Policy, 5*(6), 495–502.

John, R. (1999). Aging among American Indians: Income security and social support networks. In T. P. Miles (Ed.), *Full Color Aging.* Washington, DC: Gerontological Society of America.

Johns, M. W. (1992). Reliability and factor analysis of the Epworth Sleepiness Scale. *Sleep, 15,* 376–381.

Johnson, B. K. (1996). Older adults and sexuality: A multidimensional perspective. *Journal of Gerontological Nursing, 22*(2), 6–15.

Johnson, J., Weissman, M. M., & Klerman, G. L. (1992). Service utilization and social morbidity associated with depressive symptoms in the community. *JAMA, 267,* 1478–1483.

Johnson, J. T., Ferretti, G. A., Nethery, W. J., Valdez, I. H., Fox, P. C., Ng, D., Muscoplat, C. C., & Gallageher, S. C. (1993). Oral pilocarping for postirradiation induced xerostomia in patients with head and neck cancer. *NEJM, 329,* 390–395.

Johnson, N. W., Warnakulasuriy, S., & Travassoli, M. (1996). Heredity and environmental risk factors: Clinical and laboratory risk matters for head and neck, especially oral cancer and precancer. *European Journal of Cancer Prevention, 5,* 5–17.

Johnson, R. (in press). Relocation stress syndrome. In M. Maas, K. C. Buckwalter, T. Tripp-Reimer, M. Hardy, & M. Titler (Eds.), *Nursing Care of Older Adults: Nursing Diagnoses, Interventions, and Outcomes.* St. Louis, MO: Mosby Year-Book.

Johnson, R., Brazill, A., & Ladrigan, P. (1997). Intrainstitutional relocation: Effects on residents' behavior and psychosocial functioning. *Journal of Gerontological Nursing, 23*(4), 35–41.

Johnson, R. A., Schwiebert, V. B., & Rosenmann, P. A. (1994). Factors influencing nursing home placement decisions: The older adult's perspective. *Clinical Nursing Research, 3*(3), 269–281.

Johnson, T. E. (1990). Caenorhabditis elegans offers the potential for molecular dissection of the aging processes. In E. L. Schneider, & J. W. Rowe (Eds.), *Handbook of the Biology of Aging* (3rd ed., pp. 45–59). San Diego: Academic Press.

Joint National Committee on Prevention, Detection, Evaluation, and Treatment of High Blood Pressure. (1997). Sixth Report of the Joint National Committee on Prevention, Detection, Evaluation, and Treatment of High Blood Pressure. *Archives of Internal Medicine, 157,* 2413–2446.

Jonas, H. (1969). Philosophical reflections on experimenting with human subjects. In P. A. Freund (Ed.), *Experimentation with Human Subjects* (pp. 1–31). New York: G. Braziller.

Jones, B., & Donner, M. W. (1991). *Normal and Abnormal Swallowing: Imaging in Diagnosis and Therapy.* New York: Springer-Verlag.

Jones, P. S. (1996). *Opening the door on elder abuse: Physical and emotional abuse, caretaker neglect, financial exploitation.* Washington, DC: U.S. Department of Health and Human Services, Administration on Aging.

Judd, S., Marshall, M., & Phippen, P. (1998). *Design for Dementia.* London: Hawker Publications, LTD.

Judge, J. O., Smyers, D., & Wolfson, L. (1992). Muscle strength predicts gait measures in older adults. *Journal of the American Geriatrics Society, 40,* 20–27.

Jung, C. (1971). *The Portable Jung.* New York: Viking Publishing Company.

Kafonek, S., Ettinger, W., Roca, R., Kittner, S., Taylor, N., & German, P. (1989). Instruments for screening for depression and dementia in long-term care facility. *Journal of the American Geriatrics Society, 37,* 29–34.

Kahana, B. (1995). Isolation. In G. L. Maddox (Ed.), *Encyclopedia of Aging* (2nd ed.). New York: Springer Publishing Company.

Kahn, M. A., Herzog, C. A., St. Peter, J. V., Hartley, G. G., & Madlon-Kay, R. D. (1998). The prevalence of cardiac valvular insufficiency by transthoracic echocardiography in obese patients teated with appetite-suppresant drugs. *NEJM, 339,* 713–718.

The Kaiser Commission on Medicaid and the Uninsured. (1999). *Medicaid and Managed Care, Fact Sheet.* Menlo Park, CA: The Henry J Kaiser Family Foundation.

The Kaiser Commission on the Future of Medicaid. (1993). *The Medicaid Cost Explosion: Causes and Consequences.* Menlo Park, CA: The Henry J Kaiser Family Foundation.

Kane, R. A., Degenholz, H. B., & Kane, R. L. (1999). Adding values: An experiment in systematic attention to values and preferences of community long-term care clients. *Journals of Gerontology, 54B*(2), S109–119.

Kane, R. A., & Kane, R. L. (1981). *Assessing the Elderly: A Practical Guide to Measurement.* Lexington, MA: Lexington Books.

Kane, R. A., Kane, R. L., Illston, L. H., Nyman, J. A., & Finch, M. D. (1991). Adult foster care for the elderly in Oregon. *American Journal of Public Health, 81*(9), 1113–1120.

Kane, R. L., Ouslander, J. G., & Abrass, I. B. (1999). *Essentials of Clinical Geriatrics.* New York: McGraw-Hill, Inc.

Kannel, W. B. (1996). The demographics of claudication and the aging of the American population. *Vascular Medicine, 1*(1), 60–64.

Kanowski, S., Kinzler, E., Lehmann, E., Schweizer, A., & Kuntz, G. (1995). Confirmed clinical efficacy of Actovegin in elderly patients with organic brain syndrome. *Pharmacopsychiatry, 28*(4), 125–133.

Kaplan, E. F., Goodglass, H., & Weintnaube, S. (1983). *The Boston Naming Test.* Philadelphia, PA: Lea and Febiger.

Kaplan, S. H., Greenfield, S., Gandek, B., Rogers, W. H., & Ware, J. E. J. (1996). Characteristics of physicians with participatory decision-making styles. *Annals of Internal Medicine, 124*(5), 497–504.

Kapp, M. B. (1992). *Geriatrics and the Law: Patient Rights and Professional Responsiblities* (2nd ed.). New York: Springer Publishing.

Kapust, L. R., & Weintraub, S. (1988). The home visit: Field assessment of mental status impairments in the elderly. *The Gerontologist, 28,* 112–115.

Karagas, M. R., Lu-Yao, G. L., & Barrett, J. A. (1996). Heterogeneity of hip fracture: Age, race, sex, and geographic patterns of femoral neck and trochanteric fractures among the U.S. elderly. *American Journal of Epidemiology, 143*(7), 677–682.

Kart, C. S. (1997). *The Realities of Aging: An Introduction to Gerontology* (5th ed.). Boston, MA: Allyn and Bacon.

Kasper, C. E. (1993). Skeletal muscle atrophy. In V. Carrieri-Kohlman, A. M. Lindsey, & C. M. West (Eds.), *Pathophysiological Phenomena in Nursing: Human Responses to Illness* (2nd ed., pp. 530–555). Philadelphia: W. B. Saunders.

Kass-Annese, B. (1997). *A Total Wellness Program for Women Over 30.* New York: Springer Publishing Company, Inc.

Kato, A., Fukunari, A., Sakai, Y., & Nakajima, T. (1997). Prevention of amyloid-like deposition by a selective prolyl endopeptidase inhibitor, Y-29794, in senescence-accelerated mouse. *Journal of Pharmacology & Experimental Therapeutics, 283*(1), 328–335.

Katz, I. R., Jeste, D. V., Mintzer, J. E., Clyde, C., Napolitano, J., Brecher, M., & the Resperidone Study Group. (1999). Comparison of resperidone and placebo for psychosis and behavioral disturbances associated with dementia: A randomized double-blind trial. *Journal of Clinical Psychiatry, 60,* 107–115.

Katz, P. O. (1998). Gastroesophageal reflux disease. *Journal of the American Geriatrics Society, 46,* 1558–1565.

Katz, S. (1983). Assessing self-maintenance: Activities of daily living, mobility, and instrumental activities of daily living. *Journal of the American Geriatrics Association, 31,* 721–727.

Katz, S., Ford, R., Moskowitz, R., Jackson, B., & Jaffe, M. (1963). Studies of illness in the aged: The index of ADL, a standardized measure of biological and psychological function. *JAMA, 186,* 914–919.

Katzmann, R. (1986). *Institutional disability: The saga of transportation policy for the disabled.* Washington, DC: Brookings.

Kaufman, S. R. (1993). Values as sources of the ageless self. In J. R. Kelly *Activity and Aging* (pp. 17–24). Newbury Park, CA: Sage.

Kaye, L. W., & Kirwin, P. M. (1990). Adult day care services for the elderly and their families: Lessons from the Pennsylvania experience. *Journal of Gerontological Social Work, 15,* 167–183.

Kayser-Jones, J. (1996). Mealtimes in nursing homes: The importance of individualized care. *Journal of Gerontological Nursing, 22*(3), 26–31.

Kayser-Jones, J. S., Bird, W. F., Paul, S. M., Long, L., & Schell, E. S. (1995). An instrument to assess the oral health status of nursing home residents. *The Gerontologist, 35,* 814–824.

Kayser-Jones, J. S., & Pengilly, K. (1999). Dysphagia among nursing home residents. *Geriatric Nursing, 20*(2), 77–82.

Kayser-Jones, J. S., & Schell, E. S. (1997). The effect of staffing on the quality of care at mealtimes. *Nursing Outlook, 45,* 64–67.

Keating, J. P., Stewart, P. J., Eyers, A. A., Warner, D., & Bokey, E. L. (1997). Are special investigations of value in the management of patients with fecal incontinence. *Diseases of the Colon & Rectum, 40,* 896–901.

Keay, T. J., & Schonwetter, R. S. (1998). Hospice care in the nursing home. *American Family Physician, 57,* 491–494.

Keeth, C. K. (1996). Enteral Nutrition. In K. A. Hennessy, & M. E. Orr (Eds.), *Nutrition Support Nursing. Core Curriculum.* Silver Springs, MD: American Society for Parenteral and Enteral Nutrition. Pg 20-1-20-10.

Kegel, A. (1950). The physiologic treatment of urinary stress incontinence. *Journal of Urology, 63,* 803–814.

Keith, P. M., & Wacker, R. R. (1994). *Older wards and their guardians.* Westport, CT: Praeger Press.

Keith, S., Reiger, D. A., & Rae, D. S. (1991). Schizophrenic Disorders. In L. N. Robins, & D. A. Reiger (Eds.), *Psychiatric Disorders in America: The Epidemiologic Cachment Area Study* (pp. 33–52). New York: The Free Press.

Kelly, J. R. (1987). *Peoria winter: Styles and resources in later life.* Lexington, MA: Lexington Books, the Free Press.

Kelly, J. R. (1993). *Activity and aging.* Newbury Park, CA: Sage.

Kelly, J. R., Steinkamp, M. W., & Kelly, J. R. (1986). Later life expectancy: How they play in Peoria. *The Gerontologist, 26,* 531–537.

Kennedy, B. K., Austriaco, N. R. Jr., Zhang, J., & Guarente, L. (1995). Mutation in the silencing gene SIR4 can delay aging. *Cell, 80,* 485–496.

Kennedy, G. J., Kelman, H. R., & Thomas, C. (1990). The emergence of depressive symptoms in late life: The importance of declining health and increasing disability. *Journal of Community Health, 15,* 93–104.

Kennedy-Malone, L., Fletcher, K., & Plank, L. (2000). *Management Guidelines for Gerontological Nurse Practitioners.* Philadelphia, PA: F. A. Davis Company.

Kertesz, A. (in press). BPSD and Frontal-Lobe Dementia. *International Psychogeriatrics.*

Keyes, C. L. M., & Ryff, C. (1999). Psychological well-being in midlife. In S. L. Willis, & J. D. Reif (Eds.), *Life in the Middle* (pp. 161–180). 1999: San Diego: Academic Press.

Kim, P. K. H. (1994). *Services to the Aging and Aged: Public Policies and Programs.* New York: Garland Publishing, Inc.

King, C. (1997). Guidelines for improving assessment skills. *Generations, 21*(1), 73–75.

Kingson, E., & Berkowitz, E. (1993). *Social Security and Medicare: A Policy Primer.* Westport, CT: Auburn House.

Kinsella, K., & Taeuber, C. M. (1993). *An Aging World II. International Population Reports.* Washington, DC: U.S. Bureau of the Census.

Kirchhoff, S. (1999). Rewrite of Older Americans Act sets stage for floor fight over control of jobs program. *Congressional Quarterly Weekly Report,* September 18, 2172.

Kisner, C., & Colby, L. A. (1990). *Therapeutic Exercise: Foundations and Techniques.* Philadelphia, PA: F. A. Davis, Inc.

Kitwood, T. (1993). Person and process in dementia. *International Journal of Geriatric Psychiatry, 8,* 541–545.

Kizer, K. W. (1996). *The Guiding Principles and Strategic Objectives Underlying the Transformation of the Veterans Healthcare System.* Washington, DC: U.S. Department of Veterans Affairs.

Kjellander, D., Bongar, B., & King, A. (1998). Suicidality in borderline personality disorder. *Crisis, 19*(3), 125–135.

Kleemeier, R. W. (Ed.). (1961). *Aging and Leisure.* New York: Oxford University Press.

Klein, A. J., & Weber, P. C. (1999). Hearing aids. *Medical Clinics of North America, 83*(1), 139–151.

Klessig, J. (1992). The effect of values and culture on life-support decisions. *Western Journal of Medicine, 157*(3), 316–322.

Klugman, K. P., & Feldman, C. (1999). Penicillin- and cephalosporin-resistant Streptococcus pneumoniae: Emerging treatment for an emerging problem. *Drugs, 58*(1), 1–4.

Knies, R. C. Jr. (1996). Assessment in geriatric trauma: What you need to know. [Review] [24 references]. *International Journal of Trauma Nursing, 2*(3), 85–91.

Knight, B. G., Lutzky, S. M., & Macofsky-Urban, F. (1993). A meta-analytic review of interventions for caregiver distress: Recommendations for future research. *The Gerontologist, 33,* 240–248.

Knipling, R. R., & Wang, J. S. (1994). *Crashes and Fatalities Related to Driver Drowsiness/Fatigue.* Washington, DC: National Highway Traffic Safety Administration, U.S. Department of Transportation.

Knoll, J. (1998). Extension of life span of rats by long-term (–) deprenyl treatment. *Mount Sinai Journal of Medicine, 55,* 64–74.

Koenig, H. G., George, L. K., Cohen, H. J., Hays, J. C., Larson, D. B., & Blazer, D. G. (1998). The relationship between religious activities and blood pressure in older adults. *The International Journal of Psychiatry in Medicine, 28*(2), 180–194.

Koenig, H. G., George, L. K., & Peterson, B. L. (1998). Religiosity and remission of depression in medically ill older adults. *The American Journal of Psychiatry, 155*(4), 536–542.

Kohn, L., Corrigan, J., & Donaldson, M. (Eds.). (1999). *To Err Is Human: Building a Safer Health System. Committee on Quality of Health Care in America, Institute of Medicine.* Wahington, DC: National Academy Press.

Koplas, P. A., Gans, H. B., Wisely, M. P., Kuchibhatla, M., Cutson, T. M., Gold, D. T., Taylor, C. T., & Schenkman, M. (1999). Quality of life and Parkinson's disease. *The Gerontological Society of America, 54A*(4), M197–M202.

Korenblat, P. E., & Wedner, H. J. (1992). *Allergy Theory and Practice.* Orlando, FL: W. B. Saunders Co.

Kosloski, K., & Montgomery, R. J. V. (1993). Perceptions of respite services as predictors of utilization. *Research on Aging, 15,* 399–413.

Kosnik, W., Winslow, L., Kline, D., Rasinski, K., & Sekuler, R. (1988). Visual changes in daily life throughout adulthood. *Journals of Gerontology, 43B*(3), P63–70.

Kotlikoff, L., & Wise, D. (1990). *The Wage Carrot and the Pension Stick.* Kalamazoo: The W. E. Upjohn Institute for Employment Research.

Koval, J. K., & Zuckerman, J. D. (1989). *Fractures in the Elderly.* Philadelphia, PA: Lippincott-Raven.

Koyama, W., Koyangi, A., Mihara, S., Kawazu, S., Uemura, T., Nakano, H., Gotou, Y., Nishizawa, M., Noyama, A., Hasewaga, C., & Nakano, M. (1998). Prevalence and conditions of urinary incontinence among the elderly. *Methods of Information in Medicine, 37*(2), 151–155.

Kramer, B. J. (1997). Chronic Diseases in American Indian Populations. In M. S. Markides, & M. R. Miranda (Eds.), *Minorities, Aging, and Health* (pp. 181–204). Thousand Oaks, CA: Sage.

Kramer, B. J. (1997). Gain in the caregiving experience: Where are we? *The Gerontologist, 37,* 218–232.

Kramer, B. J. (1992). Health and aging of urban American Indians. *Western Journal of Medicine, 157,* 281–285.

Kramer, B. J. (1999). The health status of urban Indian elders. *The IHS Primary Care Provider, 24*(5), 69–73.

Krantz, G., Christenson, M., & Lindquist, A. (1998). *Assistive Products: An Illustrated Guide to Terminology.* Bethesda, MD: American Occupational Therapy Association.

Krause, N. (1999). Mental disorder in later life: Exploring the influences of stress and socioeconomic status. In C. S. Aneshensel & J. C. Phelan (Eds.), *Handbook of the Sociology of Mental Health* (pp. 183–208). New York: Kluwer Academic/Plenum Publishers.

Krause, N. (1986). Social support, stress, and well being among older adults. *Journal of Gerontology, 41,* 513–519.

Krauss, N. A., & Altman, B. M. (1998). *Characteristics of nursing home residents, 1996* [Web Page]. URL http://www.meps.ahcpr.gov/papers/99-0006/99-0006.htm [1999, August 2].

Krout, J. (1994). *Providing Community Based Services to the Rural Elderly.* Thousand Oaks, CA: Sage Publications, LTD.

Krout, J. A. (1983). Correlates of senior center utilization. *Research on Aging, 5,* 339–352.

Krout, J. A. (1983). Knowledge and use of services by the elderly: A critical review of the literature. *Inter-*

national Aging and Human Development, 17, 153–167.

Krout, J. A. (1990). The organization, operation, and programming of senior centers in American: A seven year follow-up. *Final Report to the AARP Andrus Foundation.* Fredonia, NY.

Krout, J. A. (1996). Senior center programming and frailty among older persons. *Journal of Gerontological Social Work, 26*(3–4), 19–34.

Krout, J. A. (1995). Senior centers and services for the frail elderly. *Journal of Aging & Social Policy, 7*(2), 59–76.

Krout, J. A. (1989). *Senior Centers in America.* New York: Greenwood Press.

Krowinski, W. J., & Steiber, S. R. (1996). *Measuring and Managing Patient Satisfaction.* Chicago, IL: American Hospital Publishing, Inc.

Kuehune, V. S., & Sears, H. A. (1993). Beyond the call of duty: Older volunteers committed to children and families. *Journal of Applied Gerontology, 12,* 425–438.

Kulys, R., & Tobin, S. S. (1980). Interpreting the lack of future concerns among the elderly. *International Journal of Aging & Human Development, 11*(2), 111–126.

Kytle, C. (1994). *The Now and Tomorrow of Continuing Care Communities: Trends and Issues in Organization, Governance, and Regulation.* Chapel Hill, NC: Continuing Care Community Residents of North Carolina and the Residents Association of Carolina Meadows.

La Puma, J., & Schiedermayer, D. L. (1991). Ethics consultation: Skills, roles and training. *Annals of Internal Medicine, 114*(2), 155–160.

La Rosa, J. C. (1994). Dyslipoproteins in women and the elderly. *Medical Clinics of North America, 78,* 163–180.

Lachs, M. S., Berkman, L., Fulmer, T., & Horwitz, R. I. (1994). A prospective community-based pilot study of risk factors for the investigation of elder mistreatment. *Journal of the American Geriatrics Society, 42*(2), 169–173.

Lachs, M. S., & Pillemer, K. (1995). Current concepts: Abuse and neglect of elderly persons. *NEJM, 332,* 437–443.

Lachs, M. S., Williams, C., O'Brien, S., Hurst, L., & Horwitz, R. I. (1997). Risk factors for reported elder abuse and neglect: A nine-year observational cohort study. *The Gerontologist, 37*(4), 469–474.

Lachs, M. S., Williams, C. S., O'Brien, S., Pillemer, K. A., & Charlson, M. E. (1998). The mortality of elder mistreatment. *JAMA, 280,* 428–432.

Laird, W. R., & McLaughlin, E. A. (1989). Management and treatment planning for the elderly edentulous patient. *International Journal of Prosthodontics, 2,* 347–351.

Lake, C. R., Mason, E. B., & Quirk, R. S. (1998). Psychiatric side effects attributed to phenylpropanolamine. *Pharmacopsychiatry, 21*(4), 171–181.

Lambert, M. T., & Fowler, D. R. (1997). Suicide risk factors among veterans: Risk management in the changing culture of the Department of Veterans Affairs. *Journal of Mental Health Administration, 24*(3), 350–358.

Lance, N. J., & Curran, J. J. (1993). Late onset, seropositive, erosive rheumatoid arthritis. *Seminars in Arthritis & Rheumatism, 23*(3), 177–182.

Landefeld, C. S., Palmer, R. M., Kresevic, D. M., Fortinsky, R. H., & Kowal, J. (1995). A randomized trial of care in a hospital medical unit especially designed to improve the functional outcomes of acutely ill older patients. *NEJM, 332,* 1338–1344.

Lander, S., Brazill, A., & Ladrigan, P. (1997). Intrainstitutional relocation: Effects on residents' behavior and psychosocial functioning. *Journal of Gerontological Nursing, 23*(4), 35–41.

Lang, A. E., & Lozano, A. M. (1998). Parkinson's disease: Second of two parts. *NEJM, 339*(16), 1130–1143.

Langmore, S. E., Terpenning, M. S., & Schork, A. (1998). Predictors of aspiration: How important is dysphagia? *Dysphagia, 13*(2), 69–81.

Langner, T. S., & Michael, S. T. (1963). *Life Stress and Mental Health.* New York: Basic Books.

Lanspery, S. C., & Callahan, J. J. (1994). *Naturally occurring retirement communities. A report prepared for the Pew Charitable Trusts.* Boston, MA: Policy Center on Aging, Heller School, Brandeis University.

Lapierre, S., Bouffard, L., & Bastin, E. (1997). Personal goals and subjective well-being in later life. *International Journal of Aging & Human Development, 45,* 287–303.

Lareau, S., Carrieri-Kohlmann, V., Janson-Bjerklie, S., & Roos, P. (1994). Development and testing of the pulmonary functional status and dyspnea questionnaire (PFSDQ). *Heart & Lung, 23,* 242–250.

Lareau, S., Meek, P. L., Press, D. A. J., & Roos, P. (1999). Dyspnea in patients with chronic obstructive pulmonary disease. *Heart & Lung, 28,* 65–73.

Larsen, E. B. (1998). Recognition of dementia: Discovering the silent epidemic. *Journal of the American Geriatrics Association, 46,* 1576–1577.

Larson, E. B., Kukull, W. A., Buchner, D., & Reifler, B. V. (1987). Adverse drug reactions associated with global cognitive impairment in elderly persons. *Archives of Internal Medicine, 107*(2), 169–173.

Larson, E. B., Reifler, B. V., Featherstone, H. J., & English, D. R. (1984). Dementia in elderly outpatients: A prospective study. *Annals of Internal Medicine, 100*(3), 417–423.

Lashley, F. R. C. (1998). *Clinical Genetics in Nursing Practice* (2nd ed.). New York: Springer Publishing Company.

Lasser, R. A., & Sutherland, T. (1998). Newer psychotropic medication use in nursing home residents. *Journal of the American Geriatrics Association, 46,* 202–207.

Lauque, S., Arnaud-Battandier, F., Mansourian, R., Guigoz, Y., Paintin, M., Nourhashemi, F., & Vellas, B. (2000). Protein-energy oral supplementation in malnourished nursing-home residents. A controlled trial. *Age & Ageing, 29*(1), 51–56.

Lawler, W. R. (1998). An office approach to the diagnosis of chronic cough. *American Family Physician, 58*(9), 2015–2022.

Lawton, M., VanHaitsman, K., & Klapper, J. (1996). Observed affect in nursing home residents with Alzheimer's disease. *Journals of Gerontology, 51B,* P3–14.

Lawton, M. P. (1972). Assessing the competence of older people. In D. Kent, R. Kastenbaum, & S. Sherwood (Eds.), *Research, planning and action for the elderly.* New York: Behavioral Publications.

Lawton, M. P. (1989). Behavior-relevant ecological factors. In K. W. Schaie, & C. Scholar (Eds.), *Social Structure and Aging: Psychological Processes.* Hillsdale, NJ: Erlbaum.

Lawton, M. P. (1993). Meanings of activity. In J. R. Kelly (Ed.), *Activity and Aging* (pp. 25–41). Newbury Park, CA: Sage Publications.

Lawton, M. P., Brody, E., & Saperstein, A. (1989). Effects of alternative support strategies. *Gerontologist, 29,* 8–16.

Lawton, M. P., & Brody, E. M. (1969). Assessment of older people: Self-maintaining and instrumental activities of daily living. *The Gerontologist, 9*(3), 179–186.

Lawton, M. P., Moss, M., Fulcomer, M., & Kleban, M. H. (1982). A research and service oriented multilevel assessment instrument. *Journal of Gerontology, 37*(1), 91–99.

Lawton, M. P., Moss, M., Kleban, M. H., Glicksman, A., & Rovine, M. (1991). A two-factor model of caregiving appraisal and psychological well-being. *Journals of Gerontology, 46B,* P181–189.

Lawton, M. P., Rajagopal, D., Brody, E., & Kleban, M. (1992). The dynamics of caregiving for a demented elder among black and white families. *Journals of Gerontology, 47B,* S156–S164.

Lawton, M. P., & Rubinstein, R. L. (Eds.). (2000). *Interventions in Dementia Care: Toward Improving Quality of Life.* New York: Springer Publishing Company.

Lawton, P., Moss, M., Fulcomer, M., & Kleban, M. (1982). A research and service oriented multilevel assessment instrument. *Journal of Gerontology, 37,* 91–99.

Lazarus, R. S. (1991). Progress on a cognitive-motivational-relationship theory of emotion. *American Psychologist, 46*(8), 819–834.

Lazarus, R. S. (1990). Stress, coping, and illness. In H. S. Friedman (Ed.), *Personality and Disease.* New York: John Wiley.

Lazarus, R. S., & Folkman, S. (1984). *Stress, Appraisal and Coping.* New York: Springer.

League for the Hard of Hearing. (1996). *Tips for Health Care Workers.* New York: Center for Health Care Access.

LeClerc, C. M., & Wells, D. L. (1998). Use of content methodology process to enhance feeding abilities threatened by ideational apraxia in people with Alzheimer's type dementia. *Geriatric Nursing, 19,* 261–268.

Leclerc, K. M., & Landry, F. J. (1996). Benign nocturnal leg cramps: Current controversies over use of quinine. *Postgraduate Medicine, 99*(2), 177–178, 181–184.

Lee, I. M., Paffenbarger, R. S., & Hennekens, C. H. (1997). Physical activity, physical fitness, and longevity. *Aging & Clinical Experimental Research, 9,* 2–11.

Lee, N., Wasson, D., Anderson, M. A., Stone, S., & Gittings, J. (1998). A survey of patient education after discharge. *Journal of Nursing Care Quality, 13*(1), 63–70.

Lee, P. P., Jackson, C. A., & Relles, D. A. (1998). Specialty distributions of ophthalmologists in the workforce. *Archives of Ophthalmology, 116*(7), 917–920.

Lehman, A. F., Sreinwachs, D. M., & Co-investigators of the PORT Project. (1998). Translating research into practice: The schizophrenia Patient Outcomes Research Team (PORT) treatment recommendations. *Schizophrenia Bulletin, 24,* 1–10.

Lemke, S., & Moos, R. (1989). Personal and environmental determinants of activity involvement among elderly residents of congregate facilities. *Journal of Gerontology, 44*(Supplement), S139–148.

Lemon, B. W., Bengston, V. L., & Peterson, J. A. (1972). An exploration of the "activity theory" of aging. *Journal of Gerontology, 27,* 511–523.

Lenssen, P. (1998). Management of total parenteral nutrition. In A. Skipper (Ed.), *Dietician's Handbook of Enteral and Parenteral Nutrition* (2nd ed.). Rockville, MD: Aspen Publishing.

Leon, J., Neuman, P., & Parents, S. (1997). *Understanding the growth in Medicare's home health expenditures.* Bethesda, MD: Project HOPE. Center for Health Affairs.

Leonard, R., & Kendall, K. (1997). *Dysphagia Assessment and Treatment Planning: A Team Approach*. Baltimore: Singular Publishing Group, Inc.

Lepor, H. (1995). Alpha blocked for the treatment of benign prostatic hyperplasia. *Urology Clinics of North America, 22*(2), 375–386.

LeRoy, G. L. (1997). Minority health. *The Ohio Family Physician, 49*(3), 3–9.

Lesthaeghe, R. (1999). Is low fertility a temporary phenomenon in the EU? *IPD Working Paper* Vol. 1999-1. Brussels: Vrije Universiteit Brussel.

Leszcz, M. (1987). Group psychotherapy with the elderly. In J. Sadavoy, & M. Leszcz (Eds.), *Treating the Elderly with Psychotherapy: The Scope for Change in Later Life* (pp. 325–349). Madison, CT: International Universities Press.

Levenson, S. A. (1996). *Subacute and Transitional Care Handbook*. Baltimore: Beverly Cracom Publications.

Levi, S. J. (1999). Evidence based practice: Systemic reviews, critical reviews. *Archives of Physical Medicine and Rehabilitation, 6*(5), 1–4.

Levkoff, S. E., Liptzin, B., Evans, D. A., Cleary, P. D., Lipsitz, L. A., Wetle, T. T., & Rowe, J. (1994). Progression and resolution of delirium in elderly patients hospitalized for acute care. *American Journal of Geriatric Psychiatry, 2*, 230–238.

Levkoff, S. E., MacArthur, I. W., & Bucknall, J. (1995). Elderly mental health in the developing world. *Social Science & Medicine, 41*(7), 983–1003.

Lewis, B., Lewis, D., & Cumming, G. (1995). Frequent measurement of chronic pain: An electronic diary and empirical findings. *Pain, 60*, 341–347.

Lewis, C., & Feil, N. (1996). Validation: Techniques for communicating with confused old-old persons and improving their quality of life. *Geriatric Rehabilitation, 11*, 34–42.

Lewis, C. B. (1996). Helpful hints for clinicians for improving risk of falls. *Topics in Geriatric Rehabilitation, 11*, 72–87.

Lewis, M. I., & Butler, R. N. (1974). Life review therapy: Putting memories to work in individual and group psychotherapy. *Geriatrics, 29*(165–69, 172–73).

Li, L. W., Seltzer, M. M., & Greenberg, J. S. (1997). Social support and depressive symptoms: Differential patterns in wife and daughter caregivers. *Journals of Gerontology, 52B*(4-Supplement), S200–S211.

Lichtman, S. (1998). Recent developments in the pharmacology of anticancer drugs in the elderly. *Current Opinion in Oncology, 10*(6), 572–579.

Lieberman, M., & Snowden, L. (1993). Problems in assessing prevalance and membership characteristics of self-help group participants. *Journal of Applied Behavioral Science, 29*(2), 166–180.

Lieberman, P., & Anderson, J. A. (1997). *Allergic Diseases Diagnosis and Treatment*. Totowa, NJ: Humana Press.

Liebig, P. S. (1998). Housing and supportive services for the elderly: Intragenerational perspectives and options. In J. S. Steckenrider, & T. M. Parrott (Eds.), *New Directions in Old-Age Policies*. Albany, NY: SUNY Press.

Liebskind, J. C., & Melzack, R. (1988). The international Pain Foundation: Meeting a need for education in pain. *Journal of Pain & Symptom Management, 3*(3), 131–134.

The Light House, Inc. (1999). *The World Through Their Eyes—Understanding Vision Loss: A Manual for Nursing Home Staff*. New York: Author.

Lilienfeld, D. E., & Perl, D. P. (1994). Projected neurodegenerative disease mortality among minorities in the United States, 1990-2040. *Neuroepidemiology, 13*(4), 179–186.

Lin, C. C. (1996). Patient satisfaction with nursing care as an outcome variable: Dilemmas for nursing evaluation researchers. *Journal of Professional Nursing, 12*(4), 207–216.

Lin, N., Woelfel, M. W., & Light, S. C. (1985). The buffering effect of social support subsequent to an important life event. *Journal of Health and Social Behavior, 26*, 247–263.

Lindberg, C. (1997). Implementation of in-home telemedicine in rural Kansas: Answering an elderly patient's needs. *Journal of the Medical Informatics Association, 4*(1), 14–17.

Linergan, E. (1996). *Geriatrics*. Stamford, CT: Appleton & Lange.

Lipson, J. G., Dibble, S. L., & Minarik, P. A. (1996). *Culture and Nursing Care: A Pocket Guide*. San Francisco, CA: UCSF Nursing Press.

Lithgow, G. J., & Kirkwood, R. B. L. (1996). Mechanisms and evolution of aging. *Science, 273*, 80.

Litwak, E. (1985). *Helping the Elderly: The Complementarity of Informed Networks and Formal Systems*. New York: Guilford Press.

Litwak, E., & Longino, C. F. Jr. (1987). Migration patterns among the elderly: A developmental perspective. *Gerontologist, 27*(3), 266–272.

Liu, K. M. K. G. & A. C. (1998). *Changes in Home Care Use by Disabled Elderly Persons: 1982–1994*. Washington, DC: The Urban Institute.

Livingston, G., Hawkins, A., Graham, N., et al. (1990). The Gospel Oak Study: Prevalence rates of dementia, depression and activity limitation among elderly residents in Inner London. *Psychological Medicine, 20*, 137–146.

Lo, B. (1995). Improving care near the end of life. Why is it so hard? *JAMA, 274*, 1634–1636.

Lobo, A., Saz, P., Marcos, G., Dia, G., & Camara, C. (1995). The prevalence of dementia and depression in the elderly community in a southern european population: The Zaragoza study. *Archives of General Psychiatry, 52,* 497–504.

Loeb, J., & Northrop, J. H. (1917). On the influence of food and temperature on the duration of life. *Journal of Biological Chemistry, 32,* 103–121.

Logemann, J. (1983). *Evaluation and Treatment of Swallowing Disorders.* Springfield, MA: Pro-ed, Inc.

Logemann, J. (1986). Treatment for aspiration related dysphagia: An overview. *Dysphagia, 1,* 34–38.

Loser, C., Wolters, S., & Folsch, U. R. (1998). Enteral long-term nutrition via percutaneous endoscopic gastrostomy (PEG) in 210 patients: A four-year prospective study. *Digestive Diseases & Sciences, 43*(11), 2549–2557.

Lowe, J., Barg, F., Norman, S., & McCorkle, R. (1997). An urban intergenerational program for cancer control education. *Journal of Cancer Education, 12*(4), 233–239.

Luborsky, L. (1962). Clinicians' judgements of mental health. *Archives of General Psychiatry, 7,* 407–417.

Lueckenotte, A. G. (1996). *Gerontologic Nursing.* St. Louis, MO: Mosby.

Luis, C. A., Barker, W. W., Gajaraj, K., Harwood, D., Petersen, R., Kasuba, A., Waters, C., Jimison, P., Petito, C., Diskson, D., & Duara, R. (1999). Sensitivity and specificity of three clinical criteria for dementia with Lewy bodies in an autopsy-verified sample. *International Journal of Geriatric Psychiatry, 14,* 526–533.

Lyckholm, L. J. (1998). Should physicians accept gifts from patients? *Journal of the American Medical Association, 280,* 1944–1946.

Lyder, C. H., Yu, C., Stevenson, D., Mangat, R., Empleo-Frazier, O., Emrling, J., & McKay, J. (1998). Validating the Braden Scale for the prediction of pressure ulcer risk in Blacks and Latino/Hispanic Elders: A pilot study. *Ostomy Wound Management, 44*(3A), 42S–50S.

Lye, M. (1992). Disturbances of homeostasis. In J. C. Brocklehurst, R. C. Tallis, & H. M. Fillit (Eds.), *Textbook of Geriatric Medicine and Gerontology* (pp. 675–693). New York: Churchill Livingstone Inc.

Lyon, L. J., & Nevins, M. A. (1989). Management of hip fractures in nursing home patients: To treat or not to treat. *Journal of the American Geriatrics Society, 32,* 391–395.

Lyons, K. S., & Zarit, S. H. (1999). Formal and informal supports: The great divide. *International Journal of Geriatric Psychiatry, 14*(3), 183–196.

Maas, M., Reed, D., Swanson, E. A., & Specht, J. P. (in press). Family involvement in care: Negotiated family/staff partnerships in special care units for persons with dementia. In S. G. Funk, E. M. Tornquist, M. Miles, J. Harrell, & J. Leeman (Eds.), *Key Aspects of Preventing and Managing Chronic Illness.* New York: Springer Publishing Company.

Maas, M., Swanson, E. A., Specht, J. P., & Buckwalter, K. C. (1994). A nursing perspective on SCUs. *Alzheimer's Disease & Associated Disorders, 8*(Supplement 1), S417–S424.

Maas, M. L., & Buckwalter, K. C. (1991). Alzheimer's disease. *Annual Review of Nursing Research, 9,* 19–55.

Macauley, J., & Katula, J. (1999). Physical activity interventions in the elderly: Influences on physical health and psychological function. In R. Schulz, G. Maddox, & P. Lawton (Eds.), *Annual Review of Gerontology and Geriatrics* (Vol. Interventions, pp. 111–154). New York: Springer Publishing Company.

Mace, N. L., & Rabins, P. V. (1981). *The 36-hour day: A family guide to caring for persons with Alzheimer's disease, related dementing illness and memory loss in later life.* Baltimore, MD: Johns Hopkins University Press.

MacKenzie, S., & Krusberg, J. (1996). Can I get there from her? Wayfinding systems for healthcare facilities. *Leadership in Health Services, 5,* 42–46.

Mackey, S. (1995). Time to talk about senior tax breaks? *State Legislatures,* 12–13.

Mackey, S., & Carter, K. (1994). State tax policy & senior citizens, 2nd ed. Denver: National Conference of State Legislatures.

MacLaughlin, J., & Holick, M. F. (1985). Aging decreased the capacity of human skin to produce Vitamin D. *Journal of Clincial Investigation, 76,* 1536–1538.

MacRae, P. G., Feltner, M. E., & Reinsch, S. (1994). A 1-year exercise program for older women: Effects on falls, injuries and physical performance. *Journal of Aging & Physical Activity, 2,* 127–142.

Maddox, G. L. (1963). Activity and morale: A longitudinal study of selected elderly subjects. *Social Forces, 42,* 195–204.

Maddox, G. L. (1979). Assessment of functional status in a programme evaluation and resource allocation model. In W. W. Holland, J. Ipsen, & J. Kostrzewski (Eds.), *Measurement of Levels of Health. WHO Regional Publications, European Series No 7.* Copenhagen: World Health Orgainzation.

Maddox, G. L. (1984). Mutual-help groups for caregivers in the management of senile dementia: A research agenda. In J. Werthemier, & M. Marios (Eds.), *Senile Dementia: Outlook for the Future.* New York: Alan R. Liss.

Maddox, G. L. (1994). Sociology of Aging. In W. Hazzard, E. Bierman, J. Blass, W. Ettinger, & J. Halter

(Eds.), *Principles of Geriatric Medicine and Gerontology* (3 ed., pp. 125–134). New York: McGraw-Hill.

Maddox, G. L., & Bratesman, S. Jr. (1997). Pre-assessment screening: An essential building block in LTC information systems. Durham, NC: Duke Long Term Care, Resources Program. Duke University Medical Center.

Magaziner, J., Lydick, E., & Hawkes, W. (1997). Excess mortality attributable to hip fracture in white women aged 70 years and older. *American Journal of Public Health, 87*(10), 1630–1636.

Magaziner, J., Simonsick, E. M., & Kahsner, T. M. (1990). Predictor of functional recovery one year following hospital discharge for hip fracture: A prospective study. *Journals of Gerontology, 45A,* M101–107.

Maguire, G. H. (1996). Activities of daily living. In C. B. Lewis (Ed.), *Aging: The Health Care Challenge* (3rd ed., Chap. 3). Philadelphia: F. A. Davis Company.

Mahan, K., & Escott-Stump, S. (2000). *Krause's food, nurtition and therapy.* Philadelphia, PA: W. B. Saunders.

Mahler, D. A. (1999). *Dyspnea.* New York: Marcel Dekker.

Mahon, P. Y. (1996). An analysis of the concept of patient satisfaction as it relates to contemporary nursing care. *Journal of Professional Nursing, 24*(6), 1241–1248.

Mahoney, D., Tennstedt, S., Friedman, R., & Heeren, T. (1999). An automated telephone system for monitoring the functional status of community-residing elders. *The Gerontologist, 39,* 229–234.

Maislin, G., Pack, A. I., Kribbs, N. B., Smith, P. L., Schwartz, A. R., Kline, L. R., Schwab, R. J., & Dinges, D. F. (1995). A survey screen for prediction of apnea. *Sleep, 18,* 158–166.

Maki, B., Holliday, P., & Topper, A. (1994). A prospective study of postural balance and risk of falling in an ambulatory and independent elderly population. *Journals of Gerontology, 49A,* M72–84.

Maklebust, J., & Sieggreen, M. (1996). *Pressure Ulcers: Guidelines for Prevention and Nursing Management* (2nd ed.). Springhouse, PA: Springhouse Publications Co.

Malloy, P. F., Cummings, J. L., Coffey, C. E., Duffy, J., Funk, M., Lauterbach, E. C., et al. (1997). Cognitive screening instruments in neuropsychiatry: A report of the committee on research of the American Neuropsychiatric Association. *Journal of Neuropsychiatry & Clinical Neurosciences, 9,* 189–197.

Malone, M., Rozario, N., Gavinski, M., & Goodwin, J. (1991). The epidemiology of skin tears in the institutionalized elderly. *Journal of the American Geriatrics Society, 39,* 591–595.

Man-Son-Hing, M., & Wells, G. (1995). Meta-analysis of efficacy of quinine for treatment of nocturnal leg cramps in elderly people. *BMJ, 310*(6971), 13–17.

Mann, W., & Lane, J. (1995). *Assistive Technology for Persons with Disabilities* (pp. 266, 309). Bethesda, MD: American Occupational Therapy Association.

Mannell, R. C. (1993). High-investment activity and life satisfaction among older adults: Committed, serious leisure, and flow activities. In J. R. Kelly (Ed.), *Activity and aging* (pp. 125–145). Newbury Park, CA: Sage.

Mansour-Shousher, R., & Mansour, W. N. (1999). Non-surgical management of hearing loss. *Clinics in Geriatric Medicine, 15*(1), 163–177.

Manton, K. G., Corder, L., & Stallard, E. (1997). Chronic disability trends in elderly United States populations: 1982–1994. *Proceeding of the National Academy of Sciences USA.*

Manton, K. G., Corder, L. S., & Stallard, E. (1993). Estimates of change in chronic disability and institutional incidence and prevalence rates in the U.S. elderly population from the 1982, 1984, and 1989 National Long Term Care Survey. *Journals of Gerontology, 48B,* S153–166.

Manton, K. G., & Soldo, B. J. (1985). Dynamics of the health changes in the oldest old: New perspectives and evidence. *Milbank Memorial Fund Quarterly/ Health and Society, 63,* 206–285.

Marcus, R., Feldman, D., & Kelsey, J. (1996). *Osteoporosis.* Burlington, MA: Academic Press.

Marino, S. (1996). Selected problems in counseling the elderly. In M. J. Holosko, & M. D. Feit *Social Work Practice with the Elderly* (2nd ed., Chap. 6, pp. 55–101). Toronto: Canadian Scholar's Press.

Marshall, S., & Houseman, C. (2000). *Communicating in Today's World.* Philadelphia: Lippincott Williams & Wilkins.

Marteau, T. M., & Croyle, R. T. (1998). The new genetics: Psychological responses to genetic testing. *BMJ, 316,* 693–696.

Martin, K. S. (2000). Home Health Care, Outcomes Management, and the Land of Oz. *Outcomes Management for Nursing Practice, 3*(4), 7–12.

Maslach, C., & Leiter, M. P. (1997). *The Truth About Burnout.* San Francisco, CA: Jossey-Bass Inc.

Mason, S. (1997). Type of soap and the incidence of skin tears among residents of a long-term care facility. *Ostomy Wound Management, 43*(8), 26–30.

Masoro, E. J. (1996). Possible mechanisms underlying the antiaging actions of caloric restriction. *Toxicologic Pathology, 24,* 738–741.

Mathias, S., Nayak, U. Y., & Isaacs, B. (1986). The Get Up and Go Test: A simple clinical test of balance in old people. *Archives of Physical Medicine & Rehabilitation, 67,* 387–389.

Matteson, M., McConnel, E. S., & Linton, A. D. (1997). *Gerontological Nursing: Concepts and Practice* (2nd ed.). Philadelphia, PA: W. B. Saunders Company.

Matteson, M. A. (2000). *A study of long-term care solutions for sundown syndrome.* Unpublished doctoral dissertation, University of Texas Health Science Center at San Antonio, San Antonio, TX.

Matthews, D. A., Manu, P., & Lane, T. J. (1991). Evaluation and management of patients with chronic fatigue. *American Journal of the Medical Sciences, 302*(5), 269–277.

Mattson, M. P., & Goodman, Y. (1995). Different amyloidogenic peptides share a similar mechanism of neurotoxicity involving reactive oxygen species and calcium. *Brain Research, 676,* 219–224.

Mattson, M. P., Partin, J., & Begley, J. G. (1998). Amyloid beta-peptide induces apoptosis-related events in synapses and dendrites. *Brain Research, 807*(1–2), 167–176.

May, B. J. (1999). *Home Health and Rehabilitation: Concepts of Care* (2nd ed.). Philadelphia: F. A. Davis.

May, R. (1975). *The Courage to Create* (p. 39). New York: W. W. Norton and Company.

Mayeux, R. (1993). The mental state in Parkinson's disease. In W. C. Koller (Ed.), *Handbook of Parkinson's Disease* (pp. 159–184). New York: Marcel Dekker.

McAlindon, T. E., Felson, D. T., Zhang, Y., Hannan, M. T., Aliabadi, P., Weissman, B., Rush, D., Wilson, P. W., & Jacques, P. (1996). Relation of dietary intake and serum levels of vitamin D to progression of osteoarthritis of the knee among participants in the Framingham Study. *Annals of Internal Medicine, 125*(5), 353–359.

McAlindon, T. E., Wilson, P. W., Aliabadi, P., Weissman, B., & Felson, D. T. (1999). Level of physical activity and the risk of radiographic and symptomatic knee osteoarthritis in the elderly: The Framingham study. *American Journal of Medicine, 106*(2), 151–157.

McCann, R. (1999). Lack of evidence about tube feeding—Food for thought. *Journal of the American Medical Association, 282,* 1380–1381.

McCarter, R. J. M. (1995). Energy utilization. In E. J. Masoro (Ed.), *Handbook of Physiology* (pp. 95–118). New York: Oxford University Press.

McCay, C., Crowell, M., & Maynard, L. (1935). The effect of retarded growth upon the length of life and upon the ultimate size. *Journal of Nutrition, 10,* 62–79.

McCloskey, A., & Holahan, D. (1995). *Medicaid Acute Care, (FS Number 41).* Washington, DC: American Association of Retired Persons, Public Policy Institute.

McClusky, H. Y. (1971). Education: Background and issues. *White House Conference on Aging* Washington, DC.

McCormick, W. C., & Wood, R. W. (1992). Clinical decisions in the care of elderly persons with AIDS. *Journal of the American Geriatrics Society, 40,* 917–921.

McDonald, W. M., Krishnan, K. R. R., Doraiswamy, P. M., & Blazer, D. G. (1991). The occurrence of subcortical hyperintensities in patients with mania. *Psychiatry Research, 40,* 211–220.

McDonald, W. M., & Nemeroff, C. B. (1996). The diagnosis and treatment of mania in the elderly. *Bulletin of the Menninger Clinic, 60,* 174–196.

McEvoy, C. L. (2000). Cognitive Changes in Aging. In M. Mezey (Ed.), *Encyclopedia of Elder Care.* New York: Springer Publishing Company.

McGee, S. R. (1990). Muscle Cramps. *Archives of Internal Medicine, 150*(3), 511–518.

McGill, D., Brown, K., Haley, J., & Schieber, S. (1996). *Fundamentals of Private Pensions.* Philadelphia: University of Pennsylvania Press.

McGlynn, T. J., & Metcalf, H. L. (1992). *Diagnosis and treatment of anxiety disorders: A physician's handbook.* Washington, DC: American Psychiatric Press.

McGuire, F., & Hawkins, M. (1998). Introduction to Intergenerational Programs. *Activities, Adaptation, & Aging, 23*(1), 1–9.

McIntosh, J. L., Santos, J. F., Hubbard, R. W., & Overholser, J. C. (1994). *Elder Suicide.* Washington, DC: American Psychological Association.

McKeith, I. (in press). BPSD and Lewy Body Dementia. *International Psychogeriatrics.*

McLeish, J. (1976). *The Ulyssean Adult.* Toronto: McGraw-Hill.

McLendon, C. B., & Blackistone, M. (1982). *Signage: Graphic communications in the build world.* USA: McGraw-Hill.

McWilliam, C. L., Brown, J. B., Carmichael, J. L., & Lehman, J. M. (1994). A new perspective on threatened autonomy in elderly persons: The disempowering process. *Social Science & Medicine, 38*(2), 327–338.

Medicare Payment Advisory Commission. (1998). *Report to Congress: Context for a Changing Medicare Program.* Washington, DC: Author.

Meerabeau, L. (1999). The management of embarrassment and sexuality in health care. *Journal of Advanced Nursing, 29*(6), 1507–1513.

Meiners, M. R. (1998). Public-private partnerships in long-term care. In L. C. Walker, E. H. Bradley, & T. Wetle (Eds.), *Public and Private Responsibilities in Long-Term Care: Finding the Balance*. Baltimore: The Johns Hopkins University Press.

Melamed, S., Ugarten, U., Shirom, A., Kahana, L., Lerman, Y., & Froom, P. (1999). Chronic burnout, somatic arousal and elevated salivary cortisol levels. *Journal of Psychosomatic Research, 46,* 591–598.

Melchior, M. E. W., Bours, G. J., Schmitz, P., & Wittich, Y. (1997). Burnout in psychiatric nursing: A meta-analysis of related variables. *Journal of Psychiatric Nursing & Mental Health Nursing, 4,* 193–201.

Melton, L. J., & Riggs, B. L. (1998). Epidemiology of age-related fractures. In L. V. K. S. Avioli (Ed.), *Metabolic Bone Disease and Clinically Related Disorders* (Chap. 3, pp. 45–72). New York: Grune & Stratton.

Melzack, R. (1975). The McGill pain questionnaire: Major properties and scoring methods. *Pain, 1,* 277–299.

Melzack, R., & Katz, J. (1992). The McGill Pain Questionaire: Appraisal and current status. In D. Turk, & R. Melzack (Eds.), *Handbook on Pain Assessment* (pp. 152–168). New York: Guilford.

Mentes, J., Adkins, J., Culp, K., Gaspar, P., Mobiliy, P., Rapp, C. G., Tripp-Reimer, T., Wadle, K., & Wakefield, B. (1998). *Hydration Management Research-Based Protocol*. Iowa City, IA: University of Iowa Gerontological Nursing Interventions Research Center.

Merkouris, A., Ifantopoulis, J., Lanara, V., & Lemonidou, C. (1999). Patient satisfaction: A key concept for evaluating and improving nursing services. *Journal of Nursing Management, 7*(1), 19–28.

Meshack, A. F., Goff, D. C., Chan, W., & Ramsey, D. (1998). Comparison of reported symptoms of acute myocardial infarction in Mexican Americans versus non-Hispanic whites (the Corpus Christi Heart Project). *American Journal of Cardiology, 82*(11), 1329–1332.

Meskin, L. H., Dillenberg, J., & Heft, M. W. (1990). Economic impact of dental service utilization by older adults. *Journal of the American Dental Association, 120,* 665–671.

Messerli, F. H., & Grodzicki, T. (1996). Hypertension and coronary artery disease in the elderly. *Clinics in Geriatric Medicine, 12,* 41–56.

Meston, C. M. (1997). Aging and sexuality. *Western Journal of Medicine, 167*(4), 285–290.

Mezey, M. (1995). Why good ideas have not gone far enough: The state of geriatric nursing education. In T. T. Fulmer, & M. Matzo (Eds.), *Strengthening Geriatric Nursing Education* (Chap. 1). New York: Springer Publishing Company.

Mid-Florida Area Agency on Aging. (1999). *Area Agencies on Aging* [Web Page]. URL http://www.mfaa.org/agingnetwork.html#AAA [1999, October 26].

Middleton Jr, E., Reed, C. E., & Addison, J. N. F. (1993). *Allergy Principles and Practices*. St. Louis, MO: Mosby-Year Book.

Milisom, I. (1996). Rational prescribing for postmenopausal urogenital complaints. *Drugs & Aging, 9*(2), 78–86.

Miller, B., Campbell, R. T., Furner, S., Kaufman, J. E., Li, M., Muramatsu, N., & Prohaska, T. (1997). Use of medical care by African American and White older persons: Comparative analysis of three national data sets. *Journals of Gerontology, 52B*(6), S104–112.

Miller, C. A. (1995). Medications and sexual functioning in older adults. *Geriatric Nursing, 16*(2), 94–95.

Miller, C. A. (1999). *Nursing Care of Older Adults: Theory & Practice*. 3rd ed. Philadelphia, PA: Lippincott.

Miller, J., Neelon, V., Dalton, J., Ng'andu, N., Bailey Jr., D., Layman, E., & Hosfeld, A. (1996). The assessment of discomfort in elderly confused patients: A preliminary study. *Journal of Neuroscience Nursing, 28,* 175–182.

Miller, J. F. (1992). *Coping with Chronic Illness: Overcoming Powerlessness*. 2nd ed. Philadelphia, PA: F. A. Davis, C.O.

Miller, M., & LeLieuvre, B. (1982). A method to reduce chronic pain in elderly nursing home residents. *The Gerontologist, 22*(3), 314–317.

Miller, M. E., Longino, C. F. Jr, Anderson, R. T., James, M. K., & Worley, A. S. (1999). Functional status, assistance, and the risk of a community-based move. *Gerontologist, 39*(2), 187–200.

Miller, P. A., & Bear-Lehman, J. (2000). Occupational Therapy. *Merck Manual of Geriatrics* (3rd ed.). West Point, Pa: Merck.

Miller, R. A. (1999). Kleemeir Award Lecture: Are there genes for aging? *Journals of Gerontology, 54A,* B297–B307.

Miller, S., Prohaska, T., & Furner, S. (1999). Nursing home admission for African Americans with Alzheimer's disease. *Journals of Gerontology, 54A*(7), M365–369.

Minkler, M., & Fuller-Thomson, E. (1999). The health of grandparents raising grandchildren: Results of a national study. *American Journal of Public Health, 89*(9), 1384–1389.

Minkler, M., Fuller-Thomson, E., Miller, E., & Driver, D. (1997). Depression in grandparents raising grandchildren. *Archives of Family Medicine, 6,* 445–452.

Minkler, M., & Roe, K. M. (1993). *Forgotten caregivers: Grandmothers raising children of the crack cocaine epidemic*. Newbury Park, CA: Sage Publications.

Mitchell, S. L., Kiely, D. K., & Lipsitz, L. A. (1998). Does artificial enteral nutrition prolong the survival of institutionalized elders with chewing and swallowing problems? *Journals of Gerontology, 53A*(3), M207–213.

Mittal, R. K. (1997). Hiatal hernia: Myth or reality? *American Journal of Medicine, 103*(5A), 33S–39S.

Mittleman, M. S., Ferris, S. H., Shulman, E., & Steinberg, G. (1995). The effects of a multicomponent program on spouse-caregivers of Alzheimer's Disease patients: Results of a treatment/control study. In L. L. Heston (Ed.), *Progress in Alzheimer's Disease and Similar Conditions.* Washington, DC: American Psychiatric Association Press.

Miyata, M., & Smith, J. D. (1996). Apolipoprotein E allele-specific antioxidant activity and effects on cytotoxicity by oxidative insults and beta-amyloid peptides. *Nature Genetics, 14*(1), 55–61.

Mobily, P. R., Herr, K. A., Clark, K., & Wallace, R. B. (1994). An epidemiologic analysis of pain in the elderly. *Journal of Aging & Health, 6*(2), 139–155.

Modly, D. M., Zanotti, R., Poletti, P., & Fitzpatrick, J. (1997). *Home Care Nursing Services: International Lessons.* New York: Springer Publishing Company.

Moen, P., Robison, J., & Dempster-McClain, D. (1995). Caregiving and women's well being: A life course approach. *Journal of Health & Social Behavior, 36,* 259–273.

Mollica, R. L. (1998). *State Assisted Living Policy, 1998.* Portland, ME: National Academy for State Health Policy.

Moltzer, G., van der Meulen, M. J., & Verheij, H. (1996). Psychological characteristics of dissatisfied denture patients. *Community Dentistry & Oral Epidemiology, 24,* 52–55.

Monk, A., & Cyrus, A. (1974). Predictors of voluntaristic intent among the aged. *Gerontologist, 14,* 425–429.

Monk, A., Kaye, L., & Litwin, H. (1984). *Resolving Grievances in the Nursing Home.* New York: Columbia University Press.

Montague, C. T., Farooqi, I. S., Whitehead, J. P., Soos, M. A., Rau, H., Wareham, N. J., Sewter, C. P., Digby, J. E., Mohammed, S. N., Hurst, J. A., Cheetham, C. H., Earley, A. R., Barnett, A. H., Prins, J. B., & O'Rahilly, S. (1997). Congenital leptin deficiency is associated with severe early-onset obesity in humans. *Nature, 387*(6636), 903–908.

Montefiore Medical Center, Center for Bioethics. (1999). *Making Health Care Decisions for Others: A Guide to Being a Health Care Proxy or Surrogate—A Quick Reference for Physicians* [Web Page]. URL http://www.montefiore.org/bioethics [1999, March 21].

Montgomery, R. J. V., & Borgatta, E. F. (1989). Effects of alternative support strategies. *Gerontologist, 29,* 457–464.

Moody, H. R. (1976). Philosophical presuppositions of education of old age. *Educational Gerontology, 1,* 1–16.

Moon, J. H., & Pearl, J. H. (1991). Alientation of elderly Korean American immigrants as related to place of residence, gender, age, years of education, time in the U.S., living with or without children, and living with or without a spouse. *International Journal of Aging & Human Development, 32*(2), 115–124.

Moon, M. (1993). *Medicare Now and in the Future.* Washington, DC: The Urban Institute Press.

Mooradian, A. D., & Thurman, J. E. (1999). Glucotoxicity. *Clinics in Geriatric Medicine, 15*(2), 255–263.

Moore, L. W. (1999). Living with macular degeneration: Creative strategies used by older women. *Journal of Ophthalmic Nursing & Technology, 18*(2), 68–71.

Moore, M. (1997). *Nutritional Care* (3rd ed.). St. Louis, MO: Mosby Yearbook.

Moos, R. E., & Lemke, S. (1984). Quality of residential settings for older people. In I. Altman, M. P. Lawton, & J. Wohlwill (Eds.), *Elderly People and the Environment* (pp. 159–190). New York: Plenum.

Morley, J. E., & Evans, W. J. (1999). *Nutritional Clinical Strategies in Long Term Care: Anorexia in the Elderly. Council for Nutritional Clinical Strategies in Long Term Care* [Web Page]. URL http://www.LTCnutrition.com [1999, May 21].

Morley, J. E., Miller, D. K., Perry, H. M., Guigoz, Y., & Vellas, B. (1997). Anorexia of aging, leptin and the Mini Nutritional Assessment. In B. J. Vellas, Y. Guigoz, & P. J. Garry (Eds.), *Mini Nutritional Assessment (MNA): Research and practice in elderly* Vol. 1 (pp. 67–78). Philadelphia, PA: Lippincott-Raven.

Morrell, R. W., Park, D. C., Kidder, D. P., & Martin, M. (1997). Adherence to antihypertensive medications across the lifespan. *The Gerontologist, 37,* 609–619.

Morris, J. N., Murphy, K., & Nonemaker, S. (1995). *Minimum Data Set 2. 0: Long Term Care Facility Resident Assessment Instrument User's Manual.* Baltimore, MD: Health Care Financing Association.

Morris, P. L., Robinson, R. G., Raphael, B., & Hopwood, M. J. (1996). Lesion location and poststroke depression. *Journal of Neuropsychiatry & Clinical Neurosciences, 8*(4), 399–403.

Morris, R., Lambert, C., & Guberman, M. (1964). New roles for the elderly. In *Papers on Social Welfare.* Waltham, MA: Brandeis University.

Morris, W. W., Buckwalter, K. C., Cleary, T. A., Gilmer, J. S., Hatz, D. L., & Studer, M. (1990). Refinement of Iowa Self-Assessment Inventory. *Gerontologist, 30,* 243–247.

Morrison, G., & Hark, L. (1999). *Medical Nutrition and Disease* (2nd ed.). Malden, MA: Blackwell Science.

Morrow, D. G., Leirer, V. O., & Sheikh, J. (1988). Adherence and medication instructions: Review and recommendations. *Journal of the American Geriatrics Society, 36,* 1147–1160.

Moscicki, E. K. (1997). Identification of suicide risk factors using epidemiologic studies. *Psychiatric Clinics of North America, 20*(3), 499–517.

Moseley, C. B. (1996). The impact of federal regulations on urethral catheterization in Virginia nursing homes. *American Journal of Medical Quality, 11*(4), 222–226.

Mosher, D. (1996). *The elderly population is booming. How to handle a demand for special services. The New York Cooperator: The Co-Op and Condo Monthly.* [Web Page]. URL http://www.hia.com/hia/coop/co-cover.html [2000, February 6].

Muir, A. L., Davidsen, I. A., Percy-Robb, J. W., Walsh, E. G., & Passmore, R. (1970). Physiological aspects of the Edinburgh commonwealth games. *Lancet, 2*(7683), 1125–1128.

Muldary, T. W. (1983). *Burnout and Health Professionals.* Stamford, CT: Appleton-Century-Crofts.

Mullan, M., Crawford, F., Axelman, K., Houlden, H., Lilius, L., Winblad, B., & Lannfelt, L. (1992). A pathogenic mutation for probable Alzheimer's disease in the APP gene at the N-terminus of beta-amyloid. *Nature Genetics, 1*(5), 345–347.

Munetz, M. R., & Benjamin, S. (1988). How to examine patients using the Abnormal Involuntary Movement Scale. *Hospital & Community Psychology, 39,* 1172–1177.

Munnell, A. (1992). Current taxation of qualified plans: Has the time come? *New England Economic Review,* March/April, 1–2.

Murden, R. A., McRae, T. D., Kaner, S., & Bucknam, M. E. (1991). Mini-Mental State exam scores vary with education in Blacks and Whites. *Journal of the American Geriatrics Society, 39,* 149–155.

Murer, C., & Lenhoff-Brick, D. (1997). *The Case Management Handbook.* New York: McGraw-Hill.

Murray, M. D., Birt, J. A., Manatunga, A. K., & Darnell, J. C. (1993). Medication compliance in elderly outpatients using twice-daily dosing and unit-of-use packaging. *Annals of Pharmacotherapy, 27*(5), 616–621.

Murray, M. D., Darnell, J., Weinberger, M., & Martz, B. L. (1986). Factors contributing to medication noncompliance in elderly public housing tenants. *Drug Intelligence & Clinical Pharmacy, 20*(2), 146–152.

Murrell, S. A., Meeks, S., & Walker, J. (1991). Protective functions of health and self-esteem against depression in older adults facing illness or bereavement. *Psychology and Aging, 6,* 352–360.

Myers, G. C. (1985). Aging and worldwide population change. In R. H. Binstock, & E. Shanas (Eds.), *Handbook of Aging and the Social Sciences.* New York: Van Nostrand Reinhold.

Myers, G. C., & Eggers, M. L. (1996). Demography. In J. E. Birren (Ed.), *Encyclopedia of Gerontology: Age, Aging, and the Aged.* San Diego: Academic Press.

Myers, R. J. (1993). *Social Security.* Philadelphia: University of Pennsylvania Press.

Myers, R. J. Within the system. In *My Half Century in Social Security.* Winsted, CT: ACTEX Publications.

Nadol, J. B. (1993). Hearing loss. *NEJM, 329*(15), 1092–1102.

Nagle, E. (1999). Rectal prolapse and fecal incontinence. *Primary Care, 1,* 101–111.

Nail, L. M., & Winningham, M. L. (1995). Fatigues and weakness in cancer patients: The symptom experience. *Seminars in Oncology Nursing, 11*(4), 272–278.

Naparstek, A., Biegel, D. E., & Spiro, H. (1982). *Neighborhood Networks for Human Mental Health Care.* New York: Plenum Publishing Company.

Narayanasamy, A. (1996). Spiritual care of chronically ill patients. *British Journal of Nursing, 5*(7), 411–416.

Nathan, B. P., Bellosta, L., Sanan, D. A., Weisgraber, K. H., Mahley, R. W., & Pitas, R. E. (1994). Differential effects of apolipoproteins E3 and E4 on neuronal growth in vitro. *Science, 264,* 850–852.

Nathanson, I. L., & Eggleton, E. (1993). Motivation versus program effect on length of service: A study of four cohorts of ombudservice volunteers. *Journal of Gerontological Social Work, 19,* 95–114.

Nathanson, M. (1996). *A proposal for mental health delivery to naturally occurring retirement communities. Psychiatry On-Line 1995 Version 1. 0* [Web Page]. URL http://www.publinet.it/pol/norc.html [2000, February 15].

National Alliance for Caregiving, & American Association of Retired Persons. (1997). *Family Caregiving in the U.S.: Findings from a National Survey.* Bethesda, MD: National Alliance for Caregiving & American Association of Retired Persons.

National Asian Pacific Center on Aging. (1999). *Report on Progress—Year Two: Community Based Capacity Building for Asian Pacific Islander Elders.* Seattle, WA: National Asian Pacific Center on Aging.

National Association for Home Care. (1999). *Basic Statistics About Home Care* [Web Page]. URL http://www.nahc.org/Consumer/hcstats.html [1999, October 21].

National Association for Home Care. (1999). *Basic Statistics About Hospice* [Web Page]. URL http://www.nahc.org/Consumer/hpcstats.html [1999, October 21].

National Association of Adult Protective Services Administrators. (1994). *Report of the Adult Services Task Force on the Perspective of the States on a Federal Adult Protective Services Statute*. Washington, DC: Author.

National Center for Assisted Living. (1998). *Assisted Living Sourcebook*. Washington, DC: American Health Care Association.

National Center for Family Literacy. (1994). *The Power of Family Literacy*. Louisville, KY: Author.

National Center for Health Statistics. (1997). *Health, United States 1996–1997 and Injury Chartbook*. Hyattsville, MD: Author.

National Center for Health Statistics. (1998). *Health, United States, 1998*. Hyattsville, MD: National Center for Health Statistics.

National Center on Elder Abuse. (1996). *Domestic Elder Abuse Information Series # 1, 2, 3* [Web Page]. URL http://www/gwjapan.com/NCEA [1999, November 19].

National Center on Elder Abuse. (1998). *The National Elder Abuse Incidence Study: Final Report*. Washington, DC: National Aging Information Center.

National Coalition for the Homeless (NCH). (1999). *Homelessness Among Elderly Persons: NCH Fact Sheet #15*. Washington, DC: Author.

National Commission of Sleep Disorders Research. (1994). *Wake Up America*. National Commission of Sleep Disorders Research.

National Council on Aging. (1990). *The National Institute of Senior Centers' Senior Center Self-Assessment and National Accreditation Manual*. Washington, DC: Author.

National Council on Aging. (1999). *The Senior Center Standards and Self-Assessment Workbook*. Washington, DC: Author.

National Council on Aging. (1999). *Senior Centers Self-Assessment and National Accreditation Manual*. Washington, DC: Author.

National Diabetes Data Group (NDDG). (1995). (Report No. 95-1468). National Institutes of Health, National Institute of Diabetes and Digestive and Kidney Diseases.

National Family Caregivers Association. (1999). *National Family Caregivers Association Home Page* [Web Page]. URL http://www.nfcacares.org [1999, June 8].

National High Blood Pressure Education Program Working Group. (1994). Report on hypertension in the elderly. *Hypertension, 23,* 275–285.

National Hospice Organization. (1998). *Hospice Care: A Physician's Guide, including Medical Guidelines for Determining Prognosis in Selected Non-Cancer Diseases*. Washington, DC: Author.

National Institute of Health. (1992). NIH consensus statement on impotence. *Impotence, 10*(4), 1–31.

National Institute of Health Consensus Development Conference. (1988). National Institute of Health Consensus Development Conference Statement: Geriatric assessment methods for clinical decision-making. *Journal of the American Geriatrics Society, 36,* 342–347.

National Institute of Health Consensus Development Panel on Depression in Late Life. (1992). Diagnosis and treatment of depression in late life—NIH consensus conference. *JAMA, 268,* 1018–1023.

National Institute on Aging. (1987). *Department of Health and Human Services Report to Congress*. (Report No. NIH Pub. No 87-2950). Washington, DC: Author.

National Institute on Aging. (1993). *Health and Retirement Study Press Release*. Washington, DC: Author.

National Institute on Aging. (2000). *What To Do About Flu—National Institute on Aging Page* [Web Page]. URL http://www.aoa.dhhs.gov/aoa/pages/agepages/flu.html [2000, June 28].

National Institute on Consumer-Directed Long-Term Services. (1996). *Principles of Consumer-Directed Home- and Community-Based Services*. Washington, DC: The National Council on Aging.

National Pressure Ulcer Advisory Panel. (1998). *PUSH Tool Information & Registration Form* [Web Page]. URL http://www.npuap.org/pushins.htm [1999, August 18].

National Pressure Ulcer Advisory Panel. (1998). *Stage I Assessment in Darkly Pigmented Skin* [Web Page]. URL http://www.npuap.org/positn4.htm [1999, August 18].

Naylor, M. D., Brooten, D., Campbell, R., Jacobsen, B. S., Mezey, M. D., Pauly, M. V., & Schwartz, J. S. (1994). Comprehensive discharge planning and home follow-up of hospitalized elders: A randomized clinical trial. *JAMA, 281*(7), 613–620.

Negri, E., Franceschi, S., Parpinel, M., & LaVecchia, C. (1998). Fiber intake and risk of colorectal cancer. *Cancer Epidemiology Biomarkers & Prevention, 7*(8), 667–771.

Neimeyer, R. A. (1994). *Death Anxiety Handbook: Research, Instrumentation, and Application*. New York: Taylor and Francis.

Neimeyer, R. A., & Van Brunt, D. (1995). Death anxiety. In H. Wass, & R. A. Neimeyer (Eds.), *Dying: Facing the Facts* (3rd ed., pp. 49–58). New York: Taylor and Francis.

Nelson, H. W. (1995). Long-term care volunteer roles on trial: Ombudsman effectiveness revisited. *Journal of Gerontological Social Work, 23*(3/4), 25–46.

Nemeth, L., Hendricks, H., Salaway, T., & Garcia, C. (1998). Patient Pathway Development. *Topics in Health Information Management, 19*(2), 79–87.

Nerenberg, L. (1996). *Financial Abuse of the Elderly.* San Francisco, CA: San Francisco Consortium for Elder Abuse Prevention, Mount Zion Institute for Aging.

Neuhaus, B. E. M. P. A. (1995). Status of functioning assessment in occupational therapy with the elderly. *Journal of Allied Health,* (Winter), 29–40.

Neuman, B. (1995). *The Neuman Systems Model* (3rd Edition ed.). Norwalk, CT: Appleton & Lange.

New York State Moreland Act Commission on Nursing Homes and Residential Facilities. (1975). *Regulating Nursing Home Care: The Paper Tigers.* New York: Author.

Newbern, V. B., & Krowchuk, H. V. (1994). Failure to thrive in elderly people: A conceptual analysis. *Journal of Advanced Nursing, 19,* 840–849.

Newcomer, R. J., & Benjamin, A. E. (1997). *Indicators of Chronic Health Conditions: Monitoring Community Level Delivery Systems.* Baltimore: Johns Hopkins University Press.

Newlin, P. R., Gibbs, B. A., Lonowski, L. R., & Meysers, P. J. (1996). Spanning the continuum of care: Managing a geriatric client with a CVA. In S. S. Blancett, & D. L. Flarey (Eds.), *Case Studies in Nursing Management* (pp. 354–368). Gaithersurg, MD: Aspen Publishers, Inc.

Nicolle, L. E. (1994). Urinary tract infection. In P. D. O'Connell (Ed.), *Geriatric Urology* (Chap. 16, pp. 399–412). New York: Little, Brown and Company.

Nicosia, J. F. (1994). Healing the human spirit: The healing paradigm. *Journal of Religion in Disability & Rehabilitation, 1*(3), 65–74.

Niederehe, G. (1988). TRIMS behavioral problem checklist (BPC). *Psychopharmacology Bulletin, 24*(4), 771–778.

Niedert, K. C. (1998). *Nutrition Care of the Older Adult: A Handbook for Dietetic Professionals Working Throughout the Continuum of Care.* Chicago, IL: The American Dietetic Association.

Niestadt, M., & Crepeau, E. B. (1998). Introduction to occupational therapy. In M. Niestadt, & E. B. Crepeau (Eds.), *Willard and Spackman's Occupational Therapy.* New York: Lippincott.

Nightingale, F. (1969). Notes on Nursing (Original Work Published 1860). (p. 103). New York: Dover Publications.

Nkongho, N. O., & Archbold, P. G. (1996). Working out caregiving systems in African American families. *Applied Nursing Research, 9*(3), 108–114.

Noble, J., Greene, H. L., Levinson, W., Modest, G. A., & Young, M. J. (1996). *Textbook of Primary Care.* St. Louis: Mosby Company.

Noelker, L. S., & David, D. M. (1994). Relationships between the frail elderly's informal and formal helpers. In E. Kahana, D. E. Biegel, & M. L. Wyjle (Eds.), *Family Caregiving Across the Lifespan.* Newbury Park, CA: Sage Publications.

Norris, R. J. (1992). Medical costs of osteoporosis. *Bone, 13*(Supplement), 11–16.

Northwestern University and the University of Calgary. (1990). *Your Vision: A Survey by The Vision Laboratories of Northwestern University and the University of Calgary.* Evanston, IL: Northwestern University and the University of Calgary.

Norton, M. C. et al. (1999). Telephone adaptation of the modified mini-mental state exam (3MS). The Cache county study. *Neuropsychiatry Neuropsychology & Behavioral Neurology, 12,* 270–276.

Nusbaum, N. J. (1999). Aging and sensory senescence. *Southern Medical Journal, 92*(3), 267–275.

Nyenhuis, D. L., & Gorelick, P. B. (1998). Vascular dementia: A contemporary review of epidemiology, diagnosis, prevention, and treatment. *Journal of the American Geriatrics Society, 46,* 1437–1448.

Nyman, J. A., Finch, M. D., Kane, R. A., Kane, R. L., & Illston, L. H. (1997). The substitutability of adult foster care for nursing home care in Oregon. *Medical Care, 35*(8), 801–813.

O'Rand, A. M., & Henretta, J. C. (1999). *Age and Inequality.* Boulder, CO: Westview Press.

Oboler, S. K., Prochazaka, A. V., & Meyer, T. J. (1991). Leg symptoms in outpatient veterans. *Western Journal of Medicine, 155*(3), 256–259.

Ochs, M. (1990). Surgical management of the hip in the elderly patient. *Clinics in Geriatric Medicine, 6,* 571–587.

Oktay, J. S., & Volland, P. J. (1987). Foster home care for the frail elderly as an alternative to nursing home care: An experimental evaluation. *American Journal of Public Health, 77*(12), 1505–1510.

Okun, M. (1994). The relation between motives for organizational volunteering and frequency of volunteering by elders. *Journal of Applied Gerontology, 13,* 115–126.

Okun, M. A., & Keith, V. M. (1998). Effects of positive and negative social exchanges with various sources on depressive symptoms of younger and older adults. *Journals of Gerontology, 53B,* P4–P20.

Okun, M. A., & Stock, W. A. (1987). Correlates and components of subjective well-being among the elderly. *Journal of Applied Gerontology, 6,* 95–112.

Omdahl, B. L., & O'Donnell, C. (1999). Emotional contagion, empathic concern and communicative responsiveness as variables affecting nurses' stress and occupational commitment. *Journal of Advanced Nursing, 29,* 1351–1359.

Omnibus Budget Reconciliation Act of 1987 (OBRA) Vol. U.S. Public Law 100-203, Subtitle C: Nursing Home Reform.

Omran, M. I., & Morley, J. E. (2000). Assessment of protein energy malnutrition in older persons. *Nutrition, 16*(1), 50–63.

Online Mendelian Inheritance in Man. (2000). *Amyloid Beta A4 Precursor Protein; APP* [Web Page]. URL http://www.ncbi.nlm.nih.gov/htbin-post/Omim/dispmim?604252 [2000, January 16].

Orange, J. B., & Ryan, E. B. (2000). Alzheimer's disease and other dementias: Implications for physician communication. *Clinics in Geriatric Medicine, 16.*

Orchard, T. J., & Strandness, D. E. Jr. (1993). Assessment of peripheral vascular disease in diabetes. Report and recommendations of an international workshop sponsored by the American Diabetes Association and the American Heart Association, September 18-20, 1992 New Orleans, Louisiana. *Circulation, 88*(2), 819–828.

Orem, D. (1985). *Nursing: Concepts of Practice.* New York: McGraw-Hill.

Orem, D. (1995). *Nursing: Concepts of Practice.* St. Louis, MO: Mosby.

Organisation for Economic Cooperation and Development (OECD). (1999). *A Caring World—The New Social Agenda.* Paris: Organisation for Economic Cooperation and Development (OECD).

Organisation for Economic Cooperation and Development (OECD). (1996). The development of long-term care policy. In Organisation for Economic Cooperation and Development (OECD) *Caring for Frail Elderly People—Policies in Evolution.* Paris: Organisation for Economic Cooperation and Development (OECD).

Ornduff, J. S. (1996). Releasing the elderly inmate: A solution to prison overcrowding. *The Elder Law Journal, 4,* 173–199.

Orr, D. (1994). *Earth in Mind: On Education, Environment, and the Human Prospect.* Washington, DC: Island Press.

Orr, W. C., & Sohal, R. S. (1994). Extension of life-span by overexpression of superoxide dismutase and catalase in Drosophila melanogastor. *Science, 263,* 1128–1130.

Ory, M. G. Jr., Hoffman, R. R., Yee, J. L., Tennstedt, S., & Schulz, R. (1999). Prevalence and impact of caregiving: A detailed comparison between dementia and nondementia caregivers. *Gerontologist, 39,* 177–185.

Osterweis, M., Solomon, F., & Green, M. (1984). *Bereavement: Reactions, Consequences, and Care.* Washington, DC: National Academy Press.

Othmer, E., & Othmer, S. C. (1994). *The Clinical Interview Using the DSM IV.* Washington, DC: American Psychiatric Press.

Ottenbacher, K. J., & Christainsen, C. (1997). Occupational performance assessment. In C. Christainsen, & C. Baum (Ed.), *Enabling Function and Well-Being* (p. chapter 5). Thorofare, NJ: Slack, Incorporated.

Ottenbacher, K. J., & Jannell, S. (1993). The results of clinical trials in stroke rehabilitation research. *Archives of Neurology, 50*(1), 37–44.

Ouslander, J. G., Urman, H., & Uman, G. C. (1988). Development and testing of an incontinence monitoring record. *Journal of the American Geriatrics Society, 34,* 83–90.

Owens, D., Webber, P., & Linderman, D. (1996). Dementia and depression: Concordance rates between dementia patients' self report of mood and nursing staff observer ratings of patients' mood. *Clinical Gerontologist, 17,* 21–41.

Oxman, T. E., & Berkman, L. F. (1990). Assessment of social relationships in elderly patients. *The International Journal of Psychiatry in Medicine, 20,* 65–84.

Ozawa, M. N., & Morrow-Howell, N. (1988). Missouri service credit system for respite care: An exploratory study. *Journal of Gerontological Social Work, 21,* 147–160.

Pacala, J. T., Boult, C., Reed, R. L., & Aliberti, E. (1997). Predictive validity of the Pra instrument among older recipients of managed care. *Journal of the American Geriatrics Society, 45*(5), 614–617.

Paier, G. S., & Strumpf, N. E. (1999). Meeting the needs of older adults for primary health care. In M. Mezey, & D. O. McGivern (Eds.), *Nurses, Nurse Practitioners: Evolution to Practice* (3rd ed., Chap. 16, pp. 300–315). New York: Springer Publishing Company.

Palmer, R. M. (2000). Acute hospital care of the elderly: Future directions. In T. T. Yoshikawa, & D.C. Norman (Eds.), *Acute Emergencies and Critical Care of the Geriatric Patient.* New York: Marcel Dekker, Inc.

Palmer, R. M., Counsell, S., & Landefeld, C. S. (1998). Clinical intervention trials: The ACE Unit. In R. M. Palmer (Ed.), *Clinics in Geriatric Medicine.* Philadelphia, PA: W. B. Saunders.

Palmore, E. (1972). Gerontophobia versus ageism. *The Gerontologist, 12,* 213.

Palmore, E. (1979). Predictors of successful aging. *The Gerontologist, 19,* 427–431.

Pandian, G., & Kowalske, K. (1999). Daily functioning of patients with an amputated lower extremity. *Clinical Orthopaedics & Related Research,* (361), 91–97.

Paolucci, S., Antonucci, G., Pratesi, L., Traballesi, M., Grasso, M. G., & Lubich, S. (1999). Poststroke depression and its role in rehabilitation of inpatients. *Archives of Physical Medicine & Rehabilitation, 80*(9), 985–990.

Paris, D., Town, T., Parker, T. A., Tan, J., Humphrey, J., Crawford, F., & Mullan, M. (1999). Inhibition of Alzheimer's beta-amyloid induced vasoactivity and proinflammatory response in microglia by a cGMP-dependent mechanism. *Experimental Neurology, 157*(1), 211–221.

Park, D. C., Herttzog, C., Leventhal, H., Morrell, R. W., Leventhal, E., Birchmore, D., Martin, M., & Bennett, J. (1999). Medication adherence in rheumatoid arthritis patients: Older is wiser. *Journal of the American Geriatrics Society, 47,* 172–183.

Park, D. C., & Jones, T. R. (1997). Medication adherence and aging. In A. D. Fisk, & W. A. Rogers (Eds.), *Handbook of Human Factors and the Older Adult* (pp. 257–288). San Diego, CA: Academic Press.

Parkerson, G. R., Broadhead, W. E., & Tse C-K. J. (1990). The Duke Health Profile: A 17 item measure of health and dysfuction. *Medical Care, 28,* 1056–1072.

Parkerson, G. R. Jr. (1998). *User's Guide for Four Duke Health Measures.* Durham, NC: Duke University Medical Center.

Parkes, A. J., & Cooper, L. C. (1997). Inappropriate use of medications in the veteran community: How much do doctors and pharmacists contribute? *Australian & New Zealand Journal of Public Health, 21*(5), 469–476.

Patterson, R., Gramner, L. C., Greenberger, P. A., & Zeiss, C. R. (1997). *Allergic Diseases Diagnosis and Management.* Phiadelphia, PA: J. B. Lippincott Company.

Paveza, G. J., & Hughes-Harrison, V. (1997). Financial exploitation of the elderly: A descriptive study of victims and abusers in an urban area. *The Gerontologist, 37*(Special Issue 1), 101–107.

Payne, R. Y., & Martin, M. (1990). The epidemiology and management of skin tears in older adults. *Ostomy Wound Management, 26,* 26–27.

Paynter, J., Ambrose, K., & Dolan, K. (1997). Integrating geriatric evaluation and management with a multidisciplinary care planning process. In P. L. Spath *Beyond Clinical Paths: Advanced Tools for Outcomes Management* (pp. 103–127). Chicago, IL: American Hospital Publishing, Inc.

Pearlin, L., Mullan, J., Semple, S., & Skaff, M. (1990). Caregiving and the stress process: An overview of concepts and their measures. *The Gerontologist, 30*(5), 583–594.

Pearlin, L. I. (1989). The sociological study of stress. *Journal of Health and Social Behavior, 30,* 241–256.

Pearson, A., Nay, R., Taylor, B., Tucker, C., Angus, V., & Ruler, A. (1998). (Report No. 4, Monograph Series). Adelaide, Australia: University of Adelaide.

Pearson, A., Punton, S., & Durant, I. (1992). *Nursing Beds: An Evaluation of the Effects of Therapeutic Nursing.* London: Scutari Press.

Pearson, J. L., Caine, E. D., Lindesay, J., Conwell, Y., & Clark, D. C. (1999). Studies of suicide in later life: Methodologic considerations and research directions. *American Journal of Geriatric Psychiatry, 7*(3), 203–210.

Pearson, J. L., Conwell, Y., & Lyness, J. M. (1997). Late-life suicide and depression in the primary care setting. *New Directions for Mental Health Services, (76),* 13–38.

Pearson, L. J. (1999). Annual update of how each stands on legislative issues affecting advanced nursing practice. *The Nurse Practitioner, 24*(1), 16–24.

Pedlar, D. J., & Smyth, K. A. (1999). Introduction: Caregiver attitudes, beliefs, and perceptions about service use. Special Issue. *Journal of Applied Gerontology, 18*(2), 141–144.

Pellegrino, E. D. (1994). Patient and physician autonomy: Conflicting right and obligation in the physician-patient relationship. *Journal of Contemporary Health Law & Policy, 10,* 47–68.

Pennswood Village. (1990). Summary of the survey of priority waiting list (with National Analysts, Inc., division of Booz-Allen and Hamilton).

Pennswood Village. (1994). Unpublished Data. N. Spears, Executive Director.

Peppard, N. R. (1991). *Special Needs Dementia Units: Design, Development and Operations.* New York: Springer Publishing Company.

Perle, S. M., Mutell, D. B., & Romanelli, R. (1997). Age-related changes in skeletal muscle strength and modifications through exercise: A literature review. *Sports Chiropractic and Rehabilitation, 11*(3), 97–103.

Perez, E. D. (1994). Hip fracture: Physicians take more active role in patient care. *Geriatrics, 49,* 31–37.

Perry, W. H. (1983). The willingness of persons 60 or over to volunteer: Implications for social service. *Journal of Gerontological Social Work, 5,* 107–118.

Pert, C. (1997). *Molecules of Emotion.* New York: Scribner.

Petersen, R. C., Smith, G. E., Waring, S. C., Ivnik, R. J., Tangalos, E. G., & Kokmen, E. (1999). Mild cognitive impairment: Clinical characterization and outcome. *Archives of Neurology, 56*(3), 303–308.

Peterson, D. A. (1983). *Facilitating Education for Older Learners.* San Francisco, CA: Jossey-Bass.

Petrovich, Z., & Baert, L. (1994). *Benign Prostatic Hyperplasia: Innovations in Management.* New York: Springer-Verlag.

Petty, T. L. (1998). Definitions, causes, course, and prognosis of chronic obstructive pulmonary disease. *Respiratory Care Clinics of North America, 4*(3), 345–358.

Pfeffer, R. I., Kurosaki, T. T., Harrah, C. H., Chance, J. M., & Filos, S. (1982). Measurement of functional activities on older adults in the community. *Journal of Gerontology, 37,* 323–329.

Pfeifer, E., & Davis, G. C. (1971). The use of leisure time in middle life. *Gerontologist, 11,* 187–196.

Pfeiffer, E. (1975). A short portable mental status questionnaire for the assessment of organic brain deficit in elderly patients. *Journal of the American Geriatrics Society, 23,* 433–441.

Pfeiffer, E. (1998). Why teams? In E. Siegler, K. Hyer, T. T. Fulmer, & M. D. Mezey (Eds.), *Geriatric Interdisciplinary Team Training.* New York: Springer Publishing Company.

Phillips, C. D., Hawes, C., & Fries, B. E. (1993). Reducing the use of physical restraints in nursing homes: Will it increase cost? *American Journal of Public Health, 83,* 342–348.

Phillips, C. D., Hawes, C., Mor, V., Fries, B. E., Morris, J. D., & Nennesticl, M. (1996). Facility and area variation affecting use of physical restraints in nursing homes. *Medical Care, 34,* 1149–1162.

Phillips, C. D., Morris, J. D., Hawes, C., Fries, B. E., Mor, V., Nennistiel, M., & Iannoacchione. (1997). Association of the resident assessment instrument with changes in function, cognition, and psychosocial status. *Journal of the American Geriatrics Society, 45,* 986–993.

Phillips, Y. Y., & Hnatiuk, O. W. (1998). Diagnosing and monitoring the clinical course of chronic obstructive pulmonary disease. *Respiratory Care Clinics of North America, 4*(3), 371–389.

Philp, E. B. (1997). Chronic cough. *American Family Physician, 56*(5), 1395–1404.

Picot, S. J. (1995). Rewards, costs and coping of African American caregivers. *Nursing Research, 44*(3), 147–152.

Picot, S. J., Debanne, S. A., Namazi, K. H., & Wykle, M. L. (1997). Religiosity and perceived rewards of family caregivers. *The Gerontologist, 37*(1), 89–101.

Picot, S. J., Stuckey, J., Humphrey, S., Smyth, K., & Whitehouse, P. (1996). Cultural assessment and the recruitment and retention of African Americans into Alzheimer's disease research. *Journal of Aging & Ethnicity, 1*(1), 5–16.

Pillemer, K. A., & Finkelhor, D. (1988). The prevalence of elder abuse: A random sample survey. *The Gerontologist, 28*(I), 51–57.

Pillemer, K. A., & Wolf, R. S. (1986). *Elder Abuse: Conflict in the Family.* Dover, MA: Auburn House Publishing Company.

Pines, A., & Aronson, E. (1988). *Career Burnout: Causes and Cures.* New York: Free Press.

Pinho, M., Ortiz, J., Oya, M., Panagamuwa, B., Asperer, J., & Keighley, M. R. (1992). Total pelvic floor repair for the treatment of neuropathic fecal incontinence. *American Journal of Surgery, 163,* 340–343.

Piper, B. A., Galsworthy, T. D., & Bockman, R. S. (1995). Diagnosis and management of osteoporosis. *Contemporary Internal Medicine, 7*(7), 61–68.

Pirisi, A. (2000). Antihistamines impair driving as much as alcohol. *Lancet, 355*(9207), 905.

Pitt, B., Scgal, R., & Martinez, F. (1997). Randomized trial of Lorsartan versus Captopril in patients over 65 with heart failure (Evaluation of Lorsartan in the Elderly Study, ELITE). *Lancet, 349,* 747–752.

Pixten, R. (1995). Comparing time and temporality in cultures. *Cultural Dynamics, 7,* 233–252.

Podsiadlo, D., & Richardson, S. (1991). The timed "Up & Go": A test of basic functional mobility for frail elderly persons. *Journal of the American Geriatrics Society, 39*(2), 142–148.

Ponza, M., Ohls, J. C., & Millen, B. E. (1996). *Serving Elders at Risk: The Older American's Act Nutrition Programs.* Washington, DC: U.S. Department of Health and Human Services Administration on Aging.

Popp, B., & Portenoy, R. K. (1996). Management of chronic pain in the elderly. In International Association for the Study of Pain, *Pain in the Elderly: A Report of the Task Force on Pain in the Elderly.* Seattle, WA: Author.

Port, S., Cobb, F. R., Coleman, R. E., & Jones, R. H. (1980). Effect of age on the response of left ventricular ejaculation fraction to exercise. *NEJM, 303*(20), 1133–1137.

Post, F. (1992). Affective illness. In T. Arie (Ed.), *Recent Advances in Psychogeriatrics.* London: Churchill Livingstone.

Post, L. F., Blustein, J., & Dubler, N. N. (1999). The doctor-proxy relationship: An untapped resource. *The Journal of Law, Medicine & Ethics, 27*(1), 5–12.

Post, L. F., & Dubler, N. N. (1997). Palliative care: A bioethical definition, principles and guidelines. *Bioethics Forum, 13,* 17–24.

Postrel, V. (1999). *The Future and Its Enemies: Growing Conflict over Creativity.* New York: Touchstone.

Potter, V. (1971). *Bioethics, Bridge to the Future.* Englewood Cliffs: Prentice-Hall.

Potthoff, S. J., Kane, R. L., & Franco, S. J. (1995). *Hospital Discharge Planning for Elderly Patients: Improving Decisions, Aligning Incentives.* Minneapolis, MN: University of Minnesota Institute for Health Services Research.

Pourat, N., Lubben, J., Wallace, S., & Moon, A. (1999). Predictors of use of traditional Korean healers among elderly Koreans in Los Angeles. *The Gerontologist, 39*(6), 711–719.

Pritchard, J. (1996). *Working with Elder Abuse: A Training Manual for Home Care, Residential and Day*

Care Staff. London, England: Jessica Kingsley Publishers.

Promislow, D. E. L. (1993). On size and survival: Progress and pitfalls in allometry of life-span. *Journals of Gerontology, 48,* B115–123.

Pruchno, R. (1999). Raising grandchildren: The experience of black and white grandmothers. *The Gerontologist, 39*(2), 209–221.

Prunchno, R. A., Michaels, J. E., & Potashnik, S. L. (1990). Predictions of institutionalization among Alzheimer Disease victims with caregiving spouses. *Journals of Gerontology, 45B,* S259–S266.

Quadango, J., & Hardy, M. (1996). Work and retirement. In R. H. Binstock, & L. K. George (Eds.), *Handbook of Aging and the Social Sciences* (4th ed., pp. 326–345). San Diego: Academic Press.

Quadri, P. (1998). MNA and cost of care. In B. Vellas, P. J. Garry, & Y. Guigoz (Eds.), *Mini Nutritional Assessment (MNA): Research and Practice in Elderly* (Vol. 1, pp. 141–148). Philadelphia, PA: Lippincott-Raven.

Quam, J. K., & Whitford, G. S. (1992). Adaption and age-related expectations of older gay and lesbian adults. *The Gerontologist, 32*(2), 367–374.

Quinet, R., & Winters, E. (1992). Total joint replacement of the hip and knee. *Medical Clinics of North America, 76*(5), 1235–1251.

Quinn, J. F. (1999). Retirement Patterns and Bridge Jobs in the 1990s. *EBRI Issue Brief No. 206.*

Quinn, J. F., & Burkhauser, R. V. (1994). Retirement and labor for behavior of the elderly. In L. G. Martin, & S. H. Preston (Eds.), *Demography of Aging* (pp. 50–101). Washington, DC: National Academy Press.

Quinn, M. E., Johnson, M. A., Andress, E. L., McGinnis, P., & Ramesh, P. (1999). Health characteristics of elderly personal care home residents. *Journal of Advanced Nursing, 30*(2), 410–417.

Quinn, M. J., & Tomita, S. K. (1997). *Elder Abuse and Neglect* (2nd ed.). New York: Springer Publishing Company.

Rabeneck, L., Wray, N. P., & Petersen, N. J. (1996). Long-term outcomes of patients receiving percutaneous endoscopic gastrostomy tubes. *Journal of General Internal Medicine, 11*(5), 287–293.

Rabins, P. (1997). The prevalence of psychiatric disorders in elderly residents of public housing. *Journals of Gerontology, 51A,* M319–M324.

Rader, J., & Tornquist, E. M. (1995). *Individualized Dementia Care: Creative, Compassionate Approaches.* New York: Springer Publishing Company.

Radloff, L. S. (1977). The CES-D Scale: A self-report depression scale for research in the general population. *Journal of Applied Psychological Measures, 1,* 385–401.

Radner, D. B. (1998). The retirement prospects of the baby boom generation. *Social Security Bulletin, 61*(1), 3–19.

Ragozzino, M. W. (1982). Population based study of herpes zoster and its sequelae. *Medicine, 5*(6), 310.

Raisz, L. G. (1997). The osteoporosis revolution. *Annals of Internal Medicine, 126*(6), 458–462.

Rajwani, R. (1996). South Asians. In J. G. Lipson, S. L. Dibble, & P. A. Minarik (Eds.), *Culture & Nursing Care: A Pocket Guide* (pp. 264–279). San Francisco, CA: University of California San Francisco.

Rank, M., & Hirschl, T. (1999). Examining the proportion of Americans ever experiencing poverty during their elderly years. *Journals of Gerontology, 54B*(4-Supplement), S184–S193.

Rao, S. S. C. (1999). Fecal Incontinence. *Clinical Perspectives in Gastroenterology, 5*(2), 277–288.

Rapoport, R., & Rapoport, R. N. (1975). *Leisure and the Family Life Circle.* London: Routledge & Kegan Paul.

Rappaport, H., Fossler, R., Bross, L., & Gildern, D. (1993). Future time, death anxiety, and life purpose among older adults. *Death Studies, 17,* 369–379.

Redmond, K., & Aapro, M. S. (1997). *Cancer in the Elderly: A Nursing and Medical Perspective.* Oxford, England: Elsevier Science Ltd.

Reed, D. M., Foley, D. J., White, L. R., Heimovitz, H., Bunchfiel, C. M., & Makasi, K. (1998). Predictors of healthy aging in men with high life expectancies. *American Journal of Public Health, 88,* 1463–1468.

Rees, T., Duckert, L., & Carey, J. P. (1999). Auditory and vestibular dysfunction. In W. R. Hazzard, J. P. Blass, W. H. Ettinger, J. B. Halter, & J. G. Ouslander (Eds.), *Principles of Geriatric Medicine and Gerontology* (4th ed., Chap. 44, pp. 617–631). New York: McGraw-Hill.

Regier, D. A., Boyd, J. H., Burke, J. D., Rae, D. S., Myers, J. K., Kramer, M., Robins, L. N., George, L. K., Karno, M., & Locke, B. Z. (1988). One-month prevalence of mental disorders in the United States: Based on five epidemiological catchment area sites. *Archives of General Psychiatry, 45,* 977–986.

Regier, D. A., Narrow, W. E., & Rae, D. S. (1990). The epidemiology of anxiety disorders: The Epidemiologic Cachment area (ECA) experience. *Journal of Psychiatric Research, 24*(Suppl. 2), 3–14.

Reid, J. D. (1995). Development in late life: Older lesbian and gay lives. In A. R. D. Augelli, & C. J. Patterson (Eds.), *Lesbian, Gay and Bisexual Identities Over the Lifespan* (pp. 215–240). New York: Oxford University Press.

Reinardy, J., & Kane, R. A. (1999). Choosing an adult foster home or nursing home: Residents' perceptions about decision making and control. *Social Work, 44*(5), 571–585.

Reis, M., & Nahmiash, D. (1997). Abuse of seniors: Personality, stress and other indicators. *Journal of Mental Health & Aging, 3*(3), 337–356.

Reisberg, B., Ferris, S. H., Franssen, E., & Kluger, A. (1986). Age-associated memory impairment: The clinical syndrome. *Developmental Neuropsychology, 2*, 401–412.

Reitan, R. M., & Wolfson, D. (1985). *The Halstead-Reitan Neuropsychological Test Battery*. Tempe, AZ: Neuropsychology Press.

Reiter, R. J. (1995). The pineal gland and melatonin in relation to aging: A summary of the theories and of the data. *Experimental Gerontology, 30*, 199–212.

Rejda, G. E. (1994). *Social Insurance and Economic Security*. Englewood Cliffs, NJ: Prentice Hall.

Relkin, N. R., & National Institute on Aging Alzheimer's Association Working Group. (1996). Apolipoprotein E genotyping in Alzheimer's disease. *Lancet, 347*, 1091–1095.

Reschovsky, A. (1994). *Do the Elderly Face High Property Tax Burdens?* Washington, DC: AARP Public Policy Institute.

Resnick, B. (1998). Health care practices of old-old. *American Academy Journal of Nurse Practitioners, 10*, 147–155.

Resnick, B. (1999). Reliability and validity testing of the Self-Efficacy for Functional Activities Scale. *Journal of Nursing Measurement, 7*(1), 5–20.

Resnick, B. (1999). Motivation to perform activities in daily living in the institutionalized older adult: Can a leopard change its spots? *Journal of Advanced Nursing, 29*(4), 792–799.

Reuben, D., Schnelle, J., Buchanan, J., Kington, R., Zellman, G., Farley, D., Hirsch, S., & Ouslander, J. (1999). Primary care of long-stay nursing home residents: Approaches of three health maintenance organizations. *Journal of the American Geriatrics Society,* (2), 131–138.

Reuben, D. B., Silliman, R. A., & Traines, M. (1988). The aging driver: Medicine, policy, and ethics. *Journal of the American Geriatrics Society, 36*(12), 1135–1142.

Reuben, D. B., & Siu, A. L. (1990). An objective measure of physical function of elderly outpatients the physical performance test. *Journal of the American Geriatrics Society, 38*, 1105–1112.

Reuben, D. B., Valle, M. S., Hays, R. D., & Siu, A. L. (1995). Measuring physical function in community-dwelling older adults: A comparison of self-administered, interviewer-administered, and performance-based measures. *Journal of the American Geriatrics Society, 43*, 17–23.

Review Symposium. (1982). *Aging & Society, 2*, 383–392.

Rhine, S. H. (1984). *Managing Older Workers: Company Policies and Attitudes*. New York: The Conference Board.

Rhoades, E. R. (1990). *Profile of American Indians and Alaskan Natives*. (Report No. HRS(P-DV-90-4). Washington, DC: U.S. Government Printing Office.

Rhoades, J., Potter, M. S., & Krauss, N. (1998). *Nursing Homes—Structure and Selected Characteristics* [Web Page]. URL http://www.meps.ahcpr.gov/papers/98006/98-0006.htmk [1999, August 2].

Rhodes, J. A., & Krauss, N. (1999). Nursing home trends, 1987 and 1996. *MEPS Chartbook, No. 3, AHCPR Publication No. 99-0032*. Rockville, MD: MEPS.

Rich, M. W. (1998). Therapy for acute myocardial infarction in older persons. *Journal of the American Geriatrics Society, 46*, 1302–1307.

Richards, S., & Hendrie, H. C. (1999). Diagnosis, management, and treatment of Alzheimer disease. *Archives of Internal Medicine, 159*, 789–798.

Richey, M. L., Richey, H. K., & Fenske, N. A. (1988). Aging-related skin changes: Development and clinical meaning. *Geriatrics, 43*(April), 49–64.

Richter, J. E. (1997). Diseases of the esophagus. In *Textbook of Internal Medicine* 3rd ed., (pp. 674–678). Philadelphia, PA: Lippincott-Raven.

Riedinger, J. L., & Robbins, L. J. (1998). Prevention of iatrogenic illness. Adverse drug reactions and nosocomial infections in hospitalized older adults. *Clinics in Geriatric Medicine, 14*(4), 681–698.

Riehmann, M., & Bruskewitz, R. C. (1994). Management of bladder outlet obstruction in elderly men. In P. D. O'Donnell (Ed.), *Geriatric Urology* (pp. 275–284). New York: Little, Brown and Company.

Rivara, F. P., Grossman, G. C., & Cummings, P. (1997). Injury Prevention. *NEJM, 333*, 543–548, 613–618.

Rivlin, A., & Wiener, J. M. (1988). *Caring for the Disabled Elderly: Who Will Pay?* Washington, DC: The Brookings Institution.

Rizack, M., & Gardner, D. (1998). The Medical Letter drug interaction program. *The Medical Letter, 40*(9), 1–4.

Roberge, R. T., McCandish, M., & Dorfsman, M. (1999). Urosepsis associated with vaginal pessery use. *Annals of Emergency Medicine, 33*(5), 581–583.

Robertsson, B. (1998). Assessment scales in delirium. *Dementia & Geriatric Cognitive Disorders, 10*, 368–379.

Robinson, B. (1983). Validation of a caregiver strain index. *Journal of Gerontology, 38*(3), 344–348.

Rockwood, K., Cosway, S., Stolee, P., Kydd, D., Carver, D., Jarrett, P., & O'Brien, B. (1994). Increasing the recognition of delirium in elderly patients. *Journal of the American Geriatrics Society, 42*, 252–256.

Roe, D. A. (1984). Adverse nutritional effect of OTC drug use in the elderly. In D. A. Roe (Ed.), *Drugs and Nutrition in the Geriatric Patient*. New York: Churchill Livinstone.

Roelofs, L. H. (1999). The meaning of leisure. *Journal of Gerontological Nursing, 25*(10), 32–39.

Rogers, A., Flowers, J., & Pencheon, D. (1999). Improving access needs a whole systems approach: And will be important in averting crises in the millennium winter. *BMJ, 319,* 866–867.

Rogers, J. C., & Holm, M. B. (1998). Evaluation of occupational performance areas. In H. S. Willard, M. E. Neistadt, E. B. Crepeau, & C. S. Spackman (Ed.), *Willard & Spackman's Occupational Therapy* (9th ed., Chap. 15, pp. 334–367). Philadelphia: Lippincott-Raven Publishers.

Rogers, W. A. (1997). Individual difference, aging, and human factors: An overview. In A. D. Fisk, & W. A. Rogers (Eds.), *Handbook of Human Factors and the Older Adult* (pp. 151–170). San Diego, CA: Academic Press.

Rosen, R. C. (1998). Sexual function assessment in the male: Physiological and self-report measures. *International Journal of Impotence Research, 10*(Suppl 2), S59–63.

Rosenthal, J., & Soroka, M. (1998). *Managed Vision Benefits* (2nd ed.). Madison, WI: International Foundation of Employee Benefits.

Rosswurm, M. A. (1998). *Home Care for Older Adults: A Guide for Families and Other Caregivers*. New York: Springer Publishing Company.

Roth, G. S., & Ingram, D. K. & L. M. A. (1999). Calorie restriction in primates: Will it work and how will we know? *Journal of the American Geriatrics Society, 47,* 896–903.

Rothstein, J. M., Roy, S. H., & Wolf, S. L. (1991). *The Rehabilitation Specialist's Handbook*. Philadelphia, PA: F. A. Davis.

Rovner, G., German, P., Brant, L., Clark, R., Burton, L., & Folstein, M. (1991). Depression and mortality in nursing homes. *JAMA, 265*(8), 993–996.

Rowe, J. W., & Kahn, R. L. (1998). *Successful Aging*. New York: Pantheon Books.

Royal Commission on Long Term Care. (1999). *Respect to Old Age: Long Term Care—Rights and Responsibilities*. London: HMSO.

Roye, C., & Balk, S. (1996). Evaluation of an intergenerational program for pregnant and parenting adolescents. *Maternal Child Nursing Journal, 24*(1), 32–40.

Rubenstein, L. V., Calkins, D. R., Greenfield, S., Jette, A. M., Meenan, R. F., Nevins, M. A., et al. (1988). Health status assessment for elderly patients. Report of the Society of General Internal Medicine Task Force on Health Assessment. *Journal of the American Geriatrics Society, 37,* 562–569.

Rubenstein, L. W., Josephson, K. R., Wieland, G. D., English, P. A., Sayre, J. A., & Kane, R. L. (1984). Effectiveness of a geriatric evaluation unit: A randomized clinical trial. *NEJM, 311,* 1664–1670.

Rubenstein, L. Z. (1998). Development of a short version of the Mini Nutritional Assessment. In B. Vellas, P. J. Garry, & Y. Guigoz (Eds.), *Mini Nutritional Assessment (MNA): Research and Practice in Elderly* (pp. 101–116). Philadelphia, PA: Lippincott-Raven.

Rubenstein, L. Z., Aronow, H. U., Schloe, M., Steiner, A., Alessi, C. A., Yuhas, K. E., Gold, M., Kempe, M., Raube, K., Nisenbaum, R., Stuck, A., & Beck, J. C. (1994). A home based geriatric assessment, follow-up and health promotion program: Design, methods, and baseline findings from a 3-year randomized clinical trial. *Aging Clinical & Experimental Research, 6*(2), 105–120.

Rubenstein, L. Z., Josephson, K. R., & Robbins, A. S. (1994). Falls in the nursing home. *Annals of Internal Medicine, 121*(442–451).

Rubenstein, L. Z., & Rubenstein, L. V. (1998). Multidimensional geriatric assessment. In Brocklehurst's *Textbook of Geriatric Medicine*. New York: Churchill Livingstone.

Rubenstein, L. Z., Stewart, S., Schroll, M., Bernabei, R., Bula, C., Jones, D., & Wieland, D. (1997). *In-home programs of prevention and comprehensive geriatric assessment: International Perspectives*. Presentation at the 1997 World Congress of Gerontology [Web Page]. URL http://www.cas.flinders.edu.au/iag/proceedings/proc0027.htm [2000, January 27].

Rubik, B. (1995). Energy medicine and the unifying concept of information. *Alternative Therapies in Health & Medicine, 1,* 34–39.

Rudkin, L., Markides, K. S., & Espino, D. V. (1997). Disability in older Mexican Americans. *Geriatric Rehabilitation, 12,* 38–46.

Ryan, J. W., & Spellbring, A. M. (1996). Implementing strategies to decrease risk of falls in older women. *Journal of Gerontological Nursing, 22*(12), 25–31.

Ryan, K. (1999). *Gynecology and Women's Health*. 7th ed. St. Louis: Mosby Yearbook.

Ryff, C. D. (1989). Happiness is everything, or is it? Explorations on the meaning of psychological well-being. *Journal of Personality and Social Psychology, 57,* 1069–1081.

Ryff, C. D., & Dunn, D. D. (1985). A life-span developmental approach to study of stressful events. *Journal of Applied Developmental Psychology, 6,* 113–127.

Ryff, C. D., & Essex, M. J. (1992). The interpretation of life experience and well being: The sample case of relocation. *Psychology and Aging, 7,* 507–517.

Ryff, C. D., & Keyes, C. L. M. (1995). The structure of psychological well-being revisited. *Journal of Personality and Social Psychology, 69,* 719–727.

Sabatino, C. P. (1999). The legal and functional status of the medical proxy: Suggestions for statutory reform. *The Journal of Law, Medicine & Ethics, 27*(1), 52–68.

Sadavoy, J., Lazarus, L. W., Jarvik, L. F., & Grossberg, G. T. (1996). *Comprehensive Review of Geriatric Psychiatry.* Washington, DC: American Psychiatric Press.

Saint, S., & Matthay, M. A. (1998). Risk reduction in the intensive care unit. *American Journal of Medicine, 105,* 515–523.

Salisbury, D. (1993). *Pension Tax Expenditures: Are They Worth the Cost? (Issue Brief No. 134).* Washington, DC: Employee Benefit Research Institute.

Salthouse, T. A. (1996). The processing-speed theory of adult age differences in cognition. *Psychological Review, 103,* 403–428.

Saunders, D. C., & Kastenbaum, R. (1997). *Hospice Care on the International Scene.* New York: Springer Publishing Company.

Savishinsky, J. (2000). *The Broken Watch: Retirement and Meaning in America.* Ithaca, NY: Cornell University Press.

Schaeffer, A. J. (1998). Infections of the urinary tract. In P. C. Walsh, A. B. Retik, D. Vaughan, & A. J. Wein (Eds.), *Campbell's Urology* (7th Ed. ed., pp. 601–603). Philadelphia: W. B. Saunders.

Schaie, K. W., & Pietrucha, M. (2000). *Mobility and Transportation in the Elderly.* New York: Springer Publishing Company.

Scheidlinger, S. (1994). An overview of nine decades of group psychotherapy. *Hospital & Community Psychiatry, 45,* 217–225.

Scheitel, S. (1996). Geriatric health maintenence. *Mayo Clinics Proceedings, 71*(289–302).

Schieber, S. J. (1990). *Beneits bargain: Why should we not tax employee benefits?* Washington, DC: Association of Private Pensions and Welfare Plans.

Schmidt, R., Mechtler, L., Kinkel, P. R., Fazekas, F., Kinkel, W. R., & Freidl, W. (1993). Cognitive impairment after acute supratentorial stroke: A 6-month follow-up clinical and computed tomographic study. *European Archives of Psychiatry and Clinical Neuroscience, 243,* 11–15.

Schneider, B., & Weiss, L. (1982). *The Channeling Case Management Manual.* Washington, DC: U.S. Department of Health and Human Services, Health Care Financing Administration.

Schneider, E. L., & Brody, J. A. (1983). Aging, natural death and the compression of morbidity: Another view. *NEJM, 309*(14), 854–855.

Schoenman, J., & Gardner, E. (1998). *Results of the First National Census of Optometrists.* New York: Project Hope, Center for Health Affairs.

Schofield, P. M., Bennett, D. H., Whorwell, P. J., Brooks, N. H., Bray, C. L., Ward, C., & Jones, P. E. (1987). Exertional gastro-oesophageal reflux: A mechanism for symptoms in patients with angina pectoris and normal coronary angiograms. *BMJ, Clinical Research Edition, 294*(6585), 1459–1461.

Scholtes, P. J., Joiner, B., & Streibel, B. (1996). *The Team Handbook* 2nd ed. Madison, WI: Oriel Inc.

Schommer, J. C., & Kucukarslan, S. N. (1997). Measuring patient satisfaction with pharmaceutical services. *American Journal of Health-System Pharmacy, 56*(23), 2721–2732.

The School of Sleep Medicine, Inc. (1999). *Sleep Net. Com* [Web Page]. URL http://www.sleepnet.com/ [2000, January 3].

Schopler, J., & Galinsky, M. (1993). Support groups as open systems: A model for practice and research. *Health and Social Work, 18*(3), 195–207.

Schorr, R. I., Fought, R. L., & Ray, W. A. (1994). Changes in antipsychotic drug use in nursing homes during the implementation of the OBRA-87 regulations. *JAMA, 271,* 358–362.

Schubert, M. M., & Izutsu, K. T. (1987). Iatrogenic causes of salivary gland dysfunctions. *Journal of Dental Research, 66,* 680–687.

Schulz, J. H. (1995). *The Economics of Aging* 6th ed. New York: Auburn House.

Schulz, R. (Ed.). (2000). *Handbook on Dementia Caregiving: Evidence-Based Interventions for Family Caregivers.* New York: Springer Publishing Company.

Schulz, R., & Beach, S. R. (1999). Caregiving as a risk factor for mortality: The Caregiver Health Effects Study. *JAMA, 282*(23), 2215–2219.

Schussheim, D. H., & Siris, E. S. (1998). Osteoporosis: Update on prevention and treatment. *Women's Health in Primary Care, 1*(2), 133–140.

Schwartz, G. E., & Russek L. G. (1997). Dynamical energy systems and modern physics: Fostering the science and spirit of complementary and alternative medicine. *Alternative Therapies in Health & Medicine, 3*(3), 46–56.

Schwartz, H. (1991). *Supportive Services for Elderly Residents of Government Assisted Naturally Occurring Retirement Communities in New York City.* New York City: UJA-Federation of Jewish Philanthropies.

Schwartz, N., Park, D., Knauper, B., & Sudman, S. (1999). *Cognition, Aging and Self Reports.* Philadelphia, PA: Psychology Press.

Science Panel on Interactive Communication and Health. (1999). *Wired for Health and Well-Being.* Washington, D.C.

Scinovacz, M. E., DeViney, S., & Atkinson, M. P. (1999). Effects of surrogate parenting on grandparents' well-being. *Journals of Gerontology, 54B*(6), S376–S388.

Scotch, R. K. (1984). *From Good Will to Civil Rights: Transforming Federal Disability Policy.* Philadelphia: Temple University Press.

Scott, J., Gade, G., McKenzie, M., & Venohr, I. (1998). Cooperative health care clinics: A group approach to individual care. *Geriatrics,* (5), 68–70, 76–78.

Scott, M., & Gelholt, A. R. (1999). Gastroesophageal reflux disease: Diagnosis and management. *American Family Physician, 59*(5), 1161–1169.

Seccia, M., Menconi, C., Balestri, R., & Cavina, E. (1994). Study protocols and functional results in 86 electrostimulated graciloplasties. *Diseases of the Colon & Rectum, 37,* 897–904.

Seidman, K. R. (1995). Staff instruction and training. In E. E. Faye, & C. S. Stuen (Eds.), *The Aging Eye and Low Vision: A Study Guide for Physicians* (2nd ed.). New York: The Lighthouse, Inc.

Sekuler, R., & Ball, K. (1986). Visual localization: Age and practice. *Journal of the Optical Society of America, A Optics & Image Science, 3*(6), 864–867.

Semenchuk, M. R., & Labiner, D. M. (1997). Gabapentin and lamotrigine: Prescribing guidelines for psychiatry. *Journal of Practical Psychiatry & Behavioral Health, 3,* 334–342.

Shapiro, J. P. (1993). *No Pity: People with Disabilities Forging a New Civil Rights Movement.* New York: Times Books.

Sharlin, K. S., Heath, G. W., Ford, E. S., & Welty, T. K. (1993). Hypertension and blood pressure awareness among American Indians of the Northern Plains. *Ethnicity and Disease, 3,* 337–343.

Sharp, C. (1999). *Where To Go from Here? Goal Attainment Scaling in Planning, Management and Evaluation of Services for the Aged.* Commissioned by Alan Johns, Director Validation Therapy Resource and Training Center. Melbourne, Australia: Validation Therapy Resource and Training Center.

Shashaty, A. (1999). Major Section 8 Housing losses forces out elders nationwide. *Aging Today,* (July/August), 1–2.

Shaughnessy, P., Kramer, A., Hittle, D., & Steiner, J. (1995). Quality of care in teaching nursing homes: Findings and implications. *Health Care Financing Review, 16*(4), 55–83.

Shaw, G. (1999). Family-centered care: Today, pediatrics, tomorrow the world. *Association of American Medical Colleges, 12*(September), 8–9.

Sheikh, J. L., & Yesavage, J. A. (1986). Geriatric Depression Scale (GDS): Recent evidence and development of a shorter version. *Clinical Gerontologist, 5,* 165–173.

Sheils, J. F., Rubin, R., & Stapleton, D. C. (1999). The estimated costs and savings of medical nutrition therapy: The Medicare population. *Journal of the American Dietetic Association, 99,* 428–435.

SHEP Cooperative Research Group. (1991). Prevention of stroke by antihypertensive drug treatment in older persons with isolated systolic hypertension. Final results of the Systolic Hypertension in the Elderly Program (SHEP). *JAMA, 265*(24), 3255–3264.

Shepherd, S., & Geraci, S. A. (1999). The differential diagnosis of dyspnea: A pathophysiologic approach. *Clinical Reviews, 9*(4), 52–74.

Sherwood, S., & Morris, J. N. (1983). The Pennsylvania domiciliary care experiment: I. Impact on quality of life. *American Journal of Public Health, 77*(12), 646–653.

Shiboski, C. H., Shiboski, S. C., & Silverman, A. (in press). Trends in oral cancer rates in the United States, 1973-1996. *Community Dentistry & Oral Epidemiology.*

Shih, J. (1997). Basic Beijing twenty four forms of Tai Chi exercise and average velocity of sway. *Perceptual & Motor Skills, 84,* 4–61.

Ship, J. A., Wolff, A., & Selik, R. M. (1991). Epidemiology of Acquired Immune Deficiency Syndrome in Persons Aged 50 Years or Older. *Journal of Acquired Immune Deficiency Syndromes, 4,* 84–88.

Shmotkin, D., & Hadari, G. (1996). An outlook on subjective well-being in older Israeli adults: A unified formulation. *International Journal of Aging & Human Development, 42,* 271–289.

Shoaf, L. R., Bishirjain, K. O., & Schlenker, E. D. (1999). The Gerontological Nutritionists Standards of Professional Practice for dietetics professionals working with older adults. *Journal of the American Dietetic Association, 99,* 863–867.

Shock, N. W. (1985). Longitudinal studies of aging in humans. In C. E. Finch, & E. L. Schneider (Eds.), *Handbook of the Biology of Aging.* New York: Van Nostrand Reinhold.

Shohet, J. A., & Bent, T. (1998). Hearing loss: The invisible disability. *Postgraduate Medicine, 104*(3), 81–90.

Shotkin, D. (1991). The role of time orientation in life satisfaction across the life span. *Journals of Gerontology, 46B*(P243–250).

Showers, C., & Ryff, C. D. (1997). Self-differentiation and well being in a life transition. *Personality and Social Psychology Bulletin, 22,* 448–460.

Shuman, S. K. (1999). Doing the right thing: Resolving ethical issues in geriatric dental care. *Elder Care: Journal of the California Dental Association,* 693–702.

Shumway-Cook, A., Baldwin, M., Polissar, N. L., & Gruber, W. (1997). Predicting the probability for falls in community-dwelling older adults. *Physical Therapy, 77*(8), 812 819.

Shumway-Cook, A., Gruber, W., Baldwin, M., & Liao, S. (1997). The effect of multidimensional exercises on balance, mobility and fall risk in community-dwelling older adults. *Physical Therapy, 77*, 213–218.

Siegel, J. S. (1980). On the demography of aging. *Demography, 17*, 345–364.

Siegler, E., & Whitney, F. (1994). *Nurse-Physician Collaboration: Care of Adults and the Elderly*. New York: Springer Publishing Company.

Siegler, E. L., Hyer, K., Fulmer, T. T., & Mezey, M. D. (1998). *Geriatric Interdisciplinary Team Training*. New York: Springer Publishing Company.

Silverman, C. A. (1998). Audiologic assessment and amplification. *Primary Care, 25*(3), 545–583.

Simberland, T. (1988). A new scale for the assessment of depressed mood in demented patients. *American Journal of Psychiatry, 145*, 955–959.

Simon-Rusinovitz, L. (1999). History, principles and definition of consumer-direction: Views from the aging community. Presented at *Alliance for Self-Determination Summit*, Bethesda, MD.

Simon-Rusinovitz, L., & Hofland, B. (1993). Adopting a disability approach to home care services for older adults. *The Gerontologist, 33*, 159–167.

Simon-Rusinovitz, L., Mahoney, K. J., Desmond, S. M., Shoop, D. M., Squillace, M. R., & Fay, R. A. (1997). Determining consumer preferences for a cash option: Arkansas survey results. *Health Care Financing Review, 19*(2), 73–96.

Simons, L. A., Tett, S., & Simons, J. (1992). Multiple medication use in the elderly: Use of prescription and non-prescription drugs in an Australian community setting. *The Medical Journal of Australia, 157*, 242–246.

Simonton. (1998). Career paths and creative lives. In C. E. Adams-Proce (Ed.), *Creativity & Successful Aging*. New York: Springer Publishing Company.

Singer, B. L., & Deschamps, D. (1994). *Gay and lesbian stats: A pocket guide of facts and figures*. New York: The New Press.

Singh, G., & Triadafilopoulos, G. (1999). Epidemiology of NSAID induced gastrointestinal complications. [Review] [38 refs]. *Journal of Rheumatology, 26 Suppl, 56*, 18–24.

Sjogren, T., Sjogren, H., & Lindgren, A. (1952). Morbus Alzheimer and morbus Pick: A genetic, clinical and patho-anatomical study. *Acta Psychiatrica Neurologica Scandinavica, 82*(suppl), 1–152.

Skoog, I., Kalaria, R. N., & Breteler, M. M. (1999). Vascular factors and Alzheimer disease. *Alzheimer's Disease & Associated Disorders, 13*(Suppl 3), S106–114.

Sloane, P., Blazer, D., & George, L. K. (1989). Dizziness in a community elderly population. *Journal of the American Geriatrics Society, 37*, 101–108.

Small, G. W., Rabins, P. V., Barry, P. P., Buckholtz, N. S., Dekosky, S. T., Ferris, S. H., Finkel, S. I., Gwyther, L. P., Khachaturian, Z. S., Lebowitz, B. D., McRae, T. D., Morris, J. C., Oakely, R., Schneider, L. S., Streim, J. E., Sunderland, T., Teri, L., & Tune, T. (1997). Diagnosis and treatment of Alzheimer Disease and related disorders. *JAMA, 2*, 152–158.

Smallegan, M. (1981). Decision making for nursing home admission: A preliminary study. *Journal of Gerontological Nursing, 7*(5), 280–285.

Smider, N. A., Essex, M. J., & Ryff, C. D. (1996). Adaptation to community relocation: The interactive influence of psychological resources and contextual factors. *Psychology and Aging, 11*, 362–372.

Smith, J., & Goodnow, J. J. (1999). Unasked-for support and unsolicited advice: Age and the quality of social experience. *Psychology and Aging, 14*, 108–121.

Snow, S. S., Logan, T. P., & Hollender, M. H. (1980). Nasal spray "addiction" and psychosis: A case report. *British Journal of Psychiatry, 136*, 297–299.

Snowden, J. S., Neary, D., & Mann, D. M. (1996). *Fronto-temporal Lobar Degeneration: Fronto-temporal Dementia, Progressive Aphasia, Semantic Dementia*. New York: Churchill-Livingstone.

Social Security Administration. (1999). *Annual Statistical Supplement*. Washington, DC: Government Printing Office.

Solomon, G. D., Skobieranda, F. G., & Graggs, L. A. (1993). Quality of life and well-being of headache patients: Measurement by the medical outcomes study instrument. *Headache, 33*(7), 351–358.

Somers, A. R., & Spears, N. L. (1992). *The Continuing Care Retirement Community: A Significant Option for Long-Term Care?* New York: Springer.

Soroka, M. (1997). The future of Medicare and managed care: Implications for optometry. *Journal of the American Optometric Association, 68*(3), 147–154.

Spector, A., Orrell, M., Davies, S., & Woods, B. (1998). Reality orientation for dementia: A review of the evidence for its effectiveness. *The Cochrane Library*, (4).

Spencer, J. W., & Jacobs, J. J. (1998). *Complementary and Alternative Medicine: An Evidence Based Approach*. St. Louis, MO: Mosby.

Spillman, B., Krauss, N., & Altman, B. (1997). A comparison of nursing home resident characteristics, 1987–1996. Presented at the Annual meeting of the Gerontological Society of America, Cincinnati, OH.

Ssali, T. S. (1999). African communities raise mental health concerns. *Synergy: Newsletter of the Austra-*

lian Transcultural Mental Health Network, (Summer), 3–10.

Stafford, G. D., Arendof, T., & Huggett, K. (1986). The effect of overnight drying and water immersion on dandicial colonization and properties of complete dentures. *Journal of Dentistry, 14,* 52–56.

Stall, R., & Catania, J. (1994). AIDS risk behaviors among late middle-aged and elderly Americans. *Archives of Internal Medicine, 154,* 57–63.

Stanley, M. A., & Beck, J. G. (in press). Anxiety disorders. *Clinical Psychology Review.*

Stanton, S. (1998). Consensus report on pelvic floor weakness of the elderly female. *World Journal of Urology, 16*(Suppl. 1), S44–S47.

Stark, A. J., Kane, R. A., Kane, R. L., & Finch, M. D. (1995). Effect on physical functioning of care in adult foster homes and nursing homes. *The Gerontologist, 35*(5), 648–655.

State Society on Aging of New York. (1999). *A Home Safety Checklist* [Web Page]. URL http://www.state.ny.us/text/safety/checklis.htm [1999, November 23].

Staurt, B. (1995). *Medical Guidelines for Determining Prognosis in Selected Non-Cancer Diseases.* Washington, DC: National Hospice Organization.

Steele, T. E., & Morton, W. A. (1986). Salicylate-induced delirium. *Psychosomatics, 27,* 455–456.

Steemen, E., & Defever, M. (1998). Urinary incontinence among elderly persons who live at home. *Nursing Clinics of North America, 33*(3), 441–455.

Stein, S. D., & Felsenthal, G. (1994). Rehabilitation of fractures in the geriatric population. In G. Felsethal, S. J. Garrison, & F. U. Steinberg (Eds.), *Rehabilitation of the Aging and Elderly Patient.* Philadelphia, PA: Williams and Wilkins.

Steinhauer, M. B. (1982). Geriatric foster care: A prototype design and implementation issues. *The Gerontologist, 22*(3), 293–300.

Sterns, A., Sterns, H. L., & Hollis, L. A. (1996). The productivity and functional limitations of older workers. In W. H. Crown (Ed.), *Handbook on Employment and the Elderly* (pp. 276–303). Westport, CT: Greenwood Press.

Sterns, H. L., & McDaniel, M. A. (1994). Job Performance and the Older Worker. In S. E. Rix (Ed.), *Older Workers: How Do They Measure Up?* Washington, DC: American Association of Retired Persons.

Steurle, E., & Bakija, J. M. (1994). *Retooling Social Security for the 21st Century.* Washington, DC: Urban Institute Press.

Stevens, J., Cai, J., Pamuk, E. R., Williamson, D. F., Thun, M. J., & Wood, J. L. (1998). The effect of age on the association between body mass index and mortality. *NEJM, 338,* 1–7.

Stevens, P. E. (1994). Protective strategies of lesbian clients in health care environments. *Research in Nursing & Health, 17*(3), 217–229.

Stewart, R. B., & Cooper, J. W. (1994). Polypharmacy in the aged. Practical solutions. *Drugs & Aging, 4*(6), 449–461.

Stewart, S., Vandenbroek, A. J., Pearson, S., & Horowitz, J. (1999). Prolonged beneficial effects of a home-based intervention on unplanned readmissions and mortality among patients with congestive heart failure. *Archives of Internal Medicine, 159,* 257–261.

Stillwater, B., Echavarria, V. A., & Lanier, A. P. (1995). Pilot test of a cervical cancer prevention video developed for Alaskan Native women. *Public Health Reports, 110,* 211–214.

Stoller, E. P., & Gibson, R. C. (1997). *Worlds of Difference: Inequality in the Aging Experience* (2nd ed.). Los Angeles: Pine Forge Press.

Stolnitz, G. J. (Ed.). (1992). Demographic causes and economic consequences of population aging: Europe and North America. UN/ECE Economic Studies. No. 3. Sales Publication No. GV. E. 92. 0. 4. New York: United Nations Population Fund [UNFPA].

Stone, L. O. (Ed.). (1999). *Cohort Flow and the Consequences of Populations Aging, An International Analysis and Review.* Ottawa: Minister of Industry.

Stone, M., & Stone, L. (1997). Ageism: The quiet epidemic. *Canadian Journal of Public Health, 88,* 293–294.

Storandt, M., & VandonBos, G. (1994). *Neuropsychological Assessment of Dementia and Depression in Older Adults: A Clinician's Guide.* Washington, DC: American Psychological Association.

Strawbridge, W. J., Wallhagen, M. I., Shema, S. J., & Kaplan, G. A. (1997). New burdens or more of the same? Comparing grandparent, spouse and adult child caregivers. *Gerontologist, 37*(4), 505–510.

Strayer, M. (1999). Oral health care for homebound and institutional elderly. *Elder Care: Journal of the California Dental Association,* 703–708.

Stroke Unit Trialists' Collaboration. (1997). Collaborative systematic review of the randomised trials of organised inpatient (stroke unit) care after stroke. *BMJ, 314*(7088), 1151–1159.

Stroup-Benham, C. A., Markides, K. S., Espino, D. V., & Goodwin, J. S. (1999). Changes in blood pressure and risk factors for cardiovascular disease among older Mexican Americans from 1982–1984 to 1993–1994. *Journal of the American Geriatrics Society, 47*(7), 804–810.

Struelens, M. J. (1998). The epidemiology of antimicrobial resistance in hospital acquired infections: Problems and possible solutions. *BMJ, 317,* 652–654.

Strull, W. M., Lo, B., & Charles, G. (1984). Do patients want to participate in medical decision making? *JAMA, 252,* 2990–2994.

Strumpf, N. E., Robinson, J. P., Wagner, J. S., & Evans, L. K. (1998). *Restraint-Free Care: Individualized Approaches for Frail Elders.* New York: Springer Publishing Company, Inc.

Stuart, B. (1995). *Medical Guidelines for Determining Prognosis in Selected Non-Cancer Diseases.* Alexandria, VA: National Hospice Organization.

Stuart, G., & Laraia, M. (1998). *Principles and Practice for Psychiatric Nursing* (6th ed.). St. Louis, MO: Mosby Year-Book.

Stuck, A., Aronow, H. U., Steiner, A., Alessi, C. A., Bula, C., Gold, M., Yuhas, K. E., Nisenbaum, R., Rubenstein, L. Z., & Beck, J. C. (1995). A trial of annual in-home comprehensive geriatric assessment for elderly people living in the community. *NEJM, 333,* 1184–1189.

Stuck, A. E., Siu, A. L., Wieland, G. D., Adams, J., & Rubenstein, L. Z. (1993). Comprehensive geriatric assessment: A meta-analysis of controlled trials. *Lancet, 342,* 1032–1036.

Summers, R. L., Cooper, G. J., Carlton, F. B., Andrews, M. E., & Kolb, J. C. (1999). Prevalence of atypical chest pain descriptions in a population from the southern United States. *American Journal of the Medical Sciences, 318*(3), 142–145.

Sunderland, T., Alterman, I. S., Yount, D., Hill, J.-L., et. al. (1988). A new scale for the assessment of depressed mood in demented patients. *American Journal of Psychiatry, 145*(8), 955–959.

SUPPORT Principal Investigators. (1995). A controlled trial to improve care for seriously ill hospitalized patients: The study to understand prognoses and preferences for outcomes and risks of treatments (SUPPORT). *JAMA, 274*(20), 1591–1598.

Sussman, N. (1998). Background and rationale for use of anticonvulsants in psychiatry. *Cleveland Clinic Journal of Medicine, 65,* SI7–SI14.

Suzman, R. M., Willis, D. P., & Manton, K. G. (1992). *The Oldest Old.* New York: Oxford University Press.

Swagerty, D., Takahasi, P., & Evans, J. (1999). Elder mistreatment. *American Family Physician, 59*(10), 2804–2808.

Swaim, M. W., & Wilson, J. A. (1999). GI emergencies: Rapid therapeutic responses for older patients. *Geriatrics, 54*(6), 20–22, 25–26, 29–30.

Swanger, A. K., & Burbank, P. M. (1995). *Drug Therapy and the Elderly.* Boston: Jones and Bartlett.

Swanson, E. A., Maas, M., & Buckwalter, K. C. (1993). Catastrophic reactions and other behaviors of Alzheimer's residents: Special unit compared with traditional units. *Archives of Psychiatric Nursing, 7*(5), 292–299.

Swartz, M., Landerman, R., George, L. K., Blazer, D. G., & Escobar, J. (1991). Somatization disorder. In L. N. Robins, & D. A. Reiger (Eds.), *Psychiatric Disorders in America: The Epidemiologic Catchment Area Study* (pp. 220–257). New York: The Free Press.

Szanto, K., Prigerson, H., Aouck, P., Ehrenprreis, L., & Reynolds, C. F. (1997). Suicidal ideation in elderly bereaved: The role of complicated grief. *Suicide, 27*(2), 194–207.

Tait, R., & Silver, R. C. (1989). Coming to terms with major negative life events. In J. S. Uleman, & J. A. Bargh (Eds.), *Unintended Thought.* New York: Guilford Press.

Takala, J., Ruckonen, E., Webster, N. R. N. M. S., & Zandstra, D. F. (1999). Increased mortality associated with growth hormone treatment in critically ill adults. *NEJM, 341*(11), 785–792.

Tallgren, A. (1999). The continuing reduction of the residual alveolar ridges in complete denture wearers: A mixed longitudinal study covering 25 years. *Journal of Prosthetic Dentistry, 27,* 120–132.

Tan, J., Town, T., Paris, D., Mori, T., Suo, Z., Crawford, F., Mattson, M. P., Flavell, R. A., & Mullan, M. (1999). Microglial activation resulting from CD40-CD40L interaction after beta-amyloid stimulation. *Science, 286*(5448), 2352–2355.

Tapscott, R. (1996). *The Digital Economy.* New York: McGraw-Hill.

Tariot, P. N., Podorgski, C. A., Blazina, L., & Leibovici, A. (1993). Mental disorders in the nursing home: Another perspective. *American Journal of Psychiatry, 150,* 1603–1608.

Tatara, T. Ph. D. (1996). *Elder Abuse: Questions and Answers.* Washington, DC: National Center on Elder Abuse, American Public Welfare Association.

Taylor, S. E., Helgeson, V. S., Reed, G. M., & Skokan, L. A. (1991). Self-generated feelings of control and adjustment to physical illness. *Journal of Social Issues, 47,* 91–109.

Teaff, J. D. (1990). *Leisure Services with the Elderly.* Prospect Heights, IL: Waveland Press, Inc.

Tejera, C. A., Saravay, S. M., & Goldman, E. (1994). Diphenhydramine-induced delirium in elderly hospitalized patients with mild dementia. *Psychosomatics, 35*(4), 399–402.

Teresi, J., Holmes, D., Benenson, E., Monaco, C., Barrett, V., Ramirez, M., & Koren, M. (1993). A primary care nursing model in long term care facilities: Evaluation of impact on affect, behavior, and socialization. *The Gerontologist, 33,* 667–674.

Teri, I. (1997). Assessment and treatment of neuropsychiatric signs and symptoms in cognitively impaired older adults: Guidelines for practitioners. *Seminars in Clinical Neuropsychiatry, 2*(152–158).

Teri, L., Rabins, P., Whitehouse, P., Berg, B., Sunderland, T., Eichelman, B., & Phelps, C. (1992). Management of behavior disturbance in Alzheimer disease: Current knowledge and future directions. *Alzheimer's Disease & Associated Disorders, 6*(2), 77–88.

Teri, L., & Wagner, A. (1992). Alzheimer's disease and depression. *Journal of Consulting & Clinical Psychology, 60,* 379–391.

Thom, D. (1998). Variation in estimates of urinary incontinence prevalence in the community: Effects of differences in definition, population characteristics, and study type. *Journal of the American Geriatrics Society, 46*(4), 473–480.

Thomas, C. L. (1997). *Taber's Cyclopedic Medical Dictionary* (18th ed.). Philadelphia: F. A. Davis.

Thomas, D., Goode, P., LaMaster, K., Tennyson, T., & Parnell, L. (1999). A comparison of an opaque foam dressing versus a transparent film dressing in the management of skin tears in institutionalized subjects. *Ostomy Wound Management, 45*(6), 22–28.

Thomas, T., Thomas, G., McLendon, C., Sutton, T., & Mullan, M. (1996). Beta-amyloid-mediated vasoactivity and vascular endothelial damage. *Nature, 380*(6570), 168–171.

Thomasma, D. C. (1991). Why philosophers should offer ethics consultation. *Theoretical Medicine, 12,* 129–140.

Thompson, G. (1999 March). New clinics let lesbian patients be themselves: An effort to help a group that shuns doctors. *The New York Times,* pp. 1, 6.

Thompson, L. W., Gallagher, D., & Breckenridge, J. S. (1987). Comparative effectiveness of psychotherapies for depressed elders. *Journal of Consulting & Clinical Psychology, 55*(3), 385–390.

Thompson, W. (1990). *Ageing is a Family Affair: A Guide to Quality Visiting, Long Term Care Facilities and You* (3rd ed.). Toronto, Canada: N.C. Press.

Thorson, J. A., & Davis, R. E. (2000). Relocation of institutionalized aged. *Journal of Clinical Psychology, 56,* 131–138.

Thüroff, J., Chartier-Kastler, E., Corcus, J., Humke, J., Jonas, U., & Palmtag, H. (1998). Medical treatment and side effects in urinary incontinence in the elderly. *World Journal of Urology, 16*(Suppl. 1), S48–S61.

Tideiksaar, R. (1997). *Falling in Old Age: Prevention and Management*, 2nd Edition. New York: Springer Publishing Company.

Tielsch, J. M., Javitt, J. C., Coleman, A., & Sommer, A. (1995). The prevalence of blindness and visual impairment among nursing home residents in Baltimore. *NEJM, 332*(18), 1205–1209.

Tilly, J. (1999). *Consumer-directed Long Term Care: Participant's Experiences in Five Countries.* Washington, DC: American Association of Retired Persons.

Tinetti, M., & Williams, C. (1997). Falls, injuries due to fall, and the risk of admission to a nursing home. *NEJM, 337,* 1279–1284.

Tinetti, M. E. (1986). Performance-oriented assessment of mobility problems in elderly patients. *Journal of the American Geriatrics Society, 34*(2), 119–126.

Tinetti, M. E., & Ginter, S. F. (1988). Identifying mobility dysfunctions in elderly patients. *JAMA, 259*(8), 1190–1193.

Tomer, A. (1992). Death anxiety in adult life: Theoretical perspectives. *Death Studies, 16,* 475–506.

Tomsak, R. L. (1997). An approach to acquired visual loss in adults. *Journal of Ophthalmic Nursing & Technology, 16*(5), 229–234.

Toner, J., Gurland, B., & Teresi, J. (1988). Comparison of self-administered and rater-administered methods of assessing levels of severity of depression in the elderly. *Journals of Gerontology, 43B,* P136–140.

Toner, J. A., Teresi, J. A., Gurland, B. J., & Tirumalasetti, F. (1999). The Feeling Tone Questionnaire: Reliability and validity of a direct patient assessment screening instrument for the detection of depressive symptoms in cases of dementia. *Journal of Clinical Geropsychology, 5,* 63–78.

Tordoff, C. (1996). The prevalence of HIV and AIDS in older people. *Professional Nurse, 12*(3), 193–195.

Torres-Standovik, R. (1999). *The Caregiver's Handbook* [Web Page]. URL http://www.medsupport.org [1999, November 12].

Toseland, R. W., & McCallion, P. (1998). *Maintaining Communication with Persons with Dementia: An Educational Program for Nursing Home Staff and Family Members.* New York: Springer Publishing Company.

Tregonning, M., & Langley, C. (1999). Chronic obstructive pulmonary disease. *Elderly Care, 11*(7), 21–25; quiz 26.

Tresch, D. (1997). The clinical diagnosis of heart failure in older patients. *Journal of the American Geriatrics Society, 45,* 1128–1133.

Tresch, D. D. (1998). Management of the older patient with acute myocardial infarction: Differences in clinical presentations between older and younger patients. *Journal of the American Geriatrics Society, 46,* 1157–1162.

Trzepacz, P. T. (1998). Update on the neuropathogenesis of delirium. *Dementia & Geriatric Cognitive Disorders, 10,* 330–335.

Tuckman, B. W. (1965). Developmental sequence in small groups. *Psychological Bulletin, 63*(6), 384–399.

Turner, F. (1992). *Mental Health and the Elderly*. New York: Free Press.

U.S. Administration on Aging. (1999). *Administration on Aging Homepage* [Web Page]. URL http://www.aoa.dhhs.gov [1999, October 30].

U.S. Advisory Commission on Intergovernmental Relations. (1972–1994). *Changing Public Attitudes on Government and Taxes*. (Report No. S1-S-23). Washington, DC: Author.

U.S. Bureau of Labor Statistics. (2000). Employment and Earnings. Vol. 47, Chap. 1. Washington, DC: U.S. Governement Printing Office.

U.S. Bureau of the Census. (1997). *Poverty Statistics on Population Groups, Current Population Survey*. Washington, DC: Author.

U.S. Congressional Budget Office. (1993). *Displaced Worker: Trends in the 1980s and Implications for the Future*. Washington, DC: Congressional Budget Office.

U.S. Department of Education. (1992). *Summary of Existing Legislation Affecting People with Disabilities*. Washington D.C.: U.S. Government Printing Press.

U.S. Department of Health and Human Services. (1998). *Clinician's Handbook of Preventive Services*, 2nd ed. Washington, DC: U.S. Department of Health and Human Services.

U.S. Department of Health and Human Services. (1999). *Mental Health: A Report of the Surgeon General*. Rockville, MD: Center for Mental Health Services, Substance Abuse and Mental Health Administration, and National Institute of Mental Health, National Institutes of Health, U.S. Department of Health and Human Services.

U.S. Department of Health and Human Services. (1992). *Older Americans Act Amendments*. Vol. Public Law 89-73.

U.S. Department of Health and Human Services. (1992). *Technology-Related Assistance for Individuals Covered Under the Americans with Disabilities Act of 1988*. Vol. Public Law 100-407, 89-73. 29 U.S. C. 2202.

U.S. Department of Health and Human Services, Health Care Financing Administration. (1998). *Health Care Financing Review, Medicare and Medicaid Statistical Supplement*. Baltimore, MD: Author.

U.S. Department of Health and Human Services, Health Care Financing Administration. (2000). Prospective Payment System and Consolidated Billing for Skilled Nursing Facilities—Update of Proposed Rule. *Federal Register*, 42 CFR 411 and 489 Medicare Program.

U.S. Department of Health and Human Services, National Center for Health Statistics. (1996). *Health United States, 1995*. (Report No. DHHS Pub No (PHS) 96-1232). Hyattsville, MD: US Government Printing Office, Washington DC.

U.S. Department of Health and Human Services, Public Health Service, Indian Health Service. (1996). *Indian Health Focus: Elders*. Hyattsville, MD: U.S. Government Printing Office, Washington DC.

U.S. Department of Transportation. (1997). (Report No. DOT-P10-97-01). Washington, DC: Author.

U.S. General Accounting Office. (1998). *California Nursing Homes: Care Problems Persist Despite Federal and State Oversight. GAO/HEHS-98-202*. Washington, DC: Author.

U.S. General Accounting Office. (1999). *Medicare Home Health Agencies: Closures Continue, With Little Evidence Beneficiary Access is Impaired. GAO/HEHS-99-120*. Washington, DC: Author.

U.S. General Accounting Office. (1997). *Medicare: Need to Hold Home Health Agencies More Accountable for Inappropriate Billings. HEHS-97-108*. Washington, DC: Author.

U.S. General Accounting Office. (1999). *Nursing Homes: Complaint Investigation Processed Often Inadequate to Protect Residents. GAO/HEHS-99-80*. Washington, DC: Author.

U.S. General Accounting Office. (1999). *Skilled Nursing Facilities: Medicare Payment Changes Require Provider Adjustment but Maintain Access. GAO/ HEHS-00-23*. Washington, DC: Author.

U.S. National Center for Health Statistics. (1996). *Prevalance of Selected Chronic Conditions, by Age and Sex*.

U.S. National Center for Veteran Statistics and Analysis. (1994). *Estimates and Projections of the Veteran Population in the U.S., by Age, for Years 1990 Through 2020 (from 1990 U.S. Census)*. Washington, DC: Author.

U.S. Preventive Services Task Force. (1996). *Guide to Clinical Preventive Services* (2nd). Washington, DC: U.S. Department of Health and Human Services.

U.S. Senate Special Committee on Aging. (Hearing Date, May 20, 1998). *Living Longer, Growing Stronger in America: The Vital role of Geriatric Medicine*. Washington, D.C.

U.S. Senate Special Committee on Aging. (1974). *Nursing Home Care in the United States: Failure in Public Policy, Introductory Report*. Washington, DC: U.S. Government Printing Office.

Uhlenberg, P., & Riley, M. (1996). Cohort Studies. In J. E. Birren (Ed.), *Encyclopedia of Gerontology: Age, Aging, and the Aged*. San Diego, CA: Academic Press.

UJA-Federation of Jewish Philanthropies. (1992). *Proposal: Providing supportive services to elderly resi-*

dents with complex and chronic conditions of government assisted naturally occurring retirement communities in New York City. New York, NY: UJA-Federation of Jewish Philanthropies.

Umlauf, M., Kurtzer, E., Valappil, T., Burgio, K., Pillion, D., & Goode, P. (1999). Sleep disordered breathing as a mechanism for nocturia. *Ostomy Wound Management, 45*(12), 52–60.

Unger, J., McAvay, G., Bruce, M., Berkman, L., & Seeman, T. (1999). Variations in the impact of social network characteristics on physical functioning in elderly persons: MacArthur studies of successful aging. *Journals of Gerontology, 54B,* S245–S251.3.

United Nations. (1956). The aging of populations and its economic and social implications. *Populations Studies* Vol. 26. New York: United Nations.

United Nations. (1992). Changing population age structures, 1990-2015. *Demographic and Economic Consequences and Implications.* Geneva: United Nations.

Urban Institute. (2000). *Unpublished estimates based on data from HVFA-2082 and HCFA-64 reports.* Washington, DC: Author.

Urdaneta, M. L., Saldana, D. H., & Winkler, A. (1995). Mexican-American perceptions of severe mental illness. *Human Organization, 54*(1), 70–77.

Vachon, R. A. (1987). Intervening a future for individuals with work disabilities: The challenge of writing national disability policies. In D. E. Woods, & D. Vandergoot (Eds.), *The Changing Nature of Work, Society, and Disability: The Impact on Rehabilitation Policy* (pp. 19–45). New York: World Rehabilitation Fund.

Valdez, I. H., & Fox, P. C. (1993). Diagnosis and management of salivary dysfunction. *Critical Review of Oral Biology & Medicine, 4,* 271–277.

Valway, S. E., Linkins, R. W., & Gohde, D. M. (1993). Epidemiology of lower-extremity amputations in Indian Health Services, 1982-1987. *Diabetes Care, 16*(Suppl 1), 349–353.

Van Andel, G., & Heintzman, P. (1996). Christian spirituality and therapeutic recreation. In C. Sylvester (Ed.), *Philosophy of Therapeutic Recreation: Ideas and Issues* (Vol. II, pp. 71–85). Ashburn, VA: National Recreation and Park Association.

van Casteren, V., van der Veken, J., Fafforeau, J., & Van Oyen, H. (1993). Suicide and attempted suicide reported by general practitioners in Belgium, 1990–1991. *Acta Psychiatrica Scandinavica, 87,* 451–455.

Van Schaardenburg, D., & Breedveld, F. C. (1994). Elderly-onset rheumatoid arthritis. *Seminars in Arthritis & Rheumatism, 23*(6), 367–378.

van Waas, M. A. J. (1990). Determinants of dissatisfaction with dentures: A multiple regression analysis. *Journal of Prosthetic Dentistry, 64,* 569–572.

Vanderpool, H. Y. (1995). Death and dying: Euthanasia and sustaining life. In *Encyclopedia of Bioethics* Revised ed. (pp. 554–561). New York: Simon and Schuster, MacMillan.

VandeWeerde, C., & Paveza, G. J. (1998). Self Neglect in Older Adults. A paper presented at the Annual Scientific Meeting of the Gerontological Society of America, Philadelphia, PA.

VanOrt, S., & Phillips, L. (1992). Feeding nursing home residents with Alzheimer's disease. *Geriatric Nursing—American Journal of Care for the Aging, 13*(5), 249–253.

Vaupel, J. W. (1997). The remarkable improvements in survival at older ages. *Philosophical Transactions of the Royal Society, London, B, 352,* 1799–1804.

Velthuis-te Wieric, E., van den Derg, H., Schaafsma, G., Hendriks, H., & Brouwer, A. (1994). Energy restriction, a useful intervention to retard human aging? Results of a feasibility study. *European Journal of Clincial Nutrition, 48*(2), 138–148.

Verdery, R. (1998). Failure to thrive. In E. H. Duthie, P. R. Katz, & R. Kersey (Eds.), *Practice of Geriatrics* (pp. 257–264). Philadelphia, PA: W. B. Saunders.

Vernava, A. M. III, Moore, B. A., Longo, W. E., & Johnson, F. E. (1997). Lower gastrointestinal bleeding. *Diseases of the Colon & Rectum, 40*(7), 846–858.

Vervoorn, J. M., Duinkerke, A. S. H., Luteijin, F., & van de Poel, A. C. (1988). Assessment of denture satisfaction. *Community Dentistry & Oral Epidemiology, 16,* 364–367.

Vigneri, S., Termini, R., Leandro, G., et al. (1995). A comparison of five maintenance therapies for reflux esophagitis. *NEJM, 333*(17), 1106–1110.

Villa, V., Wallace, S., & Markides, K. (1997). Economic diversity and an aging population: The impact of public policy and economic trends. *Generations, 21*(2), 13–18.

Vita, A. J., Terry, R. B., Hubert, H. B., & Fries, J. F. (1998). Aging, health risks, and cumulative disability. *NEJM, 338,* 1035–1041.

Voelkl, J. E. (1993). Activity among older adults in institutional settings. In J. R. Kelly, *Activity and Aging* (pp. 231–245). Newbury Park, CA: Sage.

Volicer, L., & Hurley, A. (1998). *Hospice Care for Patients with Advanced Progressive Dementia.* New York: Springer Publishing Company.

Volicer, L., Stelly, M., Morris, J., McLaughlin, J., & Volicer, B. J. (1997). Effects of dronabinol on anorexia and disturbed behavior in patients with Alzheimer's disease. *International Journal of Geriatric Psychiatry, 12*(9), 913–919.

Volmink, J. (1996). Treatments for posttherapetic neuralgia—a systematic review of randomized controlled trails. *Family Practice, 13*(1), 84.

Wacker, R. R., Roberto, K. A., & Piper, L. E. (1998). *Community Resources for Older Adults: Programs and Services in an Era of Change.* Thousand Oaks, CA: Pine Forge Press.

Wagner, E. (1996). The promise and performance of HMOs in improving outcomes in older adults. *Journal of the American Geriatrics Society,* (44), 1251–1257.

Waid, M. O. (1998). *Brief Summaries of Medicare & Medicaid.* Health Care Financing Administration, 19.

Waitzkin, H. (1984). Doctor-patient communication. Clinical implications of social scientific research. *JAMA, 252,* 2441–2446.

Wallace, R. K., Benson, H., & Wilson A. F. (1971). A wakeful hypometabolic physiologic state. *American Journal of Physiology, 221*(3), 795–799.

Wallace, S. P., Levy-Stroms, L., Kington, R. S., & Andersen, R. M. (1998). The persistence of race and ethnicity in the use of long-term care. *Journals of Gerontology, 53B*(2), S104–112.

Wallis, M. S., Bowen, W. R., & Guin, J. D. (1991). Pathogenesis of onychoschizia (lamellar dystrophy). *Journal of the American Academy of Dermatology, 24,* 44–48.

Walsh, K. (1999). Shared humanity and the psychiatric nurse-patient encounter. *Australian & New Zealand Journal of Mental Health Nursing, 8*(1), 2–8.

Walters, E. J. C. (1994). *Consumers' Directory of Continuing Care Retirement Communities.* Washington, DC: American Association of Homes and Services for the Aging.

Ward, R. A. (1979). The meaning of voluntary association participation to older people. *Journal of Gerontology, 34,* 438–445.

Wardlaw, J. M., Warlow, C. P., & Counsell, C. (1997). Systematic review of evidence on thrombolytic therapy for acute ischaemic stroke. *Lancet, 350*(9078), 607–614.

Ware, J., Bayliss, M., Rogers, W., Kosinski, M., & Tarlov, A. (1996). Differences in 4-year health outcomes for elderly and poor, chronically ill patients treated in HMO and fee-for-service systems. Results from the Medical Outcomes Study. *JAMA, 276*(13), 1039–1047.

Ware, J. E., & Sherbourne, C. D. (1992). The MOS 36-item Short Form Health Survey (SF-36). I. Conceptual framework and item selection. *Medical Care, 30,* 473–483.

Ware, J. E. Jr. (1996). The SF-36 health survey In B. Spilker (Ed.), *Quality of Life and Pharmacoeconomics in Clinical Trials* (2nd ed.). Philadelphia: Lippincott-Raven.

Warren, J. J., Kambhu, P. P., & Hand, J. S. (1994). Factors related to acceptance of dental treatment services in a nursing home population. *Special Care in Dentistry, 14,* 15–20.

Warshaw, G., & Moqeeth, S. (1998). Hearing impairment. In T. T. Yoshikawa, E. L. Cobbs, & K. Brummel-Smith (Eds.), *Practical Ambulatory Geriatrics* (2nd ed., Vol. 16, pp. 118–125). St. Louis: Mosby Year-Book.

Watson, D., Selkoe, D., & Teplow, D. (1999). Effects of the amyloid precursor protein Glu693Gln 'Dutch' mutation on the production and stability of amyloid beta-protein. *Biochemical Journal, 340,* 703–709.

Weaver, T., Laizner, A. M., Evans, L. K., Maislin, G., Chugh, D. K., Lyon, K., Smith, P. L., Schwartz, A. R., Redline, S., Pack, A. I., & Dinges, D. F. (1997). An instrument to measure functional status outcomes for disorders of excessive sleepiness. *Sleep, 20,* 835–843.

Wei, J. Y. (1999). Coronary heart disease. In W. R. Hazzard, J. P. Blass, W. H. Ettinger, J. B. Halter, & J. G. Ouslander (Eds.), *Principles of Geriatric Medicine and Gerontology* (4th ed., Chap. 47, pp. 661–668). New York: McGraw-Hill Publishing Company.

Weidner, A. C. (1998). Imaging studies of the pelvic floor. *Obstetrics & Gynecology Clinics of North America, 25*(4), 25–48.

Weiland, D., Kramer, B. J., Waite, M. S., & Rubenstein, L. Z. (1996). The interdisciplinary team in geriatric care. *American Behavioral Scientist, 39,* 655–664.

Weinberg, A., Minaker, K., & The Council on Scientific Affairs. (1995). Dehydration: Evaluation and management in older adults. *JAMA, 274,* 1562–1556.

Weiner, D., Pieper, C., McConnell, E., Martinez, S., & Keefe, F. (1996). Pain measurement in elders with chronic low back pain: Traditional and alternative approaches. *Pain, 67,* 461–467.

Weiner, D. K., Ladd, K. E., Pieper, C. F., & Keefe, F. J. (1995). Pain in the nursing home: Resident versus staff perceptions. *Journal of the American Geriatrics Society, 43,* SA2–5.

Weiner, J. (1994). Forecasting the effects of health reform on U.S. physicians workforce requirements: Evidence from HMO staffing patterns. *Journal of the American Medical Center, 272,* 222–230.

Weinstein, B. (1998). Disorders of hearing. In J. C. Brocklehurst (Ed.), *Geriatric Medicine and Gerontology* (5th ed.). London: Churchill Livingstone.

Weintraub, M. (1990). Compliance in the elderly. *Clinics in Geriatric Medicine, 6*(2), 445–452.

Weissert, W. G., & Henrick, S. C. (1994). Lessons learned from research on effects of community-based long-term care. *Journal of the American Geriatrics Society, 84*(11), 1813–1817.

Weitz, J. I., Byrne, J., Clagett, G. P., Farkouh, M. E., Porter, J. M., Sackett, D. L., Strandness, D. E. Jr., &

Taylor, L. M. (1996). Diagnosis and treatment of chronic arterial insufficiency of the lower extremities: A critical review. *Circulation, 94*(11), 3026–3049.

Wengel, S. P., Roccaforte, W. H., & Burke, W. J. (1998). Donepezil improves symptoms of delirium in dementia: Implications for future research. *Journal of Geriatric Psychiatry & Neurology, 11*(3), 159–161.

Wenger, G. C. (1990). Elderly carers: The need for appropriate intervention. *Aging & Society, 10,* 197–219.

Wexler, N. (1992). Clairvoyance and caution in the Human Genome Project. In D. J. Kevles, & L. Hood (Eds.), *The Code of Codes: Scientific and Social Issues in the Human Genome Project* (Chap. 3, pp. 211–243). Cambridge, MA: Harvard University Press.

Wexner, S. D., Marchetti, F., & Jagelman, D. G. (1991). The role of sphincteroplasty for fecal incontinence reevaluated: A prospective physiologic and functional review. *Diseases of the Colon & Rectum, 34,* 22–30.

Wheatley, D., & Smith, D. (Eds.). (1998). *Psychopharmacology of Cognitive and Psychiatric Disorders in the Elderly* (Chap. 6). London: Chapman and Hall.

Wheeler, J. A., Gorey, K. M., & Greenblatt, B. (1998). The beneficial effects of volunteering for older volunteers and the people they serve: A meta-analysis. *International Journal of Aging & Human Development, 47,* 69–79.

White, A. J. (1998). *The Effect of PACE on Costs to Medicare: A Comparison of Medicare Capitation Rates to Projected Costs in the Absence of PACE.* Cambridge, MA: ABT Associates.

White, B. (1994). *Competence to Consent.* Washington, DC: Georgetown University Press.

White, J. V. (1999). The utility of body mass index in predicting health risk. *The Consultant Dietitian, 24*(2), 1–7.

Whitehouse, P., Maurer, K., & Ballenger, J. B. (2000). *Concepts of Alzheimer Disease: Biological, Clinical, and Cultural Perspectives.* Baltimore: Johns Hopkins Press.

Widdicombe, J. G. (1999). Advances in understanding and treatment of cough. *Monaldi Archives for Chest Disease, 54*(3), 275–279.

Wiener, J. M. (1981). *A Sociological Analysis of Government Regulation: The Case of Nursing Homes.* Unpublished doctoral dissertation, Harvard University, Cambridge, MA.

Wiener, J. M., & Hanley, R. J. (1992). Caring for the disabled elderly: There's no place like home. In S. M. Shortell & U. E. Reinhardt, *Improving Health Policy and Management: Nine Critical Research Issues for the 1990s* (pp. 75–110). Ann Arbor, MI: Health Administration Press.

Wiener, J. M., Illston, L. H., & Hanley, R. J. (1994). *Sharing the Burden: Strategies for Public and Private Long-Term Care Insurance.* Washington, DC: The Brookings Institution.

Wiener, J. M., & Stevenson, D. G. (1998). State policy on long-term care for the elderly. *Health Affairs, 17*(3), 81–100.

Wilber, K., & Buturain, L. (1993). Developing a daily money management service model: Navigating the uncharted waters of liability and viability. *The Gerontologist, 33*(5), 687–691.

Wilber, K. H., & Reynolds, S. L. (1996). Introducing a framework for defining financial abuse of the elderly. *Journal of Elder Abuse & Neglect, 8*(2), 61–80.

Wilber, K. H., & Reynolds, S. L. (1995). Rethinking alternatives to guardianship. *The Gerontologist, 35,* 248–257.

Wilkinson, T. J., Henschke, P. J., & Handscombe, K. (1995). How should toilets be labelled for people with dementia. *Australian Journal on Aging, 14,* 163–165.

Williams, C., & Graham, J. D. (1995). Licensing standards for elderly drivers. *Consumer's Research Magazine, 78*(12), 18–23.

Winklevoss, H. E., & Powell, A. V. (1984). *Continuing Care Retirement Communities.* Homewood, IL: Irwin.

Wisocki, P. A. (1998). The experience of bereavement by older adults. In M. Hersen, & V. B. Van Hasselt, *Handbook of Clinical Geropsychology.* New York: Plenum Press.

Witta, K. M. (1997). COPD in the elderly. Chronic obstructive pulmonary disease. *Advance for Nurse Practitioners, 5*(7), 18–23.

Woerner, M. G., Alvir, J. M., Saltz, B. L., Lieberman, J. A., & Kane, J. M. (1998). Prospective study of tardive dyskinesia in the elderly: Rates and risk factors. *American Journal of Psychiatry, 155,* 1521–1528.

Wolf, A., & Colditz, G. (1996). Social and economic effects of body weight in the United States. *American Journal of Clinical Nutrition, 63*(Suppl), S466–469.

Wolf-Klein, G. (1989). Screening examinations in the elderly: Which are worthwhile. *Geriatrics, 44*(12), 36–47.

Wolf, R. S., & Pillemer, K. (1994). What's new in elder abuse programming? Four bright ideas. *Gerontologist, 34*(1), 126–129.

Wolf, S. L., Barnhart, H. X., Ellison, G. L., & Coogler, C. E. (1997). The effect of Tai Chi Quan and computerized balance training on postural stability in older subjects. *Physical Therapy, 77,* 104–107.

Wolfe, F., Zhao, S., & Lane, N. (2000). Preference for nonsteroidal antiinflammatory drugs over acetaminophen by rheumatic disease patients: A survey of 1,799 patients with osteoarthritis, rheumatoid arthritis, and fibromyalgia. *Arthritis & Rheumatism, 43*(2), 378–385.

Wolfe, M. S., Xia, W., Ostaszewski, B. L., Diehl, T. S., Kimberly, W. T., & Selkoe, D. J. (1999). Two transmembrane aspartates in presenilin-1 required for presenilin endoproteolysis and gamma-secretase activity. *Nature, 398*(6727), 513–517.

Wolfson, E., & Mower, R. S. (1994). When the police are in our bedroom, shouldn't the courts go after them? An update on the fight against sodomy laws. *Fordham Urban Law Journal, 21*(4), 997–1055.

Wolfson, L., Whipple, R., Derby, C., Judge, J., King, M., Amerman, P., Schmidt, J., & Smyers, D. (1996). Balance and strength training in older adults: Intervention gains and Tai Chi maintenance. *Journal of the American Geriatrics Society, 44,* 498–506.

Wolinksy, F. D., Fitzgerald, J. F., & Stump, T. E. (1997). The effect of hip fracture on mortality, hospitalization, and functional status: A prospective study. *American Journal of Public Health, 87*(3), 398–403.

Won, A., Lapane, K., Gambassi, G., Bernabei, R., Mor, V., & Lipsitz, L. (1999). Correlates and management of non-malignant pain in the nursing home. *Journal of the American Geriatrics Society, 47,* 936–942.

Wood, D. J., Ions, G. K., & Quinby, H. M. (1992). Factors which influence mortality after subcapital hip fracture. *Journal of Bone and Joint Surgery, 748,* 199–202.

Woods, R. T. (1999). Mental health problems in late life. In R. T. Woods (Ed.), *Psychological Problems of Aging: Assessment, Treatment and Care* (Chap. 4). London: John Wiley & Sons, Ltd.

Woodson, S. A. (1997). Sexual health across the lifespan: Teaching women about their sexuality. *AWHONN Lifelines, 1*(4), 34–39.

Woollacott, M. (1993). Age-related changes in posture and movement. *Journal of the American Geriatrics Society,* (48), 56–60.

Work Group on Research and Evaluation of Special Care Units (WGRESCU). (1996). National Institute on Aging collaborative studies: Special care units for Alzheimer's disease. Presented at the Gerontological Society of America, Preconference Program, Washington, D.C.

World Health Organization. (1999). *Model for the Disablement Process* [Web Page]. URL http://www.worldhealth.org [1999, November 22].

World Health Organization. (1999). *Aging and Health: A Global Challenge for the 21st Century,* Kobe, Japan, November 10, 1998–November 13, 1998. Geneva: Author.

World Health Organization on Aging and Health. (1999). *The Scope of the Challenge* [Web Page]. URL http://www.who.int/aging/scope.html [2000, January 17].

Wound, Ostomy and Continence Nurses Society (WOCN). (1998). *Professional Practice Fact Sheet: Medicare Part B Coverage for Support Surfaces in the Home Health Setting* [Web Page]. URL http://www.wocn.org/ [1999, August 18].

Wright, L. K., Clipp, E. C., & George, L. K. (1993). Health consequences of caregiver stress. *Medicine, Exercise, Nutrition, and Health, 2,* 181–195.

Wulf, H. (1998). Epidural analgesia in postoperative pain therapy: A review. *Anaesthesist, 47*(6), 501–510.

Wyman, J. F. (1999). Urinary incontinence. In J. T. Stone, J. F. Wyman, & S. A. Salisbury (Eds.), *Clinical Gerontological Nursing: A Clinical Guide to Practice* (pp. 203–217). Philadelphia: W. B. Saunders.

Yakoboski, P., & Dickemper, J. (1997). Increased saving but little planning: Results of the 1997 Retirement Confidence Survey. *EBRI Issue Brief, 191.*

Yalom, I. D. (1995). *The Theory and Practice of Group Psychotherapy,* 4th Edition. New York: Basic Books.

Yan, J. (1999). Petty welfare-reform limits burden on elders. *Aging Today,* (July/August), 1–2.

Yancik, R., & Ries, L. (1998). Cancer in older persons—magnitude of the problem—how do we apply what we know? In L. Balducci, G. H. Lyman, & W. Ershler (Eds.), *Comprehensive Geriatric Oncology* (2nd Vol., Chap. 5, pp. 95–103). Harwood Academic Publishers.

Yee, D. L. (1997). Can long-term care assessments be culturally responsive? *Generations, 21*(1), 25–29.

Yee, D. L., & Capitman, J. A. (1996). Health care access, health promotion, and older women of color. *Journal of Health Care for the Poor & Underserved, 7*(3), 257–272.

Yee, D. L., Sanchez, Y. N., & Shin, A. (1999). *Establishing Information Infrastructure for API Elders: A roadmap for assuring rights and protections of HCFA beneficiaries (Final Recommendations on the Vulnerable Populations Project).* Seattle, WA: National Asian Pacific Center on Aging.

Yellowitz, J. A. (1999). Providing oral cancer examinations for older adults. *Elder Care: Journal of the California Dental Association, 718*–723.

Yesavage, J. A., Brink, T. L., Rose, T. L., Lum, O., Huang, V., Adey, M., & Leirer, V. O. (1983). Development and validation of a geriatric depression screening scale: A preliminary report. *Journal of Psychiatric Research, 17,* 37–49.

Yoshikawa, T. T., Cobbs, E. L., & Brummel-Smith, K. (1998). *Practical Ambulatory Geriatrics.* St. Louis, MO: Mosby Year-Book.

748   REFERENCES

Young, J. J., French, L., & Catague, E. (1995). *Breaking Down Barriers for Pacific/Asian Elderly. Final Report*. Seattle: National Asian Pacific Center on Aging.

Young, R. C., & Klerman, G. L. (1992). Mania in late life: Focus on age at onset. *American Journal of Psychiatry, 149,* 867–876.

Ytterstad, B. (1999). The Harstad injury prevention study: The characteristics and distribution of fractures amongst elders—an eight year study. *International Journal of Circumpolar Health, 58*(2), 84–95.

Yu, B. P. (1995). Putative interventions. In E. J. Masoro (Ed.), *Handbook of Physiology* (pp. 613–631). New York: Oxford University Press.

Zahn, M. A., & Gold, S. D. (1984). *State Policy and Senior Citizens*. Denver: National Conference of State Legislatures.

Zarit, S., Reever, K., & Bach-Peterson, J. (1980). Relatives of the impaired elderly: Correlates of feelings of burden. *The Gerontologist, 206,* 649–655.

Zarit, S. H. (1997). Brief measures of depression and cognitive function. *Generations, 21*(1), 41–43.

Zastrow, C. H. (1999). *The Practice of Social Work* (6th ed.). Pacific Grove, CA: Brooks/Cole.

Zeleznick, J., Post, L. F., Mulvihill, M., Jacobs, L. G., Burton, W. B., & Dubler, N. N. (1999). The doctor-proxy relationship: Perception and communication. *The Journal of Law, Medicine & Ethics, 27*(1), 13–19.

Zinn, J. S. A. W. E., & Rosko, M. D. (1993). Variations in the outcomes of care in Pennsylvania nursing homes: Facility and environmental correlates. *Medical Care, 31*(6), 475–487.

Zinny, G. H., & Grossberg, G. T. (1998). *Guardianship of the Elderly: Psychiatric and Judicial Aspects*. New York: Springer Publishing Company.

Zisook, S., & Schucher, S. R. (1996). Grief and bereavement. In J. Sadavoy, & W. Lazarus (Eds.), *Comprehensive Review of Geriatric Psychiatry* (2nd ed.). Washington, DC: American Psychiatric Press.

Zitter, M. (1997). A new paradigm in health care delivery: Disease management. In W. E. Todd, & D. Nash *Disease Management: A Systems Approach to Improving Patient Outcomes* (pp. 1–25). Chicago: American Hospital Publishing.

Zung, W. (1965). A self-rating depression scale. *Archives of General Psychiatry, 12,* 63–69.

SUBJECT INDEX

CONTRIBUTOR INDEX